Scientific Basis for Nuclear Waste Management XXIII

MATERIALS RESEARCH SOCIETY
SYMPOSIUM PROCEEDINGS VOLUME 608

Scientific Basis for Nuclear Waste Management XXIII

Symposium held November 29–December 2, 1999, Boston, Massachusetts, U.S.A.

EDITORS:

Robert W. Smith
Idaho National Engineering and
Environmental Laboratory
Idaho Falls, Idaho, U.S.A.

David W. Shoesmith
University of Western Ontario
London, Ontario, Canada

Materials Research Society
Warrendale, Pennsylvania

Single article reprints from this publication are available through
University Microfilms Inc., 300 North Zeeb Road, Ann Arbor, Michigan 48106

CODEN: MRSPDH

Published by:

Materials Research Society
506 Keystone Drive
Warrendale, PA 15086
Telephone (724) 779-3003
Fax (724) 779-8313
Web site: http://www.mrs.org/

Library of Congress Cataloging-in-Publication Data

Scientific basis for nuclear waste management XXIII : symposium held November 29–December 2, 1999, Boston, Massachusetts, U.S.A. / editors, Robert W. Smith, David W. Shoesmith
p.cm.—(Materials Research Society symposium proceedings, ISSN 0272-9172 ; v. 608)
Includes bibliographical references and indexes.
ISBN 1-55899-516-1
ISSN 0275-0112
I. Smith, Robert W. II. Shoesmith, David W. III. Materials Research Society symposium proceedings ; v. 608
2000

Manufactured in the United States of America

CONTENTS

*Invited Paper

*Invited Paper

REPOSITORY PERFORMANCE

RADIONUCLIDE SORPTION AND TRANSPORT

*Invited Paper

CEMENT-BASED MATERIALS AND WASTE CONTAINMENT

CORROSION OF CERAMIC WASTEFORMS

RADIATION EFFECTS

NATURAL ANALOGS

*Invited Paper

WASTEFORM CHARACTERIZATION AND PROCESSING

*Invited Paper

CORROSION AND CHARACTERIZATION OF GLASS WASTEFORMS

PREFACE

The symposium entitled "Scientific Basis for Nuclear Waste Management XXIII" was held November 29–December 2 at the 1999 MRS Fall Meeting in Boston, Massachusetts. This year's symposium attracted 142 abstracts, and this proceedings volume contains 114 written contributions from the Meeting. The papers have been organized into 12 separate sections covering various topics in nuclear waste management.

The organizers of this symposium would like to acknowledge the contributions of the following individuals who chaired topical sessions and helped review the manuscripts published in this volume. Their efforts and insights into the various scientific and engineering disciplines associated with this program greatly contributed to the success of the symposium and the quality of the proceedings volume.

Tae M. Ahn
Ned E. Bibler
Andrew H. Bond
David B. Chamberlain
Ken R. Czerwinski
Darrell S. Dunn
William L. Ebert
Rodney C. Ewing
Melinda A. Hamilton
Larry C. Hull
Brian M. Ikeda
Carol M. Jantzen
Lawrence H. Johnson
Fraser King
Michelle A. Lewis
Bernard P. McGrail
William M. Murphy
Roberto T. Pabalan
Osvaldo Pensado
Swami V. Raman
George D. Redden
Donald T. Reed
Surendra P. Shah
Eric R. Siegmann
Kastriot Spahiu
Dennis M. Strachan
Lou Vance
John D. Vienna
Lars O. Werme
David J. Wronkiewicz

The efforts of many anonymous reviewers are also gratefully acknowledged. We also acknowledge the generous financial support of The Idaho National Engineering Laboratory. We would like to extend a special thank you to Margaret A. Knecht for editorial assistance, and to the staff of the Materials Research Society for essential organizational assistance.

Robert W. Smith
David W. Shoesmith

June 2000

MATERIALS RESEARCH SOCIETY SYMPOSIUM PROCEEDINGS

Volume 557— Amorphous and Heterogeneous Silicon Thin Films: Fundamentals to Devices—1999, H.M. Branz, R.W. Collins, H. Okamoto, S. Guha, R. Schropp, 1999, ISBN: 1-55899-464-5

Volume 558— Flat-Panel Displays and Sensors—Principles, Materials and Processes, F.R. Libsch, B. Chalamala, R. Friend, T. Jackson, H. Ohshima, 2000, ISBN: 1-55899-465-3

Volume 559— Liquid Crystal Materials and Devices, T.J. Bunning, S.H. Chen, L.C. Chien, T. Kajiyama, N. Koide, S-C.A. Lien, 1999, ISBN: 1-55899-466-1

Volume 560— Luminescent Materials, J. McKittrick, B. DiBartolo, K. Mishra, 1999, ISBN: 1-55899-467-X

Volume 561— Organic Nonlinear Optical Materials and Devices, B. Kippelen, H.S. Lackritz, R.O. Claus, 1999, ISBN: 1-55899-468-8

Volume 562— Polycrystalline Metal and Magnetic Thin Films, D.E. Laughlin, K.P. Rodbell, O. Thomas, B. Zhang, 1999, ISBN: 1-55899-469-6

Volume 563— Materials Reliability in Microelectronics IX, C.A. Volkert, A.H. Verbruggen, D.D. Brown, 1999, ISBN: 1-55899-470-X

Volume 564— Advanced Interconnects and Contacts, D.C. Edelstein, T. Kikkawa, M.C. Öztürk, K-N. Tu, E.J. Weitzman, 1999, ISBN: 1-55899-471-8

Volume 565— Low-Dielectric Constant Materials V, J. Hummel, K. Endo, W.W. Lee, M. Mills, S-Q. Wang, 1999, ISBN: 1-55899-472-6

Volume 566— Chemical-Mechanical Polishing—Fundamentals and Challenges, S.V. Babu, S. Danyluk, M. Krishnan, M. Tsujimura, 2000, ISBN: 1-55899-473-4

Volume 567— Ultrathin SiO_2 and High-K Materials for ULSI Gate Dielectrics, H.R. Huff, C.A. Richter, M.L. Green, G. Lucovsky, T. Hattori, 1999, ISBN: 1-55899-474-2

Volume 568— Si Front-End Processing—Physics and Technology of Dopant-Defect Interactions, H-J.L. Gossmann, T.E. Haynes, M.E. Law, A.N. Larsen, S. Odanaka, 1999, ISBN: 1-55899-475-0

Volume 569— *In Situ* Process Diagnostics and Modelling, O. Auciello, A.R. Krauss, E.A. Irene, J.A. Schultz, 1999, ISBN: 1-55899-476-9

Volume 570— Epitaxial Growth, A-L. Barabási, M. Krishnamurthy, F. Liu, T.P. Pearsall, 1999, ISBN: 1-55899-477-7

Volume 571— Semiconductor Quantum Dots, S.C. Moss, D. Ila, H.W.H. Lee, D.J. Norris, 2000, ISBN: 1-55899-478-5

Volume 572— Wide-Bandgap Semiconductors for High-Power, High-Frequency and High-Temperature Applications—1999, S.C. Binari, A.A. Burk, M.R. Melloch, C. Nguyen, 1999, ISBN: 1-55899-479-3

Volume 573— Compound Semiconductor Surface Passivation and Novel Device Processing, H. Hasegawa, M. Hong, Z.H. Lu, S.J. Pearton, 1999, ISBN: 1-55899-480-7

Volume 574— Multicomponent Oxide Films for Electronics, M.E. Hawley, D.H.A. Blank, C-B. Eom, D.G. Schlom, S.K. Streiffer, 1999, ISBN: 1-55899-481-5

Volume 575— New Materials for Batteries and Fuel Cells, D.H. Doughty, L.F. Nazar, M. Arakawa, H-P. Brack, K. Naoi, 2000, ISBN: 1-55899-482-3

Volume 576— Organic/Inorganic Hybrid Materials II, L.C. Klein, L.F. Francis, M.R. DeGuire, J.E. Mark, 1999, ISBN: 1-55899-483-1

Volume 577— Advanced Hard and Soft Magnetic Materials, M. Coey, L.H. Lewis, B-M. Ma, T. Schrefl, L. Schultz, J. Fidler, V.G. Harris, R. Hasegawa, A. Inoue, M.E. McHenry, 1999, ISBN: 1-55899-485-8

Volume 578— Multiscale Phenomena in Materials—Experiments and Modeling, D.H. Lassila, I.M. Robertson, R. Phillips, B. Devincre, 2000, ISBN: 1-55899-486-6

Volume 579— The Optical Properties of Materials, J.R. Chelikowsky, S.G. Louie, G. Martinez, E.L. Shirley, 2000, ISBN: 1-55899-487-4

Volume 580— Nucleation and Growth Processes in Materials, A. Gonis, P.E.A. Turchi, A.J. Ardell, 2000, ISBN: 1-55899-488-2

Volume 581— Nanophase and Nanocomposite Materials III, S. Komarneni, J.C. Parker, H. Hahn, 2000, ISBN: 1-55899-489-0

Volume 582— Molecular Electronics, S.T. Pantelides, M.A. Reed, J. Murday, A. Aviram, 2000, ISBN: 1-55899-490-4

Materials Research Society Symposium Proceedings

Volume 583— Self-Organized Processes in Semiconductor Alloys, A. Mascarenhas, B. Joyce, T. Suzuki, D. Follstaedt, 2000, ISBN: 1-55899-491-2
Volume 584— Materials Issues and Modeling for Device Nanofabrication, L. Merhari, L.T. Wille, K. Gonsalves, M.F. Gyure, S. Matsui, L.J. Whitman, 2000, ISBN: 1-55899-492-0
Volume 585— Fundamental Mechanisms of Low-Energy-Beam-Modified Surface Growth and Processing, S. Moss, E.H. Chason, B.H. Cooper, T. Diaz de la Rubia, J.M.E. Harper, R. Murti, 2000, ISBN: 1-55899-493-9
Volume 586— Interfacial Engineering for Optimized Properties II, C.B. Carter, E.L. Hall, C.L. Briant, S. Nutt, 2000, ISBN: 1-55899-494-7
Volume 587— Substrate Engineering—Paving the Way to Epitaxy, D.P. Norton, D.G. Schlom, N. Newman, D.H. Matthiesen, 2000, ISBN: 1-55899-495-5
Volume 588— Optical Microstructural Characterization of Semiconductors, J. Piqueras, T. Sekiguchi, M.S. Unlu, N.M. Kalkhoran, 2000, ISBN: 1-55899-496-3
Volume 589— Advances in Materials Problem Solving with the Electron Microscope, J. Bentley, U. Dahmen, C. Allen, I. Petrov, 2000, ISBN: 1-55899-497-1
Volume 590— Applications of Synchrotron Radiation Techniques to Materials Science V, S.R. Stock, D.L. Perry, S.M. Mini, 2000, ISBN: 1-55899-498-X
Volume 591— Nondestructive Methods for Materials Characterization, T. Matikas, N. Meyendorf, G. Baaklini, R. Gilmore, 2000, ISBN: 1-55899-499-8
Volume 592— Structure and Electronic Properties of Ultrathin Dielectric Films on Silicon and Related Structures, H.J. von Bardeleben, D.A. Buchanan, A.H. Edwards, T. Hattori, 2000, ISBN: 1-55899-500-5
Volume 593— Amorphous and Nanostructured Carbon, J. Robertson, J.P. Sullivan, O. Zhou, T.B. Allen, B.F. Coll, 2000, ISBN: 1-55899-501-3
Volume 594— Thin Films—Stresses and Mechanical Properties VIII, R. Vinci, O. Kraft, N. Moody, P. Besser, E. Shaffer II, 2000, ISBN: 1-55899-502-1
Volume 595— GaN and Related Alloys—1999, R. Feenstra, T. Myers, M.S. Shur, H. Amano, 2000, ISBN: 1-55899-503-X
Volume 596— Ferroelectric Thin Films VIII, R.W. Schwartz, S.R. Summerfelt, P.C. McIntyre, Y. Miyasaka, D. Wouters, 2000, ISBN: 1-55899-504-8
Volume 597— Thin Films for Optical Waveguide Devices and Materials for Optical Limiting, K. Nashimoto, B.W. Wessels, J. Shmulovich, A.K-Y. Jen, K. Lewis, R. Pachter, R. Sutherland, J. Perry, 2000, ISBN: 1-55899-505-6
Volume 598— Electrical, Optical, and Magnetic Properties of Organic Solid-State Materials V, S.P. Ermer, J.R. Reynolds, J.W. Perry, A.K-Y. Jen, Z. Bao, 2000, ISBN: 1-55899-506-4
Volume 599— Mineralization in Natural and Synthetic Biomaterials, P. Li, P. Calvert, R.J. Levy, T. Kokubo, C.R. Scheid, 2000, ISBN: 1-55899-507-2
Volume 600— Electroactive Polymers, Q.M. Zhang, T. Furukawa, Y. Bar-Cohen, J. Scheinbeim, 2000, ISBN: 1-55899-508-0
Volume 601— Superplasticity—Current Status and Future Potential, P.B. Berbon, M.Z. Berbon, T. Sakuma, T.G. Langdon, 2000, ISBN: 1-55899-509-9
Volume 602— Magnetoresistive Oxides and Related Materials, M. Rzchowski, M. Kawasaki, A.J. Millis, M. Rajeswari, S. von Molnár, 2000, ISBN: 1-55899-510-2
Volume 603— Materials Issues for Tunable RF and Microwave Devices, Q. Jia, F.A. Miranda, D.E. Oates, X. Xi, 2000, ISBN: 1-55899-511-0
Volume 604— Materials for Smart Systems III, M. Wun-Fogle, K. Uchino, Y. Ito, R. Gotthardt, 2000, ISBN: 1-55899-512-9
Volume 605— Materials Science of Microelectromechanical Systems (MEMS) Devices II, M.P. deBoer, A.H. Heuer, S.J. Jacobs, E. Peeters, 2000, ISBN: 1-55899-513-7
Volume 606— Chemical Processing of Dielectrics, Insulators and Electronic Ceramics, A.C. Jones, J. Veteran, S. Kaushal, D. Mullin, R. Cooper, 2000, ISBN: 1-55899-514-5
Volume 607— Infrared Applications of Semiconductors III, B.J.H. Stadler, M.O. Manasreh, I. Ferguson, Y-H. Zhang, 2000, ISBN: 1-55899-515-3
Volume 608— Scientific Basis for Nuclear Waste Management XXIII, R.W. Smith, D.W. Shoesmith, 2000, ISBN: 1-55899-516-1

Prior Materials Research Society Symposium Proceedings available by contacting Materials Research Society

MATERIALS RESEARCH SOCIETY SYMPOSIUM PROCEEDINGS

Fuel Cladding and Spent Nuclear Fuel

CLADDING EVALUATION IN THE YUCCA MOUNTAIN REPOSITORY PERFORMANCE ASSESSMENT

Eric R. Siegmann*, J. Kevin McCoy**, Robert Howard***
*Duke Engineering and Services, eric_siegmann@ymp.gov, ** Framatome Technologies,
*** TRW Environmental Safety Systems, (all) 1211 Town Center Drive, Las Vegas, NV 89144

ABSTRACT

The Yucca Mountain Project (YMP) 1998 Total System Performance Assessment Viability Assessment (TSPA-VA) analyzed the degradation of Zircaloy clad commercial fuel rods and the resulting exposure of the fuel in the event of a waste package failure. The cladding degradation mechanisms considered were damage before emplacement, mechanical failure from drift collapse, localized corrosion, general corrosion, delayed hydride cracking (DHC), hydride reorientation, creep rupture, and stress corrosion cracking (SCC). The potential for further cladding degradation due to cladding rupture as a result of fuel oxidation was also considered in the modeling effort. These models have been improved for use in future TSPAs.

The current cladding degradation model divides the analysis into two phases, cladding failure (perforation) and cladding unzipping (crack propagation caused by the expansion of UO_2 fuel after reaction with water). Cladding failure occurs during reactor operation, from creep strain failure during high temperature periods in dry storage or in the early periods in the repository, or localized corrosion. After a Waste Package (WP) containing spent nuclear fuel in the repository fails, moisture is assumed to enter the waste package and the failed cladding starts to unzip (tear open) from the formation of secondary uranium phases. This slowly exposes the fuel. In addition, the inventory of fission products located in the gap between the cladding and fuel pellet is rapidly released. The cladding model limits the amount of fuel that is exposed to moisture and becomes available for dissolution. As a result, the doses to the affected population are reduced (factor of 20 to 50 in TSPA-VA) from the case where cladding is not considered.

INTRODUCTION

Earlier studies have evaluated cladding degradation under repository conditions [1-3] and dry storage [4-5] conditions which are similar to early repository conditions. Experiments also measured the releases from damaged cladding. [6-7] The Yucca Mountain Project (YMP) 1998 Total System Performance Assessment Viability Assessment (TSPA-VA) included cladding degradation as part of the fuel degradation modeling. TSPA-1995, a previous analysis of repository performance, neglected the presence of cladding, as did most earlier PAs. When cladding was neglected, all the fuel in the waste package (WP) was considered available for dissolution at the speed of the intrinsic fuel dissolution rate. For some radionuclides, solubility limits were reached which controlled the rate of those radionuclides' leaving the WP. In the current TSPA model the cladding is considered an integral part of the waste form.

CLADDING MODEL

The cladding model summarizes numerous studies of cladding degradation and is incorporated into the TSPA computer model. The model describes two phases, cladding perforation and cladding unzipping. Cladding perforation is the formation of small cracks or holes in the cladding from various sources ranging from failures during reactor operation to cladding creep rupture during repository storage. Perforation permits the fuel inside the cladding

Mat. Res. Soc. Symp. Proc. Vol. 608

to begin to react with the moisture or air and leads to the cladding unzipping phase. In the unzipping phase, the cladding is torn open by the formation of secondary mineral phases on the fuel, and the radionuclides are available for release. The various components of the model are discussed below.

Cladding Condition as Received

The initial cladding condition analysis describes the condition of the commercial nuclear fuel as it is expected to be received at the YMP site. This analysis generates the initial boundary condition for the subsequent analysis of degradation of the cladding in the repository. It also evaluates the fraction of fuel rods that are perforated before emplacement and are immediately available for cladding unzipping when the WP fails. Earlier studies of cladding initial conditions have been performed [2,4,8]. The TSPA-VA used a single value for these initiating conditions but statistical distributions have since been developed.

The cladding degradation model is based on the Westinghouse 17 × 17 rod fuel design. This design represents over 30% of the PWR fuel discharged to date and also has the thinnest Zircaloy cladding. It is assumed that the BWR cladding degrades in a similar manner. This is conservative since BWR cladding is thicker and is discharged with lower burnups and stresses. In addition, most BWR assemblies are enclosed in flow channels (sheet metal boxes) which offer additional protection.

Starting with a distribution of PWR fuel burnups that are anticipated for storage at YMP, this model develops distributions for various cladding properties. Table 1 summarizes these distributions and includes the mean and upper 5% values.

Table 1 Model Results of Expected Fuel Stream into YMP

Property	**Mean Value**	**Upper 5% Value**
Burnup	44.1 MWd/kgU	63.3 MWd/kgU
Internal Pressure	4.8 MPa	7.3 MPa
Oxide Thickness	54 μm	112 μm
Peak Hydride Content	358 ppm	738 ppm
Crack Size	19 μm	57 μm
Stress (27°C)	38.4 MPa	61.8 MPa
Stress Intensity Factor, K_I	0.47 MPa-m$^{0.5}$	1.08 MPa-m$^{0.5}$

A distribution for the fraction of cladding within a WP that failed as a result of reactor operation was developed from the fraction of rods failed as a function of calendar years by assuming that the fuel assemblies are loaded into WPs in their order of discharge from the reactor. This loading sequence tends to place fuel with high failure rates (BWR fuel in 1970, also 1973-1976, and PWR fuel in 1972, 1983, and 1989) into consecutive WPs and produces larger variations in rod failure fractions than would be expected with thermal blending. A factor of four uncertainty was applied to represent the uncertainty in rod failure data to address incipient failures of the surrounding four rods in the square array assembly. The creep failure analysis included rod failure from dry storage and transportation using temperature profiles starting at 350°C. This analysis shows that a small fraction of the fuel with high stresses would fail if exposed to design basis storage and shipping temperatures. Table 2 gives the calculated percentage of rods that have failed cladding at emplacement. These fuel rods will undergo cladding unzipping and fuel dissolution when the WP fails.

Table 2 Model Prediction of Percent and Cause of Rods failed in a WP

Rod Failure Mode	Percent of Rods Failed/WP
Reactor Operation incl. Incipient Failures	0.109 (Range: 0.0 to 21.1)
Pool Storage	0.0
Dry Storage	0.045
Dry Storage & Transportation, Creep*	0.46 (Range 0.1 to 4.9)
Dry Storage & Transportation, DHC	0.0
Transportation (Vibration, Impact)	0.01
Fuel Handling	(Included in above)
Total	0.62 (Range: 0.16 to 26)

* Matsuo's creep correlation used.

Localized Corrosion

Corrosion of zirconium has been observed in fluoride-containing environments. Since fluoride is present in Yucca Mountain groundwater, fluoride corrosion may occur in waste packages. Two scenarios for fluoride corrosion have been considered. In the first (bathtub scenario), the waste package is full of water, and fluoride ions are transported to the cladding by aqueous diffusion. In the second (flow-through scenario), water enters the waste package through a breach on the top and drips out through a breach on the bottom. These two scenarios represent extremes of the rate of drainage.

The flow-through scenario is the more severe of the two. In this scenario, fluoride can be rapidly transported through the waste package by advection, whereas in the bathtub scenario it is transported by diffusion, which is a comparatively slow mechanism. In the flow-through scenario, advective flow is directed downward by gravity, so fluoride attack can be localized on a relatively small area of cladding. In contrast, diffusion does not have a preferred direction, so the fluoride can be transported to a large volume of the waste package in the bathtub scenario. Spreading the fluoride over a larger area of cladding means that more fluoride will be consumed in breaching each fuel rod. Since the flow-through scenario is more severe, the bathtub scenario was not considered further.

A bounding approach has been used to describe the flow-through scenario. It might be expected that the corrosion of zirconium is sufficiently slow and the flow of groundwater through the waste package is sufficiently fast that some fluoride will simply flow through the waste package without reacting. Credit has not been taken for this loss of fluoride. Instead, it is assumed that corrosion of the cladding is limited by the supply of fluoride.

In determining the amount of fluoride that is necessary to breach a fuel rod, it is assumed that fluoride removes all the cladding from a 10-mm length of the fuel rod by reacting to form ZrF_4. The as-manufactured thickness of the cladding may be used because, although some of the zirconium may be oxidized, the zirconium atoms remain in the products of corrosion. Fluoride attack is assumed to completely degrade one fuel rod before degradation begins on another rod. This assumption is conservative because rods breach as soon as enough fluoride is available; there is no delay in breaching one rod because fluoride is being diverted to start degrading another.

The resulting model is that the fraction of fuel rods failed by fluoride corrosion starts at zero when the waste package is breached. After breach, the fraction failed is proportional to the volume of water that has entered the package, reaching one when 2400 m^3 of water has entered the waste package. An alternative description is that the fraction of fuel rods that fail in a given year is the volume of water that enters the waste package during that year divided by 2400 m^3. Upper and lower limits are 10 times and 1/10 of the best estimate rate to represent the uncertainties in this model. This analysis makes the rod failure fraction linearly dependent on the water ingression rate (% failed = 0.0413 × m^3 water entering the WP). The water ingression into the WP increases with time as additional patches open. Rod failure rate also depends on the location of the WP group because of different drip rates apply in different repository regions. As an example, with 50 liters/year of J-13 water (2.2 ppm fluoride) entering the WP, 20% of the rods would fail by fluoride corrosion in 10,000 years.

Creep Failure

Repository design features such as backfill, drip shields, or thermal loading affect the fuel temperature. A statistical distribution of rod properties has been developed so that creep failure is included in the model. In the creep analysis, the rods were exposed to a temperature history that includes 20 years of dry storage starting at 350°C, three weeks of transportation at 350°C, and then a temperature history for the repository. The temperature profile within a WP is handled by considering rods in six zones across the WP. Matsuo's creep correlation[10] was used although more recent analyses using Murty's correlation[3] gave similar results. The strain failure criteria was a distribution based on eighteen tests of irradiated cladding[11] with a mean failure strain of 3.3%, and a range from 0.4% to 11.7%. Figure 1 gives the fraction of rods failed by strain as a function of WP surface temperature. It shows a high activation energy for creep and suggests a basis for the cladding temperature limit of 350°C, since the peak cladding temperature is about 50°C to 60°C above the WP surface temperature. The plateau on the left side of Figure 1 gives the fraction of rods that have failed during dry storage and transportation. Using Murty's creep correlation would increase the mean value from 0.46% to 2%.

Mechanical Damage

Seismic failures (fuel failure within an intact WP from seismic motion) and mechanical loading from a rubble overburden was analyzed. The seismic analysis showed that the rods failed only from very severe earthquakes (once per million year events), but then most of the rods would fail. Therefore the seismic failures are treated as disruptive events, and when such an earthquake occurs, all cladding is failed and available for unzipping. The analysis of mechanical loading from a rubble bed (after drift collapse and WP degradation) showed that the rods would fail under these conditions. The current waste package design is predicted to offer structural protection for hundreds of thousands of years and therefore, failure from this mechanism is not included. Based on the results of the seismic analysis, damage from rock fall is also unlikely.

Other Failure Mechanisms

A review of the various hydride degradation mechanisms was performed. Delayed Hydride Cracking (DHC) of existing cracks was analyzed using the distribution of stresses and crack sizes summarized in Table 1. Stress intensity factors with a mean of 0.47 MPa-$m^{0.5}$ (range 0.002 to 2.7 MPa-$m^{0.5}$) were calculated; these are below the threshold stress intensity factors, which are in the range of 5 to 12 MPa-m$^{0.5}$. Therefore, crack propagation by DHC is not expected. These stress intensities are also below those needed to produce SCC in all but the highest stressed rods. SCC is addressed in a later analysis. Hydride reorientation has not been modeled.

Cladding Unzipping and Fuel Dissolution

In TSPA-VA, the fuel in rods that were failed before emplacement was assumed to be completely exposed for dissolution while the fuel in exposed patches was available for dissolution but the remaining ends of the rods were not. This model was not necessarily conservative and is being upgraded. Fuel rods with perforated cladding are expected to remain intact until the WP fails and permits air and moisture to enter. While the humidity is low, dry unzipping could occur. Since the WP is expected to last for at least 200 years, the fuel temperatures will be too low for dry unzipping (fuel conversion to U_3O_8) to occur. Wet unzipping is modeled to start at WP failure. The fuel matrix is dissolved at the intrinsic dissolution rate and precipitates locally as metaschoepite. This secondary phase isolates most of the fuel from the moisture but the fuel in the torn cladding region continues to react, increasing volume, and forcing the tear further along the cladding. This reaction region is cone shaped and propagates along the rod at approximately 40 times the intrinsic dissolution rate. It is assumed that the perforation is in the center of the rod. This maximizes the release rate. Figure 2 gives the time to unzip a rod as a function of temperature. The unzipping time is also a function of local chemistry and pH. In addition, the gap inventory is instantly released when the cladding is perforated.

Figure 1. Creep Failures vs. WP Temperature

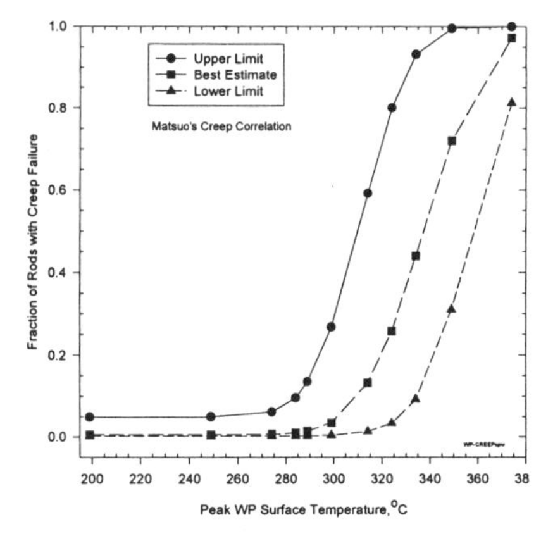

Figure 2. Rod Unzip Times vs. Temperature

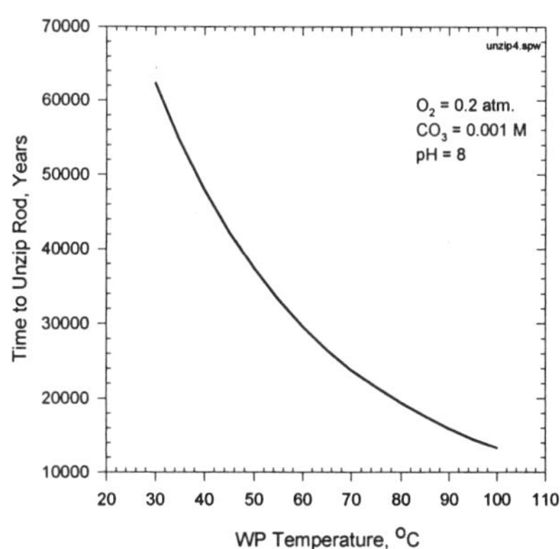

RESULTS AND CONCLUSIONS

Earlier TSPAs modeled the waste form as bare UO_2 which was available for dissolution at the intrinsic dissolution rate. Water in the WP quickly became saturated with many of the radionuclides, limiting their release rate. In the current TSPA, cladding is modeled as part of the waste form and limited the amount of fuel available at any time to dissolve. The major components of cladding perforation were failure in reactor operation, creep failure, and localized corrosion. The cladding then unzips from the production of secondary uranium phases. The cladding model limits the amount of fuel that is exposed to moisture and becomes available for dissolution. As a result, the doses to the affected population are reduced (factor of 20 to 50 in TSPA-VA) from the case where cladding is not considered.

ACKNOWLEDGEMENTS

The authors are grateful to Te-Lin Yau and Michael G. Bale for their localized corrosion studies, Hee M. Chung for the hydride analysis, William J. O'Connell for his unzipping analysis, and Steven A. Steward for the intrinsic dissolution modeling.

REFERENCES

1. T.M. Ahn, G.A. Cragnolino, K.S. Chan, and N. Sridhar, *Scientific Bases for Cladding Credit in the High-Level Waste Management at the Proposed Yucca Mountain Repository*, in Scientific Basis for Nuclear Waste Management XXII, edited by J. Lee and D. Wronkiewicz, (Mater. Res. Soc. Proc. **556**, Warrendale, PA 15086-7576, 1999).

2. S. Cohen & Associates, *Effectiveness of Fuel Rod Cladding as an Engineered Barrier in the Yucca Mountain Repository*, S. Cohen & Associates, McLean, Virginia, 1999.

3. P.J. Henningson, Cl*adding Integrity Under Long Term Disposal*, Doc. ID: 51-1267509-00, Framatome Technologies, Lynchburg, VA, 1998.

4. M.E. Cunningham, E.P. Simonen, R.T. Allemann, I.S. Levy, and R.F. Hazelton, *Control of Degradation of Spent LWR Fuel During Dry Storage in an Inert Atmosphere,* PNL-6364, Pacific Northwest Laboratory, Richland, Washington 1987.

5. M. Peehs, *Assessment of Dry Storage Performance of Spent LWR Fuel Assemblies with Increasing Burn-Up.* Erlangen, Germany: Siemens KWU-NBT. Co-ordinated Research Program (CRP) on Spent Fuel Performance Assessment and Research (SPAR), First RCM held in Washington DC-USA, April 20-24, 1998.

6. C.N. Wilson, *Results from Cycles 1 and 2 of NNWSI Series 2 Spent Fuel Dissolution Tests*, HEDL-TME-85-22, Richland, Washington: Westinghouse Hanford Company, 1987.

7. C.N. Wilson, *Results from NNWSI Series 3 Spent Fuel Dissolution Tests,* PNL-7170, Richland, Washington: Pacific Northwest Laboratory, 1990.

8. T.L. Sanders, K.D. Seager, Y.R. Rashid, P.R. Barrett, A.P. Malinauskas, R.E. Einziger, H. Jordan, T.A. Duffey, S.H. Sutherland, and P.C. Reardon, *A Method for Determining the Spent-Fuel Contribution to Transport Cask Containment Requirements,* SAND90-2406, Albuquerque, New Mexico: Sandia National Laboratories, 1992.

9. E. Hillner; D.G. Franklin, ; and J.D. Smee, . *The Corrosion of Zircaloy-Clad Fuel Assemblies in a Geologic Repository Environment.* WAPD-T-3173, West Mifflin, Pennsylvania: Bettis Atomic Power Laboratory. 1998

10. Y. Matsuo, "Thermal Creep of Zircaloy-4 Cladding Under Internal Pressure." *Journal of Nuclear Science and Technology, 24 (*2), 111-119. Tokyo, Japan: Atomic Energy Society of Japan, 1987.

11. H.M. Chung, F.L. Yaggee, and T.F. Kassner, "Fracture Behavior and Microstructural Characteristics of Irradiated Zircaloy Cladding." *Special Technical Publication, 0* (939) 775. Philadelphia, Pennsylvania: American Society for Testing and Materials, 1987.

Long Term Creep Behavior of Spent Fuel Cladding for Storage and Disposal

T. Bredel*, C. Cappelaere*, R. Limon*, G. Pinte*, P. Bouffioux**
*CEA/DRN/DMT/SEMI CE Saclay F91191 Gif sur Yvette Cedex, France
** EDF/DRD/ Pôle Industrie -Les Renardières – F77818 Moret sur Loing, France

ABSTRACT

In the framework of the 1991's French Law which defines the 3 major research lines to manage of nuclear wastes, CEA (Commissariat à l'Energie Atomique) is supporting a wide R&D program dealing with the long term behavior of spent fuel in various boundary conditions representative of interim storage and geological disposal. One major issue concerns the potential evolution with time of the thermo-mechanical properties of the irradiated cladding, the question of its integrity and whether the cladding can be considered as a first confinement barrier for radionuclides. The answer will strongly influence the design and the safety analyses of the interim storage and potentially the first stage of a geological disposal.
After irradiation, the thermo-mechanical properties of the spent fuel cladding are altered compared to those of the non irradiated material : presence of numerous irradiation defects due to the high irradiation, presence of external zirconia layers and hydrogen within the Zircaloy due to the external corrosion. Furthermore, the cladding is submitted to a relative high internal pressure field which is related to the production and release in the free volumes of fission gases and helium. Since the cladding is expected to undergo a relative high temperature field (~300 - 400°C), long term creep is expected to become a relevant deformation mechanism which can potentially lead to a breaching of the cladding.
In order to deal with this strategic issue, experimentation on irradiated samples as well as modelling work are in progress in CEA with the support of EDF and FRAMATOME. The ambitious objectives are to define and qualify the long term mechanical properties of the cladding, in particular a long term creep law and an adapted breaching criterion. Long term creep properties are studied through a stepwise approach from short term (few days) and high stress field experiments to long term (few years) and low stress field experiments. The validation of the extrapolation is ensured by complete metallurgical characterizations (HRTEM, XRD, H_2 content) performed before and after deformation. The first step of this work, medium term (~ 1 month) creep experiments on irradiated cladding will be presented.

INTRODUCTION

In the framework of the orientation law of 1991 which defines three major lines for the waste management, the CEA and EDF have implemented a wide program dealing with the long term behavior of spent fuel in various boundary conditions representative of interim storage and geological disposal.
For the long term interim storage studies, it has to be pointed out that this phase has to be considered as a provisory mean to manage long term nuclear waste while political decisions are been made [1,2]. It is therefore necessary to study the long term behavior of the cladding

Mat. Res. Soc. Symp. Proc. Vol. 608 © 2000 Materials Research Society

in order to determine if it may be considered as the primary confinement barrier for radionuclides, for how long and in which conditions.
After irradiation, the thermomechanical properties of the spent fuel cladding are altered in comparison to those of the non irradiated material : the presence of numerous irradiation defects due the high irradiation fields can be observed as well as the presence of external zirconia layers and hydrogen within the Zircaloy due to the external corrosion. Furthermore, the cladding is submitted to a relatively high internal pressure field which is related to the production and release in the free volumes of fission gases and helium. Since the cladding is expected to undergo a relatively high temperature field (~300 – 400°C), in dry storage conditions long term creep is expected to become a relevant deformation mechanism which can potentially lead to a breaching of the cladding. The prediction of the creep behavior of the irradiated cladding necessitates the establishment of a long term creep law and an adapted breaching criterion that can be extrapolated to the long term interim storage conditions.

STATE OF THE ART

At the CEA, the thermal creep of the cladding has been mostly studied in the temperature and stress ranges representative of reactor service conditions.
The following model has been developed by Soniak et al. [3,4] for the following conditions :

- Short term thermal creep on irradiated and unirradiated material for test duration from 1 h to 140 h,
- temperatures between 350 and 400°C for unirradiated materials and between 350 and 380°C for irradiated ones,
- stress between 100 and 445 MPa for unirradiated materials and between 310 and 550 MPa for irradiated ones,
- fluence between 0 and 10^{26} n.m-².

A new type of formulation has been developed for creep under constant temperature and stress conditions :

$\varepsilon = a \log(1+b(\exp(ct)-1))$

For general conditions with the strain hardening hypothesis, the following law of viscoplasticity is obtained

$$\frac{d\varepsilon}{dt} = v_s + (v_p - v_s)\exp\left(-\frac{\varepsilon}{\varepsilon_0}\right) \qquad (1)$$

In which v_p and v_s are functions of temperature, stress and fluence correspond relatively to the primary and the secondary strain rates and are linked to the a, b and c parameters by the following relationships :

$v_p = abc$ and $v_s = ac$

However, the expected duration of storage is much larger that the time domain in which the creep laws are known. Moreover, the temperature range covers only the first stage of the storage, and the lowest stress studied is about 2.5 times higher than that expected during storage. Finally, the open literature provided little information.

PROGRAM

In order to deal with this issue, a wide program of creep tests on irradiated samples associated with extensive metallurgical characterizations is in progress at CEA with the support of EDF and FRAMATOME. These creep tests are performed for more severe stress conditions than those estimated during the interim storage in order to obtain significant strain levels and even ruptures. The main objective is to establish a deformation law and a new rupture criterion that will be recommended for the long term interim storage.

CREEP TEST DESCRIPTION

The samples are part of a cladding tube, the fuel of which has been previously removed chemically. The minimal sample length is 60 mm. This length has been determined by calculation and experimentally confirmed. It has been chosen in order to obtain the best compromise between the absence of end-effect perturbation and the minimum use of irradiated material. The pressure-tightness at the two ends is obtained with adapter plugs crimped at room temperature.

The zirconia layer is systematically removed by mechanical cleaning at the two ends of the tube. The risk of cracking due to the low ductility of the irradiated material can only be minimized by a soft cutting and a mechanical polishing of this cutting to remove the surface which prevents crack initiation. After this preparation, the samples are accurately measured (diameter and thickness) in order obtain reliable stress and strain.

The CEA creep test facility holds six independent furnaces implanted in the same lead cell. Three independent heating zones for each furnace ensure a thermal gradient less than 1°C around 25 mm on each side the sample medium part. The test sample is pressurized with argon in order to obtain the stress according to the test specification. The mean stress is chosen as $\sigma=\sigma_\theta-\sigma_r=PDm/2e$ where σ_θ and σ_r are respectively the diametral and radial stresses, Dm and e are respectively the mean diameter (at mid thickness) and the tube thickness.

During the tests, a computer records and displays the displacement (two probes by sample), the pressure (one gauge by sample), the sample temperature (one thermocouple by sample), the cell and under-cell temperature.

CREEP TEST RESULTS AND COMPARISON WITH MODEL

The creep tests have been performed on samples coming from two similar rods irradiated for four cycles in a French PWR and whose burnups are in the 47200 ± 500 MWd/tU. They have been cut at the second and third grid spans ensuring therefore comparable corrosion (of about 25 μm) and average hydride content of 200 ppm. This last point is particularly relevant in regard of the impact of hydrogen on the creep behavior [5]. By this way, the fluence and the hydrogen content influence on the creep behavior can be studied by performing different series of tests on samples cut at different spans and with different doses.

The creep tests have been performed during about 40 days in the temperature range 380-420°C and stress values of 150 and 200 MPa.

The Figures 1 and 2 show creep tests performed at 380°C, 400°C and 420°C for two stress values of 150 and 200 MPa respectively.

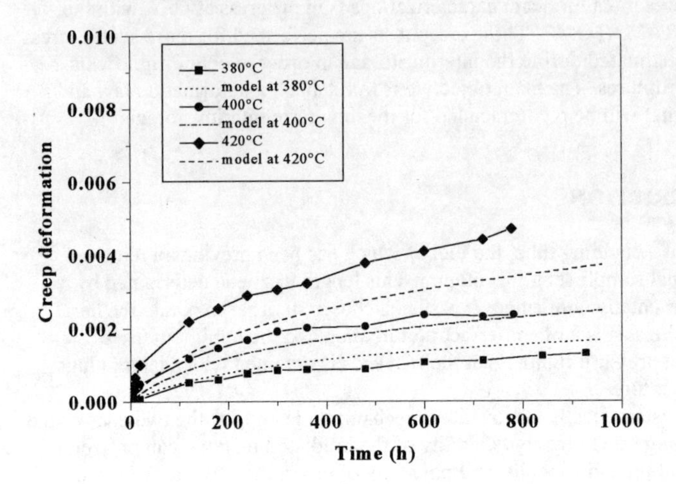

Figure 1: Influence of the temperature on creep deformation – σ = 150 MPa

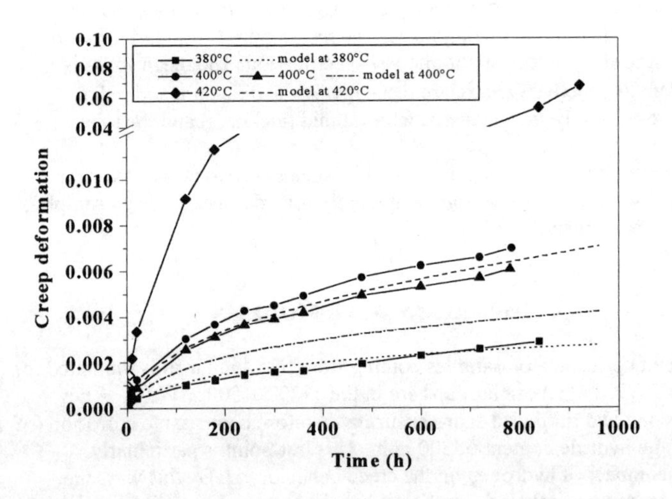

Figure 2: Influence of the temperature on creep deformation – σ = 200 MPa

The model [3] to which the tests are compared has been established on unirradiated and irradiated Zircaloy 4 cladding tubes for shorter duration and at higher stress creep tests. It clearly appears that the extrapolation of this model shows a tendency of underestimating the

creep deformation. This underestimation is more severe as the stress and the temperature increase. It might be due (at least partly) to the material recovery and annealing of irradiation defects during the test duration (up to 40 days) despite the temperature values (below 420°C). This hypothesis has to be confirmed by further metallurgical observations and microstructural characterizations.
These microstructural characterizations will also be used to understand the mechanisms that govern the creep behavior. They will also provide information on how parameters such as hydriding and irradiation affect the creep behavior and eventually lead to the development of a more physical modelling which could be used for reliably predicting the deformation.

CONCLUSION

The extrapolation of the model for high stress to the long term interim storage conditions drives to creep deformation previsions of the cladding very different for irradiated and unirradiated material. The model predicts that a creep deformation of 0.01 will be reached after 1 year for unirradiated material and after more than 300 years for irradiated materials.

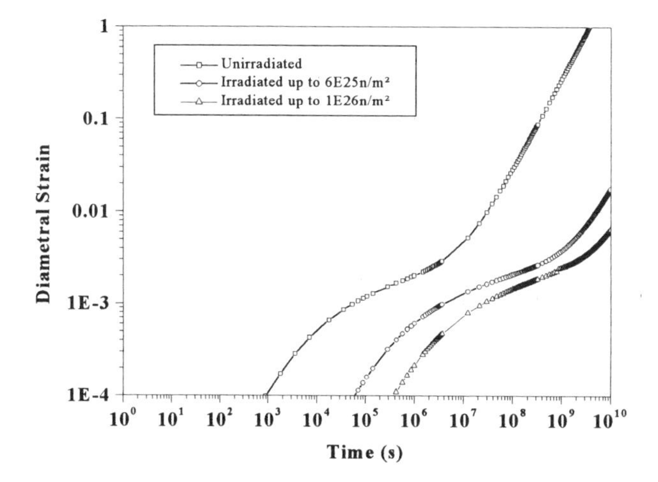

Figure 3 : Predicted thermal creep of Zircaloy 4 under internal pressure. T=350°C, σ=100 MPa (from [3]).

During interim storage, the material will have probably an intermediate behavior due to a partial recovery (at least at the beginning of the storage) and the hydrides that will probably slow down this phenomenon.
In order to take into account the parameters that may influence the creep deformation such as the irradiation temperature, the oxidation, the fluence..., it is necessary to enlarge our database with creep tests performed on samples coming from different spans and with different fluences. After reaching the limits of the model used, the next step is to realize

A more ambitious long term goal is to eventually develop a physical creep model which would predict the creep deformation more accurately and more convincingly than the empirical model.

ACKNOWLEDGEMENT

The authors want to thank their industrial partner FRAMATOME (France) for its financial support. V.Basini as well as many of our colleagues at the Laboratory have contributed to this work and their assistance is gratefully acknowledged.

REFERENCES

1. C.Poinssot, P.Toulhoat , J.M.Gras "R&S needs for accounting for the long term evolution of spent nuclear fuel in interim storage and geological disposal : objectives and programs".

2. P.Bernard, B.Barre, N.Camarcat, M.Boidron, B.Boullis, J.M.Cavedon, D.Iracane. "*Progress in R&D relative to high level and long-lived radioactive wastes management"*. Conference GLOBAL'99, Jackson Hole (USA), 29 aug-3 sept (1999).

3. A.Soniak et al., 12th ASTM Symposium on Zirconium in the Nuclear Industry (June 1998)

4 . P.Yvon, A. Soniak et al., Proceeding Enlarged Halden Programm Group Meeting, Loen, Norway, May 24-25, 1999.

5. P.Bouffioux, N.Rupa, 12th ASTM Symposium on Zirconium in the Nuclear Industry (June 1998)

HYDRIDE-RELATED DEGRADATION OF SPENT-FUEL CLADDING UNDER REPOSITORY CONDITIONS

H. M. Chung*, heechung@anl.gov
*Energy Technology Division, Argonne National Laboratory, Argonne, IL 60439

ABSTRACT

This report summarizes results of an analysis of hydride-related degradation of commercial spent-nuclear-fuel cladding under repository conditions. Based on applicable laboratory data on critical stress intensity obtained under isothermal conditions, occurrence of delayed hydride cracking from the inner-diameter side of cladding is concluded to be extremely unlikely. The key process for potential initiation of delayed hydride cracking at the outer-diameter side is long-term microstructural evolution near the localized regions of concentrated hydrides, i.e., nucleation, growth, and cracking of hydride blisters. Such locally concentrated hydrides are, however, limited to some high-burnup cladding only, and the potential for crack initiation at the outer-diameter side is expected to be insignificant for the majority of spent fuels. Some degree of hydride reorientation could occur in high-burnup spent-fuel cladding. However, even if hydride reorientation occurs, accompanying stress-rupture failure in spent fuel cladding is unlikely to occur.

INTRODUCTION

Current regulatory requirements on nuclear waste repositories specify that no more than one part in 100,000 of the radionuclides present 1000 years after closure of the repository may be released annually from the engineered barrier system for a period of 10,000 years [1]. In spent-fuel rods, structural integrity of the cladding has been identified as one of the key factors that limits the release of radionuclides and significantly influences the overall performance of a repository system. However, potential degradation of microstructure and mechanical properties of spent-fuel cladding, e.g., via hydride-related processes, can be profound. The hydride-related degradation of Zircaloy cladding can be classified into two distinct aspects, delayed hydride cracking (DHC) and hydride reorientation. Extensive hydride reorientation could exacerbates not only the susceptibility to DHC but also the potential for outright stress-rupture failure. Because hydrogen uptake and hydriding are by far more significant in PWR cladding than in BWR cladding, the present analysis is focused on degradation of Zircaloy-4-clad PWR spent-fuel cladding.

DELAYED HYDRIDE CRACKING

DHC in CANDU Pressure Tube and Laboratory Simulation Tests

Delayed hydride cracking is a proven field event for Zircaloy-2 and Zr-2.5%Nb CANDU pressure tubes. Majority of the investigations on DHC were, however, conducted on compact-tension specimens under laboratory conditions. Cracking in a hydrogen-charged precracked compact-tension specimen occurs in three stages: When stress intensity is lower than a threshold level, commonly referred to as K_{IH}, an incipient crack or flaw is stable and does not propagate. When stress intensity exceeds the critical level K_{IH} but is lower than the fracture toughness K_{IC}, a crack grows slowly at a stable rate. When stress intensity is greater than the fracture toughness K_{IC}, an unrestrained fast crack propagation occurs. Stress intensity is

Mat. Res. Soc. Symp. Proc. Vol. 608

determined by: $K_{IH} = \sigma(a\pi/2)^{0.5}$, where K_{IH} is stress intensity for DHC propagation (in MPa $m^{0.5}$), σ is stress (in MPa), and a is the crack size (in m). The stable crack growth for $K_{IH} < K < K_{IC}$ is attributed to DHC, a process that repeats a cycle in which hydrogen solutes diffuse to the crack tip, hydrides precipitate at or near the crack tip, and, subsequently, the hydrides or the metallic region near the hydrides crack under stress and thus leads to a slowly advancing crack. Coleman showed elegant examples of hydrides that form nearly parallel to an advancing crack [2]. Test conditions of most accelerated laboratory investigations of DHC are characterized by several aspects, i.e., isothermal test condition, unirradiated hydrided specimen, Mode I compact-tension specimen, little or no residual stress in the specimen, negligible driving force for hydrogen diffusion due to temperature gradient, and short duration of testing. Based on laboratory studies of this type, a threshold stress intensity K_{IH} of 5.5-8.0 MPa $m^{0.5}$ has been reported for CANDU pressure tube materials [3-5]. In contrast, Efsing and Petterson reported somewhat higher value of K_{IH} of 7.5-9.0 MPa $m^{0.5}$ for unirradiated hydrided Zircaloy-2 at ≈300°C that had 500-1000 wppm hydrogen and yield strength of 500-650 MPa [6]. Metallurgical factors that are relevant to field DHC (i.e., CANDU pressure tube failure), simulated crack growth tests in laboratory, and PWR spent-fuel cladding under repository conditions are summarized in Table 1.

Table 1. DHC-relavant factors in CANDU pressure tubes, compact-tension specimens, and spent-fuel cladding.

Factor	CANDU Reactor Pressure Tube	Compact Tension Specimen	Spent-Fuel Cladding in Repository
Material	Zr-2.5Nb or Zircaloy-2	Zr-2.5Nb, Zircaloys	Zircaloy-2, Zircaloy-4
DHC Proven	yes	yes	no
Time to Occur	many years	< several days	not known
Hydrogen Source	coolant and moisture	precharged hydrogen	H in cladding, limited
DHC Initiation at	OD hydride blisters	precrack tip	not known
Temperature Gradient	large (≈200°C), at pressure tube and calandria contact	negligible	small, not known
Stress Gradient	low	high	not known
Primary Driving Force for Hydrogen Migration	temperature gradient	stress gradient	stress gradient plus temperature gradient
Hydride Blistering	yes	no	not known
Reoriented hydrides	precursor to blistering	yes	not known
Concentrated Hydrides	yes, precursor to blistering	yes, near crack tip	some spots at high burnup
Concentrated Oxygen	OD region beneath oxide	absent or negligible	OD region beneath oxide
Applied Stress (MPa)	≈140	high	60-130 MPa
Residual Stress	some	negligible	high near OD-side oxide

The occurrence of DHC in a CANDU pressure tube cannot be predicted based on a threshold stress intensity K_{IH} of ≈5.5 MPa $m^{0.5}$. Crack initiation in failed CANDU pressure tubes occurred in fact not in association with fabrication-related flaws but in association with hydride blisters that formed during operation on the outer-diameter (OD) side of the tube [7-11]. That is, long-term microstructural evolution was the key in crack initiation in the failure CANDU pressure tubes, which can be divided into several processes: sagging of the horizontal hot pressure tube (maintained at ≈300°C), contact of the tube OD with the cold calandria tube (temperature ≈70°C), H migration to the cold contact spot, nucleation of a hydride blister, growth of the hydride blister, crack initiation in the blister, and crack growth in the blister and adjacent metal.

Potential for DHC Initiation at ID of Spent-Fuel Cladding in Repository

Most incipient cracks in spent-fuel cladding, produced as a result of pellet-cladding interaction or pellet-cladding mechanical interaction, are on the inner-diameter (ID) side. The potential for collecting H, concentrated hydride precipitation, and hydride blistering in the ID region is insignificant or negligible because the temperature of the metal is higher at the ID side than in midwall or the OD side, and hence the thermal driving force for H collection is either absent or negligible. Therefore, the conditions are similar to (or less deleterious than) those of the isothermal tests in laboratory, and the critical stress intensity factor of ≈5.5 MPa $m^{0.5}$ can be used to predict potential propagation of the incipient cracks. Oxide- or hydride-associated residual stress on the ID side is also small, and therefore, the maximum applied stress on this side will not exceed ≈130 MPa, even for a burnup as high as ≈60 MWd/kgU. Therefore, for an incipient crack on the ID side to propagate via DHC mechanism, the crack size must exceed ≈1000 μm, or greater than the thickness of PWR fuel cladding metal (i.e., 570-630 μm). Therefore, initiation and propagation of DHC on the ID side is not a concern under repository conditions.

Potential for DHC Initiation at OD of Spent-Fuel Cladding in Repository

The OD side of the cladding under repository conditions will be cooler than the ID side or the midwall; therefore, there could be some degree of driving force for H solutes to migrate over a long period of time from the hotter ID or midwall region to the metallic layer at the OD side, under either temperature or stress gradients, or both. The key process for potential DHC initiation at the OD side, then, is potential long-term microstructural evolution, i.e., nucleation, growth, and cracking of hydride blisters in the metallic region, especially on some types of localized "hydrogen-collecting spots." The "hydrogen-collecting" spots, characteristically colder and higher in stress, or both, could be locations near pellet/pellet boundaries, near pellet/pellet gaps, beneath a spacer grid, or beneath a spalled or radially cracked oxide layer. These spots are illustrated in Fig. 1. It is not clear if radial cracks in oxide, similar to those reported by Einziger and Kohli [12], could be produced as a result of long-term creep of the cladding, and if so, how they will potentially influence the susceptibilities to hydride blistering and blister cracking.

Evidence of concentrated hydrides in some high-burnup spent-fuel claddings has been reported near pellet/pellet interfaces [13,14], near pellet/pellet gaps [14,15], and beneath a spalled oxide layer [14,16]. However, all these evidences were observed only in PWR spent-fuel cladding fabricated from standard Zircaloy-4 (Sn content ≈1.5 wt.%) and operated to high burnup, e.g., >55 MWd/kgU. Oxidation and hydriding in high-Sn standard Zircaloy-4 cladding are well known to be high at high burnup. In contrast, the degree of oxidation and hydriding in present-day fuel claddings commonly used in light-water reactors, such as low-Sn Zircaloy-4, Zirlo, Zr-1Nb, optimized Zircaloy-4, and optimized Zircaloy-2, is relatively low even at high burnup. No evidence for concentrated hydrides has been reported for low or medium burnup. Therefore, formation of locally concentrated hydrides appears to be limited to a small fraction of spent-fuel cladding fabricated from standard Zircaloy-4 and operated to burnups higher than ≈55 MWd/kg. Consequently, although we cannot rule out or predict unequivocally at this time that a hydride blister forms and grows near such locally concentrated hydrides under repository conditions, the potential for initiation of DHC near such concentrated hydrides is predicted to be insignificant for the majority of commercial spent fuels.

HYDRIDE REORIENTATION

Hydride reorientation in spent fuel cladding under dry storage conditions has been investigated by Einziger and Kohli [12]. Hydride reorientation consists of dissolution of the normal circumferential hydrides as the temperature is increased higher than the heatup solubility limit, and precipitation of radial hydrides as the temperature is decreased below the cooldown solubility limit under stress. According to McMinn et al., the heatup and cooldown solubility limits differ significantly, the mechanism of which is not well understood [17]. The reorientation of hydrides is known to be strongly influenced by the stress [18-21] and texture [2] of the cladding. However, the effect of cooling rate has been reported to be insignificant [20]. Therefore, the minimum stress for reorientation applicable to spent irradiated cladding is of particular importance.

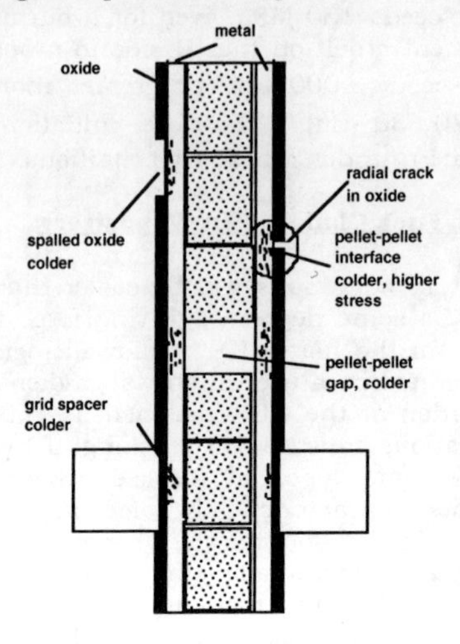

Fig. 1.
Illustration of potential spots for hydrogen concentration and hydride blister formation in spent-fuel cladding

Minimum Stress for Hydride Reorientation

The minimum stress for hydride orientation in unirradiated specimens has been reported to be ≈85-95 MPa [18-20]. In contrast, Hardie and Shanahan reported an unusually low threshold stress of only ≈35 MPa for Zr-2.5Nb pressure tube specimens [21]. However, the texture of their pressure tube specimen was entirely different from that of spent-fuel cladding, that is, the basal pole in the former material was ≈90° away from the radial direction, whereas the basal pole of the latter is ≈30° away from the radial direction. Because basal planes are nearly parallel to $\{107\}_{Zr}$ habit plane of hydride precipitation, radial hydrides in the former material are inherently easier to precipitate than in the latter material. Therefore, the minimum stress reported by Hardie and Shanahan [21] does not seem to be applicable to spent fuel-cladding. Einziger and Kohli observed reoriented (radial) hydrides in PWR spent fuel cladding (burnup ≈27-31 MWd/kgU) that was exposed to ≈323°C for up to ≈87 days and then cooled at a rate of ≈5°C/h under an applied internal stress of ≈143 MPa [12]. This level of stress exceeds significantly the minimum stress for hydride reorientation of ≈85-95 MPa that was observed for unirradiated hydrided specimens. In view of this information, reorientation of hydrides under repository conditions cannot be ruled out in some high-burnup spent-fuel cladding in which applied stress exceeds ≈90 MPa.

Failure Stress of Hydride-Reoriented Spent Fuel Cladding

From the standpoint of structural integrity, the threshold stress for stress-rupture failure of hydride-reoriented cladding is more important than the phenomenon of hydride reorientation itself. A data base for such stress, however, is not available except for the experiment of Einziger and Kohli [12]. In that experiment, the cladding remained intact under an applied stress of ≈143 MPa at ≈323°C for ≈87 days and during subsequent cooldown, despite the fact that extensive reorientation of hydrides occurred. The peak temperature is similar to that expected for spent-fuel cladding under repository conditions. Therefore, the experiment of Einziger and Kohli can be viewed as a conservative upper bound from the standpoint of the deleterious effect of hydride reorientation. Failure stress of spent-fuel cladding has been reported by Chung et al. [22] for burnup of 22-28 MWd/kgU and by Garde [23] and Smith et al. [13] for 41-64 MWd/kgU. Cladding specimens in these internal-pressurization burst tests contained few radial hydrides before test, although some high-burnup specimens may have contained localized radial hydrides, e.g., near pellet/pellet boundaries. Based on the information from these investigations, failure stress of the ≈30 MWd/kgU fuel cladding of Einziger and Kohli [12] is estimated to be ≈400 MPa at ≈323°C when hydride reorientation did not occur. Therefore, the failure stress of the hydride-reoriented cladding would have been somewhere between ≈143 MPa and ≈400 MPa. Even the lower limit of ≈143 MPa is still significantly greater than a conservative internal stress of ≈70-90 MPa that is estimated for ≈30 MWd/kg. Therefore, it can be concluded that even if hydride reorientation occurs in some medium- and high-burnup spent-fuel cladding under repository conditions, subsequent stress-rupture failure in the hydride-reoriented cladding is not likely to occur.

CONCLUSIONS

Most incipient cracks in spent-fuel cladding are located on the inner-diameter (ID) side. The driving force for hydrogen concentration and the potential for hydride blistering on the ID side are insignificant or negligible because the temperature on the ID side is higher than on the outer-diameter (OD) side or in midwall. Therefore, the condition is similar to that of isothermal laboratory simulation tests of delayed hydride cracking, and the critical stress intensity factor of ≈5.5 MPa $m^{0.5}$ can be used to predict potential propagation of incipient cracks at the inner diameter. Based on this value of critical stress intensity, crack size distribution, and maximum applied stress, delayed hydride cracking (DHC) from the ID side is unlikely to occur.

The OD side of the cladding under repository conditions will be somewhat cooler than the ID side or the midwall; therefore, there could be some degree of driving force for hydrogen solutes to migrate from the hotter ID and midwall regions to the metallic layer at the OD side, under either temperature or stress gradient, or both. The key process for potential DHC initiation at the OD side, then, is long-term microstructural evolution, i.e., nucleation, growth, and cracking of hydride blisters in the metallic region, especially on some types of localized hydrogen-collecting spots. The hydrogen-collecting spots, characteristically colder and higher in stress, or both, could be those near pellet-pellet boundaries, near pellet-pellet gaps, and beneath a spalled or radially cracked oxide layer. These local regions of concentrated hydrides are, however, limited to a small fraction of spent-fuel cladding fabricated from conventional Zircaloy-4 and operated to burnup >55 MWd/kg. Consequently, potential for initiation of DHC near such locally concentrated hydrides is expected to be insignificant for the majority of commercial spent fuels.

Texture plays an important role in reorientation of hydrides in Zr-base alloy and Zircaloy tubings. Unusually low threshold stress for hydride reorientation reported for some Zr-2.5Nb pressure tube specimens is probably not applicable to commercial spent-fuel cladding, because the texture of the pressure-tube specimens is entirely different from that of spent-fuel cladding. Some degree of hydride reorientation may occur in high-burnup spent-fuel cladding under repository conditions. However, even if hydride reorientation occurs in a limited fraction of spent-fuel cladding discharged after high-burnup operation, accompanying stress-rupture failure in the hydride-reoriented cladding is not expected to occur.

ACKNOWLEDGMENTS

This work was supported by the U.S. Department of Energy, Office of Civilian Radioactive Waste Management.

REFERENCES

1. Code of Federal Regulations 1977, Title 10 Part 60, Section 60.113, Article (a)(ii)(B), U.S. Government Printing Office, Washington DC, pp. 137.
2. Coleman, C. E. 1982, in *Zr in the Nuclear Industry: 5th Intnl. Symp.,* ASTM STP 754, D. G. Franklin, ed., ASTM, Philadelphia, p. 393.
3. Simpson, L. A.; and Puls, M. P. 1979, *Met. Trans. A,* 10A, 1093.
4. Puls, M. P.; Simpson, L. A.; and Dutton, R. 1982, in *Fracture Problems and Solutions in the Energy Industry,* Pergamon Press, New York, pp. 13-25.
5. Shi, S.-Q.; and Puls, M. P. 1996, in *Hydrogen Effects in Materials,* A. W. Thompson and N. R. Moody, eds., TMS, Warrendale, PA, p. 612.
6. Efsing, P.; and Petterson, K. 1996, in *Zr in the Nuclear Industry: 11th Intnl. Symp.,* ASTM STP 1295, E. R. Bradley and G. P. Sabol, eds., ASTM, Philadelphia, pp. 394.
7. Chow, C. K.; and Simpson, L. A. 1986, in *Case Histories Involving Fatigue and Fracture Mechanics,* ASTM STP 918, ASTM, p. 78.
8. Cheadle, B. A.; Coleman, C. E.; and Ambler, J. F. R. 1987, in *Zr in the Nuclear Industry: 7th Intnl. Symp.,* ASTM STP 939, R. B. Adamson and L. F. P. Van Swam, eds., ASTM, Philadelphia, pp. 224-240.
9. Leger, M.; Moan, G. D.; Wallace, A. C.; and Watson, N. J. 1989, in *Zr in the Nuclear Industry: 8th Intnl. Symp.,* ASTM STP 1023, L. F. P. van Swam and C. M. Eucken, eds., ASTM, Philadelphia, pp. 50-65.
10. Coleman, C. E.; Cheadle, B. A.; Causey, A. R.; Chow, C. K.; Davies, P. H.; McManus, M. D.; Rodgers, D.; Sagat, S.; van Drunen, G. 1989, ibid., pp. 35-49.
11. Moan, G. D.; Coleman, C. E.; Price, E. G.; Rodgers, D. K.; and Sagat, S. 1990, *Int. J. Pres. Vessel & Piping,* 43, 1-21.
12. Einziger, R. E.; and Kohli, R. 1984, *Nucl. Technol.* 67, 107-123.
13. Smith Jr., G. P.; Pirek, R. C.; Freeburn, H. R.; and Schrire, D. 1994, *The Evaluation and Demonstration of Methods for Improved Nuclear Fuel Utilization,* DOE/ET/34013-15, CEND-432, ABB Combustion Engineering, pp. 4-60 to 4-73.
14. Garde, A. M.; Smith, G. P.; and Pirek, R. C. 1996, in *Zr in the Nuclear Industry: 11th Intnl. Symp.,* ASTM STP 1295, E. R. Bradley and G. P. Sabol, eds., ASTM, Philadelphia, pp. 407-430.
15. Yang, R. L.; Ozer, O.; and Klepfer, H. H. 1991, in *Proc. Intnl. Topical Meeting on LWR Fuel Performance,* April 21-24, 1991, Avignon, France, ANS and ENS, pp. 258-271.
16. Guedeney, P.; Trotabas, M.; Boschiero, M.; Forat, C.; and Blanpain, P. 1991, in *Proc. Intnl. Topical Meeting on LWR Fuel Performance,* April 21-24, 1991, Avignon, France, ANS and ENS, pp. 627-638.
17. McMinn, A.; Darby, E. C.; and Schofield, J. S. 1998, in *Zr in the Nuclear Industry: 12th Intnl. Symp.,* June 15-18, 1998, Toronto, in press.
18. Marshall, R. P. 1967, *J. Nucl. Mater.* 24, 34-48.
19. Bai, J. B.; Ji, N.; Gilbon, D.; Prioul, C.; and Francois, D. 1994, *Met. and Mater. Trans. A,* 25A, 1199-1208.
20. Chan, K. S. 1996, *J. Nucl. Mater.* 227, 220-236.
21. Hardie, D.; and Shanahan, M. W. 1975, *J. Nucl. Mater.* 55, 1-13.
22. Chung, H. M.; Yaggee, F. L.; and Kassner, T. F. 1987, in *Zr in the Nuclear Industry: 7th Intnl. Symp.,* ASTM STP 939, R. B. Adamson and L. F. P. Van Swam, eds., ASTM, Philadelphia, pp. 775-801.
23. Garde, A. M. 1989, in *Zr in the Nuclear Industry: 8th Intnl. Symp.,* ASTM STP 1023, L. F. P. Van Swam and C. M. Eucken, eds., ASTM, Philadelphia, pp. 548-569.

ANALYSIS OF DRY STORAGE TEMPERATURE LIMITS FOR ZIRCALOY-CLAD SPENT NUCLEAR FUEL

T.A. Hayes*, R.S. Rosen**, M.E. Kassner*, ***, K.S. Vecchio*.
* Mechanical and Aerospace Engineering Department, Univ of California, San Diego, La Jolla, CA 92093-0411 USA
** Lawrence Livermore National Laboratory, Livermore, CA 94550 USA
*** Mechanical Engineering Department, Oregon State University, Corvallis, OR 97331 USA

ABSTRACT

Safe interim dry storage of spent nuclear fuel (SNF) must be maintained for a minimum of twenty years according to the Code of Federal Regulations. The most important variable that must be regulated by dry storage licensees in order to meet current safety standards is the temperature of the SNF. The two currently accepted models for defining the maximum allowable initial storage temperature for SNF are based on the diffusion controlled cavity growth (DCCG) failure mechanism proposed by Raj and Ashby. These models may not give conservative temperature limits. Some have suggested using a strain-based failure model to predict the maximum allowable temperatures, but we have shown that this is not applicable to SNF as long as DCCG is the assumed failure mechanism. Although the two accepted models are based on the same fundamental failure theory (DCCG), the researchers who developed the models made different assumptions, including selection of some of the most critical variables in the DCCG failure equation. These inconsistencies are discussed together with recommended modifications to the failure models based on more recent data.

INTRODUCTION

Interim dry storage of spent nuclear fuel (SNF) rods is of critical concern because of the shortage of wet storage capacity and delays in the availability of a permanent disposal repository. Safe dry storage must be maintained for a minimum of twenty years [1]. The NRC has approved two models [2,3] to determine the maximum initial temperature limit for SNF in dry storage that supposedly meet all safety criteria [1] and yield consistent temperature limits. Though these two models are based on the same fundamental failure theory, different assumptions have been made including the choice of values for material constants in the failure equation. This paper will discuss these inconsistencies as well as some of the shortcomings of the current models and suggest some modifications.

CURRENT MODELS

Currently, the maximum allowable initial temperatures for interim dry storage of SNF rods are determined using either the equations developed by Lawrence Livermore National Laboratory (LLNL) [2] or temperature limit curves developed by Pacific Northwest National Laboratory (PNNL) [3]. Both the PNNL and the LLNL models predict that cavitation failure under dry storage conditions may occur by diffusion controlled cavity growth (DCCG). Many investigators (e.g., [4]) have suggested that under conditions similar to dry storage, failure may not occur exclusively by DCCG, but rather by a coupled diffusion and power law creep mechanism. Without large far field strains, however, (as are absent in the case of zircaloy SNF [5,6]), cavity growth does not seem possible by power law creep. We therefore concluded, though tentatively, that a diffusion controlled cavity growth model is the most reasonable.

Mat. Res. Soc. Symp. Proc. Vol. 608 © 2000 Materials Research Society

LLNL used the DCCG analysis by Raj and Ashby [7] to predict the time to failure of zircaloy under dry storage conditions based on a 'limited damage' approach. The equations presented by LLNL can be used to calculate initial temperature limits for SNF for specified temperature decay profiles and stresses. PNNL predicted temperature limits using a fracture map to account for various fracture mechanisms predicted to be active over a relevant range of stresses and temperatures. For dry storage temperatures, the fracture map indicates that either DCCG (~0-110 MPa) or power law creep (~110-160 MPa) controls failure. The most relevant stresses for dry storage are less than 110 MPa, so both models predict DCCG controls failure for the bulk of SNF in dry storage.

Instead of DCCG, some have suggested (e.g., [5,8]) using a creep-strain limit approach where creep-strain is limited to 1%. This approach is based on the proposition that cavity growth and fracture only occur after significant plastic strain (much more than 1%). DCCG, however, is not strain dependent and it has been shown (e.g. [9, 10]) that cavities can nucleate and grow in various metals after very low plastic strains (much less than the strain to fracture).

Inconsistency between LLNL and PNNL models

The two models predict nearly consistent temperature limits for stresses most relevant to dry storage (from about 40-100 MPa). However, it turns out that the near coincidence of the model predictions results from a fortuitous combination of differing assumptions by PNNL and LLNL. First, PNNL (but not LLNL) used a 'recovery factor' in all failure equations. This factor is used to account for the recovery of some of the reduction in ductility of zirconium alloys from irradiation damage. Because the DCCG fracture model is *not* a function of strain, a reduced ductility would not directly affect the fracture time. It is not clear, therefore, whether a "recovery factor" should be used with DCCG. Additionally, LLNL chose a value of 0.15 (15%) "area fraction of decohesion" to correspond to failure based on estimated post-dry-storage SNF handling forces [11]. The PNNL model, however, incorporates the original form of the Raj and Ashby [7] failure equation which inherently assumes that failure occurs at an area fraction of decohesion of 0.50 (50%). Finally, PNNL and LLNL chose different values for the material constants used in the DCCG failure equation. These differences include atomic volume, Ω, grain boundary thickness, δ, cavity spacing, λ, grain boundary diffusion coefficient, D_o^{gb} and Q_{gb}, as well as other unspecified constants.

Temperature sensitivity of current models

Both models are very sensitive to changes in the temperature decay of the SNF. A calculated initial temperature thought to cause failure in 40 years essentially predicts failure in 5 years for the *same* initial temperature when using Equation 1, which PNNL suggests is reasonable. The temperature profile used by PNNL when comparing to the LLNL model is

$$T(K) = C \times [t(\text{months})]^{-0.282} \qquad (1)$$

where C can be adjusted to modify the initial temperature of the SNF. This predicts the temperature of the SNF as a function of time after removal from reactor.

The DCCG failure equation with this assumed temperature profile predicts that nearly all of the damage will occur in the first 5 years. If a power type temperature profile such as Equation 1 is the best choice, then both (PNNL and LLNL) models predict that if failures are going to occur, they will occur within the first 5 years in dry storage. Again, a failure according to the LLNL criteria implies that the material has experienced a reduced cross sectional area (reduced structural integrity), not necessarily a rupture. Thus, care must be exercised when using the DCCG failure equation to evaluate long storage times. The predicted maximum allowable temperature can be misleading because of this sensitivity, as illustrated in Figure 1. This figure shows that the 40 and 5 year failure lines are coincident and that both are very close to the 1 year failure line.

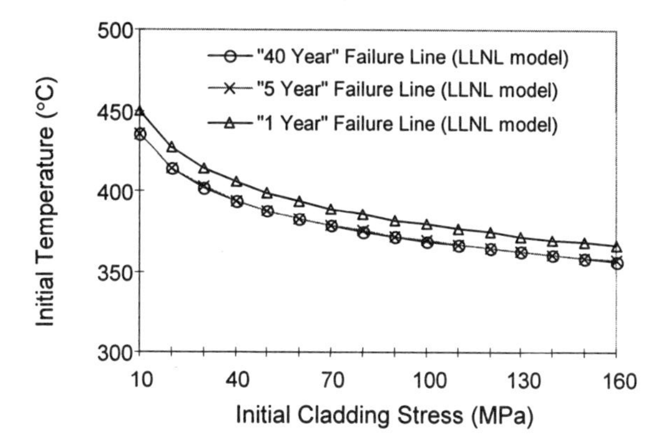

Figure 1. Comparison of 40, 5 and 1 year failure lines predicted using the LLNL model.

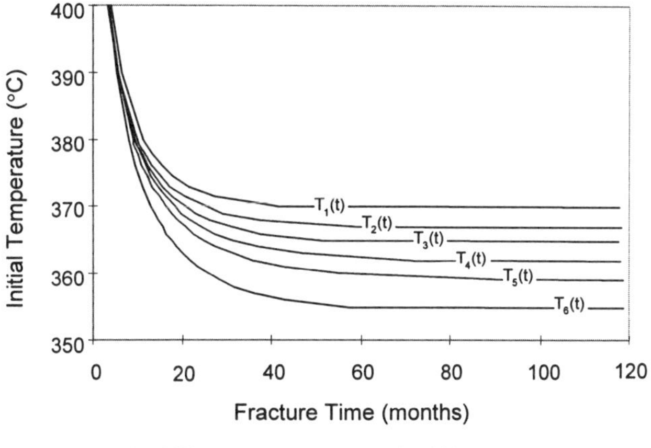

$T_1(t) = C1 \times 10^3\, t^{-0.282}$ (K), $T_2(t) = C2 \times 10^3\, t^{-0.26}$ (K), $T_3(t) = C3 \times 10^3\, t^{-0.24}$ (K)
$T_4(t) = C4 \times 10^3\, t^{-0.22}$ (K), $T_5(t) = C5 \times 10^3\, t^{-0.20}$ (K)

$$T_6(t) = \begin{cases} T_1(60) + [T_1(180) - T_1(60)]/2 - [T_1(180) - T_1(60)]/120\, t \text{ (K)} & t < 180 \\ T_1(180) \text{ (K)} & t \geq 180 \end{cases}$$

Figure 2. Fracture time as a function of initial storage temperature for various temperature decay functions for 5-year-old SNF with an initial cladding hoop stress of 100 MPa.

When using a temperature profile similar to Equation 1, small changes in the initial temperature result in large differences in the predicted failure time. Alternatively, large changes in the specified failure times result in insignificant changes in the maximum allowable initial temperature (e.g., failure times ranging from 5 years to infinity). Therefore an applicant who determines a maximum temperature for a predicted failure of 300 years has, at the same time, predicted failure after 5 years (which is *less* than the licensed period of the storage facility). Figure 2 illustrates how more conservative temperature decay profiles would affect the predicted failure time based on the LLNL model. The sensitivity of the model is reduced with each successively less concave temperature decay profile (1-5). The time over which 99% of the damage occurs similarly increases (from 5 years with profile 1 to 9 years with profile 5). This is because each successively less concave profile predicts more time at higher temperatures where the diffusion coefficients and stresses are higher and cavitation occurs more quickly. The linear profile is the most conservative. It predicts that more damage occurs (shorter fracture time) than any of the power decay profiles. It also decreases the sensitivity of the failure time to the initial temperature of the SNF compared to the temperature decay profile suggested by PNNL.

Although measured profiles seem to obey power-like temperature decay, the uncertainty involved in developing the temperature profile (see reference 12 discussion) and the fact that the predicted failure time is so sensitive to this profile suggest using a conservative profile. We therefore recommend using an accurate temperature profile measured under the proposed storage configuration or, if unavailable, using a conservative linear decay profile.

Critical material parameters

Grain boundary diffusion coefficient

Self-diffusion coefficients reported in the literature for zirconium (zircaloy not available) vary by more than four orders of magnitude (see [12] for references). Choosing a grain boundary diffusion coefficient based on these data, as PNNL and LLNL have done, is not straightforward. PNNL suggests an activation energy, Q_{gb}, of 175 kJ/mole and a value for D_o^{gb} of 3.89 x 10^{-6} m^2/sec (basis not reported). LLNL suggests using the activation energy, Q_{gb}, of 131 kJ/mole and D_o^{gb} of 5.9 x 10^{-6} m^2/sec reported by Garde et al. [13]. This yields a diffusion coefficient that is 3 to 5 orders of magnitude faster than PNNL's suggested D_{gb} at 400°C and 200°C, respectively. At the time these models were developed, no grain boundary diffusion studies had been reported for zirconium. The values chosen by LLNL and, presumably, by PNNL were based on the approximate relationship between the self-diffusion and grain boundary diffusion activation energies $Q_{gb}\approx 0.6Q_{sd}$ [14]. This is one serious limitation with these models as the more than four order of magnitude variation in reported values for the diffusion coefficient translates linearly to the predicted failure time. Since the time that these models were developed, however, grain boundary diffusion data have been reported. Vierregge and Herzig [15] measured the grain boundary diffusion coefficient of zirconium. This study gives D_{gb} as

$$D_{gb} = 4.2^{+5.9}_{-2.5} \times 10^{-13} \exp\left(\frac{-167 \pm 7\,\text{kJ/mole}}{RT}\right) \Big/ \delta \ \frac{m^2}{s} \qquad (2)$$

where δ is the grain boundary width. The predictions of Equation 2 are close to that suggested by LLNL. The error indicated in Equation 2, however, translates to a variation of more than 100°C in the maximum initial temperature when using Equation 1 for the temperature profile. Furthermore, it is not clear what effect irradiation may have on the grain boundary diffusion coefficient of zirconium alloys. Irradiation damage results in a higher concentration of vacancy type dislocation loops near grain boundaries [16,17] and also causes dispersion of Fe which could substantially increase the diffusion rate (see reference 12 discussion). The effect that alloying zirconium on grain boundary diffusion under conditions relevant to dry storage has not been established, however zircaloys are richer in iron than zirconium and iron has been shown to dramatically increase the self-diffusion coefficient of zirconium.

Cavity spacing

PNNL assumes a constant cavity spacing of 2.6 μm (basis not reported) while LLNL assumes a constant spacing of 10 μm based on data in reference 9. These two values may not be conservative. There is evidence that irradiation may lead to a cavity spacing of the order of 0.1-0.2 μm [17]. It may be reasonable, therefore, to assume a 0.1-0.2 μm cavity spacing.

ROD SURVEILLANCE TESTING

It has been suggested [5,18] that SNF stored under current dry storage conditions appears safe over extended periods of time based on rod surveillance testing. Post-storage integrity (a handling requirement after dry storage), however, has not (to our knowledge) been evaluated after long term storage. Furthermore, rod surveillance does not allow quantification of stresses and temperatures. Also, the microstructure of these fuel rods was not examined after dry storage. It is not known whether the condition of the fuel rods is really safe according to the current safety standards. The established 15% area fraction of decohesion limit may have been surpassed *without* rupture occurring in dry storage, leaving the rods unsuitable for post dry storage handling [11].

CONCLUSIONS

Although based on the same fundamental fracture theory (DCCG), the two currently accepted models (by PNNL and LLNL) include many inconsistent assumptions, including the assumed failure condition and choice of values for constants in the DCCG failure equation. Many of the inconsistent assumptions arise because both models include variables that have not been accurately measured for irradiated zircaloy (e.g., grain boundary diffusion coefficient and cavity spacing). Using the limited data available for zirconium is tenuous, as the effects of alloying on the grain boundary diffusion coefficient of zirconium and the effects of irradiation on the diffusion coefficient and the cavity spacing have not been established. These factors combined with extreme sensitivity to temperature leave the current models with marginal value. It should not be assumed that these models yield conservative temperature limits until experimental data for cavity spacing, grain boundary diffusion and post-dry storage integrity for irradiated zircaloy is obtained. A strain-based model is not applicable as long as DCCG is the assumed fracture mechanism. Long term creep testing under dry storage conditions to determine the controlling fracture mechanism and grain boundary diffusion measurements on irradiated zircaloy should be started immediately.

ACKNOWLEDGEMENTS

This work was performed under the auspices of the U.S. Department of Energy (DOE) by Lawrence Livermore National Laboratory under Contract W-7405-ENG-48, and a subcontract from LLNL to the University of California San Diego under contract B345708. The financial support from the DOE National Spent Nuclear Fuel Program and Office of Spent Fuel Management (EM-67) is greatly appreciated.

REFERENCES

1. U.S. Code of Federal Regulations, Part 72, Title 10, "Energy"

2. M. Schwartz and M. Witte, Lawrence Livermore National Laboratory, UCID-21181, September 1987

3. I. Levy, B. Chin, E. Simonen, C. Beyer, E. Gilbert, and A. Johnson, Jr., Pacific Northwest Laboratory, PNL-6189, May 1987

4. W. Nix, Mech. Behavior of Mater.-V, **2**, pp. 1383-1398 (1988)

5. M. Peehs and J. Fleisch, J. Nuc. Mater., **137**, pp. 190-202 (1986)

6. B. Chin, "Final Report, Deformation and Fracture Map Methodology for Predicting Cladding Behavior During Dry Storage," 872BC1PNLF, September 15, 1987

7. R. Raj and M. Ashby, Acta Met., **23**, pp. 653-666 (1975)

8. C. Pescatore and M. Cowgill, Brookhaven National Laboratory, EPRI TR-103949, May 1994

9. R. Keusseyan, C. Hu and C. Li, J. Nucl. Mater., **80**, pp. 390-392 (1979)

10. B. Cane and G. Greenwood, Met. Sci., **9**, pp. 55-60 (1975)

11. R. Chun, M. Witte, and M. Schwartz, Lawrence Livermore National Laboratory, UCID-21246, October 20, 1987

12. T. Hayes, R. Rosen and M. Kassner, Lawrence Livermore National Laboratory, UCRL-ID-131098, December 1999

13. A. Garde, H. Chung and T. Kassner, Acta Met., **26**, pp. 153-165 (1978)

14. D. James and G. Leak, Philos. Mag., **12**, pp. 491-503 (1965)

15. K. Vieregge and Chr. Herzig, J. Nuc. Mater., **173**, pp. 118-129 (1990)

16. M. Griffiths, J. Nuc. Mater., **159**, pp. 190-218, (1988)

17. M. Griffiths, R. Styles, C. Woo, F. Phillipp and W. Frank, J. Nuc. Mater., **208**, pp. 324-334 (1994)

18. M. McKinnon and A. Doherty, Pacific Northwest National Laboratory, PNNL-11576, June 1997

SPENT FUEL DISSOLUTION: AN EXAMINATION OF THE IMPACTS OF ALPHA-RADIOLYSIS

P.A. SMITH*, L.H. JOHNSON**
*Safety Assessment Management Ltd., 9 Penny Green, Settle, North Yorkshire, BD24 9BT, UK
**Nagra, Hardstrasse 73, CH-5430, Wettingen, SWITZERLAND

ABSTRACT

When spent fuel eventually comes into contact with groundwater, fuel matrix dissolution will be strongly influenced by redox conditions in the near field. The most significant factors influencing redox conditions at the fuel surface in a reducing environment are alpha-radiolysis of water and the presence of reductants such as Fe(II) and H_2 arising principally from canister corrosion. The radiolytic yield (G) of molecular oxidants, generally considered to be ~1 molecule of hydrogen peroxide per 100 eV for alpha-radiolysis, is expected to be considerably lower in the presence of reductants but the overall effect on the rate of matrix dissolution cannot yet be reliably quantified. We have attempted to estimate the effective yield of oxidants by examining the results from various studies of spent fuel and UO_2 dissolution, including alpha-radiolysis experiments. The analysis suggests that the effective yield (G_{eff}) is likely to be no greater than 0.01 in the repository environment. The implications of low G_{eff} values are discussed in relation to fuel dissolution rates. Some other aspects of radiolysis relevant to near-field redox chemistry are also examined, including the potential significance of alpha emitters sorbed in the repository near field in producing radiolytic oxidants.

INTRODUCTION

When a spent fuel disposal canister is eventually breached, and water contacts the fuel, rapid release of an "instant release fraction" occurs, together with the much slower release of radionuclides incorporated in the fuel matrix, as the matrix dissolves. In a reducing repository environment in which a steel canister is surrounded by a bentonite buffer, the only source of oxidants is alpha-radiolysis of water that comes into contact with the fuel. Although the maximum theoretical yield of oxidants based on mechanistic radiolysis modelling is large enough to cause relatively rapid dissolution [1], there are several lines of evidence that suggest that the impacts of radiolysis may be overstated by such models. The present study presents an approach to estimating the production rate of oxidants from alpha-radiolysis, based on a review of a number of studies of spent fuel dissolution, alpha-radiolysis and natural analogues.

PROCESSES OCCURING AFTER CANISTER FAILURE

The present analysis relates to a repository located in crystalline rock or clay, in which groundwater saturation would occur relatively rapidly after closure. The physical and chemical processes occurring subsequent to the breaching of a steel or composite copper/steel canister include:

i) the failure of the Zircaloy cladding surrounding the fuel rods by a variety of potential mechanisms, including localised corrosion and hydrogen-induced cracking, with associated implications for mass transport,

ii) the radiolysis of water that contacts the fuel surface, producing radiolytic species that may interact with the fuel matrix, and lead to radionuclide release, and

Mat. Res. Soc. Symp. Proc. Vol. 608

iii) the corrosion of steel surfaces inside the canister to produce iron oxide phases, dissolved Fe(II) and H_2 (dissolved and gaseous). The bentonite surrounding the canister ensures that all transport is diffusion controlled.

EFFECTIVE YIELDS FROM SPENT-FUEL DISSOLUTION, ALPHA-RADIOLYSIS AND NATURAL-ANALOGUE STUDIES

Spent Fuel Studies

One approach to estimating G_{eff} is to assume that alpha-radiolysis is the principal oxidant source in spent fuel dissolution experiments. This is likely to result in an overestimate of G_{eff} because other sources of oxidants exist in such experiments, e.g., dissolved oxygen and β- and γ-radiolysis of water. We begin by considering the experiments of Forsyth and Werme [2], performed under aerated conditions. The alpha dose rate at the fuel surface, d [rad s^{-1}], together with the effective G value, gives the rate of production of radiolytic H_2O_2 per gram of irradiated water, P [mol a^{-1} g^{-1}], via the expression:

$$P = \frac{d \times 6.25 \times 10^{13}\ \text{eV rad}^{-1} \times 3.15 \times 10^{7}\ \text{s a}^{-1} \times 0.01 \times G_{eff}}{6.02 \times 10^{23}\ \text{molecules mol}^{-1}} = 3.27 \times 10^{-5} dG_{eff}. \quad (1)$$

where 0.01 is the assumed maximum yield of H_2O_2 in molecules per eV. Assuming all of the H_2O_2 produced reacts with the fuel, the fractional dissolution rate that results is:

$$D = 238\ \text{g mol}^{-1} \times 10^{-9}\ \text{t}_{\text{HM}}\ \text{g}^{-1} P \frac{M}{M_{HM}}. \quad (2)$$

where M [g] is the mass of irradiated water and M_{HM} [t_{HM}] is the mass of uranium in a sample. The experiments involved 16 g samples, which, assuming a typical specific surface area for spent fuel of 2×10^{-4} m^2 g^{-1} [3], would have a geometric surface area of 3.2×10^{-3} m^2. Although there are uncertainties in surface area, because water may penetrate narrow (1 to 2 nm) grain boundaries to some degree, this does not change the results significantly because the volume of irradiated water enclosed by the grain boundaries is so small that the total yield of oxidants is unaffected. Further assuming a 30 μm α-particle range in water, the mass of irradiated water would be:

$$M = 3.2 \times 10^{-3}\ \text{m}^2 \times 30 \times 10^{-6}\ \text{m} \times 10^{6}\ \text{g m}^{-3} = 0.096\ \text{g} \quad (3)$$

16 g of fuel corresponds to a mass of uranium, M_{HM}, of:

$$M_{HM} = \frac{16\ \text{g} \times 238\ \text{g mol}^{-1} \times 10^{-9}\ \text{t}_{\text{HM}}\ \text{g}^{-1}}{270\ \text{g mol}^{-1}} = 1.41 \times 10^{-8}\ \text{t}_{\text{HM}}. \quad (4)$$

Substituting Eq. 1, Eq. 3 and Eq. 4 in Eq. 2, and rearranging:

$$G_{eff} = 1.89 \times 10^{4} \frac{D}{d}. \quad (5)$$

The measured fractional dissolution rate, D, was 1.1×10^{-4} a^{-1} for the aerated fuel dissolution experiment. The dose rate, d, for the fuel used in this experiment, at 15 years following unloading from a reactor (the age of the fuel in the original dissolution experiments of Forsyth and Werme [2]) is ~ 40 rad s^{-1}. Thus, according to this approach, the effective G value is:

$$G_{eff} = \frac{(1.89 \times 10^4) \times (1.1 \times 10^{-4})}{40} = 0.05, \qquad (6)$$

It is, however, known that the dissolution rate of UO_2 and spent fuel are very similar in such experiments, and therefore radiolysis cannot be the most important factor controlling the rate (the slightly higher dissolution rate of spent fuel may, however, be attributable to radiolysis, although β- and γ-radiolysis are likely to be more important than α-radiolysis [4,5]). Rather, dissolved oxygen is probably rate controlling under aerated conditions. Thus spent fuel dissolution experiments performed under anoxic conditions are of greater interest in relation to the analysis above. Such experiments, which give rates that are 5 times lower for LWR fuel [2] to 40 times lower for CANDU fuel in flow-through experiments [5], imply G_{eff} values in the range 0.01 to 0.001. Again, the effects of β- and γ-radiolysis may be significant, implying that the derived G_{eff} values are still likely to be overestimates.

Another study of spent-fuel dissolution that provides information on the effective G-value under anoxic conditions is that of Eriksen et al. [6]. They reported that a direct measurement of the H_2 yield for a sample of spent fuel in deaerated groundwater gave a yield of one tenth of the expected value, implying an effective G-value of 0.1.

Alpha-Radiolysis Experiments

A potentially more useful set of data that avoids the complication of simultaneous α-,β- and γ-radiolysis is that from the α-radiolysis experiments of Sunder et al.[7]. Based on a large number of measurements of corrosion potential of a UO_2 electrode exposed to a thin film of water irradiated by alpha particles, they derived corrosion rates using the electrochemical model of Shoesmith and Sunder [8]. Radiolysis occurs in the water-filled gap between an alpha source of strength S [Ci] and a UO_2 electrode. H_2O_2, generated by radiolysis at a rate P [mol a^{-1}], reacts with the UO_2 electrode, area A [m^2], giving a dissolution rate D [mol m^{-2} a^{-1}]. H_2O_2 may also be transported by diffusion, at a rate F [mol a^{-1}], to the outside of the cell, where it is removed by advection. Mass balance gives:

$$P - DA = F . \qquad (7)$$

Diffusive losses can be estimated by assuming 1-D radial diffusion out of the gap between the alpha source and the UO_2 electrode, with a zero concentration boundary condition at the exit from the gap. For these conditions, the diffusive losses are given by

$$F = 7.3 \times 10^{-3} [H_2O_2]_0 \qquad (8)$$

where $[H_2O_2]_0$ is the concentration in the gap between the source and the UO_2 electrode. The production rate is given by:

$$P = 9.7 \times 10^{-2} S G_{eff} . \qquad (9)$$

The results of the radiolysis experiments take the form of dissolution rates for different source strengths with a "best fit" function, that takes the form:

$$D = 3.16 \times 10^6 S^{4.3}, \tag{10}$$

for dissolution rates up to about 4×10^{-4} mol m^{-2} a^{-1} (0.03 μg cm^{-2} day^{-1}).

Substituting the rates of production (Eq. 9) and diffusion (Eq. 8) in the mass balance (Eq. 7), and rearranging:

$$S = \frac{7.3 \times 10^{-3}[H_2O_2]_0 + AD}{9.7 \times 10^{-2} G_{eff}}. \tag{11}$$

Substituting Eq. 11 in Eq. 10:

$$D = 3.16 \times 10^6 \left(\frac{7.3 \times 10^{-3}[H_2O_2]_0 + AD}{9.7 \times 10^{-2} G_{eff}} \right)^{4.3} = \left(\frac{36.5[H_2O_2]_0 + D}{14.93 G_{eff}} \right)^{4.3}, \tag{12}$$

for a UO_2 electrode of wetted surface area $A = 2 \times 10^{-4}$ m^2. Eq. 12 can be solved numerically to obtain D as a function of $[H_2O_2]_0$, for any assumed G_{eff} value. These solutions can be compared to the results of some further experiments, in which the alpha source was not used, but instead, dissolution rates of the UO_2 electrode were derived in H_2O_2 solutions of known concentration [8]. The comparison is made in Figure 1 across the range of D for which Eq. 12 is valid (i.e. the range for which Eq. 10 approximates the experiments). The results of the latter experiments are scattered, but follow the trend predicted by Eq. 12 and are consistent with a G_{eff} value in the range of about 0.001 to 0.004. One must view the inference of such a low G_{eff} value from the radiolysis experiments with some caution. Even though diffusive losses have been accounted for in the above analysis, the uncertainties regarding the rate of decomposition of H_2O_2 and the slow rate of reaction of the resultant O_2 with the UO_2 electrode have not been accounted for, and these factors may cause underestimation of G_{eff}. Nonetheless, the results suggest that G_{eff} is significantly less than 1 for α-radiolysis at a UO_2 surface. Christensen and Sunder [1] have noted previously that mechanistic models for α-radiolysis predict corrosion rates that are orders of magnitude higher than those derived from electrochemically determined values. The reasons for this remain unclear.

The Effect of Near-Field Reductants on G_{eff}

A further factor influencing the yield of oxidants from alpha-radiolysis is the presence of reductants such as H_2 and Fe^{2+} in the near field. High hydrogen partial pressures are expected in the near field, as a result of corrosion of steel combined with the high pressures (10 to 15 MPa) required for gas breakthrough for highly compacted bentonite [9]. Reaction of hydrogen with hydroxyl radicals leads to much reduced H_2O_2 and O_2 concentrations in solution, leading to G_{eff} values that are likely to be much less than 0.01, as shown in radiolysis modelling studies [4,10]. Similarly, the presence of Fe^{2+}, which can be expected to arise from magnetite corrosion product dissolution, results in low radiolytic yields because the Fe^{2+}/ Fe^{3+} couple acts as a recombiner [11]. The presence of Fe(II) and H_2 in uraninite deposits may well be important factors in limiting the extent of radiolytic dissolution. For example, in the case of the Cigar Lake deposit

[12], Liu and Neretnieks [13] found it necessary to invoke a G_{eff} value of 0.01 to explain the lack of evidence for radiolytic oxidation of uraninite.

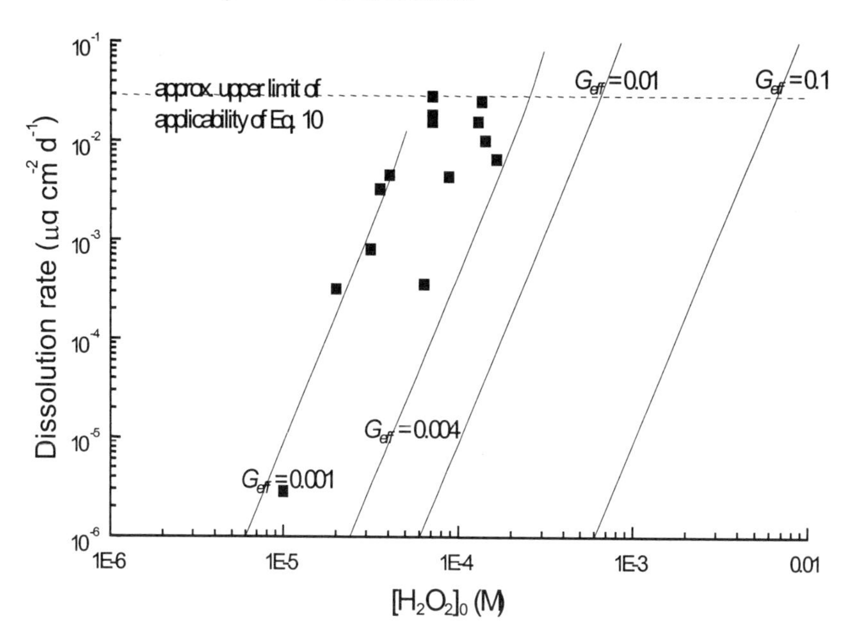

Figure 1: Dissolution rate of the UO_2 electrode as a function of H_2O_2 concentration in the cell, $[H_2O_2]_0$. The results of experiments that use a known H_2O_2 concentration are compared to the results of calculations (the solid lines) based on assumed values of G_{eff}, together with experiments that use an α-source to generate H_2O_2.

Summary and Implications of Derived G_{eff} Values

Table 1 summarises the range of G_{eff} values discussed above and the main considerations in interpreting these vales in the context of performance assessment calculations. The summary suggests that G_{eff} is unlikely to exceed 0.01 for spent fuel that has decayed sufficiently that β- and γ-radiolysis effects are small and is likely to be lower still for conditions that are anoxic, with H_2 and Fe^{2+} present in solution.

For a G_{eff} value of 0.01, the estimated integrated production of oxidants in 10^5 a for a canister with 1.6 t of spent UO_2 fuel with a burnup of 50,000 MWd t^{-1} is ~150 mol, a quantity capable, in principle, of oxidising only ~2 % of the fuel. A G value of 1 would thus lead to the possibility of complete oxidation of the fuel inventory. For mixed-oxide (MOX) fuel of a similar burnup, with its much higher ^{239}Pu content, the quantity of oxidants produced for $G_{eff} = 0.01$ is sufficient to dissolve ~ 10% of the fuel inventory in 10^5 a.

Table 1: Summary of G_{eff} values based on interpretation of spent fuel, radiolysis and natural analogue studies

G_{eff} –value	**Origin**	**Comments**
1	Primary yield in homogeneous solution.	Predicts dissolution rates approximately 10 times higher than actually measured for spent fuel in aerated water; also not consistent with Cigar Lake observations.
~ 0.1	G_{eff} derived from spent fuel dissolution rate in aerated water [2].	Overestimate because O_2 and beta/gamma radiation can explain the rate under these conditions.
0.1	Based on Eriksen et al. [6].	Actual measurement of H_2 yield for spent fuel in deaerated water; beta/gamma effects may be dominant.
~ 0.02	Derived from spent fuel dissolution rates under anoxic conditions [2].	Beta/gamma effects may be dominant.
0.01	Derived from Cigar Lake natural-analogue study [12,13].	Significant uncertainties remain.
< 0.01	From spent CANDU fuel dissolution experiments in flow-through system in deaerated water [5].	May underestimate the rate, due to transport losses from the experimental system.
~ 0.001 – 0.004	Derived from the alpha-radiolysis experiments of Sunder et al. [7].	Uncertainties in the treatment of diffusive losses from apparatus and the effect of H_2O_2 decomposition.
0.001	Consistent with mechanistic radiolysis calculations [4,10], and for Fe^{2+} acting as a recombiner [11].	Not yet experimentally supported.

OTHER RADIOLYSIS EFFECTS

The surface of spent fuel is not the only location at which radiolytic oxidant production can occur. Additional production of oxidants can originate:

i) from within the porous precipitate (i.e. UO_3 hydrates incorporating alpha emitters), that may form in the fuel fractures and in the fuel/sheath gap and, due to its porosity, which is higher than that of the fuel itself, may increase the volume of irradiated water, with only limited absorption of alpha particles in the solid hydrates,

ii) from dissolved alpha emitters in the water within the breached canister, which, in spite of their low concentration, may provide an efficient production route for oxidants since there is no energy absorption on solids and

iii) from alpha emitters that are sorbed on the buffer material, which again has a relatively high porosity (~ 40%), giving only limited absorption of alpha particles.

In the following scoping calculations, each of these mechanisms is examined to determine if their contribution to oxidant production is significant compared to that arising from radiolysis of water at the fuel surface. The calculations of the effects of precipitated alpha emitters are generic, in that they apply to any alpha emitter, on the conservative assumption that the concentrations in the precipitate are the same as those in the fuel. To simplify the calculations of the effects of dissolved and sorbed alpha emitters, only the most important alpha emitter, ^{239}Pu, is considered.

The Effect of Precipitated Alpha Emitters

As fuel dissolution proceeds, U(VI) complexes either diffuse away from the fuel surfaces, or are precipitated in the void spaces within the fuel pellets and in the gap between the fuel pellets and the cladding, eventually filling these spaces. Such precipitation has been observed in both static dissolution experiments [14] and unsaturated tests [15, 16] with spent fuel. The precipitates themselves, with the radionuclides that they incorporate, provide a source of alpha particles, but may also absorb alpha particles originating from within the fuel. These two competing effects are scoped by considering a hypothetical case where all voids are filled with precipitates. The fraction, η_1 ,of alpha radiation generated in a medium of porosity, ε, that interacts with water molecules is given by [13]:

$$\eta_1 = \frac{3\varepsilon\delta_{\alpha,f}}{8(1-\varepsilon)\delta_{\alpha,w}}, \qquad (13)$$

where $\delta_{\alpha,f}$ and $\delta_{\alpha,w}$ are, respectively, the ranges of alpha particles in fuel and water (~ 11 μm and 40 μm). For simplicity, no distinction is made between the range in fuel and the range in precipitate, which are, in any case, expected to be similar. For a fuel rod comprising pellets of radius r = 5 mm and surface area A_S = 0.15 m^2 m^{-1}, with 45μm voids (typical surface area and void aperture values for spent fuel) filled with a hydrate precipitate of porosity 0.45 [17], the overall porosity is:

$$\varepsilon = \frac{A \times 4.5 \times 10^{-5}\ \mathrm{m} \times 0.45}{\pi r^2} = 0.0387, \qquad (14)$$

giving η_1 = 0.004. This can be compared to the case where no precipitate is present, in which the fraction of particles that crosses wetted surfaces and interacts with water molecules is given by:

$$\eta_2 = \frac{3A_S\delta_{\alpha,f}}{16\pi r^2} = 0.004. \qquad (15)$$

Since $\eta_1 \sim \eta_2$, the competing effects that arise when voids are filled with precipitates, namely additional alpha-particle generation, and a reduction in the volume of water available for radiolysis, roughly cancel each other. The effects of precipitates on radiolytic oxidant production are thus not significant. It must be stressed that such a conclusion is based on the assumption that the precipitate is contained within existing cracks and voids in a fuel rod. Should the cladding fail to provide physical containment of the precipitate, oxidant production could increase significantly as the volume of porous precipitate increased.

The Effect of Dissolved Alpha Emitters

For the case of dissolved ^{239}Pu, the production of radiolytic oxidants would be limited by the concentration of Pu in solution. The maximum production rate would occur in the case where the fuel dissolution rate remained high enough to maintain Pu at its solubility limit. The oxidant production rate would thus remain constant with time. For example, the void volume of a Nagra spent PWR fuel canister is 700 litres. Assuming a Pu solubility of 10^{-8} mol/l, a conservative value for a bentonite porewater pH of 8 [18] and a G value of 1 (here we conservatively use the maximum G value), the cumulative H_2O_2 production to 10^5 years (at which time the ^{239}Pu will have largely decayed) is ~ 1 mol, which represents a very small fraction of the ~150 mol produced by radiolysis of water at the fuel surface (assuming $G_{eff} = 0.01$) over the same period.

The Effect of Sorbed Alpha Emitters

For the case of irradiation of pore water in the bentonite, we consider the following scenario. The canister is assumed to be breached along a crack of aperture W [m]. Fuel dissolution maintains the Pu concentration at its solubility limit along the crack. Neglecting the curvature of the canister/bentonite interface and the effects of the outer boundary of the bentonite, but taking into account diffusion from the crack and radioactive decay, the steady-state concentration, c, of ^{239}Pu in solution in the buffer is governed by the equation:

$$\frac{1}{r'}\frac{d}{dr'}\left(r'\frac{dc}{dr'}\right) - c = 0, \qquad (16)$$

where

$$r' = r\sqrt{\frac{R_b\lambda}{D_b}}, \qquad (17)$$

and

$$R_b = 1 + \rho_b K_d\left(\frac{1-\varepsilon_b}{\varepsilon_b}\right). \qquad (18)$$

r [m] is a radial coordinate, extending from the crack into the surrounding bentonite, with its origin at the centre of the crack, $D_b = 0.0166$ m^2 a^{-1} is the pore diffusion coefficient of bentonite, $\varepsilon_b = 0.38$ is the bentonite porosity, $\rho_b = 2760$ kg m^{-3} is its solid density, $K_d = 5$ m^3 kg^{-1} is the sorption constant for plutonium on bentonite (all data from NAGRA[19]) and λ [a^{-1}] = ln $2/t_{1/2}$, where $t_{1/2}$ [a] is the half life of ^{239}Pu (2.41×10^4 years). For times prior to steady-state being achieved, the analysis is conservative in that the amount of adsorbed ^{239}Pu will be overestimated. From the steady-state concentration of Pu as a function of r, the quantity of ^{239}Pu sorbed on the bentonite, and thus the time-integrated production of radiolytic oxidants, can be calculated. This production is given in Table 2 for $c_0 = 10^{-8}$ mol/l as a function of G and of crack aperture, W. For a G value of about 0.1 or less, the production is again considerably smaller than the ~150 mol produced by radiolysis of water at the fuel surface over the same period, and is insensitive to the crack aperture.

Table 2:
The H_2O_2 production due to radiolysis from Pu sorbed on bentonite, integrated up to 10^5 years, as a function of G and of crack aperture, W.

Crack aperture [m]	H_2O_2 production in 10^5 a [moles]		
	$G = 1$	$G = 0.1$	$G = 0.01$
10^{-3}	239	23.9	2.39
10^{-4}	174	17.4	1.74

Eriksen and Ndalamba [20] studied α-radiolysis in water-saturated bentonite and determined that $G(H_2O_2) = 0.69$, although the measurements were not at the high H_2 partial pressures that would be expected to arise from corrosion of steel, which would be expected to reduce the yield considerably. Estimates of radiolytic oxidant production in the Cigar Lake deposit by Liu and Neretnieks [13] suggest that the effective G value is ~0.01 for a fine-grained ore with a porosity of 10%. The low efficiency of oxidant production was assumed to arise from recombination of oxidants and reductants.

It is apparent that the quantity of Pu sorbed on the bentonite, and thus the H_2O_2 production, are proportional to the solubility at the crack/bentonite interface. The H_2O_2 production values in Table 2 are based on a solubility of 10^{-8} mol/l, consistent with the Pu solubility estimated by Bruno et al. [18] for pore water in compacted bentonite. An increase in Pu solubility to a value of 10^{-6} mol/l would give an integrated oxidant production to 10^5 a from α-radiolysis in the bentonite that is similar in magnitude to that at the surface of the fuel. The impact of any uncertainty in Pu solubility on potential oxidant production is thus potentially important in bounding the range of redox conditions to be expected in the near field.

CONCLUSIONS

Analysis of the results of experiments involving spent fuel dissolution and dissolution of UO_2 in the presence of oxidants produced by α-radiolysis suggests that the effective yield of H_2O_2 may be considerably lower than 1 molecule per 100 eV, the value typically assumed for this process. Of particular interest is the value derived from the α-radiolysis experiments of Sunder et al. [7], which are performed without the additional effects of β- and γ-radiolysis, which make interpretation of spent fuel dissolution experiments difficult. These data suggest that G_{eff} may be in the range 0.001 to 0.004. Such values are derived without consideration of the effects of the presence of Fe^{2+} and H_2, which would be expected to further reduce G_{eff}. In spite of these observations, significant uncertainties remain in the evaluation of the effects of α-radiolysis on spent fuel dissolution rates and on near-field redox chemistry.

ACKNOWLEDGEMENTS

The authors thank J.W. Schneider and P.Zuidema of Nagra for their comments and suggestions and D.W. Shoesmith for his suggestions regarding the interpretation of the α-radiolysis experiments.

REFERENCES

1. H. Christensen and S. Sunder, Current state of knowledge in radiolysis effects on spent fuel corrosion. Studsvik Report STUDSVIK/M-98/71 (1998).
2. R. Forsyth and L.O. Werme, Spent fuel corrosion and dissolution, J. Nucl. Mat. **190**, p. 3-19 (1992).
3. L.H. Johnson, The dissolution of irradiated UO_2 fuel in groundwater, Atomic Energy of Canada Limited Report, AECL-6837 (1982).
4. H. Christensen, Calculations simulating spent-fuel experiments, Nucl. Tech. **124** (1998), p. 165.
5. J.C. Tait and J.L. Luht, Dissolution Rates of Uranium from Unirradiated UO_2 and Uranium and Radionuclides from Used CANDU Fuel Using the Single-pass Flow-through Apparatus. Ontario Hydro Used Fuel Disposal Program Report 06819-REP-01200-0006 R00 (1997).
6. T.E Eriksen, U-B. Eklund, L.O. Werme and J. Bruno, Dissolution of irradiated fuel: A radiolytic mass balance study, J. Nucl. Mat. **227**, p. 76-82 (1995).
7. S. Sunder, D.W. Shoesmith and N. H. Miller: Oxidation and dissolution of nuclear fuel (UO_2) by the products of alpha radiolysis of water, J. Nucl. Mater. **244**, p. 66-74 (1997).
8. D.W. Shoesmith and S. Sunder: The prediction of nuclear fuel (UO_2) dissolution rates under waste disposal conditions. J. Nucl. Mat. **190**, p. 20-35 (1992).
9. S.T. Horseman, J.F. Harrington and P. Sellin, Gas migration in MX-80 buffer bentonite, in *Scientific Basis for Nuclear Waste Management XX*, edited by W.J. Gray and I.R. Triay (Mat .Res. Soc. Proc. **465**, Pittsburgh, PA 1997). p. 1003-1010.
10. J.C. Tait and L.H. Johnson: Computer modelling of alpha radiolysis of aqueous solutions in contact with used UO_2 fuel in Proceedings of the 2nd Int. Conf. on Radioactive Waste Management, Winnipeg, MB, Sept., 1986, Canadian Nuclear Society, Toronto, p. 611-615.
11. H. Christensen and E. Bjergbakke, Alpha-radiolysis of aqueous solutions, in *Scientific Basis for Nuclear Waste Management IX*, edited by L.O. Werme, (Mat. Res. Soc. Proc. **50**, Pittsburgh, PA 1985) p. 401-408.
12. J.J. Cramer and J. Smellie: Final report of the AECL/SKB Cigar Lake analog study. Atomic Energy of Canada Limited Report, AECL-10851, COG-93-147. (1994).
13. J. Liu and I. Neretnieks: Some evidence of radiolysis in a uranium ore body - quantification and interpretation, in *Scientific Basis for Nuclear Waste Management XVIII*, edited by T. Murakami and R.C. Ewing (Mat. Res. Soc. Proc. **353**, Pittsburgh, PA 1995). p. 1179-1186.
14. S. Stroes-Gascoyne, L.H. Johnson, P.A. Beeley and D.M. Sellinger, Dissolution of used CANDU fuel at various temperatures and redox conditions, in *Scientific Basis for Nuclear Waste Management IX*, edited by L.O. Werme, (Mat. Res. Soc. Proc. **50**, Pittsburgh, PA 1985), p. 317-326.
15. D. Wronkewiecz, J.K. Bates, S.F. Wolf and E.C. Buck, Ten-year results from unsaturated drip tests with UO_2 at 90 $^{\circ}C$: implications for the corrosion of spent nuclear fuel, J. Nucl. Mat. **238** , p. 78-95 (1996).

16. L.H. Johnson and P. Taylor, Alteration of spent Candu fuel in aerated steam at 150°C, Report prepared for the US DOE/YMSCO under the terms of Contract #DE-AC08-95NV11784, AECL-12003, COG-98-364 (1999).
17. F. King and J.S. Betteridge, Ex situ measurements of the porosity of precipitated corrosion products on oxidized UO_2, Ontario Hydro Report No 06819-REP_01200-0059 R00 (1998).
18. J. Bruno, E. Cera, J. De Pablo, L.Duro, S. Jordana and D. Savage. Determination of radionuclide solubility limits to be used in SR-97. Uncertainties associated to calculated solubilities. SKB Technical Report 97-33. (1997).
19. Nagra: Kristallin-I Safety Assessment Report, Nagra Technical Report 93-22, Nagra, Wettingen, Switzerland (1994).
20. T.E. Eriksen and P. Ndalamba. On the formation of a moving redox-front by α-radiolysis of compacted water saturated bentonite. SKB Technical Report 88-27 (1988).

Uranium Oxide Mass Loss Rate in Water for an Interface under Alpha-Irradiation

C. Corbel [a], J-F Lucchini [b], G. Sattonnay [a], M-F Barthe [c], F. Huet [d], P. Dehaudt [d],

C. Ardois [a], B.Hickel [a], J.L.Paul [b]

[a] CEA Saclay, DSM/DRECAM/SCM/ Laboratoire CEA de radiolyse

[b] CEA Marcoule, DCC/DRRV/SCD/LECM

[c] CERI-CNRS, Orléans

[d] CEA Grenoble, DTP/SECC

ABSTRACT

This work uses an external alpha beam to irradiate an uranium oxide/water interface and investigates the release of uranium in aerated deionized water under alpha irradiation. A high energy alpha beam delivered by a cyclotron (CERI-CNRS) goes through the oxide and emerges in the water with a 20 MeV energy. First results are reported here showing that the uranium mass loss rate increases by three orders of magnitude in aerated deionized water under high flux ($\geq 3.3 \times 10^{10}\ \alpha.cm^{-2}.s^{-1}$).

INTRODUCTION

The environmental assessment of nuclear spent-fuel (UOX) disposal requires a prediction of release rates of uranium once contact between spent fuel and groundwater is established. A first step in the modelling is to replace spent fuel by uranium oxide (UO_2) and use the release rates predicted for uranium oxide submitted to the same leaching conditions as spent fuel. The reliability of such an approach is however questionable due to the difference in the radioactivity of UOX and UO_2.

Spent fuel is a gamma, beta and alpha radioactive material with an activity depending on its burn-up and storing age. The decay time for spent fuel to recover the value of its initial alpha activity as fresh fuel is over more than one million of years. The strong gamma and beta activity of spent fuel decrease by more than three orders of magnitudes in the first few hundred years after disposal. Since the metallic container is expected to survive this period, ground water reaching spent fuel after this period will be subjected mainly to alpha radiolysis [1]. At the spent fuel//water interface, the water composition is modified by the radiolytic species produced by the emerging alpha particules and the spent fuel structure is modified by the defects produced by the alpha particules along their track. Reliable estimates of the uranium release rates due to alpha activity are still lacking although the dissolution of UO2 doped with alpha emitters [2,3] or irradiated by a low flux alpha source [4,5] has been earlier investigated.

Mat. Res. Soc. Symp. Proc. Vol. 608 © 2000 Materials Research Society

This work proposes a new approach to investigate how alpha emission from a UO_2 surface may affect the release of uranium at the UO_2/H_20 interface. The production rate of defects in UO_2 and radiolytic species in H_2O are controlled by the alpha energy, stopping power and flux. To monitor these three physical quantities at the UO_2/H_2O interface, we use an alpha-beam supplied by a cyclotron. The beam passes through the UO_2 disk and emerges into the water in contact with the disk. Depending on initial beam energy and disk thickness, the beam energy at the interface can be adjusted between 0 - 20 MeV. The beam has however a narrow energy distribution at the interface due to straggling. We report here the first results obtained for an alpha beam of 20 MeV at the UO_2/H_2O interface. They show that, at high flux, the uranium mass loss rate increases strongly in deionized aerated water.

EXPERIMENTAL PROCEDURE

Two sintered UO_2 discs, A0 and A1, of 8 mm diameter and 200 μm thickness with a well polished surface in contact with water are used in this study. At the end of the polishing, the discs have been annealed under a mixture of He/Ar gas at 1400 °C to remove polishing damage and adjust the oxygen to metal ratio to the stoichiometric value (2.01). The alpha-beam with an initial energy of 45 MeV was delivered by the cyclotron at CERI (CNRS-Orléans). After passing through Ti windows and the UO_2 discs, the beam emerged from UO_2 into water with a calculated energy of about 20 MeV and a stopping power of 20 KeV/μm which corresponds to a range in water of about 500 μm [6]. The beam was collimated so that the irradiated surface of the discs had only a 6 mm diameter (0.2827 cm^2). Only the irradiated surface of the disc was leached by a 10 ml volume of deionized water (Q millipore) with a 18 $M\Omega.cm^{-1}$ resistivity.

To follow the alteration of the UO_2 surface, sequential batch dissolution tests of one hour each were performed on the UO_2 discs, before, under and after alpha-irradiation without dismounting the discs from the leaching cell. The leaching experiments under each irradiation consisted in a sequence of three dissolutions where the alpha flux was kept constant. This sequence of three dissolutions was repeated just before and after irradiation to compare the uranium releases before and after irradiation with those measured under irradiation. In addition, sequences of more than three tests were performed before and after irradiation to follow the evolution of uranium release as a function of time. Several cycles of irradiation were performed on A0 and A1at flux varying from $(0.33 \text{ to } 3.3)\times10^{11}\ \alpha.cm^{-2}.s^{-1}$. All the tests were carried out at room-temperature using Teflon leaching cells. The uranium mass in the solutions was deduced from the concentration of uranyl ions measured after acidification by spectrofluorimetry.

RESULTS

Before each cycle of irradiation, each disc was sequentially leached until the uranium mass loss rate per hour reaches a value equal to or lower than 1 $\mu g.l^{-1}.h^{-1}$. Under alpha-irradiation at a flux of $3.3\times10^{11}\ \alpha.cm^{-2}\ s^{-1}$, the mass loss rate per hour increases by nearly three orders of magnitude as illustrated in Figure 1 for disc A1 and reaches a value of about 900 $\mu g.l^{-1}.h^{-1}$. The mass loss rate per hour remains approximately constant for each sequential irradiation of one hour. This constant value indicates that the UO_2/H_2O interface has reached a quasi-steady state under one hour irradiation. After irradiation, the mass loss rate per hour decreases progressively as a function of water renewal. Four sequential dissolutions of one hour are sufficient to recover a value of the uranium mass loss rate per hour equal to or lower than 1 $\mu g.l^{-1}.h^{-1}$. This recovery of the uranium release suggests that the altered surface layer produced by irradiation was progressively removed by the sequential dissolutions. A same behaviour is shown in Figure 2 for disc A0 irradiated with a lower flux of $3.3.x10^{10}$ $\alpha.cm^{-2}s^{-1}$. The mass loss rate per hour reaches a value of about 300 $\mu g.l^{-1}.h^{-1}$. It is lower by a factor of three than the value obtained at high flux. We can therefore conclude that the uranium mass loss rate per hour decreases with decreasing alpha flux.

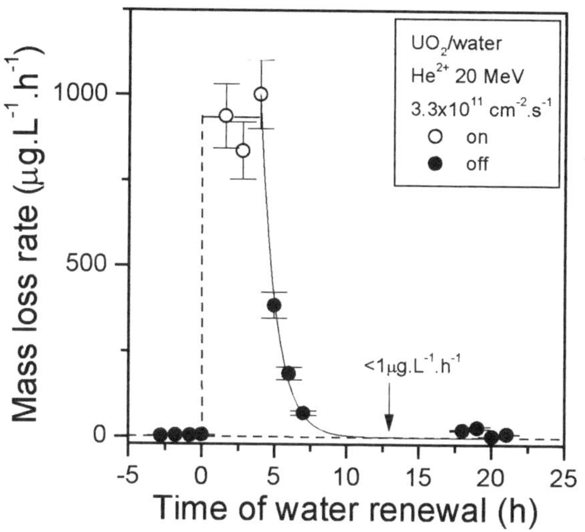

Figure 1. Uranium mass loss rate per hour in aerated deionized water before, under and after alpha irradiation. Water is renewed each hour. Before this irradiation, the oxide had cumulated a fluence of $6.3x10^{15}$ $\alpha.cm^{-2}$ in previous irradiations.

The uranium mass loss rate from uranium oxide has been earlier shown to be strongly dependent on the surface oxidation [6]. In its valence U^{VI}, the rate of dissolution of uranium is much higher than in its valence U^{IV}. Assuming that only U^{VI} is passing into solution, one can estimate that the oxidation rate of U^{IV} per alpha is about $0.27x10^{-13}$ $\mu g\alpha^{-1}$ at high flux in Figure 1 and about $0.87x10^{-13}$ $\mu g\alpha^{-1}$ at lower flux in Figure 2.

The increases of uranium mass loss rate under gamma irradiation in aerated water has been earlier related to the formation of the stable radiolytic species H_2O_2 [7]. It is much less clear that it is the main radiolytic species to play a role under alpha-irradiation. For example, there is still to elucidate the role of $OH^{\bullet}$ radicals produced at the UO_2/H_2O interface under alpha irradiation.

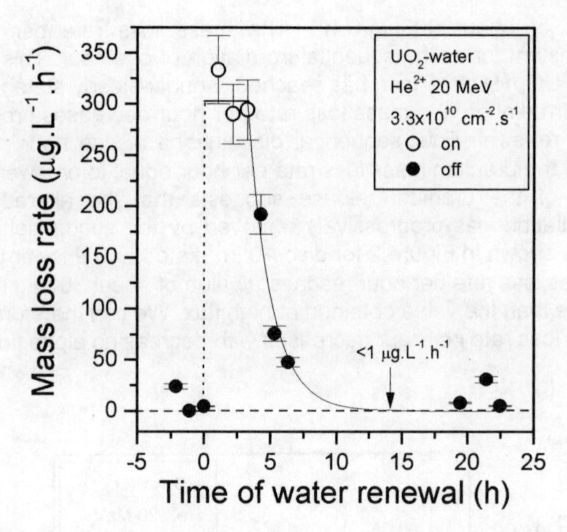

Figure 2. Uranium mass loss rate per hour in aerated deionized water before, under and after alpha irradiation. Water is renewed each hour. Before this irradiation, the oxide had cumulated a fluence of 6.3×10^{15} $\alpha.cm^{-2}$ in previous irradiations.

The irradiation conditions in the present studies are far from those in storage conditions for spent fuel. In the present experiments, the energy and flux of the α-particles, which emerge in water, are higher than those generated by α-decay in spent fuel. The flux of 3.3×10^{10} $\alpha.cm^{-2}$ s^{-1} and energy dose rate near the interface (≤30 µm) are about three orders of magnitude higher than those emitted through the surface of a high burn-up spent fuel with an initial alpha activity of 2×10^{9}Bq α/g(U). An advantage of using a high irradiation flux is that, in a short irradiation time, one can accumulate damage comparable to the one produced over several years in spent fuel. With the high flux of 3.3×10^{11} $\alpha.cm^{-2}$ s^{-1}, one obtains in one hour of irradiation a damage in UO_2 which can be compared to the one induced for about four years of storage in a spent fuel of 2×10^{9} Bq α/g (U). The discs have undergone a cycle of irradiations over several months and cumulated a fluence of about 2.5×10^{16} $\alpha.cm^{-2}$ for A0 and 2.6×10^{16} $\alpha.cm^{-2}$ for A1. At the end of the cycle, the release rate between irradiations and after removal of the altered surface was equal to or lower than 1 $\mu g.l^{-1}.h^{-1}$, indicating that the defects produced along the alpha (>20 MeV) track have little effect on the release of uranium out of irradiation.

CONCLUSION

These first results show that the impact of alpha-irradiation on uranium release in water is important in aerated conditions when the alpha flux has a high value ≥ 3.10^{10} $\alpha.cm^{-2}.s^{-1}$. This suggests that much stronger oxidizing conditions than in aerated water are produced by alpha irradiation at high flux.

ACKNOWLEDGEMENTS

The authors are grateful to M.-T. Bajard of CNRS-CERI for her help in the sample preparation and in the irradiation experiments.

REFERENCES

1. S. Sunder, Nucl. Tech. **122**, p. 211 (1998)
2. W. J. Gray PNL/SRP-6689 Effect of Surface Oxidation , Alpha radiolysis and Salt Brine Composition on Spent Fuel and UO2 Leaching Perfomance
3. V. V. Rondinella, Hj Matzke, J. Cobos and T. Wiss, Mat. Res. Soc. Symp. Proc. **556** p. 447 (1999)
4. M. G. Bailey, L. H. Johnson and D.W. Shoesmith, Corrosion Science, **25**, p.233 (1985)
5. S. Sunder, D.W. Shoesmith and N.H. Miller, J. Nucl. Mater., **244**, p.66 (1997)
6. Ziegler J.F., Biersack J.P. et Littmark U., The Stopping and Range of Ions in Solids, J. F. Ziegler (Ed.), Pergamon Press, New York, 1985.
7. D.W. Shoesmith, S. Sunder, J. Nucl. Mater. **190**, 20 (1992)

POTENTIAL INCORPORATION OF TRANSURANICS INTO URANIUM PHASES

C. W. KIM[1], D. J. WRONKIEWICZ[1], and E. C. BUCK[2]
[1]Dept. of Geology and Geophysics, University of Missouri, Rolla, MO 65409, cheol@umr.edu
[2]Argonne National Laboratory, 9700 S. Cass Avenue, Argonne, IL 60439

ABSTRACT

The UO_2 in spent nuclear fuel is unstable under moist oxidizing conditions and will be altered to uranyl oxide hydrate phases. The transuranics released during the corrosion of spent fuel may also be incorporated into the structures of secondary U^{6+} phases. The incorporation of radionuclides into alteration products will affect their mobility. A series of precipitation tests were conducted at either 150 or 90°C for seven days to determine the potential incorporation of Ce^{4+} and Nd^{3+} (surrogates for Pu^{4+} and Am^{3+}, respectively) into uranium phases. Ianthinite ($[U_2^{4+}(UO_2)_4O_6(OH)_4(H_2O)_4](H_2O)_5$) was produced by dissolving uranium oxyacetate in a solution containing copper acetate monohydrate as a reductant. The leachant used in these tests were doped with either 2.1 ppm cerium or 399 ppm neodymium. Inductively coupled plasma-mass spectrometer (ICP-MS) analysis of the solid phase reaction products which were dissolved in a HNO_3 solution indicates that about 306 ppm Ce (K_d = 1020) was incorporated into ianthinite, while neodymium contents were much higher, being approximately 24,800 ppm (K_d = 115). Solid phase examinations using an analytical transmission electron microscope/electron energy-loss spectrometer (AEM/EELS) indicate a uniform distribution of Nd, while Ce contents were below detection. Becquerelite ($Ca[(UO_2)_6O_4(OH)_6]\cdot 8H_2O$) was produced by dissolving uranium oxyacetate in a solution containing calcium acetate. The leachant in these tests was doped with either 2.1 ppm cerium or 277 ppm neodymium. ICP-MS results indicate that about 33 ppm Ce (K_d = 17) was incorporated into becquerelite, while neodymium contents were higher, being approximately 1,300 ppm (K_d = 5). Homogeneous distribution of Nd in the solid phase was noted during AEM/EELS examination, and Ce contents were also below detection.

INTRODUCTION

The UO_2 in spent nuclear fuel is unstable under moist oxidizing conditions such as those which are expected to exist at the proposed nuclear waste repository at Yucca Mountain, Nevada, Laboratory studies examining the alteration of UO_2 and spent nuclear fuel under such conditions have shown that the alteration products are primarily uranyl (UO_2^{2+}) phases [1-3]. Spent nuclear fuel typically contains 95 to 99% UO_2, up to 1% Pu, and up to 4% other actinides (e.g., Np, Am, Cm) and fission products (e.g., Sr, Cs, Tc, I, Mo, Se) [4]. The radioactivity and toxicity of long-lived actinide elements is a cause for concern if they are released into the environment. The actinides contained in the spent fuel (U, Np, Pu, Am, and Cm) may be released during the oxidation and dissolution of the UO_2 matrix. Migration of these elements may be retarded by their incorporation into the structures of uranium phases. Results of experiments on the corrosion of spent fuel indicate that actinides such as Np, and fission products such as Cs, Sr, and Mo are being incorporated into the alteration products [3,5,6].

The crystal chemistry of Ce^{4+} and Nd^{3+} may be similar to Pu^{4+} and Am^{3+}, respectively, because of the similarity of valence charges and their ionic radii (Ce^{4+}, 0.94 Å; Pu^{4+}, 0.93 Å;

Mat. Res. Soc. Symp. Proc. Vol. 608 © 2000 Materials Research Society

Nd^{3+}, 1.04 Å; Am^{3+}, 1.07 Å [7]). Thus, Ce^{4+} and Nd^{3+} may be used as appropriate surrogate elements for Pu^{4+} and Am^{3+}, respectively. Kim et al. [8] have experimentally examined the potential for actinide substitution in dehydrated schoepite by using Ce^{4+} as a surrogate for Pu^{4+}. Results indicate a Ce concentration of approximately 25 to 10 ppm, with the concentration in the solid progressively decreasing with increasing reaction time and crystal size. Ianthinite is of interest with respect to potential transuranic element retention due to the presence of both U^{4+} and U^{6+} ions into its structure. The structure of ianthinite will accommodate considerable amounts of Pu^{4+}. In corrosion experiments of UO_2 with groundwater, becquerelite is a common alteration product, and it is one of the early phases to form [1,2]. The purpose of this study is to characterize the potential for the incorporation of Pu^{4+} and Am^{3+} into uranium phases by synthesizing ianthinite ($[U_2^{4+}(UO_2)_4O_6(OH)_4(H_2O)_4](H_2O)_5$) and becquerelite ($Ca[(UO_2)_6O_4(OH)_6]\cdot 8H_2O$) doped with Ce^{4+} and Nd^{3+}. Results from this study may provide useful data for estimating realistic release rates for radionuclides from near-field environments within the repository.

EXPERIMENTAL

Ianthinite was produced in this study by dissolving uranium oxyacetate ($UO_2(CH_3COO)_2\cdot 2H_2O$) and copper acetate monohydrate ($Cu(C_2H_3O_2)_2\cdot H_2O$) as a reductant into a solution containing either Ce^{4+} or Nd^{3+}. Solutions were produced by dissolving 0.32 grams of uranium oxyacetate and 0.08 grams of Cu acetate monohydrate into a 15 ml solution containing either 2.1 ppm cerium or 399 ppm neodymium. The pH of the resulting leachant was 3.3. The solution was transferred into Teflon-lined Parr reaction bombs (Model 4749). The bombs were heated at 150°C in Lindberg/Blue M mechanical convection ovens with a temperature uniformity of ±2°C. Copper acetate, at 140°C, decomposes to Cu^{2+} and Cu^{+}, the latter which plays a role as a reducing agent for uranium [9]. After seven days, the reaction bombs were cooled to room temperature. The precipitates consisted of fine purple crystals of ianthinite, small crystals of cuprite (Cu_2O) and tenorite (CuO). The crystals that formed were rinsed three times with deionized water and then air-dried.

Becquerelite was produced in the present study by dissolving uranium oxyacetate and calcium acetate ($(CH_3COO)_2Ca\cdot H_2O$) in a solution containing either Ce^{4+} or Nd^{3+}. Solutions were produced by dissolving 0.42 grams of uranium oxyacetate and 0.35 grams of Ca acetate into a 20 ml solution containing either 2.1 ppm cerium or 277 ppm neodymium. The pH of the resulting leachant was 5.1. Test solutions were maintained in Savillex Teflon vessels (Model 561R2) that were heated to 90±2°C for seven days. After cooling to room temperature, the precipitates consisted of fine platy yellow crystals of becquerelite. The solid phase reaction products that formed were rinsed three times with deionized water and then air-dried.

An analysis of the solid phase reaction products (after dissolution in a nitric acid solution) was performed by a Perkin-Elmer ELAN-5000 inductively coupled plasma (Ar-plasma)-mass spectrometer (ICP-MS). Duplicate ICP-MS scans demonstrated that the average percent deviations were 0.5 and 6.7 for the Ce and Nd analysis, respectively. An analysis of standard solutions also indicated that the accuracy values for the Ce and Nd analysis were within ±5 % or better. A selected portion of the solid phase reaction products were also transferred onto aluminum mounts, coated with carbon, and examined in a JEOL T330A scanning electron microscope (SEM) operated at 15 kV. The SEM is equipped with an energy dispersive X-ray spectrometer (EDS). Selected particles of the reaction products were also crushed into thin particles. The resultant electron-transparent thin-sections of the particles were transferred to holey carbon-coated copper grids and examined in a JEOL 2000 FX II analytical transmission electron microscope (AEM) operated at 200 kV with a LaB_6 filament. The AEM is equipped with a Gatan 666 parallel electron energy-loss spectrometer (EELS), which has an energy resolution of about 1.6-1.8 eV. Electron diffraction patterns were taken with a charge-coupled device (CCD) camera which permits very low intensity viewing and, therefore, is ideal for electron beam-sensitive materials such as the uranium phases formed in this study. Crystallographic data were also collected from the inorganic crystal structure database produced jointly by Gmelin-Institut für Anorganische Chemie and Fachinformationszentrum FIZ Karlsruhe, so that electron diffraction patterns could be simulated with a computer utilizing Desktop Microscopist software.

RESULTS AND DISCUSSION

Ianthinite

SEM examination of the solid phase reaction products demonstrated the presence of characteristic crystals of ianthinite (Fig. 1a). The figure 1b shows an AEM image of the ianthinite particles. The computer simulations of ianthinite partially matched with the experimentally obtained CCD electron diffraction pattern from the uranium phase shown in the inset of figure 1b, but extra diffraction patterns appeared suggesting the presence of an additional phase. There are several factors that may contribute to the pattern-matching difficulties. These particles were noted to slowly and progressively change from the purple color of ianthinite to a yellowish color more characteristic of dehydrated schoepite following their removal from the test solutions. The sample that was examined by AEM may therefore have been partially oxidized to a schoepite type phase. In addition to oxidation, dehydration in the vacuum chamber during analysis could lead to structural rearrangement [10]. Finally, there is limited information on atomic parameters of ianthinite for the computer simulations.

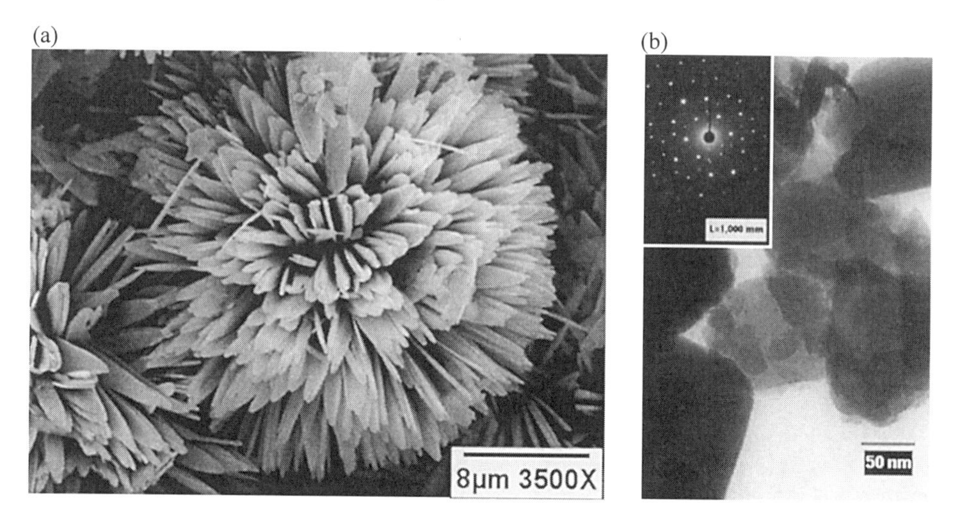

Figure 1. (a) SEM micrographs of ianthinite crystals. (b) AEM bright field image of ianthinite particles. Inset: Selected area electron diffraction pattern from ianthinite.

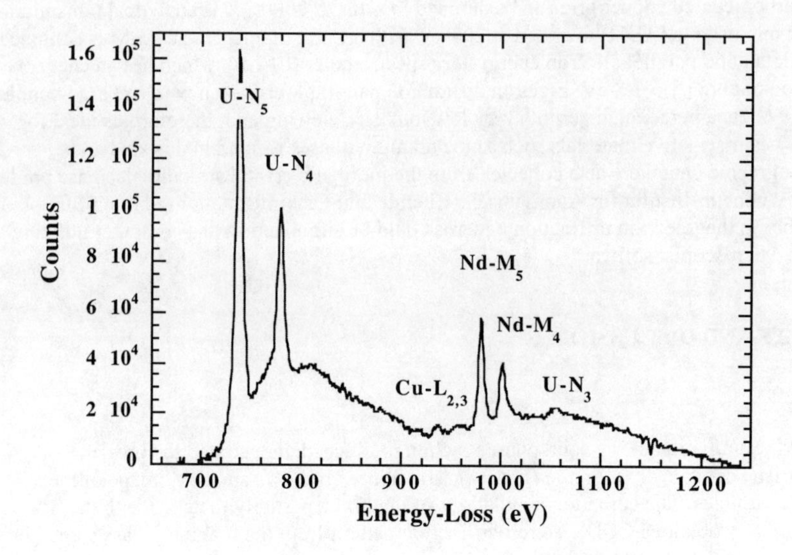

Figure 2. Electron energy-loss spectrum of Nd-bearing ianthinite. $M_{4,5}$ edges of Nd are clearly visible.

EELS analysis was performed on particles that were sufficiently thin for microanalysis. The cerium concentration in ianthinite was below the limit of detection for the AEM/EELS technique. The ICP-MS results from dissolved solids indicate that about 306 ppm Ce was present in the ianthinite, while neodymium contents were much higher, being approximately 24,800 ppm. Neodymium was also detected during the AEM/EELS examination of these samples (Fig. 2). The peak ratio of Nd/O was homogeneous (0.015 to 0.020) throughout the samples suggesting that Nd was uniformly distributed in the structure of ianthinite.

The structure of ianthinite ($[U_2^{4+}(UO_2)_4O_6(OH)_4(H_2O)_4](H_2O)_5$) consists of sheets of uranyl and uranus polyhedra connected through hydrogen bonds to interlayer H_2O groups [11]. The incorporation of Ce^{4+} into this structure is likely to occur by direct substitution for U^{4+} within the polyhedra sheet due to the similarity of valence charges and their ionic radii (Ce^{4+}, 0.94 Å; U^{4+}, 0.97 Å [7]). However, the incorporation of Nd^{3+} needs an appropriate charge-balancing substitution (Nd^{3+}, 1.04 Å; U^{4+}, 0.97 Å [7]). One possible substitution mechanism involves $OH^- \leftrightarrow O^{2-}$ [12], with the suggested charge-coupled substitution being $Nd^{3+} + OH^- \leftrightarrow U^{4+} + O^{2-}$. Therefore, structural formula of the Nd-bearing ianthinite is as follows:

$$[U_2^{4+}(UO_2)_4O_6(OH)_4(H_2O)_4](H_2O)_5 \ + \ xNd^{3+}$$

$$\Downarrow$$

$$[U_{2-x}^{4+}Nd_x(UO_2)_4O_{6-x}(OH)_{4+x}(H_2O)_4](H_2O)_5$$

where x is 0.04 in this study

Becquerelite

Examination of the solid phase reaction products with SEM indicates the presence of characteristic platy crystals of becquerelite (Fig. 3). With SEM/EDS it was not possible to detect any elements in the becquerelite other than uranium, calcium, and oxygen. The AEM image of the uranyl phase also shows the characteristic platy morphology of becquerelite (Fig. 4a). The presence of becquerelite was confirmed by electron diffraction analysis (Fig. 4b). The computer simulations of becquerelite agreed with the experimentally obtained CCD electron diffraction pattern from the uranyl phase shown in Figure 4b.

6µm 5000X

Figure 3. SEM micrographs of becquerelite crystals.

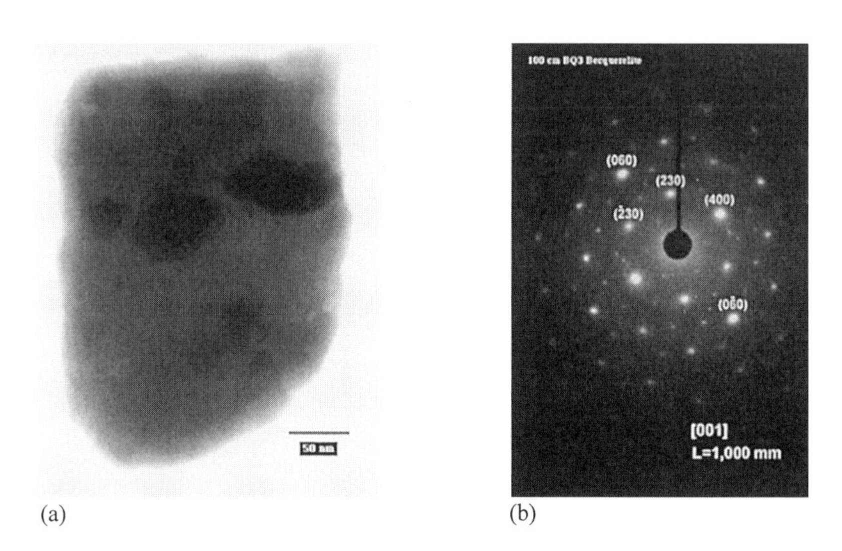

Figure 4. (a) AEM bright field image of becquerelite particle. (b) Selected area electron diffraction pattern from becquerelite.

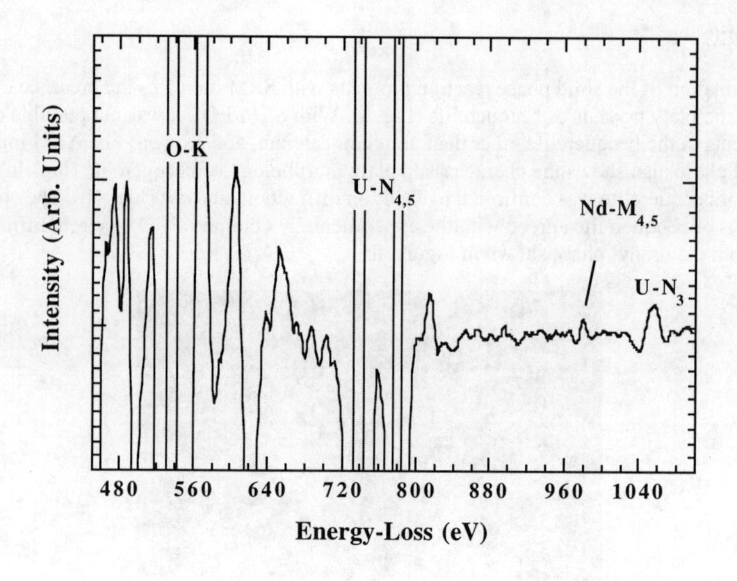

Figure 5. Electron energy-loss spectrum of Nd-bearing becquerelite. The vertical scale intensity has been increased 50x for easier viewing of the Nd peak located at 980 eV.

EELS analysis was performed on becquerelite particles that were sufficiently thin for microanalysis. The cerium concentration in becquerelite was also below the limit of detection for the AEM/EELS technique. A uniform distribution of Nd was noted during the AEM/EELS examination of these samples (Fig. 5). The ICP-MS results from dissolved solids indicate that about 33 ppm Ce was present in the becquerelite, while neodymium contents were much higher, being approximately 1,300 ppm.

The structure of becquerelite ($Ca[(UO_2)_6O_4(OH)_6]\cdot8H_2O$) consists of sheets of uranyl polyhedra connected through hydrogen bonds to interlayer Ca^{2+} and water molecules [13]. The incorporation of Nd^{3+} into this structure is likely to occur by substituting for Ca^{2+} in the interlayer, assuming that appropriate charge-balancing substitutions can occur (Nd^{3+}, 1.04 Å; Ca^{2+}, 0.99 Å [7]). One possible substitution mechanism involves $O^{2-} \leftrightarrow OH^-$. Therefore, the suggested charge-coupled substitution will be $Nd^{3+} + O^{2-} \leftrightarrow Ca^{2+} + OH^-$. This is a reverse of the charge-balancing mechanism proposed to occur during the substitution of Nd^{3+} in the structure of ianthinite. The incorporation of Ce^{4+} into this structure is likely to occur by substitution for U^{6+} within the polyhedra sheet (Ce^{4+}, 0.94 Å; U^{6+}, 0.80 Å [7]), together with local charge-balancing substitutions ($Ce^{4+} + 2OH^- \leftrightarrow U^{6+} + 2O^{2-}$).

Distribution coefficients

Table I shows the distribution coefficients, K_d values, of Ce and Nd in ianthinite and becquerelite tests from this study, and in dehydrated schoepite from the previous work [8]. The K_d value of Ce was higher than that of Nd in ianthinite test, suggesting that substitution of Ce^{4+} for U^{4+} was favored over the substitution of Nd^{3+} for U^{4+}. In the tests with becquerelite, a lower

K_d value was obtained for Nd relative to Ce, implying that substitution of Ce^{4+} for U^{6+} is favored over that of Nd^{3+} for Ca^{2+}. This occurs even though Nd^{3+} and Ca^{2+} have smaller disparity of charge and ionic radius. It is thought to be probably due to the interlayer structure of becquerelite. This has to be further studied in the future. Additional tests with Nd concentrations approximating those of Ce (2.1 ppm) are being conducted to assess differences in atomic substitution between the two lanthanide elements from solutions with similar ionic strengths.

Table I. K_d values of Ce and Nd in uranium phases. All tests at seven days.

Uranium phases	Ce in leachate (ppm)	Ce in solid (ppm)	K_d	Nd in leachate (ppm)	Nd in solid (ppm)	K_d
Ianthinite	0.3	306	1020	216	24,800	115
Becquerelite	1.9*	33	17	271	1,300	5
Dehydrated schoepite [8]	1.9	25	13	273	-	-

* Number calculated from weight of solid becquerelite and concentration of Ce in solid phase.

CONCLUSIONS

This paper presents results from a study of potential incorporation of transuranics into uranium phases. Results of this study suggest that:

1. Ce^{4+} (surrogate for Pu^{4+}) directly substitutes for U^{4+} in the structure of ianthinite in the amount of about 306 ppm (2.1 ppm in leachant, $K_d = 146$).
2. Nd^{3+} (surrogate for Am^{3+}) substitutes for U^{4+} in the structure of ianthinite, together with local charge-balancing substitutions in the amount of approximately 24,800 ppm (399 ppm in leachant, $K_d = 62$).
3. Ce^{4+} substitutes for U^{6+} in the structure of becquerelite in the amount of about 33 ppm (2.1 ppm in leachant, $K_d = 16$), assuming that local charge-balancing substitutions occur.
4. It is thought that Nd^{3+} substitutes for Ca^{2+} in the interlayer sites of becquerelite in the amount of approximately 1,300 ppm (277 ppm in leachant, $K_d = 5$), assuming that local charge-balancing substitutions occur.

ACKNOWLEDGEMENTS

This work was supported by the U.S. Department of Energy, under contract DE-FG07-97ER14820. The AEM was performed at Argonne National Laboratory, Chemical Technology Division. The authors thank to Robert J. Finch for his valuable discussions on this study. Appreciation must be given to F. Scott Miller for thoughtful discussions regarding micrographs work and Laura M. Luther for ICP-MS analysis.

REFERENCES

1. D. J. Wronkiewicz, J. K. Bates, S. F. Wolf, and E. C. Buck, J. Nucl. Mater. **238**, 78 (1996).

2. D. J. Wronkiewicz, J. K. Bates, T. J. Gerding, and E. Veleckis, J. Nucl. Mater. **190**, 107 (1992).

3. P. A. Finn, J. C. Hoh, S. F. Wolf, S. A. Slater, and J. K. Bates, Radiochim. Acta **74**, 65 (1996).

4. J. O. Barner, Pacific Northwest Laboratory Report PNL-5109 (1985).

5. E. C. Buck, R. J. Finch, P. A. Finn, and J. K. Bates in *Scientific Basis for Nuclear Waste Management XXI*, edited by I. G. McKinley and C. McCombie (Mater. Res. Soc. Proc. **506**, Davos, Switzerland, 1997) pp. 87-94.

6. E. C. Buck, D. J. Wronkiewicz, P. A. Finn, and J. K. Bates, J. Nucl. Mater. **249**, 70 (1997).

7. L. Pauling, *The Nature of the Chemical Bond,* 3rd ed. (Cornell University Press, Ithaca, 1960), p. 644.

8. C. W. Kim, D. J. Wronkiewicz, and E. C. Buck, Radiochim. Acta (submitted).

9. C. Bignand, Bull. Soc. Franç. Minér. Crist. **78**, 1 (1955).

10. E. H. P. Cordfunke, G. Prins, and P. Van Vlaanderen, J. Inorg. Nucl. Chem. **30**, 1745 (1968).

11. P. C. Burns, R. J. Finch, F. C. Hawthorne, M. L. Miller, and R. C. Ewing, J. Nucl. Mater. **249**, 199 (1997).

12. P. C. Burns, R. C. Ewing, and M. L. Miller, J. Nucl. Mater. **245**, 1 (1997).

13. M. K. Pagoaga, D. E. Appleman, and J. M. Stewart, Amer. Miner. **72**, 1230 (1987).

IN SITU LONG TERM MEASUREMENTS OF pH AND REDOX POTENTIAL DURING SPENT FUEL LEACHING UNDER STATIONARY CONDITIONS – THE METHOD AND SOME PRELIMINARY RESULTS

K. SPAHIU*, L. WERME*, J. LOW, U-B. EKLUND****
***SKB, Brahegatan 47, S- 102 40 Stockholm, Sweden.**
****Studsvik Nuclear AB, S-61182 Nyköping, Sweden.**

ABSTRACT

In order to get a better understanding for the spent fuel corrosion process, the variations of important intensive parameters such as pH and the redox potential (E_h) of the bulk solution have been continuously measured during long term sequential leaching experiments. These data may be used together with standard chemical analytical data for modeling spent fuel corrosion, especially in anoxic or reducing conditions. In order to overcome difficulties caused by the strong radiation field and the long experiment times, a method for in-situ measurements of pH and E_h using a computer controlled system was worked out. The stability of the measuring system over long time periods was then tested; pH values stable within 0.05 pH units/year in buffered systems were measured. The variations of these parameters in a variety of conditions and solution compositions were followed continuously during the spent fuel leaching process.

A discussion of the results of spent fuel leaching in anoxic conditions is presented, pointing out the difficulties to realize in laboratory near field conditions. An interesting case of calcite co-precipitation/co-dissolution is also presented. Dissolution experiments show that the calcite precipitated previously during spent fuel leaching experiments in synthetic groundwater contained considerable amounts of actinides and fission products.

INTRODUCTION

Spent fuel leaching studies have been performed in a variety of conditions and solution compositions since the early 70´s [1, 2 and references therein]. An overview of the results accumulated in SKB´s fuel leaching experimental program in the hot cells at Studsvik in recent years is provided in [3], while some more recent fuel leaching actinide data are treated in [4]. The aim of these studies is to obtain data on the spent fuel oxidation-dissolution process, which are then used to propose models on the source term evolution during long periods of time. A good understanding of the complex system of spent fuel reactions in deep groundwater is required in order to be able to make such long - term predictions. The high radiation field under which such experiments have to be performed makes difficult the use of more involved experimental setups. This is the reason why in the majority of studies the only parameters measured are the concentrations of the fission products and actinides in the leach solution at various time intervals. In a series of studies the corrosion potential of uranium (IV) oxide and spent fuel electrodes has been measured and used as a basis for the modeling of spent fuel dissolution process [1, 5, 6].

Two main intensive parameters, which influence the corrosion of spent fuel and the state of the released radionuclides, are the pH and the redox conditions under which the spent fuel leaching is carried out. Therefore, all thermodynamic modeling of the near field evolution is performed in E_h-pH coordinates. During the spent fuel oxidation-dissolution process, however, reactions that cause changes in the initial values of these parameters take place. This work describes an attempt to follow *in situ* the evolution of the proton concentration and the redox

Mat. Res. Soc. Symp. Proc. Vol. 608

potential during several months of stationary spent fuel corrosion experiments. This information, together with the analytical data on the concentrations of various radionuclides in the leach solution, may then be used to improve the modeling of the spent fuel corrosion process.

The redox conditions are extremely important for the stability of the $UO_2(s)$ spent fuel matrix. The majority of the studies in low salinity groundwaters have been performed under oxidizing conditions [3], but there are also data available from anaerobic or anoxic conditions [7]. We will present data on the variation of pH and E_h during spent fuel dissolution under oxidizing and anoxic conditions.

EXPERIMENTAL

Due to the duration of the fuel leaching experiments, measurements without liquid junction were preferred. The leach solutions contained in all cases 10 mM NaCl, in order to assure stable potentials of the Ag/AgCl reference electrode. The reference electrodes were prepared according to Brown [8] with slight modifications. In order to express the redox potentials referred to S. H. E. (Standard Hydrogen Electrode) and to calibrate the glass electrode, a preliminary Gran titration [9, 10] was performed in the presence of a hydrogen electrode. After the calibration, the glass electrode and the corresponding reference, as well as a Pt - electrode, were placed in a 250 ml Pyrex flask containing 200 ml test solution.

The anoxic conditions were assured by passing Ar or Ar+2%CO_2 gas mixtures through a system of three washing bottles. The same system was used during electrode calibration described above with hydrogen gas. The first bottle contained a Cr(II) "oxygen trap" solution, the second distilled water and the third the same solution as in the test flask. The outlet of the gas from the measuring cell was carried out through a mini-washing bottle, to make the system airtight and ensure a slight overpressure. Low flow rates of gas were used, enough only to obtain a very slow bubbling in the end flask. However, as it will be discussed further, it seems impossible to get rid of traces of oxygen only by flushing with an inert gas.

The electrode potentials were monitored each hour for a few days outside the hot cell, before the test vessel was transferred inside. The tests were performed at ambient hot cell temperature (20-25°C). The readings of the electrodes before and after transferring to the hot cell were always the same within a few tenths of a millivolt. In order to check if the high radiation field in the cell would affect the readings of the electrodes, or if radiolytic reactions would change pH (-log$[H^+]$ in our case) and Eh, the stability of the measuring system was tested during several days inside the hot cell. These data are shown in figures 1 to 3 with negative day numbers. Averages of 24 (or 12 measurements/day in later experiments) values per day are shown by each point in the figures. The typical standard deviations/day of the potentials or pH were lower than the diameter of the symbols in the figures. Day zero represents the introduction of the fuel/clad segment (~ 20 mm, 16 g. UO_2) into the test vessel. After each exposure the fuel specimens were transferred to new flasks with fresh solution. The analysis of the leachates was performed via mass spectrometry (ICP-MS) [11, 12].

RESULTS AND DISCUSSION

Spent fuel leaching and other studies concerning redox sensitive elements are reported as performed under oxidizing or anoxic experimental conditions.

With **oxidizing** conditions usually it is meant that the experiments are performed in the presence of the laboratory atmosphere, with an oxygen partial pressure of about 0.2 atm. In this case the concentration of the dissolved atmospheric oxygen in the solution is of the order of

$[O_2]_{diss} = k\ P_{O2} \approx 2.5x10^{-4}$ M. For strongly radioactive materials, such as spent fuel, one has to take into account also the oxygen produced through the radiolysis of water. In radiolysis experiments performed by Eriksen et al [13, 14] it is shown that after ~13 days the concentration of radiolytic oxygen reaches more or less stationary levels around $\approx 10^{-7}$ M. Thus the influence of radiolytic oxygen can be neglected under oxidizing conditions. Other radiolytic oxidants, such as H_2O_2 or radicals, however, may have an influence in the oxidation/dissolution of the spent fuel matrix [1].

The –log $[H^+]$ measurements in the hot cell during spent fuel leaching in carbonated solutions under **oxidizing conditions** showed no variation during several months [15]. The bicarbonate concentration of 2 mM seems to buffer the solutions sufficiently. The -log $[H^+]$ values are quite near to the value 8.46, calculated with the equilibrium constants for carbonic acid at 0.01 M NaCl [16] and atmospheric CO_2 partial pressure. The platinum electrode readings are more interesting. During more than 12 days in the hot cell, prior to the introduction of the spent fuel segment the measured potential was ~460 mV. This is quite oxidizing, but about 250 mV lower than the value calculated for equilibrium with dissolved atmospheric oxygen (see below). This is in agreement with literature data [17, 18, 19] and almost coincides with the value measured by Natarjan and Iwasaki [20] in oxygenated waters of the same pH. In the first days after the introduction of the spent fuel segment in the test vessel a quick decrease in the Pt electrode readings is observed followed by a somehow slower increase. After about 20 days the potentials level out and are followed by a slower decrease after about 80 days [15]. More analytical solution data are needed to fully understand this behavior. A possible explanation is through the role of radiolytic hydrogen in the initial decrease, since the atmospheric oxygen concentration is high as compared to the radiolytic one.

The term anoxic (or anaerobic) groundwater usually refers to the complete absence of oxygen. In general, **anoxic conditions in laboratory studies** refer to the conditions realized in a glove box or vessel, flushed with an inert gas like Ar or N_2. The following items should be considered while working in anoxic conditions. First, an equilibrium calculation would show that the redox conditions are not as believed (and some times reported) mildly oxidizing or even reducing. In fact in most cases the redox conditions are strongly oxidizing. The relevant equilibrium and constant in this case are:

$$2\ H_2O \leftrightarrow 4\ H^+ + O_2\,(g) + 4e^- \qquad \log K = -\ 82.8 \qquad (1)$$

Hence:

$$20.7 + (\log P_{O2})/4 = pH + pe$$

In a glove box the anoxic conditions refer in general to oxygen concentrations of the order of 1 ppm or less (i. e. $\log P_{O2}$= -6 to -8) and for a typical groundwater with pH = 8.2, it follows:

$$pe = 20.7 - pH - 2 = 10.5 \quad \text{or} \quad E_h \approx +\ 630\ mV$$

In this case for spent fuel studies the dissolved oxygen $[O_2]_{diss} = k\ P_{O2} \approx 10^{-9}$ to 10^{-11} M, is negligible as compared to radiolytic oxygen.

Second, a problem that we encountered during long term experiments in inert gas atmosphere and real or simulated ground waters was calcite precipitation [4] (see also below). Real or synthetic groundwaters are always in (or calculated assuming) equilibrium with calcite, thus the inert gas atmosphere caused CO_2 losses from the solution, followed by the pH increase and the precipitation of calcite. This may be avoided by adding CO_2 in the inert gas.

Sometimes small amounts of hydrogen (2-5%) are added to the inert box flushing gas, claiming that in such a way reducing conditions are assured at ambient temperature. In this case redox potential measurements at the typical groundwater pH interval result in very negative values (around –500 mV depending on the reference electrode and the calibration), apparently confirming very reducing conditions. In fact such conditions exist only in the surface of the Pt electrode, a known H_2 catalyst, but not in the bulk solution. In this case the Pt electrode is simply functioning as a hydrogen electrode and responding to the concentration of the protons in solution (this explains the low negative values corresponding at pH=8 to a contribution of –59.16 log [H] = - 473 mV). The same should hold for E_h measurements in boreholes, when measurable amounts of hydrogen are present in the ground water.

The –log $[H^+]$ variation during spent fuel leaching in **anoxic conditions** for more than 2 years duration is shown in Fig 1. The initial period was thought to give an idea of the stability

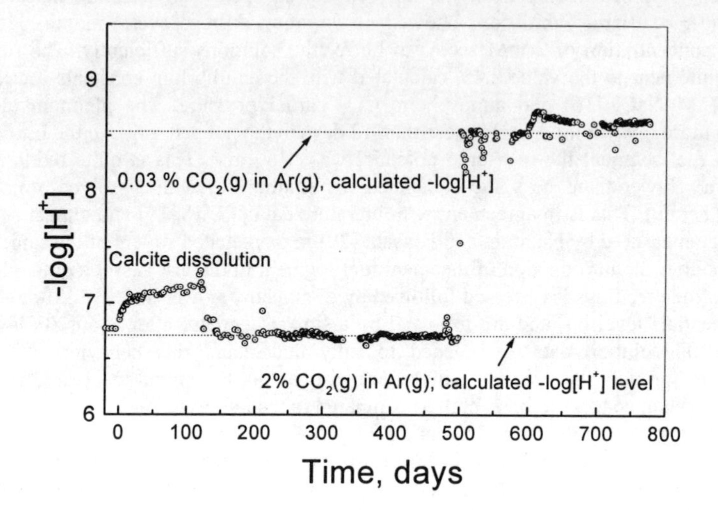

Fig. 1. The variation –log $[H^+]$ during spent fuel leaching in 10 mM NaCl, 2 mM $NaHCO_3$ solutions under anoxic conditions (Ar+2% CO_2 or Ar+0.03%CO_2 gas mixture)

of the measuring system in a pH buffered system by 2 mM $NaHCO_3$ under an Ar + 2%CO_2 atmosphere. In the initial period of more than 100 days the –log $[H^+]$ increased steadily by more than 0.5 units. This was quite unexpected in a buffered solution. The analysis of the solution showed that there were large amounts of Ca^{2+} (more than the total Ca^{2+} in 250 ml groundwater) and Mg^{2+} ions in the leach solution, even though the starting solution contained none of these ions. The source of Ca and Mg was the spent fuel segment, which had been leached in simulated Allard groundwater and under an Ar+5%H_2 atmosphere for prolonged time intervals [7]. The dissolution of calcite (confirmed by XRD) was accompanied with the release in solution of relatively large amounts of actinides and REE, which possibly had co-precipitated with calcite. The pH values measured before spent fuel introduction and after all calcite was dissolved coincided quite well with the calculated values (dotted lines in Fig. 1). A linear drift of the E^0 of the glass electrode [21] of -0.013 mV/day could be estimated for an almost 500 day period.

The variation in the redox potentials during the same experiment is shown in Fig. 2. Outside the hot cell values ~ 360 mV were measured, about 100 mV lower than those measured in oxidizing conditions. The Pt electrode was very sensitive to traces of oxygen as could be seen when a lowering of the gas flow occurred, even though the system was not opened. The pH increase caused by the change of the concentration of CO_2(g) from 2% to 0.03% is accompanied with a corresponding decrease in the redox potential. It seems that the sum of pH + pe is approximately constant under such conditions, indicating a one electron to one proton ratio in the redox determining process (e. g., in Eq. 1).

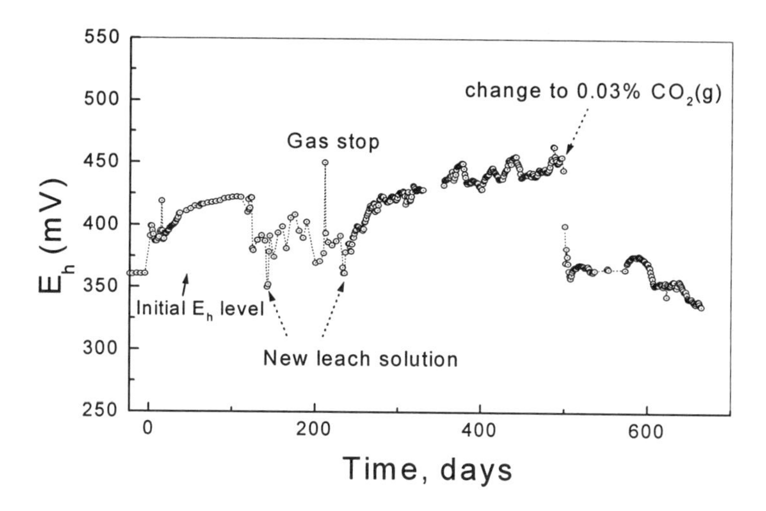

Fig. 2. The variation of the redox potential during spent fuel leaching in 10 mM NaCl, 2 mM Na HCO_3 solutions under anoxic conditions (Ar+2% CO_2 or Ar+0.03%CO_2 gas mixture)

After the test of the stability of the measuring system in buffered solutions, measurements of the pH and E_h in **un-buffered 10 mM NaCl solutions** in contact with air were performed.

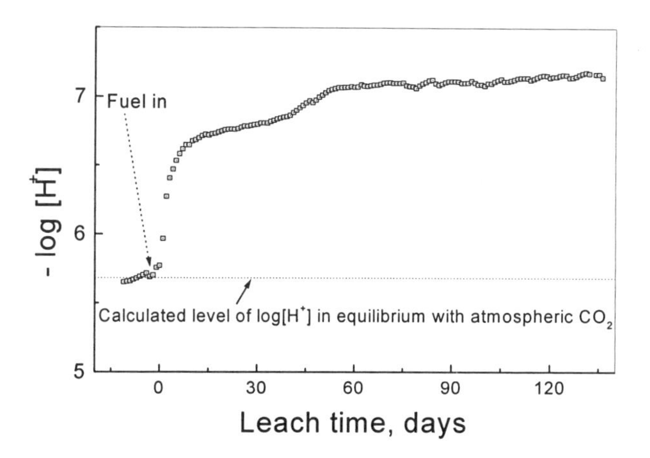

Fig 3. The variation of pH during spent fuel leaching in 10 mM NaCl solutions and oxidizing conditions (in contact with air).

These measurements would help interpret some spent fuel leaching experiments carried out in distilled water and oxidizing conditions, provided the low NaCl concentration does not influence the dissolution process (comparison of ICP-MS analyses shows that this is the case. The drift of the glass electrode, estimated for a 12 day period before introducing the spent fuel segment in such solutions, was ~0.6 mV/day. The measured log $[H^+]$ ~ - 5.7 coincides quite well with calculated value for equilibrium with atmospheric CO_2 (dotted line in Fig. 3). A considerable increase of pH was observed after introducing the spent fuel. Attempts to explain this increase in pH (corresponding, however, to μM levels of base) by dissolution of the basic oxides of some fission products (e. g. CsO and SrO) show that the increase in pH is slightly higher than what would be caused by the dissolved amount of these oxides.

REFERENCES

1. Johnson, L. H. and Shoesmith, D. W., *Spent fuel.* In: Radioactive Waste Forms for the Future, Eds. Lutze, W and Ewing, R. C., (North-Holland Physics Publishing, The Netherlands 1988).
2 Oversby, V. M., *Nuclear waste Materials,* In.: Materials Science and Technology, Eds. Cahn, R. W., Haasen, P. and Kramer, E. J., Vol. 10B, Nuclear Materials, Part 2, (edited by Frost, B. R. T., VCH Verlagsgesellschaft mbH, Federal republic of Germany 1994).
3 Forsyth, R. S. and Werme, L. O., J. Nucl. Mater. **190,** 3 (1992).
4 L.O. Werme, K. Spahiu, J. Alloys and Compounds, **271-273** 194 (1998).
5 D. Shoesmith and S.Sunder AECL Report – 10488, 1991.
6 D. W. Shoesmith, S. Sunder and W. H. Hocking, *In The electrochemistry of novel materials*, Eds. J. Lipkowski and P. N. Ross (VCH, New York, NY, 1994), pp. 297-337.
7 R. S. Forsyth, SKB Technical Report 97-25, 1997.
8 A. S. Brown, J. Am. Chem. Soc. **56** 646 (1934).
9 G. Gran, Acta Chem. Scand. **4,** 559 (1950).
10 G. Gran, Analyst, **77** 661(1952).
11 R. S. Forsyth, SKB Technical Report 97-11, 1997.
12 Forsyth, R. S. and Eklund, U-B, SKB Technical Report 95-04, 1995.
13 T. Eriksen, U-B. Eklund, L. Werme, J. Bruno, J. Nucl. Mater., **227**, 76 (1995).
14 J. Bruno, E. Cera, U-B Eklund, T. Eriksen, M. Grive, K. Spahiu, presented at Migration 99, Lake Tahoe, Nevada, September 1999. To be published in *Radiochimica Acta.*
15 K. Spahiu, L. Werme, J. Low, U-B. Eklund, presented at 1998 Spent Fuel Workshop, May 18-20, Las Vegas, Nevada, 1998 (unpublished).
16 I. Grenthe, K.Spahiu. and T. Eriksen, J. Chem.Soc. Faraday Trans. **88**, 1267 (1992).
17 D. K. Nordstrom, E. A. Jenne, J. W. Ball, In Chemical Modelling in Aqueous Systems, edited by E. A Jenne, ACS Advan. in Chem. Series 93, (Am. Chem. Soc., Wasington D.C., 1979), pp. 51-79.
18 M. Whitfield, Limnol. and Oceanog., **19** 857 (1974).
19 Garrels, R.M. and Christ, C.L.: *Solutions, Minerals, and Equilibria*, Harper & Row, New York 1965, 450 p.
20 K. A. I. Nararajan and I. Iwasaki, Minerals Sci. Eng. **6** 35 (1974).
21 J. N. Butler and R. N. Roy, in: *Activity coefficients in electrolyte solutions* 2nd ed., edited by K. S. Pitzer (CRC Press, Boca Raton, 1991), pp. 156-189.

SIMFUEL LEACHING EXPERIMENTS IN PRESENCE OF GAMMA EXTERNAL SOURCE (^{60}CO)

J.A. Serrano*, J. Quiñones*, P.P. Díaz Arocas*, J.L. Rodríguez Almazán*, J. Cobos*, J.A. Esteban**, A. Martínez-Esparza**
* Departamento de Fisión Nuclear. CIEMAT. Avda. Complutense, 22. 28040 Madrid. SPAIN
** ENRESA, C/ Emilio Vargas, 7. 28043 Madrid. SPAIN

ABSTRACT

One of the factors considered within the studies of performance assessment on spent fuel under final repository conditions is the effect of the radiation on its leaching behavior. Radiation from spent fuel can modify some properties of both solid phase and leachant and therefore it would alter the chemical behavior of the near field.

Particularizing in the effect of the radiation on the leachant, it will cause generation of radiolytic species that could change the redox potential of the environment and therefore may bring on variations in the leaching process.

In this work, the chemical analogue utilized was SIMFUEL (natural UO_2 doped with non-radioactive elements simulating fission products) and the leachants selected were saline and granite bentonite waters both under initial anoxic conditions. To emulate γ radiation field of a spent fuel, leaching experiments with external ^{60}Co sources in a irradiation facility (Nayade) were performed. Initial dose rate used was 0.014 Gy/s.

Preliminary results indicate that radiation produces an increase of the uranium dissolution rate, being the concentrations measured close to those obtained in oxic atmosphere without radiation field. In addition the solubility solid phases from experimental conditions were calculated, for both granite bentonite water and 5 m NaCl media.

On the other hand, a tentative approach to model the role of γ radiolysis in these SIMFUEL tests has been carried out as well.

INTRODUCTION

The long-term interactions between spent fuel and groundwater has to be understood in order to dispose the fuel in a safe final underground repository. With this goal in mind, and given the limitations related to the handling of irradiated fuel, spent fuel dissolution experiments combined with more detailed studies using non-irradiated chemical analogues are very useful for performance assessment (PA) studies.

One of the factors considered within the PA studies on spent fuel behavior under final underground repository conditions is the effect of the radiation on the leaching processes. The radiation emitted by spent nuclear fuel alter the properties of the solid and the leachant, in such way that it modifies the chemistry of the near field [1, 2, 3, 4 and their references].

Radiolysis of the groundwater in contact with the used fuel produces oxidants for the oxidative dissolution of the nuclear fuel. The highest initial absorbed dose rate is generated by the β and γ radiation and after a long period the α radiation will be higher than those. Experimental measurements of the effects of radiolysis on the matrix dissolution rate are complicated by the fact that the effects of α, β and γ radiolysis can not be separated.

Particularizing in the effect γ radiation, SIMFUEL leaching experiments with and without presence of external ^{60}Co sources in bentonitic groundwater (GBW) [5] and in 5 m NaCl were performed. In this paper the initial results obtained from these experiments are presented.

Mat. Res. Soc. Symp. Proc. Vol. 608 © 2000 Materials Research Society

EXPERIMENTAL PROCEDURE

Fig.1 Gamma irradiation facility, Nayade, used for SIMFUEL leaching experiments.

Characteristic of the irradiation facility

Leaching experiments were performed in a γ irradiation facility (NAYADE) situated at CIEMAT. This radioactive facility allows irradiating materials under water by γ emitter (^{60}Co).

Reactors leaching experiments were placed in a well-closed basket and this was immersed in the pool. The basket was encircled by γ sources.

Previously to start the experiments, Fricke dosimetry of the basket was performed, being the initial dose absorbed for the leaching vessel of 0,014 Gy/s [6]. This γ dose chosen is consistent with a 40 MWd/kg spent fuel after 20 years of cooling time.

Leaching experiments

SIMFUEL pellets used for these tests were supplied by AECL Research Chalk River Laboratories. The concentration of additives in SIMFUEL simulate the fission products representative of a burn up of 50 MWd/kg U. The percent of each element in UO_2 matrix and general characteristics of SIMFUEL are given elsewhere [7].

Fig.2 Detail of the irradiation basket used for leaching experiments.

Sampling and preparation of all leaching tests were performed inside of a glove box under anoxic conditions (N_2 atmosphere). The oxygen content in the glove box atmosphere was always lower than 1 ppm. SIMFUEL powder samples were conditioned as was described in previous report [8]. They presented a specific surface of 0.04 $m^2 \cdot g^{-1}$. Leachants used were synthetic groundwater GBW [5] and 5 m NaCl.

Parallel leaching experiments, inside a glove box under nitrogen atmosphere (without γ sources) and in the Nayade facility were performed. All the tests were got ready in N_2 atmosphere inside a glove box to avoid superficial oxidation before to start the experiments. Solutions were analyzed by ICP-MS and fluorimetric techniques.

RESULTS AND DISCUSSION

Table I shows the U concentration from SIMFUEL after 190 days in N_2 atmosphere with and without presence of γ emitters (labeled *Nayade* and *Glove box*, respectively) in comparison with others obtained without external sources in air atmosphere [9] for the same experimental time (labeled *Air*).

Table Error! Unknown switch argument. U concentration after 190 d of experimental time in various conditions for both GBW and 5m NaClNaCl: "Glove Box", nitrogen atmosphere; "Nayade", initial nitrogen atmosphere and presence of external γ source and "Air", air atmosphere.

Experiment	*U (mol·kg⁻¹) in GBW*	*U (mol·kg⁻¹) in 5m NaClNaCl*
Glove box	$4.6 \cdot 10^{-8}$	$3.6 \cdot 10^{-8}$
Nayade	$1.7\ 10^{-6}$	$2.2\ 10^{-6}$
Air	$3.2 \cdot 10^{-6}$	$6.7 \cdot 10^{-6}$

It is observed that uranium concentrations measured in anoxic atmosphere (N_2 atmosphere) without irradiation (Glove box) both in GBW and in NaCl are lower than those obtained in the irradiation facility, Nayade (Fig.3 and 4, respectively). From this it is reasonable to think that γ dose received by the leaching system is enough to alter SIMFUEL leaching behavior, due to generation of oxidant radiolytic products.

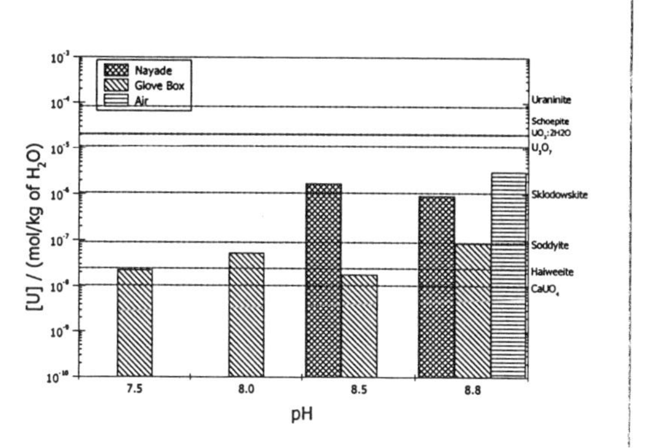

Fig.3 Uranium concentration vs. pH in GBW under irradiation, reducing and oxidizing conditions (Nayade, Glove box and air atmosphere, respectively). Lines correspond with the solubility solid phases calculated for Nayade conditions at pH 8.8.

On the other hand, uranium concentrations obtained from Nayade experiments are in the same order of magnitude that those obtained in air atmosphere.

As it was expected, uranium concentration values obtained in NaCl and GBW are very close. However the ratio between U concentration measured in the presence and absence of a γ field for tests performed in NaCl solution is almost double than in GBW. This fact could be related with the composition of the leachants, i.e. with the initial concentration of Cl^- and the subsequent generation of radiolytic species such us $ClOH^-$, ClO_2^-, etc.

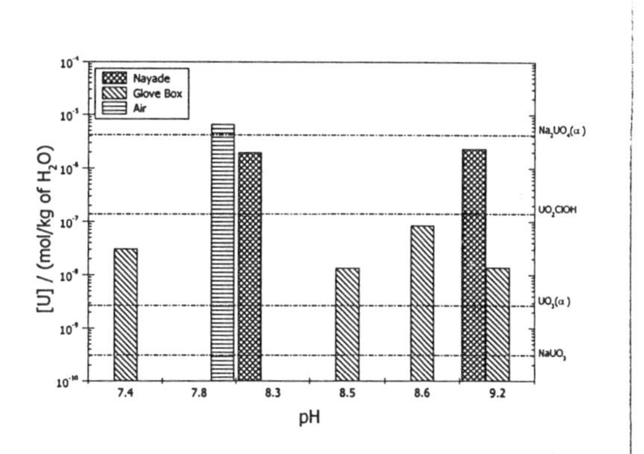

Fig.4 Uranium concentration vs. pH in NaCl under irradiation, reducing and oxidizing conditions (Nayade, Glove box and air atmosphere, respectively). Lines correspond with the solubility solid phases calculated for Nayade conditions at pH 9.2.

From experimental conditions of these tests (Eh, T, pH and leachant chemical compositions), and assuming the hypothesis that concentrations measured (Table I) reach the steady state, solubility control phases for U concentration in solution were calculated. The geochemical modeling codes used were EQ3/6 [10] for brine solutions (Pitzer data base "data0.pit.R2") and Phreeqc v.1.6 [11] for GBW tests supported by the thermodynamic data base NEA [12].

Fig.3 shows the U concentration measured versus pH for the tests performed in GBW under irradiation, reducing and oxidizing conditions (Nayade, Glove box and Air tests, respectively). In this figure, the lines represent the uranium solubility calculated for each solid phase considered for the Nayade experimental conditions at pH 8.8. As can be deduced, from this thermodynamic calculation, candidate solid phase for controlling uranium solubility in GBW under γ irradiation is Sklodowskite ($Mg(H_3O)_2(UO_2)_2(SiO_4)_2{\cdot}4H_2O$). In the case of the experiments carried out in 5m NaCl, Fig.4, the controlling solid phase calculated, also for Nayade conditions at 9.2, is Na_2UO_4 (α). In Table II, are summarized the controlling uranium phases in GBW and 5 m NaCl for the various chemical conditions studied. All of these thermodynamic calculations have to considered only as references points due to some of them are notoriously slow to form at low temperatures.

Table Error! Unknown switch argument. Solubility controls on uranium concentration in GBW and 5 m NaCl for Nayade, Glove box and Air experiments. (Bold writing correspond with the expected controlling solids).

	GBW		*NaCl*	
	Solid	*U (mol·kg⁻¹)*	*Solid*	*U (mol·kg⁻¹)*
Nayade	Shoepite	$2.049{\cdot}10^{-5}$	**Na_2UO_4(α)**	$4.2088{\cdot}10^{-6}$
	Haiweeite	$2.46{\cdot}10^{-8}$	UO_2ClOH	$1.4026{\cdot}10^{-7}$
	Skoldowskite	1.084·10-6		
	U_3O_7 (β)	$1.134{\cdot}10^{-5}$		
	$UO_3{\cdot}2H_2O$	$2.125{\cdot}10^{-5}$		
	Uraninite	$8.375{\cdot}10^{-5}$		
Air	Shoepite	$1.975{\cdot}10^{-5}$	**$NaUO_3$**	$7.1524{\cdot}10^{-6}$
	Skoldowskite	$1.931{\cdot}10^{-6}$	UO_2ClOH	$2.7987{\cdot}10^{-8}$
	Soddyite	$8.503{\cdot}10^{-8}$		
	$UO_3{\cdot}2H_2O$	$2.048{\cdot}10^{-5}$		
Glove box	Coffinite	$3.994{\cdot}10^{-9}$	Na_2UO_4	$6.6705{\cdot}10^{-7}$
	Rutherfordine	$1.771{\cdot}10^{-9}$	**UO_2ClOH**	$8.8500{\cdot}10^{-8}$
	U_4O_9	$4.216{\cdot}10^{-7}$	UO_3(α)	$1.0705{\cdot}10^{-9}$

A tentative approach to model the role of γ radiolysis in these SIMFUEL tests was carried out. The kinetic model used [13] is based on UO_2 matrix dissolution studies developed by Christensen et al. [14, 15]. Code applied for calculations was Maxim Chemist [16].

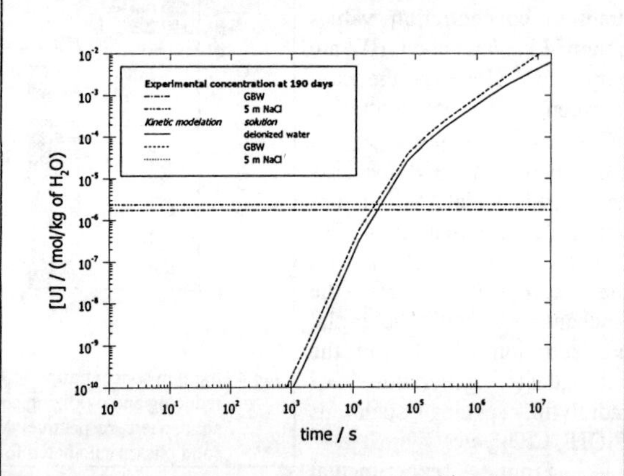

Fig.5 Evolution of U concentration from the kinetic model used in this work.

This model deems that UO_2 oxidation processes with O_2 and H_2O_2 can be reduced in two reactions taking into account experimental reaction constant values [17 and their references]. For chlorides, kinetics reactions used were taken from Kelm et al. [18, 19]. Retrodiffusion of O_2 (gas) and H_2 (gas) was considered.

Due to the special geometry of the experimental system (external γ source) it was assumed that all the radiolytic products generated in the whole leachant can react with the

SIMFUEL powder surface. In addition, diffusion liquid-gas was taken into account.

The Fig 5 shows the experimental data of U concentrations measured at 190 d for both GBW and 5 m NaCl solutions (horizontal lines) and the U concentration evolution calculated considering various leachants: deionized water, GBW and 5 m NaCl. U concentration calculated under 5m NaCl curve sit on top of GBW but is higher than the calculated in deionized water in which not chlorides reactions were considered. As can be observed in this Fig. 5, experimental U concentrations values obtained in Nayade tests are much lower than those predicted for the same reaction time for both NaCl and GBW solutions, in fact, the model achieves the experimental uranium concentrations in less than 3 hours. These high uranium concentrations calculated by the kinetic model can be related with the water layer in contact with water since, as was aforementioned, all the radiolytic products generated in the bulk leachant can react with the leachate.

Ongoing experimental work will provide data to try to obtain matrix alteration rates for various γ doses combining with groundwater chemistry by a more complex kinetic a thermodynamic model.

CONCLUSIONS

Enhanced U dissolution from SIMFUEL powder samples was observed by performing static leaching tests under anoxic conditions at room temperature in presence of external γ source in both GBW and in NaCl 5m solutions. After 190 days of leaching, the ratio between experiments performed with and without γ field was approximately 30 times for GBW and 60 times for 5 m NaCl.

Uranium concentrations obtained from the tests started under reducing atmosphere and underwent to a initial γ dose of 0.014Gy/s are comparable with others performed under oxic conditions (air atmosphere).

From thermodynamic calculations solid phases candidate for controlling uranium dissolution in the tests performed in presence of γ sources were Skoldowskite for synthetic granite bentonite groundwater and $Na_2UO_4(\alpha)$ for 5m NaCl tests.

Uranium concentrations calculated by the kinetic model were much higher than the measured values for both synthetic granite bentonite groundwater and 5m NaCl tests. These differences may be related with the water layer in contact with water since all the radiolytic products generated in the bulk leachant were assumed to be able to react with the leachate. Moreover, any restriction due to groundwater chemistry was taken into account.

A prediction of the role of γ radiolysis on SIMFUEL leaching, and in consequence for spent fuel dissolution behavior tests pass to consider a combination of thermodynamic and kinetic aspects and of course accurate experimental data. On going SIMFUEL leaching tests with various geometries and dose rates will try to provide data for future calculations.

ACKNOWLEDGEMENTS

This work has been carried out in the Dpto. of Nuclear Fission of the CIEMAT inside the agreement of the Association CIEMAT ENRESA (Annex XIV) in the area of the definitive storage of spent fuel. The authors would like to thank J. Fuentes, A. Quejido, M.J. Tomas, A. Fariñas and J.M. Altelarrea for their valuable help for performing this work.

REFERENCES

1. S. Sunder, D.W. Shoesmith, L.H. Johnson, G.J. Wallace, M.G. Bailey and A.P. Snaglewski in *Scientific Basis for Nuclear Waste Management X*, edited by J. Bates and W.B. Seefeldt (Mat. Res. Soc. Symp. Proc. **84,** Boston, 1986) 103-113.
2. H. Christensen, in *Scientific Basis for Nuclear Waste Management XVI*, edited by T. Abrajano, Jr and.L. H. Johnson (Mat. Res. Soc. Symp. Proc. **212**, Boston, 1990) 213-220.
3. A. Loida, B. Grambow, H. Geckeis and P. Dressler, in *Scientific Basis for Nuclear Waste Management XVII*, edited by T. Murakami and R. C. Ewing, (Mat. Res. Soc. Symp. Proc. **353**, Kyoto, 1994) 577-584 .
4. J. Cobos, V.V. Rondinella, Hj. Matzke, A. Martinez-Esparza, T. Wiss. "Radiolysis effects on uranium realease under spent fuel storage conditions". Presented at GLOBAL´99. 30 Agosto-2 Sept. 1999. Jackson Hole, Wyoming, USA.
5. CIEMAT specific procedure Nº PR-X8-01 (1996).
6. J. Fuentes, private communication, CIEMAT (1997).
7. P.G. Lucuta, R.A. Verall., Hj. Matzke, B.J. Palmer. J. Nucl. Mater. 178, 48 (1991).
8. J. Garcia-Serrano, J.A. Serrano, P. Diaz-Arocas, J. Quiñones, and J.L.R. Almazan, in *Scientific Basis for Nuclear Waste Management XIII*, edited by V.M. Oversby and P.W. Brown (Mat. Res. Soc. Proc. **412**, Pittsburgh, PA 1996), p. 229-240.
9. J. Serrano, J. Quiñones, PP Diaz, J.L. Rodríguez, J. Cobos, C. Caravaca. "Influence of redox potential on the dissolution behaviour of SIMFUEL".Presented at Spent Fuel 98 Wokshop. The Vegas, Nevada. USA. 18-20 May 1998.
10. T.J. Wolery, EQ3NR A computer program for geochemical aqueous speciation – solubility calculations: Theoretical manual, user's guide and related documentation (version 7.0) (1992).
11. D. L. Parkhurst "User's guide to phreeqc. A computer program for speciation, reaction-path, advective-transport, and inverse geochemical calculations" u.s. geological survey Water-Resources Investigations Report 95-4227 (1995)
12. Grenthe et al. "Chemical Thermodynamics of Uranium", edited by H. Wanner and I. Forest, NEA-OCDE (1992).
13. J. Quiñones, J.A. Serrano, P. Diaz Arocas, J.L. Rodríguez Almazán, J.A. Esteban, A. Martinez Esparza. Ciemat Informe Técnico DFN/RA-03/SP-99 (1999).
14. H. Christensen, S. Sunder, D.W. Shoesmith. J. of Alloys and Compounds **213/214**, 93-99 (1994).
15. H. Christensen . Presented at Spent Fuel Workshop 1999. Ontario Canada (1999)
16. M.B. Carver, D.V. Hanley, K.R. Chaplin. Maksima Chemist. A program for mass action kinetics simulation by automatic chemical equation manipulation and integration by using Stiff techniques. AECL-6413. (1979)
17. J. Bruno, E. Cera, L. Duro, J. Pon, J. de Pablo, T. Eriksen SKB Technical Report **TR-98-22** (1998)
18. M. Kelm private communication (1998).
19. M. Kelm Presented at Spent Fuel Workshop 1999. Ontario Canada (1999).

OXIDATIVE ALTERATION OF SPENT FUEL IN A SILICA-RICH ENVIRONMENT: SEM/AEM INVESTIGATION AND GEOCHEMICAL MODELING

Yifeng Wang*, Huifang Xu**

* Sandia National Laboratories, 4100 National Parks Highway, Carlsbad, New Mexico 88220. E-mail: ywang@sandia.gov
** Department of Earth and Planetary Sciences, The University of New Mexico, Albuquerque, New Mexico 87131. E-mail: hfxu@unm.edu

ABSTRACT

Correctly identifying the possible alteration products and accurately predicting their occurrence in a repository-relevant environment are the key for source-term calculations in a repository performance assessment. Uraninite in uranium deposits has long been used as a natural analog to spent fuel in a repository because of their chemical and structural similarity. In this paper, a SEM/AEM investigation has been conducted on a partially alterated uraninite sample from a uranium ore deposit of Shinkolobwe of Congo. The mineral formation sequences were identified: uraninite → uranyl hydrates → uranyl silicates → Ca-uranyl silicates or uraninite → uranyl silicates → Ca-uranyl silicates. Reaction-path calculations were conducted for the oxidative dissolution of spent fuel in a representative Yucca Mountain groundwater. The predicted sequence is in general consistent with the SEM observations. The calculations also show that uranium carbonate minerals are unlikely to become major solubility-controlling mineral phases in a Yucca Mountain environment. Some discrepancies between model predictions and field observations are observed. Those discrepancies may result from poorly constrained thermodynamic data for uranyl silicate minerals.

INTRODUCTION

It is proposed in the United States that spent fuel will be disposed in a geologic repository at Yucca Mountain, Nevada. In such an oxidizing and silica-rich subsurface environment, spent fuel (UO_2) will experience complex alteration processes when it contacts with incoming groundwater. The alteration will include the oxidative dissolution of UO_2 and the formation of various uranium-bearing secondary mineral phases, which usually act as a secondary waste form and directly control the mobility of uranium in a repository environment. Therefore, correctly identifying the possible alteration products and accurately predicting their occurrence in repository-relevant environments are the key for the source-term calculation in a repository performance assessment. Uraninite (UO_{2+x}) in uranium deposits has long been used as a natural analog to spent fuel in a repository because of their chemical and structural similarity [1, 2]. In this paper, we present our scanning electron microscopy(SEM) and analytical electron microscopy (AEM) observations on uraninite alteration in a sample from a uranium ore deposit of Shinkolobwe of Congo. Based on the observations, we conduct thermodynamic calculations to constrain the possible reaction paths of spent fuel alteration in a silica-rich

Mat. Res. Soc. Symp. Proc. Vol. 608 © 2000 Materials Research Society

environment. The capability and limitation of the current thermodynamic model are also discussed.

SAMPLE AND EXPERIMENTAL METHOD

A partially altered (weathered) uraninite is from a uranium ore deposit of Shinkolobwe of Congo. The surrounding rocks of the ore deposit are dolomitic siltstones. The oxidative alteration of uraninite was caused by downward meteoric water. Polished thin section was prepared for scanning electron microscopy (SEM) investigation. Crushed uraninite powder crystals were used for transmission electron microscopy (TEM) and electron energy-loss spectroscopy (EELS) analyses. Oxidation states of U were determined using a FEG TEM (Philips 420 ST) and associated Gatan PEEL system [3].

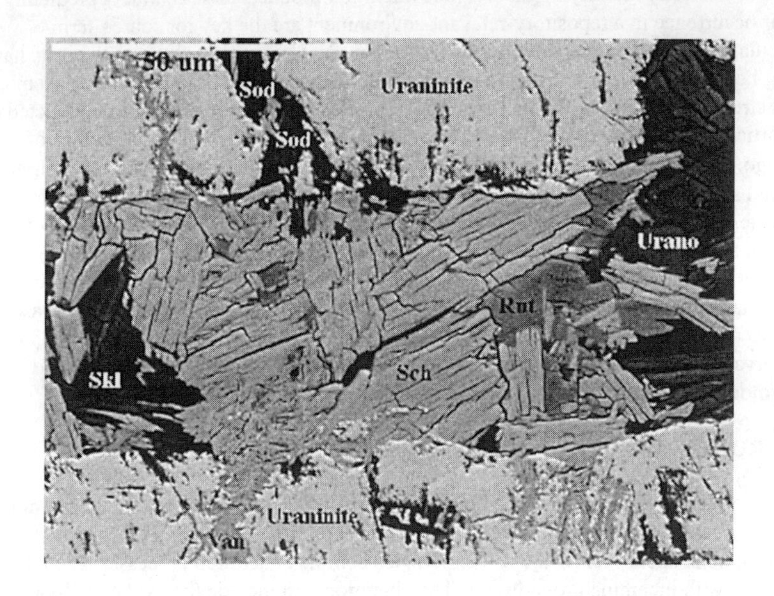

Figure 1. SEM image of a weathered uraninite ($UO_{2.5}$) sample from a uranium ore deposit of Shinkolobwe of Congo. Sod – soddyite ($(UO_2)_2SiO_4{\bullet}2H_2O$), Urano – uranophane ($Ca(UO_2)_2Si_2O_7{\bullet}6H_2O$), Sch – Schoepite ($UO3{\bullet}2H_2O$), Skl – sklodowskite ($Mg(UO_2)_2Si_2O_6(OH)_2{\bullet}5H_2O$), Van – vandendriesscheite ($PbU_7O_{22}{\bullet}12H_2O$), Rut – rutherfordine ($UO_2CO_3$).

SEM results indicate that the main alteration products are schoepite, soddyite and uranophane (Figure 1). Other U-bearing phases are sklodowskite, rutherfordine, and kasolite ($Pb(UO_2)SiO_4{\bullet}H_2O$). Silica-free U^{6+}-bearing minerals (such as schoepite) are the reaction products of uraninite with silica-depleted solutions. Alteration reactions preferentially occur along fractures in a uraninite crystal. Electron energy-loss spectroscopy results show the U^{6+}/U^{4+} ratio of uraninite is about 1/1. TEM results indicate that uraninite crystals are heterogeneous in crystallinity and composition (especially Pb concentration) (Figure 2). The alteration products of the uraninite, such as soddyite, are relatively poor in Pb with respect to the uraninite. Radiation damage caused by alpha decay mainly results in low crystallinity of uraninite (Figure 2).

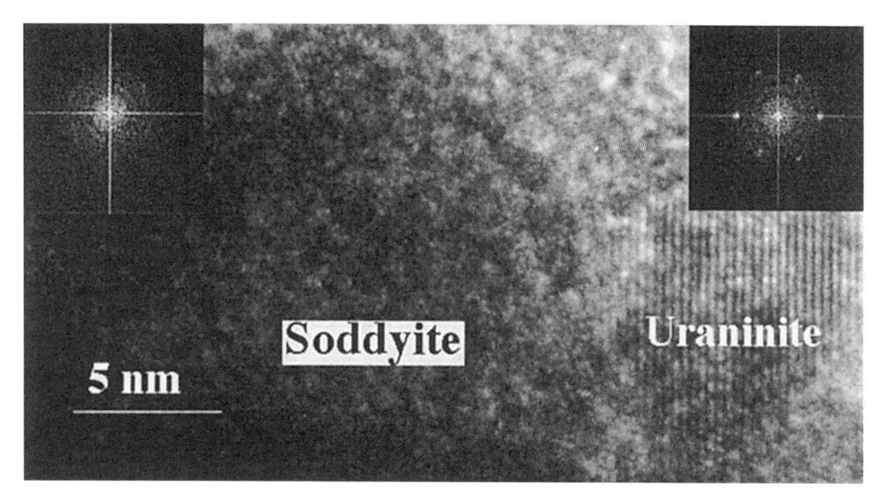

Figure 2. TEM image showing an interface between the uraninite and its alteration products, soddyite. EDS spectra indicate that the soddyite is relatively poor in Pb with respect to the uraninite. Inserted patterns on the upper corners are Fast Fourier Transform (FFT) of the image from soddyite and uraninite areas. Amorphous feature of the soddyite is caused by irradiation of high-energy electron beam (radiation damage).

The mineral formation sequence can be identified from the SEM image (Figure 1). Uraninite was directly replaced by either soddyite or schoepite. The soddyite was in turn replaced by uranophane. The early-formed schoepite could be replaced either by carbonate unranyl mineral phases such as rutherfordine or by uranyl silicates such as soddyite and uranophane. Lead, a fission product of uranium, was leached during uraninite alteration and was precipitated as vandendriesscheite next to uraninite grains. The occurrence of Mg uranyl silicate mineral (sklodowskite) and uranyl carbonate (rutherfordine) reflects the surrounding carbonate environment.

GEOCHEMICAL MODELING

Computer code EQ3/6 [4, 5] with an enhanced thermodynamic data was used to calculate the reaction path of spent fuel alteration in an oxidizing, silica-rich repository environment. In order to make our calculations more relevant to the Yucca Mountain environments, we use the composition of water from well J-13 at the Nevada Test Site. This water is often used to represent the groundwater at the proposed high level waste repository at Yucca Mountain, Nevada [6]. Two sets of reaction-path calculations were conducted. In both of them, UO_2 was titrated into one kilogram of J-13 groundwater with a fixed O_2 fugacity of 3.3 x 10^{-31} atm and CO_2 fugacity of 5.0 x 10^{-3} atm, which were calculated from the initial groundwater composition. In the first set of calculation, we assume that the groundwater is not buffered by ambient rocks. This set of calculations mimics spent fuel dissolution in the near field, that is, close to waste packages. In contrast, in the second set of calculations, we assume that the groundwater is always buffered by a silicate mineral (microcline) and carbonate mineral (calcite). The calculation results are presented in Figures 3 and 4.

As shown in Figure 3A, in the near field of a disposal room, depending on the relative rate of spent fuel dissolution to silica supply by incoming groundwater, the solution next to a waste package can become depleted in silica, resulting in the formation of uranyl hydrates ($UO_3.2H_2O$). Away from the waste package, the uranyl hydrates will then be replaced by soddyite and $CaUO_4$, depending on the initial concentrations of Si and Ca in the incoming water. Farther away from the waste package, the soddyite will be replaced by Ca-uranyl silicate (e.g., haiweeite ($Ca(UO_2)_2(Si_2O_5)_3{\bullet}5H_2O$)). Figure 3B shows that the silica concentration in the groundwater have a great impact on uranium solubility. In an oxidizing environment, the solubility of uranium in a silica-depleted groundwater is generally at least two magnitude orders higher than that in a silica-rich solution. Note that in a silica-depleted environment the oxidative dissolution of spent fuel has little impact on solution pH.

Farther away from a waste package, for instance, in the far field, the groundwater is most likely to be buffered by minerals in the host rock. Figure 4 displays the reaction path of spent fuel dissolution in the J13 groundwater buffered by both microcline and calcite. In this case, spent fuel will be directly replaced by soddyite, which will be in turn replaced by haiweeite. The formation of soddyite will induce silicate dissolution, resulting in an increase in solution pH and therefore the solubility of uranium in the solution.

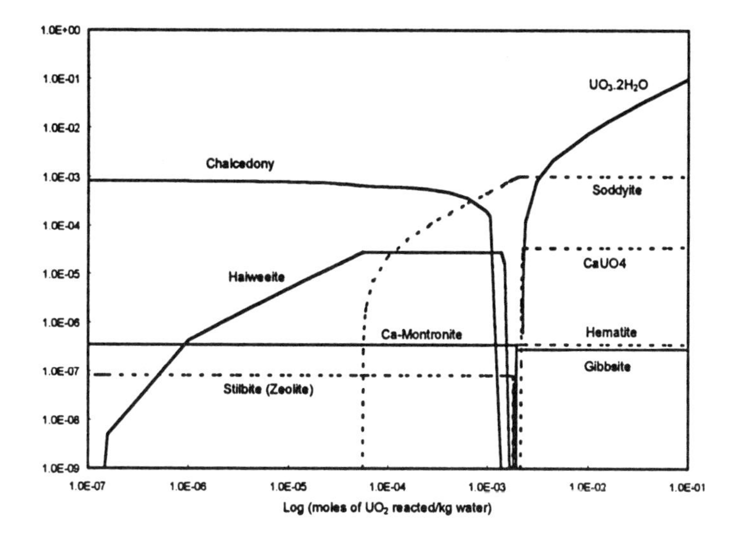

A

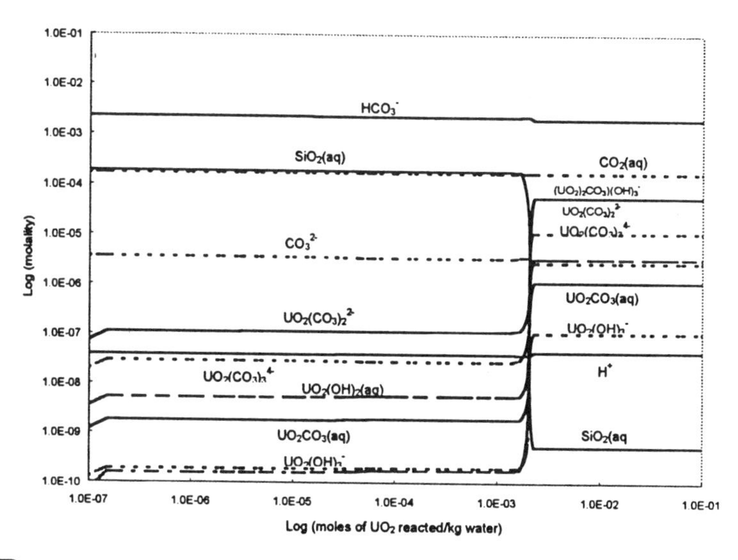

B

Figure 3. Reaction-path calculation of oxidative dissolution spent fuel in the J13 groundwater. In the calculation, the solution is not buffered by microcline and calcite.

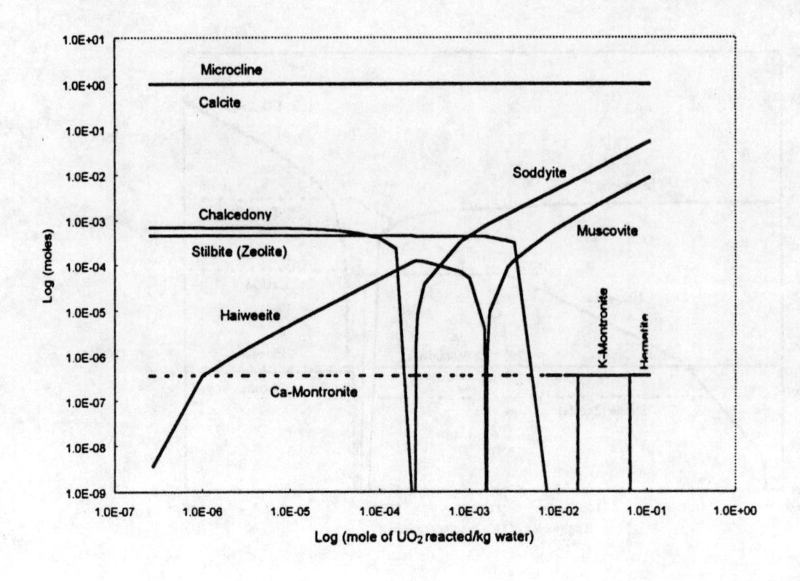

A

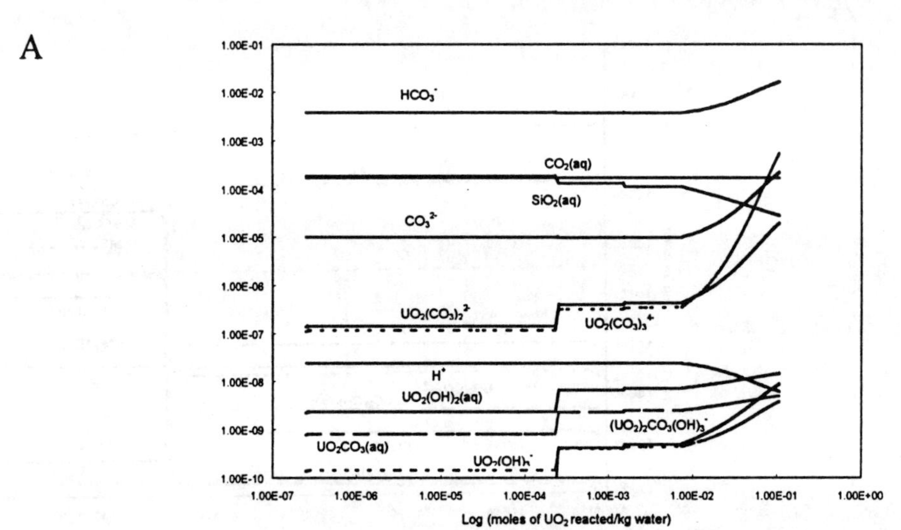

B

Figure 4. Reaction-path calculation of oxidative dissolution spent fuel in J13 groundwater. In the calculation, the solution is buffered by microcline and calcite.

Our computer simulations show that the predicted mineral formation sequence is quite stable for a given O_2 and CO_2 fugacity. The predicted sequence is in general consistent with our SEM observations on a natural analog sample. That is, in an oxidizing and silica-rich environment, the alteration of uraninite (or spent fuel) follows two reaction paths: uraninite → uranyl hydrates → uranyl silicates → Ca-uranyl silicates or uraninite → uranyl silicates → Ca-uranyl silicates. Our simulations also show that uranium carbonate minerals are unlikely to become major solubility-controlling mineral phases in Yucca Mountain environments. However, some discrepancies are observed between model prediction and field observations in detail mineral phases. For example, the SEM observation shows that a dominant Ca-uranyl silicate mineral is uranophane instead of haiweeite as model predicted. In the calculation, uranophane can precipitate only when both soddyite and haiweeite are suppressed. This discrepancy may result from the poorly constrained thermodynamic data for uranophane in the computer database.

CONCLUSIONS

A SEM/AEM investigation has been conducted on a partially alterated uraninite sample obtained from a uranium ore deposit of Shinkolobwe of Congo. The mineral formation sequences were identified: uraninite → uranyl hydrates → uranyl silicates → Ca-uranyl silicates or uraninite → uranyl silicates → Ca-uranyl silicates. Reaction-path calculations were conducted for the oxidative dissolution of spent fuel in a representative Yucca Mountain groundwater. The predicted sequence is in general consistent with the SEM observations. The calculations also show that uranium carbonate minerals are unlikely to become major solubility-controlling mineral phases in a Yucca Mountain environment. Some discrepancies exist between model predictions and field observations in terms of the occurrence of uranophane. This discrepancy may result from the poorly constrained thermodynamic data for some uranyl silicate minerals.

ACKNOWLEDGMENTS

Sandia is a multiple laboratory operated by Sandia Corporation, a Lockheed Martin Company, for the United States Department of Energy (US DOE) under Contract DE-AC04-94AL85000. This work is based upon a research conducted at the Transmission Electron Microscopy Laboratory in the Department of Earth and Planetary Sciences of the University of New Mexico, which is partially supported by NSF and State of New Mexico. H. Xu also acknowledge the NSF of China for partial support of the study.

REFERENCES

1. R. J. Finch and R. C. Ewing, J. Nucl. Mater., 190, p. 133-156 (1992).
2. M. Fayek, T. K. Kyser, R. C. Ewing, and M. L. Miller, Mat. Res. Soc. Symp. Proc., 465, p. 1201-1207 (1997).
3. H. Xu and Y. Wang, J. Nucl. Mater., 265, p. 117-123 (1999).

4. T. J. Wolery, EQ3NR, A Computer Program for Geochemical Aqueous Speciation-Solubility Calculations: Theoretical Manual, User's Guide, and Related Documentation (Version 7.0). Lawrence Livermore National Laboratory, UCRL-MA-110662 PT III (1992).
5. T. J. Wolery and S. A. Daveler, EQ6, A Computer Program for Reaction Path Modeling of Aqueous Geochemical Systems: Theoretical Manual, User's Guide, and Related Documentation (Version 7.0). Lawrence Livermore National Laboratory, UCRL-MA-110662PTIV (1992).
6. J. E. Harrer, J. F. Carley, W. F. Isherwood, and E. Raber, Report of Committee to Review the Use of J-13 Well Water in Nevada Nuclear waste Storage Investigations. Lawrence Livermore National Laboratory, UCID-21867 (1990)

Container Fabrication and Corrosion

FABRICATION AND TESTING OF COPPER CANISTER FOR LONG TERM ISOLATION OF SPENT NUCLEAR FUEL

L. O. Werme
Swedish Nuclear Fuel and Waste Management Co.(SKB), Box 5864, SE-102 40 Stockholm, SWEDEN.

ABSTRACT

In 1993 SKB launched its Encapsulation Plant Project. Within this project, SKB has:
a) designed a facility for encapsulation of nuclear fuel
b) laid down the design premises for a canister for disposal of nuclear fuel
c) tested and developed fabrication methods for copper canisters
d) evaluated the long term chemical and mechanical behaviour of the canister
e) made preliminary plans for a factory for the production of copper canister
f) constructed a canister laboratory for full scale testing of the key operations in an encapsulation plant.

The conclusions of this project were that a canister consisting of an outer layer of 50 mm copper over an insert of cast nodular iron would provide sufficient corrosion protection and would have sufficient mechanical strength. This canister can be produced by several methods such as forming from rolled plates, hot extrusion, and pierce and draw. These methods have been tested at full scale and the results from these manufacturing tests will be presented and discussed. The canister will be sealed by electron beam welding in the encapsulation plant and the integrity of the weld will be verified by ultrasonic testing and high-energy radiography. The final development work in this area will be performed in the Canister Laboratory in Oskarshamn, Sweden. The laboratory is now in operation and is equipped with a 100 kW EB-welding unit, a 9 MV digital X-ray unit and a state-of-the-art phased array ultrasonic testing unit.

INTRODUCTION

Sweden has 12 nuclear reactors in operation at 4 different sites. These reactors produce about 50% of all the electricity used in Sweden. By 2010 the current nuclear program will have produced approximately 8000 metric tons of spent nuclear fuel. In 1976, Sweden started a program for handling the nuclear waste and a comprehensive system for nuclear waste management has been developed. This system is owned and managed by the Swedish Nuclear Fuel and Waste Management Co. (SKB) and consists of a central interim storage facility for spent nuclear fuel (CLAB), an underground repository for radioactive operational waste (SFR) and a transport system for spent fuel and radioactive waste.

After 30 to 40 years of storage in the interim storage facility (CLAB), the fuel will be encapsulated in corrosion resistant disposal canisters. Encapsulation of the fuel will take place in an encapsulation plant built as an extension to CLAB. After encapsulation, the fuel will be transported to a geological repository, where the canisters will be deposited at a depth of 500 to 700 m in granitic rock. Leakage of radioactive substances from this repository will be prevented by a multibarrier system consisting of the fuel itself, a corrosion-resistant canister, a bentonite clay buffer, and the overlying rock mass. The canister will prevent all dispersal of radioactivity to the surrounding rock, as long as it is intact. The other barriers can retard and attenuate radionuclide

Mat. Res. Soc. Symp. Proc. Vol. 608 © 2000 Materials Research Society

dispersal to acceptable levels if the canister starts to leak. The barrier system is illustrated in Figure 1.

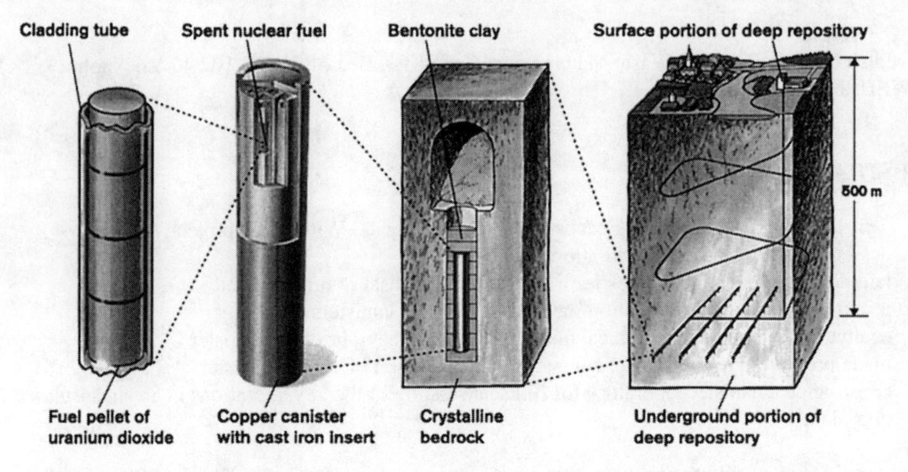

Figure 1 - The multiple-barrier system in the repository: the nuclear fuel, the canister, the bentonite buffer and the bedrock.

The function of the canister in the geologic repository is to isolate the spent fuel from the surrounding environment. This requirement is interpreted to mean that the canister should completely isolate the waste and that no known corrosion processes should be able to violate the integrity of the canister for at least 100,000 years. This requirement on canister integrity leads to requirements on:

- initial integrity,
- chemical resistance in the environment expected in the repository,
- mechanical strength under the conditions expected to prevail in the repository.

To meet the requirement that other barriers should retard and attenuate radionuclide dispersal to acceptable levels if the canister begins to leak, the requirement must also be made that the canisters must not have any harmful effect on other barriers in the repository. This imposes requirements on:

- choice of a canister material that does not adversely affect buffer and rock,
- limitation of heat and radiation dose to the near field,
- a configuration such that the fuel in the canister remains subcritical even if water enters the canister.

A detailed discussion of the design premises for the canister for disposal of spent nuclear fuel is given in ref. [1].

CANISTER DESIGN CRITERIA

Initial integrity

The canisters must be intact when they leave the encapsulation plant. This is considered to have been achieved if the canisters are fabricated, sealed and inspected with methods that guarantee that no more than 0.1% of the finished canisters will contain defects larger than the acceptance specifications for non-destructive testing.

Chemical resistance

The decay heat in the fuel is generated primarily by fission products during the first few hundred years. From a radiotoxicological viewpoint, isolation of the fuel for 1,000 years corresponds to a reduction by approximately a factor of 10, but also ensures that most of the more readily accessible radionuclides have decayed. It takes another 50,000 years for a further ten-fold reduction to be obtained.

Isolation for 100,000 years would reduce the toxicity of the fuel by a factor of more than 100, plutonium would be reduced ten-fold and the toxicity of the fuel would be comparable to that of the uranium from which it had been fabricated.

Mechanical strength

The canisters must withstand deposition at a depth of up to 700 m in granitic rock, surrounded by bentonite clay. This entails an evenly distributed load of 7 MPa of hydrostatic pressure. The bentonite clay swells upon water uptake and will, if the volume is restricted, exert a pressure on the rock and the canister. For the density (2000 kg/m^3) chosen for the bentonite clay, this pressure will be 7 MPa. These pressures are regarded as additive and isostatic. The canister should be designed for these loads with the customary safety margins.

The strength of the canister must also withstand a postulated increase in hydrostatic pressure due to glaciation. In this case, 30 MPa is added to the groundwater pressure of 7 MPa, corresponding to an ice cap approximately 3,000 m thick. The total groundwater pressure is then 37 MPa, and with the bentonite's swelling pressure the total pressure is 44 MPa. This load case is to be regarded as an extreme case for which no extra safety margins are required.

Impact on other barriers

Dissolved substances from the canister material and its corrosion products must not chemically alter the buffer so that its swelling properties, hydraulic conductivity, and diffusion resistance are appreciably impaired in the buffer as a whole or locally by creating rapid transport pathways through the buffer. The same applies to gaseous corrosion products. The growth of corrosion products on the canister surface must not lead to a pressure build-up in the buffer that could threaten the mechanical integrity of the canister.

If extensive vaporisation of groundwater takes place during the initial phase in the repository before it is water-saturated, enrichments of precipitated substances, such as salts, can form on the canister surface. Then, when the repository is saturated with water, the water chemistry in the immediate vicinity of the canister can be considerably more aggressive than otherwise foreseen. To prevent this, the canister is designed for a maximum outside temperature below 100°C.

The canister must provide sufficient radiation shielding to prevent radiation from altering the bentonite buffer or the water chemistry in the near field. This can occur by radiolysis of water or humid air prior to water saturation. The contribution of the radiolysis products to canister corrosion must be negligible in comparison with other corrosion during the life of the canister in the repository.

The canister must be designed so that the fuel remains subcritical even if water should enter the canister. This will influence the design of the interior of the canister in that the canister must either be designed so that the quantity of moderator (water) the canister can hold is limited, or suitable neutron absorbers must be incorporated in the canister.

MATERIALS SELECTION

Chemical resistance

Over a period of more than 15 years, an extensive database on groundwater chemistry has been built up. Despite the fact that the data represent widely spread geographic regions of Sweden, they present a relatively consistent picture. Groundwater in granitic rock in Sweden is oxygen-free and reducing below a depth of 100 to 200 meters. The redox potential below this depth ranges between -200 and -300 mV on the hydrogen scale and the water has a pH ranging from neutral to mildly alkaline (pH 7–9).

The range of variation is relatively limited when it comes to solute content as well. The chloride concentration in the groundwater can, however, vary within very wide limits, ranging from approx. 0.15 $mmol/dm^3$ to approx. 1.5 mol/dm^3. The chloride concentration is balanced by equivalent quantities of sodium and, particularly in more saline water, calcium. The concentrations of other substances are generally low. Moreover, reducing groundwaters have a typical concentration of dissolved sulphides of $1.5 \cdot 10^{-5}$ $mmol/dm^3$.

The chemical environment in the immediate vicinity of the canister is determined by the composition of the bentonite pore water. This is, in turn, determined by the interaction between the bentonite and the groundwater in the surrounding rock. The chloride concentration is not affected by the bentonite, which means that in the case of groundwaters with high chloride contents, equivalent values are obtained in the bentonite pore water with a corresponding increase in sodium concentration and also in calcium concentration. With the exception of chloride, the concentrations of other dominant anions will be determined for the foreseeable future by the interaction of the groundwater with the bentonite and will be on the order of millimoles per litre of bicarbonate and tens of millimoles per litre of sulphate.

Resistance to corrosion can be achieved in several ways. The canister can be fabricated using a material that is not attacked by corrosion under the foreseen conditions; in other words, the material is immune to corrosion. Alternatively, corrosion resistance can be achieved by choosing a material that is passive under repository conditions. A third alternative is to fabricate the canister of a material that corrodes in a predictable fashion and give it a corrosion allowance that guarantees the desired service life in the repository.

Materials that are immune to corrosion are the noble metals. Passive materials include titanium, titanium alloys, stainless steels, and others. Canisters with corrosion allowances could be fabricated of low-alloy carbon steels.

Corroding materials, such as low-alloy carbon steels, have corrosion rates from a tenth of a micron per year up to several microns per year under repository conditions. Over a period of 100,000 years, this corresponds to a corrosion loss of from several tens of millimetres up to several hundred millimetres. In a conservative case with a corrosion rate of several microns per year an unacceptable canister wall thickness would be required.

Passive metals have a very low corrosion rate. All passive metals are relatively new materials and very little is known about their properties over long periods of time. This is primarily true of the stability of the passive film, which is the basis of their corrosion resistance. Many of these metals, such as aluminium and titanium, are, in themselves, highly chemically unstable and would corrode very rapidly if the passive film were broken down. They are also sensitive to localised corrosion, particularly crevice corrosion, albeit to a varying degree.

Noble metals, which would be immune to corrosion under all circumstances that could occur in the repository, are really only gold and platinum. Neither of these metals is a practical alternative as a canister material. Copper, on the other hand, has a wide stability range in oxygen-free water [2], and oxygen-free conditions are expected during most of the repository performance lifetime. Dissolved sulphides in the groundwater change the situation, and copper can then corrode by formation of copper sulphide and hydrogen. The concentrations of dissolved sulphide in the near field of the canister are, however, very low and the corrosion attack on copper will, during long periods of time, be controlled by the availability of dissolved sulphides [3,4,5].

Based on corrosion considerations, copper is ranked highest among candidate materials. The corrosion properties of copper are well known, and the occurrence of native copper and copper archaeological material several thousand years old makes it possible to verify models of general corrosion and pitting in varying environments over long periods of time.

The expected corrosion behaviour of a copper canister in the repository environment has been studied thoroughly for more than 15 years, and an expected life of over 100,000 years can be achieved with adequate margins using a reasonable thickness of the canister wall [3,4,5,6]. Based on the most recent corrosion evaluation in ref. [6], the total corrosion during 100,000 years is estimated to be less than 5 mm. This is much thinner than the minimum copper thickness required for safe handling of a waste canister and the ultimate wall thickness will, therefore, be determined by other factors than corrosion shielding.

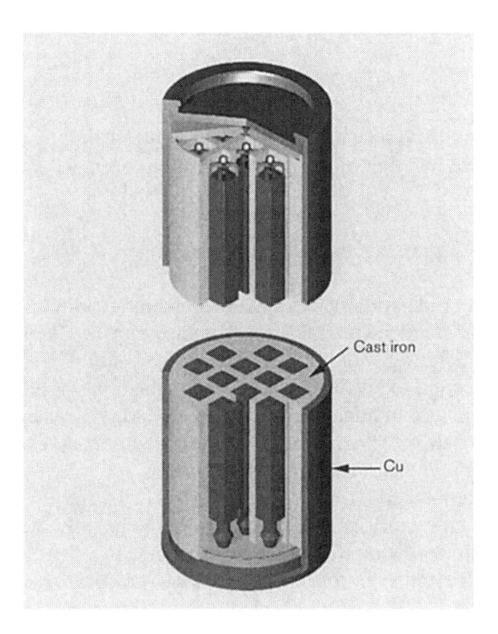

Figure 2 - Exploded view of canister for spent nuclear fuel. Version for BWR elements.

Mechanical strength

The canister must be designed for an external pressure of 44 MPa. Its strength is largely dependent on the design. The canister can be designed so that its own shell is capable of withstanding the external pressure. However, from point of view of corrosion it is desirable that the copper corrosion barrier is as stress-free as possible. Therefore, the canister has been designed so that the required mechanical strength is provided by a suitable inner structure that supports the outer copper shell. A nodular cast iron insert, as Figure 2 shows, gives this internal support.

The pressure of failure for a canister configured according to Figure 2 was calculated to be about 80 MPa for the BWR version and about 115 MPA for the PWR version.

REFERENCE DESIGN

The reference canister has an overall length of 4833 mm and an outer diameter of 1050 mm. It consists of an insert of cast nodular iron with a 50 mm spacing between the fuel channels and 50 mm minimum metal cover to the periphery. The wall thickness of the copper canister is 50 mm to satisfy a requirement of 100 mm total metal cover for radiation shielding.

The inner container or insert is cast with an integral bottom of nodular iron. The fuel channels are constructed of hot-drawn mild steel square tubes, 160x160 mm, with 10 mm material thickness in the form of a rack, which is then embedded in cast nodular iron. The spacing between nearby tubes in the rack is 30 mm. The weight of a canister with fuel is about 25 metric tons in the BWR version and about 27 metric tons in the PWR version. The total copper weight is about 7.5 metric tons.

The insert is placed inside an outer canister or shell (overpack) of copper. To enable the two components to be put together, the insert has an outer diameter that is 3.5 mm smaller than the inner diameter of the copper shell. This clearance guarantees a maximum strain in the copper of less than 4% when the copper shell is pressed against the insert as a result of the water pressure and the bentonite's swelling pressure build-up. This value is met by the requirements on the ductility of the copper material.

The copper canister is fabricated either in the form of two tube halves formed from rolled plate that are then welded together with two longitudinal welds, or of seamless extruded tubes. The longitudinal welds are made by conventional electron beam welding.

CANISTER FABRICATION

Three methods for the manufacturing of copper tubes have been tested in full scale:

- Rolling/Roll forming into tube halves welded together by longitudinal welding seams
- Extrusion of tubes
- Pierce and draw processing

The pierce and draw process is a process of pressing and pulling over mandrels to the required dimensions of the tube.

Today all these alternatives are established industrial methods used for manufacturing tubes in dimensions equivalent to the SKB canister tube dimensions. There are a number of available suppliers in Europe that can conduct test manufacturing with these methods. The major part of extrusion and pierce and draw processing production today deals with the manufacturing of tubes in more or less alloyed steel, often for demanding use in nuclear power plants or in the off-shore industry. However, before SKB initiated test manufacturing, there was no, or insignificant, experience in the test manufacturing of copper tubes in the current canister dimensions.

Copper lids and bottoms are machined from material that has been preformed by forging. The forging results in a form of material that enables a lesser amount of material to be machined off, but it also results in the material gaining a better (finer grain size) structure and also increased strength. Suppliers in the Nordic as well as other European countries can produce forged blanks for lids and bottoms.

The current welding technique for longitudinal welding of tube halves and fix-welding of bottoms used for the manufacturing of the canisters, is the electron beam welding technique (EB welding). The seal welding of the copper lids on canisters filled with spent nuclear fuel, is also planned to be performed by means of EB welding at the future encapsulation plant. The technique used for seal welding will be tested at SKB´s Canister Laboratory in Oskarshamn. The EB welding will be carried out in a welding chamber under conditions of partial vacuum or high vacuum. SKB has developed this special technique for welding of copper canisters together with TWI in England.

Roll-forming of tube halves and longitudinal welding

The basic materials used in roll forming are hot-rolled copper plates, which are formed into half-tubes. These half-tubes are then welded together with EB-welding. Figure 3 shows two half-tubes being put together for welding. After the longitudinal welding has been carried out, the tube has to be annealed for residual stresses to be removed, which otherwise could cause changes in its shape during machining. After annealing, the tube is machined to its final dimensions. Figure 4 shows the setting up of a copper tube for the internal drilling.

Figure 3 - The two half-tubes are put together for electron beam welding.

Figure 4 - Internal drilling of copper tube.

Extrusion of tubes

Extrusion is one of the two methods identified for use in seamless manufacturing of tubes. The extrusion tests were performed at Wyman-Gordon Ltd, which as far as we know is the only company with presses large enough to manage extrusion in the dimensions required. Wyman-Gordon has facilities in Scotland and USA.

Extrusion is a hot-forming process. The basic material is a cylindrical copper ingot of suitable size (about 11 metric tons, length approx. 2 m and diameter approx. 0.85 m), see Figure 5. This ingot is heated and by upset forging formed into a cylinder, which is then punched into a hollow blank for subsequent extrusion. Figure 6 is a photo of a punched blank.

The punched blank is heated to the required temperature and placed in position under a press tool. An internal mandrel determines the inner diameter. A large press tool is pushed downward and the tube is formed and pressed in a vertical upward single motion until the required dimension is reached. Figure 7 is a photo of extrusion-processed copper tubes.

Figure 5 - Copper ingot for extrusion.

Figure 6 - Photo of a punched blank ready for extrusion.

Figur 7 - Extrusion-processed copper tubes

Pierce and draw processing of copper tubes

The second method for use in seamless manufacturing of tubes is pierce and draw processing. At Vallourec & Mannesmann in Germany, SKB has been given access to a production facility in which full-scale test manufacturing can be carried out. Pierce and draw processing is, like extrusion, a hot-forming method. The basic material used is a cylindrical copper ingot of appropriate size. The material is first hot-formed by upset forging, which gives a shorter material with larger diameter, and is then placed in a special tool in which it is pierce-punched. The bottom part of the material is kept intact in order to support the piercers during the continued forming, which is done in several steps.

The forming results in the gradual increase of the inner and outer diameters of the SKB copper tubes until they have obtained the required final dimensions. This is done by successively changing the piercers and the outer ring through which the tube material is pressed. The tool measurements are adjusted to enable the gradual decrease of the wall thickness of the tube while, at the same time, the length of the tube is increased. Between each forming stage, the tube material is re-heated to the required temperature.

LID WELDING AND NON-DESTRUCTIVE EXAMINATION

With one exception, all the steps in the fabrication of canister and canister insert can be performed in a non-radioactive environment. The final step, however, the emplacement and seal welding of the canister lid as well as the non-destructive inspection of the seal weld will have to be carried out in a hot cell. In order to test these critical operations, SKB has built a Canister Laboratory in the town of Oskarshamn. In the laboratory the electron beam welding and non-destructive testing operations are being tested and developed further. The full-scale equipment is installed in three different stations: a welding station, a station for X-ray testing and a station for ultra sonic testing and machining. Other important parts of the process, which will be tested, include transport and handling equipment for canisters, canister lids, fuel assemblies and equipment in the handling cell.

Figure 8 - The top of a canister with lid before EB-welding.

Electron beam welding

Development of electron beam welding for the sealing of spent fuel copper canisters has been going on for more than 15 years at TWI, Cambridge, UK [see e.g. ref.7]. The goal of this work has now been achieved: a wide round-bottomed weld free of root defects. The EB unit at the Canister Laboratory is capable of producing electron beams with a power of 100 kW. A differential pumping system allows the gun column to be operated at atmospheric pressure in the welding chamber. The normal operating gas pressure in the welding chamber is 0.2 mbar, i.e. considerably higher the vacuum required for conventional electron beam welding. Figure 8 shows a lid about to be welded in the welding chamber at the Canister Laboratory. The picture also shows the lid lifting and positioning device (top) and the EB-gun nozzle (left).

Figure 9 - A spent fuel canister set for inspection by digital radiography.

Digital radiography

For the spent fuel canisters, conventional X-ray radiography cannot be used. The 50 mm wall thickness together with the lid design result in a total metal thickness of 100 mm that has to be inspected. This requires high energy X-rays for sufficient radiation penetration. The background radiation from the encapsulated fuel requires digital radiation detection to be used. For this purpose a 9MV digital radiography system has been installed. BIR INC., USA, supplied it. The radiation source is a Varian linear accelerator. The system is capable of detecting a density variation of less that 1 %, corresponding to 1 mm size pores in the weld. Figure 9 shows the top of a canister inside the radiography cave. The radiation source is at the top of the picture and the detector array at the bottom. The weld is inspected at an angle, so that also the position of the flaws in the weld can be determined.

Ultrasonic testing

As a complement to radiography, the welds will also be inspected ultrasonically. The surface of the weld corresponds to 150,000 mm^2 and flaws larger than 2 mm are to be detected. In order to be able to carry out such an inspection in a reasonable time, the ultrasonic scanning system has to be highly automated. The unit used at the Canister Laboratory has a 64-element phased array transducer, which allows electronic scanning of the weld. R/D Tech France has supplied the system. In parallel with the application work at the Canister Laboratory, development work is going on at the University of Uppsala [8]. The work includes studies of ultrasonic beams in copper, defect detection and methods for noise reduction.

Friction stir welding

As an alternative method for sealing copper canisters, SKB and TWI have developed a solid phase welding method. This method was invented in 1991 at TWI and is called friction stir welding. Consistent with the more conventional methods of friction welding, which have been practised since the early 1950s, the weld is made in the solid phase, that is no melting. Since its invention, the process has received world-wide attention and today two Scandinavian companies are using the technology in production, particularly for joining aluminium alloys.

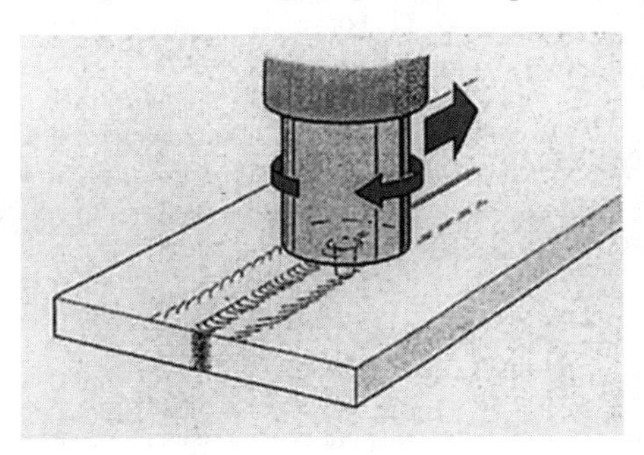

Figure 10 - The principle of friction stir welding

In friction stir welding (FSW) a cylindrical, shouldered tool with a profiled probe is rotated and slowly plunged into the joint line between two pieces of sheet or plate material, which are butted together (See Figure 10). The parts have to be clamped onto a backing bar in a manner that prevents the abutting joint faces from being forced apart. Frictional heat is generated between the wear resistant welding tool and the material of the workpieces. This heat causes the latter to soften without reaching the melting point and allows traversing of the tool along the weld line. The plasticised material is transferred from the leading edge of the tool to the trailing edge of the tool probe and is forged by the intimate contact of the tool shoulder and the pin profile. It leaves a solid phase bond between the two pieces. The process can be regarded as a solid phase keyhole welding technique since a hole to accommodate the probe is generated, then filled during the welding sequence.

SKB's work programme has demonstrated the feasibility of friction stir welding of 50 mm thick copper plate [9]. An experimental canister lid (and base) welding machine has, therefore, been designed and constructed. Test welding is now in progress and if the tests are successful, the machine will eventually be transferred to SKB's Canister Laboratory for further testing and development.

FUTURE WORK

Based on results obtained until now, the work with testing and further developing the manufacturing techniques will be continued. The work that has been carried out until now has proven that SKB has established a concept for copper canister design and manufacturing that, in all probability, can be optimized and tested during coming years to ensure that it can function efficiently in the production process. So far over 20 tubes have been manufactured and the current situation and the plans for future work can be summarized as follows:

- For the manufacturing of copper tubes, roll-forming of rolled copper plate to tube halves with longitudinal welding is a method that has now been tested on a relatively large number of tubes, which probably can be developed into a functioning production method. However, tests with seamless tube manufacturing have yielded such promising results that the immediate focus in tube manufacturing will be on continued testing of extrusion and pierce and draw processing.
- Electron beam welding of bottoms and lids has only been performed to a limited extent using the purpose-built welding equipment. Further work in this area will be performed at the Canister Laboratory, where also the NDE techniques will be further developed and tested.
- The development work with friction stir welding will continue. Repair weld trials will also be undertaken, with friction stir welding used to remove fusion weld flaws.

Figure 11 - Complete full-size canister

REFERENCES

1. L.O. Werme, "Design premises for canister for spent nuclear fuel", Technical Report TR-98-08, 1998, Swedish Nuclear Fuel and Waste Management Co. (SKB).

2. B. Beverskog and I. Puigdomènech, "SITE-94. Revised Pourbaix diagrams for copper at 5-150°C", SKI Report 95:73, 1995, Swedish Nuclear Power Inspectorate.

3. The Swedish Corrosion Institute and its reference group, "Copper as canister material for unreprocessed nuclear waste - evaluation with respect to corrosion", KBS Technical Report 90, 1978, Swedish Nuclear Fuel and Waste Management Co. (SKB).

4. The Swedish Corrosion Institute and its reference group, "Corrosion resistance of a copper canister for nuclear fuel", SKBF/KBS Technical Report 83-26, 1983, Swedish Nuclear Fuel and Waste Management Co. (SKB).

5. L. Werme, P. Sellin and N. Kjellbert, "Copper canisters for nuclear high level waste disposal. Corrosion aspects", SKB Technical report 92-26, 1992, Swedish Nuclear Fuel and Waste Management Co. (SKB).

6. P. Wersin, K. Spahiu and J. Bruno, "Kinetic modelling of bentonite-canister interaction. Long-term predictions of copper canister corrosion under oxic and anoxic conditions", SKB Technical Report 94-25, 1994, Swedish Nuclear Fuel and Waste Management Co. (SKB).

7. K.R. Nightingale, A. Sanderson, C. Punshon, L.O. Werme, "Advances in EB Technology for the Fabrication and Sealing of Large Scale Copper Canisters for High Level Nuclear Waste Burial", 6th International Conference on Welding and Melting by Electron and laser Beams, CISFFEL 6, 1998, Vol. 1, pp. 323-330.

8. P. Wu, T. Stepinski, "Inspection of copper canisters for spent nuclear fuel by means of Ultrasonic array System. Modelling, defect detection and grain noise estimation", SKB Technical Report TR-99-12, 1999, Swedish Nuclear Fuel and Waste Management Co. (SKB).

9. C-G. Andersson, R.E. Andrews, "Fabrication of Containment Canisters for Nuclear Waste by Friction Stir Welding", presented at the "1st International Symposium on Friction Stir Welding", held at the Rockwell Science Center, Thousand Oaks, California, 14-16 June 1999. The Proceedings are available as a CD-ROM from TWI, Cambridge, UK.

PASSIVE DISSOLUTION AND LOCALIZED CORROSION OF ALLOY 22 HIGH-LEVEL WASTE CONTAINER WELDMENTS

D. S. Dunn, G. A. Cragnolino, and N. Sridhar
Center for Nuclear Waste Regulatory Analyses, Southwest Research Institute,
6220 Culebra Road, San Antonio, TX 78238-5166, ddunn@swri.org.

ABSTRACT

Localized corrosion of high level nuclear waste containers is considered an important factor that will have a strong influence on the overall performance of the proposed repository at Yucca Mountain, NV. The present candidate container material, Alloy 22 [UNS N06022 (57Ni-22Cr-13.5Mo-3W-3Fe)], is highly resistant to localized corrosion. Assessing the performance of the HLW containers also requires an evaluation of localized corrosion resistance and determination of the passive corrosion rate of Alloy 22 weldments. The localized corrosion resistance of welded Alloy 22 specimens was evaluated by measuring the repassivation potential in chloride containing solutions whereas potentiostatic tests were used to determine the passive corrosion rate of the welded material. The results for the welded material are compared to those for the base metal.

INTRODUCTION

A key factor in the overall performance of the proposed Yucca Mountain (YM) repository for the disposal of high level nuclear waste is the ability of the waste packages to contain the radioactive waste for an extended period [1]. Waste package designs have gradually shifted to include thick walled containers fabricated from more corrosion resistant alloys. Recently, several enhanced design alternatives (EDA) were proposed to replace the Department of Energy (DOE) viability assessment (VA) waste package design[2]. The EDA II waste package specifications include a 5-cm thick type 316 nuclear grade (NG) stainless steel (SS) (69Fe-17Cr-12Ni) inner container surrounded by a 2-cm thick Alloy 22 (57Ni-22Cr-13.5Mo-3W-3Fe) outer container. Although additional design options in the EDA II, including a dripshield encompassing the waste packages and the use of backfill, may extend the lifetime of the waste package, this design is primarily dependent on the corrosion resistance of the outer Alloy 22 barrier.

Previous investigations have shown that Alloy 22 in the mill annealed condition is resistant to localized corrosion in the range of chloride containing environments expected at the YM site [3,4]. In addition, the uniform corrosion rate of mill annealed Alloy 22 was shown to be sufficiently low to promote long waste package lifetimes in the absence of localized corrosion [5]. Welding operations during closure of the WP, followed by prolonged exposures to elevated temperatures in the emplacement drifts, may affect the phase stability of corrosion-resistant Alloy 22. In this case, generation of short- and long-range ordered structures or formation of brittle intermetallic phases [i.e., topologically close-packed (TCP) phases] may accelerate uniform and localized corrosion processes and even affect mechanical properties [6]. Although excellent performance has been demonstrated for the mill annealed material, variations in the corrosion resistance of the container material as a consequence of fabrication and welding must be considered in order to determine the long term performance of the waste packages. The objective of this paper is to examine the effects of welding and thermal exposure on the passive dissolution rate and localized corrosion resistance of Alloy 22.

MATERIALS AND METHODS

Passive corrosion rates were measured using cylindrical specimens 6.2 mm in diameter and 48.6 mm long. Crevice corrosion repassivation potential (E_{rcrev}) measurements were performed using specimens machined with flat surfaces that allowed the attachment of polytetrafluoroethylene (PTFE) crevice forming washers to the specimens surfaces. Alloy 22 crevice specimens were machined from two 12.7-mm thick sections (heat 2277-8-3235) that were welded across the crevice area using an Alloy 622 (58Ni-21Cr-14Mo-3W-3Fe) filler rod (heat XX1045BG11). Tests were also conducted using specimens machined from hot-rolled and annealed, 12.7-mm thick plate (heat 2277-8-3175) thermally aged at temperatures of 870 °C for times ranging from 0.5 to 240 h.

Mat. Res. Soc. Symp. Proc. Vol. 608 © 2000 Materials Research Society

Compositions of the test specimens are provided in Table 1. Deaerated test solutions contained 1 to 4 M NaCl, 1.42 mM HCO_3^-, 0.20 mM SO_4^{2-}, 0.16 mM NO_3^-, and 0.10 mM F^-, added as sodium salts. The chloride concentration of the test solutions was many times greater than the chloride concentration of the groundwater in the Yucca Mountain region that is typically 2×10^{-4} molar. The higher chloride concentrations were chosen to simulate the effect of evaporation and concentration on the surfaces of the waste packages. Prior to the start of a test, all specimens were polished to a 600 grit finish, cleaned ultrasonically in detergent, rinsed in deionized (DI) water, and ultrasonically cleaned in acetone. At the completion of each test, specimens were cleaned and examined using an optical microscope and a scanning electron microscope (SEM).

Passive dissolution rates of Alloy 22 specimens were measured in 2-L glass cells that were fitted with a water-cooled Allihn-type condenser and a water trap to minimize solution loss at elevated temperatures and air intrusion. A saturated calomel electrode (SCE) was used as a reference electrode in all experiments. The SCE was connected to the solution through a water-cooled Luggin probe with a porous silica tip so that the reference electrode was maintained at room temperature. A platinum flag was used as a counter electrode. All solutions were deaerated with high-purity nitrogen (99.999 percent) for a period of at least 24 h prior to the start of the tests in order to obtain accurate anodic current density measurements at potentials ranging from −200 to 800 mV_{SCE} for a period of 2 days. Tests were conducted with a computer controlled potentiostat with a current resolution of 1.25×10^{-10} A/cm^2.

Repassivation potentials were determined on specimens with artificially formed crevices using cyclic potentiodynamic polarization (CPP) tests at temperatures ranging from 95 to 175 °C in a 316L SS autoclave with a PTFE liner following a procedure that is similar to ASTM Standard G61 [7]. The autoclave was equipped with a platinum counter electrode and an internal Ag/AgCl (0.1 M KCl) reference electrode. For comparison, all potential values were converted to the SCE scale at 25 °C. CPP testing was performed using a computer controlled potentiostat with a scan rate of 0.167 mV/s. The scans were reversed at a current density of 5 mA/cm^2. Repassivation potentials of the base alloy and thermally aged specimens were also measured in glass cells. Tests were conducted by scanning the applied potential from −600 to 900 mV_{SCE} at a rate of 0.1 mV/s to initiate localized corrosion. After a potentiostatic hold at 900 mV_{SCE} for 10 to 30 minutes, the repassivation was measured by reducing the applied potential at a rate of 0.167 mV/s.

RESULTS

Potentiostatic anodic current transients for base metal Alloy 22 specimens thermally aged 4 h at 870 °C are shown in figure 1. The temperature of 870 °C was selected because the formation of secondary phases in Alloy 22 occurs most rapidly in the range of 800 to 900 °C [8]. At −200 mV_{SCE}, the anodic dissolution current is close to 8×10^{-8} A/cm^2 at the start of the test and decreases below the resolution limit of the instrument after 20 h. Several large anodic current spikes were also observed suggesting a periodic increase in the dissolution rate for the aged specimens. A slightly lower initial anodic dissolution current density was measured when the specimen was polarized to 0 mV_{SCE}. However, after 12 h, the anodic current density sharply increased to over 10^{-5} A/cm^2. Significant intergranular corrosion was observed at the specimen/PTFE gasket. Similar results were obtained at 200 mV_{SCE} with the exception that the initiation time for intergranular corrosion was much shorter. Intergranular corrosion was only observed on a small surface area of the specimen. Note that in figure 1 the dissolution current is normalized to the entire exposed surface area. As a result, the anodic current density in the areas where intergranular corrosion occurred is actually greater than that depicted in figure 1. Similar results were obtained for specimens thermally aged for longer times. The initiation of intergranular corrosion at potentials lower than 200 mV_{SCE} was inconsistent and significant variations were observed in the measured anodic dissolution current densities among different specimens. When no intergranular corrosion was initiated, the anodic current density was similar to that measured for mill annealed specimens. Increasing the potential and the thermal aging time resulted in the appearance of numerous anodic current spikes indicating highly variable corrosion rates. Grain boundary etching and incipient intergranular corrosion were noted on these specimens, whereas severe intergranular attack was observed on all specimens where the anodic current density was greater than 10^{-5} A/cm^2.

Potentiostatic anodic current transients were also measured on welded Alloy 22 specimens. These specimens have a significantly larger surface area compared to the cylindrical specimens used for the majority of the passive corrosion tests. Approximately one-fourth of the total exposed specimen surface area (20 cm^2)

was weldment. All of the corners and edges were rounded prior to testing in order to prevent preferential attack in these areas. The results of anodic dissolution rate measurements of the welded specimens are shown in figure 2. At potentials less than 600 mV_{SCE} the anodic current density for the welded specimens are in the range of 2×10^{-8} to 4×10^{-8} A/cm^2 and are quite similar to the anodic current density measured for the base alloy. The anodic current density of the welded specimen tested in 4 M chloride adjusted to pH 2.7 is slightly greater than that measured in the alkaline solutions. At 600 mV_{SCE} the anodic current density of specimens tested in solutions adjusted to pH 2.7 and 11.0 increase substantially and at 800 mV_{SCE} the anodic current density measured in all solutions is greater than 10^{-4} A/cm^2. Post test examination of the specimen revealed preferential attack in the weld region that exposed the weld microstructure; however, no intergranular corrosion of the adjacent base metal was observed.

Comparisons of the E_{rcrev} values for welded versus the mill annealed Alloy 22 are shown in figure 3. For the welded material, the E_{rcrev} measured at 95 °C was approximately 100 mV lower than that for the mill annealed material. At 125 °C, no significant difference in the E_{rcrev} values was observed. Post test examination of the specimens revealed that preferential localized attack was observed in the welded region of specimens tested at temperatures of 95 and 125 °C; however, the severity of the attack to the welded regions increased with both temperature and chloride concentration.

Figure 4 shows the comparison of E_{rcrev} measurements for the base alloy to the thermally aged specimens. The E_{rcrev} for the base alloy in the 4 molar chloride solution was measured to be 350 mV_{SCE} whereas E_{rcrev} values ranging from -240 to -260 mV_{SCE} were measured on specimens that were thermally aged for 0.5 to 24 hours. Examination of the specimens revealed intergranular corrosion in the crevice regions of the specimens. As the aging time increased, the number of intergranular corrosion sites and the severity of attack increased. From the results shown in Figure 4, it is apparent that the E_{rcrev} is reduced by thermal aging; however, in the range from 0.5 to 24 hours the E_{rcrev} does not appear to be a strong function of aging time.

DISCUSSION

Passive corrosion rate measurements were performed on both heat treated and welded Alloy 22 because the formation of TCP phases in Ni-base materials may alter the resistance of the alloys to localized corrosion and increase the passive dissolution rate [9]. Although the microstructure of the thermally aged specimens used in this study was not characterized, thermal aging at 870 °C for 4 h is expected to result in the formation of TCP phases such as P, σ, and μ-phases [10,11]. Results obtained with the thermally aged specimens indicate the passive corrosion rate of Alloy 22 is not affected by the formation of TCP phases; however, the range of passivity is substantially reduced by thermal aging. The compositions of TCP phases are known to be Mo rich [10,11]. Although the role of Mo is not completely clear, incorporation of Mo into the TCP phases, particularly at the grain boundaries, may reduce the amount of Mo available to stabilize the passive film. As a result, intergranular corrosion of the thermally aged specimens occurred at low potentials, whereas passivity was maintained on the mill annealed material.

Specimens that were thermally aged at 870 °C had low E_{rcrev} values compared to the as received material. In addition, intergranular attack in creviced regions was also observed with the thermally aged specimens. Formation of TCP phases does not appear to alter the passive corrosion rate of Alloy 22, however, the aged material is much more susceptible to localized corrosion. The marked changed in E_{rcrev} after only 0.5 hours at 870 °C is consistent with the observations of Heubner et al. [8]who reported the precipitation of TCP phases in the grain boundary regions after 0.25 hours at temperatures ranging from 800 to 900 °C.

In contrast to the results obtained for the thermally aged specimens, for the welded specimens both the passive corrosion rate and the potential range over which the passive film was stable were much closer to that of the mill annealed material. Because the formation of P, σ, and μ-phases in Alloy 22 weldments has been identified [11], the welded regions can be expected to be less corrosion resistant than the base metal. Some preferential attack of the weld region occurred at potentials greater than 600 mV_{SCE}, but no deep intergranular penetration was observed. The relatively good performance of the weldments compared to the thermally aged specimens is consistent with the results reported by Heubner et al. [8] using ASTM G28 Method A [12] tests that consisted of boiling 42 g/L $Fe_2(SO_4)_3$ + 50 percent H_2SO_4. Heubner et al. [8] indicated that the corrosion rate of the welded material was 1.61 mm/yr or slightly more than twice that of the mill annealed material (0.73 mm/yr). On the other hand, the corrosion rate for Alloy 22 thermally aged for 100 h at 800 °C was 7.38 mm/yr. The

Table 1. Chemical compositions of the materials used in this study (in weight percent)

Material	Fe	Ni	Cr	Mo	W	Co	Mn	Si	S	C	Others
22 (2277-8-3175)	3.80	Bal.	21.40	13.60	3.00	0.09	0.12	0.030	0.002	0.004	P:0.008 V:0.15
22 (2277-8-3235)	3.94	Bal.	21.40	13.47	2.87	1.31	0.24	0.023	0.001	0.003	P:0.008 V:0.17
622 (XX1045BG11)	3.05	Bal.	20.73	14.13	3.15	0.09	0.24	0.06	0.001	0.006	Cu: 0.09 P: 0.007 V: 0.01

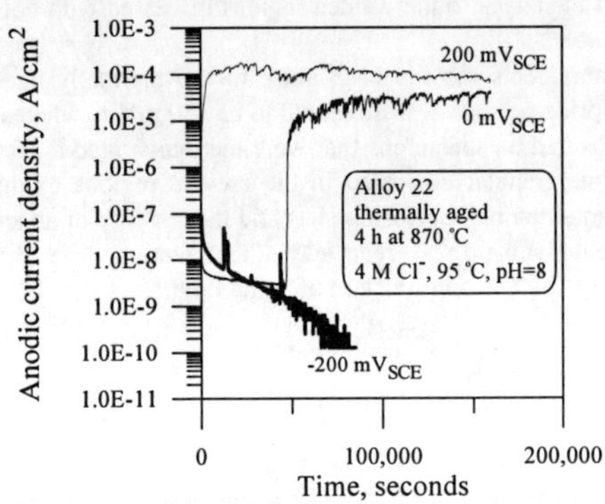

Figure 1. Anodic current transients for thermally aged base metal Alloy 22 specimens.

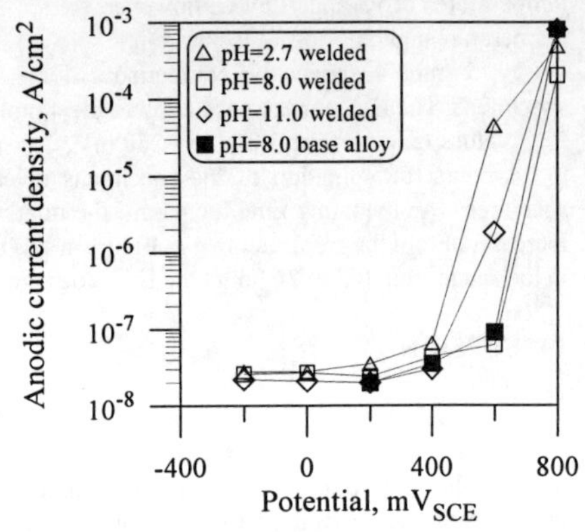

Figure 2. Steady-state anodic current densities measured for Alloy 22 in 4 M chloride solutions.

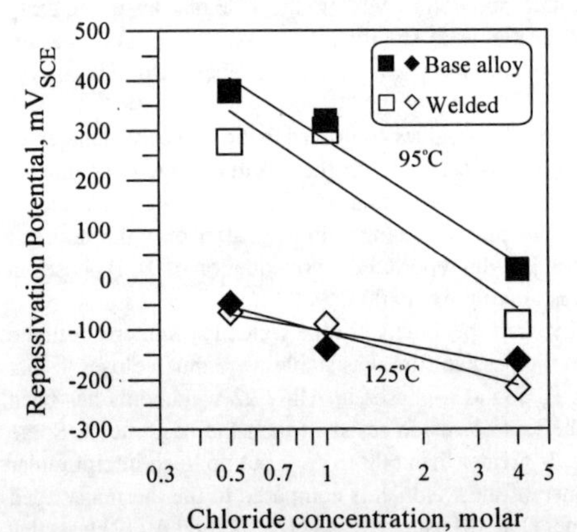

Figure 3. Repassivation potential for crevice corrosion measured on Alloy 22 specimens in chloride solutions at pH 8.

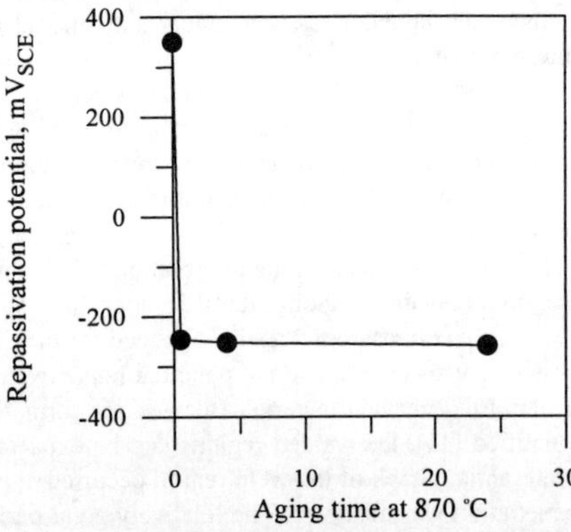

Figure 4. Crevice corrosion repassivation potential for Alloy 22 as a function of thermal aging time.

corrosion rates reported by Heubner et al. [8] were based on weight loss measurements and do not accurately reflect the severity of localized attack at the grain boundaries. No intergranular penetration was observed for the mill annealed material after the ASTM G28 Method A [12] test. However, intergranular penetration of 1 to 2 μm for the welds and penetrations of 100 to 250 μm for the thermally aged material were reported by Heubner et al. [8]. It is apparent from the comparison of the mill annealed, thermally aged, and welded materials that the welds do not contain a sufficient amount of secondary phases to substantially increase the localized corrosion susceptibility in the absence of a crevice.

Welding was observed to increase the localized corrosion susceptibility of Alloy 22. This was indicated by both lower values of E_{rcrev} for the welded specimens (figure 3) and preferential attack of the weldments. In addition, localized attack on the welded specimen exhibited deep intergranular penetrations that were not observed in the as-received material. Cieslak et al., [13] has shown that the interdendritic regions of weldments are depleted in Ni and enriched in Mo and Ti. It is apparent that segregation of the alloying elements leads to an increased crevice corrosion susceptibility of the weldments. In this study, the welded specimens were machined from plate that was 12.7-mm thick; however, the Alloy 22 container proposed in the EDA II is 20-mm thick. The additional thickness will require more weld passes and, as a result, the container will be at an elevated temperature for a longer period during the closure weld possibly increasing the amount of TCP phases present in the weldments and decreasing the localized corrosion resistance. In addition, the TCP phase precipitation kinetics may be accelerated by cold work associated with forming processes necessary to fabricate the container. Since the precipitation of TCP phases is dependent on both time and temperature, the results of this study indicate that the thermal mechanical history of the WP, including forming and cutting operations, welding, weld repairs, annealing cycles, and the temperature after emplacement are important considerations for the WP performance in the proposed repository.

CONCLUSIONS

Although the passive corrosion rates of Alloy 22 weldments were comparable to those measured on the base metal, greater susceptibility to localized and intergranular corrosion was observed with the welded specimens. Repassivation potentials of weldments were lower than the base metal, suggesting greater susceptibility to localized corrosion in oxidizing chloride environments. Because the composition of the weldments and the base metal are virtually identical, the lower repassivation potentials are believed to result from the formation of topologically close-packed phases. Thermal aging at 870 °C for 30 minutes, which may result in a microstructure similar to that of the base metal adjacent to a thick section weld, was found to increase the intergranular corrosion susceptibility of Alloy 22. These results indicate that the performance of the weldments and the heat-affected zone may reduce the life of the waste packages in the repository environment.

ACKNOWLEDGMENTS

The authors thank Walter Machowski and Jerry Sievert for assistance in conducting some of the experiments presented in this paper. This paper was prepared to document the work performed by the CNWRA for the NRC under contract No. NRC-02-97-009. This paper is an independent product of the CNWRA and does not necessarily reflect the views or the regulatory position of the NRC.

REFERENCES

1. U.S. Department of Energy, Repository Safety Strategy: U.S. Department of Energy's Strategy to Protect Public Health and Safety After Closure of a Yucca Mountain Repository, YMP/96-01 Rev.1, Washington, DC: U.S. Department of Energy, Office of Civilian Radioactive Waste Management, 1988.

2. Harrington, P. License application design selection (LADS) overview and process. *111th Meeting Advisory Committee on Nuclear Waste, Washington, DC, July 20, 1999.* Washington, DC: Nuclear Regulatory Commission. 1999.

3. D. S. Dunn, G. A. Cragnolino, and N. Sridhar Methodologies for Predicting the Performance of Ni-Cr-Mo Alloys Proposed for High Level Nuclear Waste Containers. Scientific Basis for Nuclear Waste Management. MRS Symposium Proceedings. 1999.

4. Dunn, D. S., Y-M. Pan, and G.A Cragnolino. Effects of Environmental Factors on the Aqueous Corrosion of High-Level Radioactive Waste Containers—Experimental Results and Models. CNWRA 99-004. San Antonio, TX: Center for Nuclear Waste Regulatory Analyses. 1999.

5. U.S. Nuclear Regulatory Commission, *Issue Resolution Status Report, Key Technical Issue: Container Life and Source Term,* Revision 2, Washington, DC: U.S. Nuclear Regulatory Commission, 1999.

6. Cragnolino, G.A., D.S. Dunn, C.S. Brossia, V. Jain, and K.S. Chan. *Assessment of Performance Issues Related to Alternate Engineered Barrier System Materials and Design Options*. CNWRA 99-003. San Antonio, TX: Center for Nuclear Waste Regulatory Analyses. 1999.

7. American Society for Testing and Materials. Standard test method for conducting cyclic potentiodynamic polarization measurements for localized corrosion susceptibility of iron-, nickel- or cobalt-based alloys: G61–86. *Annual Book of ASTM Standards. Volume 03.02: Wear and Erosion—Metal Corrosion*. West Conshohocken, PA: American Society for Testing and Materials: 237–241. 1999b.

8. Heubner, U.L., E. Altpeter, M.B. Rockel, and E. Wallis. Electrochemical behavior and its relation to composition and sensitization of Ni-Cr-Mo alloys in ASTM G–28 solution. *Corrosion* 45(3): 249–259. 1989.

9. Kasparova, O.V. The break-down of the passive state of grain boundaries and intergranular corrosion of stainless steels. *Protection of Metals* 34(6): 520–526. 1998.

10. Raghavan, M., R.R. Mueller, G. A. Vaughn, and S. Floreen. Determination of isothermal sections of nickel rich portion of Ni-Cr-Mo system by analytical electron microscopy. *Metallurgical Transactions* 15A: 783–792. 1984.

11. Cieslak, M.J., T.J. Headley, and A.D. Romig, Jr. The welding metallurgy of Hastelloy alloys C-4, C-22, and C-276. *Metallurgical Transactions* 17A: 2,035–2,047. 1986a.

12. American Society for Testing and Materials. Standard tests methods of detecting susceptibility to intergranular attack in wrought, nickel-rich, chromium-bearing alloys: G28. *Annual Book of ASTM Standards. Volume 03.02: Wear and Erosion—Metal Corrosion.* West Conshohocken, PA: American Society for Testing and Materials: 84–88. 1999d.

13. Cieslak, M.J., G.A. Knorovsky, T.J. Headley, and A.D. Romig, Jr. The use of PHACOMP in understanding the solidification microstructure of nickel base alloy weld metal. *Metallurgical Transactions* 17A: 2,107–2,116. 1986b.

STRESS CORROSION CRACK GROWTH IN COPPER FOR WASTE CANISTER APPLICATIONS

KJELL PETTERSSON, MAGNUS OSKARSSON
Dept of Materials Science and Engineering, Brinellvägen 23, KTH, SE-100 44 Stockholm, Sweden

ABSTRACT

Stress corrosion crack growth in pure copper has been studied with the aim of determining data which may be used in extrapolations to conditions of interest for use of copper as a canister material for long term storage of spent nuclear fuel. The canister should retain its integrity for a time of 100 000 years and the ultimate aim of the project is to show that stress corrosion cracking is not a threat to the canister integrity. The crack growth is studied as a function of applied stress intensity factor and environmental factors such as solute concentration and electrochemical potential. The approach used is to determine crack growth rates under conditions which give low but measurable rates which can then be extrapolated to times and conditions of relevance to canister integrity. The most accurate method of on-line crack growth monitoring, DC potential drop, is unfortunately in conflict with the low resistivity of copper. The results so far indicate that there is a stress intensity threshold for SCC growth in copper which is not appreciably lower than 30 MPa$\sqrt{m}$. Additional data are needed to improve the credibility of the threshold value. It is also proposed that crack growth experiments are performed in which the environment is gradually made less aggressive in order to prove that growth stops before environmental conditions of relevance to canister integrity are reached.

INTRODUCTION

Relatively pure copper has been proposed as a suitable canister material for the long term storage of high level nuclear waste. There are numerous potential problems, real or imaginary, which must be investigated before it can be considered certain that a copper canister will survive its design life of 100 000 years. One of the problems is corrosive attack on the canister from the outside. Such an attack can take several forms, a general corrosive attack, a pitting attack or an attack in the form of stress corrosion cracks. The choice of copper is based largely on the presumption that it will be very resistant to the first two forms of corrosive attack and this presumption has been confirmed by a large number of investigations and evaluations of the corrosion resistance of copper under the anticipated canister storage conditions. It is currently less clear how it can be proved that no stress corrosion cracks will nucleate and grow in the canister material under storage conditions.

One approach has been to assume that the initiation and growth of stress corrosion cracks is controlled by the electrochemical potential of the material in the canister environment. It is then presumed that the formation and growth of stress corrosion cracks are impossible below a certain potential, the critical potential. If this potential can be determined by experiments and if it can then be proved that it is impossible for the waste canister to attain this potential in the storage environment it is then proved that stress corrosion cracking can not occur. Several investigations have been carried out with the aim of determining this critical potential for copper [1–3]. These tests are however short term tests and none of them have lasted more than 1000 h, a time which differs from the canister exposure time by a factor of 10^6. The tests did, however, demonstrate the absence of a stress corrosion attack below a certain potential under the particular test conditions used, constant extension rate testing with extensive plastic

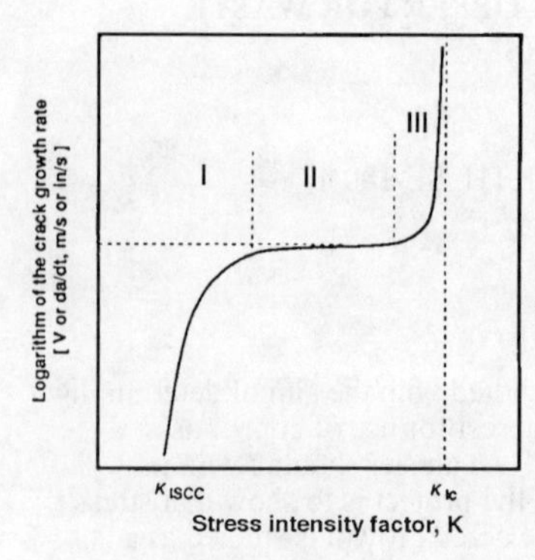

Figure 1. The three stages of stress corrosion crack growth

straining of the material. Unfortunately, none of the authors has elaborated on what this information is worth in the context of the long term integrity of the canister.

In the absence of convincing arguments for the significance of a critical potential it is reasonable to seek other approaches for the assessment of SCC susceptibility of the canister. One possibility is to accept that cracks may form but that the growth of cracks is extremely slow under storage conditions. One would then use the so called fracture mechanics approach where it is assumed that the crack growth rate is a function of the environment and the applied stress intensity factor. Figure 1 shows a relationship frequently observed for metals susceptible to stress corrosion cracking. According to Figure 1 the crack growth can be divided into three different stages, I, II and III and four stress intensity factor ranges: the ranges corresponding to the three stages and the range below K_{ISCC} in which no crack growth takes place. Stage I is characterized by a strong dependence on the stress intensity factor K_I often adequately described by the following equation:

$$\frac{da}{dt} = C \cdot K_I^n = C \cdot \left(\sigma\sqrt{\pi a}\right)^n \qquad (1)$$

where σ is the applied stress, *a* the crack length and *C* a constant. In stage II the crack growth is weakly dependent on K_I and controlled by diffusion processes in the crack environment. In stage III, finally, mechanical crack growth processes interact with the chemical process and at a high K_I-value a final purely mechanical failure occurs. In the context of the long term canister integrity stage I is the most interesting. The approach to an assessment of canister integrity could then be to determine the stress corrosion crack growth behaviour of copper over a range of stress intensity factor and a range of environmental conditions in the hope that relationships according to eq. 1 were found for the influence of stress intensity factor and some other relationships for the influence of environment. These relationships would then be used for extrapolations to canister conditions and, if the procedure was successful, extrapolated growth rates would be so low that the long term integrity of the canister was secured.

The purpose of the present project was to demonstrate the possibility of obtaining crack growth data for copper as a function of applied stress intensity factor as a pre-project to a more long term project for determining crack growth data. The reason for carrying out a pre-project was simply that since pure copper is a very soft and ductile metal it is not at all certain that the fracture mechanics approach to stress corrosion cracking will work. At the time of the start of the project some work on crack growth had been done in Finland where useful data had been obtained [4] although a few problems with branching of cracks had been observed [5]. More recently testing on sub-sized CT-specimens has been reported from Canada. [6]

EXPERIMENT

Material

Two different plates of annealed or hot rolled copper have been used in the experiments. One 30 mm thick plate contained 55 ppm P and 14 ppm Ag while the other plate with 40 mm thickness contained 35 ppm P and 10 ppm Ag. All other impurities are present only in concentrations less than 5 ppm. For the 30 mm thick plate a texture determination showed that the grain orientations were random. It is assumed that the same applies to the thicker plate. The P addition is intentional and done in order to improve the creep ductility of the material [7]. It also ensures that there will be no grain growth problems during storage [8]. The grain sizes of the materials ranges from 50–500 μm with a linear intercept average of about 130 μm. The materials have been used in the as-received condition as well as after various amounts of cold work performed by compression in the normal direction. In the annealed condition the yield strength is about 50 MPa, the uniform elongation about 40 % and the reduction of area at fracture nearly 100 %. An attempt to determine the fracture toughness on material with 20 % cold work resulted in two invalid tests from which it can be concluded that the fracture toughness is at least 130 MPa$\sqrt{m}$.

CERT-testing

A limited number of CERT tests were performed in screw-driven Instron tensile tester. The test environment was a 0.3 M solution of $NaNO_2$ in water in an open plastic container fitted to the load train so that the specimen was completely immersed in the liquid. The corrosion potential was measured versus a Hg/Hg_2SO_4 reference electrode (REF 601, Radiometer). A few tests were performed with the specimen potential controlled by a Wenking PGS 81 potentiostat. All potentials in the paper have been converted to the saturated calomel electrode scale (SCE)

Crack growth testing

For the crack growth testing was used a standard 25 mm thick CT specimen with screw-holes for attachment of current and potential leads for measurement of crack growth with the potential drop method.

The constant load testing was performed in a creep test rig where a dead weight load was applied to the specimen train through a lever arrangement. The load could be changed easily by just adding or removing weights from a weight pan. Before testing each specimen was subjected to fatigue pre-cracking so that a sharp crack of suitable depth was present before the stress corrosion crack growth test. The stress amplitude during the later parts of fatigue pre-cracking was such that the maximum stress intensity factor was well below the stress intensity planned for the stress corrosion crack growth test.

Figure 2. Specimen in grips of creep machine.

The test environment was circulated through two small plastic containers attached to the specimen one on each side. The arrangement can be seen in Figure 2. A platinum counter electrode was fixed to the inside of the flat vertical face of the container and on one of the edges a hole was drilled in which the reference electrode was mounted. With this arrangement it was possible to both measure the potential of the specimen and to control the potential of the specimen with the potentiostat.

The crack length was monitored by the reversing DC potential drop method. Due to the low resistivity of copper a current of 35 A was used in order to get a reasonably large signal. The current was only applied when the potential drop was measured, usually once every ten minutes. Typically the potential drop was 60 μV and it is thus unlikely that it can have an influence on measured corrosion potentials which are two to three orders of magnitude greater. Initially a dummy specimen identical to the active test specimen was used as a reference in order to compensate for any small variations in current and temperature. The whole measurement procedure was controlled and monitored by a computer. The relation between crack length and potential was checked with a specimen with fatigue cracks of known lengths. It turned out that the use of the correlation from ASTM E647 was sufficiently good to be used in the present case. Minor adjustments may be made post-test based on direct measurements of initial and final crack length.

RESULTS

CERT-testing

In the as-received condition testing at room temperature in $NaNO_2$ reduced the total elongation from 48 to 15% while testing at 80 °C gave a total elongation of 37 %. This indicates that the copper is less sensitive to SCC in this environment at higher temperatures. Specimens with 20% cold work were tested under potential control at room temperature. Specimens tested in air had an elongation to failure of 17 %. For potentials at or above 10 mV_{SCE} the total elongation was reduced to about 13 % in the $NaNO_2$ solution. There was no observable dependence on potential except for the existence of a threshold value somewhere below 10 mV. Without potential control the observed potential varied from -30–10 mV, in basic agreement with [3]. In all cases of an environmental effect on failure strain a post-test inspection of the specimen revealed a non-ductile appearance of the fracture surface.

Crack growth testing, phase 1

Phase 1 of the crack growth testing took place in a laboratory subject to daily variations in temperature. The first tests were performed with as-received copper and at a temperature of 80 °C. No crack growth was observed and the stress intensity factor was limited to 25 $MPa\sqrt{m}$ due to extensive plastic deformation of the specimen at higher values. In order to limit plastic deformation subsequent tests were performed on specimens with 10 % cold work. However no crack growth could be started at 80 °C for K_I-values up to 40 $MPa\sqrt{m}$ when plastic deformation was too extensive. Therefore the test temperature was reduced to room temperature at which the material had proved to be significantly more sensitive according to the CERT tests. An annealed specimen was tested with an initial K_I of 26.6 $MPa\sqrt{m}$ at which stress corrosion crack growth occurred. Growth out of the crack plane and initiation of cracks on the surface together with extensive plastic deformation made impossible any meaningful evaluations of growth rate and further tests were done on specimens with 10 % cold work.

The first such test was performed in 0.3 M $NaNO_2$ with a starting K_I of 32.2 $MPa\sqrt{m}$. Growth was monitored for 120 h after which the test was terminated in order to confirm by SEM examination that the cause of growth was SCC. It had then grown 1 mm. In a subsequent test with 0.1 M $NaNO_2$ it was impossible to start the crack to grow. In a new test with 0.3 M solution K_I had to be raised to 38 $MPa\sqrt{m}$ before the crack started to grow. It then grew steadily 5 mm for 150 hours until the test had to be terminated due to extensive plastic deformation. A dilution of the solution to 0.2 M about 100 hours into the growth phase had no effect on the growth rate. In the following tests the effect of reducing and increasing K_I was studied. An unloading by 10% generally stopped growth while reloading made the growth resume in the studied range of 40–50 $MPa\sqrt{m}$. The noise level in the potential drop signal in terms of crack length was about 0.05 mm with more long term variations with a larger amplitude. These latter variations were identified as due to daily variations in temperature of the room and it was decided to move the experiment to an air-conditioned room with a controlled temperature. The crack growth rates observed during this phase of the experiment agreed well with previous Finnish data [4, 5] and the following crack growth law was formulated on the basis of all available data:

$$\frac{da}{dt} = 5.2 \cdot 10^{-24} K_I^{11} \quad \text{mm/s} \qquad (2)$$

It is interesting to note that this law gives an allowable initial crack depth of 5 mm at 100 MPa for a 100000 year life of the waste canister.

Crack growth testing, phase 2

In phase 2 several improvements in measurement technique were made. The room had a more constant temperature and the electrochemical potential of the specimen could be measured and controlled. The noise in potential drop readings was also reduced by various means. The material used in this phase had a cold work level of 20 % in order to reduce plastic deformation at high values of K_I. All tests were run at room temperature. In the first test crack growth started at 30.6 MPa$\sqrt{m}$. The growth rate decayed to zero in about a 100 hours and did not start again until K_I was increased to 34 after 340 hours. The crack then made a jump of about 0.1 mm. Such jumps were observed frequently in later tests and may be interpreted as caused by tunnelling of the crack into uncracked material leaving uncracked ligaments behind and thus the crack is apparently stationary until the ligaments are broken by an increase in load. Unfortunately similar jumps are occasionally observed on reduction of K_I and it is then more difficult to explain the jumps. Another feature of this first test in phase 2 was that cracks initiated on the side faces of the specimen, and the recorded potential drop values are thus not entirely due to growth of the primary crack and no evaluations of crack growth rate are meaningful.

In test No. 2 potential control was possible and the test started at an open circuit potential of - 55 mV_{SCE} which increased to a steady value of 0 mV_{SCE} after some time. The starting K_I was 35.3 which gave a growth rate of 2.6×10^{-6} mm/s. A reduction to 32.2 reduced the growth rate to 2.8×10^{-7} mm/s. The potential was then fixed at 70 mV_{SCE} which increased the growth rate by a factor of 5. Similar changes were done all through the test until it was terminated after 1362 hours. All changes in crack growth rate due to changes in potential or K_I did go in the expected direction, but it should be noted that a given combination of K_I and potential did not necessarily result in the same crack growth rate. Occasionally changes in crack growth rate occurred which were not caused by any known external influence. These may however be explained by irregularities in the crack front which suddenly start to even out. An observation of particular interest was a significant crack growth rate of 4×10^{-6} mm/s at a potential of -90 mV_{SCE} at a K_I of 40.4 MPa$\sqrt{m}$. This is very close to -100 mV_{SCE} where almost no environmental influence was observed in CERT testing in both refs. 1 and 3.

Test No. 3 was performed in a similar way to test No. 2, the response to changes in potential and K_I was studied. During this test use of a reference specimen to compensate for temperature was terminated and instead the temperature on the active specimen was measured and all potential values were compensated to a 20 °C value by use of the temperature coeffiecient for the resistivity of copper. This removed the long wave variations in observed crack length but the noise remained almost the same despite the fact that the evaluated crack length now depended only on two voltage measurements instead of four. At one stage the solution of $NaNO_2$ was diluted from 0.3 M to 0.15 M at constant potential. This resulted in a reduction of crack growth rate and the test was concluded with that concentration. Basically the same response to changes in potential or K_I as before was observed. In Figure 3 is shown a compilation of crack growth data from tests No. 2 and 3. The data have been grouped according to potential, L is $E<0$ mV_{SCE}, M is $0<E<50$ mV_{SCE} and H is $E>50$ mV_{SCE}.

Test No. 3 was stopped by flushing the crack with water. The crack in the specimen was held open with a wedge to prevent closure during unloading. The crack was subsequently impregnated with an epoxy resin in order to stabilize it for preparation of cross section TEM specimens of the crack tip region. Unfortunately the crack was too branched to get any good TEM specimens. However it was possible to determine by electron diffraction that the film on the fracture surface was Cu_2O.

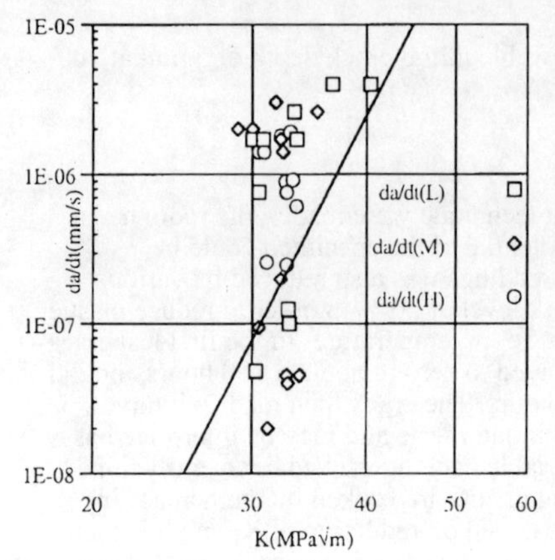

Figure 3. Crack growth data from tests No. 2 and 3. H, L, and M stands for high, medium and low potentials respectively. The straight line is drawn after eq. 2.

Test No. 4 was run from the start with 0.15 M $NaNO_2$. Initially something looking like crack growth occurred at a K_I of about 25 but after 400 hours the growth rate had decayed to zero. Polarizations to +110 mV_{SCE} during this stage had no effect. After 400 hours loadings and unloadings gave observations of positive and negative jumps in crack length but no steady growth. Many of the jumps occurred in the wrong direction, a phenomenon for which we have no explanation. A post test metallographic examination indicated blunting of the crack tip in one cross section and a grain boundary dissolution attack at the crack tip in another cross section.

In test No. 5 the environment was changed to 0.06 M ammonia, an environment in which SCC of copper has been observed previously [8]. The EC potential was initially -114 mV_{SCE} but drifted to -50 mV_{SCE} in 50 hours when a K_I of 32 $MPa\sqrt{m}$ was applied to the specimen. As in previous tests different applied potentials were used as well as different K_I values. The potential drop record indicates that crack growth occurred during the test but a post test SEM examination of the fracture surface showed only fatigue from the pre-cracking as well as from the post test fatigue cracking done before breaking up of the specimen. A possible part explanation to the observed increase in potential drop may be that the parts of specimen exposed to the ammonia solution had started to dissolve. On the face opposite the counter electrode 0.3 mm had been removed.

DISCUSSION

The data in Figure 3 strongly suggest a K_{ISCC} for copper of about 30 $MPa\sqrt{m}$. This value for K_{ISCC} of copper was also proposed in a previous report based only on the results of Phase 1 of the present experiments [10]. However the experiments were not performed with the aim of determining K_{ISCC}. If that had been the case they would have been done differently. It has been shown by Ståhle et al. that on unloading a CT-specimen during SCC growth crack closure will occur at the crack tip [11]. This may stop the crack for an indefinite time and that may have contributed to the failure to observe any growth below 30 $MPa\sqrt{m}$ in the present experiments. A more reliable way to determine a K_{ISCC} would be to gradually reduce the load on a specimen with a growing crack. Such a procedure was not possible with the creep machines used. In any case it seems unlikely that even a properly determined K_{ISCC} would be much lower than 30 $MPa\sqrt{m}$.

It appears that an analogue procedure might be used in the application of a less aggressive environment. In test No. 3 the crack growth continued after dilution of the solution to 0.15 M while in test No. 4 it was impossible to get the crack growth to start when the experiment started with the more dilute solution. In view of these observations it seems unlikely that it will be possible to start crack growth in a groundwater environment in the laboratory. Perhaps it may be a sufficently convincing demonstration that SCC is not a problem for waste storage if experiments were performed where one started growth in an aggressive environment and then gradually made the environment less aggressive. If crack growth stopped under those circumstances it seems extremely unlikely that it would ever start if the environment was less aggressive from the start of exposure. One problem with such an approach is that it is probably impossible to measure growth rates $< 10^{-9}$ mm/s in copper. Note that in the present experiments even in the best cases the noise level was 0.02 mm which is comparable to the growth in one year with 10^{-9} mm/s. However if the growth experiments are supplemented with calculations of oxidant availability at the crack tip of the type performed by King et. al. [6] which might explain the reduction in growth rate with change of environment then the case becomes stronger.

ACKNOWLEDGEMENT

This work was funded by The Swedish Nuclear Fuel and Waste Management Company, SKB. The authors would like to acknowledge the support and encouragement from SKB's project manager Lars Werme.

REFERENCES

1. L. A.Benjamin, D. Hardie, and R. N. Parkins, British Corrosion Journal, **23**, 89, (1988).

2. R. N. Parkins, Report from University of Newcastle upon Tyne, Department of Metallurgy end Engineering Materials, March 1988.

3. B. Rosborg and B.-M. Svensson, Studsvik Report STUDSVIK/M-94/73, 1994 (in Swedish).

4. S. Hietanen, Report No. VTT-MET B-232, Metalliilaboratorio VTT 1993.

5. H. Hänninen, (private communication).

6. F. King, C. D. Litke and B. M. Ikeda, Corrosion 99, Paper No. 482, NACE, 1999.

7. R. Sandström and P. J. Henderson, Mat. Sci. Eng. A, **246**, 143, (1998)

8. K Pettersson, Report No. TRITA-MAC-0594, KTH, 1996.

9. Y. Suzuki and Y. Hisamatsu, Corrosion, **21**, 353, (1981).

10. K Pettersson and M. Oskarsson, Report No. TRITA-MAC-0611, KTH, 1997.

11. P Ståhle, F. Nilsson, and L Ljungberg, Corrosion 88, Paper No. 281, NACE, 1988.

PITTING CORROSION OF COPPER . EQUILIBRIUM - MASS TRANSPORT LIMITATIONS

C. TAXÉN
Swedish Corrosion Institute, Roslagsvaegen 101, hus 25, SE-104 05 Stockholm, Sweden,
Claes.Taxen@corr-institute.se

ABSTRACT

Predictions from a mathematical model of the propagation of a corrosion pit in copper are reported. The model uses equilibrium data for solid and aqueous species to calculate local chemical and electrochemical equilibria in small volume elements. Mass transport between elements under local internal equilibrium is calculated using aqueous diffusion coefficients with the constraint of electrical neutrality. Propagation of a corrosion pit is deemed possible when the fraction of the oxidised copper that forms solid corrosion products, at the copper metal, is insufficient to completely cover the underlying metal. The effect of pH and salt concentrations in the bulk water was studied by varying the composition of the water.

Results are presented in the form of *E-log* $[Cl^-]$ diagrams where *E* is the applied potential and $[Cl^-]$ is the total bulk concentration of chloride. The *E-log* $[Cl^-]$ diagram shows two separate areas where pitting is found to be possible. One region at low chloride concentration and high potential and one region at high chloride concentration and low potential. Increased sulphate concentration is found to be detrimental with respect to pitting corrosion, particularly in the high potential region. Increased carbonate concentration is found to be beneficial, particularly in the low potential region. Pitting corrosion of copper can be described as a case of galvanic corrosion where cuprous oxide at a pH similar to that of the bulk is the cathode material for oxygen reduction and copper metal at the local, lower pH in a corrosion pit may behave as the anode.

INTRODUCTION

Most of the experience of the pitting of copper comes from studies in tap water. In tap water systems pitting corrosion has been known to cause leaks in copper pipes short time after installation. In an assessment of the integrity of a copper canister for spent nuclear waste against corrosion, the problem of pitting must therefore be addressed. The integrity of the copper canister against general corrosion has been judged more on grounds of the thermodynamic properties of copper than on experimentally determined corrosion rates. This fact together with the long time perspective has made us take a thermodynamic approach also to the problem of pitting corrosion.

THE MODEL AND THE CALCULATIONS

In the calculations of the concentration profiles, in and around a pit, we assume that there is local equilibrium within each volume element. The equilibrium concentrations for a large number of aqueous species are calculated for each element. Solids may precipitate and dissolve to satisfy the equilibrium conditions. Given the geometric frame, the equilibrium conditions, a set of specified assumptions and approximations, we can calculate the concentration profile in and around a corrosion pit. From the extent of the aqueous transport of copper we determine the domains in which pitting is possible and the domains in which copper is immune to pitting. The pitting domain is, in a water of a given composition, represented as a minimum pitting potential. For the free corrosion potential to attain a value higher than the minimum pitting potential, a cathodic process is

Mat. Res. Soc. Symp. Proc. Vol. 608 © 2000 Materials Research Society

required to take place at that potential. A site for this cathodic process is at the cuprous oxide surface outside the site of the pit. The likelihood that cuprous oxide is stable and accessible for aqueous oxygen decreases with increasing potential. We determine likely values for the maximum corrosion potentials at which cuprous oxide can act as a cathode.

A Geometric Model of The Site of a Corrosion Pit

Figure 1 illustrates schematically how we imagine the site of a corrosion pit on copper. There is a cavity in the copper metal caused by a preferential anodic dissolution at the bottom of the cavity. Porous corrosion products fill up a large fraction of the cavity volume but extend also outside the cavity. Aqueous reactants and corrosion products diffuse and migrate between the bulk of the solution outside the pit and the site of the anodic dissolution at the bottom of the pit. Chemical transformations between aqueous species take place everywhere so that local equilibrium always is attained. Figure 2 illustrates the description of the site of a corrosion pit as consisting of thin shells. The division is such that the ratio between the aqueous area and the thickness of the shell is constant between all shells. The mass transport through such shells is approximated as that through rectangular elements in one dimension. This approximation allows us to model the chemical behaviour of a corrosion pit without exact knowledge of the shape of the cavity or the shape of the crust of corrosion products outside the pit.

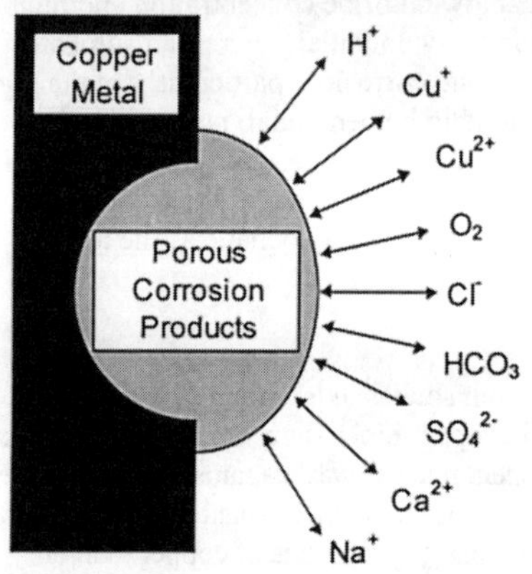

Figure 1. Schematic illustration of the site of a corrosion pit in copper with aqueous species diffusing and migrating.

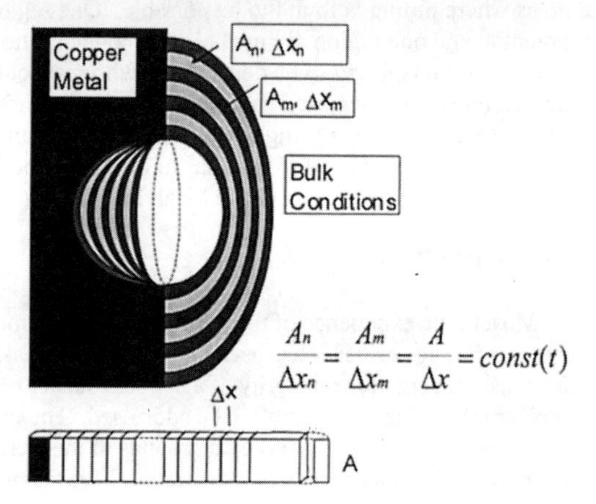

$$\frac{A_n}{\Delta x_n} = \frac{A_m}{\Delta x_m} = \frac{A}{\Delta x} = const(t)$$

Figure 2. Description of the site of a corrosion pit as consisting of thin shells.

Summary of the Model

The chemical system comprises the components Cu^+, Cu^{2+}, H^+, Na^+, Ca^{2+}, HCO_3^-, Cl^-, SO_4^{2-} and all significant combinations. All together 36 aqueous species and 16 solid compounds are considered. Thermodynamic data are taken from a compilation by Puigdomenech [1]. Diffusion coefficients for the aqueous species come from a variety of literature sources. Activity coefficients are calculated from the local ionic strength using Davies' approximation [2]. Typically 1000 to 2000 volume elements are used to describe the equilibria and the mass transport from the undisturbed bulk solution to the corroding copper metal at the growing front of a corrosion pit. Details of the model have been published elsewhere [3, 4].

Assumptions Made

The calculations of the concentration profiles in and around a corrosion pit and the determination of the pitting potentials are based on the following assumptions and approximations:

- diffusion and migration are the only modes of transport
- the changes in the pit geometry caused by the pitting process are so slow that diffusion and migration of aqueous species can be regarded as occurring in a fixed geometry.
- the growth rate of the pit is so low that the solution and porous solids in any small volume element in the pit can be considered to be in local internal chemical equilibrium.
- there is one axis in and outside the pit around which there is rotational symmetry. Perpendicular to this axis of symmetry there are surfaces along which no concentration changes. The flux of an aqueous species is always perpendicular to these surfaces.

RESULTS AND INTERPRETATION

Figures 3 to 5 show results for a corrosion pit formed at an applied potential of 288 mV (NHE), in a water with the bulk composition in table I. Figure 3 illustrates the redox conditions in and around the cavity. In the bulk of the solution outside the cavity, the redox potential is governed by the concentration of dissolved oxygen. Oxygen diffuses towards the cavity and is reduced by aqueous cuprous species diffusing and migrating outwards. At a pH slightly lower than pH 4, virtually all oxygen is consumed. This drop in oxygen concentration is manifested as a sharp drop in redox potential in figure 3. The redox potential continues to decrease until the stability region of copper metal is encountered. The local redox potential at the corroding copper metal at the bottom of the pit is reflected in a slightly higher measurable potential outside the cavity. The small slope of the line, indicating the external potential, is caused by the successive additions of the local iR-drops and diffusion potentials caused by current transport and concentration gradients.

Figure 4 shows the fraction of the oxidised copper that is transported outwards as dissolved species. In this water and at this potential we find that the transport of cuprous species is negligible and the aqueous transport of copper is almost exclusively in the form of cupric

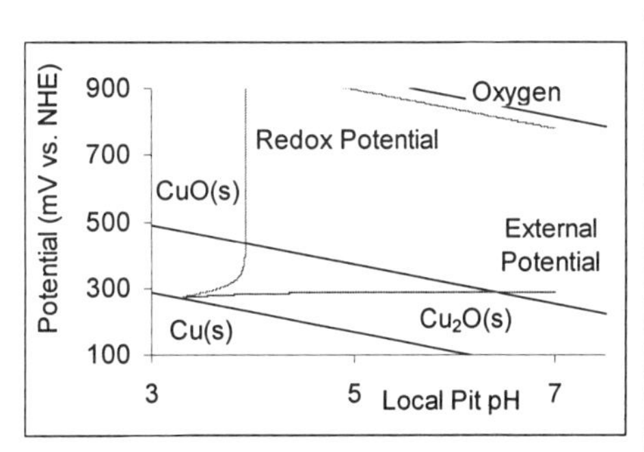

Figure 3. Conditions in a corrosion pit in copper illustrated in a potential-pH diagram. The straight lines indicate the relative stability of copper and its oxides and the reversible oxygen line at 1.0 atmosphere.

Table I. Composition of the water studied.

Temperature °C	25	
pH	7.0	
Total Concentrations	moles/litre	mg/litre
Chloride	0.0020	70.8
Sulphate	0.0040	384.0
Carbonate (as CO_3^{2-})	0.0012	69.0
Calcium	0.0005	21.6
Sodium	0.0101	231.5
Copper	0	0
Oxygen	5.0E-6	0.16

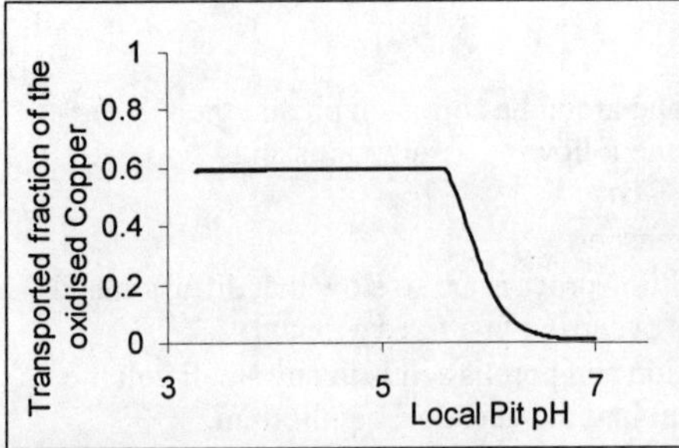

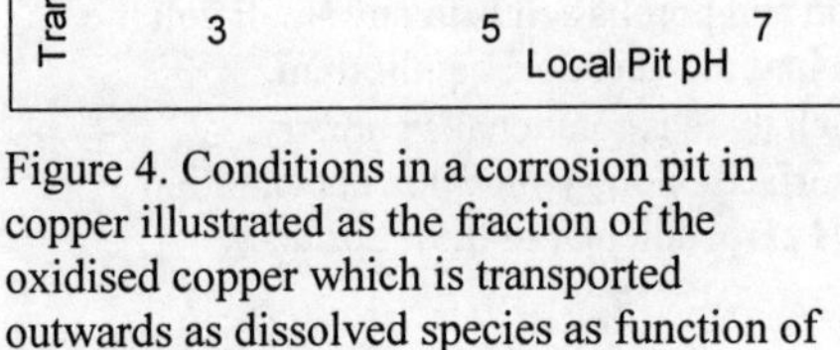
Figure 4. Conditions in a corrosion pit in copper illustrated as the fraction of the oxidised copper which is transported outwards as dissolved species as function of the local pH in the pit.

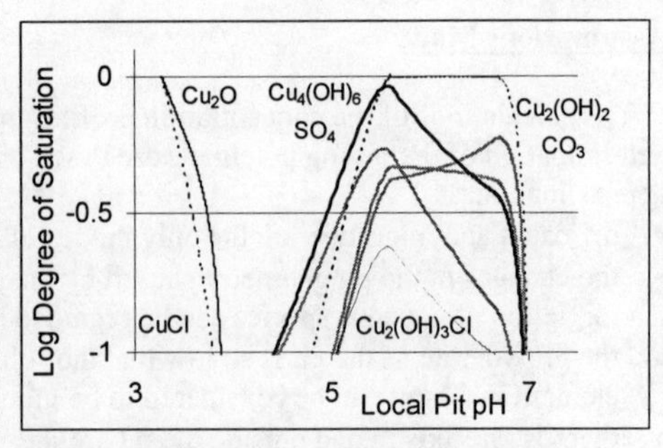

Figure 5. Log degree of saturation of the cuprous and cupric solids considered, as function of the local pH in the pit.

species. At about pH 5.5, saturation and precipitation of cupric solids occurs. This is shown by a decrease in the transported fraction as higher pH values are encountered. Figure 5 shows the degree of saturation with respect to the cuprous and cupric solids considered, as function of the local pH in the pit. In the higher pH-range, there is precipitation of malachite, $Cu_2(OH)_2CO_3(s)$. Over a narrow pH range around pH 5.5, brochantite, $Cu_4(OH)_6SO_4(s)$ is close to saturation. Atacamite, $Cu_2(OH)_3Cl(s)$, is not close to saturation under the conditions considered. Cuprous oxide, $Cu_2O(s)$, and cuprous chloride, CuCl(s), both reach saturation at the lowest pH.

Conditions in and around corrosion pits such as illustrated in figures 3 to 5 were calculated for a large number of potentials and water compositions. When the fraction of the oxidised copper that is transported away from the site of metal oxidation is plotted versus the external potential, a continuous curve is obtained. Figure 6 shows such curves for waters with the bulk composition given in table I with successive additions of sodium chloride. The chloride concentrations are indicated by the numbers at each curve. The circle at about 300 mV indicates the conditions for the corrosion pit illustrated in figures 3 to 5. The general trend in figure 6 is that increasing potentials give increasing values of the aqueous transport of copper. Increasing chloride concentrations shift the location of the curves to lower potentials. However, when cuprous chloride is formed in the pit, precipitation takes place immediately at the site of metal oxidation.

At potentials below the potential where CuCl(s) is formed, chloride assists in transporting cuprous species out from the pit whereas above this potential chloride is drawn into the pit. Mass transport of chloride from the bulk determines how much CuCl(s) can form in a corrosion pit. At low chloride concentrations precipitation of CuCl(s) causes only a small kink in the curves in figure 6. At high chloride concentrations in the bulk, mass transport can sustain the formation of large amounts of CuCl(s) in the pit and the parameter for copper transport in figure 6 drops to very low values.

Continued propagation sets a limit to the amount of solids that can be allowed to precipitate. Propagation is possible only when bare copper metal is exposed to the pit solution. When 60% of the oxidised copper forms cuprous oxide at the site of oxidation, the oxide occupies 100% of the volume. This is our boundary condition for immunity against pitting corrosion. When less than 40% of the oxidised copper is transported away, the pit cannot continue to grow because no metal is exposed to the solution and there is no room for aqueous mass transport. As shown in table II, at least 70% must be transported away when CuCl(s) is considered instead of $Cu_2O(s)$.

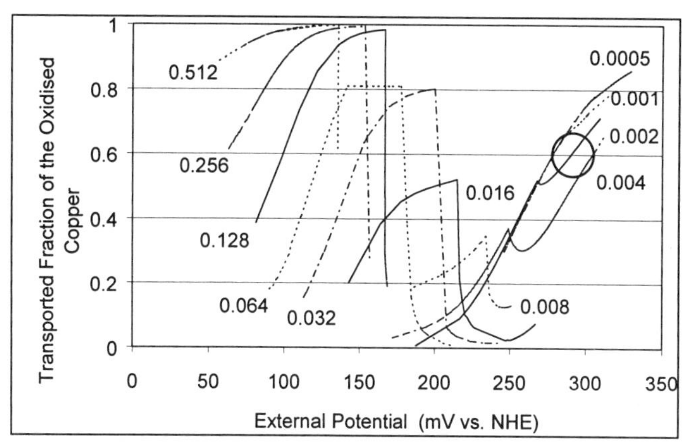

Figure 6. Transported fraction of the oxidised copper as function of the potential. Chloride concentration in the bulk varied. pH 7.0. The numbers by the curves indicate the bulk chloride concentration in moles per litre. The circle shows the location of the corrosion pit illustrated in figures 3 to 5.

If we use the limit of 40% aqueous mass transport for copper, we can use the data in figure 6 to construct a map over regions where we find pitting to be possible. Figure 7 shows such a map for 25 °C at pH 7.0. We have used the term immunity against pitting corrosion when propagation is impossible because of a complete coverage of copper metal by cuprous oxide. For lack of better words we used the term passivity for conditions when a pit cannot propagate because too much CuCl(s) would form. Figure 7 also shows the coexistence potential for cuprous oxide with the most stable cupric solid. Below this potential, conducting Cu_2O(s) is stable against oxidation and can behave as cathode material. In waters where the pitting potential is higher than this coexistence potential, the cathode material itself, Cu_2O(s) at pH 7.0, is less noble than copper metal at the pH of a corrosion pit.

Influence of Other Water Parameters

Increased sulphate concentrations are found to be detrimental with respect to pitting corrosion, particularly in the high potential region. The primary reason for this, is complex formation between sulphate and cupric ions. This allows higher aqueous copper concentrations to form at a given potential. Increased carbonate concentrations are found to be beneficial, particularly in the low potential region. The reason for this is that carbonate, in the form of carbonic acid, assists in transporting protons out from the pit. The facilitated proton transport favours the formation of Cu_2O(s) rather than the competing formation of aqueous cupric chloride complexes. The pH of the bulk has a small influence on the potential required for pitting corrosion.

Table II. Excess fraction copper when a volume of metal is converted to the same volume corrosion product.

	Cu(s)	Cu_2O(s)	CuCl(s)	Units
Molar weight	63.55	143.09	99.0	g/mol
Density	8.92	6.0	4.14	g/cm^3
Molar volume	7.12	23.85	23.91	cm^3/mol
Volume contents	0.14	0.08	0.04	mol Cu/ cm^3
Excess fraction copper		0.40	0.70	mole/ mole

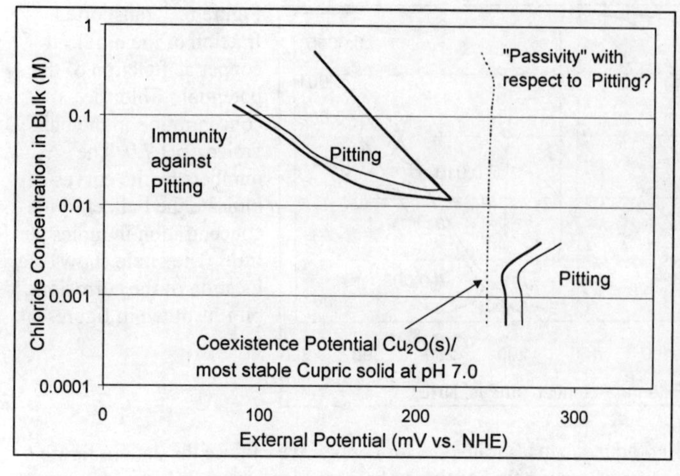

Figure 7. Potential – chloride concentration diagram with areas where we find that a corrosion pit can propagate. The thick line indicates that the fraction of the oxidised copper which is transported away from the site of the oxidation as aqueous species is equal to 0.4, the thin line indicates conditions corresponding to the transported fraction equal to 0.5.

Comparison To Experimental Results And Practical Experience.

The results show, in general, good agreement with published experimental results and practical experience from pitting corrosion of copper. The solid phases formed seem to be correctly predicted and the beneficial influence of carbonate as well as the detrimental influence of sulphate is well-established [5]. The complicated effect of the chloride concentration that we find has some support in the literature. Chloride in the concentration range of 0.01 moles per litre has been found to cause pitting corrosion [6] while additions of up to 0.002 moles per litre was found to be beneficial [7].

CONCLUSIONS

We suggest that the following conclusions are drawn from the work:

- there is a minimum potential for pitting corrosion of copper
- the value of this minimum potential is a complicated function of the water composition
- among the common anions, chloride has the greatest influence on this value
- in chloride rich waters there may exist also an upper potential limit for pitting corrosion

ACKNOWLEDGEMENT

This work was commissioned by the Swedish Nuclear Fuel and Waste Management Co. (SKB).

REFERENCES

1. I. Puigdomenech, To be publised by the Swedish Fuel and Waste Management Co.
2. W. Stumm and J.J. Morgan, in *Aquatic Chemistry* 2.nd ed. (Wiley, New York, 1981) p. 135.
3. C. Taxén, KI-rapport 1996:8E. Swedish Corrosion Institute.
4. C. Taxén, 13:th International Corrosion Conference, Nov 25-29, 1996, Melbourne Australia. Proceedings, Paper 141.
5. E. Mattson and A. .M. Fredriksson, Br. Corros. J. **3**, 246 (1968).
6. J. G. N. Thomas and A. K. Tiller, Br. Corros. J. **7**, 256 (1972).
7. M. Edwards, J. Rhering and T. Meyer, Corrosion. **50**, 366 (1994)

MECHANICAL PROPERTIES, MICROSTRUCTURE AND CORROSION PERFORMANCE OF C-22 ALLOY AGED AT 260°C TO 800°C

R. B. REBAK*, T. S. E. SUMMERS** and R. M. CARRANZA***
*Haynes International Inc., Kokomo, IN, 46901, USA, rrebak@haynesintl.com
**Lawrence Livermore National Laboratory, Livermore, CA, 94551, USA
***Comisión Nacional de Energía Atómica, 1429 Buenos Aires, Argentina

ABSTRACT

Changes in the microstructure, mechanical properties and corrosion resistance of C-22 alloy were studied systematically as a function of aging temperature and aging time. Aging was performed in the temperature range 260°C to 800°C for times between 0.5 h and 40,000 h. For aging temperatures of 600°C and higher, precipitation of tetrahedral close packed (TCP) phases in C-22 alloy induce a decrease in its mechanical properties and corrosion resistance in aggressive acidic solutions. At the lower aging temperatures, long range ordering (LRO) was observed, which did not produce changes in the chemical resistance of the alloy. Arrhenius extrapolations of the high temperature data predict that C-22 alloy will be thermally stable when exposed to temperatures in the order of 300°C for times higher than 10,000 years.

INTRODUCTION

C-22 alloy (Ni-22Cr-13Mo-3W) is a candidate material for use in the fabrication of high level nuclear waste containers that are planned for the Yucca Mountain site in Nevada. [1-2] C-22 alloy was selected because of its excellent resistance to localized corrosion (pitting corrosion, crevice corrosion and stress corrosion cracking) in chloride containing environments.[3-8] Due to the heat generated by the radioactive decay of the waste, the containers might experience temperatures as high as 250°C during their first 1,000 years of emplacement. The lifetime design of the containers is 10,000 years, and the maximum allowed temperature is 350°C. Studies of the thermal stability of C-22 alloy at 350°C for such long times are impractical in the laboratory. Previous work has shown that the mechanical and corrosion properties of C-22 alloy do not change when the alloy is aged at 427°C for up to 40,000 h.[5] However, when Ni-Cr-Mo alloys (such as C-22) are aged at temperatures above 600°C, the precipitation of detrimental second phases affect the corrosion resistance and mechanical properties of the aged alloys. [9-11]

The purpose of this work was to study systematically the thermal stability of HASTELLOY® C-22® alloy[0] (N06022) at temperatures of approximately 600°C and higher. Assuming that the aging mechanism remains the same at the lower temperatures, the thermal stability data obtained at high temperature can be used to predict the possibility that these detrimental TCP phases would form at the maximum likely temperature of the containers (250°C) for times longer than 10,000 years.

EXPERIMENTAL

Samples of C-22 alloy were aged for 0.5 to 40,000 h at 260 - 800°C (500 - 1472°F). Aging was performed in air and afterwards the samples were rapid air cooled. Samples were prepared either from a 0.25 inch (6.35 mm) or 0.5 inch (12.7 mm) thick plates. The following

[0]® HASTELLOY and C-22 are registered trademarks of Haynes International, Inc.

Mat. Res. Soc. Symp. Proc. Vol. 608 © 2000 Materials Research Society

heats were used for the aging studies: 2277-7-3173 (427°C, 343°C and 260°C for mechanical properties and corrosion studies), 2277-6-3181 (aging at 482 to 800°C for corrosion studies), heats 2277-0-3195 and 2277-3-3223 (aging at 593°C to 760°C for mechanical properties) and heat 2277-3-3223 (aging at 482°C to 760°C for mechanical properties). Metallographic and microstructure studies were carried out on samples from all of the above heats. Changes in the tensile mechanical properties and impact energy were determined according to ASTM E 8 and E 23 standards, respectively. Changes in the corrosion resistance were determined using the standard ASTM G 28 A test (boiling solution of 50% H_2SO_4 + 42 g/l of $Fe_2(SO_4)_3$) and immersion tests in boiling 2.5% HCl and 10% NaOH solutions (according to ASTM E 31). Electrochemical impedance tests were carried out according to ASTM G 106 in simulated J-13 water at 95°C.[4] J-13 water is well water from the Yucca Mountain site.[12]

RESULTS AND DISCUSSION

Microstructure

In the mill annealed (MA) condition C-22 alloy exhibits a metastable gamma (γ) phase; that is, when the alloy is exposed to high temperatures, precipitation of second phases will occur. The formation of second phases in C-22 alloy can be divided in two distinctive regimes according to the temperature range at which it occurs in a laboratory time frame. In the range of near 600°C and above, the formation of brittle TCP phases such as μ and P phases and secondary carbides can be found. In the temperature range below 600°C, only long range ordering (LRO) was observed. For example, LRO phases of a few nm in size were observed in C-22 alloy aged at 427°C for 30,000 h and phases less than 10 nm in size were observed after aging for 1,000 h at 538°C (Figure 1). LRO was not observed at 343°C even after aging for 40,000 h. Figure 1 also shows that LRO was observed at 593°C but none at 649°C; therefore, the transition temperature appears to be between these temperatures. In the high temperature range, as the aging time increases at a given temperature, the precipitation of the second phases first develops preferentially at specific grain boundaries and later starts to form at twin boundaries and finally within the grains or in the bulk (Figure 1).

Mechanical Properties

Figure 2 shows the reduction of area during tensile testing, and Figure 3 shows the impact energy (both at room temperature) for two heats of C-22 alloy that were aged at temperatures above approximately 600°C. The ductility and impact energy of the alloy decreased as the aging temperature and aging time increased. This decrease in the mechanical properties is attributed to the formation of brittle TCP phases mainly at the grain boundaries. It has been previously reported that aging of C-22 alloy at 427°C for up to 40,000 h only produced a slight decrease in the reduction in area (from 80% to 75%); that is, formation of LRO in the early stages does not seem to affect considerably the mechanical properties of C-22 alloy.[5] The impact energy of C-22 alloy aged at 427°C for 40,000 h was the same as the impact energy of MA material.

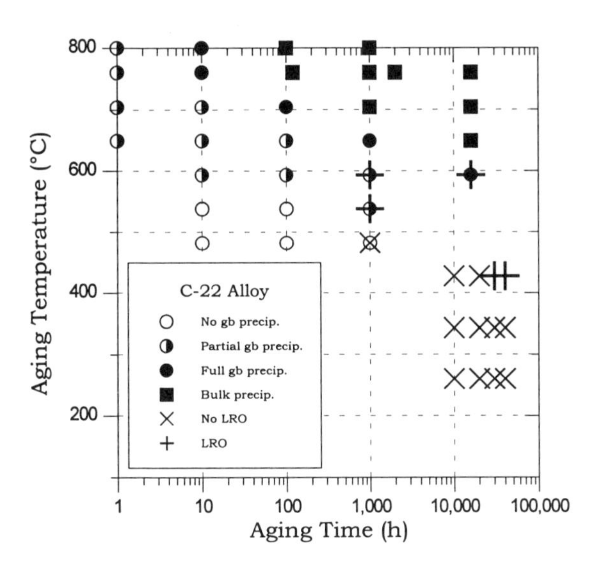

Figure 1: Time-Temperature-Transformation (TTT) Diagram for C-22 alloy.

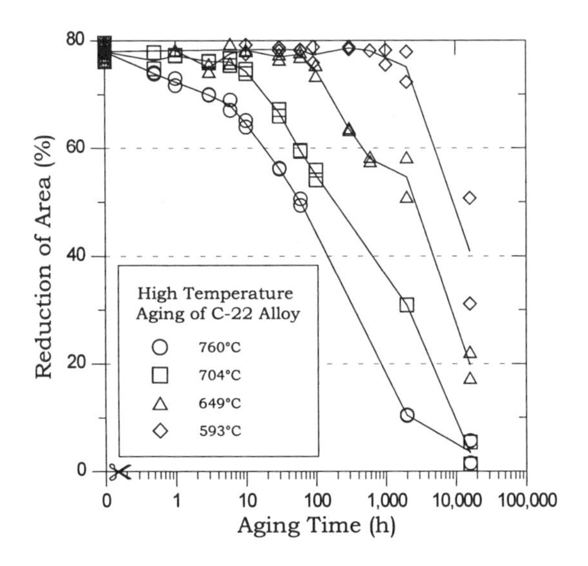

Figure 2: Changes in reduction of area for C-22 alloy with aging temperature and aging time.

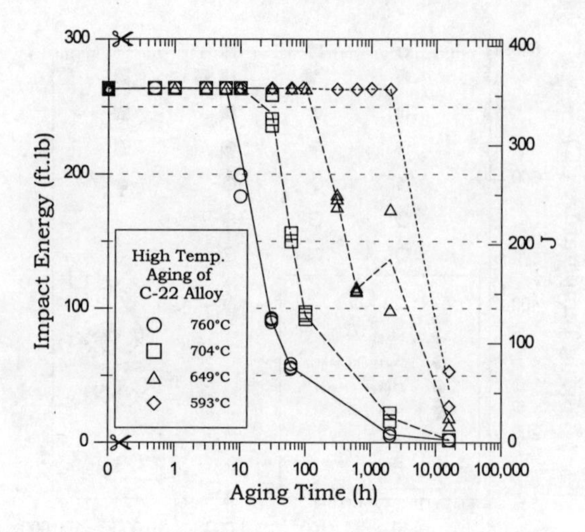

Figure 3: Changes in impact energy for C-22 alloy with aging temperature and aging time.

Corrosion

Figure 4 shows the changes in the corrosion rate of C-22 alloy in the acidic and oxidizing ASTM G 28 A solution as a function of aging temperature and aging time. The corrosion rate increased gradually as the aging temperature and aging time increased. A similar behavior was observed for the aged alloys tested in boiling 2.5% HCl acid solution (reducing conditions). The TCP phases that form principally at the grain boundaries are generally richer in chromium (Cr) and molybdenum (Mo) than in the matrix; therefore, the matrix immediately adjacent to these phases are depleted in these elements causing a localized decrease in the corrosion resistance of the alloy. A depletion of Mo causes an increase in the corrosion rate in HCl and a depletion in Cr causes an increase in the corrosion rate in the ASTM G 28 A solution. On the other hand, 96 h immersion corrosion testing in boiling 10% NaOH and 10% NaOH + 4% NaCl solutions did not produce measurable corrosion rates (below detection limit of 0.003 mm/year) in C-22 alloy both in the MA condition and as a function of aging temperature and aging time. Similarly, electrochemical impedance testing in simulated J-13 water at 95°C showed that the corrosion rate of MA C-22 alloy was in the order of 0.0002 mm/year and did not change as a function of aging temperature and aging time. It was shown before that the incipient formation of LRO was not detrimental for the localized corrosion resistance of C-22 alloy.[5]

Arrhenius Extrapolation

Thermally activated processes (such as diffusion) can be described by an Arrhenius expression of the form

$$\ln k = \ln A - E/RT \qquad (1)$$

where k is the rate of the process, A is a constant, R is the gas constant, T is the temperature in K and E is the apparent activation energy. A plot of ln(k) vs. 1/T is a straight line with a gradient of -E/R.

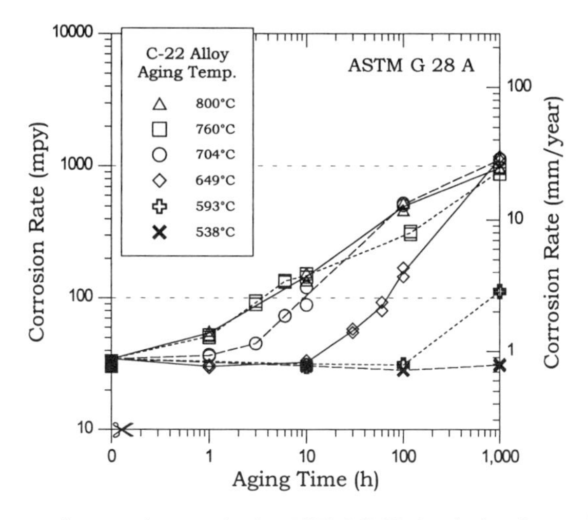

Figure 4: Changes in corrosion rate in the ASTM G 28 A solution for aged C-22 alloy.

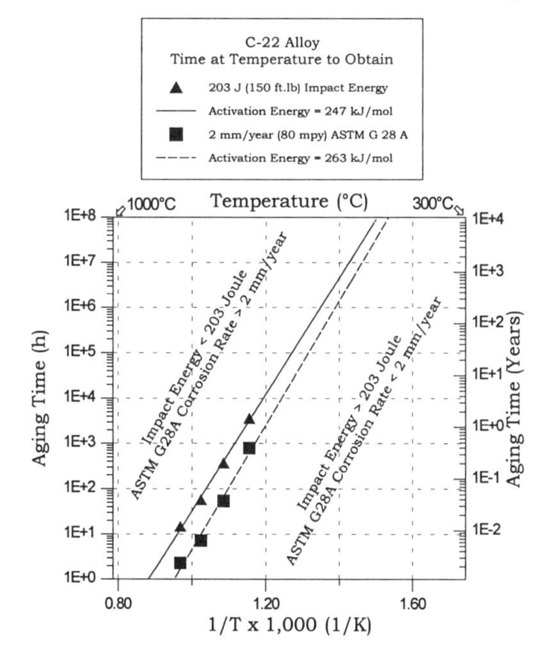

Figure 5: Long time prediction of the mechanical (impact energy) and electrochemical (corrosion resistance) behavior of C-22 alloy using Arrhenius extrapolations.

Figure 5 shows the aging time that is necessary for C-22 alloy to have an impact energy of 203 J (150 ft.lb) (from Figure 3) and a corrosion rate of 2 mm/year (80 mpy) in the ASTM G 28 A solution (from Figure 4). Figure 5 predicts that an aging time of over 10000 years would be necessary at 300°C for C-22 alloy to have an impact energy of approximately 50% the value of the MA material and to show a significant decrease on its corrosion resistance. The estimated activation energy for the impact energy (mechanical property) is 247 kJ/mol and for the corrosion resistance (electrochemical property) is 263 kJ/mol. These two values of activation energy are typical for the diffusion of relevant alloying elements in nickel (e.g. Cr). [13]

CONCLUSIONS

(1) Aging of C-22 alloy at temperatures greater than approximately 600°C produces the formation of TCP precipitates such as μ and P phases, which start to form at grain boundaries. LRO was found below 600°C and above 427°C.

(2) Aging of C-22 alloy at temperatures approximately 600°C and above causes a decrease in ductility and impact energy.

(3) Aging of C-22 alloy at temperatures greater than approximately 600°C causes a decrease in corrosion resistance in oxidizing and reducing acidic solutions but no change in its corrosion resistance in caustic solutions or J-13 water.

(4) The incipient formation of LRO after aging at 427 °C for 40,000 h was not detrimental to the localized corrosion resistance or impact energy of the alloy and only slightly detrimental for the tensile properties.

(5) Arrhenius extrapolation of the data from thermally activated processes in C-22 alloy predicted good mechanical properties and corrosion resistance of the alloy even after aging for over 10,000 years at temperatures in the range of 300°C.

REFERENCES

1. U. S. Nuclear Waste Technical Review Board, 1997 Findings and Recommendations, Report to The U. S. Congress and The Secretary of Energy, Arlington, Virginia (April 1998).
2. A. A. Sagüés, Paper QQ14.1 in Proceedings of the Symposium on Scientific Basis for Nuclear Waste Management XXII, Materials Research Society, Boston, Nov. 30- Dec. 4, 1998, **556** (In Print).
3. P. E. Manning, J. D. Schöbel, Werkstoffe und Korrosion, **37**, p. 137-145 (1986).
4. S. J. Lukezich, The Corrosion Behavior of Ni-Base High Performance Alloys in Simulated Repository Environments, MS Thesis, The Ohio State University, 1989.
5. R. B. Rebak and N. E. Koon, Paper 153, Corrosion/98, NACE International, Houston (1998).
6. K. A. Gruss, G. A. Cragnolino, D. S. Dunn, N. Sridhar, Paper 149, Corrosion/98, NACE International, (1998).
7. R. B. Rebak, P. Crook, in Proceedings of the Symposium Critical Factors in Localized Corrosion III, Fall Meeting of the Electrochemical Society, Boston, Nov. 1-6, 1998, Volume PV 98-17, p. 289-302 (1999).
8. D. S. Dunn, G. A. Cragnolino, N Sridhar, Paper QQ14.6 in Proceedings of the Symposium on Scientific Basis for Nuclear Waste Management XXII, Materials Research Society, Boston, Nov. 30- Dec. 4, 1998, **556** (In Print).
9. F. G. Hodge, Corrosion, **29**, p. 375 (1973).
10. U. L. Heubner, E. Altpeter, M. B. Rockel and E. Wallis, Corrosion, **45**, p. 249 (1989).
11. T. S. E. Summers, T. Shen, R. B. Rebak in Proceedings of the International Conference on Ageing Studies and Lifetime Extension of Materials, Oxford, UK, July 12-14, 1999.
12. J. E. Harrar, J. F. Carley, W. F. Isherwood and E. Raber, Report on the Committee to Review the use of J-13 Well Water in Nevada Nuclear Waste Storage Investigations, LLNL UCID 21867 (Univ. of California, Jan. 1990).
13. Smithells Metals Reference Handbook, Seventh Edition (Butterworth-Heinemann, Oxford, 1992).

THEORETICAL STUDY OF BULK AND SURFACE PROPERTIES OF DIGENITE $Cu_{2-\delta}S$

P. A. KORZHAVYI, I. A. ABRIKOSOV, and B. JOHANSSON
Condensed Matter Theory Group, Physics Department, Uppsala University
SE-75121 Uppsala, Sweden

ABSTRACT

In connection with the problems of mid-temperature embrittlement and sulfide corrosion of copper, we perform an *ab initio* study of intrinsic properties of copper(I) sulfide in the anti-fluorite crystal structure (digenite). The energies of the (111) and (110) non-polar surfaces of Cu_2S are calculated using the interface Green's function technique. The (111) surface is found to have the lowest energy, in agreement with the cleavage pattern of digenite mineral. The locally self-consistent Green's function method is used to obtain the formation and interaction energies of native point defects in bulk digenite. The results show that digenite exists as a non-stoichiometric compound $Cu_{2-\delta}S$ with stable (constitutional) cation vacancies, in agreement with experiment. This natural presence of constitutional cation vacancies combined with the calculated low formation energy of Frenkel defects implies a high cation mobility in $Cu_{2-\delta}S$, which is consistent with the superionic behavior of digenite.

INTRODUCTION

Corrosion of copper canister resulting in formation of copper(I) sulfide may occur under waste disposal conditions in the case of the availability of dissolved hydrogen sulfide (HS^-) in the ground waters [1], or as a result of activity of sulfur-reducing bacteria [2] at the interface between the copper canister and bentonite clay. A high rate of sulfide corrosion of copper may be expected from the measured fast self-diffusion of cations in copper(I) sulfides [3,4]. The activation energy for cation self-diffusion was found to be quite low and nearly independent of the crystal structure. Superionic conductivity observed in high-digenite [5] is a clear manifestation of the high cation mobility in copper(I) sulfides.

The mechanical strength of the copper canister under waste disposal conditions is also of importance. There is strong evidence that the embrittlement of copper at temperatures above 100-150°C is caused by grain boundary segregation of sulfur [6], most probably in the form of copper(I) sulfide [7]. The fracture surface is found to be enriched in sulfur and heavily cavitated, suggesting a low value of the surface tension [6].

The physical properties of copper(I) sulfides are therefore of practical interest. A microscopic description of these properties may be obtained using *ab initio* electronic structure calculations. In the present work we calculate the energies of two non-polar open surfaces as well as the formation and interaction energies of point defects (antisite atoms, vacancies, and self-interstitials) in the high-temperature modification of Cu_2S (high-digenite).

METHOD OF CALCULATIONS

We have performed *ab initio* total energy calculations using the interface Green's function (IGF) technique [8] to study the properties of open surfaces, and the locally self-consistent Green's function (LSGF) method [9] was used to derive the energies of point defects.

The spherical shape approximation was used for the one-electron potential whereas the non-spherical (multipole) components of the electron density were included in the expansion of the electrostatic potential and energy. This so-called ASA+M technique allows one to obtain surface energies [8] and vacancy formation energies [10] for transition and noble metals with an accuracy typical for the most precise full-potential methods. Our calculations were performed with the angular momentum cutoff $l_{max} = 2$. For the charge density, multipole components up to $l = 4$ were taken into account.

Mat. Res. Soc. Symp. Proc. Vol. 608 © 2000 Materials Research Society

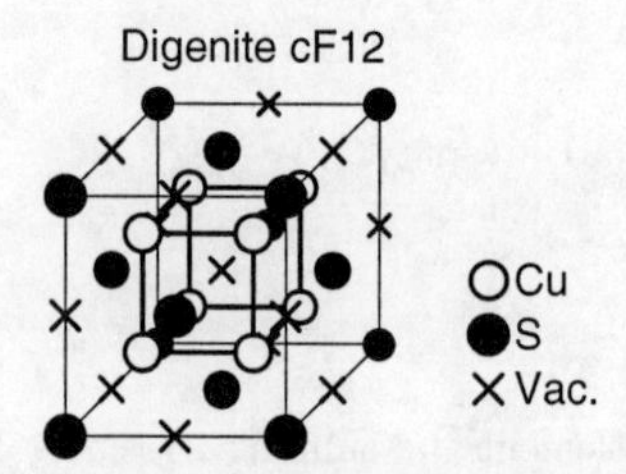

FIG. 1. Anti-fluorite unit cell of digenite.

The crystal structures of copper(I) sulfides are based on a close-packed anion sublattice with the interstitial sites partially occupied by the cations [11]. In high-digenite (anti-fluorite prototype structure) the sulfur atoms form a face-centered cubic (fcc) sublattice, the copper atoms occupy the tetrahedral interstitial sites, whereas the octahedral interstitial sites are vacant, see Fig. 1. In order for the structure to appear more close-packed in the theoretical treatment, empty spheres were put into the octahedral sites. Vacancies were also modeled by empty spheres. Equal radii for the atomic and empty spheres were used in our calculations.

We used $2 \times 2 \times 2$ supercells (128 lattice sites) based on the cubic unit cell of Cu_2S to simulate single point defects, and $2 \times 2 \times 4$ supercells (256 lattice sites) were used to extract the interaction energies of defect pairs in Cu_2S.

The total energy was calculated in the framework of the local density approximation (LDA) employing the Perdew, Burke, and Ernzerhof exchange-correlation potential [12] (without the gradient corrections). The volume of each supercell was relaxed to its calculated equilibrium value. The effect of local relaxations around the defects was not considered in the present study.

SURFACE PROPERTIES

The calculated (unrelaxed) energies of three low-index surfaces for pure fcc Cu as well as of two non-polar surfaces for Cu_2S are listed in Table I. Our calculations give the correct anisotropy of the surface energy in pure Cu. This anisotropy can be interpreted in terms of the number of nearest neighbor bonds that need to be broken in order to cleave the crystal thereby creating two surfaces with the given orientation. Although very approximate, this bond-cutting analysis may be quite useful for obtaining a qualitative description of bonding.

The surface energy of pure Cu suggests an energy of the Cu-Cu bond of about 0.24 eV. The bond-cutting analysis of the surface energy of Cu_2S yields a very strong Cu-S bond of 1.83 eV but a weakened Cu-Cu bond of only 0.17 eV in digenite. As a result, the surface energy of Cu_2S is roughly proportional to the surface density of broken Cu-S bonds.

The lowest surface energy is attained at the (111) surface where only one Cu-S bond per surface unit cell is broken. Thus, our calculations predict an octahedral equilibrium shape of digenite crystal, as well as the {111} type of cleavage, in agreement with the mineralogy data [13].

TABLE I. Calculated surface energies of pure Cu (fcc) and Cu_2S (digenite) for different surface orientations. Numbers of broken bonds, n_{Cu-S} and n_{Cu-Cu}, per formula unit (f.u.).

	Orientation	Surface energy		n_{Cu-S}	n_{Cu-Cu}
		J/m^2	eV/f.u.		
Cu	(111)	2.20	0.77		3
	(100)	2.34	0.94		4
	(110)	2.42	1.37		6
Cu_2S	(111)	1.42	1.17	1	3
	(110)	1.61	2.17	2	4

BULK PROPERTIES

The prototype structure shown in Fig. 1 is perfectly ordered. Among the the four fcc sublattices one is fully occupied by S atoms, two other (equivalent) sublattices are occupied by Cu atoms, and the fourth sublattice is vacant, i.e. occupied by vacancies (V). Site occupancy is not perfect in a real crystal where one may have constitutional and thermal point defects.

We consider totally six types of native point defects: antisite atoms (Cu_S, S_{Cu}), vacancies (V_S, V_{Cu}), and interstitial atoms (Cu_V, S_V). The notation i_α stands for alloy component $i = \{Cu, S, V\}$ occupying sublattice $\alpha = \{Cu, S, V\}$.

In Fig. 2 we show the calculated energy of formation and lattice parameter of the stoichiometric Cu_2S and of the supercells containing different types of the native point defects. Since the individual point defects are not composition-conserving, the compositions of the supercells are off-stoichiometric.

As Fig. 2a shows, the copper vacancies are stable (constitutional) defects in the S-rich $Cu_{2-\delta}S$. Moreover, Cu-rich as well as stoichiometric digenite turns out to be unstable with respect to a decomposition into the S-rich $Cu_{2-\delta}S$ (containing copper vacancies) and pure Cu. This is consistent with the Cu-S phase diagram [14] in which the Cu-rich boundary of the digenite phase field does not reach the ideal stoichiometric composition even at pre-melting temperatures.

Thermal defects in a compound with a fixed atomic composition must appear as a result of composition-conserving defect reactions. The effective formation energies of point defects may be calculated as the energies of the corresponding defect reactions per one *thermal* pont defect (V_{Cu} do not count since they are constitutional defects) assuming the reactants and the products are in an unbound state [15]. The calculated effective formation energies of native point defects in S-rich $Cu_{2-\delta}S$ as well as the defect reactions are listed in Table II.

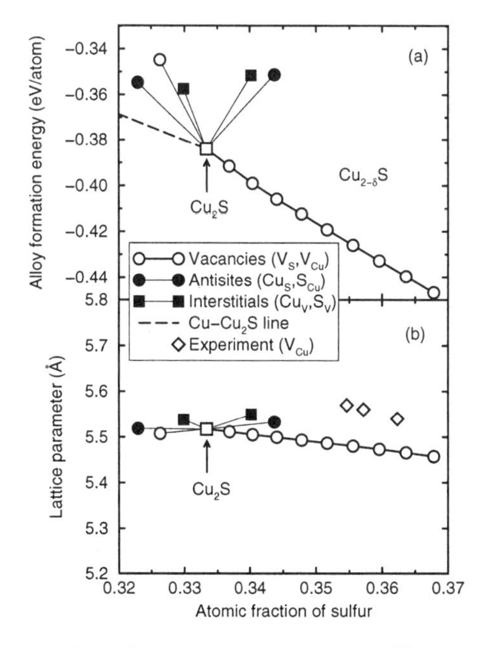

FIG. 2. Calculated formation energy (a) and lattice parameter (b) of Cu_2S containing native point defects. Experimental lattice parameters [16] are shown for comparison.

TABLE II. Calculated effective formation energies, E^{eff} (eV), of native defects in the S-rich $Cu_{2-\delta}S$. The value given in parenthesis corresponds to thermal Cu vacancies.

Defect	Reaction	E_d^{eff}	Defect	Reaction	E_d^{eff}
V_{Cu}	$0 \rightarrow Cu_S + 3V_{Cu}$	0(0.55)	V_S	$0 \rightarrow 2V_{Cu} + V_S$	2.21
Cu_S	$0 \rightarrow Cu_S + 3V_{Cu}$	0.55	S_{Cu}	$3V_{Cu} \rightarrow S_{Cu}$	5.40
Cu_V	$0 \rightarrow Cu_V + V_{Cu}$	1.81	S_V	$2V_{Cu} \rightarrow S_V$	4.65

TABLE III. Pair interaction energies, E^{int} (eV), of selected defects in Cu_2S.

Pair	Distance	E^{int}	Pair	Distance	E^{int}
$V_{Cu} - V_{Cu}$	$(1/2)a_0\langle 100\rangle$	+0.258	$V_{Cu} - Cu_S$	$(1/4)a_0\langle 111\rangle$	+0.150
	$(1/2)a_0\langle 110\rangle$	−0.014	$V_{Cu} - Cu_V$	$(1/4)a_0\langle 111\rangle$	−0.585
	$(1/2)a_0\langle 111\rangle$ [a]	+0.171	$Cu_S - Cu_V$	$(1/2)a_0\langle 100\rangle$	−0.648
	$(1/2)a_0\langle 111\rangle$ [b]	−0.007	$Cu_V - Cu_V$	$(1/2)a_0\langle 110\rangle$	+0.014
	$a_0\langle 100\rangle$	0.000	$Cu_S - Cu_S$	$(1/2)a_0\langle 110\rangle$	−0.052

[a] Interaction through a S atom
[b] Interaction through an octahedral interstitial site

According to our calculations, Cu vacancies are the primary thermal defects which may most easily be generated in the crystal together with the antisite Cu_S defects or interstitial Cu_V copper atoms. It is noteworthy that Cu atoms may substitute on the S sublattice at elevated temperatures, this possibility was so far neglected in the analysis of the diffraction data [17].

Pair interaction energies for all possible defect pairs within the first three coordination shells have been calculated. However, our analysis has shown that the defect interactions cannot significantly reduce the formation energies of the high energy defects (V_S, S_{Cu}, and S_V) whose equilibrium concentrations are expected to be negligible. Therefore, in Table III we report only on the defect interactions of the three lowest energy point defects in digenite: V_{Cu}, Cu_S, and Cu_V.

The interaction energy between two Cu vacancies shows an attenuated oscillatory behavior as a function of distance with the first minimum at $\frac{1}{2}a_0\langle 110\rangle$ that is in qualitative agreement with the experimentally observed ordering pattern of vacancies in digenite. The experimental data [17–21] suggest that Cu vacancies are arranged into planar clusters with the shortest vacancy-vacancy separation distance of $\frac{1}{2}a_0\langle 110\rangle$ in the $\langle 111\rangle$ planes.

Large negative values of the interaction energy are found for the defect pairs $V_{Cu} - Cu_V$ and $Cu_S - Cu_V$. It is clear that these interactions favor the formation of the Frenkel defect pairs, $0 \rightarrow V_{Cu} + Cu_V$, in the presence of constitutional Cu vacancies and antisite Cu atoms. The formation and annihilation of the Frenkel pairs is the mechanism of cation diffusion in $Cu_{2-\delta}S$. Therefore, the strong attractive interactions found in this work may have important consequences on the cation mobility, since they lower the activation barrier of atomic diffusion. The cooperative attractive interaction effects between interstitial atoms and vacancies are also a necessary condition of superionic conductivity [22].

We modeled the process of formation of the Frenkel pairs in the presence of Cu vacancies using direct calculations for 128-site supercells. The initial configurations contained 1 to 3 Cu vacancies. The vacancies in the supercells representing the initial as well as the final configurations were arranged into clusters surrounding the octahedral interstitial site; the separation distance between any two vacancies was $\frac{1}{2}a_0\langle 110\rangle$, see Fig. 3.

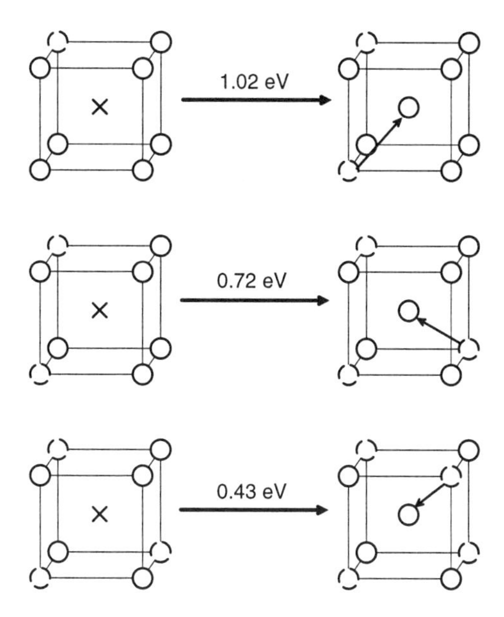

FIG. 3. Schematic representation of the initial and final defect configurations modeling formation of Frenkel pairs in the presence of constitutional Cu vacancies. Notations are the same as in Fig. 1. Copper vacancies are shown by circles with a dashed contour.

The calculations show that the effective formation energy of a Frenkel pair reduces to about 0.43 eV in the case where the interstitial Cu atom is surrounded by four Cu vacancies in the final configuration (to be compared with the formation energy of an unbound Frenkel pair of 1.81 eV in the absence of other defects). The actual formation energy of the Frenkel pair in $Cu_{2-\delta}S$ should be even lower than the presently calculated value, since we did not consider the energy of local relaxations around point defects. Nevertheless, the calculated formation energy of the Frenkel defect correlates reasonably well with the experimental activation energy of cation self-diffusion in copper(I) sulfides, 0.4 – 0.5 eV [3,4].

SUMMARY

In order to obtain a microscopic description of the physical properties of copper(I) sulfides relevant to the problems of nuclear waste storage in copper canisters, we have performed *ab initio* calculations for surfaces and point defects in high-digenite.

The surface energy of Cu_2S was found to be lower than that of pure Cu. We also found that the Cu-Cu bond is weakened in Cu_2S because of the competing chemical interaction of Cu with S. The calculated anisotropy of the surface energy of Cu_2S is consistent with the observed octahedral shape of digenite mineral as well as with the $\{111\}$ type cleavage.

Our calculations show that digenite is an off-stoichiometric phase with thermodynamically stable copper vacancies. The calculated pair interactions of copper vacancies are in qualitative agreement with the experimentally observed patterns of vacancy ordering.

At finite temperatures, two other types of native point defects may also have high concentrations in digenite: antisite Cu atoms and interstitial Cu atoms in octahedral positions. The energy of a Frenkel defect pair is shown to be sensitive to the local environment and becomes very low in the presence of constitutional vacancies. Our calculations show that this effect is due to a large negative interaction energy of an interstitial Cu atom with a Cu vacancy.

ACKNOWLEDGMENTS

This work is entirely funded by SKB AB, the Swedish Nuclear Fuel and Waste Management Company.

REFERENCES

[1] N. M. Perea, in *Scientific Basis for Nuclear Waste Management XX*, edited by W. J. Gray and I. R. Triay (Mater. Res. Soc. Proc. **465**, Pittsburgh, PA, 1997) p. 1153.
[2] M. Motamedi and K. Pedersen, this volume.
[3] R. V. Bucur and R. Berger, Solid State Ionics **76**, 291 (1995).
[4] R. Berger and R. V. Bucur, Solid State Ionics **89**, 269 (1996).
[5] J. B. Boyce and B. A. Huberman, Phys. Rep. **51**, 189 (1979).
[6] P. J. Henderson, J. O. Österberg, and B. Ivarsson, Technical Report TR 92-04 (SKB, Stockholm, 1992).
[7] P. A. Korzhavyi, I. A. Abrikosov, B. Johansson, Acta Mater. **47**, 1417 (1999).
[8] H. L. Skriver and N. M. Rosengaard, Phys. Rev. B **46**, 7157 (1992).
[9] I. A. Abrikosov, A. M. N. Niklasson, S. I. Simak, B. Johansson, A. V. Ruban, and H. L. Skriver, Phys. Rev. Lett. **76**, 4203 (1996).
[10] P. A. Korzhavyi, I. A. Abrikosov, B. Johansson, A. V. Ruban, and H. L. Skriver, Phys. Rev. B **59**, 11693 (1999).
[11] P. Villars and L. D. Calvert, *Pearson's Handbook of Crystallographic Data for Intermetallic Phases* (American Society for Metals, Ohio, 1985) Vol. 2, p. 2005.
[12] J. P. Perdew, K. Burke, and M. Ernzerhof, Phys. Rev. Lett. **77**, 3865 (1996).
[13] *The System of Mineralogy of J. D. Dana and E. S. Dana*, 7th edited by C. Palache, H. Berman, and C. Frondel (John Wiley & Sons, NY, 1944) Vol. 1, p. 180.
[14] D. J. Chakrabarti and D. E. Laughlin, in *Binary Alloy Phase Diagrams*, Second Edition, ed. by T. B. Massalski (ASM International, 1990) p. 1467.
[15] P. A. Korzhavyi, I. A. Abrikosov, and B. Johansson, in *High-Temperature Ordered Intermetallic Alloys VIII*, edited by E. P. George, M. Yamaguchi, and M. J. Mills (Mater. Res. Soc. Proc. **552**, Pittsburgh, PA, 1999) pp. KK5.35.1-8.
[16] M. A. Gezalov, G. B. Gasimov, Yu. G. Asadov, G. G. Guseinov, and N. V. Belov, Sov. Phys. Crystallogr. **24**, 700 (1979).
[17] S. Kashida and K. Yamamoto, J. Phys.: Condens. Matter **3**, 6559 (1991).
[18] C. Manolikas, P. Delavignette, and S. Amelinckx, Phys. Status Solidi **A33**, K77 (1976).
[19] C. Conde, C. Manolikas, D. Van-Dyck, P. Delavignette, J. Van Landuyt, and S. Amelinckx, Mat. Res. Bull. **13**, 1055 (1978).
[20] D. Van-Dyck, C. Conde-Amiano, and S. Amelinckx, Phys. Status Solidi **A58**, 451 (1980).
[21] J. N. Gray and R. Clarke, Phys. Rev. **B33**, 2056 (1986).
[22] Yu. Ya. Gurevich and Yu. I. Kharkats, Phys. Rep. **139**, 203 (1986).

Performance Assessment

CHEMICAL INTERACTIONS IN THE NEAR-FIELD OF A REPOSITORY FOR SPENT NUCLEAR FUEL – A MODELLING STUDY

Heikki Kumpulainen, Jarmo Lehikoinen and Arto Muurinen
VTT Chemical Technology, P.O. Box 1404, FIN-02044 VTT, Finland

ABSTRACT

The near-field chemistry of the repository for spent nuclear fuel arising from interactions between the groundwater, compacted bentonite clay, canister and the spent fuel was calculated using a three-successive-closed-systems approach. The calculations were performed for fresh granitic and saline groundwaters using the thermodynamic computer codes, HYDRAQL/CE and EQ3/6. The effects of water chemistry inside the canister as well as water radiolysis on fuel dissolution were taken into consideration. The groundwater and the three barriers of the near-field were accounted for by this approach, with particular emphasis given to the pH, Eh and actinide solubilities.

INTRODUCTION

It has been planned to dispose of spent nuclear fuel in Finland in a repository at a depth of about 500 m in crystalline bedrock. The near-field barriers in the excavated space comprise a compacted bentonite buffer surrounding a copper-lined iron canister, the canister itself and the spent fuel matrix (UO_2). These barriers interact via groundwater and evolve towards thermodynamic equilibrium with each other and with the geochemical environment.

The canister is predicted to remain intact for a very long period of time in the base case of the safety assessment. In one conceivable scenario, in which a defect in a canister has remained undetected, the groundwater could come into contact with the spent nuclear fuel. The fuel matrix (UO_2) and the iron inside the copper canister, may act as redox buffers and help to maintain some of the radionuclides in their low-solubility forms, thus contributing to lower radionuclide releases. The evolution of the near-field chemistry towards thermodynamic equilibrium depends on both the chemical reactions and mass transport phenomena.

The transport of solutes in the very slowly moving groundwater is, in practice, the only way by which the radionuclides of the spent fuel matrix can be released into the far-field. When assessing the safety of the repository, knowledge of the rates at which radionuclides will be released into the groundwater is needed. In safety assessment, the radionuclide solubilities are important parameters, which depend on the chemical conditions that develop in the near-field.

The objective of this work is to model the near-field chemistry arising from interactions between the groundwater, bentonite, canister and the spent fuel using the groundwater and bentonite considered in the Finnish nuclear waste disposal concept. The outcome of this study is the chemistry in the solution contacting the spent fuel, the actinide solubilities and the effect of the type of synthetic groundwater on them.

MODELLING AND RESULTS

The near-field chemical interactions have usually been modelled by taking into account co-existing transport phenomena and chemistry [1]. These coupled models often calculate the transport of solutes in porous media relatively accurately, but frequently oversimplify the

Mat. Res. Soc. Symp. Proc. Vol. 608 © 2000 Materials Research Society

chemical interactions of the aqueous species and their heterogeneous reactions with the solid phases. In this study, the equilibrium modelling of chemical interactions in the near-field has been realized by means of a sequential-closed-systems approach [2], which ignores solute transport but provides a detailed description of the equilibrium chemistry. The groundwater and the three barriers of the near-field were accounted for by this approach with particular emphasis given to the pH, Eh and actinide solubilities in the water in contact with spent fuel. The effects of water chemistry inside the canister as well as water radiolysis on fuel dissolution were included in the model. The groundwater (either fresh or saline) was first allowed to equilibrate with the bentonite. The resulting bentonite water was permitted to react with the canister. Finally, the canister-equilibrated water was reacted with the spent nuclear fuel.

The assumptions in the modelling of the complex near-field system are:

- a single spent-fuel canister is considered,
- reducing groundwater conditions prevail,
- only canister iron is considered (altered to magnetite), and
- a temperature of 25 °C and a pressure of 1 bar.

The bentonite system was calculated using the HYDRAQL/CE thermodynamic code (descendant of HYDRAQL [3]), which includes cation exchange, the major reaction of bentonite with water. The modelling tool used for the canister and spent fuel calculations was the EQ3/6 code, version 7.2b [4]. The aqueous speciation and saturation indices were calculated from analytical data. Reaction path modelling (EQ6) was applied for calculating the chemical and mineralogical interactions in the aqueous chemical systems. The thermodynamic database version applied was GEMBOCHS.V2-EQ8-DATA0.COM.R2 (2 Aug. 1995), and the aqueous activity coefficients were calculated by means of the B-dot equation (e.g., [4]).

Synthetic groundwaters

The compositions of the synthetic fresh and saline groundwaters used in the calculations are shown in Table I. For the model calculations, an Eh of -0.25 V was assumed for both waters.

Table I. Compositions of the synthetic groundwaters (in mol L^{-1}).

Component	Fresh	Saline
Na^+	2.26E-03	2.09E-01
K^+	1.00E-04	5.40E-04
Ca^{2+}	4.64E-04	9.98E-02
Mg^{2+}	1.90E-04	2.30E-03
HCO_3^-	1.80E-03	3.50E-05
Cl^-	1.48E-03	4.17E-01
SO_4^{2-}	1.00E-04	4.40E-05
pH (-)	9.1	8.2

Bentonite

After the repository closure, the groundwater is first equilibrated with the bentonite that surrounds the canister. In addition to the swelling montmorillonite component, bentonites contain quartz, feldspars, minor amounts of carbonates, sulphates and sulphides (e.g., [5]). Unfortunately, the thermodynamic data available for many silicates is often scarce and/or uncertain [6].

The chemical composition of the MX-80 bentonite used in the HYDRAQL/CE equilibration calculation is given elsewhere [7]. A bentonite-to-water ratio of 3.6 Mg m^{-3} was assumed. Confidence in the bentonite modelling was gained from Fig. 1, where a comparison of a previously calculated bentonite equilibration with experimental results for a simulated fresh and a saline groundwater is given (see [7]). However, the bentonite-to-water ratio considered in this work is somewhat higher than in Fig. 1.

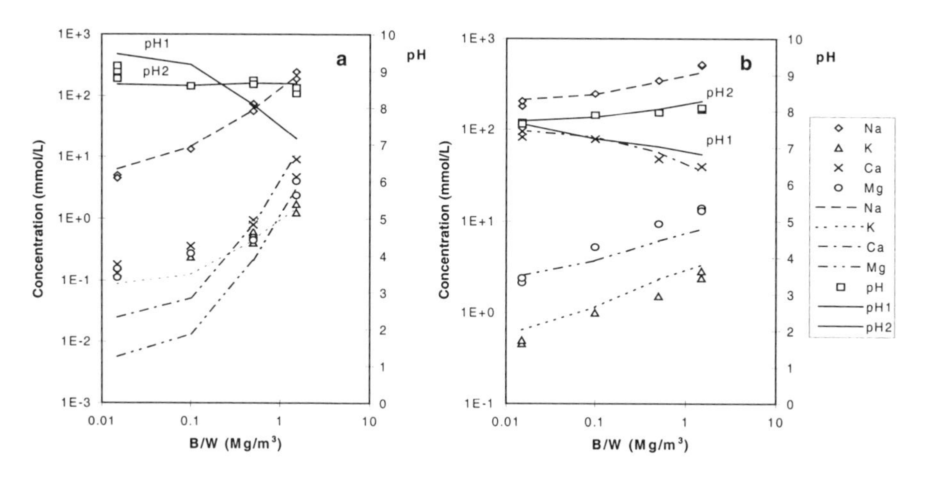

Figure 1. Measured (symbols) and calculated (lines) equilibrium concentrations of cations in the equilibrating solution as a function of B/W (bentonite-to-water ratio). Modelling of the water-bentonite interaction in a closed system (pH 1) followed by re-equilibration of the solution in $\log P_{CO_2}$ of -3.36 (pH 2) for **a)** fresh and **b)** saline water. See [7] for details.

Canister

The groundwater in the repository equilibrates with the bentonite and eventually penetrates the bentonite bed before coming into contact with the canister. The modelling of the interaction of bentonite-equilibrated water with the canister was based on a number of assumptions. Due to the stability of copper in the anoxic repository conditions, only iron was supposed to react with the bentonite-equilibrated water. In addition, it was presumed that owing to hydrolysis, the iron inside the canister had already been altered to magnetite, Fe_3O_4 [8].

Two alternative paths for the reactions with the canister were followed. In the first one, the precipitation of haematite (Fe_2O_3) was considered to be kinetically hindered [9], which is the most probable case, and also used here for the calculation of the near-field chemistry. In the second one, the precipitation of haematite was determined by its solubility product, if the conditions were favourable for its formation.

When bentonite water reacts with the canister (here, with its corrosion product, magnetite), the precipitation of haematite takes place at later stages of the reaction. During the course of the reaction, the sulphide concentration decreases, pH becomes more basic, and pyrite precipitates according to

$$2Fe_3O_4 + 13H^+ + 11HS^- + SO_4^{2-} \Leftrightarrow 6FeS_2 + 12H_2O. \qquad (1)$$

At later stages, reaction (2) lowers the sulphate content in the solution and increases the pyrite and haematite concentration as well as the pH.

$$15Fe_3O_4 + 4H^+ + 2SO_4^{2-} \Leftrightarrow 22Fe_2O_3 + FeS_2 + 2H_2O \qquad (2)$$

Spent fuel

The canister-equilibrated water (for the case not containing precipitated haematite) was allowed to react with the spent nuclear fuel. The composition of the spent fuel used in the calculations, which corresponds to a cooling time of 1000 a, is shown in Table II. The radionuclides were assumed homogeneously distributed in the spent fuel matrix and released in stoichiometric proportions to the UO_2 matrix. The oxidation reaction of UO_2 was simulated by assuming that all the reactive oxidants produced by α-radiolysis of water could be represented by O_2(aq). It was also conservatively postulated that all the available oxygen was consumed by the fuel reactions. Further, the UO_2 of the fuel matrix was supposed to be oxidized under the fixed oxygen fugacity determined by the U_3O_7/U_3O_8-equilibrium.

Table II. Spent fuel composition with a burnup of 36 MWd/(kg IHM) [in mol/(kg IHM)] (IHM = Initial Heavy Metal) [10,11].

Element	Content	Element	Content
C	7.25E-04	O	8.38E+00
Se	7.61E-04	Np	7.93E-03
Tc	8.40E-03	Pu	3.42E-02
Sn	8.59E-04	Am	1.95E-03
I	2.04E-03	Sr	4.22E-03
Cs	1.30E-02	Pd	1.60E-02
Th	5.17E-06	Sm	6.12E-03
U	4.00E+00		

Under more realistic conditions, α-radiolysis affects only a thin water layer close to the fuel surface. There are experimental indications of this kind of behaviour from leaching experiments [12]. The possible consequence is a radiolysis-controlled oxidative dissolution at the fuel surface followed by a re-precipitation of the redox-sensitive elements, since the bulk conditions will be reducing due to the presence of Fe(II) corrosion products [13].

The highly crystalline NpO_2 and PuO_2 were not assumed to precipitate. The precipitation of $CaUO_4$ was also not considered in the calculations. The aqueous species not included in the analysis were $Np(OH)_5^-$ and $Pu(OH)_5^-$, since their occurrence at pH >7 is not probable [14].

According to the calculated results, the solubility-limiting solid phases for the actinides are $AmOHCO_3$ for Am, $Np(OH)_4$ for Np and $Pu(OH)_4$ for Pu. The modelling results for the canister water-spent fuel equilibration are shown in Table III.

Table III. Calculated water chemistry and concentrations [in mol/(kg soln.)] for a few elements in solution equilibrated with spent fuel for the simulated fresh and saline groundwater case.

Element	Fresh	Saline
Am	3.91E-07	1.40E-06
Cs	1.30E-03	1.30E-03
Np	8.16E-09	1.40E-08
Pu	1.74E-09	1.74E-09
Se	9.06E-06	3.55E-06
Tc	8.40E-04	8.40E-04
Th	6.69E-15	6.57E-15
U	1.69E-05	2.26E-06
pH (-)	7.92	7.65
Eh (V)	0.098	0.115

Near-neutral pH and slightly oxidizing conditions were computed in both groundwaters. In the cases where magnetite was assumed to oxidize to haematite, virtually the same near-field chemistry with regard to solution pH and Eh developed as shown in Table III. The uranium concentration in the solution was relatively high due to the U(VI) carbonate complexation under oxidizing conditions.

DISCUSSION AND CONCLUSIONS

The effect of groundwater type on the near-field chemistry and actinide solubilities was calculated. Near-neutral and slightly oxidizing water chemistries were found for both fresh and saline groundwater. Changing from fresh to saline groundwater

- had only a minor effect on the pH and Eh of the water in contact with spent fuel,
- increased Am solubility three-fold,
- increased Np solubility two-fold,
- did not affect Pu solubility, and
- decreased U solubility.

Changes in actinide solubility occurred concurrently with changes in speciation, whereas no changes in Pu speciation were seen. The decrease in uranium solubility in the saline groundwater was probably due to the lower carbonate complex formation compared to the fresh water case.

The credibility of the model results is dependent on the model itself, its assumptions and the quality of the chemical database. It is important to keep in mind that thermodynamic equilibrium modelling does not provide information about reaction kinetics. The calculated solubility of the actinides is valid only in the thin aqueous layer assumed to form on the surface of the spent fuel, where water radiolysis may maintain oxidizing conditions. Although oxidizing conditions facilitate radionuclide dissolution, the redox-sensitive radionuclides begin to re-precipitate after migrating to the reducing environment.

ACKNOWLEDGMENTS

We cordially thank Dr. Randy Arthur, Monitor Scientific, LLC, Denver, CO, USA, for valuable discussions during the work. This work was financially supported by the Ministry of Trade and Industry of Finland (KTM) and was carried out as a part of the publicly financed nuclear waste management programme.

REFERENCES

1. H. Kumpulainen, J. Lehikoinen, A. Muurinen, and K. Ollila, Report VTT Tiedotteita 1912, 1998 (in Finnish).

2. R.C. Arthur, and M.J. Apted, Report SKI 96-31, 1996.

3. C. Papelis, K.F. Hayes, and J.O. Leckie, Dept. Civil Eng., Stanford Univ., CA, Technical Report No. 306, 1988.

4. T.J. Wolery, Report UCRL-MA-110662 Pt.1, 1992.

5. M. Müller-Vonmoos, and G. Kahr, Report NTB 83-12, 1983.

6. R. Grauer, Report NAGRA TR 86-12E, 1986.

7. A. Muurinen, and J. Lehikoinen, Eng. Geol., 54, 207 (1999).

8. A.E. Bond, A.R. Hoch, G.D. Jones, A.J. Tomczyk, R.M. Wiggin, and W.J. Worraker, Report SKB 97-19, 1997.

9. H.St.C. O'Neill, Am. Miner., 73, 470 (1988).

10. M. Anttila, Report YJT-92-03, 1992 (in Finnish).

11. M. Anttila, personal communication.

12. A. Loida, B. Grambow, H. Geckeis, and P. Dressler, in Scientific Basis for Nuclear Waste Management XVIII, edited by T. Murakami, and R.C. Ewing (Mater. Res. Soc. Proc. 353, Pittsburgh, PA, 1995), pp. 577-584.

13. L. Werme, P. Sellin, and R. Forsyth, Report SKB 90-08, 1990.

14. R.C. Arthur, and M.J. Apted, Report SKI 96-30, 1996.

RESULTS OF AN AQUEOUS SOURCE TERM MODEL FOR A RADIOLOGICAL RISK ASSESSMENT OF THE DRIGG LLW SITE, UK

J.S. SMALL, P.N. HUMPHREYS, T.L. JOHNSTONE, R. PLANT, M.G. RANDALL, AND D.P. TRIVEDI

BNFL Research & Technology, Risley, Warrington, WA3 6AS, UK.

ABSTRACT

A radionuclide source term model has been developed which simulates the biogeochemical evolution of the Drigg low level waste (LLW) disposal site. The DRINK (**DRI**gg **N**ear field **K**inetic) model provides data regarding radionuclide concentrations in groundwater over a period of 100,000 years, which are used as inputs to safety assessment calculations. The DRINK model considers the coupled interaction of the effects of fluid flow, microbiology, corrosion, chemical reaction, sorption and radioactive decay. The model simulates the development of a period of reducing conditions resulting from degradation of cellulose and steel wastes. Under these conditions U and Th remain as solubility controlling solids for periods over 30,000 years and provide an important source of daughter nuclides such as Ra. The fraction of ^{14}C is followed through all reactions involving carbon. Less than 5% of ^{14}C is present as mobile aqueous species.

INTRODUCTION

Disposals to the UK LLW site at Drigg, owned and operated by British Nuclear Fuels (BNFL), are authorised by the Environment Agency under the terms of the Radioactive Substances Act 1993. A post closure radiological safety assessment (PCRSA) is being prepared to support the safety case for the Drigg site. An integral part of the PCRSA is the DRINK source term model which simulates the biogeochemical evolution of the Drigg near surface site. DRINK provides radionuclide concentrations within the near field which are input as radionuclide fluxes to a groundwater assessment model. Simulated activities of radionuclides remaining in the near field are used for the calculation of impacts associated with potential human intrusion. In addition to quantitative model inputs, DRINK helps provide a phenomenological understanding of the behaviour of the site over extended periods of time, which is a requirement of the regulatory guidance.

The Drigg site includes two disposal systems: 1) An original system operated from 1959 to 1988 comprising a series of parallel trenches excavated into glacial clays, back filled with LLW and covered with an interim water resistant cap. 2) Current disposal of compacted waste placed in steel ISO-freight containers, with void space filled with a highly fluid cement based grout. These containers are then disposed of in a series of open concrete vaults. Figure 1 illustrates the disposition of the two disposal systems. Drigg LLW contains a large proportion of cellulosic waste together with disposed steel and contaminated soil. Radionuclides with highest activities in the inventory include ^{3}H, ^{241}Pu, ^{137}Cs, ^{234}U and ^{90}Sr. The long-lived radionuclides ^{238}U and ^{232}Th have the highest molal concentration.

This paper outlines the functionality of the DRINK source term model and gives examples of its output. The results provide a simulation of the nature and timing of the chemical changes which occur during the degradation of wastes in the Drigg near field. The varying aqueous concentration of radionuclides is calculated considering their varying solubility under the simulated conditions.

Mat. Res. Soc. Symp. Proc. Vol. 608 © 2000 Materials Research Society

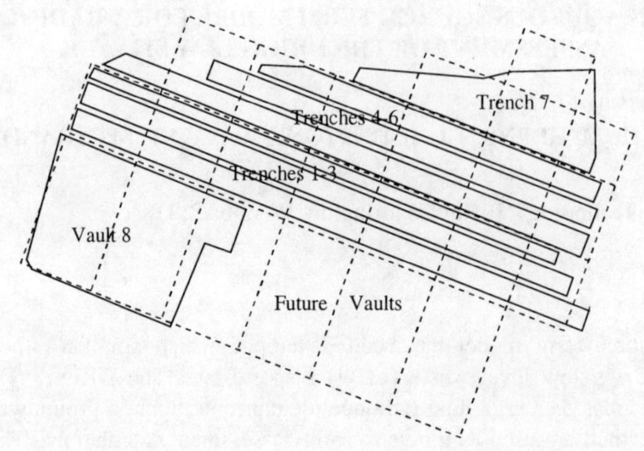

Figure 1. Plan view of the disposition of the Drigg Trench and Vault disposal systems (solid lines), and coincidence with the DRINK finite difference grid (dashed lines).

THE DRINK SOURCE TERM MODEL

The DRINK model utilises the BNFL biogeochemical reaction Generalised Repository Model (GRM) [1,2] to simulate the evolving geochemistry of the Drigg trenches and vaults. GRM considers kinetically controlled steel corrosion and microbial induced cellulose degradation reactions. The products of these processes are used to determine an evolving redox condition, taking account of kinetically controlled microbially mediated redox reactions between redox product species and species in groundwater (e.g. SO_4), and minerals in soils (e.g. $Fe(OH)_3$). Redox potential (pe) is calculated by using standard mass action equations [3] considering the most oxidising couple. The resulting pe is used as a constraint for equilibrium speciation and mineral equilibrium calculations by a routine based on PHREEQE [4], which determines the pH and master species concentrations, including those radionuclides which are solubility controlled. GRM describes the 2-dimensional lateral groundwater flow in the saturated zone by means of a finite difference solver. The discretisation of the finite difference grid used in the DRINK model is shown in Figure 1. Vertical flow is considered on a cell by cell basis and is used to simulate the release of radionuclides from the unsaturated zone to the saturated zone. In DRINK, sorption is modelled using a distribution coefficient (Kd) which is selected taking into consideration the simulated geochemical model, and the types of sorbant surfaces present in the Drigg trenches and vaults. Radioactive decay is considered on a cell basis for dissolved, sorbed and precipitated phases, and for the unsaturated zone.

BIOGEOCHEMICAL EVOLUTION OF THE DRIGG SITE

Chemical conditions in the Drigg trenches is simulated to vary over a period of around 1000 years. During this time degradation of cellulose and steel corrosion result in the establishment of conditions more reducing and acidic than the local groundwater. Figure 2 shows examples of the time evolution of concentrations of solid and dissolved species. In the DRINK model the first stage of cellulose degradation is the hydrolysis of cellulose modelled by a first order kinetic reaction with a pH dependent hydrolysis constant [2]. The computation of the cellulose concentration in Figure 2a includes the presence of hydrolysis in the saturated zone, and the transfer of cellulose and other materials from the unsaturated zone resulting from settlement. In effect cellulose in the saturated zone is replaced by that in the unsaturated zone until around 100 years when the unsaturated zone is depleted of cellulose. After this time an exponential decrease in cellulose concentration is simulated. Iron corrosion is modelled by a zero order kinetic reaction. In the saturated zone iron increases (Figure 2b) in concentration because of transfer from the unsaturated zone as a result of settlement.

Products of cellulose hydrolysis and corrosion are subject to microbial mediated redox reactions [1,2]. Glucose is metabolised by aerobic and anaerobic processes to yield CO_2, while acetate and other volatile fatty acids (VFAs) are produced by fermentation. The processes of cellulose degradation, microbial mediated redox reactions, gaseous exchange and mineral equilibration have been considered by DRINK for the computation of the rate of CO_2 production displayed in Figure 2c. The acetate concentration versus time show in Figure 2d was calculated from the model including acetate production by fermentation, and consumption by redox reactions. Acetate, and H_2 produced during corrosion are the main electron donors in the model which are responsible for the reduction of species such as SO_4 in groundwater, and Fe(III) oxyhydroxide in disposed soils.

The pH determined by the PHREEQE speciation calculation (Figure 2e) shows the generation of acidic conditions associated with the formation of CO_2 and acetate. After around 800 years, the pH returns to that of the local groundwater, which is in equilibrium with calcite at log pCO_2 = -2. pe is determined by the $[SO_4^{2-}]/[HS^-]$ couple for the first 2,000 years of the DRINK simulation. The concentrations of sulphate and sulphide species indicate a period of more reducing conditions during the first 1,000 years during which time S(II) is the dominate oxidation state (Figure 2f). The environment is significantly more reducing than the local groundwater, whose redox condition is consistent with equilibrium with siderite and Fe(III) oxyhydroxide. After periods of 1,000 to 2,000 years the trenches reoxidise as a result of the presence of NO_3^- in groundwater.

The evolving biogeochemical conditions associated with cellulose degradation and steel corrosion involve reactions between minerals considered by the PHREEQE module. A simplified mineral assemblage is considered comprising calcite and Fe(III) oxyhydroxide representing the reactive component of disposed soil. Under the acidic and reducing conditions simulated for the Drigg trenches calcite and Fe(III) oxyhydroxide dissolve during the early stages of the model. Siderite precipitates during the first 1,000 years of the model and is the main corrosion product under the simulated conditions, calcite and Fe(III) oxyhydroxide reprecipitates after 800 years as the chemical composition of leachate waters return to that of the local groundwater (Figure 2g).

pH buffering of the Drigg vaults

The chemical buffering effect of the Portland Cement based grout used as a backfill material in the Drigg vaults is simulated by the inclusion of two end-member phases to represent the calcium silicate solid-solution (CSH) [5]. Figure 2h shows the variation in concentrations of the calcium rich end member (CEMCSH), the silica end member (CEMSIO) and calcite. The variation in these concentrations is consistent with a carbonation reaction where CO_2 reacts with CSH to produce calcite and where the CSH becomes more silica-rich. Two phases of carbonation are simulated, a rapid period during the first 100 years where CO_2 generated from cellulose degradation contributes to the reaction, and a slow carbonation phase due to the presence of CO_2 in inflowing groundwater. The DRINK model simulates that it takes around 10,000 years for the calcium end member to be dissolved. During this time the pH is around 11 (at 10 °C). Despite the establishment of alkaline conditions the vaults are computed to undergo a similar pattern of biogeochemical evolution as the Drigg trenches described above. Reducing conditions are established as a result of cellulose degradation and steel corrosion reactions, reoxidation occurs over a period of 2,000 to 10,000 years. Under alkaline conditions different secondary mineral phases are produced with Fe(II) hydroxide and FeS formed as corrosion products.

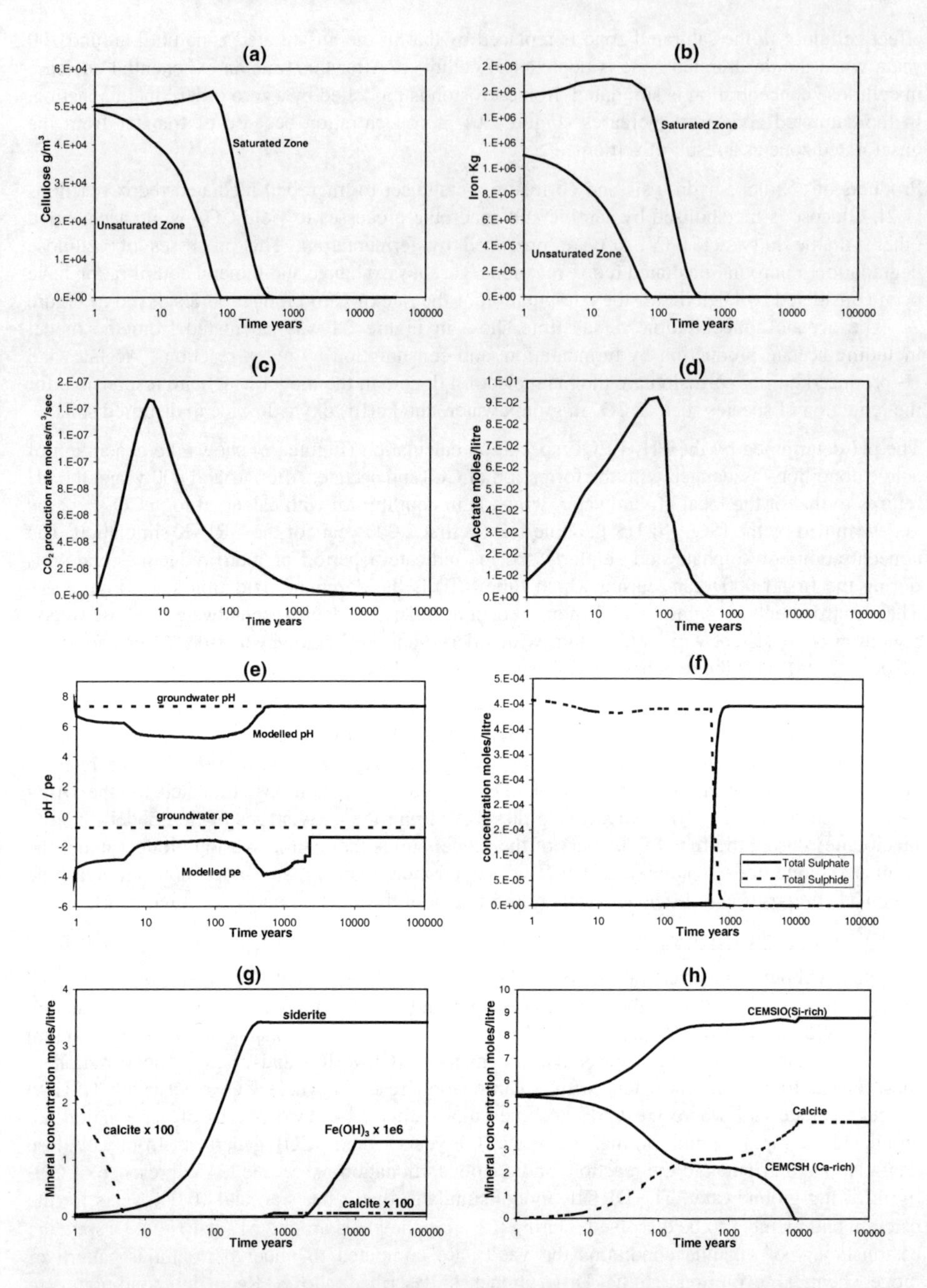

Figure 2. a-g Concentration profiles of waste materials, and degradation products simulated by the DRINK model for a representative model cell from the Drigg trenches, (h) simulated evolution of cement minerals in the Drigg vault saturated zone.

RADIONUCLIDE RELEASE MECHANISMS AND NEAR-FIELD CONCENTRATIONS

The DRINK model considers the time variation in pH, pe and aqueous speciation to determine changing solubility controls over radionuclide elements. In the DRINK model all isotopes of a particular element are combined for the purpose of determining solubility. The principal solubility controlled radionuclides are Th and U and their aqueous concentrations under representative trench conditions are illustrated in Figure 3a. Uranium solubilities are simulated to be at minimum values during the first 1,000 years as a consequence of the reducing conditions, which stabilise the U(IV) solid phase UO_2. After reoxidation of the trenches UO_2 solubility increases as consequence of U(VI) aqueous speciation. At approximately 30,000 years UO_2 completely dissolves and U concentration decreases sharply. Under the alkaline vault conditions U concentration is solubility controlled by U(IV) species within a concentration range of 1e-9 to 1e-10 moles/litre. Thorium solubility is very strongly influenced by the generation of high concentrations of acetate during the first 200 years (Figure 2d), with Th-acetate being the dominant aqueous species under the acidic conditions of the trenches. When acetate concentration is low, Th concentration is at a low and constant level because of the dominance of the $Th(OH)_4$ species which is unaffected by pH variation. Under the alkaline vault conditions Th solubility is unaffected by Th-acetate complexing. As a consequence of these solubility controls U and Th are retained as solids within the Drigg site. This has important implications for maintaining the concentration of sorption controlled daughter nuclides such as ^{226}Ra, (Figure 3a) which increase in concentration over the 1,000 to 30,000 year period. The rate of generation of ^{226}Ra is lower after 10,000 years when solid U dissolves. The behaviour of other sorption controlled radionuclides is illustrated in Figure 3b. Sorption controlled release is characterised by a washout of the radionuclide, exemplified by Tc which is assigned a minimal Kd value. Other radionuclides such as ^{90}Sr are influenced by decay while Am concentration increases as a result of ingrowth of ^{241}Am from ^{241}Pu decay.

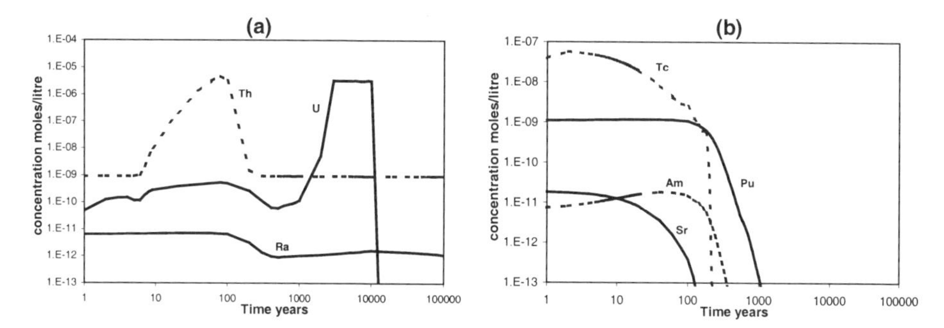

Figure 3. Variation in radionuclide elemental concentrations in the Drigg trenches influenced by (a) solubility controlled processes, and (b) sorption controlled processes.

^{14}C modelling

Carbon 14 is treated separately from other radionuclides because of the complication of considering the stable isotopes (^{12}C and ^{13}C) which behave in a virtually identical chemical manner. The approach adopted in DRINK is to assign ^{14}C to the reactive cellulose phase, and to recalculate the fraction of ^{14}C in all carbon species, minerals, gases and microbial substrates at each reaction and transport step, but neglecting isotope fractionation effects. The detailed biogeochemical modelling of cellulose degradation and mineral precipitation is therefore central to modelling ^{14}C release. Figure 4 shows the redistribution of the whole ^{14}C inventory during the

DRINK simulation. The majority of the ^{14}C is redistributed to siderite and calcite which form as secondary precipitates, approximately 10% of the inventory remains as inert organic material (humin). Less than 5% of the ^{14}C inventory is present as acetate or carbonate species, which is thus available for release from the near-field; an even smaller amount is present as gaseous CO_2.

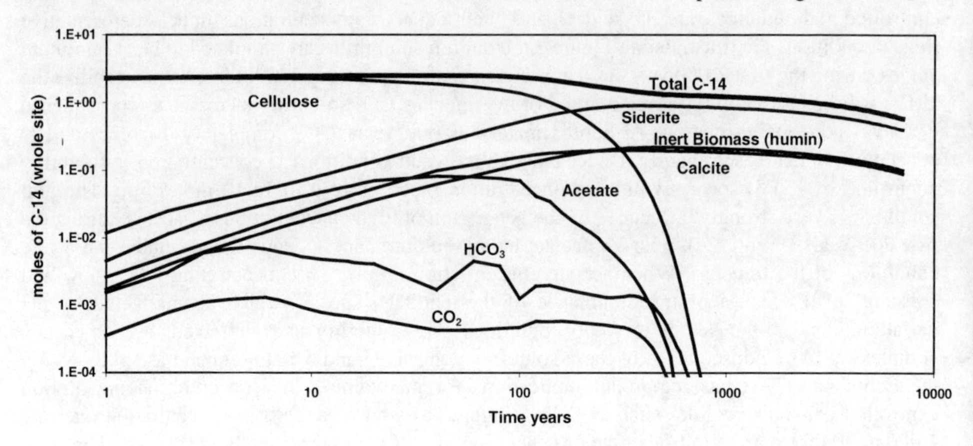

Figure 4. Variation in the distribution of C-14 between solid, aqueous and gaseous species for the whole of the Drigg site.

CONCLUSION

The DRINK model simulates that waste degradation results in the development of reducing and acidic conditions in the Drigg trenches during the first 2,000 years after site closure. Following this, the trenches reoxidise and chemical conditions return to that of the local groundwater. In the Drigg vaults a cementitious backfill maintains an alkaline environment for over 10,000 years. Using this model the varying solubility of U and Th in the trenches and vaults has been calculated over a period of 100,000 years. U and Th are retained within the site as solids and provide a source for daughter radionuclides such as Ra. The distribution of ^{14}C between aqueous species and solids minerals and microbial substrates has been calculated on the basis of the model of cellulose degradation and reactive chemical transport. Less than 5% of the ^{14}C inventory is present as mobile species. The continuously varying aqueous concentrations of radionuclides calculated by DRINK are directly input to a geosphere and biosphere groundwater risk assessment model used for the Drigg PCRSA. In addition the DRINK model supports the Drigg post closure safety case by contributing to the phenomenological understanding of the behaviour of the site.

REFERENCES

1. Manton, S., Johnstone, T., Trivedi, D. P., Hoffmann, A. and Humphreys, P. N., Radiochimica Acta, **68**, 75-79 (1995).
2. Humphreys, P. N., Johnstone, T., Trivedi, D. and Hoffmann, A., in *Scientific Basis for Nuclear Waste Management XVIII,* edited by T. Murakami and R.C Ewing. Mat. Res. Soc. Proc.., **353**, 211-218 (1995).
3. Stumm, W. and Morgan, J.J. *Aquatic Chemistry*. 2nd Ed.. Wiley, New York (1981), p449.
4. Parkhurst, D.L., Thorsteson, D.C. and Plummer, L.N. PHREEQE – A computer program for geochemical calculations, USGS Water Resour.. Invest., 80-96 (1980).
5. Berner, U.R. Waste Management **12**, p201,219 (1992).

LONG TERM PERFORMANCE ASSESSMENT FOR A PROPOSED HIGH-LEVEL RADIOACTIVE WASTE DISPOSAL SITE AT YUCCA MOUNTAIN - EXPLANATION OF THE SHAPE OF RELEASE CURVES

SITAKANTA MOHANTY, ROBERT W. RICE(consultant)
Center for Nuclear Waste Regulatory Analyses
Southwest Research Institute
6220 Culebra Rd, San Antonio, TX 78238
smohanty@swri.edu

ABSTRACT

The Nuclear Regulatory Commission (NRC) Total-system Performance Assessment (TPA) code is a tool to independently evaluate the long-term performance of the proposed Yucca Mountain (YM) high-level radioactive waste (HLW) repository by modeling processes such as the dissolution of the spent nuclear fuel (SNF) and the subsequent transport through unsaturated and saturated highly heterogeneous fractured porous media to a hypothetical release into the biosphere. The release rates of radionuclides into the biosphere corresponding to some of the SNF dissolution models show a sinusoidal trend with an overall decrease in the rate with time; for other models the sinusoidal behavior is nonexistent. This study identifies two key mechanisms contributing to these trends: the spatial discretization of the repository into subareas and radionuclide-specific sorption properties in the saturated zone alluvium. Other mechanisms which might affect the release rates, such as solubility limits, radioactive decay and ingrowth, inventory depletion, and transport properties of the unsaturated zone (UZ), do not significantly contribute to these trends.

INTRODUCTION

The NRC has the responsibility to review the license application for the HLW repository site at YM. In support of its regulatory review activities, the NRC staff has focused on detailed technical evaluation [1,2] to understand and quantify the isolation characteristics and capabilities of the proposed YM repository system. To support these technical assessments, the NRC and the Center for Nuclear Waste Regulatory Analyses (CNWRA) recently developed Version 3 of the TPA code [3].

YM is located in a semi-arid environment on the Nevada Test Site in southern Nevada, and rises several hundred meters above the surrounding land. The current design for the YM repository is to dispose of SNF in waste packages (WPs) emplaced in drifts approximately 300 m below the top of YM and 300 m above the water table. Following emplacement in the repository, the WPs will eventually fail, and the infiltrating water will contact the SNF in the WP and cause SNF dissolution. After the SNF dissolves into the contacting water, the flowing water will transport radionuclides out of the engineered barrier system (EBS), which is comprised of the WP, concrete invert that supports the WP, and possibly shields and backfill, through the hydrologically UZ and saturated zone (SZ) to a hypothetical release into the biosphere 20 km from the repository footprint.

Because of the high level of uncertainty in projecting the probable evolution of the repository, alternative conceptual models are used to evaluate the performance of the YM repository. To represent uncertainty in predicting the SNF dissolution rate, which is one of several key factors in assessing repository performance, four alternative conceptual models have been proposed [3] to characterize the nature of the time evolution of releases from the EBS. While the EBS releases show an expected decreasing trend for each of the SNF dissolution rate models, a very distinctive, yet peculiar sinusoidal behavior is observed in the calculational results of radionuclide release rates into the biosphere. To investigate this behavior, this paper first describes the general trends in the biosphere release rates and then hypothesizes

Mat. Res. Soc. Symp. Proc. Vol. 608 © 2000 Materials Research Society

several potential mechanisms. These mechanisms include: (i) solubility limits of the radionuclides, (ii) decay and ingrowth of the radionuclide inventory, (iii) inventory depletion of the SNF in the WP, (iv) transport properties of the UZ and SZ, and (v) spatial discretization of the repository. The paper guides the reader through contributions from these mechanisms to the trends exhibited in the biosphere release rates.

MODEL DESCRIPTION

To analyze radionuclide release rates into the biosphere, the repository is discretized into seven subareas representing different regions in the repository that are comprised of different thicknesses of hydrostratigraphic units, chemical sorption (K_d) and hydrologic properties, and matrix/fracture flow interactions. For each subarea, the transport of radionuclides through the UZ and SZ is modeled with a vertical column, representing the UZ, connected to a horizontal streamtube, representing the SZ. Simulations are conducted to 100,000 yr.

For the release and transport calculations, eleven radionuclides in the SNF are modeled: Am-241, C-14, Cl-36, Cm-245, I-129, Np-237, Pu-239, Se-79, Tc-99, Th-230, and U-234. These radionuclides were identified in a screening process as having the potential to significantly contribute to dose in the biosphere [4]. Two decay chains are also modeled: Cm-245–>Am-241–>Np-237 and U-234–>Th-230. The EBS release rates of these radionuclides provide a source term or time-dependent boundary condition for the UZ and SZ transport model. Using this source term, the release rates are computed at the outlet of the UZ and SZ flow path for each subarea. The outlets of the SZ flow paths correspond to the location of the pumping well that represents the interface between the geosphere and the biosphere.

The four SNF dissolution models used to compute the source term for UZ and SZ transport represent alternative conceptualizations of SNF release rates from the WP and are expected to bound the conceptual uncertainties associated with conditions that may exist in and around the WP and SNF subsequent to WP failure [3]. Model 1 estimates the SNF dissolution rate in solutions containing carbonate anions. The SNF dissolution rate in the presence of Ca and Si ions, which are found in the YM groundwater, is calculated in Model 2. Model 3 utilizes a dissolution rate derived from a natural analog. The formation of secondary minerals is considered in Model 4. Mathematical formulations of the SNF dissolution models and the associated assumptions are given elsewhere [4] and are not the focus of this paper. Rather, the modifications to the trends in the release rates at the EBS due to the transport of radionuclides through the UZ and SZ is the primary focus.

RESULTS AND DISCUSSION

The TPA simulations are conducted by exercising the four SNF dissolution rate models first to determine the release rates from the EBS. The release rate then is used as a boundary condition for transport through the UZ and SZ. The discussion describes detailed analyses of the results which provide an explanation for the release rate trends based on general factors, such as radionuclide and subarea effects, and the proposed mechanisms identified previously.

The release rates for I-129, Tc-99, Se-79, and Cl-36, which represent four of the eleven radionuclides evaluated in TPA simulations, are provided in this section. These four radionuclides are analyzed in detail because they have the largest contribution to dose [4].

The total release rates for these four radionuclides out of the EBS and into the biosphere calculated with the four SNF dissolution rate models are provided in figures 1(a) and 1(b), respectively. Figure 1(a) shows two peaks which correspond to two distinct WP failure times, and the height of each peak is proportional to the number of failed WPs. The first peak in the release rates, which occurs before 5,000 yr, is attributable to initially failed WPs (i.e., defective or damaged prior to or during emplacement), while the second peak that occurs at about 17,000 yr is from WPs failed by corrosion.

For discussion purposes, the four SNF dissolution rate models can be classified into two categories, Group1 and Group 2, where Group 1 models show undulated release rates with an overall decreasing trend in rate with time, and Group 2 models show rather smooth release rate characteristics. As shown in figure 1(a), the EBS release rates for the Group 1 models, Models 1 and 4 that exhibit fast and slow release rates, feature an increase to a peak value and thereafter decrease several orders of magnitude. The EBS release rates for the Group 2 models, Models 2 and 3 that have an intermediate release rate, reach a maximum value that remains relatively constant with time. The rapid decrease in release rates for Group 1 models is attributable to source depletion for Model 1 and to instantaneous releases from the gap inventory (i.e., portion of the radionuclide inventory assumed to be held loosely on the grain boundaries, cladding/fuel gap, and cladding) for Model 4, whereas the Group 2 models maintain an intermediate release rate which is not large enough to be dominated by source depletion or small enough to be dominated by instantaneous releases from the gap inventory.

To investigate the trends in the Group1 and Group 2 models, particularly the undulations and decrease in the overall trend in the release rates after 50,000 yr for the Group 1 models, release rates into the biosphere are provided in figure 2 for each radionuclide. Only the results from Models 1 and 2 are presented in this figure because the trends in their release rates are representative of the Group 1 and Group 2 models during the 100,000-yr simulation period. The total release in figure 2(a) for Model 1 and in figure 2(b) for Model 2 corresponds to the release rates provided in figure 1(b). Results in figure 2 show a rapid decrease in release rates for individual radionuclides for the Group 1 models compared to the Group 2 models. This rapid decrease is responsible for the decrease in the release rates shown in figure 1(b) after 50,000 yr. Additionally, the results in figure 2 illustrate that different radionuclides contribute to the undulations in the total release. The effect is more evident in figure 2(a), which shows that at early times the total release is attributable to Cl-36 and I-129, and at later times to Tc-99 and Se-79. The reasons for this behavior are discussed in detail later in this paper. However, although figure 2 shows the effect of different radionuclides on the release rate, these figures do not indicate the reasons for undulations in the release rate for the same radionuclides, such as Tc-99.

To further explore the reasons for the undulations in figure 2, Tc-99 release rates into the biosphere are presented in figure 3 for each subarea. In this figure, the Tc-99 release rates are presented, because the trends in the Tc-99 release rates are representative of the trends in the I-129, Se-79, and Cl-36 release rates. The total release of Tc-99 in figure 3(a) for Model 1 and in figure 3(b) for Model 2 correspond to the Tc-99 release rates in figures 2(a) and 2(b), respectively. From figure 3, it is noticeable that the undulations in the Tc-99 release rate is associated with contributions from different subareas. Thus, the total release rates shown in figure 3(a) explain the undulation of Tc-99 in figure 2(a) and also explains the two undulations in figure 1(b).

Thus, the results in figure 2 and 3 identify radionuclide and subarea contributions to the undulations in the release rates and the causes for the rapid decrease observed in figure 1(b) after 50,000 yr. The mechanisms proposed previously in this paper are associated with these trends. The following sections discuss the impacts of these proposed mechanisms on the release rates into the biosphere.

Solubility Limits

The solubility of a radionuclide can affect the magnitude and time of the peak EBS release rate, which may change the release rate into the biosphere. Releases that are not solubility limited are expected to exhibit a more rapid increase in the release rate compared to releases that are solubility limited which show a more gradual increase in the release rate. However, the radionuclides I-129, Tc-99, Se-79, and Cl-36 are highly soluble [3] and the only significant difference in the EBS release rates, which were analyzed but not included in this paper, is the magnitude of the releases for the Group 1 and Group 2 models. This difference is caused by the initial inventory of SNF in the WP. Consequently, solubilities do not cause the difference between the release rates for the Group 1 and Group 2 models.

Radioactive Decay and Ingrowth

The release rate of a radionuclide may be affected by radioactive decay and ingrowth. The

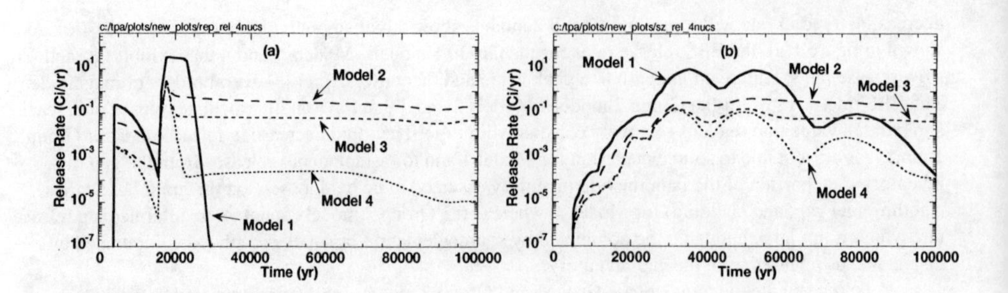

Figure 1. Total release rates for Tc-99, I-129, Se-79, and Cl-36 (a) out of the EBS and (b) into the biosphere for the four SNF dissolution rate models

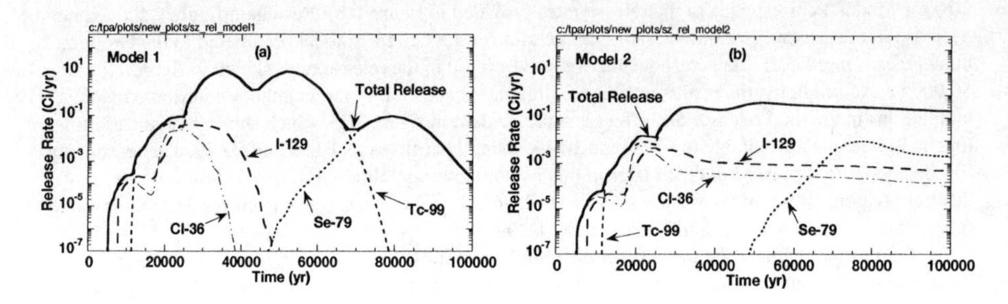

Figure 2. Release rates for Tc-99, I-129, Se-79, and Cl-36 into the biosphere using (a) Group 1 (Model 1) and (b) Group 2 (Model 2) SNF dissolution rate models

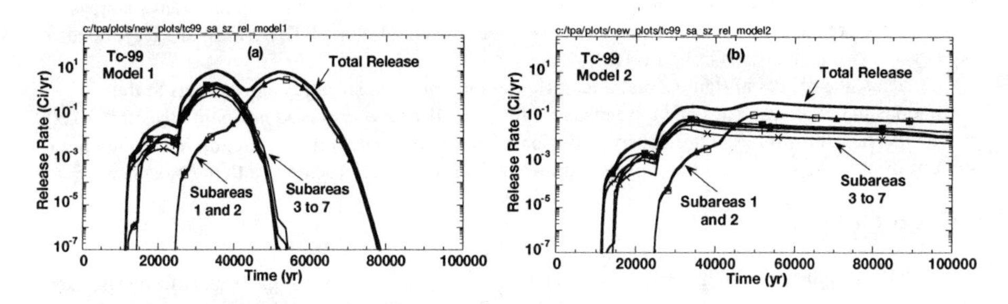

Figure 3. Release rates for Tc-99 from each subarea into the biosphere using (a) Group 1 (Model 1) and (b) Group 2 (Model 2) SNF dissolution rate models

radionuclides I-129, Tc-99, Se-79, and Cl-36 are not modeled in the TPA code with chain decay. Because the half-lives of these radionuclides, 1.57×10^7, 2.13×10^5, 6.5×10^4, 3.01×10^5 yr, are the same order of magnitude as the 100,000-yr simulation time, half-lives do not significantly decrease the inventory available for release, and thus the EBS release rates, by orders of magnitude as shown in the Group 1 model release rates in figure 1(b). Consequently, radioactive decay and ingrowth do not cause the differences observed in the Group 1 and Group 2 models in figure 1(b).

Inventory Depletion

A depletion of inventory may cause a significant change in the release rate with time because the amount of a radionuclide released from the SNF is related to the amount available for release. Except for Model 1, every SNF dissolution model shows a relatively constant release rate with time (Figure 1(b)) after the gap inventory is released. The effects of inventory depletion in Model 1 which quickly releases and depletes the WP inventory, are evident in figure 2(a). For example, the last undulation in this figure is solely caused by the depletion of Se-79 inventory. Therefore, the effects of inventory depletion on the undulation in the release rate curves are model-specific and do not explain the behavior of Group 1 models in general.

Transport Properties of the Unsaturated and Saturated Zones

The transport of radionuclides may be delayed by sorption and desorption in the UZ and SZ. However, in a separate analysis of results not presented in this paper, the UZ release rates are not significantly different from the EBS release rates. Thus, the analysis presented in this section will focus on transport in the SZ.

Figure 2 displays radionuclide-specific differences in the biosphere release rates. For the total release, Cl-36 has the largest contribution prior to about 8,000 yr. From about 8,000 to 12,000 yr, both Cl-36 and I-129 contribute the most to total release. Between about 12,000 and 68,000 yr, Tc-99 dominates total release and, thereafter, Se-79 is the major contributor to total release. The differences in the arrival time of the peak release for these radionuclides are caused by retardation in the SZ alluvium. Radionuclides with the smallest retardation factors arrive earlier at the biosphere, while later arrival times correspond to radionuclides with larger retardation factors. The SZ alluvium retardation factors are 1.0, 2.0, 5.5, and 22.4 for Cl-36, I-129, Tc-99, and Se-79, respectively [3]. Thus, for the Group 1 models, the sinusoidal behavior observed in figure 1(b) can be attributed to contributions from these various radionuclides. Sorption in the SZ alluvium does not cause a similar trend for the Group 2 models because the EBS release rates, as shown in figure 1(a), remain relatively constant.

As described in this section, the undulations of the Group 1 model release rates in figure 1(b) are caused by SZ sorption. However, repository spatial discretization also contributes to the undulations as discussed in the following section.

Spatial Discretization

The EBS release rates are computed for seven subareas in the TPA code and each subarea has a unique UZ and SZ hydrostratigraphy that could affect radionuclide transport. The UZ is previously identified as having no effect on the biosphere release rates. Therefore, only the effects of spatial discretization in the SZ will be discussed.

The Tc-99 release rates discussed earlier and provided in figure 3 for the Group 1 and Group 2 models show the same behavior. The releases from subarea 1 and 2 are delayed by about 15,000 yr when compared to the release rates from subareas 3 through 7. This delay causes two peak release rates at about 30,000 and 50,000 yr in figure 1(b). These peaks are attributable to the differences in the length of the SZ flow path and the groundwater travel time. For subareas 1 and 2, the SZ flow path length is approximately 4,000 to 7,000 m greater than the length of the SZ flow path for subareas 3 through 7 [3]. The differences in the flow path length produce an SZ groundwater travel time of about 6,000 yr for subareas 1 and 2, compared to approximately 2,800 yr for subareas 3 through 7. Consequently, spatial discretization of the repository into subareas is also responsible for the sinusoidal behavior of the Group 1 models.

SUMMARY AND CONCLUSIONS

Two categories of SNF dissolution rate models were identified in this paper: Group 1 models which exhibit release rates dominated by source depletion and instantaneous releases from the gap inventory, and Group 2 models which show release rates that are not dominated by these effects and have rather smooth release rate characteristics. The release rates into the biosphere computed using the Group 1 models show a sinusoidal nature and a relatively rapid decrease after 50,000 yr while, for the Group 2 models, the sinusoidal behavior vanishes. This paper proposes that these trends in the release rate may be caused by the effects of solubility limits, inventory depletion, radioactive decay and ingrowth, UZ and SZ transport properties, and repository spatial discretization. After analyzing the EBS and biosphere rates, an explanation for the trends in the biosphere release rates was identified.

The sinusoidal trend in the biosphere release rates calculated using Group 1 models may be attributed to two mechanisms that primarily contribute to this trend: (1) spatial discretization of the repository into subareas, and (2) radionuclide-specific sorption properties in the SZ alluvium. Furthermore, the decrease in the biosphere release rates for the Group 1 models after 50,000 yr is caused by the rapid decrease in the EBS release rate. The Group 2 models do not exhibit these trends, because the biosphere release rates are relatively constant with time.

Thus, the SNF dissolution rate model utilized in a TPA simulation can affect the trend in the calculated release rates into the biosphere. Moreover, the biosphere release rates are strongly influenced by the properties of the SZ flow path. The behavior of the biosphere release rates is important because these release rates are used to compute dose to the receptor, which is the performance measure for the HLW repository at YM.

ACKNOWLEDGMENTS

The paper was prepared to document work performed by the Center for Nuclear Waste Regulatory Analyses (CNWRA) for the Nuclear Regulatory Commission (NRC) under contract No. NRC–02–97–009. The activity reported here was performed on behalf of the Office of Nuclear Material Safety and Safeguards. The paper is an independent product of the CNWRA and does not necessarily reflect the views or regulatory position of the NRC.

REFERENCES

1. R.B. Codell, N. Eisenberg, D. Fehringer, W. Ford, T. Margulies, T. McCartin, J. Park, and J. Randall, *Initial Demonstration of the NRC's Capability to Conduct a Performance Assessment for a High-Level Waste Repository*, NUREG–1327, Nuclear Regulatory Commission, Washington, DC, 1992.

2. R.G. Wescott, M.P. Lee, N.A. Eisenberg, T.J. McCartin, and R.G. Baca, eds., *NRC Iterative Performance Assessment Phase 2*, NUREG–1464, Nuclear Regulatory Commission, Washington, DC, 1995.

3. S. Mohanty and T.J. McCartin, *Total-system Performance Assessment (TPA) Version 3.2 Code: Module Description and User's Guide*, Center for Nuclear Waste Regulatory Analyses, San Antonio, TX, 1998.

4. S. Mohanty, R. Codell, R.W. Rice, J. Weldy, Y. Lu, R.M. Byrne, T.J. McCartin, M.S. Jarzemba, and G.W. Wittmeyer, *System-level Repository Sensitivity Analysis Using TPA Version 3.2 Code,* NUREG (to be published), U.S. Nuclear Regulatory Commission, Washington, DC: 1999.

SAFETY ASSESSMENT OF BORE-HOLE REPOSITORIES FOR SEALED RADIATION SOURCES DISPOSAL

M.I. OJOVAN, A.V. GUSKOV, L.B. PROZOROV, A.E. ARUSTAMOV, P.P. POLUEKTOV, B.B. SEREBRYAKOV.
Scientific and Industrial Association "Radon", The 7-th Rostovsky Lane, 2/14 Moscow, 119121, Russia, Oj@tsinet.ru.

ABSTRACT

Bore-hole repositories (BHR) are considered to be promising for disposal of HLW and spent sealed radiation sources (SRS). A safety assessment of BHR disposal of SRS was performed using geologic environmental analysis, available parameters for BHR and SRS design and radionuclide inventory. The probabilistic calculations take into account some data uncertainties and variability. The results showed that there is practically no release of short-lived radionuclides into the environment for about 1000 years. This is completely due to the very low corrosion rate of the lead matrix in which the SRS are encapsulated. Various models were applied for more detailed numeric simulation of the repository temperature, radiation fields, and transport of released radionuclides in the geosphere. Ultra-conservative scenarios were chosen for these models. The worst case comprises both breaching of all engineered barriers and flooding of the disposal site plus eventual failure of an imperfect SRS immobilization matrix with some sources partly exposed by the breached matrix. For this extreme case, the maximum dose was found to be not higher than 55 – 75 μSv/y.

INTRODUCTION

Experience in operation of shallow ground BHR for SRS and inspection of their status have indicated the possibility that high-power ionising radiation fields could accelerate corrosion of engineered barriers, allowing radionuclides to penetrate into underground water and the surrounding soil [1]. A technological scheme to immobilize SRS in a metallic matrix directly in the BHR was therefore developed that increased the SRS BHR safety considerably. This scheme allows the use of a conventional scheme of SRS BHR disposal followed by in-place immobilisation of the SRS to provide an additional barrier – a metal layer – between the sources and environment. This decreases the potential release of radionuclides and enhances disposal safety [2].

After sufficient radionuclide decay (requiring about 500 - 1000 years), the metallic blocks containing the SRS will be removed and melted for matrix material reutilisation.

To estimate the applicability of the SRS immobilisation technology in BHR one should evaluate the long term safety of the repository based on the most likely event scenarios.

SIMULATION

For modelling we used a number of computer codes that model each scenario for potential release of radionuclides from the disposal site and transport into the environment. We considered many possible scenarios, among them the most important were:

- Lateral spreading of radionuclides;
- Transport of radionuclides in perched water tables due to flow into the repository;
- Transport of radionuclides into perched water tables due to flow of water into the repository and heat convection of this water caused by radionuclide decay heating of the repository;
- Transport of radionuclides due to the inhomogeneity of filtration factors when water floods the repository;
- Transport of radionuclides due to thermal gradients in the flooded repository, specifically heat convection caused by radionuclide decay heat;
- Transport of radionuclides due to capillary uplift of ground water.

Mat. Res. Soc. Symp. Proc. Vol. 608 © 2000 Materials Research Society

In order to assess SRS BHR safety we adopted an ultra-conservative model of event scenarios. We assumed (despite its improbability) that the SRS embedded in lead can contact the stainless steel BHR walls, which can be destroyed by the resulting physico-chemical reactions.

The calculation of population dose for evaluatiing repository safety was made with the GENII code based on recommendations given in ICRP publications 26, 30 and 60 and U. S. Department of Energy Hanford disposal site environmental models [3].

INITIAL PARAMETERS OF MODELS

The modelling results suggest that engineered barriers of concrete, carbon steel and stainless steel cannot be corroded sufficiently to destroy their integrity earlier than 77 years after emplacement. After this time significant corrosion of the lead matrix and SRS cases, which are stainless steel, begins. The corrosion rate of the lead matrix is 5×10^{-6} m per year. The corrosion is accompanied by transport of radionuclides into underground waters. We assumed that the SRS contain radionuclides both as soluble salts and as insoluble metal oxides. We assumed that the distribution factor for the metal oxides is 5, which means that only 20 % of the radioactivity can be released into the environment. The radionuclide inventory, 180,000 Ci immediately after disposal, was taken as 90% Co-60 and 10% Cs-137.

The host rock heat conductivity was 2.8 W/(m^2 °C) and is used in the calculations for all rock and engineered barriers. The lead matrix heat conductivity is 35 W/(m^2 °C) [1, 2, 4].

The volume heat capacity of the host rock, matrix and barriers is assumed to be constant and equal to 1200 kJ/(m^3 °C). We assumed that the water content of rock is equal to the amount of water that would be accomodated by the common porosity, e.g. 0.3.

The distribution coefficient for Cs-137 ion sorption by surface loamy soils was assumed to be 200; for Moscow moraine clay, 2000; for moraine sediments, 530; for concrete corrosion products and bore hole soil fill, 1.5 [5]. Co-60 exists as complex anions, which are essentially not sorbed by soil. Therefore we considered two cases: one in which the distribution coefficient K_d for unfixed Co-60 was zero, another in which this K_d for fixed Co-60 was 20. The latter K_d was used for all rocks and engineered barriers.

The filtration coefficient for surface loamy soil was taken from conservative data as 0.1 m/day; for moraine sediments, 1 m/day; for disturbed soil and products of corrosion, 0.5 m/day. The lead matrix is assumed to be impenetrable. For Moscow moraine clay the filtration coefficient is 10^{-4} m/day.

We assumed that at the instant when all barriers are destroyed (after 77 years) the BHR voids are filled with disturbed soil and crushed concrete.

We assumed that the perched water table level is 0.5 m above the Moscow moraine clay, has an infiltration flow velocity of 3×10^{-5} m/day, and this water table is constant. Consistent with Reference [6], for lateral spreading the longitudinal dispersion was taken as 50 m and the transverse dispersion as 10 m at a distance of 2 km from the BHR. To calculate the vertical transport in the case of BHR flooding, the longitudinal dispersion was taken as 0.3 m and the transverse dispersion as 0.1 m. The common porosity of soil was taken as 0.3, whereas the active porosity was taken as 0.2.

RESULTS

The calculation results show that for 180,000 Ci total radioactivity, the maximum temperature in the repository is 160 °C, which does not exceed the maximum permitted value.

According to the hydrogeologic data, the slope of the underground water horizon nearest the BHR is about 10^{-4}. Groundwater within the moraine horizons has a hydrostatic pressure of 22 m. Therefore radioactivity can not be directly transferred with the infiltrating stream to the water horizon and will migrate in a horizontal direction in the clay with a very low velocity. However the results show that the concentration of radionuclides at 2 km from the BHR (typical distance to an assumed nearest water source) will be zero. From these results we can conclude that the lateral spreading of radioactivity is not a hazard.

We also analyzed radionuclide release from the matrix into the water horizon via the perched water table and horizontal spreading of nuclides in the perched water table. We assumed that a layer of loamy soils is located 0.5 m above the matrix and extends to a radius of 10 m from the BHR. Thereafter we assumed that the infiltrating stream flows into the area of the breached concrete barrier and leaches radionuclides. Along the stream and in the region of the products of barrier degradation the filtration coefficient is constant and equal to 0.5 m/day.

We mathematically modelled the situation when the stream first flows down and then up, resulting in leaching of radionuclides into the perched water table followed by transport in the horizontal direction. The major portion of leached radioactivity is located near the repositories (Fig. 1). As one can see from Table I., only data for unfixed Co-60 can be compared with the limits determined by the regulatory document NRB-96.[See Reference 4]

Table I. Predicted concentrations of nuclides and doses for radionuclide transport into the water horizon through a perched water table.

Nuclide	Maximum concentration, Bq/l	Maximum dose, Sv/y	Maximum time, y
Co-60(K_d=0)	4.1	5.7×10^{-5}	89
Co-60(K_d=20)	7.4×10^{-8}	1.1×10^{-12}	144
Cs-137	1.1×10^{-4}	7.3×10^{-9}	465

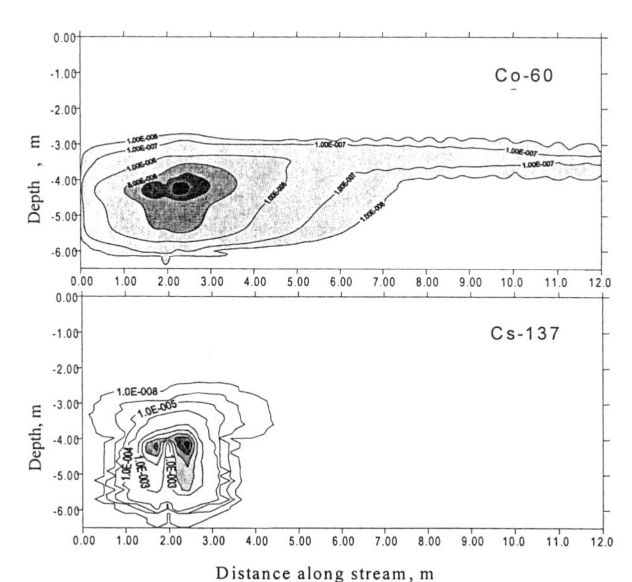

Fig. 1 Concentration of radionuclides (Ci/m^3) after 100 years for an infiltrating stream flowing through the BHR.

To simulate radionuclide transport in a perched water table while taking into account heat convection caused by BHR heating from radioactive decay, we again used the assumptions of the previous scenario. However, in the vicinity of the repository we assumed a vertical increase in the ground water level. We again mathematically modelled the situation for the stream first flowing down and then upward. This results in radionuclide leaching to the perched water table and then transport in the horizontal direction. We conclude that filtration coefficient inhomogeneities can play a more significant role than the assumed convection in transport of radionuclides in the case of a perched water table (Fig 2, Table II).

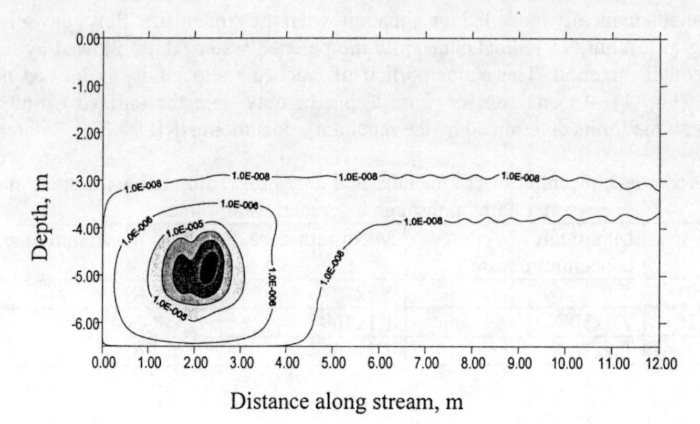

Fig. 2 Groundwater radionuclide concentration (Ci/m^3) after 100 years taking into account convection caused by heating of the BHR due to radioactive decay.

Table II. Predicted concentration of radionuclides and doses for a scenario with transport of radionuclides in the perched water table due to convection caused by heating of the repository.

Nuclide	Maximum concentration, Bq/l	Maximum dose, Sv/y	Maximum time, y
Co-60(K_d=0)	5.4	7.5 x 10^{-5}	90
Co-60(K_d=20)	3.2 x 10^{-8}	4.5 x 10^{-12}	150
Cs-137	5.0 x 10^{-5}	3.3 x 10^{-9}	460

We also analyzed potential radionuclide transport in case of full flooding of the BHR. Such flooding is possible if a confining layer (water-resistant clay) is located below and close to the bottom of the BHR, as assumed for this scenario. We assumed that the concrete barrier is not completely destroyed. Only in this case is there a possibility for descending and ascending streams to access the metal block containing the SRS. One can see (Table III) that repository flooding results in essentially no release of radioactivity.

Table III. Predicted concentration of radionuclides and doses for scenario with radionuclides transport when flooding repository.

Nuclide	Maximum concentration, Bq/l	Maximum dose, Sv/y	Maximum time, y
Co-60 (K_d=0)	4.6 x 10^{-5}	6.4 x 10^{-10}	106
Co-60 (K_d=20)	1.9 x 10^{-23}	2.7 x10^{-28}	204
Cs-137	5.6 x 10^{-4}	3.7 x 10^{-8}	220

We also analyzed the scenario for potential radionuclide transport due to heat convection caused by heating of the flooded BHR. The conditions are the same as in the previous scenario, except the water level gradient and the filtration coefficient were constant. The filtration coefficient was taken as 0.5 m/day.

One can see by the calculated results (Table IV) that BHR heating does not cause significant transport of radionuclides to the ground surface.

Table IV. Predicted concentration of radionuclides in ground water and doses for the scenario with radionuclide transport due to heat convection caused by heating of the BHR.

Nuclide	Maximum concentration, Bq/l	Maximum dose, Sv/y	Maximum time, y
Co-60 (K_d=0)	4.5×10^{-7}	6.3×10^{-12}	115
Co-60 (K_d=20)	2.6×10^{-30}	3.6×10^{-35}	152
Cs-137	2.8×10^{-9}	1.9×10^{-13}	249

Using an empirical model of the aerated zone [7] we considered potential radionuclide transport caused by ground water capillary uplift.

We assumed that the BHR was located in clay. An 0.3-m wide space between the matrix and the walls of the BHR was filled with sandstone soil. The BHR was filled with water to a height of 1 m and there was no infiltration through the clay.

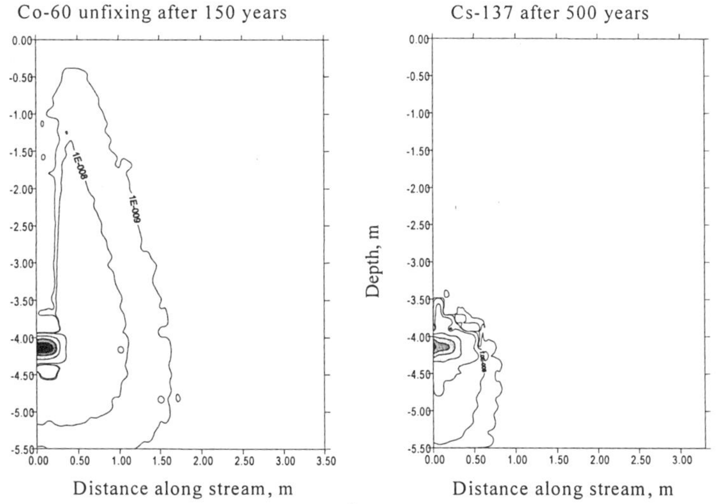

Fig. 3 Concentration of radionuclides (Ci/m^3) in groundwater due to groundwater capillary uplift.

The results of the calculations (Table V) show that only the unfixed Co-60 can be observed. From the difference between the distribution of Cs-137 and unfixed Co-60 (Fig. 3) we see that this Co-60 rises with groundwater capillary uplift while Cs-137 remains at the site of the SRS matrix. Because the unfixed Co-60 concentration is insignificant, this indicates that groundwater capillary uplift does not cause any significant surface water and soil contamination.

Table V. Predicted concentration of radionuclides and doses for scenario with groundwater capillary uplift.

Nuclide	Maximum concentration, Bq/l	Maximum dose, Sv/y	Maximum time, y
Co-60 (K_d=0)	1.5 x 10^{-2}	2.1 x 10^{-7}	147

CONCLUSION

The results of calculations show that the maximum annual dose dose to a member of the population does not exceed 5.5 - 7.5.x.10^{-5} Sv/y even in the case of complete destruction of the engineered barriers and complete flooding of the BHR. It indicates the high degree of safety of SRS disposal in a BHR when sources are immobilized by encapsulation in a metal (lead) matrix. The predicted data comply with the annual dose limit (10^{-4} Sv/y) used at the present time in the Russian Federation and confirm that theoperation of these BSR SRS repositories would be safe for population.

Due to its chemical properties, lead forms essentially only insoluble compounds with groundwater anions. The heterogeneous character of the exchange reaction fixes the lead corrosion products near the repository. In addition, as has been noted above, after a certain period of time (500-1000 years) the metallic matrix with sources would be removed and melted for material reutilisation. This is an additional means to protect the environment from potential contamination by lead corrosion products.

REFERENCES

1. Sobolev I.A., Timofeev E.M., Ojovan M.I., Arustamov A.E., Kachalov M.B., Shiryaev V.V., Mat. Res. Soc. Symp. Proc., Vol.506, 1998, p.1003-1008.
2. Arustamov A.E., Ojovan M.I, Kochalov M.B., Mat Res. Soc. Proc., v 556 (1999) pp. 961-966.Napier B.A., Peloquin R.A., Strenge D.L., Ramsdell J.V. GENII - The Hanford Environmental Radiation Dosimetry Software System. V.1: Conceptual Representation. Pasific Northwest Laboratory. Washington, 1988.
3. Arustamov A.E., Ojovan M.I., Kachalov M.B., Sobolev I.A., Shiryaev V.V. Metal Matrices for the Immobilization of Highly-Radioactive Spent Sealed Sources. – WM '98 Proceedings, Tucson, Arizona, March 1-5, 1998.
4. Derivation of Default Acceptance Criteria for Disposal of Radioactive Waste to Near Surface Facilities: Development and Implementation of an Approach. IAEA TECDOC, Vienna, 1998.
5. Savage D. The Scientific and Regulatory Basis for Geological Disposal of Radioactive Waste. John Wiley&Son. Chichester, New York, Brisbane, Toronto, Singapore, 1995.
6. Van Genuchen M. Th. A closed-form equation for predicting the hydraulic conductivity of unsaturate soil.- Soil. SCI. Am. J., 1980, v. 44, pp. 892-898.

PREDICTION OF WASTE PACKAGE LIFE FOR HIGH-LEVEL RADIOACTIVE WASTE DISPOSAL AT YUCCA MOUNTAIN

OSVALDO PENSADO AND SITAKANTA MOHANTY
Center for Nuclear Waste Regulatory Analyses (CNWRA)
Southwest Research Institute, 6220 Culebra Rd, San Antonio, TX 78238

ABSTRACT

The U.S. Department of Energy (DOE) has released a Viability Assessment (VA) of the proposed high-level waste (HLW) repository at Yucca Mountain (YM), Nevada. The proposed standard for the repository requires the evaluation of the predicted —over 10,000 yr— performance of the system. In the VA, it is argued that long waste package (WP) lifetimes are responsible for significant containment and delay in release of the HLW. This paper examines the DOE WP lifetime prediction by combining the VA parameters modeling the corrosion behavior of the WP materials with the repository performance assessment code developed by the U.S. Nuclear Regulatory Commission (NRC) and the Center for Nuclear Waste Regulatory Analyses (CNWRA). The objective of the analysis is to exercise and strengthen the NRC review capabilities in preparation for receipt of the DOE license application for the proposed HLW repository. The results indicate that during the first 10,000 yr, radionuclide release depends almost entirely on the fraction of initially defective waste packages (IDWP). Radionuclide release resulting from corrosion of the WPs occurs only beyond 10,000 yr. The long WP lifetimes (>10,000 yr), reported in the VA and in this study, primarily depend on corrosion rates developed from an expert elicitation consensus. Given the importance in the estimation of the onset of radionuclide release and the lifetime of the WP materials, it is suggested the need for a stronger technical basis to assess the fraction of IDWP and to support the low corrosion rates of the WP materials.

INTRODUCTION

The NRC, with the technical support of the CNWRA, has been developing capabilities to review the DOE license application for the proposed HLW repository. One of the main tools to be used during the review process is the Total-system Performance Assessment (TPA) code developed by the CNWRA and NRC [1]. Based on a Monte Carlo scheme, the TPA code computes the expected annual dose to the average member of the critical group in the event of failure of the WPs to isolate the radionuclides. The NRC has selected a *basecase* (defined as a particular set of models and model parameters describing the YM repository system) to perform sensitivity and uncertainty analyses and to study the general behavior of the system as simulated by the TPA code.[1]

The DOE has recently released a VA for the proposed HLW repository at YM [2]. A major objective of the DOE YM repository program is to show the feasibility of near-complete containment of radionuclides within the WPs for several thousand years [3]. The proposed standard for the repository requires the evaluation of the predicted —over 10,000 yr— performance of the system. Of interest to this paper are the Total System Performance Assessment studies [2, Vol. 3], particularly the analysis of the corrosion phenomena leading to WP failure (i.e., number of failed WPs and the time of failure). In this paper, the NRC *basecase* is combined with the parameters characterizing the corrosion phenomena reported in the VA (i.e., the NRC TPA code is used to simulate the corrosion processes as described in the VA). The objectives are to exercise and strengthen the NRC review capabilities in preparation for receipt of the DOE license application for the proposed HLW repository.

PROBLEM DESCRIPTION

The WP design proposed in the VA includes two metallic barriers: an inner corrosion resistant material composed of Alloy 22 (2 cm thick) and a corrosion allowance material (CAM) composed of ASTM

[1] The DOE has an analogous definition to the NRC *basecase*, referred to in this paper as the *DOE base case*.

Mat. Res. Soc. Symp. Proc. Vol. 608 © 2000 Materials Research Society

A516 (10 cm thick) [2, Vol. 2]. The carbon steel (ASTM A516) is meant to provide structural strength, protection against rock fall, and reduction in the intensity of alpha and gamma radiation (thus decreasing the probability of the presence of radiolytic products such as hydrogen peroxide, which are detrimental to the corrosion resistance of the WP materials) [2, Vol. 2].

The NRC models the CAM behavior by considering several corrosion modes, including dry air oxidation and corrosion under humid air and aqueous environments. The physical condition determining the corrosion mode is the relative humidity (RH). Thus, when the RH exceeds a lower threshold, the corrosion mode changes from dry oxidation to humid air corrosion, and when the RH lies above an upper threshold, aqueous corrosion is initiated (i.e., at high RH it is assumed the formation of a water film on the WP surface). Current NRC modeling does not consider the presence of dripping (i.e., the direct impingement of water on the WPs) [1]. Localized corrosion in aqueous environments (an accelerated corrosion process) is assumed to occur when the corrosion potential is above a critical potential, defined by the temperature and chloride concentration. Experimental data indicate that Alloy 22 is a material highly resistant to localized corrosion with a passive dissolution rate independent of environmental factors such as pH and chloride concentration and only slightly dependent on the temperature [4]. On the other hand, carbon steel in aqueous environments and subject to a corrosion potential above a critical potential (defined by the temperature and chloride concentration) is prone to localized corrosion [1].

The DOE corrosion model differs from the NRC model in several technical areas. In the NRC model, radionuclides are released after breaching of a site on the WP surface. On the other hand, the DOE modelers have proposed the concept of corrosion patches developing on the WPs —964 patches are assumed to exist on a WP, each with a uniform area of 310 cm^2, and an arbitrary pit density of 10 pits cm^{-2} (the maximum number of pits allowed on a patch is 32,000) [5]. The 964 patches cover the whole WP surface of 30 m^2. The choice for the patch size has been motivated by the DOE experimental program, where sample coupons with a surface area of 310 cm^2 have been used to determine corrosion rates via weight loss measurements [5]. The patches are assumed to corrode under different mechanisms; the faster mechanisms include the results of dripping. According to the DOE modeling, it is feasible for Alloy 22 to display localized corrosion [5].

In this paper, the NRC *basecase* has been modified to emulate the VA description of corrosion phenomena. Because of the difference in the models, a one-to-one mapping of the concepts is not feasible. However, careful attention was given to the appropriate selection of parameter values to provide an adequate representation of the DOE models and data. From the several cases presented by the DOE, only WP failure under dripping was selected because this case was expected to provide the highest WP failure rate. The DOE and NRC models coincide in that the carbon steel outer overpack could be breached by localized corrosion with a pit growth law of the form $d = B\,t^n$, where d is the pit depth (m), B is a constant (m yr^{-n}), n is a dimensionless constant ($0<n<1$), and t is the time. Alloy 22 seems to be affected only by passive dissolution under the environmental conditions relevant to YM [4]. From an expert elicitation consensus [5], the DOE proposed probability distribution functions (PDFs) for B, n, and the general corrosion rate for Alloy 22 (represented by r).[2] Details of the definition of such PDFs are provided in Table I. The DOE researchers have proposed cumulative PDFs for the general corrosion rate for Alloy 22 under dripping and nondripping environments at multiple pH ranges (from 2.5 to 10), equilibrium potentials (340 mV_{SHE} and 640 mV_{SHE}), and temperatures (25°C, 50°C, and 100°C) [5]. From these expert-elicitation data, a single PDF (enclosing all the pH and equilibrium potential variation) was computed for each temperature and for each dripping and nondripping case. The data considered in the present analysis correspond to the dripping case at 100°C.

Radionuclide release from the WPs caused by corrosion failure is predicted to occur after several thousand years. For example, the expected value for B is 40 mm yr^{-n}, and for n, 0.5 (see Table I). The failure time of the carbon steel computed from these numbers is 6.25 yr. Therefore, localized corrosion of carbon steel (assuming the presence of an aqueous environment and a corrosion potential above the repassivation potential) is a phenomenon leading to CAM failure in a time on the order of years (an extremely short time estimate). On the other hand, since the expected corrosion rate for Alloy 22 at 100°C under dripping

[2] General corrosion includes passive and active dissolution. It is a form of nonlocalized corrosion, also called uniform corrosion. A material affected by general corrosion with a rate, r, on a single surface is breached at a time, Δt, if $r\Delta t$ equals the initial thickness (measured across the exposed surface) of the material.

Table I: Parameters and PDFs used in the computations. The values listed on the "Range" row for the normal and log-normal PDFs correspond to the domain values where the cumulative distribution functions equal 0.01 and 99.9, respectively. In the uniform and log-uniform PDFs, the values listed correspond to the end points of the domain.

		B (mm yr^{-n})	***n***	***r*** (mm/yr)	**Fraction of IDWP**
PDF	DOE NRC	log-normal uniform	normal constant	log-normal uniform	log-uniform uniform
Range	DOE NRC	7.8×10^{-4}, 5.6×10^{2} 0.866, 8.66	0.243, 0.747 0.45	5×10^{-7}, 3×10^{-3} 6.2×10^{-4}, 2×10^{-3}	10^{-5}, 10^{-3} 10^{-4}, 10^{-2}
Expected value	DOE NRC	39.46 4.8	0.5 0.45	10^{-4} 1.31×10^{-3}	2.15×10^{-4} 5.05×10^{-3}

conditions is 10^{-4} mm/yr, the expected alloy lifetime is of the order of 200,000 yr. Thus, the main protection against radionuclide release is provided by the Alloy 22 wall.

The presence of IDWP caused by WP fabrication or handling defects leads to early release of radionuclides. DOE estimated the probability for the presence of IDWP based on a study on pressure vessels and assuming failure independence of the two WP walls [2]. The IDWP are postulated to influence the repository performance only after 1,000 yr [2]. The rationale behind this time is not clear, since the IDWP are part of the system at the time of the waste emplacement. Table I summarizes and compares the numerical values assigned by the DOE and NRC to B, n, r, and the fraction of IDWP.

RESULTS AND DISCUSSION

The NRC *basecase* was modified to incorporate the DOE definitions of B, n, r, and the fraction of IDWP. Modifications to the *basecase* and designated labels are listed in Table II. The *case labels* (bc+DOEcorr, bc+DOEjf, and bc+DOEcorrJF) are arbitrary names of the modified cases intended to facilitate the discussion. Figure 1 displays the total expected dose rate during a 100,000-yr period. Each expected dose rate is computed as an average of 250 independent Monte Carlo realizations, except for the

Table II: Modifications to the NRC *basecase* analyzed in this paper. The *case labels* are arbitrary names intended to facilitate the discussion.

Case label	**Modification to the NRC *basecase***
bc+DOEcorr	DOE PDFs for B, n, and r
bc+DOEjf	DOE PDF for the fraction of IDWP
bc+DOEcorrJF	DOE PDFs for B, n, r, and the fraction of IDWP

DOE data, which were reported in the VA. In the first 10,000 yr, the total dose rate is dominated by the IDWP. The *basecase* and the bc+DOEcorr case predict comparable dose rates because the fraction of IDWP is the same for both cases. Similarly, the bc+DOEjf and bc+DOEcorrJF cases produce comparable dose rates in the first 10,000 yr. In the *basecase* and bc+DOEcorr case, radionuclide release starts at ~1,500 yr, which is the time required for water to be available inside the WPs prior to the release of radionuclides. The onset of radionuclide release for the bc+DOEjf and bc+DOEcorrJF cases is further delayed (starts at ~2,500 yr) because of the DOE assumption that IDWP are relevant to the system only after 1,000 yr. Further basis is needed to support this assumption because of its importance in estimating the onset of radionuclide release. Both the *basecase* and the bc+DOEjf case produce similar dose rates after 20,000 yr. On the other hand, the bc+DOEcorr and bc+DOEcorrJF cases predict comparable dose rates beyond 50,000 yr. The similarity exists because corrosion is the leading radionuclide release mechanism beyond 10,000 yr. As depicted in Figure 1, corrosion is the predominant release mechanism beyond 15,000 yr according to the NRC description, and beyond 40,000 yr according to the DOE approach. Both the NRC *basecase* and the DOE base case produce

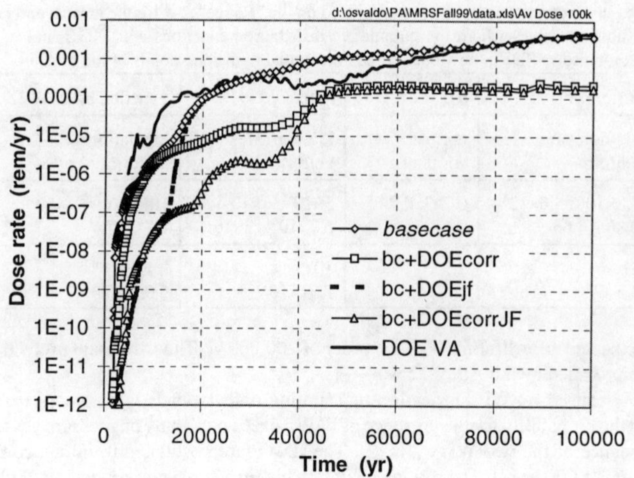

Figure 1: Expected dose rate versus time. The DOE VA data were reported as the 50th percentile case [2, Vol. 3, Figure 5-15]. The DOE 50th percentile case and the *basecase* predict comparable dose rates. In the first 15,000 yr, the bc+DOEjf and the bc+DOEcorrJF dose rates are similar. The bc+DOEjf dose rate is comparable to the *basecase* dose rate beyond 20,000 yr.

comparable dose rates; higher dose rates are predicted by the DOE base case in the initial 10,000-yr period.

Figure 2 displays the average (computed on the basis of 250 Monte Carlo realizations) fraction of WPs failed due to corrosion, which leads to the dose rate versus time in Figure 1. The DOE data in Figure 2 correspond to the DOE base case as defined in the VA.[3] During the first 10,000 yr, the five cases predict similar integrity of the WPs. As expected from the data in Figure 1, the fraction of failed WPs for the *basecase* and for the bc+DOEjf case are comparable. The bc+DOEcorr and bc+DOEcorrJF cases also produce similar results. According to the data in Figures 1 and 2, it is clear that the initial radionuclide release is controlled by the fraction of IDWP. It is only after the first thousands or tens of thousands of years that corrosion failure arises as the dominant release mechanism. The *basecase* predicts that, on average, all

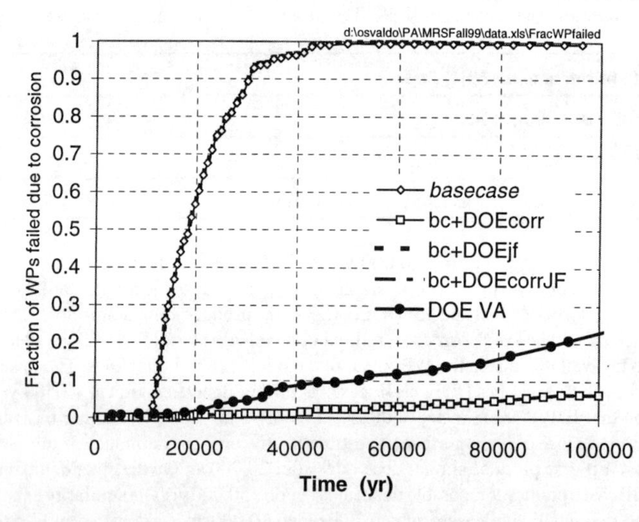

Figure 2: Average fraction of waste packages (WPs) failed due to corrosion versus time. The averages were computed from 250 Monte Carlo realizations. The DOE VA data were reported as the DOE base case data (100,000-yr period) [2, Vol. 3, Figure 4-12]. The *basecase* and the bc+DOEjf case produce comparable results (upper data). The bc+DOEcorr and the bc+DOEcorrJF also produce comparable results (lower data).

[3] The DOE general-corrosion base case is defined as the case with a 50-percent uncertainty and 50-percent variability split and the median corrosion rate taken at the 50th percentile of the uncertainty variance [2, Vol. 3].

of the WPs fail after 50,000 yr. The DOE base case predicts that about 20 percent of the WPs fail within 100,000 yr. Figure 2 also indicates that if the DOE corrosion and the fraction of IDWP descriptions are incorporated into the *basecase* (the bc+DOEcorrJF case), fewer WPs are predicted to fail due to corrosion. Thus, the analysis shows that incorporation of the DOE corrosion and the fraction of IDWP descriptions into the *basecase* leads to lower doses due to a decrease in the WP failure rate. Despite the similitude in the fraction of failed WPs for the DOE base case and the bc+DOEcorrJF case (see Figure 2), the associated dose rates vary at times by more than two orders of magnitude (see Figure 1). On the other hand, despite the remarkable difference in the fraction of failed WPs for the DOE base case and the *basecase*, the associated dose rates are comparable, never deviating by more than an order of magnitude. The difference between the *basecase* and the DOE base case, concerning the dose rate, is more evident in Figure 3, which shows the complementary cumulative distribution functions (CCDF) for the peak dose rate in 10,000 yr and 100,000 yr. Since the peak dose rates of the DOE base case are higher than those of the *basecase*, it can be concluded that the DOE analysis has embedded conservatism in other parts of the system, not analyzed in this paper. A more thorough analysis will be completed in the future, after accurate WP design information is made available by the DOE.

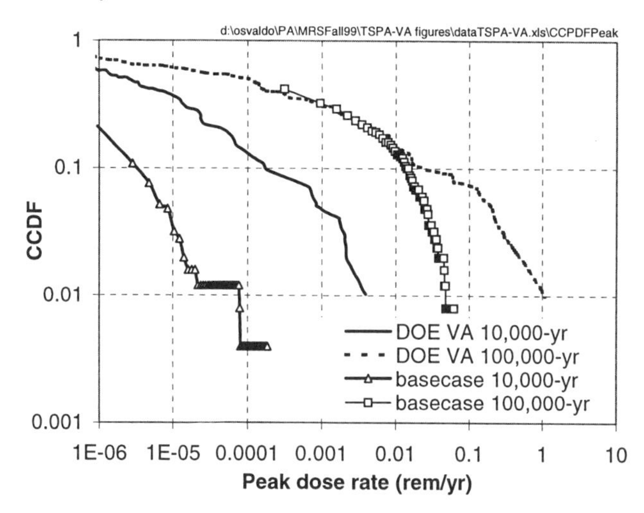

Figure 3: Complementary cumulative distribution function (CCDF) for the peak dose. The DOE VA data were reported in [2, Vol. 3, Figure 4-26].

Corrosion is a phenomenon with 100-percent certainty to occur in time. One of the objectives of the performance assessment codes developed by the DOE and NRC is to estimate this time, and link it to the release of radionuclides. The long life associated to the WP materials, Alloy 22 in particular, is the result of the assumed low corrosion rates. However, it must be pointed out that Alloy 22 does not have a long-term industrial service history. The corrosion rates herein used are estimates only partially supported by experimental data. A stronger experimental and theoretical basis is needed to support the corrosion resistance properties of this nickel-chromium alloy. For example, the authors believe that over a long time, the alloy may be depleted of some components due to preferential dissolution, as observed in Alloy C-276, anodically polarized in 0.5 N NaCl at 70°C [6]. Preferential dissolution may cause the corrosion resistance properties of nickel-chromium alloys to change with time. This possibility needs to be evaluated from a more fundamental viewpoint, since preferential dissolution may be the cause for earlier failure than that computed from current general corrosion or passive dissolution models.

CONCLUSIONS

Some capabilities of the NRC to review the DOE license application for the proposed HLW repository have been exercised. DOE data concerning the description of the corrosion phenomena and the fraction of IDWP have been combined with the NRC description of the repository system (here referred to as the *basecase*) and the results of the performance of the system (in terms of the dose to the average member of a critical group) are discussed. Use of the DOE data in the NRC models leads to decreased dose rates with respect to *basecase* data. The expected dose rate for the NRC *basecase* and for the DOE base case (data reported in the VA [2, Vol. 3]) are comparable in the 100,000-yr period.

In this paper, it is argued that the early release of radionuclides from the system is controlled by the number of IDWP. After this initial period, corrosion becomes the leading radionuclide release mechanism. The time at which corrosion begins to dominate the release mechanisms depends on the corrosion rates assigned to Alloy 22. According to the NRC description, corrosion is the main release mechanism beyond 15,000 yr, and beyond 40,000 yr according to the DOE description.

In the DOE assessment, most of the dose within the first 10,000 yr originates from IDWP. Therefore, further technical bases for the treatment of IDWP in performance assessment studies by DOE will enhance confidence that during the first 10,000 yr the expected dose will be well below compliance limits. It is recommended to complement the estimation of the fraction of IDWP with risk analysis tools and to consider that IDWP are part of the system at the time of the waste emplacement. Further experimental and theoretical studies supporting the use of general corrosion rates to assess the durability of the Alloy 22 wall will benefit the repository program. For example, potentially slow phenomena such as preferential dissolution could be explored as a mechanism likely to lead to a change in the corrosion resistance properties of Alloy 22.

ACKNOWLEDGMENTS

The paper was prepared to document work performed by the CNWRA for the NRC under contract No. NRC–02–97–009. The activity reported here was performed on behalf of the Office of Nuclear Material Safety and Safeguards. The paper is an independent product of the CNWRA and does not necessarily reflect the views or regulatory position of the NRC.

REFERENCES

1. S. Mohanty and T.J. McCartin (Coordinators), *Total-system Performance Assessment (TPA) Version 3.2 Code; Module Description and User's Guide*, (Center for Nuclear Waste Regulatory Analyses, San Antonio, TX, 1998), pp. 4-43 – 4-45.

2. U.S. Department of Energy, *Viability Assessment of a Repository at Yucca Mountain*, (DOE/RW-0508, 1998), Vol. 2: pp. 5-6 – 5-8; Vol. 3: pp. 3-81, 4-11, 4-12, 4-23, 4-65, and 5-15.

3. TRW Environmental Safety Systems, Inc., *Mined Geologic Disposal System Advanced Conceptual Design Report. Engineered Barrier Segment/Waste Package*. (TRW Environmental Safety Systems, Inc., B00000000–01717–5705–00027, Rev. 00, Las Vegas, NV, 1996), Volume III.

4. D.S. Dunn, Y.-M. Pan, and G.A. Cragnolino, *Effects of Environmental Factors on the Aqueous Corrosion of High-Level Radioactive Waste Containers—Experimental Results and Models,* (Center for Nuclear Waste Regulatory Analyses, Report 99-004, San Antonio, TX, 1999), pp. 1-4 – 5, 4-1 – 4, 4-8 – 12, and 5-1 – 2.

5. Civilian Radioactive Waste Management System, *Chapter 5, Total System Performance Assessment-Viability Assessment (TSPA-VA), Analyses Technical Basis Document* (Waste Package Degradation Modeling and Abstraction), (CRWMS, B00000000–01717–4301–00005, Rev 00, 1998), pp. 5-53, 5-68, 5-69, 5-73, 5-102, F5-13, and F5-19.

6. M.A. Cavanaugh, J.A. Kargol, J. Nickerson, and N.F. Fiore, Corrosion-NACE, **39**, pp. 144–159 (1983).

Repository Performance

Behavior of MgO as a CO_2 Scavenger at the Waste Isolation Pilot Plant (WIPP), Carlsbad New Mexico

J.L. Krumhansl, H. W. Papenguth, P-.C. Zhang, J. W. Kelly, H.L. Anderson and J. O. E. Hardesty

Sandia National Laboratories, Albuquerque, New Mexico, 87185

ABSTRACT

Transuranic nuclear wastes being disposed of in the Waste Isolation Pilot Plant (WIPP) contain large amounts of organic material that may decay producing substantial quantities of undesirable CO_2. Because of this possibility a MgO backfill has been included in the repository design. In addition to scavenging CO_2, the backfill may also hydrate forming $Mg(OH)_2$. Backfill hydration may provide a sink that may sorb much, or all, of the brine that enters the repository. The key to understanding the full implications of including the MgO backfill lies in knowing the rates of the various processes that may operate as the repository ages and evolves. Both carbonation and hydration reactions were found to occur rapidly enough to have beneficial impacts on repository performance.

INTRODUCTION

The WIPP is a repository for the disposal of defense-related transuranic wastes located in southeastern New Mexico. Wastes are currently being interred at the site for permanent disposal at a depth of 650 meters in the thick bedded rock salt of the Salado Formation. Inclusion of a MgO backfill surrounding the waste is an integral part of the overall disposal plan. The rational for its inclusion is based in an extensive set of performance assessment (PA) calculations formalized while obtaining the site operating license (1).

A large proportion of WIPP waste consists of organic materials (paper, cardboard, plastic, rubber, wood etc.) contaminated with traces of Pu. In early PA trials it appeared that abundant CO_2 from the decaying organics might lower brine pH and significantly elevate actinide solubilities. MgO was identified (2) as the optimal material to scavenge the unwanted CO_2. Later refinements in the PA package removed this concern. However, the U.S. Environmental Protection Agency (USEPA) also stipulated that before a license would be granted additional "assurance measures" would have to be included that went beyond the minimum design needed to demonstrate compliance. Inclusion of the MgO backfill is one such "assurance measure".

Although initially intended to scavenge CO_2, the impact of a MgO backfill clearly goes beyond this seemingly simple function. In fact, so much MgO is currently slated for burial with the waste (85,600 tons) that predicting its behavior is integral to understanding the overall chemical evolution of the repository. The likely evolution of the backfill is illustrated in Fig. 1. MgO hydration [producing $Mg(OH)_2$)] will be the first major change in the backfill. This will indirectly have a large impact on the timing of CO_2 generation since significant microbial degradation of the organic materials in the waste cannot commence until the waste is at least dampened by brine. With restricted brine seepage the backfill may consume all the water and the potential impacts of CO_2 generation may become moot issues.

Once CO_2 generation commences various magnesium carbonates will start forming. Magnesite ($MgCO_3$) is the most thermodynamically stable magnesium carbonate. But, this is also a classic example of how metastable phases can introduce great complexity into an otherwise simple system. In this case, the problem is the long-term stability of mixed $[Mg(OH)_2]_x \cdot [MgCO_3]_y \cdot [H_2O]_z$ phases such as hydromagnesite and nesquehonite, as

Mat. Res. Soc. Symp. Proc. Vol. 608

indicated by the upward diagonal trace of the arrow from the $Mg(OH)_2$ in Fig. 1. The timing of their appearance may affect brine availability as hydrous phases first form and then invert to anhydrous $MgCO_3$. These changes will also have a large impact on the physical properties of the backfill. Thus, putting the CO_2 scavenging ability of the backfill in its proper perspective requires consideration of all of the evolutionary steps that may lead to the final production of the end-stage magnesite.

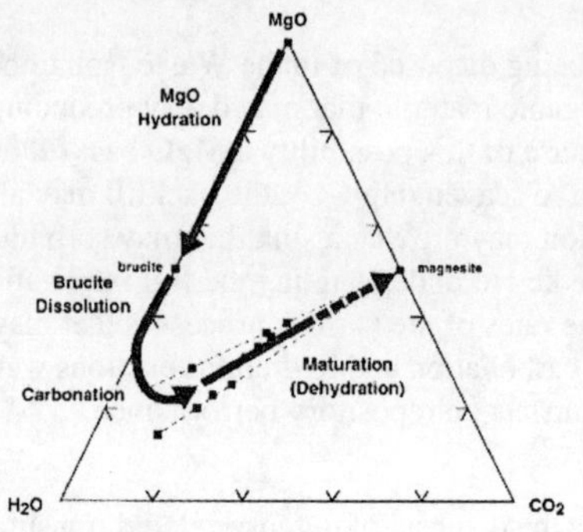

Figure 1 Chemical Evolution MgO backfill.

HYDRATION EFFECTS

Under repository conditions the hydration of MgO may occur in two ways; by contact with brine and by exposure to water vapor. Brine hydration rates are relatively well understood but the vapor-phase hydration studies are less advanced. Simpson (3) investigated MgO hydration in water vapor at temperatures from 20° to 150° C. At 20°C and 100% humidity, over reaction times up to about 80 days, only half the material was hydrated. At 60° C, the impact of particle size became noticeable, and complete hydration was observed in the 31–62.5 µm-sized fraction, but not for the 250 – 350 µm-sized fraction. Simpson also noted that the vapor pressure of the brine has a significant effect on hydration rates, although thermodynamically $Mg(OH)_2$ was the stable phase over all the brines studied. Our recent studies confirmed this by using various salt-saturated solutions to provide a range of relative humidities. At the highest relative humidities the vapor phase hydration progresses in a linear manner but for lower relative humidities about 1% converts rapidly. After this, the hydration ceased for the remaining 185 days of the test. This pronounced drop in reaction rate occurs at a humidity greater than that of a saturated NaCl solution so it will have a large effect on predicting the in situ hydration rate of the MgO backfill.

Hydration rates in brines were measured for a variety of conditions and demonstrated that hydration rates depended on several experimental factors: brine chemistry, notably the Mg content of the brine, the solid to brine ratio, and the temperature of the experiment. This latter feature was exploited in many cases to accelerate reactions that would require many decades (or centuries) at the anticipated repository temperature (about 28° C). Brine influx to the WIPP workings may either take place by the (very) slow oozing of fluid from the rock salt adjacent to the repository – or as a consequence of drilling activities that penetrate both the repository and a brine reservoir that may exist below the repository. Samples from other deep brine pockets suggest that this latter fluid would be closer to a $NaCl\text{-}Na_2SO_4$ brine and contain significantly less Mg than brines from the Salado rock salt in which the facility is located.

These studies also found that hydration in brines did not proceed linearly with time. Rather, the process could be divided into three stages. Immediately after being placed in the brine there is a waiting period during which the transformation to $Mg(OH)_2$ is not perceptible using X-ray diffraction or scanning electron microscopy (SEM). Several lines of evidence suggest that this is a time during which nucleation of the new $Mg(OH)_2$ crystals occurs. The duration of the waiting period is markedly shortened at relatively high solid:fluid ratios, indicating that a degree of supersaturation needs to build up in the solution before the reaction can progress. It was also found that high Mg brines prolong the waiting period. Finally, the phenomenon was studied at 40°, 60°, and 90° C with a "generic weep" brine such as might slowly seep into the repository from the surrounding rock salt (saturated NaCl with about 1 M Mg). An Arrhenius plot of the incubation periods at these three temperatures indicated an activation energy of about 71 kJ/mole (17 kcal/mole). This is significantly larger than expected for rates governed by mass transport to and from the reacting surface, but is consistent with what might be expected if nucleation were controlling the onset of a process. This work also suggests that waiting periods in the range from one to two months would be expected under repository conditions, which agrees with the delay implied by the high Mg brine curve in Fig. 2 (triangles). Once perceptible hydration has occurred there is a period during which about a third of the sample converts to brucite, $Mg(OH)_2$, in a relatively short time (several months). After that, conversion rates again slowed perceptibly, though tests at 60° C and higher show that full conversion will eventually be reached.

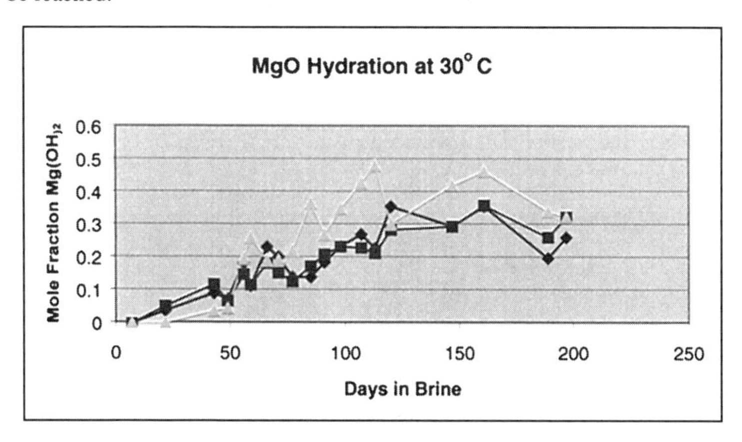

Figure 2 Hydration of MgO in deionized water (diamonds), 4M NaCl (squares) and 4M NaCl plus 1M MgCl (triangles). The large scatter has been traced to differing reactivities of individual pellets. Also note the delay in the Mg-containing brine relative to other experiments.

In addition to desiccating the repository, MgO hydration will have other long-term impacts on overall repository performance. Conversion to $Mg(OH)_2$ can potentially fill most of the pore space in the backfill and exert significant pressures on the facility. Accelerated dehydration studies using deionized water at 90° C generated pressures (30-35 MPa) when the containment vessel had no head space into which the hydrating MgO could expand. This loading exceeds the lithostatic pressure in the repository. In similar experiments using generic weep brine the pressure was just a fraction of this value (3-4 MPa). In the WIPP, however, that degree of physical confinement is unlikely since a great deal of void space will exist in, and among, the waste packages. However, backfill expansion may still serve to compact the waste, and impart mechanical strength to the backfill. Fully confined swelling tests were also observed to lower the bulk permeability of a MgO backfill to values similar to that obtained with compacted rock salt backfills ($\sim 10^{-17}$ m^2).

CO_2 UPTAKE PROPERTIES

CO_2 uptake is the second phase of backfill evolution. Mass balance calculations indicate that with the presently designed waste inventory and backfill loading about 26 mole % of the MgO would be consumed if all the CO_2 the waste could potentially produce were scavenged. Although decay is anticipated to be a slow process, and CO_2 levels in the WIPP will probably be low, the initial investigations involved accelerated tests where 100% CO_2 was bubbled through various sized pellets saturated with different WIPP brines. For pellets in the 2-4 mm diameter range it was found that within a few weeks the uptake rate slowed to a virtual halt before reaching the 26 mole percent benchmark. However, pellets having diameters less than 1 mm reached this benchmark in less than two weeks.

X-ray diffraction and SEM examinations revealed the cause for this difference. With larger particles a dense coating of needle-like nesquehonite crystals formed that almost isolated the pellet interior pellet from the surrounding solution (Fig. 3a). However, on smaller particles the dominant alteration material was fine-grained hydromagnesite (Fig. 3b). Higher magnification (Fig. 4b) revealed that hydromagnesite formes loose masses of bladed crystals with a card-house structure. This apparently provides less protection for the underlying pellets than the nesquehonite coatings. Nesquehonite transforms to hydromagnesite below a CO_2 partial pressure of about 10^{-2} atm. (4). Apparently, the greater surface area of the smaller pellets provided greater reactivity and lowered the effective CO_2 fugacity in solution below this limit.

Figure 3 A (left) Nesquehonite needles coating a 2-4 mm MgO pellet. B (right) fine grained hydromagnesite coating on 1 mm MgO pellets, note: pellet hollowed out by MgO dissolution.

Long-term testing was then undertaken at lower CO_2 partial pressures (5% and 0.5%, using 2-4 mm diameter pellets) to mimic expected WIPP conditions and assess whether hydromagnesite formation would assure a sustained ability to scavenge CO_2 past the first few weeks of testing. These studies verified that the MgO backfill maintained its long-term reactivity (Fig. 4a), and that surface coatings formed under these circumstances were hydromagnesite (Fig. 4b). However, as testing progressed it was also noted that the CO_2 uptake rate decreased significantly in spite of the presence of coatings composed primarily of hydromagnesite. Over the last analysis interval (200 – 306 days) conversion rates had slowed to 1.5% per year for 5% CO_2, and 5.1% per year in 0.5% CO_2 experiments. Further, after 306 days, only 14.7% of the MgO was converted in the 5% CO_2 experiment and 11.2% in the 0.5% CO_2 experiment. Several more years of experimentation will be needed to demonstrate the full 26 mole % conversion at low CO_2 partial pressure.

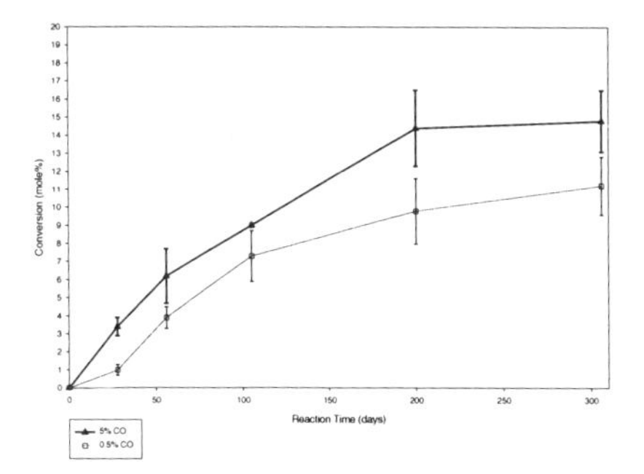

Figure 4 A (left) CO_2 uptake of 2-4 mm pellets in 5% (top curve) and 0.5% CO_2 gas (bottom curve) out to 306 days. B (right) hydromagnesite coating formed in 0.5% CO_2 on the surface of a MgO pellet immersed in generic weep brine after 1 year exposure.

MAGNESITE FORMATION

The final step to achieving a mature backfill is formation of anhydrous magnesite ($MgCO_3$) by reactions such as:

$$Mg_4(CO_3)_3(OH)_2{\cdot}3H_2O \rightarrow Mg(OH)_2 + 3H_2O + 3MgCO_3. \quad (1)$$

Again, to assess reaction rates in a reasonable time frame required accelerated testing at elevated temperatures (100° – 200° C), and again it was found that brine chemistry was an important factor (5). It was also found that a waiting period preceded the onset of perceptible transformation. An Arrhenius plot (log [time] vs. 1/T) was used to extrapolate high temperature transformation rates down to room temperature. This procedure suggests that in a simple saturated NaCl brine no transformation would be observable for about two decades. In a generic weep brine (with elevated Mg and sulfate) the waiting period would be about ten times longer. Once reactions commence the first half of the hydromagnesite would be transformed in roughly another 4 years in a simple NaCl brine, and in an additional 75 years in generic weep brine. We cannot presently estimate the time needed to transform the last half of the hydromagnesite.

CONCLUSIONS

In evaluating MgO backfills it is apparent that brine entry by several paths must be considered. Brine entry from the adjacent rock salt will be slow enough that the hydration rates reported here should prevent free brine from accumulating in the repository during most of the period of regulatory concern (10,000 years). If free brine does finally accumulate in the mine it will do so in a setting where much of the void space has already been filled by the expansion accompanying earlier MgO hydration, and where brine circulation would be significantly impeded by a low permeability backfill.

A different scenario is anticipated if the first significant amount of brine to enter the repository results from drilling operations which penetrate brine pockets below the repository. Our measured hydration rates show that several months may be needed before swelling of the pellets would fill enough of the pores to restrict brine flow. Thus, brine entry from such a source is likely to be widespread and CO_2 production might start at essentially the same time as hydration. What is unclear is how fast the microbial activity will commence. Will the MgO consume the brine before CO_2 generation is significant? Or, will microbial activity win the race and, for a time, convert $Mg(OH)_2$ to hydromagnesite as quickly as it forms? If the latter scenario prevails our data clearly demonstrate that pelletized MgO will prevent a CO_2 accumulation that would lower the pH and elevate actinide solubilities.

To summarize, by virtue of its ability to buffer pH, scavenge carbonate and act as a desiccant, the MgO backfill has the potential for significantly improving the ability of the WIPP to retain the Pu, and other actinides, disposed of there.

REFERENCES

1. US DOE, Title 40 CFR Part 191 Compliance Certification Application for the Waste Isolation Pilot Plant. DOE/CAO-1996-2184 (1996).
2. Bynum, R.V., Stockman, C., Wang, Y., Peterson, A., Krumhansl, J., Nowak, J., Cotton, J., Patchet, S.J., and Chu, M.S.Y., ICEM'97, 6th International Conference on Radioactive Waste. Management and Environmental Remediation, Singapore, October 12-16, 1997, New York, NY, American Society of Mechanical Engineers, p. 357-361 (1997).
3. D.R. Simpson, Some Characteristics of Potential Backfill Materials, ONWI-449 (1983).
4. F. Lippmann, Sedimentary Carbonate Minerals (Springer-Verlag, N.Y., N.Y., 1973), 86-88.
5. P-.C Zhang, H.L. Anderson, J.W. Kelly, J.L. Krumhansl, H.W., Papenguth, Appl. Geochem., in review (2000).

ACKNOWLEDGEMENT

This work sponsored by Sandia National Laboratories, Albuquerque New Mexico, USA, 87185. Sandia is a multiprogram laboratory operated by Sandia Corporation, a Lockheed Martin Company, for the U.S. Department of Energy [DOE] under contract: DE-AC04-94AL8500.

VADOSE ZONE MONITORING SYSTEM FOR SITE CHARACTERIZATION AND TRANSPORT MODELING

J. B. SISSON, A. L. SCHAFER, and J. M. HUBBELL
Integrated Earth Sciences Dept., Idaho National Engineering and Environmental Laboratory, Idaho Falls, ID 83415-2107, jys@inel.gov

ABSTRACT

Monitoring the vadose zone below buried waste provides an early warning of contaminate transport toward the groundwater. To quantify the transport mechanisms, vadose zone hydraulic characteristics and the physical variables need to be obtained. We have designed and implemented a Vadose Zone Monitoring System (VZMS) to monitor or sample the 3 state variables of the vadose zone, water potential, water content and chemical concentration. The state variables are monitored using an Advanced Tensiometer (AT), a borehole water content sensor (BWCS) and a vacuum lysimeter, respectively. This system was installed at the Savannah River Site (SRS) E-Area disposal site, where low level wastes have been disposed of in shallow trenches. The system has operated for several months providing nearly continuous water content and water potential data. The vacuum lysimeters were activated on a quarterly schedule. Installation details and an example data set are presented to illustrate the effectiveness of the VZMS, and demonstrate the utility of the VZMS as an indicator of contaminant transport.

Advanced Tensiometer

Surface cap
Data logger
Waste disposal site
Inner guide pipe
Outer guide pipe
Pressure transducer
Water reservoir
Porous ceramic cup
Water table

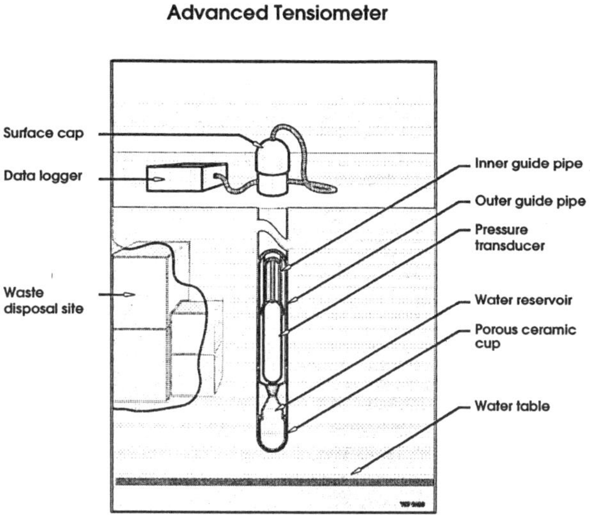

Figure 1. Components and installation of the advanced tensiometer.

INTRODUCTION

Current regulations require that groundwater be monitored for contaminant presence. Once contamination is detected in the groundwater above regulatory limits, the contaminant must be removed or contained. Achieving removal or containment becomes a combined legal and technical challenge. If the contaminant migration could be detected in the vadose zone above the aquifer before seriously impacting groundwater, engineering solutions could be designed and implemented before incurring regulatory penalties. To achieve this end, we have designed and implemented a VZMS. Its primary objective is to provide early detection and warning of contaminant transport, allowing time for engineering solutions to be evaluated, optimized, and implemented prior to contaminants impacting groundwater.

MATERIALS AND METHODS

In the absence of large temperature gradients, contaminant transport in the vadose zone is governed by 3 state variables: water potential, water content and contaminant concentration. Water potential is the amount of energy needed to remove water from the soil and the gradient in water potential determines the direction of water movement. Water content determines the fraction of the porous media available for transport, the hydraulic conductivity, internal surface area available for chemical interaction, and ultimately the pore water velocity along transport paths. The chemical transport rate is a function of chemical concentration, the internal surface area it interacts with, and soil water flux.

Monitoring waste disposal sites can require monitoring water potentials from land surface to depths of more than 200 m under climatic conditions ranging from arid to humid. Development of the AT was necessary in order to meet the wide range of hydrologic conditions beneath existing waste sites. As shown in Figure 1, the AT consists of a porous ceramic cup that acts as a semipermeable membrane in that it allows the transfer of water between the soil and the water chamber. The pressure in the water chamber is the water potential and is monitored using an electronic pressure sensor (Hubbell and Sisson, 1998) with its output continuously available to the data logger at land surface. The design of the AT allows the chamber to be serviced and refilled by raising the pressure sensor a few centimeters which allows water standing above the chamber to flow in. This is accomplished from land surface, and part of this study was to assess the required service schedule.

Vadose zone materials consist of solid material (sand, organic material etc.), air, soil water, and chemicals in the soil water. The bulk dielectric constant of vadose zone materials is determined by the dielectric constant of each constituent and the relative proportions of each. Changes in the bulk dielectric constant can be primarily attributed to changes in soil water content in situations where the chemical concentration remains relatively constant. We have developed the Borehole Water Content Sensor (BWCS) to continuously record changes in soil water content by sampling changes in dielectric content. The BWCS consists of a modified CS505 Fuel Moisture Sensor (Campbell Scientific, Inc., Logan, UT), its waveguide is placed against the borehole wall, and its electronic leads are brought to land surface.

The BWCS is a capacitance sensor and responds to small changes in the dielectric constant of the surrounding material. It differs from the TDR in that the TDR measures the time an electrical signal is resident in the waveguide. The output of the BWCS consists of a stream of pulses whose period is proportional to the capacitance and therefore of the dielectric constant of the material in contact with the waveguide. A TDR is sensitive to the surface to subsurface cable length and requires post manipulation of the data to infer water content. Calibration of the BWCS is direct, and occurs by placing the BWCS in contact with soils of varying water content and simply recording the corresponding pulse period. We calibrate the BWCS in the laboratory using soils obtained from the specific site of interest and a regression curve is fit to the data. The BWCS does not need to be recalibrated and is permanently installed. Its continued performance is evaluated by comparing it to the tensiometer data, and its long term performance in insitu installations is being evaluated in this study.

In this VZMS, we have used standard vacuum lysimeters (Soil Moisture Equipment Corp., Santa Barbara, CA) to obtain soil water solution samples. Soil water samples are obtained by evacuating the lysimeters on a quarterly schedule as opposed to having a continuous sampling or recording mechanism.

As illustrated in Figure 2, the VZMS is constructed using the AT assembly as a center backbone, with the lysimeters placed below the ceramic cup. The BWCS were attached to the AT with a lever arm used to firmly seat the BWCS against the native soil at the borehole wall. Installation of the VZMS is done in an open 8 inch hole configuration, using silica flour to backfill the borehole around the lysimeter, AT and BWCS. In a multiple depth configuration, bentonite is used to seal intervals between the combined sensors.

The VZMS system was installed at Savanna River's E Area waste disposal site, where low level wastes were discharged in shallow trenches. The primary objective for installing a VZMS at the Savanna River Site was to monitor for tritium migration at the E-Area trenches and to provide design information for a more expansive monitoring program. The design information sought included calibration and maintenance requirements, criteria for the placement of instrumentation, and reliability estimates.

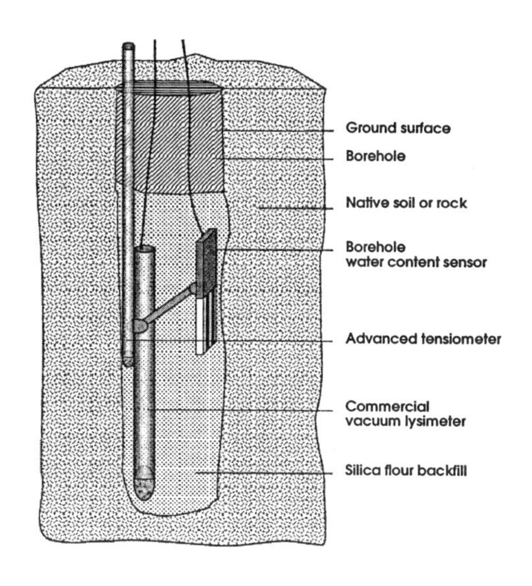

Figure 2. VZMS: showing BWCS, AT, and vacuum lysimeter.

In this pilot scale demonstration, our device set was placed at 4 different depths in each of 3 boreholes. The configuration was selected to allow determination of horizontal and vertical gradients in all three state variables. Data from the AT and BWCS were recorded hourly and the data was downloaded to the analysis team daily using cellular telephones. The lysimeters are to be sampled at 3-month intervals for laboratory analysis. Lysimeter data are not yet available.

Prior to installation, site specific soils from Savanna River were obtained to calibrate the BWCS. Soil water content was adjusted in the laboratory across a representative range of expected field values, and corresponding BWCS measurements were made. The resulting data and regression curve are shown in Figure 3. This figure illustrates increased variability with increased moisture content as the soil becomes plastic and undergoes deformation. At lower water content, this deformation is insignificant. Within the range of moisture content, a linear regression fits the data, moreover, a single calibration curve was found to adequately represent all of the moisture content sensors.

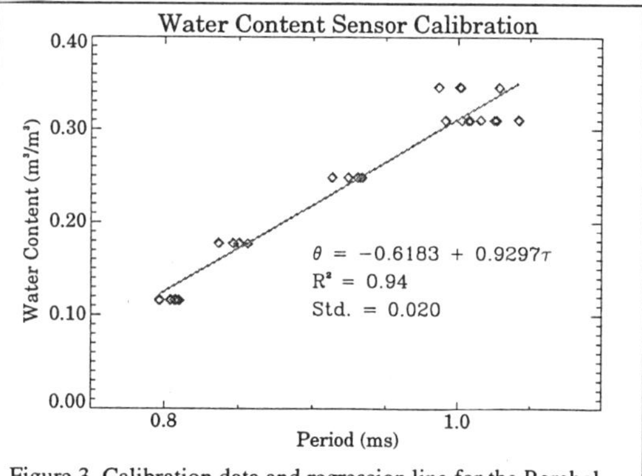

Figure 3. Calibration data and regression line for the Borehole Water Content Sensors (BWCSs).

MONITORING RESULTS

Following calibration, the VZMS was installed at the E Area Site. Six months of subsequent data collection and nearly continuous monitoring of water potential and water content has with negligible maintenance. Example water potential and water content for one well (SRS-6) is given below in Figures 4 and 5, respectively.

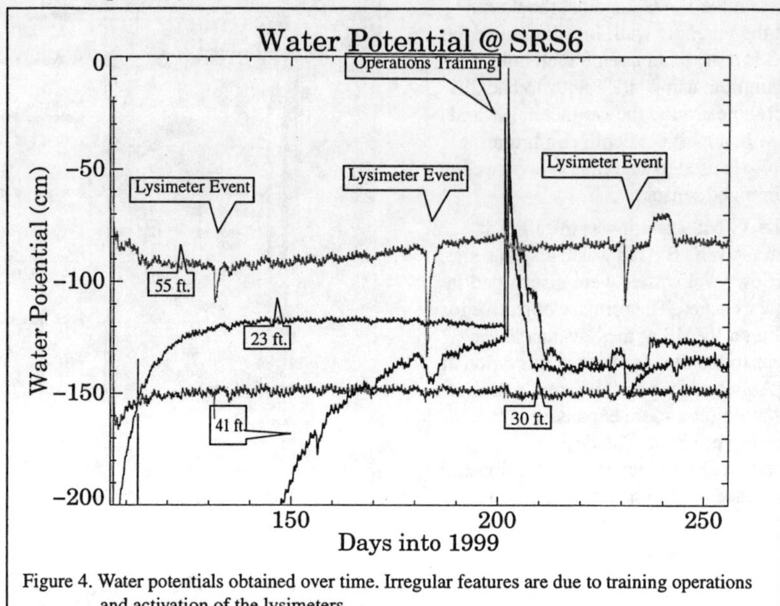

Figure 4. Water potentials obtained over time. Irregular features are due to training operations and activation of the lysimeters.

Recorded water potential indicates that a fairly long time can be required for the silica flour used as backfill to reach equilibrium with the surrounding soil. This equilibration time arises because of the borehole diameter of 20 cm which results in a relatively large volume of backfill material. The water content of the backfill material has to equilibrate with that of the native soil. Water content estimated from the BWCS indicates that since the BWCS was placed against the borehole, the response is less effected by the backfill and the equilibration time is much shorter than for the AT. Differences in the equilibration time at the various depths are primarily a function of the differences in water content of the backfill material used. More moist backfill material required less time to equilibrate suggesting that a silica flour slurry is the more optimal method of installation.

Following the equilibration period, the gradient in water potential stabilized at all depths, allowing determination of the driving force for water movement. In this well, the gradient is near unity, indicating that the soil column is draining to the groundwater at approximately 18 m (59 ft.) below land surface. The data shows that the AT is sensitive to small fluctuations in surrounding water content, as indicated by changes induced by sampling the lysimeters. These purging events are also indicated by the step changes in the water content data. The water content data is much smoother than the water potential data because the AT is sensitive enough to record the effects of barometric and earth tide fluctuations. The earth tide influence is currently being investigated.

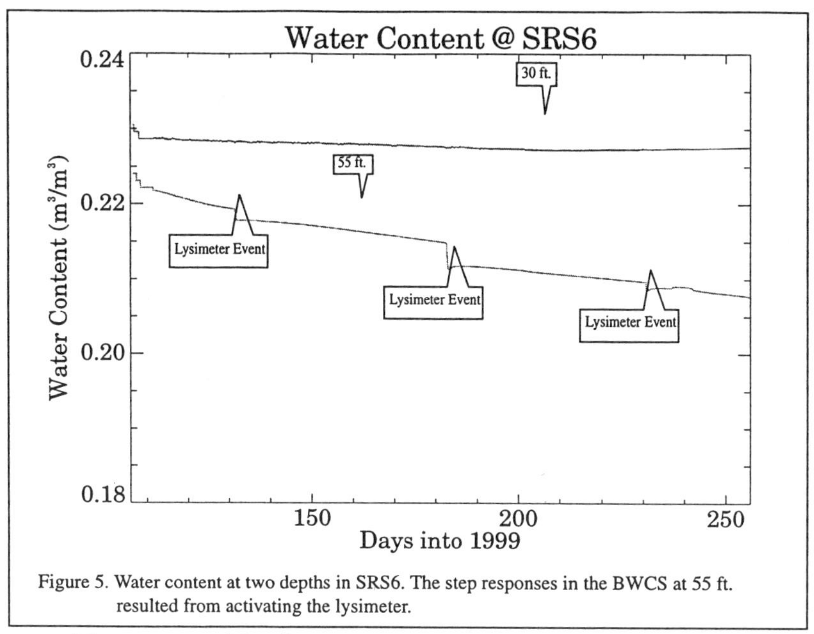

Figure 5. Water content at two depths in SRS6. The step responses in the BWCS at 55 ft. resulted from activating the lysimeter.

In general the water potential and water content were relatively constant in time, indicating that water movement was slow. Prior to and following the installation of the VZMS at the Savannah River Site rainfall has been well below normal. Without a significant rainfall event, computation of unsaturated hydraulic properties using the combined in situ moisture content and water potential can not occur. In this particular installation, the sensors were installed in sand overlain by silts and clays which were not anticipated before the installation. Percolation through the silts and clays is slow, and large changes in water content of the sand will require relatively large fluxes. To obtain improved estimates of soil water flux and pore water velocities needed for chemical transport studies, instruments will need to be placed in these flux-limited zones. Given the accuracy of the data collected thus far, it is anticipated that accurate estimates of unsaturated hydraulic properties can be obtained using the VZMS. The hydraulic properties are required for numerical models used to describe water movement and chemical transport through the vadose zone. Numerical model results are used in risk analysis to evaluate the impact of disposal operations to the environment. Current thought is that hydraulic properties obtained *in situ* are more accurate and more easily defended to regulators then laboratory obtained values.

At the time of this writing the laboratory analysis of lysimeter samples was unavailable and will not be reported here.

CONCLUSIONS

A VZMS was developed at the INEEL and deployed at the Savanna River Site for monitoring water potential, water content and concentration needed to estimate contaminant transport. This pilot scale system was installed at four depths in each of three boreholes. Data being collected with this instrumentation will be used to assess the required density of the site wide monitoring network. The network density will be established based on observed changes in vertical and horizontal moisture and will depend on the ability of natural subsurface features to control moisture redistribution.

The data and results obtained to date indicate that the BWCS and the AT are capable of operating several months in succession without maintenance. Monitored water content and water potential indicate that the system is performing within design specifications after nine months of continued operation. As a result of observations thus far, we estimate that maintenance will be required biannually. This information and the resultant data quality will be used to define operating costs and personnel scheduling for an expanded site-wide sampling network. Based on the ability to detect small scale changes in water potential and the ability to detect simultaneous changes in water content, this system will greatly enhance our ability to quantify vadose zone transport behavior.

REFERENCES

1. Hubbell, J.M. and J.B. Sisson, 1998, "Advanced tensiometer for shallow or deep soil water potential measurements," *Soil Sci.*, Vol. 163, No. 4, pp. 271-276.

2. Sisson, J.B. and J.M. Hubbell, 1998, Water potentials to depths of 30 meters in fractured basalt and sedimentary interbeds. Indirect Methods for Estimating the Hydraulic Properties of Unsaturated soils. (ed) M. Th. Van Genuchten, Riverside, CA.

THERMAL PARAMETRIC STUDIES FOR RADIOACTIVE WASTE MANAGEMENT

D. MANZONI*, S. LE BONHOMME*, B. SOULIER**, Y. GUILLOUX***
*EDF, Département TTA, 6 Quai Watier, 78401 Chatou , FRANCE
**ENS de Cachan, 61 av du Pdt Wilson, 94235 Cachan, FRANCE
***ANDRA, 1-7 rue Jean Monnet, 92298 Châtenay-Malabry , FRANCE

ABSTRACT
The objective of this study is to optimize the storage system with respect to thermal constraints. Due to a large number of 3D calculations, a simple temperature model based on an Design Of Experiment method has been developed. The most significant parameters have been exhibited by an extensive parametric study. A satisfactory agreement is observed between polynomial model and 3D approach.

INTRODUCTION

A radioactive waste disposal is characterized by various essential factors: cooling time, number of waste packages by tunnel, rock and buffer characteristics, drift spacing etc...This system have to verify some thermal limits, in particular in the buffer, so that the degradation of the clay and the canister corrosion will be moderated.
In order to confirm and optimize the Initial Option of Storage, EDF and ANDRA carried out thermal parametric studies. A Design Of Experiment method has been performed to organize the calculations and to build a simplified model of the maximum buffer temperature.

1. DESIGN DESCRIPTION

This study considers a storage concept of tunnels filled with spent fuel UOX 33 GWd/MTU Burnup. The current container design is made of 10 cm of carbon steel. The spent fuel is stored directly inside the containers. 4 PWR assemblies are disposed in the container, as shown on figure 1 :

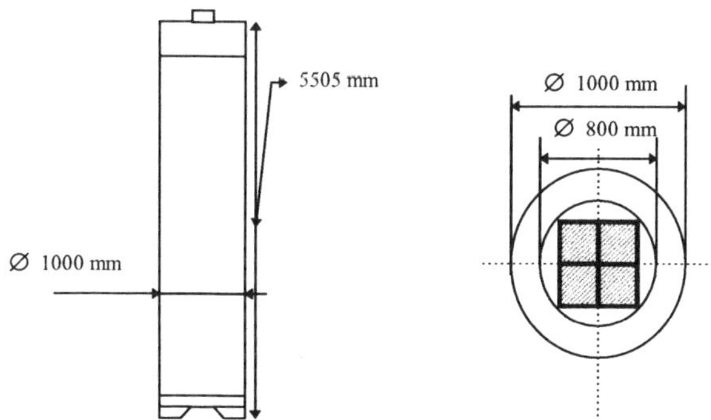

Figure 1 : Characteristics of the container

Figure 2 shows the fuel decay heat curves per container and per assembly for UOX 33 GWD/MTU Burnup.
The material properties of the different materials are given in table 1. However, temperature dependent material properties can be taken into account to improve the modeling.

Mat. Res. Soc. Symp. Proc. Vol. 608 © 2000 Materials Research Society

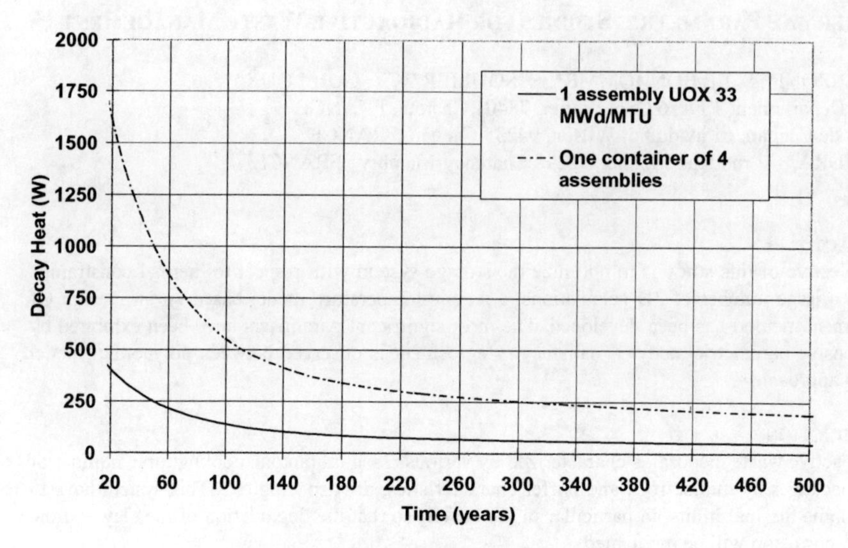

Figure 2 : Spent fuel UOX 33000 MWd/MTU

	Thermal conducticity *($W.m^{-1}.K^{-1}$)*	Density (*$kg.m^{-3}$*)	*Heat capacity ($J.kg^{-1}.K^{-1}$)*
Steel	35	7850	500
Concrete	1.75	2300	1000
Air	0.025	1.16	1017
Buffer	k = 1.3	2200	1050
Clay	$k_x = k_y = 1.8$; $k_z = 1.4$	2360	775.5

Table 1 : Thermal-Physical properties of the various materials

2. THERMAL CALCULATIONS WITH A FINITE ELEMENT MODEL

The materials temperatures are evaluated with the EDF thermal code called SYRTHES. It deals with conduction and radiation (limited to transparent medium) for transient problems in three dimensions. The solid structure is discretized by an unstructured grid made of around 300 000 nodes, and the radiation mesh is a surface grid. The grids are completely independent but still respect the same geometrical boundary. The tunnel geometry and all the heat transfers taken into account in the model are reported in figure 3.

The convection is modeled by a heat transfer coefficient h and the air temperature :

- At the surface of earth, it is assumed that $h \cong 10W/m^2.K$ and $T_{air} = 13°C$. This value of h is conventional for outdoor convection : it includes radiation.
- In the drift, in case of ventilation, the next formula is chosen : $h \cong 4.87V + 2.43W/m^2.K$ where V is the air velocity (m/s). This correlation has been established by Gillies et al. for underground mines (see [1]).

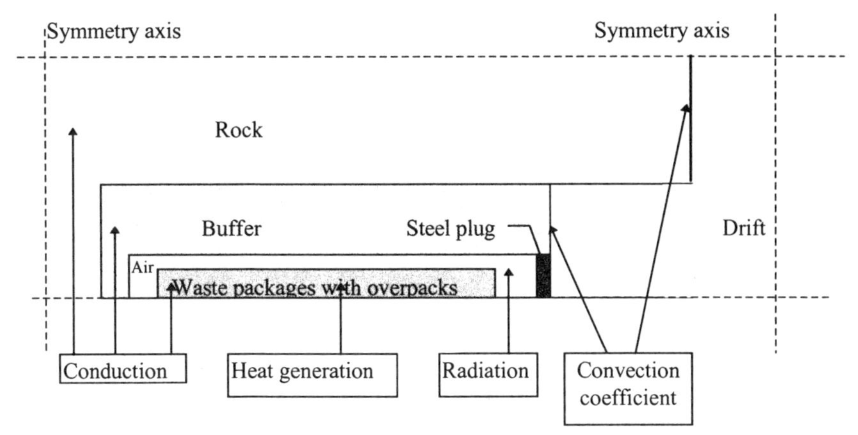

Figure 3 : Heat transfers in a tunnel repository

The heat flux is assumed to be transferred to the buffer only by radiation. The radiation is taken into account in all directions around the containers. The very fine carbon steel casing around the buffer has been neglected. The emissivity of buffer and container are chosen equal to 0.9 (it corresponds to a degraded stainless steel emissivity). The conduction is taken into account in all different materials. An initial rock temperature of 38°C has been chosen. In order to improve the accuracy of the model, an initial temperature gradient can be imposed.
Adiabatic boundary conditions are used on the sides of the domain in order to simulate an infinite number of tunnels. This is a conservative hypothesis. An uniform time-dependent heat flux is applied along the container.
The time step used in the transient calculations is adapted so that a residual convergence criterion of 10^{-5} is satisfied. Thus, a 6 hours time step is used for the first days, and a 2 years time step is used for the last years of the calculations.
Figure 4 shows the temperature field near the tunnel at the time of maximum buffer temperature (31.5 years), for a storage of 4 containers per tunnel and for 30 years initial cooling time.

3. Parametrical Study

The maximum temperature of buffer response can be approximated by a polynomial model. This type of approximation is valid if the response surface is smooth enough, without discontinuities. Eight parameters have been taken into account :

- the assemblies initial cooling time (between 30 and 100 years): **A (years)**
- the number of packages per shaft (between 1 and 10 packages): **N**
- the ventilation time before the emplacement of gravel backfill into the drift (between 15 and 100 years): **D (years)**
- the temperature of the ventilation air (between 10°C and 50°C): **T (°C)**
- the step between the tunnels (between 12m and 30m): **Px (m)**
- the distance between the drifts (between 5m and 25m for the half of the distance): **Dy (m)**
- the thickness of the buffer (between 0.3m and 1m): **e (m)**
- the thermal conductivity of the buffer (between 0.5W/m.K and 2.5W/m.K) : **k (m)**

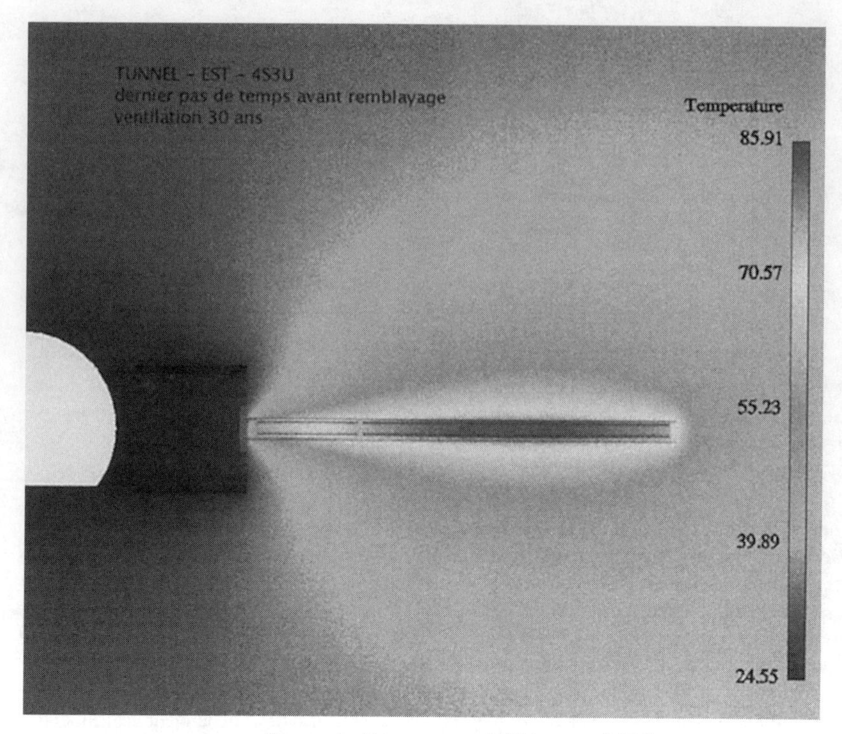

Figure 4 : Temperature field in tunnel (°C)

Variables in the polynomial model are represented by the 8 parameters and by the interactions between parameters (quadratic and linear):

$$T_{\max} = x_0 + x_1 P_1 + x_2 P_1^2 + x_3 P_2 + ... + x_i P_1 P_2 + x_{i+1} P_1 P_3 + ... + x_{44} P_7 P_8 \quad (1)$$

The coefficients of the polynomial model are identified by a linear regression from a set of calculations (equation 3).
For a N-calculations sample, the experiment problem is written in the following matricial form :

$$Y = AX + E \quad (2)$$

where Y is the vector of the N calculated responses, X is the vector of the 45 coefficients to be estimated, A is a matrix of 45 columns and N rows (constituted of 45 parameters values set for each calculations), and E denotes the modeling errors, called residuals (differences between the finite element model and the polynomial one). The coefficient vector X is estimated by a minimization of the square residual sum :

$$\hat{X} = (A'A)^{-1} A'Y \qquad \text{where A' denotes A transpose} \quad (3)$$

Assuming that the residuals can be represented by a gaussian law with a constant variance σ^2 in the computer experiment domain (hypothesis which is verified a posteriori), the covariance matrix only depends on the choice of the calculations points :

$$\text{cov}(\hat{X}) = (A'A)^{-1}.\sigma^2 \quad (4)$$

The diagonal terms represent the variance of the coefficients and the cross-coupling terms the correlations between coefficients. So it is possible to define a set of adequate calculations by minimizing the terms of the covariance matrix. This is the basis of experiment design building. Several predefined experiment designs ([2], [3], [4]) have been defined and optimized to study response surfaces. The most well known designs are the composite and the orthogonal designs. The method of central composite Design of Experiment have been chosen for this study.

For 8 parameters, only 81 3D calculations have been run to realize the polynomial model leading to parametric studies. It can be noticed that the same parametric studies with no polynomial model would have needed more than 20,000 3D calculations.

4. ANALYSES AND RESULTS

Two responses have been estimated : the maximum temperature of buffer during the ventilation time and during the backfill time. Each polynomial has 44 coefficients and a constant term. Coefficient analysis allows hierarchization of parameters. Conclusions of such a study must be strictly drawn in the experiment domain. Parameters effects are just valid in their variation range and a non significant parameter in the variation domain may become important with a wider range. Table 2 presents polynomial coefficients obtained for the two temperatures :

	Linear terms						Quadratic terms		Interaction terms
Response	A	N	Dy	Px	e	k	e*e	k*k	k*e
Tmax ventilation	-1.28	**19.92**	-0.40	-1.19	**17.86**	**-25.93**	***-3.69***	***5.94***	***-9.39***
Tmax backfilling	-0.62	**21.49**	-1.09	-4.04	***8.95***	**-17.31**	1.48	3.79	***-6.59***

Table 2 : Polynomial coefficients for response

Less significant columns are suppressed.

X	Very highly significant parameter
X	Highly significant parameter
X	Significant parameter

Numbers indicated in each cell of table 2 are the estimated polynomial coefficients. They represent the physical values of linear, quadratic or interactive effect of parameters. To evaluate if the model is accurate, the estimated values of temperature are compared to the FEM calculations for the Experimental Design points. The residuals between polynomial and FEM results for the temperature response during ventilation time are reported on figure 5. A difference of 8°C maximum is noticed. More calculations have been performed in order to compare polynomial model expectations and FEM calculations for random examples out of prime design. A difference of 5°C maximum is observed.

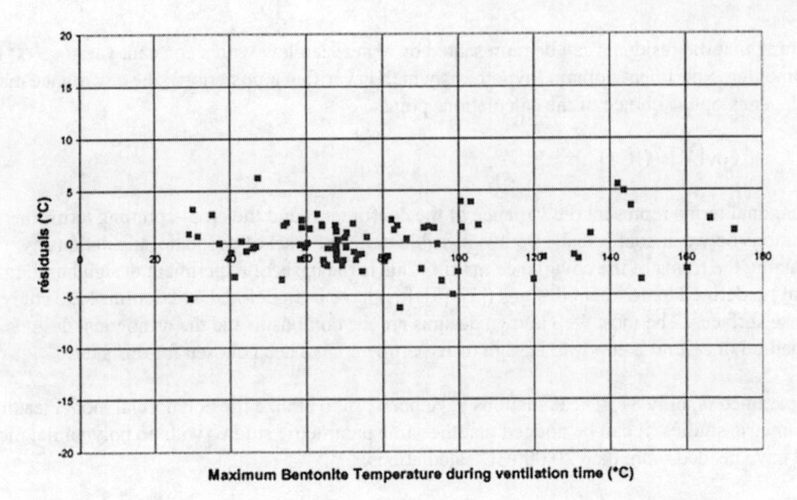

Figure 5 : Residuals between polynomial and FEM calculations for ventilation time

A global analysis of parameters effects is possible using table 2. It shows that the most important parameters are : - the number of containers per tunnel, - the thermal conductivity of buffer, - the thickness of buffer. Linear terms are the most significant. The global effect of the buffer conductivity and thickness appears in the linear, quadratic and interaction terms.
So a significant reduction of the buffer temperature is possible by using these three key parameters : reducing the number of containers by tunnel, increasing the conductivity of the buffer, reducing the buffer thickness (in case of law buffer conductivity compared to the surrounding rock conductivity).

CONCLUSION

The original use of Design Of Experiment method for numerical analysis allows to perform computer experiments in a rational way with a minimum number of calculations ; the benefit becomes greater as the number of parameters increases. The determination of the maximum of buffer temperature by a polynomial model can help to optimize the dimensions of a storage system with respect to the thermal criteria. A satisfactory agreement is noticed between the 3D calculations and the polynomial approach.

REFERENCES

1. A.D.S. Gillies, P. Creevy, G. Danko, P.F. Mousset, Proceedings of the 5th US Mine Ventilation Symposium, Ch 36, pp.288-298 (1989).
2. J. Sachs, W.J. Welch, T.J. Mitchell, H.P. Wynn, 1989, Statistical Science, Vol.4. N°4, 409-435, (1989).
3. G.E. Box, N.R. Draper, *Empirical model-building and surface response*, edited by Wiley Interscience (New York, 1987).
4. D. Bosselut, G. Regnier, B. Soulier, SMIRT Structural Mechanics In Reactor Technology (1997)

LONG TERM TEST OF BUFFER MATERIAL AT ASPO HARD ROCK LABORATORY, SWEDEN

O. KARNLAND*, TORBJÖRN SANDÉN**
*Clay Technology AB and Geol. Department University of Lund, Sweden, ok@claytech.se
**Clay Technology AB, Lund, Sweden, ts@claytech.se

ABSTRACT

Bentonite clay has been proposed as buffer material in several concepts for nuclear high level waste repositories. The "Long Term Test of Buffer Material" (LOT) series aims at validating models and hypotheses concerning physical properties in a bentonite buffer material and of related processes regarding mineralogy, microbiology, cation transport, copper corrosion and gas transport under conditions similar to those in a the Swedish KBS3 repository.
The test series comprises 7 test parcels, which will be run for 1, 5 and 20 years. The testing principle is to place parcels containing heater, central copper tube, pre-compacted clay buffer, instruments, and parameter controlling equipment in vertical bore-holes in granitic rock. The parcels are equipped with heaters in order to simulate the decay power from spent nuclear fuel at standard KBS3 conditions (90°C) and adverse condition (130°C). Adverse conditions in this context refer also to high temperature gradients over the buffer, and additional accessory minerals leading to i.a. high pH and high potassium concentration in clay pore water. Temperature, total pressure, water pressure and water content are measured during the heating period. At test termination standard chemical and mineralogical analyses, and physical testing are made. The 2 pilot tests (1-year tests) have been completed and preliminary results concerning the bentonite analyses and tests are presented. The 5 test parcels in the main test series have been started during the fall, 1999.

INTRODUCTION

Bentonite properties are closely related to the interaction between the main mineral montmorillonite and the ground-water. The interaction is affected by the ambient physico-chemical conditions. A number of laboratory test series, made by different research groups, have resulted in various buffer performance and alteration models. According to the latter, no major alteration is expected to take place in the buffer at the prevailing conditions in a KBS3 repository neither during, nor after water saturation. The models may to various degrees be checked in the present LOT field tests [1].

Objectives

Five research groups are engaged in the LOT tests and carry on research concerning clay alteration, accessory minerals, cation diffusion, bacteria survival, and copper corrosion, respectively. The objectives may be summarised in the following items:

- Check of models concerning buffer performance under quasi-steady state conditions after water saturation, e.g. swelling pressure and cation exchange capacity,
- Produce data concerning gas penetration pressure and gas transport capacity in the buffer,
- Check of models on buffer degrading processes, e.g. illitisation and salt enrichment,
- Check of models for cation diffusion in bentonite,
- Information concerning survival, activity and migration of bacteria in the buffer,
- Check of models for copper corrosion, and information regarding type of corrosion,

Mat. Res. Soc. Symp. Proc. Vol. 608 © 2000 Materials Research Society

- Information which may facilitate the realisation of the full scale test series with respect to clay preparation, instrumentation, data handling and evaluation.

Field experimental concept

The test series (Table 1) include 7 test parcels and concern realistic repository conditions except for the smaller scale and the controlled "adverse" conditions in 4 tests parcels. Adverse conditions in this context refer to high temperatures, high temperature gradients over the buffer, and additional accessory minerals leading to i.a. high pH and high potassium concentration in clay pore water.

Table 1. Lay out of the planned Long Term Test series.

Type	No.	max T, °C	Controlled parameter	Time, years	Remark
A	1	130	T, [K$^+$], pH, am	1	pilot test
A	0	120-150	T, [K$^+$], pH, am	1	main test
A	2	120-150	T, [K$^+$], pH, am	5	main test
A	3	120-150	T	5	main test
S	1	90	T	1	pilot test
S	2	90	T	5	main test
S	3	90	T	>>5	main test

A = adverse conditions, S = standard KBS3 conditions
T = temperature, [K$^+$] = potassium concentration, K-feldspar
pH = high pH from cement, ~12.5, am = accessory calcite, gypsum and feldspar

Each parcel is composed of exchangeable heater, central copper-tube, approximately 40 highly compacted bentonite (Volclay MX-80) cylinder rings, gauges, and various additives (Figure 1). The parcels were placed in percussion drilled bore-holes with a diameter of 300 mm and a depth of 4 m in diorite rock at a depth of 450 m below ground. The axially compressed cylinder rings had an initial density of 2080 kg/cm^3, outer diameter of 280 mm, and a height of 100 mm, which give a total bentonite mass of approximately 400 kg in each parcel. The final density of the bentonite clay, at full saturation and after swelling in the test holes, was calculated to be 2000 kg/m^3.

Temperature, total pressure, water pressure and water content are measured during the heating period. Each parcel (somewhat different in the two pilot tests) contain 25 thermocouples, 3 total pressure gauges, 3 water pressure gauges, 4 relative humidity sensors, 7 titanium filters, and 12 water sampling containers. The thermocouples are jacketed by cupro-nickel alloy and all other equipment is made of titanium. All sensors are connected to a standard computer and registrations are made every hour.

The lower 2 m of the heaters had a power of 600 W in the S1 parcel and of 1000 W in the A1 parcel. The radial temperature decrease from the copper tube to the rock in the lower part of the S1 parcel was from 90 to 50 C°, and from 130 to 80 C° in the A1 parcel. In the upper part of the 2 parcels the temperature was close to background and with almost no gradient from copper tube to rock.

The 2 pilot tests were terminated and extracted by overlapping core-drilling outside the original bore-hole. The entire 4.5 m long S1-parcel and approximately 20 cm rock cover were successfully lifted in one piece from the rock, whereas a central part of the A1 parcel was lost during drilling.

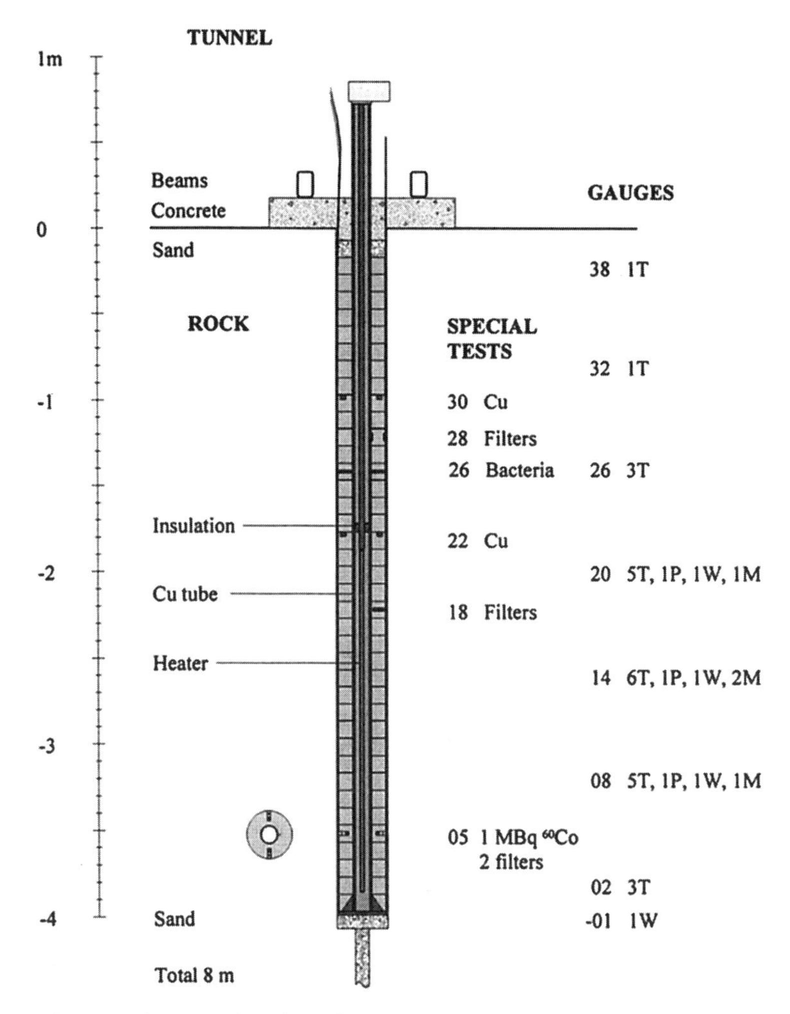

Figure 1. Cross-section view of an S-type parcel. The first figures in column denote block number and second figures denote the number of sensors. T denotes thermocouple, P total pressure sensor, W water pressure sensor, and M moisture sensor.

LABORATORY RESULTS

After uplift, selected parts were divided in order to give test material for the following physical tests (total number of tests within brackets):

- water ratio of parcel material (210),
- density of parcel material (75),
- hydraulic conductivity of reference (10) and parcel material (20),
- swelling pressure of reference (10) and parcel material (20),
- bending strength of reference (10) and parcel material (19),
- shear strength of reference (1) and parcel material (3),

and the following chemical and mineralogical analyses (total number within brackets):

- ICP-AES element analyses of total and clay fraction, reference (20) and parcel material (50),
- Cu-trien analyses of cation exchange capacity of reference (10) and parcel material (20),
- XRD analyses of total and clay fraction, reference (20) and parcel material (50),
- SEM microstructure and element analyses (spot and mapping).

All test and analyses have been completed and the following section briefly describes some techniques and compiled preliminary results.
Hydraulic conductivity and swelling pressure were determined in small (sample diameter 2 cm) swelling pressure cells. Swelling pressure was calculated from measured axial force at zero water pressure, and the hydraulic conductivity was calculated from the percolated water volume under a moderate hydraulic gradient according to Darcy´s law (Figure 2).

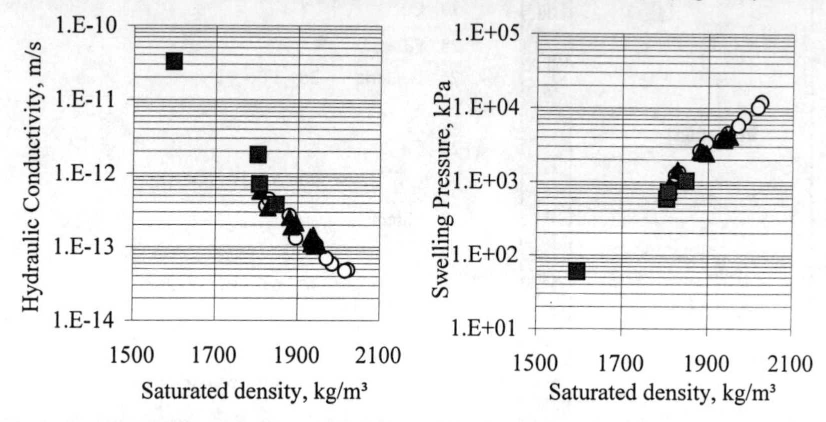

Figure 2. Measured hydraulic conductivity and swelling pressure as a function of clay sample density at full saturation. Squares represent A1 parcel material and triangles S1parcel material. Open circles represent reference material.

Total material and the clay fraction from reference samples and parcel samples were analysed by standard ICP/AES technique with respect to major element content. Table 2 shows major element results from analyses of all fractions in material from the S1 parcel. Figure 3 illustrates measured changes in silica content in the total material, and minor uptake of copper in the clay fraction.

Table 2. Compilation of results from ICP/AES analyses of main elements expressed as oxides in the total material reference material (R, 10 analyses) and in the S1 parcel (30 analyses). LOI denote loss of ignition.

	R				S1			
	Mean	Max	Min	Sdev	Mean	Max	Min	Sdev
SiO2	64.4	65.3	63.8	0.58	63.0	64.6	61.9	0.76
Al2O3	20.0	20.5	19.6	0.23	20.1	20.9	19.1	0.47
SiO2/Al2O3	3.2	3.3	3.2	0.05	3.1	3.3	3.0	0.10
CaO	1.2	1.2	1.2	0.03	1.3	1.4	1.2	0.05
Fe2O3	3.8	3.9	3.8	0.04	3.7	3.8	3.6	0.05
K2O	0.5	0.6	0.4	0.04	0.5	0.6	0.5	0.02

MgO	2.5	2.6	2.4	0.04	2.5	2.6	2.4	0.07
MnO2	0.0	0.0	0.0	0.00	0.0	0.0	0.0	0.00
Na2O	2.3	2.4	2.2	0.04	2.3	2.4	2.3	0.02
P2O5	0.1	0.1	0.0	0.01	0.1	0.1	0.1	0.00
TiO2	0.1	0.2	0.1	0.00	0.1	0.2	0.1	0.00
Sum	94.8	96.2	94.1	0.68	93.7	95.6	93.1	0.73
LOI	5.2	5.3	5.1	0.06	5.5	5.7	5.4	0.11

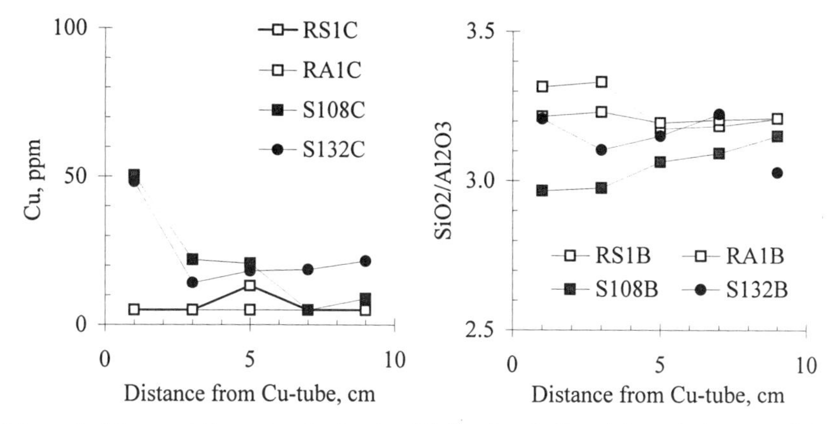

Figure 3. Measured Cu content in the clay (C) fraction (left) and the ratio between SiO_2 and Al_2O_3 (right) in the total material (B) in block 8 (high T part) and 32 (low T part) in the S1-parcel. Open symbols represent results from the reference material.

A copper trietylentetramin technique was used in order to determine the cation exchange capacity (CEC) [2]. A total number of 10 samples from the reference material and 25 samples from parcel material have been analysed. Typical results are shown in Figure 4.

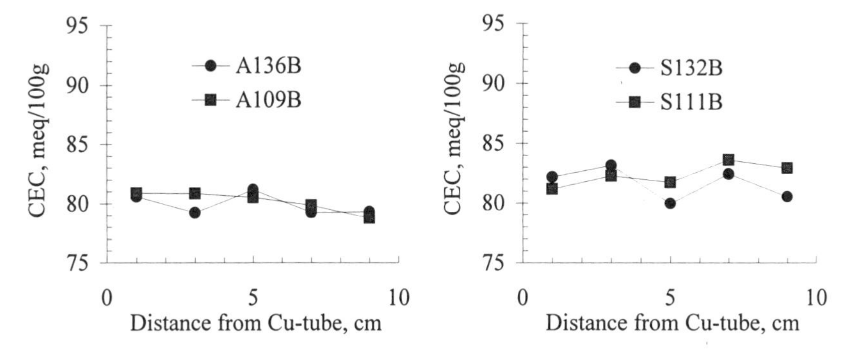

Figure 4. Measured CEC values for material from block 9 (hot) and 36 (cold) in the A1-parcel (left) and block 11 (hot) and 32 (cold) in the S1 parcel (right). Mean CEC-value from reference material is 80 meq/100g clay.

Two different preparation techniques were used for the X-ray diffraction analyses (XRD) in order to produce randomly oriented specimens, and oriented clay fraction specimens. The packed, unsorted specimens were scanned in the 2θ interval 3-65°. The oriented, clay fraction specimens were scanned in the 2θ interval 3-40°. After saturation with ethylene glycol the 2θ interval 3-15° was scanned again, in order to analyse the swelling properties. Figure 5 show typical results from a hot part of the S1 parcel.

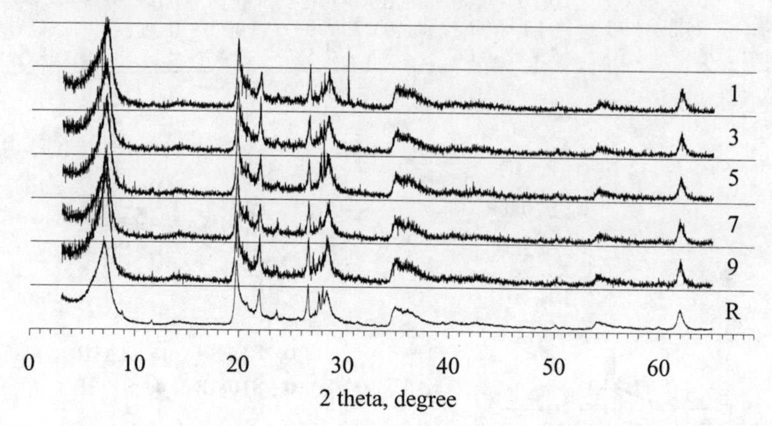

Figure 5. Recorded XRD patterns from block 8 (hot part) in the S1 parcel. Figures denote the radial distance to Cu-tube in cm, and R denotes mean result of reference material.

CONCLUSIONS

The analyses concerning physical and mineralogical conditions in reference and exposed bentonite material may be summarised in the following way:

- No significant difference was found with respect to clay mineral structure according to CEC, XRD analyses,
- Minor mineralogical differences were found with respect to exchangeable cations and reduced silica content according to ICP/AES element analyses,
- Indications of redistribution of accessory minerals were noticed in the exposed material according to SEM analyses,
- No significant changes in physical properties were found.

An overarching conclusion is that no unpredicted changes were found as a result of the exposure to 90 and 130 C° with the exception for a minor uptake of Cu into the clay matrix.

REFERENCES

1. O. Karnland, *Test Plan - Long term test of buffer material*. SKB, IPR-99-01, Stockholm 1999.

2. L.P.Meier and G. Kahr, *Determination of the cation exchange capacity (CEC) of clay minerals using the complexes of copper (II) ion with trietylenetetramine and tetraethylenepentamine,* Clays and Clay Minerals . Vol. 47, No. 3, 386-388, 1999

O_2 CONSUMPTION IN A GRANITIC ENVIRONMENT

I. PUIGDOMENECH[1], L. TROTIGNON[2], S. KOTELNIKOVA[3], K. PEDERSEN[3], L. GRIFFAULT[4], V. MICHAUD[2], J.-E. LARTIGUE[5], K. HAMA[6], H. YOSHIDA[6], J. M. WEST[7], K. BATEMAN[7], A. E. MILODOWSKI[7], S. A. BANWART[8], J. RIVAS PEREZ[9], E.-L. TULLBORG[10]

[1] Royal Institute of Technology (KTH), Inorg. Chem., Stockholm, Sweden, ignasi@inorg.kth.se
[2] CEA, Cadarache, France
[3] Univ. of Göteborg, Dept. of Cell and Molecular Biology, Göteborg, Sweden
[4] ANDRA, Chatenay-Malabry, France
[5] CNRS, Aix-en-Provence, France
[6] JNC, Tono Geoscience Center, Gifu, Japan
[7] BGS, Fluid Processes Group, Keyworth, UK
[8] University of Sheffield, Dept. of Civil & Structural Engineering, Sheffield, UK
[9] University of Bradford, Dept. of Civil & Environmental Engineering, Bradford, UK
[10] Terralogica AB, Gråbo, Sweden

ABSTRACT

The fate of O_2 in a granitic repository has been addressed by an international project: *The redox experiment in detailed scale (REX)*. The emphasis of the project was on a field experiment involving groundwater in contact with a fracture surface. To this aim a borehole, ≈20 cm in diameter, was drilled at 380 m depth in the tunnel of the Äspö Hard Rock Laboratory, Sweden. Injection pulses of molecular oxygen were performed at *in situ* temperature and pressure. Several microbial and chemical parameters were studied as a function of time: microbial counts, pH, O_2-concentration, Eh (redox potential), *etc.*

The field study has been supported by laboratory experiments to determine O_2 reaction rates and mechanisms. These laboratory studies have been performed with Äspö samples (both for inorganic and microbially mediated proccsses). A replica experiment has also been completed at CEA, France, with the other half of the fracture surface obtained in the drilling procedure of the field experiment. The aim of the replica experiment has been to duplicate as far as possible the conditions of the REX *in situ* experiment, for example by using groundwater sampled at the REX site in Sweden, shipped in special containers to France.

The data that has been collected from the O_2 injection pulses in the REX field and replica experiments have been compared with the rates of O_2 uptake determined in the laboratory experiments. These data allow an estimate of the life-times for oxygen uptake in fractures in granitoids, which is of consequence for performance assessment calculations.

INTRODUCTION

Molecular oxygen entrapped in a crystalline rock repository after closure could affect the corrosion of metal canisters. Similarly, future intrusions of oxygen-rich melt waters during a glacial event may affect the integrity of the canisters, as well as the migration of radionuclides.

This paper presents results from an international project (*The redox experiment in detailed scale: REX)* that aimed to study the fate of O_2 in the granitic environment of the underground Äspö Hard Rock Laboratory in South East Sweden.

Mat. Res. Soc. Symp. Proc. Vol. 608 © 2000 Materials Research Society

A block scale redox experiment was carried out previously in a fracture zone at 70 m depth in the entrance tunnel to Äspö[1,2]. In spite of massive surface water input, the fracture zone remained persistently anoxic. The main conclusion from that study was that the increased inflow of relatively organic-rich shallow groundwater instead of adding dissolved oxygen, it added organic compounds that acted as reductants in the subsurface environment. These conclusions are specific to that particular fracture zone, the experimental conditions, and the time scale (3 years) of the experiment, but are probably also relevant for other conductive fracture zones with similar surface conditions.

The detailed scale redox experiment (REX) was started to focus on the question of O_2 that is trapped in the tunnels when the repository is closed. The objectives of the experiment were to determine:

- How does oxygen trapped in the closed repository react with the rock minerals in the tunnel and deposition holes and in the water conducting fractures?
- What is the capacity of the rock matrix to consume oxygen?
- How long time will it take for the oxygen to be consumed and how far into the rock matrix and water conducting fractures will the oxygen penetrate?

EXPERIMENTAL CONCEPT

The emphasis of the project has been on a field experiment involving groundwater in contact with a fracture surface. To this aim a borehole, ≈20 cm in diameter, was drilled at 380 m depth in the tunnel of the Äspö Hard Rock Laboratory. Fe-carrying phases in the fracture surface were chlorite, clay minerals, epidote and pyrite. Injection pulses of O_2 dissolved in groundwater were performed at *in situ* temperature and pressure. Several microbial and chemical parameters were determined as a function of time: pH, Eh, and O_2-concentration, microbial counts and activities, structure of microbial populations, *etc*. The set-up for the REX field experiment is illustrated in Figure 1.

The field study has been supported by laboratory experiments to determine O_2 reaction rates and mechanisms with Äspö samples (both for inorganic and microbially mediated processes). A replica experiment has been performed at CEA Cadarache, France, with the other half of the fracture surface obtained in the drilling procedure of the field experiment. The aim of the replica experiment has been to duplicate as far as possible the conditions of the REX *in situ* experiment, for example by using groundwater sampled at the REX site in Sweden, shipped in special containers to France.

RESULTS

Supporting Laboratory Investigations:

Laboratory experiments performed at Bradford University have tested O_2 uptake by rock and fracture filling mineral samples collected from the Äspö tunnel. Fracture-filling minerals have also been collected from the NW-3 fracture zone using the "triple tube" technique in a 3 m long borehole (KA3065A). The core from this borehole has been characterised and the sieved fractions have also been used in the laboratory tests at the University of Bradford. Oxygen uptake rates have been determined for most of the samples[3]. The results obtained so far show that the rate of O_2 uptake depends both on the particle size of the samples and on their origin in the Äspö tunnel (variation in mineral composition, degree of alteration, *etc*). The values of the first-order rate constant obtained are in the range 1 to 140×10^{-3} L g^{-1} day^{-1}.

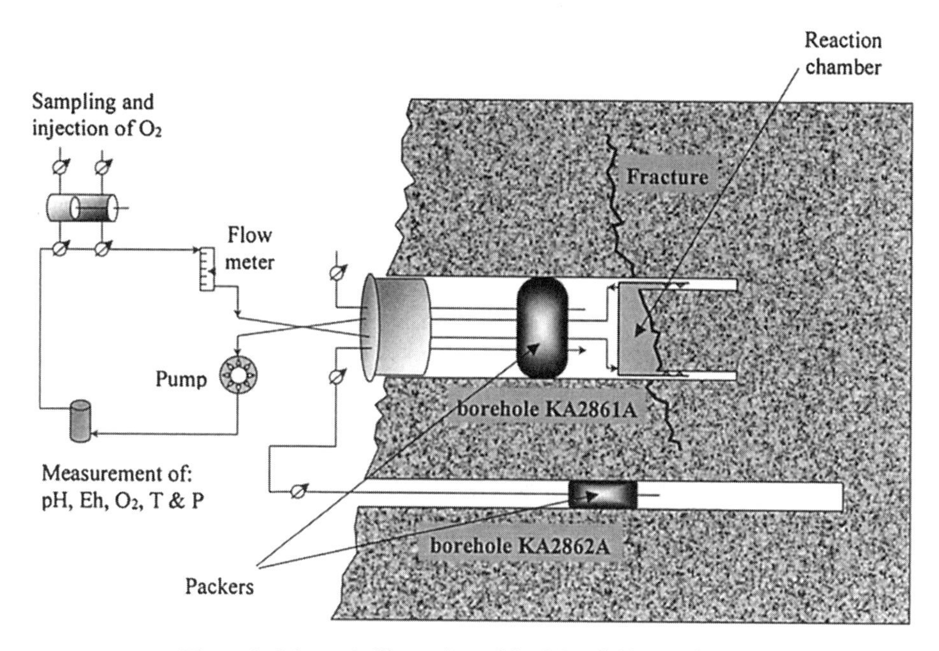

Figure 1. Schematic illustration of the REX field experiment.

Measurements of dissolved gases (CH_4, H_2, *etc*) in Äspö groundwaters have been performed[4,5]. They have been combined with the measurements of microbial O_2 reduction and CO_2 production in Äspö groundwaters. These results showed that O_2 may be consumed by methanotrophic, hydrogen-oxidising or heterotrophic bacteria in a closed nuclear waste repository. The results on bacteriological oxygen consumption experiments performed in the field at Äspö are given elsewhere[5]. Monod-type kinetic rate laws were used to calculate the time scale to consume O_2 in a closed repository[5,6]. The results showed that an initial oxygen content of $[O_2]_o$ = 8 mg L^{-1} (0.25×10^{-3} mol L^{-1}) would be consumed by microbial processes in the groundwater to levels below 0.1% in 1.0 year.

The Japan Nuclear Cycle Development Institute and the British Geological Survey jointly studied the microbial effects on redox groundwater-rock interactions using samples from Äspö. Detailed results from these laboratory investigations were reported elsewhere[7,8]. Both batch and flow experiments were performed. Biofilm development was observed in the column experiments. The columns with bacteria became clogged very rapidly, and minor amounts of smectite were observed in the reaction residues examined after this period (about two weeks). Experiments without bacteria did not clog up. These results are consistent with bacterially enhanced smectite formation being rcsponsible for the clogging of the column experiments[9].

The In-Situ Experiment:

The REX field experiment has been conducted on a single fracture at 8.81 m from the tunnel wall (borehole KA2861A, $\Phi \approx 200$ mm). The drillcore was sent to CEA (Cadarache, France) where the replica of the field experiment has been completed as described below.

The aim of the field study is to isolate the innermost part of the borehole and to monitor the uptake of O_2 as a function of time. A detailed description of the set-up for the REX field experiment has been reported[6], and it is summarised in Figure 1.

A series of O_2 injection pulses were performed in the REX field (and replica) experiments. O_2 concentrations in the range 1 - 8 mg L^{-1} were consumed in the experiments within a few days, 5 to 10 days. Data for one of the O_2-pulses is shown in Figure 2. The data could be described with a first order kinetic equation. Further refinements of the rate law were not possible owing to the limitations of the field experiment, mainly: hydraulic leakage through microfractures and o-rings, and diffusion of atmospheric O_2 into the reaction loop.

Microbial data from the field experiment showed O_2-induced aerobic microbial respiration and a succession of microbial groups in the groundwater and on mineral surfaces. It was unexpectedly found that in some cases the number of iron-reducing organisms increased during the O_2 pulses. Similar effects were observed in the replica experiment. It appears that iron-reducing bacteria developed because of the increased Fe(III) concentrations that followed the O_2 pulses.

The number of cells attached as biofilms exceeded the number of free-living cells. These biofilms would potentially lead to changes in the absorbing properties of the mineral surfaces. The O_2 uptake capacity of the rock is largely affected by the microbial films developing in fracture surfaces.

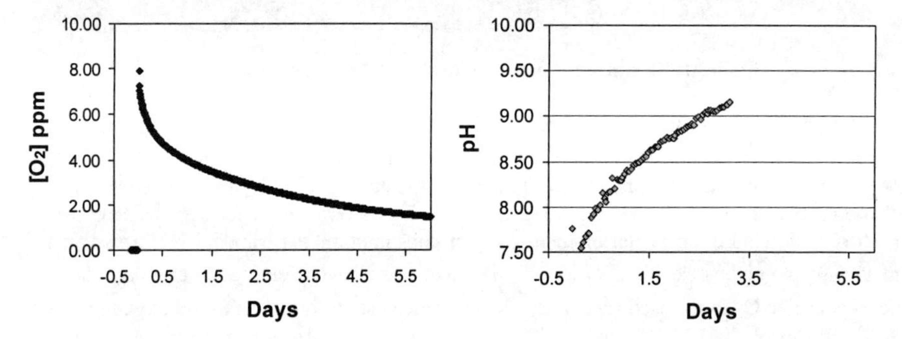

Figure 2. Results from one of the O_2-pulses in the REX field experiment.

The replica experiment

As part of the REX project, ANDRA and CEA have performed a laboratory experiment that reproduced closely the field test, *i.e.* a study in which a diorite core section would be submitted to the same succession of O_2 injections as in the field experiment. The use and purposes of such an experiment, called the *replica* experiment, are both methodologic (how to dimension and complement the *in situ* experiment) and scientific (gathering of more data and better understanding of processes). The small spatial extension of the *in situ* reaction chamber was especially suitable for this project of a replica. The replica experiment was performed on the other half of the fracture surface used in the field experiment. The replica set-up was designed in order to work at a total pressure of 1 bar; different tests and preliminary investigations guided the choice of materials for the different parts of the set-up[6] (core confinement unit, measurement unit, water equilibration unit, tubes, fittings...). The replica experiment started in June 1998 and ended in May 1999. During this period, O_2 pulses of increasing intensity were injected and different parameters were monitored (especially $O_2(aq)$, pH, Eh, solution chemistry, microbial populations). There were long anoxic periods between series of O_2 pulses in order to follow the progressive return to reducing conditions. It was observed that the oxygen uptake kinetics could be fairly simulated by a rate law combining both Monod kinetics (enzymatic kinetics) and a first order contribution (Fig. 3). The form of this rate law is:

$$\frac{d[O_2]}{dt} = -\frac{k_1[O_2]}{K+[O_2]} - k_2[O_2]$$

the values for k_1, K and k_2 were found to be respectively 5-20 μmol $L^{-1}day^{-1}$, 0.05-0.50 μmol L^{-1}, and 0.04-0.10 day^{-1}. During periods of oxygen uptake, the pH was observed to decrease regularly. After O_2 depletion, the pH increased back and a rapid drop of Eh (measured using a Pt electrode) was observed. Upon addition of successive oxygen pulses, the carbonate alkalinity of the water increased progressively. This feature was consistent with the succession of major bacterial groups (aerobes, anaerobes, iron-reducing bacteria, *etc*). During long anoxic periods, the iron and manganese concentrations in solution increased strongly, consistently with the continuous drift of Eh to lower values (typical values attained in three months were Eh = -0.05 V_{NHE} for pH about 6.2 at 15°C). At the end of anoxic periods, a strong growth of sulfate reducing bacteria was observed. The whole dataset obtained on the replica experiment is now under comparison with results from the field experiment and from other participants. A coupling between mineral/solution/bacterial interaction is strongly suggested in both cases.

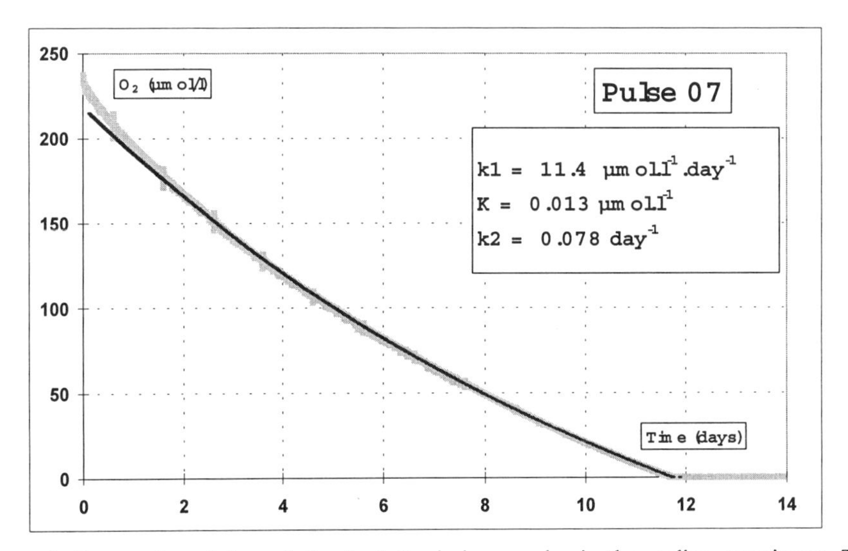

Figure 3. Temporal evolution of dissolved O_2 during a pulse in the replica experiment. The observed evolution is compared to a curve calculated using a combination of first rate law and enzymatic catalysis (Monod law for microbial processes).

CONCLUSIONS

- O_2 uptake rates in the field are comparable to laboratory results obtained in France and UK.
- Microbes have a substantial role in O_2 uptake, both in biofilms and in unattached state. The microbial processes are apparently coupled to reactions at mineral surfaces through iron or manganese cycling (*e.g.*, by fast chemical oxidation of Fe(II) in solution, and consequent microbial reduction of Fe(III) by iron-reducing bacteria).
- For performance assessment, the reducing capacity contributed by microbial activities must include the access to substrates like molecular hydrogen and methane, both originating in deep geological sources.
- Previous laboratory O_2 uptake rates by mineral samples[10], and results from the URL in Canada[11] resulted in expected time scales for O_2 depletion in typical granite fractures between 0.2 and 350 years. The data collected within the REX project results in calculated time scales for oxygen uptake in granitic fractures that are in the order of a few days.

ACKNOWLEDGEMENTS

This work has been financed by the following organisations: Swedish Nuclear Fuel and Waste Management Co. (SKB), Sweden; Agence Nationale pour la Gestion des Déchets Radioactifs (ANDRA), France; and Japan Nuclear Cycle Development Institute (JNC), Japan.

REFERENCES

1. S. Banwart, E.-L. Tullborg, K. Pedersen, E. Gustafsson, M. Laaksoharju, A.-C. Nilsson, B. Wallin, and P. Wikberg, J. Contaminant Hydrol. **21**, 115-125 (1996).

2. S. A. Banwart, P. Wikberg, and I. Puigdomenech, in *Chemical Containment of Waste in the Geosphere*, edited by R. Metcalfe and C. A. Rochelle (Special Publications, Vol. 157, Geological Society, London, 1999), pp. 85-99.

3. J. Rivas Perez and S. Banwart, Report No. SKB-PR-HRL-98-15, 1998.

4. S. Kotelnikova and K. Pedersen, FEMS Microbiology Rev. **19**, 249-262 (1997).

5. S. Kotelnikova and K. Pedersen, Report No. SKB-TR-99-17, 1999.

6. I. Puigdomenech, S. A. Banwart, K. Bateman, L. Griffault, E. Gustafsson, K. Hama, S. Kotelnikova, J.-E. Lartigue, V. Michaud, A. E. Milodowski, M. Morosini, K. Pedersen, J. Rivas Perez, L. Trotignon, E.-L. Tullborg, J. M. West, and H. Yoshida, Report No. SKB-ICR-99-01, 1998.

7. K. Bateman, J. West, K. Aoki, H. Yoshida, P. Coombs, M. R. Gillespie, P. Henney, S. Reeder, and A. E. Milodowski, Min. Mag. **62A**, 124-125 (1998).

8. H. Yoshida, K. Hama, J. M. West, K. Bateman, A. E. Milodowski, S. J. Baker, P. Coombs, V. L. Hards, B. Spiro, and P. D. Wetton, Report No. BGS-WE/99/8C and SKB-IPR-99-19, 1999.

9. K. Bateman, P. Coombs, K. Hama, V. L. Hards, A. E. Milodowski, J. M. West, P. D. Wetton, and H. Yoshida, in *Proc. 9th V M Goldschmidt Conference* (LPI Contribution Nr.971, Luna & Planetary Institute, Houston, Texas, 1999), pp. 20-21.

10. A. F. White and A. Yee, Geochim. Cosmochim. Acta **49**, 1263-1275 (1985).

11. M. Gascoyne, Hydrogeol. J. **5**, 4-18 (1997).

NUMERICAL SIMULATION OF BENTONITE EXTRUSION THROUGH A NARROW PLANAR SPACE

JOONHONG AHN, PAUL L. CHAMBRÉ, and JEROME VERBEKE
Department of Nuclear Engineering, University of California, Berkeley, California 94720-1730, ahn@nuc.berkeley.edu

ABSTRACT

A mathematical model for bentonite expansion through a narrow planar space has been developed based on Terzaghi's theory for clay deformation due to water intrusion. The bentonite expands in a radial direction through a horizontal planar gap, with a constant aperture, filled with water. The permeability and the compressibility of the bentonite are assumed functions of its void ratio. The resulting governing equation is a non-linear diffusion-like equation with void-ratio-dependent coefficients. Numerical solutions for the space-time-dependent void ratio in the expanding bentonite are obtained by applying the Finite Element Method. The finite element solution is combined with a predictor-corrector scheme for evaluations of the void ratio distribution and the location of the moving bentonite-tip boundary. A computer code has been developed for the numerical solutions. The numerical scheme is supported by comparing the results with an analytical solution.

INTRODUCTION

This paper presents results of a numerical analysis for bentonite expansion in a planar gap. A mathematical model has been developed based on Terzaghi's consolidation theory [1] for clay material and Darcy's law for water flow. Numerical solutions for the space-time-dependent void ratio in the expanding bentonite are obtained by applying the Finite Element Method. A computer code, SABRE, has been developed for the numerical solutions.

MATHEMATICAL FORMULATION OF BENTONITE SWELLING

We consider a cylindrical geometry as defined in Figure 1. A water-filled horizontal gap initially intersects a bulk bentonite of radius $R(0)$. The aperture width L is assumed constant in the gap. The expanding bentonite occurs in the region $R(0) \le r \le R(t)$ at time t.

Water flows in at the expanding tip, $R(t)$, through the expanding bentonite layer between $R(0)$ and $R(t)$, and then flows into the bulk bentonite region through the intersection of the gap with the bulk bentonite at $r = R(0)$. The bulk bentonite swells and extrudes into the gap, resulting in the advancement of the tip in the gap. To simplify, the bentonite expansion in the gap is decoupled from the bulk bentonite region by assuming an appropriate boundary condition at the intersection at $R(0)$ (Figure 1). Radial symmetry is assumed for the expansion.

The fundamental equation for the behavior of a deformable medium with incompressible fluid was obtained in [2]. By assuming that the elevation of the planar gap is fixed, and that the expanding bentonite is saturated with water, the governing equation for the space-time-dependent void ratio $e(r, t)$ [dimensionless] in the expanding bentonite in a cylindrical coordinate system is obtained as

$$\frac{1}{1+e}\frac{\partial e}{\partial t} = \frac{1}{r}\frac{\partial}{\partial r}\left(D(e) r \frac{\partial e}{\partial r} \right) \text{ on } R(0) < r < R(t) \text{ for } t > 0, \text{ where } D(e)\ [\mathrm{m}^2/s] \equiv \frac{k(e)}{\mu a_v(e)}. \quad (1)$$

The void ratio e is defined as the ratio between the void volume and the solid-phase volume of the bentonite. μ [Ns/m^2] is the viscosity of water. $k(e)$ [m^2] is the permeability of bentonite. The initial and boundary conditions are shown in Figure 1.

Mat. Res. Soc. Symp. Proc. Vol. 608 © 2000 Materials Research Society

$a_v(e)$ is the coefficient of compressibility [m^2/N], which is defined as $a_v = -de/d\sigma'$. The effective stress σ' [N/m^2] is considered to be carried by the solid skeleton of porous bentonite. For water-saturated bentonite, σ' is equal to the swelling pressure, which results from the effects of electrical repulsion and attraction between bentonite particles. A plot of e versus log σ' is approximately a straight line [3]. $a_v(e)$ and $k(e)$ can be approximated in a form of an exponential function of e [2] (See Table I for the assumed functional forms of $k(e)$ and $a_v(e)$).

From the mass conservation of water in the region $R(0) \le r \le R(t)$, shown in Figure 2, the equation for the location of the tip in the gap is written as

$$\frac{dR(t)}{dt} = \left[D(e)\frac{\partial e}{\partial r} \right]_{r=R(t)}, \quad \text{for } t > 0. \tag{2}$$

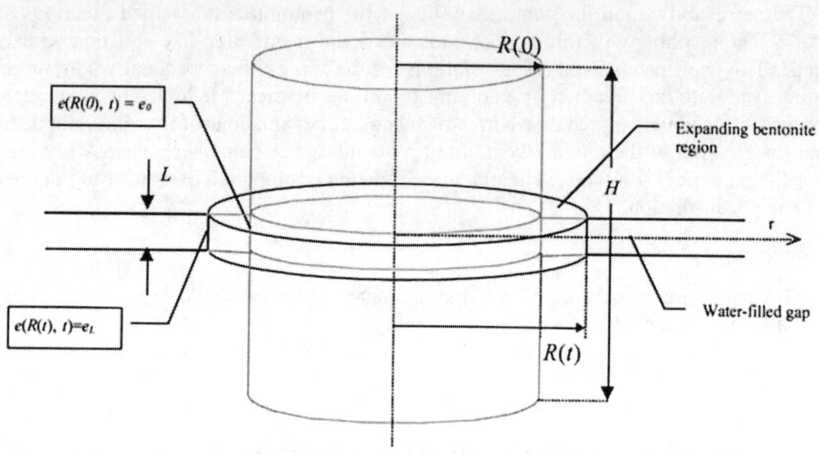

Figure 1 **Geometrical definition of the problem and boundary conditions.**

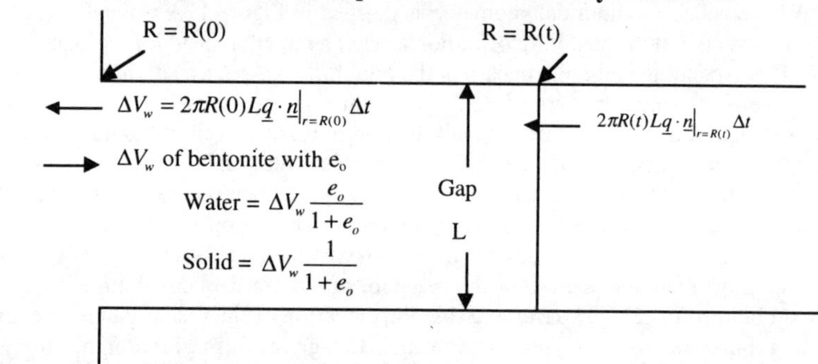

Figure 2 **Assumed mass flow at the inner boundary at $r = R(0)$ where the gap intersects with the bulk bentonite, and at the outer boundary, $r = R(t)$. $\underline{q}$ is the Darcy flux, and $\underline{n}$ is a unit normal to the boundaries at $r = R(0)$ and $r = R(t)$. $\Delta V_w(t)$ is the volume of water crossing the boundary at $r = R(0)$ from the gap to the bulk bentonite during the time interval Δt.**

NUMERICAL SOLUTION SCHEME

The numerical algorithm to calculate $e(r,t)$ is based on a fully implicit Galerkin-Finite Element technique combined with a predictor-corrector scheme. The temporal accuracy is controlled by automatic time step modification. The void ratio $e(r,t)$ is approximated by $\tilde{e}(r,t)$, a sum of N shape functions $\phi_k(r)$ weighted by N nodal values $e_k(t)$, $k = 1, 2, \ldots, N$. $e_k(t)$ satisfies the following system of non-linear, first-order ordinary differential equations in time,

$$\underline{\underline{M}}(t)\dot{\underline{e}}(t) + \underline{\underline{K}}(t)\underline{e}(t) = \underline{F}(t), \tag{3}$$

where $\underline{e}(t)$ is the nodal value vector composed of $e_k(t)$, $k = 1, 2, \ldots, N$. The dot symbol denotes the derivative with respect to time. The double underline denotes a matrix. Elements of $\underline{\underline{M}}$, $\underline{\underline{K}}$, and $\underline{F}$ are written as

$$M_{ik}(t) = \int_{R(0)}^{R(t)} \phi_i \frac{1}{1+\tilde{e}} \phi_k r dr,\ K_{ik}(t) = \int_{R(0)}^{R(t)} rD(\tilde{e}) \frac{d\phi_i}{dr}\frac{d\phi_k}{dr} dr, \text{ and } F_i(t) = \left[\phi_i rD(\tilde{e})\frac{d\phi_k}{dr} e_k\right]_{r=R(0)}^{r=R(t)}. \tag{4}$$

With linear shape functions, only the terms for $k = i$-1, i and i+1 in the matrices M_{ik} and K_{ik} in (3) are non-zero. Only the two extreme terms for $i = 1$, and N in F_i are non-zero.

The motion of the tip reads :

$$\dot{R}(t) = D(e_N)\left[\left.\frac{d\phi_{N-1}}{dr}\right|_{r=R(t)} e_{N-1} + \left.\frac{d\phi_N}{dr}\right|_{r=R(t)} e_N\right] \text{ for } t>0. \tag{5}$$

We solve (3) and (5). The scheme is described as follows:

- From the knowledge of the tip position $R(t_n)$ and the void ratio $\underline{e}(t_n)$ at time t_n, $R(t_{n+1})$ and $\underline{e}(t_{n+1})$ at time t_{n+1} are predicted by the Adams-Bashforth formula [4].
- The internal nodes are relocated by the predicted tip position obtained in the previous step.
- The nodal vector $\underline{e}(t_{n+1})$ of the void ratio on the relocated finite-element grid is computed by the implicit trapezoid rule. The nonlinear algebraic system is solved by the Newton-Raphson iterative technique with predicted values of $\underline{e}(t_{n+1})$ and $R(t_{n+1})$ as first estimates. The convergence criterion on the Newton-Raphson scheme for $\underline{e}(t_{n+1})$ and $R(t_{n+1})$ is arbitrarily set such that the norm of the relative error between two successive iterations be less than 0.0001.
- The position of the tip is corrected with the void ratio obtained in the previous step.
- The nodal vector $\dot{\underline{e}}$ and $\dot{R}$ of the tip position are computed for the next time step.
- An automatic selection of the time increment is made, based on the requirement that the relative norm of the error should be equal to or less than a pre-set input value ε. The relative norm of the error is estimated by the formula proposed by Gresho *et al.* [5]. Here ε is arbitrarily set to be 0.001.

NUMERICAL EVALUATIONS

Input data

The parameter values and the functional forms of $k(e)$ and $a_v(e)$ are summarized in Table I. The radius of the initial bulk bentonite is arbitrarily assumed to be 25 mm. The void ratio e_L at the tip is assumed arbitrarily 1.0 or 2.0 in the following simulation. The void ratio e_o at the intersection, $r = R(0)$, is based on the void ratio reported elsewhere [6] for compacted bentonite to be used in geologic disposal.

Comparison with analytical solution with constant coefficients

We modify the governing equation (1) and consider the following equation with constant coefficients,

$$\frac{1}{1+e_o}\frac{\partial e}{\partial t} = D\frac{1}{r}\frac{\partial}{\partial r}\left(r\frac{\partial e}{\partial r}\right), \tag{6}$$

subject to the necessary side conditions for Eq. (1).

The approximate solution to (6) for early times is obtained from [7]. The location, *R(t)*, of the tip is

$$R(t) - R(0) = K\sqrt{t}, \text{ where } K = \left(1+\frac{1}{e_L}\right)\sqrt{D}(e_L - e_o)\frac{2}{\sqrt{\pi(1+e_o)}}. \tag{7}$$

As Eq. (2) and Figure 2 show, the expansion is primarily controlled by the water inflow through the tip. We take the values of $k(e)$ and $a_v(e)$ at e_L to evaluate a constant D. With the assumed functions for $k(e)$ and $a_v(e)$ and the values of e_L and e_o shown in Table I, D can be calculated to be 0.1065 mm^2/hour for $e_L = 1.0$, and 0.0584 mm^2/hour for $e_L = 2.0$. With these D values, K is obtained as 0.300, and 0.500 mm/hour$^{1/2}$, respectively. In Figure 3, the results represented by the solid lines have been obtained by (7), whereas the numerical results by the SABRE code are shown by the discrete dots. With the greater void ratio at the tip, expansion is faster because more water flows in through the tip.

The agreement between the analytical and the numerical is good. This agreement indicates that the numerical scheme has been correctly implemented into the computer code.

Table I Assumed Parameter Values and Functions

Parameter	Value / formula	Note
$R(0)$ [m]	0.025	Arbitrarily chosen
e_L [1]	1.0 and 2.0	Arbitrarily chosen
e_o [1]	0.5	Based on repository condition [6].
μ [Ns/m^2]	0.001	For water at 20°C
$k(e)$ [m^2]	$k(e) = k_o \exp(3.44e),\ k_o = 7.16\times10^{-22},\ 0.5<e<1.7$	Based on the extrapolated data in [8].
$a_v(e)$ [m^2/N]	$a_v(e) = a_{vo}\exp(4.04e),\ a_{vo} = 1.33\times10^{-8},\ 0.5<e<0.8$	

Spatial distribution of void ratio in the extruding bentonite

Figure 4 shows the spatial distributions of the void ratio in the bentonite expanding in the planar gap obtained by the SABRE code for the case with $e_L = 2.0$. Profiles are shown for 1, 10, 100, and 1000 hours. For each curve, the tip is located at the point where the void ratio $e_L = 2.0$. For the assumed functions, $k(e)$ and $a_v(e)$, the coefficient, *D(e)*, is an exponentially decreasing function of *e*. The profiles show that the gradient of the void ratio at the tip is greater than at the gap intersection. The gradient at the tip decreases as time proceeds.

DISCUSSIONS

Limitations of the present model involve the approximate treatment of bentonite deformation by the one-dimensional consolidation theory. For example, effects of the friction force at the interface between the gap surface and the expanding bentonite are considered to be more prominent for a smaller gap aperture. To reproduce the aperture dependency of the bentonite expansion in a gap, we need an equation for force equilibrium relating changes in effective stress to the deformation (strain) of the bentonite skeleton in addition to one for fluid flow. This is being considered now.

In the present model, the void ratio at the expanding tip is prescribed and assumed constant. The e-dependency of $k(e)$ and $a_v(e)$ are assumed exponential. These parameters should be measured experimentally in the range between e_L and e_o.

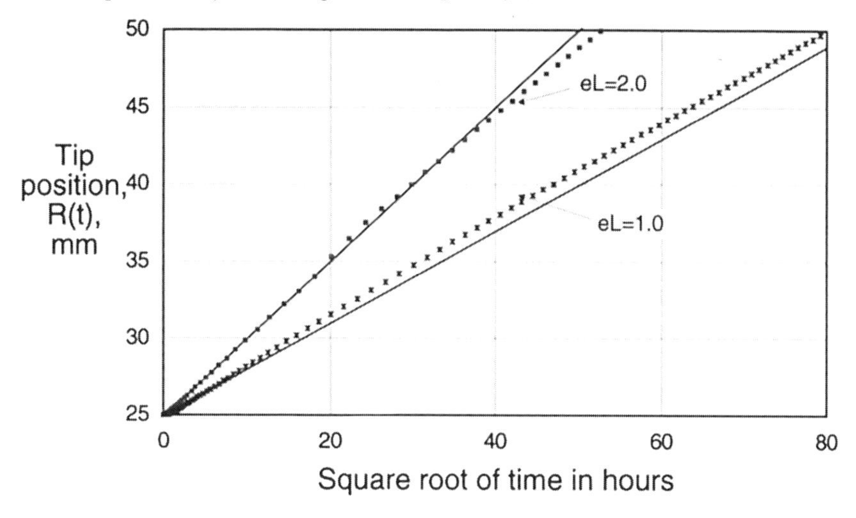

Figure 3 Comparison of numerical results (discrete dots) with the analytical solution (7) (solid lines) for different void ratios at the tip. Parameters are shown in Table I.

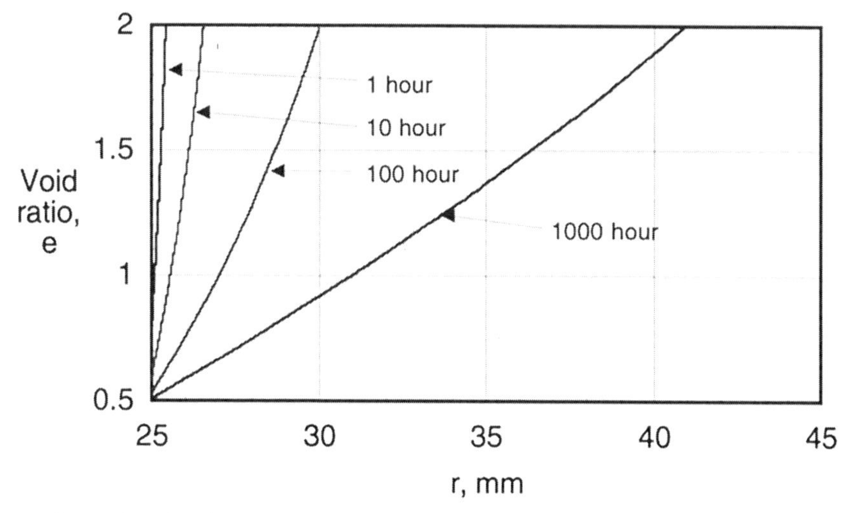

Figure 4 Spatial distributions of the void ratio in the expanding bentonite in the gap at different times for the void ratio, e_L, at the tip equal to 2.0. The inner boundary $R(0)$ is located at 25 mm.

CONCLUSIONS

A mathematical model has been established for the bentonite expansion in a planar gap. Numerical solutions are obtained for the space-time dependent void ratio in the expanding bentonite. A computer code, SABRE, has been developed for the numerical solutions. A fully implicit Galerkin's weighted residual method on deforming elements combined with a predictor-corrector scheme has been applied. The numerical results obtained by the code have been compared with the analytical solution.

The following conclusions are derived:

- The comparison with the analytical solution indicates that the governing equations (1) and (2) developed in this study have been properly implemented into the numerical code, SABRE.
- The model has been established based on a number of assumptions. To simulate real situations by this model, the void ratio e_L at the expanding tip, the permeability and the compressibility coefficient of the bentonite in the range between e_L and e_o must be prescribed.

REFERENCES

[1] K. Terzaghi, Erdbaumechanik auf bodenphysikalischer Grundlage, Leipzig und Wien, F. Deuticke, 1925.

[2] T. N. Narashimhan and P. A. Witherspoon, Numerical Model for Saturated-Unsaturated Flow in Deformable Porous Media 1. Theory, *Water Resources Research*, **13**(3) 657-664, 1977.

[3] K. Terzaghi, R. B. Peck, and G. Mesri, *Soil Mechanics in Engineering Practice*, 3rd ed., Wiley-Interscience, 1996.

[4] W. H. Pressm B. P. Flannery, S. A. Teukolsky, and W. T. Vetterling, *Numerical Recipes, The Art of Scientific Computing*, Cambridge University Press, 1986

[5] P. M. Gresho, R.L. Lee, R.L. Sani, *Recent Advances in Numerical Methods in Fluids,* Pineridge Press, 1980.

[6] Power Reactor and Nuclear Fuel Development Corporation, *Technical Report on Research and Development for Geological Disposal of High-Level Radioactive Wastes*, PNC TN 1410 92-08, 1992.

[7] H. S. Carlslaw and J. C. Jaeger, *Conduction of Heat in Solids,* 2nd Ed., Oxford University Press, 1959.

[8] Power Reactor and Nuclear Fuel Development Corporation, *Evaluation of Extrusion of Bentonite Buffer (I)*, PNC TN8410 97-313, 1997.

IMPACT OF POROSITY HETEROGENEITY IN THE DIFFUSION OF SOME ALKALI- AND ALKALINE EARTH-METALS IN CRYSTALLINE ROCK

H. JOHANSSON, J. BYEGÅRD*, M. SKÅLBERG
Dep. of Nuclear Chemistry, Chalmers University of Technology, S-412 96 Göteborg, Sweden
*byegard@nc.chalmers.se

ABSTRACT

Data from diffusion experiments using Na^{+}, Ca^{2+}, Sr^{2+}, Cs^{+} and Ba^{2+} as tracers in a crystalline rock type (Äspö diorite, originating from the Äspö Hard Rock Laboratory) have been evaluated using a heterogeneity model. The model concept consists of diffusion in a variety of different channels with different porosities. The porosity distribution used has been obtained from independently performed measurements of porosity distributions using the ^{14}C-polymethylmethacrylate impregnation method. Breakthrough curves in through-diffusion experiments as well as penetration profiles in the matrix have been evaluated using the porosity distributions. In the calculations only two parameters, the pore diffusivity (D_p) and the sorption distribution coefficient (K_d) have been varied in order to fit the experimental data to the proposed model. For the penetration profile of more strongly sorbing tracers, i.e., Cs^{+} and Ba^{2+}, a significantly better explanation of the data is obtained using a heterogeneity model compared to using a uniform porosity distribution model. The data from the through-diffusion experiments gives a better explanation of the shape at the beginning of the breakthrough curve. The implication of the proposed diffusion model is discussed, both from an *in situ* sorption experiment application and a performance assessment application.

INTRODUCTION

For a repository for spent nuclear fuel in deep bedrock, as proposed in many countries, the role of the surrounding rock acting as a barrier in retarding released radionuclides is an important parameter. The retardation process in fractured rock is due to the sorption of radionuclides on the fracture walls combined with diffusion into the fracture coatings and further into micropores of the rock, i.e., matrix diffusion [1]. By assuming a homogeneous distribution of the porosity and combining this with the distribution coefficients (K_d) determined from static batch experiments with crushed rock material, performance assessment has shown that the matrix diffusion is the far major retardation process for transport in the geosphere [2].

A parameter that has focused some interest during the recent years is the heterogeneity of the diffusion pathways. Haggerty et al. [3] has shown that the results obtained in *in situ* experiments performed in Culebra dolomite can be explained by a heterogeneous mass transfer, applying a log-normal distribution of the diffusion rates. By impregnation of samples of crystalline generic rock types with ^{14}C labelled methylmethacrylate (^{14}C-MMA) followed by polymerisation, it has been shown that the porosity of the rock is strongly heterogeneously distributed [4,5]. Bradbury and Green [6] applied a dead end pore model in order to interpret the early part of the breakthrough curve in a through-diffusion experiment. However, a general problem is that involving heterogeneity in the interpretation of experiments gives more parameters to use in the calculations. Doubts can therefore be raised whether the obtained better fitting of the results are due to involvement of a correct retardation process (in this case

Mat. Res. Soc. Symp. Proc. Vol. 608 © 2000 Materials Research Society

heterogeneity) or whether it simply is a result of more parameters used in the fitting calculations.

In this work, independently obtained porosity distributions [5] of some rock material from the Äspö Hard Rock Laboratory in Sweden, have been used to interpret the breakthrough curves of some performed through-diffusion experiments [7] and in-diffusion profile studies [5]. These results have earlier been interpreted assuming a homogeneous porosity distribution and comparisons of the two interpretation methods are discussed.

EXPERIMENTAL

Experimental procedures

The impregnation of Äspö-diorite with ^{14}C-MMA and the measurements of the total porosities and spatial porosity distributions were described by Johansson et al. 1998 [5]. The methods and results of the through-diffusion experiments using tritiated water (HTO), Na^+, Ca^{2+} and Sr^{2+} have been described by Johansson et al. 1997 [7] while the studies of penetration profiles of Cs^+ and Ba^{2+} were described by Johansson et al. 1998 [5].

Diffusion theory

The theory of the through-diffusion applied to the present experimental conditions, assuming the rock as a homogeneous medium, has been described earlier [5]. For the introduction of heterogeneity to the model, it has been assumed that the rock consists of several channels with different porosity; distributed accordingly to the distribution found in the ^{14}C-MMA impregnation studies. The diffusion equation for the through diffusion experiment is thus modified to the following solution (based on Crank 1975 [8])

$$C_r = \frac{C_2 V_2}{C_1 A_s l_s} = \sum_{\varepsilon} n_{\varepsilon} \left[\frac{(D_p \varepsilon) t}{l_s^2} - \frac{\varepsilon + K_d \rho}{6} - \frac{2(\varepsilon + K_d \rho)}{\pi^2} \sum_{n=1}^{\infty} \frac{(-1)^n}{n^2} \exp\left\{ - \frac{(D_p \varepsilon) n^2 \pi^2 t}{l_s^2 (\varepsilon + K_d \rho)} \right\} \right] \quad (1)$$

where C_r is the scaled ratio of the concentration in the measurement container, C_2, relative to that of the injection container, C_1. A_s is the geometrical surface area of the sample which has thickness l_s and V_2 is the volume of the measurement container. D_p is the pore diffusivity, ε is the porosity and n_ε is the frequency of that porosity. K_d is the sorption coefficient and ρ is the density of the rock sample.

The analytical solution to the transport equation for the in-diffusion experiments, assuming a linear sorption isotherm, constant C_0 during the experiment and a semi-infinite boundary at the sealed end of one sample, is (based on Crank 1975 [8])

$$\frac{C_x}{C_{x=0}} = \sum_{\varepsilon} n_{\varepsilon} \left[(\varepsilon + K_d \rho) \cdot \mathrm{erfc}\left\{ \frac{x}{2} \left(\frac{D_p \varepsilon t}{\varepsilon + K_d \rho} \right)^{-0.5} \right\} \right] \quad (2)$$

where C_x is the measured average concentration of a slice at distance x and time t, $C_{x=0}$ is the constant concentration in the pore water in the first thin layer at the surface of the solid, erfc is the error function complement (erfc = 1-erf). The other parameters are the same as in eq. (1).

For the case of interpreting the diffusion as a homogeneous process, only one porosity is used and the frequency factor n_ε thus becomes equal to 1. The porosity used in the homogeneous case was determined from the breakthrough of HTO in that particular rock sample. To be correct, the porosity appearing besides D_p in equation (1) and (2) should be the through-transport porosity, i.e. the fraction of the total porosity that carries the diffusion flux through the sample. This parameter is poorly known, therefore the through-transport porosity was approximated with the porosity, ε.

The pore diffusivity, D_p, and the sorption coefficient, K_d, were calculated by fitting the data to eq. (1-2). A steepest gradient method was used to minimise the error sum of squares according to the least square method. The errors of the parameters were estimated assuming normal distribution of the errors.

RESULTS AND DISCUSSION

The porosity distribution in Äspö diorite is presented in Fig 1, together with an attempt to fit the results to a log-normal distribution. The fitting is reasonable, indicating that it seems reasonable to generalise the porosity as log-normal distributed. Attempts to address heterogeneity in transport modelling as a result of log-normal distribution of diffusion rates can be found in the literature [3,9].

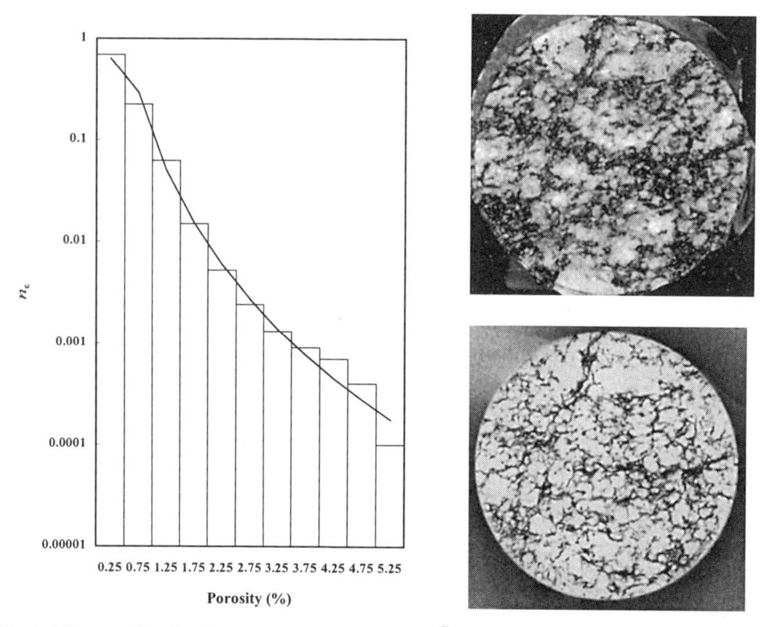

Fig 1 The porosity distribution histogram (left) for Äspö diorite, measured from autoradiographs of ^{14}C-PMMA impregnation (right, lower). A photograph of the rock sample is also shown (right, upper)

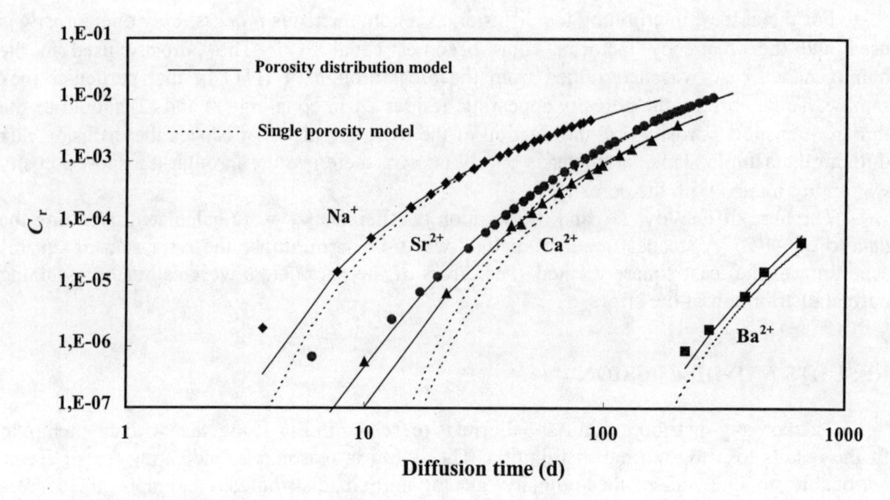

Fig 2 Experimental results and corresponding calculated breakthrough curves for the through-diffusion experiment in Äspö diorite for the different cationic tracers used.

The results of the through-diffusion experiments in the 1 cm diffusion cells, using Na^{+}, Ca^{2+}, $Sr^{2+}$and Ba^{2+} as tracers, are illustrated in Figure 2. In the figure, the fit using both a homogeneous porosity model and a heterogeneous porosity model are given. Comparing the results of a homogeneous porosity model with a heterogeneous porosity model shows that the heterogeneous porosity model can better fit the results at the first part of the breakthrough of the tracer. A general observation is that the homogeneous model can either fit the start of the breakthrough curve or fit the breakthrough when steady state is obtained; not both. The improvement obtained by introducing the independently measured heterogeneity indicates that the heterogeneity plays an important role in the diffusion and retardation of the cation exchange sorbing species.

By using the heterogeneously distributed porosity model to interpret the penetration profiles of Cs^{+} and Ba^{2+} (Fig. 3), a significant improvement of the fitting is obtained, compared to the homogeneous porosity model. In earlier work, Johansson et al. 1998 [5] managed to fit the results using a two-pathway diffusion model. This method lacks of the disadvantage of using as much as four parameters to fit each data set. By the heterogeneous porosity model and using the independently determined porosity distribution, the good fitting can be obtained, using only two fitting parameters for each data set. Johansson et al. 1998 [5] already acknowledged that their succeed in fitting the results using a two pathway model could indicate that a general heterogeneity could influence the system, and that there could be more than two different diffusion pathway types in the rock.

The heterogeneous porosity distribution used in this work is based on a determination in two dimensional slices. The uncertainties when extrapolating the distribution results to three dimensions and deeper penetration depths should therefore be acknowledged.

The calculated pore diffusivities (D_p) and sorption distribution coefficients (K_d) are presented in Table I. It has been pointed out earlier [7] that batch sorption experiments tend to overestimate the sorption coefficients; values determined in batch experiment can not properly be used to predict results in diffusion experiment for these slightly sorbing tracers. As can also

Table I Evaluated pore diffusivities, D_p, rock and sorption coefficients, K_d, in Äspö diorite for the different tracers used

Cell Nr	Model	Nuclide	Cell Size (cm)	D_p $(=D_e/\varepsilon)^C$ (m²/s)	K_d (m³/kg)	Reference	K_d (m³/kg) (Batch Exp.)^D [10]
				Through-diffusion experiment results			
3	A	$^{22}Na^+$	1	$1.8\cdot10^{-11}$	$1.5\cdot10^{-6}$	[7]	$(5\text{-}10)\cdot10^{-6}$
	B			$1.6\cdot10^{-11}$	$1.3\cdot10^{-6}$	this work	
4	A	$^{22}Na^+$	2	$1.7\cdot10^{-11}$	$2.3\cdot10^{-6}$	[7]	$(5\text{-}10)\cdot10^{-6}$
	B			$9.5\cdot10^{-12}$	$2.8\cdot10^{-7}$	this work	
10	A	$^{85}Sr^{2+}$	1	$6.4\cdot10^{-12}$	$6.9\cdot10^{-6}$	[7]	$(3\text{-}10)\cdot10^{-5}$
	B			$1.7\cdot10^{-11}$	$2.0\cdot10^{-4}$	this work	
11	A	$^{85}Sr^{2+}$	2	$5.3\cdot10^{-12}$	$2.5\cdot10^{-6}$	[7]	$(3\text{-}10)\cdot10^{-5}$
	B			$9.9\cdot10^{-11}$	$3.0\cdot10^{-5}$	this work	
15	A	$^{45}Ca^{2+}$	1	$7.3\cdot10^{-12}$	$5.5\cdot10^{-6}$	[7]	$(4\text{-}6)\cdot10^{-5}$
	B			$1.2\cdot10^{-11}$	$1.5\cdot10^{-5}$	this work	
17	A	$^{45}Ca^{2+}$	2	$1.7\cdot10^{-11}$	$4.9\cdot10^{-6}$	[7]	$(4\text{-}6)\cdot10^{-5}$
	B			$3.5\cdot10^{-11}$	$1.4\cdot10^{-4}$	this work	
19	A	$^{133}Ba^{2+}$	1	$(1\cdot10^{-13})$		[10]	$(7\text{-}40)\cdot10^{-4}$
	B			$1.2\cdot10^{-12}$	$3.7\cdot10^{-5}$	this work	
				Penetration studies experiment results			
21	A	$^{133}Ba^{2+}$	2	$7.2\cdot10^{-12}$	$2\cdot10^{-4}$	[5]^E	$(7\text{-}40)\cdot10^{-4}$
	B			$4.7\cdot10^{-12}$	$4.4\cdot10^{-4}$	this work	
25	A	$^{137}Cs^+$	2	$2.0\cdot10^{-11}$	$8\cdot10^{-4}$	[5]^E	$(8\text{-}300)\cdot10^{-3}$
	B			$1.9\cdot10^{-11}$	$2.5\cdot10^{-3}$	this work	

A- Homogeneous porosity model
B- Heterogeneous porosity model
C- Presented in [5] and [7] as D_e-values, recalculated using the porosity determined from the diffusivity of tritiated water.
D- Contact time 14 days, variations due to different size fractions used.
E- The slower of the two diffusion processes found

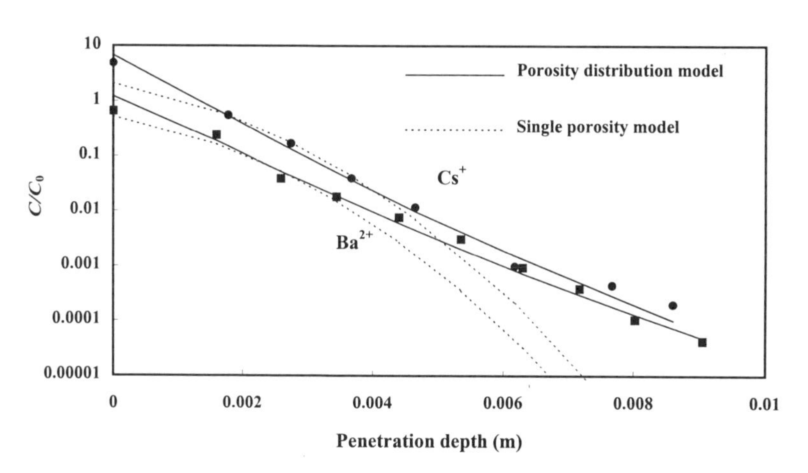

Fig. 3 Penetration profile of Cs^+ and Ba^{2+} in Äspö diorite, compared to the different model calculations.

be seen in Table I, the introduction of porosity heterogeneity gives a somewhat better consistency between the two different ways to determine the sorption coefficients. However, the heterogeneously distributed porosity model still generally gives sorption distribution coefficients under or in the lower part of the interval of the batch sorption determined K_d.

In this work, the heterogeneous model has been useful for interpreting laboratory results for short penetration depths; in this case for diffusion cells with a length of 1-2cm. However, it could be questioned if a heterogeneity model is required for longer penetration depths. Bradbury and Green [6] have earlier shown that their diffusion model approach (partly conducting porosity and dead end pores) will decrease in significance as the sample length was increased. The impact of porosity heterogeneity for performance assessment (PA) is therefore not clear. However, even if the porosity heterogeneity may have little or none effect on the repository scale migration models, it will be of importance for making a correct interpretation of laboratory results, since laboratory data is used as input data for PA.

Implementation of heterogeneity in the porosity could, however, be of importance for interpreting short-term *in situ* transport experiments (e.g, [11]), since the results may be influenced by diffusion with short penetration depth similar to the results presented in this article.

ACKNOWLEDGEMENTS

This work was founded by Swedish Nuclear Fuel and Waste Management Co (SKB) under the TRUE-programme. Mrs Marja Siitari-Kauppi, Helsinki University, Finland, is gratefully acknowledged for the ^{14}C-MMA impregnation measurements.

REFERENCES

1. I. Neretnieks, J. Geophys. Res. **85**, 4379 (1980)
2. L. Moreno, B. Gylling, I. Neretnieks, J. Contam. Hydrol. 25, 283 (1998)
3. R. Haggerty, S.W. Fleming, L. C. Meigs, S. A. McKenna, in review, submitted December 1998 to *Water Resources Research*
4. K. H. Hellmuth, S. Lukkarinen, M. Siitari-Kauppi, Isotopenpraxis Enviro. Health Stud. **30**, 47 (1994)
5. H. Johansson, M. Siitari-Kauppi, M. Skålberg, E. L. Tullborg, J. Cont Hydrol. **35**, 41 (1998)
6. M. H. Bradbury and A. Green, J. Hydrol. **82**, 39 (1985)
7. H. Johansson, J. Byegård, G. Skarnemark, M. Skålberg, Mat. Res. Soc. Symp. Proc. **465** Pittsburg PA 1997 p. 871-878
8. J. Crank, The mathematics of Diffusion, 2nd ed., Oxford University Press, New York 1975, p 50-51
9. R. Haggerty and S. M. Gorelick, Soil Sci. Soc. Am. J. **62**, 62 (1998)
10. J. Byegård, H. Johansson, M. Skålberg, E. L. Tullborg, SKB TR-98-18, Swedish Nuclear Fuel and Waste Management Co, Box 5864, SE-102 40 Stockholm, 1998
11. A. Winberg (ed) Äspö Hard Rock Laboratory International Cooperation Report 96-04, Swedish Nuclear Fuel and Waste Management Co, Box 5864, SE-102 40 Stockholm, 1999

Radionuclide Sorption and Transport

SIMULATION OF RADIONUCLIDE MIGRATION IN GROUNDWATER AWAY FROM AN UNDERGROUND NUCLEAR TEST

A.F.B. TOMPSON *¶, C.J. BRUTON *, W.L. BOURCIER *, D.E. SHUMAKER **, A.B. KERSTING †, D.K. SMITH †, S.F. CARLE *, G.A. PAWLOSKI *, J.A. RARD *
*Geosciences and Environmental Technologies Division, **Center for Applied Scientific Computing, † Analytical and Nuclear Sciences Division, Lawrence Livermore National Laboratory, Livermore, California, 94551, ¶ afbt@llnl.gov

ABSTRACT

Reactive transport simulations are being used to evaluate the nature and extent of radionuclide contamination within alluvium surrounding an underground nuclear test at the Nevada Test Site (NTS). Simulations are focused on determining the abundance and chemical nature of radionuclides that are introduced into groundwater, as well as the rate and extent of radionuclide migration and reaction in groundwater surrounding the working point of the test. Transport simulations based upon a streamline-based numerical model are used to illustrate the nature of radionuclide elution out of the near-field environment and illustrate the conceptual modeling process. The numerical approach allowed for relatively complex flow and chemical reactions to be considered in a computationally efficient manner. The results are particularly sensitive to the rate of melt glass dissolution, distribution of reactive minerals in the alluvium, and overall groundwater flow configuration. They provide a rational basis from which defensible migration assessments can proceed.

INTRODUCTION

There is increasing concern about environmental risks posed by radionuclides produced by underground nuclear tests [1,2]. These risks are dependent, in large part, on the physical and chemical mechanisms that control the extent to which radionuclides are introduced and transported in groundwater. In this paper, we review a series of groundwater flow and streamline-based reactive transport simulations designed to evaluate the impact of some of these processes at the Cambric nuclear test site at NTS. Ultimately, our results will be used as a means to (i) better understand the complex mechanisms involved in radionuclide release and retardation, (ii) design data acquisition strategies for future validation and characterization purposes, and (iii) provide better estimates of radionuclide release for transport simulations designed to assess the fate of radionuclides over longer times and larger spatial scales.

NUCLEAR TEST EFFECTS AND RADIONUCLIDE DISTRIBUTION

The detonation of an underground nuclear device releases an immense amount of energy that vaporizes the geologic and device-related materials in a local region surrounding the testing point [3-5]. This produces a cavity into which overlying formation materials eventually collapse, creating a vertical "rubble" chimney that may extend to the surface. Compressive shock waves generated by the test will fracture or alter the formation beyond the cavity wall. For tests conducted beneath the water table, groundwater will also be vaporized near the explosion point.

Mat. Res. Soc. Symp. Proc. Vol. 608 © 2000 Materials Research Society

As temperatures cool and gas pressures dissipate, components of the cavity gasses begin to condense in an order determined by their relative vapor pressures or boiling points. First among these are condensing rock vapors that accumulate into a melt glass puddle at the bottom of the cavity. Groundwater eventually refills the cavity region.

Radionuclides associated with an underground nuclear explosion are derived from the original materials in the device, nuclear reactions connected with the explosion, and activation products created in the geologic medium. Complex dynamic processes occurring milliseconds to hours after detonation will control their chemical nature and spatial distribution. Most radionuclide vapors will be retained in the immediate cavity region by rebounding compressive stresses in the formation. In some cases, small amounts of radionuclides may escape the cavity region as a result of pressure-driven "prompt injection". During cooling, heavier radionuclides with higher boiling points (such as ^{241}Am or ^{239}Pu) will condense first and largely be incorporated within the melt glass [5]. Lighter radionuclides (such as tritium, ^{3}H) tend to condense later within a "radioactive" or "exchange" volume surrounding the cavity, typically within 2 to 5 cavity radii about the testing point [3]. Other radionuclides will partially condense both within the melt and the rubble zone. Some radionuclides (such as ^{85}Kr) may exist only as noncondensible gases and move outside the immediate vicinity of the cavity/chimney system.

Little is known about how radionuclides are distributed within melt glass or exchange volume rubble, nor of their chemical state in the rubble following condensation. Some may become associated with the solids of the chimney or cavity, while others, including the noncondensibles, may become incorporated within pore waters. When groundwater infills the cavity, the "rubble" fractions may form aqueous species or solid phases consistent with the aqueous chemistry and minerals in the rubble.

THE CAMBRIC TEST

The Cambric nuclear test was conducted at Frenchman Flat in NTS in 1965. Frenchman Flat, located in the southeast corner of the NTS, is an intermountain basin formed by Tertiary-age faulting typical of the Basin and Range physiographic province. The working point and resulting test cavity are centered in Quaternary/Tertiary alluvium, approximately 70 m beneath the ambient water table and 290 m beneath the ground surface. The alluvium is composed of interbedded silts, clays, sands and gravels derived largely from silicic volcanic rocks (tuff and rhyolitic lava). Alteration minerals include clinoptilolite, calcite, smectite, illite/muscovite, and iron oxide, all of which may possess sorptive potential.

Site Data and Simulation Approach

As reviewed in [6-8], the Cambric test had a small yield (0.75 kt) which produced cavity and exchange volumes approximately 10.9 m and 18 m in radius, respectively. The melt debris is comprised of approximately 900 metric tons of glass that occupies a bulk region of 400 m^3 at the bottom of the cavity, assuming a 10% porosity [5,6]. The Cambric test was the subject of a long term radionuclide migration experiment between 1975 and 1991, from which a considerable amount of pertinent chemical and physical data have been obtained [7,8]. The experiment involved the pumping of a nearby well in order to induce the elution of relatively mobile radionuclides (such as ^{3}H, ^{36}Cl, ^{85}Kr, and ^{129}I). Our simulations were limited to a subset of radionuclides and their migration under ambient (non-pumping) flow conditions. This approach is being pursued largely as a means to develop a framework for estimating radionuclide release and migration here and at other tests at NTS [6].

Table 1: Half-life and derived inventory of selected radionuclides associated with the Cambric test, decay corrected to zero time on May 14, 1965 [6].

Radionuclide	Half Life (yr)	Derived Inventory (moles)
^{3}H	12.3	2.04
^{90}Sr	28.8	3.44×10^{-3}
^{137}Cs	30.2	1.07×10^{-2}
^{155}Eu	4.7	8.46×10^{-5}
^{239}Pu	24100	13.0
^{241}Am	432	5.19×10^{-2}

Table 1 shows the derived inventory of radionuclides considered in the current simulations, decay corrected to zero time on May 14, 1965 [6]. Because details of the initial radionuclide distribution are uncertain, these inventories were assumed to be uniformly distributed throughout the melt glass and exchange volumes according to the fractional distribution data in Table 2. All tritium is assumed to condense within the water that occupies the pores of the exchange volume and melt glass. The data in Table 2 were estimated from radiochemical diagnostics and thermodynamic properties of the elements. Prompt injection processes are not considered to be important at Cambric.

Table 2: Estimated distribution of selected radionuclides among the glass, rubble, gas, and groundwater at the Cambric test [2,6].

Radionuclide	Glass %	Rubble %	Gas %	Water %
^{3}H			2	98
^{90}Sr	25	75		
^{137}Cs	10	90		
^{155}Eu	95	5		
^{239}Pu	95	5		
^{241}Am	95	5		

Groundwater samples taken from below the cavity were assumed to be representative of water chemistry in the vicinity of the Cambric test. The ambient groundwater composition used in the simulations is given in Table 3, and is similar to that of waters from volcanic aquifers at NTS. The redox state of the groundwater was assumed to be controlled by equilibrium with atmospheric oxygen, under which conditions radionuclides tend to be most mobile, leading to conservative estimates of migration.

Table 3: Ambient groundwater chemistry used in simulations.

Constituent	Concentration (mg/L)
Na	63
K	8
Ca	16
Mg	4
Sr	0.24
HCO_3	177
Cl	16
SO_4	32
HPO_4	0.31
SiO_2	65
pH	8.0

Radionuclides distributed within the exchange volume were partitioned between pore water and mineral surfaces according to ion exchange and surface complexation reactions. Geochemical speciation calculations were used to identify potential radionuclide-bearing aqueous complexes in Cambric groundwaters, with provision for potential fluctuations in pH and bicarbonate concentration. Calculations were also used to identify the most likely radionuclide-bearing solids that might precipitate if saturation were achieved, either in the initial distribution, or during melt glass dissolution. Table 4 summarizes the aqueous species and solids considered in the simulations. Solid solutions and co-precipitation were not provided for in the models, so clinoptilolite and smectite solid solutions are represented by the calcium-rich compositional end members, Ca-clinoptilolite and Ca-beidellite, respectively.

RADIONUCLIDE RELEASE AND RETARDATION MODELS

Use of the streamline-based reactive transport approach allows the use of detailed equilibrium and kinetic models to describe chemical interactions. Release of radionuclides from melt glass was simulated using a kinetic rate law for glass dissolution. Alteration minerals were allowed to precipitate when saturation limits were reached, and allowed to redissolve if dictated by chemical conditions. Surface complexation, ion exchange and precipitation/dissolution were assumed to control release from the rubble/exchange volume and retardation in the alluvium. Although the framework mineralogy of the alluvium (e.g. quartz, feldspars) was assumed to be inert, surface-active (called reactive) minerals in the alluvium were allowed to precipitate and dissolve according to their saturation state, and alter the total sorptive capacity of the reactive minerals accordingly. See [6] and below for more details.

Co-precipitation, solid solutions, colloid-facilitated transport, and changes in redox state were not considered in these simulations, nor was sorption to carbonates.

Melt Glass Dissolution Kinetics

Melt glass is a heterogeneous brecciated mixture of vesicular and massive glass that largely retains the chemical composition of the host rock. At Cambric, the glass has a rhyolitic composition with about 75 weight % silica, similar to the alluvium underlying Frenchman Flat [6,9]. Glass also contains small amounts of radionuclides consistent with data in Tables 1 and 2.

The rate of release of radionuclides from the glass is proportional to the rate at which glass reacts with groundwater, and depends on temperature, pH and fluid chemistry. In a porous medium, this can be described by an equation of the form

$$\phi \frac{dc_j}{dt} = v_j r = v_j A_s k\left(\Pi_i^N a_i^{p_i}\right)\left(1 - \frac{Q}{K}\right), \qquad (1)$$

where c_j is the aqueous concentration of radionuclide j in groundwater, ϕ is the melt glass porosity, r is the intrinsic rate of glass dissolution per unit volume of bulk medium, and v_j is a stoichiometric coefficient describing the mole fraction of radionuclide j in the glass. The rate (r) is dependent on the specific surface area of the glass (A_s), a temperature-dependent rate coefficient (k), a dimensionless product factor dependent on the activities (a_i) of N catalytic or inhibitive aqueous species, and an affinity term (1-Q/K) that provides for a slow-down in the rate resulting from fluid saturation effects. In this work, only the effect of pH (a_{H+}) is included in the product term, and both the rate coefficient and the exponent were determined from experimental data [10]. In the affinity term, Q and K are the activity product and equilibrium constant for the glass dissolution reaction, respectively. In the following simulations, we assume Q is the activity of SiO_2 (aq), and K is the solubility product for amorphous silica. These parameters provide a conservative estimate of glass dissolution rate. The ambient groundwater was assumed to buffer the pH at 8 and the temperature was assumed constant at 25 °C.

One of the more critical parameters in (1) is the specific melt glass surface area (A_s). As discussed in [6], our nominal value of 0.5 cm^2/g (or about 118 m^2/m^3) was estimated from reactive surface area measurements of fractured glass in waste form canisters [11].

Surface Complexation and Ion Exchange

Radionuclides partitioned into the "exchange" volume (Table 2) may exist as aqueous species, sorbed species on the surfaces of reactive minerals and colloids, and as discrete solid phases and components of solid solutions. The state of the radionuclide depends on radionuclide concentrations in solution, the presence of reactive minerals, and groundwater chemistry. The initial state of a radionuclide and the chemical processes that retard it will collectively influence radionuclide "release" from the exchange volume and subsequent mobility.

Surface complexation, ion exchange, precipitation and dissolution were assumed to control radionuclide release from the exchange volume and retardation in the alluvium. Reactive minerals in the alluvium include goethite (iron oxide), clinoptilolite (zeolite), and smectite and illite/muscovite (clays). A one-site nonelectrostatic surface complexation model was used to describe Pu and Sr sorption onto goethite. Ion exchange was modeled using the ideal Vanselow convention. Exchange was considered between Sr and Ca on clinoptilolite, Ca, Mg and Sr on smectite, and Cs, Na, and K on illite/muscovite [6]. Surface areas and cation exchange capacities were taken from the literature [6] in the absence of site-specific data.

Given the lack of readily available sorption data for Eu and Am onto iron oxides and silicates at the time of this work, Eu and Am were not assumed to participate in any sorption reactions. They therefore migrated as tracers in the simulations in the absence of retarding mechanisms.

HYDROLOGIC FLOW AND TRANSPORT

A simple, steady state groundwater flow model in the vicinity of the Cambric test was developed as an initial basis to forecast radionuclide migration away from the cavity region. The model domain is comprised of a 450 m-long by 360 m-wide by 210 m-deep prismatic block "carved" out of the local alluvium just beneath the water table. It includes the cavity and chimney features of the Cambric test, and encompasses the region influenced by the long-term migration experiment [7,8]. The domain was oriented such that its longer side is collinear with the topographical gradient and principal direction of geologic deposition, as well as the apparent (ambient) horizontal hydraulic gradient (about 0.001 in magnitude [7]).

Table 4: Aqueous and solids considered in the reactive transport simulations. Tritium (^3H) was only used in a calibration test and is not included here.

H^+	OH^-	Na^+	Ca^{2+}
K^+	Mg^{2+}	$O_2(aq)$	Fe^{2+}
Fe^{3+}	$Fe(OH)_3(aq)$	$Fe(OH)_4^-$	$Fe(OH)_2^+$
Al^{3+}	AlO_2^-	$SiO_2(aq)$	$HSiO_3^-$
HPO_4^{2-}	$H_2PO_4^-$	HCO_3^-	CO_3^{2-}
SO_4^{2-}	Cl^-		
Cs^+			
Sr^{2+}			
Eu^{3+}	$EuOHCO_3(aq)$	$Eu(OH)_2CO_3^-$	$EuOH(CO_3)_2^{2-}$
Am^{3+}	$AmCO_3^+$	$Am(CO_3)_2^-$	$Am(OH)_2^+$
Pu^{4+}	$PuO_2(CO_3)_2^{2-}$	$PuO_2(CO_3)_3^{4-}$	$PuO_2CO_3^-$
$PuO_2(OH)_2(aq)$	PuO_2^{2+}	PuO_2^+	
Melt glass			
Muscovite	Ca-Clinoptilolite	Ca-Beidellite	Goethite (FeOOH)
β-Cristobalite (SiO_2)	Calcite ($CaCO_3$)		
$EuOHCO_3$	$AmOHCO_3$	$PuO_2(OH)_2 \cdot H_2O$	

Away from the test, the alluvium was represented by a series of horizontal layers with distinct hydraulic and chemical properties. Within each layer, nonuniform distributions of hydraulic conductivity were specified according to a log-normal correlated Gaussian random field model. This served to reflect observed variability in available (yet sparse) conductivity data, although uncertainty in the parametric data required for the model (ln K variance and spatial correlation scales) and its overall applicability remains and is deserving of additional experimental confirmation [6]. In general, mean conductivity values in the layers (all scalar) ranged from 0.2 to 10 m/d, with some localized extreme values reaching over 50 m/d. Correlation scales ranged from 6 m (vertical) to 24 m (horizontal in the direction of flow), while ln K variances were approximately 2.3. The porosity was assumed to be a uniform value of 0.4. Hydraulic conductivities in the cavity, chimney, and melt glass zones were assumed to be smaller than those found in the surrounding media. Since the material in the cavity and chimney represents slumped or collapsed alluvium, smaller values seemed appropriate [3,6], even though this would tend to divert flow around radioactive zones abutting the test. The permeability of the glass was assumed to be quite small (0.04 m/d), with a porosity of 0.1. The principal chemical properties of the geologic media were specified in terms of the abundance of reactive minerals in the alluvium and cavity/chimney rubble (Table 5), as well as the specific properties of the melt glass. These will be discussed further in the context of the reactive transport model (below).

Steady flow in the system was induced through specification of a small and fixed hydraulic gradient of 0.001 across the longitudinal axis of the domain, with all lateral boundaries being considered no-flow boundaries. This is consistent with conditions in Frenchman Flat. An earlier calibration of this model incorporated a pumping well to reflect the 16-year migration experiment and the subsequent capture of the tritium inventory shown in Tables 1 and 2 [6]. Remaining simulations incorporated ambient flow conditions only and focused on radionuclides other than tritium. Figure 1 shows a portion of the domain and a series of streamlines that pass through the exchange volume and melt glass zones in the model. These streamlines were integrated from the flow solution and form the basis for the reactive transport simulations discussed below.

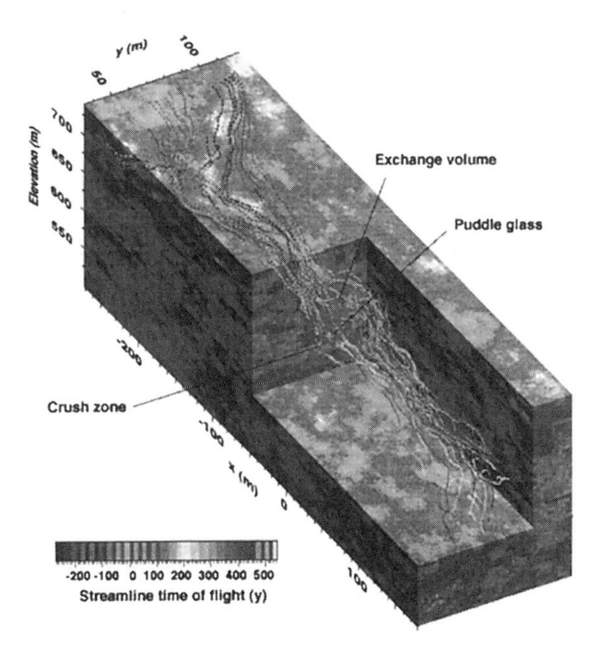

Figure 1: Portion of the flow model domain, with flow moving from the upper left to lower right. Shading indicates relative values of hydraulic conductivity (dark is high). Selected streamlines shown passing through the exchange volume and melt glass zones.

Streamline Transport Model

Reactive transport simulations were accomplished with a streamline-based transport model. In this approach, a three-dimensional transport problem is recast into a large number of independent one-dimensional reactive transport simulations that correspond in a one-to-one fashion to a large number of streamlines that have been extracted from a simulated three-dimensional flow field [6,12,13] developed on a regular, fixed grid.

The streamline mapping procedure is, in some sense, a regridding process tailored specifically for transport simulations. The procedure used to select the streamlines was constrained by two important issues: (i) a desire to use streamlines passing through the initially-contaminated regions only, as opposed to the entire domain, so as to focus all computational effort on the migrating radionuclides, and (ii) a desire to have at least one streamline passing through each grid block inside of the contaminated zone, so as to provide a numerical basis to "pick up" all of the mass comprising the initial inventory as it is introduced into the aqueous regime.

Table 5: Distribution of volume percentages used in the simulations for the alluvium/exchange volume and melt glass. Reactive or inert medium configurations are used at different spatial locations in different transport simulations.

Phase	Reactive medium	Inert medium	Melt glass
Glass	0	0	90
Inert matrix	47	60	0
Ca-Clinoptilolite	5	0	0
Ca-Beidellite	5	0	0
Calcite	1	0	0
Muscovite	1	0	0
Goethite	1	0	0
Pore space	40	40	10

This particular approach did not ensure that all grid-blocks downgradient of the initially contaminated region had streamlines passing through them. However, such grid blocks, should they exist, would likely be associated with very low conductivities and a lack of streamline resolution in these regions would be less important from an overall migration perspective. Also, this approach did not inherently allow for mass transfer between streamlines, as would occur from transverse dispersion and diffusion processes. This is recognized as an approximation, although the errors incurred would be smaller in advectively dominated problems such as this, where the resolution of the flow model is relatively large ($\Delta x_i \sim 2$ m). Altogether, our simulations were based upon 809 streamlines extracted from the flow simulation (Fig. 1).

Along each streamline, the one-dimensional simulations were based upon the GIMRT reactive transport model [14], reconfigured in a "time-of-flight" formulation [6,12]:

$$\phi\frac{\partial(u_j + u_j^{im})}{dt} + \phi\frac{\partial u_j}{d\tau} - \phi\frac{\partial}{d\tau}\left(\frac{D}{V^2}\frac{\partial u_j}{d\tau}\right) = -\sum_{m=1}^{N_m} v_{jm} r_m \quad (2)$$

In this expression, the curvilinear spatial coordinate along the streamline has been transformed into a time of flight variable (τ). The concentrations u_j and u_j^{im} represent the total mobile and immobile concentrations of radionuclide j, each being linear combinations of primary and secondary species that are related through equilibrium mass action relationships [6,12]. Local longitudinal dispersion (D) was not considered in the current work. The rate terms on the right hand side of (2) represent accumulation of radionuclides from dissolution of melt glass or other precipitates, or loss through precipitation of other solids considered in the model (Table 4). Rate laws similar to (1) are used here; surface areas and rates for processes other than melt glass dissolution are discussed in [6]. The effects of radioactive decay are incorporated in an approximate, *ex post facto* manner by appropriate adjustment of concentrations at the end of a simulation. Cumulative impacts of radionuclide decay accrued over the course of a simulation cannot be treated in this manner and must be addressed directly within the model.

Simulations along each line are initiated by assigning the initial aqueous chemistry (Table 3) and appropriate solid phase specification (see below) along the entire line. This may include the melt glass for those portions of the lines passing through the glass volume or the reactive and inert minerals elsewhere. For those lines passing through the exchange volume, the applicable radionuclide inventory is assigned to the appropriate section of the line and distributed among the liquid and solid phases as conditions dictate.

Reactive Mineral Heterogeneity

The types of reactive minerals (Table 5) are known from mineralogic descriptions of alluvium from two boreholes in the vicinity of Cambric. These descriptions clearly indicate that the mineralogic abundance is spatially variable. Based upon the observation that zeolites are in greatest abundance in a layer with the lowest hydraulic conductivity, we assumed that hydraulic conductivity distribution is inversely correlated with the reactive mineral distribution. This assumption was used in a series of sensitivity simulations to gauge the effects of the mineralogic distribution on the overall elution of radionuclides out of the model system. We also included a simulation with an increased glass dissolution rate, as might be produced by a larger melt glass surface area (A_s) or a larger rate coefficient (k) in (1).

Using the terminology of [6], Mineralogic Model 10 assigned the nominal reactive mineral specification (Table 5) uniformly to all points in the model domain. Model 11 assigned an inert mineral specification (Table 5) uniformly to all points in the model domain. Model 12 assigned the reactive specification to the 1% of the domain with the lowest conductivity values, with an inert specification elsewhere. Model 10a was similar to Model 10, although the nominal goethite surface area was reduced to 50 m^2/g from 600 m^2/g. Model 13a assigned the reduced goethite specification (as in 10a) to the 80% of the domain with the lowest conductivity values. Model 10d was the same as model 10a, except that the glass dissolution rate was increased by a factor of 100.

SOME SIMULATION RESULTS

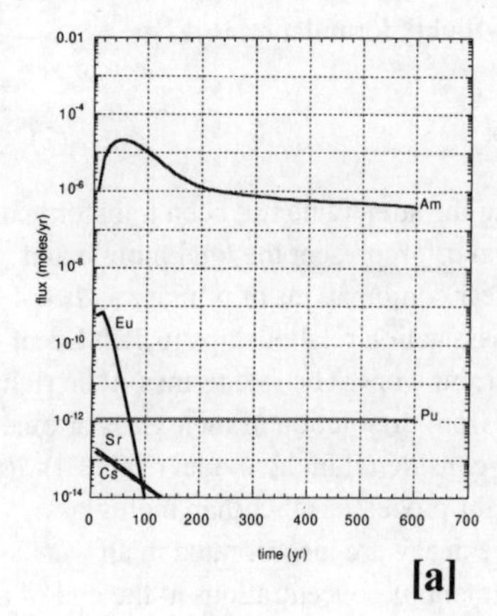

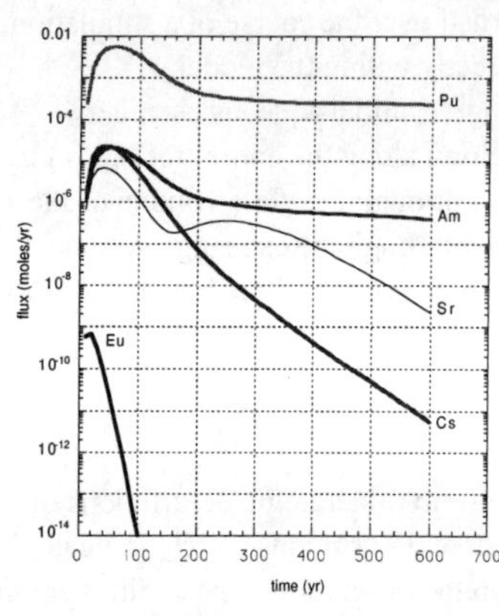

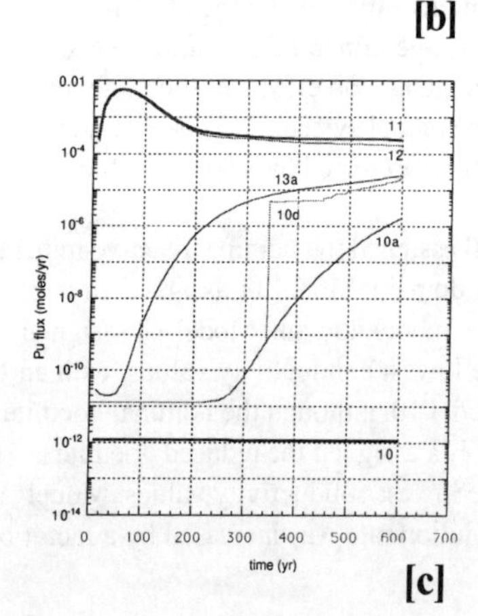

Figure 2a shows the total flux of the total Eu, Am, Pu, Sr, and Cs out of the domain over a 600 year period, as integrated over all streamlines, corresponding to Mineralogic Model 10. Because its long half live and lack of retardation, Am shows the most significant breakthrough, with the 5% exchange volume fraction eluting first and the remaining signal coming from slower dissolution from the melt glass. Although not retarded, the Eu breakthrough is much less notable, owing to radioactive decay. The Pu, Sr, and Cs fluxes should be considered as "0", since their corresponding concentrations are all below small background levels required in GIMRT. Radionuclide precipitation did not occur in this simulation.

Figure 2b shows the total flux of the same radionuclides corresponding to Mineralogic Model 11, in which no reactive minerals were present in the domain. Although the Am and Eu results are unchanged, Pu, Sr, and Cs now exhibit breakthrough owing to their increased mobility in the system. The only chemical processes affecting elution out of the domain in this simulation are melt glass dissolution and radioactive decay. Most elution curves show the early release of the exchange volume components, followed by a slower release from the glass. Pu shows the largest release rate because its inventory (Table 1) was the largest of the radionuclides considered.

Figure 2c shows the aggregate flux of total Pu for all mineralogic models considered. The small fraction of reactive minerals in Model

Figure 2: Elution profiles for total Cs, Sr, Eu, Pu, and Am out of the model domain, as based upom Mineralogic Model 10 (a) and Model 11 (b). Elution profile for Pu under all Mineralogic Models (10, 11, 12, 101, 13a) shown in (c).

12 slightly affected the latter end of the elution curve when compared to Model 11. Comparison of Model 10a to Model 10 shows breakthrough occurs with reduction of the goethite surface area, despite the fact that the reactive minerals are uniformly distributed in both models. The elution profile of Model 10a is enhanced further in Model 13a when the same reactive mineral specification is constrained to be in the 80% of the materials with the lowest hydraulic conductivity. Faster flow pathways in higher conductivity regions with no reactive minerals contribute to the increased mobility of Pu. Finally, in comparing Model 10a to 10d, a dramatic increase in the Pu elution arising from the increased melt glass reaction rate is observed. The elution profile flattens in Model 10d when $PuO_2(OH)_2 \cdot H_2O$ reaches saturation and precipitates in the melt glass, essentially throttling the aqueous Pu release.

CONCLUSIONS

Our simulations indicate that the cavity, chimney, and melt glass environments can act as significant sources of radionuclides for hundreds to thousands of years, or more [6]. Important processes controlling the rates of radionuclide introduction into groundwater include melt glass dissolution and chemical retention effects associated with sorption processes on reactive minerals in the alluvium and cavity/chimney regions. The elution results were particularly sensitive to the abundance and distribution of reactive minerals in the system and the dissolution rate of melt glass.

The streamline modeling approach proved to be an exceptionally useful and flexible technique for studying this problem. It allowed a highly resolved 3D reactive transport problem to be decomposed into a large, yet tractable, number of 1D reactive transport problems, whose results could later be recombined into a 3D solution. Solutions were assembled in a piecemeal fashion using less computer time and allowing for problematic aspects of the solution to be solved more quickly. Analysis of preliminary 1D solutions allowed for faster benchmarking of specific sensitivity and design issues, as well as for diagnosis and interpretation of particular transport and reaction behavior before the more time-consuming 3D simulations were performed.

Efforts are underway to measure glass surface areas from retrieved melt glass samples and analog volcanic glasses, as well as to better characterize the mineralogy and heterogeneity of alluvium. Data collection efforts are being planned to better characterize *in situ* radionuclide concentrations for monitoring and model validation purposes. Field data are critically important for better understanding the relevant physical and chemical processes in the system, as well as for calibrating and validating predictive models.

ACKNOWLEDGEMENTS

This work was conducted under the auspices of the U. S. Department of Energy by Lawrence Livermore National Laboratory under contract W-7405-Eng-48. This work was funded by the Underground Test Area Project, U. S. Department of Energy, Nevada Operations Office.

REFERENCES

1. U.S. DOE, *Regional groundwater flow and tritium transport modeling and risk assessment of the underground test area, Nevada Test Site, Nevada,* U. S. Department of Energy, Nevada

Operations Office, Environmental Restoration Division, Las Vegas, NV, DOE/NV--477 (1997).

2. IAEA, *The radiological situation at the atolls of Mururoa and Fangataufa. Inventory of radionuclides underground at the atolls. Interim version, Vol. 3*, International Atomic Energy Agency, Vienna (1998).

3. I. Borg, R. Stone, H. B. Levy, and L. D. Ramspott, *Information pertinent to the migration of radionuclides in ground water at the Nevada Test Site. Part 1: Review and analysis of existing information,* Lawrence Livermore National Laboratory, Livermore, CA, UCRL-52078 (1976).

4. L. S. Germain and J. S. Kahn, *Phenomenology and containment of underground nuclear explosions,* Lawrence Livermore National Laboratory, Livermore, CA, UCRL-50482 (1968).

5. D. K. Smith, Characterization of nuclear explosive melt debris, Radiochimica Acta, 69, 157-167 (1995).

6. A. F. B. Tompson, C. J. Bruton, and G. A. Pawloski, eds., *Evaluation of the hydrologic source term from underground nuclear tests in Frenchman Flat at the Nevada Test Site: The Cambric test,* Lawrence Livermore National Laboratory, Livermore, CA, UCRL-ID-132300 (1999), (http://www-ep.es.llnl.gov/www-ep/UGTA, August 2, 1999).

7. D. C. Hoffman, R. Stone, and W. W. Dudley, Jr., *Radioactivity in the underground environment of the Cambric nuclear explosion at the Nevada Test Site,* Los Alamos National Laboratory, Los Alamos, NM, LA-6877-MS (1977).

8. E. A. Bryant, *The Cambric migration experiment: A summary report,* Los Alamos National Laboratory, Los Alamos, NM, LA-12335-MS (1992).

9. L. Schwartz, A. Piwinskii, F. Ryerson, H. Tewes, and W. Beiringer, Glass from underground nuclear explosions. Journal of Noncrystalline Solids, 67, 559-591 (1984).

10. J. J. Mazer, *Kinetics of glass dissolution as a function of temperature, glass composition, and solution pH,* Ph.D. thesis, Northwestern University (1987).

11. R. G. Baxter, *Description of defense waste processing facility reference waste form and container,* Savannah River, Aiken, SC, DP-1606, rev. 1 (1983).

12. M. R. Thiele, R. P. Batycky, M. J. Blunt, and F. M. Orr, Simulating flow in heterogeneous systems using streamtubes and streamlines, SPE Reservoir Engineering, 10, 5-12 (1996).

13. S. B. Yabusaki, C. I. Steefel, and B. D. Wood, Multidimensional, multicomponent, subsurface reactive transport in nonuniform velocity fields: code verification using an advective reactive streamtube technique, Journal of Contaminant Hydrology, 30, 299-331 (1998).

14. C. I. Steefel and S. B. Yabusaki, *OS3D/GIMRT, Software for modeling multicomponent and multidimensional reactive transport, User manual and programmer's guide, Version 1.0,* Pacific Northwest National Laboratory, Richland, WA, PNL-11166 (1996).

ASSESSING THE EFFECTIVE REACTIVE SURFACE AREA IN HETEROGENEOUS MEDIA THROUGH THE USE OF CONSERVATIVE AND REACTIVE TRACERS

Robert W. Smith and Jonathan R. Ferris
Idaho National Engineering and Environmental Laboratory, Biotechnologies Department,
Idaho Falls, ID 83415-2107, rqs@inel.gov

ABSTRACT

The characteristics and abundance of reactive surfaces in aquifer media have long been recognized as key factors controlling the migration of contaminants and other dissolved constituents in groundwater. The authors have shown previously that the effective reactive surface area of a heterogeneous aquifer is a complex function of groundwater advective velocity and the correlation structures of the physical and chemical heterogeneities. Although in principle, the available surface area within an aquifer could be estimated using geostatistical techniques and laboratory BET surface area determination for individual samples, this approach is fraught with difficulties associated with inadequate sample coverage and lack of appropriate methods for scaling laboratory surface area measurements. An alternative approach (multiple-tracer approach) to the difficult process of estimating available aquifer surface area relies on the use of conservative and slightly reactive tracers to assess the integrated effective reactive surface area along flow paths. This multiple-tracer approach sacrifices detailed understanding of the fine-scale heterogeneity but can provide integrated large-scale estimates of effective reactive surface area useful for the prediction of reactive transport. The approach is demonstrated by the analysis of the breakthrough curves for paired tracers in heterogeneous media.

INTRODUCTION

Many radionuclides, metals, and other contaminants chemically react with aquifer materials through co-precipitation or adsorption reactions. These reactions result in the partitioning of radionuclides from groundwater onto the solid phase and retardation of migration. In addition, the distribution of adsorbing phases and the extent of their interactions with radionuclides coupled with the groundwater behavior result in the heterogeneous distribution of contaminants in the subsurface. Although aquifers can be composed of heterogeneous assemblages of minerals of differing grain sizes and reactive properties, the overall reactivity in an aquifer is often dominated by grain coatings rather than the bulk mineralogy [1]. Chief in importance among the grain coatings are hydrous ferric oxides because of widespread occurrence in the subsurface, large surface area (approximately 30 m^2/g to 700 m^2/g for hydrous ferric oxide [2, 3]), and a high degree of reactivity. For example, surface area measurements on naturally oxide-coated sands from a site near Oyster, Virginia, range from 0.14 m^2/g to 3.1 m^2/g and are highly correlated with the abundance of hydrous oxide coatings (Figure 1). Furthermore, surface area measurements made on the sands after stripping the oxide coatings are significantly smaller (>95%) than for the coated sands and are consistent with surface areas estimated from measured grain-size distributions. Finally, characterizations of Oyster sand and studies of uranium adsorption onto these sands [4] indicate that uranium adsorption is proportional to total surface area (Figure 1). Although the above example and other studies demonstrate the importance of the surface area of hydrous ferric oxide grain coatings in controlling biogeochemical reactions, the incorporation of the concept of reactive surface area into field-scale investigations of biogeochemical processes in advecting groundwater systems is limited.

Mat. Res. Soc. Symp. Proc. Vol. 608 © 2000 Materials Research Society

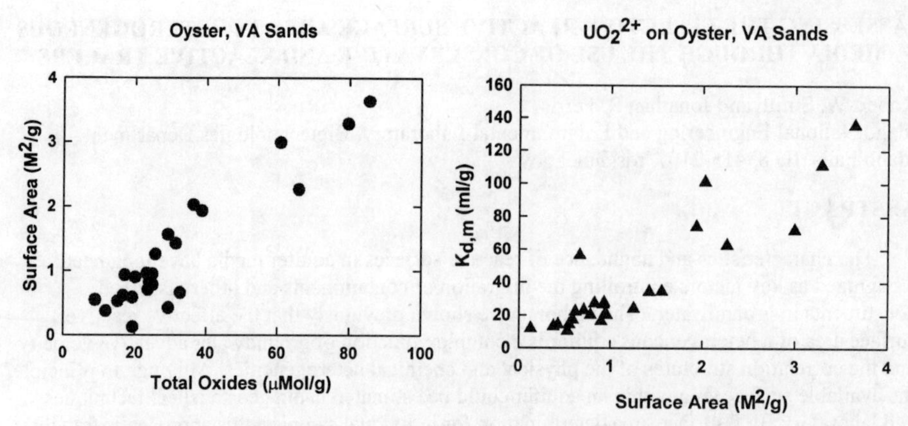

Figure 1. The relationship between reactive surface area and abundance of oxide coatings and surface area and the mass-based $K_{d,m}$ for the adsorption of UO_2^{2+} at pH 5 for sands collected from Oyster, Virginia [4].

Smith and Schafer [5] have previously proposed from theoretical considerations that surface area should be an anisotropic property in a heterogeneous porous media. However, routine static surface area characterization methods such as BET will not capture the anisotropy of the media. To measure effective reactive surface area for intact media, an approach that uses conservative and adsorbing tracers in solution to determine effective reactive surface area in advecting systems is proposed. The separate use of advecting systems and adsorption from solutions are established surface area characterization techniques with well developed theoretical basis [6]. The unique aspect of the multiple-tracer approach described here is to apply these techniques simultaneously (adsorption from an advecting solution) to intact heterogeneous porous media in both column studies and at the field scale. The in situ determination of the effective reactive surface area will allow more robust assessment and prediction of the transport of radionuclides, metals, and contaminants in the subsurface.

APPROACH

The use of tracers to estimate hydrological and reactive parameters of columns and aquifers is well established. Typically, one or more tracers with differing characteristics are introduced into the system being investigated and the breakthrough of the tracers is monitored. The multiple-tracer approach can be used to estimate the in situ effective reactive surface area (i.e., the surface area that a packet of advecting water interacts with) for intact cores and aquifers as described below.

In the case of pure advective transport of a tracer (i.e., ignoring the effects of dispersion), the velocity of a tracer exhibiting linear reversible adsorption is given by

$$V_r = \frac{V_W}{R_{f,r}} \qquad R_{f,r} = \frac{V_c}{V_r} = \frac{D/T_c}{D/T_r} = \frac{T_r}{T_c} \tag{1}$$

where $R_{f,r}$ is the retardation factor for the reactive tracer. V_r, V_W, and V_c are the velocities of the reactive tracer, water, and a conservative tracer introduced simultaneously with the reactive tracer (i.e., $V_W = V_c$), respectively. T_r and T_c are the times required for the reactive and conservative tracer, respectively, to travel distance D. Smith and Schafer [5] have previously shown that $R_{f,r}$ is related to the surface area along the flow path by

$$R_{f,r} = 1 + K_{d,m}\varrho\frac{1-\theta}{\theta} = 1 + K_{d,A}A_S \tag{2}$$

where ρ is the grain density of the medium, θ is the porosity, and $K_{d,m}$ and $K_{d,A}$ are mass-based and surface area-based distribution coefficients, respectively. Equations (1) and (2) allow in situ or effective reactive surface area to be estimated from multiple-tracer tests and the relationship

$$A_S = \left(\frac{T_r}{T_c} - 1\right)\Big/K_{d,A}\,. \tag{3}$$

Equation (3) demonstrates conceptually how the breakthrough curves of the multiple tracers can be used to assess the effective reactive surface area. In practice, inverting tracer experiments in real systems to obtain reactive surface area is complicated by the need to explicitly consider dispersion (i.e., broadening or asymmetry in the breakthrough curve) and to adequately account for the aqueous and surface complexation reactions of the tracers. Because of these complexities, the evaluation of experimental and field results to determine the effective reactive surface area requires using a biogeochemical transport code that incorporates an experimentally based chemical reaction model in the solution of the advection-dispersion transport equations. Alternatively, the relatively simple Method of Moments [7] can be used. The Method of Moments is advantageous because it requires neither making a priori assumptions about the nature of the porous media nor solving the advection-dispersion transport equations. Furthermore, it is applicable to both column studies and field tests.

For conditions of constant flow, the corrected first temporal moment, $\overline{T_i}^*$, (i.e., center of mass of the plume) for the breakthrough of a finite tracer pulse is given by

$$\overline{T_i}^* = \frac{\int_0^\infty TC_i\,dT}{\int_0^\infty C_i\,dT} - \frac{T_{iS}}{2} \tag{4}$$

where C_i is the dimensionless tracer concentration (equal to C/C_0) and T_{iS} is the duration of the tracer injection. Although the breakthrough curves should be complete, the curves may be extrapolated with an exponential function provided data are available to describe the tails of the curves.

The multiple-tracer approach outlined above is an integrated approach that provides a surface area measurement averaged over the entire flow path. This approach does not require a priori assumptions about the nature or structure of geochemical or physical heterogeneities along the flow path. However, additional spatially distributed point data (e.g., characterization samples) are often available that can be used to condition transport calculations. The joining of these two types of observations (e.g., integrated and point) can be accomplished through the analysis of the breakthrough behavior. In this case, the optimal distribution of surface area can be determined using inverse techniques that reflect both the integrated surface area along flow paths and the local measurement of this property. Results from this type of analysis provide fundamental characterization information needed for the simulation of complex biogeochemical processes in the heterogeneous subsurface.

Table 1. Material properties used for the reactive tracer simulations presented in Figure 3.

$k_1 = 1.110E\text{-}06\ cm^2$	$k_2 = 1.088E\text{-}08\ cm^2$	$R_{f,1} = 4.80$	$R_{f,2} = 39.34$
$\rho = 2.67\ g\ cm^{-3}$	$\theta = 0.40$	$A_{S,1} = 28{,}600\ cm^{-1}$	$A_{S,2} = 288{,}000\ cm^{-1}$
$K_{d,A} = 1.33E\text{-}4$ cm		Effective $R_{f,H} = 4.83$	
$m_1 = 0.9,\ m_2 = 0.1$		Effective $A_{S,H} = 28{,}900\ cm^{-1}$	
Effective $k_H = 1.000E\text{-}6\ cm^2$		Effective $R_{f,V} = 8.25$	
Effective $k_V = 1.000E\text{-}7\ cm^2$		Effective $A_{S,V} = 54{,}500\ cm^{-1}$	
$k_H/k_V = 10$		$A_{S,H}/A_{S,V} = 0.53$	

TRANSPORT SIMULATIONS

Numerical simulations in preparation for laboratory experiments were conducted to evaluate the appropriateness of the multiple-tracer approach to estimate effective reactive surface of porous media with small-scale heterogeneities. A system composed of alternating high- and low-permeability layers, as shown in Figure 2, was considered. In these simulations, it was assumed that sand grains were uniformly coated with equally reactive (on a per area basis) hydrous metal oxides. The differences in physical and reactive properties between the high- and low-permeability layers are strictly a function of the grain size, as described previously [5, 8]. The coupling physical and chemical heterogeneity results in anisotropy in both permeability and retardation factors [5]. Two-dimensional simulations for a conservative tracer (e.g., bromide) and a slightly reactive tracer (e.g., fluoride or strontium) were conducted using media properties given in Table 1. Flow directions both perpendicular to and parallel to the layering were evaluated. Simulations were conducted using the TETRAD computer code for a two-layer 50 × 50-cm system with 2,500 nodes. The total thickness of the low-permeability layer was 5 cm ($m_2 = 0.1$); ρ and θ were set to 2.67 g/cm^3 and 0.40, respectively, for both layers; and head gradients of 0.005 and 0.05 were imposed across the flow domain parallel and perpendicular to the layering, respectively. Breakthrough curves were calculated by summing the fluxes. The Method of Moments, as described above, was used to evaluate the tracer breakthrough curves and estimate retardation factors and effective reactive surface areas.

RESULTS

Figure 3 shows the breakthrough curves for conservative and reactive tracers passing through a heterogeneous layered system. Also shown in Figure 3 are the moments estimated from the discrete breakthrough data shown in the figure. As shown in this figure, the effective retardation factor for heterogeneous media is a function of the transport direction. In the direction perpendicular to the layering, the tracers are forced through the low-permeability and highly reactive layers. In the direction parallel to layering, the reactive tracers largely bypass the low-permeability zones and remain confined to the high-permeability less reactive zones. Smith and Schafer [5] have previously suggested that the effective reactive surface area for binary layered systems can be estimated from

$$A_{S,H} = \frac{(1-m_2)k_1 A_{S,1} + m_2 k_2 A_{S,2}}{(1-m_2)k_1 + m_2 k_2} \quad ; \quad A_{S,V} = (1-m_2)A_{S,1} + m_2 A_{S,2} \qquad (5)$$

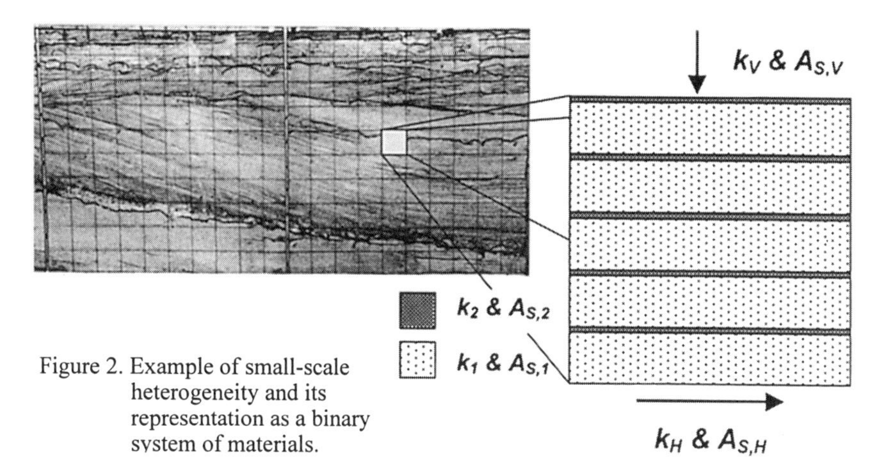

Figure 2. Example of small-scale heterogeneity and its representation as a binary system of materials.

where $A_{S,H}$ and $A_{S,V}$ are the effective surface areas parallel or perpendicular to the layering, respectively. $A_{S,i}$, k_i and m_i are the total surface area, permeability, and relative thickness, respectively, of the two materials composing the alternating bands in Figure 2. The effective reactive surface areas for the two simulations determined by the Method of Moments and Equation (3) are 29,000 cm^{-1} and 54,000 cm^{-1} parallel and perpendicular to the layering, respectively. These values compare favorably to the values predicted from Equation (5) and given in Table 1. Our simulations lead to the counter intuitive result that for heterogeneous media, reactive surface area is not a constant, but rather an anisotropic property much like permeability. For the simulations considered here, Equation (5) yields an anisotropy ratio ($A_{S,H}/A_{S,V}$) of 0.53. This value compares favorably to the value of 0.54 derived from the tracer simulations using the Method of Moments. The advantage of using the Method of Moments over an inverse method is that it eliminates the need to explicitly consider the effects of dispersion in determining the retardation times. In fact, the evaluation presented here was conducted using a spreadsheet and trapezoid rule integration. In more complex systems with multiple sites with differing reactivity, the simple Method of Moments approach would not be appropriate. However, the judicious selection of tracers (simple aqueous chemistry and linear or Langmuir adsorption) should allow the determination of the effective reactive surface area for a variety of materials under both laboratory and field conditions.

CONCLUSIONS

Reactive surface area is a key property of an aquifer that often controls the retardation of radionuclides, metals, and other contaminants in advecting groundwater. In addition, the nature and abundance of reactive surfaces control many naturally occurring biogeochemical processes. Although aquifers are composed of assemblages of minerals, coatings of hydrous metal oxides often control reactivity. Many well-established techniques are available for laboratory determination of surface area. However, their utility in defining the actual surface area of intact media that are available to react with solutes in an advecting groundwater is limited. This limitation can be overcome by the multiple-tracer approach. This approach uses advecting conservative and reactive tracer breakthrough curves and $K_{d,a}$ derived from batch experiments to determine the effective reactive surface area for the medium (either in column studies or field tests). The breakthrough curves are analyzed using the Method of Moments. The Method of

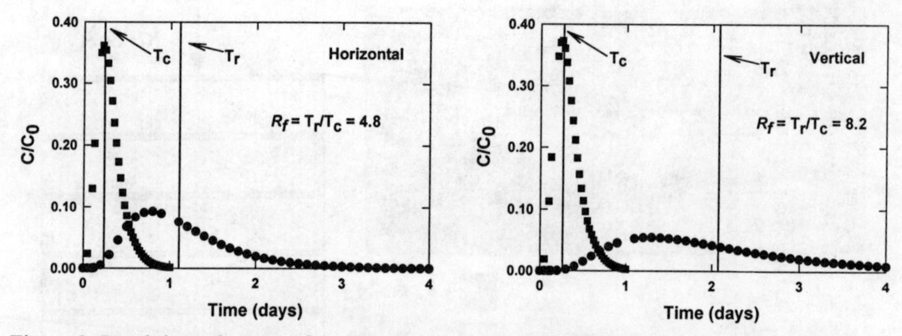

Figure 3. Breakthrough curves for a conservative (T_c) and reactive (T_r) tracer in a layered system.

Moments requires neither a priori assumptions about aquifer heterogeneity nor solving the advective-dispersion equations. Application of the Method of Moments to numerical simulations of transport in heterogeneous porous media suggests that this method can be used to accurately estimate the reactive surface area encountered by advecting tracers and that the estimated surface areas are consistent with theoretical derivations.

This research was funded by the Natural and Accelerated Bioremediation Research Program, Office of Biological and Environmental Research, U.S. Department of Energy, performed under U.S. Department of Energy Contract No. DE-AC07-99ID13727.

REFERENCES

1. E. A. Jenne, in *Symposium on Molybdenum in the Environment,* Vol. 2, edited by W. Chappell, K. Petersen (M. Dekker, Inc., New York, 1977) pp. 425-553.

2. R. W. Smith and E. A. Jenne, Environ. Sci. Technol. **25**, 525–531 (1991).

3. D. A. Dzombak and F. M. M. Morel, *Surface Complexation Modeling: Hydrous Ferric Oxide*, (Wiley Interscience, New York, 1990) p. 393.

4. J. J. Rosentreter, H. S. Quarder, R. W. Smith, and T. L. Mcling, in *Adsorption of Metals by Geomedia*, edited E. A. Jenne (Academic Press, San Diego, 1998) pp. 181–192.

5. R. W. Smith and A. L. Schafer, in *Scientific Basis for Nuclear Waste Management XXII*, edited by D. J. Wronkiewicz and J. H. Lee (Mat. Res. Soc. Symp. Proc., **556**, Pittsburgh, PA, 1999) pp. 1051-1058.

6. T. Allen, *Particle Size Measurements, Vol. 2, Surface Area and Pore Size Distribution*, 5th ed. (Chapman Hall, London, 1997) p. 251.

7. M. Jin, M. Delshad, V. Dwarakanath, D. C. McKinney, G. A. Pope, K. Sepehrnoori, C. E. Tilburg, and R. E. Jackson, Water Resour. Res. **31**, 1201–1211 (1995).

8. R. W. Smith, A. L. Schafer, and A. F. B. Tompson, in *Scientific Basis for Nuclear Waste Management XIX*, edited by W. M. Murphy and D. A. Knecht (Mat. Res. Soc. Symp. Proc., **412**, Pittsburgh, PA, 1996) pp. 693–699.

DECAY-SERIES DISEQUILIBRIUM STUDY OF IN-SITU, LONG-TERM RADIONUCLIDE TRANSPORT IN WATER-ROCK SYSTEMS

SHANGDE. LUO*, TEH-LUNG KU*, ROBERT ROBACK‡, MICHEAL MURRELL‡, TRAVIS L. MCLING¶

* Dept. of Earth Sciences, Univ. of Southern California, Los Angeles, CA 90089-0740, sluo@usc.edu

‡ MSJ 514, Los Alamos National Laboratory, Los Alamos, NM 87545

¶ Idaho National Engineering and Environmental Laboratory, Idaho Falls, ID 83415-2107

ABSTRACT

Uranium and thorium-series disequilibrium in nature permits the determination of many *in-situ* physico-chemical, geologic and hydrologic variables that control the long-term migration of radionuclides in geologic systems. It also provides site-specific, natural analog information valuable to the assessment of geologic disposal of nuclear wastes. In this study, a model that relates the decay-series radioisotope distributions among solution, sorbed and solid phases in water-rock systems to processes of water transport, sorption-desorption, dissolution-precipitation, radioactive ingrowth-decay, and α recoil is discussed and applied to a basaltic aquifer at the Idaho National Engineering and Environmental Laboratory (INEEL), Idaho.

INTRODUCTION

Performance assessment models for nuclear waste disposal generally invoke laboratory-derived distribution coefficients (K_d) for individual radioelements to simulate radionuclide migration. Questions can be raised as to the extent to which laboratory data reflect the behavior of natural geochemical systems because of K_d dependence on properties such as groundwater chemistry, aquifer mineralogy, colloidal presence, and microbial activities in the system. Natural analog studies in geological environments provide an alternative approach to model testing and validation. Application of this approach to undisturbed systems allows an in-situ assessment of long-term migration behaviors of radionuclides. Naturally occurring uranium and thorium series radionuclides are well suited for such studies because several isotopes of the same element continuously enter groundwaters and because the supply rate of many of these radionuclides can be estimated with adequate accuracy [1]. In what follows we will show that a variety of chemical, geologic and hydrologic processes that control radionuclide transport can be understood through modeling the decay-series disequilibria observed in a natural system.

MODELING THE URANIUM- AND THORIUM-SERIES DISEQUILIBRIUM

In commenting on the use of decay-series disequilibria as natural analogs for nuclide movements in groundwater system, McKinley and Alexander [2] have cited the failure of previous attempts to distinguish (1) the process of sorption from precipitation, and (2) sorbed species from those present inside mineral grains or precipitates that are impermeable to water. To address their concerns, Ku et al. [3] proposed a model that relates the steady-state distributions of decay-series nuclides in dissolved, sorbed, and solid pools of a geologic system to in-situ processes of water transport, sorption-desorption, dissolution-precipitation, radioactive ingrowth-decay, and α recoil. The model has the following mathematical expression:

$$Q + P_d + P_r + R_f^p A^p = k_p C + R_f A \qquad (1)$$

where A ($=\lambda C$; λ is the decay constant and C the concentration) is the nuclide activity in groundwater (dpm L^{-1}), with superscript p referring to its radioactive parent; k_p is the precipitation rate constant (y^{-1}); Q, P_d, and P_r are the rates (atoms L^{-1} y^{-1}) of supply by water flow, dissolution, and α recoil, respectively; and R_f is the retardation factor, expressed as:

Mat. Res. Soc. Symp. Proc. Vol. 608 © 2000 Materials Research Society

$$R_f = 1 + K = 1 + \frac{k_1}{k_2+\lambda} \quad (2)$$

where K is a dimensionless distribution coefficient between the sorbed and dissolved pools for the radionuclide that has k_1 and k_2 as its sorption and desorption rate constants (y^{-1}), respectively. The model assumes that (1) first-order kinetics govern the processes of sorption-desorption and dissolution-precipitation of radionuclides (It can be shown that dissolution is reduced to zeroth order for a constant radionuclide concentration in the solid pool) and (2) α- recoil input from the sorbed and dissolved pools to the solid pool is negligible. Implicit in assumption (1) is a linear sorption isotherm for the range of concentrations of the nuclides of interests. For R_f to be independent of nuclide concentrations, it is also assumed that decay of radionuclides on the rock surface (the sorbed pool) releases all daughter nuclides into the dissolved pool. The validity of these assumptions has been assessed by Ku et al. [3] and Murphy [4]. In this paper, the model is applied to characterize the nuclide transport in groundwaters at the Idaho National Engineering and Environmental Laboratory (INEEL), Idaho (Fig. 1).

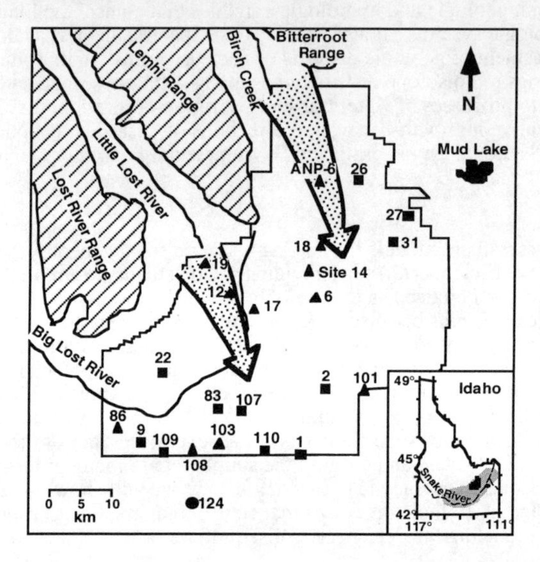

Fig. 1. Map showing the sampling locations. Groundwaters were sampled on Apr. 10, 1997 (circles), Sept. 12–17, 1997 (triangles) of 1997, and Aug. 26–Sept. 10, 1998 (squares) for measurements of U, Th, Ra and Rn isotopes [5]. The two large arrows indicate the major groundwater pathways as delineated by the modeling results. The inset shows the locations of INEEL (blackened area) and the Snake River Plain Aquifer (shaded area) in Idaho, USA. The groundwater is unconfined, contains <325 mg/L of dissolved solutes, and with typical pH of ~8.0 and Eh of ~227 mV, is mostly of the Ca-Na-bicarbonate type, relatively high in silica, and saturated in dissolved oxygen [5].

Equation (1) can be simplified by setting $Q = 0$ for isotopes of Th (^{232}Th, ^{230}Th, ^{228}Th, and ^{234}Th) and Ra (^{226}Ra, ^{228}Ra, and ^{224}Ra) because of their affinity to aquifer solids in geologic environments. Thus,

$$P_d + P_r + R_f^p A^p = k_p C + R_f A \quad (3)$$

The rate of dissolution of a nuclide from rocks (P_d) is the product of rock dissolution rate (ω, g L^{-1} y^{-1}) and the nuclide concentration (C^r, atoms g^{-1}) in rocks. The number of atoms of a decay-series radionuclide in solutions or in rocks generally decreases with increasing the nuclide's decay constant. Therefore, for short-lived radionuclides such as ^{228}Th, ^{234}Th, ^{228}Ra, ^{224}Ra and ^{222}Rn, dissolution and precipitation can be neglected, giving:

Table I. Measured decay-series isotope activities (dpm/m^3) and activity ratios in groundwater at INEEL, Idaho*

Well No.	^{238}U (×10^3)	^{232}Th	^{226}Ra	^{222}Rn (×10^3)	^{234}U/^{238}U	^{234}Th/^{238}U	^{230}Th/^{232}Th	^{228}Th/^{232}Th	^{228}Th/^{228}Ra	^{228}Ra/^{226}Ra	^{224}Ra/^{228}Ra
USGS-124	1.200	0.492	47.2	(170)	2.54	0.056	1.18	4.6	0.068	1.84	1.01
USGS-86	0.765	2.761	18.8	(860)	2.00	0.036	0.95	1.2	0.068	2.66	1.47
USGS-101	0.987	0.055	10.9	(50)	2.20	0.043	1.23	3.9	0.016	1.21	1.03
USGS-103	1.099	0.110	5.1	(117)	2.12	0.038	1.18	3.4	0.045	1.65	1.20
USGS-108	1.311	0.129	4.3	(47)	2.12	0.019	1.25	1.7	0.044	1.19	1.33
USGS-19	1.173	0.045	1.5	(346)	2.92	0.017	1.00	2.1	0.028	2.20	2.18
USGS-12	1.763	0.070	14.0	(632)	2.61	0.028	1.46	5.5	0.011	2.51	1.23
ANP-6	1.630	0.035	7.4	(1455)	2.95	0.024	1.18	14.3	0.023	2.93	4.18
USGS-17	1.288	0.686	17.4	(209)	1.89	0.015	1.01	1.5	0.031	1.94	1.14
Site 14	1.545	0.053	14.9	(112)	2.51	0.023	1.27	7.7	0.017	1.62	1.54
USGS-18	1.540	0.054	10.4	(361)	2.57	0.031	1.27	7.6	0.018	2.22	1.52
USGS-6	1.306	0.266	14.5	(114)	2.23	0.029	1.06	3.8	0.042	1.63	1.09
USGS-110	1.516	0.037	14.3	16	2.42	0.021	1.59	5.6	0.021	0.70	1.26
USGS-2	1.356	0.033	15.3	37	2.23	0.017	1.89	3.3	0.006	1.28	1.34
USGS-107	1.640	0.024	8.5	84	2.37	0.014	1.50	4.2	0.008	1.59	1.14
USGS-83	1.019	0.035	18.4	565	2.12	0.023	1.32	7.7	0.006	2.38	1.16
USGS-22	0.323	1.754	19.3	293	1.60	0.071	0.99	2.2	0.084	2.37	1.18
USGS-1	1.279	0.036	11.6	22	2.37	0.017	1.53	6.7	0.019	1.08	1.31
USGS-9	1.175	0.090	11.9	85	2.14	0.019	1.23	3.1	0.023	1.02	1.28
USGS-109	1.293	0.051	6.6	73	2.07	0.029	1.45	4.4	0.020	1.68	1.33
USGS-27	2.310	0.628	15.6	465	2.40	0.010	0.98	1.5	0.031	1.94	1.49
USGS-31	1.695	0.167	13.9	314	2.43	0.010	0.96	5.2	0.031	2.00	1.62
USGS-26	1.780	0.050	8.4	42	2.75	0.007	1.00	5.9	0.035	1.01	1.55

* Except for ^{238}U which was measured by thermal ionization mass spectrometry, all isotopes were measured by decay counting techniques [5]. The analytical errors (1-σ) derived from counting statistics were <0.5 % for ^{238}U, <5 % for ^{234}U, ^{234}Th, ^{228}Th, ^{226}Ra, ^{228}Ra, ^{224}Ra and ^{222}Rn, and about 3-10 % for ^{230}Th and ^{232}Th. Data in parentheses are interpolated values from the observed relationship between ^{228}Ra/^{226}Ra and ^{222}Rn [5].

$$P_r + R_f^p A^p = R_f A \tag{4}$$

Because $P_r \approx 0$ for β-decay products, applying eqn. (4) to ^{228}Th relates the retardation factors of ^{228}Th and ^{228}Ra to their activity ratio (A_{Ra228}/A_{Th228}) in groundwater as:

$$\frac{R_{f,Th228}}{R_{f,Ra228}} = \frac{A_{Ra228}}{A_{Th228}} \tag{5}$$

For uranium isotopes ^{238}U and ^{234}U, eqn. (1) is used to estimate Q (atoms L^{-1} y^{-1}) and the groundwater transit time (τ_w, y) is related to Q by the relationship:

$$Q = \frac{(C^i - C)}{\tau_w} \tag{6}$$

where C^i is the radionuclide concentration in recharging waters (atoms L^{-1}). Positive values of Q denote net gain due to fluid transport.

Equations (1) – (6) form the basis of our assessment of the effects of sorption-desorption, dissolution-precipitation, and advection-diffusion in the radionuclide transport in groundwater. The α-recoil input (P_r) can be determined from the groundwater ^{222}Rn activities [1]. The relationships depicted by the above equations allow us to determine retardation factors (R_f) of U, Th and Ra from measurements of short-lived Ra and Th isotopes, as well as the rock dissolution rate (ω), the precipitation rate constants (k_p) of U, Th and Ra, and the groundwater transit time (τ_w) from measurements of long-lived U, Th, and Ra isotopes.

RESULTS AND DISCUSSION

Table I shows the measurement results of isotopes of U (^{238}U, ^{234}U), Th (^{232}Th, ^{230}Th, ^{228}Th, ^{234}Th), Ra (^{226}Ra, ^{228}Ra, ^{224}Ra) , and Rn (^{222}Rn) in groundwaters collected from a basaltic aquifer at INEEL, Idaho (Fig. 1). A very large range of activities exists: from 10^4-10^6 dpm m^{-3} for ^{222}Rn, to 10^3 dpm m^{-3} for ^{238}U and ^{234}U, to 10-10^2 dpm m^{-3} for ^{226}Ra, ^{228}Ra, ^{224}Ra, and ^{234}Th, and to less than 1 dpm m^{-3} for ^{232}Th, ^{230}Th, and ^{228}Th. Radioactive disequilibria among the radionuclides occurred as a result of water-rock interaction. Applying the observed disequilibria to the model presented above, we have made estimates on rates of α recoil, sorption-desorption, and dissolution-precipitation; in-situ retardation factors of isotopes of U, Th and Ra; and water transit time in the aquifer. The major results are summarized in Table II.

Alpha-Recoil Input

The model-derived α-recoil rates show great regional variations. Given an α-recoil range of ~0.05 μm, rock density of ~2.8 g cm^{-3} and ^{238}U activity of 0.3 dpm g^{-1} in basalt at INEEL[5, 6], these rates can be translated to an effective surface area (S) of the aquifer rocks in the range $(2\text{-}1300)\times10^6$ cm^2 L^{-1}, equivalent to a fracture width of ~0.02-10 μm for a planar structure or a pore diameter of ~0.04-20 μm for a tubular configuration. Relatively large fractures occur in the southern part of INEEL.

The α-recoil rates (Table II) are significantly lower than the ^{222}Rn activities in groundwater (Table I), suggesting that the in-situ production from decay of ^{226}Ra in the dissolved and sorbed pools must contribute an important source of ^{222}Rn to groundwater. This is particularly true for those waters from the southern part of INEEL, in which the ^{222}Rn activities are low and the $P_{r,Rn222}/A_{Rn222}$ ratios are mostly close to ~0.1. The U and Th concentrations in rocks vary only slightly throughout the study area and are typical of those for basalts [6]. It is therefore the fracture size and density, not the chemical compositions, of the aquifer rocks that exert a main control on the supply rate of ^{222}Rn by α-recoil.

Retardation Factors of Ra, Th and U

Ra isotopes: In spite of their vastly different half lives, the Ra isotopes ^{226}Ra, ^{228}Ra and ^{224}Ra should have similar R_f values because rapid desorption rate of radium is inferred from the following model equation [5]:

$$P_{r,\,Ra224} + R_{f,Ra228}A_{Ra228} = R_{f,Ra224}A_{Ra224} \tag{7}$$

For most of the samples, A_{Ra224} is only slightly higher than A_{Ra228} (Table I). Equation (2) requires $R_{f,Ra228} \geq R_{f,Ra224}$. To meet these requirements, eqn. (7) shows that the recoil input of ^{224}Ra from the basaltic aquifer rocks must be small and $R_{f,Ra228} \approx R_{f,Ra224}$, i.e., the desorption rate constant (k_2) of Ra must be much greater than the decay constant of ^{224}Ra (= 0.19 day^{-1}). Thus we have: $R_{f,Ra224} = R_{f,Ra228} = R_{f,Ra226} = R_{f,Ra}$. It is also inferred that the activity of ^{224}Ra in any groundwater should not be smaller than that of ^{228}Ra . However, because of the short half life of ^{224}Ra, aging of water in well bores (e.g., if wells are not completely purged before sampling) or in large fractures may cause the ^{224}Ra /^{228}Ra ratio to become <1.

Table II. Model-derived α-recoil rate (P_r, atoms $L^{-1}min^{-1}$) of ^{222}Rn, retardation factors (R_f) and precipitation rate constants (k_p, y^{-1}) of Ra, Th, and U, rock dissolution rate (ω, mg $L^{-1}y^{-1}$), and water transit time (τ_w, y)*

Well No.	P_r×10^{-2}	R_f×10^{-4} (Ra)	R_f×10^{-6} (^{232}Th)	R_f×10^{-6} (^{228}Th)	R_f×10^{-6} (^{224}Th)	R_f×10^{-3} (U)	k_p×10^{-3} (Th)	k_p×10^{-1} (Ra)	k_p×10 (U)	ω	τ_w
USGS-124	0.02	0.36	0.62	0.14	0.006	0.34	0.04	0.22	0.29	71	39.8
USGS-86	4.56	2.14	**	0.32	0.051	1.87	0.08	3.11	3.52	736	16.2
USGS-101	0.02	0.44	1.03	0.27	0.013	0.53	0.55	0.34	0.39	98	75.3
USGS-103	0.27	1.76	1.07	0.39	0.021	0.76	0.49	1.61	0.56	176	57.5
USGS-108	0.13	0.81	0.21	0.19	0.041	0.76	0.42	1.27	0.43	176	69.4
USGS-19	2.43	6.86	**	2.43	0.051	0.67	4.95	20.57	14.48	718	0.4
USGS-12	2.16	2.97	11.50	2.73	0.123	3.27	3.32	4.51	1.20	750	15.6
ANP-6	13.02	2.07	**	0.89	0.051	0.42	3.00	8.22	2.33	344	0.6
USGS-17	0.41	0.96	0.41	0.31	0.040	0.56	0.14	0.88	0.72	310	76.9
Site 14	0.50	0.42	0.87	0.25	0.012	0.23	0.42	0.27	0.16	73	64.4
USGS-18	1.85	1.69	3.41	0.94	0.044	1.24	2.22	2.06	0.88	388	9.4
USGS-6	0.13	0.70	0.57	0.16	0.008	0.22	0.11	0.40	0.24	95	92.9
USGS-110	0.02	2.06	5.47	0.99	0.041	0.87	1.02	0.33	0.26	123	48.8
USGS-2	0.10	1.62	9.23	2.92	0.146	2.40	1.88	1.33	0.49	200	41.6
USGS-107	0.14	3.36	18.27	4.47	0.202	2.84	6.64	3.65	0.92	514	21.1
USGS-83	1.47	1.38	12.78	2.27	0.094	2.01	2.48	1.55	0.99	278	33.3
USGS-22	0.83	1.28	**	0.15	0.051	3.40	0.08	1.38	4.63	456	66.8
USGS-1	0.05	2.27	7.73	1.17	0.047	0.80	0.82	0.85	0.28	95	53.8
USGS-9	0.18	2.29	2.87	1.00	0.052	0.97	0.67	0.68	0.57	196	50.7
USGS-109	0.24	3.45	7.03	1.73	0.078	2.29	1.87	3.34	0.75	305	43.4
USGS-27	2.15	1.19	**	0.38	0.051	0.40	0.40	0.22	(-4.0)	812	(0.4)
USGS-31	1.67	1.16	0.84	0.38	0.023	0.14	0.53	1.01	0.48	285	45.1
USGS-26	0.14	2.71	4.30	0.78	0.033	0.24	1.75	0.94	0.45	282	29.6

* The model input also included 0.28 and 0.31 dpm g^{-1} for ^{238}U and ^{232}Th in the basalt, and 1.8 and 5.4 dpm L^{-1} for ^{238}U and ^{234}U in the recharging waters, respectively. The α recoil input for the Th series nuclides equals that of ^{222}Rn multiplying by 0.9. The parenthesized values at USGS-27 is anomalous due to influence of nearby agricultural activities on the groundwater U concentration [5]. The uncertainties of the estimates come largely from estimation of the α-recoil input rates which should be good to within a factor of 2 [1].

** Values are either negative or smaller than those of ^{228}Th. In these cases, the same retardation factors were assumed for ^{232}Th and ^{228}Th and the average k_1 and k_2 values, 1.08 min^{-1} and 0.58 y^{-1} respectively, were used to calculate $R_{f,Th234}$.

Estimates of $R_{f,Ra}$ for all samples average (1.6±0.9)×10^4, similar to that reported by Krishnaswami et al. [1] for a number of gravel aquifers in Connecticut. Although the two aquifers have different rock types, they are all nearly saturated in dissolved oxygen, and contain relatively low concentrations of dissolved solids. For hypersaline or more reducing waters, Ra generally has higher mobility [7]. It appears that the groundwater chemistry, not the mineralogy of the aquifer rocks, mainly controls the retardation of Ra isotopes. Our estimates may reflect the typical $R_{f,Ra}$ values for the oxygenated, low-salinity groundwaters.

We have shown that the minimum desorption rate constant of Ra is 0.19 d^{-1}. Therefore, from eqn. (2), the sorption rate constant (k_1) of Ra is estimated to have a minimum value of 2.1±1.2 min^{-1}, i.e., the sorption occurs on time scales of less than one minute, much faster than their transport by groundwater flow.

Th isotopes: The retardation factors of ^{228}Th (R_{fTh228}) were estimated using eqn. (5) to range mostly from 10^5 to 10^6, about two orders of magnitude higher than those of Ra isotopes. R_{fTh228} exhibits large spatial variations. Low R_{fTh228} values occur mainly in wells where we also found high ^{230}Th and ^{232}Th concentrations due possibly to colloidal association. The low R_{fTh228} values could also reflect such an association.

The retardation factors of Th isotopes are isotope-dependent: $^{232}Th \approx {}^{230}Th > {}^{228}Th > {}^{234}Th$ (Table II). That $R_{fTh232} > R_{fTh228}$ for most of the samples suggests that the desorption rate constant of Th must be comparable to or smaller than the decay constant of ^{228}Th (0.363 y^{-1}). From eqn. (2), we estimated the sorption and desorption rate constants (k_1 and k_2) of Th to range from 0.12 to 4.1 min^{-1} and 0.1 to 2.5 y^{-1}, respectively Whereas the sorption rates of Th and Ra are comparable, the desorption of Th is about two orders slower than that of Ra.

U isotopes: Because of the long half lives of ^{238}U and ^{234}U, their R_f values are expected to be the same and estimated to be mostly in the range of 10^2–10^3, much smaller than those for Ra and Th isotopes. However, even in this oxygenated, bicarbonate-rich groundwater of INEEL, U is moderately retarded by aquifer solids. This should serve as a cautionary note to using U isotopes as a conservative tracer for groundwater dating or mixing studies.

Since most aquifers have porosities < 0.3 and rock densities ~ 2.8 g cm^{-3}, the in-situ distribution coefficients (K_d) of radionuclides are estimated to be at least an order of magnitude smaller than the values of R_f ($\approx K$)[1], i.e., they are on the order of 10^5 for Th, 10^3 for Ra and 10^2 for U. Because R_f is site-dependent, the radionuclide migration in water-rock systems can be better understood from the in-situ K_d, rather than from the laboratory-derived K_d.

Dissolution and Precipitation

Dissolution and precipitation exhibit large spatial variations (Table II), with high rates occurring in recharge areas of the Birch Creek and Little Lost River drainages. There, the groundwater apparently is "aggressive" enough to dissolve aquifer rocks at a rate up to 800 mg L^{-1} y^{-1}, much higher than rates of <100 mg L^{-1} y^{-1} in southeastern INEEL where the groundwater is older. If the dissolution rate is assumed to be constant with time, it would take less than about one million years for the aquifer rocks to be completely weathered. However, since dissolution occurs mainly in the fractured rocks, fresh basalts of one million years old may still exist if they are not fractured. It should be noted that the above estimates represent the total-rock dissolution rates, and since >75% of the minerals in the rock are oxides of Si, Al, and Fe [6] which are mostly re-precipitated into the solid pool during dissolution, the net dissolution rates of the aquifer rock will be much smaller than the total dissolution rates as estimated above.

Precipitation occurs on the time scales of days for Th, years for Ra, and 10^2 years for U and shows large regional variations. High k_p values are generally found in areas of high dissolution rates, e.g., in recharge regions north of INEEL. High dissolution and precipitation rates, hence high weathering rates, in these areas are likely in association with the presence of abundant microfractures in aquifer rocks as suggested by the groundwater ^{222}Rn data discussed earlier.

The above estimates indicate that, for short-lived Th (^{234}Th and ^{228}Th) and Ra (^{228}Ra and ^{224}Ra) isotopes, the dissolution and precipitation rates are orders-of-magnitude lower than the rates of supply by *in-situ* production from their sorbed and dissolved parents and/or by α-recoil injection from the solid pool. This justifies us to ignore dissolution and precipitation for short-lived Th and Ra isotopes in our model. However, for long-lived U (^{234}U and ^{238}U), Th (^{230}Th and ^{232}Th), and Ra (^{226}Ra) isotopes, dissolution and precipitation have rates comparable to, or greater than, the in-situ production and/or α-recoil supply. These processes must all be taken into account in the model simulations.

Groundwater Transit Time

Estimated groundwater transit times (τ_w) range from <10 y in the recharge regions to >90 y in central INEEL, with the oldest waters occurring in the central and southeast parts of the site. For the entire Snake River Plain Aquifer, the transit time has been estimated by other methods at 200–250 y [8]. By dividing the aquifer into several compartments, Ackerman [9] showed that the travel time of regional flows in the Mud Lake compartment (which covers the INEEL area) is less than 100 y.

The spatial variation of τ_w serves to highlight the preferential flow paths in the aquifer, with flow velocity taken to be inversely related to τ_w. The results suggest two major groundwater pathways (Fig. 1). The wells ANP-6 and USGS-19 with τ_w<10 years represent the locations of recharge. The recharged water migrates from the vicinity of Birch Creek and Little Lost River via two preferential paths southward into the major aquifer. Stagnated water exists in areas southeast of the Lost River and Lemhi Ranges. A relatively young water occurs at USGS-86 in the southwestern corner of INEEL and it may originate from the Big Lost River drainage.

CONCLUSIONS

Modeling U and Th series disequilibria in groundwaters at INEEL has provided quantitative estimates for the following parameters: (1) time scales of sorption (minutes for Ra and Th), desorption (days for Ra and years for Th), and precipitation (days for Th, years for Ra, and centuries for U); (2) retardation factors due to sorption (>10^6 for ^{232}Th, ~10^4 for ^{226}Ra, and ~10^3 for ^{238}U); (3) dissolution rates of rocks (~70 to 800 mg L^{-1} y^{-1}); and (4) groundwater transit time (<10 to ~90 years). Two local north-south preferential flow pathways have been delineated from the spatial variations in groundwater transit time. It appears that groundwater chemistry and size and density of the aquifer-rock microfractures exert an important control on retardation factors and precipitation and dissolution of radionuclides in groundwater.

This study shows that the observed decay-series disequilibria in groundwater are best explained by the presence of a sorbed pool of radionuclides which are exchangeable with their dissolved counterparts through sorption/desorption. Recognizing the role of this sorbed pool cautions us on the use of batch experiments to determine distribution coefficients of radionuclides. If freshly crushed rocks were used in such experiments, it could present artificially altered surface sorption sites. In nature, such sites would normally be formed over geological time scales. The in-situ parameters derived from the decay-series disequilibria could be used in the site-dependent performance assessment models to more realistically predict the nuclear waste migrations in the far field. They may also be used for testing and validation of the performance assessment models that are based on the laboratory-determined sorption-desorption and dissolution-precipitation parameters.

Studies of radionuclide transport in geologic systems based on naturally occurring decay-series disequilibria, such as the multiple-tracer approach of the present study, have the advantage of obtaining in-situ sorption/retardation data integrated over a range of timescales. However, a detailed characterization of the systems faces the limitation of inadequate constraints on the physical, chemical, and geological processes which control the nuclide distribution among various geochemical reservoirs. For example, our treatment of the interfacial processes of dissolution-precipitation and adsorption-desorption has left room for improvement, in the sense that bulk averages rather than individual mineral surfaces are considered. The role of colloidal transport also awaits evaluation, so do the influences of water chemistry and rock mineralogy.

ACKNOWLEDGMENTS

Work sponsored by DOE Environmental Management Science Program.

REFERENCES

1. S. Krishnaswami, W. C. Graustein, K. K. Turekian, and J. F. Dowd, Water Resour. Res. **18**, 1663-1675 (1982).
2. I. G. McKinley and W. R. Alexander, J. Contaminant Hydrology **13**, 249-259 (1993); Radiochimica Acta **74**, 263-267 (1996).
3. T. L. Ku, S. Luo, B. W. Leslie, and D. E. Hammond, in *Uranium Series Disequilibrium: Applications to Earth, Marine and Environmental Sciences,* edited by M. Ivanovich and R. S. Harmon (Clarendon Press, Oxford, 1992), pp. 631-668; Radiochim. Acta **80**, 219-223 (1998).

4. W. M. Murphy, in *Sixth EC Natural Analogue Working Group Meeting, Proceedings of an international workshop held in Santa Fe, New Mexico, USA*, edited by H. von Maravic and J. Smellie (European Commission, EUR16761, 1996), pp. 233-241.
5. S. Luo, T. L. Ku, R. Roback, M. Murrell, and T. L. McLing, Geochim. Cosmochim. Acta (in press).
6. L. L. Knobel, L. Cecil, T. S. and Wood, *U. S. Geological Survey Open-File Report* 95-748, Idaho Falls, Idaho, 1995.
7. J. G. Zukin, D. E. Hammond, T. L. Ku , and W. A. Elder, Geochim. Cosmochim. Acta **51**, 1719-1931.
8. W. W. Wood and W. H. Low, Geol. Soc. Amer. Bull. **97**, 1456-1466 (1986).
9. D. J. Ackerman, *U. S. Geological Survey Water Resources Investigations Report* 94-4257, Idaho Falls, Idaho, 1995.

CHEMICAL EVOLUTION OF LEAKED HIGH-LEVEL LIQUID WASTES IN HANFORD SOILS

M. NYMAN, J. L. KRUMHANSL, P. ZHANG, H. ANDERSON, T. M. NENOFF
Sandia National Laboratories, P.O. Box 5800 MS-0710, Albuquerque, New Mexico, 87185-0710

ABSTRACT

A number of Hanford tanks have leaked high level radioactive wastes (HLW) into the surrounding unconsolidated sediments. The disequilibrium between atmospheric CO_2 or silica-rich soils and the highly caustic (pH > 13) fluids is a driving force for numerous reactions. Hazardous dissolved components such as ^{133}Cs, ^{79}Se, ^{99}Tc may be adsorbed or sequestered by alteration phases, or released in the vadose zone for further transport by surface water. Additionally, it is likely that precipitation and alteration reactions will change the soil permeability and consequently the fluid flow path in the sediments. In order to ascertain the location and mobility/immobility of the radionuclides from leaked solutions within the vadose zone, we are currently studying the chemical reactions between: 1) tank simulant solutions and Hanford soil fill minerals; and 2) tank simulant solutions and CO_2.

We are investigating soil-solution reactions at: 1) elevated temperatures (60 - 200 °C) to simulate reactions which occur immediately adjacent a radiogenically heated tank; and 2) ambient temperature (25 °C) to simulate reactions which take place further from the tanks. Our studies show that reactions at elevated temperature result in dissolution of silicate minerals and precipitation of zeolitic phases. At 25 °C, silicate dissolution is not significant except where smectite clays are involved. However, at this temperature CO_2 uptake by the solution results in precipitation of $Al(OH)_3$ (bayerite). In these studies, radionuclide analogues (Cs, Se and Re – for Tc) were partially removed from the test solutions both during high-temperature fluid-soil interactions and during room temperature bayerite precipitation. Altered soils would permanently retain a fraction of the Cs but essentially all of the Se and Re would be released once the plume was past and normal groundwater came in contact with the contaminated soil. Bayerite, however, will retain significant amounts of all three radionuclides.

INTRODUCTION

There are approximately 11 million cubic meters of liquid effluent containing radionuclides stored in large underground tanks at Hanford, Washington [1, 2]. In all, sixty seven tanks are suspected leakers and there are a great many other contaminant plumes around the sited arising from various sources; surface spills, pumping problems, early waste disposal practices, etc.[1] Near the highest temperature tanks (up to 270 °C, with many tanks having been above 100 °C for prolonged periods, [7]) the strongly basic concentrated, aluminum, phosphate and sodium-rich waste solutions[3-6] in contact with silica-rich soil minerals provide an ideal recipe for the in-situ formation of zeolites. Additionally, as cooled fluids flow away from the tanks their pH will fall as they encounter CO_2. This will initiate precipitation of hydrous Al – hydroxides. The predicted location for these reactions are shown in figure 1 (adapted from Appendix H of the GW/VZ Integration Project Specification DOE, 1998). [8]

Mat. Res. Soc. Symp. Proc. Vol. 608

Highly soluble radionuclides likely to travel in contaminant plumes include ^{79}Se, ^{99}Tc and ^{133}Cs. The upper limits of the radionuclides in the solutions are approximately 5.7 ppm ^{133}Cs, 20 ppm ^{99}Tc, and up to 70 ppm ^{79}Se.[9] The phases which form as a result of solution interactions with neutralizing media such as CO_2 or silicates must be identified in order to determine the location, speciation and phase association of these contaminants within the Vadose zone. Further, changes in soil permability as a result of alteration and precipitation processes must be assessed with regard to their impact on the flow paths of past or future leaks. These questions were addressed by initiating a suite of scoping studies intended to simulate the various changes that would occur as a leaked fluid would (DSSF-7) migrated thorugh the vadose zone toward the water table.

EXPERIMENT

Soil-Solution Reactions

Soils used in this study were obtained from the Hanford Reservation and are typical of the sandy surface cover deposited by the last great Lake Missoula flood. The fluid used in these experiments ("DSSF-7 ") was developed to be similar to that originating from rinsing operations associated with tank 101AW [3] . The major components of the DSSF-7 simulant solution, the radionuclide analogues, and the concentrations of these species are compiled in Table I. The nonradioactive analogues include Re for ^{99}Tc, and nonradioactive Cs and Se for ^{133}Cs and ^{79}Se, respectively. The soils were exposed to DSSF-7 solutions for 1 - 2 weeks at 90° and 200 °C in Parr reaction vessels and for 82 days at room temperature, 60°, and 90° C in polyethylene bottles. After completing the 90° and 200 °C experiments the liquid phase was isolated from the solid phase by filtration. Concentrations of Cs, Se and Re in both the solid phases and liquid phases were determined by ICP-Mass Spectrometry. The 82 day experimental fluids did not contain radionuclide surrogates but were used to identify solid alteration products and observe textural features reaulting from the treatment (i.e. grain cementation and zeolite morphologies).

CO_2-Solution Reactions

Two experiments were also performed to study uptake of CO_2 by the DSSF-7 test solution . First, an aliquot of DSSF-7 solution was left stirring at room temperature in contact with the air to verify that eventually even this dilute source of CO_2 could neutralize enough hydroxide to initiate precipitation of a solid phase. Second, an accelerated experiment was executed by bubbling pure CO_2 gas through DSSF-7 solution at room temperature. There was enough of this second precipitate that it could be isolated by filtration and analyzed by X-ray diffraction (phase identification) and ICP-MS (radionuclide surrogate content). The remaining DSSF-7 solution was also analyzed by ICP-Mass Spectrometry for Se, Cs and Re.

Table I. Composition of DSSF-7 Simulant Solution

Component	Concentration in solution	Component	Concentration in solution
*Cs	100 ppm	$Na_2HPO_3 \cdot 7H_2O$	0.014 mol/L
*Se	100 ppm	NaOH	3.885 mol/L
*Re	100 ppm	$Al(NO_3)_3 \cdot 9H_2O$	0.721 mol/L
$NaNO_3$	1.162 mol/L	Na_2CO_3	0.147 mol/L
KNO_3	0.196 mol/L	NaCl	0.102 mol/L
KOH	0.749 mol/L	$NaNO_2$	1.512 mol/L
Na_2SO_4	0.008 mol/L		

* Radionuclide analogue

RESULTS

Solution-Soil Reactions

Figure 2 shows the X-ray diffraction spectra of unaltered sandy Hanford soil (a), and Hanford soils exposed to DSSF-7 solution for 82 days at 25 °C (b), 60 °C (c) and 90 °C (d). The major phases of the unaltered soils are quartz, plagioclase feldspar and grains of basalt – consisting mainly of feldspar and minor amounts of pyroxene. X-ray diffraction analyses of the soil minerals show that with increasing temperature more of the sample is altered to a cancrinite-group zeolite. At 90 °C both quartz and feldspar were completely removed from the sample. Basaltic fragments, however, were relatively less effected by the treatment. The unaltered soil (consisting of quartz feldspar and basalt grains[10]) and soil altered by DSSF-7 solution at 90 °C are compared by SEM in figure 3. The unaltered soil consists of large, angular grains varying in diameter from several microns to several hundred microns. The altered soil still shows outlines of the original grains, but multiple grains are cemented together by cancrinite overgrowths. The cancrinite overgrowth consists of botryoidal masses (several microns in diameter). Soil alteration as a function of the solution: soil ratio was also studied. At 200 °C, 1 gram of soil and 10 grams of DSSF-7 solution resulted in complete alteration to cancrinite. On the other hand, 8 grams of soil and 10 grams of solution at 200 °C resulted in incomplete alteration of the minerals and quartz was still the dominant phase observed in the X-ray diffraction. In addition, a different well-formed zeolite was were observed in this sample (figure 4). This phase was tentatively identified as analcime by the crystal morphology and by several minor peaks observed in the X-ray powder diffraction pattern, .

Solution-CO_2 Reactions

The precipitate formed from bubbling CO_2 through DSSF-7 solution was identified as bayerite, $Al(OH)_3$, by X-ray diffraction. A white precipitate, presumably bayerite, was also noted in the experiment exposing DSSF-7 to air. This suggests the CO_2 in the atmosphere is present in sufficient quantity for $Al(OH)_3$ precipitation from a leaked tank fluid.

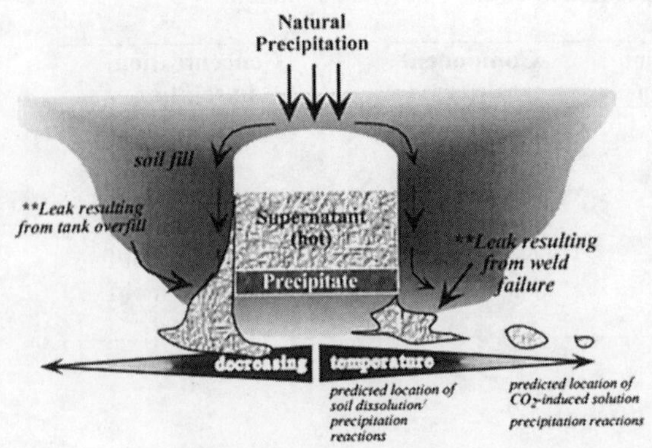

Figure 1. Schematic of a Hanford waste tank, depicting the site of the soil/solution reactions within the Vadose zone.

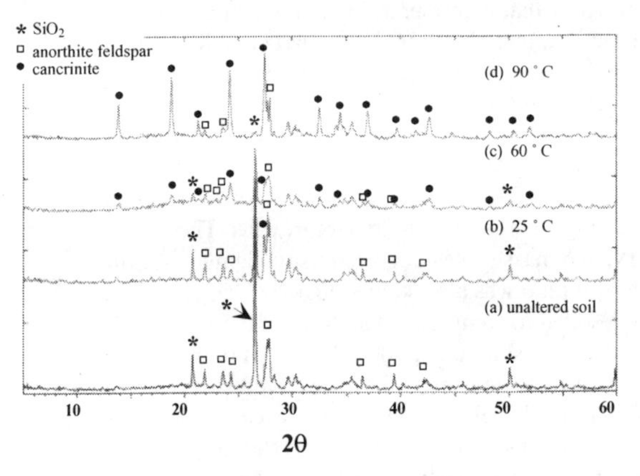

Figure 2. X-ray diffraction spectra of unaltered Hanford soil (a), and soil soaked in DSSF-7 solution for 2 weeks at 25 °C (b), 60 °C (c) and 90 °C (d).

Figure 3. Unaltered Hanford soils predominantly quartz and feldspar grains (top) Hanford soils altered by DSSF-7 solution at 90 °C; feldspar and quartz grains are overgrown and cemented with cancrinite (bottom). Magnification = 50 x.

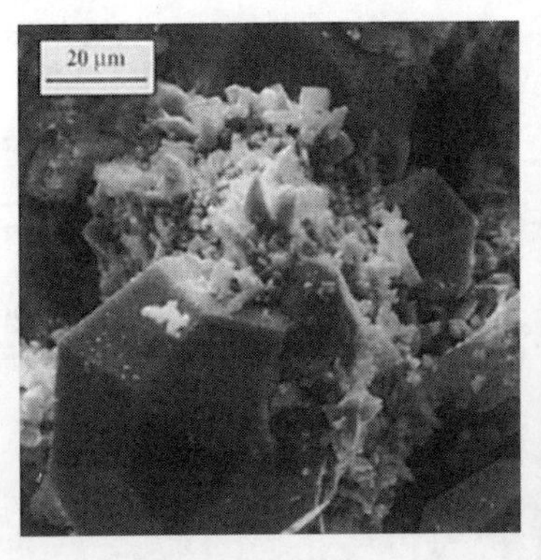

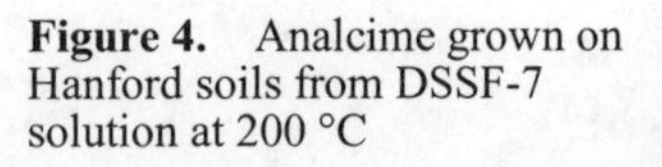

Figure 4. Analcime grown on Hanford soils from DSSF-7 solution at 200 °C

Radionuclide Analogue Sequestration

Se, Cs, and Re concentrations were monitored in experiments of DSSF-7 with both Hanford soils and CO_2 to determine the radionuclide sequestration ability of the reaction products (results compiled in Table II). Major alteration phases identified by X-ray diffraction are cancrinite and analcime (both aluminosilicate zeolites) for the soil-solution experiments, and $Al(OH)_3$ for the CO_2 precipitation experiments. The zeolites are capable of sequestering Cs^+ in micropores, and $Al(OH)_3$ may sandwich Cs^+ between octahedral layers. The Se and Re, which exist in solution primarily as SeO_4^{2-} and ReO_4^-, may coprecipitate or surface adsorb onto alteration phases.[11-14]

In general, a 10 - 20% decrease in concentration of the radionuclide analogues was observed in the reaction supernates. Higher soil : solution ratios resulted in removal of more of the radionuclide analogue from solution. There is a noticeable decrease in Cs concentration from the solution of experiment 5, as well as the relatively high concentration of Cs in the reacted solids (Table II, values in parenthesis). This is the only experiment in which analcime was identified as an alteration product. This suggests that the selectivity for Cs is greater for analcime than for cancrinite.

Further, both Re and Se were removed from the fluid during the sediment – DSSF-7 interaction studies, but were not detectable in the reacted sediments. This suggests that the process responsible for their removal could be reversed by exposing contaminated sediments to fresh groundwater once a plume had passes. It also follows that neither anion seems to have followed the more abundant nitrate, nitrite, and carbonate into the anion sites present in the cancrinite structure. Part of the Cs, however, was retained after the sample was washed which suggests it is fixed in the zeolitic alteration products. Finally, the ability of the bayerite to retain all three radionuclide is notable. Thus, such precipitates may significantly impact radionuclide mobility in the cooler, more distant, parts of the vadose zone.

Table II. Radionuclide Analogue Concentration in Experiment Supernates

Experiment Description	**PPM of radionuclide analogue in experiment supernate and in the solid reaction products ( )**		
	Se (ppm)	**Cs (ppm)**	**Re (ppm)**
1) DSSF solution initial concentration	86	96	98
2) 1:10 soil : solution 90 °C / 2 weeks	76 (>250)	74 (49)	89 (>5)
3) 1:1 soil : solution 90 °C / 2 weeks	72 (>250)	63 (38)	81 (>5)
4) 1:10 soil : solution 200 °C / 1 week	79 (>250)	87 (30)	88 (>5)
5) 1:1 soil : solution 200 °C / 1 week	71 (>250)	39 (124)	84 (>5)
6) CO_2 enhanced $Al(OH)_3$ precipitation	68 (8,390)	78 (9,000)	84 (8,020)

CONCLUSIONS

Studies have been carried out in order to: 1) model the reactions which take place between a hot alkaline tank solution and soil fill adjacent to a leaking tank at Hanford, and 2) predict the fate of ^{133}Cs, ^{99}Tc, and ^{79}Se radionuclides. Preliminary results reveal that at high temperatures the sediments will be zeolitized and at lower temperatrues precipitation of bayerite (or related hydrous Al oxides) may dominate. These studies have also shown that, at least in a batch reactor mode, the radionuclides are only partially immobilized by these changes. Better simulations involving more appropriate mass:fluid ratios and a realistic assessment of mass transport processes are necessary before these results could be applied to quantitatively predicting radionculide distributions in an actual field setting.

Secondly, the overgrowth textures, cementation of grains, and volumenous precipitation of hydrous Al-hydroxide gels all suggest that chemical reactions within a migrating plume will almost certainly alter fluid flow paths adjacent to a leaking tank. This result must be considered when developing vadose zone hydrologic models to predict the transport of radionuclides away from a leaking tank.

ACKNOWLEDGEMENTS:

This work was supported by the U.S. DOE under contract DE-AC04-94AL85000.

REFERENCES:

[1] E. K. Wilson, *Chemical and Engineering News Sept. 29, 1997* (1997) 30.

[2] D. L. Illman, *Chemical and Engineering News June 21, 1993* (1993) .

[3] S. F. Agnew, J. G. Watkin, *TWRS Characterization Program, report # LA-UR-3590*, Los Alamos National Laboratory, Los Alamos 1994.

[4] S. F. Agnew, *report # LA-UR-3860*, Pacific Northwest National Lab, Richland, WA 1996.

[5] R. J. Serne, M. I. Wood, *report # PNNL-7279, DE90013546*, Pacific Northwest National Labs, Richland, WA 1990.

[6] R. L. Weiss, *report # WHC-SD-WM-ER-309*, Pacific Northwest National Lab, Richland, WA 1988.

[7] B. D. Flanagan, *report # WHC-SD-WM-TI-591*, Pacific Northwest National Lab, Richland 1994.

[8] National Research Council, DOE, *Committee on Subsurface Contamination at DOE Complex Sites: Research Needs and Opportunities* 1998.

[9] E. M. J. Kupfer, *report # HNF-SD-WM-TI-740*, Pacific Northwest National Lab, Richland, WA 1997.

[10] R. J. Serne, J. M. Zachara, D. S. Burke, *report # PNNL-11495 UC-510*, Pacific Northwest National Laboratory, Richland 1998.

[11] L. B. Sand, F. A. Mumpton, *Natural Zeolites; Occurrence, Properties, Uses*, Pergamon Press, London 1978.

[12] A. J. Ellis, W. A. J. Mahon, *Chemistry and Geothermal Systems*, Academic Press 1977.

[13] R. M. Barrer, *Hydrothermal Chemistry of Zeolites*, Academic Press, London 1982.

[14] J. Breck, *Zeolite Molecular Sieves*, Wiley and Sons, New York 1974.

TECHNETIUM-99 CHEMISTRY IN REDUCING GROUNDWATERS: IMPLICATIONS FOR THE PERFORMANCE OF A PROPOSED HIGH-LEVEL NUCLEAR WASTE REPOSITORY AT YUCCA MOUNTAIN, NEVADA

Roberto T. Pabalan, David R. Turner, and Michael P. Miklas, Jr.
Center for Nuclear Waste Regulatory Analyses, Southwest Research Institute
6220 Culebra Road, San Antonio, TX 78238-5166 (rpabalan@swri.edu; dturner@swri.edu)

ABSTRACT

Performance assessment calculations by the U.S. Department of Energy and the Nuclear Regulatory Commission indicate that Tc-99 is a major contributor to dose to a hypothetical receptor group 20 km downgradient of a proposed high-level nuclear waste repository at Yucca Mountain, Nevada, within the first 10,000 yr after permanent closure. This result is due in large part to the high solubility and low retardation of Tc under oxidizing conditions in the Yucca Mountain environment. Recent site characterization data on the chemistry of saturated zone groundwater at Yucca Mountain and vicinity indicate the presence of locally reducing geochemical conditions, which could decrease the solubility and enhance the sorption and retardation of Tc-99. In this study, a preliminary assessment of the potential effects of reducing conditions on the transport and release of Tc-99 was conducted. Sensitivity analyses using the NRC/CNWRA Total-system Performance Assessment code (TPA Version 3.2) indicate that decreased Tc solubility and increased Tc sorption due to reduction of Tc(7+) to Tc(4+) can significantly delay the arrival of Tc-99 at the receptor group location. Decreased Tc solubility can decrease the Tc-99 dose by three orders of magnitude relative to the TPA 3.2 base case. Enhanced Tc retardation in the tuff aquifer only does not greatly decrease the calculated Tc-99 peak dose, whereas increased Tc retardation in the alluvial aquifer alone prevents Tc-99 from reaching the receptor group in 50,000 yr. The release and transport of other redox-sensitive radioelements could be affected in a manner similar to Tc. Thus, reduced groundwater conditions could significantly enhance the performance of the geologic barrier system and reduce the dose to the receptor group.

INTRODUCTION

Technetium-99 is a radionuclide of concern in safety assessments of high-level nuclear waste (HLW) repositories because of its long half-life (2.13×10^5 yr) and its relative abundance in HLW (fission yield ~ 6 percent). A redox-sensitive element, Tc exists in a heptavalent (7+) state under nonreducing conditions as the pertechnetate (TcO_4^-) anion (Fig. 1). The TcO_4^- anion does not sorb significantly onto mineral surfaces, and compounds of Tc(7+) generally have high solubility so that the concentration of Tc in groundwater is not expected to be limited to low values by solubility constraints. Conversely, under reducing conditions Tc is present in a tetravalent (4+) state. The solubility of Tc(4+) solids, such as $TcO_2 \bullet 1.6H_2O$ (Fig. 1), is much lower compared to Tc(7+) solids, and Tc(4+) aqueous species sorb more strongly onto mineral surfaces compared to Tc(7+) species. For example, sorption experiments by Lieser and Bauscher[1] using sediments, predominantly quartz, and groundwaters from Gorleben, Germany, indicate that at pH = 7 and Eh > 0.17 V, in the stability field of TcO_4^-, Tc K_d is low and relatively constant at ~ 0.2 mL/g (Fig. 2). At lower Eh, where the Tc(4+) species $TcO(OH)_2(aq)$ is the dominant aqueous form, K_d is high, with a value of ~ 1,000 mL/g. Although Lieser and Bauscher[1] indicated that precipitation of low solubility Tc(4+) solids may have contributed to Tc removal in their experiments, the effect of low Eh on sorption is significant.

As part of the United States HLW program, the Department of Energy (DOE) and the Nuclear Regulatory Commission (NRC) conducted total-system performance assessment (PA) simulations of the proposed geologic repository at Yucca Mountain, Nevada (YM).[2,3,4] The DOE and the NRC use PA to systematically analyze what can happen at the repository after permanent closure, and to calculate as a function of time the total dose affecting a hypothetical human receptor group located some distance downgradient from YM in the event of failure of waste packages (WPs) to isolate the radionuclides. The

Mat. Res. Soc. Symp. Proc. Vol. 608

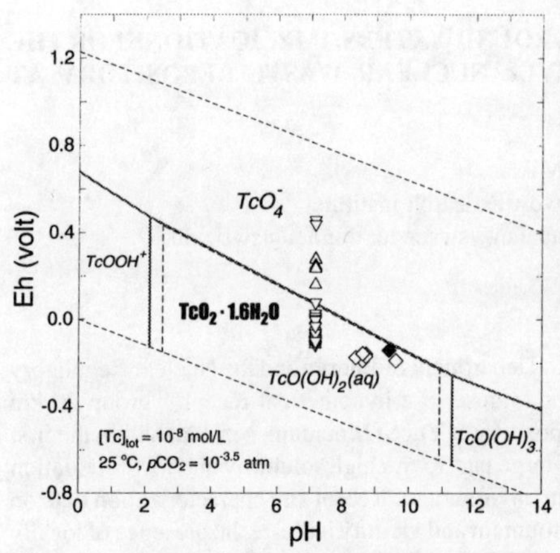

Figure 1. Eh-pH diagram for aqueous Tc species calculated using the NEA thermodynamic data.[9] Also shown are Eh measured in well UE-25 WT#17 (△,▽)[5], and on samples from wells USW H-3 (◆) and USW H-4 (◇).[6]

NRC PA activity, part of an ongoing process at the NRC to prepare for the review of a potential DOE license application for a HLW repository at YM, is designed to provide information on the potentially important isolation characteristics and capabilities of the proposed repository system based on the available information regarding the geologic setting and the natural and engineered barrier systems. An important aspect of the NRC PA is quantifying the sensitivity of the results to, and the uncertainty associated with, the input parameters.

In the DOE and NRC PA analyses, oxidizing conditions were assumed over the entire transport path from YM through both the unsaturated and saturated zones. Under these conditions, there is no solubility limiting phase for Tc-99, and the PA source term models for Tc assumed a high solubility limit such that Tc release is essentially constrained by the dissolution rate of the waste form. After mobilization from the waste form, the dominant aqueous species is assumed to be TcO_4^-, and Tc sorption or retardation, modeled using a K_d or a retardation coefficient (R_f), is typically assumed to be either very low[3,4] or zero[2] over the entire transport path length. Because of the assumed high solubility and low retardation, both the DOE and the NRC PA results indicate that Tc-99 is a dominant contributor to the dose rate during the first 10,000 yr after closure of the repository and remains as a significant contributor to the total dose at longer times.[2,3]

Some site characterization data suggest that localized reducing conditions may exist in the groundwaters at Yucca Mountain. These data include downhole measurements of Eh in well UE-25 WT#17 (Fig. 3),[5] Eh measurements on a water sample from well USW H-3 and on "thief" samples from well USW H-4,[6] and indirect evidence in the form of reduced sulfide minerals observed in cuttings from NC-EWDP-1D,[7] a well located in the alluvial zone. In this study, a preliminary assessment of the potential effects of reducing conditions on Tc-99 transport and release and on the performance of the proposed repository system was conducted using the NRC/CNWRA PA code (TPA Version 3.2).

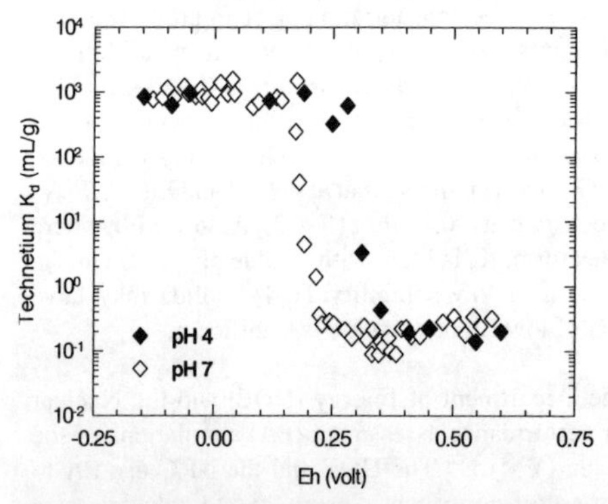

Figure 2. Technetium sorption coefficient, K_d, as a function of Eh. Values taken from Lieser and Bauscher[1]

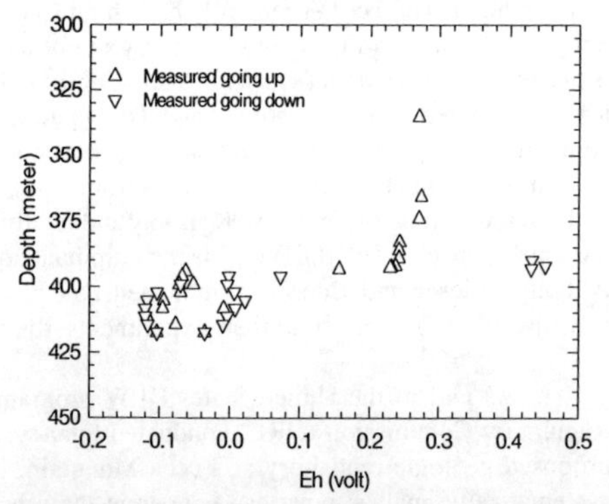

Figure 3. Eh measured downhole in well UE-25 WT#17. Data taken from DOE (1999; data tracking number LAAM831311AQ98.003)[5]

TECHNICAL APPROACH

The NRC has selected a base case (defined as a particular set of models and model parameters describing the YM repository system) to perform sensitivity and uncertainty analyses and to study the general behavior of the system as simulated by the TPA Version 3.2 code.[4,8] Because oxidizing conditions were assumed throughout the geologic barrier system at YM, previous NRC PA calculations assumed zero retardation of Tc in the fractures and matrix of the volcanic tuff in the saturated zone, and a log uniform probability distribution function, with minimum and maximum values of 1 and 30, respectively, was assumed for Tc R_f in the alluvium.[8] In addition, the solubility limit of Tc in the NRC PA calculations was set to a high value (1 mol/L). To investigate the effects of saturated zone reducing conditions on Tc-99 transport and repository performance, the Tc R_f parameters for the saturated zone and the Tc solubility limit were modified from the TPA Version 3.2 base case (oxidizing) values.

Downhole measurements of Eh and pH in well UE-25 WT#17[5] indicate an average pH of 7.14 ± 0.09 and Eh values less than 0.17 V below a depth of about 390 m (Fig. 3). In the absence of data on Tc sorption on Yucca Mountain rocks under reducing conditions, the Tc K_d of 1,000 mL/g derived by Lieser and Bauscher[1] for pH 7 and low Eh was used. For Tc-99 sorption on the matrix of volcanic tuff in the saturated zone, this K_d corresponds to an R_f of 2.6×10^4, based on porosity (ϕ) and bulk density (ρ) base case values of 1 percent and 2.65 g/cm^3, respectively.[8] Zero retardation ($R_f = 1$) of Tc in the fractures of volcanic tuff was maintained in the calculations due to uncertainty about the type and extent of fracture coatings. In TPA Version 3.2, transport through alluvium is modeled using a porous media approach, and there is no component of fracture flow and transport. A Tc R_f value of 1.7×10^5, based on a K_d of 1,000 mL/g, ϕ of 12.5 percent, and ρ of 2.47 g/cm^3, for transport through the alluvium was used in the sensitivity analyses. The TPA Version 3.2 does not have the capability to handle localized groundwater conditions, although one can specify different transport properties for the tuff and the alluvial aquifers. In this study, higher retardation factors are used (i) over the entire length of the saturated zone flow path from the repository to the receptor group location 20 km away, (ii) for the tuff aquifer only, or (iii) for the alluvial aquifer only.

The Tc solubility limit used in the sensitivity analyses, calculated using the NEA Tc thermodynamic data[9] and the chemistry of groundwater reported for UE-25 WT#17[5,10] (data tracking numbers LAAM831311AQ98-001, -002, and -003) and summarized in Table 1, is 4×10^{-9} mol/L. The solubility-limiting phase was assumed to be $TcO_2 \cdot 1.6H_2O$. Using an anhydrous phase, TcO_2, results in a much lower calculated solubility (4×10^{-13} mol/L), but $TcO_2 \cdot 1.6H_2O$ is more relevant to the formation of Tc oxides in reducing groundwaters.[11]

Table 1. Cation and anion concentrations in water samples taken from well UE-25 WT#17[5,10]

Component	Concentration (mg/L)
Sodium	21.1 ± 1.4
Silicon	20.0 ± 3.7
Calcium	11.2 ± 1.8
Potassium	3.24 ± 1.24
Magnesium	1.13 ± 0.38
Iron	0.78 ± 0.78
Manganese	0.41 ± 0.13
Sulfate	15.2 ± 7.0
Chloride	9.20 ± 4.30
Nitrate	3.80 ± 2.17
Fluoride	3.34 ± 0.56
Bromide	1.11 ± 1.22
Phosphate	<0.1
pH	7.14 ± 0.09

It is important to note that radionuclide release begins only after WP failure, thus WP lifetime significantly affects repository performance and the Tc dose to the receptor group. The WP, based on the DOE Viability Assessment design,[2] is robust. It has an outer corrosion allowance material composed of carbon steel and an inner material of Alloy C-22. The latter is highly resistant to localized corrosion, with a passive dissolution rate independent of environmental factors, such as pH and chloride concentration, and only slightly dependent on temperature.[12] Thus, for the base case, WP failure by corrosion does not occur until greater than 18,000 yr on average. To investigate the effect of container life on the sensitivity analysis results, additional TPA runs were done using the properties of Alloy 625, a less corrosion resistant material, instead of Alloy C-22 in the input file to induce earlier WP failure and focus on the performance of the geologic barrier.

Only dose from Tc-99 is discussed here, but the potential effects of reducing groundwater conditions on dose from other redox-sensitive radioelements (Pu, Np, U) may be evaluated using the same approach. All TPA runs were conducted for 50,000 yr using 250 realizations for each run. All parameters, except those for Tc and the WP material discussed previously, were maintained at the base case values listed in appendix A of the TPA Version 3.2 user manual.[8] The base case data set has 838 parameters, with 592 defined as constants and 246 sampled parameters specified with probability distribution functions.

RESULTS

The sensitivity analyses used peak dose as the measure of repository performance to eliminate the time dependency of the performance measure and simplify the interpretation. The results in Figs. 4 and 5 are presented in terms complementary cumulative distribution functions (CCDFs), which are estimates of the probability that a given consequence (in this case, peak dose) will be exceeded. For the TPA Version 3.2 base case with an Alloy C-22 corrosion resistant material, using solubility and retardation parameters for Tc based on oxidizing conditions results in peak doses to the receptor group from Tc-99 at 50,000 yr (closed circles in Fig. 4) that are less than the 25 mrem/yr peak mean total effective dose equivalent (TEDE) limit proposed by the NRC in Draft 10 Code of Federal Regulations Part 63. Using a lower solubility limit of 4×10^{-9} mol/L to simulate the potential effect of reducing conditions on Tc mobilization (but with low Tc retardation), the calculated dose from Tc-99 was decreased by about three orders of magnitude (open circles in Fig. 4). Using enhanced Tc retardation in the tuff aquifer to reflect stronger sorption of Tc(4+), but maintaining a high (oxidizing condition) solubility limit, the calculated Tc-99 dose (open diamonds in Fig. 4) decreased by less than an order of magnitude compared to the base case. On the other hand, with enhanced Tc retardation in the alluvial aquifer only or along the full length of the saturated zone, Tc-99 did not reach the receptor group in 50,000 yr, even with a high solubility limit.

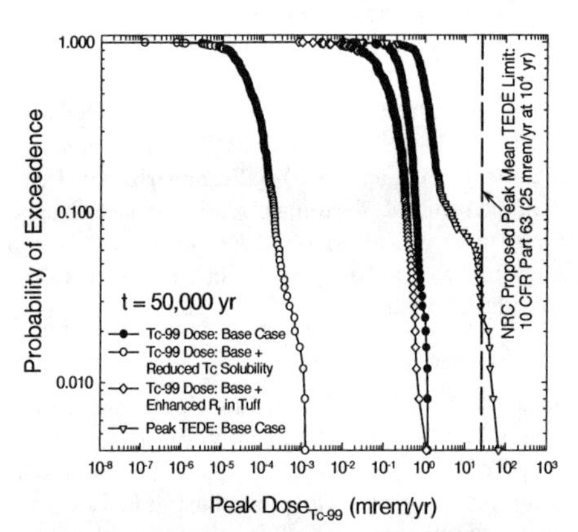

Figure 4. CCDF for Tc-99 dose to a receptor group 20 km from the repository calculated using the base case parameter set,[8] the base case parameters with decreased Tc solubility limit, or the base case parameters with enhanced Tc retardation in the tuff aquifer. Calculations with enhanced retardation along the full length of the saturated zone or in the alluvial aquifer only resulted in zero Tc-99 dose and are not plotted. Shown for reference are the peak total dose from all radionuclides and the proposed NRC 10 CFR Part 63 TEDE limit at 10,000 yr.

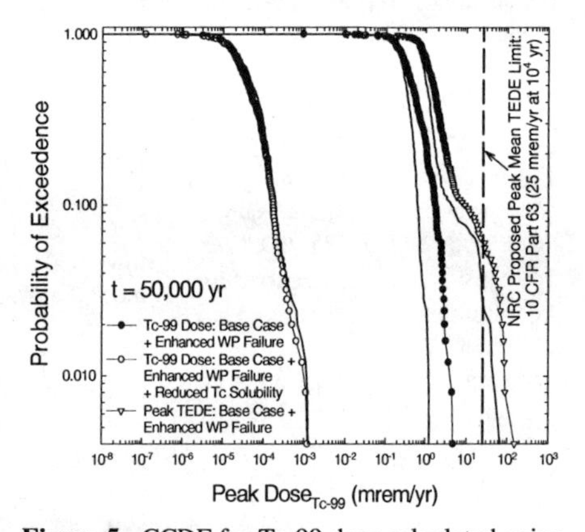

Figure 5. CCDF for Tc-99 dose calculated using Alloy 625, instead of Alloy C-22, properties to simulate earlier WP failure, and the base case parameter set[8] or the base case parameters with decreased Tc solubility limit. Calculations with enhanced retardation along the full length of the saturated zone, which resulted in zero Tc-99 dose, are not plotted. The heavy lines are the CCDFs for Tc dose and TEDE for the base case with Alloy C-22 taken from Fig. 4.

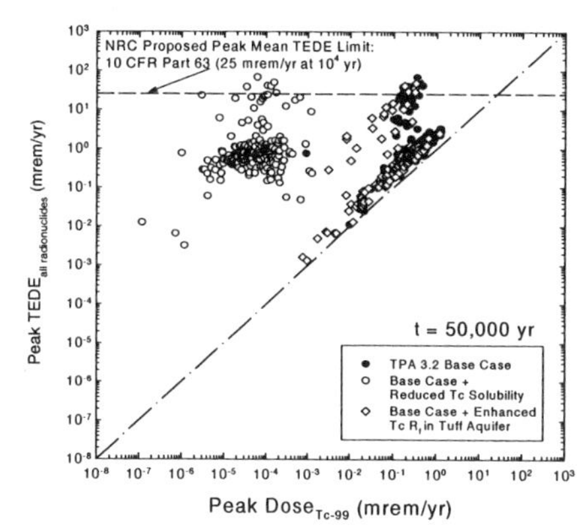

Figure 6. Calculated peak TEDE and Tc-99 dose. Values for each of 250 realizations are plotted. Symbols as in Fig. 4. Calculations with enhanced retardation along the full length of the saturated zone or in the alluvial aquifer only resulted in zero Tc-99 dose and are not plotted. The dashed-dotted line represents a 1:1 relationship between log (Peak TEDE) and log ($Dose_{Tc-99}$). Points along the line indicate that the peak dose is all due to Tc-99.

The resulting CCDFs for the cases in which the effect of decreased container life on Tc release was evaluated are shown in Fig. 5. The simplified approach of using Alloy 625 properties, instead of Alloy C-22, for the corrosion resistant material in the WP reduced the average time to WP failure to about 10,200 yr. The CCDFs indicate that, under oxidizing conditions (high Tc solubility and low Tc retardation), the Tc-99 peak dose to the receptor group at 50,000 yr is slightly higher with Alloy 625 than with Alloy C-22 due to the earlier WP failure. Lowering the Tc solubility limit decreased the Tc-99 dose by more than three orders of magnitude relative to the oxidizing condition case, although the results for Alloy 625 and Alloy C-22 are about the same. Enhancing Tc retardation in the saturated zone delayed Tc transport sufficiently such that no Tc-99 dose was received by the receptor group at 50,000 yr, even with the enhanced canister failure.

The calculated values of Tc-99 peak dose and peak TEDE are plotted in Fig. 6. Under TPA Version 3.2 base case conditions, the doses from Tc-99 calculated for 250 realizations lie close to the diagonal line, i.e., Tc-99 makes up a significant proportion of the peak TEDE at 50,000 yr. Other radionuclides, predominantly I-129, also contribute to the peak TEDE at 50,000 yr. A reduction in Tc solubility lowers the calculated dose from Tc-99 and also minimizes the contribution of Tc-99 to the peak TEDE (indicated by the greater vertical distance of the open circles from the diagonal line). Enhanced Tc retardation in the tuff aquifer alone does not greatly decrease the calculated Tc-99 peak dose, whereas increased Tc retardation in the alluvial aquifer only or along the full length of the saturated zone prevents Tc-99 from reaching the receptor group in 50,000 yr.

CONCLUSIONS

A preliminary assessment of the potential effect of reducing groundwater on the performance of a proposed HLW repository at YM was conducted using the TPA Version 3.2 code. Perhaps not surprisingly, the sensitivity analyses indicate that decreased Tc solubility and increased Tc sorption due to reduction of Tc(7+) to Tc(4+) can significantly delay the arrival of Tc-99 at the receptor group location. In general, the reduced form of redox-sensitive radioelements, such as Np, Pu, and U, are less soluble and sorb more strongly than oxidized species.[13] The release and transport of other redox-sensitive radioelements will be affected in a manner similar to Tc. Thus, reduced groundwater conditions could significantly enhance the performance of the geologic barrier system and reduce the calculated dose to the receptor group.

In performance assessment calculations for a multiple barrier geologic repository, dose to the receptor group is extremely dependent on the robustness of the engineered barrier. If it can be demonstrated with certainty that container life exceeds a significant proportion of the compliance period, then using conservative assumptions to model radionuclide release and transport through the geologic system may be sufficient to provide confidence in the total system performance of the repository. If, however, uncertainty remains in the long-term performance of the engineered materials, then additional realism in PA simulations of the potential isolation capability of the geologic barrier at YM is necessary to demonstrate whether the proposed repository can meet its performance objectives.

Under the current structure of TPA Version 3.2, it is not possible to specify localized reducing zones. Thus, the dependence of Tc release and repository performance on the relative extent of reduced conditions was not studied in detail. In these preliminary sensitivity analyses, conditions sufficient to enhance Tc sorption were assumed for the tuff aquifer only, for the alluvial aquifer only, or for the entire saturated zone transport path. The results indicate that, under the assumptions used in the TPA 3.2 calculations, the tuff aquifer plays a relatively minor role in retarding Tc-99 transport.

Only little information is available on the redox state of groundwaters in the YM area. Given that retardation in the tuff aquifer does not significantly affect the calculated dose to the receptor group, any characterization of the extent of reducing conditions in the saturated zone at YM should focus on the alluvial aquifer. Also, a revision of the NRC/CNWRA TPA code to permit simulation of localized effects is necessary to more realistically evaluate the effect of hydrochemistry on the solubility and transport of redox-sensitive radionuclides.

ACKNOWLEDGMENTS

The reviews by G. Wittmeyer, B. Sagar, and R. Smith are greatly appreciated. This paper was prepared to document work performed by the CNWRA for the NRC Office of Nuclear Material Safety and Safeguards under contract No. NRC-02-97-009. This paper is an independent product of the CNWRA and does not necessarily reflect the views or regulatory position of the NRC.

REFERENCES

1. K.H. Lieser and C. H. Bauscher, Radiochimica Acta **42**, 205 (1987); **44/45**, 125 (1988).
2. U.S. Department of Energy, *Viability Assessment of a Repository at Yucca Mountain*, DOE/RW-0508 (Office of Civilian Radioactive Waste Management, Las Vegas NV, 1998).
3. U.S. Nuclear Regulatory Commission, *NRC Sensitivity and Uncertainty Analyses for a Proposed HLW Repository at Yucca Mountain, Nevada, Using TPA 3.1. Results and Conclusions*, NUREG-1668, Vol. 2 (Nuclear Regulatory Commission, Washington, D.C., 1999).
4. S. Mohanty, R. Codell, R.W. Rice, et al., *System-Level Repository Sensitivity Analyses Using TPA Version 3.2 Code*, CNWRA 99-002 (Center for Nuclear Waste Regulatory Analyses, San Antonio, TX, 1999).
5. U.S. Department of Energy, *Downhole Eh and pH Measurements for UE-25 WT#17*, Available at http://m-oext.ymp.gov/html/prod/db_tdp/atdt/internet/TDIF307214.html (1999).
6. A.E. Ogard and J.F. Kerrisk, *Groundwater Chemistry Along Flow Paths Between a Proposed Repository Site and the Accessible Environment*, LA-10188-MS (Los Alamos Laboratory, Los Alamos, NM, 1984).
7. Nye County. *Nye County Nuclear Waste Repository Project Office: Early Warning Drilling Program.* Available at http://www.nyecounty.com/ewdpmain.htm (1999).
8. S. Mohanty and T. McCartin, *Total-system Performance Assessment (TPA) Version 3.2 Code: Module Descriptions and User's Guide* (Center for Nuclear Waste Regulatory Analyses, San Antonio, TX (1998).
9. Sandino, M. and E. Osthols (editors), *Chemical Thermodynamics of Technetium* (North-Holland, Amsterdam, in press).
10. U.S. Department of Energy, *Yucca Mountain Site Characterization Project Technical Data Management Site and Engineering Properties*, Available at http://m-oext.ymp.gov/html/prod/db_tdp/sep/internet/ default.htm (1999).
11. R.E. Meyer, W.D. Arnold, F.I. Case, and G.D. O'Kelley, Radiochimica Acta **55**, 11 (1991).
12. D.S. Dunn, Y.-M. Pan, and G.A. Gragnolino, *Effects of Environmental Factors on the Aqueous Corrosion of High-Level Radioactive Waste Containers—Experimental Results and Models*, CNWRA 99-004 (Center for Nuclear Waste Regulatory Analyses, San Antonio, TX, 1999).
13. D. Langmuir, *Aqueous Environmental Geochemistry* (Prentice Hall, Upper Saddle River, NJ, 1997).

THERMODYNAMIC INTERPRETATIONS OF CHEMICAL ANALYSES OF UNSATURATED ZONE WATER FROM YUCCA MOUNTAIN, NEVADA

LAUREN BROWNING[1], WILLIAM M. MURPHY[1], BRET W. LESLIE[2], WILLIAM L. DAM[2]
[1]Center for Nuclear Waste Regulatory Analyses, Southwest Research Institute, 6220 Culebra Rd., San Antonio, Texas 78238 USA. [2]U.S. Nuclear Regulatory Commission, Washington, DC 20555

ABSTRACT

Analytical pore water compositions from Yucca Mountain were evaluated for internal thermodynamic consistency. Significant ionic charge imbalances, unequilibrated aqueous speciation relationships, and erratic variability with depth in some species concentrations were found. Thermodynamic consistency was restored by introducing measured CO_2 gas pressure as a constraint, imposing equilibrium aqueous speciation, and adjusting pH to achieve charge balance. Reinterpreted water chemistry data were used to evaluate and interpret vertical and lateral variations in water chemistry, differences between unsaturated zone pore and perched water compositions, and water-rock equilibria.

INTRODUCTION

Understanding unsaturated zone (UZ) groundwater chemistry is necessary in predicting the long-term performance of Yucca Mountain as a possible high level nuclear waste disposal facility. The composition and evolution of UZ waters may affect corrosion of engineered barriers, waste form alteration, radionuclide release, retardation of radionuclide transport, dissolution/precipitation of minerals, and changes in porosity and permeability. Predictions of waste package failure times and radionuclide release rates are particularly sensitive to the evolution of carbonate ion concentration and pH [1]. Consequently, the quantity and chemistry of groundwater contacting the engineered barriers constitute a key element of subsystem abstraction for performance assessment. Ironically, it is difficult to obtain reliable groundwater compositions from unsaturated zone environments that may otherwise be desirable for isolation of nuclear waste.

Yang et al. [2,3] measured chemical compositions of pore water and perched water from Yucca Mountain. Perched waters were sampled from boreholes using plastic bailers and pore waters were extracted from borehole core samples using high-pressure uniaxial compression techniques. Although the accuracy of the resultant pore water chemistry data was unavoidably compromised by air drilling of core samples, pore water evaporation, and compression techniques [2,3], they provide a valuable characterization of groundwater chemistry at Yucca Mountain.

This paper describes equilibrium speciation calculations performed to characterize uncertainties in analytical pore water data [2,3]. We restore thermodynamic consistency to the data given a specific set of assumptions. Revised data were used to evaluate and interpret vertical and lateral variations in UZ groundwater chemistry in terms of water-rock equilibria and to compare unsaturated zone, saturated zone, and perched water compositions.

BACKGROUND: UNSATURATED ZONE WATER CHEMISTRY ANALYSES

Yang et al. [2,3] extracted pore waters from core samples from boreholes USW UZ-14 and UE-25 UZ#16, SD-7, SD-9, and SD-12. Stratigraphic units penetrated by the boreholes are (in descending order): the Paintbrush Group (comprising the Tiva Canyon Tuff, Yucca Mountain Tuff, Pah Canyon Tuff, and Topopah Spring Tuff), the Calico Hills Formation, and the Prow Pass Tuff. Perched water compositions from boreholes USW UZ-14, and SD-7, and extracted pore water compositions, were measured using ICP spectroscopy and ion chromatography.

Although the pore water analyses of Yang et al. [2,3] provide a valuable characterization of groundwater chemistry at Yucca Mountain, there are indications that aspects of the data are unreliable. Yang et al. [2,3] noted charge imbalances in the chemical analyses. In addition, Apps [4] concluded that measured pH values are inaccurate, based on inconsistencies of pH measurements of water from the J-13 well [5]. To correct pH, Apps [4] performed aqueous speciation calculations with EQ3 version7.2b [6] using analytical pore water chemistry from borehole USW UZ-14 to calculate pH assuming equilibrium with calcite. If these measurements and assumptions are correct, then the calculations of Apps [4] would restore charge balance

Mat. Res. Soc. Symp. Proc. Vol. 608

without assigning an adjustable charge balancing species; however, the results showed an increase in charge imbalances from the values indicated by the analytical data. The range of analytical pH for pore waters extracted from similar depths within individual boreholes appears unreasonably wide. For example, in analyses from borehole UE-25 UZ#16, pH varies from 7.0 to 8.6 across a depth interval of less than 10 meters in the Calico Hills Formation. Erratic fluctuations in pH provide additional support for assertions that measured pH values are unreliable [4]. Similar abrupt variations are present in some reported major aqueous species concentrations. For example, analytical SiO_2 (aq) concentrations vary by a factor of more than 2.5 for samples extracted from approximately the same depth in borehole UE-25 UZ#16. Potassium occurs in primary and secondary phases at Yucca Mountain and is an important component of Yucca Mountain waters, but Yang et al. [2,3] did not always report K^+ concentrations. Finally, particulate aluminum in filtered samples resulted in unreliable Al concentrations [2]. Clearly, uncertainty in pore water analyses of Yang et al. [2,3] requires further evaluation.

Gas samples collected from boreholes in the unsaturated zone at Yucca Mountain since 1983 [2,7] provide additional insight about the groundwater environment. After an initial stabilization period following borehole drilling, measured gas phase concentrations at any given depth remained remarkably constant with time. Measured CO_2(g) pressures are about four times atmospheric throughout most of the Topopah Spring Tuff, and are significantly higher in the Tiva Canyon and lower Topopah Spring Tuff. Biological activity is probably responsible for the elevated near-surface CO_2(g) pressures [7], and the higher CO_2(g) pressures near the bottom probes may be the result of drilling-fluid contamination [2].

REEVALUATION AND REVISION OF WATER CHEMISTRY ANALYSES

We performed aqueous speciation calculations using the code EQ3 version 7.2b [6] demonstrating that analytical pH and associated CO_3^{2-}/HCO_3^- ratios are inconsistent with chemical equilibrium. Figure 1 shows that analytical values for log [CO_3^{2-}/HCO_3] (molal) vs. pH do not lie on the line defined by mass action equations for the pH dependent speciation of C-bearing solutions. Moreover, many pore water analyses with high reported pH contain significant HCO_3^-, but CO_3^{2-} data are not reported. Only low pressure extractions from borehole UE25 UZ#16 were used for pH analysis, but pH values were reported for water extracted over a larger range of extraction pressures for boreholes SD-7, SD-9, SD-12, NRG-6, and NRG-7a [3]. Both techniques result in unequilibrated aqueous speciation relations. Similar problems in carbonate speciation and pH appear with perched water data of Yang et al. [2], but to a smaller degree.

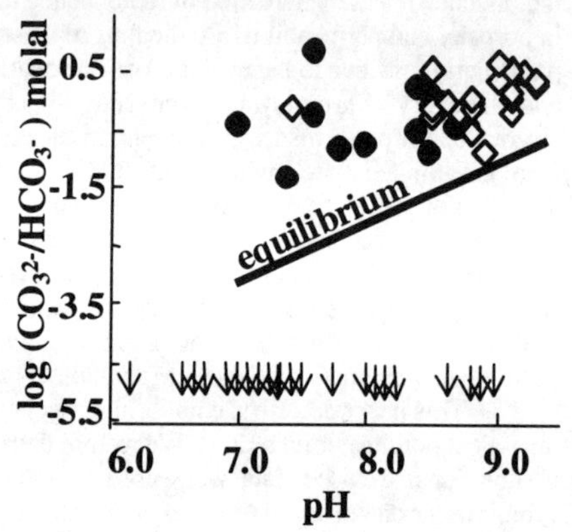

Figure 1. Analytical pore water compositions from boreholes UE-25 UZ#16 [2] (circles) and SD-7, SD-9, SD-12, NRG-6, and NRG-7a [3] (diamonds) are inconsistent with equilibrium aqueous speciation relations for the reaction $H^+ + CO_3^{2-} \Leftrightarrow HCO_3^-$. Equilibrium line calculated for 25°C. Arrows indicate all data with no reported CO_3^{2-}.

Downhole gas measurements at Yucca Mountain indicate that CO_2(g) pressures in the UZ are higher than atmospheric [2], suggesting that CO_2(g) loss during the extraction and/or chemical processes may have contributed to the thermodynamic inconsistencies demonstrated in Figure 1. However, CO_2(g) loss cannot explain either the apparent absence of CO_3^{2-} in HCO_3^--rich alkaline solutions or the lack of ionic charge balance described in [2,3], because rapid equilibration of aqueous carbonate speciation reactions is expected. A loss of CO_2(g) would shift pH to higher values along the equilibrium line. One possible explanation for data lying above the equilibrium line is that CO_2(g) loss occurred after the pH measurement, but before measurement of aqueous carbonate species concentrations. In this case, analytical carbonate species concentrations would reflect equilibration at a higher pH than was originally measured, increasing the

CO_3^{2-}/HCO_3^- concentration.

To restore thermodynamic consistency to the pore water analyses, equilibrium aqueous speciation calculations were performed by introducing measured CO_2 gas pressure as a constraint and imposing electroneutrality by adjustment of pH. A $CO_2(g)$ concentration of 0.12 percent (1.02e-3 bar) was adopted in this study to constrain carbonate chemistry equilibria. This value is consistent with measured $CO_2(g)$ concentrations in the Topopah Spring Formation from borehole USW UZ-1 [2] and a total pressure of 0.85 bar. Where K^+ data do not exist, its concentration was fixed at 14 mg/L, based on average UZ4-TP analyses from depths of 91 to 96 m [8]. Sensitivity studies showed that pH values calculated based on charge balance were insensitive to reasonable variations in estimated K^+ concentration. Temperatures used in these calculations (20°-34°C) were estimated from the geothermal gradient [9].

The main assumptions of this approach are that $CO_2(g)$ pressures were constant throughout the sampled intervals and that the major analytical ionic concentrations (except for C-bearing species) are reliable. We judge that measured gaseous CO_2 pressures more accurately constrain carbonate chemistry equilibria in unsaturated groundwaters than analytical aqueous HCO_3^-, CO_3^{2-}, or pH values. Both aqueous carbonate species and pH are difficult to measure for pore waters from unsaturated rocks and are demonstrated to be thermodynamically inconsistent. It is reasonable to expect that equilibration was achieved between gaseous CO_2 and carbonate species in pore waters extracted from high porosity zones of unsaturated nonwelded tuffs. The apparent disequilibrium between measured $CO_2(g)$ pressures from fractures and HCO_3^- pore water concentrations noted by Yang et al. [2] may be explained by inaccuracies in analytical carbonate species concentrations.

Perched water analyses were interpreted differently because these data appear more reliable than pore water analyses - likely because standard extraction procedures yield sufficient volumes of perched water for analyses. However, charge imbalances and deviations from equilibrium in the HCO_3^--CO_3^{2-}-H^+ system were corrected by minor adjustment of HCO_3^- concentrations.

RESULTS AND DISCUSSION

Revised Pore Water Compositions

Revised water chemistry differs from analytical data by maintaining theoretical thermodynamic equilibrium in terms of aqueous speciation and electrical neutrality and having different total H^+ and total carbonate concentrations. Revised solution compositions from the Calico Hills Formation exhibit a slightly larger total carbonate concentration than the analytical pore water. However, the general trend in pore water composition from calcium sulfate or calcium chloride types in the Paintbrush Group to sodium bicarbonate types in the Calico Hills Formation reported by Yang et al. [2,3] is preserved (Figure 2). Figure 3 compares the pH distribution of the analytical data, the reinterpretation of [4], and this work for borehole UE-25 UZ#16. The method used by Apps [4] shifts the average pH upward to about 8.2, but does not tighten the range in reported pH. The method used by Apps [4] also does not reduce erratic variability of pH over small vertical distances. The approach introduced in this work reduces variability of pH and shifts the average pH to approximately 8.2. If natural

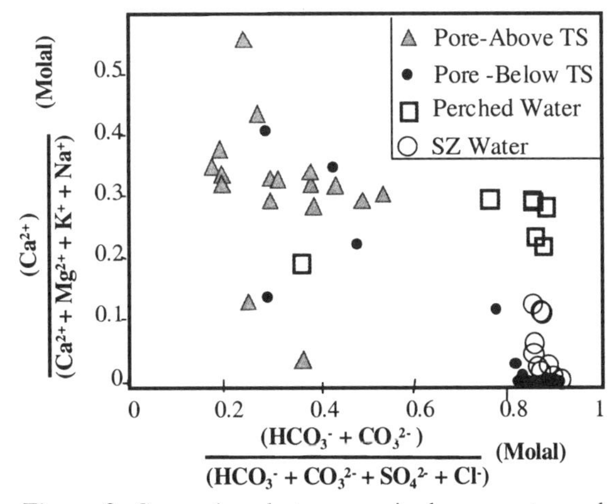

Figure 2. Comparison between revised pore water and perched water compositions from borehole USW UZ-14 [2] and saturated zone water compositions from Yucca Mountain [10]. Pore water compositions below and above the Topopah Spring (TS) have different symbols.

pH varies smoothly with depth, then these results support the assumption of constant $CO_2(g)$ used in our calculations and the presence of drilling-fluid contaminants in the lower gas probes [2].

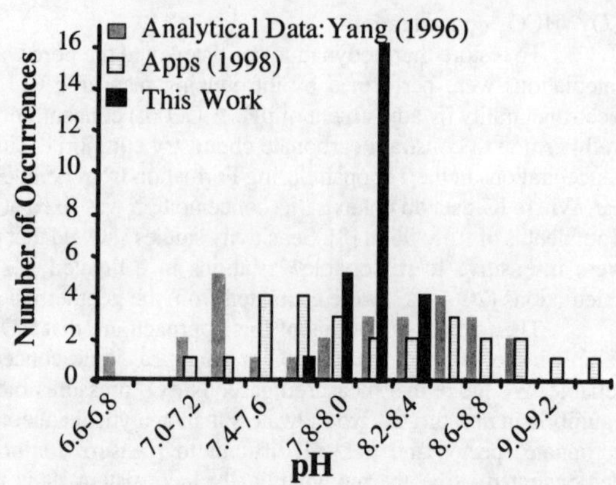

Figure 3. Distribution of pH in borehole UE-25 UZ#16.

Differences between Saturated, Perched, and Unsaturated Water Compositions

Kerrisk [10] and Murphy [11] concluded that saturated zone tuffaceous aquifer groundwaters are generally undersaturated with respect to calcite. Murphy [11] bolstered this observation using reported data for isotopic compositions of calcites and groundwaters and the absence of occurrence of calcite in rocks from groundwater producing zones in boreholes. In general, saturated zone groundwaters from the vicinity of Yucca Mountain have calculated equilibrium CO_2 fugacities (10^{-2}-10^{-3} bar) in excess of atmospheric values and in excess of values measured in the gas phase from the unsaturated zone at Yucca Mountain [10]. Measurements of pH of saturated zone groundwaters are subject to uncertainties because of the potential for degassing of CO_2 during groundwater collection, and other analytical errors, and a variety of reported pH values for these groundwaters is unsurprising. Nevertheless, interpretation of data taken from Kerrisk [10] for saturated zone waters in the vicinity of Yucca Mountain demonstrates a consistent pattern of elevated CO_2 pressure, minimal charge imbalance (when speciated at groundwater collection temperatures), and undersaturation with respect to calcite [11]. Reinterpreted perched water compositions exhibit these same characteristics.

Figure 2 compares the compositions of groundwaters from the saturated zone in the vicinity of Yucca Mountain reported by Kerrisk [10] with perched and pore water data for borehole USW UZ-14. Below the Topopah Spring Tuff, the concentrations of SO_4^{2-},Ca^{2+}, and Mg^{2+}, which are probably derived predominantly from the ground surface, are significantly diminished. The compositions of perched waters, deep pore waters, and saturated zone waters differ from near-surface pore waters by low relative proportions of Ca^{2+} and high proportions of total carbonate relative to other anions. Despite possible differences in their flow pathways and residence times, these three groundwater types have evolved to similar compositions. Pore waters are generally more concentrated than perched waters, and pore water concentrations increase with depth [2]. The Cl^- concentrations in perched waters and saturated zone waters are lower than in the adjacent pore waters, which led Yang et al. [2] to conclude that perched waters experienced less water-rock interaction and therefore must be supplied by fast fracture flow. Although perched waters may be supplied predominantly by fracture flow, Cl^- is unlikely to be affected strongly by water-rock interactions. Elevated pore water concentrations might be an artifact of the compression techniques used to extract the waters from unsaturated zone core for chemical analysis. Compression techniques were not used to collect either perched or saturated zone waters. Evaporation would cause unsaturated zone pore waters to become more concentrated, particularly in nonvolatile anions, but would have a lesser effect on adjacent perched waters replenished by fast pathway fracture flow.

Lateral and Vertical Variations in Mineral Saturation States

Thermodynamically interpreted groundwater compositions were used to infer mineralogical controls in the UZ at Yucca Mountain. Average calcite saturation states (log Q/K) are close to 0 for pore waters collected from individual boreholes - consistent with the assumption of Apps [4]. For example, log Q/K values for calcite average 0.133 in borehole UE-25 UZ#16 and –0.132 in borehole USW UZ-14. These averages, however, mask scatter in log Q/K values between +/- 0.5. Calcite is not ubiquitously present in geologic formations throughout the UZ at Yucca Mountain [12], and it is possible that pore waters undersaturated in calcite were extracted from rocks lacking calcite. Supersaturation with respect to calcite is more difficult to explain. No systematic variations occur between the saturation state of calcite and either depth or the analytical charge imbalances [2,3] which might belie our assumption of constant CO_2(g) pressures. Dissolved SiO_2 concentrations in most samples remain close to cristobalite saturation.

To explore potential relationships between vertical and lateral variations in groundwater chemistry and clinoptilolite compositions at Yucca Mountain, we estimated clinoptilolite compositions assuming equilibrium with extracted pore waters from boreholes USW UZ-14 and UE-25 UZ#16. Clinoptilolite at Yucca Mountain exhibits cation variations mainly among the major components K, Na, and Ca [13]. Unfortunately, most pore water analyses of Yang et al. [2,3] provide no information on K^+. We therefore focused on clinoptilolite-water exchange reactions among Na and Ca. Free ion concentrations of Ca^{2+} and Na^+ were used to calculate the Na to Ca ratio in clinoptilolite according to equilibrium for the reaction $Na_2Al_2Si_{10}O_{24}{:}8H_2O + Ca^{2+} <=> CaAl_2Si_{10}O_{24}{:}8H_2O + 2\ Na^+$.

Two approaches were used to calculate Na-Ca exchange in clinoptilolite. First, estimated thermodynamic data for endmember Ca- and Na-clinoptilolite [14] were used to calculate log K of the exchange reaction. These data were found reasonable in simulations of the natural geochemical system at Yucca Mountain [15]. Despite significant variations in the Ca^{2+}/Na^+ ratio of pore waters with depth, corresponding clinoptilolite compositions estimated in this manner remained close to the Ca-endmember composition. This result is inconsistent with observed clinoptilolite compositions [13], suggesting inaccuracies in the estimated thermodynamic data. An alternate approach, using the standard Gibbs free energy of ion exchange derived from experimental data at 25°C [16], produced estimated clinoptilolite compositions that more closely match observations.

Figure 4 shows estimated clinoptilolite compositions from boreholes UE-25 UZ#16 and USW UZ-14 superimposed on diagrams of log $(mCa^{2+}/m(Na^+)^2)$ versus depth. These boreholes are located on the eastern and western sides of Yucca Mountain, respectively. Clinoptilolite compositions were calculated assuming unit aqueous activity coefficients, ideal binary Ca-Na_2 (Gapon model) solid solution, and the standard Gibbs free energy of Ca-Na ion exchange consistent with data from [16]. Pore water compositions from both boreholes show a sharp decrease in $mCa^{2+}/m(Na^+)^2$ free ion ratios down to the base of the Calico Hills Formation. This trend indicates that Ca^{2+} is supplied to the pore waters from a near-surface environment and is progressively depleted in concentration with depth due to the precipitation of Ca-bearing minerals [2,15]. Estimated clinoptilolite compositions vary sympathetically, becoming increasingly sodic with depth in the Calico Hills Formation. Estimated Calico Hills clinoptilolite compositions from the eastern and western boreholes are consistent with "representative clinoptilolite compositions" in the Calico Hills Formation from the eastern and western sides of Yucca Mountain [13]. However,

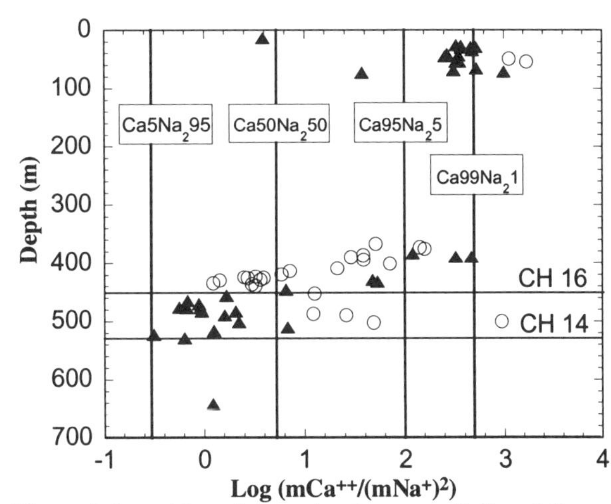

Figure 4. Logarithm of the ratio of aqueous Ca^{2+} molality to the square of aqueous Na^+ molality plotted versus depth from the ground surface for pore waters from boreholes USW UZ-14 (triangles) and UE-25 UZ#16 (circles). Vertical lines represent clinoptilolite solid solution compositions in mole percent binary Ca-Na_2 endmember components at equilibrium with solution composition ratios on the abscissa. Horizontal lines mark approximate depths of the base of the Calico Hills Formation (CH) in the two boreholes.

systematic variations in pore water composition with depth imply that clinoptilolite compositions vary significantly within individual lithologic units. This argues against the applicability of "representative" clinoptilolite compositions [13]. Our calculations show a return to more Ca-rich clinoptilolite compositions below the Calico Hills Formation on the eastern side of Yucca Mountain (UE-25 UZ#16), as observed by Broxton et al. [13]. However, Broxton et al. [13] also report a general trend in western boreholes toward more calcic clinoptilolite compositions with depth beginning at the Calico Hills Formation, which is not evident in Figure 4 (USW UZ-14). Estimated clinoptilolite compositions vary both laterally and vertically in ways that are difficult to generalize in large-scale trends.

CONCLUSIONS

Reported pore water compositions at Yucca Mountain [2,3] contain charge imbalances, unequilibrated H^+-HCO_3^--CO_3^{2-} relationships, and erratic variability in species concentrations with depth. No systematic trends were identified in the inaccuracies, suggesting that extraction, compression, and/or chemical analysis procedures change solution compositions in ways that are incompletely understood. Internal thermodynamic consistency was restored to the data by fixing the CO_2 gas pressure at measured values, imposing electroneutrality by adjustment of pH, introducing a reasonable K^+ concentration, and recalculating equilibrium aqueous speciation relations at inferred in-situ temperatures. This approach reduced erratic variability in pH values with depth and shifted pH values to approximately 8.2. Revised pore water compositions provide useful information about the UZ mineralogy at Yucca Mountain. Significant vertical and lateral variations in solution composition, even within individual lithologic units, reflect complex changes in mineral composition that could, if ignored, affect the reliability of process-level and performance assessment models. Browning et al. [17] show, for example, that mineral and infiltrating fluid compositions have an important effect on the results of reactive transport simulations of gas-water-rock interactions at Yucca Mountain under both ambient steady state and thermally transient conditions. Sensitivity of repository performance to these chemical variations, both in the ambient setting and under thermally perturbed conditions, should be addressed.

ACKNOWLEDGEMENTS

This paper was prepared to document work performed in part by the Center for Nuclear Waste Regulatory Analyses (CNWRA) for the Nuclear Regulatory Commission (NRC) under Contract No. NRC-02-97-009. This study was performed on behalf of the NRC Office of Nuclear Material Safety and Safeguards, Division of Waste Management. The paper is an independent product and does not necessarily reflect the views or regulatory position of the NRC.

REFERENCES

1. W. J. Gray, H. R. Leider, and S. A. Steward, PNL-10540m, 1995.
2. I. C. Yang, G. W. Rattray, and P. Yu, U.S.G.S. WRIR 96-4058, 1996.
3. I. C. Yang, P. Yu, G. W. Rattray, J. S. Ferarese, and R. N. Ryan, U.S.G.S. WRIR 98-4132, 1998.
4. J. A. Apps LBNL-40376, UC-814, 1997.
5. J. E. Harrar, J. F. Carley, W. F. Isherwood, and E. Raber, UCID-21867, NNA.1910131.0274, 1990.
6. T. J. Wolery, UCRL-MA-110662- Pt.3, 1992.
7. D. C. Thorstenson, E. P. Weeks, H. H. Haas, and J. C. Woodward *in Nuclear Waste Isolation in the Unsaturated Zone* (Focus '89 Proc. Las Vegas, NV, 1990) pp. 256-270.
8. I. C. Yang, A. K. Turner, T. M. Sayre, and M. Parviz, U.S.G.S. WRIR 88-4189, 1988.
9. Sass and Lachenbruch (1982) USGS-OFR-82-973.
10. J. F. Kerrisk, LA-10929-MS, 1987.
11. W. M. Murphy, in *Scientific Basis for Nuclear Waste Management XVIII*, T. Murakami and R. C. Ewing (Ed.). Mater. Res. Soc. Symp. Proc. 353, pp. 419-426, 1995.
12. D. L. Bish and S. J. Chipera, LA-11497-MS, 1989
13. D. E. Broxton, D. L. Bish, and R. G. Warren, Clays and Clay Min. 35 (2), 89 (1987).
14. J. F. Kerrisk, LA-10560-MS, 1983.
15. W. M. Murphy (Focus '93 Proc. Am. Nuc. Soc., La Grange Park, IL, 1994), p. 115-121.
16. R. T. Pabalan and F. P. Bertetti, J. Soln. Chem. 28 (4), 367 (1999).
17. L. Browning, W. M. Murphy, and D. Hughson, EOS Sup. 80 (17), Spring AGU, 1999.

RESEARCH INTO EFFECTS OF REPOSITORY HETEROGENEITY

[#]M.M. Askarieh, [*]A.V. Chambers, and [*]L.J. Gould
[*]AEA Technology, 424 Harwell, Didcot, Oxon OX11 0RA, UK.
[#]United Kingdom Nirex Limited, Curie Avenue, Harwell, Didcot, Oxon OX11 0RH, UK.

ABSTRACT

Calculations of the consequences of the geological disposal of solid radioactive waste require models that apply a simplified description of the evolution of a repository. In work carried out to date for United Kingdom Nirex Limited (Nirex), models of repository performance have assumed that the contents of the repository are homogeneous. There is no direct representation in the models, for example, of the fact that all the waste will be contained initially within waste packages, with backfill surrounding the packages.

It is important that the assumption of a homogeneous repository is demonstrated to be cautious (i.e. the calculations of radiological consequence, based on a homogeneous model, should err in the direction of slightly overestimating any release of radionuclides). This paper introduces models that simulate the impact of specific heterogeneities on radionuclide release. These have been constructed and applied, in order to assess the appropriateness of the assumption of homogeneity.

Calculations using the NAMMU computer program have examined the release of radionuclides from within waste packaging. These calculations suggest that recent performance assessments overestimate the release of short-lived radionuclides. Calculations, including the use of the INHOMOG computer program, have examined the relative placement of different categories of waste packages within repository vaults (for the Nirex disposal concept). It is concluded from the calculations that, for a few radionuclides, consideration of the distribution of some waste packages may be appropriate in future performance assessment studies.

THE SEPARATION OF WASTE AND BACKFILL

The assumption of homogeneity in the performance assessment calculations undertaken to date introduces an important bias. For example, it excludes from consideration the role that the waste packaging may have in restricting the release of radionuclides from the wastes to the surrounding backfill (see Reference [1] for a description of the Nirex disposal concept and overview of near-field evolution).

The waste containers (mainly stainless steel [2]) may provide an important barrier to radionuclide migration in the short term, although their physical structure may not play an important role in repository performance in the long term. This is because, for example, the containers may be perforated, either as a consequence of slow corrosion processes, or through the provision of vents to allow gases generated by the degradation of some of the wastes [3] to escape. These perforations may allow the release of dissolved radionuclides once the waste packages have resaturated.

Waste encapsulation grouts (cementitious materials used to immobilise ILW within the waste containers and similar in composition to the materials discussed in [4]) may also provide an important physical barrier to radionuclide release from the waste packages. Pristine waste encapsulation grouts have a low permeability and diffusivity [4], relative to that of the backfill, and may limit radionuclide release. In particular, groundwater flow may be largely excluded from low permeability waste packages, and instead may be channelled through the higher permeability backfill. However, it is unlikely that the waste encapsulation grout will remain in pristine condition in the long term. For example, corrosion of metal wastes may lead to a volume expansion, and cracking of the grout may ensue (as well as possible rupturing of the waste containers), increasing the effective permeability of the contents of the waste packages.

Mat. Res. Soc. Symp. Proc. Vol. 608 © 2000 Materials Research Society

NAMMU Calculations

Calculations can be performed to examine the role of waste packaging as a barrier to radionuclide release. A model has been constructed to represent a stack of seven identical cylindrical waste packages (i.e. drums) within a repository vault. The arrangement of drums in this waste stack is based on the specification for the disposal of 500 litre drums in stillages within the repository vaults as for the Nirex 97 assessment [2]. The model was two-dimensional, with all waste stacks assumed identical. Angular variations in the properties of the system relative to an axis running vertically through the centre of the waste drums were ignored. It was also assumed that the direction of groundwater flow in the vicinity of the waste drums was vertically upwards based on the results of hydrogeological modelling as reported in Reference [2]. A schematic of the model is provided in Figure 1(a). The model also considered the case that the containers had corroded, such that it is possible for groundwater flow to have access to the drum contents.

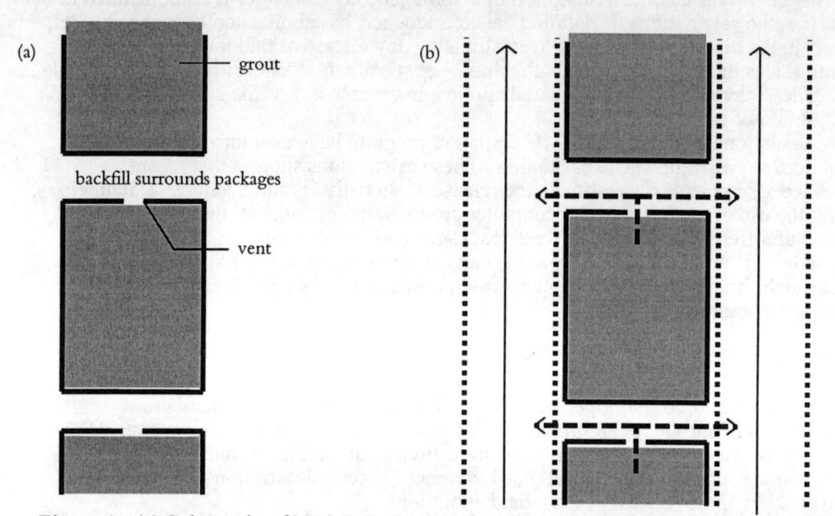

Figure 1. (a) Schematic of NAMMU model of part of waste stack (intact containers). (b) Schematic of radionuclide migration and water flow simulated using NAMMU. Broken arrows indicate path of possible radionuclide diffusion. Solid arrows represent direction and position of significant groundwater flow. Dots represent boundary of significant flow.

Simulations of radionuclide migration for this model were conducted using the NAMMU computer program [5]. NAMMU (Numerical Assessment Method for Migration Underground) is a software package for modelling groundwater flow and transport processes in porous media. It applies a finite-element method to solve the governing equations for radionuclide migration. In the present context it was applied to calculate the flux of a radionuclide from the top of a waste stack. An appropriate groundwater volume flux through the repository was selected based on the Nirex 97 assessment [2]. Also:

- for simplicity the radionuclide was taken to have a sufficiently long half-life that radioactive decay need not be considered in the calculations. It was also assumed to be non-sorbing and that the aqueous concentration was unaffected by solubility limitation;
- the concentration of the radionuclide in the system was expressed as a fraction of the initial concentration within the waste packages (hence calculated fluxes were in units of yr^{-1} rather than moles yr^{-1}). All of the waste packages were assigned the same initial inventory;

- the backfill and in-drum regions were assigned appropriate transport properties [4] but these were modified in a simple fashion to take account of the possibility of cracking. Table I provides a summary of these data;
- changes in the properties of the radionuclide, containers, backfill and waste encapsulant (e.g. through continued exposure to the groundwater) were not simulated;
- conditions within each drum were assumed to be homogeneous initially. These calculations are therefore only applicable to intimately grouted wastes [2].

Table I. Effective permeability and intrinsic diffusivity for backfill and in-drum regions applied in calculations of radionuclide release from waste packaging.

Material	Permeability / m^2	Intrinsic Diffusivity / m^2s^{-1}
Backfill region	10^{-15}	1.4×10^{-10}
In-drum region	10^{-18}	10^{-12}

Three cases were considered: (1) The waste drum material was assumed to remain completely intact. Each of the drums had a circular vent of radius 7.5 cm in its top surface, which was modelled as an open hole. (2) Identical to Case 1 except that each of the drums had a vent hole of radius 2.4 cm in its top surface (the area of the vent holes was a factor of 10 smaller than in the first case). This was done to examine in a simple fashion the impact of possible changes in future vent design, specifically the size of the vent, on the rate of release of radionuclides from the waste stack. (3) The waste containers were absent. Only the waste encapsulation material provided a barrier to release. This was done to examine the relative importance of the waste containers and the waste encapsulation material as barriers to radionuclide release.

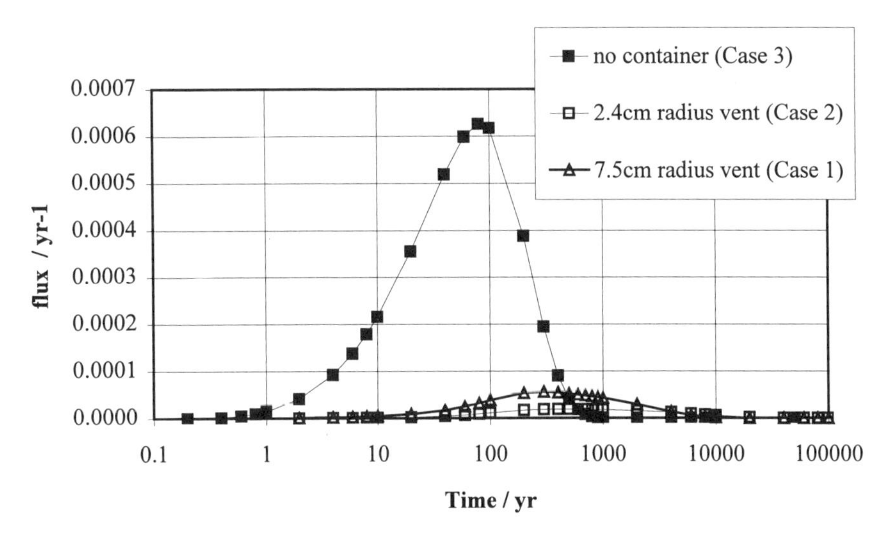

Figure 2. Radionuclide flux from top of the waste stack simulated by NAMMU

The results of the calculations are shown in Figure 2. Due to the permeability contrast between the backfill and the grout material, for Case 3 most of the groundwater flow was predicted to pass through the backfill region, rather than the waste (see Figure 1(b)). For Cases 1 and 2 the transport of the radionuclide from the waste into the backfill was predicted to be dominated by diffusion through the vents (Figure 1(b)). Once in the backfill region, the radionuclide was transported by advection and diffusion in the backfill to the top of the stack. As the cross-sectional area of the vent was reduced, the calculated peak flux became smaller and occurred later, suggesting that vent design has a small influence on the performance of the waste package as a physical barrier.

As sorption processes were not considered, the calculations are representative of the release of weakly-sorbing radionuclides such as ^{36}Cl and ^{129}I [6]. It is indicated in Figure 2 that much of the inventory of the radionuclides would be released from the waste packaging on a timescale that is short compared to the half-lives of ^{129}I (1.57×10^7 years) and ^{36}Cl (3.01×10^5 years). For short-lived radionuclides such as ^{137}Cs (30.14 years) and ^{90}Sr (28.6 years), analysis of Figure 2 (Cases 1 and 2) suggests that much of the inventory would have decayed before release. (Cases 1 and 2 are likely to be realistic at relatively early times after closure). For radionuclides that can sorb strongly onto the grout, or with concentrations within the waste packages that may be solubility-limited, retention within the waste packaging is likely to be longer than implied by these calculations.

There is a distinction to be made between the period of time over which the waste packages provide a barrier to radionuclide release and the period of time over which the waste containers remain intact. The waste containers may remain intact for timescales in excess of 10^4 years [1], but for weakly-sorbing radionuclides, the NAMMU calculations imply an effective physical barrier of duration less than 500 years. This is because of the importance of the vent described in the NAMMU calculations as a route to radionuclide release. The period of time over which the waste packages provide a barrier to radionuclide release depends on the the radionuclide considered. For strongly-sorbing radionuclides, or radionuclides whose aqueous concentrations within the drums are solubility limited, the containers may corrode away before release through the vents can occur. In such instances, the influence of the near-field porewater, and the properties of the encapsulated waste, on the corrosion of the containers as a determinant of corrosion lifetime, is important.

RELATIVE PLACEMENT OF WASTE PACKAGES WITHIN THE REPOSITORY

It is expected that heterogeneities in radionuclide distribution will occur over small length scales, in particular between waste packages in a waste stack. At present, Nirex has not specified that any wastes will be subject to controlled emplacement. That is, the placement of waste packages within a repository vault will be determined only by the order in which they are delivered to a repository vault. Furthermore, there would be heterogeneities in both the vertical and horizontal direction within a repository vault. This may have consequences for the release of radionuclides from a repository and for calculations of risk. This is because the distance between the downstream boundary between a repository and the host rock, and that of packages with a high inventory of radionuclides, will differ, depending on the arrangement of the waste packages.

The computer program INHOMOG has been developed to calculate the release of radionuclides from repository vaults as a function of the heterogeneous distribution of wastes [7]. In short, INHOMOG considers the impact on radionuclide release from the near field, or a section thereof, from a non-uniform distribution of the waste inventory. It takes into account the spatial distribution (in one dimension) of waste disposal units (e.g. waste packages, vault sections or an entire repository vault) with different chemical and physical properties. It calculates the flux and the integrated flux for key radionuclides arising as a result of this non-uniform distribution, and simulates the consequences of the interaction of waste disposal units within the distribution. The calculations assume that the disposal units are connected sequentially by groundwater transport processes, with advection, diffusion and dispersion through a one-dimensional porous medium simulated. The transport and decay equations are expressed as finite differences. Spatial variability is described by assigning different conditions (e.g. radionuclide solubility or initial amount of organic complexants etc.) to each grid point in the finite difference scheme. INHOMOG represents chemical processes very simply in terms of a solubility limit and a sorption coefficient for each radionuclide. These may be specified to vary with, for example, the elapsed time, the pH, or the concentration of any organic complexants present.

INHOMOG Calculations

A number of INHOMOG calculations were performed to determine the magnitude of the effect of heterogeneous distributions in the inventory of ^{129}I relative to the downstream boundary of the repository. The calculations assumed that groundwater flow was vertically upwards through the repository vaults and was uniformly-distributed (as in Nirex 97 [2]). Other data required by the calculations were also obtained from [2]. Three calculations were undertaken for a section of a disposal vault where either: (1) the ^{129}I inventory was distributed homogeneously through the vault section examined; (2) the waste packages with the highest ^{129}I inventory were assumed to be placed in the top rows of containers in the waste stacks within the vault section i.e. the distribution of the packages containing the highest ^{129}I inventory was biased towards the downstream boundary; (3) the waste packages with the highest ^{129}I inventory were assumed to be placed in the lowest rows of containers in the waste stacks within the vault section.

Consideration was given to the available inventory data [8] and distribution of the inventory of ^{129}I between waste packages from different waste streams, and to the design of the repository, as described in Reference [2], in defining the distribution of the ^{129}I inventory for Cases 2 and 3. A number of simplifying assumptions were also made in the calculations, in particular, as the calculations examined a one-dimensional system, dilution processes were ignored. Physical containment provided by the waste packaging was also ignored.

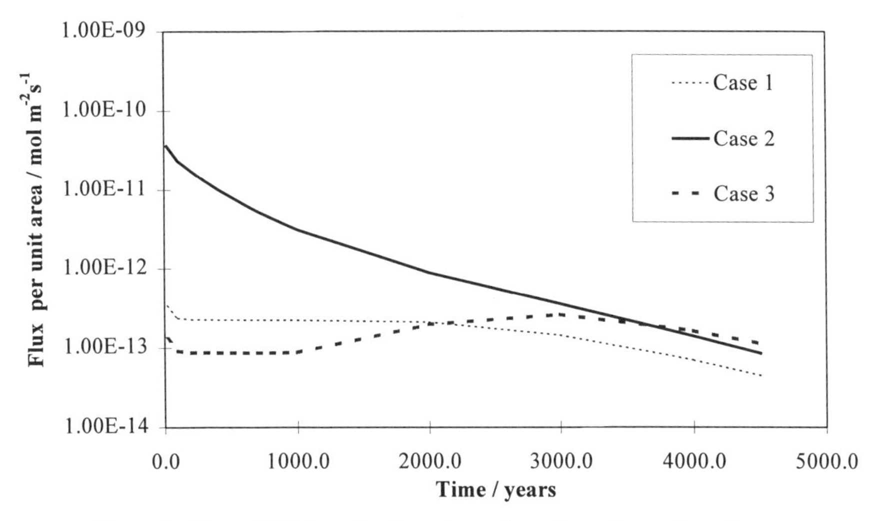

Figure 3. Flux of ^{129}I from the downstream boundary of a vault section.

For each of these cases, a calculation of the flux of the inventory of ^{129}I from the near field duc to transport is illustrated in Figure 3. Over the timescale of interest, decay is insignificant for ^{129}I. The calculations show that a 'detrimental' package arrangement (Case 2) can give rise to ^{129}I near-field fluxes significantly greater than the release calculated at early times for a homogeneous system. The time to release 90% of the ^{129}I inventory was found to have been halved in Case 2 relative to Case 1. Alternatively, the homogeneous distribution for Case 1 leads to an increased rate of release initially compared to the specific distribution of Case 3. This suggests that for some radionuclides with characteristics similar to the ^{129}I inventory, scenarios where different waste package arrangements arise may need to be considered in future performance assessments. Conversely, in the absence of interactions with other species that may have less homogeneous distributions, it is difficult to find a waste package distribution that concentrates the ^{238}U inventory in one region of the repository such that a significant deviation in performance from that of a homogeneous repository can arise.

SUMMARY AND DISCUSSION

It is important that the assumption of a homogeneous repository is demonstrated to be cautious (i.e. should err, if at all, in the direction of slightly overestimating any release of radionuclides). This paper has examined the impact of specific heterogeneities on radionuclide release.

Calculations using the NAMMU computer program have examined the release of radionuclides from waste packages and suggest that recent performance assessments overestimate the release of short-lived radionuclides by not considering the barrier to release that the waste packaging provides. For the Nirex disposal concept there is a distinction to be made between the period of time over which the waste packages provide a barrier to dissolved radionuclide release and the period of time when the containers remain intact.

Calculations using the INHOMOG computer program have examined the relative placement of different categories of waste packages within repository vaults. These show that certain heterogeneous distribution of the inventory of ^{129}I could give rise to lower near-field containment times than for a homogeneous system. In the case of the Nirex 97 assessment [2], however, such variations in the source term would not affect calculations of radiological risk to man. The duration of the release of key radionuclides to the surface is largely controlled by the performance of the host rock in which the repository is sited [2].

ACKNOWLEDGEMENT

The authors would like to thank Nirex for funding the technical work reported in this paper, and also Dr. Andy Baker, Dr. David Lever (both AEA Technology) and Dr. Alan Hooper (Nirex) for their helpful comments during the preparation of this paper.

REFERENCES

1. M.M. Askarieh, A.V. Chambers, G.E. Hickford and S.M. Sharland, in *Scientific Basis for Nuclear Waste Management XXI* (Mater.Res.Soc.Proc. **506**, Pittsburgh PA, 1998), pp465-477.

2. *Nirex 97: An Assessment of the Post-closure Performance of a Deep Waste Repository at Sellafield - Volume 3: The Groundwater Pathway*, Nirex Science Report S/97/012, 1997.

3. P.J. Agg, R.W. Cummings, J.H. Rees, W.R. Rodwell and R. Wikramaratna, *Nirex Gas Generation and Migration Research: Report on Current Status in 1994*, Nirex Science Report S/96/002, 1996

4. A.W. Harris and A.K. Nickerson, *An Assessment of the Mass-transport Properties of Cementitious Materials under Water-saturated Conditions*, Nirex Report NSS/R309, 1995.

5. L.J. Hartley, C.P. Jackson and S.P. Watson, NAMMU (Release 6.3) User Guide, AEA-ES-0138, Release 6.3, Issue 1, 1996.

6. S. Bayliss, R. McCrohon, P. Oliver, N.J. Pilkington and H.P. Thomason, *Near-field Sorption Studies: January 1989 to June 1991*, Nirex Report NSS/R277, 1996.

7. A.V. Chambers, S.J. Williams and S.J. Wisbey, *Nirex Near-field Research: Report on Current Status in 1994*, Nirex Science Report S/95/011, 1995.

8. *The Physical and Chemical Characteristics of UK Radioactive Wastes, Volumes 1 and 2; The Radionuclide Content of UK Radioactive Wastes*, Nirex Report No. 696, 1996.

SORPTION OF URANIUM AND PLUTONIUM ON BENTONITE ALTERED BY HIGHLY ALKALINE WATER

M. BROWNSWORD*, M. MIHARA** and S.J. WILLIAMS*
*AEA Technology plc, 220.32 Harwell, Didcot, Oxfordshire, OX11 0RA, United Kingdom
**Japan Nuclear Cycle Development Institute (JNC), Tokai-mura, Ibaraki, Japan

ABSTRACT

Bentonite may be used, in conjunction with cementitious materials, in repository designs for TRU wastes in Japan. Therefore, possible effects of the interaction of highly alkaline water with bentonite are of interest. In this study, the sorption of plutonium and uranium has been measured on a bentonite sample that had been altered by hydrothermal exposure to a high pH solution.

Samples of Kunipia-F® bentonite were exposed to a pH 14 solution of sodium, potassium and calcium hydroxides for 20 days at 130°C, 160°C and 200°C under air. X-Ray analysis showed that alteration at 160°C was similar to alteration at 200°C.

Batch sorption experiments were carried out at room temperature under nitrogen by contacting 200°C-altered bentonite with synthetic equilibrated water containing both uranium and plutonium in the presence of sodium dithionite as a reducing agent. The experiments were equilibrated for three months and then sampled after centrifugation, 0.45 μm filtration, and 10,000 nominal molecular weight cut-off (NMWCO) filtration. Uranium and plutonium remaining in solution were determined. The final redox potentials were between -500 to -510mV vs SHE and the final pH values were 12.2.

Sorption of both actinides showed some dependence on the solid:liquid separation method. The average R_D value for uranium sorption after centrifugation was $2 \times 10^3 cm^3 g^{-1}$ which increased to 3×10^4 $cm^3 g^{-1}$ after 0.45 μm filtration. However, the effect of a further decrease in the pore size of the filter had little additional effect; 10,000 NMWCO filtration gave an average R_D of 5×10^4 $cm^3 g^{-1}$. The sorption of plutonium after solid:liquid separation by centrifugation was very similar to uranium with an average R_D of 6×10^3 $cm^3 g^{-1}$. After 0.45 μm filtration the increase in R_D to 2×10^5 $cm^3 g^{-1}$ was more marked than for uranium. There was no significant increase in R_D value after 10,000 NMWCO filtration compared to 0.45 μm filtration.

INTRODUCTION

Bentonite may be used in conjunction with cementitious materials in repository designs for TRU wastes in Japan. Therefore, possible effects of the interaction of highly alkaline water with bentonite are of interest. The purpose of this study was to measure the sorption of plutonium and uranium on bentonite that had been altered by hydrothermal exposure to a high pH solution containing sodium, potassium and calcium hydroxides. The batch sorption technique was used and experiments were carried out at a liquid to solid ratio of 20:1 by contacting the altered bentonite with synthetic water containing both uranium and plutonium under reducing conditions at room temperature.

EXPERIMENTAL DETAILS

Altered bentonite samples for use in the sorption experiments were prepared by hydrothermal treatment [1]. Samples of compacted Kunipia-F® bentonite (montmorillonite

Mat. Res. Soc. Symp. Proc. Vol. 608 © 2000 Materials Research Society

content = 98-99 wt.% [2]) had been exposed to a solution of sodium, potassium and calcium hydroxides at approximately pH 14 under an air atmosphere at 130°C, 160°C and 200°C in order to investigate the alteration behaviour of the compacted bentonite. The dry density was 1.8 g cm^{-3} and the liquid:solid ratio was $40cm^3$:11.3g.

It was shown by X-Ray analysis that the surface of bentonite in contact with the highly alkaline solution had been altered to analcite, a zeolite, to a depth of about 2 mm after 20 days (Figure 1). X-Ray data from samples treated at 160°C and 200°C gave almost the same pattern. It was therefore decided to use the 160°C-altered samples to prepare equilibrated water for analysis and to use the 200°C-altered samples in the sorption experiments.

5g of crushed 160°C-altered bentonite were equilibrated with 250 cm^3 saturated calcium hydroxide solution in contact with a small amount of solid calcium hydroxide (approximately 0.1g) under nitrogen at room temperature. After one month the equilibrated water was filtered under nitrogen through an Amicon Diaflo™ PM-30 polysulphone filter (30,000 nominal molecular weight cut-off (NMWCO)). The pH was measured, and samples were taken for anion analysis by ion chromatography, and for cation analysis by inductively-coupled plasma mass spectrometry (ICP-MS) (following acidification with 1 mol dm^{-3} nitric acid). The results of these analyses are shown in Table 1. Based on these analytical results, a synthetic solution for use in the sorption experiments was made up to simulate the equilibrated solution.

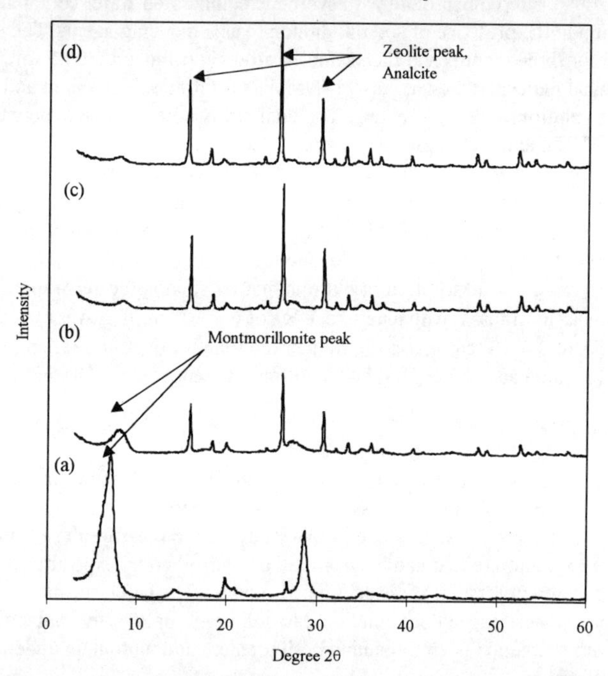

Figure 1 XRD of Bentonite Samples
(a) unaltered sample (starting material), (b) 130°C-altered sample,
(c) 160°C-altered sample, (d) 200°C-altered sample

Table 1 Water Analysis after Equilibration of Altered Bentonite with Calcium Hydroxide Solution

Chemical Species	Concentration / mol dm^{-3}
Al	4.4×10^{-4}
B	$\leq 3.7 \times 10^{-6}$
Ca	2.8×10^{-3}
Fe	$\leq 3.6 \times 10^{-7}$
K	4.3×10^{-3}
Mg	$\leq 8.2 \times 10^{-6}$
Na	8.4×10^{-3}
Si	6.4×10^{-5}
Sr	6.4×10^{-6}
CO_3^{2-}	6.7×10^{-5}
F^-	4.6×10^{-6}
Cl^-	7.0×10^{-5}
Br^-	2.4×10^{-6}
SO_4^{2-}	2.3×10^{-6}
NO_3^-	4.4×10^{-6}
NO_2^-	$\leq 1.3 \times 10^{-6}$
pH	12.2

Uncertainty = ±10%

The batch sorption experiments were carried out in triplicate by contacting 2g portions of the 200°C-altered bentonite with synthetic water containing both uranium and plutonium under reducing conditions in the presence of sodium dithionite. All operations were carried out in a recirculated nitrogen-atmosphere glovebox with an oxygen content of ≤1 ppm. The experimental methodology was as follows:

(i) small volumes of acidic solutions of uranium-233 and plutonium-238 were added to 130 cm^3 of the synthetic solution so as not to exceed the approximate solubility limits at pH 12 of 10^{-7} mol dm^{-3} and 10^{-10} mol dm^{-3} respectively [3];

(ii) sodium dithionite (sodium hydrosulphite) was then added to give a concentration of 2.5×10^{-3} mol dm^{-3} to provide reducing conditions;

(iii) the pH was measured, and re-adjusted from pH 12.04 to pH 12.2 with 1 mol dm^{-3} sodium hydroxide solution;

(iv) the solution was then filtered through a 30,000 NMWCO filter (Millipore) to ensure that the aqueous concentrations of uranium and plutonium did not exceed their respective solubility limits.

Aliquots of this filtrate (40 cm^3) were added to the three centrifuge tubes containing the altered bentonite and shaken until the bentonite was thoroughly wetted. Further aliquots were taken for the determination of the initial uranium-233 and plutonium-238 concentrations. Known activities of uranium-236 and plutonium-239 were added as yield monitors prior to electro-deposition onto stainless steel counting trays and the resulting thin sources analysed using a Canberra alpha spectrometry system. The initial uranium concentration was measured as 6×10^{-8} mol dm^{-3} and the initial plutonium concentration was 5×10^{-11} mol dm^{-3}.

The sorption experiments were equilibrated for three months with gentle agitation at room temperature. During equilibration, the pH and redox potential were monitored every two weeks, and adjusted if necessary. At the end of the equilibration, the solution was sampled after:

(i) centrifugation;
(ii) centrifugation followed by 0.45 μm filtration (Millipore Millex HV™ filters);
(iii) centrifugation followed by 0.45 μm filtration and then 10,000 NMWCO filtration (Millipore TGC™ filters).

Filters were washed and pre-conditioned prior to use based on the methodology described by Rai [4]. Samples were acidified and the final uranium and plutonium concentrations measured by alpha spectrometry as described above. The final pH and redox potentials were measured.

In order to check that sorption was predominantly onto the bentonite, rather than the vessel walls, the residual liquid and solids were poured out, and the vessel walls carefully washed with the synthetic bentonite-equilibrated solution. The vessel walls were then washed with dilute nitric acid and this wash analysed for uranium-233 and plutonium-238. The amounts of wall sorption were small (the largest value was equivalent to 1.2% of the added actinide).

The sorption distribution ratios (R_D values) for plutonium and uranium were then calculated from the decreases in solution concentrations. Because the amount of wall sorption was small, it was not necessary to correct for this when calculating the R_D values.

RESULTS

The results are given in Table 2. The measured redox potentials at the end of the experiments were in the range -500 to -510mV vs SHE and the final pH values were all 12.2.

The R_D values for uranium and plutonium sorption both show some dependence on the solid:liquid separation method. The R_D values for uranium sorption after centrifugation are 1.6×10^3 to 2.4×10^3 cm^3g^{-1} which increase to 1.7×10^4 to 5.1×10^4 cm^3g^{-1} after 0.45 μm filtration. However, the effect of a further decrease in the pore size of the filter has little additional effect; 10,000 NMWCO filtration gives R_D values in the range 3.4×10^4 to 5.1×10^4 cm^3g^{-1}.

Sorption of plutonium after solid:liquid separation by centrifugation is very similar to uranium; R_D values are in the range 5.5×10^3 to 7.2×10^3 cm^3g^{-1}. After 0.45 μm filtration the increase in R_D is more marked than for uranium and values in the range 8.2×10^4 to 2.6×10^5 cm^3g^{-1} are obtained. There is no significant increase in measured sorption compared to 0.45 μm filtration after filtration through 10,000 NMWCO filters; R_D values are between 5.3×10^4 and 3.1×10^5 cm^3g^{-1}.

Overall, the data show a general trend of an increase in R_D value for both uranium and plutonium from centrifuged results to 0.45 μm filtration but little further increase in R_D after 10,000 NMWCO filtration. It also shows that sorption of plutonium (other than after centrifugation) is greater than uranium although the difference is less than one order of magnitude.

DISCUSSION

The uranium sorption data can be compared with uranium sorption data [5] obtained for an unaltered bentonite (Kunigel VI [2]). Those experiments were also carried out under reducing conditions but at the pH of bentonite-equilibrated water (approximately pH 10). The sorption of uranium at room temperature gave R_D values between 1.7×10^4 to 3.9×10^4 cm^3g^{-1}.

Table 2 The Sorption of Uranium and Plutonium onto High pH-altered Bentonite at a Water : Solids Ratio of 20:1

Actinide	Expt No	pH[1]	Eh[2] / mV vs SHE	Distribution Ratio[3] (R_D) / cm^3g^{-1}					
				After Centrifugation	Mean of Centrifuged Values	After 0.45 µm Filtration	Mean of 0.45 µm Filtered Values	After 10,000 NMWCO Filtration	Mean of 10,000 NMWCO Filtered Values
Uranium	1	12.2	-500	$(1.9 \pm 0.2) \times 10^3$		$(1.7 \pm 0.4) \times 10^4$		$(3.4 \pm 0.5) \times 10^4$	
	2	12.2	-500	$(2.4 \pm 0.2) \times 10^3$	2×10^3	$(3.4 \pm 0.5) \times 10^4$	3×10^4	$(5.1 \pm 0.7) \times 10^4$	5×10^4
	3	12.2	-510	$(1.6 \pm 0.2) \times 10^3$		$(5.1 \pm 1.2) \times 10^4$		$(5.1 \pm 0.7) \times 10^4$	
Plutonium	1	12.2	-500	$(5.5 \pm 0.6) \times 10^3$		$(2.6 \pm 0.4) \times 10^5$		$(3.1 \pm 0.5) \times 10^5$	
	2	12.2	-500	$(7.2 \pm 0.8) \times 10^3$	6×10^3	$(8.2 \pm 1.1) \times 10^4$	2×10^5	$(1.8 \pm 1.0) \times 10^5$	2×10^5
	3	12.2	-510	$(6.0 \pm 0.7) \times 10^3$		$(1.3 \pm 0.2) \times 10^5$		$(5.3 \pm 0.6) \times 10^4$	

1 Measured pH values = ±0.2
2 Measured Eh values = ±10 mV
3 Uncertainties based solely on 2σ counting statistics

There was no difference between 0.45 μm filtration or 10,000 NMWCO filtration. These values are similar to the R_D values of 1.7×10^4 to 5.1×10^4 cm^3g^{-1} measured in this study. The plutonium data can also be compared with plutonium sorption data obtained for Kunigel VI from 0.1 mol dm^{-3} and 0.01 mol dm^{-3} $NaClO_4$ solutions [6]. Those experiments were carried out under reducing condition at room temperature in a pH range of approximately 5 to 11. The sorption of plutonium gave R_D values between 1.6×10^3 to 9.7×10^4 cm^3g^{-1} after 10,000 NMWCO filtration in 0.01 mol dm^{-3} $NaClO_4$ solution. This range of values is approximately one order of magnitude less than the R_D values measured in this study.

CONCLUSIONS

The sorption of plutonium and uranium has been measured on a sample of Kunipia-F® bentonite altered by hydrothermal exposure to a high pH solution. Sorption of uranium and plutonium both show some dependence on the solid:liquid separation method. The average R_D value for uranium sorption after centrifugation is 2×10^3 cm^3g^{-1} which increases to 3×10^4 cm^3g^{-1} after 0.45 μm filtration. However, the effect of a further decrease in the pore size of the filter has little additional effect. Filtration through 10,000 nominal molecular weight cut-off (NMWCO) filters gives an average R_D value of 5×10^4 cm^3g^{-1}. The sorption of plutonium after solid:liquid separation by centrifugation is very similar to uranium; the average R_D value is 6×10^3 cm^3g^{-1}. After 0.45 μm filtration the increase in R_D to 2×10^5 cm^3g^{-1} is more marked than for uranium. There is no significant increase in R_D value measured after filtration through a 10,000 NMWCO filter compared to 0.45 μm filtration.

A comparison of these data with those for uranium and plutonium sorption onto an unaltered bentonite suggest that the sorption of uranium and plutonium at high pH on bentonite altered by exposure to highly alkaline water, as studied here, is unlikely to be significantly reduced compared to sorption on unaltered material for an aqueous phase defined by filtration through filters with pore sizes of ≤0.45μm.

ACKNOWLEDGEMENTS

This work was funded by the Japan Nuclear Cycle Development Institute (JNC), Ibaraki, Japan, and their permission to publish these results is gratefully acknowledged.

REFERENCES

1. S. Ichige, M. Ito and M. Mihara, 1998 Annual Meeting of the Atomic Energy Society of Japan, 1998, p.611 (in Japanese).
2. Japan Nuclear Cycle Development Institute, The Draft Second Progress Report on Research and Development for the Geological Disposal of HLW in Japan, 1999, (http://www.jnc.go.jp/kaihatu/tisou/zh12/Draft_2nd/s02/pdf/04-01-02-01-02.pdf)
3. B.F. Greenfield, D.J. Ilett, M. Ito, R. McCrohon, T.G. Heath, C.J. Tweed, S.J. Williams and M. Yui, Radiochimica Acta **82**, 27 (1998).
4. D. Rai, Radiochimica Acta **35**, 97 (1984).
5. G.M.N. Baston, J.A. Berry, M. Brownsword, T.G. Heath, D.J. Ilett, C.J. Tweed and M. Yui in Scientific Basis for Nuclear Waste Management XX, edited by W.J. Gray and I.R. Triay (MRS, Pittsburgh, PA, 1997) pp.805-812.
6. Power Reactor and Nuclear Fuel Development Corporation, Annual Report of the Study on the Geological Disposal of HLW, 1996 (in Japanese).

SURFACE CHARGE AND ELECTROPHORETIC PROPERTIES OF COLLOIDS OBTAINED FROM HOMOIONIC AND NATURAL BENTONITE

T. Missana, M.J. Turrero, A. Adell.

CIEMAT - Departamento de Impacto Ambiental de la Energía
Avenida Complutense 22, 28040 MADRID (SPAIN)

ABSTRACT

Suspensions of colloids obtained from a Spanish bentonite were studied by potentiometric acid-base titrations and electrophoresis in order to analyse their surface chemical properties, which are responsible of their stability behaviour and that are very important to consider in radionuclides sorption modelling.

"Fast" titrations and *"batch - back"* titrations techniques were used to determine the contribution of the pH - dependent charge and the difference in the results obtained are discussed. Experimental data obtained by acid/base titrations were interpreted according the EDL theory. The model prediction agreed satisfactorily with the experimental data in the alkaline pH range. Protonation / deprotonation reactions of surface functional groups (SOH) appeared to be the main surface charge - determining mechanism in the alkaline pH range whereas ion - exchange type reactions, had to be taken into account over the acidic pH range. Surface potentials were calculated for different salt concentrations, from experimental data and taking into account both layer and edge sites charge contributions.

INTRODUCTION

It has been recently shown that bentonite colloids can be generated at the near/field far field interface of a radioactive waste repository [1] and can therefore affect the transport of contaminants to the biosphere. The study of the surface charge and electrophoretic properties of such colloids is necessary to improve the understanding of the mechanisms that lead to the formation of pseudo-colloids at the repository/far-field interface, as well as of their stability or sorption properties.

The surface charge at the colloid / water interface can be due to the ionisation of the surface functional groups (SOH) at the solid surface according to the following equilibria:

$$SOH + H^+ \xleftrightarrow[K_{a1}]{} SOH_2^+ \quad \textbf{E.1}$$

$$SOH \xleftrightarrow[K_{a2}]{} SO^- + H^+ \quad \textbf{E.2}$$

The main contribution to the clay mineral charge is permanent (intrinsic) and it is represented by an excess of negative charge due to isomorphic substitutions within their sheet structure. This negative charge is usually compensated by the adsorption of cations in the layer surface that, in presence of water, can be easily exchanged with other cations present in solution. The total amount of these cations is called the *cation exchange capacity* (CEC) of the material and represents the degree of charge substitution and therefore the magnitude of the intrinsic charge. [2] However, at the edge of the particles surface functional groups similar to those present on the surface of an alumina/silica particle exist. Due to these surface functional groups, an additional electrical double layer can be formed at these edges by the adsorption/desorption of protons. The properties of these surface sites have been found to be similar to those of oxi-hydroxides surfaces [3] [4], therefore they are supposed to follow the equilibria expressed by equations E.1 and E.2.

The aim of this work is to study the surface chemical properties of colloids obtained from purified and homoionised FEBEX[5] bentonite, to determine the pKs of the E.1 and E.2

Mat. Res. Soc. Symp. Proc. Vol. 608

reactions, and to calculate the surface potential at different salt concentration, considering the contributions due to both edge and structural charge. This parameter is of importance in colloidal stability predictions and in surface complexation models. The pH - dependent surface charge, both in oxy - hydroxides [6] and in clay [7] [8] is usually determined by potentiometric titrations and different methods can be used, trying to optimise the technique depending on the material to be studied [9]. The proton adsorption *vs.* pH behaviour was therefore studied using different acid/base titration techniques, namely *"fast"* and *"batch-back"* titrations in order to analyse the difference that can arise from the two different experimental approaches.
The effects of the salt impurities were studied using as - received bentonite.

MATERIALS AND METHODS

FEBEX bentonite is a clay with a smectite content greater than 90% (93±2%), with quartz (2±1%), plagioclase (3±1%), cristobalite (2±1%), potassic feldspar, calcite and trydimite as accessory minerals. More details about this clay can be found elsewhere.[10]
Bentonite colloids from the as-received material were prepared by putting in contact, for a week, the clay sieved at a fraction < 64 μm and water in an appropriate solid to liquid ratio. For electrokinetics measurements a suspension with a solid to liquid ratio 1:20 was used. The suspension was then centrifuged for 20 min at 7000 rpm and the supernatant, which contained the colloidal fraction, finally collected. A solid to liquid ratio of 1:1000 was used in titration experiments and the colloidal fraction was separated by filtration with a 0.45 μm membrane. Homoionised and purified bentonite was prepared following the method described by Bayens and Bradbury [7]. Suspensions of homoionised clay at different ionic strengths were obtained by dialysis methods and the colloidal fraction obtained by centrifuging.
ζ - potential measurements were carried out by means of the laser Doppler electrophoresis technique with a Malvern Zetamaster apparatus equipped with a 5 mW He-Ne laser (λ=633 nm).
Potentiometric titrations were performed at three different ionic strengths ($1\cdot10^{-1}$, $1\cdot10^{-2}$, $1\cdot10^{-3}$ M) of $NaClO_4$. The suspensions were stirred and purged by N_2 prior to titration. Titrations were carried out in N_2 atmosphere in order to minimise the effects of CO_2. "Fast" titrations were carried out by performing continuous additions of NaOH or HCl (5μL) each 3 minutes. The reference solution was obtained by filtering the initial suspension with a 0.01 μm filter membrane and it was titrated by the same procedure of the colloidal suspension. The difference between the total H^+/OH^- added to the suspension and those required to bring the reference solution to the same pH represented the total amount of $[H^+]$ consumed/released by the solid phase (ΔH^+).
"Batch - back" titrations were carried out by placing 40 ml of the initial suspensions in polyethylene bottles and adding a known amount of NaOH or HCl to give a series of initial pH values between 3 and 11. The sample to which neither acid or base were added was used as reference sample. After the addition of acid/base, the sample were shaken 24 hours and then filtered with a 0.01 μm Millipore membrane. The supernatant was then titrated back to the pH of the reference sample, in the same way of fast titrations. The difference between the added acid/base in batch titration and the base/acid used in back titration represented the total amount $[H^+]$ consumed/released by the solid phase.

RESULTS AND DISCUSSION

Experimental results

Figure 1 shows the ζ-potential as a function of pH of a suspensions of bentonite colloids obtained from the homoionised/purified clay and conditioned at ionic strength of $1\cdot10^{-3}$ M with $NaClO_4$. In the same figure, the ζ-potential of the colloids obtained from "natural" bentonite has been also plotted. Since the water used to prepare "natural" bentonite suspensions is deionised water (MilliQ system) the salinity of the solution only depends from the salts impurities leached from bentonite and varies with the initial solid to liquid ratio used to prepare the colloidal suspension. At an initial solid to liquid ratio 1:20, the ionic strength is 5 - 7 10^{-3} M.

The two different samples show very similar ζ-potentials, that indicating very small effects of the bentonite impurities on its electrokinetic properties. Bentonite colloids show fairly low negative potential (≈ -30 mV), almost constant in all the pH range considered. Only a few mV increase / decrease of the potential is observed at the lowest (<4) and highest (>10) pH. A quite interesting behaviour was also observed measuring the ζ-potential of the colloids as a function of the ionic strength. A clear dependence of the potential from the ionic strength could not be seen, and there was not any evidence of decreasing of potential when increasing the ionic strength (see Table 2). This result was already observed in Na-montmorillonite but not yet clearly explained [14] [11].

The results of the *fast-titration* experiments onto homoionic bentonite are shown in Figure 2. The curves of ΔH^+ as a function of pH show only a small dependence with the ionic strength and a clear intersection point between the curves is not observed. ΔH^+ is positive in the range of pH < 6.5 - 7 and negative in the range of pH > 6.5 - 7. In the alkaline range, the degree of deprotonation increases with the ionic strength, this behaviour being in agreement with the behaviour of oxy-hydroxides. In the acidic range, the dependence with the ionic strength is anomalous and this clearly indicates that other charge determining mechanisms have to be considered. In fact, ionic exchange process between Na^+ and H^+ [7][8] can play an important role at low pH. The Na^+/H^+ exchange takes place at structural-charge sites (X) and it is represented by the following reaction:

$$XH + Na^+ \underset{K_x}{\leftrightarrow} XNa + H^+ \qquad \text{E.3}$$

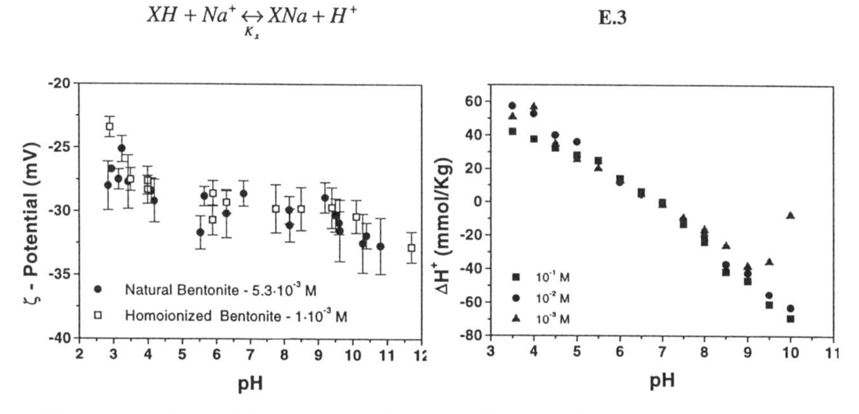

Figure 1: ζ-potential of the (□) homoionised bentonite colloids and (●) of the "natural" bentonite colloids.

Figure 2: Fast - Titration curves of the homoionised and purified bentonite colloids at (■) $1\cdot10^{-1}$; (●) $1\cdot10^{-2}$ and (▲)$1\cdot10^{-3}$M with $NaClO_4$.

Figure 3 shows the results of the batch - back titrations (compared with the results obtained from fast titration). As can be clearly observed, the two methods gave the same results in the alkaline range whereas in the acidic range the ΔH^+ values obtained with batch-back titration methods are typically higher than that obtained with fast titrations. This confirms that protonation/deprotonation reactions of the SOH sites are the main surface charge determining mechanisms in the alkaline range and that other processes, with slower kinetics, have to be included in the modelling of titration curves in the acidic pH range.

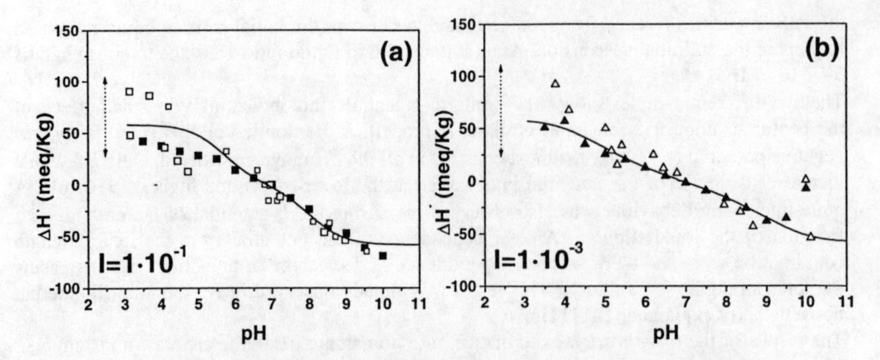

Figure 3: Comparison between the fast (solid points) and batch-back titration experiments (open points). (a) $1\cdot10^{-1}$M and (b) $1\cdot10^{-3}$M. The solid line correspond to the modelling obtained using the parameters in Table 1.

The results of batch-back titrations are usually not affected by dissolution processes which could be then discarded as responsible of the increasing of ΔH^+ values. As already reported, the number of structural sites in equilibrium with the solution could increase with time [8]: at shorter time scales (fast titrations), only the structural sites exposed to the external planes would be "active" and accessible for ion-exchange reactions whereas at longer time scales (batch-back titrations) a higher fraction of these sites could be involved in the exchange reaction. In addition, in the acidic range, the formation of a surface positive charge can act as a screen for the diffusion of positive ions from the internal layers to the solution thus producing the no immediate accessibility of all the structural sites for exchange reactions.
Both the electrophoretic measurements and titration experiments show that the main charge contribution of FEBEX bentonite colloids is due to the structural negative charge, as expected. The ζ-potential is, in fact, always negative in spite of the smaller positive variable charge that is developed at the edge sites at pH<6.5. (Fig.2 and 3). The concentration of the surface functional groups at the edges (N_s) was estimated with the FITEQL code being approximately 5 meq/100 g. Since the total CEC of FEBEX bentonite is approximately 103 meq/100 g, the contribution of the functional groups (SOH) it is only a 5 % approximately of the total charge.
The parameters used to fit the titration curves using the double layer theory are shown in Table 1. The reactions expressed in the E.1 E.2 and E.3 have been take into account simultaneously. [12] The concentration of the active cation exchange reaction sites (X_{ACT}) sites was determined by the best-fit of the fast titration curves and also represents a small fraction of the whole structural sites.
As can be observed in Figure 3, the fit is very good in the alkaline range and acceptable in the acidic range above all in the case of fast titrations. A better fit of the batch-back titrations curves could be obtained by assuming an higher concentration of "active" X_{ACT} sites, according to the kinetic effects observed.
Whereas the electrokinetic behaviour of the natural and homoionised/purified bentonite is practically identical, the titration curves were observed to be strongly affected by the presence of the bentonite impurities.
The natural bentonite titration curves showed a very small dependence with the ionic strength but the inversion of charge was observed at an higher pH (8.3) than that identified in the case of the purified material. The very high buffer capability of the as received bentonite strongly affected the titration experiments, with important kinetic effects, even if very low bentonite concentrations were used. The precipitation/dissolution of calcite impurities was observed to

be the main pH controlling mechanism, this leading to some difficulties in the interpretation of the experimental data.

Table 1

Parameters used for the modelling of titration curves

Tot. Surface Area	725 m^2/g
N_s	0.08 $\mu eq/m^2$
X_{ACT}	0.10 $\mu eq/m^2$
pK_{a1}	5.8
pK_{a2}	-7.9
pK_X	-4.5

Calculation of the double layer parameters from the experimental data

The total charge density of the particle, σ_0, is the sum of the contribution of structural charge density in the layers, σ_{INT}, that can be set equal to the CEC of the sample and the variable charge, σ_H, determined by potentiometric titrations.

In the Gouy-Chapman double layer model, $\sigma_0 + \sigma_d = 0$ and

$$\sigma_d = -0.1174\sqrt{I}\,sinh\left(\frac{ze\Psi}{2kT}\right) \qquad \textbf{E.4}$$

where σ_d, is in C/m^2 and Ψ in Volt and where k is the Boltzmann constant, T the absolute temperature, e the electron charge, z the electrolyte valence and I the solution ionic strength in mol/l. A semi - empirical relationship derived by the description of the double layer around a spherical particle [13] is often used to calculate the charge at the slipping plane (σ_ζ) using the experimental ζ-potential [14]:

$$\sigma_\zeta = \varepsilon_0\varepsilon_r \frac{kT}{ze}\kappa r\left\{2sinh(\frac{z\zeta^*}{2}) + \frac{4}{\kappa r}tanh\left(\frac{z\zeta^*}{4}\right)\right\} \quad \text{where}\ \zeta^* = \frac{e\zeta}{kT} \qquad \textbf{E.5}$$

where κ represents the inverse Debye length and r the radius of the particle.

Table 2 shows the calculations of the double layer parameters from the experimental data for three different pH and at the three ionic strength considered using E.4 and E.5. As can be seen, ζ- potential and the charge evaluated at the shear plane are much lower than the actual surface potential / charge. Surface potentials of approximately -105, -160 and -220 mV have been calculated for the ionic strength of $1{\cdot}10^{-1}$ $1{\cdot}10^{-2}$ $1{\cdot}10^{-3}$ M respectively. A feature that has to be remarked is that the variation of the surface potentials with the pH is very small (few mV), as was also previously experimentally observed for ζ potentials.

CONCLUSIONS

Results showed that, in the alkaline range, deprotonation of the surface SOH groups represents the main contribution to the edge charge. In this case, the results of the two titration techniques used (namely *fast* titrations and *batch-back* titrations) are very similar. In the acidic range, where cation-proton exchange reactions can be of importance, the adsorption/desorption of protons showed a slower kinetics which is most probably responsible of the difference found in the experimental results obtained using the two different techniques. The charge due to the adsorption/desorption of protons is, in absolute value much

less than the structural charge, as confirmed by electrophoretic measurements that showed a negative almost-constant ζ-potential over the entire pH range.
The potential calculated by means of the Gouy-Chapman theory, and considering the contribution of both the layer and the edge charge, showed only a small dependence with the pH, as occurred for the ζ potential which was experimentally determined, but much higher absolute values even at the higher ionic strengths considered.

Table 2

Calculation of the double layer parameters from the experimental data

I = 0.1 M				
pH	**ζ (mV)**	**σ_ζ (C/m^2)**	**σ_d (C/m^2)**	**Ψ (mV)**
4	-32.5	$-2.54\cdot10^{-2}$	-0.132	-102
7	-38.0	$-3.04\cdot10^{-2}$	-0.137	-104
10	-39.3	$-3.17\cdot10^{-2}$	-0.146	-107
I = 0.01 M				
4	-32.0	$-7.96\cdot10^{-3}$	-0.130	-159
7	-34.0	$-8.52\cdot10^{-3}$	-0.137	-162
10	-34.6	$-8.69\cdot10^{-3}$	-0.145	-165
I=0.001				
4	-29.0	$-2.31\cdot10^{-3}$	-0.130	-218
7	-29.5	$-2.35\cdot10^{-3}$	-0.137	-221
10	-31.0	$-2.49\cdot10^{-3}$	-0.138	-221

ACKNODLEGEMENTS

Part of this work has been funded within the frame of the European Commission's Nuclear Fission Safety Research and Training Programme (1994-1998) under contract F14-CT96-0021 and by the Spanish Ciemat Enresa.

REFERENCES

[1] T. Missana, M.J. Turrero, A. Melon, Scientific Bases for Nuclear Waste Management XXII, MAT. RES. SOC. PROC., **556** - (1999) (In press).
[2] H van Holpen, *"An Introduction to clay colloids chemistry"*, Wiley Interscience (1977)
[3] W. Stumm and J.J. Morgan in *Aquatic Chemistry* John Wiley and Sons, New York (1981)
[4] G. Sposito in *The surface chemistry of soils,* Oxford University Press (1984)
[5]The FEBEX (Full scale Engineered Barrier EXperiment) consists in an "in-situ" test, in natural conditions and at a full scale; a "mock-up" test at almost full scale and a series of a laboratory test to complement the information from the two large scale tests. The aim of this project is to study the behaviour of components in the near field of a repository of high level wastes in crystalline rock.
[6] D. Dzombak, F. Morel *"Surface Complexation Modelling: Hydrous Ferric Oxides"* N.Y. J. Wiley and Sons (1990)
[7] B. Bayens and M.H. Bradbury, PSI Bericht Nr. 95 - 10, Paul Sherrer Institute (1995)
[8] H Wanner, Y. Albinsson, O. Karland, E. Wieland, P. Wersin, L. Charlet, Radiochimica Acta, 66/67, 157-162 (1994)
[9] C.P. Shultess, D.L. Sparks, Soil, Sci, Soc. Am., 50, 1406-1411 (1986)
[10] ENRESA Technical Report " Febex bentonite: Origin, properties and fabrication of blocks 05/98 (1998)
[11] I. Sondi, J. Biscan, V. Pravdic, J. Colloids Int. Sci., 178, 514 (1996)
[12] J. van der Lee, "CHESS another speciation and surface complexation computer code" Technical Report N LHM/RD/93/39/CIG, Ecole de Mines de Paris (1993).
[13] A. L. Loeb, P.H. Wiersema, J Th G Overbeek *The electrical double layer around a spherical colloid particle*, MIT Press, Cambridge (1961)
[14] R.J. Hunter, *"Zeta Potential in Colloid Science"*, Academic Press London (1988)

DIFFUSION OF PLUTONIUM IN COMPACTED BENTONITES IN THE REDUCING CONDITION WITH CORROSION PRODUCTS OF IRON

K. Idemitsu*, X. Xia*, T. Ichishima*, H. Furuya*, Y. Inagaki*, T. Arima*, T. Mitsugashira**, M. Hara**, Y. Suzuki**
*Department of Applied Quantum Physics and Nuclear Engineering, Kyushu University, Japan;
**The Oarai Branch, Institute for Materials Research, Tohoku University, Japan

ABSTRACT

In a high-level waste repository, a carbon steel overpack will corrode after the repository is closed. This will create a reducing environment in the vicinity of the repository. Reducing conditions are expected to retard the migration of redox-sensitive radionuclides such as plutonium.

The apparent diffusion coefficients of plutonium were measured in compacted bentonites (Kunigel V1® and Kunipia F®, Japan) in contact with carbon steel and its corrosion products under a reducing condition and, for comparison, without carbon steel under an oxidizing condition. Gas bubbles were observed in some bentonite specimens which had low dry densities after contact with carbon steel for approximately two years. This observation suggests hydrogen generation during corrosion of the carbon steel. The apparent diffusion coefficients measured were approximately 10^{-14} m^2/s under the reducing condition and less than 10^{-15} m^2/s under the oxidizing condition. There was a significant effect of redox conditions on the apparent diffusion coefficients. The effects of dry density (0.8 to 2.0 Mg/m^3) and montmorillonite contents (50% for Kunigel V1 or 100% for Kunipia F), however, were not observed clearly. The chemical species of plutonium were expected to be $PuOH^{2+}$ for the reducing condition and $Pu(OH)_4$ for the oxidizing condition, respectively.

INTRODUCTION

Compacted bentonite and carbon steel are being considered as a candidate for buffer and overpack material for high-level waste disposal in Japan. To perform an adequate safety analysis, it is essential to make reliable assessments of the diffusion behavior of radionuclides in compacted bentonite. So far, most of the diffusion experiments in bentonite have been carried out under oxidizing conditions. But in a real repository a carbon steel overpack will be corroded by consuming oxygen trapped in the repository after the depository's closure. This will create a reducing environment in the vicinity of the repository. Reducing conditions are expected to retard the migration of redox-sensitive elements such as uranium[1,2], technetium[3,4], and so on[5]. Iron corrosion products are also expected to interfere with the migration of radionuclides by filling the pores in bentonite and sorbing radionuclides. In this study the apparent diffusion coefficients of plutonium were measured in compacted Na-bentonites (Kunigel V1® and Kunipia F®, Japan) contacted with carbon steel and its corrosion products under reducing conditions and, for comparison, without carbon steel under oxidizing conditions.

Mat. Res. Soc. Symp. Proc. Vol. 608 © 2000 Materials Research Society

EXPERIMENTAL PROCEDURE

Two types of sodium bentonite were used in this experiment: Kunigel V1® and Kunipia F®, Japan. Kunigel V1 and Kunipia F contain approximately 50% and 100% of montmorillonite, respectively. The mineral composition of Kunigel V1 is shown in Table I. The chemical composition of the bentonites is shown in Table II. Bentonite powder was compacted into cylinders of 10 mm diameter and 10 mm height with a dry density of 0.8 to 1.6 Mg/m^3. Each compacted bentonite was inserted into an acrylic resin column as shown in Figure 1 and saturated with deionized water for a month. In the experiments with corrosion products, each water-saturated compacted bentonite was kept in touch with a partly corroded surface of carbon steel, 12 mm diameter and 2 mm thickness, for 7 days at room temperature to let the corrosion products of iron diffuse into the bentonite. Thirty microliters of solution containing ca. 1 kBq of ^{239}Pu was put on the surface between the carbon steel and the bentonite. After a diffusion period of two years, each column was disassembled.

Then the bentonite was pushed out and was sliced in steps of 0.5 to 2 mm. Each slice was submerged in 1N HCl solution to extract Pu, and the liquid phase was separated by centrifugal method. Then 0.2 mL of the supernatant was taken to be measured with an α liquid scintillation counter. Analogous diffusion experiments without contacting carbon steel were performed for comparison.

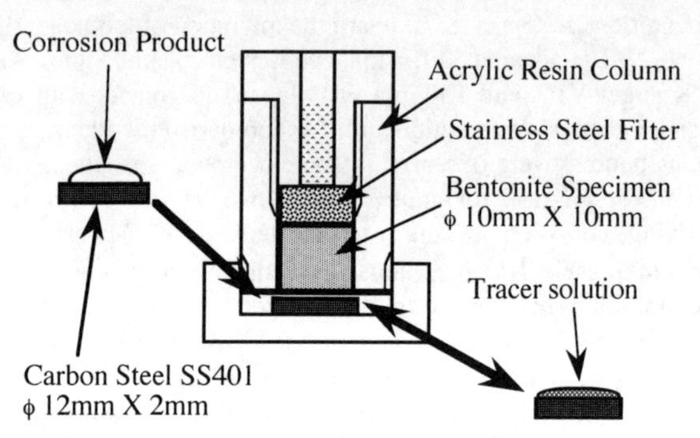

Figure 1. Schematic of experimental apparatus.

Table I. Mineral composition of 'Kunigel V1®'.

Mineral	weight %
Montmorillonite	46 to 49
Quartz	0.5 to 0.7
Chalcedony	37 to 38
Plagioclase	2.7 to 5.5
Calcite	2.1 to 2.6
Dolomite	2.0 to 2.8
Analcime	3.0 to 3.5
Pyrite	0.5 to 0.7

Table II. Chemical compositions of bentonites used in this study (wt%).

Bentonite	Kunigel V1®	Kunipia F®
SiO_2	70.7	58.36
Al_2O_3	13.8	20.36
Fe_2O_3	1.49	1.34
FeO	0.62	0.51
TiO_2	0.20	0.13
MnO	0.22	<0.01
Na_2O	2.56	2.93
K_2O	0.33	0.09
CaO	2.30	0.42
MgO	2.26	2.97
P_2O_3	0.05	<0.01
CO_2	2.20	-
S	0.29	-

RESULTS AND DISCUSSION

Observation of the Bentonite Specimens

In order to identify the reducing condition provided by the corrosion products of iron, the sliced bentonites were observed and their color photos were taken. Reducing conditions were demonstrated by the changes in color of the corrosion products of iron, from red before contact with bentonite to dark green after diffusion periods of two years. The red coloration is considered to be Fe_2O_3 and/or $Fe(OH)_3$ which resulted from the reaction between Fe and O_2 in air, while inside of the bentonites, the concentration of O_2 is extremely low so that it can be ignored. Thus, the corrosion of iron occurs inside the bentonites by the following reaction:

$$3Fe + 4H_2O = Fe_3O_4 + 4H_2 \text{ (gas)} \qquad (1)$$

In fact, the color of the corrosion product turned red again after a few minutes of exposure to the air. This means that carbon steel in bentonite maintained a reducing condition during contact with the bentonite. The colors of the bentonite slices were different from the original color of the bentonite and varied as a function of depth from the interface between the bentonite and the carbon steel. This color change suggests that corrosion products of iron diffused into the bentonite. Some small bubbles were observed in some low-density compacted bentonites in this experiment. It is considered that the bubbles were formed by H_2 gas generated by a corrosion reaction as shown in equation (1). Thus it is concluded that the environment of diffusion was a reducing condition. The bubbles prevent bentonites from contacting with iron at some point so that Pu could lose its diffusion path, thus Pu in this case diffused very slowly. The Pu concentrations in some slices were less than the detection limit (1Bq/ml) of the measurement device. Therefore, for some bentonite samples with relatively low dry density, suitable profiles were not given because of the absence of significant data.

Penetration Profiles and Determination of Apparent Diffusion Coefficients

Penetration profiles of Pu with corrosion products (reducing condition) and without corrosion products (oxidizing condition) are shown in Figure 2 in the case of Kunigel V1 with a density of 2.0 Mg/m^3. The plutonium profiles with carbon steel are higher than those without carbon steel. The concentration of ^{239}Pu in the first slice was much higher than that of the second slice. This suggests that ^{239}Pu could be precipitated at the interface of the bentonite specimen and the plutonium concentration in the bentonite pore water could be kept at a solubility of Pu at the interface. In addition, plutonium could not penetrate the other side. Thus the profiles were fitted by the solution[6] with a constant concentration boundary with the assumption of semi-infinite media,

$$C_{Bent} = C_0 \operatorname{erfc}\left(\frac{x}{2\sqrt{D_a t}}\right) \qquad (2).$$

Where C_{Bent} is the concentration of ^{239}Pu in the bentonite specimen and C_0 is a constant concentration at the interface of the bentonite specimen, or a concentration of ^{239}Pu adsorbed on the bentonite specimen equilibrium to a solubility of plutonium. There are two fitting parameters: apparent diffusion coefficients, D_a, and constant concentrations at the surface of bentonite, C_0. The fitting curves are also shown in Figure 2. The apparent diffusion coefficients obtained are shown in Table III. The apparent diffusion coefficients measured were approximately 10^{-14} m^2/s under the reducing condition and less than 10^{-15} m^2/s under the oxidizing condition. There was a significant effect of redox conditions on the apparent diffusion coefficients. It turned out contrary to the results for uranium measured by the similar experiment [2], where the apparent diffusion coefficients were 4 to 11×10^{-14} m^2/s under a reducing condition and 9 to 14×10^{-13} m^2/s under an oxidizing condition. The apparent diffusion coefficients as a function of the dry density of the bentonites are illustrated in Figure 3. The density has no remarkable effect on it.

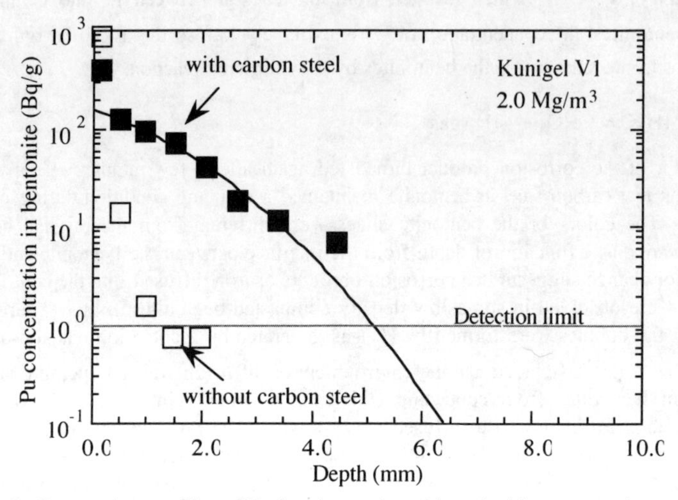

Figure 2. Penetration profiles of Pu in a bentonite with and without corrosion products.

Table III. The apparent diffusion coefficients with the dry density of bentonites both for a reducing and an oxidizing condition.

Bentonite	Dry Density (Mg/m^3)	Diffusion Coefficients (m^2/s)	
		Reducing condition	Oxidizing condition
Kunigel V1	1.0	-	$<10^{-15}$
	1.2	-	
	1.4	$(1.2 \pm 0.3) \times 10^{-14}$	
	1.6	$(1.9 \pm 1.0) \times 10^{-14}$	
	1.8	$(2.6 \pm 0.5) \times 10^{-14}$	
	2.0	$(2.6 \pm 0.3) \times 10^{-14}$	
Kunipia F	0.8	-	$<10^{-15}$
	1.0	-	
	1.2	-	
	1.4	-	
	1.6	$(1.5 \pm 0.2) \times 10^{-14}$	
	1.8	$(8.8 \pm 3.3) \times 10^{-15}$	

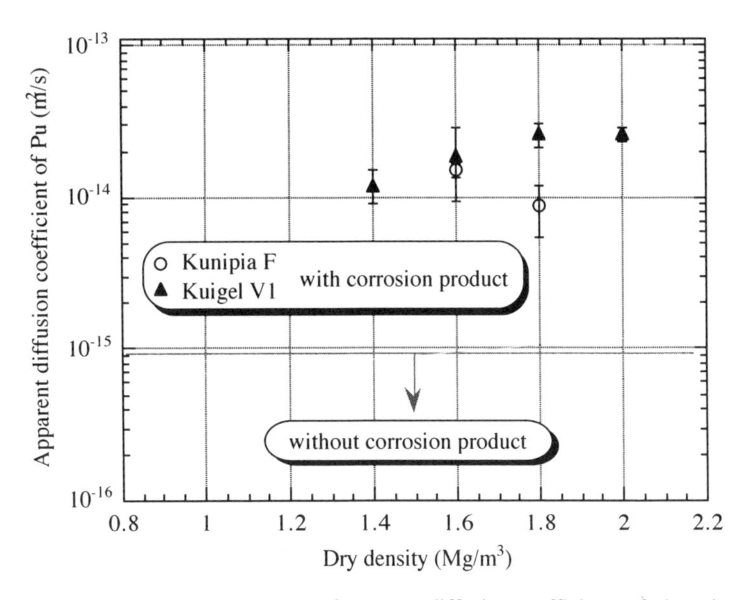

Figure 3. Dry density dependence of apparent diffusion coefficients of plutonium.

Effects of the Reducing Condition

The reducing condition, the presence of corrosion products of iron, can enhance the diffusion of plutonium in the bentonites more than one order of magnitude higher in apparent diffusion coefficients than those estimated under the oxidizing condition. This could be accounted for by the complexity of chemical forms of plutonium and the changes in chemical equilibrium depending on the redox conditions in the reaction system. From a thermodynamic calculation, the dominant species of Pu are tri-valent $Pu(OH)^{2+}$ under the reducing condition and tetra-valent $Pu(OH)_4$ under oxidizing conditions, respectively.

The water in the interlayer of montmorillonite is considered to be the dominant medium of Pu diffusion in montmorillonite. The absorption mechanisms of Pu in a bentonite would be the formation of complex compounds with surface water. $Pu(OH)_4$ could be easy to form complex compounds, and thus is adsorbed in the bentonite and hardly moves through the interlayer,. whereas $Pu(OH)^{2+}$ more easily penetrates into the interlayer space than does $Pu(OH)_4$. Consequently, the diffusion of Pu under the reducing condition was faster than that in the oxidizing condition. However, the opposite results were obtained for the diffusion of uranium in the same circumstance. In the case of uranium, the authors proposed that uranium in the bentonite pore water probably exists as a neutral hydroxide complex under the reducing conditions and as an anionic carbonate or hydroxide complex under the oxidizing conditions. Both results prove that the redox conditions can affect the chemical forms of radionuclides and further influence their migration behaviors.

On the other hand, in general, the diffusion of Pu is very slow compared with the diffusion of some typical fission products in the same bentonites. The apparent diffusion coefficients of Cs and Sr are in the range of 10^{-12} to 10^{-11} m^2/s on average and were largely influenced by the dry density of the bentonites. The differences are probably due to their different mechanisms of diffusion. For Pu, only dissolved species of Pu are suggested to possibly diffuse into the bentonites, and the water in the interlayer of montmorillonite is the dominant medium. And the solubility of Pu is normally low because it is easy to form complex compounds which hardly migrate.

CONCLUSIONS

The reducing conditions provided by the presence of corrosion products of iron can enhance the diffusion of Pu in the bentonites. The environment of repository under the deep underground is normally an environment of reducing conditions. The observed apparent diffusion coefficients of Pu were in the range of 8.8×10^{-15} to 2.6×10^{-14} m^2/s under the reducing condition, while less than 10^{-15} m^2/s under the condition without the presence of corrosion production of iron. It is two or three orders of magnitude lower than for some fission products in the same bentonites in general. Although the diffusion is very slow, it could be concluded that plutonium is very important in the safety assessment of HLW repository for very long disposal.

The dry density of the bentonites had no remarkable effect on the diffusion coefficients in this study. The dominant path of the diffusion is considered as the interlayer space of the montmorillonite, which is the major component of the bentonites.

The diffusion of Pu under reducing conditions in the bentonites is considered to be controlled by the diffusion of dissolved tri-valent species of Pu in the interlayer water of montmorillonite. It should be noted that there are two other possibilities for enhancement of the diffusion of Pu in the bentonite under the reducing conditions; alteration of bentonite by corrosion product and plug flow by migration of hydrogen gas bubbles.

ACKNOWLEDGMENT

The authors wish to thank Mr. H. Asano and Ms. Y. Iwata (Ishikawajima-Harima Heavy Industries Co., Ltd.) for their technical support.

REFERENCES

1. K. Idemitsu, H. Furuya and Y. Inagaki in *Scientific Basis for Nuclear Waste Management XVI*, edited by C.G. Interrante and R.T. Pabalan (Mater. Res. Soc. Proc. **294**, Pittsburgh, PA, 1993), pp. 467-474.
2. K. Idemitsu, H. Furuya, Y. Tachi and Y. Inagaki in *Scientific Basis for Nuclear Waste Management XVII*, edited by A. Barkatt and R.A. Van Konynenburg (Mater. Res. Soc. Proc. **333**, Pittsburgh, PA, 1994), pp. 939-946.
3. Y. Kuroda, K. Idemitsu, H. Furuya, Y. Inagaki and T. Arima in *Scientific Basis for Nuclear Waste Management XX*, edited by W.J. Gray and I.R. Triay (Mater. Res. Soc. Proc. **465**, Pittsburgh, PA, 1997), pp. 909-916.
4. B. Torstenfelt in Radiochimica Acta **39**, pp. 97-104 (1986).
5. Y. Albinsson, B. Christiansen-Sätmark, I. Engkvist and W. Johansson in Radiochimica Acta **52/53**, pp. 283-286 (1991).
6. J. Crank, *The Mathematics of Diffusion*, 2nd ed. (Clarendon Press, Oxford, 1975), pp. 11-21.

EFFECT OF IONIC CHARGE ON EFFECTIVE DIFFUSION COEFFICIENT IN COMPACTED SODIUM BENTONITE

H. Sato*
*Japan Nuclear Cycle Developement Institute, 4-33 Muramatsu, Tokai-mura, Naka-gun, Ibaraki-ken 319-1194, JAPAN, sato@tokai.jnc.go.jp

ABSTRACT

Effective diffusion coefficients(De) in bentonite were measured as a function of ionic charge to evaluate the degree of surface diffusion and anion exclusion. The De measurements for Ni^{2+}, Sm^{3+} and SeO_3^{2-} were carried out for 1.8Mg•m^{-3} by through-diffusion method. Sodium bentonite, Kunigel-V1® was used. The order of obtained De values was Sm^{3+}>Ni^{2+}>SeO_3^{2-}. These De values were compared with those reported to date. Consequently , the order of De values was Cs^+>Sm^{3+}>HTO>Ni^{2+}>anions(I^-, Cl^-, CO_3^{2-}, SeO_3^{2-}, TcO_4^-, $NpO_2CO_3^-$, $UO_2(CO_3)_3^{4-}$), showing a tendency of cations>HTO>anions. The reason that the De of Ni^{2+} was lower than that of HTO may be because the free water diffusion coefficient(Do) of Ni^{2+} is about 1/3 of that of HTO. The formation factors(FF) were in the order, Sm^{3+}>Cs^+>Ni^{2+}>HTO>anions, indicating a possibility of surface diffusion in cations and of anion exclusion in anions. In this case, the FF of Sm^{3+} was approximately 5 times greater than that of HTO. However, since the Do of Sm^{3+} is about 1/3 of that of HTO, the De of Sm^{3+} may have been a little higher than that of HTO. Based on this, it is presumed that surface diffusive effect on De in bentonite is insignificant.

INTRODUCTION

Effective diffusion coefficient(De) in bentonite is one of the key parameters for performance assessment of the geological disposal of high-level radioactive waste[1], because it controls the release rates of nuclides from buffer material. Sodium bentonite has been considered as a candidate of the buffer material so far. It is known that De values of nuclides in bentonite depend on the diffusion species[2]. It is therefore ideal that De is determined every species, but De values of nuclides obtained under repository in releavant conditions are quite limited. In particular, no De in bentonite obtained under reducing conditions has been reported.

For measurement of De in bentonite, some studies have been reported so far and they have discussed for surface diffusion and anion exclusion. Muurinen et al.[3] have measured the De values of Cs^+ and Sr^{2+} for a Na-bentonite, MX80(density 1.75Mg•m^{-3}), pointing out a possibility of surface diffusion. Chueng and Gray[4] have also measured the De values of Cs^+, I^- and Cl^- for a Na-bentonite, Avonlea bentonite(densities 1.25 and 1.75Mg•m^{-3}), discussing on surface diffusion of Cs^+ and on anion exclusion of I^- and Cl^-. However, neither the degree of surface diffusion nor anion exclusion can be quantitatively estimated in their study, because there is no datum for neutral species to be compared. For this bentonite, Oscarson and Gray[5] and Choi et al.[6] have also treated and discussed on surface diffusion based on obtained De values(HTO, Sr^{2+}, Na^+, I^- and Ca^{2+} for [5] and HTO, I^- and Sr^{2+} for [6]), concluding that surface diffusion in cations is unimportant for performance assessment. Muurinen et al.[7] have discussed a possibility of anion exclusion from the effect of ionic strength and bentonite density on De values of U and Cl. Eriksen and Jansson[8] have also obtained the De values of Cs^+, Sr^{2+} and I^- for MX80, discussing on both surface diffusion of cations and anion exclusion of I^-, and have calculated the surface diffusivities(Ds) of both cations. Furthermore, some other reports for surface diffusion in cations and anion exclusion in anions have been found.

For Japanese bentonite, Kato et al.[9, 10], and Sato and Shibutani[2] have reported De values of HTO, $^{137}Cs(Cs^+)$, $^{99}Tc(TcO_4^-)$, $^{237}Np(NpO_2CO_3^-)$, $U(UO_2(CO_3)_3^{4-})$, $^{125}I(I^-)$, $^{36}Cl(Cl^-)$ and $^{14}C(CO_3^{2-})$ for a Na-bentonite, Kunigel-V1® obtained as a function of dry density of the bentonite and the obtained De values show a tendency of cations>neutral species(HTO)>anions. It is generally familiar that De values of cations are higher than those of anions. However, no quantitative discussion for the effect of the ionic charge of diffusion species on De in bentonite has been carried out so far.

In this study, De in compacted bentonite was measured as a function of the ionic charge of

Mat. Res. Soc. Symp. Proc. Vol. 608 © 2000 Materials Research Society

diffusion species to quantitatively evaluate the degree of surface diffusion and anion exclusion.

EXPERIMENTAL

Material and Experimental Conditions

Sodium bentonite, Kunigel-V1®, which is a crude bentonite and was treated in the reference case of performance assessment in the second progress report[11], was used. Major clay mineral of the bentonite is Na-montmorillonite, which mode is 46–49wt% and chalcedony, quartz, plagioclase, calcite, dolomite, analcite and pyrite are contained as impurities. The detailed mineralogy is described in the literatures of Ito et al.[12, 13]. A simulated porewater, prepared to obtain certain concentration by dissolving NaCl, Na_2CO_3 and Na_2SO_4 in distilled water, was used in all diffusion experiments. The concentration and chemical composition was determined based on the results of bentonite leaching tests for various liquid-solid ratios. Tables I and II show the experimental conditions and the chemical composition of the simulated porewater, respectively.

Table I Experimental Conditions for Diffusin Experiments

Bentonite	Na-bentonite, Kunigel-V1® (Kunimine Industries Co. Ltd.)
Dry density	$1.8 Mg \cdot m^{-3}$
Tracer	Sm ($SmCl_3$), Se (Na_2SeO_3), Ni (Ni-63+$NiCl_2$ (carrier))
Concentration	Sm:0.01M, Se:0.001M, Ni:$2 kBq \cdot ml^{-1}$ (Ni-63)+0.001M
Porewater	simulated porewater (see Table II)
pH	Ni, Sm: 5 – 6, Se: not adjusted (monitoring)
Temperature	room temperature
Atmosphere	Se: Ar-atmosphere (O_2 concentration < 1ppm) Ni, Sm: aerobic condition
Producibility	n=2

Table II Chemical Composition of the Simulated Porewater

Ion	Concentration(M)
Na^+	0.83
Cl^-	0.0071
SO_4^{2-}	0.12
CO_3^{2-}	0.29

Experimental Procedure

The diffusion experiments were carried out by through-diffusion method[3]. Figure 1 shows the schematic view of a diffusion cell. Bentonite was dried at 105°C for over night and filled in the sample holder with the size of 20mm in diameter and 3mm in thickness with a dry density of $1.8 Mg \cdot m^{-3}$. The bentonite in the holder was saturated with simulated porewater before diffusion experiment. For Ni and Sm, the bentonite was saturated with the simulated porewater adjusted at pH5–6. The saturation was carried out in a vacuum chamber for a week after degassed an hour. The pH of the porewater was monitored and adjusted using HCl if necessary.

All experiments for Se were conducted under Ar atmosphere(O_2<1ppm). The operations for Se were performed in an atmospheric controled glove-box. The simulated porewater, degassed for over night, was injected into both cells to saturate the bentonite.

Each tracer solution was prepared by dissolving Na_2SeO_3, $NiCl_2$ and $SmCl_3$ in the simulated porewater so as to obtain cetain concentrations. The tracer solution for Se was prepared with degassed simulated porewater in the glove box. The tracer solution for Ni was prepared by dissolving a ^{63}Ni stock solution and $NiCl_2$ as a carrier in the simulated porewater so as to obtain a concentration of $2 MBq \cdot l^{-1}$(10^{-3}M for carrier). In this case, since the concentration of ^{63}Ni is equivalent to 1.5×10^{-8}M, total concentration of Ni can be regarded as the concentration of carrier. All tracer solutions were checked that no colloidal formation was all found from the results of filtration tests for 4 filters: 0.45μm membrane filter, 300,000, 30,000 and 10,000MWCO(Molecular Weight Cut Off) ultrafilters.

After the saturation of bentonite, the porewater in the tracer cell was exchanged with the tracer solution and then the experiment was started. Samples(0.5ml for Ni, 10ml for Se and Sm) were periodically taken from the measurement cell and an identical volume of the porewater was added excepting Ni. Since sampling volume was small for Ni, no addition of the porewater was carried out. A small fraction of samples were also extracted from the tracer cell. The pH and ORP of the solution were monitored. Furthermore, through-diffusion tests for sintered metal filters were carried out for each element to correct concentration gradients in the filters when De values are calculated. The samples were analyzed for Se and Sm concentrations with ICP-AES(detection limit: 0.5ppm for both). Those for Ni were analyzed for β activity(65.9keV) emitted from ^{63}Ni with a liquid scintillation counter (detection limit: $0.2Bq \cdot ml^{-1}$).

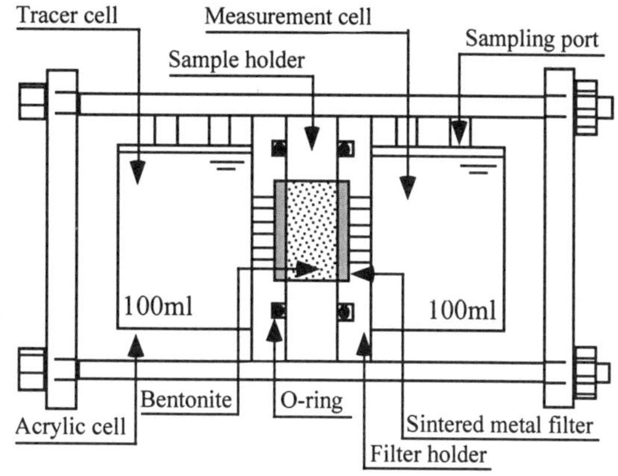

Figure 1 Schematic view of diffusion cell

The accumulative quantities of Ni, Sm and Se permeated through bentonite from the tracer cell were obtained with time based on these analyzed data. For Ni, no addition of porewater was carried out after sampling. In this case, the accumulative quantity of Ni permeated through bentonite was calculated from the following equation.

$$Qn = Cn \cdot \{V - (n-1) \cdot v\} + \sum_{i=1}^{n-1}(C_i \cdot v) \quad (n = 1, 2, 3, 4, \cdots\cdots) \tag{1}$$

Where Qn is the accumulative quantity of the tracer permeated through bentonite up to the n-th sampling(cpm), Cn is the analyzed concentration in the n-th sample($cpm \cdot ml^{-1}$), V is the solution volume in the measurement cell(ml) and v is the sampling volume(ml)(0.5ml for Ni).

On the other hand, the accumulative quantities of Sm and Se permeated through bentonite were calculated from the following equation.

$$Qn = Cn \cdot V + \sum_{i=1}^{n-1}(C_i \cdot v) \quad (n = 1, 2, 3, 4, \cdots\cdots) \tag{2}$$

Where Qn is the accumulative quantity of the tracer permeated through bentonite up to the n-th sampling(μg) and Cn is the analyzed concentration in the n-th sample(ppm).

At the end of diffusion experiment, bentonite in the holder was pushed out and cut with a knife into 0.3mm pitched slices. Each slice was immediately weighed and dried at 105°C for over night to obtain water quantity. The slices for Se were immersed in 10ml distilled water for several hours to extract Se from the slices. The slices for Ni and Sm were immersed in a 10ml HCl solution(0.1M) to extract these elements from the slices. The concentrations of Ni, Sm and Se were then analyzed and the concentration profiles in the bentonite were determined.

Diffusion Theory

The calculations of De values were based on Fickian law[14]. The diffusion equation for a one-dimensional non-steady state is generally expressed by the following equation[15].

$$\frac{\partial C}{\partial t} = \left(\frac{\phi \cdot Dp}{\alpha}\right) \cdot \frac{\partial^2 C}{\partial x^2} \tag{3}$$

Where C is the concentration of the tracer in the bentonite($kg \cdot m^{-3}$), t is the diffusing time(s), Dp is the diffusion coefficient in the porewater($m^2 \cdot s^{-1}$), ϕ is the porosity, α is the rock capacity factor($\alpha = \phi + \rho \cdot Kd$), ρ is the dry density of the sample($Mg \cdot m^{-3}$), Kd is the distribution coefficient($m^3 \cdot Mg^{-1}$) and x is the distance from the source(m).

The $\phi \cdot Dp/\alpha$ is equal to apparent diffusion coefficient(Da). The accumulative quantity of tracer permeated through bentonite up to an arbitrary time for equation(3), based on initial and boundary conditions, is written as follows:

Initial conditin

$C(t, x) = 0, t = 0, 0 \leq x \leq L$

Boundary condition

$C(t, x) = Co \cdot \alpha, t > 0, x = 0$

$C(t, x) = 0, t > 0, x = L$

$$\frac{Q(t)}{A \cdot L \cdot Co} = \frac{De}{L^2} t - \frac{\alpha}{6} - \frac{2\alpha}{\pi^2} \sum_{n=1}^{\infty} \left\{ \frac{(-1)^2}{n^2} exp\left(\frac{De \cdot n^2 \cdot \pi^2 \cdot t}{L^2 \cdot \alpha} \right) \right\} \quad (4)$$

Where Q(t) is the accumulative quantity of the tracer permeated through bentonite(cpm for Ni, μg for Sm and Se), A is the cross-section area of the sample(m^2), L is the thickness of the sample(m), Co is the concentration of the tracer in the tracer cell($cpm \cdot ml^{-1}$ for Ni, ppm for Sm and Se) and De is the effective diffusion coefficient($m^2 \cdot s^{-1}$).

At long time such as steady state, the exponentials fall away to zero. Therefore, equation(4) is approximately written by the following equation for steady state.

$$\frac{Q(t)}{A \cdot L \cdot Co} = \frac{De}{L^2} t - \frac{\alpha}{6} \quad (5)$$

The De is calculated from the slope of Q(t)/(A•L•Co) with time in steady state based on equation(5). If surface diffusion does not occur, De is expressed by the following parameters[1, 15, 16].

$$De = \phi \cdot (\delta / \tau^2) \cdot Do = \phi \cdot G \cdot Do = FF \cdot Do \quad (6)$$

Where δ is the constrictivity, τ^2 is the tortuosity, Do is the free water diffusion coefficient ($m^2 \cdot s^{-1}$), G is the geometric factor(or tortuosity factor) and FF is the formation factor.

It is familiar that Do depends on species. The Do is calculated by the Nernst expression[17].

$$Do = R \cdot T \cdot \lambda / (F^2 \cdot Z) \quad (7)$$

Where R is the gas constant($8.314 J \cdot mol^{-1} \cdot K^{-1}$), T is the absolute temperature(K), λ is the limiting ionic equivalent conductivity($m^2 \cdot S \cdot mol^{-1}$), F is the Faraday constant($96{,}493 C \cdot mol^{-1}$) and Z is the absolute value of the ionic charge.

The Do values of Ni^{2+}, Sm^{3+} and SeO_3^{2-} are calculated to be 6.61×10^{-10}, 6.08×10^{-10} and $8.13 \times 10^{-10} m^2 \cdot s^{-1}$(25°C)[18], respectively, using equation(7).

Correction of Concentration Gradient in Filter for De

The concentration gradient of tracer in the filter which was used to constrict the swelling of bentonite is also included in De calculated based on equation(5) and this concentration gradient in the filter must be corrected to calculate true De in bentonite. In this study, De was corrected by the following equation derived for steady state.

$$De = \frac{L}{(L + 2L_f)/De_t - 2L_f/De_f} \quad (8)$$

Where De_t is the effective diffusion coefficient before correction($m^2 \cdot s^{-1}$), De_f is the effective diffusion coefficient in the filter($m^2 \cdot s^{-1}$) and L_f is the thickness of the filter(m)(1mm).

RESULTS AND DISCUSSION

Diffusion Coefficients

Figure 2 shows the changes in concentrations of Ni, Sm and Se in both cells with time in through-diffusion experiments and the concentration profiles in bentonite. The concentrations of Ni and Sm in the measurement cell show non-linear curves in transient state and increase in a straight line with time in steady state. For Se, no non-linear curve, shows transient state like Ni and Sm, is found. This may be because Se is little sorptive onto bentonite[19]. Each through-diffusion experiment was carried out in duplicate and good producibility was obtained. The concentration profiles of Ni, Sm and Se in the bentonite approximately linearly decrease from the tracer cell side to the other side in all cases. This indicates that the diffusion in all cases is in steady state. The concentrations of Ni, Sm and Se in the tracer cell are approximately kept constant, although some variation in the plots is found.

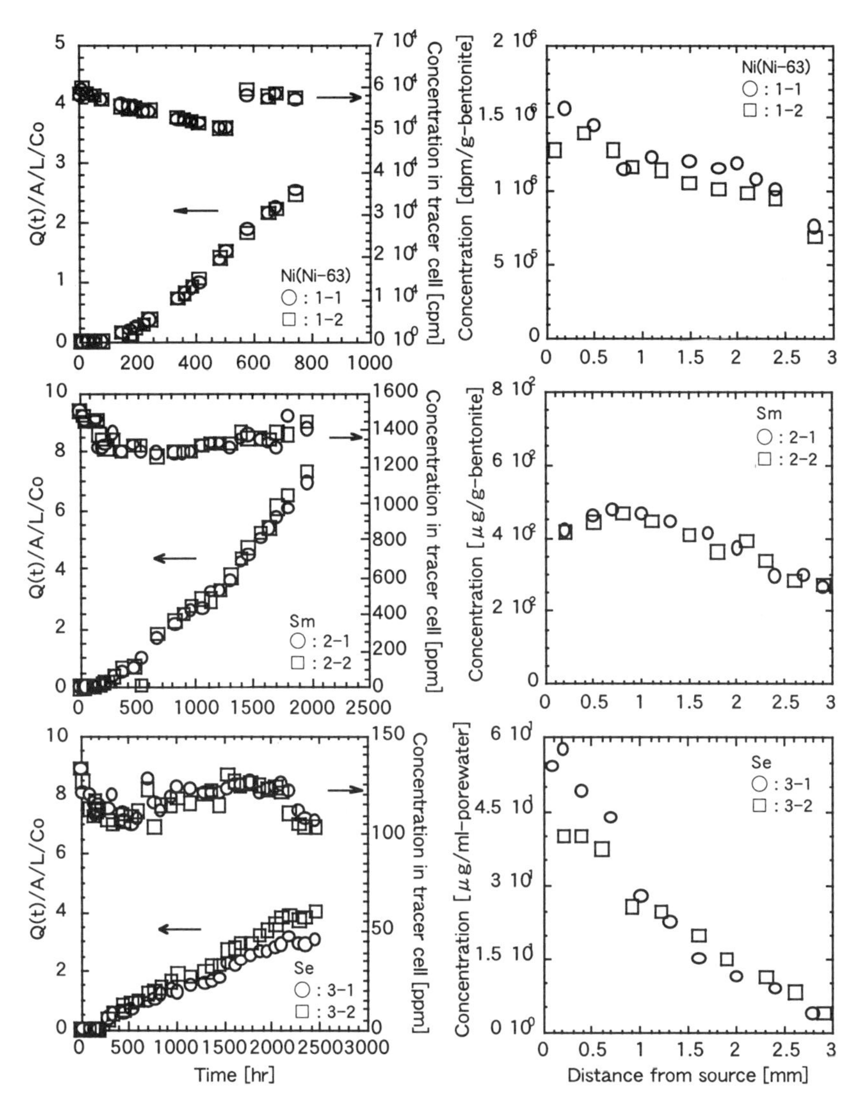

Figure 2 Changes in concentrations of Ni, Sm and Se in the tracer and measurement cells as a function of time in through-diffusion experiments(left) and the concentration profiles of these elements in bentonite(right).

Table III shows obtained De_f, De_t and De values together with pH, Eh and temperature. Both pH and Eh were approximately stable during the experiments. Although Sm is possible to form carbonate complexes such as $SmCO_3^+$ and $Sm(CO_3)_2^-$ in a high pH region[21], since the pH values were kept around 5.5, Sm^{3+} is predominant in this case. Also for Ni, hydroxide such as $Ni(OH)_2(aq)$ is formed in a high pH region[22], but Ni^{2+} is predominant in this pH condition. Similarly the dominant species of Se is estimated to be SeO_3^{2-} from Eh-pH diagrams[22].

Table III De_f, De_t and De values for Ni, Sm and Se

No.	Element (Nuclide)	Species	Effective diffusion coefficient [m2/s] Def	Det	De	pH	Eh [mV]*	Temperature [°C]
1–1	Ni (Ni-63)	Ni^{2+}	5.0x10-11	3.2x10-11	2.6x10-11	5.6±0.1		23.3±0.8
1–2			3.3x10-11	3.1x10-11	2.9x10-11	5.6±0.1		23.3±0.7
2–1	Sm	Sm^{3+}	1.4x10-11	3.0x10-11	1.3x10-10	5.3±0.3		23.5±0.5
2–2			1.5x10-11	2.9x10-11	7.3x10-11	5.4±0.2		23.5±0.5
3–1	Se	SeO_3^{2-}	4.2x10-11	9.2x10-12	6.1x10-12	11.4±0.2	65±30	22.5±1.6
3–2			6.2x10-11	1.2x10-11	8.1x10-12	11.3±0.5	68.1±23.3	22.9±1.6

De_f : effective diffusion coefficient in the filter
De_t : effective diffusion coefficient in the filter plus bentonite
De : effective diffusion coefficient in bentonite
*Eh vs. SCE, Eh vs. SHE can be calculated based on equation proposed by Ostwald; Eh=ORP+0.2415–0.00079(Tc–25)[20], where Eh is the Eh vs. SHE(V), ORP is the Eh vs. SCE(V) and Tc is the temperature(°C).

Ionic Charge and Bentonite Density Dependencies for De

Figure 3 shows the De values of Ni^{2+}, Sm^{3+} and SeO_3^{2-} as a function of ionic charge. The De values increase with increasing ionic charge. The authors[2] have measured De values of HTO, $^{137}Cs(Cs^+)$, $^{99}Tc(TcO_4^-)$, $^{237}Np(NpO_2CO_3^-)$ and $U(UO_2(CO_3)_3^{4-}$ as a function of dry density and reported that the De values were in the order, Cs^+>HTO>anions. Kato et al.[9] have obtained De values of HTO, $^{137}Cs(Cs^+)$ and $^{99}Tc(TcO_4^-)$ as a function of dry density and reported the same result. Moreover, Kato et al.[10] have obtained De values of $^{125}I(I^-)$, $^{36}Cl(Cl^-)$ and $^{14}C(CO_3^{2-})$ as a function of dry density and reported that the De values were in the order, $I^- \geq Cl^- > CO_3^{2-}$. As clear also in studies to date, De values for cations tend to be higher than those of HTO and De values for anions tend to be lower than those of HTO. This cause is generally interpreted to be due to electrostatic interaction. The surface of bentonite is negatively charged and cations are electrostatically attracted, while anions are repulsed. The retardation effect by the later is called anion exclusion. However, detailed mechanism for surface diffusion on solid-liquid interface has not been made clear. The concentration distributions of cations and anions from the surface of solid can be calculated based on electric double layer theory and some studies for the modelling of De based on this theory have been reported[2, 9, 10, 23]. Here the degree of surface diffusion and anion exclusion is quantitatively evaluated based on De data obtained in this study and data reported to date.

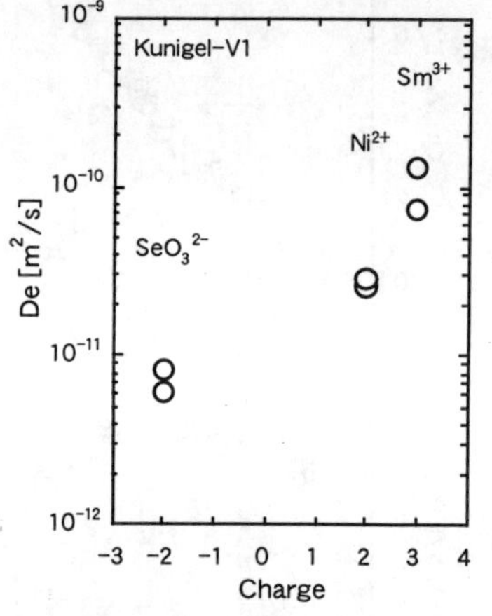

Figure 3 De values of Ni^{2+}, Sm^{3+} and SeO_3^{2-} plotted as a function of ionic charge

The obtained De values were compared with those[2, 9, 10] reported to date. The order of De values was consequently Cs^+>Sm^{3+}>HTO>Ni^{2+}>anions(I^-, Cl^-, CO_3^{2-}, SeO_3^{2-}, TcO_4^-, $NpO_2CO_3^-$ and $UO_2(CO_3)_3^{4-}$), showing a tendency of cations>neutral species>anions. Figure 4 shows De values as a function of dry density reported to date. The order of De values does not necessarily agree to the number of ionic charge. This reason may be because the Do of Ni^{2+}($6.61x10^{-10} m^2 \cdot s^{-1}$(25°C)) is about 1/3 of that of HTO($2.28x10^{-9} m^2 \cdot s^{-1}$ (25°C)[24]). Similarly, since the Do of Sm^{3+}($6.08x10^{-10} m^2 \cdot s^{-1}$(25°C)) is also about 1/3 of that of Cs^+($2.06x10^{-9} m^2 \cdot s^{-1}$ (25°C)[18], it is presumed that the De values of Sm^{3+} were a little lower than those of Cs^+.

Bentonite Density and Ionic Charge Dependencies for FF

The FF values were calculated normalizing De values by Do values. Figure 5 shows the FF values as a function of dry density. The Do values of I^-, Cl^-, CO_3^{2-} [18] and TcO_4^-[25] are calculated $2.05x10^{-9}$, $2.03x10^{-9}$, $9.23x10^{-10}$ and $1.95x10^{-9} m^2 \cdot s^{-1}$ (25°C), respectively. Although the Do values of $NpO_2CO_3^-$ and $UO_2(CO_3)_3^{4-}$ have not been measured, that of $UO_2(CO_3)_3^{4-}$ has been estimated $7.2x10^{-10}$ $m^2 \cdot s^{-1}$(25°C)[26] based on the Stokes equation[17] by the ionic radius calculated from the molecular structure. Since the hydrous radius of ion generally increases with increasing the absolute value of ionic charge, Do decreases with increasing the absolute value of ionic charge, meaning that the Do of $NpO_2CO_3^-$ is similar values to those of ions which take the same absolute value of ionic charge. The Do of HCO_3^- ($1.19x10^{-9} m^2 \cdot s^{-1}$ (25°C)[18]) was thereupon used as the analogue of $NpO_2CO_3^-$.

The FF values showed a tendency to be in the order, Sm^{3+}>Cs^+>Ni^{2+}>HTO>SeO_3^{2-} > $NpO_2CO_3^-$ for $1.8Mg \cdot m^{-3}$ and those for cations wholly showed a tendency to be high for overall density. This indicates a possibility of surface diffusion in cations. While, the FF values for all anions were wholly lower than those of HTO for overall density and the effect of ionic charge on De was clearly found. This indicates a possibility of anion exclusion in anions.

The degree of surface diffusion and anion exclusion was calculated for $1.8Mg \cdot m^{-3}$. The FF values of Sm^{3+}, Cs^+ and Ni^{2+} were approximately 5, 3 and 1.3 times greater than that of HTO, respectively. While, those of TcO_4^-, $NpO_2CO_3^-$ and SeO_3^{2-} were approximately 1/7, 1/16 and 1/5 of that of HTO, respectively. If this effect is caused by electrostatic interaction, this is affcted by the ionic strength of porewater. In this study, simulated porewater with an ionic strength of approximately 1.2 was used. Therfore, there is a possibility that more significant surface diffusive and anion exclusive effect is found for porewater with lower ionic strength.

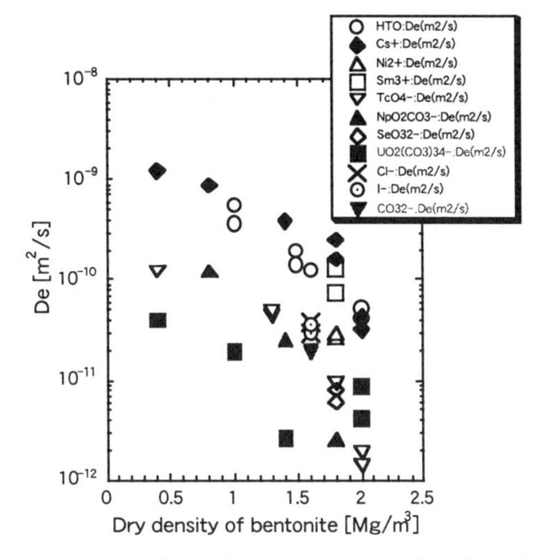

Figure 4 De values of various ions as a function of dry density of bentonite reported to date

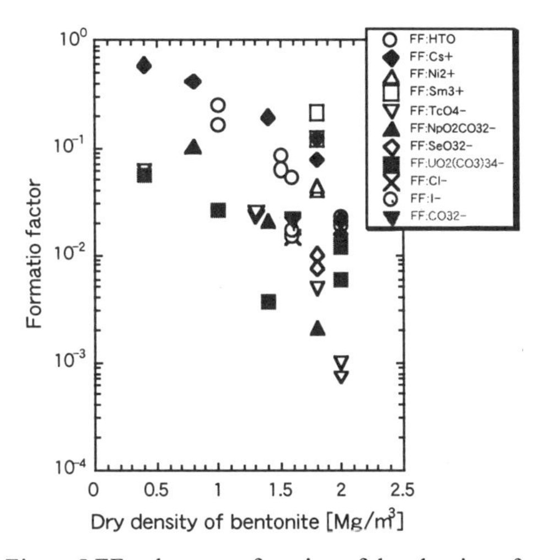

Figure 5 FF values as a function of dry density of bentonite

CONCLUSION

The De values of Ni^{2+}, Sm^{3+} and SeO_3^{2-} in bentonite for $1.8Mg \cdot m^{-3}$ were obtained. The De values, including De data reported to date, were in the order, Cs^+>Sm^{3+}>HTO>Ni^{2+}>anions for overall density, showing a tendency to be in the order, cations>neutral species>anions.

The FF values for the same density were in the order, Sm^{3+}>Cs^+>Ni^{2+}>HTO>SeO_3^{2-}

$>NpO_2CO_3^-$, showing a tendency to be in the order, cations>neutral species>anions. The degree of surface diffusive effect for a dry density of 1.8Mg•m^{-3} was approximately estimated to be 5 times for Sm^{3+}, 3 times for Cs^+ and 1.3 times for Ni^{2+} as much assuming that FF of HTO is a standard. The degree of anion exclusive effect for the same condition was approximately calculated to be 1/7 for TcO_4^-, 1/16 for $NpO_2CO_3^-$ and 1/5 for SeO_3^{2-} as much.

ACKNOWLEDGMENTS

The author would like to thank Messrs. S. Ueta and H. Kato of Mitsubishi Materials Corporation for the performance of diffusion experiments.

REFERENCES

1. H. Sato, PNC TN8410 98-097, 1998.
2. H. Sato and T. Shibutani, PNC Technical Review No.91, PNC TN8410 94-284, 1994 (in Japanese).
3. A. Muurinen, P. Pentilä-Hiltunen, and J. Rantanen, in *Scientific Basis for Nuclear Waste Management X*, edited by J. K. Bates and W. B. Seefeldt (Mater. Res. Soc. Proc. **84**, Pittsburgh, PA, 1987) pp. 803–811.
4. S. C. Chung and M. N. Gray, in *Scientific Basis for Nuclear Waste Management XII*, edited by W. Lutze and R. C. Ewing (Mater. Res. Soc. Proc. **127**, Pittsburgh, PA, 1989) pp. 677–681.
5. D. W. Oscarson and M. N. Gray, Clay and Clay Minerals **42**, 534 (1994).
6. J. W. Choi, D. W. Oscarson, and M. N. Gray, J. Contaminant Hydrology **22**, 189 (1996).
7. A. Muurinen, P. Pentilä-Hiltunen, and K. Uusheimo, in *Scientific Basis for Nuclear Waste Management XII*, edited by W. Lutze and R. C. Ewing (Mater. Res. Soc. Proc. **127**, Pittsburgh, PA, 1989) pp. 743–748.
8. T. E. Eriksen and M. Jansson, SKB 96-16, 1996.
9. H. Kato, M. Muroi, N. Yamada, H. Ishida, and H. Sato, in *Scientific Basis for Nuclear Waste Management XVIII*, edited by T. Murakami and R. C. Ewing (Mater. Res. Soc. Proc. **353**, Pittsburgh, PA, 1995) pp. 277–284.
10. H. Kato, T. Nakazawa, and S. Ueta, in *Scientific Basis for Nuclear Waste Management XXII* (Mater. Res. Soc. Proc. **556**, in press).
11. Japan Nuclear Cycle Development Institute, JNC TN1400 99-010, 1999.
12. M. Ito, M. Okamoto, M. Shibata, Y. Sasaki, T. Danbara, K. Suzuki, and T. Watanabe, PNC TN8430 93-003, 1993 (in Japanese).
13. M. Ito, M. Okamoto, K. Suzuki, M. Shibata, and Y. Sasaki, J. Atomic Energy Soc. Japan, **36** (11), 1055–1058 (1994)(in Japanese).
14. J. Crank, *The Mathematics of Diffusion*, 2nd ed. (Pergamon Press, Oxford, 1975).
15. K. Skagius and I. Neretnieks, KBS TR82-12, 1982.
16. H. Sato, T. Shibutani, and M. Yui, J. Contaminant Hydrology **26**, 119 (1997).
17. R. A. Robinson and R. H. Stokes, *Electrolyte Solutions*, 2nd ed. (Butterworths, London, 1959). p. 317.
18. Y. Marcus, *Ion Properties* (Marcel Dekker, Inc., New York, 1997), pp. 168–170.
19. T. Shibutani, M. Yui, and H. Yoshikawa, in *Scientific Basis for Nuclear Waste Management XVII*, edited by A. Barkatt and R. A. Van Konynenburg (Mater. Res. Soc. Proc. **333**, Pittsburgh, PA, 1994) pp. 725–730.
20. S. Tajima, *An Introduction to Electrochemistry*, 3rd ed. (Kyoritsu, Tokyo, 1986), p.102 (in Japanese).
21. S. Shibutani, PNC Technical Review, No.97, PNC TN8410 96-011, 1996 (in Japanese).
22. D. G. Brookins, *Eh-pH Diagrams for Geochemistry* (Springer-Verlag, Berlin, 1988).
23. H. Sato and M. Yui, in *Scientific Basis for Nuclear Waste Management XVIII*, edited by T. Murakami and R. C. Ewing (Mater. Res. Soc. Proc. **353**, Pittsburgh, PA, 1995) pp. 269–276.
24. Chemical Society of Japan, *Chemical Handbook*, 4th ed. (Maruzen, Tokyo, 1993), p. II-61 (in Japanese).
25. H. Sato, M. Yui, and H. Yoshikawa, J. Nucl. Sci. Tech., **33** (12), 950–955 (1996).
26. T. Yamaguchi, PNC TN1100 96-010, 156–160, 1996 (in Japanese).

ADSORPTION PARAMETERS FOR RADIOACTIVE LIQUID-WASTE MIGRATION

L. C. HULL, M. N. PACE, G. D. REDDEN
Idaho National Engineering and Environmental Laboratory, Idaho Falls, ID 83415

ABSTRACT

Proton titration experiments have been conducted at the Idaho National Engineering and Environmental Laboratory on synthetic goethite and soil in an effort to develop adsorption parameters that will help predict migration of radioactive liquid waste. This is the initial step in a reactive transport project to understand contaminant migration in a system characterized by strong chemical gradients. For this stage, two levels of pretreatment were applied to the soil to remove carbonate minerals and soluble salts to focus on the remaining mineral fraction. Without some sort of treatment or conditioning, native soil has a large buffer capacity that interferes with proton titration experiments. In this report, results are presented from the initial stages of the project.

INTRODUCTION

One of the most significant problems in the application of geochemistry to fate and transport modeling is the lack of an approach to adsorption that can account for changes in the subsurface geochemical environment such as changes in pH, ionic strength, and competing cations. An essential requirement to advance the study of geochemistry is to integrate mechanistic ion exchange and surface complexation into reactive transport models to allow practical application for performance and risk assessment.

At the Idaho National Engineering and Environmental Laboratory (INEEL), liquid radioactive waste from reprocessing spent nuclear fuel is 1 *M* nitric acid with milligram/liter levels of uranium. Past spills of the liquid waste to the soil were rapidly neutralized by calcite in the soil matrix, with concurrent large pH changes, generation of carbon dioxide gas, strong chemical gradients, and adsorption of uranium onto soil minerals. Adsorption parameters that can accommodate associated changes in the subsurface geochemical environment are needed to understand the migration potential of the waste.

A reactive transport project is under way at the INEEL to study the mechanisms controlling contaminant transport in the vadose zone. The working hypothesis of the reactive transport project is that uranium transport in the vadose zone is primarily controlled by surface complexation reactions on mineral oxide surfaces and requires knowledge of uranium partitioning on individual soil components under the influence of the carbonate system. Many laboratory experiments have been conducted on the adsorption of uranium on single, synthetic minerals [1, 2, 3]. However, little work has been done to expand these studies to adsorption on mixtures of minerals or on natural soil materials. Proton titration experiments are being conducted on single minerals and on soil to quantify surface charge and to evaluate buffer capacity. Future experiments will measure adsorption of uranium on these materials.

EXPERIMENT

Goethite Preparation

Goethite (α-FeOOH) was prepared under CO_2-free conditions following a method similar to that of Atkinson, Posner, and Quirk [4]. Sodium nitrate salts in the goethite suspension were removed by decanting and dialysis exchange with deionized water to yield a final solvent conductivity of less than 20 μ S. The product was stored at 3°C to prevent microbial growth. Analysis of the solid by x-ray diffraction (XRD) yielded a spectrum consistent with 100% goethite.

Mat. Res. Soc. Symp. Proc. Vol. 608 © 2000 Materials Research Society

Soil Characterization and Pretreatment

Soil from the spreading areas at the INEEL, which is typical of soil from waste disposal areas, was obtained and sieved through a 2-mm sieve and characterized. Bulk mineralogy was determined by XRD analysis. Organic matter was removed with hydrogen peroxide. An absence of CO_2-generation suggests that little organic matter was present initially. Next, the soil was treated with dithionite—citrate—bicarbonate to quantify reducible crystalline and amorphous oxides [5]. A nominal surface area for the soil sample was measured using the Brunauer, Emmett, and Teller (BET) method [6].

For this stage of the project, the soil was treated to remove calcite. Soil was placed in dialysis tubing and immersed in a sodium acetate/acetic acid buffer solution (pH 5.0) to dissolve carbonates and other soluble salts [5]. Once CO_2 production was complete, the soil was dialyzed against deionized water until the conductivity was less than 20 μ S. The soil was oven-dried at 30°C and sieved through a 2-mm sieve. A riffle splitter was used to obtain uniform subsamples of treated soil used for characterization and adsorption experiments. An XRD analysis of the treated and untreated soil samples showed that the calcite was removed.

After the initial titration experiments on soil treated with the acetate buffer solution at pH 5.0, it was concluded that a more aggressive conditioning of the soil was necessary to remove soluble salts and exchangeable cations because pH stability was difficult to establish during titration experiments. The method used for conditioning the soil was taken from Baeyens and Bradbury [7]. Approximately 0.5 g of soil was added to 50-mL centrifuge tubes. Thirty mL of 0.01 *M* NaCl solution, adjusted to pH 3.5 with HCl, were also added to each tube. The tubes were mixed for 30 minutes on a rotating wheel then taken off and centrifuged. The supernatant was decanted and the pH was measured. This process was repeated until the pH of the supernatant was approximately 3.5. Then the soil in each tube was washed with 0.01 *M* NaCl (without adjusting the pH) following the same process until the pH was approximately equal to the initial pH of the neutral NaCl solution. The conditioned soil samples were stored in a refrigerator at 3°C with 30 mL of 0.01 *M* NaCl.

Titration Experiments

Custom automated titrators were used to conduct titrations of soil and mineral samples. Solutions of hydrochloric acid and CO_2-free sodium hydroxide were prepared. The NaOH solution was calibrated with potassium hydrogen-phthalate (KHP). The HCl solution was calibrated against the calibrated NaOH. A titration was performed using the background solution (0.01 *M* NaCl) to check for internal consistency of the acid and base reagents.

Proton stochiometry titrations of the solids for estimating surface charge were performed under nitrogen. Goethite titrations were performed using 1-g/L suspensions in 0.01 *M* NaCl. Soil titrations were performed on an approximately 1.7-g/L solution of soil in 0.01 *M* NaCl. Both solutions had a surface area concentration of about 57 m^2/L. The pH of the slurry was lowered to 5.5 and allowed to equilibrate overnight while purging with nitrogen to ensure that the CO_2 was removed. The pH was then lowered to 3.0 and allowed to stabilize. Once stable, the slurry was titrated with NaOH to pH 11.0 and then back to pH 3.0 with HCl while constantly purging with nitrogen.

RESULTS

Scanning electron microscopy (SEM) was used to look for crystalline and amorphous iron oxide on the soil. Iron produces higher electron backscattering than most other elements in soil. Iron-containing phases can be seen as lighter areas in Figure 1. Fully crystallized grains of iron minerals, such as magnetite, are not apparent in the soil. Iron-containing phases occur as small particles adhering to the surfaces of mineral grains.

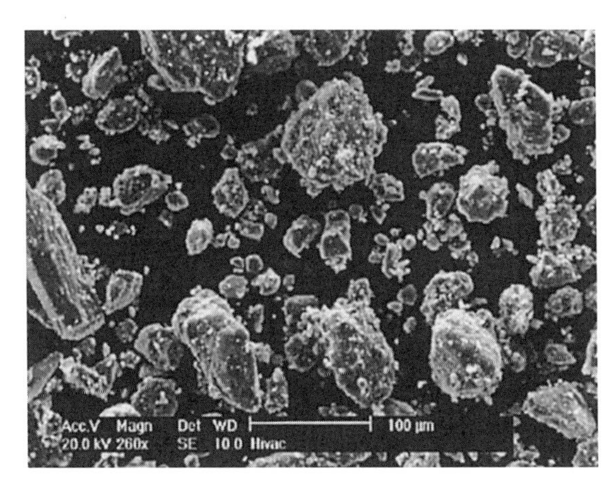

Figure 1. Photograph of soil taken by scanning electron microscopy.

The mineralogy of the soil determined by XRD is shown in Table I. Analysis of the specific experimental material has not been completed, but the table gives a general indication of the material composition. Although iron phases were identified by SEM, no specific iron or manganese minerals have been identified in XRD analysis of bulk soil samples. The clay minerals consist of illite, smectite, and mixed layer illite-smectite [8].

Table I. Mineralogy in percent by weight of soil typical of INEEL waste disposal areas [9].

Mineral	Minimum	Maximum	Median	Mean	Samples
Quartz	23	44	34	35	6
Plagioclase feldspar	10	16	14.5	14	6
Potassium feldspar	11	16	13	14	6
Calcite	0	21	3	5	6
Pyroxene	0	12	10	8	6
Dolomite	0	3	0.5	1	6
Detrital mica and clays	14	28	25.5	23	6

The surface area concentration of the untreated soil was determined to be 30.3 $m^2/g \pm 0.19$ % and that of the treated soil 34.2 $m^2/g \pm 0.08$ %. The density of oxide adsorption sites for the soil was calculated using the following equation [10].

Surrogate sites = (site density * surface area * molecular weight * mole ratio of surrogate oxide to soil oxide * extractable metal) / 10^6

At this point in the process, this information is being used primarily as a guide to identify the oxide phases that will be most important in affecting adsorption. The calculated site densities for the soils are listed in Table II. Iron oxide and manganese oxide provide the greatest number of adsorption sites per gram of soil according to the calculations.

Table II. Estimated oxide site densities for soil.

Soil Oxide	Extractable Metal (μmol/g)	Surrogate Oxide	Site Density[a] (μmol/m^2)	Surface Area[a] (m^2/g)	Molecular Weight (g/mol)	Surrogate Sites (μmol/g)
Iron	23.8	α-FeOOH	27.2	48	89	2.79
Manganese	0.906	δ-MnO_2	370	74	87	2.16
Aluminum	4.10	γ-Al_2O_3	13	117	102	0.32
Silicon	12.7	am-SiO_2	8.3	170	58	1.04

a. Data for the surrogate oxides were taken from Kent et al. [10, Table 3-1].

Acid and base titrations were performed to quantify the pH-dependent surface charge of goethite and soil and to compare the behavior of the soil with a well-studied and well-characterized model mineral. Goethite titrations show rapid response and stability during pH changes. Experiments performed in this study showed that the amount of acid needed to reach a target pH value is greater for treated soil than for goethite (see Figure 2). Even after the target pH value was reached for the treated soil, additional acid was needed to stabilize the pH. Possible explanations for this include the presence of exchangeable cations, diffusion of hydrogen ions into porous grains, or dissolution reactions. The treated soils were then further conditioned with 0.01 *M* NaCl adjusted to pH 3.5 before experimentation on the titrator. The conditioning process is intended to remove adsorbed cations and other soluble salts as well as saturate all exchange sites with sodium. The conditioned soil showed lower buffer capacity than the treated, unconditioned soil, taking less acid to stabilize at a target pH value (Figure 2).

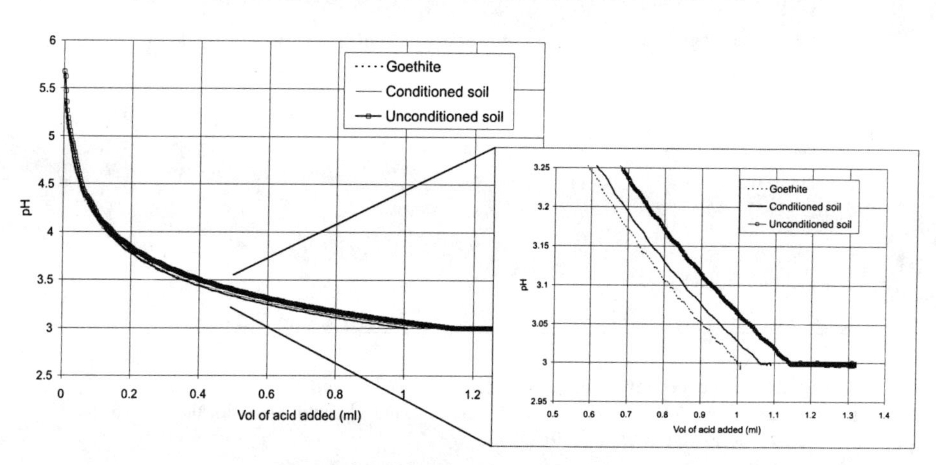

Figure 2. Stability of conditioned soil.

Surface charges were calculated from the titration curves for goethite and conditioned soil. Figure 3 illustrates the relation between surface charge and pH produced from a titration of 1-g/L goethite in 300 mL 0.01 *M* NaCl. Figure 4 illustrates the relation between surface charge and pH from a titration of approximately 1.7-g/L conditioned soil in 300 mL of 0.01 *M* NaCl. The titration curves for goethite are not reversible suggesting that some dissolution may have occurred. The surface charge for the conditioned soil (Figure 4) shows reversibility in pH ranges of 10-7. This reversibility suggests that dissolution of mineral phases in the soil is not a significant problem in that pH range. However, the titration curve for the conditioned soil is not reversible at pH values <7 suggesting that dissolution of mineral phases may occur.

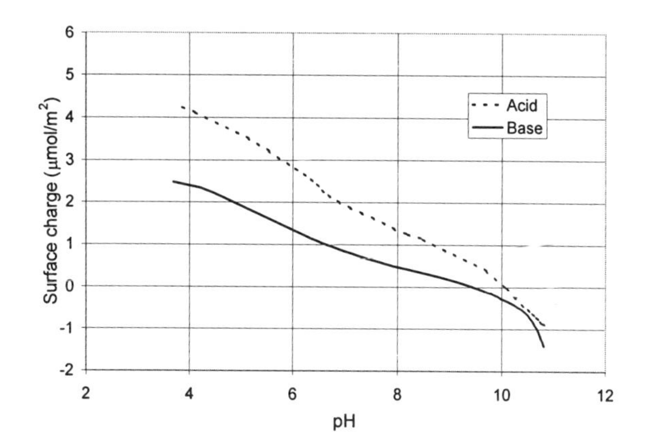

Figure 3. Proton stochiometry titration of goethite (IS = 0.01 M).

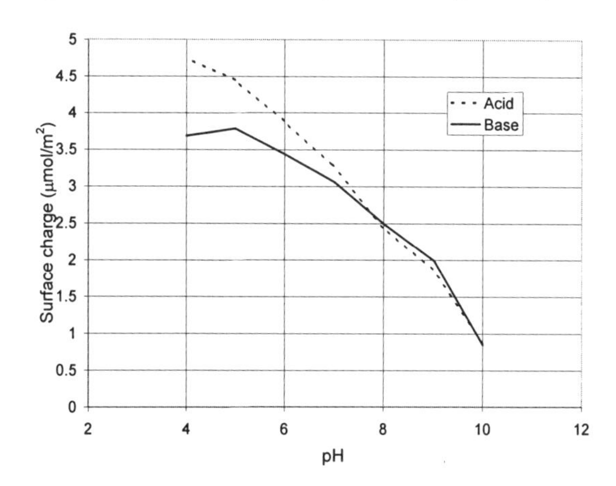

Figure 4. Proton stochiometry titration of conditioned soil (IS = 0.01 M).

CONCLUSIONS

The reactive transport project at the INEEL is developing and evaluating an approach to mechanistic adsorption for natural soils. Surface charge titration experiments with native soils indicate that soluble salts and adsorbed cations interfere with titrations. The soluble salts and adsorbed cations can be removed by aggressive conditioning of the soil at pH 3.5. Conditioning of the soil can improve the experimental performance of the surfaces for titrations, but the effects of soil conditioning on adsorption and the ability to predict adsorption onto nonconditioned surfaces based on the measured adsorption isotherms are yet to be determined.

ACKNOWLEDGEMENTS

This work was conducted by the Idaho National Engineering and Environmental Laboratory for the U.S. Department of Energy under Contract No. DE-AC07-99ID13727.

REFERENCES

1. C. D. Hsi and D. Langmuir, Geochim. et Cosmochim Acta 49, 1931 (1985).

2. G. Redden, J. Li, and J. Leckie, Adsorption of Metals by Geomedia, Academic Press, San Diego, CA, p. 291 (1998).

3. T. E. Payne, G. R. Lumpkin, and T. D. Waite, Adsorption of Metals by Geomedia, Academic Press, San Diego, CA, p. 75 (1998).

4. R. J. Atkinson, A. J. Posner, and J. P. Quirk, J. Phys. Chem. 71, 550–558 (1967).

5. A. Klute, Methods of Soil Analysis, Part 1, Physical and Mineralogical Methods, 2nd ed., Soil Science Society of America, Inc., Madison, Wisconsin, 1986.

6. S. Brunauer, P. H. Emmett, and E. Teller, J. Am. Chem. Soc. 60, 309 (1938).

7. B. Baeyens and M. H. Bradbury, J. of Contam. Hydrology 27, 199 (1997).

8. C. T. Rightmire, Description and Hydrogeologic Implications of Cored Sedimentary Material from the 1975 Drilling Program at the Radioactive Waste Management Complex, (Water-Resources Investigations Report 84-4071, U. S. Geological Survey 1984).

9. R. C. Bartholomay, L. L. Knobel, and L C. Davis, Mineralogy and Grain Size of Surficial Sediment from the Big Lost River Drainage and Vicinity (Open-File Report 89-384, U.S. Geological Survey 1989).

10. D. B. Kent, V. S. Tripathi, N. B. Ball, J. O. Leckie, and M. D. Siegel, Surface Complexation Modeling of Radionuclide Adsorption in Subsurface Environments (NRC FIN A1756, Division of High-Level Waste Management, Office of Nuclear Material Safety and Safeguards, U.S. Nuclear Regulatory Commission 1988).

EFFECTS OF SURFACTANT MODIFICATION ON THE SORPTION BEHAVIOR OF NATURAL ZEOLITES FOR STRONTIUM(2+) AND URANIUM(6+)

James D. Prikryl, F. Paul Bertetti, and Roberto T. Pabalan
Center for Nuclear Waste Regulatory Analyses, Southwest Research Institute
6220 Culebra Road, San Antonio, TX 78238-5166
jprikryl@swri.edu; pbertetti@swri.edu; rpabalan@swri.edu

ABSTRACT

Natural zeolite minerals have strong ion exchange selectivity for alkali and alkaline earth radionuclides, such as strontium and cesium, but only limited sorption affinity for actinides such as uranium. In a previous study, surfactant-modification of the zeolite surface was shown to enhance uranium(6+) sorption at solution pHs greater than about 6. In this study, experiments were conducted to evaluate the ability of surfactant-modified zeolite (SMZ) to sorb both strontium(2+) and uranium(6+). A surfactant-modified zeolite was prepared by treating specimens composed mainly of clinoptilolite with the cationic surfactant hexadecyltrimethylammonium-bromide (HDTMA). The surfactant-modified clinoptilolite was reacted with strontium solutions having a range of strontium concentration, and with a mixed solution of strontium and uranium. Experiments were conducted as a function of pH and in equilibrium with atmospheric CO_2(g). Surfactant-modification of clinoptilolite decreases the ion exchange selectivity of clinoptilolite for strontium at acidic to near-neutral pH. Strontium exchange data were compared to previously developed model predictions of strontium ion exchange behavior on clinoptilolite. The presence of both uranium and strontium in solution had little effect on either the ion exchange selectivity of the surfactant-modified clinoptilolite for strontium or on the enhanced sorption of uranium by SMZ at pH greater than 6.

INTRODUCTION

Zeolites, such as clinoptilolite, possess a net negative structural charge resulting from isomorphic substitution of cations in the crystal lattice. This net negative charge results in the favorable ion exchange selectivity for many cations, such as Cs^+, Sr^{2+}, and NH_4^+, which makes natural zeolites of interest for use in the treatment of wastewaters and the remediation of contaminated sites [1–4]. However, the negative charge also causes natural zeolites to have little or no affinity for oxyanions of toxic heavy metals, such as chromate (CrO_4^{2-}) and selenate (SeO_4^{2-}), and zeolite sorption affinity for actinides such as uranium, which sorb primarily through a surface complexation mechanism, is limited [5,6]. Recent studies have shown that treatment of clinoptilolite with cationic surfactants yields sorbents (surfactant-modified zeolites or SMZs) that have a strong affinity for selenate and chromate [1,7], as well as nonpolar organics, and that also retain their ion exchange selectivity for heavy metal cations, such as Pb^{2+} [8–10]. Enhanced sorption of actinides and other radionuclides would expand the potential application of SMZs in the treatment and remediation of contaminated water. Results of studies investigating actinide sorption on SMZs indicate that surfactant-modification enhances clinoptilolite's ability to sorb U^{6+}, particularly at pHs greater than six where U^{6+} sorption on unmodified zeolite is typically low due to formation of anionic U^{6+} aqueous carbonate complexes [11]. The enhanced sorption of U^{6+} is interpreted to be due to anion exchange with counterions on the external portion of a surfactant bilayer or admicelles [11]. The anion exchange sites form when positively charged surfactant head groups are presented to the surrounding solution where they are balanced by counterions, which can be replaced by other

Mat. Res. Soc. Symp. Proc. Vol. 608

anions in solution [7,12].

In this study, experiments were conducted to evaluate the ability of surfactant-modified zeolite to sorb both Sr^{2+} and U^{6+} when exposed to a mixed radionuclide solution. The surfactant-modified clinoptilolite was reacted with strontium solutions having a range of strontium concentration, and with a mixed solution of strontium and uranium.

EXPERIMENTAL METHODS

Material Preparation

Clinoptilolite used in this study was obtained from Minerals Research Company (Clarkson, New York) and originated as a clinoptilolite-rich tuffaceous rock from Death Valley Junction, CA. The material was prepared as described in Prikryl and Pabalan [11] and Pabalan [13]. The internal and external cation exchange capacities (CECs) of the clinoptilolite were measured using the method of Ming and Dixon [14] as modified by Haggerty and Bowman [1]. The external CEC was determined to be 0.11 meq/g (milliequivalents per gram), while the internal CEC was determined to be 1.86 meq/g, for a total calculated CEC of 1.97 meq/g of clinoptilolite.

Surfactant-modified zeolite (SMZ) was prepared by reacting the clinoptilolite samples with reagent-grade HDTMA (Fisher Scientific). A previous study [1] suggested that anion sorption on the SMZ is maximized when the external CEC of the zeolite is fully satisfied by HDTMA. Therefore, the clinoptilolite samples were treated with a quantity of HDTMA approximately equal to 100 % of the zeolite external CEC. However, recent investigations have concluded that maximization of SMZ sorption capacity may require additional loading and is influenced not only by the initial concentration of HDTMA in solution, but also the time allowed for reorganization of HDTMA on the zeolite surface [15,16]. In this study, 10 g of clinoptilolite powder was treated with 120 mL of a 0.016 M HDTMA solution. The mixture was mechanically shaken for 24 hr at 25 °C, centrifuged, and the supernatant decanted. The solid was then rinsed with deionized water several times, filtered, air-dried, and stored in plastic containers prior to sorption experiments.

Batch Sorption Experiments

Batch sorption experiments were conducted by reacting weighed amounts of clinoptilolite or SMZ with Sr^{2+} solutions or Sr^{2+}- and U^{6+}-bearing solutions over a range of radionuclide concentration and pH (Table 1). Experimental solutions containing Sr^{2+} were prepared by mass from reagent grade $Sr(NO_3)_2$. Uranium-bearing solutions were prepared by diluting a purchased ^{233}U standard (Isotope Products, Inc., Burbank, CA). The pH of experimental solutions was varied in ~0.5 unit increments, typically in the range from 2.5 to 9, by addition of HNO_3 or $NaHCO_3$. Solution ionic strength depended on pH adjustment and varied from 0.003 to 0.009 M.

The experiments were conducted in loosely capped polycarbonate bottles in equilibrium with air and at a solid mass to solution volume (M/V) ratio of 4 g/L. The mixtures were allowed to react for 14 days to allow equilibrium with respect to the sorption reaction(s) and atmospheric CO_2(g) to be attained [1,11,16,17]. The amount of Sr^{2+} or U^{6+} retained on the clinoptilolite was determined from the difference between initial and final radionuclide concentrations in experimental solutions. Strontium concentrations in experimental solutions were determined by spiking solutions with ^{90}Sr (100-pCi/g solution) and then analyzing (Cerenkov counting) the activity of ^{90}Sr and ^{90}Y after secular equilibrium had been reached in the aqueous samples. The ratio of initial to final ^{90}Sr activity was then used in conjunction with the initial total Sr concentration to calculate the final Sr concentration. The concentration of ^{233}U was analyzed by counting its alpha activity. All radioactivity measurements were made using a Packard 2505 TR/AB liquid scintillation analyzer.

Table 1. Summary of experimental conditions for Sr^{2+} and U^{6+}+ Sr^{2+} sorption using Death Valley Junction clinoptilolite (CDV) with and without surfactant modification. Each experiment consisted of a series of solutions from pH 2–9. SMZ: surfactant-modified zeolite.

Experiment Series	Initial Sr^{2+} conc. (ppm)	Initial U^{6+} conc. (ppb)	Substrate	Solid mass (g)	pH range	Soln. vol. (mL)
PCDV-4B, 4G, 4K, 4D, 4E	0.5, 25, 100, 180, 600	-	SMZ	0.1	2–9	25
PCDV-4H, 4J, 4I	25, 100, 180	-	clinoptilolite	0.1	2–9	25
PCDV-4	100	100	clinoptilolite	0.1	2–9	25
PCDV-4A	100	100	SMZ	0.1	2–9	25

RESULTS

Sr^{2+} Sorption Experiments

Results of the Sr^{2+} sorption experiments on HDTMA-modified clinoptilolite at initial Sr^{2+} solution concentrations of 0.5, 25, 100, 180, and 600 ppm are plotted in Figures 1 and 2. In all the experiments, the amount of Sr^{2+} removed from solution generally increases with increasing pH up to pH~7. The capability of the solid to remove Sr^{2+} from solution at acidic to neutral pH is, in general, related to the initial Sr^{2+} solution concentration—the lower the initial Sr^{2+} solution concentration, the greater the percentage of Sr^{2+} removed from solution. Above pH~7 sorption of Sr^{2+} begins to decrease (most easily observed in the K_D plots for the 0.5 and 25 ppm experiments). However, a dramatic increase in Sr^{2+} removal (over 98% Sr^{2+} removed) was observed in experimental solutions with initial Sr^{2+} concentrations of 180 ppm and 600 ppm at pHs greater than 8.5 and 7.5, respectively. Examination of these solutions revealed the formation of a white precipitate in the sample containers. In fact, in most experiments with initial Sr^{2+} concentrations of 100, 180, and 600 ppm, with or without HDTMA, a white precipitate was observed forming in sample containers at alkaline pHs (above pH of 7.5). In each series of experimental solutions where a precipitate was observed, the amount of precipitate increased with increasing pH.

Results of the Sr^{2+} sorption experiments on untreated clinoptilolite and SMZ for initial Sr^{2+} concentrations of 25, 100, and 180 ppm are plotted in Figure 3. Comparison of Sr^{2+} removal by clinoptilolite and HDTMA-modified clinoptilolite shows that untreated clinoptilolite is more effective in removing Sr^{2+} from solution at acidic and neutral pHs. At neutral to alkaline pHs, the magnitude of sorption of Sr^{2+} by untreated and HDTMA-modified clinoptilolite is similar.

U+Sr Sorption Experiments

U+Sr sorption experiments were conducted to determine the effectiveness of HDTMA-modified clinoptilolite in removing multiple radionuclides from solution. The percentages of U^{6+} sorbed on the solids are plotted versus equilibrium pH in Figure 4. Sorption of U^{6+} on untreated and HDTMA-modified clinoptilolite is similar up to a pH of about 6. For both solids, U^{6+} sorption increases with increasing pH to about pH 6 and then decreases at higher pH. However, more U^{6+} is retained on the HDTMA-modified clinoptilolite at higher pH values. It should be noted that an increase in the amount of U^{6+} removed from solution also occurs in association with the formation of the precipitate (Figure 4).

Sr^{2+} sorption behavior in the U+Sr experimental solutions is similar to its sorption behavior in solutions without U^{6+} present for both untreated clinoptilolite and SMZ (Figures 5 and 3).

DISCUSSION

At acidic to neutral pHs, where Sr^{2+} ions are the dominant aqueous species, ion exchange between Sr^{2+} and cations in the clinoptilolite structure is the predominant mechanism for removal of Sr^{2+} from solution [17]. The amount of Sr^{2+} removed from solution through ion exchange is a function not only of the amount of clinoptilolite in the system, but also the ionic strength of the solution, the concentration of Sr^{2+} and competing cations such as Na^+ or H^+, the selectivity or preference of the clinoptilolite for the cations in solution, and the composition of the clinoptilolite [13,17]. The clinoptilolite used in this study contains about 4.8 and 2.8 percent by weight Na_2O and K_2O, respectively [13]. Since the selectivity of clinoptilolite is $K^+ > Sr^{2+} > Na^+$, it is not unexpected that, at the concentrations used in this study, not all Sr^{2+} in solution will be removed by the clinoptilolite, even at concentrations well below the CEC. Indeed, the Sr^{2+} will tend to primarily replace Na^+ in the zeolite, especially at lower concentrations, while larger concentrations of Sr^{2+} are needed to effectively displace the K^+ in the zeolite. This is illustrated in Figure 6, which depicts predicted ion exchange isotherms of Sr^{2+} with a pure Na- and K-clinoptilolite in Sr^{2+} solutions of similar ionic strength to those used in this study. Data from the 25, 100, and 180 ppm Sr^{2+} experiments plot between the Na and K isotherms. The change in Sr^{2+} sorbed by ion exchange with change in pH is due partly to variations in ionic strength caused by pH adjustments, which affect the ion activity ratios of cations in solution, but is primarily due to the competition of H^+ for exchange sites in the zeolite.

Comparison of Sr^{2+} removal by untreated and HDTMA-modified clinoptilolite at acidic to neutral pHs shows that untreated clinoptilolite is more effective at removing Sr^{2+} from solution. This phenomenon might be anticipated since HDTMA occupies at least the ECEC portion of the possible exchange sites in the zeolite. Using data from the 100-ppm experiment, the difference in Sr^{2+} sorbed at pH~4 is about 20%, or 500 μg, of Sr^{2+} in solution. This is equivalent to 0.11 meq/g of zeolite, which is the measured ECEC for the purified clinoptilolite used in this study. Additionally, the difference in uptake of Sr^{2+} between the SMZ and clinoptilolite appears to lessen as pH increases to near neutral values. This might indicate that the sorbed HDTMA, which is stable at low pH [10], enhances the competitive effect of H^+ in the ion exchange process. Therefore, it seems likely that the reduction of Sr^{2+} uptake at acidic to near-neutral pHs in the SMZ is caused by a loss of ECEC and other exchange sites due to the presence of sorbed HDTMA.

The appearance of a white precipitate at alkaline pH in some experimental solutions is not unexpected. Thermodynamic calculations suggest that at 25 °C the experimental solutions become saturated with respect to strontianite ($SrCO_3$) above a pH of 7.2. Strontianite has low solubility and is likely the white precipitate, which is formed when Sr^{2+} reacts with CO_3^{2-} in solution at alkaline pH. Although Sr^{2+} remains the dominant species in solution, in carbonate-containing waters, Sr^{2+} may also be present as carbonate and hydroxy species at alkaline pHs. Unfortunately, though sorption of Sr-bearing species on the HDTMA-modified clinoptilolite may take place at these pH values, it is difficult to determine whether or not Sr^{2+} sorption is occurring given the presence of the precipitate. However, in experiments conducted at 0.5 and 25 ppm Sr^{2+} there was a distinct reduction in the sorption of Sr^{2+} at pH values above 7.

Figure 4 compares U^{6+} sorption on clinoptilolite and SMZ, with and without Sr^{2+} present. At pH values above 6, the SMZ is more effective at retaining U^{6+} than non-modified clinoptilolite. The increase in U^{6+} sorption associated with the appearance of the strontianite precipitate is most likely due to coprecipitation of U^{6+} within the strontianite or due to enhanced sorption of U^{6+} on the Sr-carbonate mineral surface. Comparison of the results of the U+Sr sorption experiments with results of earlier experiments involving the sorption of U^{6+} on SMZ [11] indicates that the presence of Sr^{2+}

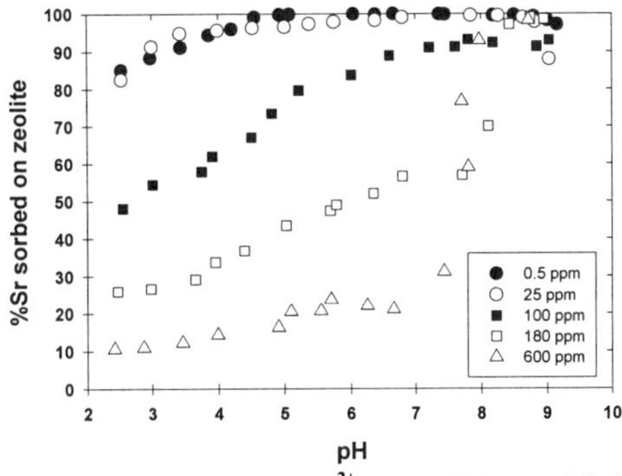

Figure 1. Sorption of Sr^{2+} on HDTMA-modified clinoptilolite at various initial Sr^{2+} concentrations.

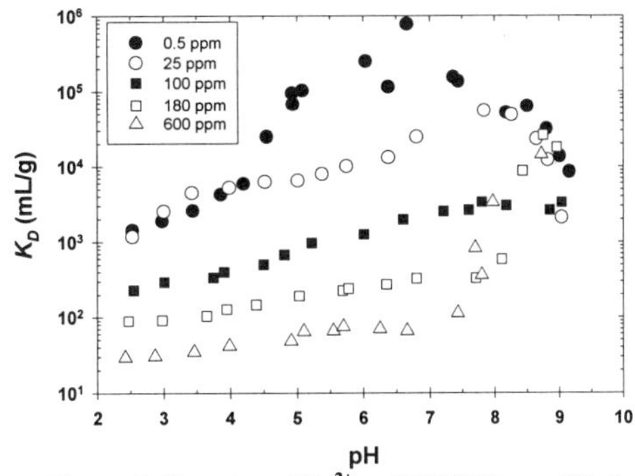

Figure 2. Sorption of Sr^{2+} on HDTMA-modified clinoptilolite at various initial Sr^{2+} concentrations plotted in terms of K_D.

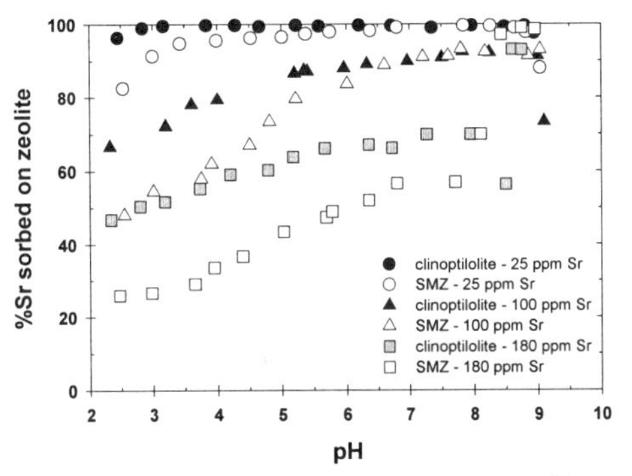

Figure 3. Comparison of sorption of Sr^{2+} on HDTMA-modified clinoptilolite (SMZ) and untreated clinoptilolite at 25, 100, and 180 ppm initial Sr^{2+} concentrations.

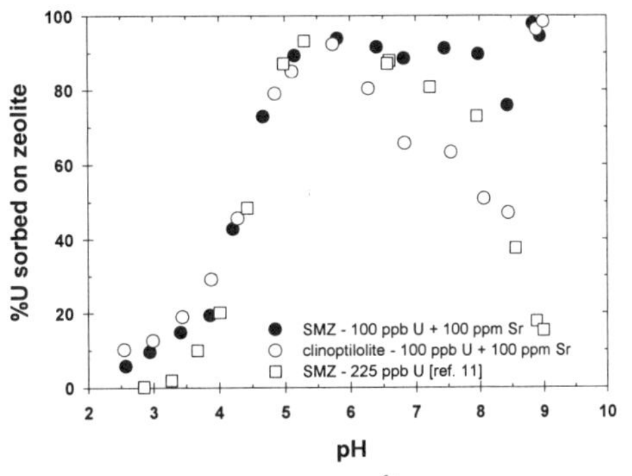

Figure 4. Sorption of U^{6+} on SMZ and untreated clinoptilolite in the presence of Sr^{2+}. Also shown is sorption of U^{6+} on SMZ in the absence of Sr^{2+} [11].

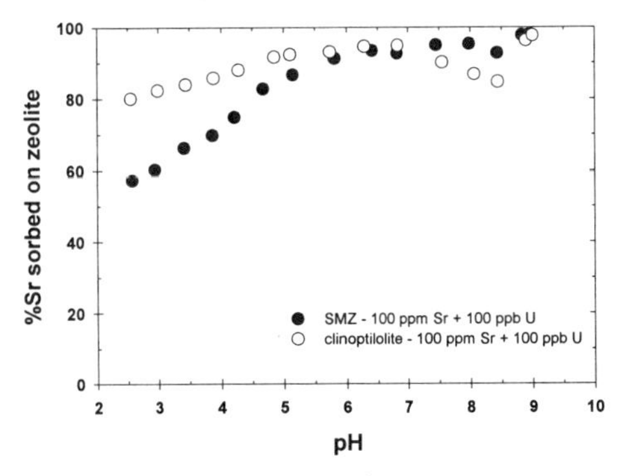

Figure 5. Sorption of Sr^{2+} on untreated and HDTMA-modified clinoptilolite (SMZ) in the presence of 100 ppb U^{6+}.

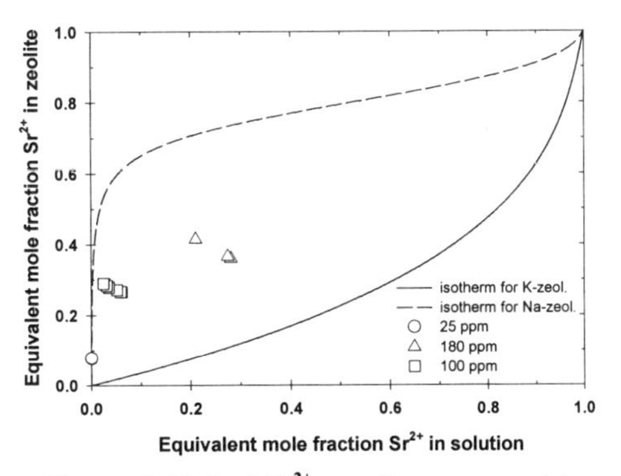

Figure 6. Plot of Sr^{2+} sorption compared to predicted ion exchange isotherms for Na-form and K-form unmodified clinoptilolite.

in solution may affect sorption of U^{6+} by the SMZ at alkaline pHs (Figure 4). However, differences in U^{6+} concentration and the possibility of $SrCO_3$ precipitation preclude concluding that the U^{6+} sorption differences are due to the presence of Sr^{2+}.

CONCLUSIONS

Surfactant-modification of clinoptilolite decreases the ion exchange selectivity of the zeolite for Sr^{2+} at acidic to near-neutral pH. The reduction in sorption is not entirely attributable to the occlusion of the external exchange sites on the zeolite. The effects of pH and counterion interaction with HDTMA require further study. Moreover, recent insights into the behavior of HDTMA and other surfactants with respect to sorption on clinoptilolite [e.g., 7,15,16] need to be considered in future studies if the ion exchange and sorption behavior of SMZs is to be fully characterized and modeled. The presence of both U^{6+} and Sr^{2+} in solution had little effect on the ion exchange selectivity of the SMZ for Sr^{2+}, and possible precipitation of $SrCO_3$ in experimental solutions limits conclusions regarding the impact of Sr^{2+} on increased sorption of U^{6+} by the SMZ at alkaline pHs.

ACKNOWLEDGMENTS

This study was funded by the Southwest Research Institute Internal Research and Development Program.

REFERENCES

1. G.M. Haggerty and R.S. Bowman, *Environ. Sci. Technol.* **28**, p. 452–458 (1994).
2. S. Robinson, T. Kent, and W. Arnold, in *Natural Zeolites '93: Occurrence, Properties, Use*, edited by D. Ming and F. Mumpton (International Committee on Natural Zeolites, Brockport, NY 1995), p. 579.
3. D. Leppert, *Geology* **50**, p. 140 (1988).
4. E. Valcke, B. Engels, and A. Cremers, *Zeolites* **18**, p. 205 (1997).
5. R.T. Pabalan, D.R. Turner, F.P. Bertetti, and J.D. Prikryl, in *Adsorption of Metals by Geomedia*, edited by E. Jenne (Academic Press, San Diego, CA 1998), p. 99–130.
6. F.P. Bertetti, R.T. Pabalan, and M. Almendarez, in *Adsorption of Metals by Geomedia*, edited by E. Jenne (Academic Press, San Diego, CA 1998), p. 131–148.
7. E.J. Sullivan, D.B. Hunter, and R.S. Bowman, *Environ. Sci. Technol.* **32**, p. 1948 (1998)
8. D. Neel and R. Bowman, in *Proc. 36th Annual New Mexico Water Conf., 7-8 November 1991*, Las Cruces, NM 1992, p. 57.
9. M. Flynn, *Sorption of Chromate and Lead onto Surface-Modified Zeolites*. New Mexico Institute of Mining and Technology, Masters Thesis, 1994.
10. R. Bowman, G. Haggerty, R. Huddleston, D. Neel, and M. Flynn in *Surfactant-enhanced Remediation of Subsurface Contamination*, edited by D. Sabatini, R. Knox, and J. Harwell (ACS Symposium Series **594**, Washington, D.C. 1995), p. 54.
11. J.D. Prikryl and R.T. Pabalan in *Scientific Basis for Nuclear Waste Management XXII*, edited by D. Wronkiewicz and J. Lee (Mater. Res. Soc. Proc. **556**, Warrendale, PA 1999), p. 1035–1042.
12. Z. Li and R.S. Bowman, *Environ. Sci. Technol.* **31**, p. 2407–2412 (1997).
13. R.T. Pabalan, *Geochim. Cosmochim. Acta* **58**, p. 4573 (1994).
14. D.W. Ming and J.B. Dixon, *Clays Clay Min.* **35**, p. 463–468 (1987).
15. E.J. Sullivan, J.W. Carey, and R.S. Bowman, *J. Colloid Interface Sci.* **206**, p. 369–380 (1998).
16. Z. Li, *Langmuir* **15**, p. 6438–6445 (1999).
17. R.T. Pabalan and F.P. Bertetti, *J. Soln. Chem.* **28**, p. 367–393 (1999).

EXPERIMENTAL STUDY OF THE NICKEL SOLUBILITY IN SULPHIDIC GROUNDWATER UNDER ANOXIC CONDITIONS

T. CARLSSON, U. VUORINEN, T. KEKKI, H. AALTO,
VTT Chemical Technology, P.O.Box 1404, FIN-02044 VTT, Finland
torbjorn.carlsson@vtt.fi, ulla.vuorinen@vtt.fi

ABSTRACT

Recent literature recognizes a lack of proper chemical data that may be necessary for the calculation of realistic values of the nickel solubility in sulphidic groundwaters. Consequently, in order for the performance assessment of radioactive waste repositories to be conservative, solubility values that clearly exceed those obtained from modelling and experiments have been used to compensate for uncertainty. The safety analysis of decommissioning waste from the Olkiluoto NPP indicates that ^{59}Ni, together with ^{94}Nb, will dominate the dose rates in the waste after about 65 000 years. However, the estimated solubility limit of nickel is uncertain, and more studies are needed.

This paper presents the results from an experimental study of the nickel solubility in sulphidic groundwater under anoxic conditions. The waters used were natural groundwater from Olkiluoto and synthetic saline water, to which sodium sulphide had been added. Two sulphide concentrations, 0.1 and 3 mg/L, were used.

INTRODUCTION

The radionuclides ^{59}Ni and ^{94}Nb are expected to be dominant in the dose rates in the decommissioning waste from the Olkiluoto Nuclear Power Plant after $6.5 \cdot 10^4$ years [1]. The estimated solubility limit of Ni used in the analysis was based on both hydrogeochemical modelling and experimental results, but it was associated with considerable uncertainties. In order for the safety analysis to be conservative, solubility values that clearly exceed those obtained from modelling and experiments have been used to compensate for uncertainty.

The main objective of the work was to determine the solubility of Ni in two sulphidic groundwaters under anoxic conditions. The waters used were natural groundwater from Olkiluoto, Finland, and synthetic saline water. Sodium sulphide was added to both the waters, yielding sulphide concentrations of 0.1 and 3 mg/L. In addition, the effect of the ferrous iron in the waters was studied by adding small amounts of ferrous chloride.

In short, the nickel solubility in the waters was studied by adding nickel chloride spiked with ^{63}Ni to the samples and comparing the initial and final nickel activities using liquid scintillation counting. The initial nickel concentrations were 10^{-6} M and 10^{-3} M. All the experiments were carried out in low carbon-dioxide nitrogen atmosphere at ambient room temperature. The duration of the experiments varied between 1 week and 2 months.

MATERIALS

In this study, two different groundwaters were used: a natural, fresh groundwater from the groundwater station PVA2 at the Olkiluoto VLJ-repository, and a synthetic Olkiluoto saline reference groundwater. In the following, the two waters are referred to as PVA2 and OL-SR, respectively. The natural groundwater was collected on site in a glass bottle flushed with N_2

Mat. Res. Soc. Symp. Proc. Vol. 608

Table I. The main chemical composition of the filtered natural groundwater PVA2 (measured values) and the complete composition of the synthetic groundwater OL-SR (calculated values) before adding nickel, ferrous iron and sulphide. The pH was measured under N_2. N.a. = not analyzed. N.d. = concentration is below the given detection limit.

Component	Unit	PVA2	OL-SR
pH		8.4	8.3
Conductivity	mS/m	60	3 500
B	mg/L	N.a.	0.9
DOC	mg C/L	12	0
DIC	mg C/L	57	0
HCO_3	mg/L	283	0
F	mg/L	N.a.	1.2
Na	mg/L	140	4 800
Mg	mg/L	3.4	55
Al	mg/L	N.d. (< 0.2)	0
Si	mg/L	5.6	0
SO_4	mg/L	26.5	4
S(-II)-tot	mg/L	0	0
Cl	mg/L	48.4	14 600
K	mg/L	5.2	21
Ca	mg/L	9.6	4 000
Fe^{2+}	mg/L	~ 0.02	0
Fe^{3+}	mg/L	~ 0.01	0
Ni	mg/L	N.d. (< 0.01)	0
Sr	mg/L	N.a.	35
Br	mg/L	N.a.	105
I	mg/L	N.a.	0.9

(grade 6.0) and brought to the laboratory where it was immediately taken into a glove box with N_2 atmosphere ($O_2 \leq 1$ ppm, ~28 °C). The PVA2 water was then filtered in the glove box (0.45 μm Gelman AquaPrep capsule) using a slight N_2 pressure and then stored in the glove box until later. The synthetic groundwater was prepared in the glove box. Both of the waters were stored in the ambient glove-box conditions. Table I shows the chemical composition of the two waters. For the natural filtered groundwater, only the measured values for the chemically relevant elements are given, while for the synthetic water the complete theoretical composition is presented. Details concerning the synthetic groundwater and its preparation are found in [2].

EXPERIMENTAL

Briefly, the nickel solubility experiments were performed under N_2 atmosphere ($O_2 \leq 1$ ppm, ~28 °C) in the following way: 1) Four stock solutions with varying concentrations of nickel and ferrous iron were made by dissolving proper amounts of $NiCl_2{\cdot}6H_2O$ (BAKER ANALYSED®, reagent) and $FeCl_2 \cdot 4H_2O$ (MERCK, p.a.) in de-ionized water. The stock solutions were spiked with small amounts of $^{63}NiCl_2$ solution (from Amersham). 2) Two sulphide stock solutions were made by dissolving proper amounts of $Na_2S \cdot 9H_2O$ (BAKER ANALYSED®, reagent) in de-ionized water. 3) The samples were prepared by first adding 0.01 mL of a nickel/ferrous iron stock solution and subsequently 0.01 mL of a sulphide stock solution to 1.50 mL of either PVA2 or OL-SR. The mixing took place in 2.0-mL Eppendorf Safe-Lock centrifuge tubes and the samples were subsequently stored in the sealed centrifuge tubes at the ambient glove-box temperature (~28 °C) until further analysis. 4) In order to determine the ^{63}Ni activity at zero time, a few activity measurements with liquid scintillation counting were taken immediately after sample preparation without any centrifugation. 5) The samples that were to be analyzed after a certain storage time were centrifuged at 23 500 G for 20 min., and aliquots of 0.50 mL were transferred to liquid scintillation vials for subsequent activity measurement. The ^{63}Ni activity was determined with a Wallac scintillation counter 1415. The activity measurements followed, with minor deviations, the procedure described in [3].

The total initial sample concentrations of nickel, ferrous iron and sulphide were as follows:
- $[Ni^{2+}]$: 0.059 mg/L ($1 \cdot 10^{-6}$ M) and 59 mg/L ($1 \cdot 10^{-3}$ M),
- $[Fe^{2+}]$: 0.1 mg/L ($1.8 \cdot 10^{-6}$ M) and 1.0 mg/L ($1.8 \cdot 10^{-5}$ M),
- $[S^{2-}]$: 0.1 mg/L ($3.1 \cdot 10^{-6}$ M) and 3.0 mg/L ($9.4 \cdot 10^{-5}$ M).

The above notation refers to the initial conditions and is not associated with the true speciation of the ions. For example, the sulphide ion S^{2-} is at the actual pH values mainly present as hydrogen sulphide, HS^-.

The sample assemblage included all the eight possible combinations of the above concentrations. Thus, in some cases, it was *a priori* foreseeable that precipitation of, e.g., sulphide might occur (see below). In addition to the above samples, a small number of blank samples were prepared and analyzed in order to determine whether sorption on walls, etc. could disturb the experiments.

PRELIMINARY RESULTS AND DISCUSSION

The two groundwaters used exhibited some notable differences in the water chemistry. In short, the synthetic OL-SR water contained relatively high amounts of Na, Ca, and Cl, but no inorganic or organic carbon. On the other hand, PVA2 water contained only small amounts of Na, Ca and Cl, but some inorganic and organic carbon. However, the two waters exhibited rather similar pH values, see Table I.

The results from an SEM examination show that the PVA2 groundwater may contain a microbial population. It should be noted that the fact that the PVA2 water was passed through 0.45-μm filters at the outset of experiments, does not guarantee that all the particulates are smaller than the nominal 0.45 μm for longer periods of time. Larger particles may form as a result of, e.g., agglomeration. Previous studies by the authors have shown that 4–10 % of the nickel with an initial concentration of $1 \cdot 10^{-5}$ M in PVA2 water was associated with the colloidal phase [4]. Figure 1 shows the nickel concentrations in the PVA2 and OL-SR samples over a 55-day period. In the PVA2 samples with an initial nickel concentration of $1 \cdot 10^{-3}$ M, the concentration drops by about 60 % as a result of precipitation (to approximately $4 \cdot 10^{-4}$ M), while for OL-SR samples the nickel seems to be approximately constant and equal to the initial value of $1 \cdot 10^{-3}$ M over the whole period. These findings agree well with observations made in a previous study [4], in which the solubility of nickel was studied in, among other things, groundwater taken from the same PVA2 groundwater station about 1.5 years earlier. The

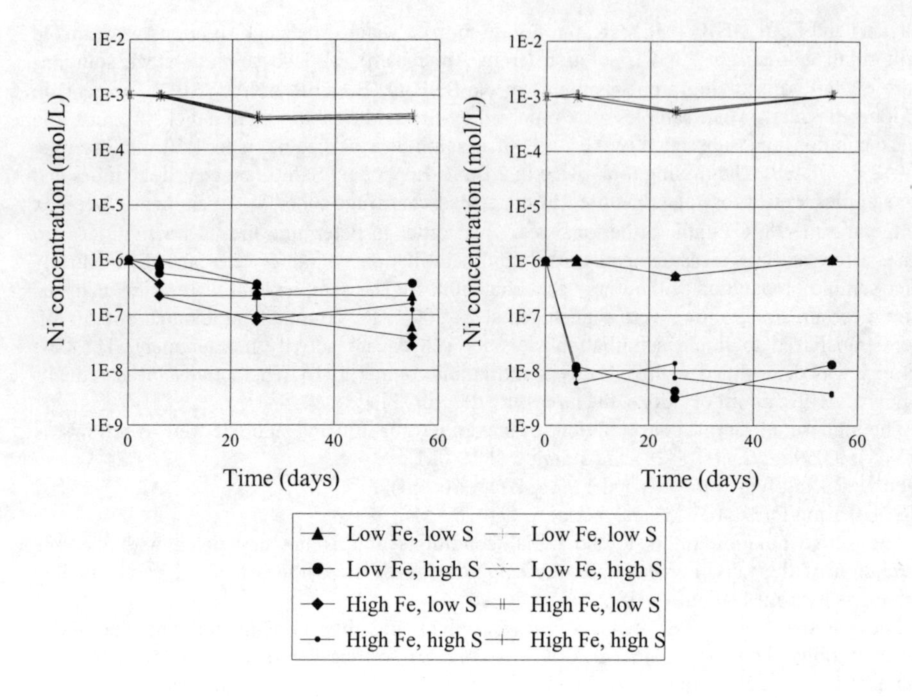

Figure 1. The final nickel concentration as a function of time in natural PVA2 groundwater (*left*) and synthetic OL-SR groundwater (*right*).

chemistry of the PVA2 water used in the two studies was similar, although slight differences were observed mainly due to handling of the groundwater after collection. The reader is referred to [4] for further details. One of the findings in [4] was that an initial nickel concentration of $1 \cdot 10^{-3}$ M in PVA2 water, to which no Fe nor S had been added, remained constant over a 24-week period.

The results in Figure 1 also indicate that the nickel solubility in the PVA2 and OL-SR samples with high initial nickel concentrations were almost unaffected by the different concentrations of ferrous iron and sulphide used. Observations from 100-mL parallel samples in this work showed that light-grey, dark or black precipitates were formed in all these cases. The iron and sulphide concentrations were varied by a factor of 10 and 30, respectively, and the higher initial concentrations for these elements were roughly in the same order of magnitude as that for nickel. Despite these facts, the formation of precipitates did not significantly affect the nickel concentration, which means that nickel was not adsorbed on or coprecipitated with the precipitates to any appreciable extent.

The results differ somewhat from those above when nickel is present at the trace concentration $1 \cdot 10^{-6}$ M. In the OL-SR samples with low sulphide content, nickel remains at the initial level during the whole 55-day period, irrespective of whether the ferrous iron concentration is high or low. In addition, in this case it seems that the possible formation of solids does not influence the amount of dissolved nickel. However, the higher sulphide concentration caused the dissolved nickel concentration to drop to about 1 % of its initial value, which suggests the formation of a nickel sulphide and/or iron/nickel sulphide. However, since the decrease in the nickel concentration is the same at both the high and low iron content, the possible interaction between nickel and iron sulphide seems to be insignificant.

The behaviour of trace nickel in the PVA2 water is clearly different from that in the OL-SR water. In PVA2, the trace concentration of nickel is seen to decrease in all the cases, but does not reach the level of $1 \cdot 10^{-8}$ M found for the OL-SR waters with the high sulphide content. A possible explanation for this is the nickel association with the organic material, although no clear evidence exists so far.

The differences between the various water samples are also manifested in the development of pH over time. Table II shows the measured pH values in PVA2 and OL-SR after 55 days. In the case of PVA2, the pH seems to be governed almost completely by the nickel chemistry; the lower nickel concentration yields a pH of about 9.0, while the higher nickel concentration yields a pH of 8.2. The pH values in PVA2 also seem to be independent of, within the experimental limits, the iron and sulphide content.

The pH values in OL-SR, on the other hand, are clearly different from those measured in the corresponding PVA2 waters. In OL-SR with the lower nickel concentration, the pH is found to decrease to about 7.7–8.2, while in the OL-SR samples with a high nickel concentration the pH drops to 7.3–7.8. According to Table II, the relationships between pH and the various concentrations of nickel, iron and sulphide seem to be more complex in the OL-SR water than in the PVA2 water. However, a thorough discussion of such relationships has to be postponed until further data are at hand.

Modelling of the water chemistry using the geochemical computer code EQ3/6 [5], indicated that the PVA2 and OL-SR samples were over-saturated with regard to several nickel and iron compounds, most of which were sulphides. Although modelling is often a more or less straightforward task, there is a great risk of producing merely meaningless results when modelling the chemistry of nickel sulphide systems. The difficulties associated with modelling nickel in sulphidic groundwaters have recently been thoroughly discussed by Thoenen [6]. Briefly, the problems emanate from large uncertainties associated with the data or even a complete absence of data that may be important in calculating, e.g., nickel concentrations in sulphidic groundwaters. Two monosulphides are likely to form, as according to our modelling millerite (NiS) and pyrrhotite (FeS) are supersaturated in the PVA2 and OL-SR waters.

Table II. Measured pH values in the PVA2 and OL-SR samples after 55 days. Initial pH values of PVA2 and OL-SR: 8.4 and 8.3, respectively.

Total concentrations (mg/L)			Measured pH	
Ni	Fe(-II)	S(-II)	PVA2	OL-SR
0.059	0.1	0.1	8.9	7.7
0.059	0.1	3.0	9.0	8.2
0.059	1.0	0.1	9.0	7.8
0.059	1.0	3.0	9.0	8.1
59	0.1	0.1	8.2	7.8
59	0.1	3.0	8.2	7.7
59	1.0	0.1	8.2	7.3
59	1.0	3.0	8.2	7.4

Unfortunately, the widely used solubility constants for NiS seem to be crude estimates only and accurate solubility data are still missing [6]. In the case of FeS formation, the possible nickel co-precipitation with the solid might be seen in a corresponding decrease in the nickel concentration, but present thermodynamic databases and geochemical models do not allow for the proper modelling of coprecipitation. Therefore, attempts to produce precise, quantitative modelling of the nickel chemistry in the studied waters will not be fruitful at the present stage. In order to attain a more detailed picture of the nickel chemistry in the studied groundwaters, further experimental data have to be collected. Plans for the near future involves XRD studies of the solid phases formed and chemical analyses of the final water chemistry.

SUMMARY

This work is part of an ongoing study on nickel solubility in a natural and a synthetic groundwater containing added amounts of ferrous iron and (hydrogen) sulphide. The solubility experiments were performed under anoxic conditions during 55 days. In cases with an initial nickel concentration of 10^{-3} M, this value was found to be constant in the synthetic water but it dropped about 60 % in the natural water, due to the precipitation of a yet unidentified substance. In synthetic water with an initial trace concentration of nickel (10^{-6} M), the high sulphide content leads to the precipitation of, probably, a nickel sulphide, while the nickel concentration is constant and equal to its initial value, when the sulphide content is low. No effects from ferrous iron are seen on the nickel solubility.

In the natural groundwater, all the nickel concentrations drop to intermediate values. In this case, further data are needed in order to discover any relationships between the nickel content and the water chemistry.

ACKNOWLEDGEMENTS

Teollisuuden Voima Oy and Posiva Oy are gratefully acknowledged for their financial support. Valuable discussions with M. Snellman (Posiva Oy) are appreciated.

REFERENCES

1. T. Vieno, F. Mészáros, H. Nordman, and V. Taivassalo, Nuclear Waste Commission of Finnish Power Companies, Report YJT-93-27, (1993).
2. U. Vuorinen and M. Snellman, Posiva Oy, Helsinki, Working Report 98-61, (1998).
3. T. Carlsson and H. Aalto, VTT Research Notes 1793, (1996).
4. T. Carlsson, H. Aalto, and U. Vuorinen, Mat. Res. Soc. Symp. Proc. 556, Pittsburgh, PA, (1999).
5. T. J. Wolery, Lawrence Livermore National Laboratory, Report UCRL-MA-110662 PT III.
6. T. Thoenen, Nuclear Technology, Vol. 126, (1999), p. 75-87.

EFFECT OF CARBONATE CONCENTRATION ON THE SORPTION OF PLUTONIUM ONTO GEOLOGICAL MATERIALS

G.M.N. BASTON, J.A. BERRY, M. BROWNSWORD, D.J. ILETT, C.M. LINKLATER, C.J. TWEED AND M. YUI*
AEA Technology plc, 220 Harwell, Didcot, Oxfordshire, UK
*Japan Nuclear Cycle Development Institute, Tokai Works, Ibaraki, Japan

ABSTRACT

This paper describes the most recent work in a programme of generic experimental and modelling sorption studies undertaken to increase confidence in the performance assessment for a potential high-level radioactive waste repository in Japan. The sorption of plutonium onto three rock samples was studied as a function of carbonate concentration under reducing conditions. Geochemical modelling was used to assist with experimental design and interpretation of results.

INTRODUCTION

The work described here forms part of the Japan Nuclear Cycle Development Institute (JNC) programme of generic studies undertaken to increase confidence in the performance assessment for a potential Japanese high-level radioactive waste (HLW) repository. In a continuation of work reported previously [1–3], batch sorption experiments have been carried out to study the sorption of plutonium at different carbonate solution concentrations onto samples of basalt, mudstone and sandstone (representing generic rock types from Japan).

One of the aims of the JNC programme is to generate data suitable for use in repository performance assessment calculations. Experimental conditions were therefore chosen in order to simulate, as closely as possible, those anticipated in the vicinity of a potential HLW repository in Japan. Thermodynamic chemical modelling was carried out to help to plan the experiments and to aid interpretation of the results.

EXPERIMENTAL

Batch sorption experiments were carried out in duplicate at room temperature (21±3°C). Sorption was studied from synthetic rock-equilibrated de-ionised water under a nitrogen atmosphere and also under nitrogen atmospheres containing 0.4% and 2.0% carbon dioxide. In addition, sorption from synthetic rock-equilibrated seawater was studied under a nitrogen atmosphere and with a nitrogen / 2.0% carbon dioxide atmosphere. Reducing conditions, to simulate those expected *in-situ*, were obtained by adding sodium dithionite.

Samples of the geological materials and mineralogical descriptions were supplied by JNC [4]. The composition of the basalt was mainly plagioclase, feldspar and quartz. The mineralogy of the mudstone was dominated by quartz, muscovite, calcite and carbonaceous material, while that of the sandstone was predominantly quartz, plagioclase and chlorite. The methodology for the preparation of rock samples, and synthetic rock-equilibrated de-ionised water and seawater has been described previously [1-3]. The liquid : solid ratio was 5:1 (100 cm^3 : 20g). The de-ionised water solutions were of low ionic strength, with carbonate concentrations of ~3 x 10^{-4} M in sandstone- and mudstone-equilibrated water and below detection limit for basalt-equilibrated water (<1 x 10^{-4} M). The seawater solutions were similar to the initial composition of the

Mat. Res. Soc. Symp. Proc. Vol. 608 © 2000 Materials Research Society

TABLE I: PREDICTED CARBONATE CONCENTRATIONS UNDER MIXED NITROGEN / CARBON DIOXIDE ATMOSPHERES

Solution	% CO_2 in atmosphere	Predicted Carbonate Concentration / M
Rock-equilibrated de-ionised water	0.4	8.07×10^{-4} - 1.29×10^{-3}
Rock-equilibrated de-ionised water	2.0	1.35×10^{-3} - 1.85×10^{-3}
Rock-equilibrated seawater	2.0	1.65×10^{-2} - 1.68×10^{-2}

synthetic seawater (carbonate concentration ~5×10^{-4} M). Thermodynamic chemical modelling was used to predict the changes to the solution carbonate concentration and pH values for solutions in equilibrium with the 0.4% and 2.0% carbon dioxide atmospheres. The experimental solutions were adjusted to the predicted higher carbonate concentrations and lower pH values. These values for the three rocks were in the ranges shown in Table I. pH values subsequently remained stable throughout the course of the experiments, indicating constant carbonate concentrations. A predicted risk of precipitation of carbonates in the seawater experiments led to the use of lower calcium concentrations and the omission of magnesium in these cases.

At the end of a three-month equilibration period for each experiment, three phase-separation techniques were employed (centrifugation, and filtration through 0.45μm and 10000MWCO filters as described previously [1–3]). The radionuclides used were plutonium-238 for the sorption experiments, and plutonium-239 as a yield tracer in the preparation of α-spectrometry counting trays. Initial plutonium concentrations were in the range 5×10^{-11} M to 3×10^{-10} M.

TABLE II: PLUTONIUM SORPTION ONTO BASALT, MUDSTONE AND SANDSTONE FROM SYNTHETIC ROCK-EQUILIBRATED DE-IONISED WATER

				R_D / m^3kg^{-1}		
Rock	% CO_2 in atmosphere	Final pH	Final Eh / mV vs S.H.E.	Centrifuged	0.45 μm filtered	10000 MWCO filtered
Basalt	0.0	10.8	-550	0.39 ± 0.04	21 ± 2	120 ± 20
	0.0	10.8	-550	0.36 ± 0.04	27 ± 3	86 ± 9
	0.4	7.4	-510	2.6 ± 0.3	86 ± 13	200 ± 40
	0.4	7.3	-510	3.0 ± 0.4	10 ± 2	130 ± 20
	2.0	6.6	-370	9.2 ± 1.1	580 ± 140	1000 ± 200
	2.0	6.6	-380	8.0 ± 0.9	680 ± 80	1000 ± 100
Mudstone	0.0	9.1	-370	2.4 ± 0.2	25 ± 3	52 ± 7
	0.0	9.1	-390	1.3 ± 0.1	21 ± 2	39 ± 5
	0.4	7.1	-380	0.9 ± 0.1	13 ± 2	110 ± 20
	0.4	7.2	-370	1.6 ± 0.2	23 ± 3	40 ± 9
	2.0	6.4	-360	0.58 ± 0.07	11 ± 1	14 ± 2
	2.0	6.4	-360	1.5 ± 0.2	44 ± 4	47 ± 6
Sandstone	0.0	10.0	-590	0.71 ± 0.07	16 ± 2	37 ± 4
	0.0	10.0	-550	0.92 ± 0.09	17 ± 2	40 ± 6
	0.4	7.1	-330	3.8 ± 0.5	240 ± 40	260 ± 60
	0.4	7.1	-330	5.3 ± 0.7	190 ± 40	300 ± 90
	2.0	6.4	-380	8.3 ± 1.0	650 ± 70	>310
	2.0	6.3	-370	3.6 ± 0.4	210 ± 20	260 ± 30

RESULTS AND DISCUSSION

Results of sorption experiments using equilibrated de-ionised water and seawater are shown in Tables II and III respectively. Uncertainties on R_D values shown in these Tables are $\pm 2\sigma$ and are based on counting statistics alone. R_D was defined by:

$$R_D = V/M \times (C_o - C_t)/C_t$$

where V = volume of solution, M = mass of rock, C_o and C_t = initial and final concentrations of plutonium in solution.

Most of the measured R_D values after filtration are very high (>10 m^3kg^{-1}). In all cases, strong sorption is maintained in the presence of carbonate. To assist with interpretation of thc experimental data, thermodynamic modelling was carried out. The geochemical speciation program HARPHRQ [5] was used, along with the associated HATCHES thermodynamic database (version NEA11) [6] and additional plutonium data supplied by JNC.

In developing the model, there were a number of key uncertainties to consider. Firstly, there is uncertainty as to the thermodynamic stability of key plutonium(III)/(IV) species in solution and the role that carbonate plays in influencing the relative stability of these oxidation states. Secondly, within the literature, there is an incomplete understanding of the effect of carbonate on plutonium sorption. Available data give conflicting results; some suggest high carbonate concentrations are associated with a reduction in sorption [7], whereas other data (AEA Technology unpublished results) suggest sorption is either unaffected or enhanced. As a result, the approach has been to derive a thermodynamic sorption model that is chemically plausible. This provides a valuable interpretative tool with some predictive capability for this complex natural system. The model is based on haematite. This iron oxide phase is of similar crystallinity to the magnetite observed in some of the samples and has an independently parameterised sorption dataset. The authors are not aware of any such datasets for iron(II) oxide mineral phases. Also, the role of iron-bearing minerals such as chlorite and biotite could not be fully assessed in this work due to lack of appropriate data.

TABLE III: PLUTONIUM SORPTION ONTO BASALT, MUDSTONE AND SANDSTONE FROM SYNTHETIC ROCK-EQUILIBRATED SEAWATER

				R_D / m^3kg^{-1}		
Rock	% CO_2 in atmosphere	Final pH	Final Eh / mV vs S.H.E.	Centrifuged	0.45 µm filtered	10000 MWCO filtered
Basalt	0.0	8.4	-520	0.84 ± 0.09	130 ± 20	240 ± 40
	0.0	8.4	-520	0.63 ± 0.06	120 ± 20	97 ± 3
	2.0	7.5	-400	5.7 ± 0.9	96 ± 13	150 ± 20
	2.0	7.5	-410	7.8 ± 1.1	120 ± 20	890 ± 740
Mudstone	0.0	8.4	-470	1.3 ± 0.1	3.5 ± 0.7	3.6 ± 0.6
	0.0	8.3	-480	1.4 ± 0.1	7.3 ± 0.9	140 ± 30
	2.0	7.6	-390	4.7 ± 2.4	1900 ± 500	5800 ± 900
	2.0	7.5	-400	6.6 ± 1.8	790 ± 190	1200 ± 200
Sandstone	0.0	8.4	-510	1.4 ± 0.1	30 ± 5	36 ± 4
	0.0	8.4	-510	2.1 ± 0.2	22 ± 2	23 ± 4
	2.0	7.5	-430	4.3 ± 0.5	11 ± 1	19 ± 2
	2.0	7.5	-430	3.9 ± 0.5	7.0 ± 0.8	9.8 ± 1.2

TABLE IV: PREDICTED AQUEOUS SPECIATION OF PLUTONIUM IN ROCK-EQUILIBRATED DE-IONISED WATER AND SEAWATER / %

	0% CO_2		0.4% CO_2	2.0% CO_2	
	DI Water	Seawater	DI Water	DI Water	Seawater
Basalt					
$Pu(OH)_4$	58				
Pu^{3+}		1	2	9	
$Pu(OH)^{2+}$		48	11	11	2
$Pu(OH)_2^+$	36	25	5		
$Pu(OH)_3$	5				
$Pu(CO_3)^+$		22	79	78	61
$Pu(CO_3)_2^-$		2	3		29
$Pu(CO_3)_3^{3-}$					4
Mudstone					
$Pu(OH)_4$	37				
Pu^{3+}		1	3	18	
$Pu(OH)^{2+}$	2	27	15	13	2
$Pu(OH)_2^+$	51	52	4		
$Pu(CO_3)^+$	5	16	75	63	56
$Pu(CO_3)_2^-$	3		1		34
$Pu(CO_3)_3^{3-}$					6
$PuSO_4^+$			1	6	
Sandstone					
$Pu(OH)_4$	3				
Pu^{3+}		1	4	17	
$Pu(OH)^{2+}$		25	17	14	
$Pu(OH)_2^+$	91	49	4		
$Pu(OH)_3$	2				
$Pu(CO_3)^+$		20	74	66	62
$Pu(CO_3)_2^-$	2	2			29
$Pu(CO_3)_3^{3-}$					4

Values of pH and Eh used in the calculations are those given in Tables II and III.

The predicted aqueous speciation is shown in Table IV. No experimental technique is presently capable of measuring the speciation or oxidation state of plutonium in neutral or alkaline pH solutions because of the very low solubility of plutonium compounds under these conditions. The predicted speciation under the experimental conditions was a mixture of Pu(III) and Pu(IV). A preliminary model based on an independently parameterised Pu(IV) sorption dataset did not adequately reproduce experimental trends. A model based on the sorption of plutonium(III) hydrolysis products and carbonate species was therefore developed. No independent data for plutonium(III) sorption onto haematite were available. The dataset was thus derived by fitting to the de-ionised water data obtained in the current work.

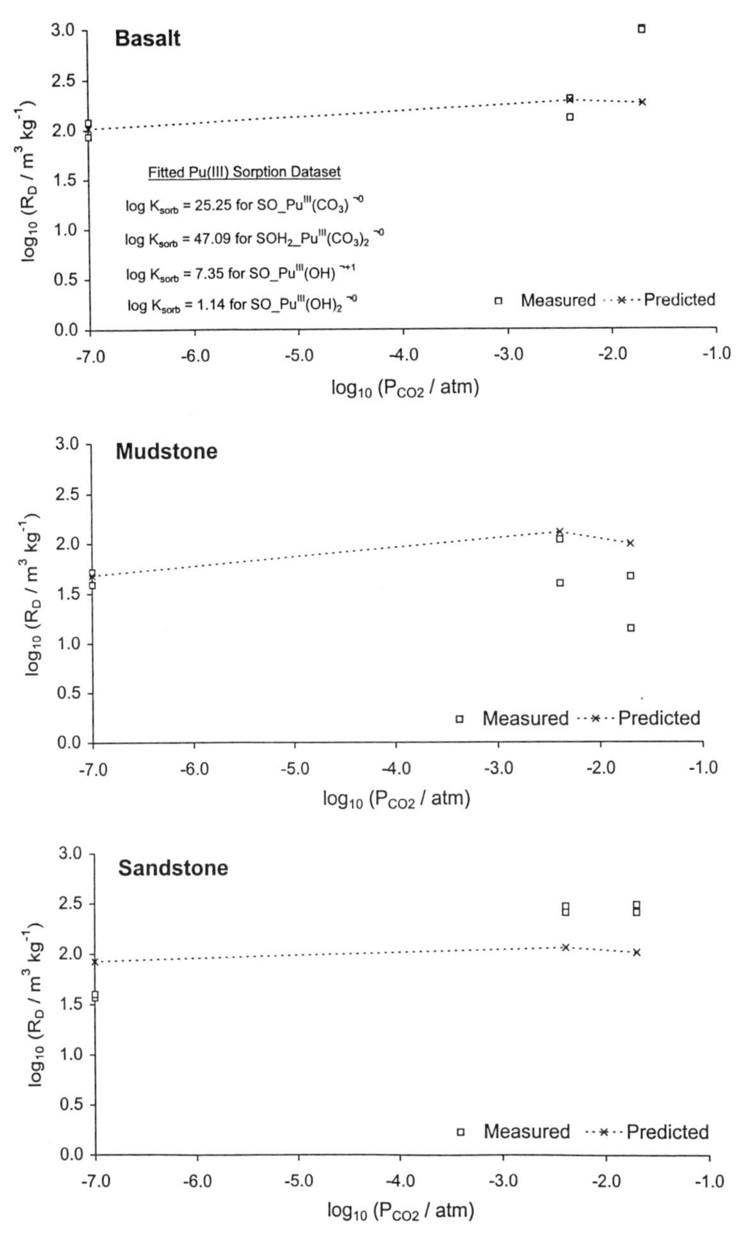

Figure 1 Comparison of the Predicted and Observed Sorption of Plutonium onto Basalt, Mudstone and Sandstone from rock-equilibrated deionised water in the presence of carbonate.

Modelled results for the de-ionised water experiments are compared with the experimental data in Figure 1. The model gave generally good agreement with observed results. The ability of the model to simulate the experimental trends for all three rocks supports the assumption that a similar phase dominates sorption in all three cases. Similar agreement was obtained for the seawater experiments; however in the case of sandstone, the model does not reproduce the observed decrease in sorption as the carbonate concentration is increased. Interestingly, there was no need for rock-specific model refinements e.g. to account for different surface areas or amounts of iron oxide at the surface for each rock type. The modelling illustrates that the observed sorption can be interpreted in terms of strong sorption of both plutonium(III) hydrolysis products and carbonate species onto iron oxide surfaces present in the rocks.

CONCLUSIONS

Results from batch sorption experiments have shown that under reducing conditions, plutonium is generally strongly sorbed onto samples of JNC basalt, mudstone and sandstone. This strong sorption is maintained in the presence of carbonate. The successful application of an iron oxide-based model illustrates that the observed sorption can be interpreted in terms of strong sorption of both plutonium(III) hydrolysis products and carbonate species.

ACKNOWLEDGEMENT

This work was funded by the Japan Nuclear Cycle Development Institute (JNC), Tokai, Japan. Their permission to publish the results is gratefully acknowledged.

REFERENCES

1. G.M.N. Baston, J.A. Berry, M. Brownsword, M.M. Cowper, T.G. Heath and C.J. Tweed, in *Scientific Basis for Nuclear Waste Management XVIII*, edited by T. Murakami and R.C. Ewing (Mater. Res. Soc. Proc. **353**, Pittsburgh, PA, 1995) pp. 989-996.

2. G.M.N. Baston, J.A. Berry, M. Brownsword, T.G. Heath, D.J. Ilett, C.J. Tweed and M. Yui, in *Scientific Basis for Nuclear Waste Management XX*, edited by W. Gray and I. Triay (Mater. Res. Soc. Proc. **465**, Pittsburgh, PA, 1997) pp. 805-812.

3. G.M.N. Baston, J.A. Berry, M. Brownsword, T.G. Heath, D.J. Ilett, C.J. Tweed and M. Yui, in *Scientific Basis for Nuclear Waste Management XXII,* edited by J. Lee and D. Wronkiewicz (Mater. Res. Soc. Proc., Pittsburgh, PA, 1999) in the press.

4. T. Shibutani, private communication (1998).

5. A. Haworth, T.G. Heath and C.J. Tweed, HARPHRQ: A Computer Program for Geochemical Modelling, Nirex Report NSS/R380 (1995).

6. J.E. Cross and F.T. Ewart, Radiochim. Acta **52/53**, 421 (1991), and subsequent releases of this database available from NEA.

7. A.L. Sanchez, J.W. Murray and T.H. Sibley, Geochim. Cosmochim. Acta **49**, 2297 (1985).

TEM Investigation of U^{6+} and Re^{7+} Reduction by *Desulfovibrio desulfuricans*, a Sulfate-Reducing Bacterium

Huifang Xu [a], Larry L. Barton [b], Pengchu Zhang [c], and Yifeng Wang [c]

[a] Department of Earth and Planetary Sciences, The University of New Mexico, Albuquerque, New Mexico 87131. E-mail: hfxu@unm.edu

[b] Department of Biology, The University of New Mexico, Albuquerque, New Mexico 87131.

[c] Sandia National Laboratories, 115 North Main Street, Carlsbad, New Mexico 88220, USA.

ABSTRACT

Uranium and its fission product Tc in aerobic environment will be in the forms of UO_2^{2+} and TcO_4^-. Reduced forms of tetravalent U and Tc are sparingly soluble. As determined by transmission electron microscopy, the reduction of uranyl acetate by immobilized cells of *Desulfovibrio desulfuricans* results in the production of black uraninite nanocrystals precipitated outside the cell. Some nanocrystals are associated with outer membranes of the cell as revealed from cross sections of these metabolic active sulfate-reducing bacteria. The nanocrystals have an average diameter of 5 nm and have anhedral shape. The reduction of Re^{7+} by cells of *Desulfovibrio desulfuricans* is fast in the media containing H_2 electron donor, and slow in the media containing lactic acid. It is proposed that cytochrome in these cells has an important role in the reduction of uranyl and Re^{7+} that is a chemical analogue for one uranium fission product Tc^{7+} through transferring electron from molecular hydrogen or lactic acid to the oxyions of UO_2^{2+} and ReO_4^-.

INTRODUCTION

The solubility of uranium and Tc is dependent on their oxidation state. U^{4+} and Tc^{4+} are sparingly soluble. Under oxidizing environments, uranium will exist in the form of uranyl (UO_2^{2+}) ion that complexes with carbonate (CO_3^{2-}) and organic ligands. When carbonate is present in aerobic solutions, uranyl forms highly soluble metal complexes of $UO_2CO_3^0$, $UO_2(CO_3)_2^{2-}$, and $UO_2(CO_3)_3^{4-}$ [1]. Waste water from the nuclear industries are important because oxidative dissolution of nuclear waste and uranium fission product Tc results in the formation of very soluble oxy-ions, such as of UO_2^{2+}, TcO_4^-, and PuO_2^{2+}. In order to immobilize the uranium in water, uranium may be reduced to insoluble uraninite (UO_2). Recent studies have demonstrated that *Desulfovibrio desulfuricans* and other sulfate-reducing bacteria are capable of reducing uranyl and other metals by enzyme-mediated reactions [2-8]. This biological processe involving reduction of heavy metals can effectively remove from solution. The detoxification process is known as dissimilatory reduction and occurs when electrons from molecular hydrogen (H_2) or lactate are transferred to oxidized metal ions such as selenate [9], or UO_2^{2+} [5]. In at least

Mat. Res. Soc. Symp. Proc. Vol. 608 © 2000 Materials Research Society

one instance, the dissimilatory reduction of uranyl ion is coupled to the growth of a sulfate-reducing bacterium, *Desulfotomaculum reducens* [10].

It is very difficult to Immobilize Tc in aerobic environment, because Tc^{7+} in the form of TcO_4^- is very stable in solution. In this report, we also use sulfate-reducing bacteria of *Desulfovibrio desulfuricans* to reduce ReO_4^-, a chemical analogue for TcO_4^- using electron donors of lactate and hydrogen gas. Based on the similarity of redox potentials of ReO_4^- and TcO_4^- [1], the bacteria that can reduce ReO_4^- may be able to reduce TcO_4^-.

CULTURE, SAMPLE PREPARATION, AND EXPERIMENTAL METHODS

Culture for Uranium Reduction

Bacteria used in the experiments were *Desulfovibrio desulfuricans* DSM 642 which was grown in the following medium as described earlier [7]: 1000 ml distilled water; Na lactate 4 ml of a 65% solution; NH_4 Cl: 2.0 g; Na_2SO_4: 4.0 g; $MgSO_4$: $7H_2O$: 2.0 g; K_2 HPO_4. $7H_2O$: 0.5 g; yeast extract: 1.0g; $FeSO_4$: 20 mg; pH adjusted to 7.4 with 20% KOH. To maintain cultures, 10 ml of growth medium was added to 13 x 125 mm anaerobic tubes fitted with rubber closures. Tubes were autoclaved and flushed with purified nitrogen before they were inoculated. Using anaerobic transfer techniques, 0.1 ml of a two-day culture was introduced into the stopped culture tubes. Incubation was at 35 °C .

To cultivate *D. desulfuricans* for uranium reduction, 1.5 liters of growth medium was placed in a 2-liter anaerobic flask and after autoclaving was flushed with purified nitrogen gas. The inoculum for the 1.5 liters of medium was 200 ml of *D. desulfuricans*. After 2 days of incubation at 35°C, cells were removed by centrifugation at 5,800 g for 30 minutes and washed with sterile, degasses bicarbonate buffer (1.25 g/L). The biomass was resuspended in 20 ml of sterile, deonized water supplemented with 25% (w/v) of acrylamide monomer, 0.25 g (w/v) solution of potassium persulfate, and crosslinkage was accelerated by adding 2.5 ml of a 5% (w/v) solution of 3-dimethylaminopropionitrile. The cell suspension was refrigerated at 4°C until the acrylamide polymerized [8]. The gel was cut into 3 mm cubes, washed with sterile water and placed in a 125 ml serum bottle along with 100 ml of degassed solution containing 5 mM uranium (U^{6+}) acetate (Electron Microscopy Sciences, Fort Washington, PA), 30 mM $NaHCO_3$, and 50 mM Tris-HCl , pH 7.6. Unless specified, chemicals were from Sigma Chemical Co., St. Louis, MO. The bottles were flushed with purified H_2 for 15 minutes and incubated at 35 °C. After 5 days, the reduction of yellow uranium (U^{6+}) was suggested from the appearance of a black film on the surface of the gel [6-8]. The specimens were prepared for TEM experiment by touching C-coated formvar Cu grids to the surface of the wet acrylamide block that had black precipitates and drying the wet Cu grids in air.

Culture for Re Reduction

The culture of *Desulfovibrio desulfuricans* DSM 642 was grown in a lactate-sulfate medium that contained the following in a liter of distilled water: 4 ml of Na lactate, 5 g tryptone (Difco Co., Detroit, MI), 4 g yeast extract (Difco Co., Detroit, MI), 1.5 g Na_2SO_4, 1.6 g $MgSO_4$, and 0.5 g cysteine. HCl. To maintain the culture, monthly transfers were made using 10 ml of medium in 13 x 150 ml anaerobic tubes. For reduction of the element rhenium, 4 liters of a culture of *D. desulfuricans* was grown in anaerobic bottles until stationary phase had been reached and cells were harvested by centrifugation. Cells were suspended in 200 ml anaerobic bottle containing 10 ml of a 50 mM Tris-HCl buffer at pH 7.6 and flushed with purified N_2 for 15 min. To evaluate the reduction of rhenate by cells of *D. desulfuricans*, a Warburg apparatus was used as previously described [11]. The 2 ml reaction mixture in Warburg flasks consisted of 50 mM Tris-HCl, pH 7.6 and 20 mg of cell protein, as determined by Lowry protein analysis, and the gas phase (electron donor) was purified H_2. The reaction was initiated by addition of 2.9 mg of potassium perrhenate (VII) (Aldrich Chem Co., Milwaukee, WI) in 0.2 ml to give a final concentration of 5 mM rhenate in the reaction. Upon completion of the hydrogen oxidation reaction was stopped after 60 minutes, the solution became black. At this time a sample was removed from the reaction mixture and placed on C-coated formvar Cu grids.

All transmission electron microscopy (TEM) and X-ray energy dispersive spectroscopy (EDS) examinations were carried out with a JEOL 2010 HRTEM with an Oxford Link ISIS EDS system. A Li-drifted Si detector with ultrathin window was used for collecting EDS spectra. Point-to-point resolution of the HRTEM is 0.19 nm, an accelerating voltage was 200 keV.

RESULTS AND DISCUSSIONS

Uranyl Reduction

Black uranium precipitates formed on the surface of the PAG after the culture was incubated for 2 days. The uranyl solution changed from yellow to colorless, and it was apparent that uranyl ion was removed from solution [6-8]. TEM images indicate that the black precipitates are nanocrystals of uranium oxide which exist on the surface of the bacterial cells (Figs. 1, 2). Electron diffraction pattern from the nanocrystals indicates the nanocrystals were uraninite (Fig. 2). The average size of the nanocrystals is about 5 nm, and the crystals are anhedral. High-resolution TEM image of the uranium oxide nanocrystals indicates lattice fringes of the uraninite end product of uranyl reduction (Fig. 1). Local areas of the cell surface are rich in uraninite nanocrystals (Fig. 2).

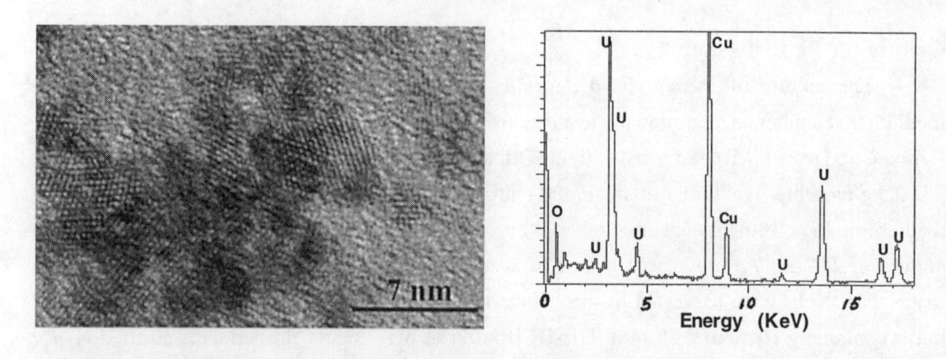

Fig. 1 HRTEM image of the reduced uraninite nanocrystals with lattice fringes (left) and EDS spectrum from the uraninite nanocrystal (right). Cu peaks result from Cu grid holding the specimen.

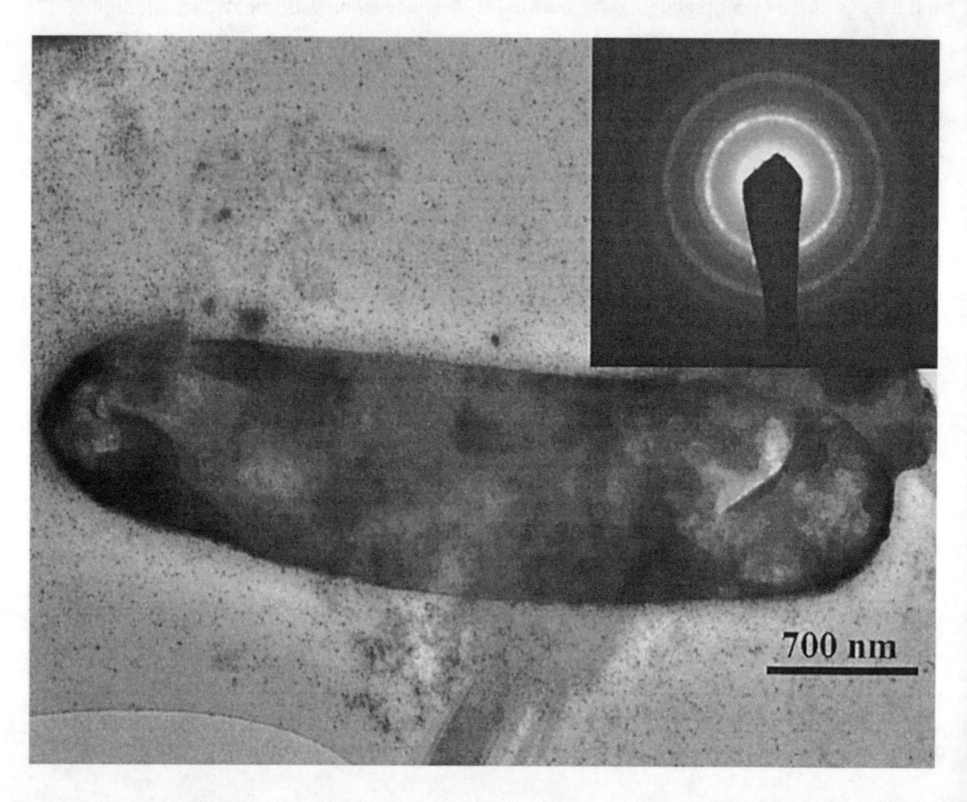

Fig. 2 A bright-field TEM image of the bacteria and the reduced uraninite crystals. Inserted SAED pattern (upper right corner) is from the uraninite nanocrystals.

ReO_4^- Reduction

Black precipitates are dominated by amorphous Re-oxide that is found coated on the surface of the *D. desulfuricans* cells (Fig. 3). The reduced Re may be in the amorphous forms of ReO_2 and $Re(OH)_4$. We also tried electron donor of lactic acid for the reduction of Re. After 5 days, there were only small amount of reduced amorphous Re as black precipitates. The reduction reaction with lactic acid as electron donor is much slower than with H_2.

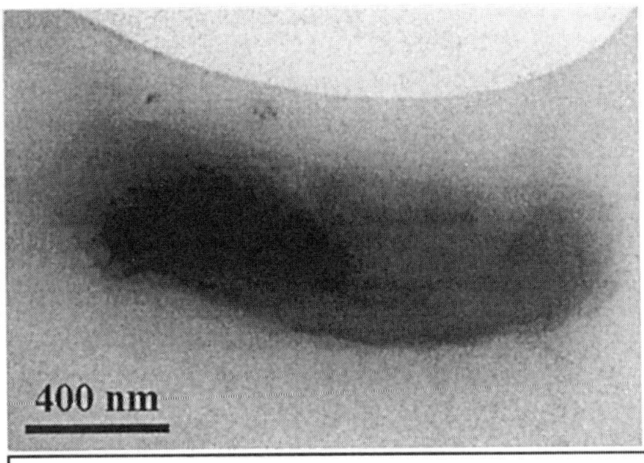

Fig. 3 A TEM image of the one bacterial cell with the reduced amorphous Re-oxide on its surface.

It is proposed that the reduction of uranyl is associated with cytochrome in the cell through metabolism of the cell. The function of cytochrome in the outer membrane would be to participate in detoxification and in this case would be reduction of uranyl through transferring electrons from the organic molecules to uranyl ions. Possible overall reactions occur in the uranyl reduction are:

$$\langle CH_2O \rangle + 2UO_2^{2+} \xrightarrow{\text{Cells of } D.\ desulfuricans} HCO_3^- + 2UO_{2\,(S)} + 5H^+$$

$$H_2 + UO_2^{2+} \xrightarrow{\text{Cells of } D.\ desulfuricans} 2UO_{2\,(S)} + 2H^+ .$$

The basic reaction of Re reduction is similar to that of U reduction. The reaction may be written as:

$$3H_2 + 2ReO_4^- \xrightarrow{\text{Cells of } D.\ desulfuricans} 2ReO_{2\,(S)} + 6H^+ .$$

In above reactions, $\langle CH_2O \rangle$ represents organic acids such as lactic acid or pyruvic acid that may serve as electron donors. Sulfate-reducing bacteria have several different molecular forms of c-type cytochromes. *D. desulfuricans* has a tetraheme (c_3) c-type cytochrome in the periplasmic space between the outer membrane and plasma membrane [12]. *D. desulfuricans* has two additional soluble c-type cytochromes that are either hexaheme or dodecaneheme proteins. Cytochromes have been reported in the outer membrane of *D. vulgaris* [13] and it would be consisted to assume that cytochromes are present in the outer membrane of *D. desulfuricans*. Hydrogenase is found in the periplasmic space of sulfate-reducers [12] and would provide electrons for cytochromes in the periplasmic space with the surface of cell, reductions occurs with the accumulation of uraninite nanocrystals on the cell surface.

CONCLUSIONS

As a conclusion, polyacrylamide gel containing immobilized *D. desulfuricans* cells may be used to recover uranium as uraninite from uranyl-bearing solutions and waste water. The products of uranyl reduction by *D. desulfuricans* are uraninite which occurs as nanocrystals outside the cell and associated with the cell outer membrane. The crystals are anhedral with an average size of 5 nm. The bacteria of *D. desulfuricans* are also able to reduce ReO_4^-. The reduced forms of tetravalent Re could be amorphous ReO_2 and $Re(OH)_4$.

REFERENCES

[1] Brookins, D. (1988). Eh—pH diagrams for geochemistry. pp. 155-157. Springer, New York.

[2] Fude, L., Harris, B., Urrutia, M. M., and Beveridge, T. J. (1994). Appl. Environ. Microbiol., 60, 1525-1531.

[3] Lovley, D. R., and Phillips, E. J. P. (1992a). Appl. Environ. Microbiol., 58: 850-856.

[4] Lovley, D. R., and Phillips, E. J. P. (1992b). Appl. Environ. Microbiol., 60: 726-728.

[5] Lovley, D. R. (1993) Dissimilatory metal reduction. Ann. Rev. Microbiol., 47: 263-290.

[6] Tucker, M. D., Barton, L. L., and Thompson, B. M. (1996). Appl. Microbiol. Biothechnol., 46: 74-77.

[7] Tucker, M.D., Barton, L. L., and Thomson, B. M. (1998a). J. Ind. Microbiol. Biotechnol. 20:13-19.

[8] Tucker, M.D., Barton, L. L., and Thomson, B. M. (1998b). Biotechnology and Engineering, 60:88-96.

[9] Tomei, F.A., Barton, L. L., Lemanski, C. L., Zocco, T. G., Fink, N. H., and Sillerud, L. O. (1995). Journal of Industrial Microbiology 14:329-336.

[10] Tebo, B. M., and Obraztsova, A. Y. (1998). FEMS Microbiol. Lett., 162, 193-198.

[11] Barton, L.L., J. LeGall, J.M. Odom and H.D. Peck, Jr. (1983). 153: 867-871.

[12] Fauque, G., Le Gall, J., and Barton, L. L. (1991) Sulfate-reducing and sulfur-reducing bacteria. In: J. M. Shively and L. L. Barton eds. "Variations in Autotrophic Life." pp. 271-339, Academic Press, London.

[13] Van Ommen Kloeke, F., Bryant, R. D., and Laishley, E. J. (1995). Anaerobe, 1, 351-358.

Cement-Based Materials and Waste Containment

INTERACTIONS OF URANIUM AND NEPTUNIUM WITH CEMENTITIOUS MATERIALS STUDIED BY XAFS

E. R. SYLWESTER[1], P. G. ALLEN[1], P. ZHAO[1], B. E. VIANI[2]
[1]Glenn T. Seaborg Institute for Transactinium Science, Lawrence Livermore National Laboratory, P.O. Box 808, MS L-231, Livermore, CA 94551, sylwester1@llnl.gov
[2]Geosciences & Environmental Technologies Division, Lawrence Livermore National Laboratory

ABSTRACT

We have investigated the interaction of U(VI) and Np(V) actinide ions with cementitious materials relevant to nuclear waste repositories using X-Ray Absorption Fine Structure (XAFS) Spectroscopy. The actinide ions were individually loaded onto untreated as well as hydrothermally treated cements. The mixtures were then equilibrated at varying pH's for periods of 1 month and 6 months.

In all cases uranium was observed to remain in the initial UO_2^{2+} form in the Near Edge (XANES) spectra. The uranium samples show evidence of inner-sphere interactions on both treated and untreated cements at all pH's, with the uranyl complexing with the mineral surface via sharing of equatorial oxygens. On treated cement near-neighbor U-U interactions are also observed, indicating the formation of oligomeric surface complexes or surface precipitates.

Neptunium was observed to undergo a reduction from the initial NpO_2^+ to Np^{4+}. Calculated % reduction showed ca. 15% of Np(V) is reduced to Np(IV) after a 1 month equilibration time. After 6 months higher % reduction of between 40% and 65% was observed. No Np-Np interactions were observed in the EXAFS spectra, which suggests that surface precipitation of NpO_2 is an unlikely mechanism for sorption.

INTRODUCTION

Cementitious materials comprise a large fraction of the near-field environment of US high-level nuclear waste repositories. Interactions of the waste radioisotopes with concrete can be expected to occur after primary containment is breached but before migration to the far field. Concrete has been shown to be a strong sorber of some actinides [1, 2]. However, it is expected that during storage times >1,000 years the infrastructure of these repositories will be subjected to radiative heating as well as variations in humidity. These conditions may alter the chemical or structural characteristics of the cement and the materials' ability to sorb waste radionuclides.

We have used X-Ray Absorption Fine Structure (XAFS) spectroscopy to study samples of Uranium and Neptunium sorbed onto both treated and untreated portland cement in order to determine changes in the speciation (oxidation state and structure) of the radionuclides as a function of cement treatment, sample pH, and time.

EXPERIMENT

Sample Preparation

Samples of crushed (<53μm) treated and untreated concrete were reacted with 0.01 M NaCl containing U or Np at mass/volume ratio of 0.4 g/L at room temperature (23±2 °C). Initial oxidation states were confirmed using UV/VIS spectrometry prior to sample

Mat. Res. Soc. Symp. Proc. Vol. 608 © 2000 Materials Research Society

preparation. Samples were then equilibrated for ca. 5-6 months at pH 9-12 [3]. The aqueous phase was drawn off and the remaining wet paste samples packaged for XAFS analysis. One "fresh" sample each of U(VI) and Np(V) sorbed to treated concrete at pH 10.2 was also prepared with one month equilibration time prior to XAFS measurement.

XAS Spectroscopy

U and Np L_{III}-edge X-ray absorption fine structure (XAFS) spectra were collected at the Stanford Synchrotron Radiation Laboratory (SSRL). The samples were studied on beamline 4-1, using a Si(220) double-crystal monochromator. All spectra were collected at room temperature in fluorescence mode using a Ge solid state detector developed at Lawrence Berkeley National Laboratory[4]. The spectra were calibrated by simultaneous measurement of UO_2 or NpO_2 references, defining the first inflection point at 17166.0 eV and 17606.2 eV for the U and Np L_{III}-edges, respectively.

The XANES and EXAFS data were extracted from the raw absorption spectra by standard methods described elsewhere [5] using a suite of programs, EXAFSPAK [6]. Modeling of the back scattering phases and amplitudes of the individual neighboring atoms was based on FEFF7.2 [7]. Input files for FEFF7.2 were prepared using the structural modeling code ATOMS 2.46b [8], based on the model compound α-$UO_2(OH)_2$ [9]. Microsoft EXCEL™ was used to perform principal component analysis (PCA) on the spectra using standard methodology [10]. The relative contributions from the two oxidation states present in the Np L_{III} XANES spectra were determined with nonlinear least-squares curve fitting using the program PEAKFIT™.

RESULTS

Uranium

Normalized U L_{III}-edges for all samples showed a primary absorption peak at 17.17 keV with a shoulder at 17.18 keV. Both of these features indicate a UO_2^{2+} moeity [10], with the primary absorption peak associated with the $2p_{3/2} \rightarrow 6d$ transition and the shoulder associated with MS resonances from the linear uranyl $[O=U=O]^{2+}$ structure.

Figure 1 shows the raw k^3-weighted EXAFS data and fourier transforms (FTs) for all samples along with the corresponding nonlinear least-squares fits over the k-range 3-13. Table I gives the coordination numbers (N), bond lengths (R), and Debye-Waller disorder factors (σ^2) calculated for all fits shown in Figure 1. There is a clear difference in the raw spectra between the samples on treated concrete and the samples on untreated concrete. Both are dominated in the low-k region by a low frequency oscillation arising from backscattering of the uranyl oxygen atoms. Spectra of the samples equilibrated on treated concrete have an additional high frequency oscillation that appears at k ~ 9, which can be attributed to the presence of larger, more distant backscatterers (i.e. uranium). These features are also reflected in the fourier transformed (FT) spectra. The Fourier transforms represent a pseudo-radial distribution function of the uranium near-neighbor environment, where peaks representing the near neighbor atoms appear at lower R values relative to their true distance from the central atom depending on the phase shift of the backscattering atom. All of the FT spectra show the preservation of the uranyl structure indicated by the presence of a U-O peak at 1.3 Å. All samples also show a split equatorial shell (see also Table I). This bond heterogeneity is consistent with surface adsorption and/or precipitate formation. The samples equilibrated on treated concrete also show a U-U interaction at 3.96 Å indicative of oligomer formation or surface precipitation.

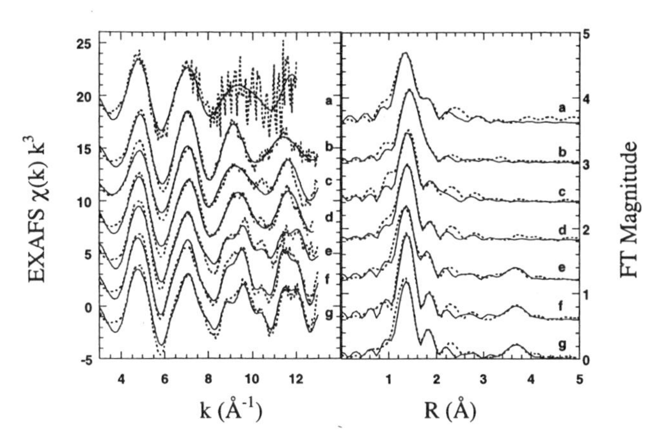

Figure 1: U L_{III}-edge EXAFS data for samples on (a) treated concrete, pH 10.2, "fresh," (b) untreated concrete, pH 12.3, (c) untreated concrete, pH 11.3, (d) untreated concrete, pH 10.3, (e) treated concrete, pH 11.3, (f) treated concrete, pH 10.3, (g) treated concrete, pH 9.3. The dashed line is the experimental data, and the solid line represents the best theoretical fit to the data. LEFT: raw k^3-weighted data. RIGHT: Fourier transforms.

Table I. U EXAFS curve-fitting results.

Sample	Shell	N[a]	R (Å)[a]	σ^2 ($Å^2$)[b]
Treated concrete, pH 9.3	U-O_{ax}	2.00	1.81	0.00201
	U-O_{eq1}	4.33	2.30	0.00776
	U-O_{eq2}	1.82	2.48	0.00776
	U-U	0.64	3.96	0.00076
Treated concrete, pH 10.3	U-O_{ax}	2.00	1.82	0.00115
	U-O_{eq1}	3.38	2.28	0.00606
	U-O_{eq2}	2.22	2.44	0.00606
	U-U	1.07	3.96	0.00338
Treated concrete, pH 11.3	U-O_{ax}	2.00	1.81	0.00268
	U-O_{eq1}	2.90	2.26	0.00330
	U-O_{eq2}	2.16	2.42	0.00330
	U-U	1.51	3.96	0.00603
Untreated concrete, pH 10.3	U-O_{ax}	2.00	1.82	0.00184
	U-O_{eq1}	3.33	2.28	0.00687
	U-O_{eq2}	1.70	2.45	0.00687
Untreated concrete, pH 11.3	U-O_{ax}	2.00	1.83	0.00242
	U-O_{eq1}	4.56	2.29	0.00967
	U-O_{eq2}	1.36	2.53	0.00967
Untreated concrete, pH 12.3	U-O_{ax}	2.00	1.83	0.00216
	U-O_{eq1}	2.54	2.23	0.00163
	U-O_{eq2}	1.44	2.39	0.00163
"Fresh," treated concrete, pH 10.3	U-O_{ax}	2.00	1.81	0.00255
	U-O_{eq1}	3.40	2.30	0.00418
	U-O_{eq2}	1.69	2.49	0.00418

[a]The 95% confidence limits for the bond lengths (R) and coordination numbers (N) for each shell are: U-O_{ax}: ±0.001 Å; U-O_{eq1}: ±0.003 Å and ±0.3; U-O_{eq2}: ±0.01 Å and ±0.2; U-$U_{(1-3)}$: ±0.005 Å and ±0.3, respectively.

[b]Debye-Waller factors for U-O_{eq1} and U-O_{eq2} were linked in the fits.

Neptunium

Figure 2 shows the normalized Np L_{III}-edges for three representative samples; the "fresh" sample on treated concrete at pH 10.2, and equilibrated samples on treated concrete at pH 9.3 and untreated concrete at pH 10.3. Also shown for comparison are absorption spectra for NpO_2^+ and NpO_2 reference compounds. The experimental samples initially appeared to contain a mix of oxidation states or species. PCA analysis confirmed the presence of two components and target testing on four spectra representing Np^{3+}, Np^{4+}, NpO_2^+, and NpO_2^{2+} species resulted in a positive fit for Np^{4+} and NpO_2^+, and a negative fit for Np^{3+} and NpO_2^{2+}. The percentage of NpO_2^+ was subsequently determined with PEAKFIT using linear combination least-squares fitting of pure Np^{4+} and NpO_2^+ absorption spectra to the sample spectra. These results are shown in Table II.

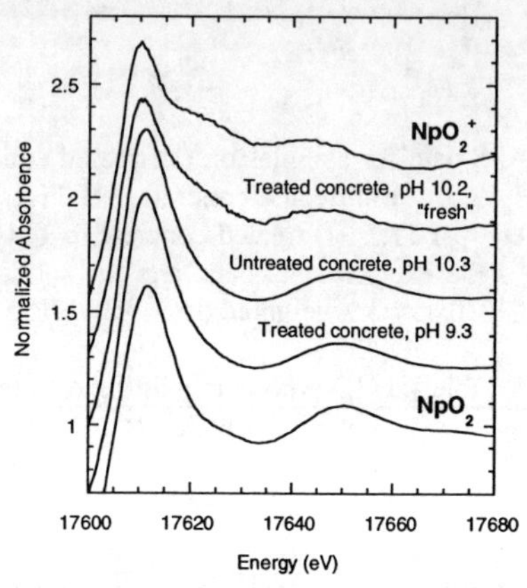

Figure 2: Comparison of the normalized Np L_{III} XANES for three of the samples with XANES spectra for pure NpO_2 and NpO_2^+.

Table II. Np EXAFS curve-fitting results and % NpO_2^+.

Sample	Shell	N[a]	R (Å)[a]	σ^2 (Å^2)	% Np(V) XANES[b]	% Np(V) EXAFS[b]
Treated concrete, pH 9.3	Np-O_{axial}	0.90	1.82	0.00300	36.3	45.0
	Np-O	4.4	2.39	0.00915		
Treated concrete, pH 10.3	Np-O_{axial}	1.2	1.86	0.00300	47.6	62.0
	Np-O	5.6	2.39	0.01535		
Treated concrete, pH 11.3	Np-O_{axial}	0.85	1.85	0.00300	44.9	43.0
	Np-O	4.2	2.39	0.00971		
Untreated concrete, pH 10.3	Np-O_{axial}	0.71	1.84	0.00300	37.8	36.0
	Np-O	4.6	2.38	0.00940		
Untreated concrete, pH 11.3	Np-O_{axial}	0.74	1.84	0.00300	38.8	37.0
	Np-O	4.7	2.40	0.00902		
Untreated concrete, pH 12.3	Np-O_{axial}	1.0	1.86	0.00300	45.9	50.0
	Np-O	3.4	2.39	0.00421		
Fresh, treated concrete, pH 10.2	Np-O_{axial}	1.6	1.86	0.00300	78.9	77.0
	Np-O	3.4	2.44	0.00792		

[a]The 95% confidence limits for the bond lengths (R) and coordination numbers (N) for each shell are: Np-O_{axial}: ±0.004 Å and ±0.05; Np-O: ±0.003 Å and ±0.3, respectively.
[b]The 95% confidence limits for % Np(V) are ±2.5% for determination by EXAFS Np-O_{axial} fraction and ±7% for determination by XANES component fitting.

Figure 3 shows the Np L_{III} raw k^3-weighted EXAFS and FT spectra. The FTs show only a single broad peak containing axial and equatorial O interactions from NpO_2^+ as well as Np^{4+}-O interactions. This FT peak moves to higher R as a function of sample equilibration time indicating a change in the Np(V):Np(IV) ratio with longer equilibration times. Curve fits were performed using the axial Np(V)-O shell expected to occur at ~1.85 Å and a Np-O shell at ~2.4 Å representing a combination of equatorial oxygens bonded to Np(V) and oxygens bonded to Np(IV).

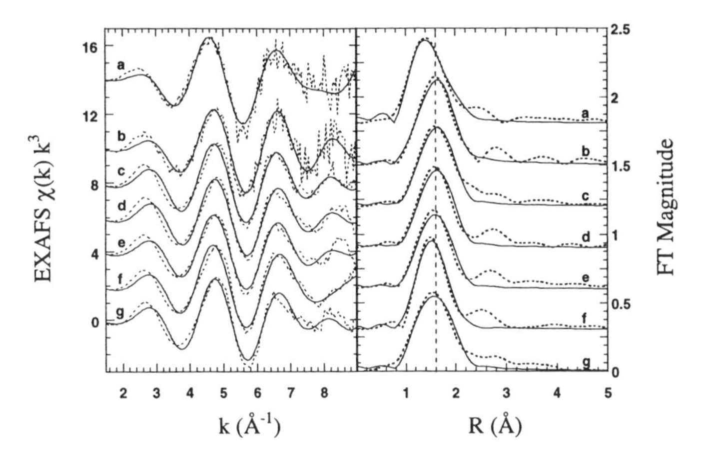

Figure 3: Np L_{III}-edge EXAFS data for samples on (a) treated concrete, pH 10.2, "fresh," (b) untreated concrete, pH 12.3, (c) untreated concrete, pH 11.3, (d) untreated concrete, pH 10.3, (e) treated concrete, pH 11.3, (f) treated concrete, pH 10.3, (g) treated concrete, pH 9.3. The dashed line is the experimental data, and the solid line represents the best theoretical fit to the data. LEFT: raw k^3-weighted data. RIGHT: Fourier transforms.

To compare the EXAFS data with the quantitative results from the XANES analysis, % Np(V) was calculated based on the coordination number of the Np(V)-O_{ax} shell. The results of both methods agree within error for six of the seven samples and confirm that reduction is correlated to equilibrium time but not pH or hydrothermal treatment.

Reduction of Np(V) in solution followed by precipitation of Np(IV) mixed oxides has been proposed as an alternate mechanism to sorption for Np partitioning to the solid phase in our experiments. However, our EXAFS spectra show no evidence of NpO_2 precipitate, which would be indicated by Np-Np interaction at R ~ 4 Å.

Preliminary batch solubility studies as a function of pH show the observed solubility to be independent of solution-cement equilibration time. Furthermore, solubility does not match calculated solubility curves for NpO_2 for $\log\beta_1$ values between 2.6-5.4 modeled in equilibration with calcite. This preliminary evidence agrees with the XAS data and supports a mechanism where Np(V) is sorbed onto the surface, followed by the kinetically slow reduction to Np(IV) while in contact with the surface.

CONCLUSIONS

XANES analysis of uranium sorbed onto concrete showed that U remains as UO_2^{2+}. Uranium EXAFS analysis showed that all samples contained a split equatorial shell, which is evidence for inner-sphere surface complexation. Spectra of uranium on hydrothermally treated concrete also contained U-U interactions, signifying oligomer formation or the formation of a surface precipitate. These results suggest that there is a significant change in U-surface structure with concrete hydrothermal treatment.

The Np XANES showed reduction from NpO_2^+ to Np^{4+}, with the amount of reduction observed to increase as a function of equilibrium time. This reduction was determined quantitatively to be ~15% in the fresh sample and ~50% in the aged samples. The amount of reduction was observed to be independent of either pH or hydrothermal treatment of the concrete. The neptunium EXAFS also showed a reduction of NpO_2^+ to Np^{4+}, and yielded quanitiative results in good agreement with the more rigorous analysis of the XANES region. The absence of Np-Np interactions in the EXAFS spectra provides evidence that neptunium partitioning is controlled by sorption of the Np(V), followed by reduction, and not by reduction followed by formation of Np(IV) surface precipitates. Although the mechanism for Np reduction is unclear, there is no evidence that hydrothermal heating of concrete affects Np speciation.

ACKNOWLEDGEMENTS

This work was performed under the auspices of the U.S. Department of Energy by Lawrence Livermore National Laboratory under Contract W-7405-Eng-48. This work was partly supported by the Yucca Mountain Site Characterization Project at LLNL, and was done (partially) at SSRL, which is operated by the Department of Energy, Division of Chemical Sciences. Additional XAFS experimental support was provided by D. Caulder and W. Lukens and the Actinide Chemistry Group at LBNL

REFERENCES

1. T. G. Heath, D. J. Ilett, C. J. Tweed, *Mat. Res Soc. Symp. Proc.* (1996).
2. B. F. Greenfield, D. J. Ilett, M. Ito, R. McCrohon, T. G. Heath, C. J. Tweed, S. J. Williams, M. Yui, *Radiochim. Acta* **82**, 27-32 (1998).
3. M. Atkins, F. P. Glasser, *Mat. Res Soc. Symp. Proc.* (1990).
4. J. J. Bucher, N. M. Edelstein, K. P. Osborne, D. K. Shuh, N. Madden, P. Luke, D. Pehl, C. Cork, D. Malone, P. G. Allen, *Rev. Sci. Instr.* **67** (1996).
5. R. Prins, D. E. Koningsberger, Eds., X-Ray Absorption: Principles, Applications, Techniques for EXAFS, SEXAFS, and XANES (Wiley-Interscience, 1988).
6. G.N. George, I.J. Pickering, "EXAFSPAK, a suite of computer programs fro analysis of X-ray absorption spectra." (SSRL, Stanford, CA, 1995).
7. J. Mustre de Leon, J. J. Rehr, S. Zabinsky, R. C. Albers, *Phys. Rev. B* **44**, 4146-4156 (1991).
8. B. Ravel, "ATOMS, a program to generate atom lists for XAFS analysis from crystallographic data." (University of Washington, Seattle, WA, 1996).
9. J. C. Taylor, H. J. Hurst, *Acta Cryst. B* **27**, 2018-2022 (1971).
10. E. R. Malinowski, Factor Analysis in Chemistry (Wiley-Interscience, 1991).
11. P. G. Allen, D. K. Shuh, J. J. Bucher, N. M. Edelstein, C. E. A. Palmer, R. J. Silva, S. N. Nguyen, L. N. Marquez, E. A. Hudson, *Radiochim. Acta* **75**, 47-53 (1996).

The leaching behavior differs from element to element. Nickel displays a curve that is almost proportional to $t^{1/2}$ whereas cobalt and chromium show a declining tendency. Manganese has an extremely fast leaching phase during the first 24 hours, which almost stops thereafter and is followed by a constant slope.

DISCUSSION

The time-dependent variation of the leaching in the tests shows no uniform behavior of the different elements, but a very uniform behavior of the leaching from the different concrete types of a specific substance. The results for nickel and manganese relate very well to the diffusion theory and models, but others do not, indicating the importance of knowledge of each specific element's binding characteristics and behavior. Various physical and chemical interactions can alter the leaching behavior and give the leaching curves a different appearance. The most important interactions are changes of pH, structural changes and the effect of carbonation.

When a cement-based material is leached, the pH -value will start to decrease, affecting the leaching rate of species whose solubility is pH -dependent. The drop of pH is mainly caused by leaching of two alkalis (Ka^+ and Na^+) and the different hydrates in the cement matrix. The alkalis are not firmly bound to the matrix and will therefore cause the first decrease in pH, from pH = 14-13.5 down to pH = 12.6 [9].

Sodium and potassium hydroxides entail the greatest and most rapid effect on pH. At pH = 12.6, the leaching of $Ca(OH)_2$ starts, causing a very slow dissolution of the cement matrix. The rate depends on the solubility, but also on the diffusivity of interfering ions, e.g. chloride ions. Still, it has been shown [9] that the dissolution front of $Ca(OH)_2$ follows the square root of time. All the hydrates will successively dissolve and a dissolution front will be formed for each hydrate. The leaching of calcium is possible to calculate with the diffusion as well as the shrinking core model, but the latter can also be used to calculate the pH -profile. As a decrease of pH would have a dissolving effect on pH-sensitive metals, the effect would be a continuous increase of the leaching rate.

The leaching of calcium and its hydrates may also alter the structure of the material, being part of the constitutive phases of the material. Leaching of each of the hydrates implies an increase of the porosity, following the different dissolution fronts. An additional factor is cracks in the solid. With a width > 1 μm, they can separate highly leached areas from almost unleached zones, while increasing the permeability and facilitating the entry of leachant [10]. The result will be a locally increasing $D(x,t)$. In leaching tests, it is unlikely that structural changes would have any significant consequences, but in real constructions they are likely to have a big impact.

The penetration of carbon dioxide is present in a moistened material as well as a dry material. The ingress rate is about 1/5 to a dry material. The reaction gives rise to formation of calcium carbonate, with several effects on the leaching. The pH in the pore water will decrease to pH < 9. Further, $CaCO_3$ -formation will change the porosity. Additionally, the uptake of carbonate during saturated conditions gives rise to precipitation of $CaCO_3$, which may block the transport of dissolved substances [10]. An increase of CO_2 also appears to increase the severity of cracking. Carbonation is in other words giving effects that both can increase and decrease the leaching rate, and it is not obvious which of the effects have the biggest impact.

Most of these interaction are diffusion governed processes which implies that the deviation from the proposed models would be linear to $t^{1/2}$ as well, or in other words, that a leaching curve taking these effects into account, would still be proportional to $t^{1/2}$. However, the test results in Figure 4 show a slope for cobalt and chromium that is not constant. This could imply one of two possibilities: There could be interactions that are not diffusion governed, or the limit for the total amount that is available for leaching at that time, could be reached. This leads to an important discussion on the definitions of the available concentration; the concentration that creates the potential for diffusion governed leaching.

To be able to quantify the diffusion coefficients, a correct estimation of the total available fraction for leaching must be made. The value of the availability in NEN 7341 is derived from a concrete which has lost its immobilization effect completely. To use such a value gives a diffusion coefficient for a material that does not function any longer. The question is whether this availability is relevant. Our judgment has been that those values are not relevant since the condition in the availability test simulates an extremely long-term stage, which is not relevant for a concrete during its lifetime in a building. Furthermore, the physical and chemical interactions described earlier may occur in reality and/or during an experiment. Therefore, it can be difficult to make long-term predictions based on diffusion coefficients from a short-term test.

CONCLUSIONS

1. The leaching behavior depends to a large extent on the binding characteristics of the substance.
2. The diffusion models can be used as a basis to predict the leaching for some metals.
3. The concrete type does not change the leaching behavior significantly.
4. Much more research is needed on the understanding and quantification of the binding of substances in cement-based materials.

ACKNOWLEDGMENTS

This work was supported by MISTRA, the Swedish Foundation for Strategic Environmental Research, as part of the research program Sustainable Building. Professor Johan Claesson is greatly acknowledged for contributing the mathematical background and analytical solutions in Equations (1)-(5).

REFERENCES

1. I. Hohberg, G. J. De Groot, A. M. H van der Veen, W. Wassing in *Waste Materials in Construction: Putting Theory into Practice* (Proc., Elsevier Science, 1997), pp. 217-228.

2. P. Schiessl, I. Hohberg, (Proc. of the Mario Collepardi Symp, Rome, Italy, 1997, pp. 27- 48.

3. Å. Andersson in *Utilizing ready-mixed concrete and mortar,* edited by R K Dhir and M. C. Limbachiya (Proc., Dundee, 1999), pp. 345-355.

4. NEN 7341 02.95. Leaching tests. Determination of the leaching availability of inorganic compounds, 1995.

5. NEN 7345 03.95. Leaching tests. Determination of the leaching behaviour of inorganic components from building monolithic waste materials with the diffusion test, 1995.

6. European Commission. Report EUR 17869 EN, 1997, 68 pp.

7. P. Moszkowicz, F. Sanchez, R. Barna, J. Méhu, J. Talanta, 46 (1998).

8. P. Moszkowicz, R. Barna, F. Sanchez, H. R. Bae, J. Méhu in *Waste Materials in Construction: Putting Theory into Practice* (Proc., Elsevier Science, 1997), pp. 491-500.

9. B. Gerard and L Tang in *Material Science and Concrete Properties* (Proc., Toulouse, 1998) pp. 215-222.

10. M. Andac and F. P. Glasser, J. Cement and Concr. Res. 29, pp. 179-186 (1999).

DIMENSIONAL ANALYSIS OF IONIC TRANSPORT MECHANISMS IN CEMENT-BASED MATERIALS

R. BARBARULO[1-2], J. MARCHAND[1†] AND S. PRENE[2]
(1) Centre de Recherche Interuniversitaire sur le Béton,
Université Laval, Sainte-Foy, Canada, G1K 7P4
(2) Division Recherche et Développement – Département EMA
Électricité de France, Les Renardières, 77818 Moret-sur-Loing, France

1. ABSTRACT

In order to investigate the validity of the local equilibrium assumption, a dimensional analysis of various ionic transport problems currently encountered in civil engineering has been performed. The dimensional analysis allows comparing, on a theoretical basis, the rate of ionic transport to the rate of chemical reaction. This approach has been applied to various practical cases. The analysis clearly shows that the local equilibrium assumption is verified in most practical cases.

2. INTRODUCTION

In most models recently developed to predict the service-life of construction materials, the influence of on-going chemical reactions on the mechanisms of transport is usually taken into account by simply assuming the existence of a local chemical equilibrium [1]. According to this hypothesis, the rate of precipitation (or dissolution) of the various species in solution should be intrinsically much faster than the rate of transport. The validity of this assumption rests on the observations that, in most degradation cases involving construction materials, chemical reactions usually progress as fronts originating from the external surfaces of the solid [2].

In order to provide more information on the subject, a comprehensive investigation of the mechanisms of ionic transport in construction materials was performed [3]. The validity of the local chemical equilibrium assumption was studied from a theoretical point of view using the dimensional analysis technique. This method allows to study the relative influence of various phenomena on the local chemical equilibrium. After a brief description of the technique, the main results of this study will be summarized.

3. COUPLING TRANSPORT AND CHEMICAL REACTIONS IN CEMENT-BASED MATERIALS

Consider a saturated porous material of porosity Φ made of a inert matrix plus a soluble phase φ. This pure phase is constituted of two compounds X_a and X_b in the stoechiometric proportions ν_a and ν_b. The liquid phase in the saturated material contains the ionic species X_a and X_b in the respective concentrations c_a and c_b. In a first approximation, chemical activity effects can be neglected and the conditions of equilibrium can therefore be expressed in terms of the concentration of each species in solution:

† Current address: Building Materials Division – NIST, Mail Stop 8621, Gaithersburg, MD 20899, U.S.A.

Mat. Res. Soc. Symp. Proc. Vol. 608 © 2000 Materials Research Society

$$c_a^{\nu_a} \times c_b^{\nu_b} = K \tag{1}$$

where K is the equilibrium constant of phase φ.

According to equation (1), precipitation – dissolution reactions can occur in the system to restore the chemical equilibrium. For instance, ions can precipitate to form more solid phase φ. When the solid phase φ is dissolved, its components X_a and X_b are released in solution.

The precipitation – dissolution reactions are generally induced by the transport of ions within the material pore structure. In a saturated medium, the diffusion of ions X_i (i = a or b) in the liquid phase can be described by the Nernst-Planck equation [1, 4]. The unidimensional version of this equation (homogenized over a representative elementary volume of the material) is given by:

$$\frac{\partial c_i}{\partial t} = D_i \frac{\partial}{\partial x}\left(\frac{\partial c_i}{\partial x} + \frac{z_i F}{RT} c_i \frac{\partial V}{\partial x}\right) - \frac{(1-\Phi)}{\Phi}\left(\frac{\partial c_{i-s}}{\partial t}\right)_\varphi \qquad i = a \text{ or } b \tag{2}$$

where c_i is the concentration of the ionic species X_i in the liquid phase, z_i its valence, D_i the diffusion coefficient of this species in the porous environment, $c_{i\text{-}s}$ the concentration of compound X_i in the solid phase. V is the so-called diffusion potential, which can be calculated using the Poisson equation:

$$\frac{\partial^2 V}{\partial x^2} + \frac{\Gamma F}{\varepsilon}\left(z_a c_a + z_b c_b\right) = 0 \tag{3}$$

where Γ is the tortuosity of the liquid phase and $\varepsilon = \varepsilon_0 \varepsilon_r$ the dielectric constant of the medium.

Chemical reactions are described by the term $\left(\frac{\partial c_{i-s}}{\partial t}\right)_\varphi$ that quantifies the exchange between the solid phase and the liquid phase of a given ionic species X_i. A general way of describing the dissolution – precipitation reaction is to assume that the speed of the dissolution – precipitation reaction will increase as the system deviates more and more from its equilibrium condition. Equation (2) thus becomes:

$$\frac{\partial c_i}{\partial t} = D_i \frac{\partial}{\partial x}\left(\frac{\partial c_i}{\partial x} + \frac{z_i F}{RT} c_i \frac{\partial V}{\partial x}\right) + \nu_i . k . \left(K - c_a^{\nu_a} . c_b^{\nu_b}\right) \qquad i = a \text{ or } b \tag{4}$$

where k is the kinetic coefficient of dissolution – precipitation of the phase φ. Any variation in the concentrations of the various ionic species within the reactive porous medium can be described on the basis of equations (3) and (4).

4. DIMENSIONAL ANALYSIS

The main advantage of using equation (4) is that it does not rely on the local equilibrium assumption. Furthermore, the kinetic nature of the chemical reaction is clearly underlined by the last term of the right-hand side of the expression. The rate of reaction can therefore be compared to the rate of ionic transport. Such a comparison can be done by using a special analytical technique called the dimensional analysis.

4.1. Dimensionless equations

In order to perform the dimensional analysis, all equations have to be first rewritten using dimensionless terms such as $\bar{t} = \frac{t}{\tau}$, $\bar{x} = \frac{x}{L_0}$, $\bar{c} = \frac{c}{c_0}$ and $\bar{V} = \frac{V}{V_0}$, where τ is a time, L_0 a length, c_0 a concentration and V_0 an electrical potential to be defined further on. By introducing these terms in equations (3) and (4), one finds:

$$\frac{\partial^2 \bar{V}}{\partial \bar{x}^2} = \left(\frac{c_0 L_0^2 \Gamma F}{\varepsilon V_0} \right) \left(z_a \bar{c}_a + z_b \bar{c}_b \right) \tag{5}$$

$$\frac{\partial \bar{c}_i}{\partial \bar{t}} = \left(\frac{\tau D_i}{L_0^2} \right) \frac{\partial}{\partial \bar{x}} \left(\frac{\partial \bar{c}_i}{\partial \bar{x}} + z_i \left(\frac{F V_0}{RT} \right) \bar{c}_i \frac{\partial \bar{V}}{\partial \bar{x}} \right) + \nu_i . \left(k \tau c_0^{\nu_a + \nu_b - 1} \right) \left(\frac{K}{c_0^{\nu_a + \nu_b}} - \bar{c}_a^{\nu_a} \bar{c}_b^{\nu_b} \right) \tag{6}$$

$$i = a \ or \ b$$

According to Buckingham's π theorem, if a system is described by n independent dimensional quantities, and if m fundamental units are necessary to express these quantities, all relations between the n quantities can be expressed by using $(n - m)$ dimensionless numbers (or groups) [5]. In the present case, the system is described with twelve dimensional quantities: c_0, τ, L_0, V_0, D_a, D_b, k, K, T, F, R, ε. The six units necessary to the description of the system are: length, time, mass, quantity of matter, temperature, and current intensity. Therefore, according to the theorem, the system can be fully described on the basis of the six (12-6) dimensionless numbers appearing in equations (5) and (6):

$$\left(\frac{\tau D_a}{L_0^2} \right), \left(\frac{D_b}{D_a} \right), \left(\frac{F V_0}{RT} \right), \left(k \tau c_0^{\nu_a + \nu_b - 1} \right), \left(\frac{K}{c_0^{\nu_a + \nu_b}} \right), \left(\frac{c_0 L_0^2 \Gamma F}{\varepsilon V_0} \right).$$

It should be emphasized that it is always possible to create additional dimensionless numbers by combining any of the six previous one. This might be useful, for instance, to investigate the influence of a specific phenomenon on the behavior of the entire system [6].

4.2. Values for τ, L_0, c_0 and V_0

The quantities τ, L_0, c_0 and V_0 appearing in equations (5) and (6) are called the characteristic dimensions of the system. For example, τ is the time taken by a given ionic species to diffuse along a typical L_0 distance. L_0 is the characteristic length of the physical system. The quantities c_0 and V_0 should be defined from their typical values in the system [3]. On the basis of the quantities τ, L_0, c_0 and V_0, one can easily determine the value of the dimensionless numbers describing the system.

4.3. Physical meaning of the dimensionless numbers

A dimensionless number generally represents the relative importance of a given phenomenon with respect to another. For instance, as can be seen in Table I, the number $\frac{k.c_0^{(\nu_a+\nu_b-1)}.L_0^2}{D_a}$, noted as *KinDiff*, compares the kinetics of dissolution (or precipitation) (k) to the rate of diffusion (D_a). Information on the physical meaning of the five remaining dimensionless numbers is given in Table I.

Table I – Physical meaning of the various dimensionless numbers

Dimensionless number	Phenomena involved
$\frac{c_0.L_0^2\Gamma F}{\varepsilon.V_0}$	Diffusion potential vs an externally applied potential
$\frac{F.V_0}{RT}$	Transport of ions by migration vs. Transport by diffusion
$\frac{D_b}{D_a}$	The rate of diffusion of species b vs that of species a
$\frac{c_0^{(\nu_a+\nu_b)}}{K}$	This number allows to verify if the system is any close (or not) to its equilibrium conditions
$\frac{k.c_0^{(\nu_a+\nu_b-1)}.L_0^2}{D_a}$	Kinetics of reaction vs. the rate of diffusion

As can be seen, the dimensionless numbers can be used to analyze, for any specific problem, the relative influence of two competing phenomena. In the following section, a typical example of the application of the dimensional analysis to a concrete durability problem will be given. The kinetics of calcium hydroxide ($Ca(OH)_2$) dissolution will be compared to the rate of ionic diffusion.

5. PRACTICAL CASES

5.1. Calcium hydroxide leaching

Calcium hydroxide is one of the major (and most soluble) phases produced by the hydration of cement with water. Its dissolution contributes to locally increase the porosity of concrete. From the standpoint of durability, the study of the mechanisms of portlandite dissolution is of interest. For this particular case, the phase φ previously defined thus corresponds to $Ca(OH)_2$ and, $X_a = Ca^{2+}$, $X_b = OH^-$, $\nu_a = 1$ and $\nu_b = 2$. The kinetic constant was estimated by conductimetry measurements to be $k = 1.3\times 10-5\ mol^{-2}.m^6.s^{-1}$.

Let us consider a mortar sample immersed in pure water. The pore solution of the material is initially in equilibrium with calcium hydroxide. The equilibrium constant of $Ca(OH)_2$ is $K = 10^{3.75}\ mol^3/m^9$, so that $c_0 = 33\ mmol/L$ [3]. The diffusion coefficients of the hydroxide and calcium ions are respectively equal to $D_{OH} = 9.0 \times 10^{-12}\ m^2/s$ and $D_{Ca} = 1.4 \times 10^{-12}\ m^2/s$. Temperature is $T = 300K$. Knowing these quantities [7], one can compute the values of the dimensionless numbers. Figure (1) shows the evolution of the number *KinDiff* at the distance L_0 from the surface of the sample.

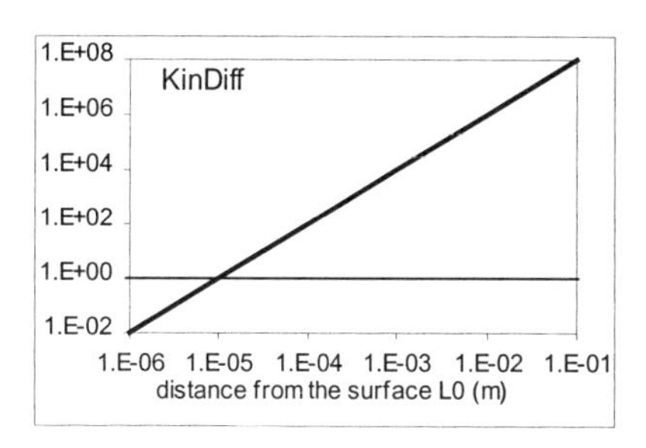

Figure (1): Evolution of *KinDiff* as a function of the distance L_0 from the sample surface.

As can be seen, at a depth of 10 μm or more, the number $KinDiff > 1$. Thus, at 10 μm from the surface, the kinetics of dissolution of $Ca(OH)_2$ is much faster than the kinetics of transport by diffusion. This conclusion validates, in a theoretical way and for the particular case of portlandite, the hypothesis of the local chemical equilibrium frequently assumed for modeling the degradation of cement-based materials.

5.2. Other applied cases

The same analysis can be performed for other cases where transport and dissolution – precipitation phenomena are involved. For instance, the analysis was applied to study the transport of ions in unsaturated concrete. In this particular case, the transport of ions by diffusion is coupled to the transport of water by capillary suction. Other solid phases critical to concrete durability, such as C-S-H, ettringite and chloroaluminate were taken into account. These calculations clearly indicate that, in most practical cases, the assumption of local equilibrium is valid. However, calculations also indicate that the local equilibrium is probably not verified when ions are accelerated by the application of external electrical potential to the system (such as in a migration test). This conclusion is in good agreement with the observations of Gérard [8] and those of Castellote [9].

6. CONCLUSION

The method of dimensional analysis appears to be an interesting technique for comparing the different physical phenomena occurring in the chemical degradation of concrete. The development of the method necessitated proposing a simple kinetic law of dissolution – precipitation.

This work allowed us to come to the conclusion that in the case of the dissolution of portlandite, the local chemical equilibrium hypothesis is validated theoretically and reinforces the experimental observations already done. This confirms the idea that the local equilibrium hypothesis can be formulated in order to model the chemical degradation of cement-based materials. The kinetics of degradation will therefore be controlled by the kinetics of ionic diffusion.

7. ACKNOWLEDGEMENTS

This work was financially supported by Electricité de France – Division Recherche et Développement – Département Etude des Matériaux.

8. REFERENCES

1. J. Marchand, B. Gérard, A. Delagrave, « Ion transport mechanisms in cement-based materials, Materials Science of Concrete, Vol. V, 1998, Edited by J. Skalny and S. Mindess, American Ceramic Society, pp. 307-400.

2. F. Adenot, « Concrete durability: Modeling the chemical and physical damage processes », Ph.D. Thesis, Université d'Orléans, France, (in French), 1992.

3. Barbarulo, J. Marchand, « Dimensional analysis of ionic transport mechanisms in cement-based materials », Cement and Concrete Research, proposed for publication, 1999.

4. É. Samson, J. Marchand, J.J. Beaudoin, « Describing ion diffusion in cement-based materials using the homogeneisation technique », Cement and Concrete Research, Vol. 29, N° 8, pp. 1341-1345.

5. H.L. Langhaar, « Dimensional analysis and theory of models », pp. 55-58, Ed. John Wiley & Sons, New York, 1951.

6. W.J. Duncan, « Physical similarity and dimensional analysis, An elementary treatise », pp. 87-97, Ed. Edward Arnold and Co., London, 1955.

7. The values of the physical constants are the following: Faraday $F = 96485\ C/mol$, the ideal gas constant $R = 8.32\ J/K.mol$ and the dielectric constant of the environment (water) $\varepsilon = 80\ m^{-3}.kg^{-1}.s^{4}.A^{2}$.

8. B. Gérard, « Influence of various coupled phenomena on the long-term behavior of concrete structures for nuclear waste storage facilities », Ph.D. Thesis, E.N.S. Cachan, France, Laval University, Quebec, Canada, (in French).

9. M. Castellote, C. Andrade, C. Alonso, «Chloride-binding isotherms in concrete submitted to non-steady-state migration experiments», Cement and Concrete Research, Vol. 29 (in press).

COUPLING BETWEEN DIFFUSIVITY AND CRACKS IN CEMENT-BASED SYSTEMS

C. TOGNAZZI*,***,+, J-M. TORRENTI**, M. CARCASSES***, J-P. OLLIVIER***
*DESD/SESD, CEA Saclay, 91 191 Gif sur Yvette, France
**ENPC, 6-8 av. Blaise Pascal, 77455 Marne la vallée cedex 2, France
***LMDC, INSA-UPS, Complexe Scientifique de Rangueil, 31 007 Toulouse Cedex 4, France
+Now at LaSAGeC, ISA du BTP, Allée du Parc Montaury, 64 600 Anglet, France

ABSTRACT

In this paper, experimental results and modelling of the coupling between diffusivity and cracks in cement-based systems are presented.

The cracks were generated in samples by means of a compression test. In this test, displacement is controlled, thus controlling the opening of the cracks. Diffusion tests on samples obtained at three different compression levels corresponding to three cracked states were then performed using tritiated water. These tests showed diffusivity was affected by the existence of cracks and the main effect was observed when cracks crossed the whole specimen (e.g. in the post-peak regime of the compression test).

Finally the diffusion test was modelled in a simple way, assuming that the cracks were circumscribed by parallel planes. With this model, an equivalent opening of the cracks, which was of the same order as the opening estimated using strain measurements or image analysis, was obtained.

INTRODUCTION

Concrete structures always present cracks, either from service loading for reinforced concrete or from shrinkage in all cases. These cracks are a potential flow channel for aggressive agents, for example. Nevertheless, concrete is used for radioactive waste disposal. In this application, diffusivity is an essential parameter. The goal of this paper is to investigate the coupling between mechanical cracking and diffusivity in cement-based materials. Mortar cylinders were loaded to different levels of compressive stress, then tested by the diffusion test. The technical experiments and the results are presented in the first part. Then, a modelling of the relation between diffusivity and cracks is discussed.

EXPERIMENT

Compression tests

The behaviour of mortar or concrete in compression follows 3 phases : in the first, the relation between compressive stress and strain is almost linear ; in the second, it becomes non-linear with positive hardening ; then, after the peak load capacity, it softens. The non-linearity reflects the appearance of microcracks which are localised in a shear band at the maximum compressive capacity and then become organised into macrocracks [1]. A compressive test with a constant rate of circumference expansion allows the sample to go past the maximum compressive capacity with a stable softening curve [2], [3]. Consequently, it creates considerable cracking across the whole specimen. Actually, in the softening phase, lateral strains (perpendicular to the loading) become very large and continue to increase (cf. figure 1).

For the study of the influence of cracks on diffusivity, the material was mortar [4]. The compressive tests were controlled by the lateral strain measured with a chain and a displacement

Mat. Res. Soc. Symp. Proc. Vol. 608 © 2000 Materials Research Society

sensor placed around the middle of the cylindrical sample (diameter 113 mm and height 220 mm) [4]. The three levels of loading were (cf. figure 1) :

- level 0 : no loading
- PEAK level : sample loaded up to a circumference strain almost equivalent to the maximum compressive capacity ($\varepsilon_2 = \varepsilon_3 = 1{,}15.10^{-3}$).
- POSTPEAK level : sample loaded up to a circumference strain such that the peak load capacity was passed with appearance of a localised macrocrack running across the specimen from side to side, visible to the naked eye on both sides. The POSTPEAK level was obtained after 4 cycles, so the control of the test was good ($\varepsilon_2 = \varepsilon_3 = 3{,}76.10^{-3}$).

Reproducibility in global behaviour (strain-stress law) was good, although cracking features were not exactly similar : softening behaviour was structural but the cracks spread with a diffuse damage zone at PEAK level whereas a macrocrack ran across every sample at POSTPEAK level. Figure 2 shows an example of the crack patterns (microcracks and macrocrack) observed by image analysis in PEAK and POSTPEAK samples.

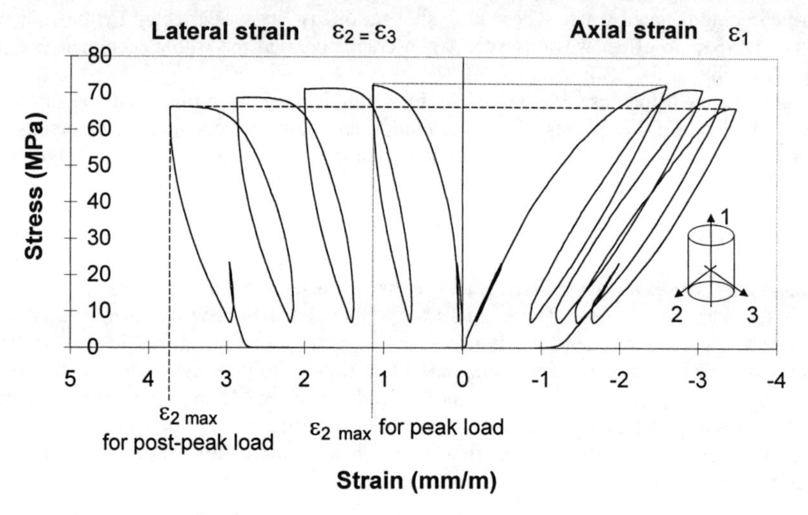

Figure 1. The three levels of loading in compression : level 0 corresponds to a virgin sample, PEAK level to the maximum compressive capacity and POSTPEAK level to the softening phase. For the POSTPEAK level, 4 cycles of loading and unloading were controlled by circumferential expansion (the first one up to the maximum compressive capacity –if the unloading was complete, the level was PEAK-, and the three others up to the softening phase).

From an experimental point of view, it is useful to reduce the frictional end confinement to obtain cracking that is as uniform as possible, parallel to the loading over the full height of the specimens. To do this, a simple technique consists of inserting a material with the same elastic properties as the concrete being tested between the sample and the steel loading platens [2]. Reactive Powder Concrete [5] was used in intermediate layers so that fairly similar cracking was obtained over the whole height of the samples.

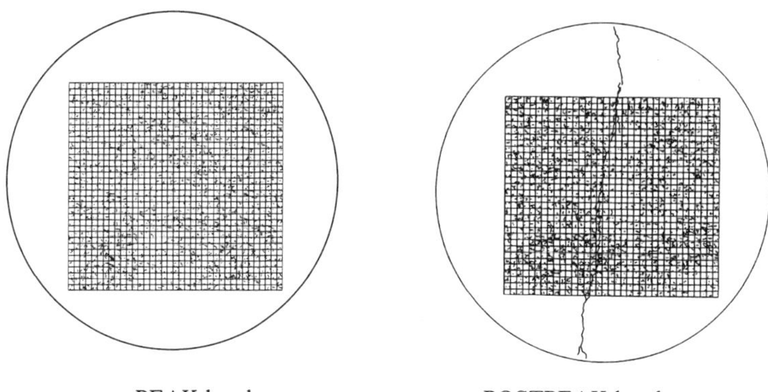

PEAK level POSTPEAK level

Figure 2. Images analysis of cracked mortar in compression (from [6])

With the strain measurements, the order of magnitude of the mechanical opening of cracks was estimated on the sample unloaded after the compressive test. Actually, the measure of circumferential strain is linked to the section expansion of the sample. Furthermore, if the change in cross-sectional area is presumed to result from only a single diametrical crack parallel to the loading, experimental extensions after unloading give an order of magnitude of 225 μm for the mechanical opening of the macrocrack for POSTPEAK level samples [4]. This opening overestimates the real width of cracks because it includes the effect of microcracks and macrocracks in a single equivalent crack. Furthermore, the length of the crack is said to be one diameter even though this underestimates the real length. So this value is a maximum for the real macrocrack. However, this value appears to be of the same order of magnitude as the opening of the macrocrack estimated by image analysis, i.e. 250 μm [6].

Diffusion tests on cracked samples

Compressive load-induced cracks ran across the disc sample, which was no longer isotropic. The transfer properties of these cracked discs were measured perpendicular to plane faces, in the loading direction.

The diffusion properties of the cracked discs perpendicular to plane faces were assessed using a simple diffusion cell apparatus similar to the one described in [7]. The test specimens (10 mm thick, 113 mm in diameter) were cut into cylinders and subjected to a compressive test to the three different levels. They were mounted on the diffusion cells and then saturated with deionized water. The upstream compartment was filled with a lime-saturated solution containing 21 MBq/l of tritiated water. This C_0 concentration remained practically constant throughout the experiment. At regular intervals, the concentration of tritiated water in the downstream compartment was measured and the solution was renewed so that the concentration remained almost nil.

Tritiated water has negligible chemical interaction with the cement paste hydrates [7].

The effective tritiated water diffusion coefficients D_e and the time-lag T_i were obtained from the steady-state regime according to Fick's first law of diffusion and the mass balance equation in one dimensional flux [8], [7].

When t tends to infinity, $\frac{Q(t)L}{C_0} = D_e(t - T_i)$ with $T_i = \frac{pL^2}{6D_e}$ (1)

where t is the time, L the thickness of the disc, C_0 the concentration in the upstream compartment, p the porosity, and Q(t) the cumulative quantity of tritiated water having passed into the downstream compartment by diffusion at time t.

Two duplicate specimens were tested for each load level. The experiment lasted approximately 5 months. Figure 3 shows the increase of QL/C_0 with time. According to equations 1, the effective diffusion coefficient D_e is the linear asymptote slope, and the time-lag T_i is the intersection point of the linear asymptote with the abscissa. The dispersal between the two duplicate specimens is small.

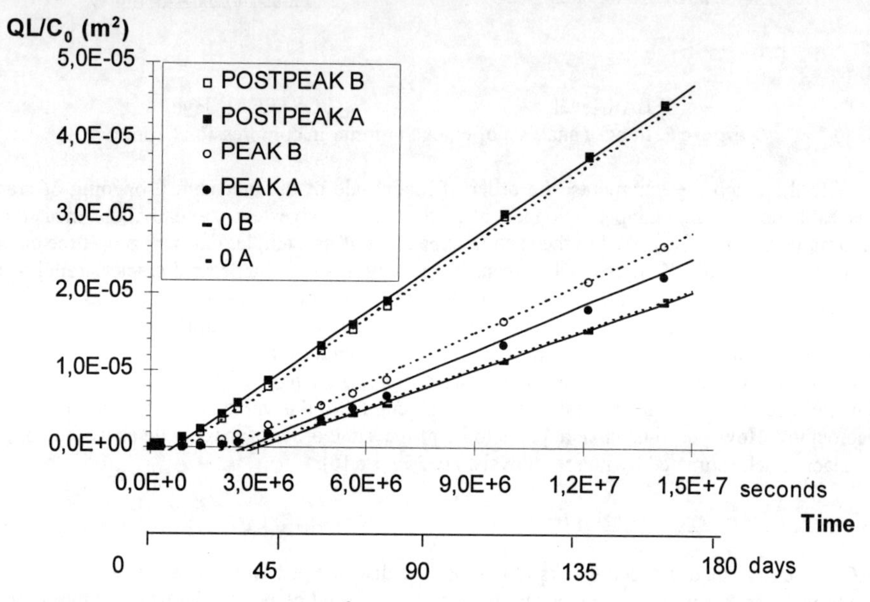

Figure 3. Variation of QL/C_0 (m^2) in tritiated water in the downstream compartment in relation to time in the diffusion tests

RESULTS

To characterise the different levels of cracking, an crack equivalent to those made mechanically was defined with a simple model of transfer in parallel. This equivalent crack was presumed to run across the sample, and to be delimited by parallel planes, perpendicular to the disc faces. Its length was taken as one diameter. Consequently the same hypotheses were used for crack geometry as for the width evaluated from strain measurements at POSTPEAK level .

The sound part of the sample, of cross-sectional area A, was in parallel with the part containing the equivalent crack, of cross-sectional area S_f (figure 4). The tritiated water fluxes J add together in a steady-state regime. As $A >> S_f$, this gives :

$$A.J_{X\ \text{cracked mortar}} = A.J_{X\ \text{sound mortar}} + S_f.J_{X\ \text{crack}} \qquad (2)$$

So, it was possible, from diffusion tests in a steady-state regime on sound or cracked samples, to calculate the cross-sectional area S_f of the equivalent crack in relation to diffusion for each level of loading.

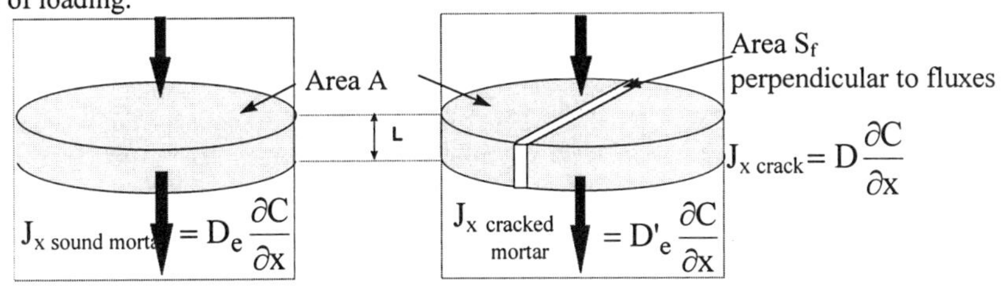

Figure 4 : Formula for a parallel model for diffusion transfer

$$S_f = \frac{A(D'_e - D_e)}{D} \quad (3)$$

where D_e is the effective diffusion coefficient measured on the sound disc, D'_e the effective diffusion coefficient on the cracked disc, and D the effective diffusion coefficient in water.

To test this simple model, a diffusion test was carried out on two samples with a "model crack" of known width and diametric length. These samples were obtained by sawing a disc (70-mm in diameter) into two parts, and coating the edges with waterproof glue. The opening measured under an optical microscope with a magnification of 50 was about 40 μm. The diffusion coefficients measured on the two different samples were $3.6.10^{-12}$ $m^2.s^{-1}$ and $4.3.10^{-12}$ $m^2.s^{-1}$. The equivalent openings in relation to the diffusion calculated from formula 3, using $2.2.10^{-9} m^2.s^{-1}$ for the tritium diffusion coefficient in water were 40 μm and 55 μm respectively. So, the order of magnitude was correct.

The model was then used to estimate the equivalent opening of mechanical cracks. Table 1 shows the measured diffusion coefficients (dispersal of 12 % - cf. [7]), estimated time-lags, equivalent crack cross sections, S_f, and openings, w, calculated from formula 3.

Loading level	Sample number	Effective diffusion coefficient (in $m^2.s^{-1}$)	D'_e / D_e	Time-lag Ti (days)	Equivalent crack section S_f (mm^2)	Equivalent crack width w (μm)
0	A	$1.7.10^{-12} \pm 2.10^{-13}$	-	28.6	0	0
0	B	$1.7.10^{-12} \pm 2.10^{-13}$	-	33.0	0	0
PEAK	A	$2.2.10^{-12} \pm 3.10^{-13}$	1.3	25.1	1.8	20
PEAK	B	$2.0.10^{-12} \pm 2.10^{-13}$	1.2	30.7	1.1	12
POSTPEAK	A	$3.3.10^{-12} \pm 4.10^{-13}$	1.9	7.6	5.7	64
POSTPEAK	B	$3.3.10^{-12} \pm 4.10^{-13}$	1.9	10.7	5.7	64

Table 1. Tritiated water diffusion test results

The three loading levels are distinguishable, but with a very small cracking effect on PEAK samples. The diffusion coefficient of the cracked sample at POSTPEAK level is twice the diffusion coefficient of mortar. This increase is significant. The time-lag decreases with mechanical loading, as with cracking, as already seen [9] for temperature cracked samples. For POSTPEAK level, the time-lag is divided by 3, which is significant. Furthermore, this time-lag decrease at a higher ratio than the diffusion coefficient increase, confirms the validity of the model of transfer in parallel : mechanically-induced cracks at POSTPEAK level do not act as additional porosity but as preferential short flow channels [4], so the time-lag apparently decreases.

The variation of diffusion was significant, but not very large, so the equivalent openings were small, under 100 µm. For example, it was only 64 µm for POSTPEAK cracking. This value combines the small effect of all microcracks, in particular those made at the peak, and the effect of the macrocrack. It is lower than the equivalent opening estimated by strain measurements, but it remains on the same scale of some hundreds of µm. This is logical because the macrocrack had a certain tortuosity and the microcracks were not all connected, since bridges of matter blocked transfer.

CONCLUSION

Displacement-controlled compressive tests make it possible to control the behaviour and cracking of mortar, which is a brittle material, and to control crack opening. Every sample at POSTPEAK level had a macrocrack across it.

The effect of cracking on diffusion parallel to loading was significant, either in terms of time-lag or diffusion coefficient : the diffusion coefficient was multiplied by two for cracks of about a hundred µm in width. With a simple model of transfer in parallel across the cracked materials, validated by tests on samples with a "model crack", an equivalent opening of the crack was calculated. It was of the same order as the opening estimated using strain measurements or image analysis, but was smaller. Actual, cracks are not straight and do not run across the whole specimen.

Lastly, this cracking, which has an effect on transfer also affects chemical degradation : the coupling between crack and degradation is presented in [4] and modelled in two dimensions with a simple model of chemical degradation in [4] and [10].

ACKNOWLEDGMENTS

The present investigation was carried out with financial support from ANDRA, EDF and GEO.

REFERENCES

1. J.M. Torrenti, E.H. Benaija, C. Boulay, , J. Eng. Mech., ASCE, vol. 119, n°12, (1993).
2. S.P. Shah, R. Sankar, ACI Mat. J., vol. 84, pp. 200-212, (1987).
3. R. Gettu, B. Mobasher, S. Carmona, D.C. Jansen, Adv. Cem. Based Mat., 3, pp. 54-71, (1996).
4. C. Tognazzi, Ph. D. Thesis, INSA Toulouse, France, 1998 (in French).
5. P. Richard, M. Cheyrezy, Cem. Conc. Res., vol. 25, n°7, pp. 1501-1511, (1995).
6. A. Ammouche, D. Breysse, in *La dégradation des bétons*, Ed. Hermès, 1999, pp. 75-87, (in French).
7. C. Richet, Ph. D. Thesis, Université Paris XI Orsay, France, 1992 (in French).
8. J. Cranck, *The Mathematics of diffusion*, 2nd Ed., Oxford Science Publications, 1979.
9. J-P. Bigas, J-P. Ollivier, Proceedings du 2ème congrès universitaire de Génie Civil, Poitiers, 6-7 Mai 1999, pp. 82-91 (in French).
10. M. Mainguy, C. Tognazzi, J-M. Torrenti and F. Adenot, « Modelling of leaching in pure cement paste and mortar », accepted for publication in Cem. Conc. Res., 1999.

CLAY-BASED GROUTING INTO THE EDZ FOR THE VAULT SEALING

Y. Sugita*, T. Fujita*, K. Masumoto** and N. Chandler***
*Japan Nuclear Cycle Development Institute (JNC), Ibaraki, Japan, sugita@tokai.jnc.go.jp
**Kajima corporation, Chofu, Tokyo, Japan
***Atomic Energy of Canada Limited (AECL), Pinawa, Manitoba, Canada, R0E 1L0

ABSTRACT

In the Japanese concept for the disposal of the high-level radioactive wastes (HLW), the potential pathways for radioactive contaminant transport would be sealed by a combination of tunnel plug, backfilling and grouting. The candidate material for these engineered barriers would be bentonite or a bentonite-based mixture taking into consideration long-term stability of the seals. Two tests of bentonite grouting for sealing an excavation damage zone (EDZ), that exists in the rock in the immediate vicinity of the tunnel, were conducted in the granitic rock at Atomic Energy of Canada Limited's (AECL's) Underground Research Laboratory (URL). One test was the trial for the development of grouting procedure and the evaluation of grouting effectiveness, and the second test was a demonstration of the grouting around the clay-block bulkhead of the Tunnel Sealing Experiment (TSX). In the trial, the injection proportion of 4.0% of grout slurry was the most efficient. The result of the seepage test showed that grouting resulted in a reduction of hydraulic conductivity of the EDZ in the floor of the tunnel. In the demonstration, although the hydraulic pulse test didn't indicate that the grouting significantly reduced the rock hydraulic conductivity, the test was a useful site-scale demonstration of bentonite grout injection for the purpose of EDZ sealing around a tunnel bulkhead.

INTRODUCTION

Clay-based mixtures and concrete are considered as candidate materials of tunnel and shaft plugs in the present sealing concept for disposal of the HLW in Japan [1]. The clay-based plug will be constructed to isolate a fracture zone. For this purpose, it is necessary not only to excavate a key for emplacement of the plug but also to reduce the hydraulic conductivity of the EDZ adjacent to the plug. The TSX is conducting in the granitic rock of the Pre-Cambrian Canadian Shield at 420 level in the URL to examine the sealing performance of two full-scale bulkheads under a high hydraulic gradient with the seepage through and around these plugs being monitored. The TSX is jointly managed by AECL, JNC, ANDRA (France) and WIPP (USA) (see Figure-1, 2) [2]. Bentonite grouting is one of the effective methods to reduce the hydraulic conductivity of fractured rock. Trial and demonstration bentonite grouting tests were conducted at the URL for the purpose of developing a grouting procedure and the evaluation of

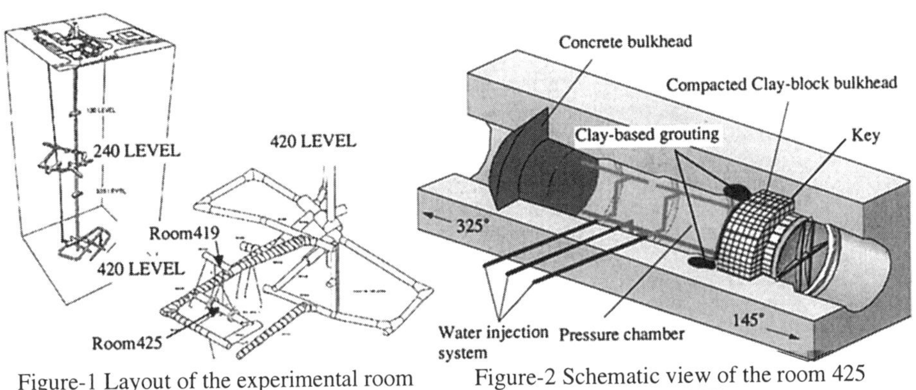

Figure-1 Layout of the experimental room of the TSX at the URL

Figure-2 Schematic view of the room 425 of the TSX

Mat. Res. Soc. Symp. Proc. Vol. 608 © 2000 Materials Research Society

grouting effectiveness for the EDZ around the clay-block bulkhead. These tests were carried out at Room 419 (395m below the ground surface) and Room 425 (420m below the ground surface) as a part of the TSX.

TRIAL BENTONITE GROUTING TEST IN ROOM 419

With regard to the effectiveness of the grouting, it is difficult to evaluate only bentonite grouting in the TSX since the EDZ is interrupted by the combination of the key, the bulkhead and the grouting (see Figure-2). In those terms, the trial bentonite grouting test was conducted in Room 419 to estimate the sealing effectiveness of only the bentonite grouting into the EDZ. In Room 419, a trial key, 1m deep by 2m wide, was excavated for the examination of the feasibility of construction of a square-shaped key in the TSX (Figure-3). The EDZ is induced both by the blasting during the excavation and by the stress redistribution around the tunnel [3]. Room 419 has the same orientation and geometry as Room 425. Therefore, it was expected that the EDZ developing around the trial key had characteristics similar to the EDZ around the clay bulkhead in the TSX tunnel (Room 425). The procedure for the trial bentonite grouting test included drilling boreholes, grout injection and borehole filling. A seepage test was conducted to estimate the effectiveness of grouting by comparing the seepage before and after grouting. Figure-3 shows the layout of the boreholes for this test. The boreholes were 76mm-diameter with shallow inclinations. Boreholes Gt1, Gt2, Gt3 were inclined at 20° with a nominal length of 1.0 m and boreholes Gt4, Gt5 were inclined at 10° with a nominal length of 0.7 m. The appropriate shallow drilling angle allowed the borehole to intersect the EDZ for a longer distance. The collar of borehole Gt2 was situated at the direction of the minimum principal stress in the key surface. The other boreholes were drilled around this borehole, collared on the vertical face of the key. The concrete dam was constructed for the seepage collection for the connected hydraulic conductivity test. To prevent direct seepage through the concrete-rock interface, two lines of bentonite strips were placed across the floor under the dam. An overflow pipe was put at a height of 1.0m on the dam to keep a constant water level in the reservoir. To collect and monitor the water seepage and outflow of the grout slurry, V-shaped trays were inserted into saw cuts on the

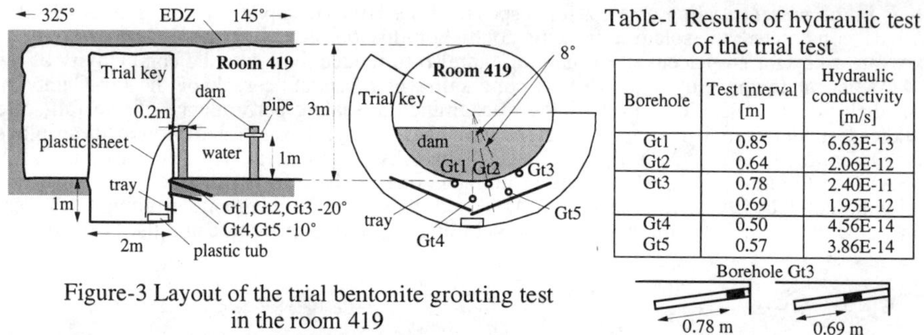

Figure-3 Layout of the trial bentonite grouting test in the room 419

Table-1 Results of hydraulic test of the trial test

Borehole	Test interval [m]	Hydraulic conductivity [m/s]
Gt1	0.85	6.63E-13
Gt2	0.64	2.06E-12
Gt3	0.78	2.40E-11
	0.69	1.95E-12
Gt4	0.50	4.56E-14
Gt5	0.57	3.86E-14

vertical face of key and a plastic tub was placed below the bottom of this tray to collect the accumulated seepage. This tub and the rock face were covered with a plastic sheet to prevent condensation or evaporation at this site. Table-1 shows the test interval and the results from the hydraulic pulse test before grouting in each of the five boreholes. This hydraulic test was a pulse recovery test [4], with a 15 cm-long mechanical packer placed into each borehole. The test interval was between the end of borehole and the packer. Gt3 had two test intervals to compare the hydraulic conductivity of the deeper test interval with of the shallower one. The results show that the rock near the floor of tunnel had a higher hydraulic conductivity because of the fractures in the EDZ. A grout injection system was shown in Figure-4. A constant pressure pump was used to eliminate pressure pulses. Grout was injected into all five boreholes at the same time. The actual maximum pressure in this trial test was set at 500kPa to avoid the possibility that the floor of the tunnel might break out when grout was injected, especially in the shallow boreholes. Grout injection progressed from a diluted slurry (0.2% by weight) to a concentrated one (8.0%). The injection was carried out at 0.2, 0.5, 1.0, 2.0, 4.0, 6.0 and 8.0% (see Table-2) bentonite by weight based on the results of the tests at Kamaishi Mine [5]. At each change of the proportion, the injection boreholes were flushed at a pressure of 100kPa to completely replace the previous slurry with the new slurry. The full injection pressure was applied only after the electrical conductivity of the slurry collected from the outlet valves (flushing slurry) became almost the same as that of the injected slurry (see Table-2).

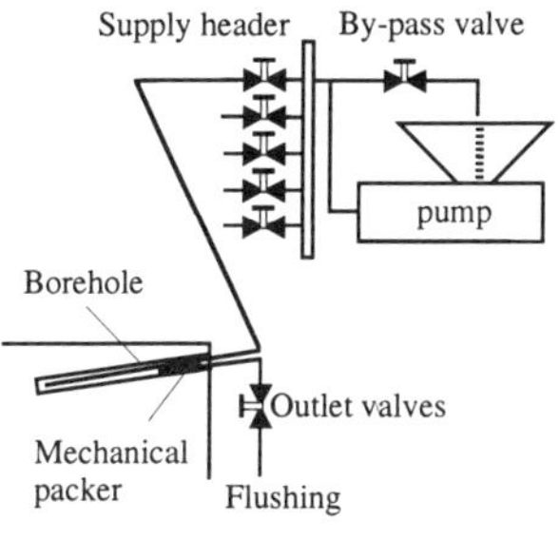

Figure-4 Grout injection system

Table-2 Results of grout injection of the trial test

Proportion [%]	Injection of the slurry			Electric conductivity [mS/cm]	
	Spent time to 500 kPa [min.]	Injection time at 500 kPa	Total injection volume [L]	Inflow	Outflow
0.2	47	4hr	53.2	0.823	0.800
0.5	20	3hr-20min	60.5	0.836	0.807
1.0	16	1hr-33min	29.8	0.886	0.848
2.0	12	1hr-42min	30.4	1.002	0.966
4.0	16	1hr-39min	17.6	1.206	1.142
6.0	20	1hr-50min	12.5	1.388	-
8.0	16	1hr-52min	4.1	1.587	0.977
reservior water				0.235	

RESULTS OF THE TRIAL BENTONITE GROUTING TEST

Measurement of the inflow and outflow rate of the slurry indicated that the outflow rate is almost same as the inflow rate as shown in Figure-5. In other words, most of the slurry flowed out from the vertical surface of the key or from the interface between the dam and the floor of tunnel. Due to this leakage it was difficult to evaluate the grouting injection technique by only the inflow rate of the grout slurry. However, the measured electric conductivity of the outflow and the inflow (injected slurry) showed the possibility of theoretical analysis of the efficiency of the grouting (see Table-2). The water from reservoir had a very low conductivity compared with the grout slurry. Since the correlation between electric conductivity and the concentration of the grout slurry is nearly linear, it was possible to calculate the rate of direct seepage from the reservoir. The outflow included fresh water seeping from the reservoir, thus diluting the grout outflow and lowering the electric conductivity (See Figure-6). The outflow rate is therefore defined by,

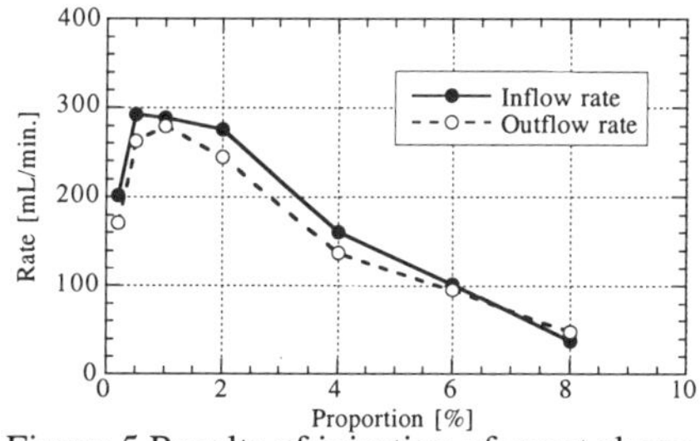

Figure-5 Results of injection of grout slurry

Figure-6 Seepage from reservior

[outflow rate] = [inflow rate] - [infiltration rate into rock] + [seepage rate from reservoir]

In this equation, the infiltration rate into the rock (mL/min) times the proportion of grout slurry (%) is equivalent to the efficiency of grout injection [6]. Figure-7 shows the bentonite infiltration rate (the infiltration rate based on the calculated seepage (mL/min) times the proportion of grout slurry (%)) as a function of grout injection proportion. This Figure indicates that 4.0% was the most efficient proportion in this test condition. The results of the connected hydraulic conductivity test were used to evaluate the effectiveness of grouting. The rate of seepage from the reservoir decreased when the grout was injected indicating a reduction in the hydraulic conductivity of the EDZ around the edge of the key. Figure-8 shows the seepage rate after pouring water into the reservoir during the entire grouting trial and summarizes the change in

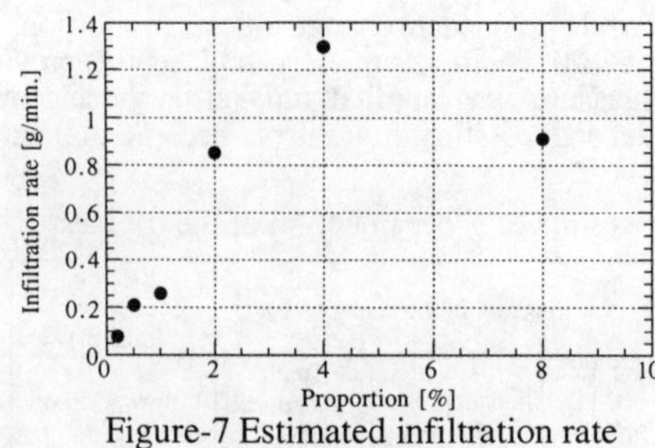

Figure-7 Estimated infiltration rate of bentonite mass

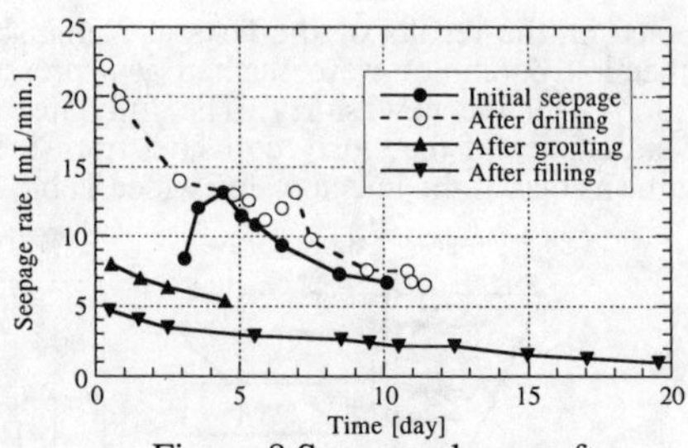

Figure-8 Seepage change after each activities

seepage rates for each activity. After filling of the reservoir, the seepage rate decayed gradually with time as the bentonite strips along the concrete-rock interface started to swell upon contact with water. The initial seepage, before drilling the boreholes, was 6.7 mL/min. The disturbance of the hydraulic condition by each activities caused a temporary increase in seepage rate, but the seepage rate decayed and became nearly constant after each activity. After drilling the boreholes, the seepage rate increased to more than 20 mL/min as a result of cutting part of bentonite strips during drilling of Gt2 and Gt3. Prior to grouting injection, the seepage rate became stable at 6.5 mL/min. After grouting, the seepage rate was 5.3 mL/min, and filling the boreholes with compacted sand-bentonite plugs accelerated this decrease because of the absorption of seepage by the bentonite and sealing the grout boreholes due to swelling. Since the rate of seepage after grouting is less than the seepage rate before grouting, it can be concluded that the grouting activities, including filling boreholes, were successful in reducing the hydraulic conductivity.

DEMONSTRATION BENTONITE GROUTING TEST IN ROOM 425

For the demonstration test in Room 425, grouting boreholes were drilled at 4 points in the EDZ at locations of maximum and minimum (extensional) stress concentrations, as shown in Figure-9. Each location has one injection borehole. The boreholes for grouting were 76mm-diameter with shallow inclinations. All boreholes (Gd1, Gd2, Gd3 and Gd4) are inclined at 15° with a nominal length of 1m. The center of the collar of the borehole is 15cm away from the edge of the bulkhead key. The applicability of grouting injection technique used in this demonstration was validated from the results of the trial test in room 419. The grouting injection procedure in the Room 425 demonstration was the same as that used for the trial (see Figure-4). An observation borehole was located 0.3m away from Borehole Gd3. Injection pressure was increased gradually with increases in the proportion of grout slurry to avoid the outflow of grout slurry. Injection pressure was 200kPa (at 0.2 and 0.5%), 300kPa (at 1.0 and 2.0%) and 500kPa (at 4.0, 6.0 and 8.0%). At each change of the proportion, the injection boreholes were flushed to completely replace the previous slurry with the new slurry, and this was confirmed by the measurement of the electric conductivity. At 0.2, 0.5, 2.0 and 6.0% grout slurry, a preparatory injection was applied to reduce the requirement for a long injection time in the tight schedule of the TSX. At first the new grout slurry was injected into the boreholes at 200kPa, then the all valves were shut.

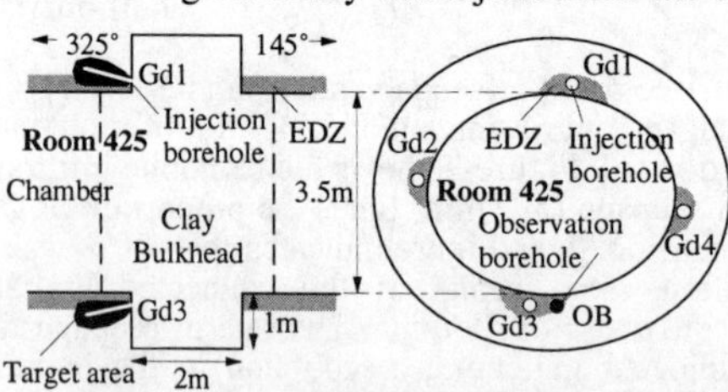

Figure-9 Layout of the demonstrational bentonite grouting test in the room 425

Table-3 Results of injection of grout slurry of the demonstrational test

Proportion [%]	Injection pressure [kPa]	Preparatory injection time at 200 kPa	Injection time at injection pressure	Total volume [mL]
0.2	200	17hr-5min	4hr	3,377
0.5	200	18hr-27min	1hr-5min	2,599
1.0	300	-	2hr	2,480
2.0	300	18hr-12min	1hr-55min	1,763
4.0	500	-	2hr	2,941
6.0	500	17hr-27min	1hr-50min	7,392
8.0	500	-	2hr-10min	2,724

Each preparatory injection time is listed in Table-3.

RESULTS OF THE DEMONSTRATION BENTONTIE GROUTING TEST

Table-3 shows the results of grout injection in each proportion of grout slurry. Figure-10 shows the injection volume of grout slurry in each proportion. Grouting injection volume increased dramatically with increase of injection pressure at all proportions of grout slurry. Reduction of the injection pressure while changing proportion is considered to have caused an increase in injection volume during the next pressurization. The rate of grout injection for all proportions of grout slurry increased slowly after setting the injection pressure. Figure-11 volume of bentonite slurry and bentonite mass. The total injection volume of grout slurry decreased when the bentonite proportion was increased from 0.2 to 2.0% at an injection pressure of 200kPa. It was considered that filling of fractures was in progress. The tendency for the decrease of injection volume of grout slurry at 300kPa, with the proportion increased from 1.0 to 2.0% was almost the same as the tendency for decreased injection volume at 200kPa. However, in case where both injection pressure and proportion of grout slurry were increased, the tendency for the decrease of injection volume of bentonite was much less. While at an injection pressure of 500kPa and with the injection proportion increased from 4.0 to 6.0%, the volume of injected slurry and bentonite mass increased with increasing proportion of grout slurry. This change in tendency from decreasing injection volume to increasing injection volume was considered to prove that increasing the injection pressure contributed to the opening of a fracture and that injected slurry was flushed out. Both injection volume of grout slurry and mass of bentonite reached a peak at the injection proportion of 6.0%. However, as shown in Figure-10, injection volume of grout slurry was still increasing before the injection was stopped at the injection proportion of 4.0%. This shows that injection of grout slurry was incomplete. The injection volume of grout slurry, at 500kPa injection pressure and injection proportion of 6.0%, became almost constant. At the injection proportion of 8.0%, injection volume of grout slurry decreased steeply. It was concluded that filling of fractures was

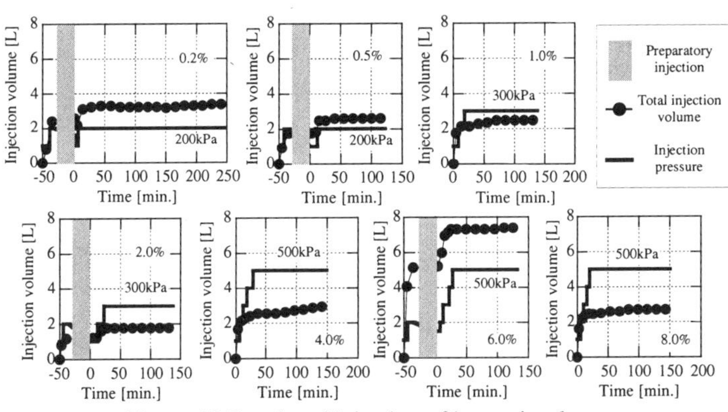

Figure-10 Results of injection of bentonite slurry

Table-4 Results of hydraulic test of the demonstrational test

Borehole	Hydraulic conductivity [m/s] before grouting	Hydraulic conductivity [m/s] after grouting
Gd1	7.36E-13	3.34E-13
Gd2	1.75E-13	3.50E-13
Gd3	1.54E-11	1.31E-11
Gd4	3.22E-11	7.53E-12

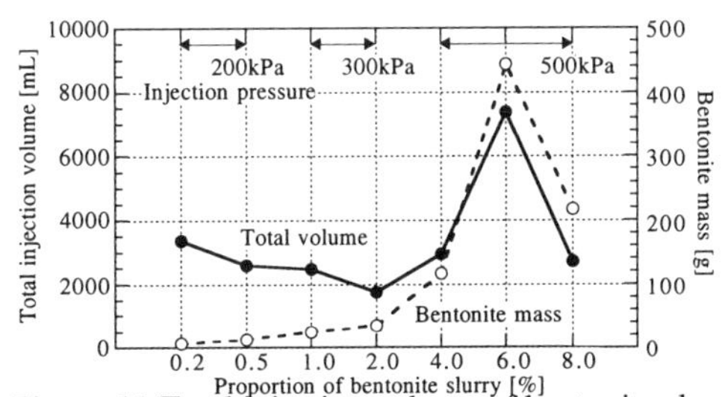

Figure-11 Total injection volume of bentonite slurry and bentonite mass

in progress. Outflow to the surface of the rock mass was not observed during injection of grout slurry. However, a zone of wetness on the face of the key, having a radius of 10 to 20cm, was observed around the collar of the injection boreholes. Near borehole Gd1 there was a wet zone on the surface of the excavation along the roof of the tunnel, especially. Some bubbles were observed around the collar of the injection borehole Gd3 during the injection proportion from 4.0 to 6.0%. These bubbles disappeared at the end of the injection at 6.0%. It was concluded that filling of a fracture was in progress during injection at 6.0%. There was no visible grout inflow into the observation borehole near borehole Gd3. Therefore, at the bottom of the drift, the grout infiltration zone radius was less than 30cm. Before and after grout injection hydraulic pulse tests were conducted to evaluate the efficiency of the grouting. As shown in Table-4, the hydraulic conductivity of borehole Gd4 before grouting was of the order of 10^{-11} m/s. The hydraulic conductivity decreased by one order of magnitude after grouting. For borehole Gd3 there was no clear efficiency of grouting. The initial hydraulic conductivity around boreholes Gd1 and Gd2 was of the order of 10^{-13} m/s. Due to such low hydraulic conductivity before grouting, it was difficult to observe any apparent efficiency of grouting. It was considered that near these boreholes, the actual EDZ was smaller than the predicted one.

CONCLUSIONS

Bentonite grouting was conducted to reduce the hydraulic conductivity of the EDZ developing adjacent to the key for the clay-based bulkhead in Room 419 and Room 425 at AECL's URL. Grout injection progressed from a diluted slurry (0.2%) to a concentrated one (8.0%) using a maximum injection pressure of 500kPa. The injection proportion of 4.0% bentonite by weight was the most efficient in the Room 419 trial. Based on the results of seepage measurements, the seepage rate from the reservoir through the rock near the floor of tunnel decreased after grouting activities. The demonstration of the clay-based grouting into the EDZ for vault sealing was successful based on the evaluation of hydraulic conductivity of the EDZ. These results indicated that the bentonite grouting caused a reduction of hydraulic conductivity of the EDZ adjacent to the clay-based plug, although the EDZ at this site was very small and the initial hydraulic conductivity was very low.

ACKNOWLEDGEMENTS

The authors wish to express their gratitude to Mr. Keiji Hara for his advice on experimental plan, to Mr. Paul Thompson for his support of grouting injection system, to Mr. Edward Kozak for his support of hydraulic test, to Mr. Masaru Toida for his support of grouting technique, and to all technicians for their excellent work at the URL.

REFERENCES

1. Japan Nuclear Cycle Development Institute, *H12 Project to Establish Technical Basis for HLW Disposal in Japan - Project Overview Report 2 -, the Draft Second Progress Report on Research and Development for the Geological Disposal of HLW in Japan*, JNC TN1400 99-010, 1999.
2. N. Chandler, D. Dixon, M. Gray, K. Hara, A. Cournut and J. Tillerson, *An in situ Demonstration of Technologies for Vault Sealing*, Proc. 19th Annual Conference of Canadian Nuclear Society, 1998.
3. N. Chandler, E. Kozak and C. Martin, *Connected Pathways in the EDZ and the potential for flow along tunnels*, Proc. EDZ Workshop, Int. Conf. Deep Geological Disposal of Radioactive Waste, Canadian Nuclear Society, 1996, p. 25-34.
4. H. J. Ramey, R. G. Agarwal and I. Martin, *Analysis of ëSlug Testí or DTS Flow Period Data*, J. Can. Petroleum Technology, vol.14, 1975, p. 37-47.
5. Y. Sugita and T. Fujita, *Clay Grouting Experiment at the Kamaishi Mine*, Proc. 27th Symposium on Rock Mechanics, 1996, p. 276-280. (in Japanese)
6. K. Masumoto, Y. Sugita, T. Fujita and N. Chandler, *Trial of Bentonite Grouting into the EDZ at AECL's Underground Research Laboratory*, Proc. 29th Symposium on Rock Mechanics, 1999, p. 36-42.

Study on Effects of Hydraulic Transport of Groundwater in Cement

M. Toyohara*, M. Kaneko*, F. Matsumura**, N. Mitsutsuka**, Y. Kobayashi*** and M. Imamura***

*Nuclear Engineering Lab., Power Systems & Services Company, Toshiba Corp., 4-1 Ukishima, Kawasaki-ku, Kawasaki, 210-0568, Japan(masumitsu.toyohara@toshiba.co.jp)
**Advanced Energy Design & Engineering. Dept., Toshiba Corp.
***Low Level Waste Management Dept., Japan Nuclear Fuel Ltd., 2-2-2, Uchisaiwai, Chiyoda-ku, Tokyo 100-0011, Japan

ABSTRACT

This paper discussed the effects of solution velocity thorough the cementitious materials on formation of secondary minerals. These minerals were produced by the reaction of hydrates in cement and chemicals in groundwater. The chemicals estimated were $NaHCO_3$, Na_2SO_4 and NaCl. Calcite yielded by the reaction of carbonate ion and Ca^{2+} was found to cause a change in the porosity volume of cement, and thus, to decrease the flow rate. The existence of sulfate ion did not affect the flow rate. However, in the case of a solution containing both chloride and carbonate ions, the flow rate increased because Ca dissolution from hydrates was induced.

INTRODUCTION

Cement is used as a material for the immobilization of radioactive waste due to its attractive property. Also, backfill, a certain kind of mortar, is used to fill the vacancies around containers of radioactive waste. These materials are thought to cause the pristine groundwater to become alkaline within the repository [1], and this change reduces solubility and enhances sorption for certain species relevant to radionuclides [2,3]. Hence, their concentrations in pore water are reduced, which contributes to reduction of the radionuclide release from repository. However, cement is not thought to control the velocity of groundwater passing through the repository, because microfractures or voids are generated gradually in cement structure. Hence, radionuclides with poor sorption to cementitious material move easily with invading groundwater after leaching from waste, and then pass out from the repository. In Japan, since groundwater velocity is presumed to be high, the retention of these radionuclides, such as ^{14}C, ^{36}Cl and ^{129}I, is a current problem, because they easily migrate with groundwater in natural barrier.

In recent years, studies have shown that filling the pore due to the secondary minerals changed the radionuclide transport or velocity of groundwater in cement materials[4,5]. The results of these studies indicate that groundwater flow was affected by the relation between the amounts of hydrates dissolved in groundwater and precipitation of calcite ($CaCO_3$). Calcite was a secondary mineral formed by the reaction of Ca ions released from hydrates and carbonate ions (CO_3^{2-}) in groundwater. If the same reactions occur under the disposal conditions in Japan and the formation of this mineral reduces the groundwater flow, it is estimated that the migration of the radionuclides with poor sorption will be suppressed, thereby simplifying the engineering barrier concept, and consequently, reducing the cost of radioactive waste disposal. The results reported in the studies alluded to above motivated us to study the relation between water velocity and formation of secondary minerals. The reactions are considered to depend both on the kinds of chemicals in groundwater and their concentrations. Thus, we initiated a basic study on the reactions that occur in pore water in cement, using powder of hydrated cement in order to accelerate the reactions. In this paper, the effects of chemicals in groundwater on hydraulic flow in cement are clarified, and

Mat. Res. Soc. Symp. Proc. Vol. 608 © 2000 Materials Research Society

then the possibility of reducing groundwater velocity in a repository composed of cement materials is evaluated.

EXPERIMENTAL

Portland cement and water were mixed sufficiently and then cured for 28 days at room temperature. The mixing ratio of water and cement was 40:100 by weight. The hardened cement was ground and the resulting powder was sieved to a particle size of 75 to 150 μ m. The powder was filled in the column and then set up in a glove box. The contents of the solution inlet into this powder were determined base on a consideration of the groundwater in Japan. In the light of reports on HLW disposal concept [6], chemicals were selected which might react with those from hydrated cement to form insoluble precipitates or with hydrates in cement. The chemicals and the concentrations are listed in Table I. Three concentrations of carbonate were selected. Sulfate and chloride ions were also studied because aluminate hydrates in cement incorporated them. These solutions were prepared from the reagent of $NaHCO_3$, Na_2SO_4 and NaCl.

Table I Solution contents and their concentration

Solution contents	Concentration
Pure water (Carbonate-free)	---
$NaHCO_3$	$1.8X10^{-2}$
$NaHCO_3$	$0.9X10^{-2}$
$NaHCO_3$	$1.8X10^{-3}$
$NaHCO_3$ NaCl	$1.8X10^{-2}$ $5.5X10^{-1}$
$NaHCO_3$ Na_2SO_4	$1.8X10^{-2}$ $1.3X10^{-3}$

Unit; mole/dm^3

Experiments were conducted using the column method. The equipment setup is shown in Fig.1. The column was cylindrical and had an inner diameter of 2 cm and a length of 10 cm. The filter was installed at the bottom of the column to prevent the powder flowing out. The powder was filled to 5cm height from the bottom of the column. Apparent vacancy was estimated to be 50 vol.%. The glove box was filled with nitrogen gas with a purity of 99.9999 vol. % so as to preclude reaction of the cement or the chemicals in solution with carbon dioxide in air. Using suction to draw solution through the powder in the column, the solution flow rate was controlled. The differential pressure between the glove box and the outlet in the column was 170 mm Hg, which was estimated based on the consideration of a hydraulic condition in Japan. At the beginning of the experiment, carbonate-free solution was inlet for 20 hours in order to homogenize the powder. The weight of discharged solution was measured to estimate its flow rate. The flow rate in this study was defined by the following equation:

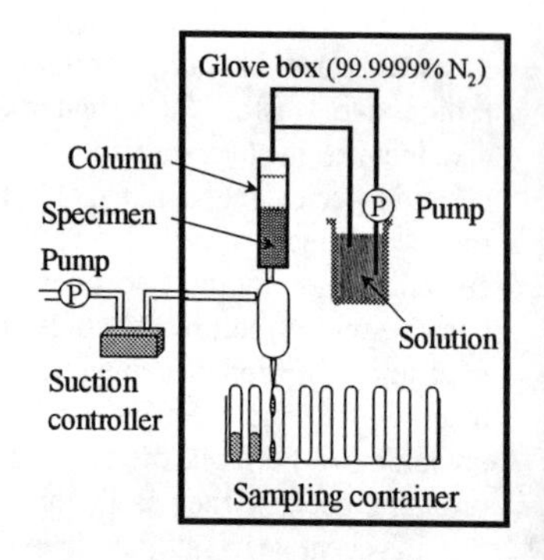

Fig. 1 Equipment setup for flow-through experiment using powdered cement

Flow rate (m/sec) = Solution weight during a period / density / area of column / time of a period.

The experiments were conducted until termination of solution flow or observation of the constant flow rate. The lump of powder in the column was removed and then was analyzed using the X-ray diffraction method (XRD) in order to study the hydrates.

Results and discussion

1. Effect of carbonate concentration in solution on flow rate

Figure 2 shows the relation between the flow rate of solution discharged and time. In the case of carbonate-free solution, the initial flow rate was around 10^{-5} m/s and then decreased rapidly. But after 25 hours it reached the constant of around $6X10^{-6}$ m/s. On the other hand, the flow rates of solutions with $NaHCO_3$ were found to decrease to one third (less than $2X10^{-6}$ m/s) of that of the carbonate-free solution. In the case of a $NaHCO_3$ concentration of $1.8X10^{-2}$ mole/dm^3, the flow

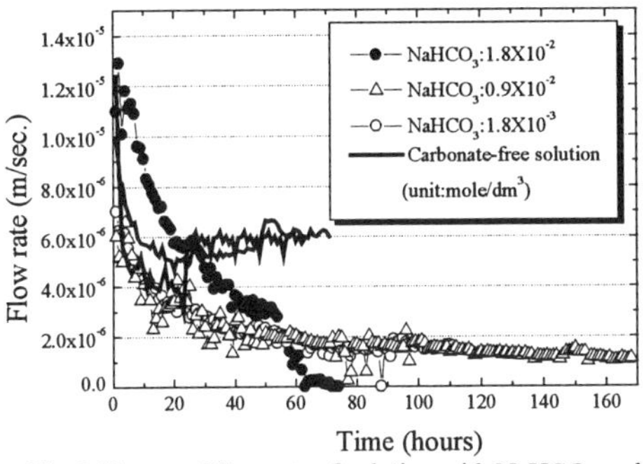

Fig. 2 Change of flow rate of solution with $NaHCO_3$ and carbonate-free solution passed through the column filled with powder of hydrated cement

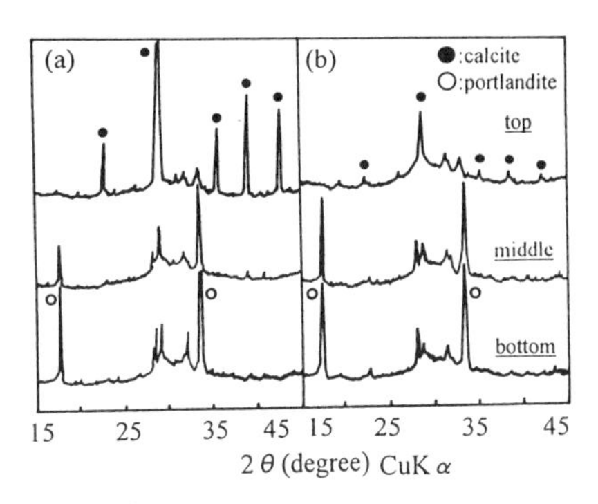

Fig. 3 X-ray diffraction pattern for powder of hydrated cement after solution with $NaHCO_3$ passed through
(a) $1.8X10^{-2}$ mole/dm^3, (b)$1.8X10^{-3}$ mole/dm^3

rate dramatically decreased to under $1X10^{-6}$ m/s and could not be measured after 70 hours. The apparent flow rate was $8.8X10^{-10}$ m/s based on the lower limit measurement of solution volume. On the other hand, at $0.9X10^{-2}$ and $1.8X10^{-3}$ mol/dm^3 of $NaHCO_3$ concentration, both flow rates decreased rapidly, but finally the flows remained constant. Figure 3 shows the results of XRD observation of residual lumps of powder in the column after experiments. Calcite was formed at the top of the lump which was always immersed in the pristine solution. In the case of solution with a $NaHCO_3$ concentration of $1.8X10^{-2}$ mol/dm^3, the diffraction for calcite was strong at that point, indicating formation of a considerable amount of this mineral, but it gradually weakened in the direction toward the bottom of the lump. Conversely, the diffraction for portlandite ($Ca(OH)_2$) was not seen at the top of the lump, but gradually strengthened in the direction toward the bottom. In the case of the concentration of $1.8X10^{-3}$ mole/dm^3, the diffraction for calcite was weak even at the top of the lump. Figure 4 shows the profiles of Ca^{2+} concentrations of discharged solutions. The

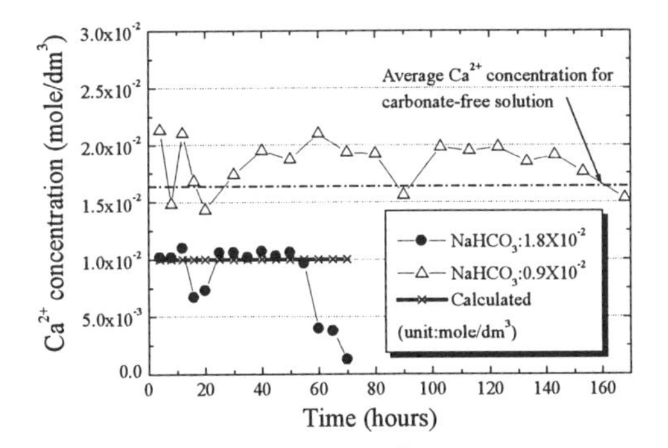

Fig. 4 Concentration profiles of Ca^{2+} in discharged solution from column filled with powder of hydrated cement

concentration of Ca^{2+} for carbonate-free solution was around $1.6X10^{-2}$ mole/dm^3 and did not change during inlet of solution. Further, a similar profile was seen in the case of a $NaHCO_3$ concentration of $0.9X10^{-2}$ mol/dm^3. But in the case of $1.8X10^{-2}$ mol/dm^3, Ca^{2+} concentration was around $1X10^{-2}$ mole/dm^3 at the initial stage of the experiment. In this case, pH of discharged solution was 12.4. Further Ca^{2+} concentration decreased dramatically after 50 hours.

From these results, calcite was presumed to decrease the volume of discharged solution passed through the column filled with the powder of hydrated Portland cement. The flow rates depended on the carbonate concentrations of pristine solution. Hence, it is considered that calcite formed by the reaction of Ca^{2+} and CO_3^{2-} might decrease the solution flow path in the powder of hydrated cement. This consideration was deduced from the results of XRD. Consequently, the carbonate concentration of pristine solution was considered to play a vital role in decreasing the flow rate. The main reactions that occurred in the pore water with a pH of around 12.4 are considered to be described by the following:

$$Ca(OH)_2 = Ca^{2+} + 2OH^-$$
$$Ca(OH)_2 + 2Na^+ + 2HCO_3^- = CaCO_3 + 2Na^+ + CO_3^{2-} + 2H_2O$$

The calculated Ca concentrations in discharged solutions based on the pH of solution are shown in Fig. 4. The equilibrium constants for solution and mineral phase are referenced from [7]. These values and the results of experiments agreed well. From these results, in the case that a certain amount of carbonate is present, these reactions are presumed to proceed simultaneously to form calcite precipitation. Further, the existence of Na^+ ion results in suppression of the dissolution of calcium hydroxide. In the case of a $NaHCO_3$ concentration of $1.8X10^{-2}$ mole/dm^3, both formation of calcite and suppression of the dissolution of portlandite decreased the pore volume in cement. On the other hand, since the concentrations of Na and carbonate ion are low in the case of $NaHCO_3$ concentration of less than $1.8X10^{-2}$ mole/dm^3, the amount of calcite formed is small and a considerable amount of calcium hydroxide may dissolve into the solution. In addition, the reason for rapid decrease of Ca concentration is presumed to increase the concentration of alkaline metals that leached from cement, due to the residence time of solution in cement.

From these results, it was concluded that the formation of calcite governed the change of porosity volume in the hydrated Portland cement. In the typical case, the decrease of porosity volume would terminate groundwater flow in it. This was confirmed to occur under the condition of a $NaHCO_3$ concentration of $1.8X10^{-2}$ mole/dm^3. In the case of lower concentration, the termination of flow was not observed, but the flow rate was significantly lower than in the case of carbonate-free solution.

2. Effects of chloride ion and sulfate ion

Figure 5 shows the effect on the flow rates of solutions having a Na_2SO_4 concentration of $1.8X10^{-3}$ mole/dm^3 or a NaCl concentration of 0.55 mole/dm^3. Each solution contains $NaHCO_3$ whose concentration is $1.8X10^{-2}$ mole/dm^3. The flow rate for solution containing Na_2SO_4 decreased to less than under $1X10^{-7}$ m/s, which was similar to that of the solution containing a $NaHCO_3$ concentration of $1.8X10^{-2}$ mole/dm^3, but the flow was not terminated. On the other hand, the reduction of flow rate was small in the case of NaCl solution. The final flow rate was around $5X10^{-5}$ m/s. Figure 6 shows the results of XRD for the lumps of powder after inlet of solution. In the case of solution containing Na_2SO_4, both calcite and ettringite ($3CaO \cdot Al_2O_3 \cdot 3CaSO_4 \cdot 32H_2O$) were found at the top of the lump, but a certain amount of portlandite also remained in the lump.

On the other hand, it was confirmed that a considerable amount of calcite formed over the lump in the case of NaCl solution. Figure 7 shows the Ca^{2+} concentration profiles for both solutions. Calcium concentration for solution containing Na_2SO_4 was found to be about $1.0X10^{-2}$ mole/dm^3, a value similar to that for carbonate-free solution, and decreased rapidly after 70 hours. However, for NaCl solution, Ca^{2+} concentration was high throughout the experiment.

From these results, the existence of sulfate ion was found to have a minor effect on the flow rate in powder of hydrated cement. The flow rate of the solution containing both $NaHCO_3$ and Na_2SO_4 also decreased in a manner similar to that in the case of solution with only $NaHCO_3$, but

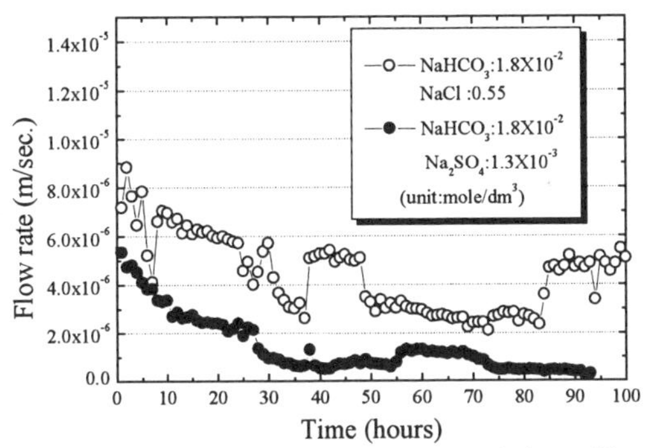

Fig 5 Changes of flow rate of carbonate solutions with Na_2SO_4 or NaCl passed through the column filled with powder of hydrated cement

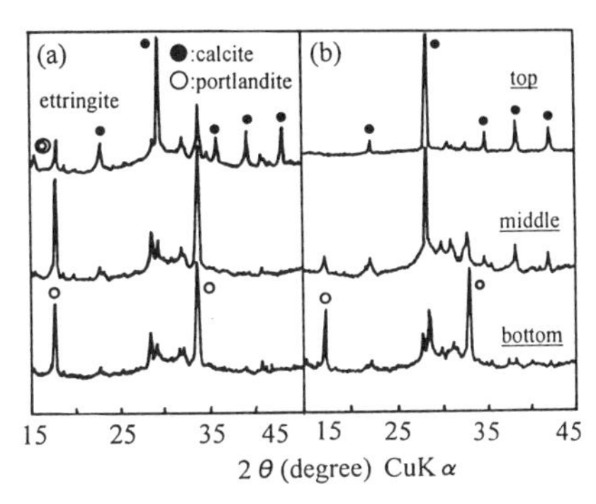

Fig. 6 X-ray diffraction pattern for powder of hydrated cement certain solutions passed through.
(a)$NaHCO_3$:$1.8X10^{-2}$, Na_2SO_4:$1.8X10^{-3}$(mole/dm^3)
(b)$NaHCO_3$:$1.8X10^{-2}$, NaCl :0.55 (mole/dm^3)

remained constant at around $1.0X10^{-6}$ m/sec after 60 hours. This reduction was presumed to yield calcite and ettringite at the top of the lump and these secondary minerals might reduce the porosity at that part. On the other hand, the existence of chloride ion in carbonate solution induced the increase in flow rate. In this case, the formation of calcite was confirmed over the entire lump. The diffraction for calcite was strong at the top of the lump and was observed even at the bottom, indicating formation of a considerable amount of this mineral. Further Ca^{2+} concentrations of discharged solution were much higher than that in the case of carbonate-free solution. Consequently, the existence of NaCl was considered to activate the dissolution of certain hydrates in the lump of hydrated cement and this suppressed the reduction in porosity brought about by the formation of calcite in the lump. The details of the dissolution reaction were not clarified but we

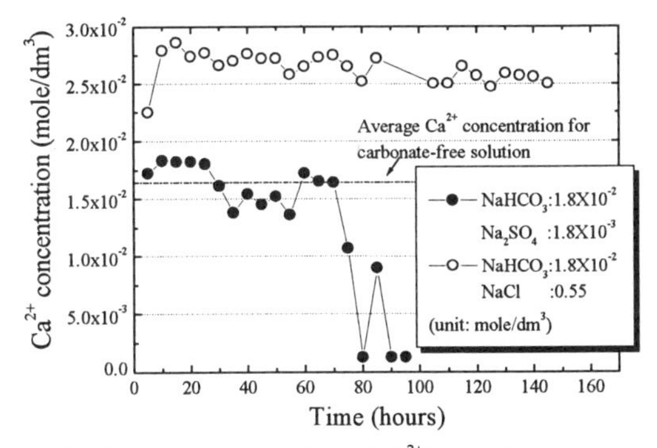

Fig. 7 Concentration profiles of Ca^{2+} in solutions passed through the column filled with powder of hydrated cement

surmised that dissolution of CSH gel was activated.

3. Summary of the relation between the flow rate and chemicals in solution

Table II shows the results of this study on the relation between the hydraulic conductivity and chemicals. The carbonate ion plays an important role in reducing the porosity in the powder. The hydraulic conductivity depends on the concentration of carbonate ions in pristine solution. From these results, a significant low hydraulic conductivity was presumed to occur even in the case of solution flow in powdered cement. The existence of chloride ion affects the porosity reduction brought about by the calcite precipitation. However, more detail studies are required in order to introduce into the disposal concept the hydraulic transport properties of solution in cement obtained by this study.

Table II Summary of hydraulic conductivity for various solutions obtained by pass-through experiments

Solution contents	Concentration ($mole/dm^3$)	Hydraulic conductivity (m/sec)
Pure water	---	$1.3X10^{-7}$
$NaHCO_3$	$1.8X10^{-2}$	$\sim 1.9X10^{-11}$
$NaHCO_3$	$0.9X10^{-2}$	$2.4X10^{-8}$
$NaHCO_3$	$1.8X10^{-3}$	$3.5X10^{-8}$
$NaHCO_3$ NaCl	$1.8X10^{-2}$ $5.5X10^{-1}$	$1.3X10^{-7}$
$NaHCO_3$ Na_2SO_4	$1.8X10^{-2}$ $1.3X10^{-3}$	$5.0X10^{-9}$

Conclusion

(1) In the case that carbonate was present, it was confirmed that flow rate of solution decreased more than in the case of carbonate-free solution. In the case of a $NaHCO_3$ concentration of $1.8X10^{-2}$ $mole/dm^3$, the flow rate dramatically reduced and was terminated. The hydraulic conductivity was around $1.9X10^{-11}$m/s.
(2) The reason for reduction of flow rate is precipitation of calcite by the reaction of Ca^{2+} and carbonate in groundwater. Calcite effects the change of porosity volume in hydrated cement.
(3) The effects of other chemicals were also studied. The flow rate of solution with both Na_2SO_4 and $NaHCO_3$ was similar to that of solution with $NaHCO_3$ only. The existence on NaCl in solution increases the flow rate because the dissolution of Ca^{2+} from hydrates is enhanced.

References

[1] Atkinson, A., AERER11777, (1985)

[2] Berner, U. R., Evolution of Pore Water Chemistry during Degradation of Cement in a Radioactive Waste Repository Environment, Waste Management, 1992, **12**: p.201-219

[3] Brownsword, M., et al., The Solubility and Sorption of Uranium(VI) in a Cementitious Repository, Mater. Res. Soc. Symp. Proc. 1990, **176**:p.577-582

[4] Sarott, F. A., et al., Diffusion and adsorption studies on hardened cement paste and the effect of carbonation on diffusion rates. Cement and Concrete Research, 1992. **22**:p.439-444

[5] Berner, U. R., Geochemical Modelling of Repository Systems: Limitation of Thermodnamic Approch. Radiochim. Acta **82**, p.423-428 (1998)

[6] PNC TN 1410 92-081, 1992, p.4-35

[7] Reardon, E. J., Problems and approaches to the prediction of the chemical composition in cement/water systems. Waste Management, 1992, **12**: p.221-239

Corrosion of Ceramic Wasteforms

GLASS/CERAMIC INTERACTIONS IN THE CAN-IN-CANISTER CONFIGURATION FOR DISPOSAL OF EXCESS WEAPONS PLUTONIUM

B. P. MCGRAIL, P. F. MARTIN, H. T. SCHAEF, C. W. LINDENMEIER, A. T. OWEN
Applied Geology and Geochemistry Department, Pacific Northwest National Laboratory, Richland, Washington 99352, pete.mcgrail@pnl.gov

ABSTRACT

A can-in-canister waste package design has been proposed for disposal of pyrochlore rich ceramics containing excess weapons plutonium. The can-in-canister configuration consists of a high-level waste (HLW) canister fitted with a rack that holds minicanisters containing the ceramic. The HLW canister is then filled with glass. The pressurized unsaturated flow (PUF) technique was used to investigate waste form/waste form interactions that may occur when water penetrates the waste containers and contacts the waste forms. Volumetric water content was observed to increase steadily from accumulation of water mass as waters of hydration associated with alteration phases formed on the glass surface. Periodic excursions in effluent electrical conductivity and pH were monitored and correlated with secondary phases formed during the test. Plutonium exited the PUF system primarily as filterable particulates. However, effluent Pu and Gd concentrations were found to decrease with time and remained at near detection limits after approximately 250 days, except during transient pH excursions. These results indicate that both Pu and Gd will be retained in the can-in-canister waste package to a very high degree.

INTRODUCTION

The U.S. Department of Energy Office of Fissile Material Disposition has indicated that the ceramic can-in-canister design is the preferred option for the disposal of surplus weapons Pu in the proposed, mined geologic repository at Yucca Mountain, Nevada [1]. The long-term release of ^{239}Pu, ^{235}U, and neutron absorbers from the ceramic in this can-in-canister configuration must be understood before a credible safety analysis can be conducted.

At some time in the distant future, water is expected to penetrate the waste containers and contact the waste forms. Although several scenarios of water contact are possible, the most credible contact mode from a hydraulics viewpoint is the slow percolation of water through the waste packages under conditions of partial hydraulic saturation. Gravity and capillary forces drive flow through the containers. Because the can-in-canister waste packages are expected to contain approximately 85 vol% of HLW glass and 15 vol% of Pu-bearing ceramic, any water percolating through the containers will react chemically with both high-level waste (HLW) glass and the Pu ceramic. Although exact flow paths through a waste package are not known, the substantially higher amount of glass compared with ceramic suggests that on average, water will contact glass before contacting ceramic. The glass is also expected to react faster than the ceramic and to contain a significant amount of exchangeable cations, which are subject to diffusive and advective mass transport. For all these reasons, the glass is expected to dominate the chemistry of any water percolating through a can-in-canister waste package. Consequently, it is important to understand how the glass/water reaction impacts, if at all, corrosion of the ceramic and the release and transport behavior of Pu and the neutron absorbers. In this paper, we summarize the results from a 17-month experiment utilizing a pressurized unsaturated flow (PUF) system at Pacific Northwest National Laboratory (PNNL).[2]

Mat. Res. Soc. Symp. Proc. Vol. 608 © 2000 Materials Research Society

EXPERIMENTAL METHODS

Because the purpose of this test is to determine if the chemical changes in water that has reacted with HLW glass alters the corrosion rate of the ceramic or release and transport of Pu and the neutron absorbers, a "sandwich" configuration, as illustrated in Figure 1, was selected. To match the approximate volumes of glass and ceramic in a can-in-canister configuration, 85% of the volume was allocated for glass (42.5% in the top and bottom beds) and 15% with ceramic. Details on the glass and ceramic materials used are given in the Materials section. Details on the PUF system are discussed next.

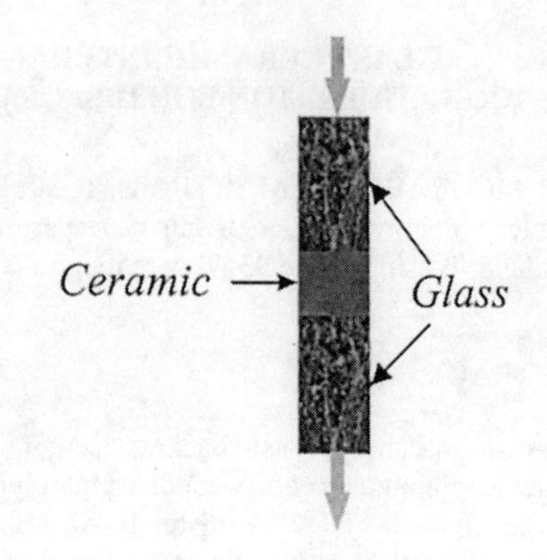

Figure 1. Schematic of PUF Interaction Test

PUF System

The PUF system has been discussed in detail elsewhere [3,4], but is briefly described here. The basic test apparatus consists of a column packed with particles of the test material or materials of a known size and density, and a computer data acquisition and control system. The column is fabricated from a chemically inert material (polyetheretherketone) so that dissolution reactions are not influenced by interaction with the column. A porous titanium plate with nominal pore size of 0.2 μm is sealed in the bottom of the column to ensure an adequate pressure differential for the conductance of fluid while operating under unsaturated conditions [5]. When water saturated, the porous plate allows water but not air to flow through it, as long as the applied pressure differential does not exceed the air entry relief pressure, or "bubble pressure," of the plate. A computer control system logs test data to disk from several thermocouples, pressure sensors, inline sensors for effluent pH and conductivity, and column weight from an electronic balance to accurately track water mass balance and saturation level. The column is also fitted with a "PUF port," which is an electronically actuated valve that periodically vents the column gases at an interval specified by the operator. The purpose of column venting is to prevent reduction in the partial pressure of important gases, especially O_2 and CO_2 that are participants in a variety of homogeneous and heterogeneous chemical reactions.

Materials

The ceramic used in this test (CPSF5) was produced at Lawrence Livermore National Laboratory and shipped to PNNL. The target composition (see Table 1) was 80 mass% zirconolite with 15 mass% Ba-hollandite and 5 mass% rutile, 16 mass% PuO_2, and doped approximately equimolar with Gd. The simulated HLW glass used in the PUF experiment was received from the Savannah River Technology Center (SRTC) labeled WP-14. Staff at SRTC transmitted a chemical analysis of the glass, which is given in Table 2.

Crushed glass and ceramic were prepared following the procedure detailed by the American Society for Testing and Materials [6]. The materials were ball milled and sieved to separate the 850 to 212 μm (-20+70 mesh) size fraction. The crushed and sieved material was then

Table 1. Target Composition for CPSF5

Oxide	Mass%
ZrO_2	17.91
TiO_2	40.85
Gd_2O_3	8.83
CaO	8.15
Al_2O_3	8.15
BaO	3.40
PuO_2	16.11

washed ultrasonically with demineralized water and ethanol to remove fines, and then dried. The specific surface area of the crushed samples was estimated by assuming the particles to be spheres having radii equal to the average opening of the sieves [7]. The Archimedes method was used to measure bulk density; the WP-14 glass was measured to be 2710 ±50 kg/m^3, giving a specific surface area of 0.0042 m^2/g. The density of the ceramic was measured to be 4480 ±110 kg/m^3 giving a specific surface area of 0.0025 m^2/g.

Table 2. Analyzed Composition of WP-14 Glass

Oxide	Mass%	
Al_2O_3	4.71	±0.042
B_2O_3	6.91	±0.058
CaO	1.10	±0.028
Cr_2O_3	0.33	±0.026
CuO	0.36	±0.006
Fe_2O_3	12.40	±0.135
K_2O	1.99	±0.049
Li_2O	4.48	±0.044
MgO	1.49	±0.017
MnO_2	3.23	±0.070
Na_2O	7.82	±0.136
Nd_2O_3	0.35	±0.019
NiO	1.05	±0.029
PbO	0.16	±0.005
SiO_2	50.92	±0.406
TiO_2	0.24	±0.004
ZnO	0.09	±0.002
ZrO_2	1.19	±0.025
Total	98.83	±1.100

PUF Test Procedure

The PUF column with an internal volume of 21.72 cm^3 was packed first with ½ the total required crushed and cleaned WP-14 glass, then with the crushed and cleaned CPSF5 ceramic, and finally with the remaining WP-14 glass. The mass difference between the full and empty column was used to calculate the initial porosity of approximately 0.46 ±0.02. Mass change and bed volume were also tracked during packing of each layer to compute the porosity for each layer. Individual bed porosity was within the reported measurement error above. The column was then vacuum saturated with water at ambient temperature. A temperature controller was programmed to heat the column to 90°C in approximately 1 h (1°C/min). The column was allowed to desaturate initially during heating by gravity drainage and was also vented periodically to maintain an internal pressure less than the bubble pressure of the porous plate. After reaching 90°C, the influent valve was opened, and influent set to a flow rate of 1 mL/d. Column venting was set to occur once an hour. Effluent samples were collected in a receiving vessel that was drained nominally on a 5-day interval but sometimes more often when changes in the effluent pH and conductivity were detected. The samples were drained into tared vials from which samples were extracted and acidified for elemental analysis with ICP-MS. Liquid scintillation counting was used for the analysis of alpha and beta emitting isotopes. During the course of the experiment, several effluent samples were also collected and filtered through Centricon 30 ultrafiltration cones (30 000 molecular weight cutoff, or approximately 1.8 nm particle size). Some of these samples were collected and filtered at as close to 90°C as possible, although some cooling (probably 10 to 20°C) occurred during sample transfer and centrifugation.

After a total run duration of 11 523 h, the PUF experiment was terminated. A comprehensive run termination and sample analysis plan was developed and executed. The column was split lengthwise with the aid of a miter box. One half of the column was subsampled as found (loose and moist particles) and the other half was allowed to air-dry in preparation for resin infiltration. For the half column with loose particles, 15 subsamples (3 subsamples contained the bulk of the Pu-doped ceramic) were obtained at 5 mm intervals. From these samples the following analyses were conducted:

1. Moisture content determination by drying in glass vials at room temperature in a sealed can with $CaSO_4$ desiccant. Samples were dried until a constant mass was obtained.
2. SEM, TEM, and XRD analyses of selected samples from the glass and ceramic layers.

After drying for 48 h, the other half column was impregnated with "LR White" (London Resin Co.) resin. The resin-infiltrated half-core was epoxied to a set of aluminum mounting blocks so that sections could be cut with a low-speed metallographic saw.

RESULTS AND DISCUSSION

Test Metrics

The results from the computer-monitored test metrics are shown in Figure 2a. The indicated volumetric water content becomes a less accurate indicator of pore saturation in long duration experiments because water mass may accumulate in alteration phases as waters of hydration. A check of the pre- and post-test column mass indicated a pore saturation of not more than 33.6%. This measurement was used to determine a global correction factor for the indicated pore saturation, which was applied to the volumetric water content data. The dotted lines in Figure 2a indicate periods (about 8 h) where data acquisition was interrupted because of a software error in the control programming. A variable used in a timer subroutine overflowed when reaching 2^{32} milliseconds, which corresponds to approximately 50 days. Data acquisition was interrupted for periods of about 8 hours. Fluid flow, however, was only affected twice, once at 133 d and the other at 233 d. The software bug was discovered and corrected after the interruption at 233 d.

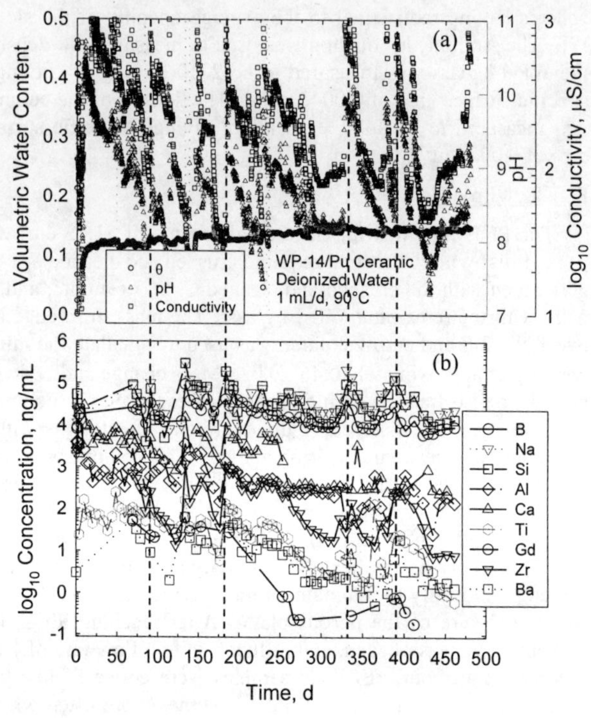

Figure 2. Computer-Monitored Test Metrics and Analyzed Effluent Composition as a Function of Time in Interactive PUF Test with Pu Zirconolite-dominant Ceramic

The data in Figure 2 indicate that the WP-14 glass, and perhaps the CPSF5 ceramic, have undergone periodic excursions in reaction rate that are manifested as excursions in effluent pH and electrical conductivity. However, the ceramic contains elements that are sparingly soluble under these test conditions, so it is highly unlikely that a dissolution rate acceleration of the ceramic could cause the transients in pH and electrical conductivity. The excursions in the effluent pH and conductivity continued after the software correction, and so appear to be unrelated to this problem as well. We have observed similar periodic and rapid increases in pH and electrical conductivity in PUF tests with a number of different glass compositions. These excursions have always correlated with the formation of secondary phases, especially Na-Ca aluminosilicate zeolites. Additional discussion regarding possible cause(s) of the transients observed in the effluent chemistry is provided in the Discussion section.

Solution Chemistry

Results from the ICP-MS analyses of effluent samples are provided in Figure 2b. There is a good correlation between the observed excursions in effluent pH and electrical conductivity with peak concentrations of constituents released from the WP-14 glass and CPSF5 ceramic. There is also a general trend of decreasing concentrations with increasing reaction time. Unfortunately, there were no elements incorporated in the ceramic to permit unambiguous determination of the dissolution rate as each element is subject to adsorption, ion-substitution, or precipitation in secondary phases. Of the elements, however, Ba is the only element present in the ceramic that is not present in the glass where these effects are minimized. Gadolinium is not present in the glass either but because of its extremely low solubility under the conditions of these tests, it is unlikely to be reliable as a dissolution rate indicator element. Consequently, Ba was used to estimate the ceramic corrosion rate. This is discussed in the Ceramic Corrosion Rate section.

Figure 3 shows the results for total Pu concentration in the effluent as a function of time. The data are highly scattered. However, there is a good correlation between spikes in the effluent pH and Pu concentration. The general trend in the data shows a decrease in Pu release as a function of time. Effluent samples between 300 and 450 d were all near or below the detection threshold, except during the transient pH excursions. A comparison of the total measured Pu concentration with the measured concentration after filtration showed that for samples with Pu significantly above the analytical detection threshold, i.e., concentrations between 10^{-10} to 10^{-8} M Pu, a large percentage (>80%) was present in filterable form.

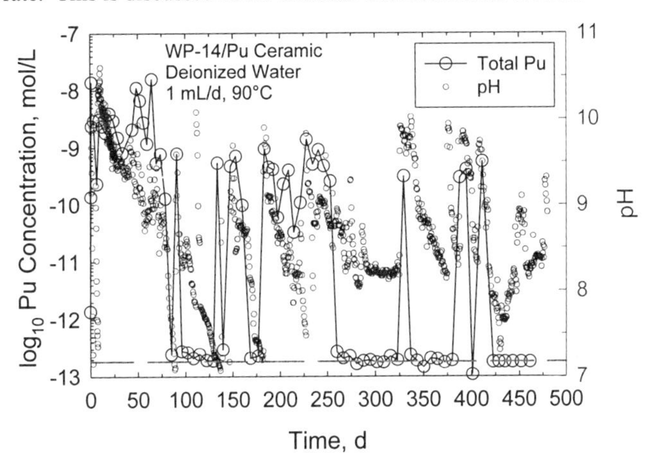

Figure 3. Effluent Pu Concentration as a Function of Time

Reacted Solids

An SEM photograph of a sample of WP-14 glass removed from the bottom of the PUF column is shown in Figure 4. Numerous deposits with a clay-like morphology were found. The morphology here though is unusual and striking in that it is studded with hundreds, if not thousands of small spherical particles. Some of these are embedded, or more likely growing out of the gel. Drying of the sample and placing it under high vacuum has clearly ruptured most of the particles. Presumably, sorption of Pu onto these particles is responsible for much of the filterable Pu release observed in the column effluents.

Milligram quantities of dried particles taken at approximately 5 mm intervals were deposited on tape and then analyzed for ^{239}Pu content with alpha energy analysis (AEA). Identification and quantification of ^{239}Pu and ^{241}Am was performed using the 5.156MeV peak and 5.486MeV peak, respectively. Results from the AEA are shown in Figure 5.

Despite the highly advection dominant system, the steep ^{239}Pu concentration gradient downstream from the ceramic bed indicates a strong attenuation of Pu in the glass bed with the concentration dropping by two orders of magnitude over a distance of 25 mm. Retention of Pu may be occurring by adsorption from solution, trapping of Pu colloids, or both. It is impossible to distinguish these mechanisms from the AEA data alone but because most all of the Pu exiting the column was found to be colloidal, colloid filtration is probably the more significant retention mechanism. Interestingly, the ^{241}Am data reverse positions with respect to ^{239}Pu on the upstream versus downstream side of the ceramic. Americium is expected to be more soluble and mobile than Pu under these test conditions and the downstream data are consistent with this expectation. The upstream data are more consistent with the original ratio in the ceramic. Because 146 days elapsed from when flow and heat were turned off to the column and when the column was dissected, the upstream profile most likely represents a diffusion profile into the glass bed. The mean diffusion penetration depth in 1-D is given by $\sqrt{\pi D_e t}/2$, where D_e is the effective diffusion coefficient. Neglecting chemical retardation, an effective diffusion coefficient of 6 x 10^{-7} cm^2/s was estimated from the ending volumetric water content in the test (0.15) and the unsaturated diffusive transport curve of Conca, Apted, and Arthur [8]. The mean unretarded penetration depth is then ≈24 mm. Consequently, the upstream data in Figure 5 are consistent with a diffusion profile where the Pu and Am are significantly retarded. The upstream concentration profile was fit with the linear adsorption model in the code CXTFIT [9] and a retardation factor of 156 provided the best fit to the ^{239}Pu data.

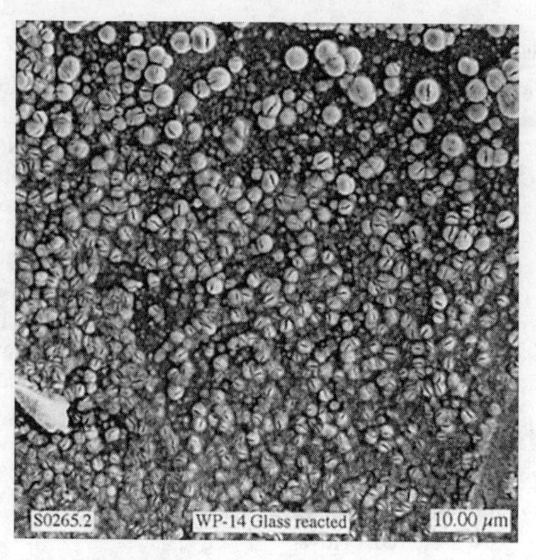

Figure 4. SEM Photo of Reacted WP-14 Glass Extracted From the Bottom of the PUF Column.

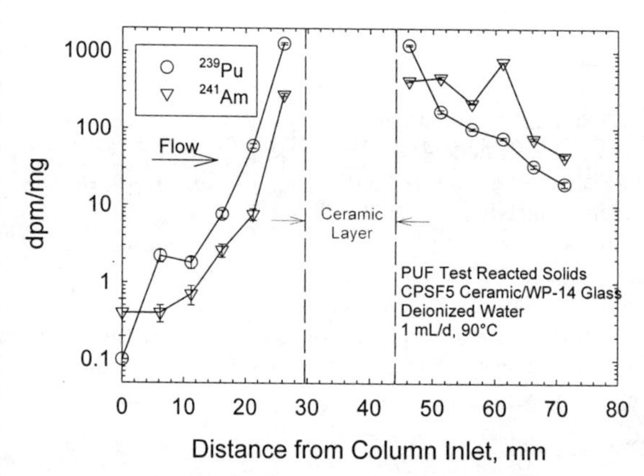

Figure 5. Alpha Energy Analysis of Glass Particles Removed from PUF Column after 480 d of Reaction. For reference, the measured count rate for the CPSF5 ceramic is 1.24 x 10^7 dpm/mg for ^{239}Pu and 1.77 x 10^6 dpm/mg for ^{241}Am.

DISCUSSION

Corrosion rate calculations were performed [2] assuming a constant water content value of 0.15, equivalent to the final measured pore saturation of 33.6%. The calculated ceramic corrosion rate as a function of time, using Ba release as the indicator element, is given in Figure 6 along with the computer-monitored pH data. The corrosion rate of WP-14 glass was calculated from the B concentration data. The results show that the release rate from the ceramic is strongly coupled to the pH excursions observed during the test, as is the glass corrosion rate. Because the pH excursions are controlled by the glass/water reaction, the data show that the performance of the ceramic is directly coupled to the chemical environment imposed by the HLW glass form. However, the overall trend for both forms is a decreasing release rate as a function of time. Moreover, the estimated ceramic corrosion rate from the Ba release is quite low, less than 10^{-5} g/(m²d) during the last 200 days of the experiment, except during the pH excursions. The calculated corrosion rate of the WP-14 glass is consistent with rates determined from long-term product consistency tests with SRL-202 glass [10].

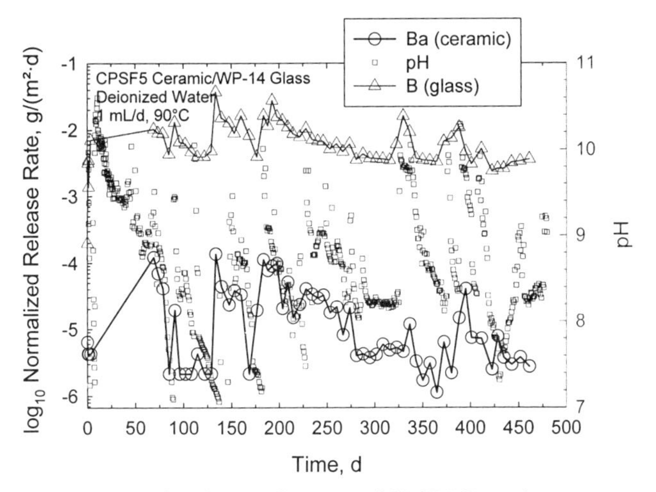

Figure 6. Calculated Corrosion Rate of CPSF5 Ceramic as a Function of Time in Interactive PUF Test

As we have observed in many other PUF tests with low-activity waste glasses, precipitation of zeolitic alteration phases causes a dramatic acceleration in the corrosion rate of these glasses. The acceleration is detectable exsitu by a rapid increase in effluent pH and electrical conductivity. In some cases, the corrosion rate has been observed to increase autocatalytically back up to the forward rate of reaction of the glass. However, because the WP-14 glass has 2.5 times less Na than LAW glass, we speculate that not only does zeolite precipitation slow as the pore fluid pH rises after precipitation begins but in fact the precipitation may cease altogether. This should cause the effluent pH to drop back to a nominal value balanced by a much slower rate of glass corrosion. Furthermore, if the high pH pore water composition becomes undersaturated with respect to the zeolite, then the zeolite will begin to dissolve. Dissolution of the zeolite will consume H^+ from the fresh pore water passing through the system. Consequently, the rate of relaxation to a lower pH, lower ionic strength pore fluid composition is dictated by the dissolution rate of the zeolite. Once the pore water pH and composition adjust to where precipitation of the zeolite becomes favorable once again, the cycle repeats.

CONCLUSION

In a 17-month long PUF experiment, a simulated HLW glass and a zirconolite-dominated ceramic for Pu immobilization have been allowed to react in a manner similar to the way they might react in a repository. The results from this experiment have provided several important insights into the long-term release behavior of Pu and neutron absorbers from the ceramic. The results show a strong coupling between the chemistry of the water percolating through a porous bed of the test materials and the corrosion rate of the ceramic. Consequently, should the HLW glass undergo a sustained acceleration in its corrosion rate due to secondary phase formation, the resulting excursion in pH could significantly impact the corrosion rate of the of the ceramic. The 2 to 3 pH unit transient excursions in pH exhibited over the entire course of this PUF experiment are probably caused by the transient acceleration in the glass corrosion rate from the formation of alteration phases. These excursions cause as much as a two order of magnitude increase in the release rate of Ba from the ceramic during these periods. However, in spite of these momentary increases, the overall rate of dissolution of the ceramic remains low.

For the first 200 days and during pH excursions, Pu was found to be exiting the PUF system primarily as filterable particulates. However, effluent Pu concentrations were found to decrease with time and remained at near detection limits after approximately 250 days, except during transient pH excursions. Similar behavior was observed for the neutron absorber, Gd, with concentrations decreasing to the detection threshold (<0.5 ppb), again except during transient pH excursions. These results indicate that both Pu and Gd will be retained in the can-in-canister waste package to a very high degree. Primarily, this is caused by the formation of highly insoluble secondary phases and/or sequestration by incorporation into or adsorption onto the hydrated surface layers formed on the HLW glass.

REFERENCES

1. S. G. Cochran, W. H. Dunlop, T. A. Edmunds, L. M. MacLean, and T. H. Gould, *Fissile Material Disposition Program Final Immobilization Form Assessment and Recommendation*, Lawrence Livermore National Laboratory, UCRL-ID-128705, Livermore, California (1997).
2. B. P. McGrail, P. F. Martin, J. P. Icenhower, H. T. Schaef, V. L. Legore, and R. D. Orr, *Evaluation of the Long-Term Performance of Titanate Ceramics for Immobilization of Excess Weapons Plutonium: Results From Pressurized Unsaturated Flow and Single Pass Flow-Through Testing*, PNNL-12240, Rev. 0, Pacific Northwest National Laboratory, Richland, Washington (1999).
3. B. P. McGrail, C. W. Lindenmeier, P. F. C. Martin, and G. W. Gee, *Trans. Am. Ceram. Soc.* **72**:317-329 (1996).
4. B. P. McGrail, P. F. Martin, and C. W. Lindenmeier, *Mat. Res. Soc. Symp. Proc.* **465**:253-260 (1997).
5. P. J. Wierenga, M. H. Young, G. W. Gee, R. G. Hills, C. T. Kincaid, T. J. Nicholson, and R. E. Cady, *Soil Characterization Methods for Unsaturated Low-Level Waste Sites*, PNL-8480, Pacific Northwest Laboratory, Richland, Washington (1993).
6. American Society for Testing and Materials, *Standard Test Methods for Determining Chemical Durability of Nuclear Waste Glasses: The Product Consistency Test (PCT)*, Standard C1285-94, Philadelphia, Pennsylvania (1995).
7. B. P. McGrail, W. L. Ebert, A. J. Bakel, and D. K. Peeler, *J. Nuc. Mat.* **249**:175-189 (1997).
8. J. L. Conca, M. Apted, and R. Arthur, *Mat. Res. Soc. Symp. Proc.* **294**:395-402 (1993).
9. J. C. Parker, and M. Th. Genuchten, *Determining Solute Transport From Laboratory and Field Tracer Experiments*, Virginia Agricultural Experiment Station, Bull. No. 84-3, Virginia Polytechnic Institute and State University, Blacksburg, Virginia (1984).
10. W. L. Ebert, and J. K. Bates, *Nuc. Tech.* **104**:372-384 (1993).

AQUEOUS DURABILITY OF TITANATE CERAMICS DESIGNED TO IMMOBILISE EXCESS PLUTONIUM

K. P. Hart§, Y. Zhang§, E. Loi§, Z. Aly§, M. W. A. Stewart§, A. Brownscombe§, B. B. Ebbinghaus‡ and W. Bourcier‡.

§ Australian Nuclear Science and Technology Organisation, PMB 1, Menai 2234, Australia (kph@ansto.gov.au)

‡ Lawrence Livermore National Laboratory, Livermore, CA, USA

Abstract

Aqueous durability has been assessed for fourteen pyrochlore- and zirconolite-rich titanate waste forms designed for the immobilisation of excess Pu. The ceramics used in this study contained about 12 wt% Pu and about 15 wt% of Hf and Gd oxides as neutron absorbers and were fabricated by cold-pressing and sintering or hot isostatic pressing.

Total release rates (i.e. unfiltered solution + vessel wall inventory) have been measured with the MCC-1 test method at 90 °C in deionised water. For all samples, 7-day release rates of Pu are between 4×10^{-5} and about 10^{-3} g.m^{-2}.d^{-1}, reducing to between about 8×10^{-6} and 3×10^{-5} g.m^{-2}.d^{-1} after more than 300 days of leaching. Release rates of U from the baseline ceramic were found to decrease to values between 6×10^{-4} and 1×10^{-5} g.m^{-2}.d^{-1} after 200 days. Hf leach rates were generally $< 5 \times 10^{-6}$ g.m^{-2}.d^{-1} after more than 7 days. Further, the addition of several % of chemical impurities to the baseline sintered formulations was found to reduce U and Gd releases by about a factor of 10 after 200 days.

Introduction

The corrosion behaviour of two types of ceramic waste forms for the immobilisation of excess Pu, zirconolite-rich and pyrochlore-rich formulations, have been studied extensively[1-3]. However, only relatively short-term test results have been reported to date. For the immobilisation of actinide-rich wastes, long-term durability data need to be obtained, given the long half lives of the actinide elements and the importance of understanding their long-term release relative to the neutron absorbers. Test data for periods of up to 300 days are reported in this paper for both zirconolite-rich and pyrochlore-rich formulations.

Experimental

The ceramic formulations studied in this work are summarised in Table 1. Pyrochlore-rich formulations have a baseline composition (wt%) of CaO (9.95), Gd_2O_3 (7.95), HfO_2 (10.65), PuO_2 (11.89), UO_2 (23.69) and TiO_2 (35.87). Hf and Gd were present as neutron absorbers to address criticality concerns. This baseline composition was designed to give a ceramic containing 95 wt% of $Ca_{0.89}Gd_{0.22}Pu_{0.22}U_{0.44}Hf_{0.23}Ti_2O_7$ (pyrochlore) plus 5 wt% of Hf-bearing rutile. Details of the formulations, many of which contain additional process impurities, and their development are given in Vance *et al* (1999)[1]. Other batches, viz. Pu97, Pu99, Pu101, and Pu103, were prepared to provide information on a wider composition field than the baseline formulation. The compositions of these latter batches are given in Table 2. Two zirconolite-rich materials[1] containing Pu, Hf and Gd were also studied.

Mat. Res. Soc. Symp. Proc. Vol. 608 © 2000 Materials Research Society

Corrosion testing was carried out at 90 ˚C in deionised water, as described elsewhere[2]. Analysis of inactive elements was carried out by ICP/MS but for Pu, the level of activity was determined by alpha spectrometry carried out on 0.1 mL of leachant evaporated onto a stainless steel planchette. This method allowed the contribution of ^{241}Am to be separated from that of the Pu[1], unlike in our previous studies[2].

Results

Long-term leaching of Pyrochlore-rich and Zirconolite-rich Formulations

The test results for Pu releases for all of the formulations listed in Table 1 are shown in Figure 1. The variation between the Pu release rates after 7 days is about a factor of 20 with the difference reducing to about 4, for the smaller number of samples studied so far, after about 300 days. Overall, the release rates decreased by about one order of magnitude between 3.5 days (i.e. 0 to 7 day period) and 324 days. Data for the effect of impurity addition on U, Pu, Gd, and Hf releases are shown in more detail in Table 3.

Figure 2 shows the U release rates measured for all of the sintered formulations. With the exception of Pu97, Pu99 and Pu101, the U release rates vary over about one and a half orders of magnitude for the 0 to 7 day period and decrease by about one to two orders of magnitude after 150 days. Release rates measured for Gd, for all of the batches listed in Table 1, remained relatively constant between 10^{-3} and 10^{-4} g m^{-2}d^{-1}. Gd release results showed a wider variation than for other elements, about an order of magnitude between triplicates, and this has contributed to scatter in the data shown in Table 3. Unfortunately space limitations do not allow for all of the Gd data to be shown in this paper. Hf concentrations were close to the detection limit in most leachants, i.e. 10^{-7} g.L^{-1}, and are generally similar to or lower than those of Pu. (Development work is being carried out to improve the detection limit of the ICP/MS for Hf and Gd.)

Although the Pu, Gd and Hf release rates were similar within about one order of magnitude for all of the ceramics listed in Table 1, the U release rate from batches Pu97, Pu99 and Pu101 has evidently been affected by the additions made to the baseline formulations. The different behaviours for U may indicate that in these batches some of the U is present at higher valence states than in the baseline formulations and as a consequence is more leachable. SEM examination of these ceramics did not identify any obvious U-bearing minor phases arising from the compositional changes relative to the baseline.

For batches 68 and 75, which were prepared and tested together, it is possible to compare the effects of impurity additions on the long-term releases of Pu, U and the neutron absorbers (Gd and Hf) from the matrix (see Table 3). Pu release rates, with the exception of those in the last leaching period, are very similar (and the formulation containing the impurities tends to have lower Pu release rates). Overall the Pu release rates are about a factor of 2 to 5 lower than those of U and about 10 to 20 times lower than those of Gd. The release rates of Hf tend to be close to the detection limit of the ICP/MS but generally appear to be similar to or lower than

---

[1] The Pu used in this study consisted of (wt%): 93% ^{239}Pu, 5% ^{240}Pu, 0.5% ^{241}Pu and 1.5% ^{241}Am. From alpha spectrometry the contribution of ^{239}Pu could not be separated from ^{240}Pu and hence, the release rates of Pu are based on both isotopes.

those of Pu (see comments above). Results reported[4] previously for fully-dense, Pu-doped Synroc-C. showed that total Pu release rates were constant at about 10^{-5} $g.m^{-2}.d^{-1}$ after about 100 days at 70 °C in deionised water. These values are very similar to those obtained for sintered, pyrochlore-rich ceramics and HIPed, zirconolite-rich ceramics in this study.

TABLE 1

CERAMIC FORMULATIONS

Batch	Precursor	Generic composition description	Preparation conditions
Pu49	Oxide	Zirconolite-rich	HIPed at 1250 °C
Pu49C	Oxide	Zirconolite-rich	Batch 49 re-HIPed at 1280 °C
Pu68	Oxide	Baseline	Sintered 1350 °C in Ar
Pu69	Oxide	Baseline‡ + 2.5 wt% impurities	Sintered 1350 °C in Ar
Pu70	Oxide	Baseline‡ + 1 wt% impurities	Sintered 1350 °C in Ar
Pu71	Alkoxide	Baseline‡ + 1.5 wt% impurities	Sintered 1350 °C in Ar
Pu72	Alkoxide	Baseline‡ + 14 wt% impurities	Sintered 1350 °C in Ar
Pu73	Alkoxide	Baseline‡ + 5.4 wt% impurities	Sintered 1300 °C in Ar
Pu75	Oxide	Baseline‡ + 1.5 wt% impurities	Sintered 1325 °C in Ar
Pu97	Oxide	Zirconolite-rich version of the baseline composition	Sintered 1350 °C in Ar
Pu99	Oxide	Brannerite-rich version of the baseline composition	Sintered 1350 °C in Ar
Pu101	Oxide	Ca-rich version of the baseline composition	Sintered 1350 °C in Ar
Pu103	Oxide	Baseline with 10 wt% phosphate addition	Sintered 1350 C in Ar
Pu118	Oxide	Baseline‡ + 1.5 wt% impurities	Sintered 1350 C in Ar

‡ Further details are given in Vance *et al.* (1999)

TABLE 2

COMPOSITION (wt%) OF BATCHES Pu97, Pu99, Pu101, AND Pu103

Component	Baseline	Pu97	Pu99	Pu101	Pu103
CaO	9.95	9.24	5.84	12.07	11.61
Gd_2O_3	7.95	6.60	6.32	7.66	8.50
HfO_2	10.65	24.21	9.92	10.88	9.69
PuO_2	11.89	9.57	12.98	12.36	13.03
UO_2	23.69	15.40	27.96	20.83	21.35
TiO_2	35.87	33.84	36.99	36.19	32.48
Al_2O_3		1.13			
P_2O_5					3.34

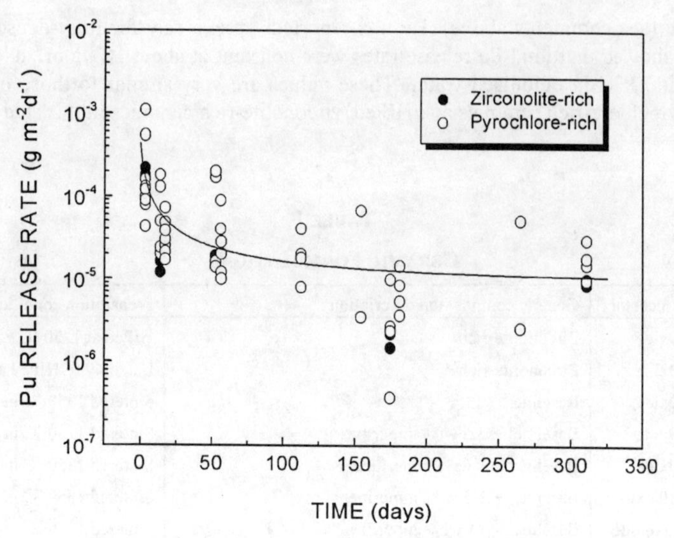

Figure 1. Total, normalised Pu release rates from zirconolite-rich (HIPed) and pyrochlore-rich (sintered) formulations measured at 90˚C in deionised water. For convenience, all the different batches of sintered material have been given the same symbol and the line shown on the graph is intended to be indicative only.

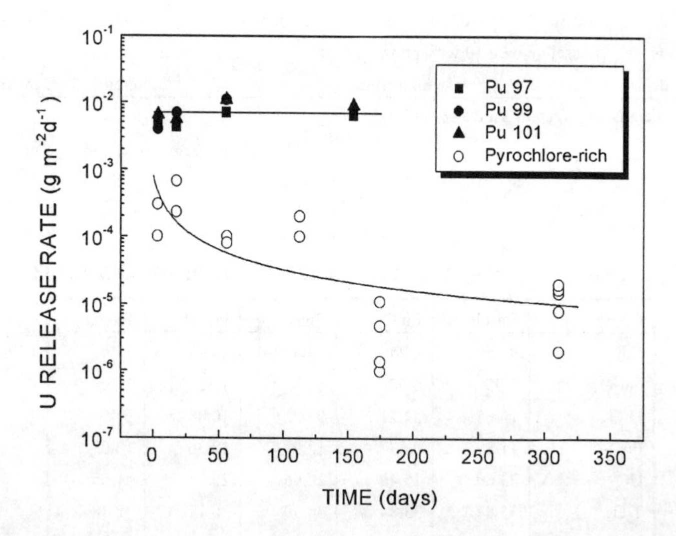

Figure 2. Total U release rates from Pu batches 69 to 73 and 97 to 118, measured at 90 ˚C in deionised water. For convenience, all the different batches of baseline sintered material have been given the same symbol and the line shown on the graph is intended to be indicative only.

TABLE 3

TOTAL RELEASE RATES OF Pu, U AND Gd from Batches Pu68 and 75 (based on median values from triplicate, polished to 6 μm samples‡– measured at 90 °C in deionised water)

	Release Rate ($g.m^{-2}.d^{-1}$)			
Test period (days)	0 to 7	7 to 28§	28 to 84	84 to 224
Baseline formulation – Pu68				
U	5×10^{-4}	1×10^{-4}	5×10^{-5}	1×10^{-4}
Pu*	1.0×10^{-4}	2.5×10^{-5}	8.9×10^{-5}	6×10^{-5}
Gd	7×10^{-4}	3×10^{-4}	1.6×10^{-3}	1×10^{-3}
Hf	2×10^{-4}	$< 5 \times 10^{-6}$	$< 4 \times 10^{-6}$	$< 2 \times 10^{-6}$
Pu75 — with impurities				
U	7×10^{-4}	2×10^{-4}	2×10^{-5}	1×10^{-5}
Pu	1.3×10^{-4}	4.8×10^{-5}	4.0×10^{-5}	3.5×10^{-6}
Gd	4×10^{-4}	1×10^{-4}	5×10^{-4}	1×10^{-4}
Hf	8×10^{-5}	$< 1 \times 10^{-5}$	$< 4 \times 10^{-6}$	$< 2 \times 10^{-6}$

‡ Maximum variation between triplicates was generally less than 50 %.
§ Average of duplicate samples
* Results based on the combined activity of ^{239}Pu and ^{240}Pu

Conclusions

Overall the major findings of the studies carried out in this work are that;

- Initial (0 to 7 day), Pu release rates of fourteen separate pyrochlore-rich and zirconolite-rich ceramics measured at 90 °C in deionised water are between 4×10^{-5} and about 10^{-3} $g.m^{-2}.d^{-1}$. After about 300 days, the release rates have declined by a factor of about 10 to 100 from the initial values.
- Release rates of U from the baseline ceramic were found to decrease by about one order of magnitude albeit slowly to values between 6×10^{-4} and 1×10^{-5} $g.m^{-2}.d^{-1}$ after more than 200 days.
- Gd release rates showed typically an order of magnitude variation between triplicate samples. Within this scatter, the release rates remained constant at about 1×10^{-3} $g.m^{-2}.d^{-1}$.
- After more than 7 days, Hf release rates were generally $< 5 \times 10^{-6}$ $g.m^{-2}.d^{-1}$.
- Longer-term releases of U and the neutron absorbers bracket those of Pu, suggesting that these elements may have similar source terms from the ceramic matrices within a repository environment.
- The addition of several percent of chemical impurities to the baseline sintered formulations was found to reduce U and Gd releases by about a factor of 10 after 200 days.

Acknowledgments

The authors wish to acknowledge the partial support of this work by the USDOE in conjunction with LLNL and thank Drs E. R. Vance and D. Shoesmith for their critical review of the manuscript.

References

1 Vance, E.R., Jostsons, A., Moricca, S., Stewart, M.W.A., Day, R. A., Begg, B. D., Hambley, M. J., Hart, K. P. and Ebbinghaus, B. B., Ceramic Transactions, **93**, 323-329, 1999.

2 Hart, K. P., Vance, E. R., Stewart, M. W. A. Stewart, Weir, J., Carter, M. L, Hambley, M., Brownscombe, A., Day, R. A., Leung, S., Ball, C. J., Ebbinghaus, B., Gray, L. and Kan, T. (1998) Mat. Res. Soc. Symp. Proc. **506**, 161-168.

3 Jostsons, A., Vance, Lou (E. R.) and Ebbinghaus, B., Immobilization of Surplus Plutonium in Titanate Ceramics, Paper presented at Global 99, Jackson Hole, Wyoming, USA, August 29 to September 3, 1999.

4 K. L. Smith, G. R. Lumpkin, M. G. Blackford, M. Hambley, R. A. Day, K. P. Hart and A. Jostsons (1997), Proc. Res. Soc. Symp. Proc., **465**, 1267-1272.

COMPOSITION, GEOCHEMICAL ALTERATION, AND ALPHA-DECAY DAMAGE EFFECTS OF NATURAL BRANNERITE

GREGORY R. LUMPKIN, S.H.F. LEUNG, AND M. COLELLA

Materials Division, Australian Nuclear Science and Technology Organisation, Private Mail Bag 1, Menai, NSW 2234, AUSTRALIA

ABSTRACT

To investigate the long-term alteration behavior of brannerite, we have undertaken a study of twelve natural samples from a range of geological environments. Our results indicate that seven of the samples exhibit only minor alteration, usually within veinlets or around the rim of the sample. The remaining five samples consist of variable amounts of unaltered and altered brannerite. SEM-EDX analyses of unaltered areas indicate that the chemical formulae may deviate from the ideal stoichiometry. The U content ranges from 0.45 to 0.88 atoms per formula unit (pfu). Maximum amounts of the other major cations on the U-site are 0.48 Ca, 0.22 Th, 0.14 Y, and 0.07 Ln (lanthanide = Ce, Nd, Gd, Sm) atoms pfu. The Ti content ranges from 1.86 to 2.10 atoms pfu. Maximum values of other cations on the Ti-site are 0.15 Fe, 0.14 Si, 0.09 Al, 0.06 Nb, 0.04 Mn, and 0.04 Ni atoms pfu. Altered regions of brannerite contain significant amounts of Si and other elements incorporated from the fluid phase, and up to 40-90% of the original amount of U has been lost as a result of alteration. SEM-EDX results also provide evidence for TiO_2 phases, galena, and a thorite-like phase as alteration products. Electron diffraction patterns of all samples typically consist of two broad, diffuse rings that have equivalent d-spacings of 0.31 nm and 0.19 nm, indicating complete amorphization of the brannerite. Many of the grains also exhibit weak diffraction spots due to fine-grained inclusions of a uranium oxide phase and galena. Using the available age data, these samples have average accumulated alpha-decay doses of 2-170 x 10^{16} alphas/mg. Our results indicate that brannerite is subject to amorphization and may lose U under certain P-T-X conditions, but the overall durability of the titanate matrix remains high.

INTRODUCTION

Brannerite, ideally UTi_2O_6, is a common accessory phase in Synroc formulations designed for the encapsulation of actinide-rich, Pu-bearing nuclear wastes [1]. Even though brannerite is a minor phase in these ceramics, the pure end-member composition contains 62.8 wt% UO_2, therefore it may account for a significant fraction of the total amount of actinides in the waste form. To ensure that the presence of brannerite does not compromise the integrity of the waste form, a combination of laboratory experiments and natural analogue studies have been undertaken to assess both the aqueous durability and radiation damage effects of this phase [2,3]. In this report, we provide preliminary compositional and structural results for a suite of twelve brannerite samples from several different localities, covering a range of geological ages and host rock environments. Samples have been characterized using optical microscopy, X-ray diffraction (XRD), scanning electron microscopy and microanalysis (SEM-EDX), and transmission electron microscopy (TEM).

EXPERIMENTAL PROCEDURES

SEM-EDX work was carried out on polished sections using a JEOL JSM-6400 SEM equipped with a Noran Si(Li) microanalysis system and operated at 25 kV for microanalysis and 15 kV for secondary and backscattered electron imaging. EDX spectra were acquired for 500 seconds and processed using Noran software that produces results normalized to 100 wt%. Spectra were obtained from unaltered brannerite, altered brannerite, and associated mineral inclusions or alteration products. TEM was performed on crushed fragments dispersed on holey carbon grids using a JEOL 2000FXII TEM equipped with a Link ISIS Si(Li) microanalysis system and operated at 200 kV. The instrument was calibrated for selected area diffraction (SAD) work over a range of objective lens currents with a gold standard. The chemistry of the brannerite fragments and mineral inclusions were checked by EDX. Spectra were processed with the Link software package TEMQuant using previously established procedures [4].

Mat. Res. Soc. Symp. Proc. Vol. 608 © 2000 Materials Research Society

SAMPLE DESCRIPTION

The samples used in this study range in age from approximately 20 Ma to 1580 Ma and occur in a variety of host rocks (see Table 1). Optical microscopy and SEM-EDX work revealed that seven out of twelve samples are either unaltered or exhibit only minor alteration, usually within narrow veinlets or around the rim of the sample (Fig. 1a,b). The remaining samples consist of variable amounts of unaltered and altered brannerite. Alteration typically occurs along microfractures into the interior of the brannerite (Fig. 1c,d). In an advanced stage of alteration, a large proportion of the brannerite is affected and other secondary phases may be present (Fig. 1e,f). The darker contrast of the altered areas shown in SEM images is consistent with U loss and possible hydration effects (see the following section).

Table 1. A summary of the localities, age data, and alpha-decay doses of natural brannerite.

Locality	Samples	t (Ma)	Host rock	Structure	D (α/mg)
Crocker's Well, SA	B1, 5, 6	1580	granite	metamict	1.7×10^{18}
Cordoba, Spain	B2, 10	~ 400	U ore deposit	metamict	~ 6×10^{17}
Ticino, Switzerland	B3, 9, 12	20-25	granite pegmatite	metamict	$2\text{-}3 \times 10^{16}$
W. Province, Zambia	B4	~ 200	U ore deposit	metamict	~ 2×10^{17}
San Bernardino Co., CA	B7	~ 65	quartz veins	metamict	~ 7×10^{16}
Bou-Azzer, Morocco	B8	~ 100	quartz veins	metamict	~ 1×10^{17}
Stanley, Idaho	B11	~ 30	placer deposit	metamict	~ 3×10^{16}
Binntal, Switzerland*	ref. [5]	11	dolomite marble	crystalline	1×10^{16}

CHEMICAL COMPOSITION

Average compositions of relatively unaltered areas of each brannerite sample are given in Table 2 (assuming all U is U^{4+}). The following composition ranges of the unaltered brannerite were determined based on 5-10 analyses of each sample: 36-42 wt% TiO_2, 30-57 wt% UO_2, 0-15 wt% ThO_2, 0-7 wt% CaO, and 0-7 wt% PbO. Additional minor constituents include up to 1.8 wt% Nb_2O_5, 2.3 wt% SiO_2, 1.2 wt% Al_2O_3, 4.2 wt% Y_2O_3, 3.0 wt% Ln_2O_3 (Ln = Ce, Nd, Sm, Gd), 0.6 wt% MnO, 2.6 wt% FeO, and 0.7 wt% NiO. Na_2O is consistently near or below the detection limit of approximately 0.1 wt%.

The chemical formula of relatively unaltered, natural brannerite may deviate considerably from the ideal UTi_2O_6 stoichiometry. The U content ranges from 0.45 to 0.88 atoms pfu. Maximum amounts of the other major cations on the U-site are 0.48 Ca, 0.22 Th, 0.14 Y, and 0.07 Ln atoms pfu. The Ti content ranges from 1.86 to 2.10 atoms per 6 oxygens. Maximum values of the other cations on the Ti-site are 0.15 Fe, 0.14 Si, 0.09 Al, 0.06 Nb, 0.04 Mn, and 0.04 Ni atoms pfu. Total cations commonly exceed the ideal value of 3.00 when normalized to 6.00 oxygens, indicating that all of the Fe is in the 3+ state and that a significant amount of the U must also be in a higher valence state than the assumed 4+ value.

Average compositions of altered areas of brannerite are given in Table 3. Considering the ranges of altered compositions documented thus far, in the most heavily altered samples, up to 75-90% of the original amount of UO_2 was lost as a result of alteration. The observed U loss is compensated in part by incorporation of large amounts of Si (up to18 wt% SiO_2) and other elements from the attending fluid phase, including Al, P, Fe, As, and possibly Pb. During alteration, Y is also typically removed from the solid brannerite, but the behavior of Ca is more erratic and may be either lost or gained.

The chemical U-Pb age of unaltered areas of samples from Crocker's Well, South Australia, range from 530 to 810 Ma, consistent with previous work [6]. These ages are much less than the 1580 Ma U-Th-Pb isotopic age determined on zircon from the host rock and may be due to loss of radiogenic Pb from the brannerite. Younger samples from Ticino, Switzerland, give chemical U-Pb ages consistent with the known ages of the pegmatite host rocks of the region.

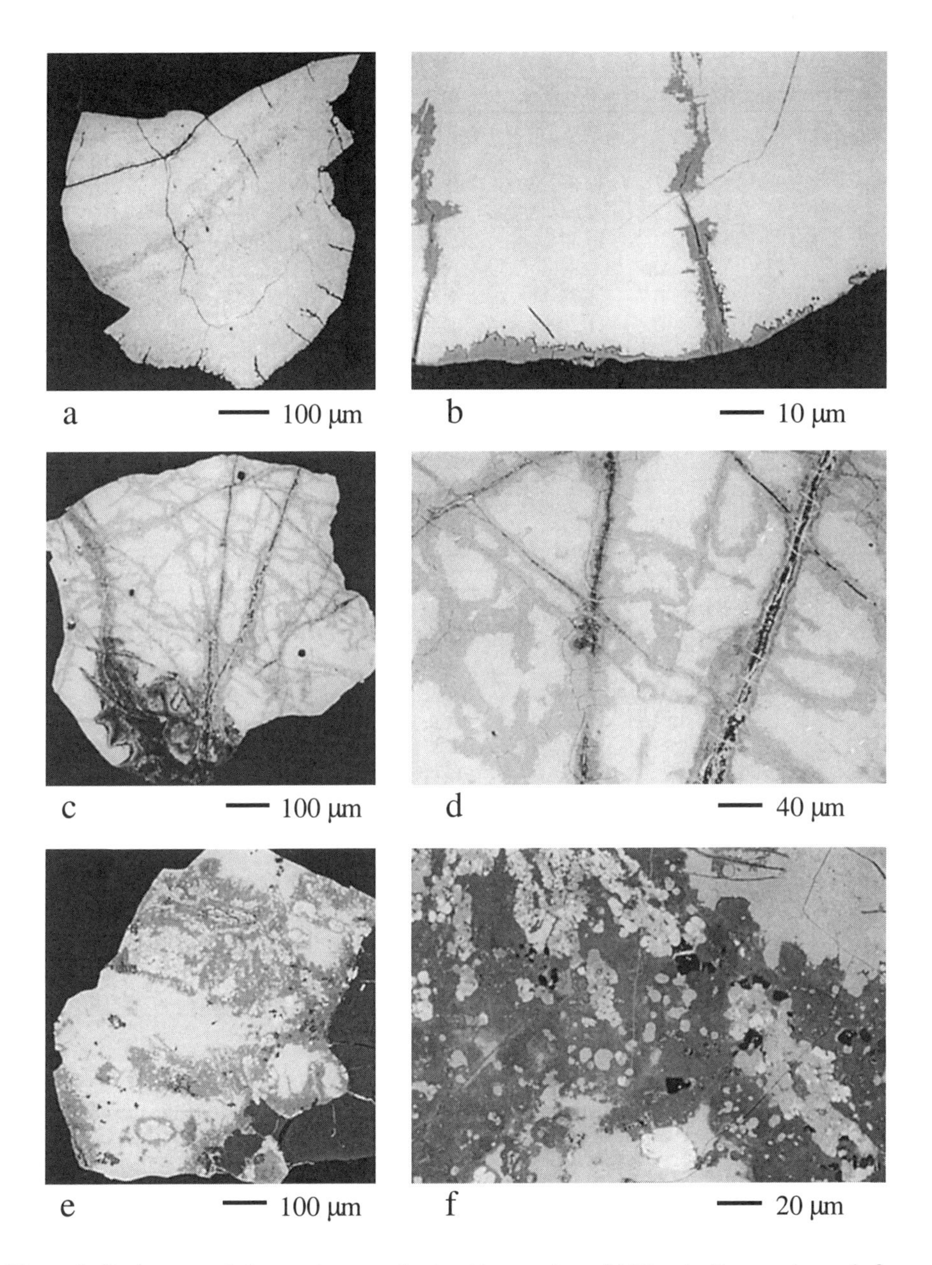

Figure 1. Backscattered electron images of natural brannerite. a,b) Chemically zoned sample from the El Cabril district, Spain. Note minor alteration (darker contrast) along rim and microfractures. c,d) This sample from Ticino, Switzerland, exhibits classic fracture controlled alteration (darker contrast). e,f) Heavily altered brannerite from Crocker's Well, South Australia. The altered areas (darker gray) contain secondary rutile or anatase (black) and thorite (white).

Table 2. Average compositions of relatively unaltered areas of twelve brannerite samples.

	B1	B2	B3	B4	B5	B6	B7	B8	B9	B10	B11	B12
Nb_2O_5	0.0	0.0	0.0	0.0	0.0	0.0	0.9	0.0	0.0	0.0	0.8	0.0
SiO_2	0.5	0.1	0.1	1.6	0.4	1.0	0.3	0.6	0.0	0.0	0.1	0.0
TiO_2	39.3	39.8	37.9	39.8	37.8	40.1	40.8	39.9	40.8	37.4	39.0	39.8
ThO_2	7.5	2.0	2.2	2.6	12.2	11.6	9.0	0.4	1.8	1.1	6.3	1.7
UO_2	40.2	55.3	50.0	45.1	34.5	36.7	38.4	50.6	55.9	52.0	46.5	56.5
Al_2O_3	0.2	0.2	0.9	0.8	0.3	0.2	0.3	0.2	0.1	0.3	0.2	0.0
Y_2O_3	1.7	0.8	0.7	1.7	1.2	1.3	3.1	1.1	0.0	0.5	2.2	1.1
Ln_2O_3	2.2	0.4	0.3	1.2	1.6	1.5	1.5	1.1	0.2	0.4	1.1	0.0
CaO	1.5	0.5	3.5	3.2	6.5	2.3	3.4	2.8	0.5	4.1	1.4	0.0
MnO	0.1	0.0	0.3	0.1	0.1	0.1	0.1	0.5	0.0	0.1	0.0	0.0
FeO	1.5	0.6	1.5	0.9	2.1	1.5	0.7	1.3	0.4	1.0	1.8	0.7
NiO	0.0	0.0	0.0	0.0	0.0	0.6	0.0	0.5	0.0	0.0	0.0	0.0
PbO	5.3	0.2	2.5	3.0	3.2	3.2	1.5	1.1	0.2	3.2	0.6	0.2

Table 3. Average compositions of altered areas of twelve brannerite samples.

	B1	B2	B3	B4	B5	B6	B7	B8	B9	B10	B11	B12
P_2O_5	0.0	0.0	0.0	0.0	0.0	0.0	0.9	0.0	0.0	0.0	0.8	0.0
As_2O_5	0.0	0.0	0.0	0.0	0.0	0.0	0.9	0.0	0.0	0.0	0.8	0.0
Nb_2O_5	0.0	0.0	0.0	0.0	0.0	0.0	0.9	0.0	0.0	0.0	0.8	0.0
SiO_2	0.5	0.1	0.1	1.6	0.4	1.0	0.3	0.6	0.0	0.0	0.1	0.0
TiO_2	39.3	39.8	37.9	39.8	37.8	40.1	40.8	39.9	40.8	37.4	39.0	39.8
ThO_2	7.5	2.0	2.2	2.6	12.2	11.6	9.0	0.4	1.8	1.1	6.3	1.7
UO_2	40.2	55.3	50.0	45.1	34.5	36.7	38.4	50.6	55.9	52.0	46.5	56.5
Al_2O_3	0.2	0.2	0.9	0.8	0.3	0.2	0.3	0.2	0.1	0.3	0.2	0.0
Y_2O_3	1.7	0.8	0.7	1.7	1.2	1.3	3.1	1.1	0.0	0.5	2.2	1.1
Ln_2O_3	2.2	0.4	0.3	1.2	1.6	1.5	1.5	1.1	0.2	0.4	1.1	0.0
CaO	1.5	0.5	3.5	3.2	6.5	2.3	3.4	2.8	0.5	4.1	1.4	0.0
MnO	0.1	0.0	0.3	0.1	0.1	0.1	0.1	0.5	0.0	0.1	0.0	0.0
FeO	1.5	0.6	1.5	0.9	2.1	1.5	0.7	1.3	0.4	1.0	1.8	0.7
NiO	0.0	0.0	0.0	0.0	0.0	0.6	0.0	0.5	0.0	0.0	0.0	0.0
PbO	5.3	0.2	2.5	3.0	3.2	3.2	1.5	1.1	0.2	3.2	0.6	0.2

RADIATION DAMAGE EFFECTS

Electron diffraction patterns of relatively unaltered areas of all of the brannerite samples typically consist of two broad, diffuse rings characteristic of amorphous materials (Fig. 2a,c,e). The diffuse rings have equivalent d-spacings of 3.1Å and 1.9Å, similar to many other metamict oxides and certain silicate minerals [7-10]. Bright field images of these grains are typically featureless (Fig. 2b), consistent with the absence of long-range periodicity. Some of the grains in sample B4 from the Western Province of Zambia were found to contain 10-200 nm sized spherical voids (Fig. 2d), similar to previous observations on metamict zirconolite and columbite [8,10]. These voids have been attributed to the accumulation of radiogenic He in the sample over time. Many of the samples examined thus far also exhibit weak diffraction spots in SAD patterns taken from certain grains (Fig. 2e). In most cases, the diffraction spots appear to be due to the presence of fine grained (generally 5-100 nm sized) inclusions of a uranium oxide phase and galena (Fig. 2f). Altered areas of the brannerite samples have yet to be analyzed by TEM.

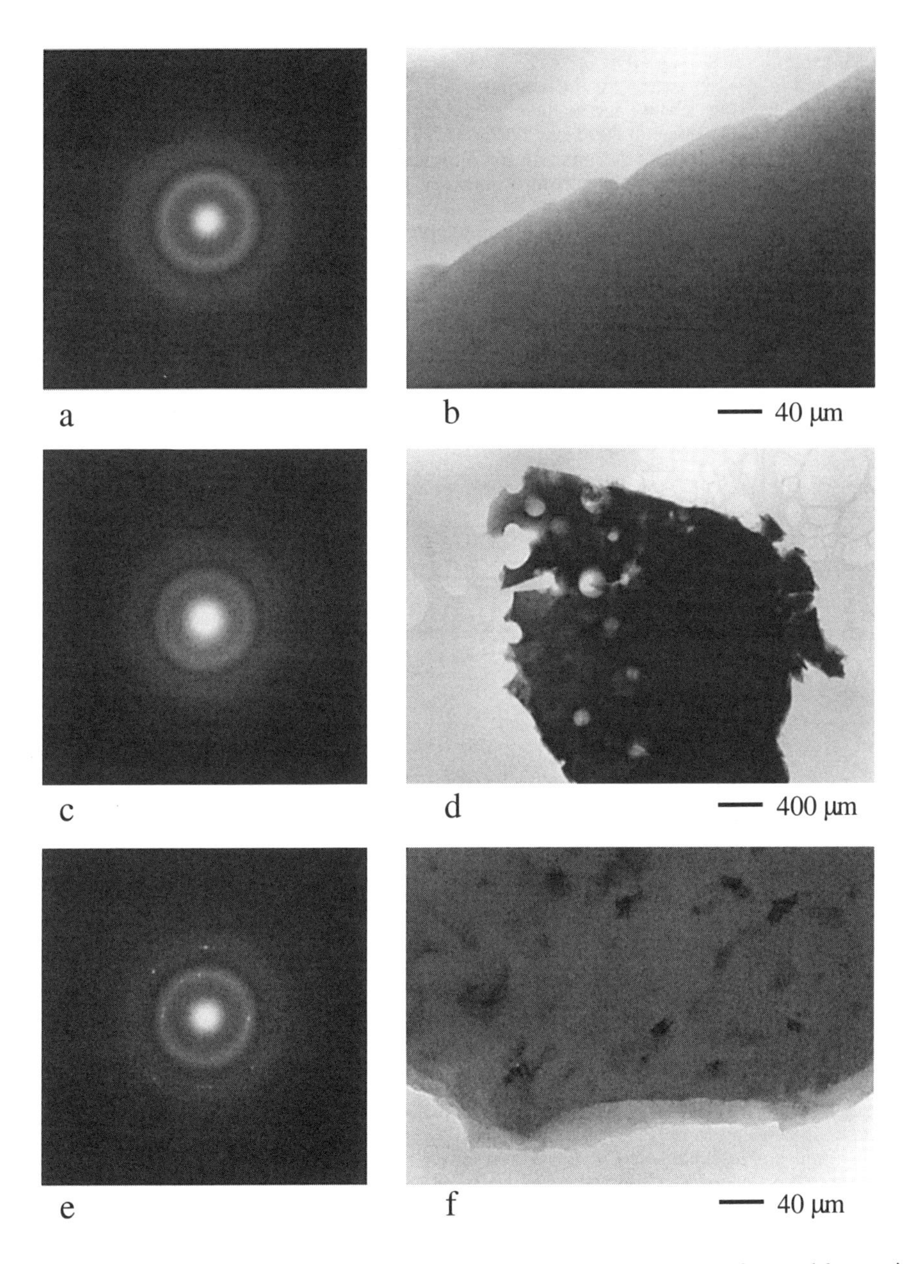

Figure 2. TEM bright field images and selected area diffraction patterns of natural brannerite samples. a,b) Fully amorphous sample from Ticino, Switzerland. c,d) Fully amorphous sample from Zambia. Note the presence of 50-200 nm sized voids (possibly due to He accumulation over time). e,f) Another example of brannerite from Zambia. Note the weak diffraction spots in (e) and the presence of 5-40 nm sized inclusions in (f).

At present, reliable age data are only available for the samples from Crocker's Well, South Australia and to a lesser extent for the samples from Ticino, Switzerland. Based on the Th and U contents and either the known age or the chemical U-Pb age determined by SEM-EDX (Table 1), the brannerites have average alpha-decay doses of 2-170 x 10^{16} α/mg. The critical amorphization dose (D_c) cannot be determined from these samples; however, using literature data given in Table 1 for a partially crystalline brannerite from Binntal, Switzerland [5], the critical dose appears to be close to 2 x 10^{16} α/mg. For comparison (see Fig. 3), natural pyrochlores and zirconolites with ages of 100 Ma or less become amorphous at doses of approximately 1 x 10^{16} α/mg [11,12].

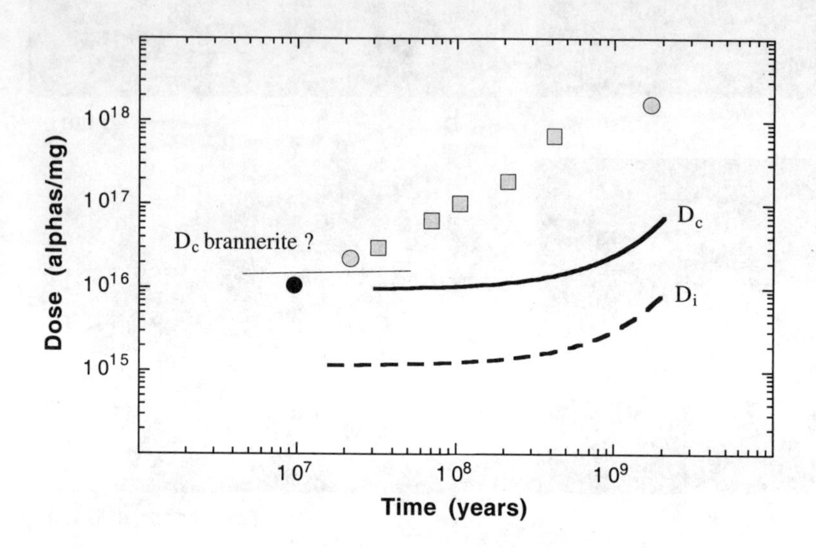

Figure 3. Dose-age plot for natural brannerite and zirconolite (see [12], D_i = onset dose, D_c = critical dose). Circles represent samples with well-established ages. Squares represent samples with estimated ages. Shaded symbols represent fully amorphous samples. Filled circle represents partially crystalline brannerite from Binntal, Switzerland [5].

DISCUSSION AND CONCLUSIONS

The compositions of relatively unaltered areas of brannerite determined in this work are generally consistent with older chemical analyses reported in the literature [13,14]. Our results demonstrate that the U content of natural brannerite is quite variable and that U can be replaced by significant amounts of Ca, Th, Y, and lanthanides. We also found that the Ti content is less variable, with minor replacement of Ti by Fe, Si, Al, Nb, Mn, and Ni. Calculated formulae based on 6 oxygen atoms suggest that some of the U may be present in a higher valence state. These results are generally consistent with previous chemical analyses [13,14] and recent synthesis work which shows that up to 0.3 atoms per formula unit of Ca or Gd can be substituted for U in brannerites fired in air or Ar at 1350-1450°C [2]. Diffuse reflectance spectroscopy carried out by Vance et al. [2] indicates that some U is present as U^{5+}, thereby providing a charge-balancing mechanism for incorporation of Ca and Gd. Similar mechanisms may be operative in the natural samples.

Geochemical alteration effects of brannerite have been poorly documented in previous work. The SEM-EDX and TEM results presented in this paper demonstrate that brannerite is usually amorphous due to alpha-decay processes and may be chemically altered along grain boundaries and microfractures. Analytical data show that the universal alteration mechanism is preferential loss of

U, compensated in part by uptake of Al, Si, P, Fe, As, and other minor elements from the fluid phase. Additionally, we have observed secondary alteration products including TiO_2 phases and thorite at an advanced stage of alteration. Nevertheless, total dissolution of the brannerite matrix appears to have been minimal. This result is consistent with studies of the processing of U ores which show that brannerite is highly resistant to dissolution in acidic fluids [15]. Under alkaline conditions, on the other hand, brannerite may be less resistant to dissolution [16].

The new data for natural brannerite allow us to establish a relative ranking of the durability of several actinide host phases in natural systems. In terms of total matrix dissolution, studies of natural samples indicate that the durability of the minerals increases in the order perovskite $\ll$ brannerite $\leq$ pyrochlore $\leq$ zirconolite. The alteration of perovskite to anatase and other phases in natural systems is well known [17-19]. Although generally highly resistant to dissolution, brannerite is susceptible to chemical alteration by ion exchange with the fluid phase and typically loses U in the process. Previous work has shown that pyrochlore may exhibit chemical alteration characterized by the loss of Na, Ca, and F. Altered pyrochlores are typically hydrated, but there is only limited evidence for loss of Th and U [11,20,21]. In contrast, chemical alteration of natural zirconolite is relatively uncommon and the Th and U contents remain more or less constant [11].

ACKNOWLEDGMENTS

This work was performed at the Australian Nuclear Science and Technology Organisation, partly funded under contract with Lawrence Livermore National Laboratory as part of the Plutonium Immobilization Project.

REFERENCES

1 B.B. Ebbinghaus, R. Van Konynenburg, F.J. Ryerson, E.R. Vance, M.W.A. Stewart, A. Jostsons, J.S. Allender, T. Rankin, and J. Congdon, presented at Waste Management '98, Tucson, AZ, 1998 (unpublished).
2 E.R. Vance, J.N. Watson, M.L. Carter, R.A. Day, G.R. Lumpkin, K.P. Hart, Y. Zhang, P.J. McGlinn, M.W.A. Stewart, and D.J. Cassidy, presented at the American Ceramic Society Annual Meeting, Indianapolis, IN, 1999 (unpublished).
3 Y. Zhang, G.R. Lumpkin, K. Hart, R. Day, S. Leung, Z. Aly, and M. Carter, presented at the
4 G.R. Lumpkin, K.L. Smith, M.G. Blackford, R. Gieré, and C.T. Williams, Micron **25**, 581 (1994).
5 S. Graeser and R. Guggenheim, Schweiz. Mineral. Petrogr. Mitt. **70**, 325 (1990).
6 K.R. Ludwig and J.A. Cooper, Contrib. Mineral. Petrol. **86**, 298 (1984).
7 T.J. Headley, R.C. Ewing, and R.F. Haaker, Nature **293**, 449 (1981).
8 R.C. Ewing and T.J. Headley, J. Nucl. Mater. **119**, 102 (1983).
9 G.R. Lumpkin and R.C. Ewing, Phys. Chem. Minerals **16**, 2 (1988).
10 G.R. Lumpkin, J. Nucl. Mater. **190**, 302 (1992).
11 G.R. Lumpkin, K.P. Hart, P.J. McGlinn, T.E. Payne, R. Gieré, and C.T. Williams, Radiochim. Acta **66/67**, 469 (1994).
12 G.R. Lumpkin, K.L. Smith, M.G. Blackford, R. Gieré, and C.T. Williams, in *Scientific Basis for Nuclear Waste Management XXI*, edited by I.G. McKinley and C. McCombie (Mater. Res. Soc. Proc. 506, Pittsburgh, PA, 1998) pp. 215-222.
13 A. Pabst, Amer. Mineral. **39**, 109 (1954).
14 F. Bianconi and A. Simonetti, Schweiz. Mineral. Petrogr. Mitt. **47**, 887 (1967).
15 R.O. Ifill, W.C. Cooper, and A.H. Clark, CIM Bulletin **89**, 93 (1996).
16 J.T. Szymanski and J.D. Scott, Can. Mineral. **20**, 271 (1982).
17 A.N. Mariano, in *Geochemistry and Mineralogy of Rare Earth Elements*, edited by B.R. Lipin and G.A. McKay (Mineralogical Society of America, Washington, D.C., 1989) pp. 309-348.
18 J.F. Banfield and D.R. Veblen, Amer. Mineral. **77**, 545 (1992).
19 R.H. Mitchell and A.R. Chakhmouradian, Can. Mineral. **36**, 939 (1998).
20 G.R. Lumpkin and R.C. Ewing, Amer. Mineral. **80**, 732 (1995).
21 G.R. Lumpkin and R.C. Ewing, Amer. Mineral. **81**, 1237 (1996).

THERMODYNAMIC STABILITY OF ACTINIDE PYROCHLORE MINERALS IN DEEP GEOLOGIC REPOSITORY ENVIRONMENTS

Yifeng Wang*, Huifang Xu**
* Sandia National Laboratories, 4100 National Parks Highway, Carlsbad, New Mexico 88220. E-mail: ywang@sandia.gov
**Department of Earth and Planetary Sciences, The University of New Mexico, Albuquerque, New Mexico 87131. E-mail: hfxu@unm.edu

ABSTRACT

Crystalline phases of pyrochlore (e.g., $CaPuTi_2O_7$, $CaUTi_2O_7$) have been proposed as a durable ceramic waste form for disposal of high level radioactive wastes including surplus weapons-usable plutonium. In this paper, we use a linear free energy relationship to predict the Gibbs free energies of formation of pyrochlore phases ($CaMTi_2O_7$). The Pu-pyrochlore phase is predicted to be stable with respect to PuO_2, $CaTiO_3$, and TiO_2 at room temperatures. Pu-pyrochlore is expected to be stable in a geologic repository where silica and carbonate components are absent or limited. We suggest that a repository in a salt formation be an ideal environment for disposal of high level, pyrochlore-based ceramic wastes. In such environment, adding CaO as a backfill will make pyrochlore minerals thermodynamically stable and therefore effectively prevents actinide release from these mineral phases.

INTRODUCTION

Crystalline phases of pyrochlore (e.g., $CaPuTi_2O_7$, $CaUTi_2O_7$) have been proposed as a durable ceramic waste form for disposal of high level radioactive wastes including surplus weapons-usable plutonium [1, 2]. Pyrochlore phases typically have a derivative fluorite structure and have a stoichiometry of $CaMTi_2O_7$, where M represents tetravalent cations such as Zr, Hf, U, Pu and other actinides. The existence of large polyhedra (with coordination numbers ranging from 7 to 8) in mineral structure allows pyrochlore phases to accommodate a wide range of radionuclides (e.g., Pu, U, Ba, etc.) as well as neutron poisons (e.g., Hf, Gd) [3].

Since most of the existing work on pyrochlore minerals has been focused on fabrication, structural characterization and leaching tests, the thermodynamic data of these mineral phases are still lacking. In this paper, we will use a linear free energy relationship to predict the Gibbs free energies of formation for various pyrochlore phases. Based on the predicted Gibbs free energies, the stability of actinide pyrochlore minerals in deep geologic repository environments will be discussed.

LINEAR FREE ENERGY RELATIONSHIP

Sverjensky and Molling have developed a linear free energy relationship that correlates the Gibbs free energies of formation of an isostructural family of inorganic

Mat. Res. Soc. Symp. Proc. Vol. 608 © 2000 Materials Research Society

solids with chemical properties of aqueous free cations [4]. This free energy relationship has been successfully applied to a wide variety of mineral phases [4, 5, 6, 7]. The linear free energy relationship for an isostructural family of minerals can be expressed as [4, 5]:

$$\Delta G^0_{f,\,MX} = a_{MX}\, \Delta G^0_{n,\,MZ+} + b_{MX} + \beta_{MX}\, r_{MZ+}\,. \tag{1}$$

where the coefficients a_{MX}, b_{MX}, and β_{MX} characterize a particular structural family of minerals MX; r_{MZ+} is the ionic radius of cation M^{Z+}; the parameter $\Delta G^0_{f,\,MX}$ is the standard Gibbs free energies of formation of solid MX; and the parameter $\Delta G^0_{n,\,MZ+}$ is the standard non-solvation energy from a radius-based correction to the standard Gibbs free energy of formation of the aqueous tetravalent cation M^{4+} [4]. The parameter $\Delta G^0_{n,MZ+}$ can be calculated by:

$$\Delta G^0_{n,\,MZ+} = \Delta G^0_{f,\,MZ+} - \Delta G^0_{s,\,MZ+} \tag{2}$$

where $\Delta G^0_{s,\,MZ+}$ is the standard Gibbs free energy of solvation of cation M^{Z+}, which can be calculated from conventional Born solvation coefficients [4].

The terms of $\Delta G^0_{n,\,MZ+}$ and r_{MZ+} are known for free metal cations. The coefficients of a_{MX}, b_{MX}, and β_{MX} can be determined by fitting the equation to limited experimental data of $\Delta G^0_{f,\,MX}$ in an isostructural family of solids. The obtained equation can be then used to predict the unknown $\Delta G^0_{f,\,MX}$ for other solids within the same family.

PREDICTION OF GIBBS FREE ENERGIES OF FORMATION OF ACTINIDE PYROCHLORE MINERALS

Since no $\Delta G^0_{f,\,MX}$ data are currently available for pyrochlore phases, Equation (1) can not be directly applied to these mineral phases. Instead, the coefficients in Equation (1) for pyrochlore phases are extrapolated from zirconolite data. For the isostructural family of $CaMTi_2O_7$ with a zirconolite structure, the Gibbs free energies of formation have been determined for the $CaZrTi_2O_7$ and $CaHfTi_2O_7$ phases [8, 9]. In order to apply the above linear free energy relationship, the coefficient a_{MX} or β_{MX} needs to be estimated independently. The coefficient β_{MX} is related to the effect of nearest neighbors or coordination number (CN) of cation [4]. In polymorphs, a structure family with a small CN (e.g., CN = 6 in calcite structure family) has a higher value of β_{MX} than a family with a big CN (e.g., CN = 9 in aragonite structure family) does [4]. The value of β_{MX} for the MO_2 family with a fluorite structure (CN of M atom is 8) is 32.0 (kcal/mole·Å) (Table 1). The value of β_{MX} for the $MSiO_4$ family with a zircon structure (CN of M atom is 7) is 64.83 (kcal/mole·Å) [5. 6]. As a first-order approximation, we use β_{MX} value of 65 (kcal/mole·Å) for the zirconolite family (CN = 7). Using the Gibbs free energies of formation of $CaZrTi_2O_7$ and $CaHfTi_2O_7$ phases [8, 9], the coefficients of a_{MX} and b_{MX} for zirconolite family can be calculated to be 0.5717 and -1024.06 kcal/mole, respectively.

Based on the results from other oxide and silicate families, the coefficient a_{MX} seems only related to the stoichiometry of solids [4]. Values of the coefficient a_{MX} are very close for all polymorphs [4]. Therefore, the a_{MX} value of 0.5717 is applied to pyrochlore phases. Similarly, we apply the β_{MvX} value of 32 (kcal/mole·Å) obtained for the MO_2 family with fluorite structure (CN = 8) to the pyrochlore family. The main difference between zirconolite and pyrochlore with the same stoichiometry of $CaMTi_2O_7$ is the coordination number of Zr. The coordination number of M atoms in the zirconolite structure and pyrochlore structure are 7 and 8, respectively. The difference in Gibbs free energies between calcite ($CaCO_3$) and aragonite polymorphs is about 0.2 (kcal/mole) [4]. The difference in Gibbs free energies between Al-Si ordered low-albite (CN of Na atom is 7) and Al-Si disordered high-albite (CN of Na atom is 9) polymorphs is about 2.0 (kcal/mole) [10]. However, the contribution from Al-Si odering in tetrahedral sites that can be calculated from the Gibbs free energies of formation of Al-Si ordered microcline and Al-Si disordered sanidine is about 1.6 (kcal/mole) [10]. Thus, the free energy contribution from the difference in coordination number of low-albite and high albite is about 0.4 (kcal/mole). Therefore, we postulate that the Gibbs free energy difference between the zirconolite and pyrochlore structures is within the range of 0.2 — 0.4 kcal/mole. It is proposed here that the Gibbs free energy of formation for Zr-pyrochlore ($CaZrTi_2O_7$) is about 0.3 (kcal/mole) higher than that of Zr-zirconolite ($CaZrTi_2O_7$). The coefficient b_{MvX} for pyrochlore phases is thus -997.67 (kcal/mole). The predicted standard Gibbs free energies of formation for other phases in the pyrochlore family with a stoichiometry of $CaM^{4+}Ti_2O_7$ are listed in Table 1.

THERMODYNAMIC STABILITY OF PYROCHLORE MINERALS IN DEEP GEOLOGIC REPOSITORY ENVIRONMENTS

Based on the predicted Gibbs free energies of formation of pyrochlore and Gibbs free energies of formation of perovskite and rutile [11], we can calculate the Gibbs free energy change of the following reaction ($\delta\Delta G_{rxt}$) of the reaction at a room temperature:

$$\underset{}{MO_2} + \underset{\text{perovskite}}{CaTiO_3} + \underset{\text{rutile}}{TiO_2} = \underset{\text{pyrochlore}}{CaMTi_2O_7}. \tag{3}$$

The calculated Gibbs free energy changes ($\delta\Delta G_{rxt}$) are listed in Table 1. The Ce-pyrochlore has been synthesized by sintering oxides of CeO_2, $CaTiO_3$, and TiO_2 [12, 13]. As the annealing temperature decreases (from 1300 °C to 1140 °C), the proportions of Ce-pyrochlore increases by consuming oxides of CeO_2 and $CaTiO_3$ phases [13]. This experimental observation is consistent with our prediction of the negative Gibbs free energy change across reaction (3) for Ce-pyrochlore. In contrast, Th-pyrochlore phase will be unstable with respect to ThO_2, $CaTiO_3$, and TiO_2, To our knowledge, no successful synthesis of Th-pyrochlore phase has been reported. It can be seen from Table 1 that the Gibbs free energy change in reaction (3) for Pu-pyrochlore (3) is more negative than that of Ce-pyrochlore (Table 1), indicating that the synthesis of Pu-pyrochlore from oxides is thermodynamically feasible.

Table 1. Ionic radii, thermodynamic data for aqueous cations, and predicted standard Gibbs free energies of formation (kcal/mole) [5]

M^{4+}	$r_{M^{4+}}$ (Å)	$\Delta G_s M^{4+}$	$\Delta G_f M^{4+}$	$\Delta G_n M^{4+}$	MO_2 (Exper.)	MO_2 (Calculated)	ΔG_f Pyrochlore $CaMTi_2O_4$	$\delta\Delta G_{rxt}$ Reaction (3)
Zr	0.79	-373.11	-141.00	232.11	-249.23	-249.21	-839.70	-1.67
Hf	0.78	-374.41	-156.80	217.61	-260.09	-259.24	-848.30	-0.24
Ce	0.94	-354.23	-120.40	233.79	-244.40	-243.28	-833.93	-1.83
Th	1.02	-344.65	-168.50	176.13	-279.34	-279.35	-864.34	3.83
U	0.97	-350.60	-124.40	226.20	-246.62	-247.40	-837.31	-1.09
Np	0.95	-353.02	-120.20	232.82	-244.22	-243.61	-834.17	-1.74
Pu	0.93	-355.45	-115.00	240.49	-238.53	-239.11	-830.42	-2.49
Am	0.92	-356.68	-89.20	267.48	-220.72	-221.35	-815.31	-5.14
Po	1.10	-339.98	70.00	409.98		-121.39	-729.36	-11.54

The negative Gibbs free energy change in reaction (3) predicted for Pu-pyrochlore implies that this mineral phase is stable with respect to PuO_2, $CaTiO_3$, and TiO_2 at a room temperature. A reaction-path calculation using computer code EQ3/6 [14] shows that PuO_2 and TiO_2 are both stable in a Yucca Mountain environment. Considering that the Yucca Mountain repository is relatively oxic, we expect PuO_2 to be stable in most deep repository environments. Therefore, out of three minerals in the left side of reaction (3), at least two of them are stable. The stability of Pu-pyrochlore thus depends on the stability of perovskite ($CaTiO_3$).

The stability of perovskite ($CaTiO_3$) is controlled by the fugacity of CO_2 and the activity of silica. In the presence of silica or CO_2, perovskite ($CaTiO_3$) is unstable in a low temperature environment [15], because of the following reactions:

$$\underset{\text{perovskite}}{CaTiO_3} + H_4SiO_4(aq) = \underset{\text{sphene}}{CaTiSiO_5} + 2H_2O \qquad \log K = 8.9 \qquad (4)$$

$$\underset{\text{perovskite}}{CaTiO_3} + CO_2(g) = \underset{\text{calcite}}{CaCO_3} + \underset{\text{rutile}}{TiO_2} \qquad \log K = 8.5 \qquad (5)$$

where K is the equilibrium constant of the reaction and is calculated using the data from [11]. However, in the absence of both silica and CO_2, $CaTiO_3$ will be stable in an aqueous solution, because the following reaction is thermodynamically not favorable:

$$\underset{\text{perovskite}}{CaTiO_3} + H_2O = \underset{\text{portlandite}}{Ca(OH)_2} + TiO_2 \qquad \log K = -4.3 \qquad (6)$$

Consequently, Pu-pyrochlore will also become stable. Based on this argument, we suggest that a repository in a salt formation be an ideal environment for disposal of high level, pyrochlore-based ceramic wastes. The stability of pyrochlore and perovskite in this environment can be further ensured by adding CaO as a backifll. CaO will sequester any carbonate originally present in groundwater through reaction:

$Ca(OH)_2$	+	CO_2	=	$CaCO_3$	$\log K = 12.8$	(7)
Bruicte				calcite		
(hydrated CaO)						

and, therefore, will keep both pyrochlore and perovskite minerals thermodynamically stable in the repository.

Even for a repository located in a silicate rock, depending on groundwater flow rates, silica and carbonate may become depleted in the near field due to chemical reactions, and then pyrochlore and perovskite may be still able to approach a stable state. In short, to assess the performance of pyrochlore ceramic waste, silica and carbonate concentrations in the near field must be carefully evaluated.

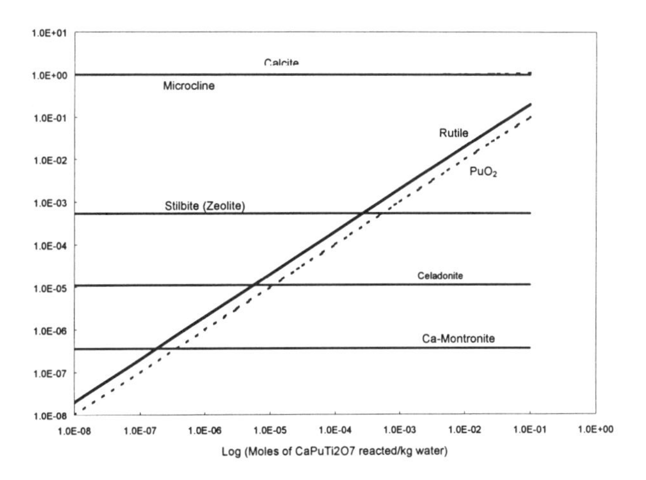

Figure 1. Reaction path calculation for Pu-pyrochlore dissolution in a Yucca Mountain repository environment. The fugacities of oxygen and carbon dioxide are maintained at 3.3×10^{-30} atm and 5.0×10^{-3} atm, respectively.

CONCLUSIONS

A linear free energy relationship has been used to predict the Gibbs free energies of formation of pyrochlore phases ($CaMTi_2O_7$). The coefficients in Equation (1) for pyrochlore phases are estimated to be: $a_{MX} = 0.5717$, $b_{MX} = -997.67$ (kcal/mole), and $\beta_{MX} = 32$ (kcal/mole·Å). The Pu-pyrochlore phase is predicted to be stable with respect to PuO_2, $CaTiO_3$, and TiO_2 at room temperatures. Pu-pyrochlore is expected to be stable in a geologic repository where silica and carbonate components are absent or limited. A repository in a salt formation is proposed as an ideal environment for disposal of high level, pyrochlore-based ceramic wastes. In such environment, adding CaO as a backfill will make pyrochlore minerals thermodynamically stable and therefore effectively prevents actinide release from these mineral phases.

ACKNOWLEDGMENTS

Sandia is a multiprogram operated by Sandia Corporation, a Lockheed Martin Company, for the United States Department of Energy under Contract DE-AC04-94AL8500.

REFERENCES

1. Dosch, R. G., Headley, T. J., Northrup, C. J., and Hlava, P. F., Sandia National Laboratories Report, Sandia 82-2980 (1982).
2. Ringwood, A. E., Kesson, S. E., Reeve, K. D., Levins, D. M., and Ramm, E. J., Synroc. In W. Lutze and R. C. Ewing eds., "Radioactive Waste Forms for the Future." North-Holland, Amsterdam, 1988, p. 233 (1988).
3. Vance, E. R., MRS Bulletin, vol. XIX, p. 28 (1994).
4. Sverjensky, D. A., and Molling, P. A. Nature, 358, 310 (1992).
5. Xu, H., and Wang, Y. J. Nucl. Mater., 275, 216 (1999).
6. Xu, H., Wang, Y., and Barton, L. J. Nucl. Mater., 273, 343 (1999).
7. Xu, H., and Wang, Y. Radiochim. Acta, in press (1999).
8. Putnam, R. L., Navrotsky, A., Woodfield, B. F., Shapiro, J. L., Stevens, R., and Boerio-Goates, J. In David Wronkiewicz ed. “Scientific Basis for Nuclear Waste Management XXII,” in press (1999).
9. Putnam, R. L., Navrotsky, A., Woodfield, B. F., Boerio-Goates, J., and Shapiro, J. L. J. Chem. Thermodynamics, 31, 3, 229 (1999).
10. Berman, R. G. J. Petrology, 29, 445 (1988).
11. Robie, R. and Hemingway, B. S. Thermodynamic Properties of Minerals and Related Substances at 298.15 K and 1 Bar (105 Pascal) Pressure and at Higher Temperatures. U. S. Geological Survey Bulletin, No. 2131, 461 pp. (1992).
12. Xu, H., and Wang, Y. In “Radioactive Waste Management and Environmental Remediation—ICEM99,” (in press) (1999).
13. Putnam, R. L. Personal communication (1999).
14. Wolery, T. J. EQ3/6, A Software Package for Geochemical Modeling of Aqueous Systems (Version 7.0), Lawrence Livermore National Laboratory, UCRL-MA-110662 PT 1-4 (1992).
15. Nesbitt, H. W., Bancroft, G. M., Fyfe, W. S., Karkhanis, S. N., and Nishijima, A. Nature, 289, p. 358-362 (1981).

DISSOLUTION KINETICS OF TITANIUM PYROCHLORE CERAMICS AT 90°C BY SINGLE-PASS FLOW-THROUGH EXPERIMENTS

J. P. ICENHOWER, B. P. MCGRAIL, H. T. SCHAEF, AND E. A. RODRIGUEZ
Applied Geology and Geochemistry Department, Pacific Northwest National Laboratory, P.O. Box 999, K6-81, Richland, WA 99352, jonathan.icenhower@pnl.gov.

ABSTRACT

Corrosion resistances of titanium-based ceramics are quantified using single-pass flow-through (SPFT) experiments. The materials tested include simple pyrochlore group ($B_2Ti_2O_7$, where B=Lu^{3+} or Gd^{3+}) and compositionally complex pure phase pyrochlore (PY12) or zirconolite- and brannerite-bearing pyrochlore-dominated (BSL3) ceramics. Experiments are conducted at 90°C over a range of pH-buffered conditions with typical duration of experiments in excess of 120 days. Apparent steady-state dissolution rates at pH=2 determined on the $Gd_2Ti_2O_7$ and $Lu_2Ti_2O_7$ samples indicate congruent dissolution, with rates of the former ($1.3x10^{-3}$ to $4.3x10^{-3}$) slightly faster than the latter ($4.4x10^{-4}$ to $7.0x10^{-4}$ g m^{-2} d^{-1}). Rates for PY12 materials into pH=2 solutions are $5.9x10^{-5}$ to $8.6x10^{-5}$ g m^{-2} d^{-1}. In contrast, experiments with BSL3 material do not reach steady-state conditions, and appear to undergo rapid physical and chemical corrosion into solution. Dissolution rates of PY12 display a shallow amphoteric behavior, with a minimum ($2.7x10^{-5}$ g m^{-2} d^{-1}) near pH values of 7. Dissolution rates display a measurable increase (~10X) with increasing flow-through rate indicating the strong influence that chemical affinity exerts on the system. These results step towards an evaluation of the corrosion mechanism and an evaluation of the long-term performance of Pu-bearing titanate engineered materials in the subsurface.

INTRODUCTION

Current disposal strategy for excess fissile materials calls for incorporation of plutonium into titanium-bearing ceramic waste forms. To provide data for credible performance analysis, the corrosion resistance of the ceramics must be thoroughly investigated. Single-pass flow-through (SPFT) experiments provide one method for assessing corrosion behavior. Flow-through rates in SPFT experiments are faster than those expected in the repository, but the utility of this method lies in determining dissolution rate parameters that will be used to estimate long-term behavior of titanate waste forms in the subsurface. Our experiments are being conducted over a pH range of 2 to 12 at 90°C over a range of flow-through rates on monolithic and powdered ceramic samples. This report summarizes the results from our experiments on titanate ceramics to date.

EXPERIMENT

Materials

The materials used in these experiments can be subdivided on the basis of mono- or multi-phase compositions. Simple pyrochlore-type materials are represented by the formula $B_2Ti_2O_7$, where B=Lu^{3+} or Gd^{3+}. The compositionally complex samples include two Ti-bearing ceramic materials that contain pyrochlore-group (betafite), zirconolite, and brannerite isostructures. Betafite isostructures are represented by the formula $A_{0-1}B_{1-2}Ti_2O_7$ (A = Ca^{2+} and Gd^{3+} and B = Gd^{3+}, Ce^{4+}, and Hf^{4+}) with an anion-deficient, fluorite structure. Zirconolite can be represented by the formula $ABTi_2O_7$ with A=Ca^{2+} and B=Hf^{4+}. The brannerite isostructure is represented by the formula $(A^{4+})Ti_2O_6$, with Hf^{4+} the likely dominant cation in surrogate-doped materials. Pyrochlore-12 (or PY12) is dominated volumetrically by the betafite minerals whereas Baseline-3 (or BSL3) contains a higher proportion of zirconolite and brannerite. In addition to the CeO_2,

Mat. Res. Soc. Symp. Proc. Vol. 608 © 2000 Materials Research Society

Gd_2O_3, and HfO_2 dopants, the ceramics also contain small concentrations (≤0.1 wt%) of Mo^{6+}, which acts as a tracer of matrix dissolution. Chemical compositions of the multiphase samples (PY12 and BSL3) tested are listed in Table 1.

Table 1. Target Composition of Ti-ceramics

Baseline-3 (BSL3)

Oxide	Mass%
CaO	11.47
CeO_2	26.84
Gd_2O_3	9.16
HfO_2	11.12
MoO_3	0.09
TiO_2	41.33
Total	100.00

Pyrochlore-12 (PY12)

Oxide	Mass%
CaO	8.23
CeO_2	35.38
Gd_2O_3	13.31
HfO_2	3.10
MoO_3	0.10
TiO_2	39.89
Total	100.01

The Lu- and Gd-bearing pyrochlore samples were prepared utilizing the sol-gel route. Following calcination at 700°C, the resulting material was pressed into a pellet and sintered at 1,200°C for 12 hours. The pellets were crushed and re-fired at 1,500°C for 30 hours. Final formation of the ceramics was accomplished by hot isostatic pressing (HIP) in Ar at 1,500°C for 2 hours at 200 MPa pressure. Monoliths were cut from the pellets, polished, and checked for homogeneity and consistency by a combination of secondary electron microscopy (SEM), back scattered electron imaging (BSEI), X-ray diffraction (XRD) and electron microprobe analysis (EMPA) methods [1]. The surface areas of the monolith samples were determined by measuring the dimensions with an electronic caliper.

Ceramic materials PY12 and BSL3 were synthesized by mixing reagent-grade oxide, carbonate, hydroxide, and halide powders that were mixed and crushed in a Teflon vessel. Slurried powders were calcined at 750°C for 1 hour and then pressed in three separate steps. The pellets were subsequently fired at 1,350°C for 4 hours. The resulting pellets were crushed, re-ground and re-fired in a die to yield a dense, sintered pellet. Porosity of BSL3 material processed in this manner is ~13%. Pressing in lieu of a second round of crushing produced a second generation of denser pellets (porosity = ~2%). The pellets were then crushed in a ball mill and sieved to separate the size fractions of interest, in this case the 150 to 75 μm and 75 to 40 μm (-100, +200 and –200, +325 mesh) size fractions, respectively. Surface area analyses of the cleaned samples were performed by Kr BET methods.

SPFT System

Table 2. Chemical Compositions of Buffer Solutions

Buffer #	pH (25°C)	Composition
1	2	0.013 M HNO_3
2	5.7	deionized water
3	7	0.01 M THAM + 0.0466 M HNO_3
4	8	0.01 M THAM + 0.029 M HNO_3
5	9	0.01 M THAM + 0.0057 M HNO_3
6	10	0.01 M THAM + <0.150 mL HNO_3
7	11	0.01 M LiCl and 0.001 M LiOH
8	12	0.01 M LiCl and 0.0107 M LiOH

Teflon reactor vessels (60 mL volume) house the sample powders and the reactor-sample configuration is kept at the temperature of interest by a calibrated constant temperature oven. In the case of the pyrochlore monoliths, the samples are situated in a grated Teflon cradle within the reactor, thus exposing the top and bottom of the monolith to solution. Solution input and effluent output lines are connected via twin ports at the top of the reactor. Computerized syringe or infusion pumps precisely control the flow of aqueous solutions into the reactors with flow-through rates set at 2, 5, or 10 mL d^{-1}. Effluent samples are collected continuously and aliquots are periodically retained for analysis. Precision ICP-AES (Ca) and ICP-MS (Mo, Ce, Gd, Hf, and Ti) methods are used to determine element concentrations in the effluent. Input solutions are made by combining deionized water with nitric acid, THAM (tris hydroxymethyl aminomethane), or lithium hydroxide + LiCl to yield solutions buffered from pH = 2 to 12 (Table 2). All input solutions are continuously sparged by nitrogen gas to prevent deviations from initial pH values. Experiments run continuously for 60 to >120 days.

Three to four blank solutions are collected for each experiment before the samples are added to the reactors. In addition, we ran continuous blanks for each input solution. The continuous blanks have the same experimental configuration as the 'real' experiments, complete with pumps transferring solutions at the temperature of interest into a reactor and effluent samples collected at the output. In this way, we can determine changes in background concentrations of elements of interest over the duration of the experiments.

Determination of Apparent Rates

Apparent normalized dissolution rates are calculated from steady-state concentrations of the elements of interest in the effluent by the expression:

$$\mathrm{rate}_i = \frac{(C_i - C_b)\nu}{f_i A} \tag{1}$$

where C_i is the concentration of the element of interest, C_b is the background concentration of the element of interest, ν is the flow-through rate, f_i is the fraction of the element in the ceramic, and A is the total surface area of the sample. As discussed below, background concentrations of some elements can be problematic, if not properly addressed. Because the rate expression above is not specific to any element, an apparent dissolution rate can be calculated from a suite of elements in the effluent solution, in this case, Ca, Ce, Gd, Hf, Mo, and Ti. As will be shown below, the dissolution rates based upon concentrations of Mo, Ce, Gd, and Ca in the effluent yield self-consistent results for samples at specified experimental conditions, with few exceptions.

RESULTS

Acidic Solutions

Experiments at 90°C with solutions at pH=2 were conducted for the four samples discussed above. Experiments were allowed to run until concentrations became invariant with respect to time, at which time they were terminated.

In experiments with titanate ceramics at pH=2, the rare earth elements Lu and Gd are soluble in aqueous solutions. Figures 1A and B illustrate the variation of apparent dissolution rate with time for experiments with the simple pyrochlore compositions, as the system approaches steady-state conditions (note the scale difference on the ordinate of the two diagrams). In general, the samples appear to dissolve congruently into solution, as indicated by the overlap in respective Gd-Lu and Ti rates. The exception to this is one $Gd_2Ti_2O_7$ sample (Gd-1), which exhibits an unexplained faster apparent rate compared to the others. It also appears that the $Gd_2Ti_2O_7$ dissolves more quickly (1.3×10^{-3} to 4.3×10^{-3}) than the $Lu_2Ti_2O_7$ (4.4×10^{-4} to 7.0×10^{-4} g m^{-2} d^{-1}) monoliths. It is not clear at this juncture if the difference in rates between Gd- and Lu-bearing ceramics is related to intrinsic differences in durability or experimental artifacts.

Concentrations of rare earth elements (Ce and Gd) were also used to determine apparent dissolution rates for the PY12 and BSL3 materials. Concentrations of Mo in the effluent solutions fall to the detection threshold at pH=2 shortly after initiation of the experiments. The MoO_4^{2-} complex is the dominant species for the pH range that we investigated and, coupled with the positive surface charge on the surface of the ceramic grains, suggests that molybdenum sorbs onto the surface of the ceramic grains after release into solution. Other cations, such as Ti and Hf, are apparently solubility limited at this pH and therefore cannot be used to estimate dissolution rates. In contrast, concentrations of Gd and Ce are not solubility limited and the resultant

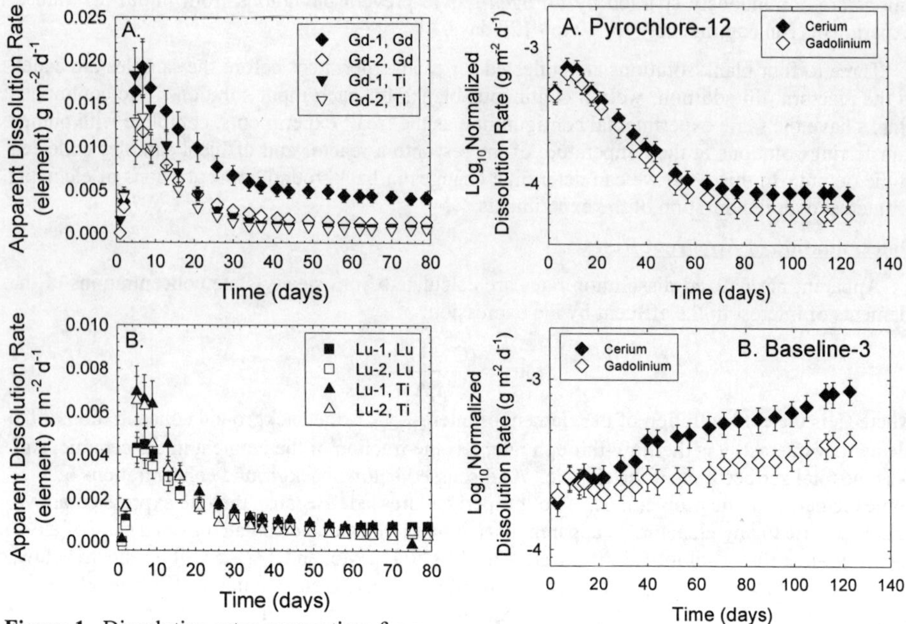

Figure 1. Dissolution rates versus time for pyrochlore ceramic monoliths at pH=2, 90°C temperature and flow-through rates of 5 mL d^{-1}. (A) $Gd_2Ti_2O_7$ (B) $Lu_2Ti_2O_7$

Figure 2. Plot of $\log_{10}$ dissolution rate versus time for experiments with Pyrochlore-12 and Baseline-3, pH=2, 90°C, 2 mL d^{-1}.

dissolution rates are discussed below. Figure 2A displays the variation in $\log_{10}$ apparent dissolution rate with time for experiments with the PY12 composition powders. Note that rates based on concentrations of Ce and Gd become invariant with time after ~90 days. Experiments with the PY12 composition at a flow-through rate of 2 mL d^{-1} yielded rates 5.9×10^{-5} to 8.6×10^{-5} g m^2 d^{-1} based on Gd and Ce concentrations, respectively. Experiments with BSL3 compositions yield enigmatic results. Concentrations of Ce, Ca and Gd in the effluent continue to rise over the course the experiments (Figure 2B), which may indicate a loss of physical integrity of the ceramic grains. We are currently examining the reacted solids to determine whether this is actually the cause of the observed rate increase.

Dissolution rates at pH = 2 for experiments with PY12 at flow-through rates of 10 mL d^{-1} are roughly 10 times faster (2 to 4×10^{-4} g $m^{-2}d^{-1}$), indicating the importance that flow-through rates exert on dissolution rates. Calcium concentrations in these experiments yield similar dissolution rates, but exhibit greater variability (4.6×10^{-4} to 1.3×10^{-3} g m^2 d^{-1}).

It is interesting to note that Bourcier [2] did not observe steady-state behavior in experiments with similar Ti-ceramics. In contrast to the results reported here for $Gd_2Ti_2O_7$, $Lu_2Ti_2O_7$ and PY12 compositions at pH=2, he noted that concentrations of U decreased continuously over the course of his long-term (>200 day) experiments. He interpreted these results to indicate that the release of elements into solution are governed by diffusion through a surface reaction layer, rather than by dissolution based on rupture of Ti—O bonds at the mineral/solvent interface. Although our data do not definitively verify either model, the steady-state rates discussed above for

TiO_2-like materials are similar in behavior to SiO_2 materials [3]. Until reliable temperature-dissolution rate data can be obtained, there is no way at present to distinguish between diffusion and surface reaction controlled mechanisms.

Neutral to Basic Solutions

The corrosion resistance of PY12 and BSL3 were determined over the range of pH values from ~6 to 12. Because Gd and Lu are solubility limited over this pH range, we did not attempt to determine the dissolution rates of the simple pyrochlore monoliths. The apparent dissolution behavior of the PY12 and BSL3 composition ceramics were determined at two flow-through rates, 2 and 10 mL d^{-1}. As described below, the rates are strongly dependent on the ratio of the flow-through rate divided by the total surface area of the sample.

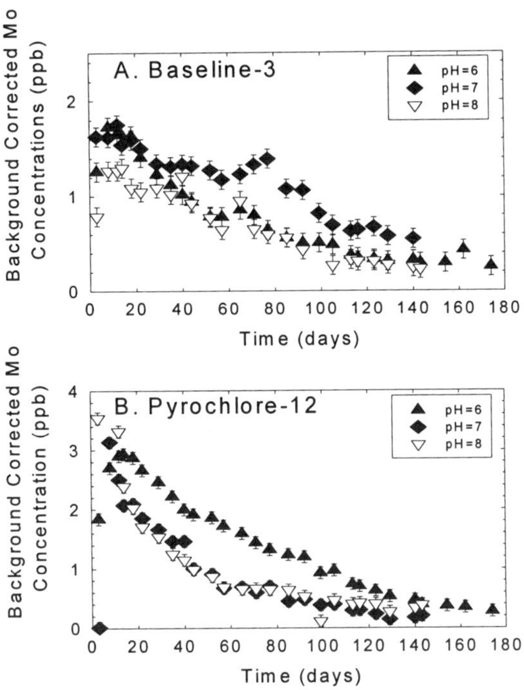

We calculated background corrected concentrations of Mo in the effluent solutions from experiments with PY12 and BSL3 by subtracting out the background concentration of Mo determined in the blank samples. We found that the concentration of Mo in the THAM-based input solutions (pH=7, 8, 9, and 10) depends on the strength of buffered solution, implying that THAM is a source of Mo contamination. Accordingly, we adjusted the strength of the THAM buffers from 0.05 M to 0.01 M, which decreased the background Mo concentration to near the detection threshold. In solutions collected from continuously run blanks, we found that concentrations of Mo rise slowly over time, due to evaporation of the input solution. We ameliorated this situation by replacing input solutions with fresh batches after three to four weeks.

Figure 3. Plot of background corrected Mo concentrations versus time for experiments at pH 6, 7, and 8, 90°C temperature, ~2 mL d^{-1}. (A) Baseline-3 (B) Pyrochlore-12. Note the change in scale on the ordinate between the two illustrations.

Even with low flow-through rates of 2 mL d^{-1}, background corrected concentrations of Mo in the effluent solutions are low, but remained well above the detection threshold (0.05 ppb). Experiments with BSL3 composition powders display a close approach to steady-state concentrations with respect to Mo (Figure 3A). Results for experiments with PY12 powders indicate a similar trend (Figure 3B). Because concentrations of Mo are approaching time-invariant conditions, a plot of concentration versus square root of time is non-linear (not shown). Consequently, interpretation of the time-dependent trends via diffusion theory does not appear to be warranted.

The near steady-state Mo concentrations yield apparent dissolution rates of 5.8×10^{-5} to 2.7×10^{-5} (g m^{-2} d^{-1}) for both the PY12 and BSL3 samples over the range of pH values studied (6 to 8) at this flow-through rate. The apparent dissolution rates at mid-pH values are therefore slower on average than the rates at pH=2. Therefore, on a plot $\log_{10}$ dissolution rate versus pH, the rates display a weak pH dependence (note scale break on diagram). Our preliminary findings seem to suggest a similar weak pH-dependence on the alkaline side of the pH scale. A more definitive statement concerning the dependence of dissolution rates on solution pH awaits further results from experiments with more alkaline (pH 9-12) solutions.

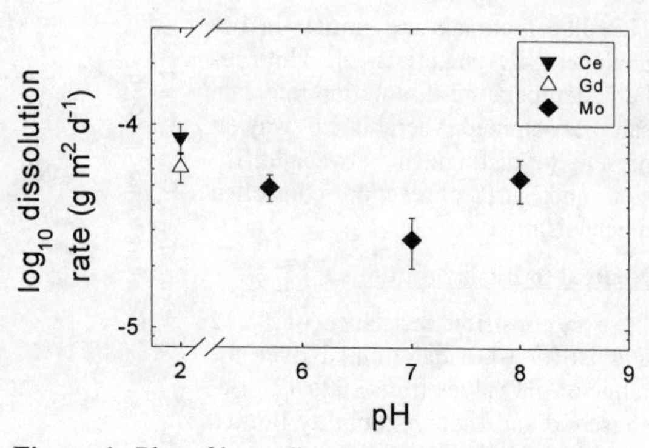

Figure 4. Plot of $\log_{10}$ dissolution rate versus pH for experiments with Pyrochlore-12, 90°C, 2 mL d^{-1}

CONCLUSIONS

We have quantified the dissolution kinetics of pyrochlore-group ceramic samples at 90°C over a range of pH values (2-12) by use of single-pass flow-through (SPFT) experiments. For simple pyrochlore samples, $Lu_2Ti_2O_7$ and $Gd_2Ti_2O_7$, dissolution appears to be congruent into pH=2 solutions. Dissolution of $Gd_2Ti_2O_7$ monoliths appears to be faster than their lutetium counterparts, although this finding has yet to be established with certainty. Powdered multiphase ceramics (PY12 and BSL3) display different behavior compared to one another in pH=2 solutions. Concentrations of Gd and Ce from experiments with PY12 reach steady-state conditions, whereas concentrations of these two elements in experiments with BSL3 materials continue to rise over time. These results indicate that BSL3 is vulnerable to rapid physical and chemical corrosion, obliging us to examine the long-term behavior of this material in other solution compositions. Dissolution rates show a weak dependence on pH values of the solutions, with a minimum near neutral conditions (pH=7). Dissolution rates of PY12 and BSL3 increase with increasing flow-through rates. Rates are faster by as much as 10X, as shown by comparing experiments at 10 and 2 mL d^{-1} flow-through rates. The faster dissolution rates at higher flow-through rates are likely a consequence of the system residing at a higher chemical affinity, as discussed by Aagaard and Helgeson [4]. Collectively, the experiments establish the baseline from which the effects of radiation damage on the dissolution rates of pyrochlore ceramic waste forms can be quantitatively assessed.

REFERENCES

1. B.D. Begg, W.J. Weber, R. Devanathan, J.P. Icenhower, S. Thevuthasan, and B.P. McGrail, *Am. Ceram. Soc. Trans.* (in press).
2. W.L. Boucier, unpublished LLNL report, 1999.
3. P.M. Dove and D.A. Crerar, Geochim. Cosmochim. Acta, **54**, 955-969 (1990).
4. P. Aagaard and H.C. Helgeson, Amer. J. Sci., **282**, 237-285 (1982).

SURFACE ALTERATION OF TITANATE CERAMICS IN AQUEOUS MEDIA

E. R. Vance, N. Dytlewski, K. E. Prince, K. P. Hart, and E. Loi
ANSTO, Menai, NSW 2234, Australia

ABSTRACT

From elastic recoil detection analysis (ERDA) of 2 MeV He ions and secondary ion mass spectroscopy (SIMS), exposure of Synroc-C to D_2O at 150°C for ~ 30 days produced surface deuteration products of a few nm in thickness, with surface roughness after polishing down to 0.25 μm diamond finish not being of critical importance in the thickness determination. Reaction at 250°C produced more extensive deuteration and general surface alteration, over depths of about a micron. SIMS did not show any surface enhancement of rare earths or Zr on Synroc-C surfaces reacted at 90°C for up to 336 days. Pu-doped Synroc-C exposed to deionised water at 70°C showed surface depletion of Pu by alpha-spectroscopy. Zirconolite-rich Synroc showed less surface deuteration than Synroc-C after reaction for 3 weeks at 150°C in D_2O. Admixtures of 0.001 M of fluoride ions to dilute HCl (pH = 2) produced deposits of anatase, ~ 20 μm thick, on perovskite after a few weeks at 90°C; these deposits were much thicker than those produced by the dilute HCl without the fluoride ions being present.

INTRODUCTION

Synroc-C [1,2] is a dense, fine-grained titanate ceramic designed for the immobilisation of Purex-type high-level nuclear waste (HLW) from spent nuclear fuel reprocessing. Thousands of tests over the last 20 years have shown the very high resistance of Synroc to aqueous dissolution, in terms of extraction of Synroc matrix components and waste ions into solution. Initial (1-day) extraction rates are < 0.1 g m^{-2} d^{-1} at boiling water temperatures for the most soluble elements, and the rates decrease significantly with increasing time, even with frequent changes of water to avoid saturation effects, to values of $< 10^{-4}$ g m^{-2} d^{-1} after periods of $> 10^3$ days. But to increase our understanding of the detailed mechanisms of its dissolution behaviour in aqueous media, surface alteration studies are necessary.

Significant knowledge of the secondary alteration products when Synroc and perovskite are reacted in aqueous media exists already [3-5]. Secondary ion mass spectroscopy (SIMS) measurements have indicated the preferential extraction of Ca, Mo, Sr, and Cs ions, together with the buildup of Ti on the surface of leached Synroc [3]. Alpha-recoil spectroscopy on sintered Synroc-C doped with Np and small quantities of Pu, Am and Cm showed actinides to build up on the surface after exposure to water [4], presumably due to dissolution followed by surface resorption. Elastic Recoil Detection Analysis (ERDA) of 2 MeV He ions showed after reacting Synroc at temperatures of 120-190°C in D_2O for a few weeks, that D appears to penetrate a few tens of nm into the surface, forming deuterated and/or deuteroxylated material as alteration products [5,6]. However evaluation of the penetration depth was potentially affected by the surface roughness of the samples.

The dissolution of perovskite, the most reactive of the Synroc phases in terms of aqueous durability in water at 70°C over periods of months has been found to be fairly insensitive to pH, with the dissolution rate increasing by only a factor of about 10 as the pH is decreased from 13 to 2 [7]. In the same work, the dissolution rate of zirconolite was found to be at least a factor of 10 lower than that of perovskite. Apparently anomalous results were however obtained by Mitamura *et al.* [8] who found for Ce/Nd and Cm/Pu-doped perovskite in 0.013 M HCl (pH = 2) at 90°C, that the Ca and dopant extraction rates first increased with increasing time between 7 and 28 days, before decreasing again between 28 and 56 days. The Ca extraction rates rose from ~1 g m^{-2} d^{-1} for 0-7 days to values as high as 10 g m^{-2} d^{-1} after 28 days. These were much higher than the values of < 0.1 g m^{-2} d^{-1} at pH = 2 found by

Mat. Res. Soc. Symp. Proc. Vol. 608 © 2000 Materials Research Society

McGlinn *et al.* [7]. After 56 days the samples were covered in a tens-of-microns thick layer of anatase [8]. Similar results were obtained on inactive samples of perovskite doped with Nd/Ce. However no such layers were evident in the work of McGlinn *et al.*[7].

The aims of the present work were to (a) extend the ERDA work [5,6] on Synroc-C reacted in D_2O by using smoother surfaces and higher temperatures, as well as to compare ERDA data with SIMS results; (b) use SIMS to see whether rare earths built up on Synroc surfaces exposed to water, as reported for sphene glass-ceramics [9]; (c) use alpha-spectroscopy to check if actinides build up on Synroc surfaces after reacting with water; (d) study zirconolite-rich versions of Synroc designed for actinide immobilisation, and (e) explore the idea that the anomalous dissolution of the doped perovskite [8] derived from traces of fluoride in the solution.

EXPERIMENTAL

Synroc-C samples were made by the standard hot-pressing route [2] and the samples for dissolution studies were polished to a 0.25 μm diamond finish. Some samples were ground further with amorphous silica for several days and the surface roughness was reduced from ~ 25 nm to ~ 5nm, as studied by a Tencor Alpha-step 200 profilometer. Samples of zirconolite-rich Synroc [13], designed for PuO_2 immobilisation but containing CeO_2 substituted on a molar basis for PuO_2, were prepared by standard Synroc-C methods, with final consolidation effected by either graphite-die hot-pressing at 1250°C, or by sintering in air at 1300°C. The corresponding dissolution samples were polished to a 0.25 μm diamond finish. The perovskite samples were made in polycrystalline form by similar means.

X-ray diffraction measurements were performed with a Siemens D-5000 instrument. SEM was conducted with a JEOL JSM-6400 machine run at 15 kV and fitted with a Tracor Northern MICRO-ZII X-ray detector and a Series II TN5502 system to carry out energy-dispersive elemental analysis.

D_2O treatment on Synroc and Synroc phases was carried out for periods of up to 30 days in Teflon vessels at 150°C and in steel pressure vessels at the higher temperatures, up to 250°C. Measurements in deionised water as well as dilute HCl (pH = 2), with or without an admixture of 0.001 M F^-, were made by modified MCC-1 methods at 90°C in Teflon containers. Solution analyses were carried out by ICP-MS.

SIMS measurements were carried out for D_2O and rare earths. For D_2O analysis, 10 keV Cs^+ primary ions were used, monitoring negative secondary ions of ^{16}O, ^{18}O and mass 18.01. The primary ion beam was focussed to a spot of around 40 μm in diameter, corresponding to a current of ~ 15 nA and raster scanned over an area of 250 x 250 μm. Secondary ions from an area 8 μm in diameter within the centre of the sputtered area were admitted into the spectrometer for analysis. The instrument was operated with a mass resolution of ~ 2000 to separate ^{18}O from mass 18.01. The depth profiles obtained for mass 18.01 reflect the distribution of $(H_2{}^{16}O+{}^{16}OD+{}^{17}OH)$ within the samples.

SIMS analyses for rare earths were performed using 12.5 keV O^- primary ions, monitoring positive secondary ions of ^{140}Ce and ^{146}Nd. The primary ion beam was focussed to a spot of around 120 μm in diameter, corresponding to a current of ~150 nA, and raster scanned over an area of 250 x 250 μm. Secondary ions from an area 60 μm in diameter within the centre of the sputtered area were admitted into the spectrometer for analysis. Moderate energy filtering was used to suppress molecular and atomic interferences on the masses of interest.

The uptake of deuterium in Synroc was also measured using ERDA using a ^4He ion beam produced from a 3 MeV Van der Graaff accelerator. A beam energy of 1.8 MeV was used, with the ion beam incident at a glancing angle of 15° to the sample surface and with the hydrogen and deuterium recoils being measured at a forward angle of 30°. The recoiling hydrogen and deuterium ejected from the near surface region were detected by a silicon surface barrier detector, in front of which was placed a 9.5 μm mylar foil to filter out the elastically scattered ^{4}He ions. A low ion beam current density of 1 nA/mm^2 was used to avoid any loss of surface deuterium that may arise due to ion beam induced sample heating.

Studies were carried out at 70°C in 20 mL of deionised water on (a) 0.76 wt% ^{239}Pu and (b) 0.0004 wt% ^{244}Cm-doped Synroc-C samples, of about 10 mm diameter and 1 mm thickness. In each test, an actinide-free Synroc-C sample was also placed ~ 1 mm away from the actinide-doped disk. After rinsing, the samples were counted in an alpha-spectrometer, using previously described methods [10].

RESULTS AND DISCUSSION

(a) Synroc-C samples with smoother surfaces and comparison of ERA and SIMS data for D_2O reaction.

A sample polished to a 0.25 μm diamond finish was polished with amorphous silica for 2 days and the surface roughness was reduced from ~ 20 to ~ 5 nm. However the ERDA profile after reaction with D_2O for 30 days at 150°C (see Fig. 3 below) was very similar to that obtained in [6], so it seems clear that after treatment in D_2O for 30 days at 150°C, polishing to a 0.25 μm diamond finish does not significantly affect the deduced deuterated layer thickness, at least from the geometrical aspects of the surface finish - strains induced by the polishing treatment are of course another matter. Whether or not there was a singularity in the dissolution rate of Synroc-C in water at some temperature between 120 and 190°C [6] was investigated by carrying out reaction at 250°C in D_2O. Also, although 250°C considerably exceeds likely repository temperatures it is of interest to look at the effect of elevated water temperatures, especially as early data indicated that Synroc is resistant even after exposure to water at 900°C [1]. After 21 days of exposure to D_2O at 250°C, it could only be deduced from ERDA that the D_2O penetration exceeded 200 nm. SEM examination showed that the TiO_2-rich alteration layer was ~ 1 μm thick.

It was concluded that the reaction rate of Synroc-C in water generally increases with temperature and that the apparently anomalous result that the intrusion of D into Synroc-C was greater at 120°C than 190°C [6] was due either to only a minor change in the reaction mechanism or due to polishing strains being larger in the sample leached at 120°C.

Synroc-C samples were reacted for 21 days at 150°C and their surfaces (0.25 μm diamond finish) were subsequently studied by SIMS as well as ERDA [5]. Fig. 1 shows the results, from which it can be concluded that both methods yielded a penetration depth of ~ 20 nm.

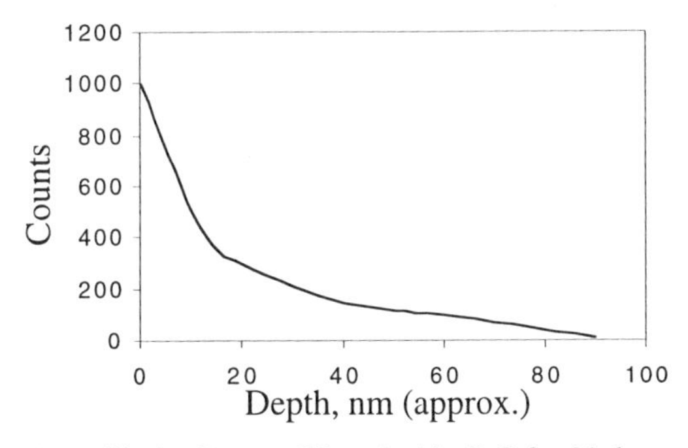

Fig. 1. SIMS depth profile for Synroc-C Leached in D_2O for 21 days at 150°C; mass number 18.007.

(b) SIMS to study rare earths at Synroc-C surfaces reacted in water

In studies of sphene glass-ceramics [9], rare earths were surface enriched after reaction with water and we studied this possibility with Synroc-C. However SIMS studies of Synroc-C samples leached for 365 days in H_2O indicated that Nd did not build up at the surface [11]. In the present work, samples exposed to deionised water at 90°C [12] were studied and there was no surface buildup of Nd or Ce after leaching for 7, 28 or 336 days. No surface buildup of Zr was noted either.

(c) Alpha-spectroscopy of Synroc surfaces exposed to water

An alpha-spectrum of the inactive disk adjacent to the Pu-doped Synroc sample is shown in Fig. 2, (leach period 84 days) together with the spectrum from 0.5 mL of a solution (0.147 Bq of ^{239}Pu/mL) of Pu source material evaporated onto the surface of an inactive Synroc disk. The similarity of the spectra suggests that the activity on the inactive sample exposed to water is all on the surface. From the known activity of the standard, the Pu activity on the disk is 0.06 +/- 0.02 Bq on either side, and from the counting efficiency the ^{241}Am activity on each side of the disk is 0.01 Bq. The calculated depletion of Pu from the surface of the active disk was 0.2 μm, from the observed decrease in the maximum alpha energy.

The inactive disk has clearly allowed actinides to simply sorb on its surface. Comparison of the source and disk spectra shows that there is relatively more Am on the surface than Pu, showing either that there is more Am in the solution than Pu or that Am is selectively sorbed by the disk. Future experiments will distinguish between these possibilities by studying the amounts of each nuclide in the solution, taking account of the tendency of the Pu to also plate out preferentially on the container walls.

In the 28-day experiment with the Cm-doped Synroc, the activity of ^{244}Cm on the inactive disk was 0.06 Bq, whereas the unfiltered leachant activity was 1.2 Bq, a factor 20 higher. The countrate for the surface of the active disk exposed to water was 95% of the unleached value and there was no enhancement/depletion of Cm activity observable at the surface, consistent with uniform extraction (~ 5%) over a depth of at least 1 μm. Nor was there any change in the maximum alpha energy.

(d) Zirconolite-rich Synroc

Zirconolite-rich Synroc samples consisted of ~ 80 wt% zirconolite and ~10 wt% each of Ba-hollandite and rutile [13]. Leaching for 30 days at 150°C in D_2O gave narrower D and H peaks than for Synroc-C (see Fig. 3), corresponding to a depth of ~ 5 nm. The hot-pressed and sintered samples gave very similar ERDA results. SEM of the samples did not reveal any secondary alteration products.

(e) Dissolution of Perovskite in Neutral and Acid solutions containing fluoride impurities

From optical and SEM studies, inactive samples of Mitamura *et al.* [8] showed considerable cracking and inhomogeneity, with regions of rare-earth-rich and Ca-rich material being common. The materials exposed to the pH = 2 solution had a secondary alteration layer of anatase on them, about 0.25 mm in thickness.

The JAERI pretreatment of Teflon containers, using the MCC method [14], involved boiling the vessels in strong nitric acid and HF, and may have liberated F^-. This was thought to possibly affect the leaching behaviour of perovskite. We prepared perovskite of the same stoichiometry by the alkoxide-nitrate route and hot-pressing in a graphite die for extended periods, viz. 10 h at 1350°C. To prevent pore formation [15], the pressure was maintained after pressing until the sample had cooled to ~ 900°C. After confirming the generally homogeneous and dense nature of the product by SEM, we prepared samples for dissolution at 90°C in (i) pH = 2 HCl solution; (ii) pH 2 + 1 mMol/L of F^-; and (iii) 1 mMol/L NaF (pH = 6).

Being the most soluble element and therefore less likely to be subject to reprecipitation effects, the Ca extraction rate is taken as a conservative measure of sample durability and the results are plotted in Fig. 4. They decreased with increasing time in the pH = 6 dilute NaF solution, and were broadly similar to values reported previously for deionised water [7]. The short-term data (3-20 days) found here at pH = 2 seemed anomalously low, relative to previous data [7], but after 56 days, there was little difference. At this stage, this sample showed a purplish coloration, due to optical interference in a very thin (< 1 μm) surface alteration layer of anatase, as shown by SEM. The sample reacted in the F^--bearing acid solution showed a white coating which SEM examination revealed as a ~ 20 μm -thick deposit of TiO_2 (XRD showed it to be anatase) and yielded higher Ca extractions than those for the pH = 2 solution alone. Thus the results were broadly similar to those obtained by Mitamura *et al.* [8], suggesting that F^- contamination played a role in producing their anomalous results.

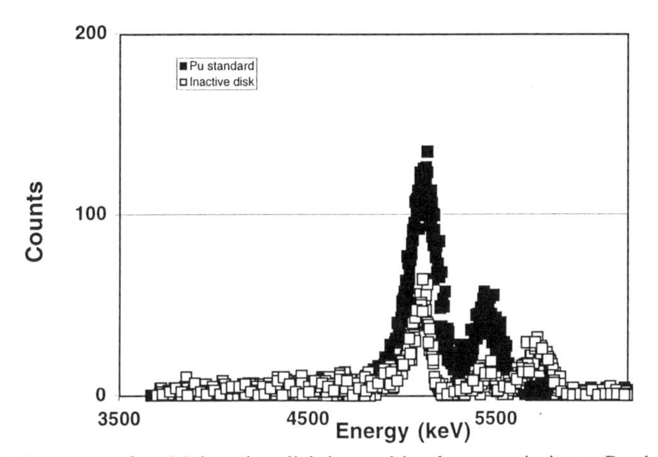

Fig. 2. Alpha spectra for (a) inactive disk located in close proximity to Pu-doped sample in aqueous solution; (b) Pu source material.

CONCLUSIONS AND FINAL REMARKS

For samples reacted in D_2O at 150°C, the SIMS profiles at mass 18.01 gave a similar penetration depth for D-bearing species to that found by ERDA previously for D [6]. Penetration of deuterated species into zirconolite-rich ceramics after reaction with D_2O at 150°C was less than that in Synroc-C. A Synroc-C sample displayed enhanced reaction with D_2O when the temperature was raised from 150 to 250°C. After exposure to deionised water at 90°C for up to 336 days, Synroc-C did not exhibit a buildup of rare earths on its surface. Alpha-spectroscopy showed sorption of actinides onto an inactive Synroc sample reacted in water in conjunction with actinide-bearing samples and depletion of actinides near the surfaces of the actinide-bearing samples. Traces of fluoride in acidic solutions had significant effects on $CaTiO_3$ surface alteration.

ACKNOWLEDGMENTS

We wish to thank G. R. Lumpkin, K. L. Smith and A. Jostsons for numerous discussions.

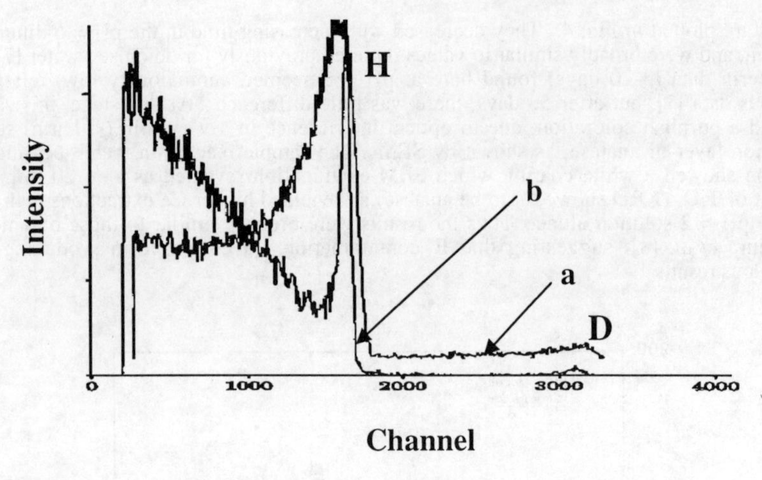

Fig. 3. ERDA depth profiles for H and D from samples reacted for 30 days in D_2O at 150°C; (a) Synroc-C; (b) zirconolite-rich ceramic hot-pressed in stainless steel bellows at 1250°C. Curves offset for clarity. Note weak D peak near channels 3100-3200.

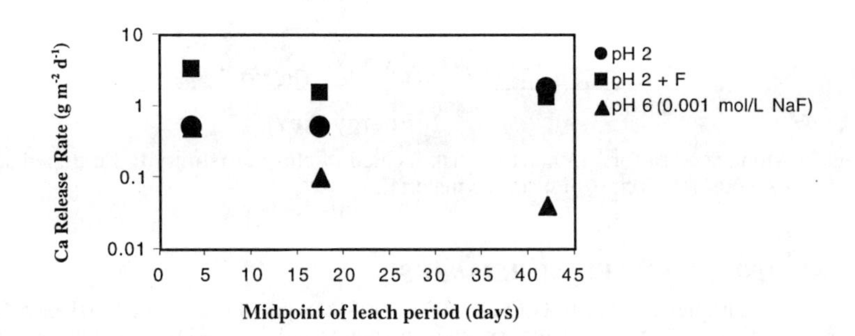

Fig. 4. Ca extraction rates with time at 90°C for polycrystalline $CaTiO_3$ reacted in (a) HCl (pH = 2); (b) HCl + 0.001 Mol/L F^- (pH = 2); (c) 0.001Mol/L NaF solution (pH = 6).

REFERENCES

1. A. E. Ringwood, S. E. Kesson, N. G. Ware, W. Hibberson and A. Major, Nature (Lond.) **278**, p. 219 (1979).
2. A. E. Ringwood, S. E. Kesson. K. D. Reeve, D. M. Levins and E. J. Ramm, in Radioactive Waste Forms for the Future, Eds. W. Lutze and R. C. Ewing (Elsevier, Amsterdam, 1988), p. 223.
3. A. Jostsons, K. L. Smith, M. G. Blackford, K. P. Hart, G. R. Lumpkin, P. McGlinn, S. Myhra, A. Netting, D. K. Pham, R. St.C. Smart and P. S. Turner (1990), Description of Synroc Durability: Kinetics and Mechanisms of Reaction, NERDDP Report.
4. H. J. Matzke, E. Toscano, C. T. Walker and A. G. Solomah, Adv. Ceram. **3**, p. 285 (1988).
5. N. Dytlewski, E. R. Vance and B. D. Begg, J. Nucl. Mater., **231**, p. 257 (1996).
6. H. J. Matzke, G. Della Mea, F. L. Friere Jr. and V. Rigano, Nucl. Instrum. Meth. **B45**, p. 194 (1990).
7. P. J. McGlinn, K. P. Hart, E. R. Vance and E. Loi, Scientific Basis for Nuclear Waste Management XVIII, edited by T. Murakami and R. C. Ewing (Materials Research Society, Pittsburgh, PA, USA, 1995), p. 847.

8. H. Mitamura, S. Matsumoto, T. Tsuboi, E. R. Vance, B. D. Begg and K. P. Hart, ibid. p. 1405.
9. P. J. Hayward, W. H. Hocking, F. E. Doern and E. V. Cecchetto, in Scientific Basis for Nuclear Waste Management-V, edited by W. Lutze (Elsevier, Netherlands, 1982), p. 319.
10. K. P. Hart, E. R. Vance, R. Stajenovic and R. A. Day, Alpha-spectroscopy and leach testing of Synroc doped with Actinide elements, in Scientific Basis for Nuclear Waste Management XXII, edited by D. J. Wronkewicz (Materials Research Society, Pittsburgh, PA, USA), in press.
11. K. L. Smith, G. R. Lumpkin, M. G. Blackford, R. A. Day and K. P. Hart, in Scientific basis for Nuclear Waste Management XIV, Eds. T. A. Abrajano, Jr. and L. H. Johnson (Materials Research Society, Pittsburgh, PA, USA, 1992), p.167.
12. G. J. Thorogood, M. App. Sc. Thesis , University of Technology, unpublished (1995).
13. E. R. Vance, B. D. Begg, R. A. Day and C. J. Ball, Zirconolite-rich Ceramics for Actinide Wastes, in Scientific Basis for Nuclear Waste Management XVIII, edited by T. Murakami and R. C. Ewing (Materials Research Society, Pittsburgh, PA, USA, 1995), p. 767.
14. Materials Characterization Center, "Nuclear Waste Materials Handbook Test Method (Rev. 7):MCC-1P Static Leach Test Method, DOE/TIC-11400, PNL, Richland, WA (1986).
15. E. R. Vance, M. L. Carter, R. A. Day and D. J. Cassidy, unpublished report

THE LONG-TERM CORROSION BEHAVIOR OF TITANATE CERAMICS FOR Pu DISPOSITION: RATE-CONTROLLING PROCESSES

A. J. Bakel, C. J. Mertz, M. C. Hash, and D. B. Chamberlain
Argonne National Laboratory, Chemical Technology Division, 9700 S. Cass Avenue, Argonne, IL 60439

ABSTRACT

The aqueous corrosion behavior of a zirconolite-rich titanate ceramic was investigated with the aim of describing the rate-controlling process or processes. This titanate ceramic is similar to SYNROC and is proposed as immobilization materials for surplus Pu. The corrosion behavior was described with results from MCC-1 and PCT-B static dissolution tests. Three important observations were made: a) Ca is released at a constant rate [$7x10^{-5}$ g/(m^2 day)] in PCT-B tests for up to two years; b) the leachates from PCT-B tests are saturated with respect to both rutile and anatase, and c) the release rates for Pu and Gd increase with time (up to two years) in PCT-B tests. The first observation suggests that the ceramics continue to corrode at a low rate for at least 2 years in PCT-B tests. The second observation suggests that the approach to saturation with respect to these TiO_2 phases does not limit the corrosion rate in PCT-B tests. The third observation suggests that the release rate of Pu and Gd are controlled by some unique process or processes, i.e., some process or processes that do not affect the release rate of other elements. While these processes cannot be fully described at this point, two possible explanations, alteration phase formation and grain boundary corrosion are forwarded.

INTRODUCTION

The dismantlement of nuclear weapons and the cleanup of weapons production sites have generated large quantities of surplus weapons-grade Pu, contaminated Pu stock, and Pu scrap in the United States [1]. The U.S. Department of Energy is currently considering two options for the disposition of surplus Pu. In the first option, the Pu would be incorporated into a mixed oxide fuel (MOX) for use in commercial reactors. Second, the Pu would be immobilized and disposed of in a stable titanate ceramic material. The titanate ceramic described and tested in this study represents an early, zirconolite-rich formulation of the proposed immobilization material. The currently preferred ceramic composition is a pyrochlore-rich formulation. We expect that the overall corrosion behavior of zirconolite-rich ceramics to be similar to that of pyrochlore-rich ceramics.

The goals of this study are to describe the long-term corrosion of a zirconolite-rich titanate ceramic and to evaluate several possible rate-controlling steps for the corrosion of these titanate ceramics. We believe that any conclusion regarding the corrosion of this zirconolite-rich ceramic, will be generally applicable to the corrosion of pyrochlore-rich ceramics.

TEST METHODS

Tests in this study were conducted according to the standard MCC-1 [2] and PCT-B [3] procedures. For the MCC-1 tests, the ceramic samples were prepared as wafers (about 10 mm in diameter and 1 mm in thickness) and the surfaces were ground to a 240-grit finish. All samples were ultrasonically cleaned in deionized water (DIW) and ethanol. Tests were conducted by sealing one ceramic wafer in a 22ml stainless steel (Type 304L) vessel with about 18 ml of DIW, leading to a ratio of the geometric ceramic surface area to leachant volume (S/V) of about 10 m^{-1}. For the PCT-B tests, the ceramic was crushed and sieved to isolate the -100+200 mesh fraction. The powder was ultrasonically cleaned in DIW and ethanol. The crushed ceramic was sealed in a 22ml stainless steel vessel with enough water to yield an S/V ratio of about 10,000 m^{-1}.

Upon the completion of a test, the leachate was removed and analyzed for pH and cations with inductively coupled plasma-mass spectrometry (ICP-MS). The test vessel was then filled with 1% HNO_3, and placed in a 90°C oven for at least 8 hours to remove any material fixed to the vessel wall. This "acid strip" solution was also analyzed with ICP-MS. The normalized mass loss [NL(i)] values presented here represent the amount of the ceramic dissolved based on the amount of an element in the leachate and the acid strip solutions. Experimental blanks were conducted, and background concentrations were subtracted from concentrations measured for test solutions.

Mat. Res. Soc. Symp. Proc. Vol. 608 © 2000 Materials Research Society

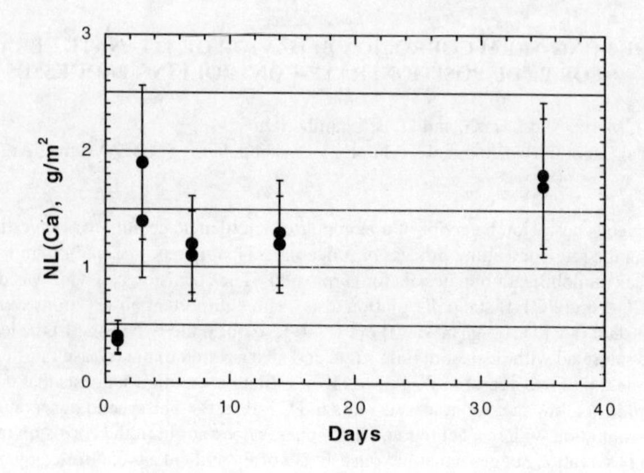

Figure 1. Normalized mass loss, based on Ca release for the zirconolite-rich ceramic in MCC-1 tests. The error bars represent 35% relative uncertainty.

TEST MATERIALS

The zirconolite-rich ceramic was fabricated at Lawrence Livermore National Laboratory. The precursor material was ground, dried, pulverized and calcined in air at 600°C for 1 hour. The calcined material was then broken up and mixed with the PuO_2. The resulting powder was cold pressed and sintered at 1325°C for about 4 hours. Further details of the fabrication method are described by Buck et al. [4].

The chemical composition of the zirconolite-rich ceramic was determined by dissolving it at 150°C in mineral acids, and analyzing the resulting solution with ICP-MS. The composition of the ceramic was measured to be: 8.0 mass % Al_2O_3, 7.3% CaO, 0.37% Cr_2O_3, 0.27% Fe_2O_3, 0.10% NiO, 0.26% ZnO, 0.21% CuO, 0.32% Ga_2O_3, 3.6% BaO, 0.31% CeO_2, 9.5% Gd_2O_3, 37% TiO_2, 16% ZrO_2, 0.35% HfO_2, and 14%PuO_2. The phase composition of the ceramic is 60-70 vol. % zirconolite, about 30% rutile, less than 5% perovskite and brannerite, and less than 1% PuO_2.

This ceramic was developed several years ago at the beginning of the testing program. Long-term tests were initiated with this material. Newer formulations are composed primarily of pyrochlore. Because zirconolite and pyrochlore are chemically and crystallographically similar, we expect that the tests described here will provide insight into the long-term corrosion behavior of the newer formulation.

RESULTS AND DISCUSSION

Previous corrosion studies with this zirconolite-rich ceramic, based on MCC-1 test data, have shown an initial rapid release of material followed by a much slower release [5]. The NL(Ca) data in Figure 1 show rapid corrosion in the first three days and slower corrosion after three days. Similar changes were observed in the release rates of other elements found in this ceramic [5], and similar behavior has been observed with SYNROC ceramics [6].

One explanation for the decrease in the corrosion rate over time is that the leachate solutions approach saturation with respect to some solid phase. We would expect on this basis that the corrosion rate would decrease as the concentration of the element or elements contained in the phase increased. Possible controlling solid phases for the dissolution of this titanate ceramic include TiO_2 and ZrO_2.

Another explanation for the observed decrease in the corrosion rate over time is the formation of protective, Ti-rich alteration layers [6]. Such continuous TiO_2 layers are often observed on corroded Ti metal surfaces [7]. Titanium-rich alteration phases have been observed on corroded perovskite and hollandite in SYNROC ceramics [8, 9]. In the same studies, however Ti-rich alteration layers were not observed on corroded zirconolite. Therefore, we do not expect that continuous, protective Ti-rich alteration layers are present on the corroded ceramics in this study.

In addition, the high initial Ca release rate shown in Figure 1 could result from the rapid corrosion of fine-grained material present on the prepared surface of the ceramic monolith. This material would presumably dissolve quickly, and would have little impact on the release rates observed in longer-term tests. Finally, the corrosion of particularly soluble material from grain boundaries could account for the corrosion behavior observed. The results of several types of tests are discussed below, and are interpreted in terms of rate-controlling processes.

Ca Release

Figure 2 shows the NL(Ca) as a function of test duration, for a series of PCT-B tests with the zirconolite-rich ceramic. The NL(Ca) increases linearly with time at a rate corresponding to about 7×10^{-5} g/(m^2 day); the R value for this linear fit is 0.997. The bounding rate estimated using Bourcier's pH and temperature dependence model [10] at 90°C and pH of 8 is about 1×10^{-4} g/(m^2 day).

The similarity of the corrosion rates measured in our PCT-B tests and the single pass flow through (SPFT) tests suggests that the corrosion mechanisms is the same during these two tests. On the other hand, the corrosion rates have been shown to decrease with time during SPFT tests [10]. This decrease has been taken as evidence of a transport-controlled rate-limiting process in SPFT tests. In contrast, the data from PCT-B tests indicate that transport is not rate limiting.

These Ca release data indicate that the corrosion of the ceramic continues in these tests for up to two years, suggesting that no protective layer formed on the phase(s) controlling the release of Ca during these tests. Also, the solutions were not approaching any limiting saturation condition.

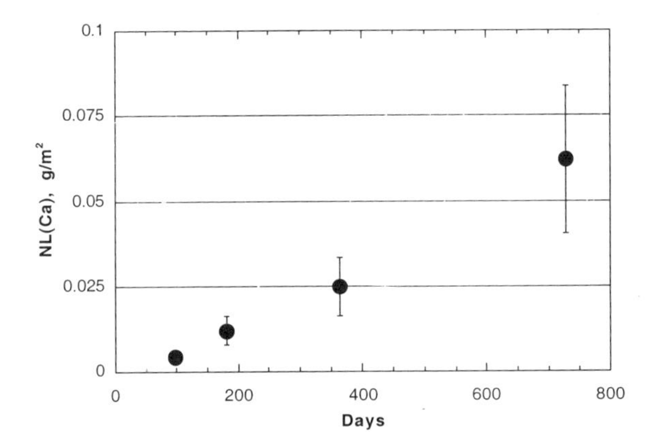

Figure 2. Normalized mass loss, based on Ca release for the zirconolite-rich ceramic in PCT-B tests. The error bars represent 35% relative uncertainty.

Ti Release

Figure 3 shows the concentrations of Ti in the leachate solutions at 90°C as a function of pH. The solubility of rutile at 100°C [11] and our analytical detection limit are also shown. The solubility of anatase (a polymorph of TiO_2 that is stable at low temperatures) is about an order of magnitude lower than that of rutile [12, 13]. Most of the leachate solutions from MCC-1 tests appear to be approximately saturated with respect to anatase. This suggests that the corrosion of the ceramic in MCC-1 tests might be controlled by the concentration of Ti in the leachate. On the other hand, the solutions from PCT-B solutions are oversaturated with respect to both rutile and anatase. These data suggest that, on this time scale, the precipitation of TiO_2 is slow relative to the corrosion of the ceramic in PCT-B tests, and that the approach to the TiO_2 solubility limit does not appear to be the rate-controlling process in PCT-B tests.

Gd and Pu Release

In contrast to the release of Ca (Figure 2), the NL(Pu) and NL(Gd) do not increase linearly with time in the PCT-B tests (Figure 4). Instead, the release rates increase with time. The data in Table 1 show that the release rates of Pu and Gd are higher in the later intervals than in the early interval in this series of tests. At the same time, the release rates of Ca are similar in all of the intervals (Table 1). These results show that the release behaviors of Gd and Pu differ from the release behavior of Ca, and that some process or processes affect the release of Gd and Pu that do no affect the release of Ca. This is expected, because the ceramic contains several different phases that may well have different corrosion behaviors.

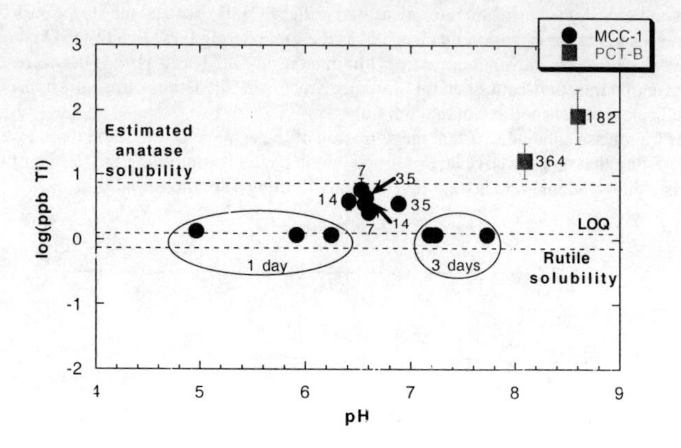

Figure 3. Ti concentrations from test leachates as a function of pH. The error bars represent 22% relative uncertainty. The numbers adjacent to the data points represent the test duration of in days. The limit of quantitation (LOQ) is shown along with the rutile and anatase solubility [11].

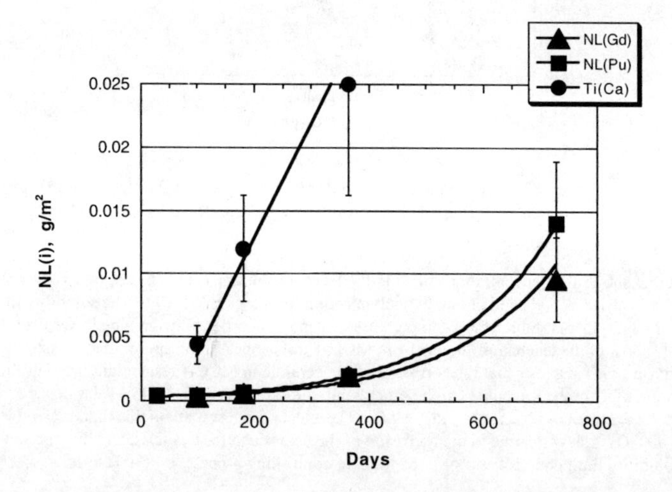

Figure 4. Values of NL(Ca), NL(Gd) and NL(Pu) for the zirconolite-rich ceramic in PCT-B tests as a function of time. The error bars represent 35% relative uncertainty.

Table 1. Interval release rates, in g/(m^2 day) for Ca, Pu and Gd for PCT-B tests with the zirconolite-rich ceramic.

	98-182 days	**182-364 days**	**364-428 days**
NR(Ca)	7×10^{-5}	9×10^{-5}	6×10^{-5}
NR(Pu)	7×10^{-6}	1×10^{-5}	7×10^{-5}
NR(Gd)	3×10^{-6}	8×10^{-6}	2×10^{-5}

We propose two possible processes that could lead to increasing release rates with time. First, the formation of alteration phases has been shown to accelerate the corrosion rate of borosilicate glasses [14]. Precipitation of alteration phases presumably alters the leachate chemistry, leading to an increase in the corrosion rate. Previous studies of corroded titanate ceramics have shown only small amounts of observable alteration products on the corroded ceramic surface [4]. We do not expect that their precipitation would affect the corrosion of the ceramic as observed. However, colloidal alteration phases might be responsible for the observed changes in corrosion behavior. The nature and effects of colloidal material are being investigated.

Second, if the surface area of the reacting ceramic increases throughout the course of the reaction (for example, by grain boundary corrosion), then the release rate would increase. An increase of at least 100-fold would be needed to explain the change in release rate of Pu/Gd between 98 and 728 days. Such an increase appears reasonable if corrosion occurs at grain boundaries, which could significantly increase the reactive surface area. However, an increase in surface area would be expected to increase the release rate of all elements.

The increase in the release rates of Pu and Gd observed in these tests is an interesting result. However, since they are based on a single set of tests, any detailed conclusions regarding the long-term corrosion behavior of this ceramic should be considered tentative. Similar long-term tests are underway with pyrochlore-rich ceramics. In addition, higher temperature PCT-B tests are planned. We expect that increasing the temperature will serve to accelerate the corrosion.

SUMMARY

The observations made in this study lead to several important observations regarding the corrosion of titanate ceramics. First, the release of Ca in these tests continues at a constant rate for up to two years. The measured Ca release is consistent with corrosion rates previously measured in SPFT tests. Second, the leachates from the PCT-B tests are oversaturated with respect to TiO_2, suggesting that Ti concentration would not affect the corrosion of this ceramic. Third, the release rates of Pu and Gd (but not Ca) increase with time in long-term PCT-B tests. Two processes that could lead to the observed rate increases, formation of alteration phase and grain boundary corrosion, are discussed.

Due to the complexity of the ceramic and the observed incongruent dissolution, it may be difficult to derive a simple equation to describe its dissolution. We believe that the results from PCT-B tests are valuable in describing the long-term behavior of this ceramic. It is interesting to note that our results are not in all cases predicted by the current model of titanate ceramic corrosion.

ACKNOWLEDGMENTS

This work was supported by the U.S. Department of Energy Office of Materials Disposition under contract W-31-109-ENG-38. Assistance was provided by Mike Nole, Steve Wolf and Kevin Quigley. Edgar Buck provided a particularly helpful review. AJB would like to dedicate this work to Mary Beth, Mercy, and Elfie at Children's Memorial Hospital.

REFERENCE

1. B. R. Myers, G. A. Armantrout, C. M. Jantzen, A. Jostons, J. M. McKibben, H. F. Shaw, D. M. Strachan, and J. D. Vienna, Technical Evaluation Panel Summary Report: Ceramic and Glass Immobilization options, Lawrence Livermore National Laboratory, Livermore, CA, Report UCRL-ID-129315 (1998)
2. ASTM, *Standard Test Method for Static Leaching of Monolithic Waste Forms for Disposal of Radioactive Waste,* ASTM Standard C1220-98, American Society for Testing and Materials, Philadelphia, PA (1998).
3. ASTM, *Standard Test Methods for Determining Chemical Durability of Nuclear Waste Glasses: The Product Consistency Test (PCT)*, ASTM Standard C1285-94, American Society for Testing and Materials, Philadelphia, PA (1994).
4. E. C. Buck, B. B. Ebbinghaus, A. J. Bakel, and J. K. Bates, "Characterization of a Plutonium-Bearing Zirconolite-Rich Ceramic," Mater. Res. Soc. Symp. Proc. 465, 1259-1266 (1997).
5. A. J. Bakel, E. C. Buck, C. J. Mertz, D. B. Chamberlain, and S. F. Wolf, "Corrosion Behavior of a Zirconolite-rich Ceramic", Lawrence Livermore National Laboratory, Livermore, CA, Report PIP 99-045 (1999).
6. A. E. Ringwood, V. M. Oversby, S. E. Kesson, W. Sinclair, N. Ware, W. Hibberson, and A. Major, "Immobilization of High-Level Nuclear Reactor Wastes in Synroc: A Current Appraisal," Nucl. Chem. Waste Manage. 2, 287-305 (1981).
7. M. G. Fontana, *Corrosion Engineering*, 3rd ed., McGraw-Hill, New York (1986).
8. K. L. Smith, M. Colella, G. J. Thorogood, M. G. Blackford, G. R. Lumpkin, K. P. Hart, K. Prince, E. Loi, and A. Jostsons, "Dissolution of Synroc in Deionized Water at 150°C," Mater. Res. Soc. Symp. Proc. 465, 349-354 (1997).
9. G. R. Lumpkin, K. L. Smith, and M. G. Blackford, "Electron Microscope Study of Synroc Before and After Exposure to Aqueous Solutions," J. Mater. Res. 6, 2218-2233 (1991).
10. W. L. Bourcier, "Interim Report on the Development of a Model to Predict Dissolution Behavior of the Titanate Waste Form in a Repository," Lawrence Livermore National Laboratory, Livermore, CA, UCRL-ID-135363 (1999).
11. K. G. Knauss, M. J. Dibley, W. L. Bourcier, and H. F. Shaw, "Ti(IV) hydrolysis Constants Derived from Rutile Solubility Measurements made from 100°C to 300°C," Lawrence Livermore National Laboratory, Livermore, CA, UCRL-JC-135165. (1999).
12. R. A. Robie, B. S. Hemingway, and J. R. Fisher, U. S. Geological Survey Bulletin 1452, U.S. Government Printing Office, Washington, DC (1979).
13. M. M. Lencka and R. E. Ritman, "Thermodynamic Modeling of Hydrothermal Synthesis of Ceramic Powders," Chem. Mater. 5, 61-70 (1993).
14. A. J. Bakel, W. L. Ebert, and J. S. Luo, "Long-Term Performance of Glasses for Hanford Low-Level Waste", in "Environmental Issues and Waste Management Technologies in the Ceramic and Nuclear Industries", Ceramic Transactions, Vol. 61, Eds., V. Jain and R. Palmer, Am. Ceram. Soc. (1995).

CHEMICAL DURABILITY OF YTTRIA-STABILIZED ZIRCONIA FOR HIGHLY CONCENTRATED TRU WASTES

Hajime Kinoshita *, Ken-ichi Kuramoto **, Masayoshi Uno *, Shinsuke Yamanaka *, Hisayoshi Mitamura ***, Tsunetaka Banba ***
*Department of Nuclear Engineering, Osaka University, Yamada-oka 2-1, Suita, Osaka 565-0871 JAPAN, khaji@nucl.eng.osaka-u.ac.jp
**Department of Nuclear Energy System, Japan Atomic Energy Research Institute, Ibaraki 319-1195 JAPAN
***Department of Environmental Sciences, Japan Atomic Energy Research Institute, Ibaraki 319-1195 JAPAN.

ABSTRACT

Neptunium has an extremely long half-life and will be the main toxic element in the later stage of disposal. Yttria-Stabilized Zirconia (YSZ), with a fluorite structure, samples doped with Np-237 in high concentration (20, 30, 40 mol %) were fabricated (sintered in Air or Ar, at 1773 K, for 80 hours), and their leaching test (in deionized water, at 423 K, for 84 days) was carried out. The results indicated that the obtained leaching rates were much smaller than those of the Synroc and glass waste form, and that the increase in Np content did not cause any drastic changes in the leaching rates of Zr, Y, or Np. They were also compared to the results of YSZ doped with Ce or Nd used as surrogates for actinides. The work showed that YSZ doped with Np in high concentration has an excellent chemical durability.

INTRODUCTION

Isolation of hazardous radionuclides is one of the most important issues in nuclear waste management. In particular, TRU elements should be treated with special concern because their half-lives are substantially longer than those of other radionuclides. The most common and practical strategy is to fix these hazardous radionuclides into a suitable solid matrix with high stability and high resistance to ground water attack, in order to retain them in underground repositories.

Yttria-stabilized zirconia (YSZ) with a fluorite structure has been studied for use in the immobilization of partitioned and highly concentrated TRU waste [1]. The biggest advantage of YSZ lies in the fact that it can form a solid solution with tri- and/or tetra-valent TRU elements and can maintain the cubic system [2-4]. Cubic-zirconia is known to have a higher resistance to α-decay damage than other ceramics such as Al_2O_3 [5]. Possible advantages from the YSZ waste form are, therefore: (i) the accommodation of TRU elements in high concentration; and (ii) the production of very stable waste forms. In addition, it has been proved that YSZ has an excellent stability, as an inert matrix, to the irradiation of neutron and fission fragments [6]. The resultant high quality waste form therefore could be utilized in other applications, such as TRU burning [7].

The general purpose of the present work is to study the applicability of YSZ as a host matrix for the waste form of highly concentrated TRU. The study was conducted focusing on the chemical durability of YSZ doped with Np. Neptunium has an extremely long half-life (2.14 million years) and will be the main toxic element in the later stage of the disposal. The chemical durability was evaluated by a leaching test. The present evaluations were also conducted using data of Ce or Nd-doped YSZ for comparison. Cerium and Nd are often used as surrogates for tetra- and tri-valent actinides, respectively [1-4].

EXPERIMENT

Sample Preparation

In the present study, YSZ samples doped with Np-237 in high concentration (20, 30, 40 mol %) were fabricated. The procedure is: (1) an Np solution (Np in nitric acid: 3 or 6 MBq/ml) was pipetted into Al_2O_3 crucibles containing YSZ powder (TZ-8Y; Toso Co., Ltd, Tokyo, Japan; ZrO_2 doped with $YO_{1.5}$ in 14.3 mol%) that had been prepared earlier by weight to a specified composition; (2) the sample mixtures were calcined under streaming air at a linear heating rate of 0.25 K/min to 523 K, then at 5 K/min to 1173 K, maintained at 1173 K for 1 h, and cooled to obtain powder mixtures of YSZ and Np dioxide [8]; (3) the mixtures were mechanically mixed using a YSZ ball mill and pelletized at 130 MPa pressure; (4) the green pellets (two for each composition were prepared) were sintered in an electric resistance furnace at 1773 K for 80 h under streams of either air or 3 % H_2-Ar.

Using basically the same procedure but sintered only under the streams of air, plain YSZ and Ce or Nd-doped YSZ specimens were fabricated. The crystalline phases in YSZ specimens were identified by X-ray diffraction using a RAD-IIC type spectrometer (Rigaku Co., Ltd., Tokyo, Japan) with CuK_α radiation. It was confirmed that all the specimens have a fluorite structure phase.

Leaching Test

The leaching test of YSZ samples doped with Np was performed in accordance with the procedure of MCC-2 [9]. The procedure is; (1) the prepared pellets were crushed and sieved under 75 μm beforehand; (2) surface area of the powder samples were determined via BET adsorption technique using a FLOWSORB-II2300 (Shimadzu Co., Ltd., Kyoto, Japan) apparatus; (3) powder samples were adjusted to have 0.5 m^2 surface area and immersed into 50 ml of deionized water at 423 K for 30 days in teflon vessels – this process was performed, as a preliminary leaching, to remove the extremely small and active particles in the samples which cause the excessive leaching [1]; (4) after the preliminary leaching, the powder samples were recovered, and the vessels were washed in 0.01 M HNO_3 and successively in deionized water, at 423 K for a day for each; (5) the recovered powder samples were washed with deionized water and, as the main leaching procedure, immersed into 50 ml of deionized water at 423 K for 84 days in washed teflon vessels; (6) the leachates were separated by ultrafiltration with 0.45μm milipore filters; (7) the teflon vessels were washed with 0.01 M HNO_3 to strip the elements attached to the vessel wall and the HNO_3 solutions were recovered; (8) the concentration of Zr and Y, and Np (both in the separated leachates and the recovered HNO_3 solutions) were measured by ICP and liquid scintillation counting, respectively; (9) the obtained results from the separated leachates and the recovered HNO_3 solutions were added for each element and the leaching rates for each element were calculated.

The leaching test of YSZ samples doped with Ce or Nd was performed basically through the same procedure in accordance with the MCC-2, and is detailed elsewhere [1]. The pH of the deionized water used for the leaching test was not directly measured, but was assumed at about 5.6 which was the value used for a similar study [10]. The tested samples and their compositions are indicated in Table I. The YSZ, indicated as the matrix in the table, always has the same molar ratio of ZrO_2: $YO_{1.5}$ (ZrO_2: $YO_{1.5}$ = 85.7: 14.3 mol%).

Table I. The tested samples and their compositions.

	Matrix(mol %)	Dopants (mol %)		
	YSZ	NpO_2	CeO_2	$NdO_{1.5}$
Np-YSZ$^{(O)}$ No.1	80	20	---	---
No.2	70	30	---	---
No.3	60	40	---	---
Np-YSZ$^{(R)}$ No.1	80	20	---	---
No.2	70	30	---	---
No.3	60	40	---	---
Ce-YSZ No.1	88.72	---	*11.28	---
No.2	72.38	---	27.62	---
No.3	51.84	---	*48.16	---
Nd-YSZ No.1	90.64	---	---	*9.355
No.2	85.48	---	---	14.52
No.3	81.15	---	---	*18.85

The * indicates the previously obtained data [1].
The $^{(O)}$ and $^{(R)}$ represent different sintering atmospheres, air and 3 % H_2-Ar, respectively.
The YSZ matrix always has same molar ratio of ZrO_2: $YO_{1.5}$ (ZrO_2: $YO_{1.5}$ = 85.7: 14.3 mol %).

RESULTS

Chemical durability of Np-doped YSZ

To evaluate chemical durability, normalized leaching rate (R_i; g $cm^{-2}day^{-1}$) of Zr, Y, Np, Ce, Nd were calculated using the following equation

$$R_i = A_i \cdot V /(F_i \cdot S \cdot T) \qquad (1)$$

where A_I is the concentration of element i in the leachate (g/ml); V is the volume of the leachate (50 ml); F_i is the mass fraction of the element i in the prepared sample; S is the initial surface area of the sample (0.5 m^2); and T is the leaching time (day). The results from a single test for Np-YSZ samples are indicated in Fig. 1. The average of uncertainties for the leaching rates of Zr, Y and Np are $2x10^{-12}$, $3x10^{-11}$ and $1x10^{-11}$ ($g/cm^2/day$), respectively. The detection limit of ICP is considered approximately $1x10^{-12}$ ($g/cm^2/day$), so that the data for Zr less than $1x10^{-12}$ ($g/cm^2/day$) (those in the circle) should not be taken into account.

Although Np-doped YSZ samples were prepared by sintering in two different atmospheres, air and 3 % H_2-Ar, the obtained leaching rates of elements did not show apparent differences between samples sintered in air (Np-YSZ$^{(O)}$ series) and those sintered in 3 % H_2-Ar (Np-YSZ$^{(R)}$ series). It was therefore considered that the difference in sintering atmosphere does not affect the chemical durability of Np-doped YSZ in the condition of this study. Hereafter, Np-YSZ$^{(O)}$ and Np-YSZ$^{(R)}$ samples are treated as the same series of samples and denoted as Np-YSZ. The figure also indicates the general level of leaching rate of Zr from a synroc [10], and that of Np from a synroc and a borosilicate glass [11, 12] for comparison. It is recognized that the obtained leaching rates of Zr and Np from Np-doped YSZ samples are much smaller than those from the synroc and borosilicate glass. Another important fact this result indicates is that the increase in Np content in the samples did not cause any drastic changes in the leaching rates of Zr, Y, or Np from Np-doped YSZ. It was therefore shown that Np-doped YSZ has an excellent chemical durability, and that there is a possibility of YSZ to be applied as a host matrix for the waste form of highly concentrated TRU elements.

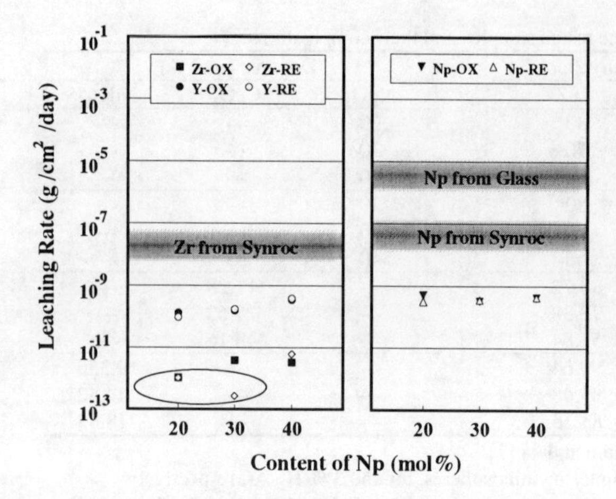

Fig. 1 Leaching rate of Zr, Y and Np from Np-doped YSZ samples in deionized water at 423 K for 84 days, and those from borosilicate glass and Synroc in the similar conditions [11, 12]. The data for Zr less than $1x10^{-12}$ ($g/cm^2/day$) (those in the circle) should not be taken account because of unreliability.

<u>Comparison of Np-doped YSZ with Ce or Nd-doped YSZ</u>

The result of Np-doped YSZ samples were also compared to those of YSZ doped with Ce or Nd. The obtained leaching rates of elements are shown in Fig. 2. The unreliable data mentioned in the former section are not indicated in this figure. It can be seen that the general leaching behavior of the elements are different depending on the kind of dopants.

Although the leaching rate of Zr did not show any significant trend in Np, Ce, or Nd-doped YSZ samples, the leaching behavior of Y in these samples is very different depending on the kind of dopants. For Np or Nd-doped YSZ, the leaching rate of Y tends to increase as the content of dopant increases. On the other hand, for Ce-doped samples, the leaching rate of Y tends to decrease as the content of dopant increases. This fact shows that the leaching behavior of Y is affected by the other constituents of the solid solution, namely by dopants.

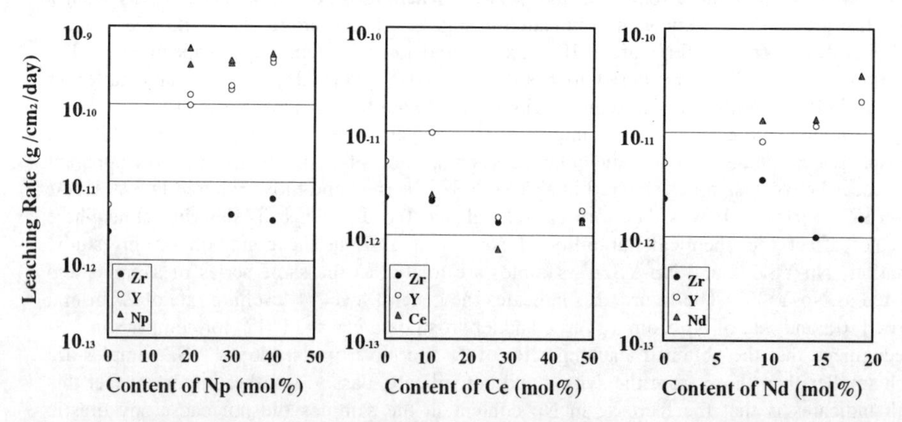

Fig. 2 Leaching rate of Zr, Y and dopants (Np, Ce, Nd) from Np, Ce, Nd-doped YSZ samples in deionized water at 423 K for 84 days (Np-doped YSZ) and for 120 days (Ce or Nd-doped YSZ)[1].

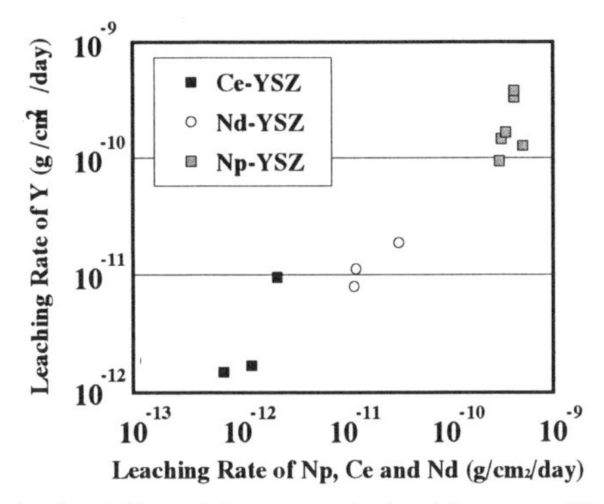

Fig. 3 Effect of dopant on the leaching rate of Y.

All the samples tested in this study consist of YSZ (ZrO_2 and $YO_{1.5}$) and a dopant (NpO_2, CeO_2 or $NdO_{1.5}$), and the leaching behavior of Zr is similar in all the samples. It is therefore natural to consider that the difference in dopants led to the difference in the leaching behavior of Y. To confirm the effect of dopants on the leaching behavior of Y, the leaching rate of Y from Np, Ce, and Nd-doped YSZ samples are plotted as a function of the leaching rates of dopants in Fig. 3. The figure shows a clear correlation between the leaching rate of Y and those of dopants (Np, Ce, and Nd). The fact that the leaching behavior of Y is affected by those of dopants implies that all the constituents of YSZ forms (for example Np, Y, and Zr for Np-doped YSZ) participate in the leaching behavior of the YSZ forms by affecting each other. A reason this effect was not prominent in the leaching behavior of Zr but in that of Y was the molar fraction of those elements. The contents of Zr in all YSZ samples were always large, but that of Y were always small, so that Zr was not affected by the other elements very much but Y was.

This effect is also recognized in their concentration in the leachates as shown in Fig. 4. The figure also indicates that the concentrations of Np and Ce in the leachates have a similar dependency on their (Np or Ce) contents; the concentrations of Np and Ce tend to increase with their content increase in the samples. Differing from Np and Ce, the concentration of Nd has a drastic increase with its content increase. A reason for this drastic increase in concentration can be the difference in the ionic radii. The ionic radii of Zr^{4+}, Np^{4+}, Ce^{4+}, Y^{3+} and Nd^{3+} are 0.084, 0.098, 0.097, 0.1019 and 0.1109 nm, respectively [13]. Therfore, Nd^{3+} causes more stress in the crystalline structure than Np^{4+} or Ce^{4+} does.

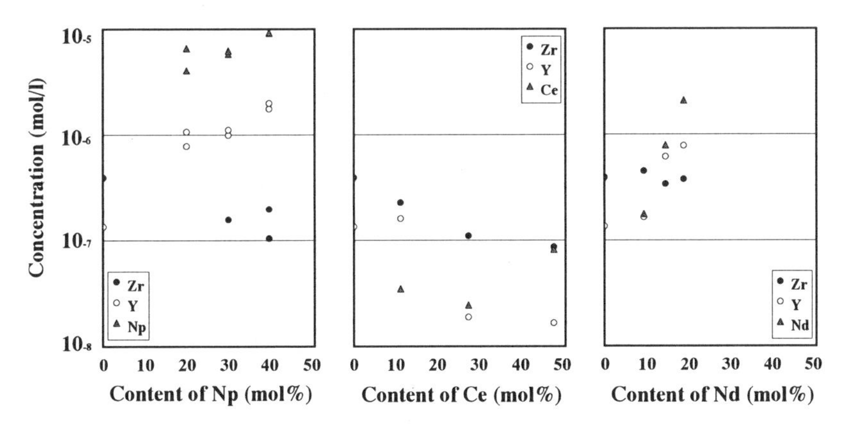

Fig. 4 Concentration of Zr, Y, and dopants (Np, Ce, Nd) in the leachates of the leaching test in deionized water at 423 K for 84 days (Np-doped YSZ) and 120 days (Ce or Nd-doped YSZ) [1].

CONCLUSIONS

Yttria-Stabilized Zirconia (YSZ), with a fluorite structure, doped with Np-237 in high concentration (20, 30, 40 mol %) were fabricated, and their leaching test, in deionized water at 423 K for 84 days, was carried out. The results indicated that the obtained leaching rates of Np and Zr were much smaller than those of the synroc and borosilicate glass, and that the increase in Np content in the samples did not cause any drastic changes in the leaching rates of Zr, Y, and Np.

By the comparison of Np-doped YSZ with Ce or Nd-doped YSZ in the leaching rates of elements, it was clarified that the leaching behavior of Y is affected by those of the other elements, especially by dopants. This fact suggests that all the constituents of YSZ forms affect each other in their leaching behavior.

Overall, the work showed that YSZ doped with Np in high concentration has an excellent chemical durability, and that there is a possibility of YSZ to be applied as a host matrix for the waste form of highly concentrated TRU elements.

REFERENCES

1. K. Kuramoto, Y. Makino, T. Yanagi, S. Muraoka, and Y. Ito in GLOBAL'95 (Proc. Int. Conf. Evalu. Emerg. Nucl. Fuel Cycl. Syst., Versailles, France, 1995), p. 1838.
2. Y. Hinatsu, and T. Muromura, Mat. Res. Bull., 21, p. 1343 (1986).
3. T. Muromura, and Y. Hinatsu, J. Nucl. Mater., 137, p. 227 (1986).
4. T. Muromura, and Y. Hinatsu, J. Nucl. Mater., 151, p. 55 (1987).
5. R. M. Berman, M. L. Bleiberg, and W. Yeniscavich, J. Nucl. Mater. 2, No.2, p. 129 (1969).
6. T. Ohmichi, JAERI-Review 96-008 (1996).
7. T. Iwasaki, and N. Hirakawa, J. Nucl. Sci. Technol., 31[12], p. 1255 (1994).
8. J. A. Fahey, R. P. Turcotte, and T. D. Chikalla, J. Inorg. nucl. Chem., 38, p. 495 (1975).
9. D. M. Strachan, B. O. Barnes, and R. P. Turcotte, Mat. Res. Soc. Symp. Proc., Vol. 3, p. 347 (1980).
10. I. Hayakawa, and H. Kamizono, J. Nucl. Mater., 202, p. 163 (1993).
11. T. Banba, H. Kamizono, S. Nakayama, and S. Tashiro, JAERI-M 89-110 (1989).
12. D. M. Levins, K. D. Reeve, W. J. Buykx, R. K. Ryan, B. W. Seatonberry and J. L. Woolfrey in Spectrum 1986 (Proc. Int. Topical Meeting on Waste Management, Am. Ceram. Soc., Pittsburgh, PA), p. 1137.
13. R. D. Shannon, Acta Cryst., A32, p. 751 (1976).

Structure and Characterization of Ceramics

EXAFS AND XANES ANALYSIS OF PLUTONIUM AND CERIUM EDGES FROM TITANATE CERAMICS FOR FISSILE MATERIALS DISPOSAL

J. A. Fortner, A. J. Kropf, A. J. Bakel, M. C. Hash, S. B. Aase, E. C. Buck, and D. B. Chamberlain
Argonne National Laboratory, Chemical Technology Division, Argonne, IL 60439

ABSTRACT

We report extended x-ray absorption fine structure (EXAFS) spectra from the plutonium L_{III} edge and x-ray absorption near edge structure (XANES) from the cerium L_{II} edge and plutonium L_{III} edge in prototype titanate ceramic hosts. The titanate ceramics studied are based upon the hafnium-pyrochlore and zirconolite mineral structures that will serve as an immobilization host for surplus fissile materials. Our samples approximate the composition envelope expected for production materials, which will contain as much as 10.5 weight % fissile plutonium and 21 weight % (natural or depleted) uranium. Three ceramic formulations were studied: one employed cerium as a "surrogate" element, replacing both plutonium and uranium in the ceramic matrix, another formulation contained plutonium in a "baseline" ceramic formulation, and a third contained plutonium in a formulation representing a high-impurity plutonium stream. The cerium XANES from the surrogate ceramic clearly indicates a mixed III-IV oxidation state for the cerium. In contrast, XANES analysis of the two plutonium-bearing ceramics shows that the plutonium is present almost entirely as Pu(IV) and occupies the calcium site in the zirconolite and pyrochlore phases. The plutonium EXAFS real-space structure shows a strong second-shell peak, clearly distinct from that of PuO_2, with remarkably little difference in the plutonium crystal chemistry indicated between the baseline and high-impurity formulations.

INTRODUCTION

We demonstrate use of XANES and EXAFS for characterizing the oxidation state and coordination environment of plutonium and cerium in prototype samples of the immobilization ceramic. Three versions of the ceramic were examined, representing pure and impure feedstocks of the fissile material, plus a cerium surrogate version containing no plutonium or uranium. The chemical formulations of the baseline ("A0") and high-impurity ("A9") titanate ceramics studied are provided in Table 1. The surrogate ceramic had a chemical composition similar to A0, with cerium substituted for plutonium and uranium on a molar basis. The A0 and A9 samples were both calcined in air (stagnant, 750 °C) and sintered in flowing argon at 1350 °C for 4 hours. The cerium surrogate sample was calcined and sintered in stagnant air (750 °C and 1350 °C, respectively). The major phases in these ceramics are pyrochlore [$A_2Ti_2O_7$], zirconolite [$ABTi_2O_7$], Hf-bearing rutile (TiO_2), and brannerite [BTi_2O_6], where A = Ca, actinides (ACT), and rare earth elements (REE), and B = ACT, REE, Zr, and Hf. Additional minor phases may occur depending on waste loading; these include uranium oxides and glassy phases. The glassy phases (and rutile) fill interstices between the major phases [1]. Both Hf and Gd are added to the ceramic formulation as neutron absorbers in order to satisfy a defense-in-depth concept for the waste form.

EXPERIMENT

The EXAFS measurements were made at the Advanced Photon Source (APS) using the Materials Research Collaborative Access Team (MRCAT) undulator beam line. Measurements were made in fluorescence mode with the incident intensity ionization chamber optimized for maximum current with linear response (~10^{10} photons detected /sec). The fluorescence ionization chamber was filled with xenon gas and produced a signal of ~10^8 photons/sec above the absorption edge. A double crystal Si (111) monochromator with resolution of better than 4 eV at 20 keV was used in conjunction with a Pt-coated mirror to minimize the presence of harmonics.

The counting time for the data ranged from 2 to 8 seconds per point in the EXAFS region. The experimental spectra were fitted using the program *FEFIT* from the University of Washington package, and *FEFF*, version 8.00, to generate the scattering paths [2].

Table 1. Ceramic compositions in weight percent.

Compound	Specimen Type		
	A0	A9	surrogate
CaO	9.95	9.44	14.52
TiO_2	35.87	34.04	39.76
HfO_2	10.65	10.11	11.82
Gd_2O_3	7.95	7.54	8.80
UO_2	23.69	22.48	-
PuO_2	11.89	11.28	-
CeO_2	-	-	25.10
Al_2O_3	-	0.50	-
MgO	-	0.44	-
$CaCl_2$	-	0.66	-
Ga_2O_2	-	0.57	-
Fe_2O_3	-	0.15	-
Cr_2O_3	-	0.08	-
NiO	-	0.13	-
CaF_2	-	0.44	-
K_2O	-	0.32	-
Na_2O	-	0.14	-
MoO_2	-	0.28	-
SiO_2	-	0.46	-
Ta_2O_5	-	0.19	-
B_2O_3	-	0.17	-
WO_2	-	0.49	-
ZnO	-	0.07	-

Standards for cerium were Ce(III)PO_4 and Ce(IV)O_2, while high-fired PuO_2 served as a plutonium standard. The Ce-L_{II} XANES from the cerium surrogate ceramic was clearly located between the Ce(III) and Ce(IV) standards (Figure 1). Fitting the edge structure with a linear combination of spectra from the Ce(III) and Ce(IV) standards yielded an oxidation state of 3.7, consistent with earlier studies on cerium-loaded Synroc [3]. Remarkably, the entire edge structure could be reasonably fit, including the edge position, the location and relative heights of the split peaks in the L_{II} spectrum, and the relative height of the continuum that occurs after the white line (Figure 2).

The near-edge structure (XANES) contains information on the formal oxidation state of the absorbing species. The XANES was fitted by a methodology described elsewhere that uses the sum of a Lorentzian and a hyperbolic tangent function [4]. The analysis of the plutonium edge in the prototype ceramics (Figure 3) demonstrates that it is almost entirely in the Pu(IV) oxidation state. The XANES was nearly identical for the A0 and A9 samples, both having the edge position and qualitative edge structure much like those of the PuO_2 standard [4,5].

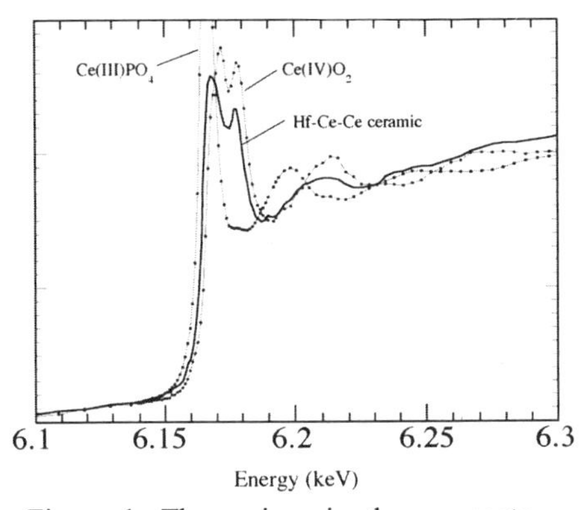

Figure 1. The cerium in the surrogate ceramic (solid line) has a mixed oxidation state, as illustrated by this comparison with XANES standards (lines with points). For clarity, the intense white line of the Ce (III) standard has been cut off in the figure.

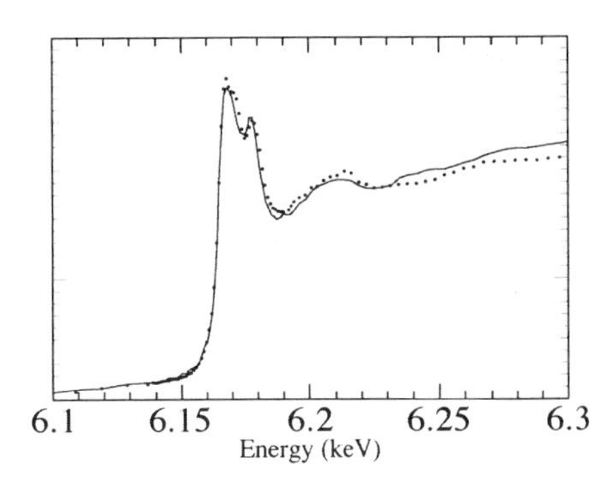

Figure 2. The cerium XANES from the ceramic (solid line) was fitted with a linear combination of 70% $Ce(IV)O_2$ and 30% $Ce(III)PO_4$ standards (dotted line).

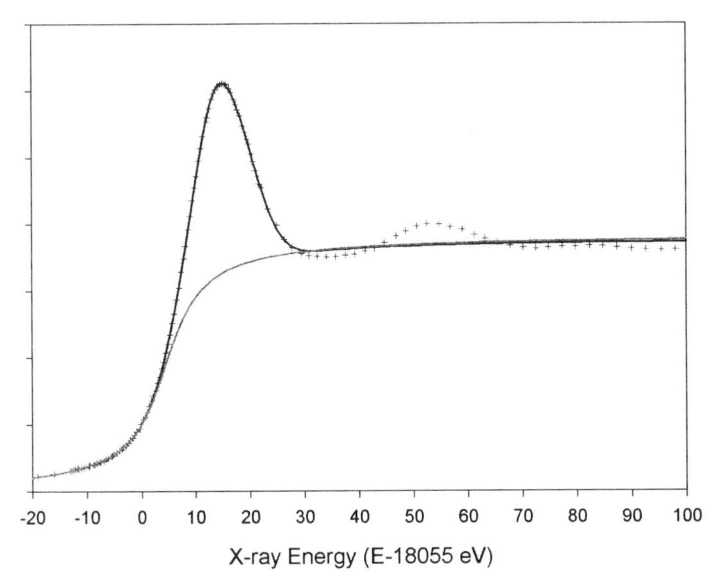

Figure 3.The plutonium XANES from the pyrochlore ceramic has an oxidation state very near Pu(IV), by comparison with PuO_2 reference data [4]. Shown is the data (hatches) with the fit generated as a sum (dark gray line) of a Lorentzian and hyperbolic tangent (light gray line) functions. For clarity, the Lorentzian function is not separately shown.

The EXAFS functions (Figure 4) were nearly identical for the A0 and A9 samples, indicating a robust preferred crystal chemical environment for the plutonium in the ceramic. The real-space analysis of the A0 ceramic in Figure 5 demonstrates a coordination environment for plutonium in the ceramic that is distinct from that in PuO_2 (Figure 6). Notably, the first coordination shell of oxygen occurs at slightly reduced bond length, with a second shell coordination environment due to titanium. Residual scattering beyond the titanium shell at 3.59 Å is due to higher coordination shells, most likely heavy elements such as Hf, Gd, U, or Pu. The data were fitted over the k-space range from 3.0 to 10.5 $Å^{-1}$, using k^3 weighting. Fitted parameters R (bond length), N (coordination number), and σ^2 (Debye-Waller and static disorder parameter) obtained are listed in Table 2. The values of the bond lengths in Table 2 differ from the apparent bond lengths in Figures 5 and 6 owing to a scattering phase shift. The phase shift was incorporated into the *FEFFIT* fitting of Table 2 but not in the Fourier transform displayed in Figures 5 and 6. The obtained structural information is consistent with Pu occupying the Ca site in the pyrochlore structure. Note that for the ideal (cubic) pyrochlore structure, the calcium and zirconium (hafnium) sites are indistinct, in contrast with previous studies on zirconolite [5]. The mechanism for charge compensation is likely cation vacancies, which are considered more energetically favorable than interstitial oxygen in these materials [3].

Table 2. Coordination results from Pu-L_{III} EXAFS analysis

Sample	R 1 (Å)	N1 (O)	σ^2	R 2 (Å)	N2 (Ti)	σ^2
A0	2.30 ± 0.05	9.5 ± 2.2	0.020 ±0.004	3.59 ± 0.02	9.0 ± 1.5	0.013 ±0.004
A9	2.30 ± 0.05	10.3 ± 2.5	0.022 ±0.004	3.59 ± 0.02	13.0 ± 2.5	0.018 ±0.004

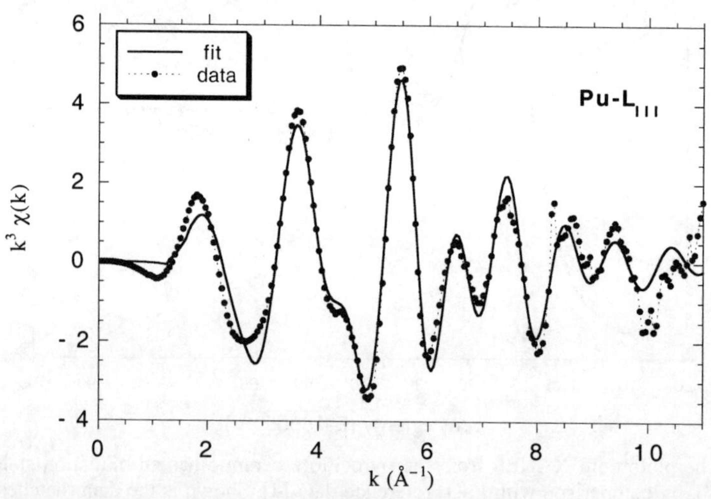

Figure 4. Plutonium L_{III} EXAFS data (dotted line) with fit obtained from *FEFF* 8 (solid line) for the A0 sample.

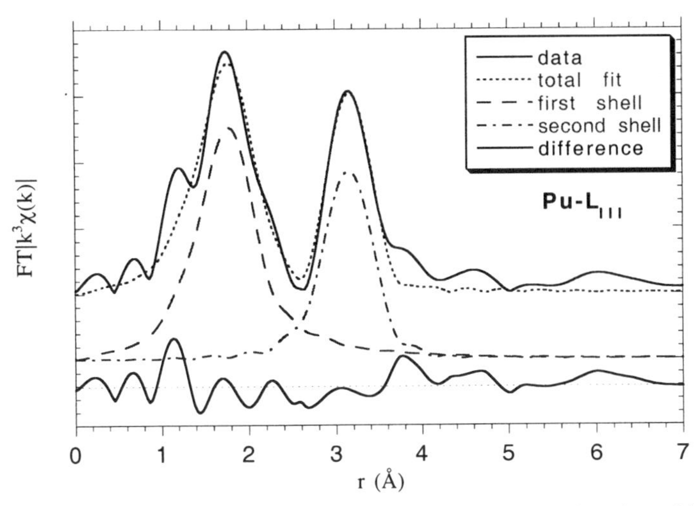

Figure 5. Plutonium EXAFS radial function with fits for the A0 sample. The residual contains substantial structure beyond the titanium shell originating from coordination by heavy elements. The individual shell fits and the residual are offset for clarity.

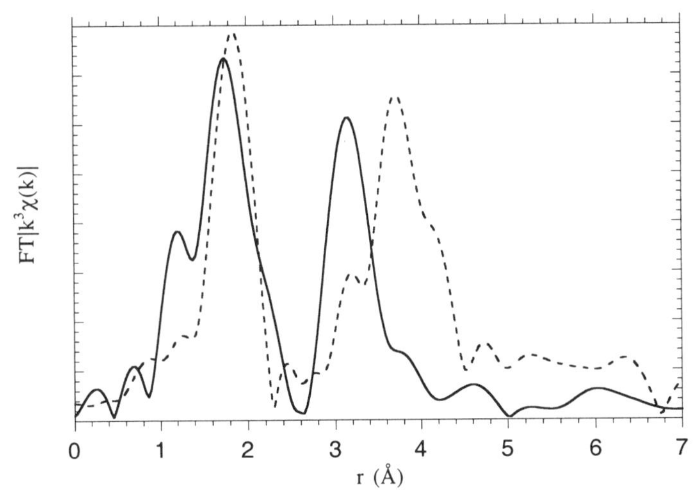

Figure 6. The EXAFS of the ceramic (solid line) shows a local structure distinct from that of PuO_2 [4] (dashed line).

CONCLUSIONS

We have demonstrated the use of EXAFS/XANES to establish the crystal chemistry of plutonium and cerium in a prototype immobilization ceramic. Plutonium is found to be nearly all Pu(IV) and to occupy a calcium site in the pyrochlore structure. Cerium, when substituted for both plutonium and uranium in the ceramic, has a formal valence of approximately 3.7, as determined by a fit of the Ce-L_{II} XANES with standards. Charge compensation for these substitutions is likely achieved by cation vacancies [3], although the present data cannot exclude other mechanisms (e.g., interstitial oxygen or formation of Ti(III)). The valence and coordination of uranium in these ceramics will be the topic of a future study [6].

REFERENCES

1. M. C. Hash, J. K. Basco, V. N. Zyryanov, and D. B. Chamberlain, Proceedings Paper for the 101st Am. Ceram. Soc. Annual Meeting, Indianapolis, 1999, *Environmental Issues and Waste Management Technologies in the Ceramic and Nuclear Industries V* (in press, due spring 2000).
2. Available from FEFF Project , c/o Todd Case, Department of Physics, Box 351560 University of Washington, Seattle WA, 98195, USA.
3. B. D. Begg, E. R. Vance, and G. R. Lumpkin, *Mat. Res. Soc. Symp. Proc.* **506**, 79 (1998).
4. A. J. Kropf, D. T. Reed, and S. B. Aase, submitted to *Journal of Synchrotron Radiation*.
5. S. R. Conradson, *Applied Spectroscopy* **52** (7), 252A (1998).
6. J. A. Fortner, A. J. Kropf, R. J. Finch, M. C. Hash, S. B. Aase, and D. B. Chamberlain, manuscript in preparation.

This work was supported by the U.S. Department of Energy Office of Fissile Materials Disposition. Use of the Advanced Photon Source was supported by the U.S. Department of Energy, Basic Energy Sciences, Office of Energy Research (DOE-BES-OER), under Contract no. W-31-109-Eng-38. The MRCAT beamlines are supported by the member institutions and the U.S. DOE-BES-OER under contracts DE-FG02-94ER45525 and DE-FG02-96ER45589.

PHASE COMPOSITIONS AND ELEMENTS PARTITIONING IN TWO-PHASE HOSTS FOR IMMOBILIZATION OF A RARE EARTH-ACTINIDE HIGH LEVEL WASTE FRACTION

S.V. STEFANOVSKY*, S.V. YUDINTSEV**, B.S. NIKONOV**, A.V. OCHKIN***, S.V. CHIZHEVSKAYA***, N.E. CHERNIAVSKAYA*
* SIA Radon, 7th Rostovskii per., 2/14, Moscow 119121 RUSSIA, itbstef@cityline.ru
**Institute of Geology of Ore Deposits, Russian Academy of Sciences, Staromonetni 35, Moscow RUSSIA
*** D. Mendeleev University of Chemical Technology, Miusskaya, 9, Moscow RUSSIA

ABSTRACT

In two-phase matrices based on pyrochlore-oxide, perovskite-oxide, and pyrochlore-zirconolite assemblages as much as 40 wt.% of the product can consist of an incorporated rare earth – actinide fraction of high level wastes (HLW). In zirconolite, with a nominal stoichiometry of $A^{VIII}B^{VII}C^{V-VI}{}_2O_7$ the actinide and rare earth ions occupy VIII- and VII-fold coordinated sites. Charge compensation is achieved by replacement of Ti^{4+} with lower valence ions such as Al^{3+}, $Fe^{2+/3+}$, Mg^{2+}, etc, that have similar radii. The highest lanthanide incorporation (45 wt.% ΣREE_2O_3) was exhibited by a calcium free zirconolite containing 8 wt.% Al_2O_3 or 0.7 formula units (f.u.). In pyrochlore ($A^{VIII}{}_2B^{VI}{}_2O_{7-x}$) actinides and rare earths occupy A-sites. Their incorporation does not require simultaneous ionic substitution in the B-sites, for example Al^{3+} or $Fe^{2+/3+}$ for Ti^{4+}, to achieve charge compensation; this simplifies the required composition of the system. Pyrochlore ceramics are suitable for immobilization of wastes with relative elevated actinide content. For a high zirconia, waste ZrO_2-based or two-phase pyrochlore-dioxide, perovskite-dioxide or pyrochlore-zirconolite ceramics are preferable. All these wasteforms may be produced via melting.

INTRODUCTION

High level wastes (HLW) contain numerous chemical elements with different properties that create problems when the waste is processed to obtain a waste form for dispodal. To facilitate treatment, HLW partitioning has been proposed in which the HLW would be partitioned to a short-lived Cs/Sr fraction and a long-lived REE-actinide rare earth element (REE)-actinide fraction that contains mainly lanthanides (La-Gd), actinides (U-Cm), and zirconium [1]. The REE-actinide fraction also can be partitioned to separate the actinides from the stable rare earth elements [2]. Compositions of the REE-actinide fraction produced in reprocessing of spent fuel from commercial and defense reactors are (in wt.%) 50-60 ΣLn_2O_3 (Nd>Ce> La≈Pr>Sm>Eu>Gd), 20-30 ZrO_2, 10-20 ΣAnO_2 (U>Np>Pu≈Am>Cm) [3] and 55 ΣAnO_2, 40 ΣLn_2O_3, 5 ZrO_2 [2], respectively.
Since actinide elements have long half-lives, they must be stored for period up to millions of years. This requires waste forms with excellent long-term stability. The best-known materials for actinide immobilization are zirconia- and titanate-based ceramics [4-14]. Zirconia-based solid solution with a fluorite-type lattice (space group *Fm3m*) is stabilized by heavy lanthanides (Gd-Lu) and yttrium [7], which are totally miscible in the system $(Zr,Ln,Y)O_{2-x}$-PuO_2 and other systems with actinide oxides with the same crystal structure. If light lanthanides, which predominate in HLW, are present in high concentration however, this structure type is unstable and rare earth and actinide zirconates with a pyrochlore-type lattice (space group *Fd3m*) are formed. Incorporation of high amount of actinides (U) and rare earths in the titanate ceramics results in complication of their phase composition with formation of extra structural varieties of zirconolite, pyrochlore, perovskite, oxides, etc. [9-16].
Formulations of most of the compounds suggested for actinide fixation may be described by the general system: CaO-TiO_2-ZrO_2-$(REE_2O_3$+$AnO_2)$. However, even ternary system CaO-TiO_2-ZrO_2 contains numerous phases, such as zirconium oxides, zirconolite, calzirtite, perovskite, zirconium titanate, and others [11]. Due to its complex elemental composition, studies of the basic system require numerous labor extensive experiments. Alternative partial systems can be studied whose compositions are specified by consideration of the nominal stoichiometry of the phases potentially capable of incorporating actinides. Varying

Mat. Res. Soc. Symp. Proc. Vol. 608 © 2000 Materials Research Society

relationships between the limits for major and minor elements, limits for the content of actinides or their surrogates in the phases, as well as phases formed outside these limits may be determined. In this way the solubility of actinides (uranium and thorium) and REEs in cubic zirconia, zirconolite, pyrochlore, and perovskite has been investigated [12-16]. In these studies solid solutions with only one additional element were studied. In the present work perovskite-, pyrochlore-, and zirconolite-based ceramics containing an REE-actinide fraction surrogate were examined. The goal of this work is to study the behavior of elements in ceramics composed of oxide and titanate phases with additives simulating the composition of the REE-actinide fraction of HLW. These data are required to optimize compositions of actinide wasteforms.

EXPERIMENTAL

Batches of the ceramics to be studied were prepared from oxides ground in an agate mortar. The formulations are given in Table I. Powders were mechanically milled and activated in an apparatus with a rotating magnetic field (linear induction rotator). The resulting powders were pressed into pellets at 300 MPa. The pellets were melted in air in alumina crucibles in a resistive furnace at 1550 ^{0}C for 20 minutes (except for Sample #7, which was held at this temperature for 180 minutes), then the temperature was reduced to 1100 ^{0}C at a rate of 10 ^{0}C/min to crystallize the melts, the furnace was then turned-off and the samples cooled to room temperature.
Ceramics containing of 20 to 50 wt.% of a surrogate of the REE-actinide fraction from spent fuel reprocessing at the Russian nuclear plant "Mayak" were produced. Initial formulations were assuming that Ce exists as Ce^{3+}. In Samples ## 1–7 Nd, Sm, and Eu were replaced by Gd, which is also a neutron absorber and especially introduced into some ceramic formulations for excess weapon plutonium disposition [17].
The samples were examined by scanning and transmission electron microscopy (SEM and TEM, respectively) equipped with energy dispersive system (EDS) using a JSM 5300 + Link ISIS unit, and X-ray diffraction (XRD) using a DRON-4 diffractometer (Cu K_α radiation). In the phase formulae calculated from the SEM/EDS data, the Ce valence state was assumed to be Ce^{3+} for pyrochlore and perovskite and both Ce^{3+} and Ce^{4+} for fluorite-structured oxides. For Sample #8, Eu and Mn are assumed to be divalent.

RESULTS

After heat-treatment of the pellets at 1550 ^{0}C Samples ## 1–3 were found to be sintered but not melted. Sample #4 was partially melted but maintained its pellet form. Samples ## 5–8 were fully melted.
Formulations of Samples ## 1 and 2 were predicted to form an REE-perovskite – U,Zr-dioxide assemblage (Table I). As seen from the XRD data (Figure 1), these samples are composed of REE-perovskites with the general formula $REEAlO_3$ (REE = La, Ce, Gd) and a fluorite-structured (space group *Fm3m*) oxide $(Zr,U,Ce)O_2$ with unit cell dimension a = 5.26 Å and a = 5.23 Å for Samples #1 and #2, respectively.
In Samples ## 3, 4, and 8, a complex zirconium–uranium oxide and pyrochlore-structured REE-calcium-uranium titanate phase have been found (Figure 1). A cubic fluorite-structured oxide (*Fm3m*) is responsible for the major reflections (in Å: d_{111} = 3.00, d_{200} = 2.61, d_{220} = 1.85, and d_{311} = 1.58) on the XRD patterns that correspond to unit cell dimension of a = 5.20 Å. The second phase in Sample #4 is a complex pyrochlore-structured oxide (*Fd3m*) responsible for the reflections d_{111} = 5.87, d_{311} = 3.09, d_{222} = 2.94, d_{400} = 2.56, d_{440} = 1.815, and d_{622} = 1.551. This yields a = 10.184 Å, which is very close to the value for pure $Gd_2Ti_2O_7$ (a = 10.186 Å). Moreover, Sample #8 has been determined to be perovskite with a presumed cubic structure (*Pm3m*). The major reflections are d_{110} = 2.69, d_{200} = 1.91, d_{211} = 1.556, d_{111} = 2.20, and d_{220} = 1.349. The simple set of reflection on the XRD pattern testifies in favor of a cubic rather than a rhombic (*Pnma*) lattice for this phase.
The relative high porosity of the samples (Figure 2) is probably due to a short melting time that was insufficient to remove gas bubbles from the melt. The analytical sum of components is close to 100 wt.%. Cerium in "pyrochlore" and as the oxide probably exists as Ce^{3+}. The phase formulations for Samples #3 and #4 were recalculated as $(Gd_{1.26}Ce_{0.33}La_{0.21}Zr_{0.18}U_{0.02})(Ti_{1.50}Zr_{0.50})O_{7.26}$ ("pyrochlore") and $(Zr_{0.46}Gd_{0.23}Ce_{0.17}Ti_{0.08}U_{0.04}La_{0.02})O_{1.88}$ (oxide). The oxide phase is enriched with Zr and U, whereas the pyrochlore-type phase accumulates predominantly Ti, La, and Gd.
The composition of Sample #5 was tailored to obtain $REEZrTiAlO_7$ formulation where rare earths

Table I. Chemical and phase compositions of the samples

	Oxides	1	2	3	4	5	6	7	8
Calculated oxide content, wt.%	CaO	-	-	-	-	-	7,7	7,7	11,0
	MnO	-	-	-	-	-	-	-	6,0
	Al_2O_3	17,0	10,8	-	-	11,8	11,4	11,4	3,0
	La_2O_3	7,3	7,9	6,6	7,3	5,1	5,2	5,2	-
	Ce_2O_3	14,0	15,0	12,6	13,9	9,7	10,0	10,0	16,0*
	EuO	-	-	-	-	-	-	-	5,0
	Gd_2O_3	36,3	39,1	33,1	36,6	25,3	25,7	25,7	6,0
	TiO_2	-	-	24,7	16,6	19,0	21,9	21,9	37,0
	ZrO_2	21,8	23,5	19,8	21,9	26,6	15,6	15,6	5,0
	UO_2	3,6	3,9	3,3	3,7	2,5	2,6	2,6	11,0
Phase composition	Specified	$REEAlO_3$ (P), $(Zr,U)O_2$ (O)		$REE_2Ti_2O_7$ (Py), $(Zr,U)O_2$		$REEZrTiAlO_7$ (Z)	$REEAlO_3$, $Ca(Zr,U)Ti_2O_7$		Py
	Actual	$REEAlO_3$ (P), $(Zr,U)O_2$ (O)		$REE_2Ti_2O_7$ (Py), $(Zr,U)O_2$ (O)		Z-2M, O, Py, P	Py, Z, P		Py, P, H, O

H – hibonite, O – cubic oxide, P – REE-perovskite, Py – "pyrochlore", Z – zirconolite, *CeO_2.

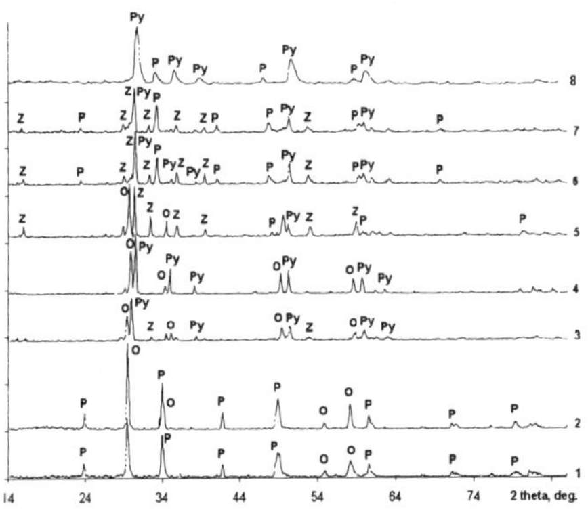

Figure 1. XRD patterns of the samples.
O – cubic oxide $(Zr,U,Ce)O_2$, P – REE-perovskite, Py – pyrochlore-structured phase, Z – zirconolite.

substitute for calcium with simultaneous incorporation of Al instead of part of Ti by scheme: $Ca^{2+} + Ti^{4+} = REE^{3+} + Al^{3+}$. The sample is composed of approximately equal amounts of the cubic and monoclinic phases (Figures 1 and 2, and Table II). The analytical sums of the components are close to 100 wt.% and the average formulae are $(Zr_{0.57}Gd_{0.22}\ Ce_{0.10}Ti_{0.08}\ U_{0.02}La_{0.01})O_{1.89}$ and $(Gd_{0.90}La_{0.10})(Zr_{0.91}\ Ce_{0.13})(Ti_{1.20}\ Al_{0.69})O_{7.00}$ respectively. The monoclinic phase is zirconolite with typical reflections: d_{311} = 3.24, d_{221} = 2.94, d_{004} = 2.76, d_{402} = 2.505, d_{404} = 2.276. Individual grains enriched with cerium and with noticeable mass imbalance (Table II) also occur. This imbalance is probably due to existence of a significant fraction of the Ce in trivalent form.

The TEM pattern of the monoclinic phase (Figure 2) is the same as the TEM pattern of zirconolite-2M variety including those with high rare earths content [18]. This phase is characterized by absence of calcium whose sites are occupied with rare earths. In the reference data on formulations of synthetic zirconolites, CaO content was as low as 7 wt.% (0.55 f.u.) [15]. In nature, a minimum amount of Ca in zirconolites was found in zirconolites from lunar basalts (average and lowest values were 0.35 and 0.21 f.u. respectively) [19]. These zirconolites also contain the highest amount of rare earths (lanthanides and yttrium) (average and the highest values are 0.51 and 0.79 f.u. respectively). Ca and REE ions occupy the same sites in the crystal lattice, therefore their content in the zirconolite is inversely correlated. This provides a basis to suggest that the last member incorporated in the zirconolite solid solutions is $REEZrTiMe^{3+}O_7$ formulation [16,19] in which the isomorphism scheme $Ca^{2+} + Ti^{4+} = REE^{3+} + Me^{3+}$ (Me = Fe, Al) is realized. Experimental studies of compounds with the general formula $Ca_{1-x}(Ce,Y)_xTi_{2-x}Fe_xO_7$ within the range $0 \leq x \leq 1$ has shown that in the cerium-bearing system a single phase zirconolite ceramic is formed at $x \leq 0.5$ [16]. At $x > 0.5$, an extra cerianite phase (CeO_2) occurred. Wide variations of solid solution compositions occur when Y substitutes for Ce, up to formation of a phase with $YZrTiFe^{3+}O_7$ formulation which is isostructural with zirconolite.

The compositions of Samples ## 6 and 7 are predicted to result a two-phase $REEAlO_3$ – $Ca(Zr,U)Ti_2O_7$ assemblage. The actual phase composition is close to the predicted one (Figure 1). A longer melting time

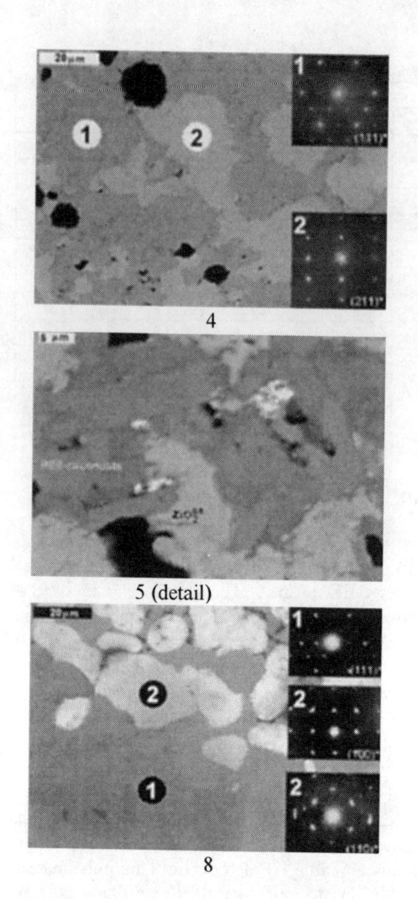

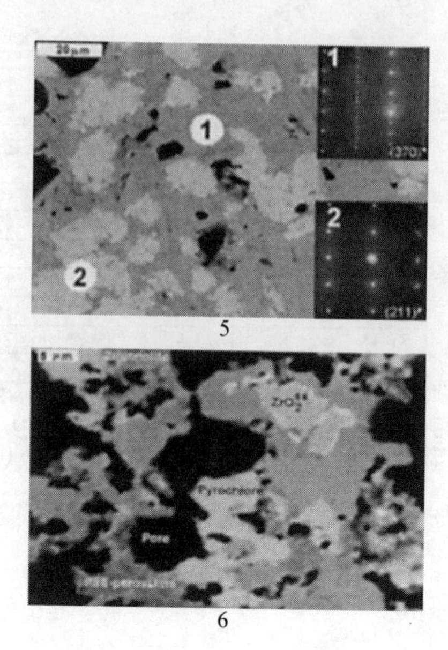

Figure 2. SEM images in backscattered electrons of the ceramic samples ## 4, 5, 6, and 8.

1 – pyrochlore (## 4 and 8) or zirconolite (#5), 2 – oxide. Inserts are TEM (electron diffraction) patterns.

did not affect phase composition (Figure 1 and Table I). In Sample #6 "pyrochlore", zirconolite, and REE-perovskite were major phases and a zirconia-based cubic solid solution enriched with Ce, Gd, and U was a minor phase. The total analyzed amount of the components in each phase differs noticeably from 100 wt.% due to the occurrence of Ce in both oxidation states Ce(III) and Ce(IV) (Table II).

Sample #8 has a predicted composition (Table I) corresponding to pyrochlore stoichiometry: $A_2B_2O_{7-x}$ (x=0.4, A = Ca, Mn, REE, U, Zr; B = Ti, Al, Mn). The melted sample is composed of "pyrochlore" and perovskite as well as a minor fluorite-structured oxide and calcium aluminotitanate (hibonite). The "pyrochlore" and perovskite compositions were found to be $(Ca_{0.66}Ce_{0.31}Mn_{0.37}U_{0.45}Eu_{0.10}Gd_{0.12})(Ti_{1.73}Zr_{0.15}Al_{0.11})O_{6.76}$ and $(Ca_{0.28}Ce_{0.20}Mn_{0.14}Zr_{0.10}U_{0.07}Eu_{0.06}Gd_{0.06})(Ti_{0.88}Al_{0.03})O_{3.00}$, respectively. The SEM images of these phases are very similar (Figure 2) which makes it difficult to characterize their distribution and amounts in the sample. Unlike the previous samples, in Sample #8 the oxide phase is enriched with U and depleted in Zr. The average oxide and hibonite formulae are $U_{0.41}Ce_{0.22}Ca_{0.12}Gd_{0.06}Eu_{0.05}Zr_{0.05}Al_{0.05}Ti_{0.02}Mn_{0.02}O_{1.75}$ and $(Ca_{0.92}Ce_{0.24}U_{0.03})(Al_{9.46}Ti_{1.14}Mn_{0.87}Zr_{0.08})O_{19.00}$, respectively. The total amount of these two phases in the sample doesn't exceed a few percent each.

The TEM patterns of the inverse lattices of these phases correspond to fluorite and pyrochlore structure types for Samples ## 4 and 8 and fluorite and zirconolite structure types for Sample #5 (Figure 2). As noted above, the electron diffraction pattern for zirconolite demonstrates a monoclinic symmetry of the crystal lattice which is assigned to zirconolite-2M. The characteristic of the electron diffraction pattern of the oxide phase in Sample #8 is the diffusive character of the reflections with elements of a circular structure (Figure 2). The Sample #8 grains are probably aggregates of the finest randomly oriented

Table II. Chemical compositions of co-existing phases in Samples

Oxides	#4		#5			#6					#8[1]			
	Py	O	Z	Zr-O	Ce-O	Py	P	Z	Zr-O	O	P	Py	Ce,U-O	H
Al_2O_3	-	-	7.8	0.3	0.1	3.0	3.0	6.0	0.7	0.1	0.8	1.3	1.5	64.4
CaO	-	-	-	-	-	7.7	21.8	6.0	1.4	0.7	8.5	8.4	3.7	6.9
TiO_2	22.9	4.4	21.3	4.6	5.7	27.3	42.0	25.7	3.5	0.7	38.7	31.4	0.8	8.2
MnO	-	-	-	-	-	-	-	-	-	-	5.4	6.0	0.9	8.3
ZrO_2	16.1	39.1	24.9	49.6	19.1	7.1	0.9	32.3	77.5	13.9	6.5	4.3	3.0	1.1
La_2O_3	6.6	2.1	3.6	1.0	5.7	4.7	4.8	0.4	0.9	1.0	-	-	-	-
Ce_2O_3	10.2	18.9[2]	5.6	11.5[2]	36.7[2]	11.4	7.5	4.4	8.1[2]	32.6[2]	18.6	12.1	20.3[2]	5.8
Gd_2O_3	43.8	28.5	36.3	29.2	12.6	31.0	17.5	21.9	9.4	23.4	5.7	4.7	5.6	<0.6
EuO	-	-	-	-	-	-	-	-	-	-	5.7	4.0	5.0	<0.7
UO_2	0.8	6.7	0.4	4.4	16.7	2.3	0.2	1.1	3.3	20.3	10.1	27.8	59.2	1.3
Total	100.4	99.7	99.9	100.6	96.6	94.5	97.7	97.8	104.8	92.7	100.0	100.0	100.0	100.0

[1] – analytical sum of components are recalculated to 100%. [2] – recalculated to CeO_2.

crystallites. Similar electron diffraction patterns are observed for partially metamictized phases having domain structure [6].

DISCUSSION

In cubic oxides with a fluorite structure and a general formula $[Zr,Ce^{4+},An^{4+},(An,REE)_x^{3+}]^{VIII}O_{2-0.5x}$, major isomorphic exchanges are $Zr^{4+} = An^{4+}$, $Zr^{4+} = Ce^{4+}$, and $Zr^{4+} = 2REE^{3+} + O^{2-}$. Ti and Ca are also present but in minor concentrations due to the significant difference in the radius of Zr^{4+} compared with that of Ti^{4+} and Ca^{2+} ions. Variations in the unit cell dimensions of the oxide phase in the samples (5.20-5.26 Å) are probably due to the differences in the composition. The unit cell dimension increases as heavy ion (Ce, Gd, U) content grows.

In zirconolite with a generalized formula $(Ca,REE,An)^{VIII}(Zr,An,REE)^{VII}(Ti,Al)^{IV-VI}{}_2O_7$, major isomorphic substitutions are $Zr^{4+} = An^{4+}$, $Ca^{2+} + 2Ti^{4+} = An^{4+} + 2Al^{3+}$, and $Ca^{2+} + Ti^{4+} = (An,REE)^{3+} + Al^{3+}$. We have synthesized a zirconolite in which all eight-coordinated sites are filled with rare earth ions. The number of Al ions in the formula is less than one. Therefore, along with the exchange $Ca^{2+} + Ti^{4+} = REE^{3+} + Al^{3+}$, one more scheme of isomorphic exchange – $Ca^{2+} + Zr^{4+} = 2REE^{3+}$ – should be assumed to agree well with the zirconolite formula calculated from the analytical data.

The pyrochlore structural type is characteristic of compounds formed in the system: $CaUTi_2O_7$-$REE_2Ti_2O_7$-$REE_2Zr_2O_7$. In general "pyrochlore" formula $A^{VIII}{}_2B^{VI}{}_2O_7$ actinides and rare earths are positioned in A-sites. The major isomorphic exchange is $Ca^{2+} + U^{4+} = 2REE^{3+}$. Since actinides and REEs occupy the same sites in the crystal lattice, an inverse correlation of their concentrations occurs. Unlike zirconolite, incorporation of rare earths into a pyrochlore lattice does not require a balance Ti replacement by lower charge cations, for example Al^{3+} or Fe^{3+}. Therefore, the Al content in pyrochlores is lower than in zirconolites. The $Zr^{4+} = Ti^{4+}$ exchange is responsible for the incorporation of Zr in pyrochlore.

Perovskite with a general formula $A^{XII}B^{VI}O_3$ (A = Na, Ca, Sr, REE, An; B = Ti, Nb, Al) can accumulate up to 40 wt.% ΣREE_2O_3 and 10 wt.% AnO_2. The perovskite $CaTiO_3$ crystal lattice has rhombic symmetry. A high REEs content stabilizes the cubic lattice. Major isomorphic substitutions in the perovskite structure are $Ca^{2+} + Ti^{4+} = REE^{3+} + (Ti,Al)^{3+}$, $3Ca^{2+} = 2\ REE^{3+}$ + vacancy, $Ca^{2+} = (Eu,Mn)^{2+}$, and $Ti^{4+} = (Ce,Zr,U)^{4+}$. A synthetic perovskite $Ce_2Ti_3O_{8.7}$, in which trivalent cerium completely replaces Ca also exists (JCPDS 20-272). A feature of the perovskite in Sample #8 is the high content of U, Zr, and REEs, and the low Ca concentration. A perovskite with a similar formulation $Ca_{0.14}Nd_{0.49}Zr_{0.10}Ti_{0.94}Al_{0.02}O_{3.00}$ was found earlier in titanate ceramics having a zirconolite-perovskite-pseudobrookite phase compositions produced by cold crucible melting [19].

The formula for hibonite is $A^{XII}B^{IV-VI}{}_{12}O_{19}$. The A and B sites are occupied by Ca^{2+} and Al^{3+} ions, respectively. Rare earths and titanium substituting for calcium and aluminum are present as minor components. Actinide incorporation is negligible due to the absence of suitable structural positions [9]. On the whole, the given phase is depleted of waste elements and its role in ceramic is insignificant.

We have calculated atomic relationships ΣREE:Zr:ΣAn for two of simulated waste compositions with different An and Zr contents (Table III). As seen from these data, the atomic relationship in the fluorite-structured zirconia-based phase (Sample #4) is closest to the relationship in Waste #1 whereas the atomic relationship in the pyrochlore-structured phase (Sample #8) is closest to the relationship in Waste #2 with high actinide content. The atomic ratio in zirconolite is strongly shifted in favor of zirconium. Therefore, as shown earlier [8,13-15,21] actinide oxides and pyrochlore are formed upon immobilization of actinide waste and excess plutonium in zirconolite matrices. Pyrochlore-based ceramics have been selected at LLNL for excess weapons plutonium disposition [17].

Table III. Atomic ratios ΣREE:Zr:ΣAn in REE-actinide fraction and various phases

Composition o the REE -An-Zr fraction	60% REE - 30% Zr - 10% An [3]	1.0 : 0.75 : 0.11
	40% REE - 5% Zr - 55% An [2]	1.0 : 0.18 : 0.92
Cubic oxide	#4	1.0 : 1.10 : 0.10
	#5	1.0 : 1.73 : 0.06
	#8	1.0 : 0.15 : 1.24
Zirconolite	#5	1.0 : 3.17 : -
"Pyrochlore"	#4	1.0 : 0.38 : 0.02
	#8	1.0 : 0.28 : 0.85
Perovskite	#8	1.0 : 0.30 : 0.22

CONCLUSIONS

Ceramics for immobilization of REE-actinide fraction of HLW composed of two major phases: REE-perovskite – Zr-U dioxide, "pyrochlore" – Zr-U dioxide, "pyrochlore – REE perovskite, and zirconolite – Zr-U dioxide have been calculated and produced via melting. The oxide phase is enriched with Zr and U whereas the pyrochlore-structured phase and the zircinolite concentrate Ti, La, Ce, and Gd. In the "pyrochlore" – REE- perovskite ceramic uranium is concentrated in the pyrochlore-type phase, while rare earths are partitioned among both the phases.

REFERENCES

1. E.G. Drozhko, A.P. Suslov, V.I. Fetisov in *High Level Radioactive Waste and Spent Fuel Management* (ASME, 1993), **2**, p.17-20.
2. V.N. Romanovski, I.V. Smirnov, A.Yu. Shadrin in *Spectrum '98, Proc. Int. Conf.* (ANS, La Grange Park, 1998), **1**, p. 576-580.
3. C.G. Sombret in *The Geological Disposal of High Level Radioactive Wastes*, (Theoph. Publ., Athens, 1987), p. 69-159.
4. R.B. Heimann, T.T. Vandergraaf, J. Mat. Sci. Lett., **7**, 583 (1988).
5. E.R. Vance, B.D. Begg, R.A. Day, C.J. Ball, Mat. Res. Soc. Symp. Proc. **353,** 767 (1995).
6. R.C. Ewing, W.J. Weber, F.W. Clinard, Progr. Nucl. Energy. **29**, 63 (1995).
7. E.R. Maddrell, Mat. Res. Soc. Symp. Proc. **412**, 353 (1996).
8. H. Yokoi, T. Matsui, H. Ohno, K. Kobayashi, Mat. Res. Soc. Symp. Proc. **353**, 783 (1995).
9. P.E. Fielding, T.J. White, J. Mater. Res. **2**, 387 (1987).
10. S.S. Shoup, C.E. Bamberger, Mat. Res. Soc. Symp. Proc. **412**, 379 (1996).
11. D. Swenson, T.G. Nieh, J.H. Fournelle, Mat. Res.Soc. Symp. Proc. **412**, 337 (1996).
12. O.A. Knyazev, S.V. Stefanovsky, S.V. Yudintsev, et.al., Mat. Res. Soc. Symp. Proc. **465**, 401 (1997).
13. S.V. Yudintsev, B.I. Omelianenko, S.V. Stefanovsky, et.al., J. Adv. Mat. *1*, 91 (1998).
14. S.V. Stefanovsky, S.V. Yudintsev, A.V. Ochkin, et.al., Mat. Res. Soc. Symp. Proc. **506**, 261 (1998).
15. E.R. Vance, D.K. Agrawal, Nucl. Chem. Waste Manag. **3**, 229 (1982).
16. S.E. Kesson, W.J. Sinclair, A.E. Ringwood, Nucl. Chem. Waste Manag. **4**, 259 (1984).
17. T. Gould, B. Myers, L. Gray, T. Edmunds, in *Proc. 3-rd Top. Meet. on DOE Spent Nuclear Fuel and Fissile Materials Management* (Charleston, SC, 1998), p. 366-373.
18. T.J. White, Amer. Mineral. **69**, 1156 (1984).
19. R. Giere, C.N. Williams, G.R. Lumpkin, Schweiz. Miner. Petrol. Mitt. **78**, 433 (1998).
20. T. Advocat, C. Fillet, J. Marillet, G. Leturcq, et.al., Mat. Res. Soc. Symp. Proc. **506**, 55 (1998).
21. A. Jostsons, E.R. Vance, R.A. Day in *Spectrum '96, Proc. Int. Conf.* (ANS, La Grange Park, 1996), p. 2032-2039.

A CORRELATION BETWEEN RELATIVE CATION RADIUS AND THE PHASE STABILITY OF ZIRCONOLITE

D. SWENSON*, P. TRIYACHAROEN
Department of Metallurgical and Materials Engineering, Michigan Technological University
1400 Townsend Drive, Houghton, MI 49931
*dswenson@mtu.edu

ABSTRACT

The maximum solubilities of several different cations in zirconolite ($CaZrTi_2O_7$) were investigated using X-ray diffraction and electron probe microanalysis. A parameter termed the relative radius ratio, defined as the ratio of the average radius of all atoms substituting for Ca and Zr to the average radius of all atoms substituting for Ti, was calculated for each chemical system, using coordination-dependent ionic radii obtained from the literature. It was found that with the possible exception of systems containing Al^{3+}, regardless of the chemical system studied, and for additional chemical systems described in the literature, the relative radius ratio for zirconolite is 1.59, with a less than one percent standard deviation, at its solubility limit. Similar strong correlations were found between relative radius ratio and phase stability for crystallographically related rhombohedral and pyrochlore structures that also appear in these systems. These results suggest that the phase stability of zirconolite is generally governed by geometry and that chemical effects are of secondary importance.

INTRODUCTION

As the most durable phase in Synroc, and with a significant capacity to incorporate actinide and rare earth elements into solid solution, zirconolite ($CaZrTi_2O_7$) has received much interest as a potential wasteform for the immobilization of excess weapons grade Pu produced from the dismantling of nuclear weapons[1]. Several studies have focused on ascertaining the solubility limits of various cationic species in zirconolite, including Ce as a Pu surrogate and various additional elements which may serve as neutron absorbers that will help in avoiding criticality upon heavy loading (see, for example, [1-8] and references contained therein). While much has been learned in these studies, it has not yet proven possible, based on available information, to devise a scheme for predicting the maximum solubility of an arbitrary assemblage of substituting cationic species in zirconolite, although such an ability would prove invaluable for practical situations where many different cationic species would be incorporated into zirconolite simultaneously. One logical starting point for such an endeavor is to consider geometric factors. It is well known (*e.g.*, predictions based on Pauling's rules) that the relative sizes of cations and anions in an ionically bonded material, in addition to charge neutrality requirements, affect the coordination of each ionic species and this in term will have an impact on the crystal structure adopted by a material.

Most of the phases formed upon complete substitution in zirconolite ($CaCeTiO_7$, $Sm_2Ti_2O_7$, and $Gd_2Ti_2O_7$) possess the pyrochlore structure[9,10]. It is well known that the zirconolite and pyrochlore structures are crystallographically related[11]. Pyrochlores possess the general stoichiometry $A_2B_2O_7$, and those most relevant to the present study have the formula $A_2^{3+(VIII)}Ti_2^{4+(VI)}O_7$, where the Roman numerals associated with each type of cation denote its coordination number within the structure. (By contrast, zirconolite has the formula $Ca^{2+(VIII)}Zr^{4+(VII)}Ti_2^{4+(VI)}O_7$. Actually, a certain small fraction of the Ti have a coordination number of five, which is neglected in the present study[11,2,3].) Owing to this crystallographic similarity, prior observations on the phase stability of the relatively common pyrochlore structure might provide insight into the phase stability of zirconolite, or for that matter any other crystallographically related structure, such as the rhombohedral phase referred to above. Subramanian *et al.*[12] have compiled most data available on the pyrochlores through 1982. They observed that for the general pyrochlore formula $A_2B_2O_7$, the radius ratio of A to B (which will be denoted here as $RR = r^{A(VIII)}/r^{B(VI)}$) of all known pyrochlores falls within the range $1.463 \leq RR \leq$

Mat. Res. Soc. Symp. Proc. Vol. 608 © 2000 Materials Research Society

1.783, as calculated using the tables of ionic radii compiled by Shannon[13]. For the titanate pyrochlores, $A_2^{3+(VIII)}Ti_2^{4+(VI)}O_7$, the observed range of RR is substantially more narrow, extending from 1.615 for $Lu_2Ti_2O_7$ to 1.783 for $Sm_2Ti_2O_7$.

In the present study, this general geometric correlation for pyrochlore phase stability has been modified, extended and applied to zirconolite. Phase diagram samples were made to determine the compositions of the phase boundaries of zirconolite in various chemical systems. The radius ratios at these phase boundaries were calculated using Shannon's tables of ionic radii for various coordination numbers and experimentally measured compositions at the phase boundaries. For a given phase, these limiting radius ratios were then compared across chemical systems.

The general correlation scheme for zirconolite phase stability will be to determine the ratio of the average radius of all cations that lie on the $Ca^{2+(VIII)}$ and $Zr^{4+(VII)}$ lattice sites (weighted according to composition) to the average radius of all cations that lie on the $Ti^{4+(VI)}$ sites. Three basic equations have been derived, each corresponding to a specific type of substitution scheme.

For a direct substitution of a tetravalent cation (such as Ce^{4+} or Hf^{4+}),

$$A^{4+(VII)} \longrightarrow Zr^{4+(VII)} \qquad [1]$$

This corresponds to a general formula $CaA_xZr_{1-x}Ti_2O_7$, (where $0 < x < 1$) and the equation for radius ratio may be expressed in terms of x and the radius of A^{4+} in seven-fold coordination:

$$RR = \frac{0.19 + (r_A^{4+(VII)} - 0.078)x}{0.121} \qquad [2]$$

For a coupled substitution of the type

$$A^{3+(VIII)} + A^{3+(VII)} \longrightarrow Ca^{2+(VIII)} + Zr^{4+(VII)} \qquad [3]$$

sample compositions correspond to the formula $Ca_{1-x}A_{2x}Zr_{1-x}Ti_2O_7$ $(0 < x < 1)$, and it is assumed that there is a one-to-one substitution of A^{3+} for Ca^{2+} and Zr^{4+} in order to maintain electroneutrality.

In the present study, this substitution scheme applied to Sm^{3+} and Gd^{3+}, but would also be relevant to Ce^{3+}. The equation for radius rato depends not only on x but on the cationic radius of A^{3+} in both 7- and 8-fold coordination:

$$RR = \frac{0.19 + (r_A^{3+(VIII)} + r_A^{3+(VII)} - 0.19)x}{0.121} \qquad [4]$$

Finally, for charge-compensated coupled substitution (applying here to $Ce^{3+} + Al^{3+}$)

$$A^{3+(VIII)} + B^{3+(VI)} \longrightarrow Ca^{2+(VIII)} + Ti^{4+(VI)} \qquad [5]$$

In this scheme, the A^{3+} cation substitutes for Ca^{2+}, while at the same time a B^{3+} cation replaces a Ti^{4+} cation in order to preserve electroneutrality. The radius ratio equation then becomes:

$$RR = \frac{0.19 + (r_A^{3+(VIII)} - 0.112)x}{0.121 + (r_B^{3+(VI)} - 0.0605)x} \qquad [6]$$

More elaborate equations may be constructed for simultaneous substitution of multiple cation types, using the basic definition of radius ratio as described above. Two will be given explicitly here. The first is a combination of schemes one and two (corresponding to Eqns. [1] and [3]). There is a coupled substitution of A^{3+} for Ca^{2+} and Zr^{4+}, and additionally a direct substitution of a tetravalent cation, B^{4+}, for Zr^{4+}. In the present study, this scheme applied to the simultaneous coupled substitution of a trivalent neutron absorbing element (such as Sm^{3+} or Gd^{3+}) plus the surrogate material for tetravalent Pu, Ce^{4+}:

$$A^{3+(VIII)} + A^{3+(VII)} + B^{4+(VII)} \longrightarrow Ca^{2+(VIII)} + 2Zr^{4+(VII)} \qquad [7]$$

Under the special restriction that the number of moles of A and B cations substituted must be equal, this corresponds to the formula $Ca_{1-x}A_{2x}B_{2x}Zr_{1-3x}Ti_2O_7$ $(0 < x < 1/3)$. The radius ratio equation is

$$RR = \frac{0.19 + (r_A^{3+(VIII)} + r_A^{3+(VII)} + 2r_B^{4+(VIII)} - 0.346)x}{0.121} \qquad [8]$$

Finally, for the coupled substitution of A^{3+} for Ca^{2+} and additional charge compensated coupled substitution of B^{3+} and C^{3+} for Ca^{2+} and Ti^{4+} (a modification of the third scheme described above):

$$A^{3+(VIII)} + B^{3+(VIII)} + 2C^{3+(VI)} \longrightarrow 2Ca^{2+(VIII)} + 2Ti^{4+(VI)} \qquad [9]$$

This would correspond to the situation where trivalent Pu surrogate material (Ce^{3+}) was being added simultaneously with a trivalent neutron absorbing element (Sm^{3+} or Gd^{3+}) with additional substitution of charge compensating elements (in this case, Al^{3+}) in order to maintain charge neutrality. Its formula, under the condition of addition of equal numbers of moles of A^{3+} and B^{3+}, is $Ca_{1-2x}A_xB_xZrTi_{2-2x}C_{2x}O_7$), where $0 < x < 1/2$. The appropriate equation for radius ratio is

$$RR = \frac{0.19 + (r_A^{3+(VIII)} + r_B^{3+(VIII)} - 0.224)x}{0.121 + (2r_C^{3+(VI)} - 0.121)x} \qquad [10]$$

EXPERIMENTAL PROCEDURE

Starting materials were commercially produced powders including Al_2O_3 (99.98% purity, less than one micron in size), $CaCO_3$ (99.95% purity, unspecified particle size) CeO_2 (99.9%, less than 5 microns), $Ce_2(C_2O_4)_3{\cdot}9H_2O$ (99% purity, unspecified particle size), Gd_2O_3 (99.9%, less than 10 microns in size), HfO_2 (98%, less than 1 micron in size), Sm_2O_3 (99.9%, unspecified particle size), TiO_2 (99.5%, nanocrystalline), ZrO_2 (greater than 95% purity, nanocrystalline). Samples were prepared to correspond to several different substitution schemes. Nominal sample weights were three grams. All powders were mixed by hand in a mortar and pestle and then calcined in air at 900 °C for one day. The powders were then pressed into 1.9 cm diameter discs using an hydraulic cold press. They were placed in a furnace on a platinum covered alumina hearth plate and annealed at 1300 °C for times ranging from one to three months. Additionally, many of the samples were removed from the furnace weekly and ground, pressed again into pellets and placed back in the furnace in order to promote homogeneity and facilitate reaching equilibrium.

After heat treatment, each sample pellet was divided into two parts. One part was ground into a fine powder and analyzed using X-ray diffraction (XRD). A Scintag automated diffractometer and CuK_α radiation were employed. Scans were run from 10 to 130 ° 2θ, and using an 0.03° step size and a count time of 10 s. Lattice parameters were refined using software provided with the diffraction unit, after major peaks were manually indexed. The remaining half of each sample was metallographically cross-sectioned, and the phases present were analyzed for chemical composition using electron probe microanalysis (EPMA) and wavelength-dispersive spectroscopy (WDS). A JEOL JXA-8600 Superprobe was used, operating at a beam current of 20 nA and an accelerating voltage of 15 kV. Standards used were Al_2O_3, $CaWO_4$, CeO_2, Gd_2O_3, Hf, Sm, TiO_2 and ZrO_2. Experimentally determined k-ratios for the cationic species were obtained and corrected using the ZAF routine. Oxygen content was determined indirectly, under the assumption that each type of cation was accompanied by a certain number of oxygen anions based on the cation's valence. Additionally, microstructural images were obtained using a JEOL JSM-35C scanning electron microscope (SEM) operating at 20 kV and generally employing backscattered electron imaging to obtain maximum atomic number contrast.

RESULTS AND DISCUSSION

More than 40 samples were prepared and analyzed. Many gave redundant information or were in single phase regions, and therefore only a few are reported here (primarily those that were two phase, which yielded the limiting solubilities of various substituting cations in zirconolite). The sample names, gross compositions, phases present as determined by XRD and phase compositions as measured by EPMA are given in Table 1. Many of these samples contained small amounts of a third phase, which was generally present in the amount of about 5 vol.%. Figure 1 shows the microstructure of sample Sm1, which, neglecting very small amount of perovskite, lies in a two phase region zirconolite + rhombohedral phase (an intermediate phase found in all systems not containing Al^{3+} which appears to be the same phase as the hexagonal phase reported by Rossell[3] in studies involving Gd substitution in zirconolite, but whose hexagonally indexed Miller indices obey the hexagonal rules for rhombohedral structures). This microstructure and the presence of a two phase zirconolite + rhombohedral phase region were typical of samples studied in the present investigation, as may be seen in Table 1. Figure 2 shows a sample that contains tetravalent Ce in addition to Sm. It also contains zirconolite and a rhombohedral phase, but the amount of perovskite is substantially higher than in the previous case (about 10 vol.%). Most samples containing Ce^{4+} were found to contain 10-15 vol.% perovskite, and this may be due to the coexistence of a mixture of Ce^{3+} and Ce^{4+} in the samples, which were annealed in air at relatively high temperatures. A change in Ce valence from 4+ to 3+ would remove the gross composition of the sample from the stoichiometry M_4O_7, which would most likely lead to the formation of additional phases such as perovskite.

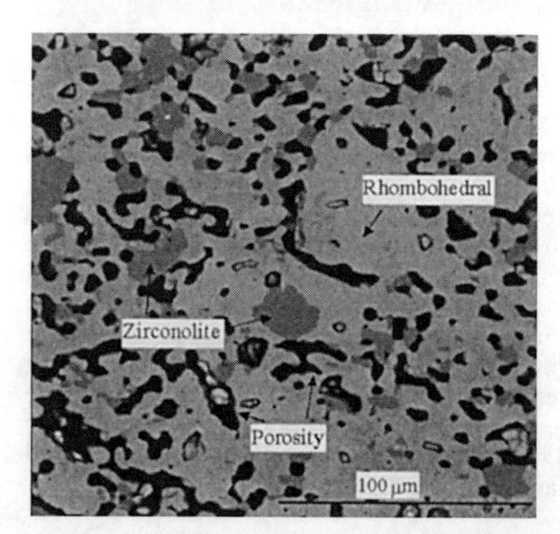

Figure. 1. SEM micrograph of Sample Sm1 depicting a two phase mixture of zirconolite and a rhombohedral phase, plus a small amount of perovskite.

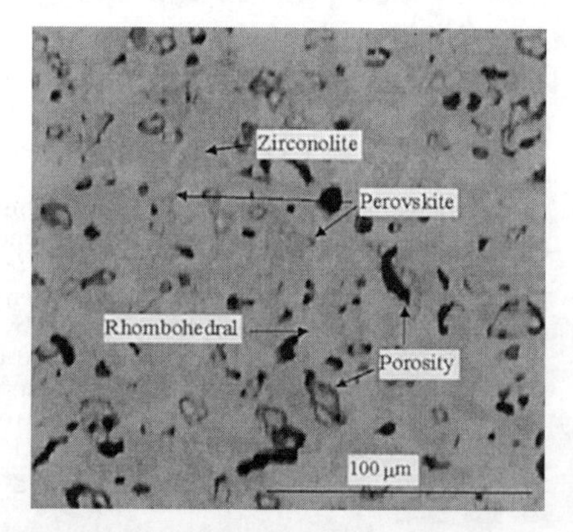

Figure. 2. SEM Micrograph of Sample Ce(IV)Sm, comprising both Ce and Sm in solution. It contains zirconolite, a rhombohedral phase and about 10 vol.% perovskite.

Sample	Gross Composition	Phase by XRD	Phase Composition by EPMA	x	RR
Table 1. Gross Sample Compositions, Phases Present and Compositions of Phases					
Sm1[3a]	$Ca_{0.75}Sm_{0.50}Zr_{0.75}Ti_2O_7$	Z[b]	$Ca_{0.84}Sm_{0.30}Zr_{0.93}Ti_{1.93}O_7$	0.15	1.60
		R	$Ca_{0.74}Sm_{0.55}Zr_{0.74}Ti_{1.97}O_7$	0.28	1.67
		Pv	$Ca_{0.71}Sm_{0.25}Zr_{0.01}Ti_{1.03}O_3$	-	-
Gd1[3]	$Ca_{0.75}Gd_{0.50}Zr_{0.75}Ti_2O_7$	Z	$Ca_{0.83}Gd_{0.31}Zr_{0.95}Ti_{1.91}O_7$	0.15	1.59
		R	$Ca_{0.71}Gd_{0.60}Zr_{0.72}Ti_{1.97}O_7$	0.30	1.67
		Pv	$Ca_{0.81}Gd_{0.14}Zr_{0.01}Ti_{1.04}O_3$	-	-
Ce(IV)1[1]	$CaCe_{0.25}Zr_{0.75}Ti_2O_7$	Z	$Ca_{0.96}Ce_{0.14}Zr_{0.93}Ti_{1.97}O_7$	0.14	1.59
		R	$Ca_{0.94}Ce_{0.47}Zr_{0.66}Ti_{1.93}O_7$	0.47	1.67
		Pv	$Ca_{0.77}Ce_{0.18}Zr_{0.01}Ti_{1.04}O_3$	-	-
Ce(III)1[5]	$Ca_{0.80}Ce_{0.20}ZrTi_{1.8}Al_{0.20}O_7$	Z	$Ca_{0.836}Ce_{0.22}Zr_{0.97}Ti_{1.79}Al_{0.19}O_7$	0.20	1.59
		Zi	$Ca_{0.01}Ce_{0.05}Zr_{0.81}Ti_{0.12}O_2$	-	-
Ce(III)2[5]	$Ca_{0.80}Ce_{0.30}ZrTi_{1.8}Al_{0.20}O_7$	U	$Ca_{0.74}Ce_{0.32}Zr_{0.96}Ti_{1.69}Al_{0.29}O_7$	0.30	-
		Zi	$Ce_{0.09}Zr_{0.80}Ti_{0.11}O_2$	-	-
Ce(IV)Sm[7]	$Ca_{0.925}Ce_{0.15}Sm_{0.15}Zr_{0.775}Ti_2O_7$	Z	$Ca_{0.92}Ce_{0.08}Sm_{0.13}Zr_{0.91}Ti_{1.96}O_7$	0.07	1.59
		R	$Ca_{0.84}Ce_{0.29}Sm_{0.29}Zr_{0.65}Ti_{1.93}O_7$	0.14	1.67
		Pv	$Ca_{0.76}Ce_{0.11}Sm_{0.11}Zr_{0.01}Ti_{1.01}O_3$	-	-
Ce(IV)Gd[7]	$Ca_{0.925}Ce_{0.15}Gd_{0.15}Zr_{0.775}Ti_2O_7$	Z	$Ca_{0.92}Ce_{0.08}Gd_{0.13}Zr_{0.90}Ti_{1.97}O_7$	0.07	1.59
		R	$Ca_{0.85}Ce_{0.29}Gd_{0.26}Zr_{0.66}Ti_{1.94}O_7$	0.14	1.67
		Pv	$Ca_{0.80}Ce_{0.10}Gd_{0.06}Zr_{0.02}Ti_{1.02}O_3$	-	-
Ce(III)Sm[9]	$Ca_{0.70}Ce_{0.15}Sm_{0.15}ZrTi_{1.70}Al_{0.30}O_7$	Z	$Ca_{0.72}Ce_{0.16}Sm_{0.19}Zr_{0.98}Ti_{1.67}Al_{0.28}O_7$	0.18	1.60
		Zi	$Ca_{0.02}Ce_{0.03}Sm_{0.01}Zr_{0.82}Ti_{0.11}Al_{0.01}O_2$	-	-
Ce(III)Gd[9]	$Ca_{0.66}Ce_{0.17}Gd_{0.17}ZrTi_{1.66}Al_{0.34}O_7$	Z	$Ca_{0.68}Ce_{0.18}Gd_{0.19}Zr_{1.00}Ti_{1.63}Al_{0.32}O_7$	0.18	1.60
		Zi	$Ce_{0.05}Gd_{0.01}Zr_{0.85}Ti_{0.09}O_2$	-	-

[a]The superscript numeral denotes the equation number of the substitution scheme employed for that sample.
[b]Z = Zirconolite, R = Rhombohedral, Pv = Perovskite, U = Unknown, Zi = Zirconia

By inspecting the far righthand column of Table 1, it may be noted that for each chemical system, the radius ratio of zirconolite at its solubility limit is 1.59. The deviation from this value is surprisingly small. Only for the samples containing Al^{3+} in solution is there any potential discrepancy. These samples (Ce(III)1, Ce(III)Sm and Ce(III)Gd) possess RR's which are right at or slightly exceed 1.59, but are nominally single phase. In particular, given the narrow range of RR's at the phase boundary in other systems, one might expect samples Ce(III)Sm and Ce(III)Gd to lie in a two-phase region. Further study must be performed in order to determine whether these systems follow the radius ratio rule of the remaining systems.

It is noteworthy as well that at its lower solubility limit, RR for the rhombohedral phase is also remarkably consistent at a value of 1.67. While not discussed here, the upper solubility limits of the rhombohedral phases and the lower solubility limits of the pyrochlore phases all held to rather distinct RR values, which corresponded to 1.68 and 1.70, respectively, for the rhombohedral structure and the pyrochlore structure.

Assuming that, as the data suggest, the RR for zirconolite is constant at its solubility limit for any assemblage of cations, Eqns. 2, 4, 6 and 8 may be utilized to predict the maximum solubility limits for such an assembly. This may be accomplished by setting each equation equal to 1.59 and solving for x. The only requirement for predictive capability is that the type of occupancy of each cationic site must be known, since the radius of a cation varies with its coordination.

As a test of the predictive capabilities of the equations, the solubilities of several cations in zirconolite as reported in the literature (Th^{4+}[3], U^{4+}[3], La^{3+}[14] and Y^{3+}[14]) were compared with their values predicted from the appropriate equations (Eqn. 2 for the tetravalent species and Eqn. 4 for the trivalent cations). The experimental and predicted values of x are given in Table 2.

Table 2. Experimental and Predicted Limiting Solubilities of Cationic Species in Zirconolite

Species	Substitution Scheme	Solubility Limit (Exp)	Solubility Limit (Calc)
Th^{4+}	$CaTh_xZr_{1-x}Ti_2O_7$	0.1	0.11
U^{4+}	$CaU_xZr_{1-x}Ti_2O_7$	0.17	0.14
La^{3+}	$Ca_{1-x}La_{2x}Zr_{1-x}Ti_2O_7$	0.07	0.05
Y^{3+}	$Ca_{1-x}Y_{2x}Zr_{1-x}Ti_2O_7$	0.2	0.24

The agreement is quite good, especially considering that the solubility limits for La and Y were estimated from XRD data and not determined directly.

CONCLUSIONS

The relative radius ratio correlation shows great promise for helping to predict the solubility limits of cations in zirconolite and its crystallographically related phases. As more experimental data become available pertaining to the simultaneous solubilities of multiple cationic species in zirconolite, it should prove possible to ascertain how generally valid this correlation is. In the mean time, the results of the present study suggest that it may prove fruitful to attempt similar correlations for the other phases present in Synroc.

REFERENCES

1. E. R. Vance, MRS Bull. **19**, 28 (1994).
2. S. E. Kesson, W. J. Sinclair and A. E. Ringwood, Nucl. Chem. Waste Man. **4**, 259 (1983).
3. H. J. Rossell, J. Solid State Chem. **99**, 38 (1992).
4. F. W. Clinard Jr., D. L. Rohr and R. B. Roof, Nucl. Instrum. Methods **B1**, 581 (1984).
5. E. R. Vance, P. J. Angel, B. D. Begg and R. A. Day, Mater. Res. Soc. Symp. Proc. **333**, 293 (1994).
6 E. R. Vance, B. D. Begg, R. A. Day, and C. J. Ball, Mater. Res. Soc. Symp. Proc. **353**, 67 (1995)
7. E. R. Vance, A. Jostsons, R. A. Day, C. J. Ball, B. D. Begg and P. J. Angel, Mater. Res. Soc. Symp. Proc. **412**, 41 (1996).
8 D. Swenson, T. G. Nieh and J. H. Fournelle, Mater. Res. Soc. Symp. Proc. **412**, 337 (1996).
9. R. S. Roth, J. Nat. Bur. Stand. **56**, 17 (1956).
10. R. A. McCauley and F. A. Hummel, J. Solid State Chem. **33**, 99 (1980).
11. F. Mazzi and R. Munno, Amer. Mineral. **68**, 262 (1983).
12. M. A. Subramanian, G. Aravamudan and G. V. Subba Rao, Prog. Solid State Chem. **15**, 55 (1983).
13. R. D. Shannon, Acta Cryst. **A32**, 751 (1976).
14. E. R. Vance and D. K. Agarwal, Nucl. Chem. Waste Man. **3**, 229 (1982).

SYNTHESIS AND STUDY OF ^{239}Pu-DOPED GADOLINIUM-ALUMINUM GARNET

B.E. BURAKOV, E.E. ANDERSON, M.V. ZAMORYANSKAYA, M.A. PETROVA
V.G.Khlopin Radium Institute, 28, 2-nd Murinskiy ave., St.Petersburg, 194021, Russia,
fax: (7)-(812)-346-1129; e-mail: burakov@riand.spb.su

ABSTRACT

Garnet solid solutions, $Y_3Al_5O_{12}$-$Gd_3Al_5O_{12}$-$Gd_3Ga_5O_{12}$ (YAG-GAG-GGG), are being considered as prospective durable host phases for the immobilization of actinide-containing waste with complex chemical compositions. Garnet samples with the suggested simplified formula: $(Gd,Ce,\ldots)_3(Al,Ga,Pu,\ldots)_5O_{12}$ containing from 3.4 to 5.3 wt.% ^{239}Pu and 3.6-5.5 wt.% Ce have been synthesized through melting of oxide starting materials in air using a hydrogen torch. Calcium and Sn were added to increase the Pu incorporation into the garnet lattice through ion charge and size compensation for Pu^{4+}. Polycrystalline materials obtained in the experiments consist of garnet, perovskite and other phases and were studied by scanning electron microscopy (SEM) and powder X-ray diffraction (XRD). Our results confirmed that the use of compensating elements such as Ca and Sn allow for significant incorporation of Pu and Ce (not less than a few wt.%) into the garnet structure. The preliminary conclusions thus so far indicate that garnet solid solution compositions may incorporate simultaneously trivalent and tetravalent actinides in significant quantities because they occupy different positions in the garnet structure.

INTRODUCTION

Plutonium waste residues originate from nuclear weapons production and may contain more than 50 % of chemical impurities including such elements as Am, Al, Mg, Ga, Fe, La, Na, Mo, Nd, Si, Ta, Ce, Ba, W, Zn, C and Cl. While for some of these residues, direct conversion to traditional glass or ceramic waste forms may be difficult, ceramic waste forms based on durable actinide host-phases are preferred for Pu, Am and other actinide immobilization. Previous work has demonstrated that garnet/perovskite ceramics $(Y,Gd,\ldots)_3(Al,Ga,\ldots)_5O_{12}/(Y,Gd,\ldots)(Al,Ga,\ldots)O_3$ are prospective durable materials which may allow the incorporation of Pu, Am and most elements listed above in the form of solid solution [1-4]. Under the same synthesis conditions without the addition of compensating elements (i.e., ion size and charge), GGG incorporates up to 6 wt.% Ce, but less than 0.1 wt.% U. Pure YAG incorporates not greater than 0.5 wt.% Ce [2]. Also, it has been demonstrated that the use of Sn and Ca as compensating (of ion size and charge) admixtures significantly increases the incorporation of U into garnet lattice [3]. However, Ce as well as U can not be considered as acceptable surrogates for Pu. Therefore, experimental studies of garnet and perovskite host-phases for Pu incorporation using Pu and not surrogate elements are required. The main goal of this work was to confirm the possibility of obtaining garnet solid solution with significant amounts of Pu through melting in air of oxide starting materials.

Mat. Res. Soc. Symp. Proc. Vol. 608 © 2000 Materials Research Society

EXPERIMENTAL

To reduce the cost of experiments with Pu, we opted to make small samples of 0.5-1.0 gram of starting materials and using melting the oxide reagents in air as the synthesis method. In order to carry out melting of the cold pressed powders, a hydrogen torch with flame temperature up to 2100°C was placed in a glove box. Previous experience has shown that melting of oxides in a hydrogen torch does not allow for equilibrium conditions. This causes the formation of two, or more, phases such as garnet, perovskite and other phases in the same sample prepared from the starting materials for garnet and perovskite. To synthesize of Pu-doped garnet we have used the following precursors:

- A – for perovskite stoichiometry, $(Gd_{0.80}Ce_{0.08}Ca_{0.12})(Al_{0.88}Ga_{0.05}Pu_{0.06}Sn_{0.01})O_3$;
- B – for garnet stoichiometry, $(Gd_{1.60}Ce_{0.30}Ca_{1.10})(Al_{3.69}Ga_{1.20}Pu_{0.10}Sn_{0.01})O_{12}$;
- C – for garnet stoichiometry, $Gd_3(Ce_{0.30}Ca_{0.40}Pu_{0.20}Ga_{0.20}Al_{0.60}Sn_{0.20}X_{0.10})Al_3O_{12}$, where X is designed empty position providing vacancies in the final crystalline structure of the garnet.

Powdered starting materials consisted of Gd, Al, Ca, Pu, Ce, Ga and Sn-oxides and were cold pressed to produce pellets 5 mm in diameter. The pellets were then melted in air in a hydrogen torch for 5-8 minutes. The temperature of the hydrogen torch was not measured precisely but was estimated to be approximately as 1900-2100°C. The samples obtained were labeled: A, B and C (Fig.1, Table I) and studied by SEM, XRD and electron microprobe methods.

RESULTS AND DISCUSSION

Results of XRD and electron microprobe analyses (Fig.1, Table I) allowed us to clearly identify the main phases in the experimental run products and consisted of:

- A – perovskite matrix (A-1) and second phase, garnet (A-2);
- B – garnet matrix (B-1) and second phase B-2 (its identification was not completed);
- C – perovskite matrix (C-1) and second phases C-2 and C-3 (their identification was not completed).

Garnet phases in samples A and B incorporated 5.5 and 3.5 wt.% Pu, respectively. Perovskite phases are characterized of 6.5 wt.% Pu contents for sample A and 3.6 wt.% Pu for sample C. In the same experiment containing both host phases, garnet and perovskite, incorporated a significant amount (several wt.%) of Ce (Table I). It was assumed that Pu and Ce occupy different positions into the garnet lattice and in the perovskite lattice: Ce^{3+} accompanies Gd^{3+}, but Pu^{4+} substitutes partly Al^{3+} (or Ga^{3+}). However, additional study of Pu and Ce valence state into the structures of garnet and perovskite structures is needed, in particular because Ce^{3+} and Ce^{4+} may co-exist in the same crystalline structure. In that case, for the ceramic based on garnet and perovskite Ce^{3+} can be considered as Am^{3+} surrogate and Ce^{4+} as a surrogate for Np^{4+} and Pu^{4+}.

Formation of garnet and perovskite from the melt (Table I) were characterized by the redistribution of Ca and Ga mainly into the garnet matrix. Previous work [3] has demonstrated that Al/Ga ratio in a melt of garnet stoichiometry can effect the yield of final phases produced: mainly garnet (at higher Ga contents) or mainly perovskite (at higher Al contents). It is also possible that admixtures of Ca and Ga can stabilize the formation of Pu-doped gadolinium-aluminum garnet which should be confirmed from the future experiments under equilibrium conditions.

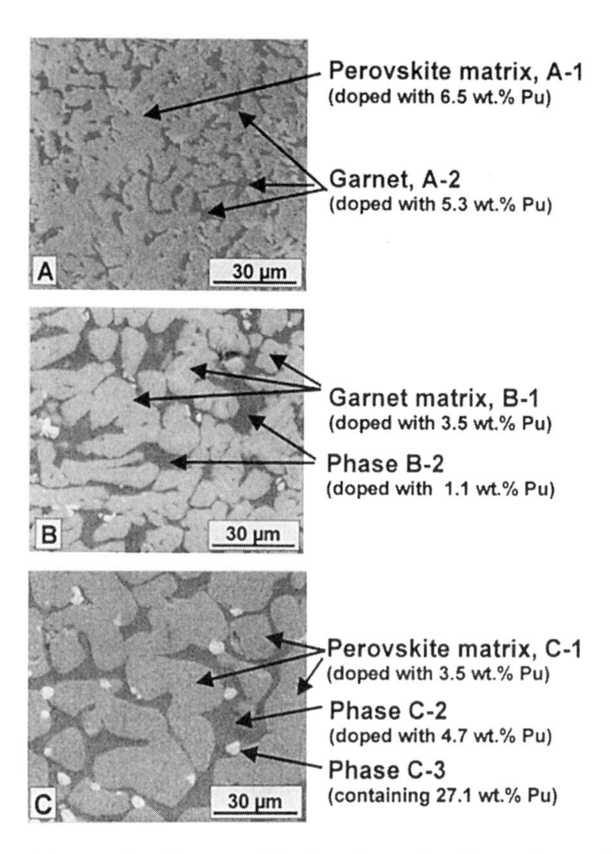

Fig.1 Back-scattered electron SEM images of Pu-doped samples of garnet/perovskite ceramic obtained through melting in air of oxide starting materials.

Table I. Chemical composition of crystalline phases in ceramic matrices (based on normalized data of microprobe analysis).

Phase (see Fig.1)	**Elements in wt.%**							
	Gd	**Ce**	**Al**	**Ga**	**Ca**	**Sn**	**Pu**	**O**
A-1, perovskite	57.8	4.0	10.0	0.4	1.0	0.2	6.5	20.1
A-2, garnet	36.1	5.5	12.3	10.3	6.0	0.1	5.3	24.4
B-1, garnet	44.4	3.6	11.9	9.9	3.3	0.1	3.4	23.4
B-2, ?	5.3	9.4	23.2	19.3	6.4	0.2	1.1	35.1
C-1, perovskite	60.0	3.3	10.8	0.5	1.0	0.2	3.6	20.6
C-2, ?	28.7	6.5	20.7	7.0	4.0	1.3	4.7	27.1
C-3, ?	43.5	11.1	1.0	-	2.4	0.3	27.1	14.6

CONCLUSIONS

Obtained results allow us to make the following conclusions:

1) It was experimentally confirmed that a Pu host-phase such as gadolinium-aluminum garnet has significant lattice capacity (not less than 5.3 wt.%) for incorporation of Pu. This ceramic material can be obtained through melting of oxide starting materials in air;
2) Gadolinium-aluminum perovskite which accompanies formation of garnet under non-equilibrium conditions is also characterized of high lattice capacity for Pu incorporation (not less than 6.5 wt.%);
3) Formation of garnet and perovskite phases in the melt under non-equilibrium conditions were characterized by redistribution of Ca and Ga primarily into the garnet lattice.
4) It is assumed that tri-and tetra-valent actinides should occupy different positions in the garnet structure, however, additional study of Pu valence states into garnet and perovskite crystalline structures is required.

ACKNOWLEDGEMENTS

Authors are very grateful to Mrs. M. A. Yagovkina and Mr. V. M. Garbuzov from V. G. Khlopin Radium Inst. for the help on sample synthesis and examination.
This work presented in this paper was supported by International Scientific and Technological Center (ISTC), Project #1063. The authors acknowledge Dr. J. M. Hanchar of George Washington University, Washington DC, US, for technical comments and correction of the English.

REFERENCES

1. B.E,Burakov, E.E,Strykanova, Proceedings of International Symposium Waste Management-98, CD-ROM version, sess34 - 05, (1998).

2. B.E,Burakov, E.B,Anderson, D.A,Knecht, in Environmental Issues and Waste Management Technologies IV, 349-356 (1999).

3. B.E,Burakov, E.B. Anderson, D.A,Knecht, M.V,Zamoryanskaya, E.E,Strykanova, M.V,Yagovkina, Mat. Res. Soc. Symp. Proc. Proceedings Scientific Basis for Nuclear Waste Management XXII, (1999) in press.

4. B.E,Burakov, E.B,Anderson, Proceedings of the 2nd NUCEF International Symposium NUCEF'98, 16-17/11/98, Hitachinaka, Ibaraki, Japan, JAERI-Conf.99-004 (Part I), 295-306 (1998).

CHARACTERIZATION OF A GLASS-BONDED CERAMIC WASTE FORM LOADED WITH U AND PU

W. SINKLER, T. P. O'HOLLERAN, S. M. FRANK, M. K. RICHMANN* AND S. G. JOHNSON
Argonne National Laboratory - West, P. O. Box 2528, Idaho Falls, ID 83403
*Argonne National Laboratory - East, 9700 South Cass Ave., Argonne, IL 60439

ABSTRACT
This paper presents microscopic characterization of four samples of a ceramic waste form (CWF) developed for disposal of actinide-containing electrorefiner salts. The four samples were prepared to investigate the influence of water content and the Pu:U ratio on CWF microstructure and performance. While the overall phase content is not strongly influenced by either variable, the presence of water in the initial zeolite has a detectable effect on CWF microstructure. It is found to influence the distribution of the major actinide host phase, a $(U,Pu)O_2$ mixed oxide.

INTRODUCTION
Argonne National Laboratory (ANL) has developed a spent fuel treatment process for sodium-bonded metallic spent nuclear fuel from the EBR II fast breeder reactor. The process involves electrometallurgical separation of the spent fuel in a KCl-LiCl molten salt bath. The three products are a) cladding hulls remaining at the anode, along with elements noble to the electrorefinement, b) uranium metal which is deposited on the cathode, and c) the electrorefiner salt bath, in which Na, Pu, U, rare earth, alkali and alkaline earth fission products accumulate.

Disposal of the electrorefiner salt is the most challenging aspect of the electrometallurgical spent fuel treatment. In order to immobilize the water soluble chlorides, the initial step in disposing of the salt is blending with zeolite 4A in a v-mixer at 500°C. This results in a reduction of the free chloride content to less than 1% of the initial quantity. Subsequently, the salt-blended zeolite is mixed with a borosilicate glass frit and hot isostatically pressed (HIP'ed). During HIPing, the zeolite 4A transforms to sodalite, a mineral aluminosilicate phase which contains chlorine and alkali elements. The resulting material is a glass-ceramic in which the primary phases are glass and sodalite, with minor constituents nepheline ($NaAlSiO_4$), $(Pu,U)O_2$ mixed oxide, halite (NaCl) and a rare earth silicate containing some Pu. The micro-structure and leach behavior of this material has been presented in several previous papers [1-3].

Part of the qualification process for acceptance of the electrometallurgical treatment is to investigate the effect on the CWF of variations in processing and compositional parameters over ranges within which they may vary during operations. This work addresses two important compositional parameters. The first of these is the water content. Water is introduced into the processing in slightly varying amounts due to the inevitable presence of a small quantity of water in the zeolite 4A. The specification zeolite water content is 0.5 wt%. The second variable addressed is the ratio of U to Pu in the electrorefiner salt. In the case of driver fuel processing, this ratio is typically large, on the order of 2:1. In the case of blanket fuel processing, which is planned at a later stage in the fuel treatment, the U:Pu ratio will be lower, on the order of 1:2.

The issue of water content is of particular interest for the CWF. The motivation for using zeolite for salt disposal is the well-known ability of zeolites to occlude salt ions [4]. While blending with zeolite dramatically reduces the overall free chloride quantity, previous studies of the CWF have suggested that occlusion may not occur in the case of the actinide salts. Instead, the U and Pu chlorides have been found to transform to a separate oxide phase during the salt/zeolite blending step. When water is present, the formation of $(U,Pu)O_2$ is explainable as a direct reaction of actinide chlorides with water, which is strongly favored thermodynamically.

Mat. Res. Soc. Symp. Proc. Vol. 608 © 2000 Materials Research Society

However, the observed formation of a separate oxide phase rather than occlusion raises an important basic question as to whether zeolite occlusion of actinide salts can occur if the competing reaction with water is suppressed by using sufficiently dry zeolites, and/or artificially large actinide contents. The present study was in part designed to address this question.

EXPERIMENTAL

Four laboratory scale samples of CWF were produced as a test matrix shown in Table 1. At ANL-East, salts of composition given in Table 2 were blended with wet or dry zeolite 4A for 20 h at 500°C under argon. The salt/zeolite blend had 10.5 wt% salt.

Table 1. Overview of experimental matrix showing sample designations

zeolite H_2O(w%)	U:Pu Ratio 3:1	U:Pu Ratio 1:3
0.12	1	2
3.5	3	4

Table 2. Approximate composition of actinide bearing 300 driver electrorefiner salt (w%)

LiCl-KCl†	NaCl	CsCl	$BaCl_2$	$RECl_3$	$AcCl_3$	others
65.0	12.7	2.1	1.0	7.9	10.1	1.2

†Eutectic compositon with 47.1 wt%KCl. RE: Rare earths; Ac: Pu and U in 3:1 or 1:3 ratio

Following blending, the salt-loaded zeolite was mixed with a borosilicate glass frit in 1:3 proportion by weight (glass:blended zeolite), and sealed in small 1" diameter evacuated HIP cans. At ANL-West, the samples were HIPed at 850°C for 1 h under 100 MPa pressure. Powder x-ray diffractometry (XRD), scanning and transmission electron microscopies (SEM and TEM) were performed using Scintag X1 (Cu Kα), Zeiss DSM 960 and JEOL 2010 instruments, respectively. TEM sample preparation was performed using standard dimpling and ion milling techniques.

RESULTS

a) X-Ray Diffractometry

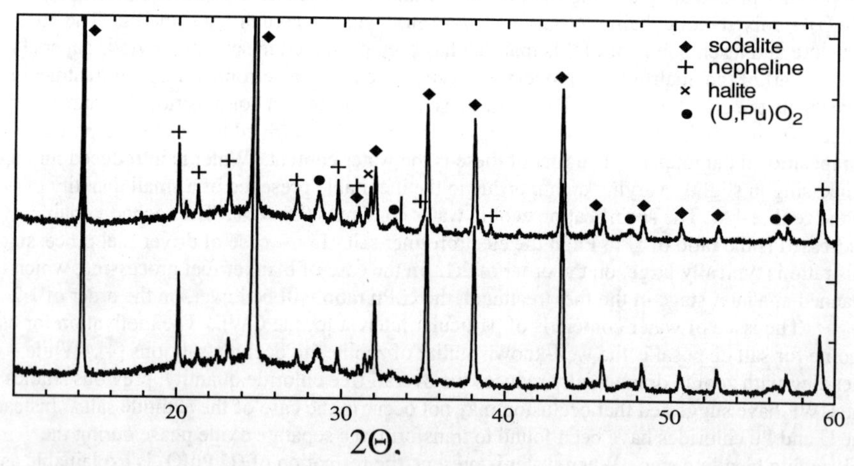

Fig. 1. XRD scans of samples 4 (top) and 2 (bottom).

Fig. 1 shows XRD scans of samples 2 and 4. The phases identified in the scans are sodalite, nepheline, the mixed actinide oxide $(U,Pu)O_2$ phase with fluorite structure, and halite. In addition, the broad rise in the background intensity near 25° 2θ is consistent with the presence of a glass phase. As can be seen, the traces are nearly identical, indicating that the influence of water on the phase content of the final CWF is minor. The only significant variation of phase content caused by the presence of water is an increase in the peaks associated with halite for high water content. For changing U:Pu ratio, the only discernable influence in the XRD trace was on the lattice parameter of the mixed oxide $(U,Pu)O_2$ phase. This varied in a way consistent with solid solutions with U:Pu of 3:1 or 1:3.

b) Scanning Electron Microscopy
Fig. 2 shows the typical appearance of wet and dry zeolite in SEM, backscattered electron (BSE) mode. The predominant microstructure of the CWF consists of polycrystalline sodalite regions joined by a glass phase. The lightest features in the images are actinide and rare-earth bearing regions. Sodalite regions appear as a more continuous light color and glass regions appear diffuse and darker. In the case of sample 1 made with dry zeolite, the actinide species appear to be more homogeneously distributed within the sample, giving the BSE image a slightly noisy appearance. In contrast, actinide-rich regions in the BSE image of the sample made with wet zeolite are more strongly clustered. No significant difference was found in SEM between samples made with high and low U:Pu ratios.

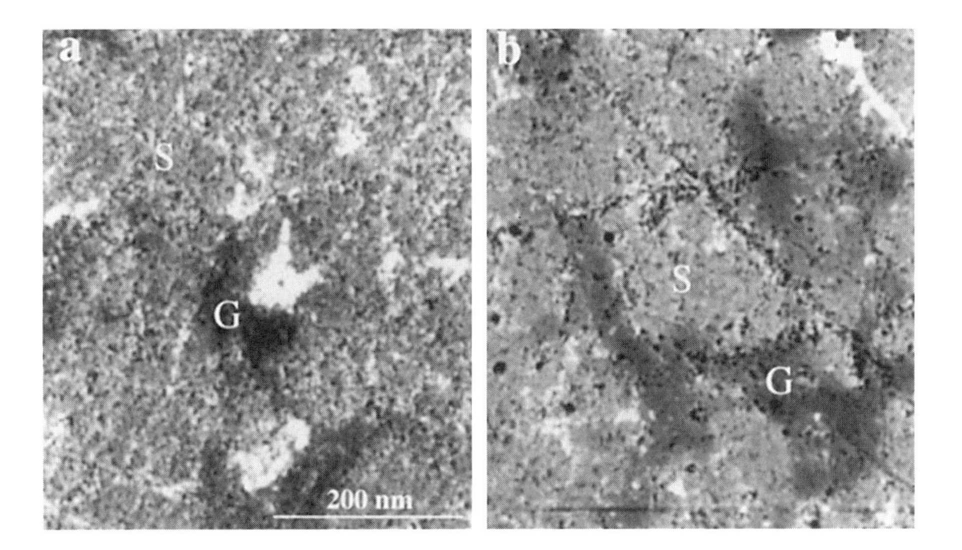

Fig. 2. SEM BSE micrographs of samples 1 (a) and 3 (b). G: Glass; S: Sodalite; bright particulate phase is actinide oxide.

c) Transmission Electron Microscopy
Fig. 3 shows a typical low-magnification microstructure image from sample 1. The major glass and sodalite phases, as well as halite and the mixed oxide phase, are all visible in the image. Energy dispersive spectroscopy (EDS) spectra showing the elemental compositions from the glass and sodalite phases do not differ significantly from those presented previously [3]. In particular, the actinide content in these phases is negligible. The major actinide-bearing phase in

the CWF is the mixed oxide. Both this phase as well as halite tend to be found within the glass near the glass/sodalite boundary, as shown in Fig. 3.

Fig. 4 shows a cluster of mixed oxide crystals in sample 3, which is identical to sample 1 except that it was made using wet zeolite. There is a clear difference between the distribution of mixed oxide particles in Figs. 3 and 4, which was consistent in all observations of these samples. Mixed oxide clusters in the samples made with wet zeolite tended to have well-defined boundaries with a roughly constant density of fine mixed oxide particles within these boundaries.

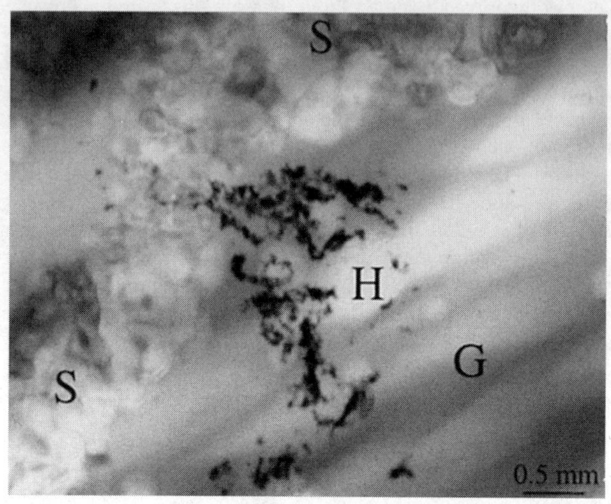

Fig. 3. Microstructure image of sample 1. S: Sodalite, G: Glass, H: Halite. The fine-grained dark phase is $(U,Pu)O_2$, found in clusters of grains with approximately 20 nm crystallite size.

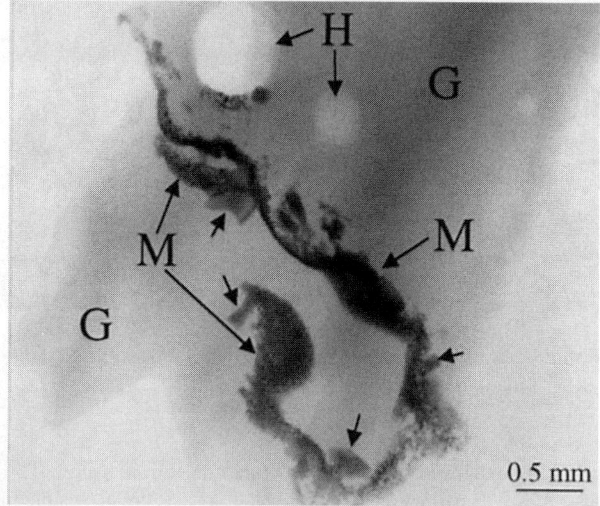

Fig. 4. Image from sample 3 showing an actinide-rich region in the glass phase. Short unlabeled arrows show examples of the rare earth based silicate phase. M: mixed oxide clusters.

In contrast to this, use of dry zeolite tended to result in large but rather incoherent and ill-defined agglomerations of the mixed oxide phase, such as that seen in Fig. 3. While a clear micro-structural difference was detected between samples made with high and low water contents, no difference among the samples as a function of the U:Pu ratio was detected using TEM.

In addition to the different morphology of clusters of mixed oxide particles, two other distinctions were noted between samples made with wet and dry zeolite. In samples made with dry zeolite, there were a significant number of larger single crystals of the mixed oxide phase, on the order of 100 nm up to sizes in excess of one micron. In contrast, such crystals were rare in the case of the samples made with wet zeolite. Finally, in a few instances, very fine mixed oxide crystals were found to have formed in the interior of sodalite grains or at sodalite grain boundaries in samples made with dry zeolite 4A. An example of this is shown as Fig. 5.

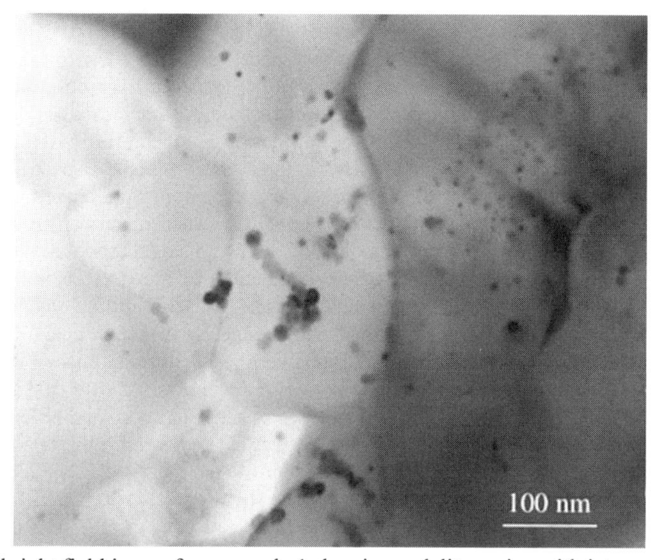

Fig. 5. TEM bright field image from sample 1 showing sodalite grains with intragranular precipitates of the mixed oxide phase (small dark inclusions).

While the majority of actinides were contained in the mixed oxide phase, a rare earth based silicate phase was found in all four samples. Examples of several rare earth based silicate crystals are seen in Fig. 4. This phase forms faceted crystals with somewhat larger sizes than the mixed oxide phase. They are commonly found within or at the boundaries of mixed oxide clusters (see Fig. 4). EDS spectra showed strong rare earth lines (primarily Nd and Ce), as well as a minor actinide component. The ratio of Pu to U was consistently higher than in any nearby mixed oxide, judging from significantly stronger intensity of the Pu Lα characteristic x-ray relative to U Lα. Because of the small quantity of this phase, its role as an actinide host is nevertheless small. The rare earth based silicate could not be indexed to any rare earth or actinide silicate phase found the JCPDS database. An orthorhombic unit cell was determined to which all diffraction patterns could be indexed. The cell parameters are:

a=22.9 Å; b=9.9 Å; c=7.2 Å

The errors are approximately 2% on the relative axis lengths, and 5% on the absolute lengths (due to uncertainty in the microscope camera length). Evidence for a b glide plane parallel to a and an a glide plane parallel to c was found in systematic absences. In addition, reflections with index l odd showed a pronounced streaking in the a-axis direction, indicating that there is significant stacking disorder along the a-axis.

DISCUSSION

While the phase content of the glass bonded sodalite CWF is relatively independent of zeolite water content and U:Pu ratio within the present ranges, a microstructural effect of high water levels was detected by TEM and SEM. The present results suggest that the water content of the zeolite 4A may influence the kinetics of actinide oxide formation. In particular, the strong clustering of actinide oxide crystals seen in Fig. 4 suggest that this region originated from the decomposition of a particle of an initial chloride phase containing actinides and rare earths by reaction with water. The reaction with water occurs during salt/zeolite blending, since it is known that the much of the actinide oxide forms during this processing step. After formation of the actinide-bearing clusters, the mobility of actinides is greatly reduced. The observation of tightly constrained clusters with well defined morphology in the case of samples made with wet zeolite 4A thus suggests that these clusters represent the outlines of initial chloride particles.

In the case of samples made with dry zeolite, the clusters of actinide-bearing crystals are less well defined. This is consistent with a more sluggish decomposition, occurring perhaps at higher temperature, thus allowing the salt particles to melt and disperse via diffusion prior to transformation to oxides and subsequent immobilization. More sluggish actinide oxide phase formation is also consistent with the observed greater quantity of large (>100 nm diameter) actinide oxide crystals in the samples made with dry zeolite 4A. The formation of larger crystals is aided by more controlled conditions of growth and a reduced thermodynamic driving force. In the case of a reaction not with water but with oxygen deriving from other phases in the CWF (the glass or zeolite), the thermodynamic driving force would be reduced. This would reduce the probability of nucleation and favor growth of existing crystals. The common observation of halite crystals near to clusters of the mixed oxide phase (see Fig. 3) supports a transformation of the actinide chlorides to oxides by ion exchange with sodium.

Finally, the present work suggests that when the water content is insufficient for reacting all of the actinide chloride to oxide, competing mechanisms such as actinide salt occlusion into the zeolite 4A may become viable. Fig. 5 illustrates the presence of intragranular and intergranular actinide oxide precipitates, which are occasionally found in sodalite regions in samples made with dry zeolite 4A. This suggests that the zeolite from which the sodalite has formed initially absorbed actinides. Precipitation of the actinides intergranularly as oxides may then have occurred during the transformation of zeolite 4A to sodalite.

ACKNOWLEDGMENTS

Argonne National Laboratory is operated for the U. S. Department of Energy by the University of Chicago. This work was supported by the Department of Energy, Nuclear Energy Research and Development Program, under contract No. W-31-109-ENG-38. Helpful discussions with J. S. Luo and W. L. Ebert are gratefully acknowledged.

REFERENCES

1. S. M. Frank, K. Bateman, T. DiSanto, S. G. Johnson, T. Moschetti, M. Noy, and T. P. O'Holleran, in *Phase Transformations and Systems Driven Far from Equilibrium, MRS Symp. Proc.,* Vol. **481**, E. Ma, P. Bellon, M. Atzmon, and R. Trivedi, Editors. MRS, Pittsburgh. (1998).
2. T. L. Moschetti, T. P. O'Holleran, S. M. Frank, S. G. Johnson, D. W. Esh, and K. M. Goff, *Ceramic Transactions* **94**, to be published, 1999.
3. W. Sinkler, D. W. Esh, T. P. O'Holleran, S. M. Frank, T. L. Moschetti, K. M. Goff, and S. G. Johnson, *Ceramic Transactions* **94**, to be published, 1999.
4. D. W. Breck, *Zeolite Molecular Sieves*, Wiley, New York (1974) , p. 529 ff.

SOLID SOLUBILITIES OF Pu, U, Hf AND Gd IN CANDIDATE CERAMIC PHASES FOR ACTINIDE WASTE IMMOBILISATION

E. R. VANCE, M. L. CARTER, B. D. BEGG, R. A. DAY and S. H. F. LEUNG, Materials Division, ANSTO, Menai, NSW 2234, Australia, erv@ansto.gov.au

ABSTRACT

Solid solubility limits of U, Pu, and the neutron absorbers Hf and Gd have been measured for zircon ($ZrSiO_4$), monazite ($CePO_4$), titanite ($CaTiSiO_5$), perovskite ($CaTiO_3$), apatite ($Ca_{10}(PO_4)_6O$), in almost all cases where these limits were not known beforehand. The method used was to oversaturate the host phase with the dopant, using a nominated substitutional scheme, and then establish the dopant content of the host phase by microanalysis/scanning electron microscopy. Tetravalent U has limited solid solubilities in titanite, perovskite and apatite. X-ray absorption near-edge and diffuse reflectance spectroscopies were used to show that U was tetravalent in U-doped perovskite prepared in both argon and hydrogen-nitrogen atmospheres, with different charge compensating schemes. Tetravalent Pu has solubilities of 0.13 and 0.02 formula units (f.u.) in perovskite and titanite respectively. Trivalent Pu has a solubility of 0.05 f.u. in titanite. Pu^{3+} dominates tetravalent Pu in monazite fired in air at 1400°C. At least 0.5 and < 0.1 f.u. of Hf are soluble in titanite and monazite respectively.Hf solubility in apatite is estimated as < 0.1 f.u. Approximately 0.3 and < 0.1 f.u. of Gd are soluble in titanite and zircon respectively.

INTRODUCTION

High-level actinide-bearing wastes are prevalent world-wide and attention has recently been focussed on the disposition of excess weapons Pu. To minimise criticality problems with Pu in the geological immobilisation option, appreciable incorporation of neutron absorbers is necessary and the aqueous leaching behaviour of the absorbers should ideally be similar to those of Pu. The leach rates themselves should also be very small.

The essential aim of the current work is to define solid solubility limits of Pu, U, and the neutron absorbers Hf and Gd in well-studied candidate ceramic phases for actinide immobilisation. These phases are monazite, titanite, perovskite, zirconolite, zircon, pyrochlore and apatite.

Table 1. Known candidate ceramic phases richest in U and Pu, in formula units (f.u.).

Phase	Composition	U^{4+}	Pu^{3+}	Pu^{4+}
Monazite,	$REPO_4$	0.5 [1]	1 [2]	0.25#
Zircon,	$ZrSiO_4$	1 (coffinite)	0.1 [3]	1 [4]
Pyrochlore,	$RE_2Ti_2O_7$	1 [5]	2 [6]*	1 [5]
Zirconolite,	$CaZrTi_2O_7$	0.7 [7]		0.7 [7]
Titanite,	$CaTiSiO_5$	0.05 [8]; 0.02#	0.05#	0.02#
Apatite,	$Ca_2RE_8(SiO_4)_6O$	0.02[9];0.5#		0.05 [9]
Perovskite	$CaTiO_3$	0.01[10];>0.1#	1 [11]	0.1[12];0.13#

*Forms monoclinic structure[6]; # Present results

Many relevant data are already known for U and Pu solid solubilities in the above phases, and these are indicated in Table 1, together with values obtained in the present work. In oxide systems, applicable U and Pu valence states lie in the ranges +4 to +6 and +3 to +6

Mat. Res. Soc. Symp. Proc. Vol. 608 © 2000 Materials Research Society

respectively, but here we are looking in the first instance at only U^{4+} and tri- and tetravalent Pu. Since solid solubilities of the different U and Pu valence states are not necessarily the same, a further aim was to define the valence states in appropriate systems by looking at the charge compensation schemes deduced from the microanalysis and in one case diffuse reflectance and near-edge X-ray absorption spectroscopies.

Data also exist for Hf and Gd incorporation into the candidate phases -see Table 2, and values obtained in the present work are shown.

Table 2. Known candidate ceramic phases richest in Gd and Hf, in f.u.

Phase	Gd	Hf
Monazite	1	< 0.01#
Zircon	< 0.1#	1
Pyrochlore	2	0.3[13]
Zirconolite	1.4[13]	1
Perovskite	1 [14]	1
Apatite	8[15]	< 0.1#
Titanite	0.3#	0.5#

Values obtained in present work.

EXPERIMENTAL

Samples were generally prepared by the alkoxide-route (see e.g. [12]), in which alkoxides and aqueous nitrate solutions are mixed thoroughly, stir-dried, calcined at ~ 750°C in an appropriate atmosphere (argon or air), wet-milled using ZrO_2 media in polystyrene containers, pelletised, and finally sintered at high temperatures in an appropriate atmosphere. Silicon and P were introduced as a 40 wt% colloidal solution (Ludox) and 85% H_3PO_4 respectively.

The main characterisation tool was scanning electron microscopy, using a JEOL 6400 instrument run at 15 keV, and fitted with a NORAN Instrument Voyager IV X-ray microanalysis System (EDS) which utilised a comprehensive set of standards for quantitative work, giving a high degree of accuracy[16]. Powder X-ray diffractometry was carried out with a Siemens D500 instrument, using Co Kα radiation.

X-ray absorption near edge spectroscopy (XANES) was carried out on Line 4-2 at the Stanford Synchrotron Research Laboratory. UTi_2O_6 and $CaUO_4$ were used as valence standards.

Diffuse reflectance spectroscopy was carried out using finely powdered samples in a Cary 5 instrument, and the data were then transformed to Kabelka-Munk plots of absorbance vs. wavelength.

PREVIOUS WORK, RESULTS AND DISCUSSION

U and Pu solubilities in Perovskite, Titanite, and Apatite

U in the Ca site of perovskite, $CaTiO_3$. Although tetravalent U can be accommodated in the B site of perovskites such as $BaUO_3$, actinide ions only occupy the A site when all the cations are smaller in size. Before exploring solid solubility limits of U in $CaTiO_3$, it was important to define the U valence states occurring under various conditions of charge compensation and annealing atmospheres. While U^{3+} is not known in oxide systems, but can occur in halides, earlier work [12] on Np in $CaTiO_3$ suggested that the +3 actinide valence was preferred under reducing conditions, relative to equivalent attempted substitutions in zirconolite. So there seemed a possibility that $CaTiO_3$ might be an oxide system in which U^{3+} could be formed if a reducing atmosphere such as 3.5% H_2/N_2 was used. Hence a sample of nominal $Ca_{0.9}U_{0.1}Ti_{0.9}Al_{0.1}O_3$ stoichiometry was synthesised in which only enough Al was added in the Ti site to compensate for U^{3+}. Firing in H_2/N_2 for 16 h at 1400°C produced a single phase, which was consistent with the U^{3+} model.

XANES was employed to study the U valence more directly. The results (Fig. 1) indicated that the position and symmetry of the U L_{III}-edge from the perovskite annealed in 3.5% H_2/N_2 was identical to that from the tetravalent U compound UTi_2O_6, indicating that the U in the perovskite was tetravalent. Given the $Ca_{0.9}U_{0.1}Ti_{0.9}Al_{0.1}O_3$ stoichiometry of the perovskite, electroneutrality would dictate that 0.1 f.u. of Ti^{3+} had formed to compensate for the presence of U^{4+} in the Ca site. DRS gave no detectable reflectance, probably because of strong absorbance from Ti^{3+} - Ti^{4+} charge transfer.

To determine whether U^{4+} would still be stable under an argon atmosphere, a second sample was so annealed at 1400°C/16 hours. The L_{III} edge position observed for this sample was also coincident with that of the tetravalent U standard, although slight asymmetric broadening on the high-energy side of the white line was evident which would suggest that a small proportion of higher-valent U may be present. This asymmetric broadening has been marked in Figure 1 with an arrow and is very pronounced in hexavalent U compounds, such as $CaUO_4$.

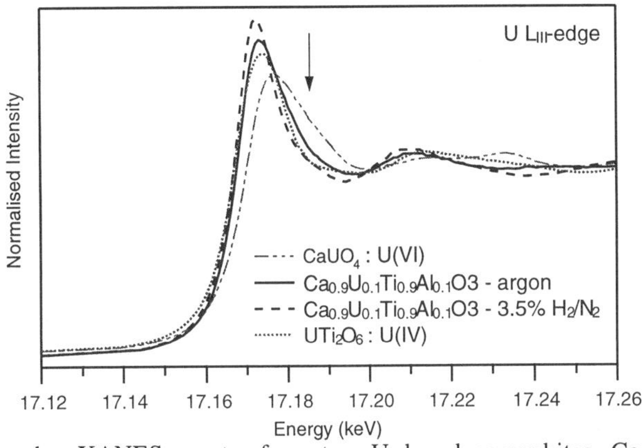

Figure 1. U L_{III}-edge XANES spectra from two U-doped perovskites, $Ca_{0.9}U_{0.1}Ti_{0.9}Al_{0.1}O_3$, annealed in argon and 3.5% H_2/N_2 along with the tetra- and hexavalent U standards, UTi_2O_6 and $CaUO_4$ respectively.

XRD of the sample annealed in argon indicated that a mixture of perovskite and pyrochlore were present. This would be expected on the basis of both the U and the Ti being in the tetravalent state. The calculated phase assemblage was :
$Ca_{0.9}U_{0.1}Ti_{0.9}Al_{0.1}O_3$ --> 0.9 $\{Ca_{0.945}U_{0.055}Ti_{0.889}Al_{0.111}\}$ + 0.05 $\{CaUTi_2O_7\}$ and this agreed well with the SEM observations. Further work is under way to determine the solubility limit of U^{4+} in perovskite, which is clearly significantly greater than the value of 0.01 f.u. reported previously [10] as being in solid solution, albeit not the solubility limit.

Pu in the Ca site of $CaTiO_3$. Begg et al.[12] showed that 0.1 f.u. of Pu^{3+} or Pu^{4+} could be readily incorporated in the Ca site of perovskite, using 0.1 (for Pu^{3+}) or 0.2 f.u. of Al (Pu^{4+}) as charge compensators, with the Pu^{3+} and Pu^{4+} samples being hot-pressed in graphite dies at 1250°C or fired in air at 1400°C respectively. The incorporation of greater amounts of Pu was then attempted using 0.3 f.u. of Pu in samples of $Ca_{0.7}Pu_{0.3}Ti_{0.7}Al_{0.3}O_3$ (Pu nominally as Pu^{3+}) or $Ca_{0.7}Pu_{0.3}Ti_{0.4}Al_{0.6}O_3$ (Pu^{4+}) stoichiometries. A sample of the former stoichiometry, fired in argon for 24 h at 1400°C, yielded essentially single-phase perovskite of the designated composition. This was expected since the endmember $PuAlO_3$ also has the perovskite structure[11]. The sample of the latter stoichiometry when fired in air at 1400°C for 24 h, yielded perovskite containing a maximum of 0.13 f.u. of Pu along with 0.33 f.u. of Al. Within experimental error, the Al content of the perovskite was twice that of the Pu, as would be expected if the Pu was present as Pu^{4+}. Other phases present were PuO_2 and $CaAl_2O_4$.

U and Pu in the Ca site of titanite. Early XRD work [8] suggested that ~ 0.05 formula units of U^{4+} would substitute in the Ca site of titanite, using Al as the charge compensator in the Ti site, i.e. a stoichiometry of $Ca_{(1-x)}U_xTi_{(1-2x)}Al_{2x}SiO_5$. Samples with x = 0.05 and 0.1 were fabricated by firing in argon at 1300°C. From SEM, these samples mainly consisted of titanite containing 0.02 f.u. of U with 0.04 f.u. of Al as compensation, plus UO_2 which contained a small amount of Ca. Traces of plagioclase ($CaAl_2Si_2O_7$) were also present. In the x = 0.1 sample, material of approximate $CaUTi_2O_7$ stoichiometry was also present. The U^{4+} solubility found here is somewhat lower than found previously[8], i.e. a range of ~ 0.03 and ~ 0.06 f.u. depending on the heat-treatment in the range of 1200-1300°C, so a more systematic study of U solubility as a function of firing temperature and cooling rate seems warranted.

A sample of $Ca_{0.9}Pu_{0.1}Ti_{0.8}Al_{0.2}SiO_5$ stoichiometry was fired in air at 1250°C and SEM showed the solubility limit of Pu in the titanite was 0.02 f.u. The average Al content of the titanite phase was 0.07 f.u., a value somewhat in excess of the expected value (0.04 f.u.) for the Pu being present as Pu^{4+} in the Ca site. The reason for the discrepancy is not presently clear. Additional anorthite ($CaAl_2Si_2O_8$) and PuO_2 were present. A sample of $Ca_{0.7}Pu_{0.3}Ti_{0.7}Al_{0.3}SiO_5$ stoichiometry fired in argon at 1250°C melted, but when fired at 1200°C for 16 hours, SEM showed the solubility limit as 0.05 f.u. of Pu. The Al content of the titanite was 0.09 f.u., and again PuO_2 and anorthite were observed as extra phases. The increased solubility of Pu in titanite in the argon-fired material could well indicate that at least some of the Pu in the titanite was present as Pu^{3+}.

U and Pu in apatite Ouzegane et al.[17] have found that about 0.15 f.u. of Th and minor U are present in Algerian britholites (apatite structures containing several f.u. of rare earths) and ~ 0.4 f.u. of Th are present in a Quebec britholite[18]. It would be expected that Th and U^{4+} would have approximately similar solubilities.

0.33 f.u. of U was notionally substituted as U^{4+} on the Ca site, with 0.66 f.u. of P being replaced by Si; an extra 0.33 f.u. of Gd was also substituted on the Ca site, with Si substituting for P, to give a nominal stoichiometry of $Ca_9Gd_{0.33}U_{0.33}(PO_4)_5(SiO_4)O$. After sintering at 1300 or 1400°C in argon, more whitlockite than apatite had formed, with residual fluorite-structured $(U,Ca,Gd)O_{2-x}$ being left also. Whereas the U, Gd and Si contents of the whitlockite were < 0.01, 0.02 and 0.3 f.u. respectively, the respective contents in the apatite phase were 0.5, 0.5 and 2.5 f.u. Although the apatite content after sintering at both temperatures was only ~ 20%, the solubility of U in the apatite has been exceeded, as indicated by the presence of the excess U-bearing fluorite. The derived solubility of U in apatite was close to the Th content of the Quebec britholite.

Pu^{4+} in monazite. Here the aim was to investigate the possible formation of a Pu^{4+} analogue of $Ca_{0.5}U_{0.5}PO_4$ [1]. Firing a sample of the appropriate stoichiometry in air at 1400°C yielded a material giving a monazite XRD pattern, plus a very small amount of PuO_2. However SEM showed that the monazite composition was given by $Ca_{0.6}Pu_{1.4}(PO_4)_2$, which is consistent with a mixture of 0.6 and 0.8 f.u. of Pu^{4+} and Pu^{3+} respectively. Minor whitlockite, containing ~ 0.15 f.u. of Pu (valence unknown) was also observed by SEM, with ~ 0.1 f.u. of Ca vacancies acting as charge compensators.

Neutron absorbers in titanite, monazite, zircon and apatite

Titanite. Previous work [8] had shown that ~ 0.2 f.u. of La would enter the Ca site of titanite, using Al compensators in the Ti site. In the present work, Gd was similarly substituted in the Ca site, i.e. $Ca_{(1-x)}Gd_xTi_{(1-x)}Al_xSiO_5$. Samples were fired in air at 1300°C, not far below the melting point of pure titanite at 1382°C. From SEM, the limiting solubility was found as ~ 0.3 f.u., a slightly higher value than that deduced for La. This discrepancy is attributed to a closer fit for Gd^{3+} in the Ca site than La^{3+}. The Al compensators were distributed fairly evenly between the Si and Ti sites. The additional phases, once the solid solubility limit was exceeded, were $(Gd,Ca)_2SiO_5$ and perovskite-structured Gd-rich titanate.0.5 f.u. of Hf was substituted in the Ti site and complete solid solution was observed. No such solid solution was observed after substituting 0.75 or 1 f.u. of Hf for Ti.

Monazite. Hf was substituted similarly to the +4 actinides, i.e. $Ce_{(1-2x)}Hf_xCa_xPO_4$. For samples fired in air at temperatures up to 1400°C, in which x = 0.05 and 0.1, the maximum solubility of Hf in the monazite was < 0.01 f.u. Additional phases observed were $CaHf_2(PO_4)_3$ and whitlockite, $Ca_3(PO_4)_2$.

Zircon. It is difficult to synthesise phase-pure zircon by ceramic techniques, even by the standard alkoxide route, although more elaborate sol-gel methods[19] and even wet-milling of oxide powders can be employed successfully[20]. Various attempts in the present alkoxide route work to incorporate rare earths in zircon via the coupled $RE^{3+} + P^{5+} \longleftrightarrow Zr^{4+} + Si^{4+}$ substitution led to the formation of fine-grained mixed zircon/xenotime products on extended firing at temperatures up to 1500°C, with little evidence of significant solid solution. The estimated solid solution limit is put at < 0.1 f.u. with further work being necessary to accurately define it.

Apatite. Hf was substituted similarly to U in apatite, as described earlier, and samples were fired in air at temperatures of 1100, 1200, 1300 and 1400°C. In all cases, the major phase was whitlockite, with some tens of percent of apatite forming, plus HfO_2. The apatites were all too fine grained to analyse, but the widespread presence of HfO_2 needles showed the Hf content of the apatite to be much less than the 0.33 f.u. in the starting material, even allowing for the apatite phase being of less abundance than whitlockite. It is conservatively estimated that there are < 0.1 f.u. of Hf in the apatite.

CONCLUSIONS AND FINAL REMARKS

The main purpose of this work was to collect information on the solubilities of actinides and neutron absorbers in the candidate ceramic phases, rather than to make judgments on their candidacy. However some remarks can be made. Apatite has high Gd solubilities, but very limited solid solubilities for U and Hf. Titanite also has low solubilities of U and Pu, although the use of argon firing atmospheres favors Pu solubility, presumably at least partly because of forming Pu^{3+} instead of the Pu^{4+} expected to be formed in air, although relatively large accommodations of neutron absorbers were observed. Monazite has large solubilities for Pu, U and Gd but not Hf. Zirconolite and pyrochlore have extensive solubilities of the actinides and the neutron absorbers, while perovskite has a high solubility for neutron absorbers and trivalent Pu but limited solubilities for tetravalent Pu and U.

The valence state of U in the Ca site of $CaTiO_3$ was shown to be +4 (solubility limit >0.1 f.u.) after firing in both argon and hydrogenous atmospheres at 1400°C, from XANES work as well as deductions made from SEM investigation.

ACKNOWLEDGEMENTS

This work was supported in part by the Environmental Management Science Program of the US Department of Energy. The XANES work was conducted at the Stanford Synchrotron Research Laboratory which is supported by the Office of Basic Energy Sciences, US Department of Energy. We wish to thank A. Brownscombe, D. Caudle and S. Krismer for assistance in sample preparation, and S. Thomson and V. Luca for help with the diffuse reflectance measurements.

REFERENCES

1. G. J. McCarthy, J. G. Pepin and D. D. Davis, in Scientific Basis for Nuclear Waste Management, Volume 2, edited by C. J. M. Northrup (Plenum, New York and London, 1979) p. 297-306.
2. C. J. Bjorklund, J. Amer. Chem. Soc., 79, p. 6347 (1958).

3. B. D. Begg, N. J. Hess, W. J. Weber, S. D. Conradson, M. J. Schweiger and R. C. Ewing, J. Nucl. Mater., in press.
4. C. Keller, Nukleonik, 5, p. 41 (1963).
5. F. J. Dickson, K. D. Hawkins and T.J. White, J. Solid State Chem., 82, p. 146 (1987).
6. S. Shoup and C. E. Bamberger, in Scientific Basis for Nuclear Waste Management XIX, edited by W. J. Gray and I. R. Triay (Materials Research Society, Pittsburgh, PA, USA, 1997), p. 379-86.
7. F. W. Clinard, Jr., L. W. Hobbs, C. C. Land, D. E. Peterson, D. L. Rohr and R. B. Roof, J. Nucl. Mater., 105, p. 248 (1982).
8. E. R. Vance and D. K. Agrawal, Nucl. Chem. Waste Manage., 3, p. 229 (1982).
9. W. J. Weber, J. Amer. Ceram. Soc., 65, p. 544 (1982).
10. E. R. Vance and K. K. S. Pillay, Rad. Effects, 62, p. 25 (1982).
11. C. Keller and K. H. Walter, J. Inorg. Nucl. Chem., 27, p. 1247 (1965).
12. B. D. Begg, E. R. Vance and S. D. Conradson, J. Alloys and Compounds, 271-3, p. 221 (1998).
13. E. R. Vance, M. W. A. Stewart, R. A. Day, K. P. Hart, M. J. Hambley and A. Brownscombe, ANSTO Report R97m030 to Lawrence Livermore National Laboratory (1997).
14. G. J. McCarthy, R. Roy and W. B. White, Mater. Res. Bull. 4, p. 251 (1969).
15. G. J. McCarthy, Nucl. Technol., 32, p. 92 (1977).
16. E. R. Vance, R. A. Day, Z. Zhang, B. D. Begg, C. J. Ball and M. G. Blackford, J. Solid State Chem., 124, p.77 (1996).
17. K. Ouzegane, S. Fourcade, J-R. Kienast and M. Javoy, Contr. Min. Petr., 98, p. 277 (1988).
18. M. R. Hughson and J. E. Sen Gupta, Amer. Mineral., 49, p. 937 (1964).
19. R. C. Ewing and R. F. Haaker, J. Amer. Ceram. Soc., 64, p. C-149 (1981).
20. D. R. Spearing and J. Y. Huang, J. Amer. Ceram. Soc., 81, p. 1964 (1998).

CATHODOLUMINESCENCE OF Ce, U AND Pu IN A GARNET HOST PHASE

M.V. ZAMORYANSKAYA and B.E. BURAKOV.
V.G.Khlopin Radium Institute, 28, 2-nd Murinskiy ave., St.Petersburg, 194021, Russia,
fax: (7)-(812)-346-1129; e-mail: burakov@riand.spb.su

ABSTRACT

Ceramic materials based on garnet, $(Y,Gd,An,\ldots)_3(Al,Ga,An,\ldots)_5O_{12}$ and perovskite $(Y,Gd,An,..)(Al,Ga,An,\ldots)O_3$ structures have been proposed for the immobilization of weapons-grade actinide-containing waste materials with complex chemical compositions. Cathodoluminescence (CL) images and emission spectra of synthetic garnet and perovskite crystals containing Ce, U and Pu were studied. It was determined that Pu^{3+} incorporated into the garnet matrix has characteristic CL emission bands at 1.9 and 1.6 eV. The loading capacity of the garnet lattice for Pu^{4+} incorporation is significantly higher than for Pu^{3+} and similar to U^{4+}. The maximum amount of Pu^{3+} that may be incorporated into the garnet structure experimentally is 0.3 wt.% in comparison with 5.3 wt.% for Pu^{4+}. The CL emission spectra of Ce^{3+} in different materials is a characteristic property and can be used for identification of garnet and perovskite phases into the multiphase ceramic matrices simultaneously with microprobe analysis.

INTRODUCTION

Ceramics based on garnet, $(Y,Gd,An,\ldots)_3(Al,Ga,An,\ldots)_5O_{12}$ and perovskite $(Y,Gd,An,..)(Al,Ga,An,\ldots)O_3$ have been proposed for the immobilization of weapons-grade actinide-containing wastes of complex chemical compositions. Crystalline phases such as garnet and perovskite are chemically and phisically extremely durable and thus desirable for the incorporation of Pu, Am, and other actinide and non-radioactive elements found in waste streams [1-3]. The lattice capacity of these materials, however, depends on the valence state and the ionic radius of the substituted ions [3]. Previous work has demonstrated that cathodoluminescence (CL) allows the determination of the following ions from the characteristic CL emission of those elements in different materials: Ce^{3+}, U^{6+} Cr^{3+}, Gd^{3+} [3]. The main features of the CL emission of ions depends on the crystalline structure of the host phase and the ion(s) emitting the CL. In some cases, CL may be also used for phase identification.
The main goal of this work was to study the Pu valence state in Pu-activated garnet using CL. No information was found in the literature concerning CL emission of the Pu-ion into the crystalline matrices. An additional purpose was to use CL for phase identification of Ce and Pu-doped garnet/perovskite ceramic materials.

EXPERIMENTAL METHODS

Powdered starting materials of different stoichiometry consisting of Gd, Al, Ca, Pu, Ce, Ga, Sn-oxides were been cold pressed in order to obtain pellets 5 mm in diameter. The pellets were then melted in air using a hydrogen torch for 5-8 minutes. The temperature of the hydrogen torch was not measured precisely but was estimated to be approximately 1900-2100°C. In comparison to the flame melting temperatures for all samples were approximately 300-400°C less. One sample

Mat. Res. Soc. Symp. Proc. Vol. 608 © 2000 Materials Research Society

of pure monocrystalline $Y_3Al_5O_{12}$ (YAG) was cut by using a diamond blade to obtain a sample with dimensions of 2 x 2 x 25 mm. Then, PuO_2 powder was spread not homogeneously on the edge of the crystal. Plutonium-activated YAG was finally obtained through melting of this edge in the hydrogen torch at a temperature of approximately 2000°C.
All samples from the experiments were examined using CL, scanning electron microscopy, and the electron microprobe. Identification of garnet and perovskite phases was confirmed by powder X-ray diffraction analysis. Acquisition of CL spectra, and electron microprobe analyses were carried out using the following parameters: accelerated voltage of 15kV and a beam current of 10-50nA depending on the objective.

RESULTS AND DISCUSSION

Analysis of Pu valence state

The cathodoluminescence image of Pu-activated YAG is not uniform indicating heterogeneity of Pu in that material. In comparison with pure YAG which has bright blue CL emission the Pu-activated YAG is characterized by the dark and light blue zones (Fig.1).

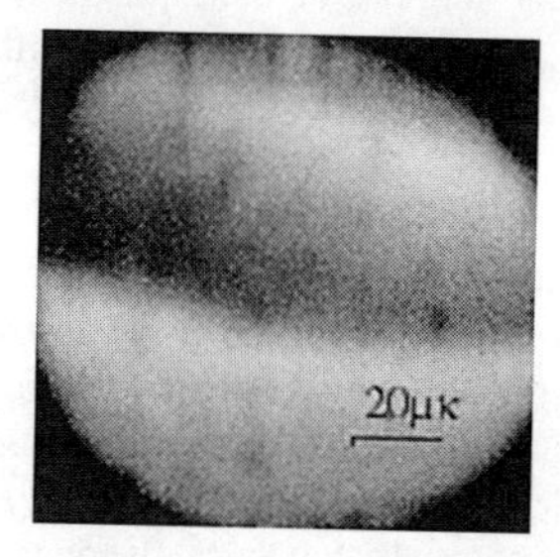

Fig.1. CL image of YAG activated with Pu^{3+}.

Results of microprobe analyses show that the concentration of Pu in dark zones (0.1-0.2 wt.% element) is higher than in the light ones (0.01-0.05 wt.% element). Two groups of narrow bands are observed in the CL spectrum of dark zones at 1.9 and 1.6 eV (Fig.2). Their intensity is strongly correlated with the Pu content in those zones. The CL spectrum of the Pu-free zones in the same sample is similar to the CL spectrum of pure standard YAG and it does not contain those bands. The CL spectrum of PuO_2 used as a standard of Pu^{4+} does not show any characteristic band emission. The author's interpretation of the CL data is that the narrow bands at 1.9 and 1.6 eV in the CL spectrum of the Pu-doped YAG crystals are caused by 5*f* electronic transitions of the Pu^{3+} ions. These results are original and published for the first time.

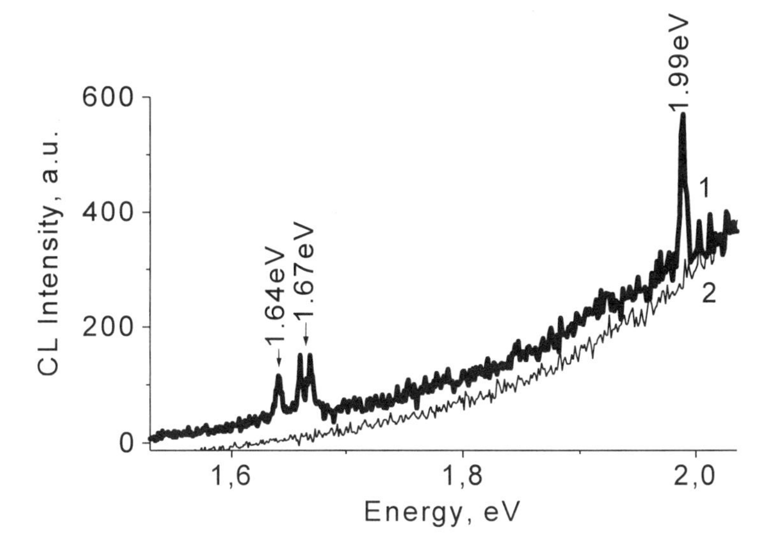

Fig.2 CL spectra of YAG: 1 – YAG activated with Pu^{3+}; 2 – pure YAG

Phase identification using CL of Ce^{3+}

The CL spectra of Ce^{3+} in garnet with compositions: $Y_3Al_5O_{12}$; $(Y,Gd)_3(Al,Ga)_5O_{12}$; $(Gd,Ca)_3(Al,Ga)_5O_{12}$ and perovskite with compositions: $YAlO_3$; $GdAlO_3$; $(Gd,Ca)(Al,Ga)O_3$ are shown in Fig.3.

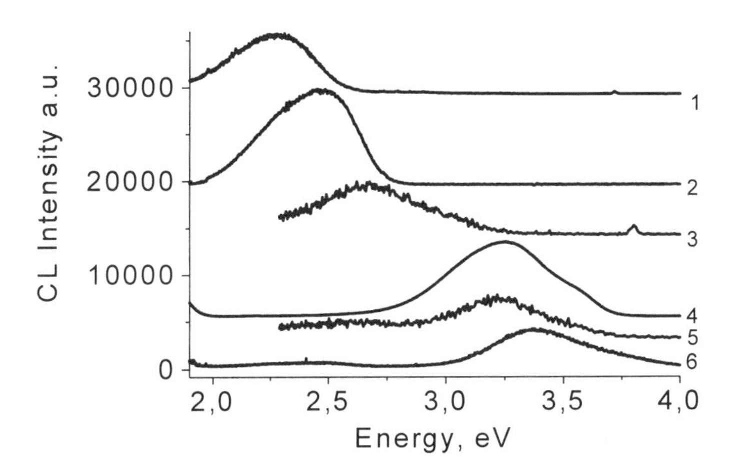

Fig.3 CL spectra of Ce^{3+} of garnet and perovskite: 1 - $Y_3Al_5O_{12}$; 2 - $(Y,Gd)_3(Al,Ga)_5O_{12}$; 3 - $(Gd,Ca)_3(Al,Ga)_5O_{12}$; 4 - $YAlO_3$; 5 - $GdAlO_3$; 6- $(Gd,Ca)(Al,Ga)O_3$.

An intensive CL emission from the Ce^{3+} ion is observed in garnet - at 2.3-2.6 eV and for perovskite at 3.1-3.5 eV. Most of the Ce in the Ce-doped garnet crystals with complex chemical composition (containing Ca, Sn [3] and other impurity elements) is in the 4+ oxidation state and is not characteristic of the CL emission. Tetravalent Ce, however, does not suppress the CL emission of Ce^{3+}. Therefore, the CL spectra of Ce^{3+} can be effectively used for the phase identification of garnet and perovskite into the multiphase matrices.

CL study of Pu-doped garnet/perovskite ceramics

Garnet and perovskite phases of the following simplified formulas :

- $(Y,Gd,Pu)_3(Al,Cr,Pu)_5O_{12}$ – *garnet 1*;
- $(Gd,Ce,Ca,Pu)_3(Al,Ga,Pu,Sn)_5O_{12}$ – *garnet 2*;
- $(Gd,Ce,Ca,Pu)(Al,Ga,Pu,Sn)O_3$ - *perovskite*

were investigated in samples doped with 5 wt.% ^{239}Pu. Their CL emission spectra are shown in Fig.4. Emission of Cr^{3+}, Gd^{3+} and Pu^{3+} are observed in the CL spectrum of *garnet 1* and Gd^{3+} and Ce^{3+} in the CL spectrum of *garnet 2*. Microprobe analyses reveal that *garnet* samples *1* and **2** contain 0.3 and 5.3 wt.% Pu, respectively. Therefore, higher Pu incorporation into *garnet 2* is correlated only with Pu^{4+}. This situation is similar to our experience with U incorporation in the garnet structure: the highest U contents (3-4 wt.%) were observed for only where the 4+ valence state was dominant. The CL emission of U^{6+} however, was observed in the garnet containing less than 0.05 wt.% U [3].

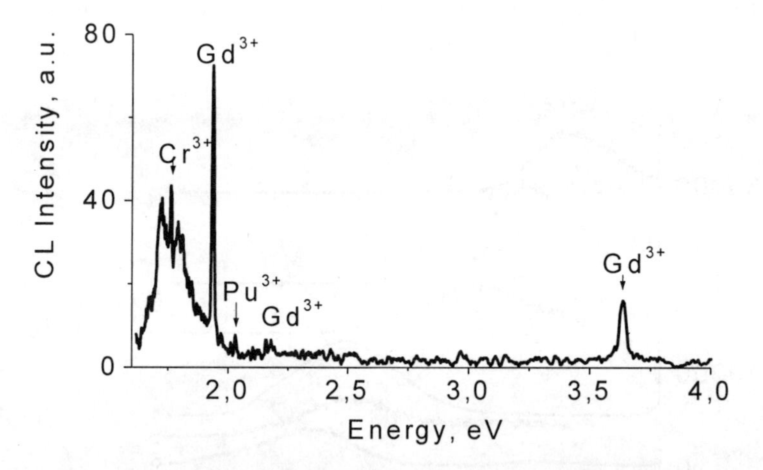

Fig.4 CL spectrum of *garnet 1* $(Y,Gd,Pu)_3(Al,Cr,Pu)_5O_{12}$

The CL spectrum of *perovskite* contains an intense emission band from Ce^{3+} at 3.2 eV in comparison with Ce^3+ at 2.6 eV for *garnet 2* in the matrix of the same sample. The peak shifts for the Ce^{3+} ion in garnet and perovskite may be attributed to differences in the crystal fields of those two host phases. This may be used as a diagnostic property with which to characterize these materials (Fig.5).

The study of Pu CL into perovskite phase is currently in progress.

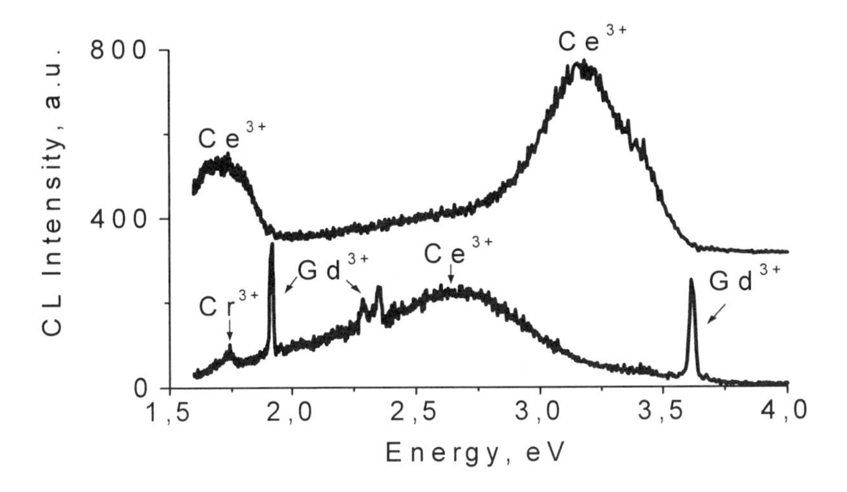

Fig.5 CL spectra of 1- $(Gd,Ce,Ca,Pu)_3(Al,Ga,Pu,Sn)_5O_{12}$ – *garnet 2*;
2- $(Gd,Ce,Ca,Pu)(Al,Ga,Pu,Sn)O_3$ - *perovskite*

CONCLUSIONS

The results obtained in this study allow us to make the following conclusions:

1) The Pu^{3+} ion incorporated into the garnet matrix has characteristic CL emission bands at energies 1.9 and 1.6 eV. No CL bands were observed for Pu^{4+} in the region 1.5-4.0 eV ;
2) The loading capacity of the garnet lattice for Pu^{4+} incorporation is significantly higher than for Pu^{3+}. The maximum amount of Pu in the valence state 3+ that could be incorporated into the garnet structure experimentally was 0.3 wt.% in comparison with 5.3 wt.% for Pu^{4+};
3) The CL spectra of Ce^{3+} can be used for a simple characteristic identification of garnet and perovskite phases into the multiphase ceramic matrices and done simultaneously with electron microprobe analyses.

ACKNOWLEDGEMENTS

Authors are very grateful to Mrs. M. A. Yagovkina, Mrs. M. A. Petrova and Mr. V. M. Garbuzov from V. K. Khlopin Radium Inst. for their participation in results interpretation and for the help on sample synthesis and examination. This work was supported by International Scientific and Technological Center (ISTC), Project #1063. Presentation of this paper was sponsored by MRS'99 Organizing Committee (Symposium QQ). Authors appreciate personally support from Symposium QQ Organizers, Drs. R.W. Smith and D.W. Shoesmith, and to Prof. J.M. Hanchar from The George Washington University, Profs. R.C. Ewing from The University of Michigan and Werner Lutze from The University of New Mexico.

REFERENCES

1. B.E, Burakov, E.E, Strykanova, Proceedings of International Symposium Waste Management-98, CD-ROM version, sess34 - 05, (1998).

2. B.E, Burakov, E.B, Anderson, D.A, Knecht, in Environmental Issues and Waste Management Technologies IV, 349-356 (1999).

3. B.E, Burakov, E.B, Anderson, D.A, Knecht, M.V, Zamoryanskaya, E.E, Strykanova, M.V, Yagovkina, Mat. Res. Soc. Symp. Proc. Proceedings Scientific Basis for Nuclear Waste Management XXII, (1999), Vol 556, 55-62.

Al_2O_3-DOPED TiO_2 CERAMIC WASTE FORMS PRODUCED BY MELTING METHOD

Masayoshi Uno, Hajime Kinoshita, Shinsuke Yamanaka,
Department of Nuclear Engineering, Osaka University, Yamada-oka 2-1, Suita, Osaka 565-0871 JAPAN

ABSTRACT

We studied the phase stability and melting temperature of alumina-doped alkaline earth and rare earth -Ti-O system with the composition variation. Melting of the mixture of Nd_2O_3, CeO_2, SrO, TiO_2 and Al_2O_3 at 1673 K in 1 hour produced $RE_2Ti_3O_9$ phase, and the chemical formula of the oxide was assumed to be $(Ce,Nd,Sr)_2(Ti,Al)_3O_9$. A Differential Scanning Calorimetry (DSC) measurement showed that the melting temperature of this compound was 1646 K. Al and Sr was considered to contribute to the reduced melting temperature of the oxides in the system. The hardness comparing with ceramic wastes and the leach rate superior to glass waste forms were confirmed.

INTRODUCTION

Solidification of low level waste by melting has been already done in nuclear plants for convenience and for a high volume reduction. It is of great interest to make waste forms of high level waste by a simple melting procedure like vitrification. However, the complex oxides of rare earth element (RE) or transuranium element (TRU) and Zr, as well as Al or Ti (a host material) usually have much higher melting temperatures than those of glass waste forms. We found that complex oxides of some alkaline earth and rare earth elements (AE,RE) and Ti with an ideal composition $(AE,RE)_2Ti_3O_9$ melt up to 1673 K by doping a small amount of alumina. The phase diagrams of the systems containing more than two RE elements or Al are not available, and the melting temperature of the oxides in the systems is not reported. We therefore, studied the phase stability and melting temperature of alumina-doped RE-Ti-O systems with compositional variation. In addition, important properties as a waste form such as density, hardness and chemical durability, of alumina doped $(AE,RE)_2Ti_3O_9$ were measured.

EXPERIMENTAL

The starting materials were commercially supplied Al_2O_3, ZrO_2, CeO_2, Nd_2O_3, SrO, TiN (from Nacalai Tesque) and TiO_2 (from Kanto Kagaku), all in powder form and of purity exceeding 99.9%.

The starting mixture mixed in an agate mortar was put in a ZrO_2 crucible and heated by a Nagano-Keiki carbon heater furnace. The temperature of samples was raised to 1673 K in 5 hours, maintained at the temperature for 1 hour and then cooled to room temperature in 5 hours in flowing argon.

The reaction products were identified by X-ray diffraction. The X-ray diffraction was performed with Cu-Kα radiation on a Rigaku RINT-2000 diffractometer equipped with a curved graphite monochrometer. The distribution of elements in the reaction products was studied by Electron probe micro analysis (EPMA) using Topkon MINI-SEM 100 and Horiba EMAX-8000 units.

The melting temperature induced on a 10mg sample by heating at a rate 25K/min in flowing argon was monitored by differential scanning calorimetry (DSC), using MAC Science DSC-3000 system.

Condensed sample after cooling was cut using a diamond impregnated saw and polished with emery papers and a diamond paste had a geometric surface area of 540-600 mm^2. Chemical durability was estimated by the Soxhlet leach test(MCC-5) method [1] followed by the analysis of the leachate by Inductively Coupled Plasma Spectrometry (ICP-ES) method by Shimazu ICP - 7500. The leaching test was performed in distilled water at 337 K for 7 days, in accordance with the procedure of MCC-5. After 7 days, the sample and the leachate were separated. The sample was washed with deionized water, and was dried to constant mass in a 383 K oven and weighed.

Density (d) was measured by the method of Archimedes using Wardon Pycnometer supplied by Shibata Scientific Tec. ltd. The Vickers hardness measurement was carried out using MHT- 1 micro Vickers hardness tester supplied by Matsuzawa Seiki co. ltd.

RESULTS AND DISCUSSION

As it is not clear that aluminum atom dissolves in RE site or Ti site runs Nos.1-7 shown in Table 1 were performed. The identified phases, matching of the identification, occurrence of melting and melting temperature measured DSC are also summarized in Table 1. In run No.1, only perovskite type Nd_2TiO_3 was obtained, but this compound did not melt at 1673K. In runs Nos. 2-4, the atomic ratio, (Al+Nd)/Ti of the starting mixture was fixed the products melted, but were not perovskite phases. Although the broadening of the X-ray diffraction peaks suggested the low crystallinity but the products might be the mixture of orthorhombic Nd_2TiO_5 and monoclinic $Nd_2Ti_2O_7$ both of which containing Al. The reactions of the staring mixtures which contained the fixed atomic ratio, Nd/(Ti+Al) (runs No. 5-7) produced the perovskite $Nd_2(Ti,Al)_3O_9$. However these products did not melt at 1673K. Since the atomic ratio, Nd/(Ti+Al)=2/3 caused the formation of the perovskite type $Nd_2(Ti,Al)_3O_9$ it is suggested that the replacement of a Ti atom by an Al atom causes the formation of perovskite ($Nd_2Ti_3O_9$) containing Al.

In runs Nos.8-12, CeO2 and SrO were added in the system. X-ray diffraction patterns showed that all the products were perovskite phases with high crystallinity with a trace of $Al_2Ti_7O_{15}$. The compositions of the products shown in table 1 were derived from the fact that they were only perovskite type $A_2B_3O_9$ phase. The product in run No. 8 without Sr and Al did not melt at 1673K but the product in run No. 9 with Sr did. Thus, the addition of Sr is found to contribute to the reduction of melting temperature of the oxide. It is also found that addition of Al contributes to the reduction of melting temperature from the measured melting temperature of the products in runs Nos.11 and 12. The melting of the products in runs Nos. 2-4 arose from the dissolution of Al since the melting temperatures of these phases are larger than 1673K according to the phase diagram [2].

Lenouv et al. [3] first reported the existence of $RE_2Ti_3O_9$ type compounds in La-Ti-O and Nd-Ti-O systems as a secondary phase, and they suggested that the crystal structure was perovskite. Richard et al. [4] later reported that $Nd_2Ti_3O_9$ was produced as an intermediate phase by the dehydration of layered perovskite $Na_2Nd_2Ti_3O_{10}$ and that the structure of the compound was I4/mmm with a=3.8334 and c=24.363 Å. However they also suggested that the real symmetry was less than I4/mmm.

Table.1 Composition of the samples(mol%) and the phase of the products

Run No	TiO_2	Al_2O_3	Nd_2O_3	CeO_2	SrO	Expected Compound	Identified compound	Matching	Melting
No.1	75	0	25			$Nd_2Ti_3O_9$	$Nd_2Ti_3O_9$	Good	X
No.2	75	3	22			$(Al_{0.3}Nd_{1.7})Ti_3O_9$			O
No.3	75	6	19			$(Al_{0.5}Nd_{1.5})Ti_3O_9$	Nd_2TiO_5 and $Nd_2Ti_2O_7$	Poor	O
No.4	75	9	16			$(Al_{0.7}Nd_{1.3})Ti_3O_9$			O
No.5	70	4	26			$Nd_2(Ti_{2.7}Al_{0.3})O_9$	$Nd_2(Ti_{2.7}Al_{0.3})O_9$		X
No.6	65	8	27			$Nd_2(Ti_{2.4}Al_{0.6})O_9$	$Nd_2(Ti_{2.4}Al_{0.6})O_9$	Good	X
No.7	57	14	29			$Nd_2(Ti_2Al)O_9$	$Nd_2(Ti_2Al)O_9$		X
No.8	67	-	11	22	-	$(CeNd)Ti_3O_9$	$(CeNd)Ti_3O_9$	Excellent	O
No.9	60	-	8	16	16	$(Sr_{0.7}Ce_{0.7}Nd_{0.7})Ti_3O_9$	$(Sr_{0.7}Ce_{0.7}Nd_{0.7})Ti_3O_9$		O
No.10	56	6	8	15	15	$(Sr_{0.7}Ce_{0.7}Nd_{0.7})(Ti_{2.7}Al_{0.3})O_9$	$(Sr_{0.7}Ce_{0.7}Nd_{0.7})(Ti_{2.7}Al_{0.3})O_9$		X
No.11	53	11	7	15	15	$(Sr_{0.7}Ce_{0.7}Nd_{0.7})(Ti_{2.5}Al_{0.5})O_9$	$(Sr_{0.7}Ce_{0.7}Nd_{0.7})(Ti_{2.5}Al_{0.5})O_9$	Excellent	1646K
No12	46	23	6	12	12	$(Sr_{0.7}Ce_{0.7}Nd_{0.7})(Ti_{2.2}Al_{0.8})O_9$	$(Sr_{0.7}Ce_{0.7}Nd_{0.7})(Ti_{2.2}Al_{0.8})O_9$		1613K

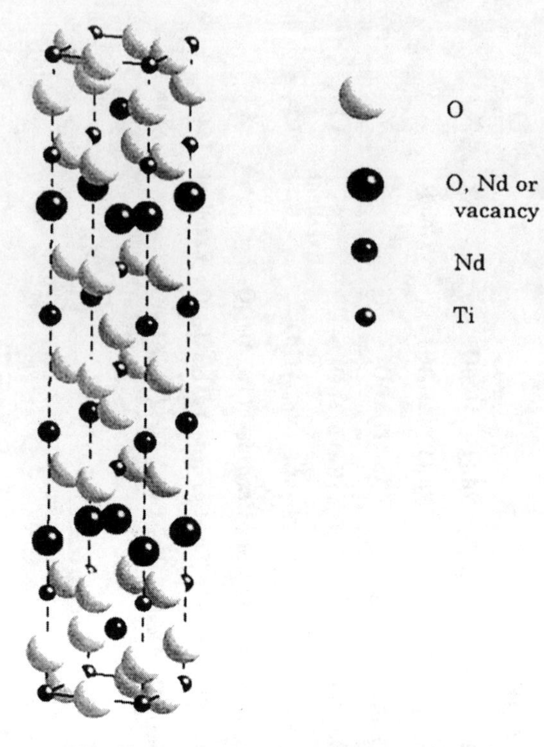

Fig. 1 Crystal structure of $Nd_2Ti_3O_9$[4]

Rietvelt analysis by Richard et al elucidated that some 1/3 Nd atom in the Nd site and 1/6 Nd or oxygen atom in the Nd and oxygen site were replaced by vacancies as shown in Fig. 1. This may cause the degradation of the properties as a waste. Some properties of the alumina doped $(AE,RE)_2Ti_3O_9$ (run No.11) as a waste were measured. The relative density calculated from the composition, lattice constant and measured density was 90.7%. Vickers hardness was 6.9 GPa. This value compares with that of Synoc or the ceramic waste[5]. The normalized leach rate of Ti was $2.24x10^{-3}$ kg/m^2. The value of Sr was below the detection limit ($0.01x10^{-3}kg/m^2$). Al was not detected from the leachate. The value for Sr of the sample was one order of magnitude smaller than that of borosilicate glass waste[6], but the value for Ti was slightly higher than that of ceramic wastes or Synroc. The results may be caused by using geometric surface of melted samples containing several pores in Soxhlet leach test in the present study. The properties of alumina doped $RE_2Ti_3O_9$, produced by melting at 1673K was confirmed to be superior to those of glass waste form.

Table 2 Properties of the alumina doped $(AE,RE)_2Ti_3O_9$ as a waste

Density	Relative	Hardness	Leachability(kg/m^2 x 10^{-3})		
(g/cm^3)	density (%)	(kg/mm^2)	Sr	Ti	Al
4.89	90.7	690	<0.01	2.24	<0.01

Perovskite phase as a host material for nuclear waste is not a new ceramic. The perovskite phase in Synroc, nominally $CaTiO_3$ was the primary host for Sr, trivalent rare earths and U^{4+}[5]. Alumina and titania base perovskite phases containing U and Pu(Ce) as a waste were studied as a host phase or a co-existing phase and reported to have capacity for immobilization of actinides [7,8,9,10]. Formation of ABO_3 type perovskite requires the larger ionic radius of A cation than that of B. No formation of perovskite phase in runs No.2-4 shows that $A_2B_3O_9$ type perovskite also satisfies this requirement. The crystallographic similarity suggests the possibility the formation of $A_2B_3O_9$ type perovskite containing actinides.

The low melting temperature that realizes the production of the waste by the convenient melting method may vary with actinide content. Perovskite type wastes containing actinides reported in the above literatures were produced by heating powder samples at 1673K-2073K for 20-80 hours. Some Synroc [11,12] and mineral-like ceramics[13,14,15] were produced by melting methods. They were produce by heat treatments at 1473-1873K. Some actinide bearing phases existed in the melted ceramics[13], but detailed information on the melting temperature of the phase and its composition variation did not reported. Alkaline elements or alkaline earth elements, which usually exist in Synroc or HLW, may contribute to low melting temperature. The effect of Sr was found in the present study. The same effect of Al content is also important since the increase in alkaline or alkaline earth element may cause increase in the leach rate of the wastes. The present perovskite phase has also possibility of host materials for immobilizing actinides produced by a convenient melting method.

CONCLUSIONS

Melting of the mixture of Nd_2O_3, CeO_2, SrO, TiO_2 and Al_2O_3 at 1673K for 1 hour produced one $(AE,RE)_2(Ti,Al)_3O_9$ phase compound. The melting temperature of this compound was 1646K. Al and Sr may contribute to the reduction of the melting temperature of the oxides in the system.

Vickers hardness compared with those of ceramic wastes and chemical durability was superior to those of glass wastes. Alumina doped $RE_2Ti_3O_9$ produced by the convenient melting method is an excellent host material for immobilizing rear earth elements although alumina this structure is considered to contain some vacancies. This material is considered to have the possibility of a host material for immobilizing actinides by melting method from consideration of the crystal structure and composition variation of melting temperature.

REFERENCES

1. Materials Characterization Center, 9-30-81.
2. R. S. Roth, J. R. Dennis and H. F. McMuride, Phase Diagrams for Ceramists, vol. 3 (American Ceramic Society, 1975).
3. A. I. Lenov et al. Bull. Acad. Sci.,5, 756 (1966).
4. M. Richard et al., J. Solid State Chem., 112, 345 (1994).
5. Radioactive Waste Forms for the Future, edited by W. Lutze and R. C. Ewing (Elsevier Science Publishing, New York, 1988).
6. K. Kawamura and J. Ohuchi, Mat. Res. Soc. Proc. 353, pp87-93(1995).
7. B. D. Begg, E. R. Vance, R. A. Day, M. Hambley and S. D. Conradson, Mat. Res. Soc. Proc. 465,pp325-332(1997).
8. B. D. Begg and E. R. Vance, Mat. Res. Soc. Proc. 465, pp333-340(1997).
9. S. V. Stefannovsky, S. V. Yudintsev, B. S. Nikonov, B. I. Omelianenko, A. I. Gorshokov, A. V. Sivtsov, M. I. Lapina and R. C. Ewing, Mat. Res. Soc. Proc., 556, pp27-33(1999).
10. S. V. Stefanovsky, S. V. Yudintsev, B. S. Nikonov, B. I. Omelianenko, and A. G. Ptashkin, Mat. Res. Soc. Proc., 556, pp121-128(1999).
11. I. A. Soblev, S. V. Stefanovsky, B. I. Omelianenko, S. V. Ioudintsev, E. R. Vance, A. Jostsons, Mat. Res. Soc. Proc., 465, pp371-378.
12. A. V. Kudrin, B. S. Nikonov and S. V. Stefavovsky. Mat. Res. Soc. Proc., 465, pp417-423(1997).

13. I. A. Soblev, S. V. Stefanovsky, S. V. Ioudintsev, B. S. Nikonov, B. I. Omelianenko and A. V. Mokhov, Mat. Res. Soc. Proc., 465, pp363-370(1999).
14. T. M. Smelova, N. V. Krylova, I. N. Shestoperov, Mat. Res. Soc. Proc., 465, pp425-431(1997).
15. S. V. Stefanovsky, S. V. Yudintsev, B. S. Nikonov, B. I. Omelianenko, A. G. Ptaskin, Mat. Res. Soc. Proc., 556, pp121-128(1999).

MELTING SIMULATED HIGH-LEVEL LIQUID WASTE WITH ADDITION OF TiN AND AlN

Masayoshi Uno, Hajime Kinoshita, Shinsuke Yamanaka,
Department of Nuclear Engineering, Osaka University, Yamada-oka 2-1, Suita, Osaka 565-0871 JAPAN,.

ABSTRACT

Calcined simulated high level liquid waste (HLLW) with a desired amount of TiN and AlN mixture was heat-treated at 1673-1873 K. It was revealed that the mixture of TiN and AlN (the atomic ratio of Al to Ti is 1:9) caused the melting of the specimen at 1673 K and the separation of the elements into two groups: alloy phase and oxide phase. The analysis of the oxide phase showed that the compounds in it could be divided into four phases, and that all fission product elements formed the complex oxides with Ti and Al. It is considered that Al and Zr dissolution in each phase contribute to the melting of the oxide phase at 1673K. A 30-days Soxhlet leach test showed that the chemical durability of the oxide phase as a waste form was superior to that of glass waste form[1].

INTRODUCTION

In order to develop a simpler and more rational solidification method than vitrification we have been studying a high temperature method by which HLLW is dry-treated in a simple process[2,3,4,5]. As shown in Fig. 1, this method consists of 4 processes. Water and nitric acid in HLLW first vaporize at 973 K in the calcination process. Cs and Rb then vaporize up to 1273K by heating calcined HLLW and would be recovered by a cold trap for the attenuation storage. Further heating with a desired amount of a reducing agent, a mixture of TiN and AlN causes reduction and melting of the sample. Elements with higher standard free energy of oxide formation (platinum metals and other transition metals) than the reducing agent are reduced and form the alloy phase. Alkaline earth elements, rare earth elements, Zr, actinides and the metal elements of the reducing agent form the oxide phase of complex oxides. With the suitable reducing agent and treatment temperature the alloy phase and the oxide phase melt in the process and the both phases are obtained separately after cooling.

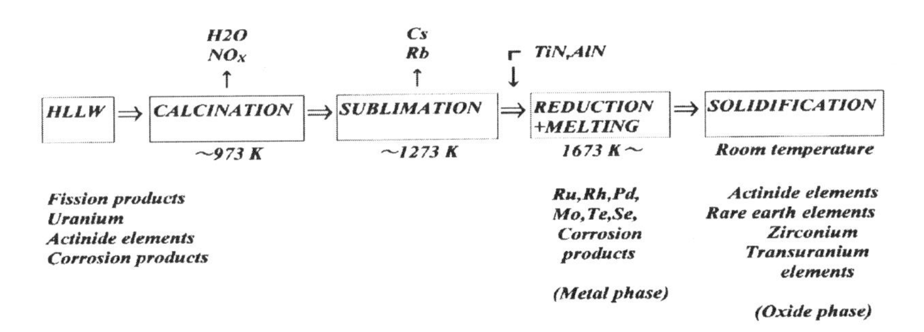

Fig.1 The process of the super high temperature method.

Mat. Res. Soc. Symp. Proc. Vol. 608 © 2000 Materials Research Society

Previous studies using simulated calcined HLLW have elucidated that the alloy phase melts up to 1673K by alloying the refractory platinum metals or Mo with corrosion products (Fe,Cr,Ni)[6], while the oxide phase melted around 1873K by the formation of complex oxides of fission product(FP) elements and Ti with addition of only TiN[7,8] and could successfully melt at 1673K by adding the mixture of TiN and AlN with the suitable composition.[9].

In the present study, the obtained oxide phase was identified and the mechanism of decrease in melting temperature was discussed. Chemical durability of the oxide phase obtained from 30 days Soxhlet leach test was also described.

EXPERIMENTAL

The composition of simulated calcined HLLW, as shown in Table 1 contains 27 FP elements, and 3 corrosion products (CP; Fe, Cr and Ni). This composition is calculated based on spent fuel of 45 GWd/t after 5 years of cooling time. It is composed of elements exist beyond amount of 1g/Mg-U in spent fuel, and contains Re and Ce instead of Tc and Pm, respectively, but contains no actinides.

Table 1 Composition of the simulated calcined HLLW

ELEMENT	CONTENT (at%)	OXIDE	CONTENT (mol%)
Mo	9.263	MoO_2	12.182
Ru	5.373	RuO_2	7.067
Pd	3.359	PdO	4.418
Re	1.999	ReO_2	2.629
Te	0.986	TeO_2	1.297
Rh	1.075	Rh_2O_3	0.707
Cd	0.254	CdO	0.334
Se	0.191	SeO_2	0.251
Ag	0.167	Ag_2O	0.110
Sn	0.189	SnO_2	0.248
Sb	0.040	Sb_2O_3	0.026
In	0.005	In_2O_3	0.003
Fe	25.827	Fe_2O_3	16.984
Cr	6.751	CrO	8.879
Ni	2.658	NiO	3.496
Cs	4.567	Cs_2O	3.003
Rb	1.122	Rb_2O	0.738
Ba	3.447	BaO	4.533
Sr	2.391	SrO	3.144
Ce	4.498	CeO_2	5.916
Nd	7.471	Nd_2O_3	4.913
La	2.334	La_2O_3	1.535
Pr	2.108	Pr_2O_3	1.386
Y	1.399	Y_2O_3	0.920
Sm	1.352	Sm_2O_3	0.889
Eu	0.233	Eu_2O_3	0.153
Gd	0.225	Gd_2O_3	0.148
Tb	0.005	Tb_2O_3	0.003
Dy	0.002	Dy_2O_3	0.001
Zr	10.711	ZrO_2	14.087

The amount of the reducing agent was determined such that it reduces all platinum metal oxides, other transition metal oxides and CP oxides, which form the metallic phase. The weight ratio of the reducing agent to the simulated calcined HLLW was about 1:3. The atomic ratio of metal elements of the reducing agent to FP elements in oxide phase was about 6:4. Al_2O_3 was also used instead of AlN since previous study[7] showed that some platinum metal oxides were thermally reduced without the reducing agent. The composition of the starting mixture is shown in table 2.

The reducing agent powder and the simulated calcined HLLW powder only mixed in a agate mortar were put in a ZrO2 crucible and heated to 1873 K or 1673K in 3 hours in flowing Ar by a Nagano-Keiki carbon heater furnace. After holding at the temperature for 1 hour the sample was furnace cooled.

The reaction products were identified by X-ray diffraction. The X-ray diffraction was performed with Cu-K α radiation on a Rigaku Rint-2100 diffractometer equipped with a curved graphite monochromator. The distribution of elements in the reaction products was studied by Electron probe micro analysis (EPMA) using JSM-5800LN and JED-2110 units (JEOL).

Samples were cut using a diamond-impregnated saw and polished with emery papers and a diamond paste had a geometric surface area of 540-600 mm^2. Chemical durability was estimated by the Soxhlet leach test (MCC-5) method[10] followed by the analysis of the leachate by Inductively Coupled Plasma Spectrometry(ICP-ES) method by Shimazu ICP - 7500. The leaching test was performed in distilled water at 337 K for 30 days, in accordance with the procedure of MCC-5. After 30 days, the sample and the leachate were separated. The sample was washed with deionized water and was dried to constant mass in a 383 K oven and weighed.

RESULTS AND DISCUSSION

Table 2 shows the results of melting of the products. Sample No.1 heat-treated with only TiN did not melt at 1673K but melted up to 1873 K. Sample No.2, where 5mol% TiN was replaced by AlN did not melt at 1673K and the separation of the alloy phase and the oxide phase did not occurred. However, by further replacement of TiN by AlN, samples Nos. 3 and 4 melted at 1673K and the alloy and oxide phases were obtained separately after cooling.

Since several small spheres of metal phase existed in the oxide phase of sample No.4, that contains twice as much Al as sample No.3, the extent of melting and separation of sample No.4 was considered to be insufficient. Therefore, the composition of the additive in sample No.3 (Ti:Al=9:1) was regarded as the best composition. Addition of Al_2O_3 instead of AlN made no difference in the results.

Table 2 Composition of the starting mixture and results of the products

Sample No.	HLLW:Additive*	Additive (Ti:Al)	Treatment temperature(K)	Result**
1	4:6	———	1873	O
2	4:6	9.5:0.5	1673	X
3	4:6	9:1	1673	O
4	4:6	8:2	1673	O

* The ratio of FP metal elements coming to the oxide phase to metal elements of the additives.

**O:Samples melted and separated.

X:Sample did not melt and separated.

The X-ray diffraction pattern for the oxide phase of sample No.3 is shown in Fig. 2. Assigning these peaks by ASTM cards, the following compounds were assumed to exist in the product; a series of AE-RE-Ti-O compounds, $BaLa_2Ti_4O_{12}$, $BaNd_2Ti_4O_{12}$, and $Ba_4La_8Ti_{17}O_{50}$, which were all orthorhombic, monoclinic $BaTi_5O_{11}$, and $Sr_5Al_8O_{17}$. The formation of solid solution may make it difficult to identify the products by only X-ray diffraction patterns. No compound containing Zr was identified in the product. The product, thus further analyzed by SEM and EPMA analysis.

The chemical analysis by EPMA suggested that the product contained 4 phases. The representative SEM and EPMA results are shown in Fig. 3 and the analyzed compositions are summarized in Table 3. The compositions are the average values of three positions in each area in Fig.3. From table 3, Al-rich phase (region (1)) may correspond to the phase assigned as $Sr_5Al_8O_{17}$ in the X-ray diffraction pattern, and the other phases (region (2)-(4)) may correspond to AE-RE-Ti-O compounds or $BaTi_5O_{11}$. From these results, the compounds in the product can be divided the following 4 phases;

(1) $AE_5Al_8O_{17}$ phase which contains La, Ti and a small amount of Zr,

(2) AE-RE-Ti oxide which contains Al and Zr,

(3) RE-Ti oxides which contains Al and Zr,

(4) RE-Ti oxides which contains a larger amount of Zr than the phase (3).

It is found, thus that all FP elements form the complex oxides of Ti and Al.

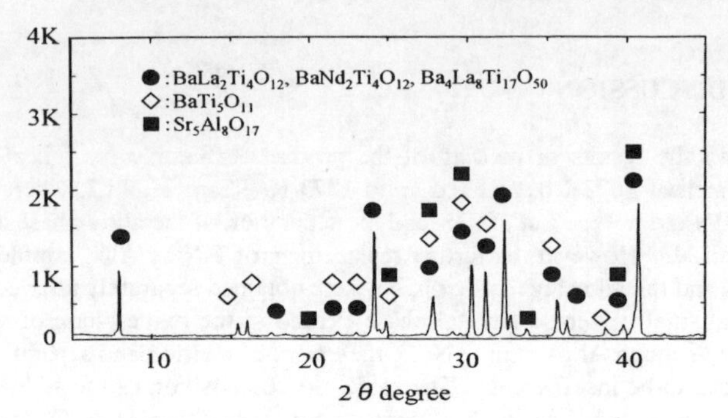

Fig. 2 The X-ray diffraction pattern for the oxide phase of sample No.3

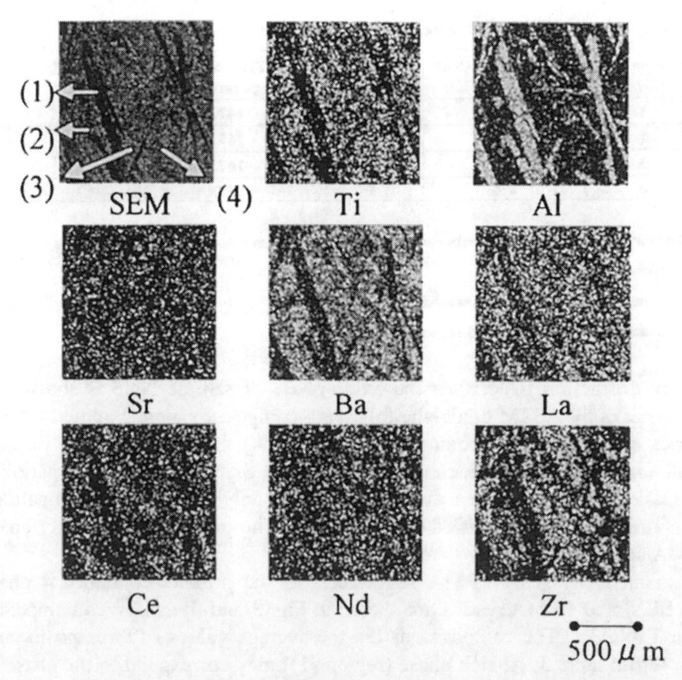

Fig. 3 SEM and EPMA results for the oxide phase of Sample No.3

Table 3 Composition of the phases in Sample No.3(atomic %)

Phase No.	Ti	Al	Sr	Ba	La	Ce	Nd	Zr
(1)	29	63	4.6	1.5	2.4	0	0	0.3
(2)	91	1.3	0.6	4.2	0.3	1.3	0	1.8
(3)	80	8.5	0	0	2.1	3.2	2.7	4.4
(4)	77	3.1	0	0	2.4	0.6	11	22

All of the binary oxides, constituting these complex oxides have melting points higher than the melting temperature of the product (1673K). Formation of complex oxides sometimes decreases the melting temperature of the system. Since AE-containing complex oxides have lower melting temperatures than the other systems[11,12,13,14], AE might play an important role in melting of the product. Phase diagrams for Sr-Ba-Al-O system are not available, but a mixture containing Al_2O_3 and SrO with the mole ratio of about 3:7 melts at 1705K according to a published SrO-Al_2O_3 pseudobinary phase diagram[9]. The $Sr_5Al_8O_{17}$ phase containing Ba, Ti and Zr may melt as high as 1673K due to the dissolution of Ba. The phases (3) and (4) may melt below 1673K, possibly due to Al dissolution. The AE-RE-Ti-Al-O phase diagrams are also unavailable, but all of the reported pseudoternary AE-RE-Ti oxides and pseudobinary RE-Ti oxides do not melt at 1673K. Any mixture of titanium oxide, RE oxides, and/or AE oxides do not melt, according to some studies[11,12,13,14]. Since Sample No.1 heated without AlN did not melt at 1673K, the lower melting temperatures of phases (2) –(4) are probably due to dissolution of Al. The study of the melting temperature of $(AE,RE)_2Ti_3O_9$ phase to confirm the effect of Al on the melting temperature is presented in another paper[15]. The lowest melting temperature in the ZrO_2-TiO_2-Al_2O_3 is 1853K[12]. Thus Zr dissolution to each phase mentioned above may contribute to the changes in melting temperatures of the products as well as dissolution of Al.

The phases identified at room temperature can transform at high temperature. The compositions of phase (3) and (4) strongly suggest the transformation. The high temperature X-ray diffraction is planned for elucidation of the more detail melting mechanism.

The chemical durability of the produced material has been examined. In the previous study[9], 7 day Soxhlet leach tests showed that concentrations of all elements in the leachate were below the detection limit of the Inductively Coupled Plasma Spectory (ICP-ES) method. 30 day tests in the present study showd that the release of Sr and Ba was $0.1x10^{-3}g/m^2$, which is one order higher than the value for glass waste form. The other elements were not detected by ICP-ES. Thus, the chemical durability of the oxide phase obtained in the method is found to be superior to that of glass waste form[1].

CONCLUSIONS

Calcined simulated high level liquid waste (HLLW) with a desired amount of TiN and AlN mixture was heat-treated at 1673-1873 K. In this treatment, it was revealed that the mixture of TiN and AlN (the atomic ratio of Al to Ti is 1:9) caused the melting of both phases at 1673 K. The analysis for the oxide phase showed that the compounds in it could be divided into four phases; (1) $AE_5Al_8O_{17}$ phase which contains La, Ti and a small amount of Zr; (2) AE-RE-Ti oxide which contains Al and Zr; (3) RE-Ti oxides which contain Al and Zr; and (4) RE-Ti oxides which contain larger amount of Zr than phase (3). It is found that all fission product elements formed the complex oxides with Ti and Al. It is thought that Al and Zr dissolution in each phase contributes to the melting of the oxide phase at 1673K. A 30-day Soxhlet leach test showed that the chemical durability of the oxide phase as a waste form was superior to that of glass waste form.

REFERENCES

1. K. Kawamura, and J. Ohuchi in Scientific Basis for Nuclear Waste Management XVIII Part 1, edited by T. Murakami and R. C. Ewing (Mater. Res. Soc. Symp. Proc. Vol. 353, Kyoto, Japan, 1995) pp. 87-93.
2. M. Horie, Trans. Amer. Nucl. Soc. 62, 111 (1990).
3. M. Horie in ENC'90 (Proc. of ENS/ANS-Foratom Conference vol. IV, 1990), p. 2281.
4. M.Horie and C.Miyake in RECOD'94 (Proc. of The forth international Conference on Nuclear Fuel Reprocessing and Waste Management vol.II, 1994), Session 9A-4.
5. M.Uno, Y.Kadotani, C.Miyake and M.Horie, Mat. Res. Soc. Symp. Proc., vol. 353, 1339 (1995).
6. M. Uno, Y. Kadotani, H. Kinoshita, C. Miyake and M. Horie, J. Nucl. Sci. Technol., 33, 973-980 (1996).
7. M. Uno, H. Kinoshita, C. Miyake and M. Horie, J. Nucl. Mater., 274, 191-196 (1997).
8. M. Uno, Y. Kadotani, C. Miyake and M. Horie, J. Nucl. Sci. Technol., 33, 879-885 (1996).
9. M. Uno, H. Kinoshita, E. Sakai, A. Ikeda, Y. Matsumoto and S. Yamanaka in NUCEF'98 (Proc. of the 2nd NUCEF Int. Symp., 1999) pp. 545-554.
10. Materials Characterization Center, 9-30-81.
11. E. M. Levin, C. R. Robbins and H. F. McMurdie, Phase Diagrams for Ceramists, vol. 2 (American Ceramic Society, 1969).
12. E. M. Levin and H. F. McMurdie, Phase Diagrams for Ceramists, vol. 3 (American Ceramic Society, 1975).
13. R. S. Roth, J. R. Dennis and H. F. McMurdie, Phase Diagrams for Ceramists, vol. 6 (American Ceramic Society, 1987).
14. M. A. Clevinger, K. M. Hill and C. Cedeno, Phase Diagrams for Ceramists, vol. 9 (American Ceramic Society, 1991).
15. M. Uno, H. Kinoshita and S. Yamanaka, presented at the 1999 MRS Fall Meeting, Boston, MA, 1999.

ISOMORPHIC CAPACITY OF SYNTHETIC SPHENE WITH RESPECT TO Gd AND U

S.V. STEFANOVSKY[1], S.V. YUDINTSEV[2], B.S. NIKONOV[2], B.I. OMELIANENKO[2], M.I. LAPINA[2]
[1] SIA Radon, 7th Rostovskii per. 2/14, Moscow 119121 RUSSIA
[2] Institute of Geology of Ore Deposits, Mineralogy, Petrography and Geochemistry RAS, Staromonetnii per. 35, Moscow 109017 RUSSIA

ABSTRACT

Phase relations in the system: CaO-TiO_2-SiO_2-(Na_2O,Al_2O_3,Gd_2O_3,UO_2) were studied. This system is of interest due to the formation of sphene, perovskite, and other phases potentially suitable for immobilization of high level waste (HLW) elements. Along with sphene, other phases found in the samples were rutile, chevkinite, anorthite, crystobalite, and pyrochlore-structured phases. Sphene is able to incorporate up to 21.5 wt.% Gd_2O_3 and 9.3 wt.% UO_2 or, in formula units: 0.25 Gd^{3+} and 0.07 U^{4+}.

INTRODUCTION

Currently high level wastes (HLW) are incorporated in aluminophosphate and borosilicate glasses [1,2] for long-term storage. Due to the thermodynamic instability of the vitreous state, glasses can not guarantee safe isolation of long-lived radionuclides, such as actinides, for the periods necessary to reduce their activity to natural background levels. Therefore more stable matrices are required for the long-term storage of actinide waste.

Promising matrices for this purpose are sphene-based glass ceramics [3-5]. These can be produced using existing vitrification technologies, which are able to incorporate various waste elements (fission products, corrosion products, and fuel components), distributing them among vitreous and crystalline phases, and which have high stability under the conditions of underground disposal.

However, an uncertainty remains with respect to the limiting amounts of actinides and rare earths in the sphene structure because the sphene composition with respect to these elements was investigated by an indirect route – on the basis of comparison of powder X-ray diffraction (XRD) patterns for the samples prepared with differing dopant concentrations. The number of direct determinations of the sphene compositions are limited to a few analyses and, as a rule, data on the content of individual elements in the mineral rather than complete chemical compositions were reported. The reason is that the extremely fine grain size of the synthetic sphene makes their investigation very complicated. At the same time such data on solid solution limits (isomorphic capacity) for elements in the sphene structure is necessary to specify the composition for the initial batch to be used to provide the required phase relations. This work describes compositions of the synthetic sphene obtained from crystallization of melts in the systems CaO-TiO_2-SiO_2-(Na_2O,Al_2O_3,Gd_2O_3,UO_2) and relationships between the sphene and other phases in these systems.

Natural sphene is a common accessory mineral of magmatic and methamorphic rocks. Its maximum its amount (about 0.3%) is observed in intrusive rocks intermediate in silica content – granodiorites [6]. In most cases the sphene compositions are close to ideal stoichiometry. The most typical impurities are rare earths, aluminum, and iron [7]. Maximum amounts of total yttrium and rare earths in the natural sphene reach 3-4 wt.% [7,8]. Actinide (uranium and thorium) content is lower by one order of magnitude and does not exceed 0.3-0.4 wt.%. [8]. Rare earth and actinide ions occupy theCa sites with the necessary charge compensation being achieved by substitution of Al^{3+} and Fe^{3+} cations for Ti^{4+} cations. The sphene variety that is the richest in Y and rare earths and contains up to 12 wt.% $(Y,REE)_2O_3$, is known as yttrotitanite [7,9]. It is suggested in some works [9,10] that some portion of the high rare earth content determined by chemical analysis is erroneously attributed to sphene, whereas in reality these high values correspond to other titanosilicates, in particular, minerals of the perrierite-chevkinite group having similar crystal lattice symmetry. Therefore, correct identification of the mineral requires simultaneous structural investigation, especially if analysis shows a high rare earth content. On the whole, the data on maximum rare earth content in natural sphene are inconsistent and they can not be used as the basis for estimation of the isomorphic capacity of this phase with respect to rare earths. Aluminum and iron oxide content in

Mat. Res. Soc. Symp. Proc. Vol. 608 © 2000 Materials Research Society

sphene varies widely and can occasionally exceed 10 wt.%. The principal mechanism of incorporation of these elements in the phase structure is due to the reaction: $Ti^{4+} + O^{2-} = Al^{3+} + F^{-}$. Increased pressure facilitates this exchange, therefore, maximum Al_2O_3 content (up to 14 wt.%) was found in the sphene of metamorphic rocks formed at pressures greater than 1000 MPa [11].

EXPERIMENTAL

We investigated 9 samples with compositions corresponding to sphene formulae taking into account isomorphic exchanges: $2Ca^{2+} = Na^{+} + Gd^{3+}$ and $2Ca^{2+} = U^{4+}$ + vacancy (Table I). The samples were prepared from Ca, Ti, Si, Gd, and U oxides and sodium nitrate (all the reagents were "chemically pure"-grade – Russian standard). Reagents were crushed and intermixed in an agate mortar. Batches in alumina crucibles were placed in a resistive furnace, heated to melting temperature (1400-1500 °C) for 3-3.5 hours, kept at melting temperature for 30 minutes for melt homogenization, cooled to 900 °C for 4 hours for crystallization, and cooled to room temperature overnight in the furnace overnight after turning off the furnace heater.

Table I. Calculated chemical compositions, formulae and phase compositions for the samples studied.

Oxides	S	Gd-1	Gd-2	Gd-3	Gd-4	U-1	U-2	U-3	U-4
	Oxide content, wt.%								
Na_2O	-	0.8	1.5	2.2	2.9	-	-	-	-
CaO	28.6	25.1	21.7	18.6	15.5	24.7	21.2	17.9	14.8
Gd_2O_3	-	4.5	8.9	12.9	16.8	-	-	-	-
UO_2	-	-	-	-	-	6.6	12.7	18.4	23.7
TiO_2	40.8	39.8	38.8	37.9	37.0	39.2	37.8	36.4	35.1
SiO_2	30.6	29.8	29.1	28.4	27.8	29.5	28.3	27.3	26.4
Ions	Formula units								
Na^{+}	-	0.05	0.10	0.15	0.20	-	-	-	-
Ca^{2+}	1.00	0.90	0.80	0.70	0.60	0.90	0.80	0.70	0.60
Gd^{3+}	-	0.05	0.10	0.15	0.20	-	-	-	-
U^{4+}	-	-	-	-	-	0.05	0.10	0.15	0.20
Ti^{4+}	1.00	1.00	1.00	1.00	1.00	1.00	1.00	1.00	1.00
Si^{4+}	1.00	1.00	1.00	1.00	1.00	1.00	1.00	1.00	1.00
O^{2-}	5.00	5.00	5.00	5.00	5.00	5.00	5.00	5.00	5.00
Phases	S>>R>Cr	S>>R	S>R>P>C	R>P>S>C	P>>R	S>>R>P>A	R>P	R>P>A	P>R>A

A – anorthite, C – chevkinite, Cr – crystobalite, P – pyrochlore-structured phase, R – rutile, S – sphene. In all the samples a vitreous phase is also present.

Materials produced were examined with an XRD (DRON-4 diffractometer, Cu Kα radiation). Two samples were investigated in more detail with scanning electron microscopy/energy dispersive spectrometry (SEM/EDS) using a JSM-5300 + "Link ISIS" system, and transmission electron microscopy (TEM) using a JEM-100c device.

RESULTS

Phase compositions of the samples are given in Table I. Along with the crystalline phases glass is also present In an amount that does not exceed 20 vol.%. As follows from the XRD data (Figure 1), sphene is the major phase in Sample S (nominal sphene composition), which does not contain waste elements, as well as in Samples Gd-1, Gd-2, and U-1, which have relatively low Gd and U content. With an increase in Gd and U concentration, peak intensities of the sphene decrease and finally disappear. Simultaneously, the sphene peaks become broader, probably due to an increase in the number of defects in the crystalline structure when Gd^{3+} and U^{4+} substitute for Ca^{2+} ions. XRD data demonstrate that in the Gd-containing system sphene remains stable to higher elemental concentrations as compared to the U-containing system.

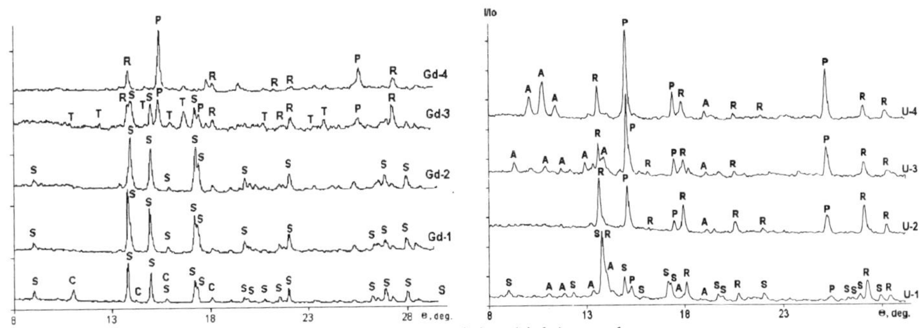

Figure 1. XRD patterns of the Gd- (left) and U-containing (right) samples.
A – anorthite, C – crystobalite, P – "pyrochlore", R – rutile, S – sphene, T – perrierite-chevkinite.

in the Gd-containing system sphene is present in Sample Gd-3 (12.9 wt.% Gd_2O_3) but it has not been observed in Sample Gd-4 (16.8 wt.% Gd_2O_3). In the U-containing system the UO_2 concentration limit that still allows sphene to crystallize ranges between 6.6 wt.% (0.05 f.u. U) and 12.7 wt.% (0.1 f.u. U). At higher Gd_2O_3 and UO_2 concentrations the stable phase is a complex oxide with a pyrochlore-type structure and ideal stoichiometry $(Ca,Gd)_2Ti_2O_7$ or $(Ca,U)_2Ti_2O_{7-x}$. Moreover, all the samples contain rutile, and some of them also contain a phase with a perrierite-chevkinte structure, whose idealized formula is $Ca(Na,Ca,REE)_4Ti_4(Si,Al)_2O_{22}$, and anorthite ($CaAl_2Si_2O_8$). The occurrence of Al in the samples and formation of anorthite and aluminosilicate glass are due to partial dissolution of the alumina crucibles.

Samples Gd-3 and U-1 have the most complicated phase compositions among the samples prepared. They contain both sphene and other (Gd,U)-containing phases and are contaminated with Al_2O_3 as mentioned above. The Al content in different parts of the samples depends on the melting duration and the Al diffusion rate in melts, and decreases from the crucible wall to the core. This allows one to study the effect of Al_2O_3 on the amount and composition of the phases by analyzing different parts of the same sample progressing from the surfaces to the core.

Sample Gd-3 consists of two zones (edge and core). Each zone is composed of at least three crystalline phases and glass. The edge contains chevkinite, sphene, and rutile. The core is composed of sphene, rutile, and pyrochlore-structured Ca,Gd-titanate. The largest crystals (200-300 μm in length and 20-50 μm in width) consist of rutile. The other phases form grains that are a few microns in size (Figure 2). Chevkinite and sphene form the needle-shaped grains. Pyrochlore is observed as skeleton crystals (Figure 2, a,c,d). Interstitial glass is located between grains of the crystalline phases and appear as dark inclusions in the SEM images. Compositions of the phases and their chemical formulae are given in Table II.

Compositions of the phases were recalculated to formulae with 2 (rutile), 5 (sphene), 7 (pyrochlore-type), and 22 (chevkinite) oxygen ions. There is an excess of silicon, as compared to ideal stoichiometry, in the sphene and chevkinite analyses. This is probably due to the small grain size causing the electron probe to capture interstitial glass. Occurrence of silicon in the pyrochlore- type phase may be caused for the same reason. Sodium, calcium, aluminum, and silicon oxides are predominant in the glass. The compositional relationship in the glass is close to Ca,Na- plagioclase stoichiometry (Table II). The reliability of the sphene and chevkinite identification is confirmed from the TEM data (Figure 2). The electron diffraction patterns correspond to standard patterns for the above-mentioned minerals.

Sphene composition in different parts of the sample is markedly varied. For example, Gd concentration in the sphene associated with chevkinite exceeds the Gd concentration in the sphene associated with the pyrochlore-type phase by more than two times. Gd content in the glass is approximately one order of magnitude lower than in the sphene, and almost 60 times lower than in the pyrochlore-type phase.

Sample U-1 also consists of four zones: edge (interacted with the alumina crucible), intermediate zone, core, and rim (top surface of the sample in the crucible). Their compositions are given in Table III. The edge is composed of rutile, anorthite, and pyrochlore-structured Ca-Ti-U-oxide grains (Figure 2, d). The grain size ranges between 10 and 50 μm for the rutile and pyrochlore-type phases, and between 50 and 100 μm for anorthite.

Table II. Chemical compositions and formulae of the phases in Sample G-3.

Oxides	Edge			Core			
	Sphene	Chevkinite	Rutile	Sphene	Pyrochlore	Rutile	Glass
	Oxide content, wt.%						
Na_2O	2.6	0.8	0.7	3.1	0.8	<0.4	6.1
CaO	16.5	12.9	0.3	19.1	5.5	<0.3	13.9
Al_2O_3	5.1	2.8	<0.3	7.9	<0.3	<0.3	25.1
Gd_2O_3	21.5	43.0	<0.3	9.5	52.8	<0.3	0.9
TiO_2	22.5	11.4	98.5	20.9	41.3	99.5	4.3
SiO_2	32.7	28.4	0.4	37.6	1.2	0.5	47.0
Total	100.9	99.3	99.9	98.1	101.6	100.0	97.3
Ions	Formula units						
Na^+	0.19	0.27	0.02	0.20	0.12	<0.01	0.57
Ca^{2+}	0.64	2.64	<0.01	0.69	0.43	<0.01	0.72
Al^{3+}	0.22	0.66	<0.01	0.31	<0.01	<0.01	1.42
Gd^{3+}	0.25	2.73	<0.01	0.11	1.27	<0.01	0.01
Ti^{4+}	0.61	1.63	0.97	0.53	2.25	0.99	0.16
Si^{4+}	1.17	5.50	0.01	1.26	0.09	0.01	2.26
Total	3.08	13.43	1.00	3.10	4.16	1.00	5.14
O^{2-}	5.00	22.00	2.00	5.00	7.00	2.00	8.00

An intermediate zone is located between the edge and the core. Within this zone the pyrochlore-type phase content decreases and the sphene and glass content increase, whereas the relative fractions of rutile and anorthite remain the same. The core, being the predominant part of Sample U-1, consists of fine-grade bulk with distributed rutile crystals (Figure 2, e). Rutile grains are observed in cross-section as parallelograms or prisms with sizes ranging between 10 and 50 μm. It is seen at higher magnification that the bulk is composed of needle-shaped sphene crystals a few microns in length and glass (Figure 2, e). Such a texture results in the non-uniformity of color on the photographs obtained with back scattered electrons: lighter for sphene and darker for glass. Within the core, the sphene grain sizes decrease from the portion near the bottom to the rim. The rim consists of vitreous bulk with inclusions of rutile grains (Figure 2, f). The maximum uranium content (45.8 wt.% UO_2) was found to be in the pyrochlore-type phase (Table III). The amount of UO_2 in the sphene is about 9 wt.%, which is close to its concentration in the glass. A feature of the sphene composition is its low Ti and high Al and Si content.
This effect is caused by the capture of glass in the analyses of fine sphene crystals. The phase compositions in the different zones is strongly varied. The main difference is the occurrence in the edge zone of the pyrochlore-structured Ca-Ti-U oxide rather than sphene. Sphene instability in this zone is probably due to melt contamination with alumina from the crucible resulting in higher Al_2O_3 content in this zone. Taking into account the compositions of individual phases (Table III), replacement of sphene by a pyrochlore-type phase may be represented as: 4.2 Sphene + 2.0 Alumina $\rightarrow$ 0.6 "Pyrochlore" + 2.7 Anorthite + 1.0 Rutile. This is confirmed by the increase of anorthite and rutile in the edge zone of the sample.

DISCUSSION

Experimental data demonstrate the high isomorphic capacity of the sphene with respect to Gd and U. Maximum elemental concentrations were found to be 0.25 and 0.07 f.u. for Gd and U, respectively (21.5 wt.% Gd_2O_3 and 9.3 wt.% UO_2). These maximum concentrations are reached when the sphene co-exists with Gd titanates (Sample Gd-4) or Ca-U titanates (Sample U-1) with a pyrochlore-type structure. These elements substitute for Ca in the sphene lattice. To maintain electric neutrality substitutions in other structural sites: $2Ca^{2+} = Gd^{3+} + Na^+$ and $Ca^{2+} + Ti^{4+} = Gd^{3+} + Al^{3+}$ in the Gd-containing sample Gd-4, and $Ca^{2+} + 2Ti^{4+} = U^{4+} + 2Al^{3+}$ in the U-containing sample U-1 are simultaneously realized. Maximum Gd and U atomic concentrations(0.25 and 0.07 respectively) found are somewhat higher than reported in ref. [5] (0.20 f.u. for Gd and 0.05 f.u. for U). One of the most likely reasons for this is the indirect nature of the

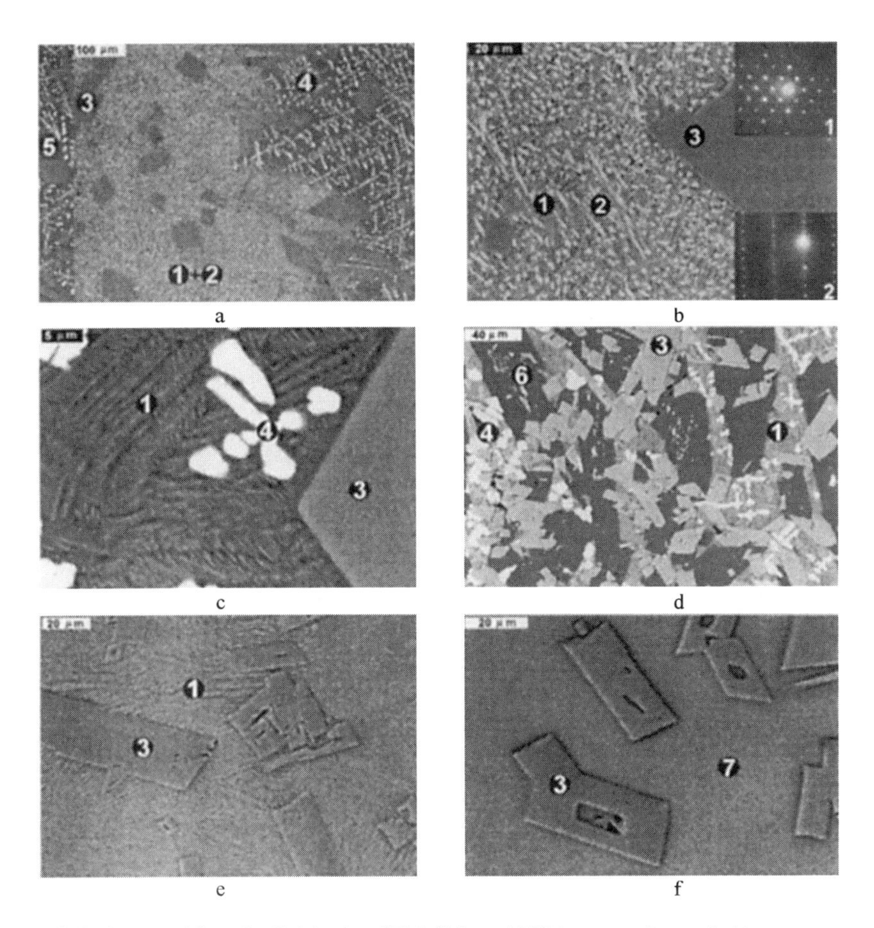

Figure 2. SEM images of Samples G-3 (a-c) and U-1 (d-f), and TEM patterns (inserts in b).
1 – sphene, 2 – perrierite-chevkinite, 3 – rutile, 4 – "pyrochlore", 5 – glass, 6 – anorthite, 7 – fine crystalline (amorphous?) bulk.

composition evaluations in the cited work based on XRD data. Another explanation is that the higher (by 200-300 °C) synthesis temperature in our experiments extends the sphene solid solution limit with an increase in solubility of the elements. The values obtained determine the maximum amount of REE-actinide HLW REE-actinide fraction elements in the sphene-based materials. When this level is exceeded in the synthetic products, the other REE and actinide host phases will form in addition to sphene or instead of it. It has been shown earlier that titanate, titanosilicate and silicate phases with the structures of pyrochlore, apatite-britholite, and chevkinite-pierrierite are stabilized at high actinide and lanthanide concentrations in ceramics [2-5].

Minimum concentrations required to reveal the pyrochlore-structured phase are 8.9 wt.% Gd_2O_3 and 6.6 wt.% UO_2 in the Gd- and U-containing systems, respectively. Chevkinite-pierrierite is present in Sample Gd-2 with 8.9 wt.% Gd_2O_3 but it is absent at higher Gd concentrations. Sphene is not observed on the XRD patterns containing 16.8 wt.% Gd_2O_3 (Sample Gd-3) and 12.7 wt.% UO_2 (Sample U-2). Lanthanide and actinide ions occupy the same Ca-sites in the sphene structure resulting in competition between them for a site in the crystalline lattice. Therefore, an important characteristic of the sphene matrix is the limit

Table III. Chemical compositions and formulae of the phases in Sample U-1

Oxides	Bottom zone			Intermediate zone			Core		Rim
	Pyrochlore	Anorthite	Rutile	Sphene	Anorthite	Rutile	Sphene	Glass	Glass
	Oxide content, wt.%								
CaO	18.3	20.5	0.3	26.7	20.3	0.4	22.6	20.8	23.7
Al_2O_3	0.4	35.5	0.6	6.6	35.6	0.6	8.5	14.1	11.9
SiO_2	0.4	40.7	0.4	39.1	42.7	0.4	39.5	51.1	41.5
TiO_2	33.3	1.3	97.4	19.2	0.9	97.1	18.1	4.1	14.8
UO_2	45.8	0.4	1.2	9.3	0.4	1.3	9.3	9.4	7.7
Total	98.2	98.0	99.9	100.9	99.5	99.8	98.0	99.5	99.6
Ions	Formula units								
Ca^{2+}	1.50	1.05	<0.01	0.94	1.01	<0.01	0.81	0.70	0.83
Al^{3+}	0.03	1.99	0.01	0.26	1.96	0.01	0.34	0.52	0.45
Si^{4+}	0.03	1.94	0.01	1.29	1.99	0.01	1.32	1.60	1.33
Ti^{4+}	1.91	0.05	0.98	0.48	0.03	0.98	0.45	0.10	0.36
U^{4+}	0.78	<0.01	<0.01	0.07	<0.01	<0.01	0.07	0.06	0.05
Total	4.25	5.03	1.00	3.04	4.99	1.00	2.99	2.98	3.02
O^{2-}	7.00	8.00	2.00	5.00	8.00	2.00	5.00	5.00	5.00

ing amount of the REE-actinide fraction that will not result in the formation of phases other than sphene. It can be expected from this study that the total amount of these elements should range between 10 and 20 wt.%. This corresponds to the value of 15 wt.% reported in ref. [4]. It also is necessary to take into account the effect of alumina on the phase composition of the synthetic products. As shown above, sphene is unstable at high alumina content, where it is replaced by a "pyrochlore"-anorthite-rutile assemblage.
It should be also noted that the distribution coefficients between sphene and glass are high for Gd (~10) and low for U (~1). These values are close to the data obtained earlier on the characteristics of elements partitioning among co-existing phases in sphene-based glass-ceramics [2,4].

CONCLUSION

In cooled melted samples of the system $CaO-TiO_2-SiO_2-(Na_2O,Al_2O_3,Gd_2O_3,UO_2)$, sphene, chevkinite, anorthite, rutile, and pyrochlore-structured phase were found. Within the series $Ca_{1-x}Na_{0.5x}Gd_{0.5x}TiSiO_5$ sphene is still present at x=0.3 whereas within the series $Ca_{1-x}U_{0.5x}TiSiO_5$ the limiting x value is 0.1. Sphene co-existing with extra phases (chevkinite, pyrochlore, rutile, anorthite) is able to incorporate up to 21.5 wt.% Gd_2O_3 and 9.3 wt.% UO_2 or, in formula units: 0.25 Gd^{3+} and 0.07 U^{4+}.

REFERENCES

1. N.P. Laverov, B.I. Omelianenko, S.V. Yudintsev, B.S. Nikonov, I.A. Sobolev, and S.V. Stefanovsky, Geology of Ore Deposits, 39, 179 (1997).
2. E.R.Vance, F.G. Karioris, L. Cartz, and M.S. Wong, Advances in Ceramics, 8, 238 (1984).
3. P.J. Hayward, E.R. Vance, C.D. Cann, and S.L. Mitchell, Ibid., 291.
4. P.J. Hayward, Glass-ceramics, in *Radioactive Waste Forms For The Future*, edited by W. Lutze and R.C. Ewing, (Elsevier Science Publishers, Amsterdam, 1988) p.427.
5. E.R. Vance and D.K. Agrawal, Nucl. Chem. Waste Manag., 3, 229 (1982).
6. V.V. Lyakhovich, *Accessory Minerals in Granitoids of the Soviet Union* (Nauka, Moscow, 1967).
7. W.A.Deer, R.A.Howie, J.Zussman, *Rock-forming minerals* (Longmans, London, 1962).
8. F.C. Hawthorne, L.A. Groat, M. Raudsepp, Amer. mineral., 76, 370 (1991).
9. I.T. Alexandrova, A.I. Ginsburg, I.I. Kupriyanova, *Rare Earth Silicates* (Nedra, Moscow, 1966)p.220.
10. S.E. Haggerty, A.N. Mariano, Contrib. Mineral. Petrol., 84, 365 (1983).
11. P.A. Pletnev, I.M. Kulikova, and E.M. Spiridonov, Notes of the All-Russian Miner. Soc. 1, 69 (1999).

MICROSTRUCTURE AND COMPOSITION OF SYNROC SAMPLES CRYSTALLIZED FROM a $CaCeTi_2O_7$ CHEMICAL SYSTEM: HRTEM/EELS INVESTIGATION

Huifang Xu [a], Yifeng Wang [b], Robert L. Putnam [c], Jose Gutierriez [d], and Alexandra Navrosky [d]

[a] Transmission Electron Microscopy Laboratory, Department of Earth and Planetary Sciences, The University of New Mexico, Albuquerque, New Mexico 87131. E-mail: hfxu@unm.edu

[b] Sandia National Laboratories, 115 North Main Street, Carlsbad, New Mexico 88220 E-mail: ywang@sandia.gov

[c] Los Alamos National Laboratory, Los Alamos, New Mexico 87545

[d] Dept. of Chemical Engineering and Materials Science, University of California at Davis, One Shields Avenue, Davis, CA 95616-8779

Abstract

Ce-pyrochlore, $CaCeTi_2O_7$ is a chemical analogue for $CaPuTi_2O_7$, which is a proposed ceramic endmember waste form for the disposition of excess weapon-usable plutonium in geological repositories. Ce-pyrochlore was synthesized by firing and annealing in air a mixture of CeO_2, TiO_2, and $CaCO_3$ with a stoichiometry of $CaCeTi_2O_7$. The annealed products contain Ce-pyrochlore, Ce-bearing perovskite, CeO_2, and minor CaO. The mixture annealed at a temperature of 1140 °C contains more pyrochlore phase than that annealed at a higher temperature (1300 °C), indicating that a low temperature condition favors the formation of the Ce-pyrochlore. The Ca/Ce ratio of the pyrochlore is slightly lower than the ideal ratio (one). Electron energy-loss spectroscopy results show that there is a small fraction of Ce^{3+} present in the pyrochlore. Ce present in perovskite is dominated by Ce^{3+}. High-resolution TEM images show that the boundary between pyrochlore and perovskite is semi-coherent. No glassy phases were observed at the grain boundary between pyrochlore and perovskite, nor between CeO_2 and pyrochlore. It is postulated, based on the presence of trivalent Ce in the Ce-pyrochlore, that neutron poisons such as trivalent cation Gd would be incorporated into the $CaPuTi_2O_7$ phase.

Introduction

Pyrochlore phases have been considered as a durable crystalline waste form for hosting weapon-usable Pu. A Ce-pyrochlore phase, $CaCeTi_2O_7$ is a chemical analogue for the proposed waste form based on $CaPuTi_2O_7$. Pyrochlore and zirconolite phases with the stoichiometry of $CaMTi_2O_7$ (or, $MCaTi_2O_7$) can be considered as a distorted structural derivative structure of fluorite structure, where M represents tetravalent cations such as Zr, Hf, U, Pu and other actinides. The

Mat. Res. Soc. Symp. Proc. Vol. 608 © 2000 Materials Research Society

existence of large polyhedra (with coordination numbers ranging from 7 to 8) in the structures allows pyrochlore and zirconolite phases to accommodate a wide range of radionuclides (e.g., Pu, U, Ba, etc.) as well as neutron poisons (e.g., Hf, Gd) [1-15]. Most previous analogue studies focus on zirconolite phase and Hf-zirconolite phase. However, the actual Pu-loaded waste form will have pyrochlore structure due to the relatively larger size of Pu. In this paper, we describe microstructure and chemistry of a Ce-pyrochlore phase.

Samples and Experimental Methods

The Ce-pyrochlore samples were synthesized from mixture of $CaCO_3$, CeO_2, and TiO_2 (anatase) fired at 1300 °C and 1140 °C in air for 96 and 340 hours, respectively. The composition of the mixture was targeted to the stoichiometry of $CaCeTi_2O_7$, that is Ca:Ce:Ti = 1:1:2. The final product fired at 1300 °C displays a black color, and the product fired at 1140 °C displays a dark green color. Double-side polished petrographic thin sections were prepared for SEM investigations. The samples for TEM studies were selected from the thin sections and then ion milled. All high resolution transmission electron microscopy (HRTEM) and EDS results were carried out with a JEOL 2010 HRTEM and Oxford Link ISIS EDS system. Point-to-point resolution of the HRTEM is 0.19 nm. Electron energy-loss spectroscopy (EELS) studies were carried out with JEOL 2010F HRTEM with a GIF system.

Results and Discussions

SEM

Backscattered electron images (Figure 1) show heterogeneous textures of ceramic samples crystallized from the mixture. The samples contain Synroc phases of Ce-pyrochlore, cerianite, and perovskite. The sample annealed at 1140°C contains more Ce-pyrochlore and less cerianite and perovskite than the sample annealed at 1300°C does (Fig. 1). The pyrochlore and Ce-bearing perovskite formed at 1300 °C contain more Ce than those formed at 1140 °C. However, cerianite in both samples are almost pure CeO_2. Based on electron microprobe analyses, atomic ratio of Ca:Ce:Ti for the pyrochlore and perovskite formed at 1300 °C are 1:1.23:2.36, and 1:0.53:1.71, respectively; atomic ratio of Ca:Ce:Ti for the pyrochlore and perovskite formed at 1140 °C are 1:1.05:2.13, and 1:0.18:1.25, respectively. Composition of the pyrochlore formed at 1140 °C is close to the ideal stoichiometry of $CaCeTi_2O_7$. High-temperature Ce-pyrochlore will contain more trivalent Ce. It is proposed that stoichiometric Ce-pyrochlore may occur at low-temperature in air. The formation of the stoichiometric Ce-pyrochlore from the Ce^{3+}-bearing pyrochlore may be expressed by:

Ce^{3+}-rich pyrochlore + Ce^{3+}-bearing perovskite + Cerianite + O_2
= Ce-pyrochlore + Ce^{3+}-poor perovskite.

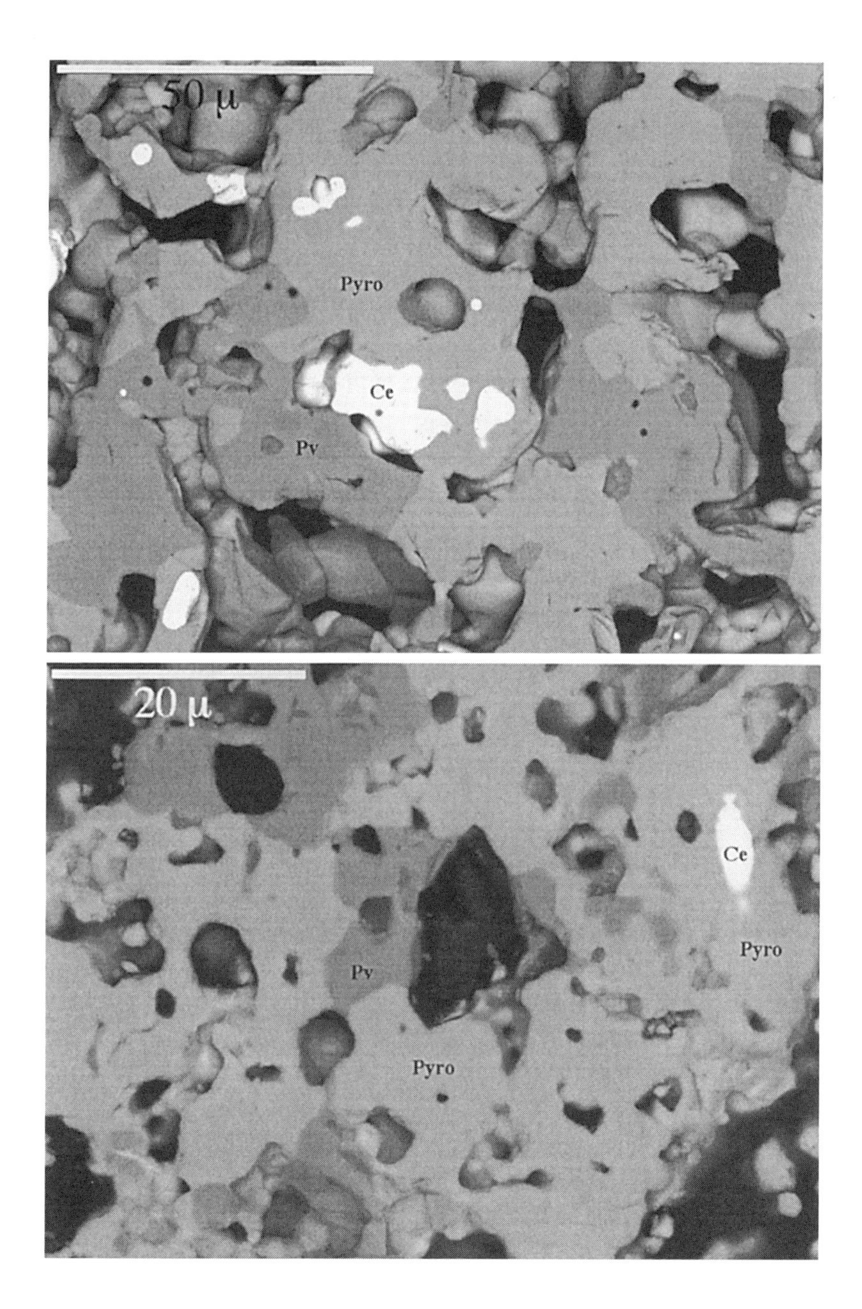

Fig. 1 Back scattered electron images of the Synroc samples annealed at 1300 °C (upper) and 1140 °C (lower) respectively. Perovskite = Pv, Cerianite = Ce, Pyrochlore = Pyro.

TEM and EELS

Transmission electron microscopy (TEM) results indicate the presence of Ce-pyrochlore, Ce-bearing perovskite, and cerianite. There are inclusion-like perovskite, pyrochlore, and minor CaO crystals within large crystals of cerianite. High-resolution TEM images show the boundary between cerianite and perovskite (Fig. 1) and the boundary between cerianite and Ce-bearing perovskite are semi-coherent. There are preferred orientation between cerianite and Ce-pyrochlore and between cerianite and perovskite. There are no glassy phases at the grain boundaries between pyrochlore and cerianite, and the boundary between CeO_2 and perovskite. There are no $CeTiO_4$ intermediate phase between cerianite and Ce-pyrochlore. This is different from a $CaZrTi_2O_4$ system, in which a $ZrTiO_4$ phase is observed between zirconolite and zirconia in the [16].

Based on EELS results, there are only Ce^{4+} in cerianite, both Ce^{4+} and Ce^{3+} in pyrochlore, and only Ce^{3+} in perovskite. The ratio of $Ce^{4+}/\Sigma Ce$ in the pyrochlore formed at 1300 °C and 1140 °C are 0.8, and 0.92, respectively. According to the atomic ratios of Ca:Ce:Ti for pyrochlore and perovskite and Ce^{4+} contents in pyrochlore, the structural formula of the Ce-pyrochlore formed at 1300 °C and 1140 °C can be calculated (Table 1).

Table 1: Composition of Ce-pyrochlore and Ce-bearing perovskite.

Sample condition	Ce-pyrochlore	Ce-bearing perovskite
1300 °C for 96 hours	$(Ca_{0.87}Ce^{3+}{}_{0.20}Ce^{4+}{}_{0.87}Ti_{0.05})Ti_2O_7$	$Ca_{1.15}Ce^{3+}{}_{0.61}Ti_{1.97}O_6$
1140 °C for 340 hours	$(Ca_{0.96}Ce^{3+}{}_{0.08}Ce^{4+}{}_{0.92}Ti_{0.04})Ti_2O_7$	$Ca_{1.59}Ce^{3+}{}_{0.29}Ti_{1.99}O_6$

The Ce-bearing perovskite formed at 1300 °C and 1140 °C have orthorhombic (pseudo-cubic) symmetry. The results indicate that a high-temperature annealing condition will result in a small amount of trivalent Ce in the pyrochlore and perovskite. This indicates that Ce-pyrochlore and Pu-pyrochlore can incorporate trivalent Gd and tetravalent Hf into the crystal structure as neutron poisons. The structural formula of Gd- and Hf bearing Pu-pyrochlore may be written as: $(Ca_{1-x}Gd_{2x}Pu_{1-x-y}Hf_y)Ti_2O_7$.

It is also reported that Pu-zircon is thermodynamically not favorable to be synthesized [17]. However, Pu-pyrochlore is more stable than Ce-pyrochlore with respect to their oxides and perovskite. According to the synthesized Ce-pyrochlore, it can be postulated that Pu-pyrochlore is a thermodynamically feasible and stable ceramic waste form for hosting Pu in silica-poor and carbonate-poor solutions, such as WIPP repository environment [18].

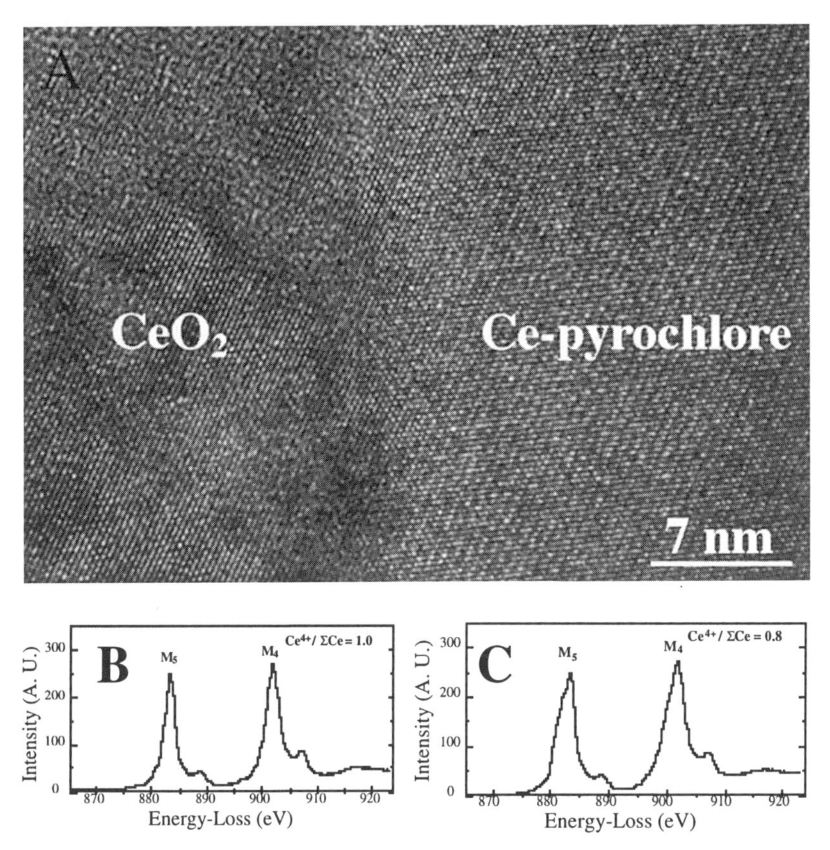

Fig. 2 (A): High-resolution TEM image of the Ce-pyrochlore dominated Synroc sample indicating bounded grain boundary between cerianite and Ce-pyrochlore. (B): EELS spectrum from the cerianite. (C): EELS spectrum from the Ce-pyrochlore.

Acknowledgment: This work is based upon a research conducted at the Transmission Electron Microscopy Laboratory in the Department of Earth and Planetary Sciences of the University of New Mexico, which is partially supported by NSF and State of New Mexico.

REFERENCES

[1] Dosch, R. G., Headley, T. J., Northrup, C. J., and Hlava, P. F., Sandia National Laboratories Report, Sandia 82-2980, 84pp (1982).

[2] Ringwood, A. E., Kesson, S. E., Reeve, K. D., Levins, D. M., and Ramm, E. J., Synroc. In W. Lutze and R. C. Ewing eds., "Radioactive Waste Forms for the Future." North-Holland, Amsterdam, p. 233 (1988).

[3] Jostsons, A., Vance, E. R., Mercer, D. J., Oversby, V. M., In T. Murakami and R. C. Ewing eds. "Scientific Basis for Nuclear Waste Management XVIII." MRS, Pittsburgh, p. 775 (1995).

[4] Ewing, R.C., Weber, W. J., and Lutze, W., In E. R. Merz and C. E. Walter eds., "Disposal of Excess Weapons Plutonium as Waste." NATO ASI Series, Kluwer Academic Publishers, Dordrecht, p. 65 (1996).

[5] Weber, W. J., Ewing, R. C., and Lutze, W., In W. M. Murphy and D. A. Knecht eds., "Scientific Basis for Nuclear Waste Management XIX." Materials Research Society, Pittsburgh, p. 25 (1996).

[6] Bakel, A. J., Buck, E. C., and Ebbinghaus, B., In "Plutonium Future — The Science." Los Alamos National Laboratories, p. 135 (1997).

[7] Begg, B. D., and Vance, E. R., In W. J. Gray and I. R. Triay eds. "Scientific Basis for Nuclear Waste Management XX." MRS, Pittsburgh, p. 333 (1997).

[8] Begg, B. D., Vance, E. R., Day, R. A., Hambley, M., and Conradson, S. D. In W. J. Gray and I. R. Triay eds. "Scientific Basis for Nuclear Waste Management XX." MRS, Pittsburgh, p. 325 (1997).

[9] Buck, E. C., Ebbinghaus B., Bakel, A. J., and Bates, J. K., Characterization of a plutonium-bearing zirconolite-rich Synroc. In W. J. Gray and I. R. Triay eds. "Scientific Basis for Nuclear Waste Management XX." MRS, Pittsburgh, p. 1259 (1997).

[10] Vance, E. R., MRS Bulletin, vol. XIX, 28 (1994).

[11] Vance, E. R., Jostsons, A., Stewart, M. W. A., Day, R. A., Begg, B. D., Hambley, M. J., Hart, K. P., and Ebbinghaus, B. B., In "Plutonium Future — The Science." Los Alamos National Laboratories, p. 19 (1997).

[12] Vance, E. R., Hart, K. P., Day, R. A., Carter, M. L., Hambley, M., Blackford, M. G., and Begg, B. D., In W. J. Gray and I. R. Triay eds., "Scientific Basis for Nuclear Waste Management XX." MRS, Pittsburgh, p. 341 (1997).

[13] Putnam, R. L., Navrotsky, A., Woodfield, B. F., Boerio-Goates, J., and Shapiro, J. L., J. Chem. Thermochem., 31, 229-243 (1999).

[14] Woodfield, B. F., Boerio-Goates, J., Shapiro, J. L., Putnam, R. L., and Navrotsky, A., J. Chem. Thermochem., 31, 245 (1999).

[15] Putnam, R. L., Navrotsky, A., Woodfield, B. F., Shapiro, J. L., Steven, R. and Boerio-Goates, J., Scientific Basis for Nuclear Waste Management XXII. MRS, Pittsburgh. p. 11 (1999).

[16] Xu, H., and Wang, Y., Scientific Basis for Nuclear Waste Management XXII. MRS, Pittsburgh. p. 47 (1999).

[17] Xu, H., and Wang, Y., Journal of Nuclear Materials, 275, 216 (1999).

[18] Wang, Y., and Xu, H., Scientific Basis for Nuclear Waste Management XXIII. MRS (Same as this volume, (2000), Pittsburgh.

Radiation Effects

ACCELERATED ALPHA RADIATION DAMAGE IN A CERAMIC WASTE FORM, INTERIM RESULTS

Steven M. Frank, Stephen G. Johnson, Tanya L. Moschetti, Thomas P. O'Holleran, Wharton Sinkler, David Esh, K. Michael Goff, Argonne National Laboratory-West
P.O. Box 2528
Idaho Falls, Idaho 83403-2528

ABSTRACT

Interim results are presented on the alpha-decay damage study of a ^{238}Pu-loaded ceramic waste form (CWF). The waste form was developed to immobilize fission products and transuranic species accumulated from the electrometallurgical treatment of spent nuclear fuel. To evaluate the effects of α-decay damage on the waste form, the ^{238}Pu-loaded material was investigated by electron microscopy for microstructure characterization, x-ray diffraction for bulk phase properties, bulk density measurement, and the product consistency test (PCT) for waste form durability. While the predominate phase of plutonium in the CWF, PuO_2, shows the expected unit cell expansion due to α-decay damage, currently no significant change has occurred to the macro- or microstructure of the material. The major phase of the waste form is sodalite and contains very little Pu, although the exact amount is unknown. Interestingly, measurement of the sodalite phase unit cell is also showing very slight expansion; again, presumably from α-decay damage.

INTRODUCTION

The ceramic waste form (CWF) used for this study was developed to immobilize fission products and actinide materials that accumulate during the electrometallurgical treatment of spent nuclear fuel at Argonne National Laboratory in Idaho in support of the ANL/DOE Spent Fuel Demonstration Project [1]. Fission products and transuranic actinides concentrate in a molten, LiCl-KCl eutectic salt during electrometallurgical treatment of spent fuel. The contaminated eutectic salt is removed from the electrorefiner and processed into the CWF for final disposal in a geological repository. The waste form produced during fuel conditioning will contain roughly 0.2 weight percent Pu, primarily of ^{239}Pu isotopic composition. Of regulatory importance to a geological repository is the waste form durability [2]. To investigate the potential long-term effects of α-decay damage on CWF durability, but in a much shorter period, tests were initiated using a CWF containing surrogate fission products and ^{238}Pu as opposed to ^{239}Pu. The use of ^{238}Pu, with its high specific activity of 3.8×10^{13} decays/min g and short half-life of 88 years, allows significant α-decay dose in a much shorter time as compared to ^{239}Pu. Thus the term accelerated alpha damage. The data presented concludes the first year of a four-year study. The current cumulated α-decay dose is approximately 5×10^{17} α-decays/gram of material. The total accumulated dose at the end of the four-year study will be approximately 2×10^{18} α-decays/gram of material. To acquire an equivalent dose using ^{239}Pu would require approximately 1100 years.

Many literature reviews have been compiled on radiation damage to materials [3]. In particular, α-decay studies of crystalline materials or glass materials containing actinide host phases reveal that the crystalline material may become amorphous due to accumulation of dislocated matrix atoms [4, 5]. Amorphization of crystalline phases leads to volume increases that may in turn lead to microcracking. Additionally, after prolonged exposure, He or other gas bubbles may develop also resulting in swelling and cracking of the waste form. This swelling

Mat. Res. Soc. Symp. Proc. Vol. 608 © 2000 Materials Research Society

and cracking of the material usually has a detrimental affect on the durability of the waste form [6]. Some crystalline materials, however, such as PuO_2, UO_2 and ZrO_2 for example, show little damage to the crystalline structure after prolonged radiation exposure [3]. This study wishes to address these issues relating to the CWF.

EXPERIMENTAL

A detailed description of the processing of the CWF is found elsewhere [7]. The process first involves occluding the salt in dry (< 0.5 wt% moisture) zeolite 4A. The zeolite is then mixed with a glass binder and consolidated at a temperature of 1100 K and a pressure of 34 MPa. During this processing, the zeolite converts to sodalite. The composition of the pre-processed mixture is shown in Table 1. The fission product surrogate salt contains KBr, KI and the chlorides of Na, Rb, Sr, Y, Cs, Ba, La, Ce, Pr, Nd, Sm and Pu. In this study, the CWF was loaded to approximately 2.5 wt% ^{238}Pu. This Pu loading is roughly 3 to 20 times the actual Pu (elemental) loading of the CWF produced during the fuel treatment process. After contacting with the zeolite, the Pu converts to the oxide. This conversion is presumably due to the reaction of Pu with oxygen from residual water in the zeolite.

Table 1. Weight percent of each component in the pre-processed mixture.

Component	LiCl/KCl Eutectic Salt	Fission Product Surrogate Salt	Zeolite 4A	Glass binder	Elemental Pu
Weight %	4.1	1.7	64.7	25.0	2.5

The ^{238}Pu-loaded CWF is being analyzed on a periodic basis to evaluate α-decay damage to the waste form. The testing schedule is planned for a minimum of 4 years. The methods used to study the extent of α-decay damage on the CWF are separated into the following sections:

1) Waste form microstructure and elemental distribution is being investigated by scanning electron microscopy (SEM) in conjunction with energy and wavelength dispersive spectroscopy (EDS/WDS). Transmission electron microscopy (TEM), in conjunction with electron diffraction (ED) and EDS, is also being performed.
2) Powder x-ray diffraction (XRD) is used to monitor bulk phase composition and changes to major phase lattice parameters.
3) Density measurements on the CWF are being performed by an immersion method. Density measurements provide information on macroscopic swelling as a function of cumulative dose.
4) Durability of the ^{238}Pu-loaded CWF, with cumulated α-decay dose, is compared to the non-radioactive CWF reference material using the PCT-A leach method [8]. The leach test uses a crushed material with a –100 to +200 mesh size fraction, demineralized water for the leachant, and a surface area to volume ratio of 2000 m^{-1}. Elemental determination of the leachate is performed by inductively coupled plasma-mass spectrometry.

RESULTS

Microscopy

The microstructure (Figures 1a and b) of the ^{238}Pu-loaded CWF is very similar to the reference (non-Pu loaded) CWF and consists of 5 to 20-micron diameter sodalite grains surrounded by the glass binder. Plutonium is found primarily as PuO_2 in the glass intergranular boundaries between sodalite grains. Figure 1a shows the microstructure shortly after the material was produced. Figure 1b shows the microstructure after one year of cumulated alpha damage. As can be seen, the two micrographs are indistinguishable with no visible indication of α-decay effects. A second Pu containing phase has been identified by TEM as a tetragonal aluminosilicate material (Figure 2). This tetragonal phase has not been observed in the CWF containing 0.2 wt% Pu and its presence is possibly due to the greater Pu loading used in this study. In addition, a strongly faceted, needle-like structure containing Pu has been observed as shown in Figure 2.

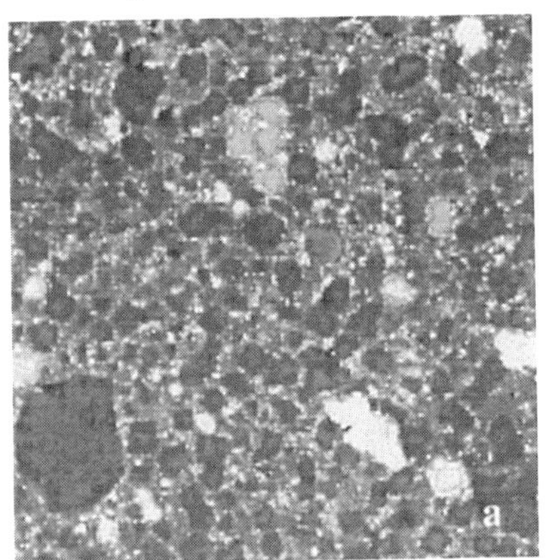

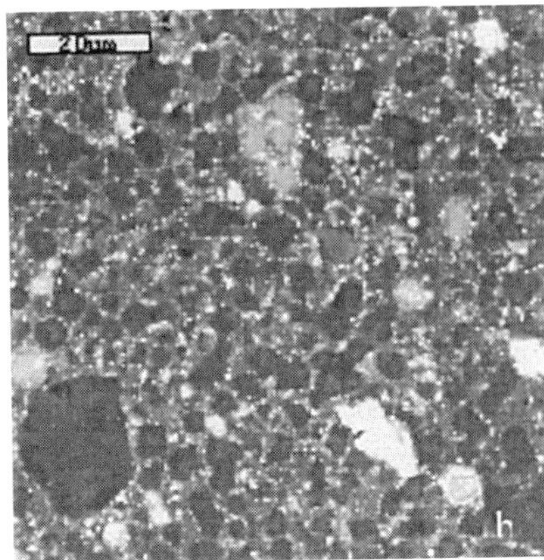

Figure 1. a) Back scattered electron image of ^{238}Pu-loaded CWF shortly after the material was prepared. This image is considered representative of the overall sample. Brightest regions are PuO_2, darkest regions are sodalite, medium dark regions are glass, and medium light regions are high Pu aluminosilicate. b) Same sample region after approximately one year (5×10^{17} α-decay/g) dose accumulation.

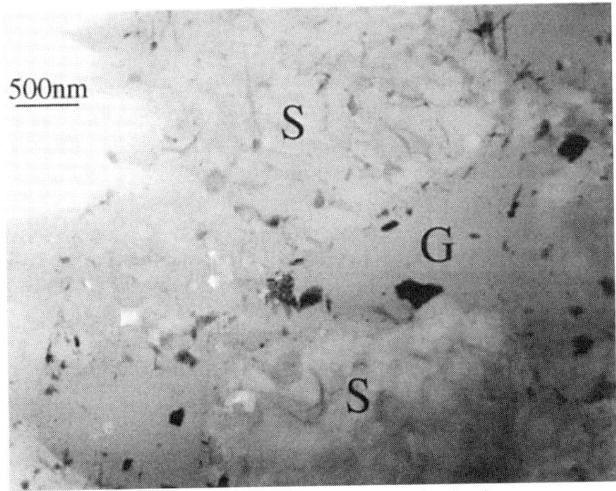

Figure 2. Low-magnification bright field image showing typical microstructure of ^{238}Pu-loaded CWF after one-year α-dose accumulation. The two main phases are labeled S: sodalite phase and G: glass phase. Plutonium containing phases are seen as the dark objects. These include PuO_2 fluoride structures, tetragonal Pu aluminosilicates and the strongly faceted, needle-like, Pu-containing material observed in the upper right of the micrograph.

X-ray Diffraction

The waste form is comprised of four crystalline components as indicated by x-ray diffraction. The weight percent of the four crystalline phases is shown in Table 3 and compares well with the reference (non-Pu loaded) CWF. The amorphous content of the sample was measured to be 22 percent, which corresponds to the target quantity of added glass binder of 25 weight percent within experimental uncertainty. The plutonium containing silicate and aluminosilicate phases observed by TEM are not observed by XRD, probably due to the low relative amounts of the phases present in the waste form. Regardless the majority of the Pu is in the oxide form.

Table 3. Crystalline phase composition of ceramic waste form in weight percent as determined by XRD. LaB_6 was added to the sample as an internal standard for refinement purposes.

Crystalline Phase	Sodalite $Na_8Al_6Si_6O_{24}Cl_2$	PuO_2	Nepheline $NaAlSiO_4$	Halite NaCl	LaB_6 NIST SRM 660
Crystalline Weight Percent found in Waste Form	84	3	7	3	3

X-ray diffraction patterns were acquired from the ^{238}Pu-loaded sample on days 2, 21, 72, 97, 146, 240 and 365 from the time the pellet was produced. For structural refinement, the LaB_6 standard was used to determine instrumental constants. The LaB_6 standard is not expected to sustain damage from radiation so the lattice parameter should not change with time. The average of the seven LaB_6 lattice parameter values is 0.415695 ± 0.000006 nm and matches very well with the NIST certified value. This excellent agreement of the measured lattice parameter of the LaB_6 standard to the NIST certified value adds confidence to the measured lattice parameters of PuO_2 and sodalite found in the sample. After pattern refinement, the change in unit cell volume (determined from the measured lattice parameters) of PuO_2 and sodalite were compared to previous determinations. Unit cell expansion of the PuO_2 phase is clearly shown in Figure 3 and reached an overall expansion of 0.6% after approximately 100 days. This expansion is less than the 1% volume expansion that has been reported elsewhere [3], and is possibly less due to initial crystal imperfections. The unit-cell expansion of PuO_2 follows a profile observed by other investigators [9]. The authors of ref. [9] have postulated that the profile maximum results from a supersaturation of defects in the lattice followed by lattice contraction due to defect migration to sink regions. If one assumes an uncertainty for the PuO_2 phase unit-cell measurement similar to that of the LaB_6 standard (at the 2σ level), than the error bars associated with each PuO_2 measurement shown in Figure 3 would be of similar magnitude to the symbols used in the Figure. Thus, for an uncertainty of this magnitude, the change in the PuO_2 profile shown in Figure 3 is greater than the measurement error. After the decrease in lattice parameter, the expansion increases gradually with the in-growth of $^{234}UO_2$ in the solid solution. In the case of the sodalite phase, a unit-cell expansion with time response similar to PuO_2 is possibly observed (also shown in Figure 3). This expansion of approximately 0.1% is questioned, however, considering the small quantity of Pu in that phase, as determined by EDS, and the greater uncertainty in the sodalite lattice parameter measurement.

Density Measurements

The density of a ^{238}Pu-loaded CWF sample has remained unchanged during the testing period to date. The density of the ^{238}Pu-loaded CWF, as determined by immersion in water, is 2.42 ± 0.01 g/cm^3. This density compares to an average density of 2.35 ± 0.02 g/cm^3 for non-Pu CWF material. The higher density of the ^{238}Pu-loaded CWF is attributed to the high Pu loading in the sample. Increase in density of the sample provides the most direct evidence of volume expansion due to α-decay damage, yet as mentioned none has been observed at the dose level of 5×10^{17} α-decays/gram of material.

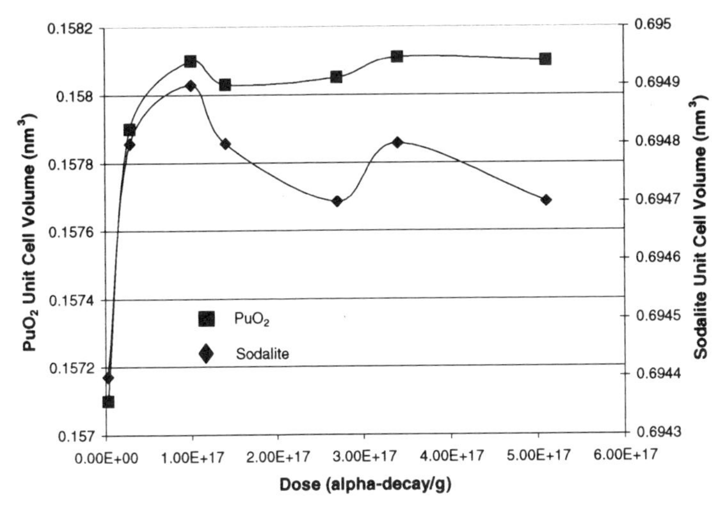

Figure 3. Unit cell volume change over time of PuO_2 and sodalite in CWF.

Durability Testing

Results of the short-term, PCT-A durability tests performed on the ^{238}Pu-loaded CWF are shown in Table 4. The PCT is used to evaluate any changes in the dissolution behavior of the CWF with time as a result of α-decay damage. These results from the first PCT, determined shortly after the material was produced, indicate that the ^{238}Pu-loaded CWF degrades in a manner similar the reference (non-Pu loaded) CWF and other high level waste forms for the elements measured [10, 11]. Two more PCT analyses will be performed on similar material that was produced at the same time as the material tested in Table 4 to determine degradation in durability after increased dose accumulation.

Table 4. Normalized elemental release in g/m^2 from a PCT (7-day) leachate solution of Pu-238 loaded CWF. The normalized release is the average value of triplicate tests. Also shown are the uncertainties associated with instrumental analysis.

Element	Normalized Release (g/m^2)	Relative Uncertainty (±1σ) (percent)	Element	Normalized Release (g/m^2)	Relative Uncertainty (±1σ) (percent)
Li	0.439	10	Al	0.035	10
K	0.090	80	Ce	0.001	30
Rb	0.165	10	Nd	0.001	20
Cs	0.148	10	Pu-238	0.001	25
Sr	0.016	40	Cl	0.376	10
Ba	0.009	25	Br	1.79	70
B	0.228	10			

CONCLUSIONS

The results of the first year of cumulated α-decay damage to the CWF indicate no microscopic or macroscopic alteration to the waste form as a whole. Only the PuO_2 phase has shown the expected unit cell volume increase resulting from lattice defects induced by alpha decay. The sodalite unit cell may also be expanding, but to much less extent than the PuO_2 phase. Because Pu is primarily concentrated to the intergranular glassy regions, the majority of decay damage is expected to occur locally in these regions. Due to the concentration of Pu in the radiation damage resistant PuO_2 fluorite structure, very little α-decay damage to the total waste form is expected. The one exception to this may be the formation and accumulation of He bubbles resulting from alpha decay. Bubble formation would most likely accumulate in the intergranular regions as well and may be a source for swelling. The greatest magnification of such swelling may be in an increase of elemental release rates (decrease material durability) observed during leach testing. Interim results from bulk density measurements, which have remained constant, and waste form durability, which has performed similar to the reference CWF, indicate no adverse effects of alpha radiation damage to the waste form at an α-decay dose of 5×10^{17} decays/gram of material.

ACKNOWLEDGMENTS

Argonne National Laboratory is operated for the U.S. Department of Energy by the University of Chicago. This work was supported by the the Department of Energy, Nuclear Energy Research and Development Program, under contract No. W-31-109-ENG-38. Thanks to Mr. Paul Hanson for sample preparation.

REFERENCES

1. J.P. Ackerman, S.M. McDeavitt, C. Pereira, L.J. Simpson, ANS Proceedings "Third Topical Meeting DOE Spent Nuclear Fuel and Fissile Materials Management", Charleston, SC, 699 (1998).
2. Office of Civilian Radioactive Waste Management, Waste Acceptance System Requirements Document, E00000000-00811-1708-00001, rev. 2 (1996).
3. W.J. Weber,R.C. Ewing, C.R.A. Catlow, T. Diaz de la Rubia, L.W. Hobbs, C. Kinoshita, Hj. Matzke, A.T. Motta, M. Nastasi, E.K.H. Salje, E.R. Vance, S.J. Zinkle, J. Mater. Res., Vol. 13, pp.1432-1484 (1998).
4. R.C. Ewing, W.J. Weber, F.W. Clinard, Jr., Prog. Nucl. Energy **29** (2), 63 (1995).
5. K.A. Boult, J.T. Dalton, J.P. Evans, A.R. Hall, A.J. Inns. J.A.C. Marples, E.L. Paige, *The preparation of fully active synroc and its radiation stability-Final Report, Oct. 1988.* AERE-R-13318 (Harwell Laboratory, Harwell, UK, 1988).
6. A.G. Solomah, H. Matzke, *Scientific Basis for Nuclear Waste Management XII,* W. Lutze, R.C. Ewing, Eds., Mat. Res. Soc. Symp. Proc., Vol. 127, pp. 241-249 (1989).
7. S.M. Frank, D.W. Esh S.G. Johnson, M.H. Noy and T.P. O'Holleran, "Production and Characterization of a Plutonium-238 Loaded Ceramic Waste Form for the Study of the Effects of Alpha Decay Damage," to be published in *Scientific Basis for Waste Management XXII,* Edited by D.J. Wronkiewicz, Materials Research Society (1999).
8. ASTM C1285-97, ASTM Philadelphia, PA, (1998).
9. T.D. Chikalla, R.P. Turcotte, Rad. Eff. **19**, 93-98 (1973).
10. S.M. Frank, K.J. Bateman, T. DiSanto, S.G. Johnson, T.L. Moschetti, M.H. Noy and T.P. O'Holleran, *Phase Transformations and Systems Driven Far From Equilibrium,* Eds., E. Ma, M. Atzmon, P. Bellon and R. Trivedi, Mat. Res. Soc. Symp. Proc. Vol. 481, pp. 351-356 (1998).
11. N.E. Bibler, J.K. Bates, *Scientific Basis for Nuclear Waste Management XIII,* Eds., V. Oversby and P. Brown, Mat. Res. Soc. Symp. Proc. Vol. 176, pp. 327-338 (1990).

FORMATION OF PEROVSKITE AND CALZIRTITE DURING ZIRCONOLITE ALTERATION

J. MALMSTRÖM*, E. REUSSER*, R. GIERE**, G.R. LUMPKIN***, M.G. BLACKFORD***, M. DÜGGELIN****, D. MATHYS****, R. GUGGENHEIM****, D. GÜNTHER*****
*Institute of Mineralogy and Petrography, ETH-Zentrum, 8092 Zürich, Switzerland (malmi@erdw.ethz.ch)
**Department of Earth and Atmospheric Sciences, Purdue University, West Lafayette, IN 47907-1397, USA
***Australian Nuclear Science and Technology Organisation (ANSTO), PMB 1, Menai, NSW 2234, Australia
****SEM-Laboratory, University of Basel, Bernoullistrasse 32, 4056 Basel, Switzerland
*****Institute of Inorganic Chemistry, ETH-Zentrum, 8092 Zürich, Switzerland

ABSTRACT

Synthetic zirconolites doped with rare earth elements (REEs) are corroded in a closed system at elevated temperature and pressure for various fluid compositions. Together with previous studies, the results indicate only a weak corrosion below 250°C at 50 MPa. Above that temperature and up to 500°C zirconolite ($CaZrTi_2O_7$) displays more rapid rates of corrosion and may be covered by various secondary phases, depending on the fluid composition. Above 500°C in Na-rich fluids, zirconolite is replaced by perovskite ($CaTiO_3$) and calzirtite ($Ca_2Zr_5Ti_2O_{16}$), but the REEs and Hf (as actinide analogues and/or neutron absorbers) are incorporated into the secondary phases. Perovskite and calzirtite exhibit an unusual crystal chemistry as determined by transmission electron microscopy and microanalysis.

INTRODUCTION

Zirconolite is a major constituent of the SYNROC nuclear waste form, where it is a principal host phase for actinides and certain fission products [1-3]. The aim of this study is to investigate the stability of zirconolite in hydrothermal fluids at high temperatures and pressures, under conditions similar to those expected for deep boreholes (33 MPa/km depth; 30-60°C/km plus self heat up to 300°C [4]). During this study, the corrosion of zirconolite has been investigated using a wide range of analytical techniques.

EXPERIMENTAL PROCEDURES

All corrosion experiments were carried out in small gold capsules with 25 µl of fluid. Primary starting materials were single phase, polycrystalline zirconolite-2M (Figure 1), $Ca_{0.8}Nd_{0.2}Zr_{1.0}Ti_{1.8}Al_{0.2}O_7$ and $Ca_{0.85}Gd_{0.1}Ce_{0.1}Zr_{0.85}Hf_{0.1}Ti_{1.9}Al_{0.1}O_7$, where REEs and Hf were used as actinide analogues and/or neutron absorbers. The ratio of zirconolite surface area to fluid volume was 4-6 cm^{-1} when the samples were loaded. Pure gold capsules were welded to obtain a closed system and placed in externally heated pressure vessels. Preliminary experiments were conducted over 21, 63 and 189 days at elevated temperatures (150-700°C) and pressure (50 MPa) for various fluid compositions (NaOH, HCl, deionized water) and molalities (0.1 - 0.001). Additional experiments were made in H_3PO_4-, CO_2- and SiO_2-rich fluids. More detailed experiments were carried out in 0.1 M NaOH at the specific conditions of 550°C and 50 MPa. These experiments were performed over several different run times (one hour to 21 days) to determine the kinetics of the corrosion process [5].

Solid starting material and solid run products were both characterized by environmental scanning

Mat. Res. Soc. Symp. Proc. Vol. 608 © 2000 Materials Research Society

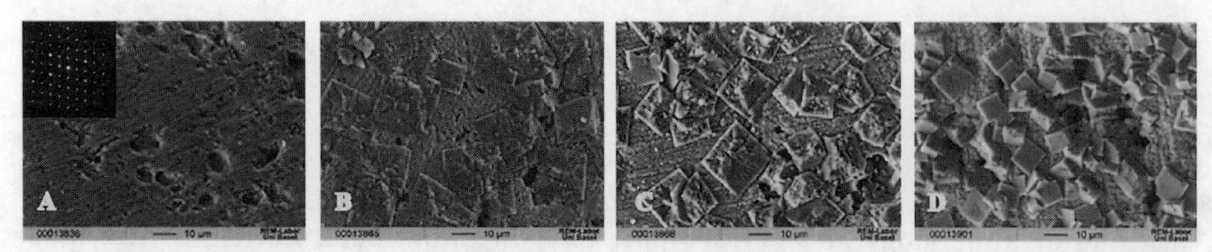

Figure 1: Secondary electron images (by ESEM in low vacuum and wet mode) of polished Nd-doped zirconolite surfaces before corrosion experiments (A), and after corrosion in 0.1 M NaOH at 550°C and 50 MPa for 12 hours (B), 24 hours (C), and 21 days (D). Top left shows a TEM selected area diffraction pattern [010] of zirconolite-2M.

electron microscopy (ESEM), electron probe microanalysis (EPMA), X-ray diffraction (XRD), transmission electron mictroscopy (TEM/EDX), infrared-spectroscopy (IR) and micro RAMAN-spectroscopy. The fluid run products were extracted from the capsule and diluted in 3 ml of 5% HNO_3 to stabilize the solution. Rhodium was added as internal standard to each sample to obtain a concentration of 10 ppb Rh in solution. Normal solution nebulization inductively coupled plasma mass spectroscopy (ICP-MS) was used in combination with a micro concentric nebulizer with desolvation unit (designed for limited sample volumes). The sample uptake was adjusted to 50 µl/min. Calibration was carried out with synthetic standards in a concentration range of 10-100 ppb and was optimized for the analysis of Ca, Zr, Ti, Nd, Al, Na, Au and Rh. [6]. Unfortunately the detection limits for Ca are too high for the concentrations of interest.

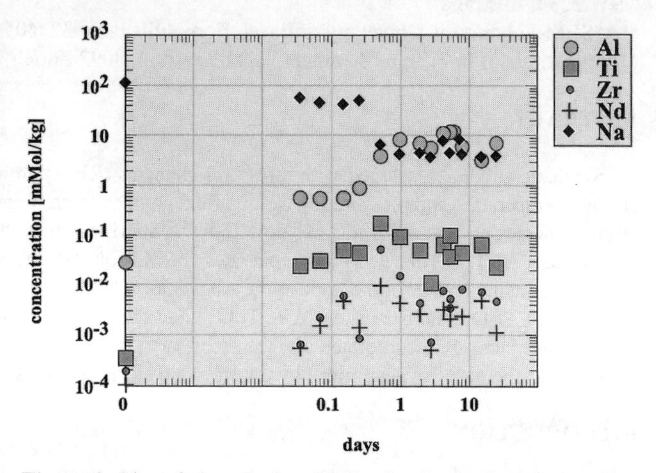

Figure 2: Plot of elemental concentration in the fluid determined by ICP-MS versus experimental run time (see corresponding ESEM images in Figure 1).

RESULTS AND DISCUSSION

Preliminary results indicate that zirconolite is corroded by dilute acidic and basic fluids at 50 MPa and temperatures above 250°C. The experiments conducted in 0.1 M HCl above 500°C produced various TiO_2 phases with variable morphologies; because of their small size and number, however, their exact composition is unknown [5]. The fate of Nd, originally present in zirconolite, is also unknown at the present time. The equivalent experiments conducted in 0.1 M NaOH formed perovskite ($CaTiO_3$) and calzirtite ($Ca_2Zr_5Ti_2O_{16}$). Qualitatively, the extent of reaction was found to be strongly related to temperature, molality of the fluid, and time [5]. From the additional experiments with other fluids we know that HCl and NaOH are more corrosive than H_3PO_4, CO_2 and SiO_2.

The time dependent studies in 0.1 M NaOH at the specific conditions of 550°C and 50 MPa

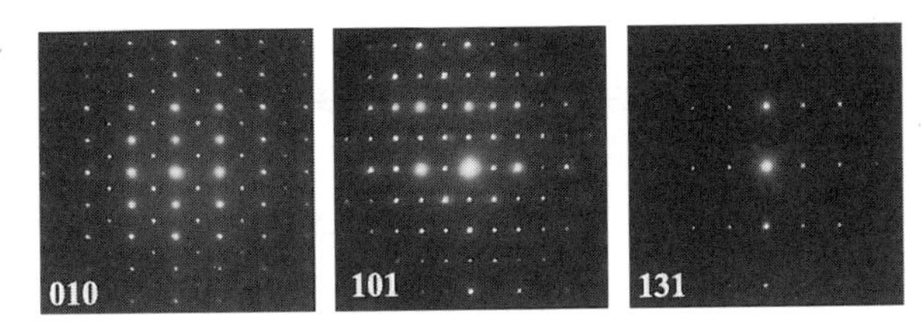

Figure 3: Selected area diffraction patterns of perovskite showing a simple orthorhombic perovskite with no high-order superstructure, which could reveal a cation-vacancy ordering mechanism.

show that reaction started after one hour with the crystallization of perovskite. Crystallization of calzirtite probably starts several hours later. After 2 to 4 days, the zirconolite surface is completely covered by perovskite and calzirtite. After the initial period of perovskite growth (24 hours) the grain size of perovskite was observed to increase with increasing time (Figure 1). Similar results are observed for the Gd-Ce-Hf-doped zirconolite.

These time dependent experiments together with the analyzed leachate data (Figure 2) reveal that the corrosion process begins with dissolution of zirconolite and a general concentration increase of the elements Zr, Ti, Nd, Al in the fluid phase. Almost immediately (observed after only 1 hour) the crystallization of perovskite begins, combined with a decrease of the Na concentration in the fluid, followed later by the crystallization of calzirtite, as reflected by the decreasing Zr concentration in the fluid from 12 hours to three days. The Al concentration increases to a high level in the fluid because it is not incorporated by the secondary phases. Elemental release rates appear to decrease with time.

The TEM diffraction patterns of these secondary phases demonstrate a simple orthorhombic perovskite with no high-order superstructure (Figure 3) and a tetragonal calzirtite with the fluorite subcell tripled along the a- and b-axes and doubled along the c-axis (Figure 4).

The observed perovskite has an unusual average chemical composition for the two different starting materials as determined by thin-film EDS analyses (Table 1):

$$Ca_{0.40}Na_{0.33}Nd_{0.13}Zr_{0.15}Ti_{0.96}O_3$$
$$Ca_{0.31}Na_{0.43}Gd_{0.04}Ce_{0.05}Zr_{0.16}Hf_{0.01}Ti_{0.98}O_3.$$

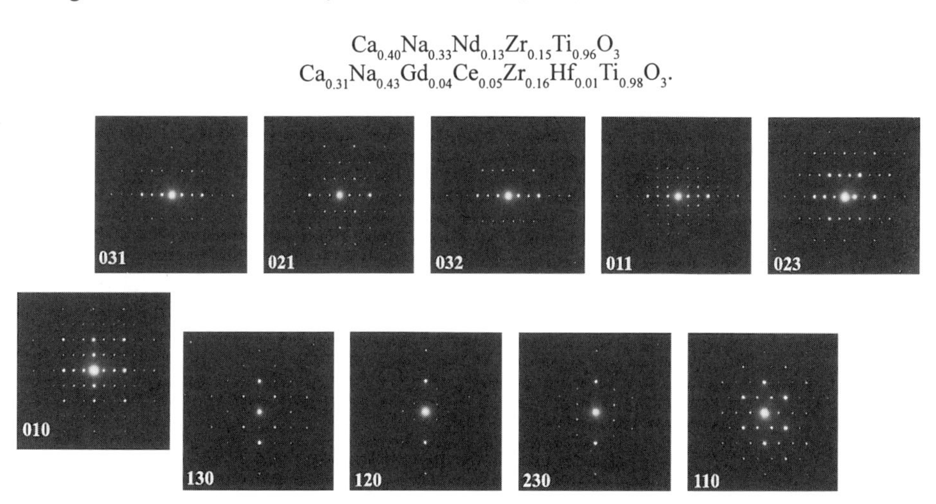

Figure 4: Selected area diffraction patterns of calzirtite. The diffraction pattern [010] (left side) shows the cubic fluorite subcell (100-type pattern) tripled along a and doubled along c. A tilting experiment around the c-axis for 45 degrees shows all zone axes up to order three (lower row). [110] shows again the fluorite subcell (110-type pattern). The equivalent tilting experiment around the a-axis shows that this calzirtite is tetragonal. The fluorite subcell is now visible on [023] (as a 110-type pattern, upper row).

Table 1: TEM energy dispersive X-ray (EDX) analysis of secondary perovskite and calzirtite resulting from the two different starting materials (standard deviation (s.d.) calculated as σ(n-1)).

	perovskite-(Nd) (n=14)		perovskite-(Ce) (n=19)		calzirtite-(Nd) (n=13)		calzirtite-(Gd) (n=3)	
wt%	mean	s. d.	mean	s. d.	mean	s. d.	mean	s. d.
Na_2O	6.75	1.00	8.99	1.52	1.05	0.30	1.17	0.21
Al_2O_3	0.01	0.05	0.02	0.10	0.14	0.40	0.08	0.13
CaO	15.08	1.38	11.70	2.18	8.11	0.44	7.67	0.79
TiO_2	51.44	2.15	52.67	1.60	17.21	0.75	17.38	0.15
FeO	0.52	0.23	0.23	0.18	0.53	0.27	0.66	0.41
ZrO_2	12.04	2.42	13.63	2.52	68.60	1.28	58.09	0.39
Ce_2O_3			5.79	1.67			0.42	0.09
Nd_2O_3	13.91	3.65			3.03	1.07		
Gd_2O_3			5.13	1.18			1.35	0.19
HfO_2	0.26	0.18	1.88	0.52	1.33	0.26	13.19	0.62
sum	100.02		100.04		100.00		100.01	
	Formula based on 3 oxygens				**Formula based on 16 oxygens**			
Na	0.33	0.05	0.43	0.07	0.31	0.09	0.36	0.06
Al	0.00	0.00	0.00	0.00	0.02	0.07	0.01	0.02
Ca	0.40	0.04	0.31	0.06	1.32	0.06	1.30	0.13
Ti	0.97	0.03	0.98	0.02	1.96	0.08	2.07	0.01
Fe	0.01	0.00	0.00	0.00	0.07	0.03	0.09	0.05
Zr	0.15	0.03	0.16	0.03	5.07	0.10	4.48	0.03
Ce			0.05	0.02			0.02	0.01
Nd	0.12	0.03			0.16	0.06		
Gd			0.04	0.01			0.07	0.01
Hf	0.00	0.00	0.01	0.00	0.06	0.01	0.60	0.03
sum	1.98		2.01		8.97		8.99	
Σ REE	0.12	0.03	0.10	0.02	0.16	0.06	0.10	0.01
Σ Zr+Hf	0.15	0.03	0.17	0.03	5.13	0.10	5.07	0.04

Three working hypotheses were considered to explain these unusual perovskite compositions. In the first approach, all Zr is allocated to the Ti site which results in a large number of Ca-site vacancies (~30%) and oxygen vacancies. However, no high-order superstructures (typical for many Ca-site vacancies) or OH^- (by IR-spectroscopy) were observed. In the second case, all Zr is assigned to the Ca site, and as third hypothesis, Zr is distributed between both sites. A [110] zone axis electron channeling experiment tested these two latter hypotheses. Results indicate that Na, REEs, and most of the Zr occupy the 8-fold coordinated Ca site, consistent with the substitutions:

$$REE^{3+} + Na^{+} \Leftrightarrow 2\,Ca^{2+}$$
$$Zr^{4+} + 2\,Na^{+} \Leftrightarrow 3\,Ca^{2+}.$$

Correlations exist between Ca and REE respectively Na as well as between Zr and Na, which are consistent with these two substitutions. A combination of these two substitutions thus best explains the overall variability on the Ca site, with all excess Zr on the Ti site.

$$Ca^{2+} + REE^{3+} \Leftrightarrow Zr^{4+} + Na^{+}.$$

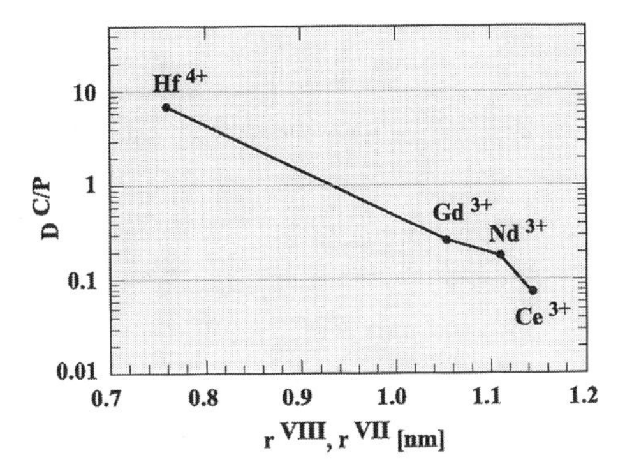

Figure 5: Average partitioning coefficient ($D^{C/P}$: average concentration in calzirtite divided by the average concentration in perovskite in wt%) of Ce^{3+}, Nd^{3+}, Gd^{3+} and Hf^{4+} between the secondary calzirtite and perovskite plotted versus their effective (REEs eight coordinated, Hf seven coordinated) ionic radii [7].

The chemical composition of calzirtite is much closer to its ideal composition. Thin-film analyses (Table 1) demonstrate, however, that the calzirtite formed in these experiments incorporates significant amounts of Na, REEs, and possibly contains up to 5% vacancies on the 8-fold coordinated Ca site:

$$Ca_{1.33}Na_{0.29}Nd_{0.14}Zr_{5.15}Ti_{1.94}O_{16}$$
$$Ca_{1.30}Na_{0.36}Gd_{0.07}Ce_{0.02}Zr_{4.48}Hf_{0.60}Ti_{2.07}O_{16}.$$

Three assumptions were made in the calculation of the average partitioning coefficients of individual REEs and Hf (Figure 5). Firstly, we have assumed that all REEs are trivalent. Secondly, we have assumed that all the REEs occupy the Ca site of perovskite and calzirtite whereas Hf occupies both sites in perovskite (as described for Zr) and only the Zr site in calzirtite. Finally, we considered the Ca site of perovskite as eight coordinated by ignoring four of the 12 oxygens which lie at distances of up to 0.32 nm away from the central cation (0.25 nm mean distance for eightfold coordination) [8]. This facilitates a direct comparison with the Ca site of calzirtite. Using ionic radii for Ce^{3+}, Nd^{3+}, Gd^{3+} in eight fold coordination and for Hf^{4+} in seven fold coordination [7], the data from Figure 5 show that elemental partitioning between the secondary phases is mainly controlled by their size. The polyhedral volumes of the two eight coordinated Ca sites are significantly different (site in calzirtite is approximately 30 % smaller than in perovskite) and thus, the large cations (trivalent REEs) have a preference for the large Ca site in perovskite ($D^{C/P}<<1$). On the other hand, the smaller cations (Hf^{4+}) prefer the smaller sites in calzirtite ($D^{C/P}>1$); here, the Zr site has a polyhedral volume that is approximately 50 % smaller than that of the Ca site in perovskite.

Lumpkin [9] described the same distribution of these elements between zirconolite and perovskite in Synroc. Therefore calzirtite and zirconolite exhibit similar behavior with respect to elemental fractionation in Na-rich systems.

High-resolution TEM images reveal an epitaxial overgrowth of secondary perovskite on primary zirconolite. The Ti polyhedra layers of zirconolite are perpendicular to c* and have a measured d-spacing of 5.32 Å (Figure 6). The Ti polyhedra layers of perovskite parallel to the (120) layers have a d-spacing of 2.64 Å (Figure 6) which is about one half of the d-spacing between the layers in zirconolite. The orientation of the boundary between the two phases is parallel to $(80\bar{2})$ in zirconolite and parallel to $(\bar{1}1\bar{6})$ in perovskite.

CONCLUSIONS

These data provide essential information about the interaction between fluids and zirconolite under closed system conditions. Above 500°C in a Na-rich fluid, zirconolite corrosion is fast and proceeds according to the reaction

$$\text{Zirconolite} + \text{NaOH} \Leftrightarrow \text{Perovskite} + \text{Calzirtite} + \text{Al(aq)}.$$

The REE and Hf initially present in zirconolite are preferentially incorporated into the solid secondary phases with respect to the fluid and distribute between the phases according to their ionic radii. This shows that zirconolite as a primary waste form breaks down leaving a surface layer of perovskite and calzirtite which act as a secondary waste form and retain the major portion of the actinide analogue and/ or neutron absorber elements released from the zirconolite.

Figure 6: High-resolution TEM image of the boundary between perovskite and zirconolite. Also shown are selected area diffraction pattern of perovskite and calzirtite.

ACKNOWLEDGEMENTS

We thank E. Vance, K. Hart and K. Smith for helpful discussions. M. Colella and D. Attard kindly prepared the TEM samples. This study is supported by a grant from ETH (Nr. 31706).

REFERENCES

[1] A.E. Ringwood, S.E. Kesson, K.D. Reeve, D.M. Levins, E.J. Ramm, in Radioactive Waste Forms for the Future, edited by W. Lutze and R.C. Ewing (Elsevier, New York, 1988), p.233.
[2] E.R. Vance, A. Jostsons, R.A. Day, C.J. Ball, B.D. Begg, P.J. Angel, in Scientific Basis for Nuclear Waste Management XIX, edited by W.M. Murphy and D.A. Knecht (Mater. Res. Soc. Proc. 412, Pittsburgh, PA, 1996), pp. 41-47.
[3] E.R. Vance, K.P. Hart, R.A. Day, B.D. Begg, P.J. Angel, E. Loi, J. Weir, V.M. Oversby, in Scientific Basis for Nuclear Waste Management XIX, edited by W.M. Murphy and D.A. Knecht (Mater. Res. Soc. Proc. 412, Pittsburgh, PA, 1996), pp. 49-55.
[4] R.C. Ewing, W.J. Weber, F.W. Clinard, Radiation effects in nuclear waste forms for high-level radioactive waste, in Progress in Nuclear Energy (Vol.29, No. 2, 1995), pp.63-127.
[5] J. Malmström, E. Reusser, R. Gieré, G.R. Lumpkin, M. Düggelin, D. Mathys, R. Guggenheim, in Scientific Basis for Nuclear Waste Management XXII, edited by D. J. Wronkiewicz and J. H. Lee (Mater. Res. Soc. Proc., Boston, MA, 1999), pp. 165-172.
[6] D. Günther, R. Frischknecht, H.J. Müschenborn, C.A. Heinrich, Direct liquid ablation: a new calibration strategy for laser ablation-ICP-MS microanalysis of solids and liquids, (Fresnius J Anal Chem 359, Springer, 1997), pp 390-393.
[7] R.D. Shannon, Acta Crystallografica (A32, 1976), pp. 751-767.
[8] P.E. Fielding, J.J. White, J. Mater. Res. 2 (1987), p. 387.
[9] G.R. Lumpkin, K.L. Smith, M.G. Blackford, in Journals of Nuclear Materials 224 (1995), pp. 31-42.

MOLECULAR DYNAMICS STUDY OF THE INFLUENCE OF A SURFACE ON A SIMPLIFIED NUCLEAR GLASS STRUCTURE AND ON DISPLACEMENT CASCADES

A. Abbas*, J.M. Delaye**, D. Ghaleb**, Y. Serruys*, G. Calas***

*Commissariat à l'Energie Atomique (CEA), Centre d'Etude Nucléaire de Saclay, DTA/CEREM/DECM/SRMP, 91191 Gif-sur-Yvette Cedex, France.
**Commissariat à l'Energie Atomique (CEA), Centre d'Etude Nucléaire de la Vallée du Rhône, DCC/DRRV/SCD/LECM, BP171, 30207 Bagnols-sur-Cèze Cedex, France.
***Laboratoire de Minéralogie-Cristallographie, Universités Paris VI-VII et Institut de Physique du Globe de Paris, UMR-CNRS-7590, France

ABSTRACT

Irradiation effects which occur in nuclear waste glasses are often reproduced by external irradiations. Interesting questions arise about the application of observed phenomena to real waste glasses. For a better understanding of the role of the surface, we have simulated a surface on a 7 oxides glass using Molecular Dynamics. Modifications of the glassy structure, limited to the first 4 Å, have been observed. We also report in this paper on the influence of surface on the behaviour of displacement cascades, characterised by an increased number of displaced atoms, and also by a larger displacements of sodium atoms in the surface region.

INTRODUCTION

It is important to understand and predict the radiation effects of fission products and actinides in real waste glass during the storage. External irradiations [1,2] (heavy and light ions, electron, γ) have been widely used to simulate effects produced by self-irradiation on glass structure and on macroscopic properties [3-6].
However possible artifacts related to external irradiations must be known before applying the results to the real case. Indeed, glass is an insulator and irradiations with charged particles induce migration of mobile elements like alkali (e.g., sodium) due to the surface potential created [7]. Moreover incident ions penetrate into the glass in one direction while the self-irradiation is isotropic in real glass. Defects relaxation can thus provoke alkali migration on long distance due to channels formation as proposed in references [8,9], while in real glasses defects relaxation is insufficient [10] to bring out such a percolation phenomenon.
Finally, the presence of a surface itself can influence radiation effects in three ways. First, the glassy structure near the surface can be different from the bulk structure [11]. Second by modifying the defect formation and migration energies or the atomic displacement energies, the surface can play a role in the structural relaxation. Finally, phenomena like sputtering of atoms or volume expansion near a surface can modify the cascade morphology [12].
The aim of this paper is to investigate the influence of a surface on ballistic displacement cascades using molecular dynamics. In a first step, we present the structural modifications observed when a surface is created. Then, displacement cascades calculated in simulation cells with and without a surface are compared. Molecular dynamics (MD) allow to describe effects produced by ballistic processes [13]. To identify the surface influence, we have preferred to simulate small energy cascades (around 1keV) to have larger statistics.

EXPERIMENT

Glass preparation ; Surface creation

The simulations were performed on glasses with 6 and 7 oxides which represent the basic matrix of nuclear glass [14]: 60% SiO_2, 16% B_2O_3, 12% Na_2O, 2% ZrO_2, 4% Al_2O_3, 6% CaO with additional UO_2 in the 7 oxides system. U-O interactions are modeled by the Lindan & Gillan potential [15]. Pair interactions are represented by a Born-Mayer-Huggins (BMH) potential:

$$\Phi(r_{ij})=A_{ij}\exp(-r_{ij}/\rho_{ij}) + \mathrm{erfc}(r_{ij}/r_c)\, q_i q_j/r_{ij} \qquad (1)$$

The first term corresponds to atomic repulsion and the second term to a screened coulombic interaction. Only the Ewald sum term in real space is calculated [16].

Mat. Res. Soc. Symp. Proc. Vol. 608 © 2000 Materials Research Society

A three body term is applied on the O-Si-O, Si-O-Si, O-Al-O and O-B-O angles to improve the local environment around formers. Formulae and parameter values of the three body terms are given in reference [14].

We prepared two different cubic cells with a side length equal to 40 Å or 80 Å, and containing respectively 5184 and 41472 atoms.
First, simulation cells with no surface are calculated by a previously described procedure [17]. The surface is then created by removing the periodic boundary conditions (PBC) in the z direction. So we introduce two parallel surfaces. The two cells with and without surface are relaxed during 10 ps at ambient temperature.
Structural changes are determined from atomic concentration profiles, angular distributions, coordination number evolution versus depth and Voronoï volumes.

Cascade initiation

Cascade initiation is performed by accelerating one atom in the cell. Primary knock-on atom (PKA) energy is limited to keep the damage volume inside the simulation box and avoid PBC artifacts. BMH potential is not able to reproduce energetic interactions which occur during collision events. So a Ziegler-Biersack-Littmark potential (ZBL) [18] is added for interatomic distances less than 0.9 Å. The connection between ZBL and BMH potentials is performed between 0.9 and 1.0 Å (1.1 and 1.3 Å for U-O interactions) by a polynomial expression.

The time step used for cascade simulations varies from 10^{-17}s at the beginning to 10^{-15}s at the end. This time step adjustment is introduced to ensure precise dynamics. At the beginning, atomic velocities can be large and a small time step is required. The progressive increase of the time step is controlled by the largest atomic velocity. The maximum atomic displacement between two successive time steps is thus limited to 0.01 Å.
The system is embedded in a thermal bath to absorb the thermal agitation created by the cascade (i.e. outward layers up to 3 Å are maintained at ambient temperature by periodically rescaling the atomic velocities).

RESULTS

Structure

Radial distribution function

The first characterisation is done by calculating the radial distribution functions (RDF) to enlighten the surface influence on the local atomic environments. No clear modifications of RDF have been observed between RDF in cells with or without surface, even if we only take into account near-surface atoms (i.e. the first 5 Å).

Concentration profile

A direct visualisation of the simulated surface profile shows a glass expansion and sodium accumulation on the surface. This observation is confirmed by plotting concentration profiles of atomic species (fig. 1). The origin corresponds to the initial location of the created surface. Each point represents the concentration of the atomic species in a 1.4 Å thick slice. The sodium profile shows an accumulation in outward layers with a depletion in layers just under the surface. We can also note that sodium concentration in the bulk is not modified. So this phenomenon is localised in the surface region (i.e. the first 4 Å).

Voronoï Volume

Voronoï volume calculation is performed far from surface to avoid artifacts due to surface vicinity (i.e. the first 4 Å are not considered). A comparison with the system without surface is done. We can distinguish three categories: formers (i.e. silicon, boron, aluminium and zirconium) characterised by a smaller volume, modifiers (i.e. calcium and sodium) characterised by a larger volume, indicating that they belong to small density areas, and finally oxygen atoms.

The evolution of Voronoï volumes versus depth shows that near the surface, we have an increase associated to modifiers and oxygen atoms, consistent with volume expansion observed from the concentration profiles. No significant evolution is observed for formers.

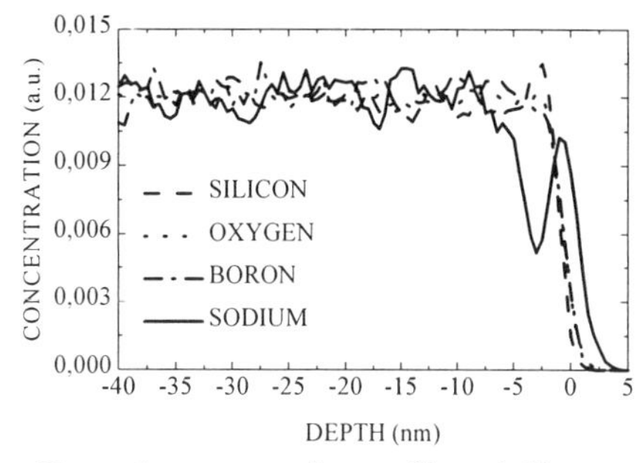

Figure 1: concentration profiles of silicon, oxygen, boron and sodium atoms.

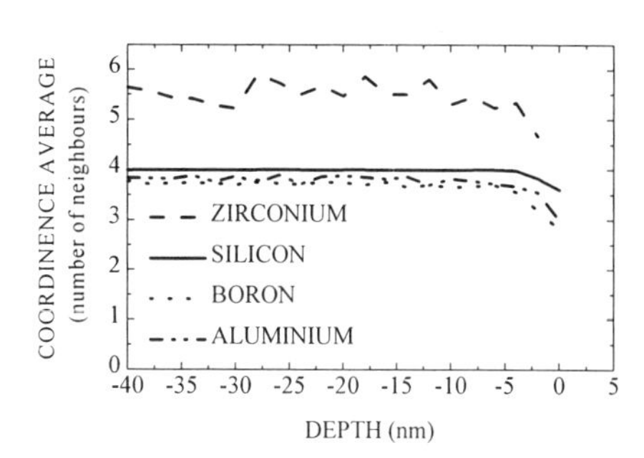

Figure 2: evolution of coordinence of different chemical species versus depth.

Coordination evolution

Creation of surface induces bond breaking, so it is interesting to see whether there are changes in atomic coordinations. The evolution of the average former coordinations with depth is plotted on figure 2. We can notice that the average coordination of formers decreases near the surface.

Angular distributions

Evolution of coordinations previously observed can be correlated to the angular distributions. Angular distributions are plotted on figure 3 for O-Si-O angles in the region near the surface and in the bulk. In the bulk, the angular distribution is centered around 109° which corresponds to 4-coordinated silicon atoms in a tetrahedral environment. There is a slight shoulder near 120° which corresponds to 3-coordinated silicon atoms. This shoulder increases in the surface vicinity.
So we confirm with the O-Si-O angular distribution the mixing of 3- and 4-coordinated Si atoms underlined by the coordination analysis.

Cascades

Cascades in small cell (5184 atoms)

A series of cascades have been simulated by accelerating an oxygen or zirconium atom with energies less than respectively 500eV and 700eV. To describe surface influence on cascade morphology, three different ways have been explored:

- the first one consists in accelerating the PKA from the surface towards the bulk in the z direction (1st),
- in the two other ones, the PKA is accelerated in the middle of the simulation cell (i.e. in the bulk) in the xy plan (parallel to the surface). Calculations in systems with (2nd) and without (3rd) surface are performed.

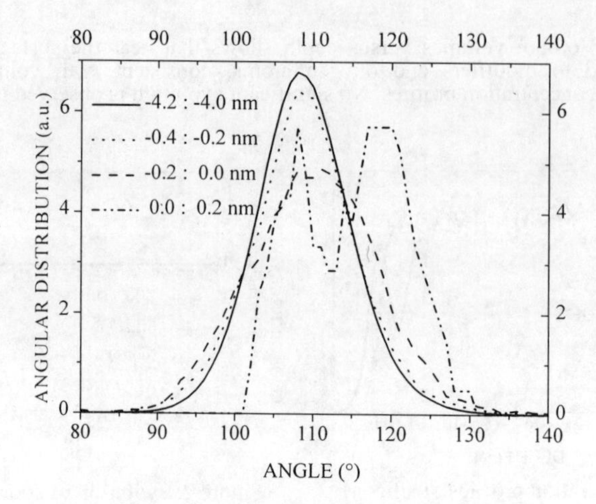

Figure 3: evolution of O-Si-O angle distribution for different depths where location of the middle of the cell is 0 nm and of the surface is 4.2 nm.

To investigate structural modifications induced by the cascades, we consider arbitrarily that an atom is displaced when the distance between its initial and final positions is larger than 1 Å. Location of displaced atoms gives a first indication of the global cascade morphology and damaged volume. Figure 4 shows two cascades initiated with the same oxygen atom accelerated in the xy plan. We can notice that the two morphologies are quite similar. At the beginning, the PKA follows roughly the same trajectory. Differences in trajectories appear only at the end of the cascade where PKA seems to go up to surface. These simulations show a general trend towards a larger number of displaced atoms in systems with surface.
The depolymerization index is an appropriate tool to analyze the structural evolution during the cascade. This index is defined as the number of chemical bonds broken between the initial and the instantaneous structure. The evolutions of depolymerisation of the glass during the cascade show no obvious influence of glass surface on relaxation. Sometimes, it favours structure restoration, but the contrary is observed as well.

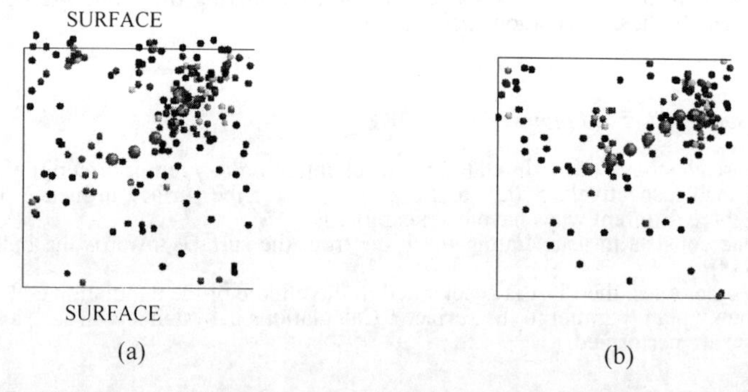

Figure 4: displaced atoms during cascades initiated by an oxygen atom of 500eV in the bulk of systems with (a) and without (b) surface.

Cascade in large cell (41472 atoms)

We have performed cascades in large cells (containing 41472 atoms). An uranium atom is accelerated at 1000 eV. Figure 5 shows two examples of cascade morphologies. The same PKA is accelerated in the same direction in the xy plan in systems with and without surface. Without surface, the cascade is located in the center of the simulation box. As previously observed [19], a cloud of displaced sodium atoms surrounds the cascade.
With surface, we found a larger number of displaced atoms which is consistent with the previous simulations in small cells. Sodium atoms seem to react in a different way. There are clearly more displaced sodium atoms near the surface and a kind of cone connecting the surfaces to the core of the cascade appears. A small number of displaced atoms are observed outside this cone. It seems that the atomic displacements appear along a preferential way from the cascade location towards the surfaces.

DISCUSSION

The results presented here indicate an influence of the surface on displacement cascades. The major surface effect on the structure is to create a 4Å deep zone depleted with sodium atoms and containing atomic environments with lower coordination. In the same time, Na migration on the surface occurs. We have to keep in mind that the cooling rate used to create the glass is very large compared to a real cooling rate. So, the average atomic diffusion length remains short, and this depletion, rather localized in our case, might occur on a longer range in the real case. Moreover, in the real case, an hydrated liquid or gas is in contact with the surface. So a structural relaxation leading to silanol bonding formation occurs, which is not represented here.
The two characteristics of the simulated surface are thus a less polymerized network with a smaller density and a sodium migration to the surface, consistent with previous observations [20].

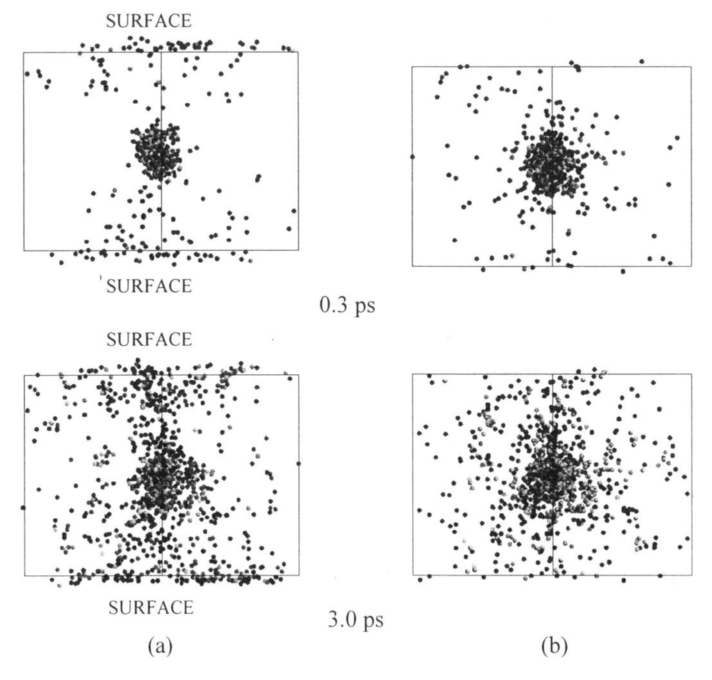

Figure 5: displaced atoms during cascades initiated by an uranium atom of 1000 eV in the bulk of systems with (a) and without (b) surface for two different times.

This simulated surface has an influence on cascade morphology. Two noticeable facts can be underlined. First, the number of atomic displacement is larger in presence of a surface, and second, the locations of the displaced sodium atoms are different. The reason is probably the additional degree of freedom for the relaxation process allowed by the surface. A study of surface influence on the migration energies is necessary but the present observations suggest that migration energies decrease for displacements near the surface.
The relaxation of elastic constraints by swelling is allowed in only one direction. This anisotropy introduced by the surface could induce the observed anisotropy of displacement paths.
An other interesting observation just evoked in this paper is the similarity of the projectile trajectories regularly noticed in the two cells with and without surface. Let us remind that starting from an unique structure, in one case a relaxation without surface is performed, in the other case a relaxation with surface is performed. So there are strong similarities between the structures with and without surface. Nevertheless, the atomic velocity distributions are different. The similarity of the projectile trajectories suggests a non-chaotic behaviour of the cascade dynamics. Permanent characteristics, which should deserve a deeper understanding, remain whatever the initial atomic velocity distribution.

RÉFÉRENCES

[1] J.-C. Dran, Y. Langevin, J.-C. Petit; Nucl. Instr. & Meth. in Phys. Res. B1 (1984) 557.
[2] G. Battaglin, G.N. Arnold, G. Mattei, P. Mazzoldi, J.-C. Dran; J. of Appl. Phys., 85 (1999) 8040.
[3] Hj. Matzke, E. Vernaz; J. Nucl. Mater. 201 (1993) 295.
[4] N. J. Hess, W.J. Weber, S.D. Conradson; Materials Res. Soc. Symp. Proceedings, 506 (1998) 169.
[5] Y. Inagaki, H. Furuya, K. Idemitsu, T. Banba, S. Matsumoto, S. Muraoka; Scientific Basis for Nuclear Waste Management XV, ed. by C.G. Sombret, 257 (1992) 199.
[6] W. Weber, R.C. Ewing , C.A. Angell, G.W. Arnold, A.N. Cormack, J.-M. Delaye, D.L. Griscom, L.W. Hobbs, A. Navrotsky, D.L. Price, A.M. Stoneham, M.C. Weinberg; J. Mater. Res., 12 (1997) 1946.
[7] P. Regnier, Y. Serruys, A. Zemskoff; Phys. Chem. of Glasses, 27 (1986) 185.
[8] A. Miotello, P. Mazzoldi; Radiation Eff. 98 (1986) 39
[9] A. Miotello, F. Toïgo; Nucl. Instr. Meth. Phys. Res. B32 (1988) 258-263
[10] M. Yamashita, Hj. Matzke; Modifications induced by irradiation in Glasses: Proc. of the Symp. F on Chemical and Physical Modifications induced by Irradiation in Glasses, E-MRS Fall Conference, November 5-7 1991, Strasbourg, France, ed. by P. Mazzoldi.
[11] D.M. Zirl, S.H. Garofalini; J. Am. Ceram. Soc., 75 (1992) 2353.
[12] R.S. Averback, T. Diaz de la Rubia; Sol. St. Phys. 51 (1998) 281.
[13] J.-M. Delaye, D. Ghaleb; J. of Nucl. Mat., 244 (1997) 22.
[14] J.M. Delaye, D. Ghaleb; Mat. Sci. & Eng. B37 (1996) 232
[15] P.J.D. Lindan, M.J. Gillan; J. Phys. Condens. Mat. 3 (1991) 3929
[16] M.J.L. Sangster, M. Dixon; Adv. Phys. 25 (1976) 247.
[17] J.-M. Delaye, D. Ghaleb; J. Non-Cryst. Sol. 195 (1996) 239.
[18] J.F. Ziegler, J.P. Biersack, U. Littmark; The stopping and range of ions in solids, edited by J.F. Ziegler, Pergamon Press Inc. 1985
[19] J.-M. Delaye, D. Ghaleb; Rad. Eff. & Def. In Sol. 142 (1997) 471.
[20] S.H. Garofalini, S.M. Levine; J. Am. Ceram. Soc. 68 (1985) 376.

TEMPERATURE DEPENDENCE OF ION IRRADIATION INDUCED AMORPHISATION OF ZIRCONOLITE

K.L. Smith*, M.G. Blackford*, G.R.Lumpkin* and N.J. Zaluzec†

* Materials Division, Australian Nuclear Science and Technology Organisation, P.M.B. 1, Menai, NSW 2234, AUSTRALIA.
† Materials Science Division, Argonne National Laboratory, 9700 South Cass Ave, Argonne, Il 60439, USA

ABSTRACT

The critical dose for amorphisation, D_c, of two end-member zirconolites ($CaZrTi_2O_7$) with different stacking fault densities, was measured as a function of irradiation temperature from 20 K to 623 K using the HVEM-Tandem Facility at Argonne National Laboratory (ANL). Below 473 K, the D_c values of both samples are identical within experimental error, showing only a small increase in D_c from (2.5 to 4.6) x 10^{18} ions m^{-2} between 20 K and 473 K. At temperatures above 473 K, the data for the zirconolite containing many stacking faults is bracketed by two data sets from almost crystallographically perfect end-member zirconolites: one collected in this study and one collected in a previous study. The raw D_c versus temperature data from the zirconolites in this and a previous study suggest that the critical temperature above which samples cannot be amorphised and/or recrystallisation is complete, T_c, is between 600 and 1000 K. The data sets collected in this study are discussed in relation to a current model.

INTRODUCTION

Zirconolite is one of the major host phases for actinides in various wasteforms, particularly Synroc [1], for immobilising high level radioactive waste (HLW). Over time, the periodic crystalline matrix is damaged by α-particles and energetic recoil nuclei resulting from α-decay events. The cumulative damage caused by these particles results in amorphisation. Data from natural zirconolites suggest that radiation damage partly anneals over geologic time and is dependent on the thermal history of the material [2]. Proposed HLW containment strategies rely on both a suitably water-resistant wasteform and geologic isolation. Depending on the waste loading, depth of burial, and the repository-specific geothermal gradient, burial could result in a wasteform being exposed to temperatures of between 373-523 K. Consequently, it is important to assess the effect of temperature on radiation damage in synthetic zirconolite.

Zirconolite containing wasteforms are likely to be hot pressed at ~ 1473 K or sintered at ~ 1623 K [3]. Zirconolite fabricated at temperatures below 1523 K (1250°C) contains many stacking faults and twins [4]. As there have been various attempts to link radiation resistance to structure [5 - 7], it is also pertinent to assess the role of original crystallographic perfection in radiation resistance.

In this study, we simulated α-decay damage in two zirconolite samples by irradiating them with 1.5 MeV Kr^+ ions using the High Voltage Electron Microscope-Tandem User Facility (HTUF) at Argonne National Laboratory (ANL) and measured the critical dose for amorphisation (D_C) at several temperatures between 20 and 623 K. One of the samples had a high degree of crystallographic perfection; the other contained many stacking faults on the unit cell scale. Previous authors proposed a model for estimating the activation energy of self annealing in zircon [8]. We will discuss our results and earlier published data in relation to that model.

EXPERIMENTAL PROCEDURE

Two zirconolite samples were prepared via the alkoxide route [9]. One was hot-pressed at 1473 K for 2 hours (1200 sample) and the other was sintered at 1723 K for 1 week (1450 sample). TEM specimens were prepared by crushing material under ethanol then passing holey carbon coated copper grids through the suspension and collecting fine particles on the

Mat. Res. Soc. Symp. Proc. Vol. 608 © 2000 Materials Research Society

carbon film. All grids were cleaned in an Ar plasma for 5 minutes using a South Bay Technology PC 150 Plasma Cleaner. This treatment minimises specimen heating and electrical charging during ion irradiation.

In addition to zirconolite, the samples fabricated for this study also contained small amounts of rutile, $ZrTiO_4$ and perovskite. Therefore the positions of zirconolite grains on grids were mapped on secondary or transmitted electron images collected using either a JEOL 2000FX TEM (equipped with an ASID scanning attachment, Tracor Northern EDX detector and Link ISIS system) or a JEOL 2010F FEG/TEM (equipped with an ASID scanning attachment, Link EDX detector and EmiSpec ES Vision system). In situ ion irradiation of the TEM specimens was performed using a 1.2 MeV modified Kratos/AEI EM7 electron microscope (operated at 300kV) interfaced with a NEC ion accelerator in the HVEM-Tandem User Facility at Argonne National Laboratory. At all temperatures, apart from room temperature, temperature variation was no more than 3 K. During irradiation at room temperature, the specimen holder temperature rose by 28 K. Grains selected for ion irradiation showed many maxima in their selected area electron diffraction (SAD) patterns. Specimens were irradiated with 1.5 MeV Kr^+ ions using the procedure described by Smith et al. [10]. The average of the dose at which all Bragg reflections had disappeared and the dose immediately prior to that dose was taken to be the critical dose for amorphisation, D_c. Most of the D_c values quoted here are the average of the data from 4 or more grains. Two of the D_c values at low temperatures were the average of only two grains. The quoted errors represent the spread of experimental results or the size of the dose steps, whichever was larger.

RESULTS

The 1450 sample basically has the zirconolite-2M polytype structure with a high level of crystallographic perfection. The 1200 sample also predominantly has the zirconolite-2M polytype structure but contains many stacking faults and twins on the scale of the unit cell.

Our data for D_c as a function of temperature (T) are shown in Table 1 and Figure 1. The errors in D_c are largest at temperatures near the critical temperature for amorphisation, T_c (the temperature above which a sample cannot be amorphised and/or recrystallisation is complete). This shows the importance of estimating the error of each data point rather than measuring the error of some data and assuming that other data will show similar variation.

The D_c versus T data for the 1200 and 1450 zirconolites can be divided into two regimes. At temperatures between 20 K and ~473 K (Regime I), the D_c values of both samples are identical within experimental error, showing only a small increase in D_c from (2.5 to 4.6) x 10^{18} ions m^{-2} between 20 K and 473 K. At temperatures above ~475 K (Regime II) , there are differences between the D_c values for the 1200 and 1450 samples, that are consistent with the 1450 sample having a higher T_c value that the 1200 sample.

Table 1. D_c versus T data for zirconolite irradiated with high energy Kr^+ ions in this and a previous study. Values in parentheses are the estimated errors.

Data collected in this study. (1.5 MeV Kr^+ ions)				Data collected by Wang et al. [11]. (1.0 MeV Kr^+ ions)			
1200 sample		1450 sample		$CaZrTi_2O_7$		$Ca_{0.8}Ce_{0.2}ZrTi_{1.82}Al_{0.2}O_7$	
T (deg.K)	Average D_c (x 10^{18} ions m^{-2})	T (deg.K)	Average D_c (x 10^{18} ions m^{-2})	T (deg.K)	Average D_c (x 10^{18} ions m^{-2})	T (deg.K)	Average D_c (x 10^{18} ions m^{-2})
20	2.5 (0.3)	50	2.9 (0.3)	25	3.0	295	2.9
150	3.5 (0.3)	150	3.4 (0.4)	99	3.0	471	2.9
303	3.7 (0.1)	303	4.0 (0.4)	200	3.1	569	3.9
400	4.3 (0.2	400	3.9 (0.2)	297	3.8	672	4.5
473	4.7 (0.8)	473	4.6 (1.1)	372	6.0	766	8.1
523	5.5 (0.9)	573	5.7 (0.3)	471	6.1	868	34
548	6.4 (0.9)	623	12.0 (4.4)	571	15		
573	10.0 (1.1)			620	31		

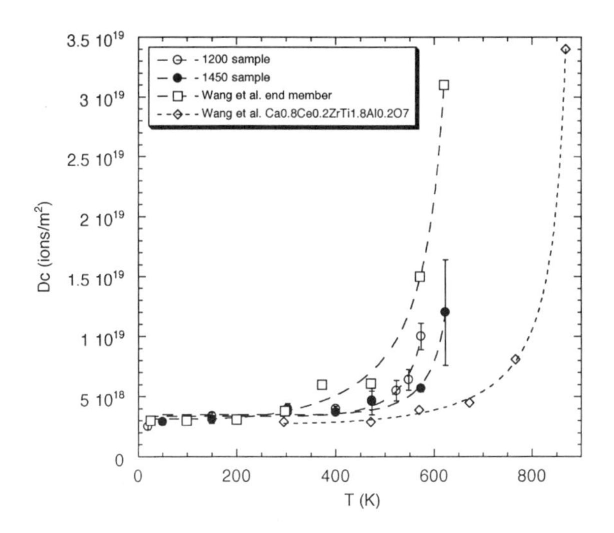

Figure 1. D_c versus temperature data from this study and that of Wang et al. [11]. The dashed curves are least squares fits of each data set.

DISCUSSION

Wang et al.[11] measured the critical dose of 1.0 MeV Kr^+ ions for amorphisation of six zirconolites of various composition, polytype and twin density. At low temperatures, their data for all samples are essentially identical to ours within experimental error, except that their data show Regime I ends at ~350 K for end-member zirconolite and ~425 K for their other zirconolites, which included Ce, Nd and Al-doped materials. Two data sets from Wang et al. are plotted in Figure 2 for comparison with our data.

The end-member zirconolite data sets in this study and that of Wang et al. all indicate that the T_c of end-member zirconolite is ~600-700 K. Furthermore, our data set for the 1200 sample (which contains many stacking faults and twins) is intermediate between the data sets for the 1450 sample and Wang et al.'s end-member zirconolite, both of which are almost crystallographically perfect. This suggests T_c is not significantly affected by stacking fault density.

In Regime I, the D_c values of all zirconolites at any given temperature are similar (see for example Table 1 and Figure 1) indicating that at temperatures below ~350 K, the radiation resistance of zirconolite is not affected by composition and stacking faults. At high temperatures (in Regime II), the data sets for different zirconolites diverge. However all the data sets in this and Wang et al.'s study indicate T_c values between 600 K and 1000 K. Our data sets for the 1200 and 1450 samples are intermediate between Wang et al.'s data for $CaZrTi_2O_7$ and $Ca_{0.8}Ca_{0.2}ZrTi_{1.8}Al_{0.2}O_7$ (see Figure 1 and Table 1), both of which have the 2M structure and are almost crystallographically perfect.

White et al. [12] measured the critical dose of 1.0 MeV Kr^+ ions for amorphisation of zirconolite at 20, 300 and 475 K and found it to be 7.1, 10 and 340 respectively in units of 10^{18} ions m^{-2}. These values are much greater than both ours and those of Wang et al. [7] (see Table 1). One possible reason for the difference is that their samples had poor thermal connection to the TEM stage and consequently experienced much more heating during irradiation than our samples or those of Wang et al.

Clinard [13] found that alpha radiation damage caused Pu-substituted zirconolite specimens to swell and that the final volume to which a specimen swelled depended on the temperature at which they were stored. At room temperature, Pu-substituted zirconolite specimens swelled by a maximum of 5 percent after ~150 days. At 575 K, specimens swelled by a maximum of 4 percent after ~350 days, while at 875 K specimens showed <0.5 percent swelling even after 600 days. In agreement with the findings of Clinard [13], the T_c values calculated both in this study and by Wang et al. suggest that significant self annealing will occur at temperatures above ~ 600 K.

Weber et al. [8] suggested that the temperature dependence of D_c can be expressed by the following equation

$$D_c = \frac{D_o}{1-\exp\left[\frac{E_a}{k}\left(\frac{1}{T_c}-\frac{1}{T}\right)\right]} \qquad \text{.............................. (1)}$$

where D_c is the critical dose for amorphisation, D_o is the amorphisation dose extrapolated to 0 K, E_a is the activation energy for self annealing, T_c is the critical temperature above which recrystallisation is complete over the entire cascade volume, k is Boltzmann's constant and T is temperature (T and T_c are in degrees Kelvin).

The D_c versus T data for some materials clearly show two plateaux and two rises (eg. the zircon data of Weber et al. [8] or the hafnon data of Meldrum et al. [14]). This has led previous authors to postulate that annealing is a two-stage process and analyse their data accordingly. Meldrum et al. also analysed their huttonite data assuming a two-stage process, even though their experimental data suggest a single stage process because huttonite is structurally similar to hafnon and zircon (which show two-stage dependencies of D_c on T).

Our data suggest a single-stage dependency of D_c on T. However the end-member zirconolite D_c versus T data of Wang et al. [11] suggest a two-stage annealing process. Consequently, we have analysed our data both as a single-stage process and a two-stage process. We designate the activation energies, "critical temperatures" and D_o values of the two regimes as E_{aI} and E_{aII}, T_{cI} and T_{cII}, and D_{oI} and D_{oII} respectively. We also tested the applicability of the model of Weber et al. by plotting ln $(1 - D_o/D_c)$ versus 1/kT. If the model is suitable, this plot should be linear.

Single-stage analysis involved least-squares fitting of equation (1) to all the data points for each sample, allowing D_o, E_a, and T_c to vary. Calculated values of D_o, E_a, and T_c for both the 1200 and 1450 samples are listed in Table 2. We also analysed the end-member zirconolite data of Wang et al. [11] on the basis of a single stage dependency of D_c on T, allowing D_o, E_a, and T_c to vary during the least-squares fit (see Table 2). Comparison of the D_o, E_a, and T_c values from our data versus those we calculate from the data of Wang et al. show that:

a) the T_c values of the 1450 sample and the end-member zirconolite of Wang et al. are comparable;
b) they are both larger than the T_c values of the 1200 sample;
c) the E_a values of the 1200 and 1450 samples are comparable and
d) they are both larger than the E_a value of the end-member zirconolite of Wang et al.

The difference between our E_a values and that calculated from the data of Wang et al. is due to the fact that as T_c is approached the gradients of our D_c versus T curves change more quickly than the gradient of Wang et al.

If one allows that our 1200 and 1450 data samples exhibit a two-stage dependence of D_c on T, then the transitions between Stage I and Stage II are at ~150 K and ~300 K respectively (see Figure 1 and Table 1). Calculated values of D_{oII}, E_{aII}, and T_{cII} (D_o, E_a, and T_c values for Stage II) for both the 1200 and 1450 samples are listed in Table 2. Insufficient data were collected from the 1200 sample to determine the D_{oI}, E_{aI}, and T_{cI} values of Stage I. For the 1450 sample we collected only 3 data points in Stage I, therefore D_{oI} was fixed at 2.8×10^{18} ions m^{-2} (estimated by extrapolation) and E_{aI} and T_{cI} were allowed to vary during the least squares analysis. Results are shown in Table 2. In combination, the values we calculate from our data assuming a single-stage and two-stage dependence of D_c on T suggest our data support a single-stage model. Specifically, E_a and E_{aII} are within error and T_c and T_{cII} are within error.

Table 2. Least squares fitted values of D_o, E_a and T_c. Values in parentheses are the estimated errors.

	Data included	Values that were allowed to vary	D_o (x 10^{18} ions m^{-2})	E_a (eV)	T_c (deg. K)
Assuming a single-stage dependence of D_c on T					
1200 sample	All	D_o E_a T_c	3.4 (0.3)	0.27 (0.08)	620 (13)
1450 sample	All	D_o E_a T_c	3.5 (0.3)	0.33 (0.10)	660 (11)
End-member zirconolite data collected by Wang et al. [11]	All	D_o E_a T_c	3.1 (0.4)	0.08 (0.02)	667 (5)
Assuming a two-stage dependence of D_c on T					
1200 sample Stage I	Not enough data				
1200 sample Stage II	≥ 150 K	D_o E_a T_c	3.7 (0.2)	0.36 (0.07)	612 (7)
1450 sample Stage I	≤ 303 K	E_a T_c	Fixed at 2.8	0.3 (1113)	339 (149410)
1450 sample Stage II	≥ 303 K	D_o E_a T_c	4.0 (0.2)	0.47 (0.12)	653 (7)

Plots of ln $(1 - D_o/D_c)$ versus 1/kT for both the 1200 and 1450 samples do not show straight lines (see Figure 2). This suggests that various defects and/or annealing mechanisms (with different activation energies) apply over the range of temperatures we investigated and/or the model suggested by Weber et al. is inadequate.

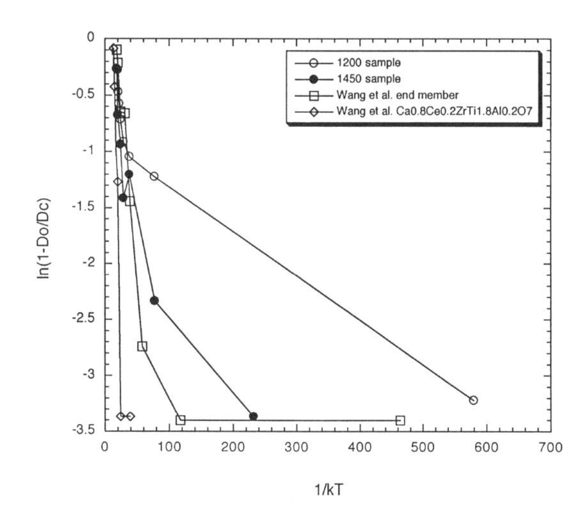

Figure 2. Plots of $\ln(1 - D_o/D_c)$versus 1/kT for the 1200 and 1450 samples in this study; and the $Ca_{0.8}Ce_{0.2}ZrTi_{1.8}Al_{0.2}O_7$ and end-member zirconolite samples of Wang et al. [11]. Continuous lines join data from individual samples.

An alternative amorphisation model to that of Weber et al. [8] has recently been developed by Meldrum et al. [14]. This new model also permits E_a and T_c values for two or more stages to be developed from D_c versus T data. Discussion of the application of this model, its merits and limitations are beyond the scope of this paper.

CONCLUSIONS

Data have been presented showing the temperature dependence of D_c for two end-member zirconolites ($CaZrTi_2O_7$), one of which is almost crystallographically perfect, while the other contains many stacking faults and twins on the unit cell scale. The plots of D_c versus T in this and a previous study [11] can be divided into two Regimes: 20 K to ~400 K and >~400K. In Regime I, D_c at any temperature is unaffected by composition or stacking fault density and D_c varies only by a factor of ~2. In Regime II, the D_c values of individual zirconolites vary greatly with temperature and there are significant differences in D_c between different zirconolites. However, the raw data from the zirconolites in this and a previous study [11] all suggest T_c values between 600 and 1000 K. The data for end-member zirconolite containing many stacking faults (the 1200 sample) lie between two data sets from almost crystallographically perfect end-member zirconolites (the 1450 sample) and the end-member data of Wang et al. [11]. The model suggested by Weber et al. [8] does not adequately describe the data collected in this study.

ACKNOWLEDGEMENTS

The authors are grateful to Adam Jostsons for extensive editorial comment, Mike Colella for assistance in producing STEM maps of specimens, Sammy Leung for SEM data and Warrick Payten for mathematical assistance. The authors also thank the HVEM-Tandem Facility staff at Argonne National Laboratory for assistance during ion irradiations, in particular we thank Charlie Allen, Ed Ryan, Stan Ockers, Tony McCormick, Pete Baldo and Lauren Funk. The Facility is supported as a User Facility by the U.S. DOE, Basic Energy Sciences, under contract W-31-109-ENG-38.

REFERENCES

[1] R.C. Ewing, W.J. Weber and F.W. Clinard Jr., Progress in Nuclear Energy, 29 [2], 63-127 (1995).

[2] G.R. Lumpkin, K.L. Smith, M.G. Blackford R. Giere and C.T. Williams, Mat. Res. Soc. Symp. Proc., 506, 215-222 (1998).

[3] A. Jostsons, E.R. Vance and B. Ebbingaus (1999) Immobilisation of surplus plutonium in titanate ceramics, presented at Global'99 "Nuclear Technology - Bridging the Millenia", Jackson Hole, Wyoming, USA, Aug. 29 – Sept. 3 1999.

[4] E.R. Vance, C.J. Ball, M.G. Blackford, D.J. Cassidy and K.L. Smith, J. Nucl. Mater., 175, p.58-66 (1990).

[5] P.K. Gupta, J. Am. Ceram. Soc., 76(5), 1088-1095 (1993).

[6] L.W. Hobbs, A.N. Seeram, C.E. Jesurum and B.A. Berger, Nucl. Instruments and Meths. in Phys Res., B116, 18-25 (1996).

[7] S.X. Wang, L.M. Wang and R.C. Ewing, Nuc. Instruments and Meths. in Phys. Res., B127/128, 186-190 (1997).

[8] W.J. Weber, R.C. Ewing and L.M. Wang, J. Mater. Res., 9(3), 688-698 (1994).

[9] A.E. Ringwood, S.E. Kesson, K.D. Reeve, D.M. Levins and E.J. Ramm (1988) Synroc, in *Radioactive Waste Forms for the Future*, edited by W. Lutze and R.C. Ewing, Elsevier, p.233-334.

[10] K.L. Smith, G.R. Lumpkin and N.J. Zaluzec, J. Nucl. Materials, 250, 36-52 (1997).

[11] S.X. Wang, G.R. Lumpkin, L.M. Wang and R.C. Ewing, in Radiation Effects in Insulators, REI-10, July 18-23, 1999, Jena, Germany (1999).

[12] T. J. White, H. Mitamura, K. Hojou and S. Furuno, Mat. Res. Soc. Symp. Proc. Vol. 333, 227-232 (1994).

[13] F.W. Clinard, Jr.,Am. Ceram. Soc. Bull., 65, 1181-87 (1986).

[14] A. Meldrum, S.J. Zinkle, L.A. Boatner and R.C. Ewing, Phys Rev. B, 59(6) 3981-3992 (1999)

RADIATION AND THERMAL EFFECTS IN ZEOLITE-NaY

Binxi Gu*, Lumin Wang*, Phil A. Simpson**, Leah D. Minc** and Rodney C. Ewing*
*Department of Nuclear Engineering and Radiological Sciences,
**Michigan Memorial Phoenix Project, University of Michigan, Ann Arbor, MI 48109.

ABSTRACT

Zeolite-NaY is susceptible to both irradiation- and thermally-induced amorphization. Amorphized zeolite-NaY loses approximately 95 percent of its ion exchange capacity for cesium due to the loss of exchangeable cation species and /or the blockage of access to exchangeable cation sites. A secondary phase was formed during the ion exchange reaction with cesium. The Cs-exchanged zeolite-NaY phase has a slightly higher thermal stability than thc unexchanged zeolite-NaY. A desorption study indicated that the amorphization of cesium-loaded zeolite-NaY enhances the retention capacity of Cs due to the closure of structural channels.

INTRODUCTION

Reversible cation exchange is one of the most important properties of zeolites. This is the basis for using zeolites in the selective removal of radionuclides, such as cesium, strontium, rare-earths, and actinides, from the high-level liquid nuclear waste [1]. Zeolite is also a potential waste form and back-fill material in nuclear waste repositories. As zeolites are intended to retain radionuclides, they will receive considerable radiation doses over time, and the cumulative doses in zeolites utilized as exchange media, waste forms, or near-field back-fill can be substantial. A number of studies have reported that zeolites are susceptible to various types of radiation damage. At room temperature, analcime, natrolite and zeolite-Y, become completely amorphized at an ionizing dose in the range of 3.2×10^{10} to 1.6×10^{11}Gy, and the dose required for amorphization decreases with increasing temperature due to the thermal instability of the zeolites [2]. Based on these results, zeolites in the near-field of a waste repository may be amorphized within 1,000 years after waste emplacement. Temperature also affects the structure and properties of zeolites. Most zeolites become dehydrated upon heating and many undergo the crystalline-to-amorphous transition at temperatures above 400~900°C, depending on the type of zeolite. In the case of nuclear waste, significant heating is possible after waste emplacement in a repository. Heat generated from decay the of fission products can result in an initial storage temperature as high as 600°C, and the temperature may still be as high as 300°C after 100 years of storage [3].

The cation exchange behavior of zeolite depends on the nature of the cation species, size, and charge, as well as the structure of the zeolite. Radiation- or thermally-induced amorphization dramatically changes the structure, which in turn leads to a redistribution of cations and a change in the framework charge. The collapse of the framework structure may also reduce the number of accessible cation sites by blocking the open channels and reducing the effective surface area. Consequently, the ion exchange and retention behavior of zeolites may be affected significantly by the amorphization process. Extensive experimental studies of ion exchange of radionuclides in zeolites have been performed over the last few decades [4,5]. However, these studies have been focused only on the effects of time, temperature, solution chemistry and pH. The effects of radiation- or thermally-induced amorphization on ion exchange capacity have not been considered in these studies. In the present study, the radiation and thermal effects on the structure of zeolite-NaY were investigated and the changes in the ion-exchange and retention capacity of Cs as a result of the structure change were studied.

Mat. Res. Soc. Symp. Proc. Vol. 608 © 2000 Materials Research Society

EXPERIMENTAL

Materials

Zeolite-NaY (Si/Al=2.55) that contained 13.0 %(wt) Na_2O was supplied by Zeolyst International Company in the form of powder with the size less than one micron. The surface area of the zeolite powder was 900 m^2/g. The chemical composition of the zeolite-NaY is: $NaAlSi_2O_6 \cdot nH_2O$. Cesium-loaded zeolite was prepared by ion exchange. Cesium content in the Cs-exchanged zeolite was analyzed using neutron activation analysis, for which the sample was first irradiated in the Ford Nuclear Reactor (FNR) at University of Michigan, and then analyzed using a gamma counter. The concentration of Cs was 21 wt. % before heat treatment. Both zeolite-NaY and the cesium-exchanged zeolite-NaY were heated in air at elevated temperature (800~1000°C) to form the amorphous structure.

Experimental Procedures

The atomic structures of zeolite-NaY and Cs-exchanged zeolite-NaY, before and after thermal treatment, were examined by powder X-ray diffraction (XRD). The XRD spectra were obtained using Cu-Kα radiation. The structures of zeolite-NaY and Cs-loaded zeolite were also investigated by transmission electron microscopy (TEM) using a JEM 2000FX microscope. Zeolite powders were crushed and suspended on a holey-carbon film supported by copper grids. The electron energy was 200 keV, and all the samples were irradiated at a dose rate of 2.5×10^{21} $e^-/m^2 \cdot s$ at room temperature.

The ion-exchange experiments used 10 mg of crystalline zeolite or 100 mg of amorphized zeolite powder in contact with 25 ml CsCl solution. The mixtures were continuously stirred on a Lab-Line shaker with an orbital speed of 150 rpm. All the ion exchange experiments were performed at room temperature. The pH values of the solution were measured before and after the exchange. The ion exchange capacities of zeolite-NaY, with and without thermal treatment, were measured as a function of time in a 1 mM CsCl solution. After various preset time intervals, centrifugation at 10,000 rpm was applied to separate the liquid phase from the solid. Samples of supernatant were analyzed for Cs^+ content by atomic absorption spectrophotometry.

A 10 mg zeolite sample loaded with cesium was put in contact with 25 ml of deionized water for 1, 2, 4, 7, 14, and 25 days, respectively. The mixtures were continuously stirred on a Lab-Line orbital shaker with the speed controlled at 150 rpm. Sample temperature was maintained at room temperature over the entire desorption period. The cesium content in the liquid phase was analyzed by atomic absorption to determine the amount of Cs released. The same experiment was also conducted with cesium-loaded samples that were heated at 900 and 1000°C in air for 30 minutes to form partially and fully amorphized phases, respectively.

RESULTS AND DISCUSSION

Structure Change due to Thermal Treatment and Electron Beam Irradiation

The atomic structures of the zeolite-NaY and ion-exchanged zeolite were investigated using XRD and TEM. The powder diffraction data obtained for zeolite-NaY are in agreement with the results in the literature [6]. The XRD patterns in figure 1 show the transition from crystalline-to-amorphous state due to thermal treatment. When heated at 800°C for 40 min followed by cooling in air, the zeolite showed a reduced peak intensity in the XRD spectrum; however, the position and relative intensity of individual peaks remained constant. Zeolite-NaY was completely amorphized at 900°C. A further increase in temperature to 1000°C resulted in melting and the formation a glass. The XRD pattern for zeolite-NaY treated at 1000°C has the same profile as that obtained at 900°C. The zeolite sample that was thermally treated at 900°C was used in the subsequent ion exchange experiment.

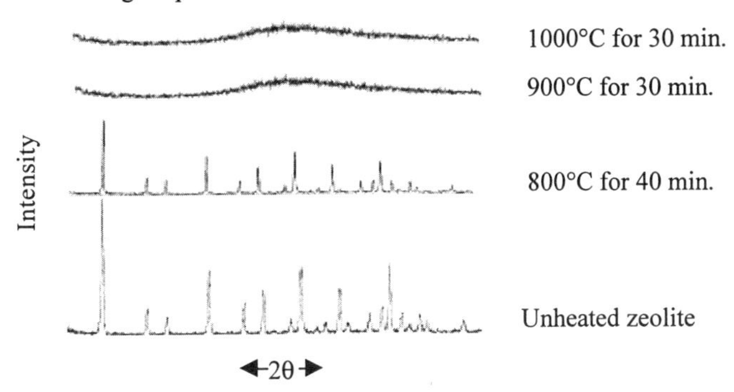

Figure 1. X-ray diffraction spectra for zeolite-NaY, unheated and heated at 800°C, 900°C, 1000°C.

The crystalline-to-amorphous transformation has also been observed for zeolite-NaY during the TEM experiment. Figure 2 is a sequence of electron diffraction patterns of zeolite-NaY taken at different cumulative electron doses. The 200 keV electron beam interacted with the zeolite sample at a dose rate of approximately 2.5×10^{21} $e^-/s{\cdot}m^2$. A gradual loss of diffraction spots in the diffraction patterns indicated that the crystalline structure was destroyed progressively with the increase in radiation dose. Full amorphization was achieved at a dose of 4.0×10^{23} e^-/m^2 at room temperature. The error involved in the measurement of amorphization dose was ±30% and may be attributed to inhomogeneous particle size or sample thickness. Several studies have confirmed that the damage to zeolite by the electron beam is dominated by a radiolytic decomposition mechanism [2]. Radiolytic damage involves the transfer of energy from the incident electron to the electrons in the specimen, which may cause bonds to break. Treacy [7] has proposed that ionization of structural water in zeolite could produce radicals that weaken bonds in the aluminosilicate framework, and this initiates collapse of the framework.

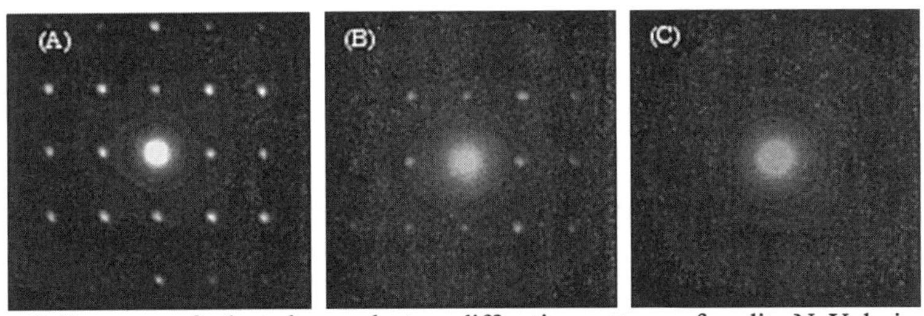

Figure 2. Sequence of selected-area electron diffraction patterns of zeolite-NaY during 200 KeV electron beam irradiation: (A) unirradiated; (B) 3.0×10^{23} e^-/m^2; (C) 4.0×10^{23} e^-/m^2.

The XRD spectra of cesium-exchanged zeolite before and after thermal treatment are shown in figure 3. The loaded Cs concentration is 21 wt. %. The exchange of Cs for Na in the zeolite structure results in reduced peak intensity, as compared with zeolite-NaY. This change is primarily due to the change in x-ray diffraction structure factors caused by introducing Cs into the crystalline structure. In addition to the peaks corresponding to the original zeolite-NaY, extra peaks were found to be present in the XRD spectrum, indicating that a secondary phase may have formed when zeolite-NaY was exchanged with Cs. After heating at 900°C for 30 minutes, the diffraction maxima that correspond to zeolite-NaY at lower values of 2θ disappeared. Increasing the temperature to 1000°C resulted in the disappearance of all remaining peaks and a fully amorphized structure. This result indicates the existence of a secondary crystalline phase that has a higher thermal stability than the original structure. The formation of two solid phases has been observed in the ion exchange reaction of zeolite-X with strontium [6]. A strontium-rich expanded zeolite phase was identified by optical microscopy. The nature of the exchanged cations was also found to be important in determining the stability and dehydration behavior of zeolite [6]. As the size of univalent cation increases, the temperature at which the water is lost increases. This phenomenon may be due to the filling of void space in the zeolite structure by the large cations after the removal of water.

Cs-loaded zeolite was also found to undergo the crystalline-to-amorphous transition when it was exposed to the electron beam at a similar dose rate as used in the zeolite-NaY case. The critical accumulated dose at which the material became fully amorphous was 5.0×10^{23} e^{-}/m^{2}. This value is slightly higher than that obtained for zeolite-NaY, indicating that Cs-exchanged zeolite is more stable under irradiation. However, the difference may also be attributed to the error (±30%) involved in the measurements, because the amorphization dose may be affected by the particle size or the thickness of the sample exposed to the electron beam.

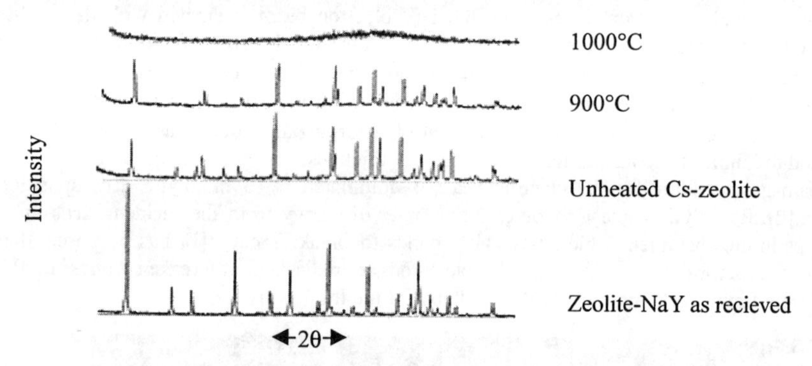

Figure 3. X-ray diffraction spectra for Zeolite-NaCsY, unheated and heated at 900°C, 1000°C. The XRD pattern of the original zeolite-NaY is shown for comparison.

Ion Exchange

The ion exchange capacity of crystalline and amorphous zeolite-NaY with cesium was measured as a function of time in 1 mM CsCl solution. The pH value of the original CsCl solution was 7.5, and no significant change in pH value (within ± 0.5) was observed after ion exchange. The Cs-Na exchange reaction on crystalline zeolite reached equilibrium quickly (in less than 5 minutes) under continuous stirring conditions, and approximately 50% of the sodium was replaced by cesium. The fast reaction rate and incomplete exchange indicated that the majority of the ion exchange occurred only in supercages where the free aperture is large enough for cesium ions to reach the exchangeable cation sites. Similar results have been reported for zeolite-X and zeolite-A under similar conditions [8]. As shown in figure 4, amorphized zeolite-NaY has a much lower exchange capacity for Cs as compared with the undamaged, crystalline zeolite. In fact, the ion exchange capacity drops to almost zero after the zeolite is fully amorphized. This result implies that zeolite has lost almost all the exchangeable cation sites upon the collapse of crystalline structure to amorphous state. In the zeolite structure, the fundamental unit is a tetrahedrally coordinated small cation, such as Si^{4+}. The substitution of aluminum for silicon produces a deficiency in electrical charge that must be locally balanced by the presence of an additional positive ion within the interstices of the structure. These interstitial sites form the basis of the exchangeable cation sites. When the framework structure collapses, the chemical bonds between silicon and oxygen, or aluminum and oxygen, are broken and local charge balance is lost. As a result, the exchangeable cation sites no longer exist. Another possible explanation for the low exchange capacity of amorphized zeolite is the closure of the open channels leading to exchangeable cation sites.

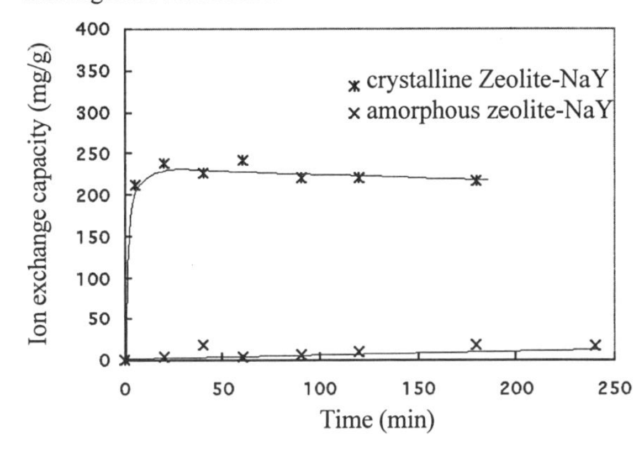

Figure 4. Comparison of ion exchange capacities of crystalline and amorphous zeolite-NaY. The amorphous phase was obtained by thermal treatment at 900°C. The experiment was conducted in 1 mM CsCl solution at room temperature.

Desorption

Cesium-exchanged zeolite-NaY, without thermal treatment as well as thermally treated at 900°C and 1000°, has been studied in a desorption experiment. Figure 5 shows the change of desorption rate of cesium in deionized water as a function of time. The sample without thermal treatment gives the highest desorption rate among all three of Cs-exchanged zeolite-NaY samples, while the fully amorphized zeolite obtained at 1000°C exhibits the lowest desorption rate. This phenomenon can be explained by the closure of open channels in the structure during the crystalline-to-amorphous transformation. Apparently, the collapse of channel openings either by thermal treatment or irradiation can block the release of the adsorbed cesium ions. Another possible explanation for the reduced desorption of Cs from the thermally treated zeolites is that the exchanged Cs ions may have migrated into the small sodalite cage when the zeolite was heated at a temperature above 900°C. As demonstrated by Noby et al. [9], the Cs ions in the Cs-exchanged zeolite-NaY can migrate from the supercage into the sodalite cage when the material is dehydrated at temperatures above 500°C. Because the diameter of Cs ion (0.34 nm) is large, as compared with the 6-ring aperture (0.22 nm) in the sodalite cage, Cs will remain trapped inside the small sodalite cages. The change of desorption behavior of ion-exchanged zeolite upon amorphization is of special interest in a nuclear waste disposal repository where zeolites may be used as waste forms or as a sorption medium for the retardation of radionuclide transport. Since radiation- and thermally-induced amorphization can easily be achieved in a nuclear waste repository due to the high radiation dose and high temperature, the retention capacity of adsorbed radionuclides may be enhanced.

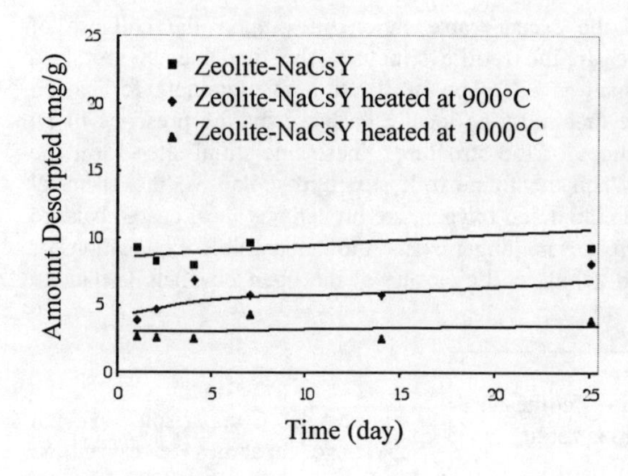

Figure 5. Desorption of cesium ions from unheated zeolite-NaCsY and zeolite-NaCsY thermally treated at 900°C and 1000°C. The experiment was conducted at room temperature in deionized water.

CONCLUSIONS

Both zeolite-NaY and Cs-exchanged zeolite-NaY undergo a crystalline-to-amorphous transformation when exposed to electron irradiation with an energy of 200 keV at a cumulative dose of ~4.0×10^{23} e^-/m^2. A fully amorphized zeolite-NaY phase was also obtained by heating the sample up to 900°C in air. At room temperature, the amorphized zeolite-NaY showed a significantly reduced ion exchange capacity as compared with the original crystalline zeolite. The reduction in ion exchange capacity is attributed to the loss of exchangeable cation species and/or the closure of access to exchangeable sites. For Cs-exchanged zeolite-NaY, complete amorphization was achieved when the sample was heated at a higher temperature (1000°C) than the original zeolite-NaY. An XRD study of Cs-exchanged zeolite indicated that a secondary phase formed during the Cs-Na exchange reaction. A desorption study indicated that the thermal treatment of cesium-loaded zeolite-NaY may enhance the retention capacity of loaded Cs ions by trapping these ions inside the resulting amorphous structure.

ACKNOWLEDGEMENTS

This work is supported by the U.S. DOE/EMSP through grant DE-FG07-97ER45652.

REFERENCES:

1. H. S. Sherry, *Molecular Sieve Zeolite*, Adv. Chem. Ser., 101, American Chemical Society, Washington, D. C., (1971) p.350.
2. S. X. Wang, L. M. Wang and R. C. Ewing, J. Nucl. Mater., **278**, p. 233 (2000).
3. R. C. Ewing, W. J. Weber, J. F. W. Clinard, Prog. Nucl. Ener., **2**, p. 63 (1995).
4. G. V. Tsitsishvili, T. G. Andronikashvili, G. N. Kirov and L. D. Filizova, *Nature Zeolite*, Ellis Horwood, New York, 1992.
5. P. Sylvester, A. Clearfield, Solvent Extraction and Ion Exchange, **16**, p. 1527 (1998).
6. D. W. Breck, *Zeolite Molecular Sieves—Structure, Chemistry and Use*, Krieger Publishing Company, Malabar, Florida, 1984.
7. M. M. Treacy, J. M. Newsam, Ultramicroscopy, **23,** p. 411 (1987).
8. P.K.Sinha, P.K.Panicker and R. V. Amalraj, Waste Manag. **15**, p. 149 (1995).
9. P. Norby, F. I. Poshni, A. F. Gualtieri, J. C. Hanson, C. P. Grey, J. Phys. Chem. B **102**, p. 839 (1998).

RADIATION EFFECTS IN CHLORIDE MOLTEN SALT COMPOSITIONS

S.V. IGNATIEV, D.V. PANKRATOV, E.I. YEFIMOV
Department of Special Nuclear Power Units, Institute of Physics and Power Engineering
Obninsk 249020, Russia, ignatiev@ippe.obninsk.ru

ABSTRACT

Radiation effects on the composition of NaCl (70 mole %) and $PbCl_2$ (30 mole %) salts irradiated by a high energy accelerator proton beam were analyzed. In the temperature interval from ~50 to ~1200 °C this composition can be considered as radiation-resistant material, that can be recommended for use as target material for accelerator-driven transmutation systems.

INTRODUCTION

At present a system consisting of a tungsten neutron generating target along with a molten salt subcritical blanket cooled by molten actinide salts is being studied as a variant of an accelerator driven transmutation system for nuclear waste destruction [1]. A system in which molten actinide chlorides are directly irradiated by high energy protons is also being considered.

The purpose of this work was to study radiation effects on the composition of NaCl (70 mole %) and $PbCl_2$ (30 mole %) salts irradiated by a high energy proton beam. The cylindrical target was 20 cm in diameter and 200 cm long. Isotopic composition and radiation parameters versus target thickness were analyzed for target irradiated by secondary neutrons, generated in the target and surrounding blanket with different flow densities and energies, by proton irradiation.

Using U. S. Los Alamos National Laboratory (LANL) LANSCE accelerator conditions (proton energy ~800 MeV, current ~1 mA), the secondary neutron flux densities and spectra of protons and secondary neutrons were calculated using the MARS-10 code. The effect of high-energy neutrons on molten salt isotopic composition and radiation parameters was studied. Radiation effects on decomposition and recombination in the molten salt components were considered.

STATIONARY NEUTRON IRRADIATION

The radioisotope composition of a NaCl+$PbCl_2$ salt target of ~3.27g/cm^3 density irradiated for 30 days by a thermal neutron flux of ~10^{12}n/cm^2·s was calculated. The analogous calculations were performed for stationary spatially uniform irradiation by a fission neutron spectrum of the same flux density.

Lead of 99.992% purity was assumed in both cases in accordance with the Russian Standard GOST-3778-65. The elements Bi, Ag, Cu, Zn, As, Sb, Mg, Fe, Sn, Cd, Na, Ca, Ni, Tl, Hg, Al, and In were considered to be the main impurities. The sodium composition was taken to be the same as that in the BN-800 reactor coolant in accordance with GOST-6-01-788-73, so that its percentage was >99.8%, while the main impurities were K, Fe, Ca, Sn, N, C, and Cl.

Under these conditions the concentrations of the major target components and most important impurities were calculated. The main nuclear reactions of the major target components and their impurities were analyzed using a compilation of nuclear reaction cross sections for fast and thermal neutron induced reactions [2]. The (n, 2n), (n, 3n), (n, p), (n, α) and (n, γ) non-elastic scattering nuclear reactions were considered for the fission neutron spectrum, while the (n, γ), (n, p) and (n, α) reactions were considered for the thermal neutrons. The activities of the radionuclides produced were calculated for the instant just after irradiation ended and after

Mat. Res. Soc. Symp. Proc. Vol. 608 © 2000 Materials Research Society

cooling intervals of 1, 10, 100 hours, 1 month and 1 year. For thermal neutron irradiation the following nuclides are of major importance just after irradiation ends: Cl-36 ($1.5\cdot10^5$ Bq/cm^3), Na-24 ($8.3\cdot10^9$ Bq/cm^3), S-35 ($2.1\cdot10^9$ Bq/cm^3), P-32 ($1.1\cdot10^6$ Bq/cm^3), K-42 ($0.9\cdot10^6$ Bq/cm^3), Cu-64 ($1.9\cdot10^5$ Bq/cm^3), As-76 ($2.4\cdot10^5$ Bq/cm^3), Ag-110m ($4.1\cdot10^3$ Bq/cm^3), Sb-122 ($1.1\cdot10^5$ Bq/cm^3), Sb-124 ($2.0\cdot10^4$ Bq/cm^3), Hg-197 ($1.8\cdot10^4$ Bq/cm^3) and Pb-209 ($9.7\cdot10^8$ Bq/cm^3).

After a 1 year cooling interval the target activity was determined mainly by S-35 ($1.15\cdot10^8$ Bq/cm^3), Cl-36 ($1.5\cdot10^5$ Bq/cm^3), Ca-45 ($2.4\cdot10^2$ Bq/cm^3), Fe-55 ($6.9\cdot10^2$ Bq/cm^3), Zn-65 ($1.4\cdot10^3$ Bq/cm^3), Ag-110m ($1.5\cdot10^3$ Bq/cm^3), Sb-124 ($3\cdot10^2$ Bq/cm^3) and Tl-204 ($1.6\cdot10^2$ Bq/cm^3).

For irradiation by the fission neutron spectrum, among the major radioactivity contributors just after irradiation are Na-24 ($3.6\cdot10^6$ Bq/cm^3), K-42 ($3.7\cdot10^3$ Bq/cm^3), Ar-41 ($1.3\cdot10^3$ Bq/cm^3), Ar-37 ($1.1\cdot10^4$ Bq/cm^3), Mn-54 ($5.4\cdot10^2$ Bq/cm^3), Mn-56 ($1.7\cdot10^3$ Bq/cm^3), Co-58 ($7.4\cdot10^3$ Bq/cm^3), Cu-64 ($3.8\cdot10^2$ Bq/cm^3), Sb-122 ($2\cdot10^3$ Bq/cm^3), Sb-124 ($3.7\cdot10^2$ Bq/cm^3), Pb-203 ($1.5\cdot10^5$ Bq/cm^3), Pb-209 ($3.3\cdot10^6$ Bq/cm^3), and Po-210 (10^2 Bq/cm^3).

The Ar-39 ($2.1\cdot10^2$ Bq/cm^3), Mn-54 ($2.4\cdot10^2$ Bq/cm^3), and Co-58 ($2.1\cdot10^2$ Bq/cm^3) were dominant after one year of cooling.

INFLUENCE OF SECONDARY NEUTRONS

The secondary neutron flux densities generated by the primary proton beam in the target in question were also calculated for the conditions of the LANSCE accelerator (proton energy E_p=800 MeV, current I_p=1 mA, Gaussian radial distribution of the proton current with sigma ~1.60 cm) using the MARS-10 code. These densities (in $n\cdot cm^{-2}\cdot s^{-1}$), averaged on the radius «R» versus the target length «Z» (in cm), are $Z_{10}=1.56\cdot10^{13}$, $Z_{20}=1.8\cdot10^{13}$, $Z_{50}=1.25\cdot10^{13}$, $Z_{70}=9.3\cdot10^{12}$, $Z_{90}=5.6\cdot10^{12}$, $Z_{110}=2.5\cdot10^{12}$, $Z_{130}=3.1\cdot10^{11}$, $Z_{150}=1.0\cdot10^{11}$, $Z_{170}=4.0\cdot10^{10}$, and $Z_{190}=2.0\cdot10^{10}$. The proton flux density was ~40 times less than the neutron density.

The spectra for secondary neutrons and protons leaking through the lateral surface of the target, as well as in the target center, in a cell with ΔZ-100-110 cm and ΔR=0-2.5 cm were also calculated. The results are presented in Tables I and II. The total number of neutrons leaking through the target lateral surface was ~8.1 per primary proton, or ~$5.0\cdot10^{16}$ n/s, while the leakage through the total target surface and planes was as much as ~8.6 neutrons per proton. The total number of leaking protons per primary proton was ~0.2.

Table I. Proton spectrum in the target

#	Energy interval, MeV	Neutron fraction, %	
		Leakage	Target center
1	10-20	-	0.1
2	20-30	2.6	0.1
3	30-50	6.2	0.8
4	50-100	20.0	38.0
5	100-200	30.0	37.0
6	200-400	40.0	23.0
7	400-600	1.2	1.0

Table II. Neutron spectrum in the target

#	Energy interval, MeV	Neutron fraction, %	
		Leakage	Target center
1	≤ 0.001	0.003	~0
2	0.001-0.01	0.025	0.03
3	0.01-0.1	0.80	0.92
4	0.1-0.2	2.20	1.70
5	0.2-0.4	2.60	3.60
6	0.4-0.8	6.50	6.60
7	0.8-1.4	13.8	13.8
8	1.4-2.5	16.8	16.5
9	2.5-4.0	13.2	13.2
10	4.0-6.5	11.5	7.7
11	6.5-10.5	7.20	3.9
12	10.5-14.5	6.60	4.2
13	14.5-20	3.80	1.60
14	20-30	3.60	3.0
15	30-50	3.40	4.0
16	50-100	3.20	7.9
17	100-200	2.70	7.40
18	200-400	1.3	3.0
19	400-600	0.70	0.70
20	600-800	0.072	0.25

Energy deposition in the target due to cascades generated by the primary protons were calculated for a beam power of ~800 kW. The energy deposition distribution for target «disks» with thickness $\Delta Z=Z_2-Z_1=10$ cm (radius R=10 cm) versus the coordinate of the first disk plane Z_1 are presented in Table III.

The total energy deposition in the target was ~437 kW or ~54% of the beam power. Practically all of the energy deposition occurred in the first half of the target, i.e. in the zones Z from 0 to 110 cm. The average energy deposition density was ~12.63 W/cm^3.

Table. III Energy deposition in the target

Z_1, cm	Energy deposition, kW	Z_1, cm	Energy deposition, kW
0	76.5	100	14.9
10	73.0	110	0.177
20	62.4	120	0.108
30	51.0	130	0.065
40	41.9	140	0.0414
50	33.5	150	0.026
60	27.1	160	0.02
70	22.2	170	0.013
80	18.3	180	0.0074
90	15.5	190	0.0044

INFLUENCE OF HIGH-ENERGY NEUTRONS

Irradiation by secondary high-energy protons resulted in the formation of radionuclides, including lead nuclide fission products, that accumulated in the molten salt target. Calculation of the secondary neutron fluxes and spectra showed that in the center of the target the neutron fractions versus energy were as follows: 80% for E_n<50 MeV and 20% for E_n>50 MeV.

The average neutron flux density over the whole target volume was ~$4.8 \cdot 10^{12}$n/cm^2·s and ~$1.2 \cdot 10^{12}$n/cm^2·s, respectively, at these energy intervals. The maximal total neutron flux density was observed near the target planes and proved to be as much as ~$2 \cdot 10^{13}$n/cm^2·s.

The formation reactions for radionuclides with half-life $T_{1/2}$ > 10 hours and atomic mass A close to that of lead were analyzed and their formation cross sections (FCS) were calculated using the Silbergberg-Tsao empiric formula [3]. The results are presented in Table IV. It was found that the non-elastic neutron scattering reactions (n, nx), leading to the formation of Pb-200, Pb-202 and Pb-203, are of primary importance.

Table IV. Formation cross sections for lead nuclides.

Target	% in natural mixture	Nuclide	$T_{1/2}$*)	FCS in mbarn versus the proton energy E_p in MeV			
				E_p=800	E_p=400	E_p=100	E_p=50
Pb-204	1.4	Pb-202	$5.3 \cdot 10^4$ y	104.0	151.0	186.0	200.0
Pb-204	1.4	Pb-200	21.54 h	170.0	210.0	240.0	250.0
Pb-206	24.1	Pb-203	51.88 h	66.0	88.0	104.0	110.0
Pb-206	24.1	Pb-202	$5.3 \cdot 10^4$ y	116.0	143.0	163.0	170.0
Pb-206	24.1	Pb-200	51.54 h	189.0	200.0	208.0	210.0
Pb-207	22.1	Pb-200	51.54 h	200.0	195.0	180.0	120.0
Pb-207	22.1	Pb-202	$5.3 \cdot 10^4$ y	123.0	140.0	150.0	155.0
Pb-207	22.1	Pb-203	51.88 h	70.0	86.0	95.0	95.0
Pb-208	52.4	Pb-203	51.88 h	74.0	84.0	90.0	92.0
Pb-208	52.4	Pb-202	$5.3 \cdot 10^4$ y	130.0	137.0	142.0	145.0
Pb-204	1.4	Hg-203	46.6 d	6.8	10.7	11.0	-
Pb-204	1.4	Pt-197	10.34 h	32.6	31.8	20.0	-
Pb-207	22.1	Pt-200	12.54 h	25.8	25.2	20.0	-

*) y, d, and h mean years, days, and hours.

The average activities A_{av} of some of the nuclides formed are presented in Table V for the instant of beam shut-off after one year of target irradiation operation.

Table V. Average radionuclide activities after one year of target irradiation operation.

Nuclide	Pb-200	Pb-202	Pb-203	Hg-203	Pt-197	Pt-200	TOTAL
A_{av},Bq/cm^3	$4.8 \cdot 10^8$	$8.8 \cdot 10^3$	$4.2 \cdot 10^8$	$4.1 \cdot 10^5$	$7.7 \cdot 10^5$	$1.2 \cdot 10^7$	~10^9

As many as ~150 products with $T_{1/2}$ > 10 hours, formed by fissioning of lead nuclides, were analyzed. At a proton energy of E_p=800 MeV the FCS range proved to be as much as from ~10^{-3}

mbarn to ~10^{-1} mbarn. The total value of the FCS was ~12 mbarn at E_p=800 MeV, and ~0.5 mbarn at E_p=400 MeV. It was found that at the instant of beam shut-off the average saturated activity of fission products did not exceed ~$5\cdot10^5$ Bq/cm^3, i. e. ~0.05% of the activity of the lead radionuclides. Table VI gives target volume averaged activities A_{av} of some volatile fission products after one year of target irradiation.

Table VI. Volatile fission product activities after one year of target irradiation operation.

Nuclide	$T_{1/2}$	σ_{800}, mbarn	A_{av}, Bq/cm^3
Kr-79	35.04 h	0.26	~10^4
Kr-81	$2.1\cdot10^5$ y	0.076	0.01
Kr-85	10.72 y	0.0053	14.0
I-124	4.18 d	0.056	$2.3\cdot10^3$
I-125	60.14 d	0.039	$1.6\cdot10^3$
I-126	13.02 d	0.02	$8.3\cdot10^2$
I-129	$1.57\cdot10^7$ y	0.0051	~10^5
I-131	8.04 d	0.0019	80.0
I-133	20.8 h	0.0007	30.0
Xe-127	36.4 d	0.039	$1.3\cdot10^3$
Xe-129m	8.899 d	0.014	$5.8\cdot10^2$
Xe-138m	11.9 d	0.0052	$2.2\cdot10^2$
Xe-133	5.24 d	0.002	80.0
Xe-133m	2.188 d	0.002	80.0

The average flux density of neutrons with energy less than ~50 MeV in a target without a blanket was ~$4.8\cdot10^{12}$n/cm$^2\cdot$s. Assuming similarity of the thermal neutron spectra to the fission neutron spectra, the activities of some important radionuclides after one year of proton irradiation were calculated. The total activity was ~$3.5\cdot10^7$ Bq/cm^3, i.e. ~3.5% of the total activity of the lead nuclides formed by the inelastic scattering of high-energy neutrons by lead.

For a target installed in a molten salt blanket with a specific energy deposition of ~50 MW/m^3, the fission neutron flux was ~$5\cdot10^{13}$n/cm$^2\cdot$s, i. e. ~10 times higher, than in the case of a «bare» target (no blanket). Thus, the total radionuclide activity will be ~$3.5\cdot10^8$, i.e. 35% of the total activity of lead radionuclides. After one year of cooling, the target activity was accounted for mainly by Pb-202 and equaled ~$8.8\cdot10^3$Bq/cm^3. At the same time the total activity of Ar-39, Mn-54, and Co-58 formed in the reactions induced by low-energy neutrons was as much as ~$1.4\cdot10^4$ Bq/cm^3 for a target installed in a blanket, and ~$1.4\cdot10^5$ Bq/cm^3 for a «bare» target, i.e. ~1.6 and ~16 times higher than the activity of Pb-202.

RADIATION DECOMPOSITION AND RECOMBINATION

The radiation strength (RS) of a molten chloride salt target irradiated by neutrons, gamma-rays, and fast electrons is an important target parameter. Thus the following were analyzed:

- Radiolysis in the system and possible appearance of new phases;
- Chloride compound strength relatively to lead strength;
- Radiolysis product influence on the process of structural material corrosion;
- Radiolysis due to residual irradiation.

The analysis was carried out in the same way as developed in Reference 4. Nearly pure chlorine was released in the radiolysis process. As the salt temperature was increased to its melting point, off gassing of chlorine decreased to ~0, evidently because the radiolysis products recombination rate increased. The total chlorine off gassing from chloride salt compositions in the temperature interval from 600 to 1200 °C was limited to ~$4.25 \cdot 10^{-6}$ molecule per 100 eV. In comparison, for a solid fuel-containing composition at a temperature of ~50 °C, off gassing was fully defined by radiolysis and was as much as ~$1.75 \cdot 10^{-2}$ molecule per 100 eV; which is a considerably higher yield than that for water used as a moderator and coolant in nuclear reactors. Because formation rates and fission product yields decrease substantially at high kinetic energy it is expected that the chloride target RS would be higher than that of the fuel-containing salt.

On the whole, in the temperature interval from ~50 to ~1200 °C, the chloride fuel composition can be considered to be a radiation-resistant material that can be recommended for use as a prospective target material for the accelerator-driven systems in question.

CONCLUSIONS

1. Due to inelastic neutron scattering by lead nuclides a comparatively high level of radioactivity is accumulated in the target. Its average value at the instant of beam shut-off after one year of irradiation operation at a beam power of 1 MW is ~10^9 Bq/cm^3. The accumulated radioactivity is accounted for by the Pb-200 and Pb-203 contributions.
2. The contribution from lead nuclide fissioning induced by high-energy neutrons does not exceed ~0.1% of the total activity of Pb-200 and Pb-202.
3. The total activity of volatile nuclides of krypton, xenon, and iodine formed in the process of lead nuclide fissioning is ~$2 \cdot 10^{14}$ Bq/cm^3 at the instant of beam shut-off.
4. At the instant of beam shut-off after one year of target irradiation operation, the total activity of the products of low-energy neutron induced fissioning is equal to ~3.5% of the total activity of Pb-200 and Pb-203 for the no-blanket condition, while for a target installed in a molten salt blanket this value is ~35%.
5. After one year of target cooling the activity induced by low-energy neutrons is 1.6 and 16 times higher than that of the accumulated Pb-202, for blanket-off and -on conditions, respectively.
6. On the whole, the influence of high-energy particles on target radiation parameters is considerable, and thus should always be taken into account.
7. In a temperature range of up to ~ 1200 °C, chloride fuel compositions can be considered to be radiation resistant materials.

REFERENCES

1. S.V. Ignatiev, D.V Pankratov, and E.I. Yefimov, Radioactive Halogen Yields From the Tungsten Target Irradiated by High-Energy Protons, Proc. of XXIII Meeting of the International Collaboration on Advanced Neutron Sources, Paul Scherrer Institute, Villigen PSI, Switzerland, Vol. II, p.p. 487-494 (1996).
2. K. Noda, Proc. of 1994 Symposium on Nuclear Data, JAERI-Conf 95-008, p. 112 (1995).
3. Silbergberg R., Tsao C.H., Partial Cross Sections in High Energy Nuclear Reactions and Astrophysical Applications, Astrophysical Journal Supplement, Series 1220 (I and II), Vol. 25, p.315 and p.335 (1973).
4. V.M. Novikov, Molten Salt Nuclear Power Units: Prospects and problems, Moscow, Energoatomizdat (1990).

HYDROGEN GAS EVOLUTION FROM WATER DISPERSING NANOPARTICLES IRRADIATED WITH GAMMA-RAY

S. SEINO, R. FUJIMOTO, T. A. YAMAMOTO, M.KATSURA,
S. OKUDA*, K. OKITSU**, R. OSHIMA***
Department of Nuclear Engineering, Osaka University,
2-1 Yamadaoka, Suita, Osaka 565-0871, Japan. sseino@nucl.eng.osaka-u.ac.jp
*) Institute of Scientific and Industrial Research, Osaka University,
8-1 Mihogaoka, Ibaraki, Osaka 567-0047, Japan
**) Department of Materials Science and Engineering, Nagasaki University,
1-14 Bunkyo, Nagasaki 852-8521, Japan.
***) Research Institute for Advanced Science and Technology, Osaka Prefectural University,
1-2 Gakuen, Sakai, Osaka 599-8570, Japan.

ABSTRACT

Hydrogen gas evolution from water dispersing nanoparticles under ^{60}Co γ-ray irradiation was investigated under various conditions. Dispersion of TiO_2, Al_2O_3, ZnO, ZrO_2 and CeO_2 nanoparticles with average sizes of 6 nm - 11 μm showed hydrogen yields much higher than that obtained by pure water radiolysis, and the yields seemed to depend on their particle size and shape rather than their chemical species. Hydrogen yield increased with increasing γ-dose but not linearly, which implies the occurrence of a kind of reversing reaction. Some of the nanoparticles supported with Pt, Au and Pd were also examined. The yields were strongly affected by these noble metals, which indicates an important role of the particle surface in the total mechanism of the present hydrogen evolution.

INTRODUCTION

It has been reported that hydrogen gas evolution from water irradiated with γ-ray is drastically enhanced by adding a little amount of nanoparticle made of a semiconductor, e.g. TiO_2[1]. In addition to such photocatalytic materials we have found that Al_2O_3 also enhances the hydrogen evolution when its nanoparticles are dispersed in water and irradiated [2]. The γ-ray is a photon creating electron-hole pairs in substances by photoelectric effect, so that these pairs may induce redox reactions on surface of these particles on the analogy of the photocatalytic mechanism. In semiconductor particles suspended in aqueous solution, UV-photon activates an electron and a hole which migrate towards solid-solution interface to reduce and oxidize species existing in the solution [3]. Photon energy of γ-ray is far higher than that of UV-photon, which might have induced such redox reactions on particles surface of Al_2O_3 generally classified as insulator with a bandgap energy too large to be activated by UV-photons. This γ-ray's energy is, however, in turn large enough to form radiochemical species such as radicals, hydrated electrons, etc. [4], which are so reactive and would participate in total phenomena. The mechanism of this enhancement experimentally pointed out is not clear at all in this stage and worth studying from the viewpoints not only of science but also of application. Its application to hydrogen gas production has been indeed proposed and discussed for making full use of γ-rays persistently emitted from the high level radioactive waste, which should be stored for several dozens of years before the geological disposal [1]. Hydrogen evolution yields so far obtained were still too low for such practical uses, but potentiality of this hydrogen evolution process has not been exploited yet. From scientific and

Mat. Res. Soc. Symp. Proc. Vol. 608 © 2000 Materials Research Society

technological viewpoints, there are many aspects to be examined. In this paper, we report on hydrogen evolution yields measured at various conditions of dose effect, surface situation, chemical species of particle and particle size. Their results are discussed taking account of the photocatalytic and radiological mechanisms. They would contribute to understanding of this phenomenon and search for more favorable particle materials for the practical use in the future.

EXPERIMENTAL

Materials examined in this work were particles of average size from 6 nm to 11 μm made of TiO_2, Al_2O_3, ZnO, ZrO_2 and CeO_2, which were provided by several suppliers. Materials produced by the physical vapor synthesis method were obtained from Nanophase Tech. Corp. (NanoTek) for all these chemical species. Their particle sizes and shapes were investigated with a transmission electron microscope to find that their particle sizes obey log-normal distribution determined from sizes measured on hundreds particles. In Table I, their geometric average sizes (D_g) and geometric standard deviations (σ_g) are tabulated. Three of them, TiO_2, Al_2O_3, ZnO, were found to have similar size distributions as shown in Fig.1. These TiO_2, Al_2O_3, were in spherical shape [5], while ZnO in parallelepiped with round corners. Another three kinds of alumina particle materials, AO-802, AO-809 and Asahi were also examined. The former two of them were supplied from Admatechs Co. Ltd. and their nominal particle sizes are 0.7 μm and 11 μm, respectively. According to the supplier's information, both the materials share a source material and a processing method, and their shapes are almost spherical. The nanoparticles of Asahi was obtained from Asahi Chemical Ltd. and of 14 nm, which involves an agglomeration though it has the smallest primary particle size of examined Al_2O_3 in this work. XRD measurements indicated that the materials of NanoTek and Asahi were of γ-Al_2O_3 and that of AO's were of mixture of θ-Al_2O_3 as the major component and α-Al_2O_3 and δ-Al_2O_3 as the minor components. Another titania particle material also examined was P25 supplied from Aerosil Nippon Ltd., which has a nominal average size of 21 nm and mainly of anatase and a few percentage of rutile.

The particle surface of TiO_2, Al_2O_3, and CeO_2 were supported with Pt, Au and Pd by sonochemical method [6] and impregnation method to investigate its effect on hydrogen evolution yield. Aqueous solution of K_2PtCl_4, $NaAuCl_4$ and $PdCl_2$ were used as metal sources, and those metal ions, e.g. Pd (II), were reduced by H atoms and organic radicals formed by an ultrasonic wave irradiation in the sonochemical method, and by H_2 gas at 473 K in the

Table I Hydrogen gas evolution yields and $G(H_2)$ from NanoTek nanoparticles with their D_g and σ_g.

Particle	D_g(nm)	σ_g	H_2 yield (μmol)	$G(H_2)$
water radiolysis	---	---	3	0.03
TiO_2	19	1.7	31	0.3
Al_2O_3	21	1.6	30	0.3
ZnO	28	1.4	32	0.3
ZrO_2	11	1.8	15	0.1
CeO_2	6	1.9	32	0.3

(^{60}Co γ-ray, 1.2 kGy/h, 18 hours, water; 50 g , particle; 0.25 g)

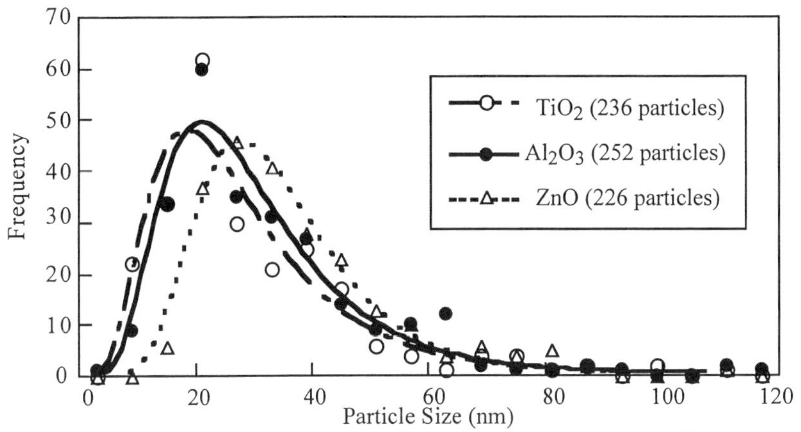

Fig.1 Size distribution of NanoTek TiO_2, Al_2O_3, and ZnO nanoparticles. Curves are calculated for log-nominal distribution with D_g and σ_g in Table I.

impregnation method. The amount of the noble metal supported on the surface was 5 wt% of the base nanoparticles.

These particles were dispersed in distilled water in a glass vial. In advance to γ-ray irradiation argon gas purged gaseous species in the free space in the vial and dissolving in water, except for experiments with AO-802 and AO-809. Hydrogen gas yield without adding any particle was measured in each series of irradiation experiment, because γ-ray irradiation induces water radiolysis to produce hydrogen gas which contributes to hydrogen evolution yields concerned in this work. Ar gas purged the air in the pure water as well.

The particles of 0.1 to 15 g was dispersed in pure water of 50 ml, to which ^{60}Co γ-ray was irradiated at room temperature. Dose rate of the γ-ray was 1.2 kGy/h and irradiation time was 18 - 72 hours. During the irradiation the dispersion was shaken by rotating a disk on which the vials were mounting, as shown in Fig. 2. After the irradiation, hydrogen gas evolved and confined in the free space of the vial was sampled with a syringe and analyzed by a gaschromatograph. From the hydrogen yields, *G*-values of hydrogen gas were calculated.

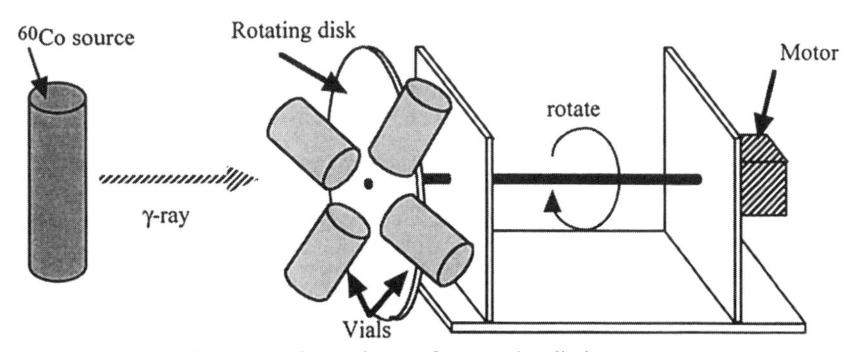

Fig. 2 Experimental setup for γ-ray irradiation.

Dose absorbed by the sample dispersion was evaluated as those absorbed by the pure water, because mass of the particle is far smaller than that of water.

RESULTS AND DISCUSSION

Figure 3 shows hydrogen gas evolution yields measured with 0.5-g TiO_2 (NanoTek) nanoparticles in 50-g water plotted against γ-ray dose together with hydrogen yields due to the pure water radiolysis. Curves are drawn only for guiding eye. It is noticed that the hydrogen yields increased with increasing γ-ray dose and that addition merely of 1 wt% of titania to water drastically enhances the yield. The increasing behavior with γ-dose seems non-linear and involving a convex curvature, implying some reversing reaction. The yield increased with the amount of the particles and saturates at large amount region [5]. The present amount, 0.5-g TiO_2, is within this saturation region. The increases either with γ-dose and additive amount indicate that the present enhancement occurs when the particles exists and γ-ray is irradiated.

Table II shows influence on hydrogen gas yield of the noble metals supported on the particle surface, in which each yield is expressed as a value normalized to those with as-received particles. The supported noble metals strongly affected the hydrogen yields, though enhancement and suppression were both observed. Platinum and gold on titania enhanced, but these on γ-Al_2O_3 (Asahi) suppressed the yield. It is remarkable that hydrogen evolution is quenched by platinum on γ-Al_2O_3 (Asahi), and palladium on the same one and three kinds listed there. Palladium's quenching effect was made sure by two supporting methods, sonochemical and impregnation. Although palladium can absorb much hydrogen by making interstitial solid solution [7] the amount of hydrogen evolved from nanoparticles without palladium were larger than palladium metal can absorb by 2 times in magnitude. These facts indicate, though implicitly, the concerned enhancement by the nanoparticles is deeply involved with some processes occurring on the particle surface.

Table I shows hydrogen gas evolution yields measured with 0.25-g of the as-received NanoTek material of TiO_2, Al_2O_3, ZnO, ZrO_2, or CeO_2 together with calculated *G*-values.

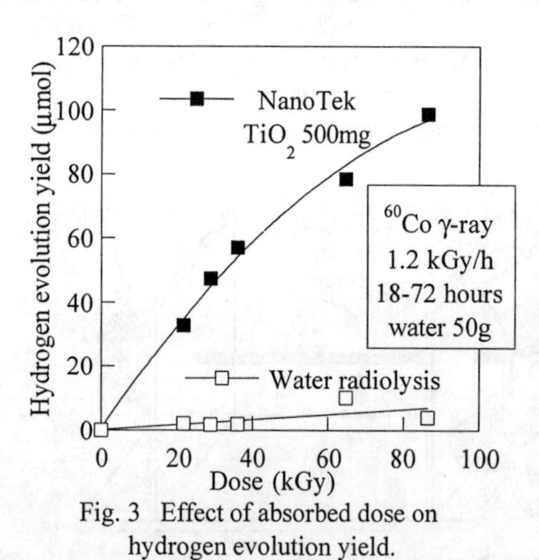

Fig. 3 Effect of absorbed dose on hydrogen evolution yield.

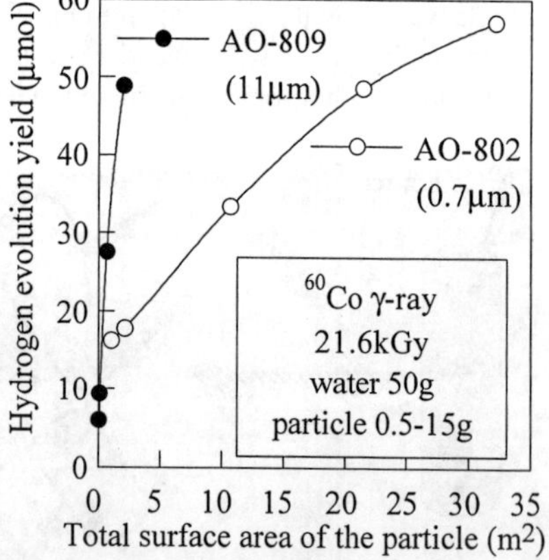

Fig. 4 Effect of total particle surface area on hydrogen evolution yield.

Table II Hydrogen gas evolution yields with nanoparticles supported with 5 wt% noble metals, which were normalized to as received nanoparticles

Particles	Particle/H_2O (wt. ratio)	H_2 evolution yields with supported noble metals None	Pt	Au	Pd*	Pd#	γ-ray Dose (kGy)
TiO_2 (P25)	0.25 / 75	1.0	2.2	2.7	---	---	32.4
γ-Al_2O_3 (Asahi)	0.25 / 75	1.0	0.0	0.4	0.0	---	57.6
TiO_2 (NanoTek)	0.10 / 50	1.0	---	---	0.0	0.0	21.6
Al_2O_3 (NanoTek)	0.10 / 50	1.0	---	---	0.0	0.0	21.6
CeO_2 (NanoTek)	0.10 / 50	1.0	---	---	0.0	---	21.6

---: not examined *: Sonochemical method #: Impregnation method

Contributions of the pure water radiolysis are given in the first line to be compared with those with the nanoparticles. For pure water radiolysis, obtained G (H_2), 0.03, was far smaller than the reported value, 0.45, possibly because Ar gas has purged the O_2 in the present aqueous phase [4]. The hydrogen evolutions were enhanced about 10 times of that from pure water radiolysis except for the case of ZrO_2, and the obtained G-values of hydrogen gas were 0.3. As already shown in Fig. 1 three of these five kinds, TiO_2, Al_2O_3, ZnO obtained from NanoTek have similar size distributions and former two ones are of spherical shape [5]. Other two, ZrO_2, and CeO_2 were found to involve nonspherical and irregular shapes. It should be pointed out here that these three kinds exhibited almost same hydrogen evolution yield regardless of chemical species. This founding implies a hypothesis worth being considered, that the enhancement concerned is dominated not by chemical species but size and shape. Although it is necessary to measure yields by varying sizes with the same shape and processing method in order to examine the hypothesis strictly. We could not obtain such a material series in nanometer scale, we could obtain such a material set only in micron scale. Figure 4 shows hydrogen evolution yields from two alumina particles, AO-802 and AO-809, with nominal average sizes of 0.7 and 11 μm, respectively. Hydrogen yields were plotted against the total surface area of the particle dispersed in water, assuming the spherical shape and density of the bulk alumina, 4 g/cm^3. A size effect is clearly indicated, that the larger particle enhances the yield much more than the smaller. Of course it is too hasty to conclude that there is such size effect even in the nanometer scale, then further study is required.

To discuss what occurs in the present heterogeneous system, responses of solid and aqueous phases to the γ-ray radiation should be recalled. In the aqueous phase, water radiolysis generates H and OH radicals which may participate in the hydrogen gas generation [5]. Even without nanoparticles hydrogen gas is generated from two H radicals in the aqueous phase, and on the contrary OH radicals reduce these H_2 molecules [4].

$$H_2O \rightarrow H, OH, \text{etc.} \quad (1)$$
$$H + H \rightarrow H_2 \quad (2)$$
$$OH + H_2 \rightarrow H_2O + H \quad (3)$$

In the particles, γ-photons and secondary electrons incident from the aqueous phase generate Compton photons, electrons with lower energies, and holes in core levels with short life times. These holes change into fluorescence X-rays and/or sets of electrons and holes via Auger process. After such cascades, there would be free electrons (e^-) and holes (p^+) with significant life times. This may be a situation similar to that assumed in the photocatalytic mechanism, which postulates that these electrons and holes migrate to the surface to occur reduction of H^+ and oxidation of OH^- [3]. H_2 molecule is generated from two H radicals.

$$H^+ + e^- \rightarrow H \tag{4}$$

$$OH^- + p^+ \rightarrow OH \tag{5}$$

$$H + H \rightarrow H_2 \tag{6}$$

Indeed there is an optimal size, ca. 10 nm, of the photocatalytic particle, which has been explained in terms of (i) probability of electron-hole recombination, (ii) migration distance from the generated point to the surface, and (iii) penetration depth of incident UV-photon [8].

It is reasonably inferred that both the kinds of irradiation effect have participated in the presently observed enhancement of hydrogen evolution. If processes in the aqueous phase were dominant, size effect could never occur. Species appearing in both series of reactions, (1)-(3) and (4)-(6), could have linked them together and controlled the whole phenomenon. OH radical appears in eq. (3) and (5) and works to reducing hydrogen yield, which might have played an important role, especially in the quench effect observed with particles supported with palladium. The H radical must be an important species in studying this phenomenon because it is available from the both branches via (1) and (4). It is considered that the particles not only provides reaction site, i.e. surface, for these common species but also drive and control the reactions by supplying holes and electrons. The size effect, if any, could be ascribed to generation and migration processes of them.

To summarize, we have measured hydrogen gas evolution yield from water dispersing a kind of nanoparticle irradiated with ^{60}Co γ-rays by varying γ-ray dose, surface situation, chemical species of the particle and particle size. The yield was found to be enhanced by addition of a little amount of the particles in comparison with the yield due to water radiolysis. It was also suggested that radicals generated by water radiolysis and holes and electrons formed in the particles react on the particle surface. Some data implied a kind of size effect but much more data should be accumulated to confirm it.

ACKNOWLEDGEMENTS

The authors thank Mr. H. Nishizawa for his help in taking TEM micrographs. This work was partially supported by Kansai Research Foundation for Technology Promotion.

REFERENCES

1. Y. Wada, K. Kawaguchi and M. Myochin, Progress in Nuclear Energy 29, 251-256 (1995)
2. T. A. Yamamoto, S. Seino, M. Katsura, K. Okitsu, R. Oshima and Y. Nagata, Nanostructured Materials 12, 1045-1048 (1999)
3. Y. LI and L. Wang, in *Semiconductor Nanoclusters Physical, Chemical, and Catalytic Aspects*, edited by P. V. Kamat and D. Meisel (Elsevier, Amsterdam, 1996) 391-415
4. I. G. Draganic and Z. D. Draganic, *The Radiation Chemistry of Water*, (Academic Press, New York and London, 1971), p. 9-14, p. 140
5. S. Seino, R. Fujimoto, T. A. Yamoamoto, M. Katsura, S. Okuda and R. Oshima, Radioisotopes, (*to be accepted*)
6. K. Okitsu, Y. Mizukoshi, H. Bandow, Y. Maeda, T. A. Yamamoto and Y. Nagata, Ultrasonic Sonochemistry, 3, S249 (1996)
7. F. A. Lewis, *The Palladium Hydrogen System*, (Academic Press, London New York, 1967), p. 13-42
8. H. Harada and T. Ueda, Chemical Physics Letters 106, 229-231 (1984)

DIAMOND DETECTORS FOR ALPHA MONITORING IN CORROSIVE MEDIA FOR NUCLEAR FUEL ASSEMBLY REPROCESSING

P. Bergonzo, F. Foulon, A. Brambilla, D. Tromson, C. Mer, B. Guizard, S. Haan
LETI (CEA-Technologies Avancées)/DEIN/SPE, CEA/Saclay, F-91191 Gif-sur-Yvette
pbergonzo@cea.fr

ABSTRACT

Recent advances in the synthesis of diamond films led to the fabrication of new devices that enable the direct measurement of the alpha activity of a corrosive liquid source. Such devices can be directly immersed in radioactive liquid solutions for the in-situ and real time measurement of their alpha activity. We report here on the fabrication of diamond from the chemical vapour deposition (CVD) technique and its use for the monitoring of the alpha activity of a plutonium 239 radioactive source diluted in concentrated nitric acid. Tests have demonstrated the excellent reliability of the CVD diamond detectors, that would give great benefit to the monitoring of the nuclear fuel assembly re-processing, and in particular for corrosive wastes identification and treatment.

INTRODUCTION

Early developments in the seventies had demonstrated that diamond could be used in order to monitor the activity of corrosive radioactive solutions [1]. In fact, diamond is known to withstand prolonged immersion in all commonly used acid solutions and even at high temperatures. However, as a result of relying on the use of mono-crystalline natural diamonds, these corrosion hard detection devices were extremely limited in size and their cost prevented industrial applications. Over the last decade, new techniques have emerged that enable the fabrication of diamond from a technique based on the plasma dissociation of a gaseous precursor. Such methods, so called the chemical vapour deposition techniques, now enable the growth of thin diamond layers on monocristalline silicon substrate from the dissociation of methane in a microwave exited plasma [2]. Diamond combines unique properties that make it a very attractive material for radiation detection applications. It is a semiconductor that has a high band gap (5.5 eV) and that can be operated at high temperatures. It exhibits high values of resistivity (> 10^{14} Ω.cm), mobility (2000 $cm^2.V^{-1}.s^{-1}$) and electrical breakdown field (10^7 $V.cm^{-1}$). It is extremely resilient to radiations and corrosion, allowing its application to nuclear particle detection in hostile environment [3, 4]. Recent progress in CVD diamond preparation techniques led to the fabrication of materials of remarkable quality for these applications, and recently X and γ ray detectors were made as well as particle detectors [5], and especially α particles [6, 7]. Such detectors can be operated under high levels of ionising radiation, high temperatures, and withstand immersion in corrosive media.

Needs have been identified within the environmental monitoring, as well as within the control and nuclear waste recycling in nuclear industries, which require a detector element capable of monitoring the nuclear activity of liquid solutions. One of the direct application is the measurement of the alpha activity of waste solutions, where radioactive residues are dissolved in concentrated acid solutions. Elements present in the active media consist of actinides from nuclear fuel processing and are mostly isotopes from plutonium, americium and uranium. The energies of the corresponding alpha particles are typically in the 5 MeV range. Because of the

Mat. Res. Soc. Symp. Proc. Vol. 608 © 2000 Materials Research Society

short range of alpha particles in the liquid phase, (e.g., < 30 µm in concentrated HNO_3 [8]), techniques generally rely on the evaporation of liquid samples and then consist of the measurement on condensed solid materials. Also, it is clear that the measurement implies the handling, transportation, processing of the radioactive liquid sample and therefore is time consuming and expensive. As such, it comes that the fabrication of devices that could simply be immersed in the liquid media and particularly when they are corrosive can be extremely attractive for real time process control. Diamond is the unique candidate for such an application since it is known to combine all properties required.

DIAMOND SYNTHESIS AND DEVICE MANUFACTURING

Large area (Ø 5 cm) diamond layers were grown on silicon from the plasma assisted chemical vapour deposition technique. From the dissociation of a gaseous precursor mixture consisting of methane diluted in hydrogen, it is possible to grow at high temperatures (typ. 800°C) a polycristalline material on carbide forming materials such as silicon. Of the commonly used energy source applied to the gaseous mixture, microwave excitation is known to result in the production of materials with the best electronic properties and here a 2.45 GHz source was used to enable the growth of diamond on 5 cm in diameter silicon substrates. The material obtained has a polycrystalline structure with grain size of the order of 10% of the layer thickness, and Raman analysis showed one intense peak at 1332 cm^{-1} and no non-diamond carbon structure. We have grown diamond using optimised growth conditions in order to obtain material with good electronic properties, that enable its use for detection applications. Typical growth rates are in the order of 0.5µm/h for good quality films exhibiting the best electronic properties and detection performances. After growth, a series of annealing steps and chemical treatments are performed in order to significantly improve the properties of the films [9].

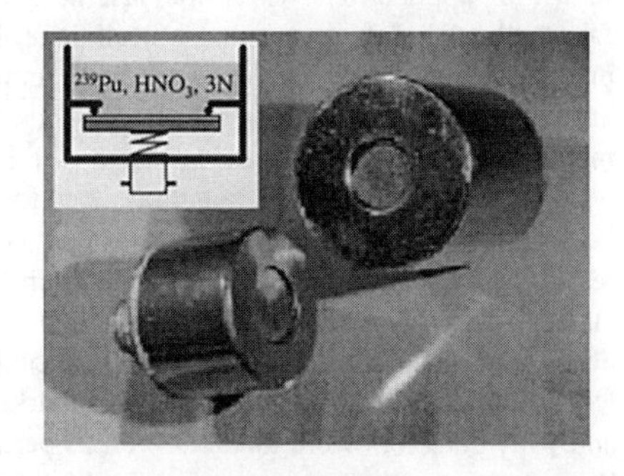

Fig. 1. (right) Photo of two detection devices that enable alpha activity measurements in corrosive liquid solutions. The device consists of a 20 µm thick CVD diamond layer, that is in direct contact with the active media. Alpha particles emitted in the vicinity of the diamond surface are detected in order to determine the activity of the solution.

Figure 1 shows a photo of detection devices, with a schematic representation of the configuration set-up. The active part consists of a ionisation chamber formed from the resistive volume of the intrinsic diamond that is polarised through electrical contacts. The device is mounted in an electrically polished stainless steel 304 case, and the diamond layer is glued using a corrosion resistant component that renders the seal between the case and the diamond layer leak proof. The silicon substrate onto which was grown the active layer forms the back contact of the detection device, and is used to bring the DC bias (V_{pol}) to the diamond. The front electrical contact is directly ensured by the liquid solution in contact with the diamond film. In fact, the liquid offers a high conductivity with respect to the diamond, and thus provides a grounded electrical contact to the diamond layer via the case of the detector. With such a configuration, the diamond is directly in contact with the solution, thus avoiding the use of any protective material that could both degrade with time and create an energy loss for low energy heavy particles. It is to be noted that because of the extremely high resistivity of the diamond layer (above 10^{14} Ω.cm), no electrochemistry could occur in the active solution from the extremely low currents involved during use. The free carriers generated by the ionising particles interacting in the diamond volume drift towards the electrodes and the induced charge is measured in the external circuit. 5 MeV alpha particles are stopped within a 12 μm depth in diamond [8]. A simple model based on Hecht's theory [10] predicts the expression of the charge collection efficiency, which is defined as the ratio of the induced charge (Q_{ind}) to the generated charge (Q_0) and is given by :

$$Q_{ind}/Q_0 = \mu\tau E/L \qquad (1)$$

where μ and τ are the carrier mobility and lifetime, respectively, E the mean internal electric field, and L the thickness of the detector. It comes that higher collection efficiencies can be obtain by limiting the thickness L to the order of the range of the particles to be detected. Another advantage of using low thicknesses is that it give rise to a low sensitivity to other ambient radiations (β, γ), thus increasing the detection selectivity. In the present case, we used a 20 μm thick diamond detectors with a $40mm^2$ active window.

MEASUREMENT IN CORROSIVE MEDIA

The alpha activity of a radioactive corrosive solution was monitored using a 37 kBq/g radioactive plutonium 239 source diluted in nitric acid, 3N. This isotope emits 5.1 MeV alpha particles. The device was biased at typically 100 V and was connected to a charge preamplifier and amplifier typical of nuclear spectrometry measurements The amplifier was set with a 2 μs shaping time. The signal was recorded using a multichannel analyser. A typical detection spectrum from the measurement in the radioactive solution can be observed in figure 2. The spectrum corresponds to the histogram of the pulse heights observed for every single alpha particle interaction. Since there is auto-absorption in the liquid source, the spectral representation corresponds to a distribution of energies from a threshold level that corresponds to the electronic noise, up to higher amplitude pulses. In fact, part of the alpha particles emitted have to travel through several microns of liquid, thus loose energy, and therefore deposit very little energy in the detector. On the other hand, some particles emitted in the close vicinity of the surface will deposit a much greater energy in the diamond, thus giving rise to higher energy counts. The result is the broad continuum observed. The insert in figure 2 also shows the evolution of this spectrum during 70 hours and per 6 hour time intervals. It clearly demonstrates the reliability of the devices since no deviation is observed. In fact, deep level defects have always been observed in both natural and CVD diamonds and are assumed to be caused by the presence of impurities as well as crystal structure defects. The progressive filling of those deep traps located in the material may result in a progressive modification of the internal electric field and thus of the detection characteristics. Various approaches can be used to address this aspect, and further to optimising the growth conditions, it can be minimised by optimising the thickness of the device to the range of the particles to be detected. Other treatments were studied and rely on the filling of the deep traps from irradiating the material under X or γ ray photons [11].

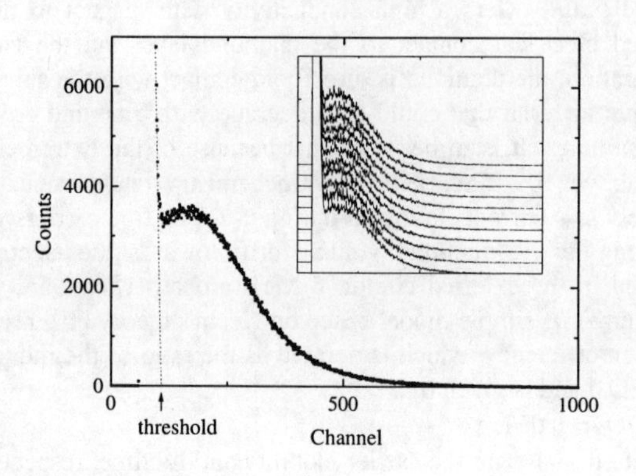

Fig. 2. (right) Typical detection pulse height spectrum measured on a ^{239}Pu diluted in HNO_3, 3N liquid source. The shape is caused by the self absorption of the alpha particles by the source. Insert shows the spectra obtained every 6 hours (shifted vertically for clarity).

ALPHA ACTIVITY MONITORING

From the spectrum given in figure 2, it comes that all interactions detected above the electronic detection threshold represent alpha particles emitted from the solution and detected in the diamond layer. Since the range of α particles in the source is short, the detector will only monitor the activity of a liquid layer that is in contact with the diamond surface. The depth of this layer is a geometrical factor that represents in theory the mean free path of the alpha particles in the liquid media. However, it is known that charge collection efficiency is generally lower than unity in diamond. The consequence here is that low energy alpha particles will not be detected since they would give rise to output signals below the electronic detection threshold. This can be considered as a further decrease in the depth of the solution probed, and simply adds up to the geometrical factor to become a constant detection ratio for the detector. This value can be evaluated before use from a calibration measurement with a solution of known activity.

By diluting the 37 Bq/g ^{239}Pu source in nitric acid, it was possible to vary the activity of the radioactive source. Figure 3 shows the evolution of the counting rate as a function of the solution activity. It demonstrates the perfect linearity of the detector response. The slope of the curve also gives the geometrical factor of the detector which is here close to $2.5\ 10^{-4}$, or that 10 counts/s corresponds to a 40 kBq/g solution.

One inherent drawback with the use of a small detector thicknesses is a resulting increase of the capacitance of the device, that results in the increase of the noise level as measured by a charge sensitive preamplifier. We have appreciated the gain that would be achieved while increasing the active area of the device. From earlier measurements [12] the variation of the electronic noise as a function of the capacitance of the device can be related to the gain in counting rate that would result from the increase in the surface area probed, for a thickness kept constant at 20 micrometer. This is shown on figure 4. It appears that a small increase in the overall counting efficiency could be achieved by scaling up the device to an active area of 80 mm^2, that results in a 40% increase of the geometrical factor of the detector. However, further to detector size considerations, part of the information on the detection spectrum shape would have been lost and a compromise was adopted in using a 40mm^2 active area for the present measurements.

Fig. 3. (right) counting rate as a function of the source activity. It demonstrates the linear behaviour of the detector.

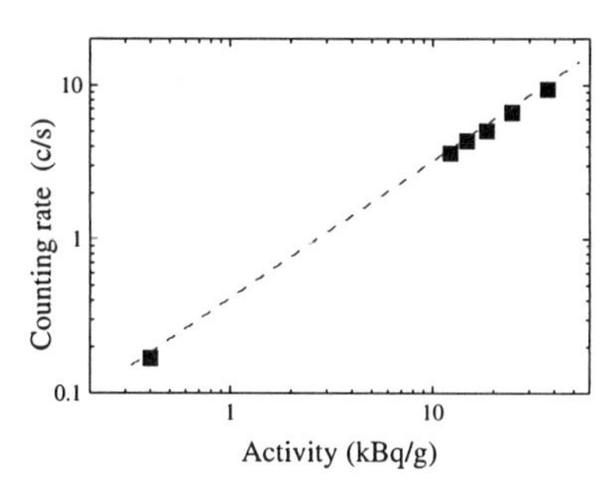

This corresponds to a limited sensitivity of these devices to typically 10 counts per second at 40 kBq/g (approx 1 mCi). Other configurations could then be used when greater sensitivities are needed, such as for example the use of multiple detection devices associated with multichannel acquisition readout circuits (e.g., Castor VLSI 32 channel analog-digital circuit [12]). The sensitivity will be improved according to the number of detection channels used.

One other important aspect that needs to be considered is the non contamination of the detector. In fact, if part of the active species in the solution tends to stick onto the diamond layer, it is clear that the overall counting rate would progressively rise during use and render the measurement erroneous. However, this remains low because of the naturally hydrophobic behaviour of diamond. Also, it has been verified that the detector could be cleaned completely using a simple rinse in nitric acid, 3N. The detector was then submitted to a wipe test and the level measured was below 0.04 Bq/cm^2, limit that states that the device is not contaminated. Longer experiments are required to verify the reliability of the detectors during months of continuous use, and current investigations are considering the installation of such devices in the French nuclear fuel re-processing plant of La Hague.

CONCLUSION

The fabrication of corrosion resistant alpha particle detectors was demonstrated. They rely on the use of thin layers (20 μm) of CVD diamond grown on silicon substrates. The devices have been tested using Plutonium isotopes dissolved in nitric acid. They appeared extremely reliable and exhibit a linear response as a function of the solution activity across the 3 explored decades. The response was stable under continuous immersion over 70 hours, and previsions would suggest they could be stable during months of continuous use. One problem that rises when devices are immersed in nuclear liquid solutions is their inherent contamination. Due to the high surface tension of CVD diamond, we found that the devices are hardly contaminated after use, and that they can be completely cleaned using a simple acid rinsing procedure. First demonstrators were based on the use of thin diamond layers, and thus were limited in size and in sensitivity because of induced capacitance noise effects. Further improvements can be considered, and in particular the use of multiple detectors with a multichannel acquisition readout circuit. These promising detectors can bring significant benefit to process control in nuclear fuel assembly reprocessing.

Fig. 4. (right) Variation of the counting rate in a 12.3 kBq/g ^{239}Pu solution as a function of the active area of the detection device. Contribution of the geometrical scale up to the increase of the noise level as measured by a charge sensitive preamplifier shows that a maximum sensitivity could be achieved at 80 mm^2. In practice, an active area of 40 mm^2 was used (see text).

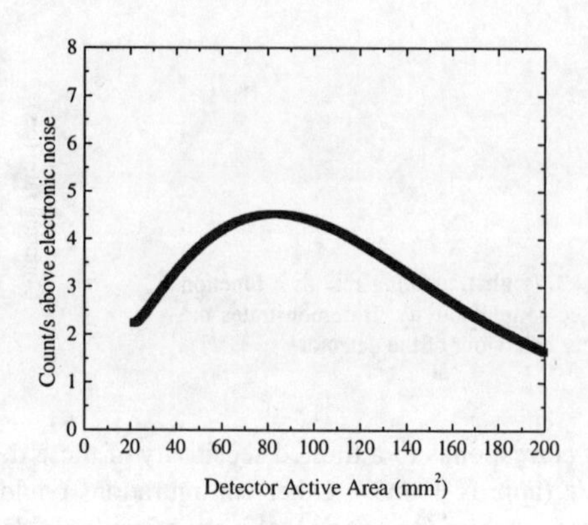

ACKNOWLEDGEMENTS

This work has been co-financed by COGEMA and CEA.

REFERENCES

[1] S.F. Kozlov, E.A. Konorova, M.I. Krapivi, V.A. Nadein, V.G. Yudina, IEEE Tran. Nucl. Sci., **24** (1977) 242

[2] E. Sevilliano and B. Wiliams, in *Diamond Films and Technology*, **Vol.8, N°2** (1998) MYU (Tokyo)

[3] M. Franklin, A; Fry, K.K. Gan, S. Han, H. Kagan, S. Kanda, D. Kania, R. Kass, S.K. Kim, R. Malchow, F. Morrow, S. Olsen, W.F. Palmer, L.S. Pan, F. Sannes, S. Schnetzer, R. Stone, Y. Sugimoto, G.B. Thomson, C. White, S. Zhao, *Nucl. Inst. Methods*, **A315**, (1992) 39

[4] D.R. Kania, L.S. Pan, P. Bell, O.L. Landen, H. Kornblum, P. Pianetta, M.D. Perry, *J. Appl. Phys.*, **65**, (1990) 124

[5] F. Foulon, P. Bergonzo, A. Brambilla, C. Jany, B. Guizard, R.D. Marshall, MRS Soc. Symp. Proc. **487** (1998) 591

[6] P. Bergonzo, F. Foulon, R.D. Marshall, C. Jany, A. Brambilla, R.D. McKeag, R.B. Jackman, IEEE Trans. Nucl. Sci., **45 (3)**, (1998) 370

[7] E.K. Souw, R.J. Meilunas, Nucl. Instr. And Meth., **A400**, (1997) 69-86

[8] Evaluated using TRIM Monte-Carlo code

[9] C. Jany, P. Bergonzo, F. Foulon, A. Tardieu, A. Gicquel, (1998) Proc. of the 5th Int. Symp. on Diamond materials, 97-32, edited by Davidson J.L. Brown W.D., Gicquel. A, Spitsyn B.V., Angus J.C., Pennington, NJ : Electrochem. Soc., Inc., 208

[10] K. Hecht, Z. Phys., **77** (1932) 235

[11] D. Tromson; P. Bergonzo; A. Brambilla; C. Mer; F. Foulon; V.N. Amosov, accepted for publication in J. Appl. Phys.

[12] C. Hordequin, F. Foulon, W. Dulinski, R. Truchetta, G. Claus, A. Brambilla, P. Bergonzo, presented at the Nuclear Science Symposium, Seattle (1999), to be published in IEEE Trans. Nucl. Sci.

Natural Analogs

GROWTH AND ALTERATION OF URANIUM-RICH MICROLITE

R. GIERÉ*, R. J. SWOPE*, E. C. BUCK**, R. GUGGENHEIM***, D. MATHYS***, and E. REUSSER****
*Department of Earth and Atmospheric Sciences, Purdue University, West Lafayette, IN 47907-1397, USA; giere@purdue.edu
**Chemical Technology Division, Argonne National Laboratory, Argonne, IL 60439, USA
***SEM Laboratory, University of Basel, CH-4056 Basel, Switzerland
****Institute of Mineralogy and Petrography, ETH Zentrum, CH-8092 Zürich, Switzerland

ABSTRACT

Uranium-rich microlite, a pyrochlore-group mineral, occurs in 440 Ma old lithium pegmatites of the Mozambique Belt in East Africa. Microlite exhibits a pronounced growth zoning, with a U-free core surrounded by a U-rich rim (UO_2 up to 17 wt.%). The core exhibits conjugate sets of straight cracks (cleavage planes) which provided pathways for a late-stage U-enriched pegmatitic fluid which interacted with the U-free microlite to produce a distinct U enrichment along the cracks and led to the formation of the U-rich rim. Following the stage of U incorporation into microlite, a second generation of hydrothermal fluids deposited mica along the cleavage planes. Subsequent to these two hydrothermal stages, the host rock was uplifted and subjected to intense low-temperature alteration during which Na, Ca and F were leached from the microlite crystals. This alteration also led to a hydration of microlite, but there is no evidence of U loss. These low-temperature alteration effects were only observed in the U-rich rim which is characterized by a large number of irregular cracks which are most probably the result of metamictization, as indicated by electron diffraction images and powder X-ray patterns.

INTRODUCTION

Pyrochlore-based ceramic waste forms are currently under development for the immobilization of excess weapons plutonium [1]. These ceramics are polyphase crystalline waste forms consisting of mainly pyrochlore and subordinate amounts of brannerite (UTi_2O_6), zirconolite ($CaZrTi_2O_7$), and rutile [2]. Minerals of the pyrochlore group conform to the simplified general formula

$$A_2B_2X_6Y \bullet nH_2O,$$

where A = Ca, Na, actinides, rare earth elements (REE), Ba, Sr, Bi, Pb; B = Nb, Ta, Ti, Zr, Sb, W, Fe; X = O, OH; and Y = O, OH, F. On the basis of the B-site cations, three subgroups are commonly distinguished within the pyrochlore group [3]: (1) pyrochlore sensu stricto (Nb-rich), (2) microlite (Ta-rich), and (3) betafite (Ti-rich). A fourth subgroup, roméite (Sb-rich), was recently suggested by Brugger et al. [4]. Synthetic pyrochlore-group phases are also prominent actinide hosts in polyphase crystalline and certain glass-ceramic waste forms designed for the immobilization of high level nuclear waste [5, 6].

Pyrochlore and microlite are strategically important Nb and Ta ores. Pyrochlore and betafite are typical accessory minerals in carbonatites, and occur in various pegmatites, syenites and related alkaline rocks, and in some hydrothermal veins. Microlite is found primarily in granitic pegmatites, whereas roméite typically occurs in metamorphosed Mn deposits and hydrothermal veins [4, 7, 8]. The pyrochlore-group minerals provide excellent natural analogues for pyrochlore-based nuclear waste forms, because samples of variable age and with high actinide contents are available.

We have studied samples of microlite samples from Mozambique with scanning electron microscopy (SEM), electron probe microanalysis (EPMA), transmission electron microscopy (TEM), and powder X-ray diffractometry (XRD) to further our understanding of the growth and alteration of this actinide-bearing natural analogue mineral.

Mat. Res. Soc. Symp. Proc. Vol. 608 © 2000 Materials Research Society

EXPERIMENTAL PROCEDURES

The microlite specimens were first characterized optically with a polarizing petrographic microscope. Subsequently, polished thin sections (30 µm thick) were studied with an Environmental Scanning Microscope (Philips XL30 ESEM) using a field emission gun as electron source.

A portion of one of the large zoned microlite crystals was broken into 0.25 mm sized fragments which were carefully sorted by color and morphology into two samples: a core sample containing only fragments from the U-free center of the crystal, and a rim sample with only fragments from the U-rich rim. Each sample was finely powdered, mounted on a zero background Si plate (using acetone), and analyzed on a Scintag X2 automated powder X-ray diffractometer equipped with a Peltier solid-state detector using Cu-Kα radiation at 45 kV and 40 mA. Data were collected from 2-70° 2θ in step scan mode (0.02° steps at 10 s/step).

Electron probe microanalysis was performed using a Cameca SX-50 microprobe which is equipped with four wavelength dispersive spectrometers and was operated at 20 kV and a beam current of 20 nA measured on a Faraday cage (beam size ≈2 µm). Samples and standards were coated with 200 Å of carbon. Synthetic oxides were used as standards for all elements except for Al and Fe, which were calibrated with mineral standards. For most elements, data collection time was 30 s on the peaks, and 15 s on background positions above and below the peaks. The raw data were corrected on-line by the PAP correction procedure [9].

For the transmission electron microscopy, microlite fragments were crushed between two clean glass slides. A small amount of alcohol was added and the particle-containing solution was pipetted onto lacy-carbon nickel grids. The samples were examined in a JEOL 2000 FXII TEM operated at 200 kV with a LaB_6 filament. Electron diffraction patterns were taken with a charge coupled device (CCD) camera which allows imaging at resolutions comparable to photographic film. Compositional analysis was performed using an IXRF Iridium II digital pulse processor with an attached Noran Instruments detector.

RESULTS

Microlite is a characteristic mineral in the early Paleozoic (approximately 500-450 Ma) lithium pegmatites of the Mozambique Belt in East Africa, where it occurs in association with albite, spodumene, lithian mica, potassium feldspar and quartz [10, 11]. These pegmatites are strongly altered as a result of tropical weathering. However, microlite is resistant to weathering and is preserved as euhedral crystals in a clay mineral matrix which consists mainly of kaolinite.

We have examined Th-free microlite from an altered pegmatite which intruded into a greenschist country rock in the Mutala area, Alto Ligonha (Mozambique). Here, the euhedral microlite crystals have a vitreous luster and exhibit a pronounced growth zoning (Fig. 1), characterized by a light green U-free core which is surrounded by a brown U-rich rim containing up to 17 wt.% UO_2 (Fig. 2). The core exhibits conjugate sets of straight cracks which represent the octahedral cleavage of microlite. These cleavage planes provided pathways for a late-stage U-enriched pegmatitic fluid. This hydrothermal fluid interacted with the U-free microlite to produce a distinct U enrichment along the cleavage planes (up to 5 wt.% UO_2; see Fig. 3) and led to the formation of the U-rich rim (500-600 µm thick; see Fig. 1). Following the stage of U incorporation into microlite, late-stage hydrothermal fluids also deposited mica along the cleavage cracks. Subsequently, the crystals were further fractured, most probably as a result of decompression during uplift. During weathering under tropical conditions, groundwater penetrated these fractured crystals and led to the deposition of clay minerals along fractures and cleavage planes.

The excellent correlation between PbO and UO_2 for all analyses representing unaltered microlite ($PbO = 0.033 + 0.051*UO_2$; $n = 218$, $r^2 = 0.959$, root-mean square residual = 0.060) indicates that Pb is almost entirely radiogenic and that the data can be used to estimate the age of formation. The calculations yielded an approximate chemical U-Pb age of 440 Ma which is in good agreement with the chemical ages obtained from uraninite and monazite inclusions. These chemical ages allowed to calculate the cumulative alpha particle dose experienced by the microlite

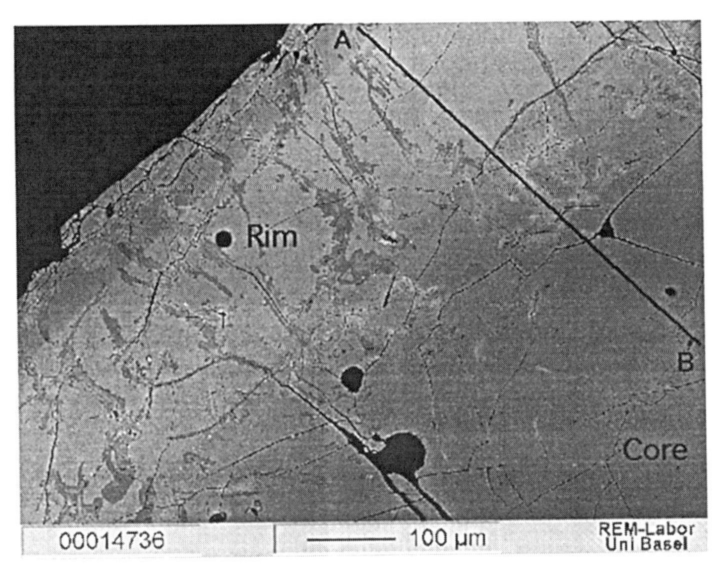

Figure 1: Backscattered electron image of microlite from Alto Ligonha, Mozambique. Note the abundant fractures which, in the rim area, are accompanied by dark alteration zones. Line A-B shows the trace of the quantitative profile shown in Figure 2.

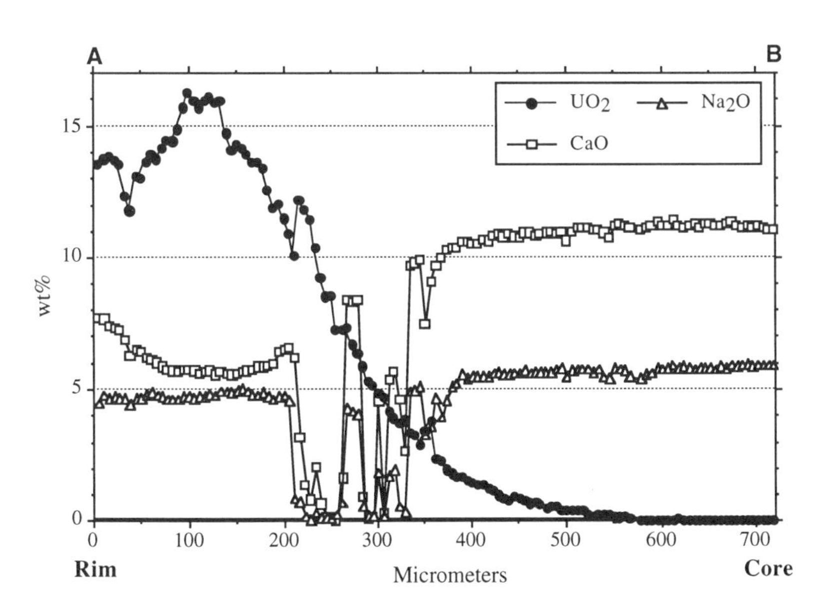

Figure 2: Concentration profile (EPMA data) across the rim into the core of microlite (profile trace shown in Fig. 1). Note that the trend for UO_2 is smooth across the two alteration zones which are marked by the depletion in Na_2O and CaO.

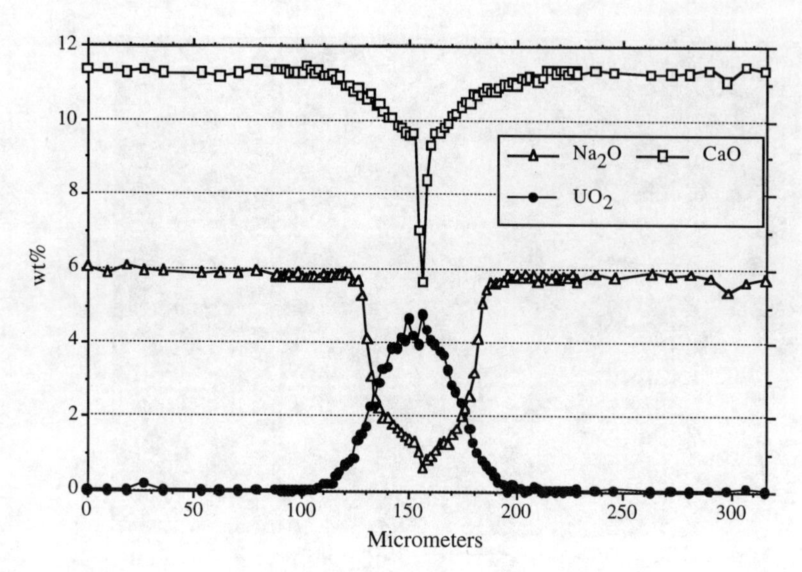

Figure 3: Concentration profile (EPMA data) across one of the cleavage cracks in the core of microlite. Note the symmetry with respect to the crack axis (located at 156 micrometers).

crystals as a result of the radioactive decay of U. In the U-rich rim, the maximum dose was approximately $2*10^{17}$ alphas/mg, suggesting that most of the U-rich rim should be completely metamict in the absence of annealing. This conclusion is based on previous studies which have shown that pyrochlore-group minerals become amorphous at a dose of > 10^{16} alphas/mg [6, 12], and is in agreement with preliminary TEM and XRD results. Our TEM investigations revealed the presence of diffuse rings in electron diffraction images, and these rings could be due to radiation damage microstructures resulting from alpha recoil collisions. The XRD patterns obtained for pulverized samples of the microlite core and rim areas demonstrate that the U-free core is completely crystalline, whereas the rim is almost entirely metamict (Fig. 4). The existence of some crystallinity (max. 5% crystallinity) in the studied rim sample (Fig. 4) is likely the result of partial annealing [12].

Further evidence for the existence of the metamict state in the rim is provided by the occurrence of abundant microfractures (Fig. 5) which are observed in the U-rich rim only. Microfracturing commonly takes place as a result of the crystalline-to-amorphous transition which is associated with an isotropic volume increase of ~ 5% in pyrochlore-group minerals [6, 12]. The fine cracks observed in the rim are arcuate and unevenly distributed, being particularly abundant around small inclusions of quartz and potassium feldspar. Backscattered electron images reveal that the U-rich rim of the studied microlites is strongly altered along these microfactures (Figs. 1, 5). Quantitative EPMA data demonstrate that this geochemical alteration is characterized by hydration (documented by low totals) and by a pronounced leaching of Na, Ca and F. At the same time, the data show that U was quantitatively retained during this alteration process (see Fig. 2) which most likely occurred under tropical weathering conditions. The chemical data demonstrate that the rate of Na, Ca and F ion exchange of metamict microlite greatly exceeds the rate of total dissolution. These low-temperature alteration effects were not observed along the pervasive fractures in the U-free crystalline core, suggesting that the susceptibility of microlite to low-temperature alteration is directly related to the microstuctural damage caused by the alpha decay of U.

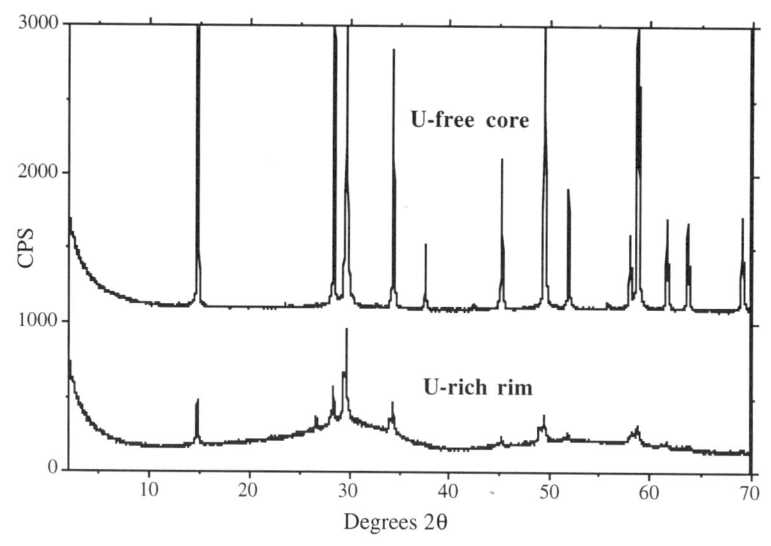

Figure 4: Powder X-ray diffraction patterns of the U-free core and the U-rich rim of microlite.

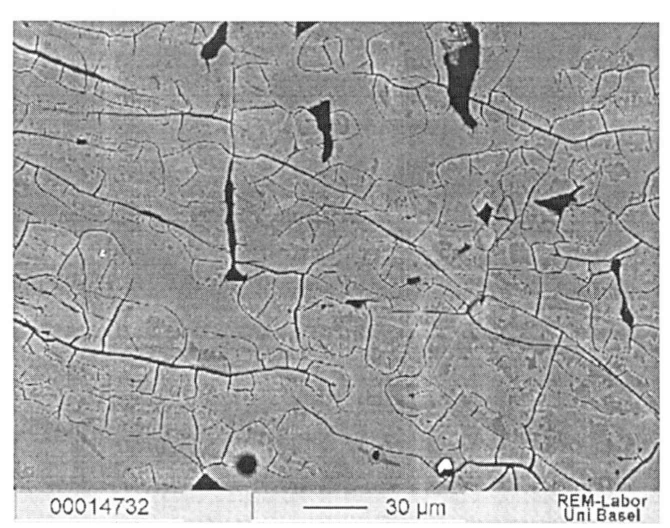

Figure 5: Backscattered electron image of an area within the U-rich rim. Note that microlite is altered (darker gray areas) along the abundant microfractures.

DISCUSSION

The results of this investigation demonstrate that 440 Ma old crystals of U-rich microlite retain a high degree of physical integrity in temperature and pressure conditions that range from values typical of igneous environments to ambient (weathering) conditions. However, the chemical composition, does change in response to changing physical and chemical conditions.

Alteration processes can be divided into primary (hydrothermal) and secondary (weathering) types [13-15]. These studies have shown that hydrothermal alteration takes place at relatively high temperatures (during late magmatic stages and the subsequent cooling) and probably before metamictization, whereas the weathering process usually occurs after the metamictization. The alteration process generally leads to dramatic changes in the A-site composition, but the concentration of actinides often remains remarkably constant. The results presented here are in perfect agreement with these observations. Additionally, our results demonstrate that only the metamict rim was significantly affected by the low-temperature alteration which had the following geochemical effects on the microlite composition: (1) hydration, (2) release of Na, Ca and F to the environment, (3) localized redistribution of radiogenic Pb, and (4) increase in the number of A-site vacancies. However, despite the adverse conditions of tropical weathering, U was retained during the alteration.

CONCLUSIONS

This natural analogue study documents that microlite is not only able to effectively scavenge U from a fluid at elevated temperatures (pegmatite stage) but also to retain the element during low-temperature alteration under tropical weathering conditions.

Viewed from a different perspective, these results are, in our opinion, necessary to fully assess the long-term performance of pyrochlore-based nuclear waste forms. Our investigation emphasizes the importance of natural analogues for evaluating the durability of waste form materials over geologic time spans and under natural conditions.

REFERENCES

1. E.R. Vance, C.J. Ball, R.A. Day, K.L. Smith, M.G. Blackford, B.D. Begg, and P.J. Angel, J. Alloys Comp. **213/214**, 406 (1994).
2. E.C. Buck, D.B. Chamberlain, and R. Gieré, in *Scientific Basis for Nuclear Waste Management XXII*, edited by D.J. Wronkiewicz and Lee. J.H. (Mater. Res. Soc. Proc. **556**, Pittsburgh, PA 1999), p. 19-26.
3. D.D. Hogarth, Am. Mineral. **62**, 403 (1977).
4. J. Brugger, R. Gieré, S. Graeser, and N. Meisser, Contrib. Mineral. Petrol. **127**, 136 (1997).
5. A.B. Harker, in *Radioactive Waste Forms for the Future*, edited by W. Lutze and R.C. Ewing, North-Holland, Amsterdam, 1988, pp. 335-392.
6. R.C. Ewing, W.J. Weber, and F.W. Clinard, Progr. Nucl. Energy **29(2)**, 63 (1995).
7. P. Cerny and T.S. Ercit, in *Lanthanides, tantalum and niobium*, edited by P. Möller, P. Cerny, and F. Saupé, Springer, Berlin, 1989, pp. 27-79.
8. G.R. Lumpkin and R.C. Ewing, in *Scientific Basis for Nuclear Waste Management VIII*, edited by C.H. Jantzen, J.A. Stone, and R.C. Ewing (Mater. Res. Soc. Proc. **44**, Pittsburgh, PA 1985), p. 647-654.
9. J.L. Pouchou and F. Pichoir, *Rech. Aérosp.* **1984-3**, 167 (1984).
10. J.M. Correia Neves, J.E. Lopes Nunes, and D.B. Lucas, Rev. Ciênc. Geol., Lourenço Marques, **Série A 4**, 35 (1971).
11. O. Von Knorring and A. Fadipe, Bull. Minéral. **104**, 496 (1981).
12. G.R. Lumpkin and R.C. Ewing, Phys. Chem.Minerals **16**, 2 (1988).
13. G.R. Lumpkin and R.C. Ewing, Am. Mineral. **77**, 179 (1992).
14. G.R. Lumpkin and R.C. Ewing, Am. Mineral. **80**, 732 (1995).
15. G.R. Lumpkin and R.C. Ewing, Am. Mineral. **81**, 1237 (1996).

CRANDALLITES AND COFFINITE: RETARDATION OF NUCLEAR REACTION PRODUCTS AT THE BANGOMBÉ NATURAL FISSION REACTOR

K.A. Jensen[*,**], J. Janeczek[***], R.C. Ewing[**], P. Stille[****], F. Gauthier-Lafaye[****], S. Salah[****]

[*] University of Aarhus, Department of Earth Sciences, Aarhus, Denmark, keld@geo.aau.dk
[**] University of Michigan, Department of Nuclear Engineering and Radiological Sciences & Department of Geological Sciences, Ann Arbor, Michigan, USA
[***] University of Silesia, Faculty of Earth Sciences, Sosnoweic, Poland
[****] Centre National de la Recherche Scientifique, Centre de Geochimie de la Surface, Strasbourg Cedex, France.

ABSTRACT

Various REE-Sr-(Pb)-crandallites, uraninite, and coffinite in the near-field of the 2 Ga old supergene-altered Bangombé U-deposit and its natural fission reactor (RZB) have been examined. The crandallite minerals may have formed during syncriticality host-rock alteration, continous alteration of phosphates, episodic Pb-loss and/or supergene weathering. Coffinitization with P_2O_5 and SO_4-substitution has occurred immediately below RZB and resulted in extensive loss of U (≤ 46%) and enrichment of Ce (≤ 190%) and Nd (≤ 780%). Additional loss of U during coffinitization also may have occurred due to dissolution. Current alteration under oxidizing conditions has resulted in partial dissolution of uraninite and coffinite and the formation of uranyl phases. Despite supergene alteration, the hydrogeochemistry (3.09 ppt U [$^{235}U/^{238}U$ = 0.7012 to 0.7019%], 4.96 ppt Ce, and 1.92 ppt Nd) suggests a remarkable retardation of lanthanides and depleted uranium by REE-Sr-(Pb)-crandallites, uraninite, coffinite, and uranyl phases at RZB.

INTRODUCTION

Chemical alteration of uraninite and coffinitization ($UO_{2+x} + H_4SiO_4 \rightarrow USiO_4.nH_2O + 2\text{-}nH_2O + 0.5xO_2$) are known to occur under reducing [1,2,3,4]. Both of these alteration mechanisms may be associated with loss of U and other impurities in the uraninite. Incorporation into the secondary phases may an effective means of retardation of radionuclides in uranium deposits, mine tailings, as well as during the alteration of spent nuclear fuel in a geological repository.

At natural fission reactor 13 at Oklo, SE-Gabon, Dymkov et al. [5] have proposed that a REE-Sr-crandallite, nominally $(Ca,REE,Pb,Sr)(Al,Fe)_3[(Si,P,As)O_4]_2(OH)_{5\text{-}6}$, formed during dissolution and chemical alteration of uraninite due to regional heating and emplacement of a 10 m distant 755±83 Ma old dolerite dyke. However, a florencite $(La_{0.38},Ce_{0.35}Nd_{0.06},Sm_{0.01},Ca_{0.03},Sr_{0.17})(Al_{2.98},Fe_{0.02})_3[(P(O,OH)_4]_2(OH)_6)$ with 30% fissiogenic Nd and 71% fissiogenic Sm has also been observed in the shallow (~ 12 m) and supergene weathered Bangombé reactor zone (RZB), 25 km SE of Oklo, where the nearest dolerite dyke is observed at approximately 100 m distance [6]. Based on petrography, the timing of florencite formation in RZB was inconclusive, but it has been proposed that the florencite formed either at the time of reactor criticality or in connection with chemical alteration and partial coffinitization of uraninite during the regional heating, as suggested at RZ13. Though mainly during supergene weathering, the uraninite in RZB has been subjected to dissolution; whereas, only minor coffinitization is observed [4,7,8,10]. Uraninite in the sandstone immediately below RZB has, in contrast, been subjected locally to extensive coffinitization [8,9].

Mat. Res. Soc. Symp. Proc. Vol. 608

In this paper we present a detailed analysis of the geochemical behavior of the most abundant elements in uraninite during coffinitization immediately below RZB. Despite the presence of lead and other chemical differences between natural uraninite ($U^{4+}_{1-x-y-z-u-v}U^{6+}_{x}R^{4+}_{y}R^{3+}_{z}R^{2+}_{u}[]_{v}O_{2+x-z/4-u/2-4v}$) and spent nuclear fuel [11], the chemical alteration behavior of U, Th, Zr, and lanthanides is expected to be comparable. We also summarize the mineralogy of crandallite group minerals that appear to be the major lanthanide, Sr- and P-bearing mineral in the weathering horizon above RZB. The degree of U, Sr, and lanthanide retardation is evaluated from hydrogeochemical data.

GEOLOGY

The 2 Ga old Bangombé uranium deposit is located 25 km NW of Franceville. The uranium deposit contains a natural fission reactor, RZB, situated at 10.40 – 11.95 m depth between the FA-sandstone host-rock and black shales from the FB-formation in the Franceville Series [10,12]. RZB is cut by three drill-cores: BA145, BAX03 and BAX08. In these drill-cores the reactor core is 5 to 10 cm thick and overlain by a 0.5 m thick chlorite-rich hydrothermal alteration halo, *argile de pile* [10,13]. The maximum ^{235}U-depletion was observed in a specimen from BAX03 with a $^{235}U/^{238}U$-ratio of 0.590% [10] as compared with the normal ratio of 0.725%. During criticality, actinides, as well as Ln's, Zr, and Mo fission products were partially redistributed and migrated at least 2 m into the sandstone and more than 5.4 m into the overlying FB-formation [10,14]. The present-day lateritization has resulted in the formation of an iron-rich horizon at 8.25 to 9.30 m depth in BAX03, 2.2 m above RZB, and partial dissolution of uraninite in RZB. Due to syncriticality migration, alteration, and weathering, isotopic equilibrium between U and the fissiogenic lanthanides is only observed in the core of the RZB. Above the reactor core, a general LREE enrichment is observed in the weathered black shales [13] and overall the retention appears to be U < Sm < Nd < Th [10].

ANALYTICAL METHODS

Petrographic analysis was completed by optical microscopy and backscattered electron imaging (BSEI). Analytical transmission electron microscopy was performed on 2 μm clay fraction specimens suspended on carbon-coated Cu-grids using a JEM 2000FX electron microscope with an energy dispersive spectrometer. The acceleration voltage of the electron beam was 200 kV. BSEI and quantitative electron microprobe analysis was conducted on a JEOL JXA-8600 Superprobe operated at 20 kV and 20 nA. Thirteen-element analyses were conducted on uraninite and coffinite by wave dispersive spectrometry. The elements were calibrated on the following standards: UO_2 (U); $ThSiO_4$ (Th); PbS (Pb); albite (Si, Al); $CaTiO_3$ (Ca, Ti); olivine (Mg); hematite (Fe); troilite (S); zircon (Zr); and phosphates for P, Y, Ce, and Nd. The elements were analyzed for at least 5 sec and up to 40 sec or until the standard deviation reached a value of 1%. The background intensity was counted 20 sec on each position. Elemental correction factors were calculated by the ZAF-program, and quantification was performed by the PRZ-program in the Tracor Northern operator system. Contributions from peak-overlaps and background interference were adjusted by empirical correction factors calculated as the average of 10 analyses on suitable standards for each oxide, respectively (see [8]).

RESULTS AND DISCUSSION

Petrographic and chemical analysis have shown that uraninite is the main source of U, Th, Zr, and lanthanides in the reactor core of RZB and the FA-sandstone immediately below. In the *argile de pile*, the overlying black shales, and the weathering zone, the lanthanides appears mainly retained by REE-Sr-(Pb)-crandallites (Fig. 1a). Uraninite immediately below the reactor core has been subjected to extensive phosphatian coffinitization associated with various degrees of dissolution. In both BAX03:12.10-12.20 and BAX08:12.38-12.40, the coffinitization was particularly advanced in regions abundant in microfractures (Fig. 1b). In BAX08:12.38-12.40, coffinitized uraninite was associated with a Fe-uranyl phosphate hydroxide hydrate (Fe-UPHH) in a fracture (Fig. 1b). In other thin-sections, goethite was the major mineral in this fracture. Fe-UPHH was also observed in fractures and uraninite dissolution voids in BAX03:12.20-12.30. Other uranyl phases observed at Bangombé include fourmarierite [16], torbernite [13], françoisite [9], uranopelite [15], and unidentified uranyl sulfate hydroxide hydrates [4,7,8]. Françoisite was observed in one thin-section only, cut from the bulk of BAX03:12.20-12.30, and is the only uranyl phase observed to contain significant amounts of lanthanides. The oxidizing groundwater in BAX03 (Eh=143 mV; pH=5.96) and the association between coffinite with mainly U^{4+} and the Fe-UPHH with U^{6+} in BAX08 shows the dynamic regime in which RZB is altered. Based on the U-Th-age of torbernite at 15 m depth in the 4-5 m distant BAX15 drill-core, oxidizing conditions may have existed for 76,500±6,800 years [13].

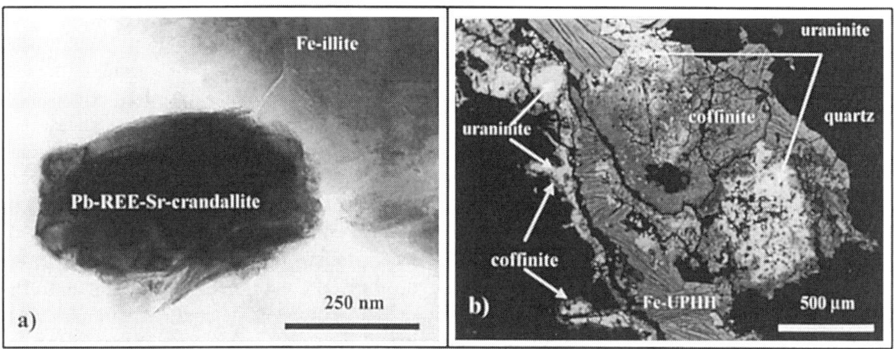

Figure 1: a) Transmission electron micrograph of crandallite in the 2 µm clay fraction of BAX03:11.10-11.21. b) Extensively coffinitized uraninite in BAX08:12.38-12.40. The coffinite is partially dissolved and associated with a Fe-uranyl phosphate hydroxide hydrate (Fe-UPHH) emplaced in a cross-cutting fracture.

Crandallite group minerals

Various REE-Sr-(Pb)-crandallites were observed in the 2 µm clay fraction of the *argile de pile* and the FB-formation in BAX03 (Table 1). Most crandallite grains occurred as up to 1000 nm anhedral to subhedral grains associated with illite (Fig. 1a). Energy dispersive spectrometry showed that Sr, La, Ce, and Nd were always among the major elements on the A-site, $AB_3(XO_4)_2(OH)_{5\text{-}6}$. Lead was also present in several of the grains and the major element on the A-site in crandallite from BAX03:10.13-10.35. The B-site was mainly occupied by Al and minor Fe, but in BAX03:11.10-11.21, Fe dominated over Al. The X-site was occupied by P and Si in samples from BAX03:6.65-8.81 to BAX03:10.55-10.75 m depth. Crandallite in the latter sample also contained minor As. Below BAX03:10.55-10.75, P was the only detactable element on the X-site. Based on chemistry, the observed crandallite minerals suggest an extensive solid solution between the pure end-members: crandallite ($CaAl_3(PO_4)(OH)_6$), goyazite ($SrAl_3(PO_4)(OH)_6$), florencite ($REEAl_3(PO_4)(OH)_6$), and plumbogummite ($PbAl_3(PO_4)(OH)_6$).

Table 1: Composition of REE-Pb-Sr-crandallites from BAX03 as determined by analytical transmission electron microscopy. The relative order of the cations, $AB_3(XO_4)_2(OH)_{5-6}$, is determined by the relative peak height on the energy dispersive spectra. The last row shows the result from electron microprobe analysis of florencite [6] in the Bangombé reactor zone. Specimens listed in italics are from the Bangombé reactor zone.

drill-core BAX03	composition
BAX03:6.65-8.81	$(Ce,Sr,Nd,Ca,K,Pb)(Al,Fe)_3[(P,Si)O_4]_2(OH,F)_{5-6}$
BAX03:10.15-10.35	$(Pb,Ce,Sr,Nd,La,Ca)(Al,Fe)_3[(P,Si)O_4]_2(OH,F)_{5-6}$
BAX03:10.55-10.75	$(Ce,Nd,La,Sr,Ca,Th)(Al,Fe)_3[(P,Si,As)O_4]_2(OH,F)_{5-6}$
BAX03:11.10-11.21	$(Sr,Ce,Nd,Pb,Ca,La)(Fe,Al)_3[PO_4]_2(OH,F)_{5-6}$
BAX03:11.21-11.41	$(Sr,Ce,Pb,Ca,Nd,La,Th)(Al,Fe)_3[PO_4]_2(OH,F)_{5-6}$
BAX03:11.62-11.90	$(Ce,Nd,Sr,Ca,Pb,La)(Al,Fe)_3[PO_4]_2(OH,F)_{5-6}$
BAX03:11.70(5D2B)	$(La_{0.38},Ce_{0.35}Nd_{0.06},Sm_{0.01},Ca_{0.03},Sr_{0.17})(Al_{2.98},Fe_{0.02})_3[(P(O,OH)_4]_2(OH)_6$

Accessory florencite, occasionally associated with lanthanide-rich plumbogummite, has also been observed in the FA-sandstone [6,17]. The Pb may come from uraninite that experienced massive lead-loss 1,000 to 700 Ma ago [2,12]. However, accessory monazite-huttonite in the FA-sandstone is also typically altered [18] during which radiogenic lead and lanthanides also may be released. Florencite was locally observed in association with monazite [6]. However, REE and Pb-rich crandallite also occur in late fracture coatings close to the surface at Bangombé, suggesting their recent formation in the weathering horizon [20]. This is consistent with numerous observations around the world [20,21]. Florencite in RZB may have formed at the expense of apatite and monazite with fissiogenic lanthanides that occur in less altered reactor zones at Oklo [12]. However, the florencite in RZB is of great age [6] and can have formed under hydrothermal conditions [20,22]. This is further supported by a relatively constant concentration of Sr (120-150 ppm) in the weathering horizon and the FB-formation, and a distinct increase in Sr (264 ppm) at the top of the *argile de pile* in BAX03 [10] reflecting syn-criticality redistribution of primordial Sr from the reactor core (48 ppm Sr). Hence, crandallite minerals may have formed continuously at Bangombé due to the various conditions (diagenetic, hydrothermal, and supergene) under which they can precipitate [20,21,22,23]. Combined with their crystal chemistry, this suggests that crandallite minerals have great potential for the long-term retardation of Sr, La, Ce, Pr, and Pb in the geosphere [22,24,25].

Coffinitization of uraninite

Table 2 lists the average composition of the uraninite, representative analysis of the coffinite with the highest concentrations of SiO_2 and P_2O_5, as well as the total compositional range of the partially coffinitized rims in BAX03:12.20-12.30 and BAX08:12.38-12.40, respectively. The uraninite hosts are relatively pure consisting of 86.39±1.94 wt.% UO_2 and 4.15±1.94 wt.% PbO in BAX03:12.20-12.30 and 89.68±0.88 wt.% UO_2 and 4.59±0.74 wt.% PbO in BAX03:12.38-12.40, respectively. The major impurities are Si, Ti, Fe, Ca, and P with oxide concentrations between 0.52±0.46 wt.% (SiO_2) and 1.02±0.34 wt.% (FeO) (Table 2). The Ti is attributed to Ti-oxide inclusions in the uraninite; whereas, Si and P may be the result of incipient coffinitization or nano-sized impurity-rich regions in the uraninite [2,4]. The minor components are ThO_2, ZrO_2, Ce_2O_3 and Nd_2O_3 that occur in concentrations below 0.42±0.20 wt.%. Thorium is almost absent (0.01±0.02 wt.% ThO_2) in the uraninite from BAX08:12.38-12.40.

Table 2: Chemistry of uraninite and coffinite, combined with the geochemical budget (Δ) in moles for coffinitization of uraninite. The geochemical budget was calculated for the transformation of 10 moles of uraninite into 10 moles of coffinite after subtracting the maximum possible PbS contamination in the analysis. Ti and Al atoms were not considered in the calculations.

	BAX03: 12.20-12.30						BAX08: 12.38-12.40					
	uraninite n = 32		coffinite n = 18				uraninite n = 10		coffinite n = 6			
wt.%	average	σ	1	2	min	max	average	σ	3	4	min	max
UO_2	86.39	1.94	69.28	66.04	66.04	85.73	89.68	0.88	71.33	79.13	70.22	79.13
ThO_2	0.12	0.03	0.42	0.27	0.25	0.59	0.01	0.02	0.08	0.16	n.d.	0.16
SiO_2	0.52	0.46	5.01	6.31	0.36	6.31	0.68	0.17	5.88	5.18	1.71	5.88
TiO_2	0.57	0.23	0.63	0.35	0.35	2.09	0.63	0.30	0.18	0.33	0.18	0.82
ZrO_2	0.17	0.12	0.12	0.47	0.12	0.91	0.42	0.20	1.73	1.83	1.73	2.05
PbO	4.15	1.05	1.56	0.82	0.82	3.64	4.59	0.74	2.06	2.06	2.06	3.70
FeO	1.03	0.34	0.55	1.39	0.48	1.39	0.84	0.22	0.71	0.85	0.71	1.42
CaO	0.92	0.35	1.06	0.95	0.25	1.06	0.93	0.32	0.85	0.80	0.50	0.85
Al_2O_3	0.21	0.23	0.51	0.42	0.25	0.74	0.10	0.01	0.16	0.15	0.07	0.16
P_2O_5	0.71	0.73	8.00	7.93	1.56	8.00	0.57	0.29	5.70	5.27	3.80	5.70
SO_3	0.10	0.11	2.50	3.17	0.15	3.77	<0.01	0.01	0.48	0.68	0.48	1.71
Ce_2O_3	0.19	0.09	0.73	0.77	0.28	0.77	0.22	0.08	0.62	0.50	0.34	0.62
Nd_2O_3	0.16	0.10	1.87	1.82	0.26	1.87	0.08	0.06	0.73	0.53	0.31	0.73
Total	95.23	1.74	92.24	90.71	85.75	95.34	98.75	1.11	90.51	97.47	87.13	97.47
% PbS	1.82	-	2.34	1.15	-	-	0.06	-	3.57	3.44	-	-
mole	average	σ	Δ	Δ	-	-	average	σ	Δ	Δ	-	-
U	8.190	0.184	-3.271	-3.775	-	-	8.189	0.080	-2.832	-2.352	-	-
Th	0.012	0.003	0.019	0.007	-	-	0.001	0.002	0.005	0.011	-	-
Si	0.221	0.195	1.377	1.675	-	-	0.280	0.070	1.705	1.438	-	-
Zr	0.034	0.024	-0.015	0.035	-	-	0.084	0.041	0.200	0.211	-	-
Pb	0.446	0.084	-0.446	-0.446	-	-	0.506	0.080	-0.440	-0.491	-	-
Fe	0.367	0.121	-0.219	-0.018	-	-	0.288	0.075	-0.088	-0.052	-	-
Ca	0.419	0.160	-0.056	-0.113	-	-	0.410	0.140	-0.102	-0.126	-	-
P	0.257	0.265	1.903	1.760	-	-	0.198	0.101	1.431	1.281	-	-
S	0.000	0.000	0.464	0.649	-	-	0.000	0.000	0.000	0.000	-	-
Ce	0.029	0.013	0.056	0.056	-	-	0.032	0.012	0.044	0.028	-	-
Nd	0.024	0.015	0.188	0.171	-	-	0.012	0.009	0.076	0.051	-	-
Total	10.000	1.065	-	-	-	-	10.000	0.610	-	-	-	-

Nominally, coffinitization results in the transformation of UO_2 into $U[SiO_4].nH_2O$. However, in the FA-sandstone below RZB, the alteration rinds also contain P (≤ 8.00 wt.% P_2O_5) and S (≤ 3.77 wt.% SO_3). The incorporation of P can be explained by solid solution between coffinite and ningoyite, $(U,Ca,Ce)(PO_4)_2.1\text{-}2H_2O$ [9]. Sulfur is normally not observed in coffinite, but $P_2O_5+SO_3$ shows a strong correlation (R^2=0.8) with $SiO_2+P_2O_5+SO_3$, similar to the correlation between SiO_2 and $SiO_2+P_2O_5$. This suggests that both $[SO_4]^{2-}$ and $[PO_4]^{3-}$ substitute for $[SiO_4]^{4-}$ in the coffinite.

Compared with the uraninite, coffinitization has resulted in a significant chemical disturbance. In addition to a more or less complete loss of Pb; the Ca and U were also partially lost; whereas, the Th, Zr, Ce, and Nd generally were enriched by 8 to 12 times their oxide content in the uraninite host (Table 2; Fig 2). Iron is partially lost, but a few analyses indicate Fe-enrichment. Particularly in BAX08 (Fig. 2b), where Fe-UPHH and goethite also occurs in the fracture. CaO and PbO appear to be the elements that suffered most from the coffinitization and occur with only 0.2 to 0.3 times their original oxide concentration. The UO_2 content in the coffinite is typically ~ 0.8 times the concentration in the uraninite. Except from CaO-enrichment, the composition of the sulfur-free phosphatian coffinite from BAX03 [9] show a similar trend when compared to the uraninite compositions in BAX03:12.20-12.30 and BAX08:12.38-12.40 determined in this project (Fig. 2).

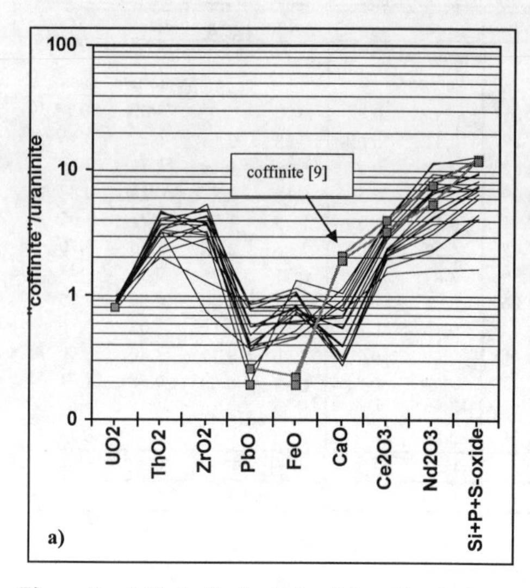

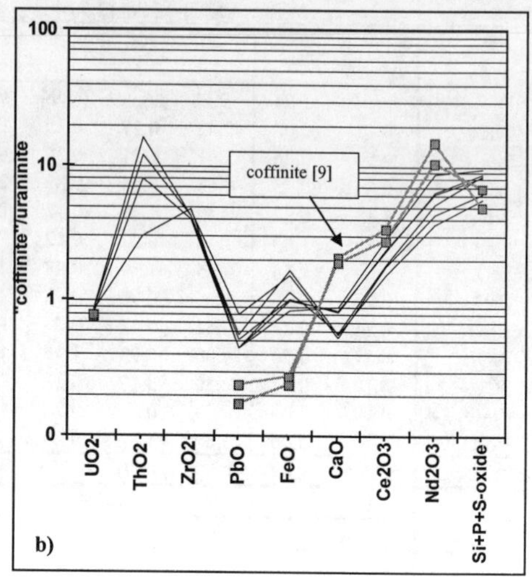

Figure 2: a) Plot of selected oxide-ratios between partially coffinitized uraninite "coffinite" and the average composition of their uraninite host in BAX03:12.20-12.30. b) Plot of selected oxide-ratios between partially coffinitized uraninite "coffinite" and the average composition of the uraninite host in BAX08:12.38-12.40. In both a and b, the ratio between phosphatian coffinite in [9] are shown with square symbols connected by a thick gray line (note Th and Zr was not analyzed).

A quantitative assessment of the geochemical effect of coffinitization was calculated by analyzing the transformation of 10 mole uraninite into 10 mole of coffinite with the observed compositions, after subtracting the amount of Pb and S that may be caused by contamination from galena (Table 2). The molar calculations confirmed that coffinitization has resulted in a massive loss of U (35 to 46%) and Pb (87 to 100%). Notably, the coffinitization in both BAX03 and BAX08 shows preferential enrichment of Nd (633 to 783%) as compared with Ce (137 to 193%) (Fig. 2; Table 2). Particularly for Nd, the enrichment is several orders of magnitude beyond the expected enrichment at the observed uranium loss. Similar, yet smaller, deviations were also observed for Zr, Ce, and Th.

In scatter plots, both Nd_2O_3 and Ce_2O_3 contents show a linear increase as function of increased contents of P_2O_5 and P_2O_5+SO_3 and a large scatter when plotted as function of SiO_2+P_2O_5+SO_3. However, above ~ 5 wt.%, P_2O_5 the increase in Nd_2O_3 is greater than that of Ce_2O_3. This suggests that either: 1) the Ce/Nd-ratio was low in the coexisting fluids during coffinitization; 2) Nd preferentially forms solid state complexes with PO_4 and SO_4 during coffinitization and/or its subsequent oxidative alteration; or 3) Ce was incorporated into other phases. The second and third option could occur if the fluids originated from an oxidized environment where Ce may precipiate as CeO_2. However, availability of P (and S) was necessary for the formation of phosphatian coffinite. Phosphorus could have been supplied at any time during continuous alteration of monazite-huttonite in the Franceville basin. However, these minerals typically have a Ce/Nd-ratio of 3 to 5 as compared with normally less than one in uraninite and coffinite. If released in equal amounts, a decrease in the Ce/Nd-ratio may occur during formation of REE-crandallite that show preferential incorporation of Ce > La ≈ Pr > Nd > Sm etc. [22]. Because crandallite can form under both reducing and oxidizing conditions, this model can always satisfy the first hypothesis. In fact, higher concentration of P_2O_5 in the black shale (0.10 to 0.25 wt.%) as compared with the FA-sandstone (0.03 wt.%), sulfate in the phosphatian coffinite and the associated Fe-UPHH suggests a link between phosphatian coffinitization and supergene weathering. Currently, however, the groundwater in BAX03 has higher concentrations of dissolved Ce (4.96 ppt) than Nd (1.92 ppt) [26]. Moreover, the groundwater in BAX03 is only slightly depleted ($^{235}U/^{238}U$ = 0.7012 – 0.7019% [27]) as compared with the average $^{235}U/^{238}U$-ratio (0.593%) in the drill-core between the packers (11.80-12.45 m). This suggests slow dissolution of uranium- and lanthanide bearing phases in the reactor zone and that the groundwater chemistry is controlled by the surrounding host-rock. This is further supported by the presence of less than 3% fissiogenic Nd and Sm in the groundwater from BAX03 [26].

CONCLUSIONS

- REE-Sr-Pb-crandallites have formed at various times during the more than 2 Ga history of the Bangombé uranium deposit and its reactor zone. One type of crandallite contains fissiogenic lanthanides and most likely formed during criticality or lead-loss and geochemical alteration of uraninite and/or phosphates 1,000 to 700 Ma ago. Crandallites also formed during continuous alteration and supergene weathering of phosphates. The broad range of stability conditions for crandallite shows its potential for long-term retardation of Sr, lanthanides, and Pb in the geosphere.

- Phosphatian coffinitization (±S) may have occurred in connection with continuous alteration of phosphates and/or leaching during supergene weathering. The coffinitization resulted in a massive uranium loss (≤ 46 %) and enrichment of Zr ≤ Th ≤ Ce << Nd. In the oxidized zone of supergene alteration, uranyl phosphate – and uranyl sulfate hydroxide minerals have formed.

- The groundwater at 11.90-12.45 m depth in BAX03 contains remarkably low concentrations of U, Ce, Nd and Sr [26] indicating that the mineral assemblage in the supergene weathering at the Bangombé reactor has a high retention capacity of U, Sr and lanthanides.

ACKNOWLEDGMENTS

This study was supported by the Swedish Fuel and Waste Management Company AB (SKB), Sweden. Partial support was provided by the Environmental Science Program of DOE (DE-FG07-97 ER14816) The work was completed under the European Union research program (CCE n° FIW-CT96-0020): Oklo Natural Analogue – Phase II, organized by EU and the French Atomic Energy Agency (CEA). The electron microprobe at the University of Aarhus is funded by the National Natural Science Research Council (SNF), the Carlsberg Foundation, and the Aarhus University Foundation. Transmission electron microprobe analyses were conducted at the Electron Microbeam Analysis Facility of the Department of Earth and Planetary Sciences, University of New Mexico and supported by the National Science Foundation, National Aeronautics and Space Administration, US Department of Energy, and the State of New Mexico.

REFERENCES

[1] J. Janeczek and R.C. Ewing, in Scientific Basis for Nuclear Waste Management XV, edited by C. Sombret (Mater Res. Symp. Proc., **257**, 1992), p. 497-504.

[2] J. Janeczek and R.C. Ewing, Geochim Cosmochim Acta, **59**, p. 1917-1931 (1995).

[3] M. Fayek and T.K. Kyser, *in* Uranium: mineralogy, geochemistry, and the environment, edited by P.C. Burns and R. Finch, Rev. in Min, **38**, p. 181-220 (1999).

[4] K.A. Jensen and R.C. Ewing, *in* Oklo Working Group, Proceedings of the second EC-CEA workshop on the Oklo-natural analogue Phase II project held in Helsinki, Finland, from 16-18 June 1998, edited by D. Louvat, V. Michaud, and H. von Marevic (Nucl. Sci. and Tech., **EUR 19116 EN**, 1999), p. 61-91.

[5] Y. Dymkov, P. Holliger, M. Pagel, A. Gorshkov, A. Artyukhina, Miner. Dep., **32**, p. 617-620 (1997).

[6] J. Janeczek and R.C. Ewing, Am. Min., **81**, p. 1263-1269 (1996).

[7] K.A. Jensen, R.C. Ewing, F. Gauthier-Lafaye, *in* Scientific Basis for Nuclear Waste Management XX, edited by W. Gray and I. Triay, (Mater. Res. Soc. Proc., **465**, 1997), p. 1209-1218.

[8] K.A. Jensen and R.C. Ewing, *in* OKLO Working Group - Proceedings of the first EC-CEA workshop on the Oklo-natural analogue Phase II project held in Sitjes, Spain, from 18 to 20 June 1997, edited by D. Louvat and C. Davies (Nucl. Sci. and Tech., **EUR 18314 EN**, 1998), p. 139-159.

[9] J. Janeczek and R.C. Ewing, Min. Mag., **60**, p. 665-669 (1996).

[10] R. Bros, F. Gauthier-Lafaye, P. Larque, J. Samuel, and P. Stille, *in* Scientific Basis for Nuclear Waste Management XVIII – Part 2, edited by T. Murakami and R.C. Ewing (Mater. Res. Soc. Proc., **353**, 1995), p. 1187-1194.

[11] J. Janeczek, R.C. Ewing, V.M. Oversby, L.O. Werme, J. Nucl. Mat., **238**, p.121-130 (1996).

[12] F. Gauthier-Lafaye, P. Holliger, P.L. Blanc, Geochim Cosmochim Acta, **60**, p. 4831-4852.

[13] R. Bros, P. Andersson, P. Roos, S. Claesson, E. Holm, J. Smellie, *in* OKLO Working Group - Proceedings of the first EC-CEA workshop on the Oklo-natural analogue Phase II project held in Sitjes, Spain, from 18 to 20 June 1997, edited by D. Louvat and C. Davies (Nucl. Sci. and Tech., **EUR 18314 EN**, 1998), p. 187-195.

[14] R. Bros, T. Ohnuki, and N. Yanase, *in* Geol. Soc. of Am. Annual Meeting and Exposition, October 25-28, 1999, Denver, Colorado, Abstracts with Programs, p. A68. (1999).

[15] D. Cui, private communication.

[16] J. Janeczek, H. Hidaka, J., Ewing, F. Gauthier-Lafaye, K.A. Jensen, W. Lapot, W., Salah, S., (in prep.).

[17] L. Perez-del-Villar, J.S., Cozar, J. Pardillo, M. Pelayo, A.J. Quijido, M.A. Labajos, *in* Oklo Working Group, Proceedings of the second EC-CEA workshop on the Oklo-natural analogue Phase II project held in Helsinki, Finland, from 16-18 June 1998, edited by D. Louvat, V. Michaud, and H. von Marevic (Nucl. Sci. and Tech., **EUR 19116 EN**, 1999), p. 141-162.

[18] R. Matheiu, L. Zetterström, M. Cuney, F. Gauthier-Lafaye, Chem. Geol. (submitted).

[19] L. Perez-del-Villar, J.S., Cozar, J. Pardillo, M. Pelayo, M.A. Labajos, *in* Oklo Working Group, Proceedings of the second EC-CEA workshop on the Oklo-natural analogue Phase II project held in Helsinki, Finland, from 16-18 June 1998, edited by D. Louvat, V. Michaud, and H. von Marevic, Nucl. Sci. and Tech., **EUR 19116 EN**, 1999), p. 163-181.

[20] Rare Earth Minerals. Chemistry, origin and ore-deposits, edited by A.P. Jones, F. Wall, and C.T. Williams, Min. Soc. Series, **7**, 372 p. (1996).

[21] J.F. Banfield and R.A. Eggleton, Clays and Clay Minerals, **37**, p. 113-127 (1989).

[22] R.G. Schwab, H. Herold, Chr. Götz, N. Pinto de Oliveira, N. Jb. Miner. Mh., **H.6**, p. 241-254 (1990).

[23] B. Rasmussen, Am Jour. Sci., **296**, p. 601-632 (1996).

[24] J.O. Nriagu, Geochim. Cosmochim. Acta, **38**, p. 887-898 (1974).

[25] C. Frondel, Science, **128**, p. 1623-1624 (1978).

[26] P. Stille, F. Gauthier-Lafaye, K.A. Jensen, P. Gomez, R.C. Ewing, D. Louvat, (in prep.).

[27] C. Skårman, C. Degueldre, M. Laaksoharju, B. Thomas, L. Tohler, *in* OKLO Working Group - Proceedings of the first EC-CEA workshop on the Oklo-natural analogue Phase II project held in Sitjes, Spain, from 18 to 20 June 1997, edited by D. Louvat and C. Davies (Nucl. Sci. and Tech., **EUR 18314 EN**, 1998), p. 227-243.

NATURAL ANALOGS AND PERFORMANCE ASSESSMENT FOR GEOLOGIC DISPOSAL OF NUCLEAR WASTE

WILLIAM M. MURPHY
Center for Nuclear Waste Regulatory Analyses, Southwest Research Institute,
San Antonio, TX 78228-0510 USA

ABSTRACT

The use of natural analog studies in performance assessments has been widely discussed and debated, but its accomplishment has been limited. Given recognized uncertainties and challenges, scientific contributions to performance assessments and support for the validity of performance assessment models are valuable from all possible sources, including natural analog studies. The conceptual basis for geologic disposal of nuclear waste and for performance assessments relies on scientific expertise based largely on studies of natural systems analogous to possible repository systems, i.e., natural analogs. Natural analog studies offer contributions to model validation based both on inductive and deductive reasoning. The utility of analog studies as a deductive tool in performance assessment is enhanced by specificity of the analog system to the repository system. As geologic sites are selected and repository designs detailed, the use of analog data in supporting deductive performance assessments should increase. Consideration of Yucca Mountain for the proposed US high level nuclear waste repository affords site specificity conducive to applications of natural analog data in performance assessment. The primary use of Peña Blanca natural analog data in recent Yucca Mountain performance assessments stems from observations of mineral products formed by alteration of natural uraninite, an analog of spent fuel. Alternate performance assessment source term models based on the Peña Blanca oxidation rate model and the schoepite solubility model yield lower, yet comparable estimated doses than the base case model in the NRC performance assessment for Yucca Mountain.

INTRODUCTION

The use of natural analog studies in performance assessments has been widely discussed and debated, but its accomplishment has been limited. The object of this paper is to illustrate how natural analog contributions to performance assessments advance the scientific basis for nuclear waste management.

Natural analog studies for nuclear waste disposal focus on numerous scientific issues, including the properties of natural materials as analogs of waste forms [1-7], the properties, longevity, and isolation of archaeological materials and their alteration products [8-10], alteration of repository host rocks due to potential hydrothermal effects or interactions with engineered materials [11-13], speciation and geochemical transport phenomena in geologic media [14-17], and the probability and consequences of volcanic activity [18]. In principle, analog studies can be used to develop conceptual models, provide or confirm parameters and functional relations used in models, test model assumptions, evaluate predictive capabilities, and define uncertainties in predictions for performance assessments.

Performance assessment is a safety assessment and management methodology used in the context of nuclear waste disposal (e.g., [19]). Various approaches to performance assessment have been taken, e.g., deterministic vs. probabilistic modeling, generic vs. specific sites, overall system performance vs. specific subsystem or multiple barrier performance. The process of performance assessment typically involves a description of the system, a specification or screening of scenarios, an estimation of consequences, an evaluation of uncertainties in the estimations, and a determination of acceptability based on safety criteria [19]. The technical complexities and uncertainties of long term geologic disposal of nuclear waste make performance assessment a difficult task and motivate diverse lines of inquiry and scientific support. Given recognized uncertainties and challenges, scientific contributions to performance assessments and support for the validity of performance assessment models are valuable from all possible sources, including natural analog studies.

Mat. Res. Soc. Symp. Proc. Vol. 608 © 2000 Materials Research Society

Performance assessment practitioners have been skeptical of the role of natural analog contributions. Eisenberg [19] noted that analog studies had "not been used to validate performance assessment methods in a decisive fashion." Sagar and Wittmeyer [20] concluded that "while [natural analog] data may be suitable for testing the basic laws governing flow and transport, there is insufficient control of boundary and initial conditions and forcing functions to permit quantitative validation of complex, spatially distributed flow and transport models." Alexander and McKinley [21] observed that "... the value of analogues in testing or validating models of radionuclide transport is widely recognized and many studies have been carried out with this aim in mind. In general, however, such studies have tended to be rather poorly focussed with little relevance to the models used in performance assessment." These challenges have restricted applications of natural analog studies in performance assessments and provided motivation to focus natural analog studies to promote more effective contributions.

Recognizing these challenges, this paper offers a current perspective on the use of natural analogs in performance assessments for geologic disposal of nuclear waste. Some philosophical aspects of the effective application of natural analog studies in performance assessments are addressed, and applications of natural analog studies in recent performance assessments are reviewed with an emphasis on the proposed US high level nuclear waste repository at Yucca Mountain, Nevada.

CONCEPTUAL BASIS FOR GEOLOGIC DISPOSAL OF NUCLEAR WASTE

The conceptual basis for geologic disposal of nuclear waste and for performance assessments relies on scientific expertise regarding the stability of geologic environments, the potential stability of waste containment structures and waste forms in geologic media, and the migration of radionuclides in geologic systems. This geologic expertise is based largely on studies of natural systems analogous to possible repository systems, i.e., natural analogs. The tie to natural systems and materials has been recognized since early considerations of nuclear waste management. For example, in 1953, Hatch [22] appealed to studies of natural materials as potential waste forms for long term isolation of nuclear wastes: "... an investigation has been under way ... to develop a method suitable for the ultimate disposal of radioactive wastes. The immediate purpose is to utilize certain desirable properties of one of the natural clays, *montmorillonite*, in such a way that the radioactive element can be combined within the clay mineral structure, and there fixed in place." Montmorillonite was recognized as a promising material for a variety of reasons among which "It is a product of weathering of volcanic rock; hence in its raw state it would be considered one of the highly stable substances of the earth." Geologic materials characteristics and stability were invoked in the 1957 National Academy of Sciences Committee on Waste Disposal recommendations regarding nuclear waste disposal: "The most promising method of disposal of high level waste at the present time seems to be in salt deposits. The great advantage here is that no water can pass through salt. Fractures are self-sealing. Abandoned salt mines or cavities especially mined to hold waste are, in essence, long-enduring tanks" [23]. The viability of this perspective on the role of geologic systems expertise in nuclear waste disposal is represented in the summary by Miller et al. [10]: "... the whole conceptual basis of geologic disposal continues to be based on the natural evidence for the long-term stability and sluggish evolution of deep geological environments."

Although engineered components of systems for geologic disposal of nuclear waste are initially well designed and controlled, their long term evolution is uncertain. In contrast, observations of natural systems that have evolved over time periods relevant to isolation of nuclear waste (e.g., $10^3 - 10^6$ years) can permit relatively reliable long term predictions. Birchard and Alexander [24] noted in 1983 that "Laboratory or engineering analysis has limited value for long-term prediction because important processes that occur on time scales much longer than a human lifetime may not be measurable on a laboratory time scale. The predictions that we can be the most confident in, will be those of natural processes which have a long (thousands of years or more), well-defined, and stable record." Stable characteristics contribute also to reliability of predictions for repository sites as well as for applicability of data from analog systems. Miller et al. [10] summarized the challenge and the essential role of natural systems studies in the context of long term performance predictions: "Despite the fact that explaining the behaviour of the natural environment unambiguously is extremely difficult owing to the many potentially important parameters involved and the ill-constrained boundaries of the system under study, interest has grown steadily in the use of analogues in performance assessment. Although such poorly-controlled 'experiments' are somewhat distasteful to the theoretical or engineering instincts of many performance assessors, it has become increasingly clear that there are really no alternative tests of long-term predictions." Indeed, natural analog systems, with their inherent uncertainties, may provide the best data for understanding the long term behavior of certain engineered materials. For example, native copper deposits (e.g., [25]) and archeological materials (e.g., [10]) provide data for the long term behavior of engineered materials in geologic media.

The study of natural analogs has developed in part as a method to provide confidence in the validity of methods and models for predictions of long term repository performance. The complexities of geologic repository systems and the long time scales of model predictions require studies at the frontiers of experimental capabilities and theoretical extrapolations. Natural analogs offer glimpses of the evolution of complex systems analogous to geologic repository systems on appropriate time scales. Although time scales for evaluation of geologic disposal of nuclear waste are long relative to laboratory or engineering time scales or human history, they are short relative to geologic time, requiring fresh perspectives from both engineers and geologists. Expectations for validation of performance assessment models for long term geologic isolation of nuclear waste must be tempered by realistic evaluations of uncertainties. Convergence of multiple lines of evidence, including those from natural analog studies, provides a scientific basis for performance assessment model validation.

Analog studies in support for repository modeling is recognized by the Nuclear Regulatory Commission in its original high level nuclear waste regulation (10 CFR Part 60) [26] and in its proposed regulation specific to the proposed repository site at Yucca Mountain (10 CFR Part 63) [27]. "Analyses and models that will be used to assess performance of the geologic repository shall be supported by using an appropriate combination of such methods as field tests, in-situ tests, laboratory tests that are representative of field conditions, monitoring data, and natural analog studies." "Any performance assessment used to demonstrate compliance ... shall ... provide the technical basis for models used in the performance assessment such as comparisons made with outputs of detailed process-level models and/or empirical observations (e.g., laboratory testing, field investigations, and natural analogs)" [27].

INDUCTIVE AND DEDUCTIVE MODEL VALIDATION USING NATURAL ANALOGS

Natural analog studies offer contributions to model validation based both on inductive and deductive reasoning. The logic of scientific discovery, observations of nature, generalizations regarding the long term stability of geological systems and geological materials, and conceptualization of systems with the potential for geologic isolation provide an inductive approach to support performance assessments for geologic disposal of nuclear waste. Eisenberg [19] subdivided analog contributions to validation of performance assessment models among component studies and system studies and noted that research focused principally on components, e.g., waste forms, engineered barriers, and sites. Eisenberg observed that use of analog data in performance assessment was largely in conceptual model development and identification of potential processes. These contributions are principally inductive, drawing on the spectrum of geologic expertise and numerous specific observations of nature to develop and validate general conceptual models for geologic disposal. Ewing and Jercinovic [28] liken the use of analogy to inductive reasoning as processes of collection and correlation of observations. Whereas analogies provide inferences of relationships, complete inductions confirm relationships based on understanding of causality. Arguments based on analogy are strengthened by the pertinence of the variables used to characterize analogous phenomena [28]. Although analogies do not provide proofs, which are recognized as impossibilities for long term performance assessments for geologic disposal of nuclear waste, analogies between natural and repository model systems contribute to confidence in models developed for performance assessments. Identification of relevant processes by observations of examples in natural systems and recognition of geologic sites and materials with sufficient stability to provide confidence in long term isolation beyond the temporal limits of human experience generally provide qualitative contributions to model validation. These inductive applications of analog studies offer an intuitive and appealing approach to address recognized and potentially daunting problems posed by the time scale and complexity of geologic disposal of nuclear waste.

Quantitative estimates of repository performance measures, which are commonly required to meet regulatory criteria of safety assessments, depend on a deductive logic of argumentation. Commonly, sets of premises and supporting data are assembled, and consequences are deduced quantitatively using rules of mathematics and statistics, permitting quantitative evaluations of uncertainties and sensitivities. Although desirable, the quantitative use of natural analog data in performance assessment has proven to be challenging. Natural system and site characterization data are abundant, but these data are typically difficult to apply directly in calculations of projected repository performance. Difficulties arise because of lack of specificity of data to repository systems and the incomplete geologic record including uncertain initial and boundary conditions for geologic processes. Nevertheless, quantitative applications of natural analog data in performance assessment can and should be made, and are a focus of the following discussion. Finally, comparison of performance assessment deductions to natural analog data can provide a check on realism or conservatism of the performance assessment results.

SPECIFICITY IN NATURAL ANALOG RESEARCH

The utility of analog studies as a deductive tool in performance assessment is enhanced by specificity of the analog system to the repository system. General principles of geologic isolation can be gleaned from observations and generalizations regarding many systems. In contrast, use of analog data for deductive applications depends on similarity of the analogous system to the system of interest. In this sense the analog system is a natural experiment providing data and quantitative relations that can be used directly in performance assessment predictions. Experimental studies are most useful when they represent closely the system of interest or when they are designed to derive fundamental parameters. If they can be identified, natural analog systems that are highly specific to geologic repository systems are particularly well suited to support performance assessment because they capture effects of the complexity of natural systems and the coupling of processes over long time periods which are inaccessible in the laboratory. Closely analogous natural systems may provide more certainty in understanding the coupled, long term evolution of a complex system than provided by consideration of experimental studies which are better constrained but which fail to reflect complexities.

The scope of natural analog studies for nuclear disposal is broad and many contributions are possible. However, many studies of natural phenomena have little or no relevance to practical aspects of waste materials, containment materials, geologic repository sites, or performance assessments for nuclear waste disposal, and are unjustified as natural analog studies for deductive performance assessments. Many natural analog studies have focused on natural uranium deposits (e.g., Oklo, Alligator Rivers, Cigar Lake, Peña Blanca, Poços de Caldas, Palmottu; see summaries in [10,29]). The stability and alteration behavior of uranium and thorium minerals and the hydrochemical transport of these elements and their radioactive decay series products are of general interest in nuclear waste disposal. The Oklo deposits are of unique geophysical and geochemical interest because these concentrated deposits of ^{235}U rich ore resulted in natural fission reactions about 2 billion years ago, an extraordinary natural analog to fission reactors and their products. These unique natural occurrences permit study of the long term behavior of nuclear activation and fission products (and their daughters) in geologic settings over geologic time. However, because of specificity, the results may be only of general interest and not amenable to deductive performance assessment applications. Petit [30] addresses this problem, stating forcefully: "The extreme heuristic utility of Oklo for geochemists should not disguise the fact that this site presents numerous differences with a real disposal site (age, materials, rock type, thermal history, etc.). All direct and global comparisons are therefore illusory, even delusive" (author's translation). Oversby [31] expands on these observations regarding applications of data from Oklo. Although natural systems studies provide an essential basis for development of inductive conceptual models for performance assessment, if natural systems are not specifically analogous, their relevance for deductive uses in performance assessment is limited or impossible. Clearly, specification of the characteristics of a repository site and engineered repository system enhances the possible utility of analog studies. As repository systems acquire definition and as analog studies are conducted that address specific technical aspects of waste isolation in those systems, data and models derived in the interpretation of natural analog systems can be used increasingly in deductive, quantitative performance assessments as a complement to laboratory based data and models. Uncertainties in interpretations of natural systems are not the only limitation in applications of natural analog studies to performance assessment. Ill defined repository sites and designs also restrict potential contributions to performance assessments from natural analog studies.

POTPOURRI OF ANALOG DATA IN PERFORMANCE ASSESSMENTS

Miller et al. [10] summarized applications of natural analog studies in performance assessment for several European programs from the period of the 1980s to early 1990s concluding that little acknowledged use of analog data had been made. They identified one study in which the copper container corrosion pitting factor, i.e., the ratio of pit depth to general corrosion depth, was determined for the Swedish KBS-3 performance assessment by observations of a variety of materials including analogous copper artifacts.

Solubility and speciation thermodynamic data bases with potential applications in performance assessment have been tested using data from sites studied as natural analogs (e.g., [32]). Generally, this type of exercise falls in the category of natural systems studies. Many thermodynamic data used in performance assessments have some basis in comparisons to natural system results.

Activity in analogous volcanic fields [33,34] have been used to test volcanic probability models for the proposed Yucca Mountain repository site [35], which have been used in performance assessments. Performance assessment models used to estimate dose due to volcanic eruptions at the proposed repository at Yucca Mountain [18] rely on data and parameter distributions gathered at active cinder cones analogous to volcanoes in the vicinity of Yucca Mountain [36].

Natural analog data have been considered, but generally discounted or neglected, in determining matrix diffusion parameters for performance assessments. At best, analog data for matrix diffusion have been used to justify conservatism of parameters based on other analyses (e.g., [10, 37]). Similarly, rates of alteration of natural bentonites have been used to justify conservatism in values used in safety analyses of backfill material [e.g., [37].

PEÑA BLANCA ANALOG INFORMATION IN PERFORMANCE ASSESSMENTS

Selection of Yucca Mountain as the only geologic site under consideration for the proposed US high level nuclear waste repository affords site specificity conducive to applications of natural analog data to performance assessment. Yucca Mountain is composed of silicic tuffs which are variably welded, nonwelded, bedded, fractured, vitric, devitrified, and zeolitized. The proposed waste emplacement horizon is in fractured welded tuffs above the water table in the unsaturated zone where the geochemical environment is oxidizing. Pore water constitutes about ten percent of the unsaturated rock by volume. Water chemistry from the unsaturated zone at Yucca Mountain is dilute, oxidizing, silica rich, and dominated by Ca-Cl and Ca-SO_4 above the proposed waste emplacement horizon and Na-HCO_3 below this horizon (see interpretation in [38]). Irradiated uranium dioxide fuel is the predominant nuclear waste form proposed for disposal at Yucca Mountain.

The Nopal I uranium deposit at Peña Blanca, Chihuahua, Mexico, is recognized as a natural analog of the proposed nuclear waste repository at Yucca Mountain, Nevada (e.g., [14,39]). Geologic, climatic, hydrologic, and geochemical characteristics of the Nopal I site are remarkably similar to Yucca Mountain, and a natural deposit of uraninite (an analog of uranium dioxide fuel) and its alteration products are exposed in the oxidizing, unsaturated zone at the site. Natural analog studies at Nopal I have focused primarily on issues of the source term for radionuclide release [7,40,41] and on the timing and mechanisms of radionuclide transport [16,42-44].

Several series of performance assessments for the proposed repository at Yucca Mountain have been conducted by the US Department of Energy (DOE), the US Nuclear Regulatory Commission (NRC), and the Electric Power Research Institute (EPRI). The latest in these series of performance assessments [45-

50] all draw on information from natural analog studies at Peña Blanca. A summary of use of Peña Blanca information in these studies illustrates current applications of natural analog data in repository performance assessments in the US high level waste disposal program.

Mineral Paragenesis

The primary use of Peña Blanca natural analog data in Yucca Mountain performance assessments stems from observations of mineral products formed by alteration of natural uraninite. This mineral paragenesis corresponds closely to that observed in long term laboratory tests of the alteration of uranium dioxide designed to simulate conditions representative of the proposed repository [51-53] (Figure 1). Laboratory test results on a time scale of years and natural analog data for a time scale of millions of years bracket the relevant time period for isolation of high level nuclear waste. Correspondence of results from these two independent sources, both closely analogous to proposed repository conditions, at bracketing time scales, provides confidence in a general model for uranium dioxide waste form alteration in the Yucca Mountain environment and on a time scale relevant to performance assessment.

The DOE Total System Performance Assessment - Viability Assessment (TSPA-VA) [45] and its supporting technical documents [46] make qualitative reference to the occurrence of uranyl (oxidized uranium) minerals as the dominant form of uranium at Peña Blanca in support of repository performance modeling. The observation of uranium mineralogy is used primarily to sustain the relevance of experimental studies of spent fuel dissolution. The role of the properties of secondary minerals in repository performance is noted particularly in the TSPA-VA because crystallographic theory and

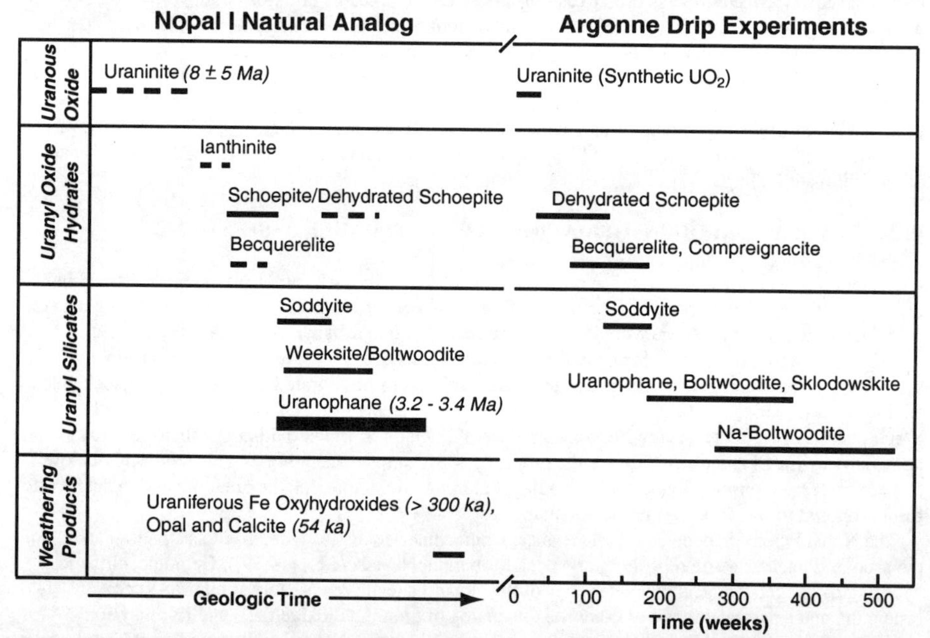

Figure 1 Paragenesis of uranium minerals at the Nopal I uranium deposit at Peña Blanca (modified from [7]) and in 90°C laboratory dripping experiments conducted at Argonne National Laboratory [53]. Dashed, thin, and thick lines for Nopal I represent minor, abundant, and very abundant present occurrences, respectively. Nopal I geochronology is provided by: chemical uranium-lead age of uraninite [7]; uranium-lead isotope age for late forming uranophane [44]; uranium decay series age limit for Fe oxyhydroxides [64]; uranium decay series ages for opal and calcite [7].

experimental data indicate that neptunium is incorporated in the structures of these relatively stable minerals [54,55]. Neptunium is consistently a major contributor to predicted radioactive releases at long times in DOE performance assessments for the proposed Yucca Mountain repository [45,56,57].

In the Technical Basis Document for the DOE TSPA-VA [46], the paragenetic development of secondary phases at Peña Blanca was compared to that observed in experimental studies of synthetic UO_2 and spent fuel (i.e., the relations illustrated in Figure 1). Based in part on these observations, the secondary uranyl minerals schoepite, soddyite, uranophane, and Na-boltwoodite are included in reactive transport simulations for TSPA-VA to predict the evolution of the chemistry of water resulting from geochemical interactions in emplacement drifts [46]. Also, a degree of confidence is provided for use of reaction rates adopted in reactive transport models for radionuclide release that are based in part on the aqueous solubility and kinetic properties of secondary uranyl minerals observed to occur in analogous natural settings such as Peña Blanca [46]. "The good match in paragenetic sequence between simulation results and laboratory and field observations suggests that the model qualitatively reproduces the replacement relations among uranyl minerals. It suggests the thermochemical and kinetic data used in the model can correctly reflect the phase relations, at least qualitatively" [46]. Similarity of secondary mineral development between the laboratory and natural field occurrences led to the conclusion that "conditions in the experimental tests have replicated an environment that may be representative of that occurring during uraninite alteration in natural oxidized systems." It was concluded that "These comparisons provide confidence that the degradation of these waste forms over time is well represented by the models, and that the long-term prediction of alteration rates is reasonably realistic" [46].

In the recent EPRI performance assessment [50] it was noted that dissolution of nuclear waste forms in the Yucca Mountain environment will lead to precipitation of more stable secondary uranyl minerals, citing Peña Blanca data as an example. Stability of these minerals was used as an argument in the performance assessment to propose lower source term concentrations of released radionuclides than provided by experimental solubility studies.

Source Term Modeling

In the recent NRC Total-system Performance Assessment (TPA) for the proposed repository at Yucca Mountain [47-49] several alternate source term models were provided. One model was based on the assumption that radionuclides from the spent fuel matrix become incorporated in oxidized secondary uranium phases represented by schoepite [41]. In the absence of distribution data, proportions of radionuclides in fuel matrix and schoepite were postulated to be equal, which may be unconservative for some species. Radionuclide releases were then modeled to occur in proportion to the solubility limited dissolution of schoepite. Schoepite solubility was calculated in the model as a function of temperature, pH, and aqueous carbonate and uranyl speciation [58]. The selection of secondary schoepite as a likely source term controlling phase was justified in part by its common occurrence in natural geologic settings analogous to the Yucca Mountain site such as Peña Blanca.

For another alternate source term model in the NRC TPA, the release rate of matrix components from spent nuclear fuel was assumed to be limited by the maximum average oxidation rate of uraninite estimated for the Nopal I uranium deposit at Peña Blanca scaled to the proposed repository inventory [41]. Radionuclide releases based on the maximum average oxidation rate of uraninite were assumed to place a conservative upper limit on the average release rate of matrix components from the waste form because sequestration of radionuclides in secondary, relatively stable uranyl minerals were disregarded in this model. Peña Blanca data indicate that uraninite oxidation occurred rapidly relative to the rate of migration of uranium out of the system [40], which suggests that the maximum average oxidation rate is a conservative limit to release rates of matrix components. The maximum average oxidation rate of uraninite at the Nopal I deposit at Peña Blanca was estimated based on reported measurements of the mass of oxidized uranium remaining at the site, the minimum period of oxidation based on geochronological dating of late forming oxidized uranium minerals, and a maximum estimate of the amount of uranium that has been oxidized and subsequently removed from the uranium deposit by aqueous transport [40,59].

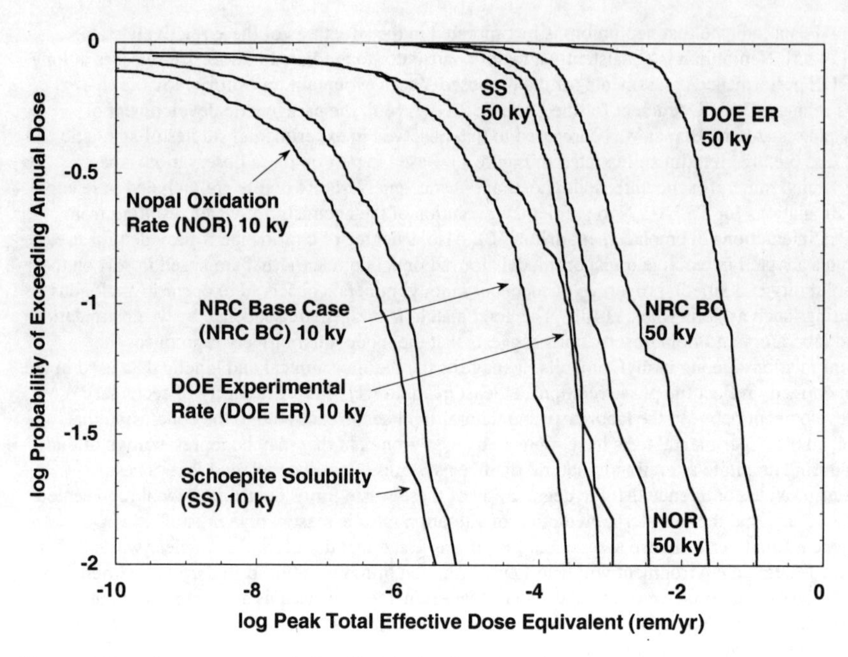

Figure 2 Complementary cumulative distribution functions for peak doses to a receptor population located 20 km from the proposed repository emplacement site over periods of 10,000 years and 50,000 years after repository closure (based on [41]). Curves represent the probability of exceeding the annual peak total effective dose equivalent given by the abscissa. Curves labeled SS, NOR, NRC BC, and DOE ER correspond to model results for the schoepite solubility, Nopal I oxidation rate, NRC base case, and DOE experimental rate source term models, respectively.

For the alternate NRC TPA source term models based on schoepite solubility and Nopal I oxidation rate, doses (with associated probabilities) were calculated for a critical group located 20 km from the waste emplacement site for periods of 10,000 and 50,000 years after repository closure [41] (Figure 2). For comparison, doses were also calculated using the same sets of assumptions, except for the matrix source term release, using the NRC base case model and the DOE model, both of which are based on laboratory data. All other aspects of the performance calculations were modeled in a fashion equivalent to that of the NRC base case source term model. Alternate source term formulations based on the Peña Blanca oxidation rate model and the schoepite solubility model yield lower doses than the base case model in the NRC TPA analyses. In the recent report on NRC performance assessment sensitivity and uncertainty analyses, comparisons were made of performance assessment results using the natural analog oxidation rate model to results computed using several other alternate conceptual models [49]. These comparisons showed generally that the natural analog oxidation rate model led to predictions of dose effects that are low relative to most alternate models considered, and concluded that "taking credit for phenomena such as ... lower release rates based on observations of natural analogs could result in considerably smaller doses" [49]. Lower estimated doses provide some confidence that base case model results are conservative. Comparable results for alternate source term models (i.e., within a few orders of magnitude) are a measure of convergence of alternate approaches to making long term estimates [60].

Transport Phenomena

Although it is suggested in the DOE TSPA-VA that information from Peña Blanca may bear on radionuclide transport issues, there is no apparent use of Peña Blanca transport data in the VA.

The importance of flow and radionuclide transport in fractures is receiving increasing recognition for performance of the proposed Yucca Mountain system (e.g., [46,47]) and is supported by the distribution of uranium surrounding the Nopal I deposit at Peña Blanca [16,43]. In the NRC review [61] of the DOE TSPA-1995 [57], the distribution of uranium in and around fractures at Nopal I [16] was used, along with other arguments, to question the effectiveness of matrix diffusion as a retardation mechanism invoked in performance modeling of radionuclide transport.

Sorption/coprecipitation on minor phases such as transition metal oxides is receiving recognition in performance modeling for the Yucca Mountain system [62] and is supported by Peña Blanca data. The association of uranium with fracture filling iron oxyhydroxides at Peña Blanca [43] supports laboratory evidence of sorption on oxyhydroxides [62,63], and strengthens arguments for the importance of sorption on oxide mineralization in fractures as a retardation mechanism for radionuclide transport.

Radionuclide transport at Peña Blanca has received extensive study and is recognized to be analogous to potential radionuclide transport in the Yucca Mountain system [14,16,43,52]. These studies contribute to conceptual modeling for performance assessments for Yucca Mountain. However, direct use of radionuclide transport data from the Peña Blanca analog system in performance assessments are absent at present. Studies of transport at Peña Blanca using uranium decay series isotopes [44,64], which provide a geologic clock, reflect unrealized analog potential for applications in performance assessments.

Dissimilarities Between Yucca Mountain and Peña Blanca

Despite remarkable similarities in geology, hydrology, climate, and waste analog material between the Peña Blanca analog and the proposed Yucca Mountain repository, and the value of investigating processes related to geologic isolation for nuclear waste on a geologic time scale, there are important differences between the natural system at Peña Blanca and the proposed repository system, and uncertainties remain concerning the evolution of the Peña Blanca system. Peña Blanca uranium deposits probably had a water saturated episode of ore genesis, sulfur species were important components during and after the period of primary mineralization at Nopal I, and aqueous conditions became acidic due to pyrite oxidation (e.g., [7,59]). Such conditions are unlikely to develop in the unsaturated and sulfide poor environment of the proposed repository. In contrast to the Peña Blanca system, the proposed repository will have engineered containment facilities and strong radiation effects. In comparison, scarce remaining old (e.g., 8±5 Ma) uraninite persists at Nopal I probably because of its physical isolation from the oxidizing environment by natural silica cementation. In addition, minor and trace components and included phases in spent fuel are dissimilar to those in primary uraninite from Peña Blanca.

In contrast to performance models for the evolution of the proposed repository at Yucca Mountain, the evolution of the Peña Blanca system was apparently characterized by episodic hydrogeochemical events as indicated by limited radiometric dating. Primary uraninite deposition occurred at approximately 8 Ma [7]; an oxidation event is indicated by dates on uranophane at approximately 3 Ma [44]; data presented by Murrell et al. [64] indicate chemical transport occurring at greater than 300 ka based on uranium decay series ages of Fe oxyhydroxides; stages of mobilization, precipitation, and remobilization over the past hundreds of thousands of years are indicated by uranium decay series data from Pickett and Murphy [44]; and a late stage mobilization of uranium and precipitation of secondary calcite and opal is indicated by uranium decay series dating of these phases at approximately 54 ka [7] (Figure 1). Dates on selected samples, which cannot represent the complete geologic history, and definition of the time scale(s) of episodicity are problematic. Nevertheless, recognition of episodicity is an inductive conclusion from the analog data which may be beneficial in constructing future performance assessments. Performance assessment models and interpretations of natural analog data that neglect the episodic nature of hydrogeochemical processes can lead to estimates of average rates and average doses far below maximum rates and doses.

CONCLUDING REMARKS

Natural analog studies are widely recognized in performance assessments for geologic disposal of nuclear waste. Applications have been primarily qualitative, providing support for conceptual models and justification for the relevance of experimental results. Quantitative applications of analog data in performance assessments exist (e.g., [41]). Although these applications are based on quantifiable geologic and geochemical data, they are attended by approximations and uncertainties associated with understanding natural system evolution over long periods of time. Similarly, extrapolations of laboratory studies and theoretical predictions have inherent uncertainties and limitations. Uncertainties in use of analog data may be no greater than those accumulated in performance assessment models using data from well constrained laboratory systems that may fail to capture the complexity, coupling, and long term effects relevant to geologic isolation of nuclear waste.

No system is perfectly analogous to a geologic repository. However, specificity in definition of repository systems and similarity of analog systems to repository systems strengthen applicability of analog data to performance assessments. As repository sites and designs are defined and specific analog systems identified, the use of natural analog data in performance assessments should increase. Confidence in predictive models and in evaluations of the safety of geologic disposal of nuclear waste should be enhanced through multiple lines of investigation, including studies of natural analogs.

ACKNOWLEDGMENTS

Invitation to present this paper by organizers of the Scientific Basis for Nuclear Waste Management XXIII is gratefully acknowledged. R.B. Codell provided assistance in performance calculations. J.D. Prikryl helped with petrographic interpretations. Constructive reviews by D.A. Pickett, M.P. Miklas, Jr., R.C. Ewing, B.W. Leslie, and B. Sagar are noted appreciatively. This paper was prepared to document work performed by the Center for Nuclear Waste Regulatory Analyses for the Nuclear Regulatory Commission (NRC) under Contract No. NRC-02-97-009. The study was performed on behalf of the NRC Office of Nuclear Material Safety and Safeguards, Division of Waste Management. The paper is an independent product and does not necessarily reflect the views or regulatory position of the NRC.

REFERENCES

[1] R.C. Ewing, in *Scientific Basis for Nuclear Waste Management* (Mater. Res. Soc. Proc. **1**, Pittsburgh, PA, 1979) 57-68.

[2] B. Grambow, M.J. Jercinovic, R.C. Ewing, and C.D. Byers, in *Scientific Basis for Nuclear Waste Management* (Mater. Res. Soc. Proc. **50**, Pittsburgh, PA, 1985) pp. 263-272.

[3] W. Lutze, G. Malow, R.C. Ewing, M.J. Jercinovic, and K. Keil, Nature **314**, 252 (1985).

[4] R.C. Ewing, B.C. Chakoumakos, G.R. Lumpkin, T. Murakami, R.B. Greegor, and F.W. Lytle, Nucl. Instru. Meth. Phys. Res. **B32**, 487 (1988).

[5] R.J. Finch and R.C. Ewing, Radiochim. Acta **52/53**, 395 (1991).

[6] J. Janeczek and R.C. Ewing, in *Scientific Basis for Nuclear Waste Management* (Mater. Res. Soc. Proc. **257**, Pittsburgh, PA, 1992) pp. 497-504.

[7] E.C. Pearcy, J.D. Prikryl, W.M. Murphy, and B.W. Leslie, Appl. Geochem. **9**, 713 (1994).

[8] I.J. Winograd, USGS Open-File Report 86-136 (1986).

[9] I. Neretnieks, in *Natural Analogue Working Group First Meeting, Brussels, November 1985*, edited by B. Côme and N.A. Chapman (CEC Nucl. Sci. Tech. EUR 10361, 1986) pp. 32-36.

[10] W. Miller, R. Alexander, N. Chapman, I. McKinley, and J. Smellie, *Natural Analogue Studies in the Geological Disposal of Radioactive Wastes* (Elsevier, Amsterdam, 1994) 395 pp.

[11] S. Levy, in *Scientific Basis for Nuclear Waste Management VII*, edited by G.L. McVay (Mater. Res. Soc. Proc., Pittsburgh, PA, 1984) 959-966.

[12] H.N. Khoury, E. Salameh, I.P. Clark, P. Fritz, A.E. Milodowski, M.R. Cave, W. Bajjali, and W.R. Alexander, J. Geochem. Explor. **46**, 117 (1992).

[13] W. Matyskiela, Geology **25**, 1115 (1997).
[14] P. Ildefonse, J.-P. Muller, B. Clozel, and G. Calas, Engin. Geol. **29**, 413 (1990).
[15] J. Suksi, T. Ruskeeniemi, A. Lindberg, and T. Jaakkola, Radiochim. Acta **52/53**, 367 (1991).
[16] E.C. Pearcy, J.D. Prikryl, and B.W. Leslie, Appl. Geochem. **10**, 685 (1995).
[17] J.A.T. Smellie and F. Karlsson, Engin. Geol. **53**, 193 (1999).
[18] M.S. Jarzemba, Nucl. Tech. **118**, 132 (1997).
[19] N.A. Eisenberg, Chem. Geol. **55**, 189 (1986).
[20] B. Sagar and G.W. Wittmeyer, in *Workshop on the Role of Natural Analogs in Geologic Disposal of High-Level Nuclear Waste*, edited by L.A. Kovach and W.M. Murphy (Nuclear Regulatory Commission, NUREG/CP-0147, Washington, DC, 1995) pp. 21-28.
[21] W.R. Alexander and I.G. McKinley, in *Proceedings of the Fourth Natural Analogue Working Group Meeting*, edited by B. Côme and N.A. Chapman (CEC EUR 10314, 1991) pp. 119-151.
[22] L. Hatch, Amer. Sci. **41**, 410 (1953).
[23] National Academy of Sciences, *The Disposal of Radioactive Waste on Land* (NAS-NRC Pub. 519, Washington, DC, 1957).
[24] G.F. Birchard and D.H. Alexander, in *Scientific Basis for Nuclear Waste Management* (Mater. Res. Soc. Proc. **15**, Pittsburgh, PA, 1983) pp. 323-329.
[25] D.P. Crisman and G.K. Jacobs, *Native Copper Deposits of the Portage Lake Volcanics, Michigan: Their Implication with Respect to Canister Stability for Nuclear Waste Isolation in the Columbia River Basalts Beneath the Hanford Site, Washington* (Rockwell International, Hanford, WA, RHO-BW-ST-26 P, 1982).
[26] Code of Federal Regulations 10, Part 60, US Government Printing Office, Washington, DC (1998).
[27] Nuclear Regulatory Commission 10 CFR Part 19 et al. Federal Register 64, #34, 8640-8679 (1999).
[28] R.C. Ewing and M.J. Jercinovic, in *Scientific Basis for Nuclear Waste Management X*, edited by J.K. Bates and W.B. Seefeldt (Mater. Res. Soc. Proc. **84**, Pittsburgh, PA, 1987) pp. 67-83.
[29] E.C. Pearcy and W.M. Murphy, *Geochemical Natural Analogs Literature Review* (Center for Nuclear Waste Regulatory Analyses, CNWRA 90-008, San Antonio, TX, 1991).
[30] J.-C. Petit, *Le Stockage des Déchets Radioactifs: Perspective Historique et Analyse Sociotechnique*, Doctoral Thesis, Ecole Nationale Supérieure des Mines de Paris, 1993.
[31] V.M. Oversby, in *Scientific Basis for Nuclear Waste Management XXIII* (Mater. Res. Soc. Proc., 2000) this volume.
[32] J. Bruno, J.E. Cross, J. Eikenberg, I.G. McKinley, D. Read, A. Sandino, and P. Sellin, *Testing of Geochemical Models in the Poços de Caldas Analogue Study* (SKB Technical Report TR 90-20 SKB, Stockholm, 1990).
[33] C.D. Condit and C.B. Connor, GSA Bull. **108**, 1225 (1996).
[34] C.B. Connor and F.M. Conway, in *Encyclopedia of Volcanoes* (Academic Press, 2000) pp. 381-393.
[35] C.B. Connor and B.E. Hill, J. Geophys. Res. **100**(B6), 10,107 (1995).
[36] B.E. Hill, C.B. Connor, M.S. Jarzemba, P.C. La Femina, M. Navarro, and W. Strauch, GSA Bull. **110**, 1231 (1998).
[37] I.G. McKinley, in *Risk Analysis in Nuclear Waste Management*, edited by A. Saltelli et al. (ECSC, EEC, EAEC, Brussels, 1989) pp. 359-396.
[38] L. Browning, W.M. Murphy, B.W. Leslie, and W.L. Dam, in *Scientific Basis for Nuclear Waste Management XXIII* (Mater. Res. Soc. Proc., 2000) this volume.
[39] W.M. Murphy, E.C. Pearcy, and P.C. Goodell, in *Fourth Natural Analogue Working Group Meeting and Poços de Caldas Project Final Workshop*, edited by B. Côme and N.A. Chapman (Commission of European Communities, EUR 13014 EN, 1991), pp. 267-276.
[40] W.M. Murphy and E.C. Pearcy, in *Scientific Basis for Nuclear Waste Management XV*, edited by C. Sombret (Mater. Res. Soc. Proc. **257**, 1992) pp. 521-527.
[41] W.M. Murphy and R.B. Codell, in *Scientific Basis for Nuclear Waste Management XXII*, edited by D.J. Wronkiewicz and J.H. Lee (Mater. Res. Soc. Proc. **556**, Warrendale, PA, 1999) pp. 551-558.
[42] J.-P. Muller, B. Clozel, P. Ildefonse, and G. Calas, Appl. Geochem. Sup. Issue **1**, 205 (1992).

[43] J.D. Prikryl, D.A. Pickett, W.M. Murphy, and E.C. Pearcy, J. Contam. Hydro. **26**, 61 (1997).
[44] D.A. Pickett and W.M. Murphy, in *Seventh EC Natural Analogue Working Group Meeting*, edited by H. von Maravic and J. Smellie (European Commission, EUR 17851 EN, 1997) pp. 113-122.
[45] Department of Energy, *Viability Assessment of a Repository at Yucca Mountain Total System Performance Assessment* (DOE/RW-0508, v. **3**, Department of Energy, Las Vegas, NV, 1998).
[46] TRW, *Total System Performance Assessment - Viability Assessment (TSPA-VA) Analyses Technical Basis Document* (TRW Environmental Safety Systems Inc., Las Vegas, NV, B00000000-01717-4301-00004 Rev. 01, 1998).
[47] CNWRA, *Total-system Performance Assessment (TPA) Version 3.2 Code: Module Descriptions and User's Guide*, 1998.
[48] Nuclear Regulatory Commission, *NRC Sensitivity and Uncertainty Analyses for a Proposed HLW Repository at Yucca Mountain, Nevada, Using TPA 3.1 Volume 1: Conceptual Models and Data*, (NUREG-1668, **1**, 1999).
[49] Nuclear Regulatory Commission, *NRC Sensitivity and Uncertainty Analyses for a Proposed HLW Repository at Yucca Mountain, Nevada, Using TPA 3.1 Results and Conclusions* (NUREG-1668, **2**, 1999).
[50] Electric Power Research Institute, *Alternative Approaches to Assessing the Performance and Suitability of Yucca Mountain for Spent Fuel Disposal* (TR-108732, EPRI, Palo Alto, CA, 1998).
[51] W.M. Murphy and E.C. Pearcy, in *Fifth CEC Natural Analogue Working Group Meeting and Alligator Rivers Analogue Project (ARAP) Final Workshop*, edited by H. von Maravic and J. Smellie (Commission of the European Communities, EUR 15176 EN, 1994) pp. 219-224.
[52] B.W. Leslie, E.C. Pearcy, and J.D. Prikryl, in *Scientific Basis for Nuclear Waste Management XVI*, edited by C.G. Interrante and R.T. Pabalan (Mater. Res. Soc. Proc. **294**, Pittsburgh, PA, 1993) pp. 505-512.
[53] D.J. Wronkiewicz and E.C. Buck, in *Uranium: Mineralogy, Geochemistry and the Environment*, edited by P.C. Burns and R. Finch (Min. Soc. Am. Rev. Min. **38**, 1999) pp. 475-497.
[54] P.C. Burns, R.C. Ewing, and M.L. Miller, J. Nucl. Mater. **245**, 1 (1997).
[55] E.C. Buck, R.J. Finch, P.A. Finn, and J.K. Bates, in *Scientific Basis for Nuclear Waste Management XXI* edited by I.G. McKinley and C. McCombie (Mater. Res. Soc. Proc. **506**, Warrendale, PA, 1998) pp. 87-94.
[56] M.L. Wilson et al. *Total-System Performance Assessment for Yucca Mountain - SNL Second Iteration (TSPA-1993)* (Sandia National Laboratories, SAND93-2675, 1994).
[57] TRW, *Total System Performance Assessment - 1995: An Evaluation of the Potential Yucca Mountain Repository* (TRW Environmental Safety Systems Inc., Las Vegas, NV. B00000000-01717-2200-00136, Rev. 01, 1995).
[58] W.M. Murphy, in *Scientific Basis for Nuclear Waste Management XX*, edited by W.J. Gray and I.R. Triay (Mater. Res. Soc. Proc. **465**, Pittsburgh, PA, 1997) pp. 713-720.
[59] W.M. Murphy, E.C. Pearcy, and D.A. Pickett, in *Seventh EC Natural Analogue Working Group Meeting*, edited by H. von Maravic and J. Smellie (European Commission EUR 17851 EN, 1997) pp. 105-112.
[60] W.M. Murphy, D.A. Pickett, and E.C. Pearcy, in *8th EC Natural Analogue Working Group Meeting*, edited by H. von Maravic and J. Smellie (European Commission, 2000) in press.
[61] R.G. Baca and M.S. Jarzemba (editors) Detailed Review of Selected Aspects of Total System Performance Assessment - 1995. CNWRA Letter Report to NRC, 1997.
[62] E.L. Hardin, *Near Field/Altered Zone Models Report* (Lawrence Livermore National Laboratory, Livermore, CA, UCRL-1D-129179, 1998).
[63] C.-K.D. Hsi and D. Langmuir, Geochim. Cosmochim. Acta **49**, 1931 (1985).
[64] M.T. Murrell, S.J. Goldstein, and P.R. Dixon, in *8th EC Natural Analogue Working Group Meeting*, edited by H. von Maravic and J. Smellie (European Commission, 2000) in press.

USING INFORMATION FROM NATURAL ANALOGUES IN REPOSITORY PERFORMANCE ANALYSIS: EXAMPLES FROM OKLO

V. M. OVERSBY,
VMO Konsult, Karlavägen 70, SE-114 59 Stockholm, Sweden

ABSTRACT

The Oklo Natural Analogue Project, Phase II was conducted by an international consortium of researchers under the sponsorship of the European Union. An important objective of the project was to examine the potential for extracting quantitative information from the Oklo natural reactors that could be used in performance assessment calculations for high level nuclear waste repositories. This paper discusses the uses and limitations of natural analogues with respect to performance assessment of repositories, the types of data needed in performance assessment calculations, the particular example of spent fuel dissolution rates, and the types of data available from Oklo studies that might contribute to understanding spent fuel behavior in a geologic setting. The main conclusions of this assessment are that important qualitative information has been obtained from studies at Oklo that gives us confidence that major processes of importance to the long-term behavior of a high level waste repository have not been overlooked in previous performance assessment calculations, but that the Oklo natural analogues are very unlikely to provide quantitative data to use as input to PA calculations.

INTRODUCTION

The Oklo Phase II Project included as one of its activities an effort to integrate the field and laboratory studies and the project modelling efforts with the Performance Assessment needs of the project supporters' organizations. This was done through the Performance Assessment Interface Group, which had three meetings during the course of the project. At the first meeting, held in conjunction with the first annual project meeting, the Technical Committee and a performance assessment representative from ENRESA discussed ways in which the project information might be better structured to communicate with PA workers. In December 1997, a second meeting of the PA Interface Group discussed three specific focus areas: spent fuel stability, radionuclide behavior at the near-field to far-field interface, and radionuclide retardation mechanisms through sorption and secondary mineral formation.

The most recent meeting of the PA Interface Group was held in Helsinki in conjunction with the 2nd annual Oklo Project meeting and included a wider group of participants, including interested parties who were from countries not participating actively in the Oklo Project. These meetings have opened a dialog between the project participants and those involved in the work of long term assessment of the behavior of a geologic repository. They have increased the understanding between the two groups of workers, but a great deal remains to be done before results from the Oklo Project can be included quantitatively in PA analyses.

In the discussion that follows, data related to the Oklo natural analogues have been taken from the summary report for the Oklo Phase I project [1], from the reports presented at the annual meetings of the Phase II project [2, 3], and from the Phase II project final report[4].

Mat. Res. Soc. Symp. Proc. Vol. 608 © 2000 Materials Research Society

POTENTIAL FOR USE OF RESULTS FROM OKLO IN PERFORMANCE ASSESSMENT OF RADIOACTIVE WASTE REPOSITORIES

Uses and Limitations of Natural Analogues

Research and development activities related to nuclear waste repositories have benefited from pre-existing knowledge in the fields of geology, geochemistry, and engineering. Geology and geochemistry were used to evaluate rock types for general suitability as repository host rocks, especially with respect to the interaction of ground water with the rock. The response of different rock types to heat and fluid circulation was known through studies of burial of sediments, diagenesis, regional metamorphism, and local thermal events such as volcanic intrusions. Mining engineering contributed information concerning the response of rock bodies to excavation. In this sense, the evaluation of natural geologic systems represented the start of "performance assessments" for waste repositories. We should, therefore, not be too surprised if further studies of geologic occurrences of specific minerals or ores fail to produce startling revelations.

Laboratory studies of potential waste products, container materials, and other engineered components that might be used in a radioactive waste repository form the framework of studies that will enable us to evaluate the potential future behavior of a waste disposal system. Even these studies have limitations, since we have to predict the future based on our understanding of past behavior of materials under conditions that only approximate those that will pertain in the repository. Our extrapolations are based on our understanding of the equilibrium chemical and physical behavior of the materials and on the kinetic restraints that may prevent equilibrium from being achieved.

In the laboratory, we can control the physical and chemical conditions of experiments rather well. In a natural setting, especially where we are using existing geologic evidence as our area of investigation, we have considerable uncertainties about fundamental parameters. We can, sometimes with difficulty, establish the present temperature, chemical compositions of solids and liquids, and flow rates - if any - of fluids in the system. But we are always left with the problem of trying to "read the record" and estimate how long the present conditions have been constant, and whether conditions in the past have been radically different. The older the natural material that we investigate is, the more uncertain we are concerning the integrated chemical/physical conditions to which it has been exposed. It is precisely these uncertainties that make it difficult, if not impossible, to provide quantitative information from geologic studies that can be used in waste repository performance assessment. This is without considering the even larger problem of finding a natural occurrence that is truly analogous to a waste disposal system or subsystem.

With all of these difficulties, why do we try to study "natural analogues" for waste disposal systems? There are two main reasons. First, we want to convince ourselves, as well as others, that we have not overlooked an important process that will affect the ability of our waste disposal system to function as designed. By studying as many natural systems as possible, we gain confidence that we have not overlooked some essential process. The particular area where natural systems provide extremely useful information is that of kinetics of slow processes with low activation energies. In the laboratory, we try to speed up processes by increasing the temperature. This may move us into a regime that has a different equilibrium chemical assemblage, but we can usually discover this and not be misled. A more difficult problem occurs when there is a process that has a very low activation energy. In this case, over thousands of years the low activation energy process will dominate, but we will only see high activation energy processes in our laboratory "accelerated" experiments. The second area where natural analogues are particularly helpful is with communication to non-specialists, both the general public and scientists whose area of expertise is somewhat to very far from those of the repository developer and performance assessor. Seeing a "natural analogue" of a waste form that is more than some tens to hundreds of millions of years old gives one a comfortable feeling that the earth in the regions within a few kilometers of the surface is not such an aggressive environment after all.

Performance Assessment Needs

The structure of a performance assessment analysis or a safety assessment of a nuclear waste repository will depend on the local geologic conditions, the particular engineered barriers selected for the repository, the choice of scenarios for analysis, and the particular regulatory framework in which repository licensing will occur. Despite these potential differences, most repository performance assessment calculations will use quantitative information concerning the behavior of the waste products and their interaction with water. If we break down the types of quantitative information that are likely to be used, we find (for repositories designed for spent nuclear fuel disposal)

Rate of dissolution of spent fuel matrix
Rate of release of fission products and actinides from spent fuel matrix
Rate of transport of radioactive materials after release from the matrix
Effect of processes that can trap or retard transport of radioactive materials.

Depending on the repository design and on the analysis tactics chosen, the importance of each of these types of information will vary. For example, it is possible to imagine a case based only on transport of material after release from the matrix. For that analysis case, the first two types of rate information would not be needed.

Spent Fuel Matrix Dissolution Rates

Many laboratory studies attempt to measure parameters that can be used to determine the rate of dissolution of spent fuel under repository disposal conditions. These studies use either spent fuel itself, uranium dioxide as an "analogue" for spent fuel, or uraninite as a "natural analogue". Some studies have also been done on a synthetic material, SIMFUEL, that attempts to duplicate the chemical and physical condition of spent fuel without using radioactive materials. In all cases, the data from these experiments must be extrapolated from the conditions used in the laboratory to the conditions expected in the repository. The key extrapolations include estimates of water chemistry under repository conditions, water flow rate, geometry of water-waste form contact, and redox conditions.

There are four basic sets of conditions for water-waste form interaction. They are oxidizing conditions with stagnant water, oxidizing conditions with flowing water, reducing conditions with stagnant water, and reducing conditions with flowing water. The complexity of the situation becomes apparent when one realizes that there is a continuous gradient in oxygen fugacity between the extremes of "oxidizing" and "reducing" conditions, and a continuous range in flow rates for water between the extremes of zero (stagnant water) and unrestricted flow. Laboratory studies, and field studies of natural analogues of waste forms, must first be located within the two continua of oxygen fugacity and water flow rate, and then the data must be extrapolated from the conditions of measurement to those expected in the repository. An additional level of uncertainty is introduced, of course, because the actual conditions in the repository cannot be precisely defined.

Depending on the location of experimental or field conditions in the continuum of flow rate, one may measure the initial rate of dissolution at infinite dilution (very large flow rate), a dissolution rate that is much slower because the species of interest is present in solution in considerable abundance (low flow rate), or the solubility of the material itself or of some secondary alteration product of the material being examined. When one is dealing with uraninites, both in the laboratory and in natural settings, one must also be aware that the presence of minor phases and of substantial amounts of lead, formed as the long-term decay product of uranium, must also be taken into account.

It is also tempting to try to estimate spent fuel dissolution rates using the release of highly soluble materials such as Cs, I, or Tc under oxidizing conditions. There are two additional considerations if this type of analysis is attempted. First, the material analyzed may not be homogeneously distributed in the fuel matrix - e.g., Tc. Second, even at equilibrium, where no net dissolution of the fuel matrix would occur, there will be a dynamic process of dissolution and reprecipitation at the fuel surface, which may release materials that are present in solid solution in the matrix. Once these materials are released, if they have a solubility higher than the fuel matrix, they will not reprecipitate. Use of the change in concentration of such an element in solution to infer the matrix dissolution rate would lead to an erroneous overestimate of the matrix dissolution rate.

Oklo as an Analogue for Spent Nuclear Reactor Fuel

The reactor zones at Oklo, Okélobondo, and Bangombé are the only places where nuclear chain reactions have been identified as occurring in natural uranium ore deposits. The reactor zones are similar to spent commercial reactor fuels in that the amount of ^{235}U at the time of reactor operation was about 3.5%, which is similar to the enrichment of ^{235}U in commercial reactor fuel. The reactor zone uraninites contain as solid solutions in the uraninite matrix many of the original fission products or the isotopes produced after decay of the original fission products. In all cases except for the current near-surface environment of the Bangombé reactor zone, the redox conditions have been such that UO_2 remained stable.

Despite the similarities between Oklo reactor zones and spent commercial reactor fuels, there are important differences between these materials that may lead to differences in their long-term behavior under geologic conditions. The first difference is that the Oklo reactor zones "burned" very slowly over periods of hundreds of thousands of years, with the ores most likely passing into and out of the critical state many times. This would lead to lower average temperatures for the Oklo materials than for the commercial fuels that are burned out over a period of 3 to 4 years. The Oklo uraninites contain Pb, produced as the long-term decay product of both ^{235}U and ^{238}U. Almost all Oklo uraninites have experienced Pb-loss during their long geologic history on at least one occasion. The Pb may now be present in galena (PbS) crystals near and within uraninite grains, or may have completely escaped from the vicinity of its parent U. The presence of large amounts of radiogenic lead - up to 20% by wt. - within the uraninite will affect its chemical stability in a different way than would the presence of a few % of fission products in solid solution in spent commercial reactor fuel. A final complication at Oklo is the presence of large amounts of organic matter in the near vicinity or intimately mixed with uraninite. This organic matter is likely to influence the chemical behavior, especially the redox stability, of the uraninite.

Because of the important differences between the uraninite at Oklo and commercial nuclear reactor fuels, the reactor zones at Oklo should not be considered as a natural example of a spent fuel repository. In addition, there are other features of a repository that are missing at Oklo, such as the canister into which spent fuel assemblies will be placed and the cladding that surrounds fuel pellets in the commercial reactor fuels. We will learn most from Oklo if we consider it as a location to examine processes that might take place after disposal of spent fuel in a geologic setting rather than as a direct material analogue for spent fuel after disposal. In this way, we can avoid the pitfalls of trying to stretch the analogy beyond its limits and, thereby running the risk of losing credibility for the concept of "natural analogues".

The most important feature at Oklo that can help us understand possible events in a repository comes from the fact that portions of the ore deposit sustained nuclear fission reactions while nearby regions of rich ore did not. This allows one to determine the necessary and sufficient conditions for criticality to occur and be sustained in a geologic environment. The behavior of the Oklo uraninites after the nuclear reactions can be compared with that of nearby uranium dioxide that did not have fission reactions. Through such comparisons we can conclude that the portions of the ore that were once reactor zones did not behave significantly differently from other parts of the ore body once the fission reactions were finished. Indeed, this is probably the reason that the discovery of the reactor zones occurred through isotopic analysis of the ore rather than through examination of the ore during prospecting and mining operations.

If we examine the evidence from reactor zone 2, we find that the original rock material, which was approximately 70% quartz, 20% uranium dioxide, and 10% clay, now consists only of uraninite and clay. During the course of the nuclear reactions, water dissolved the quartz and subsequently transported it out of the reactor zone to nearby locations, where the quartz reprecipitated. During this dissolution and transport, little of the U was removed, indicating that the redox conditions remained within the stability field of U(IV) rather than the more soluble U(VI) species. The local environment was able to counteract the effects of radiolysis arising during and after the nuclear fission reactions and prevent significant oxidative dissolution and transport of U from the system after the reactions had proceeded to the extent that rare earth element fission products were present in moderate amounts.

Another important conclusion that can be drawn from the Oklo reactor zones is that criticality in a natural setting is difficult to achieve and, if it occurs at all, it is a self-limiting process. This is an important point that is very relevant to performance assessment in geologic repositories, especially if waste forms rich in fissile isotopes are to be included in the inventory.

Finally, we find that reactor zone 2 seems to have been preserved for nearly 2 billion years, despite a period of tectonic activity at about 900 million years ago, as evidenced by numerous dikes in the Oklo vicinity. The thermal pulse accompanying the dyke intrusions and the fluid circulations that are evidenced by trapped fluid inclusions in secondary mineral phases, were not sufficiently powerful to homogenize the uraninites on the scale of a few centimeters, let alone dissolve and transport away significant amounts of the uranium and fission product relicts. While this does not give us quantitative data for use in performance assessment calculations, it does provide excellent qualitative information to support the expected stability of UO_2 in environments that maintain redox potentials that are within the U(IV) stability region.

While acknowledging the important positive evidence that can be obtained from reactor zone 2, we should be careful to remember that there are some things we cannot learn from studies at Oklo. Most importantly, we cannot determine the rate of dissolution of uraninite in reactor zone 2 conditions. We do not know how much water has flowed through the ore after reactions stopped, nor at what temperatures water may have contacted the ore and for how long. We also cannot determine unambiguously whether it is the organic matter, the ferrous iron in clay minerals, or the uraninite itself that controls the local redox potential.

DISCUSSION AND CONCLUSIONS

In addition to the use of Oklo reactor zones as an analogue for spent commercial reactor fuel, we may use other aspects of the Oklo deposits to learn about processes that may be important for nuclear waste disposal. In addition to the information obtained concerning criticality in a natural environment, we can learn about the effects of radiation-induced radiolysis on materials by examining minerals from reactor zones and from nearby ore samples that did not sustain reactions. We can study the effects of different materials, such as organic matter, on the preservation of the ore, and at Bangombé, we can examine the effects of near-surface weathering in a tropical climate on the dissolution behavior of uraninite.

Through the presence of excess ^{99}Ru, we have evidence for retention of ^{99}Tc even when there was still significant fluid circulation occurring that removed Mo and other soluble fission products from the reactor zones. This suggests that Tc may be stabilized in the metal alloys formed by fission products and may not be as easily oxidized to a soluble form as has been assumed in many performance assessment calculations.

Studies of secondary minerals at Oklo have shown incorporation of fission products and Pu, which is evidenced by small excesses of ^{235}U in some apatite and clay samples. These minerals formed before the decay of ^{239}Pu occurred, and thus at rather high temperatures (several hundred degrees C), so they cannot be used to infer trapping of fission products and actinides under conditions expected for long-term behavior of a waste repository. Sorption studies using materials from Oklo can, however, be used to understand the importance of sorption for retardation during recent fluid circulation, especially in the active weathering environment at Bangombé.

Finally, while we cannot extract quantitative information concerning uraninite dissolution rates and fission product transport rates at Oklo and Bangombé, we have learned something very important. We have seen no evidence for processes occurring in the reactor zones after the reactions were finished and the ore returned to ambient (albeit rather high) temperatures that are different from processes that have been seen in other U ore bodies that have not undergone sustained nuclear chain reactions. That means that any uranium occurrence in a natural setting would be an equally good analogue for spent fuel disposal. This allows us to examine U deposits, or even small occurrences of U in rocks that are similar to the intended host rock for the repository of interest and not have to worry about whether the absence of fission reactions in the materials might make the analogy less valid than the Oklo reactor zone analogues.

REFERENCES

1. P.-L. Blanc, Oklo-Natural analogue for a radioactive waste repository. Volume 1. Acquirements of the project. EUR16857/1 EN, European Commission, Nuclear Science and Technology, 1996.
2. D. Louvat and C. Davies, Oklo Working Group, Proceedings of the first joint EC-CEA workshop on the OKLO-natural analogue Phase II project held in Sitges, Spain, from 18 to 20 June 1997, EUR 18314 EN, European Commission, Nuclear Science and Technology, 1998.
3. D. Louvat, V. Michaud, and H. von Maravic, Oklo Working Group, Proceedings of the second joint EC-CEA workshop on the OKLO-natural analogue Phase II project held in Helsinki, Finland, from 16 to 18 June 1998, European Commission, Nuclear Science and Technology, in press.
4. Final report of the Oklo Phase II project, to be published by European Commission, Nuclear Science and Technology, in preparation.

CLAYSTONE CONSTRAINTS ON MODELS OF THE LONG-TERM CHEMICAL EVOLUTION OF BUFFER POREWATERS

R. C. ARTHUR*, J. WANG**
* Monitor Scientific, LLC, Denver, CO 80235, rarthur@monitorsci.com
** Beijing Research Institute of Geology, Beijing, PR CHINA, radwaste@public.bta.net.cn

ABSTRACT

Geochemical models of clay-water interaction are evaluated using data characterizing the hydrochemical, isotopic and mineralogical properties of a natural claystone that is similar to bentonite buffers that have evolved over long periods of time in a deep geologic repository for nuclear wastes. Model predictions in general compare favorably with the mineralogy and hydrochemistry of this formation if increases in pH due to partial $CO_2(g)$ exsolution from porewater samples is accounted for. This suggests the models can be used with enhanced confidence to accurately predict the chemistry of pore fluids in bentonite buffers, enabling more defensible estimates to be made of radioelement solubilities that partially define the source term in performance assessments. Modifications are needed, however, to improve the accuracy of thermodynamic data supporting the models and to account for potential effects on porewater compositions of reactions that are kinetically inhibited from attaining equilibrium over experimental time scales.

INTRODUCTION

Solubility-limited radioelement concentrations, which partially define the source term in performance assessments of disposal concepts for HLW, are often calculated based on the chemistry of porewaters in the bentonite buffer of the EBS [1]. For this reason a number of models of bentonite-water interaction have been proposed to predict the chemical evolution of buffer porewaters over time scales of several millions of years. Equilibrium is assumed in all the models, which differ primarily in their treatment of the solid-solution properties of the smectite clays. From simplified to comprehensive, the models account for stoichiometric, ion-exchange, surface-chemical, ideal-mixing or ideal site-mixing reactions among the smectites and an aqueous phase. Model parameters are calibrated based on the results of short-term experiments (ion-exchange; surface-chemical), inferred stability relations among coexisting clay minerals and groundwaters (stoichiometric; ideal-mixing) or semi-empirical observations of activity-composition relations among smectite clays and illite (ideal site-mixing). The accuracy of the models is untested, however, under conditions expected in the buffer over time scales considered in performance assessments. Here we evaluate the validity of the equilibrium basis of several models, and accuracy of associated model parameters, using as constraints data characterizing the hydrochemical, isotopic and mineralogical properties of the London Clay, a natural claystone analogous in many respects to bentonite buffers that have evolved over long periods of time.

MINERALOGY AND HYDROCHEMISTRY OF THE LONDON CLAY

In 1987 the British Geological Survey drilled six boreholes through a sequence of mudrocks near the city of Bradwell, Essex on the east coast of England [2]. The mudrocks comprise 40-60 m of London Clay and 3-35 m of the Lower London Tertiaries sands, silts and clays. Pore fluids were extracted from a total of 45 drillcore samples of London Clay obtained from 25 depth intervals in one borehole (B101) and 12 depth intervals in a second borehole (B102). The porewaters were extracted using compression rigs capable of squeezing solutions from drillcore samples at loads up to 70 MPa [2]. The resulting database is unique because: 1) porewater compositions are determined in a series of samples representing the entire depth profile of a natural claystone, 2) the porewaters and clay minerals remain in contact until the porewaters are extracted for analysis, and 3) borehole locations have experienced different recharge histories, ranging from fresh water (B101) to seawater (B102).

The mineralogy and physical properties of the London Clay are similar to predicted properties of EBS buffers that have evolved over extended periods of time. It contains dominant

Mat. Res. Soc. Symp. Proc. Vol. 608 © 2000 Materials Research Society

smectite (montmorillonite) and illite, and minor amounts of kaolinite, chlorite, quartz and feldspar [2]. Calcite is detectable in small amounts at depths between 5 and 12 m, dolomite is present between 4 and 20 m, and siderite is present between 4 and 35 m. Pyrite is absent in upper portions of the claystone, but trace amounts occur below 16 m. The average cation-exchange capacity is 28 meq $100g^{-1}$, and cation occupancies are Ca = 11 meq $100g^{-1}$, Mg = 12 meq $100g^{-1}$, and Na = 5 meq $100g^{-1}$. Moisture contents vary from 29 to 42% at saturated densities averaging 1.9 g cm^{-3}.

Chemical analyses of London Clay porewaters are documented in reference [2]. Here we note that the concentrations of some aqueous constituents, particularly K^+ and Na^+, were observed to vary as a function of the amount of porewater extracted, and thus as a function of the applied load utilized in the sampling apparatus. The cause of this fractionation behavior is unknown, but it has been suggested that high loads cause water from interlayer sites in smectite to move more freely through the clay [2]. This complicates efforts to define a single representative composition of porewater in a given sample of the London Clay. In following discussions, we consider all reported porewater analyses because it is not possible to decide if any one sample is more or less representative than another.

Stable oxygen- and hydrogen-isotope analyses of London Clay porewaters reveal maximum depletions corresponding to meteoric recharge in the colder climate prevailing in southern England during the last glacial maximum, about 20,000 years ago [2]. Diffusional mixing of groundwater from geologic units above and below the London Clay has occurred up to the present time.

Estimates of the *In-Situ* pH of London Clay Porewaters

Preliminary assessments of mineral-water equilibria in the London Clay reveal severe inconsistencies with respect to the mineralogy of this formation. Calculated mineral stability relations in the CaO-MgO-Al_2O_3-SiO_2-H_2O system are shown in Fig. 1a, for example, where the compositions of all porewaters extracted from B101 and B102 drillcores (a total of 105 samples) are represented by symbols. As can be seen, the porewaters plot in the stability field of dolomite, which is consistent with the observation that dolomite is present in this formation, at least in detectable amounts between 4 and 20 m. All but a few solutions plot in the stability fields of Mg-saponite and Ca-saponite, however, which conflicts with the observation that the London Clay consists of dioctahedral montmorillonites rather than trioctahedral saponites.

We speculate that the discrepancies stem from increases in pH resulting from partial $CO_2(g)$ exsolution from porewater samples. Loss of CO_2 from groundwater samples is a common problem [3], and may have occurred during storage for 18 months of B101 and B102 drillcores prior to porewater extraction. The *in-situ* pH is therefore estimated using a *back-titration* model [3] to simulate the reverse of the exsolution process, assuming that no other heterogeneous reactions occur during the titration and that the titration end-point is reached when the saturation index of dolomite, or calcite, equals zero (referred to below as the *dolomite* or *calcite* models, respectively). Equilibrium of the corrected solution composition with respect to dolomite is a credible constraint because ratios of the aqueous activities of Mg^{2+} to Ca^{2+} are consistently greater than unity in all London Clay porewaters, indicating that calcite is unstable relative to dolomite [4]. The rate of dolomite precipitation is extremely slow [4], however, and it is therefore possible that pore solutions equilibrate with calcite and are metastably supersaturated with respect to dolomite.

Results based on the dolomite model indicate *in-situ* pH values are between 5.3 and 7.8. These values are generally much lower than those measured in B101 pore fluids (7.8–11.5) and bracket the pH range measured in B102 solutions (6.5-7.7) [2]. Representative effects on mineral stabilities are illustrated in Fig. 1b, where it can be seen that the *in-situ* pH estimates are consistent with the mineralogy of the London Clay insofar as both Mg-montmorillonite and dolomite are predicted to be at equilibrium with pore solutions. Results using the calcite model (not shown in the figure) are less consistent because many porewaters plot in the Mg-saponite field. Calculated amounts of $CO_2(g)$ lost from pore solutions are moderate, averaging 25% and 47% of estimated initial concentrations in B101 and B102 samples, respectively, indicating that

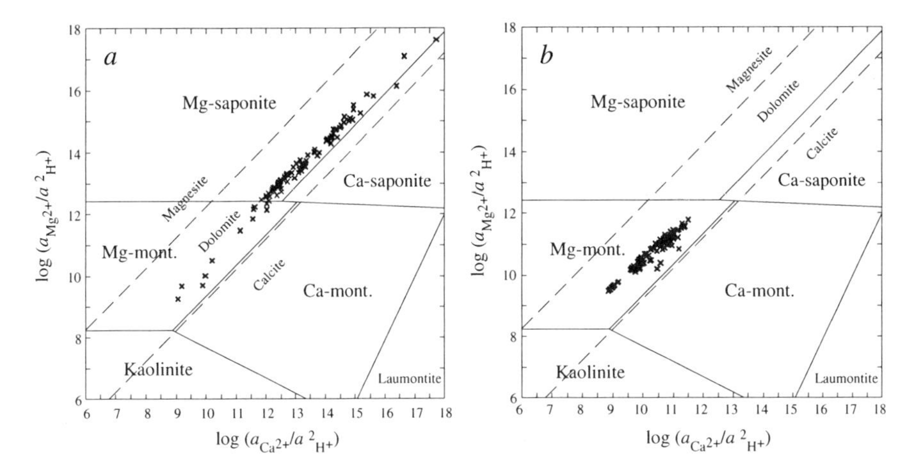

Figure 1. Stability diagram for the CaO-MgO-Al_2O_3-SiO_2-H_2O system at 25°C, assuming solubility equilibrium with respect to quartz. The compositions of London Clay porewaters (symbols) are based on pH measurements [2] (*a*) or estimates of the *in-situ* pH (*b*). The stability fields of the saponites and montmorillonites do not vary significantly over a range of $SiO_2(aq)$ concentrations bounded approximately by quartz and chalcedony solubilities (see Fig. 3).

even partial $CO_2(g)$ exsolution significantly affects the pH measurements. The estimated *in-situ* compositions of porewaters are adopted provisionally in evaluations described below.

ASSESSMENTS OF MINERAL-WATER EQUILIBRIA IN THE LONDON CLAY

Mineralogical and hydrochemical constraints derived from the study of the London Clay are used in this section to evaluate whether mineral-water equilibria considered in models of bentonite-water interaction are reasonable, and if associated thermodynamic data are accurate. We use the Geochemist's Workbench [5] and its supporting thermodynamic database [6] to carry out associated aqueous-speciation calculations. Analytical charge imbalances, generally less than ±4% [2], are corrected by adjusting concentrations of major cations or anions accordingly. The *in-situ* temperature is assumed to be 25°C.

Stoichiometric Model

Typical results of this type of model (*e.g.*, [7]) are illustrated in Fig. 1b, where it is assumed that the clay minerals are stoichiometric phases of fixed composition. The figure is drawn assuming equivalent concentrations in the respective clay minerals of exchangeable Ca^{2+} or Mg^{2+}, which as noted above closely coincides with the measured relative concentrations of these cations in the London Clay. The porewater compositions plotted in Fig. 1b should therefore lie near the boundary between the stability fields of Ca- and Mg-montmorillonite. As can be seen, however, although the distribution of data points parallels this boundary, all the data lie well within the Mg-montmorillonite stability field. This suggests that minor adjustments are needed to improve estimates of equilibrium constants for hydrolysis reactions involving Mg-montmorillonite and/or Ca-montmorillonite, which define the location of this boundary.

Ion Exchange Model

Equilibrium is assumed in this type of model (*e.g.*, [8]) for reactions such as,

$$2NaX + H^+ + CaCO_3(c) \leftrightarrow CaX_2 + 2Na^+ + HCO_3^-, \quad (1)$$

where X refers to an "exchanger phase" of fixed, but unspecified, composition representing octahedral and tetrahedral sites in montmorillonite. For conditions in the London Clay, where calcite is apparently unstable relative to dolomite, the appropriate reaction is given by:

$$NaX + H^+ + 1/2CaMg(CO_3)_2(c) \leftrightarrow 1/2CaX_2 + Na^+ + 1/2Mg^{2+} + HCO_3^-. \quad (2)$$

We use the following procedure to test presumptions of equilibrium for both reactions. The pH is calculated using the corresponding mass-action equations, and:

- the average concentrations of exchangeable cations in the London Clay noted above (converted to units of mol kg^{-1} assuming an average moisture content of 34% [2]);
- concentrations of Na^+ and Mg^{2+} in porewaters extracted from B101 and B102 drillcores;
- *in-situ* dissolved carbonate concentrations estimated using the calcite or dolomite back-titration models for reactions (1) or (2), respectively; and
- equilibrium constants for reactions (1) and (2) that are calculated using a value of the equilibrium constant for the ion-exchange reaction from reference [8].

Results, referred to as *model* pH values, are compared with corresponding estimates of the *in-situ* pH in Fig. 2. Trends among the data paralleling the dashed line representing agreement between *in-situ* and model pH are partly an artefact of the calculation procedure because *in-situ* values of dissolved carbonate concentrations are used to calculate model pH. The trends are also determined in part, however, by the aqueous activities of Na^+ and Mg^{2+}, which span a range of roughly two orders of magnitude among all B101 and B102 samples [2]. Agreement between model and *in-situ* pH is better when dolomite rather than calcite is assumed to be stable. This suggests that equilibrium with respect to reaction (2) is consistent with estimates of *in-situ* pH and total carbonate concentrations, concentrations of exchangeable cations measured in the London Clay, aqueous concentrations of Na^+ and Mg^{2+}, and thermodynamic data that define the equilibrium constant for this reaction. Slight disagreement between *in-situ* and model pH values are apparent, however, and may be due to errors in any one or several of these parameters.

Ideal Site-Mixing Model

In these models (*e.g.*, [9]) the activities of thermodynamic components of clay-mineral

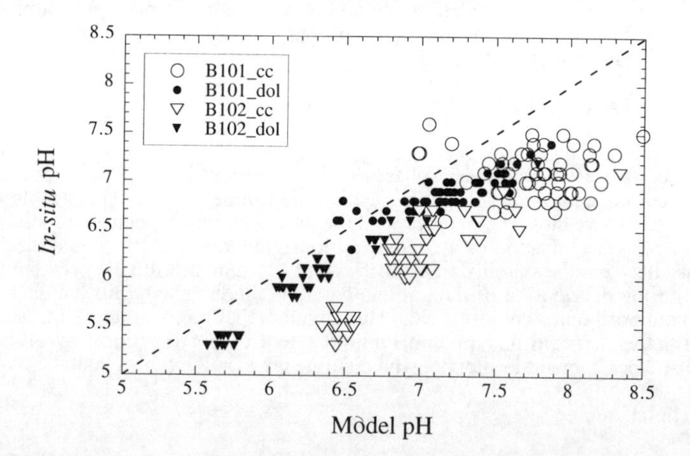

Figure 2. Comparison of *in-situ* pH and model pH values. The latter are calculated assuming pore solutions equilibrate with montmorillonite and either calcite (cc) or dolomite (dol).

solid solutions are used as descriptive variables enabling stability fields of these minerals to be rendered on conventional stability diagrams. The stability diagram for the K_2O-Al_2O_3-SiO_2-H_2O system is shown in Fig. 3, where ranges in the activities of components that are stoichiometrically equivalent to muscovite (a_{mu}) and pyrophyllite (a_{pyr}) [10] define stability fields of illite and montmorillonite solid-solutions, and an intervening "mixed-layer" field in which both illite and montmorillonite are stable. The compositions of most London Clay porewaters, represented by symbols in the figure and calculated using *in-situ* pH values estimated using the dolomite model, plot in the mixed-layer or illite stability fields. Porewaters plotting in the mixed-layer field may be consistent with the mineralogy of the London Clay, where kaolinite, illite and montmorillonite coexist, but to confirm this the compositions of illite and montmorillonite solid solutions must be determined to enable calculation of component activities. Porewaters plotting outside the mixed-layer field are inconsistent with the mineralogy of this formation. Experimental fractionation of K^+ concentrations (decreasing with increasing applied load) and $SiO_{2}(aq)$ concentrations (increasing with increasing applied load) [2] are not sufficient to explain the discrepancy between calculated and observed mineral stabilities for the data plotting outside the mixed-layer field. Estimates of *in-situ* pH assuming equilibrium with respect to calcite rather than dolomite shift porewater compositions upward toward the illite-K-feldspar boundary, generating poorer agreement with the observed mineralogy.

CONCLUSIONS AND IMPLICATIONS FOR PERFORMANCE ASSESSMENT

Models of mineral-water equilibria applied to the London Clay are generally consistent with

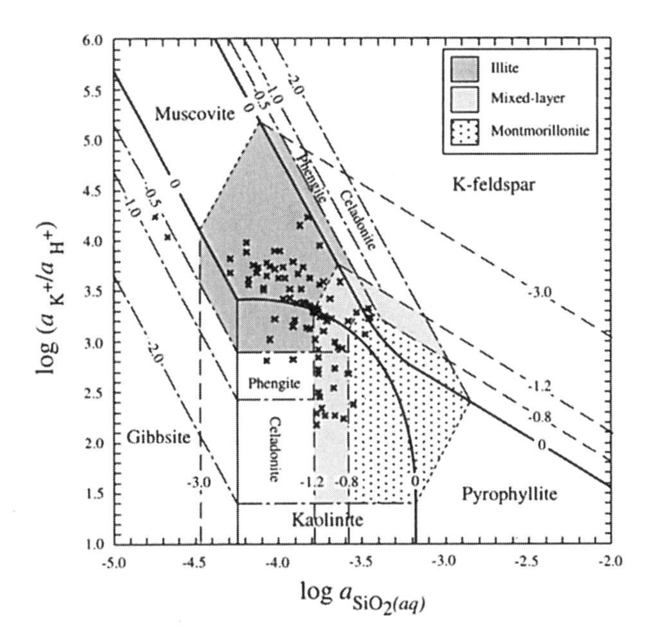

Figure 3. Stability diagram for the K_2O-Al_2O_3-SiO_2-H_2O system at 25°C, including dioctahedral clay minerals belonging to the muscovite-pyrophyllite solid-solution series. The long-dash lines refer to values of log a_{pyr} as labelled. Dash-dot-dash lines denote values of log a_{mu} as labelled. Solid lines refer to unit activity of a_{pyr} or a_{mu}, except where they are curved. The curvature corresponds to the loci of points generated by tie lines (short dash) between the stability fields of kaolinite and K-feldspar representing equilibrium of an aqueous solution and clay minerals of fixed composition with respect to a_{pyr} and a_{mu}.

the mineralogy and hydrochemistry of this formation, if increases in pH due to partial $CO_2(g)$ exsolution from porewater samples is accounted for. This conclusion suggests enhanced confidence is warranted in the use of such models to predict the chemistry of pore fluids in bentonite buffers, enabling more defensible estimates to be made of radioelement solubilities that partially define the source term in performance assessments.

Improvements to the models are needed, however. Models developed solely on the basis of short-term experiments may fail to account for the effects on solution chemistry of kinetically controlled heterogeneous reactions. This is exemplified in the present study by the apparent stability of dolomite rather than calcite in London Clay porewaters. Dolomite does not precipitate over laboratory time scales and is not presently considered in ion-exchange models of buffer-porewater interactions. Thermodynamic data supporting stoichiometric and ion-exchange models appear to be slightly, but significantly, in error. An ideal site-mixing model suggests London Clay porewaters equilibrate with kaolinite, illite and montmorillonite, but exceptions are common. A more rigorous test of this model requires quantitative data characterizing the compositions of illite and montmorillonite solid solutions.

Improvements also are needed in methods to extract pore fluids from compacted argillaceous media. Squeezing using compression rigs generates variable fractionation of solute concentrations depending on the applied load, thus complicating interpretations of representative porewater compositions. Alternative methods such as ultra-centrifugation or *in-situ* measurements using microanalytical techniques may help to clarify such interpretations.

Additional field studies similar to that carried out at Bradwell, but focussing on claystones that are similar to less altered buffers, would shed light on the geochemical evolution of these materials over intermediate time scales in a repository environment. Large and undisturbed bentonite deposits exist in China, and several are excellent candidate sites for such studies. The Gaomiaozi deposits in Xinhe County, Inner Mongolia Autonomous Region, and the Xintan deposit in Anhui Province, are of particular interest because reaction zones are well developed between original Na-montmorillonite and Ca-montmorillonite (Gaomiaozi), or illite (Xintan).

ACKNOWLEDGEMENTS

Support for this study from the Swedish Nuclear Power Inspectorate is gratefully acknowledged. H. Sasamoto (Japan Nuclear Cycle Development Institute) and D. Savage (Quintessa, Ltd.) contributed insightful comments on approaches to modeling clay-water reactions.

REFERENCES

1. I. G. McKinley and D. Savage, in *Fourth International Conference on the Chemistry and Migration Behavior of Actinides and Fission Products in the Geosphere* (R. Oldenbourg Verlag, München, 1994), p. 657-665.
2. A.H. Bath, C.A.M. Ross, D. Entwisle, M.R. Cave, K.A. Green, S. Reeder and M. Fry, *Hydrochemistry of Porewaters from London Clay, Lower London Tertiaries and Chalk at the Bradwell Site*, Rep. Brit. Geol. Surv. WE/89/26, Keyworth, U.K. (1989).
3. F.J. Pearson, J.L. Lolcama and A. Scholtis, *Chemistry of Waters in the Boettstein, Weiach, Riniken, Schafishheim, Kaisten and Leuggern Boreholes: A Hydrochemically Consistent Data Set*, Nagra TR 86-19, Nagra, Wettingen, Switzerland (1989).
4. W. Stumm and J.J. Morgan, *Aquatic Chemistry*, 3rd ed., Wiley Interscience, New York, 1996, pp.393-394.
5. C. M. Bethke, *Geochemical Reaction Modeling*, Oxford Univ. Press, New York, 1996, 397p.
6. data0.3245r46, compiled by the Geochemical Modeling Group, Lawrence Livermore National Laboratory, Livermore, CA, and accessible by ftp to *s122.es.llnl.gov*.
7. R.M. Garrels, Clays Clay Miner., 32(6), p.161-166 (1984).
8. H. Wanner, P. Wersin and N. Sierro, *Thermodynamic Modeling of Bentonite-Groundwater Interaction and Implications for Near-Field Chemistry in a Repository for Spent Fuel*, SKB TR 92-37, Swedish Nuclear Fuel and Waste Management Co., Stockholm, Sweden (1992).
9. P. Aagaard and H.C. Helgeson, Clays Clay Miner., 31(3), p.207-217 (1983)
10. W.F. Giggenbach, Chem. Geol., 49, p. 231-242 (1985)

AN ARCHEOLOGICAL SITE AT AKROTIRI, GREECE, AS A NATURAL ANALOG FOR RADIONUCLIDE TRANSPORT: IMPLICATIONS FOR VALIDITY OF PERFORMANCE ASSESSMENTS

D.L. HUGHSON, L. BROWNING, W.M. MURPHY, R.T. GREEN
Center for Nuclear Waste Regulatory Analyses, Southwest Research Institute, San Antonio, TX 78238-5166, dhughson@swri.org

ABSTRACT

Natural analog studies provide a means to test assumptions in performance assessment models. Data on the spatial distribution of metals derived from archeological artifacts at Akrotiri, Greece, were used to evaluate performance assessment model assumptions of radionuclide transport through unsaturated volcanic tuffs hosting the proposed repository at Yucca Mountain, Nevada. Processes controlling transport of trace elements from 3600 year old metallic artifacts at Akrotiri are analogous to processes controlling radionuclide transport at Yucca Mountain. Archeological information temporally and spatially constrains the source and chemistry of these trace elements, providing a comparison basis for model validation efforts. Despite these constraints and characterization of the Akrotiri site, the data are open to different interpretations as the extent of the plume emanating from the Minoan artifacts, transport pathways, heterogeneities in parameters, and boundary conditions are uncertain. Rather than validating a conceptual model of flow and transport, the Akrotiri site data indicate that 1-D models of aqueous phase transport in the unsaturated zone are conservative. Different conceptual models, apparently matching site data, produce different predictions illustrating that predictive flow and transport modeling is uncertain due to under-determined parameters and boundary conditions.

INTRODUCTION

The Akrotiri archeological site on the island of Santorini, Greece, was selected as a natural analog for supporting performance assessment modeling of the proposed high-level nuclear waste (HLW) repository at Yucca Mountain, Nevada, due to the presence of a well-constrained source of trace metals over a several thousand year time period and similarities in climate and geology. Both sites are in oxidizing, unsaturated, silicic volcanic rock in relatively arid climates. Around 1600 BC the Minoan eruption buried Akrotiri under several meters of ash. Bronze utensils on a ground floor subsequently became a temporally and spatially well-constrained source of trace metal contaminants. Sampling and testing activities at the Akrotiri archeological site were conducted to characterize its hydrological and geochemical features and to look for evidence of a plume of trace metals emanating from the artifacts [1,2]. These data showed subtle evidence for a plume of Cu, Pb, and Zn.

The approach taken in modeling trace metal transport at the Akrotiri archeological site [2] was patterned after that typically employed in performance assessments. Site characterization information, such as permeability and moisture retention measurements, and reasonable estimates (e.g., infiltration rate, solubility, and sorption coefficient) were used in a numerical model of flow and transport through porous media to predict concentration distributions of trace metal species over time. Input parameters were varied to investigate effects of uncertainty and to attempt to bound estimates. Model results predicted steady state concentration profiles from source to water table in less than the 3600 yr time frame, contrary to the field data which indicated an apparent concentration transient. This discrepancy persisted over parameter intervals of, for example, a factor of 36 for saturated hydraulic conductivity and 66% increase or reduction in infiltration flux. Some qualitative agreement between model results and observations was noted, however the discrepancy between model results and the data regarding trace metal distribution was not reconciled.

In this study we employ alternative hypotheses in an attempt to explain the distribution of trace metals observed near the location where the artifacts were removed. This modeling effort should be viewed as heuristic in contrast to the previous, predictive modeling work. We examine two alternatives suggested previously [2], variations in infiltration flux and spatial heterogeneity in model parameters. Effects of boundary flux assumptions are modeled in 1-D and potential effects of 3-D geological and anthropogenic features are simplified to 2-D with variable boundary flux assumptions. Natural analog sites, such as the

Mat. Res. Soc. Symp. Proc. Vol. 608 © 2000 Materials Research Society

archeological excavation at Akrotiri, offer opportunities for ascertaining performance model predictive reliability that are otherwise unavailable. Successful modeling of the natural analog would help to increase confidence in the model predictions. Likewise failure in modeling the natural analog site may lead to iterative model improvements and quantitative assessment of model prediction uncertainties.

MODELING STUDY

Recap of Previous Work

Rock and soil samples were collected from 8 boreholes to a depth of approximately 0.5 m, all within approximately 1 m of the artifacts' location, and analyzed for Cu, Pb, Ag, Co, and Zn by selective extraction. A contour plot of total Cu, extracted sequentially from cation-exchangeable sites, associated with carbonates, and with reducible Fe and Mn oxides is shown in Figure 1. Borehole, sample, and artifact locations are shown schematically in these plots. The contour plot indicates lateral spreading of the Cu plume away from the artifact location in a southerly direction with minimal vertical transport apparent except near borehole VII which is adjacent to a discrete vertical fracture. Parameters treated as constant in this study are given in Table 1. Saturation is noted as S, α and n are parameters of the van Genuchten moisture retention function, and k is permeability.

The model domain tested [2] was a 1-D column consisting of 1.5 m of Minoan ash overlying a 4 cm packed earth layer overlying 22.5 m of Cape Riva tuff to the water table. With a flux boundary condition at the top representing estimated infiltration of 15 cm/yr [2], the hydrologic properties given in Table 1, and no sorption to the solid phase, the dissolved aqueous species derived from the artifacts are predicted to achieve a steady state profile within 30 yr. Including equilibrium sorption onto solids by means of a linear model with estimated [2] distribution coefficients $K_d = 0.45$ m^3/kg for packed earth and $K_d = 0.0027$ m^3/kg for Cape Riva tuff and a bulk density for the Cape Riva tuff of $\rho_b = 1.74E+6$ g/m^3, the plume takes over 1000 years to reach a uniform distribution to the water table. This corresponds to a retardation coefficient of about 30, as calculated from the relationship

$$R = 1 + \frac{\rho_b K_d}{\theta} \tag{1}$$

where K_d is the distribution coefficient, ρ_b is bulk density, and θ is moisture content.

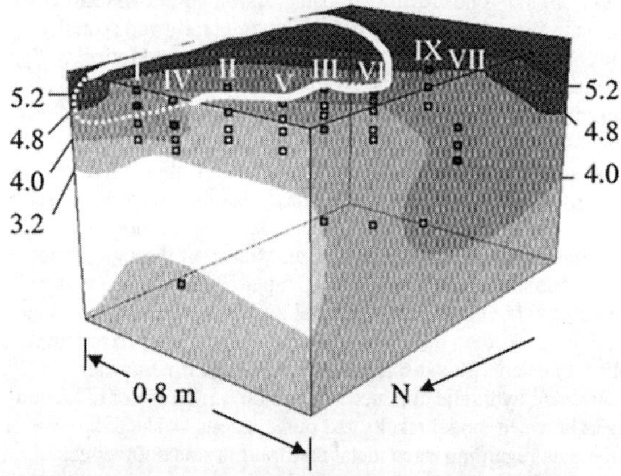

Figure 1. 3D perspective of sample locations (small squares), boreholes (Roman numerals), and Cu concentration contours. Numbers on the sides indicate ppm Cu associated with solids. Sides and top are 2D slices through solid contours. White line shows the approximate location of the base of the artifacts, dotted where extending beyond plotted range.

Alternative Hypotheses

1. Precipitation and Infiltration

Murphy et al. [2] estimated runoff using U.S. Bureau of Reclamation methods and applied a mean net infiltration of 15 cm/yr varying sinusoidally from 20 cm infiltration in the winter to 5 cm evapotranspiration in the summer. Data from the Hellenic National Meteorological Service (Figure 2) show the average monthly precipitation over the period 1974-1991. The average rainfall over this period is 32 cm/yr. However potential evapotranspiration, estimated using the empirical method of Hamon [3], is exceeded by precipitation only 4 months out of the year, keeping the soil surface dry during summer months. Steady state flow in the 1-D model, using properties

Table 1. Model parameters from [2] treated in this study as constants.

Material	Porosity	VG n	VG α (Pa^{-1})	k (m^2)	Residual S
Minoan ash	0.60	1.4	5.01×10^{-5}	1.945×10^{-11}	0.02
Packed Earth	0.30	1.13	2.76×10^{-4}	1.144×10^{-14}	0.10
Cape Riva	0.25	1.7	7.02×10^{-5}	1.716×10^{-11}	0.02

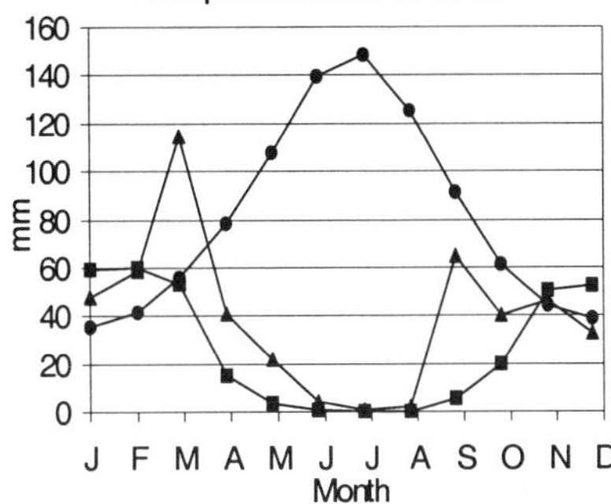

Figure 2. Precipitation data in mm for Latitude 36.25° N, Longitude 25.26° E, averaged over the period 1974-1991.

given in Table 1 and a prescribed boundary condition at the surface of S = 0.1, results in an upward flux of water from the water table of about 4 mm/yr, assuming a constant temperature of 20°C.

Redistribution of moisture from precipitation infiltration is complicated by evapotranspiration and soil heterogeneity. A low permeability packed earth layer on the floor of room Δ3 at the Akrotiri archeological site may have been part of a more extensive low permeability horizon compacted by human activity or may have been limited to the area of the Δ3 room, possibly diverting infiltration and creating a dry shadow below the artifacts. Remnants of the flagstone floor of the collapsed building may also have diverted infiltration above, creating a dry shadow surrounding the artifacts. Figure 3 illustrates the temporal effect of an areally extensive low permeability layer on moisture redistribution during and following a hypothetical storm. The model is a 1-D soil column of 1.5 m Minoan tuff overlying 0.04 m of packed earth overlying 22.46 m of Cape Riva tuff, with a water table boundary at the bottom and a constant flux boundary at the top, so the lateral extent of the low permeability layer is effectively infinite. The hypothetical storm lasts two days during which saturation at the surface is maintained at 0.85. Following the storm the soil surface dries out and is maintained at a saturation of 0.2. Initially the soil surface is at a saturation of 0.4 which is the hydrostatic condition.

This simulation suggests that, in the presence of a low permeability layer, an isolated 2 day precipitation event would contribute very little to percolation flux in the Cape Riva formation. Comparison of the average monthly precipitation with the average maximum precipitation in a 24 hour period during the month (Figure 2) suggests that such isolated storms are the norm at Akrotiri. Murphy et al. [2] performed sensitivity analyses reducing the flux boundary condition by 66% to 5.1 cm/year. However, referring to the discussion above, further sensitivity analyses with boundary flux reduced by a factor of 10 or more seem warranted. For the conservative solute, reducing flux by a factor of 10 scales the time required to reach a uniform concentration distribution to the water table from approximately 30 to 300 yr. However, the reduced infiltration results in a lower moisture content, except near the water table, which increases the retardation factor. Thus, as shown in Figure 4, the retarded plume, with 1.5 cm/yr infiltration flux and sorption to the solid phase as described by the distribution coefficients given above, has not reached the

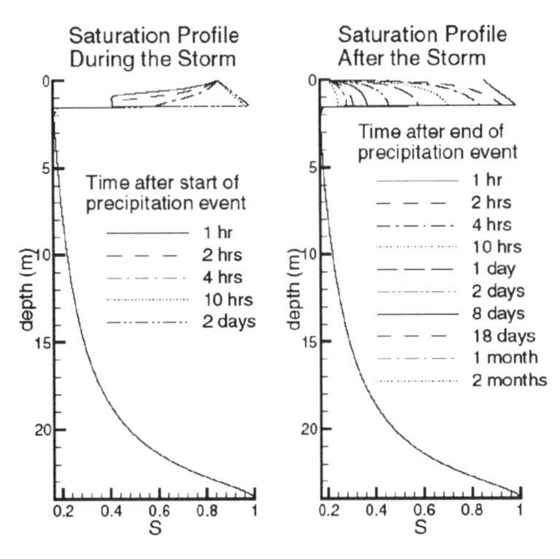

Figure 3. Saturation as a function of depth and time during a simulated storm.

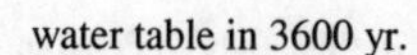

water table in 3600 yr.

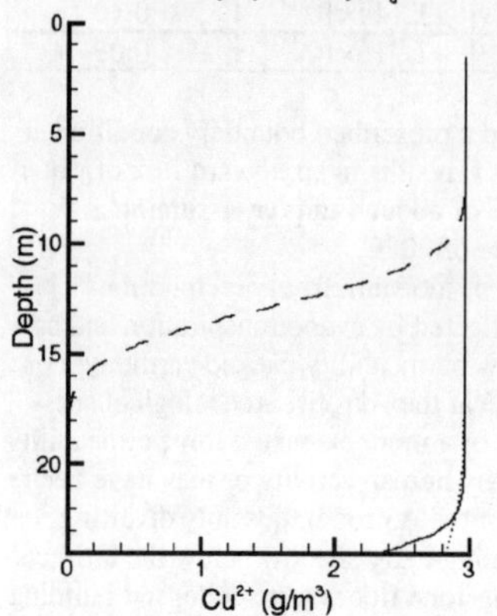

Figure 4. 1-D simulations illustrating the combined effects of plume retardation and low infiltration.

2. Heterogeneity

Parameters such as permeability commonly vary over several orders of magnitude in geological formations. Large discrete vertical fractures were observed in the Cape Riva formation within the vicinity of the artifacts' location. Borehole VII was located specifically to sample one such feature. Cu analyses of samples from borehole VII appear to indicate that more vertical transport occurred near this feature (Figure 1) than was observed from the other borehole data. The hydrological properties of the packed earth floor were not characterized. It was observed to be compacted and to have a high clay content suggesting low permeability. The floor was not level but dipped approximately 10 degrees in a southerly direction. Figure 5 shows the development of an aqueous phase Cu^{2+} plume from an atacamite ($Cu_2Cl(OH)_3$) source located above a thin low permeability layer extending laterally for 1.4 m. The entire model geometry of Figure 5 dips 10 degrees to the right with respect to gravity so the plume is asymmetric. A uniform infiltration flux of 1.5 cm/yr is applied along the top boundary and the resulting streamlines are shown in Figure 5. Infiltration flux diverges above the packed earth layer due to its low permeability and converges below due to capillarity in the Cape Riva formation. Sorption to the solid phase was not considered in this simulation so solute advection corresponds to water velocity. A molecular diffusion coefficient of 7.5E-6 cm^2/s was assumed. Model simulated saturation in the Minoan ash layer surrounding the source term is about 0.3 with 1.5 cm/yr infiltration. The umbrella effect of the low permeability layer, however, creates a dry zone in the Cape Riva formation immediately beneath the artifact location where flux vectors are less than 1 mm/yr and saturation is reduced to the range of 0.06 to 0.09. This corresponds to a moisture content in the range of 0.03 to 0.05 and a corresponding retardation factor of 95 to 158, assuming a distribution coefficient of K_d = 0.0027 m^3/kg. Distribution coefficients in [2] were determined from isotherm data for sorption of Cu on quartz [4] and kaolinite [5] to represent values for tuff and packed earth, respectively, using a water chemistry model [2]. We calculated 4.67E-5 molar concentration of Cu^{2+} in equilibrium with acatamite at pH = 6 using the geochemical speciation code EQ3. Taking an approximate average of 3 ppm of Cu sorbed to the solid phase from the data of Murphy et al. [2] gives an estimate of K_d =0.001 m^3/kg which reduces these retardation coefficients to the range of 36 to 59. At the high end of the K_d estimate, however, we have a plausible conceptual model for the observed Cu distribution. Lower infiltration flux due to evapotranspiration and a low permeability layer results in smaller flux velocities, lower moisture content, and increased retardation. Less solute transport may have occurred through the packed earth due to its clay content and possibly larger distribution

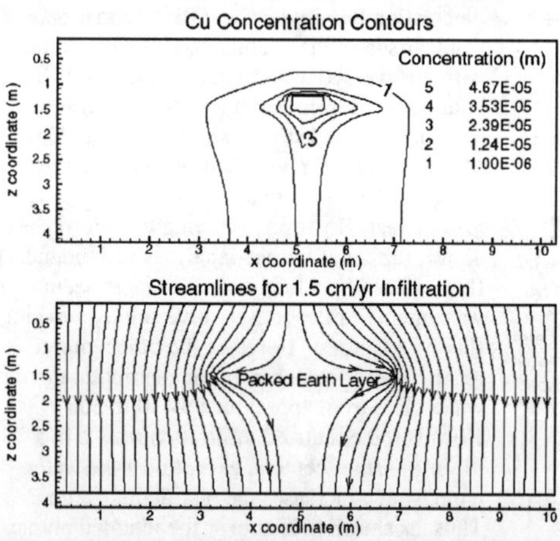

Figure 5. Distribution of aqueous phase Cu^{2+} at 3600 yr and infiltration flux of 1.5 cm/yr (without sorption) resulting from the presence of a low permeability layer underneath the source.

coefficient. Assuming a distribution coefficient $K_d = 0.45\ m^3/kg$ in the packed earth layer and a bulk density of 1300 kg/m^3 gives retardation coefficients in the range of 975 to 1085.

The effect of the large discrete vertical fracture intersecting borehole VII on unsaturated flow and transport is difficult to evaluate because it may act as a barrier to unsaturated flow under low infiltration flux and as a conduit for flow during a precipitation event. This transition is not easily handled using a porous media continuum model. The fracture was simulated by increasing the permeability five orders of magnitude over the permeability of the Cape Riva tuff in a column of elements in the grid. This may exaggerate the tendency of the fracture to act as a conduit for flow because moisture retention properties were the same as the surrounding Cape Riva tuff. The effect of a hypothetical high permeability vertical feature located at the downslope end of the packed earth layer at 3600 yr is shown in Figure 6 for 1.5 cm/yr infiltration and in Figure 7 for 15 cm/yr infiltration. The more widespread concentration distribution of aqueous Cu^{2+} at 1.5 cm/yr infiltration than at 15 cm/yr infiltration can be explained by mixing and dilution. The streamlines in Figure 6 show the path of advective transport below the packed earth layer. Transport through the packed earth layer is primarily by diffusion. With 1.5 cm/yr of infiltration the liquid velocities are low enough to allow for diffusive spreading of the plume. At 15 cm/yr, though, advection dominated flow below the packed earth layer sweeps the solute into the fracture where it is diluted by flux converging on this highly permeable zone and is transported out the bottom boundary of the model. Even though the plume is more widespread at 1.5 cm/yr infiltration, significantly more mass has been transported to the water table with 15 cm/yr infiltration flux. With 1.5 cm/yr infiltration 44 g of the mineral atacamite source term has been dissolved by 3600 yr (nominal width of the 2-D model is 0.2 m). Increasing the infiltration rate to 15 cm/yr resulted in 341 g of atacamite dissolved in 3600 yr, although the resulting contamination plume was less widespread.

These simulations did not include sorption. Considering a retardation coefficient for the packed earth layer as above in the transport model with 1.5 cm/yr infiltration would result in a sorbed phase concentration distribution in the rock similar to that observed from borehole data (Figure 1). These simulation results tend to support the hypothesis that infiltration is low as the artifacts removed were observed to be only slightly corroded with surface ornamentation of the bronze pieces essentially intact [2].

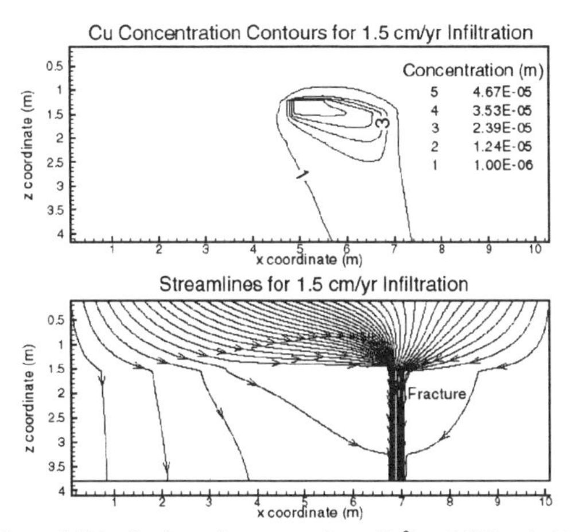

Figure 6. Distribution of aqueous phase Cu^{2+} at 3600 yr (without sorption) resulting from the presence of both a low permeability layer underneath the source and a vertical high permeability zone representing a fracture.

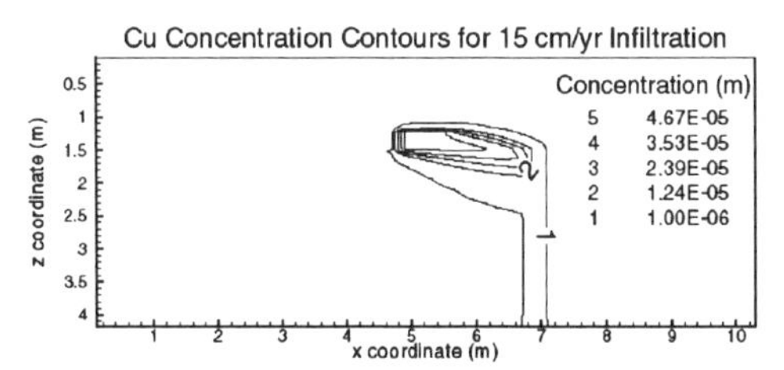

Figure 7. Identical model geometry as that shown in Figure 6 but with the higher 15 cm/yr infiltration rate.

Implications for Performance Assessments

Problems inherent in validation of groundwater flow and solute transport models, particularly verification of model predictions as bases for public policy decisions, are widely recognized [6]. The uncertainty in model predictions due to model non uniqueness is illustrated here by showing the effect of two alternative conceptualizations on solute transport predictions. Alternative models, apparently matching field data, yield different predictions for long-term release of trace metals to the water table by varying only infiltration flux and heterogeneity. Despite the good temporal and spatial constraints on the source term and the small size of the modeling domain, many degrees of freedom still exist. One goal of performance assessments is to bound the range of possibilities. Clearly the 1-D model with the higher flux rate of 15 cm/yr and no retardation predicts an early arrival of aqueous trace metal species at the water table. Mass flux from the vertical 1-D model is per unit horizontal area. For an area equivalent to the atacamite source term in the 2-D model (Figures 5, 6, and 7) the 1-D model with no retardation calculates 2 moles of Cu arriving at the water table in 3600 yr. This corresponds to dissolution of 427 g of atacamite which agrees reasonably well with the 341 g of atacamite dissolved in the 2-D model over the same time period. Including sorption to solids and assuming a lower flux rate of 1.5 cm/yr, however, results in a dramatically different prediction. In that case none of the dissolved Cu is transported to the water table within 3600 yr.

CONCLUSIONS

Two factors, spatially heterogeneous hydraulic properties and sorption to solids with reduced infiltration, could potentially explain the field data of Cu distribution at the Akrotiri archeological site. Transport models incorporating either of these factors, however, make divergent predictions within the 3600 yr time frame. In the case of heterogeneity and high infiltration, the plume arrived at the water table in about 30 yr and several hundred grams of the source mineral were removed by 3600 yr. In the case of low infiltration combined with sorption and retardation, no Cu reached the water table and little source mineral was removed by 3600 yr. Determining which of these models (if either) is more representative of transport processes at the site is difficult without more field work. Site characterization and modeling alone may result in unrealistic predictions. However, if more field work indicated that heterogeneity is a major factor affecting transport at the site but no evidence of an extensive plume was found, it would remain uncertain whether transport did not occur or if the sampling holes did not intersect the plume. So another conclusion is that the 1-D model with high infiltration and no retardation conservatively bounds the predictions. The non-conservative bound would be that very little transport has occurred.

ACKNOWLEDGMENTS

This report was prepared to document work performed by the Center for Nuclear Waste Regulatory Analyses for the Nuclear Regulatory Commission under Contract No. NRC 02-97-009. The report is an independent product of the CNWRA and does not necessarily reflect the views or regulatory position of the NRC.

REFERENCES

1. W.M. Murphy and E.C. Pearcy in *Scientific Basis for Nuclear Waste Management XIX*, edited by W.M. Murphy and D.A. Knecht (Mater. Res. Soc. Proc. **412**, Pittsburgh, PA, 1996) pp. 817-822.
2. W.M. Murphy, E.C. Pearcy, R.T. Green, J.D. Prikryl, S. Mohanty, B.W. Leslie, A. Nedungadi, J. Cont. Hydr. **29**, 245 (1998).
3. W.R. Hamon, Int. Assoc. Sci. Hydrol. Publ. **63**, 52 (1963).
4. P.W. Schindler, P. Liechti, J.C. Westall, Neth. J. Agric. Sci. **35**, 219 (1987).
5. W. Salomons and U. Förstner, *Metals in the Hydrocycle* (Springer, New York, 1984).
6. N. Oreskes, K. Shrader-Frechette, K. Belitz, Science **263**, 641 (1994).

Wasteform Characterization and Processing

GRAPHITE PROCESSING WITH CARBON RETENTION IN A WASTE FORM

M.I. OJOVAN, O.K. KARLINA, V.L. KLIMOV, G.Yu. PAVLOVA
Scientific and Industrial Association "Radon",
7th Rostovsky Lane, 2/14, Moscow, 119121, Russia, Oj@tsinet.ru

ABSTRACT

Conversion of waste graphite into a stable waste form acceptable for long term storage and disposal was considered both theoretically and experimentally. A self-sustaining transformation process of graphite composited with suitable precursors was studied. The powdered precursors that were used were used were: Al+SiO_2 (1), Al+TiO_2 (2) and Ti+SiO_2 (3).

Numeric thermodynamic simulation was performed. Equilibrium temperatures and chemical compositions of reaction products were determined for a wide range of component ratios in the source mixtures. The highest temperatures (up to 2300 K) were observed for precursor type (2). Precursor type (3) demonstrated a minimal rise of temperature of up to 1900 K.

Regions of compositions with complete binding of all chemical elements as well as production of stable final products were found to be rather narrow. About 10 – 13 wt.% of carbon can be processed in composition with given precursors. The gas phase reaction products were studied to minimize carry over of radionuclides. Carbon monoxide was shown to be the main component of the gas phase.

The self-sustaining synthesis process was conducted in ceramic crucibles at ambient pressure in an air atmosphere. Batch masses ranged between 0.1 – 1 kg. Best results were obtained for processing of graphite composited with Al and TiO_2. XRD analysis has identified titanium carbide and corundum in the waste form produced. These experiments confirmed that carbon can be converted completely into a stable waste form.

INTRODUCTION

Considerable amounts of high-level graphite waste containing fragments of fuel and fission products were accumulated during operation of uranium–graphite reactors. There is evidence that the ^{14}C-content in reactor graphite may be as much as 1 wt.% [1]. For safe disposal and long term storage, such waste must be properly processed into chemically stable materials.

At present known technologies for radioactive graphite processing are :

1. Graphite incineration in an oxygen-containing atmosphere, whichis the primary processed used [2]. However, this method does not provide radionuclide immobilization in a stable substance. Moreover, ^{14}C from the graphite passes into the gaseous phase as carbon dioxide. As a result, the necessity inevitably arises for additional technological operations that extremely complicate the waste treatment process.

2. The Savannah River Technology Center has developed an immobilization process for mixture of graphite fines with calcium fluoride (CaF_2, 15 wt.%) and plutonium oxide (PuO_2, 12 wt.%) [3]. The mixture is heated at 700 ^{o}C with a sodium borosilicate glass frit resulting in a waste form acceptable for disposal. Evidently this process consumes externally applied heat.

3. Recently attempts were made to apply self-sustaining high-temperature synthesis (SHS) to graphite waste processing [4,5]. SHS-reactions proceed in mixtures prepared by addition of a powdered metal fuel (aluminum or titanium) as an energy carrier and titanium or silicon dioxide

Mat. Res. Soc. Symp. Proc. Vol. 608 © 2000 Materials Research Society

to radioactive graphite waste: C(graphite) +Al + SiO_2 (1), C(graphite) +Al + TiO_2 (2), C(graphite) + Ti + SiO_2 (3).

When combustion of the mixture is initiated, metallothermic oxide reduction and interaction of elements reduced by graphite occur. As a result, graphite is chemically bound into stable metal carbides, which form a matrix for radionuclide oxide immobilization. High temperatures are developed from the metallothermic oxide reduction and carbide formation reactions. As a consequence, partial evaporation and dissociation of the reaction products and deviation from the stoichiometric equilibrium inevitably occur.

The detailed thermochemical analysis performed in this paper for reactions in mixtures (1) – (3) in item 3 above is aimed at determining the reaction temperature and phase composition of the reaction products. The technological aspects are not discussed.

THERMODYNAMIC SIMULATION OF THE THERMOCHEMICAL TREATMENT PARAMETERS

Thermodynamic simulation is, in essence, a numerical experiment and, at present, is widely used for prediction and analysis of the characteristics of high-temperature processes and reactions in multicomponent systems, including thermochemical treatment of radioactive waste (see, e.g., [6,7]). For thermodynamic simulation, the complex program ASTRA.4 [8] involving a database of thermodynamic characteristics of chemical compounds, comprising extensive domestic and foreign reference data [9,10], was used.

In the thermodynamic calculations, the formation of the following compounds due to reactions in mixtures (1)-(3) was considered: in the gas phase: O, O_2, C, C_2, C_3, CO, CO_2, C_2O, C_2O_3, Al, Al_2, AlO, AlO_2, Al_2O, Al_2O_2, Al_2O_3, AlC, AlC_2, Al_2C_2, Ti, TiO, TiO_2, Si, Si_2, Si_3, SiO, SiO_2, SiC, SiC_2, Si_2C, Si_2C_2, and Si_3C; in the condensed phase: C, Al, Al_2O_3, Al_4C_3, Si, SiO_2, SiC, Ti, TiO, TiO_2, Ti_2O_3, Ti_3O_5, Ti_4O_7, TiC, TiSi, $TiSi_2$, and Ti_5Si_3.

The calculation results are shown in Figures 1-3 as ternary diagrams.

Equilibrium reaction temperature as a function of mixture composition is plotted in Fig. 1. The highest temperatures of about 2200 K and a wide area of 2000 K were observed in mixture (1) (Fig. 1a). A maximum temperature of 2300 K was generated in the mixtures (2), and for these mixtures, 2200 K is seen in a very wide region of compositions (Fig. 1b,d). The mixtures (3) develop the lowest temperatures of the three mixtures - the maximum temperature is only 1900 K (Fig. 1c). However, the real temperature might be much lower than the calculated equilibrium temperature because of reaction incompleteness.

Earlier, it was shown that the mixtures (2) produce the best results [11]. Therefore we shall only further consider further the phase composition of Al–C–TiO_2 mixtures.

The equilibrium phase composition of the combustion products of Al-C-TiO_2 mixtures is shown on Fig. 2. As can be seen, the region of mixture formulations that forms stable chemical compounds such as TiC, Al_2O_3 and Ti_xO_y, is rather wide (area 12 on the diagram). Area 12 is shown in detail on Fig. 2b, which also shows four subareas – 13, 14, 15 and 16. Only the stoichiometric area 13 contains Al_2O_3 and TiC with no titanium oxides.

The mixture formulations corresponding to diagram areas 1 - 11 are unsuitable for thermochemical treatment of graphite due to the presence of unbound elements (C, Al, and Ti) and easily hydrolyzed aluminium carbide Al_4C_3 in the end reaction products.

It should be noted that one must not overlook the temperature conditions corresponding to a given mixture (Fig. 1) when treating the calculation results analytically on phase composition of the end products. There could be a situation when the optimum (from the viewpoint of the best end product phase composition) mixture compositions would not react in a self-sustaining reaction due to the calculated temperature being too low.

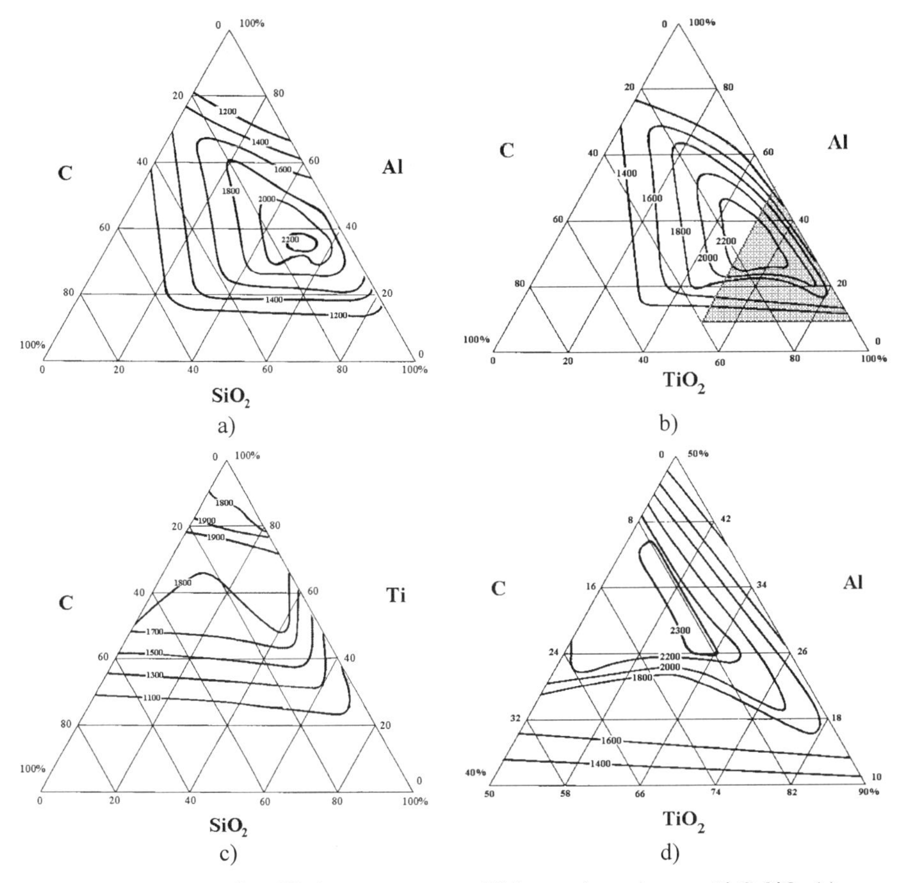

Fig.1. Isotherms of equilibrium temperature (K) in reacting mixtures Al-C-SiO_2 (a), Al-C-TiO_2 (b, d – corresponds to shaded area in b) and Ti-C-SiO_2 (c) at pressure p=0,1 MPa.

For reaction temperatures of 2000-2300 K, the main noncondensing components of the gas phase are carbon monoxide and carbon dioxide. Fig. 3 demonstrates the isolines of equilibrium concentrations of carbon monoxide (a) and dioxide (b) in the gas phase reaction products for mixture (2). The CO_2 content is always a few orders of magnitude lower than the CO content. The concentrations of CO and CO_2 are given on Fig. 3 in kg per 1 kg of the total mixture of combustion products (gas and condensed phases).

A correlation of the Fig. 2 data with those of Fig. 3 shows that for the most interesting (with respect to end product composition) areas of the phase diagrams the CO content approaches 0.01 kg/kg (Fig. 3a). And thus, in view of [1], ^{14}CO and $^{14}CO_2$ releases can reach 10^{-4} and 10^{-7} kg/kg of the batch, respectively.

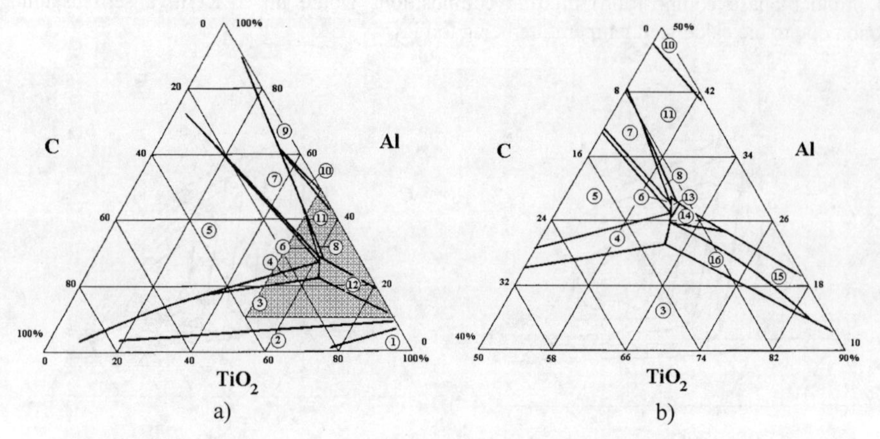

Fig.2. Equilibrium phase composition of condensed reaction products in reacting Al-C-TiO_2 mixtures. a) – all mixtures, b) –mixtures from the highlighted area on a).

Phase areas: 1 – C, Al_2O_3, TiO_2, Ti_4O_7; 2 – C, Al_2O_3, Ti_4O_7, Ti_2O_3; 3 – C, Al_2O_3, Ti_2O_3, TiC; 4 – C, Al_2O_3, TiC; 5 – C, Al_2O_3, Al_4C_3, TiC; 6 – Al_2O_3, Al_4C_3, TiC; 7 – Al, Al_2O_3, Al_4C_3, TiC; 8 – Al, Al_2O_3, TiC; 9 – Al, Al_2O_3, Ti, TiC; 10 – Al, Al_2O_3, Ti, TiO, TiC; 11 – Al, Al_2O_3, TiO, TiC; 12 – Al_2O_3, Ti_xO_y, TiC; 13 – Al_2O_3, TiC; 14 – Al_2O_3, TiO, TiC; 15 – Al_2O_3, TiO, Ti_2O_3, TiC; 16 – Al_2O_3, Ti_2O_3, TiC.

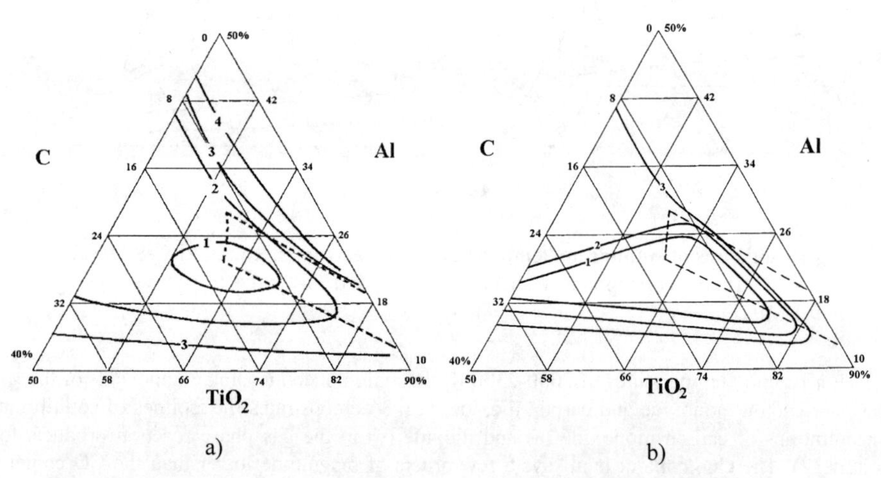

Fig. 3. Isocurves of carbon monoxide (a) and carbon dioxide (b) equilibrium concentration in reacting mixtures of Al-C-TiO_2. (Dashed lines confine area 12 on Fig. 2a). The numbers denote concentrations - on Fig. 3a: 1 – $0.5\cdot10^{-1}$, 2 – 10^{-2}, 3 – 10^{-5}, 4 – 10^{-8} kg of CO/kg; - on Fig. 3b: 1 – 10^{-5}, 2 – 10^{-6}, 3 – 10^{-7} kg of CO_2/kg.

EXPERIMENT

We did not aim to perform exhaustive experimental studies, however, we carried out some tests on mixtures (1)-(3). For test simplification, we thoroughly excluded the steps of preliminary batch compaction and hot compaction of the end product from the sum total of mechanical operations.

Standard graphite, aluminum, titanium, titanium dioxide and silicon dioxide powders were used as source components. Mixing was performed manually. Mixture portions of 0.1 – 1.0 kg were incinerated in a ceramic crucible. Special precautions were not applied to exclude the effects of ambient air.

Under such conditions, we could not achieve stable and complete combustion of Al-C-SiO_2 mixtures. The reaction ceased in a few seconds after the initiator burnout that proceeds on the surface of the source mixture.

The combustion of Ti-C-SiO_2 mixtures ran better. The whole mixture continued to glow for a few minutes after initiator burnout.

However, the best results were achieved for Al-C-TiO_2. In this case the end product was a porous cylinder which took the shape of the crucible. According to XRD data, the reaction products of the mixture, which was composed of Al – 28, C – 9.5 and TiO_2 – 62.5 wt.%, consisted of titanium carbide, alumina and titanium oxide (Fig. 4). This data agrees with the results of the thermodynamic simulation data (see Fig. 2b).

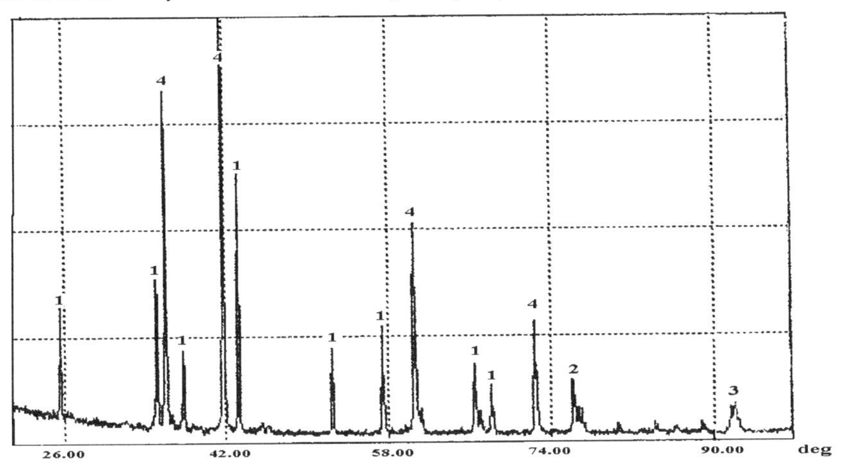

Fig. 4. The XRD-analysis data of the end reaction products for Al-C-TiO_2 mixtures. 1 – Al_2O_3, 2 – TiC, 3 – TiO, 4 – TiC, TiO.

CONCLUSIONS

Using the numeric thermodynamic simulations:

- The reactions of graphite conversion into chemically stable compounds (titanium and silicon carbides) were studied
- The ranges of temperatures supporting the most effective conditions for carbide-formation reactions were established in ternary diagrams of the source mixtures

- In the ternary diagrams of the source mixtures, areas were established which are characterized by complete chemical binding of graphite and by the absence of unstable compounds in the end products
- It was established that, under conditions of high reaction temperature, certain source mixtures produce some amounts of undesirable carbon monoxide.

In conclusion, it would not be out of place to make the following remark. When the data on thermodynamic simulation are treated analytically, one must bear in mind that when the calculation results are closer to reality, the process temperature is higher and the duration of the process is longer.

ACKNOWLEDGMENT

The authors are grateful to Dr. N. Penionzhkevich for XRD analysis and N. Manyukova for computer graphs.

REFERENCES

1. V.I. Bulanenko, V.V. Frolov, A.G. Nikolayev, Atomic Energy. **81**, 304 (1996) (in Russ.)

2. Ya.V. Natanzon, V.V. Tokarevski, V.A. Kremnev, V.Ya. Petrishchev, and V.P. Titov, SU Patent No. 1 718 277 (31 May 1989).

3. T.S. Rudisill, J.C. Merra, and D.K. Peeler in *Abstracts of the Materials Research Society 1998 Fall Meeting* (Boston, Massachusetts, 1998) pp.714-715.

4. A.G. Merzhanov, I.P. Borovinskaya, N.S. Mahonin, et al., Russ. Patent No. 2 065 220 (18 March 1994).

5. E.Ye. Konovalov, O.V. Starkov, M.P. Muishkovski, L.S. Gudkov, and F.D. Lisitsa, Atomic Energy. **84**, 239 (1998) (in Russ.).

6. V.V. Kropochev, M.I. Ojovan, I.A. Sobolev in *Proceedings of the 1997 Intern. Conf. on Incineration and Thermal Treatment Technologies* (Oakland, California, 1997) pp. 207-212.

7. I.A. Sobolev, M.I. Ojovan, G.A. Petrov, V.L. Klimov, and V.L. Tarasov in *Proceedings of the 1998 Intern. Conf. on Incineration and Thermal Treatment Technologies* (Salt Lake City, Utah, 1998) pp.311-313.

8. B.G. Trusov, *Simulation of Chemical and Phase Equilibrium at High Temperatures: Computer Code,* State Reg. No. 920 054 (31 March 1992).

9. *JANAF Thermochemical Tables,* J. of Phys. and Chem. Ref. Data, **14**, Suppl. No.1 (1985).

10. Gurvich L.V., Veits I.V., Medvedev V.A., et al., *Thermodynamic Properties of Individual Substances,* 4th ed. Vol.1–4. (Hemisphere Publ. Corp. & CRC Press, New York, 1989–1994).

11. V.L. Klimov, M.I. Ojovan, O.K. Karlina, and G.Yu. Pavlova in *Proceedings of the 1999 Intern. Conf. on Incineration and Thermal Treatment Technologies* (Orlando, Florida, 1999).

LONG-TERM BEHAVIOR OF BITUMEN WASTE FORM

I.A. SOBOLEV, A.S. BARINOV, M.I. OJOVAN, N.V. OJOVAN, I.V. STARTCEVA, Z.I. GOLUBEVA

Scientific and Industrial Association 'Radon', the 7-th Rostovsky Lane 2/14, Moscow, RF, Fax: (095) 248 1941, E-mail: Oj@tsinet.ru

ABSTRACT

Waste blocks were produced in a bituminization plant using NPP-operational and other liquid wastes of low and intermediate level activity and tested under laboratory and near-surface wet disposal conditions. Leach rates of radioactive and non-radioactive waste components and depths of radionuclide penetration into the host loamy soil were estimated. Bituminized waste seems to occupy a middle position between cemented and vitrified waste forms in terms of radionuclide retention ability. For certain samples of the bitumen waste form, the testing covers a period of more then a quarter of a century that is of great importance for prediction of the waste form behavior over the required time period of several hundred years.

INTRODUCTION

The waste matrix is considered the first and the most important barrier to radionuclide release in a multi-barrier waste disposal concept. For over two decades, SIA 'Radon' has been investigating three main types of matrices for low and intermediate level waste (LILW) conditioning [1, 2]. Cemented, bituminized and vitrified waste blocks were produced in industrial and pilot scale facilities and disposed of at the testing area consisting of an open testing site and experimental shallow-ground repositories. Bitumen has been used as a matrix material for encapsulating liquid radwaste (sludges from evaporation or precipitation, etc.) since 1968. Two industrial bituminization plants started operation in 1974 and 1978: a continuously operating, two-stage facility of high capacity and one more based on the usage of the rotary thin film evaporator/ mixer. Three types of bitumen were used in the laboratory and field experiments (Table I).

Table I. Fractional compositions and standard physical properties of bitumens.

Type of bitumen		BN-II	BN-III	BN-IV
Fractions, wt %	oils (aliphatic hydrocarbons)	55.5	54.5	50.0
	resins	17.5	15.0	11.0
	asphaltenes	27.0	30.5	39.0
Depth of the hardening zone at 25°C, mm		81-120	41-80	21-40
Softening temperature, °C		40	45	70
Ignition temperature (in an open crucible), °C		200	200	230

This paper presents results from laboratory and long-term field tests performed on samples/blocks of bituminized waste. Institutional and real NPP-operational wastes were used.

EXPERIMENT

The scope of laboratory work involved research of interactions in the system *waste form - water* (the impact of waste content, bitumen type and bitumen admixtures on radionuclide retention

Mat. Res. Soc. Symp. Proc. Vol. 608 © 2000 Materials Research Society

properties of bituminized waste) and in the *water - soil* system (the influence of waste form on backfill and soil parameters), as well as estimation of waste product radiation stability.

Leach resistance was measured according to E. D. Hespe [3] on samples with waste loadings of 30, 35, 40, 50, and 60 %. Two leach stages were identified. At the initial stage of high leach rates (on the order of $1 \cdot 10^{-1}$ to $1 \cdot 10^{-3}$ cm/day), the dissolution of salts from the outer layer of the waste product occurs, for 15-50 days. After this stage, the leach rate rapidly decreases with time. The next stage is characterized by a slow decrease in leach rate to values of about $8 \cdot 10^{-5}$ cm/day (for waste loadings of up to ~40 %). This is indicative of a diffusion mechanism for waste component release. Normalized leach rates for waste macrocomponents are somewhat higher than those for radionuclides (after 15 days). Some leach trends were revealed for elements and species encountered in wastes. Cations and anions may be arranged in order of decreasing leach rate as follows:

$$Na^{+} > NO_3^{-} > K^{+} > Cs^{+} > Ca^{2+} > Mg^{2+} > Cl^{-} > SO_4^{2-}.$$

Rates of radionuclide release from bitumen compound increase with the rise in waste loading. For waste loadings of up to 25-30 wt.%, only a slow growth in radionuclide leach rates was observed. For higher waste content, this growth is more pronounced as a concequence of pore formation resultingn in enhanced ingress of water to the interior of the bitumen compound. Waste loading of 50 wt.% is considered a threshold level.

Waste samples prepared using soft bitumen BN-II proved to be the most leach resistant. Samples using BN-IV, characterized by a high content of asphaltens, displayed the worst leach resistance properties.

It was found that waste component leaching from the bitumen compound into the loamy soil (host rock of a repository site) depends on soil humidity. So, radionuclide leach rates increased by a factor of about eight with an increase in loamy soil humidity from 5 to 30 %. Other changes in physical, physico-chemical and physico-mechanical properties of argillaceous ground caused the increase of Sr and Cs migration rates by a factor of 10-14.

Parallel with the laboratory studies on bitumen samples containing real radwastes or simulators, field experiments are ongoing to address the issue of how effective the waste bitumen matrix is as a part of an engineered barrier system. Long-term behavior of bituminized waste under near-surface repository conditions is very dependent on equilibrium in the system soil - water - bitumen waste form. Two groups of factors affecting the longevity of bitumen waste products may be distinguished: *intrinsic factors* including conditioned product quality (homogeneity, water content, etc.), type of bitumen used, admixture presence, waste content and its composition, ets.; and *external factors*, such as the ambient temperature, sorption properties of the soil, leachant chemistry, water flow rate, etc.

Real NPP-operational (Kursk NPP) and decontamination wastes generated in the facility itself ('Radon') were used. The principal radioactive waste components were ^{137}Cs, ^{134}Cs, ^{90}Sr, and ^{60}Co [2]. Radwastes represented low alkalinity solutions with nitrates of alkaline and alkaline-earth metals as dominating non radioactive components (up to 87 wt.% of the total salt content). So, waste from Kursk NPP (reactor RBMK) was characterized by a salt content of about 350 g/l, of which sodium nitrate constituted approximately 86 %.

Three types of commercial paving and construction oil bitumen were used: BN-II, BN-III and BN-IV. Their fractional compositions and some physical parameters are given in Table I. An admixture of K-Ni ferrocyanide as a leach retarder was used in preparation of certain test samples.

Waste blocks were produced in the bituminization plant of SIA 'Radon'. Before solidification, liquid wastes were concentrated in an evaporator/dryer to about 60 wt % dry matter. Preheated waste sludge and bitumen were fed into an evaporator/mixer. The temperature at the casting stage did not exceed 135 °C. Hot mixtures were poured into moulding cars or into carbon steel containers with a wall thickness of 1.5-2.5 mm. Containers for open site tests were of a few liters or a few tens of liters in volume with the upper part open to allow contact of solidified waste and the atmosphere. The main parameters of the waste blocks tested were described in [2]. The testing site consists of an open site and two isolated shallow ground repositories. The total area is approximately 1200 m^2. A complete description of the testing sites is given elsewhere [1, 4].

Long-term testing of waste forms is being performed mainly by monitoring contamination of the contact water (precipitation, groundwater) and soil. Water samples are routinely collected, usually twice a month, for chemical, radiochemical and radiometric analyses. The same analyses have been applied to soil drill cores obtained through drilling the soil surrounding the waste blocks.

RESULTS AND DISCUSSION

Radionuclide leach rates were determined for samples which differed in waste loading, type of bitumen used, isotope composition of the initial waste, and by the presence of leach retarder.

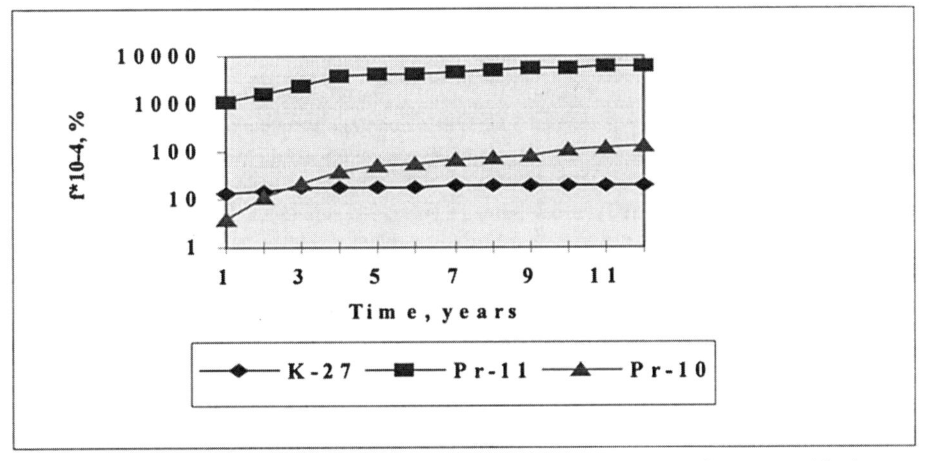

Fig.1. Radioactivity fractions leached by precipitation and groundwater from waste blocks as functions of time. Pr-11 (open site) and K-27 (repository) are analogues. Pr-10 contains K-Ni ferrocyanide admixture.

^{137}Cs was the main radioactive contaminant of the contacting water (up to 96 % of the total leached radioactivity). Therefore, radioactivity release rates were strongly affected by the presence of ferrocyanide admixture (see Pr-10 in Fig.1), especially for waste loadings ≥ 50 %. Data generated from laboratory, open-site and repository tests showed that the leach behavior of the waste product is notably dependent on type of bitumen used. Soft bitumen (BN-II) is the best as a waste matrix in terms of leach resistivity. The major part of the leached radioactivity, 75 to 86 %, is released from waste blocks during the first year of the testing period. Leach parameters of samples are given in Table II.

Table II. Leach parameters of bitumen waste blocks

Test conditions		Open area		Shallow-ground repository			
Parameter	Time(yr)	Pr-10	Pr-11	K-27	K-4	K-5	K-6
Specific	1	26.74	1310	32.43	91.73	125.11	106.11
radioactivity,	11	92.80	696	10.72	56.91	66.08	41.29
Bq/l	24	62.73	-	-	50.47	44.52	23.45
Leach	1	$1.02 \bullet 10^{-7}$	$7.29 \bullet 10^{-5}$	$7.11 \bullet 10^{-7}$	$2.19 \bullet 10^{-6}$	$4.10 \bullet 10^{-5}$	$1.85 \bullet 10^{-7}$
rate,	11	$2.63 \bullet 10^{-7}$	$4.56 \bullet 10^{-5}$	$1.32 \bullet 10^{-7}$	$8.66 \bullet 10^{-7}$	$1.52 \bullet 10^{-5}$	$5.94 \bullet 10^{-8}$
$g/cm^2 \cdot day$	24	$2.57 \bullet 10^{-7}$	-	-	$4.92 \bullet 10^{-7}$	$8.07 \bullet 10^{-6}$	$3.43 \bullet 10^{-8}$
f,	1	0.0004	0.11	0.001	0.004	0.070	0.0002
%	11	0.01	0.64	0.002	0.005	0.082	0.0003
	24	0.02	-	-	0.007	0.084	0.0004

Data on leaching from bituminized samples prepared from the same source mixture (see Fig.3) and tested under open site and repository conditions indicate the essentially better leaching behavior of bituminized waste throughout the shallow-ground disposal testing. Therefore the data of Table II may be considered as conservative estimates of the radionuclide containment properties of the various waste forms.

Two blocks of bituminized waste, S-I and S-II, had been placed directly into bore pits. Their parameters were: waste loading, 29 % and 5 %; specific radioactivity, $6.66 \cdot 10^5$ Bq/kg and $1.85 \cdot 10^7$ Bq/kg; weight, 500 kg and 300 kg , resp. The space around the waste blocks in the pits had been filled by a host loamy soil. Therefore the waste matrix was the only barrier to radionuclide release into the environment. After 11 years, measurements of soil contamination have been carried out. Maximum soil radioactivity was registered underneath the waste blocks and reached $2.81 \cdot 10^4$ Bq/kg for S-I and $2.66 \cdot 10^6$ Bq/kg for S-II. Doubled background values for the disposal site host rock, $1.85 \cdot 10^3$ Bq/kg, were observed at a maximum distance of 40 cm below S-I and 26 cm below S-II. Hence, the average annual rates of radionuclide migration into the host rock were estimated to be approximately 3.6 cm/yr and 2.4 cm/yr, respectively. They indicate the salt content of bituminized products may have a stronger effect on radioactivity release and soil contamination compared with the activity concentration in these products (due to competition of Na+ and Cs+ for absorption and cation-exchange sites of the soil). Results show quite good radionuclide containment properties for the bitumen matrix as well as for thehost rock of the disposal site.

As a result of the interaction of groundwater, precipitation, and the bitumen compound, changes in the chemistry of the leachate water occur. Results of long-term groundwater chemistry monitoring are provided in Table III.

Table III. Leachate groundwater chemistry data, mg/l.

	Time, years	K-27	K-4	K-5	K-6	Initial data
pH	1	8.03	7.54	7.54	7.39	7.69
	12	8.08	7.85	7.67	7.66	
	27	-	8.09	7.96	8.09	
TDS	1	845	1430	4131	7250	261
	12	646	1772	3632	3193	
	27	-	1790	3202	1980	
Na^+	1	186	293	1134	1603	19.6
	12	167	402	934	546	
	27	-	469	833	452	
Ca^{2+}	1	54				39.8
	12	40	42	49	52	
	27	-	24	39	31	
NO_3^-	1	174	841	2626	1935	3.2
	12	37	534	1829	330	
	27	-	427	1521	164	
HCO_3^-	1	505				230
	12	542	1349	766	622	
	27	-	970	653	753	

The salinity of the contact groundwater increased from 0.9 to 11 g per litre. The initial Ca-hydrocarbonaceous type of groundwater changed to HCO_3^- - NO_3^- - Na^+. The initial pore water chemistry, represented by HCO_3^- - Ca^{2+}- - Na^+ with a salinity of 0.07 g/l, changed to NO_3^- - HCO_3^- - Ca^{2+} - Na^+ after contact with the bitumen blocks. Pore water salinity reached 0.5 g/l.

PREDICTION OF LONG TERM BEHAVIOR

It is assumed that the mechanism of radionuclide leaching will remain the same over the entire time of storage. After some initial period of time (less than 1 year) this mechanism is assumed to be governed by the diffusion of radionuclides. Therefore one can consider that the leached radioactivity fraction consists of two terms:

$$f=f_0+f_1,$$

where f_0 is the leached radioactivity fraction due to initial dissolution processes, and f_1 is given by the formula:

$$f_1=(S/V)[\Sigma q_{0i}(D_i/\lambda_i)\Phi(\sqrt{\lambda_i t})]/\Sigma q_{0i},$$

where S is the interface surface, V is the volume of the waste block, q_{0i} is the initial specific radioactivity of the i-th radionuclide in the block, D_i is the diffusion coefficient, $\Phi(\sqrt{\lambda_i t})$ is the error integral, λ is the decay constant, and t is time.

Fig.2 shows the prediction of leaching behavior of a block (K-27) during storage for 300 years. We took f_0 from experimental data after one year of storage - f_0=0.00138%. The experimental data can be seen in the beginning of the plot.

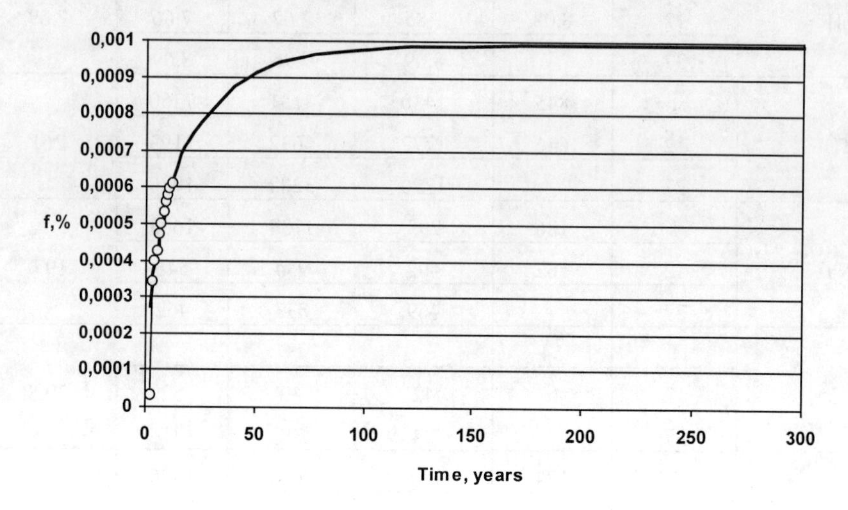

Fig.2. The leached radioactivity fraction of a block of bituminized radioactive waste (f_1).

One can see that the total (and maximum) leached radioactivity fraction will be reached by the time the value $f_{max} = f_0 + 0.001\% = 0.00238\%$.

CONCLUSION

Bitumen may be considered a suitable host material for immobilizing certain low-level-activity wastes, primarily for wastes having poor moisture absorption in a dry state and for placement in a near-surface repository whose design requires the introduction of several engineered barriers.

LITERATURE

1. I.A. Sobolev, A.S. Barinov and M.I. Ozhovan, Soviet Atomic Energy **69** (5), 950-953 (1990).

2. I.A. Sobolev, A.S. Barinov, M.I. Ojovan, N.V. Ojovan, I.V.Startceva, and Z.I. Golubeva, in *Proc. Int. Workshop Radwaste Bituminization'99*, Rez -Prague, Czech., June 6 - July 2 1999

3. E.D. Hespe, Atomic Energy Rev. **9**, (91) 195-207 (1971).

4. I.A. Sobolev, A.S. Barinov, M.I. Ojovan, N.V. Ojovan, G.N.Chuikova, and I.V. Startseva: in *Proc. Int. Conf. Waste Management*, Tucson, AZ, February 28 - March 4 1999, CD-ROM.

CHARACTERIZATION OF A CERAMIC WASTE FORM ENCAPSULATING RADIOACTIVE ELECTROREFINER SALT

T.L. Moschetti, W. Sinkler, T. DiSanto, M.H. Noy, A.R. Warren, D. Cummings, S.G. Johnson, K.M. Goff, K.J. Bateman, and S.M. Frank
Argonne National Laboratory-West, P.O. Box 2528, Idaho Falls, ID 83403-2528

ABSTRACT

Argonne National Laboratory has developed a ceramic waste form to immobilize radioactive waste salt produced during the electrometallurgical treatment of spent fuel. The first ceramic waste forms that immobilize the radioactive waste salt have been produced. This study presents the first results from electron microscopy and durability testing of a ceramic waste form produced from that radioactive electrorefiner salt. The waste form consists of two primary phases: sodalite and glass. The sodalite phase appears to incorporate most of the alkali and alkaline earth fission products. Other fission products (rare earths and yttrium) tend to form a separate phase and are frequently associated with the actinides, which form mixed oxides. Seven-day leach test results are also presented.

INTRODUCTION

Argonne National Laboratory processes EBR-II spent fuel by an electrometallurgical treatment [1]. This treatment creates a radioactive salt waste stream containing transuranics and fission products. To immobilize the radioactive salt, a ceramic waste form has been developed [2-3]. The ceramic waste form (CWF) is produced by blending salt and zeolite 4A at approximately 773 K resulting in a salt loading of about 10.5 wt%. A composition of the electrorefiner salt incorporated into the waste form is given in Table I. The salt-loaded zeolite is then mixed with glass (in a weight ratio of 3:1) and hot isostatically pressed to obtain a consolidated, durable waste form. The maximum temperature and pressure within the hot isostatic press were 1123 K and 100 MPa. For the CWF, the uranium content is 0.5 wt% while the plutonium content is 0.02 wt% and neodymium, the dominant rare earth, is 0.05 wt%.

Table I. Measured Composition of Electrorefiner Salt Incorporated in the Ceramic Waste Form.

Element	Measured wt %	Element	Measured wt %
Li	5.87	La	0.21
Na	1.9	Ce	0.42
K	21.3	Pr	0.21
Cs	0.66	Nd	0.70
Sr	0.15	Sm	0.21
Ba	0.25	Np	0.0265
Y	0.12	U†	7.207
Fe	0.066	Pu†	0.291
Cl‡	57.63		

†Isotopic compositions are approximately 60% ^{235}U, 40% ^{238}U and 99% ^{239}Pu, rest ^{240}Pu and ^{238}Pu.

‡Calculated assuming anionic composition is 100% Cl^- and standard chloride stoichiometries for all elements (3 chlorides for all rare earths and actinides). Analyses for iodide and bromide were not performed but are expected to be 350 μg/g and 45 μg/g, respectively.

Mat. Res. Soc. Symp. Proc. Vol. 608 © 2000 Materials Research Society

EXPERIMENTAL

Both transmission electron microscopy (TEM) and scanning electron microscopy (SEM) were performed. The SEM was done with a Zeiss DSM 960A scanning electron microscope (Thornwood, NY). Energy dispersive and wavelength dispersive x-ray spectroscopy (EDS/WDS) was done by interfacing the detectors and instrumentation of the microscope to an Oxford ISIS series 300 x-ray analysis system, software version 3.2, and the Oxford software Winspec, version 1.3 (Oxford, UK). The TEM was performed with a JEOL 2010 transmission electron microscope (Peabody, MA) operated at 200 kV, and equipped with a LaB_6 filament and EDS detector. The instrument also has electron diffraction (ED) capabilities.

A seven-day Product Consistency Test (PCT-B) was performed according to ASTM C1285-94 [4]. The samples were first crushed and sieved to obtain particle sizes between 75-150 μm. Then they were placed in de-mineralized water at 90°C. The only difference from PCT-A is that 11.2 ml instead of 10 ml of de-mineralized water was added for each gram of the CWF sample. This resulted in a surface area-to-volume ratio (SA/V) of 2000 m^{-1} for the CWF. Upon completion of the PCT, chloride concentrations in the leachate were measured with an ion specific electrode and the pH with a pH meter. Other elemental concentrations in the leachate were determined with inductively coupled plasma-mass spectroscopy or -atomic emission spectroscopy. The normalized mass loss (NML) for various elements was calculated using

$$NML = \frac{C_i * V}{f_i * SA} \quad (1)$$

where C_i is the concentration of the ith element in the leachate and f_i is the mass fraction of the ith element in the unleached sample. Samples of the CWF were tested in triplicate along with a triplicate of the ARM-1 waste glass reference material [5] and a duplicate of blanks.

MICROSCOPY RESULTS

A typical backscattered electron (BSE) micrograph of the first ceramic waste form produced from radioactive electrorefiner salt is shown in Figure 1. The microstructure of the waste form generally consists of sodalite and glass regions, 50-150 μm in size. Glass sometimes penetrates into the sodalite regions, though on a much finer scale. Actinide/rare earth-containing phases and halite are minor phases that tend to be found along grain boundaries. Two fission products, Cs and Sr, were below the detection limit of EDS/WDS. This microstructure corresponds well with that of previously studied waste forms, which were loaded with surrogate fission product salt, plutonium, or uranium [6-8].

Rare earths, actinides and yttrium form separate phases in the waste form and often are closely associated with one another. The composition of these phases varies but uranium is predominant throughout. This is unsurprising since the uranium content is at least four times greater than the content of plutonium, rare earths, or yttrium. Due to their small size, the chemical form (such as oxide or silicate) of these elements could not be determined by SEM/EDS. Previous studies on surrogate fission product- and actinide-doped waste forms suggest that plutonium and rare earths form silicates as well as mixed oxides [7-8]. However, TEM/ED results on the electrorefiner-salt-loaded waste forms show that the uranium, plutonium, and rare earths are predominantly solid solutions of mixed oxides. The plutonium and rare earths may have formed silicates in these other studies because of higher concentrations of plutonium and rare earths in the waste forms.

Figure 2 is a typical bright field image of the CWF obtained by TEM, where examples of phases are marked. The four phases observed in the sample are as follows: (1) mixed oxide $(Pu,U)O_{2-x}$ with fluorite structure confirmed by ED, (2) sodalite, (3) glass and (4) halite (NaCl). Occasionally rare earths were detected with the mixed oxide phase as well.

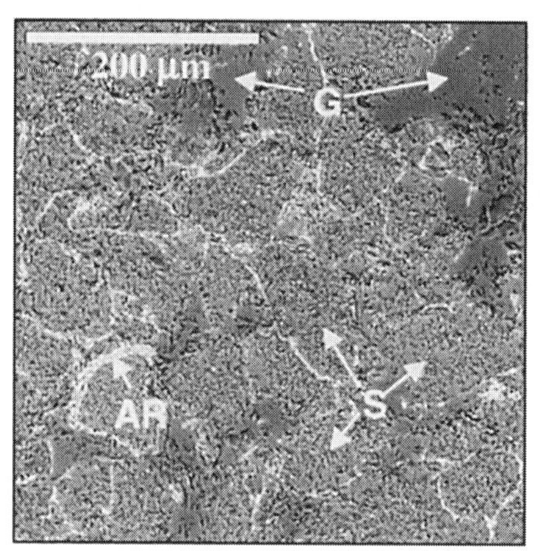

Figure 1. Typical Micrograph of the First CWF Produced with Radioactive Electrorefiner Salt. The CWF generally consists of glass (G) and sodalite (S) granules that are 50-150 µm across. The high contrast inclusions containing various amounts of actinides and rare earths (AR) tend to concentrate along the sodalite grain boundaries.

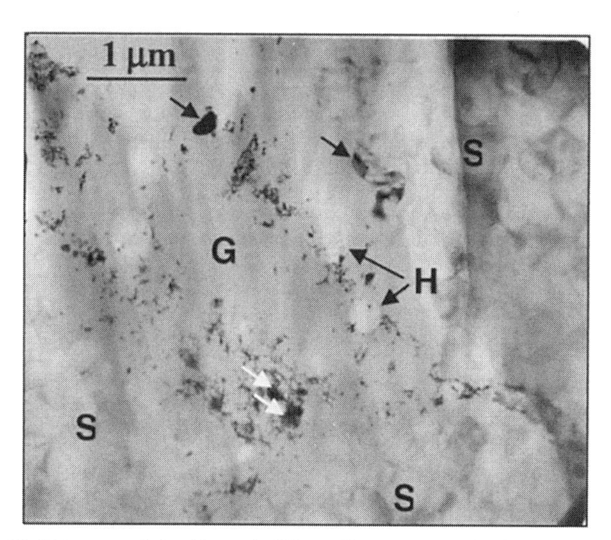

Figure 2. Bright Field Image of the Ceramic Waste Form. The principal phases found by TEM are sodalite (S), glass (G), mixed oxides containing varying amounts of U, Pu, and rare earths, and spherical inclusions of halite (H). Unmarked arrows indicate larger mixed oxide particles.

The microstructure determined by TEM for the CWF loaded with the radioactive salt, whose composition was indicated in Table I, was typical of many CWF samples. As was the case with examinations of surrogate and U/Pu-doped CWFs [7-8], the microstructure of the radioactive salt-loaded CWF is dominated by large regions of glass and sodalite. The halite phase is present as spherical inclusions within the glass, often near the glass/sodalite phase boundary. The actinide-bearing mixed oxide phase is primarily present as fine unfaceted crystals approximately 20 nm in diameter. In addition to these very fine crystals, a few larger mixed oxide crystals are also present, interspersed within regions containing the smaller grains. The mixed oxide crystals are situated within the glass phase, or occasionally sandwiched between two sodalite regions. The actinide contents of the glass and sodalite phases were near detection limits for EDS. This, plus the existence of a separate actinide-based phase in the CWF, in spite of the actinides being a relatively minor component of the salt, is indicative of a negligible quantity of actinide dissolved in either the glass or sodalite phases.

PCT RESULTS

The results for selected elements of the PCT are given in Table II. In the table, the CWF is also compared with a glass similar to the Defense Waste Processing Facility (DWPF) glass [9] and the Environmental Assessment (EA) glass [5]. The normalized mass loss is given in both references for a PCT-A, which is similar to the test described in the Experimental section except 10 ml of leachant were used per gram of sample. This still results in SA/V ratio of about 2000 m^{-1} for the glass samples so the results should still be comparable. As can be seen from Table II, the normalized mass loss for elements in the radioactive CWF is comparable to the DWPF glass and is much smaller than the EA glass.

The mixed oxide phase appears to be very durable. This is clear from the lower NML of actinides, rare earths, and yttrium. The release of Pu is also lower than that from material similar to DWPF glass, given in Table II.

Table II. Normalized Mass Loss for Selected Elements in the 7-Day PCT. Units are in g/m^2.

Element	Radioactive CWF[†]	DWPF glass [9] 200R	DWPF glass [9] 165/42	EA glass [5]
Li	0.71	0.485	0.275	5.0
Na	0.26	0.495	0.18	6.9
K	0.14	0.285		
Cs	0.22	0.10	0.060	
Ba	0.058			
Sr	0.052		0.0095	
Si	0.068	0.225	0.165	4.3
Al	0.082	0.15	0.16	
B	0.71	0.55	0.205	8.3
Cl	2.18			
Y	0.020			
Ce	0.018			
Nd	0.0077			
Pu	0.013	0.049[‡]	0.065[‡]	
U	0.010			

[†]Errors at two sigma for concentrations used to calculate values are 5% for Al, Li, K, Na, and Si and 10% for all others. [‡]Mass loss includes all alpha emitters, primarily Pu-238 and Cm-244.

SUMMARY

The microstructure of the radioactive CWF is very similar to the surrogate and U/Pu-doped CWFs studied previously [7-8]. It consists of large sodalite and glass regions with actinide/rare earth/yttrium-containing phases and halite decorating the boundaries between these regions. The uranium, plutonium, yttrium and rare earth fission products are observed in a mixed oxide phase. On a finer scale, this phase tends to be displaced into the glassy regions near the sodalite/glass region boundaries. Cesium and strontium were not observed, but they may be uniformly distributed in the waste form.

Based on the 7-day PCT results, the durability of the radioactive CWF is comparable to the reference materials similar to the DWPF glass and is much better than the reference EA glass. More durability testing is planned for similar waste forms in other hot isostatically pressed canisters of the waste form material and will be presented at a later date.

ACKNOWLEDGEMENTS

The authors gratefully acknowledge Clay Brower, Julie Colborn, Christal Mason and Ronda Elliott for their assistance in sample preparation and analyses. This work was supported by the U.S. Department of Energy, Nuclear Energy Research and Development Program, under contract no. W-31-109-ENG-38.

REFERENCES

1. J.P. Ackerman, T.R. Johnson, L.S.H. Chow, E.L. Carls, W.H. Hannum and J.J. Laidler, Prog. Nucl. Energy 31:141-154 (1997).
2. C. Pereira, M.A. Lewis and J.P. Ackerman, Spent Nuclear Fuel-Treatment Technologies, Reno, NV, 1996.
3. K.M. Goff, R.W. Benedict, K. Bateman, M.A. Lewis, C. Pereira, C.A. Musick, *International Topical Meeting on Nuclear and Hazardous Waste Management*, Spectrum, Seattle, WA, pp. 2436-2443 (1996).
4. ASTM C1285-94, Standard Test Methods for Determining Chemical Durability of Nuclear Waste Glasses: The Product Consistency Test (PCT), *Annual Book of ASTM Standards*, (American Society for Testing and Materials, Philadelphia, 1995), pp. 797-814.
5. C.M. Jantzen, N.E. Bibler, D.C. Beam, C.L. Crawford, and M.A. Pickett, Westinghouse Savannah River Co. report WSRC-TR-92-346, revision 1 (1993).
6. S.M. Frank, K.J. Bateman, T. DiSanto, S.G. Johnson, T.L. Moschetti, M.H. Noy, and T.P. O'Holleran in *Phase Transformations and Systems Driven Far from Equilibrium*, edited by E. Ma, P. Bellon, M. Atzmon, and R. Trivedi (Mat. Res. Soc. Symp. Proc. 481, Warrendale, PA 1998), pp. 351-356.
7. T.L. Moschetti, T.P. O'Holleran, S.M. Frank, S.G. Johnson, D.W. Esh, and K.M. Goff, in *Environmental and Waste Management Technologies in the Ceramic and Nuclear Industries V*, edited by G.T. Chandler and X. Feng (Ceramic Transactions 107, tentative publication date: April 2000); W.S. Sinkler, T.P. O'Holleran, S.M. Frank, T.L. Moschetti, S.G. Johnson, D.W. Esh, K.M. Goff, *ibid.*
8. W. Sinkler, T.P. O'Holleran, S.M. Frank, M.K. Richmann and S.G. Johnson, presented at the 1999 MRS Fall meeting, to be published in *Scientific Basis for Nuclear Waste Management XXIII*, edited by R.W. Smith and D.W. Shoesmith, 1999.

9. N.E. Bibler and J.K. Bates in *Scientific Basis for Nuclear Waste Management XIII*, edited by V. Oversby and P. Brown (Mat. Res. Soc. Symp. Proc. 176, Pittsburgh, PA 1990), pp. 327-338.

TEM CHARACTERIZATION OF CORROSION PRODUCTS FORMED ON A STAINLESS STEEL-ZIRCONIUM ALLOY

J.S. Luo* and D.P. Abraham
Chemical Technology Division, Argonne National Laboratory, Argonne, IL 60439
*E-mail address: Luo@cmt.anl.gov

ABSTRACT

The corrosion products formed on a stainless steel-15Zr (SS-15Zr) alloy have been characterized by transmission electron microscopy (TEM) and energy dispersive x-ray spectroscopy (EDS). Examination of alloy particles that were immersed in 90°C deionized water for two years revealed that different corrosion products were formed on the stainless steel and intermetallic phases. Two corrosion products were identified on an austenite particle: trevorite ($NiFe_2O_4$) in the layer close to the metal and maghemite (Fe_2O_3) in the outer layer. The corrosion layer formed on the intermetallic was uniform, adherent, and amorphous. The EDS analysis indicated that the layer was enriched in zirconium when compared with the intermetallic composition. High-resolution TEM images of the intermetallic-corrosion layer interface show an interlocking metal-oxide interface which may explain the relatively strong adherence of the corrosion layer to the intermetallic surface. These results will be used to evaluate corrosion mechanisms and predict long-term corrosion behavior of the alloy waste form.

INTRODUCTION

Stainless steel-zirconium (SS-Zr) alloys have been developed as waste forms to immobilize and retain fission products separated during the electrometallurgical treatment of spent nuclear fuel [1, 2]. The baseline waste form is a stainless steel-15 wt% zirconium (SS-15Zr) alloy, which is prepared by melting appropriate amounts of Type 316 stainless steel (SS316) and high-purity zirconium. The SS-15Zr alloy displays a eutectic microstructure that contains the stainless steel-type phases, ferrite and austenite, and $ZrFe_2$-type Laves intermetallic polytypes C36 (dihexagonal, $MgNi_2$-type) and C15 (cubic, $MgCu_2$-type) [3]. Many fission products, such as Nb, Pd, Rh and Ru, are preferentially incorporated into the $ZrFe_2$–type intermetallics, whereas others such as Tc and Mo are present in all phases of the alloy waste forms [2]. The actinide elements do not form separate actinide-rich phases, but appear as locally enriched regions within the $ZrFe_2$–type intermetallics [2].

A series of tests is being conducted at Argonne National Laboratory to evaluate the corrosion behavior of the SS-15Zr metal waste form. In this paper, we present preliminary results from the microstructural characterization of corrosion layers formed on SS-15Zr alloy phases using transmission electron microscopy (TEM) and energy dispersive x-ray spectroscopy (EDS). Determining the corrosion products formed on SS-15Zr alloy phases is critical to delineating the mechanisms of radionuclide release from the metal waste form.

Mat. Res. Soc. Symp. Proc. Vol. 608

EXPERIMENTAL

Specimens of SS-15Zr alloy, crushed to 75 to 150 μm size fraction, were immersed in 90°C deionized water for two years. The solution volume was such that the sample surface area-to-leachant volume ratio was ~2000 m^{-1}. On completion of the test, individual particles of the stainless steel and intermetallic phases were selected on the basis of scanning electron microscopy (SEM), embedded in resin, and sectioned by a Reichert-Jung Microtome to yield ~50-nm thick samples for TEM examination. Transmission electron microscopy was performed with a JEOL 2000FX unit operating at 200 kV and a PHILIPS CM30 unit operating at 300 kV, both equipped with an energy dispersive x-ray spectrometer. High-resolution TEM was carried out with a JEOL 4000EX microscope operating at 400 kV with a point-to-point resolution of 1.65 Å.

RESULTS AND DISCUSSION

Corrosion Products on an Austenite Particle

Figure 1 is a TEM image of corrosion products formed on an austenite particle from the reacted SS-15Zr alloy. The corrosion layer was largely detached from the metal, probably due to the mechanical force applied during ultramicrotomy. However, it was still possible to establish a spatial relationship between the metal surface and the corrosion layer. At least two distinct corrosion products formed on the surface. The corrosion product that appears to have been in immediate contact with the stainless steel surface (i.e., B in Fig. 1) exhibited a relatively dense and uniform microstructure, whereas the corrosion product on the outer surface of the layer (C in Fig. 1) was more porous. Assuming that no materials were lost during sample handling and preparation, the total layer thickness was estimated to be between 0.5 and 1.0 μm.

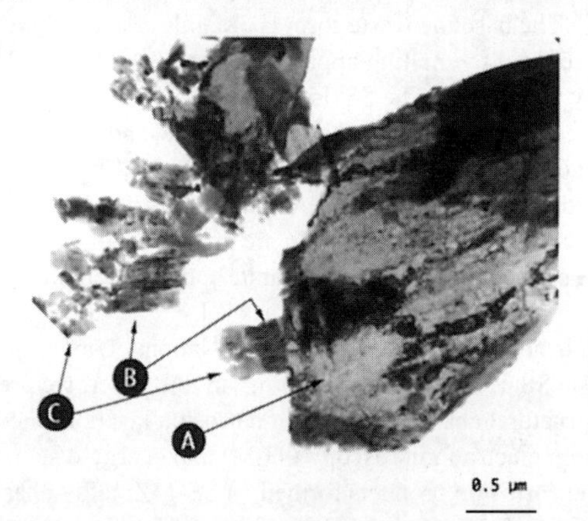

Figure 1. Bright-field TEM micrograph of an austenite–corrosion layer interface. At least two corrosion products (B and C) have formed on the stainless steel (A) surface, based on their distinctive microstructural characteristics.

Electron diffraction analyses indicated that both corrosion products were crystalline. The corrosion product B was indexed to match the cubic trevorite structure (JCPDS-ICDD 10-325, nominal formula $NiFe_2O_4$); the outer product C was indexed to match the face-centered cubic maghemite-c structure (JCPDS-ICDD 39-1346, nominal formula Fe_2O_3). The $NiFe_2O_4$ spinel was observed by Nakayama et al. [4, 5] in the passive film of 18Cr-8Ni austenitic stainless specimen that was heated in 300°C deoxygenated water for 24 h. This spinel structure was unaffected even after the sample was heated for 3 h at 1000°C. The presence of nickel in a spinel-type lattice was also reported by Castle and Clayton [6] in the oxide layer of a stainless steel alloy heated in 200°C water. These studies show that the spinel structure can be stabilized by nickel ions under appropriate experimental conditions.

Figure 2 compares the EDS spectra obtained from the austenite phase A, and the corrosion products B and C. The austenite spectrum has a minor zirconium peak, probably resulting from the small zirconium solubility in the phase. The corrosion product B consists mainly of Fe, Ni, and O, apparently consistent with the nominal composition of trevorite; $NiFe_2O_4$. The product C, which is part of the outer corrosion layer, appears to incorporate a small amount of Ni in the maghemite structure. Chromium was not observed in either corrosion product. This finding is somewhat surprising since the corrosion resistance of stainless steels is usually attributed to the presence of chromium in the passivation layer.

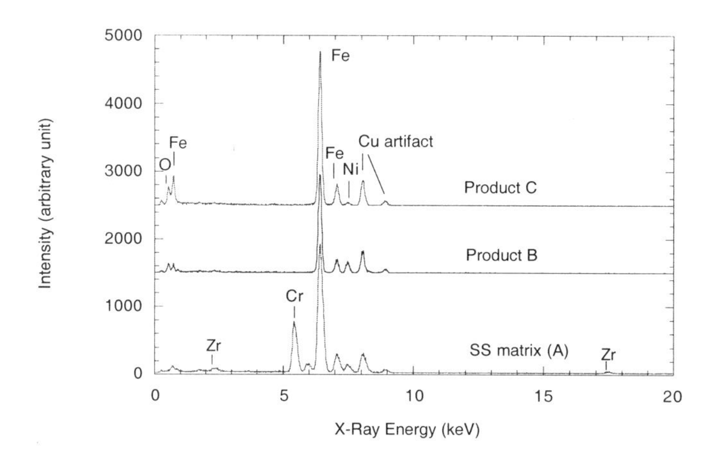

Figure 2. EDS spectra from the stainless steel (SS) matrix (austenite), and the corrosion products B and C. The corrosion products are enriched in Fe and Ni, and free of Cr.

Corrosion Products on an Intermetallic Particle

The corrosion layers formed on intermetallic particles were small and ranged in thickness from 10 to 100 nm. A bright-field TEM image of a typical corrosion layer is shown in Fig. 3; the layer is uniform and well adherent to the metallic surface. Electron diffraction patterns obtained from the intermetallic particle and corrosion layer are shown in Fig. 4a. The intermetallic was identified as C15, the cubic Laves polytype. The corrosion layer exhibited broad diffuse rings (Fig. 4a), indicating that the layer was largely amorphous. The EDS analysis showed that the layer was enriched in Zr and Cr and depleted in Fe and Ni, when compared with the metallic matrix (Fig. 4b).

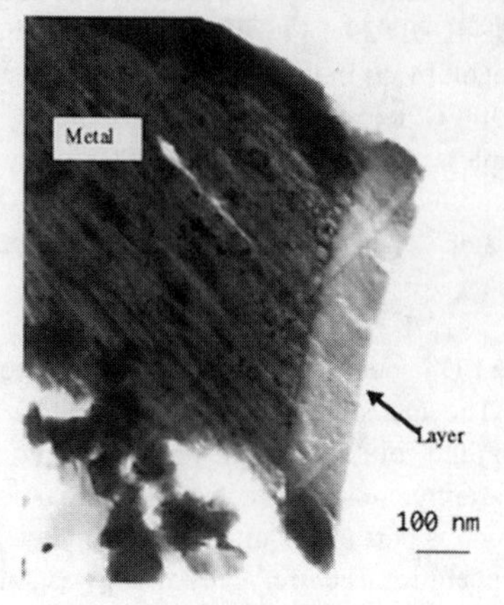

Figure 3. Bright-field TEM image of the corrosion layer formed on an intermetallic particle.

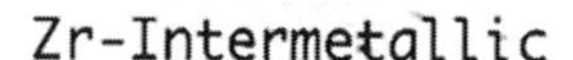

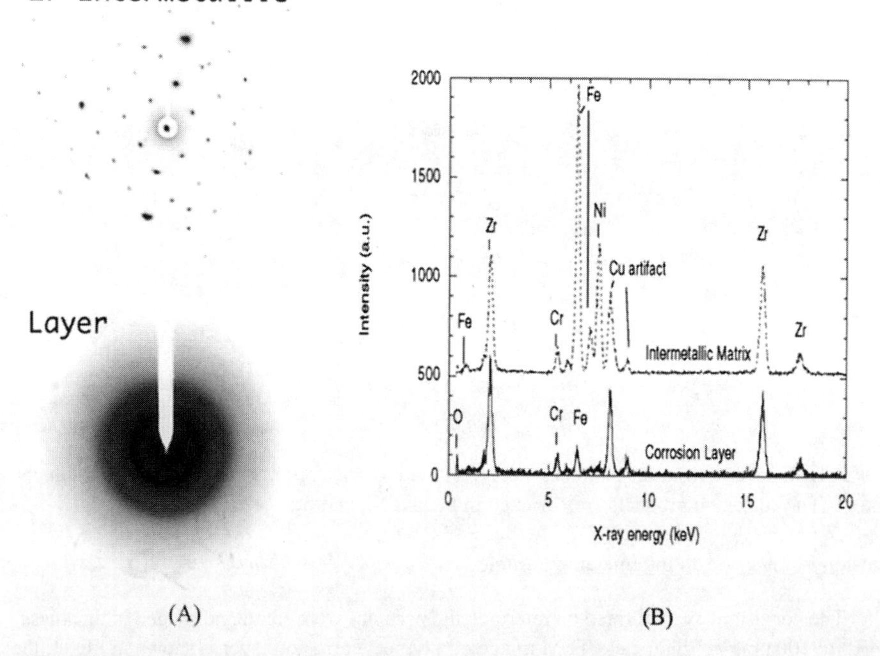

Figure 4. Electron diffraction patterns (A) and EDS spectra (B) obtained from the intermetallic and the corrosion layer formed on its surface.

Figure 5 is a high-resolution TEM image of the intermetallic-corrosion layer interface. The crystalline intermetallic lattice was imaged along the [110] direction, and its (220) plane with a d-spacing of 0.25 nm was clearly resolved. The corrosion layer exhibited the typical contrast of a disordered lattice, which is consistent with the electron diffraction data. It is evident from Fig. 5 that the interface is not flat on the atomic scale. Domains of disorder are observed on the intermetallic side of the image, and small crystalline regions are detected in the amorphous corrosion layer. The interfacial configuration observed appears to be the result of a gradual transformation of the intermetallic into an amorphous corrosion product, probably due to the ingress of oxygen into the metal. Such an interlocking metal-oxide interface may explain the relatively strong adherence of the corrosion layer to the intermetallic surface.

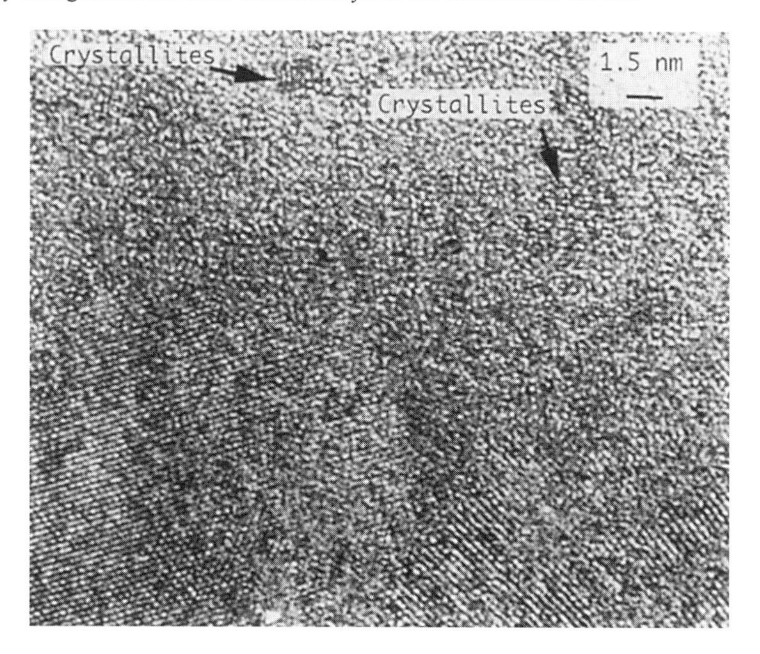

Figure 5. High-resolution TEM image of the intermetallic-corrosion layer interface. The image was taken along the [110] direction of C15, the cubic $ZrFe_2$ polytype. Crystallites observed in the amorphous corrosion layer (disordered zones) are shown by arrows.

The thinness of the corrosion layer indicates that the intermetallic is very resistant to corrosion. This may be because the corrosion layer is amorphous. Investigations in the Fe-Cr system have shown that alloy corrosion resistance increases with chromium content, with a dramatic improvement observed at ~12 wt% Cr [7, 8]. This result has been correlated to the increasing tendency of the oxide layer to become more disordered as the alloy chromium content increases, the suggestion being that an amorphous oxide is more resistant to breakdown than a crystalline oxide. Furthermore, amorphous oxides may be more protective because the mobility of ions may be smaller than in crystalline structures containing defects and grain boundaries [9]. In any case, the thin corrosion layers and the adherence of the layer to the intermetallic suggest that fission products and actinide elements that are present in the oxide layer of the metal waste form could be retained on the alloy surface at least during the early stage of corrosion.

CONCLUSIONS

Corrosion layers formed on stainless steel–15Zr alloy particles that were reacted in 90°C deionized water for two years have been characterized. The corrosion products formed on an austenite particle were identified as trevorite ($NiFe_2O_4$) and maghemite-c (Fe_2O_3) with a small Ni content. Corrosion layers on intermetallic particles were uniform, adherent, amorphous, and enriched in Zr when compared with the intermetallic composition. High-resolution TEM images of the intermetallic-corrosion layer interface showed an interlocking metal-oxide, which suggests oxide formation due to oxygen ingress into the metal. The corrosion resistance of the intermetallic particles may be attributed to these amorphous oxide layers that effectively retard oxygen diffusion into the metal.

ACKNOWLEDGMENTS

This research was supported by the U.S. Department of Energy under Contract W-31-109-ENG-38. The authors thank R. Finch for assistance in sample preparation, and R. Cook and R. Csencsits for assistance in TEM operations.

REFERENCES

1. S.M. McDeavitt, D.P. Abraham, and J.Y. Park, J. Nucl. Mat. 257, 21-34 (1998).

2. D.P. Abraham, D.D. Keiser, Jr., and S.M. McDeavitt, Proc. Intl. Conf. on Decommissioning and Decontamination and on Nuclear and Hazardous Waste Management, Vol. 2, American Nuclear Society, LaGrange Park, IL p. 783 (1998).

3. D.P. Abraham, J.W. Richardson, Jr., and S.M. McDeavitt, Mat. Sci. Eng. A239-240, 658 (1997).

4. A. T. Nakayama and Y. Oshida, Corrosion 24, 336 (1968).

5. J. E. Castle and C. R. Clayton, Passivity of Metals, R. P. Frankenthal and J. Kruger, eds., The Electrochemical Society Inc, Princeton, NJ p.714 (1978).

5. G. Aronwitz and N. Hackerman, J. Electrochem. Soc. 110, 633 (1963).

6. C. L. McBee and J. Krueger, Electrochim. Acta 17, 1337 (1972).

9. T. P. Hoar, Palladium Medal Address, J. Electrochem. Soc. 117, 17c (1970).

LEACHING CHARACTERISTICS OF THE METAL WASTE FORM FROM THE ELECTROMETALLURGICAL TREATMENT PROCESS: PRODUCT CONSISTENCY TESTING

S. G. Johnson, D. D. Keiser, S. M. Frank, T. DiSanto, A. R. Warren and M. Noy
Argonne National Laboratory-West
P.O. 2528
Idaho Falls, ID 83403

ABSTRACT

Argonne National Laboratory has developed an electrometallurgical treatment for spent fuel from the experimental breeder reactor II. A product of this treatment process is a metal waste form that incorporates the stainless steel cladding hulls, zirconium from the fuel and the fission products that are noble to the process, i.e., Tc, Ru, Nb, Pd, Rh, and Ag. The nominal composition of this waste form is stainless steel/15 wt% zirconium/ 1-4 wt% noble metal fission products /1-2 wt % U. Leaching results are presented from crushed sample immersion tests on simulated metal waste form samples. The test results will be presented and their relevance for waste form product consistency testing discussed.

INTRODUCTION

The electrometallurgical (EM) treatment process of spent fuel at Argonne National Laboratory (ANL) yields three product streams: a uranium metal product, a contaminated salt waste stream and an irradiated cladding hull waste stream [1]. The stainless steel cladding hulls are consolidated into a metal waste form (MWF) ingot with a nominal composition of stainless steel components (80 wt%), zirconium (15 wt%), fission products noble to the process (1-5 wt%) and residual actinides (1-5 wt%). The last two categories include the elements: Tc, Ru, Ag, Rh, Pd, Nb, U and Pu [2]. The topic of waste form process qualification has been addressed for borosilicate glass [3] and the plutonium immobilization product [4]. The work discussed in this paper is part of an effort to qualify processing oa a MWF for disposal in a geologic repository.

The main premise of this paper is that a method of performing assessing the consistency of the product produced from the metal waste fabrication for the electrometallurigal treatment process at Argonne is required. The fundamental objectives for a product consistency or product quality test are to: 1) detect an out-of-control process, 2) it must be operational in a remote (hot cell) environment and 3) it should be scalable to accommodate larger metal waste form ingots planned in the future. This paper examines using immersion testing with crushed material as a means of evaluating the consistency of the MWF. Other techniques are also being evaluated for this purpose and it is likely that a small number of tests will be used to monitor product consistency. These other methods may include the following: density measurements, hardness testing, microscopy, monolithic immersion tests, electrochemical corrosion measurements, elemental analysis and specific resistance measurements.

Borosilicate glass high level waste relies on input process stream analysis, a short-term immersion test (ASTM C1285-97), and determination of crystalline content to establish the basis of a well-controlled process. The metal waste form resulting from the EM process can similarly rely on input stream analysis, although this information is not always timely and the issue of a statistical sample of a large collection of cladding hulls is unresolved. The analysis of the final product to complement this input stream information would provide for a better degree of process control. The method and results presented in this paper involve an immersion test that has been applied to an alloy sample that represents the target composition of major constituents of the MWF.

Further mention regarding the method of production of the metal waste form from the EM process is appropriate here so that the motivation for devising and applying tests to it is clear. The product specifications for the metal waste form state that it shall consist of between 5 and 20 wt% zirconium, with the target being 15 wt%, with small amounts of noble fission products depending on the irradiation history of the fuel being processed and actinides with the balance being stainless steel constituents. The cladding hulls are stainless steel, whereas the remaining constituents are present as elements plated onto the hulls in the electrorefiner. It has been demonstrated in actual hot operations that additions of both stainless steel and zirconium are necessary to facilitate the formation of an ingot with the appropriate composition that is well consolidated. It is planned in the future that the addition of stainless steel, which is used to provide a molten pool of material to encourage the remainder of the material to form a coherent ingot, would consist of irradiated reactor hardware. It is

Mat. Res. Soc. Symp. Proc. Vol. 608 © 2000 Materials Research Society

not clear, given the number of operations that will occur in a hot cell, that relying solely on input stream control will provide the appropriate degree of confidence in the quality of the product. It is because of this uncertainty that tests are being considered for the MWF product to confirm that the process is under control. This paper describes and presents results from a single test, the crushed sample immersion test, for testing the metal waste form product.

EXPERIMENTAL

The material used for this investigation was a 316SS/15 wt % Zr alloy ingot. This ingot contained no noble metal fission product elements or actinides and is therefore a non-radioactive surrogate metal waste form. Two methods of sample preparation were investigated, these are detailed below, and the resulting material was then subjected to an immersion test modeled after ASTM C1285-97 [5] with some modifications due to the differing density of the metal waste form. The sample fraction used for the test was sieved and the –100 mesh to +200 mesh fraction was used. The test duration was 14 days and the temperature was 90 °C. The leachants used were either demineralized water (DW) or simulated ground water (SGW) of the J-13 variety, please see ref. [6] for the exact chemical composition of this common leachant. The leach vessels used were Teflon. The surface area-to-volume of leachant ratio was 2000 m^{-1}. The density of the metal material used was 7.7 g/cm^3 and therefore the ratio of crushed metal to leachant was adjusted to 2.8 gram of metal to 10.0 ml of leachant to provide the surface area to volume ratio of 2000 m^{-1}. The surface area of the material used was calculated using the same spherical approximation used for glass. The two sample preparation methods are listed and described below:

1) A bandsaw was used to generate material which was then introduced into the micromill for further processing before being sieved and used for the test.
2) A drill press equipped with a coring bit was used to obtain shavings that were than further processed using a micromill. The drill press was operated at a slow speed to minimize heating.

The micromill portion of the preparation was performed in the following fashion. The material was milled for 30 seconds and then sieved using an ultrasonic siever and the material of size such that it was retained on the 100 mesh screen was re-milled for 30 seconds, then resieved. This was repeated until enough material was gathered for the test.

The test solution from the 14 day immersion test was analyzed after being filtered through a 0.45 μm filter. The crushed metal material was then removed from the test vessel and the vessel rinsed with water to remove all visible solid material. An aliquot of 2 vol. % HNO_3 equal to the original volume of leachate present in the vessel was then added and the vessel placed into a 90 °C oven for 18 hours. This solution is the acid strip and was analyzed separately.

Although two methods of sample preparation were used to generate test material for this work scanning electron microscopy results clearly indicate that the two methods yield material that is indistinguishable; this is presented in the results section of this paper.

Analysis Methods

The elemental analyses for the solutions generated from the two immersion tests described above were performed using an Inductively Coupled Plasma Mass Spectrometer. The scanning electron microscopy (SEM) work was performed with a Zeiss DSM 960A digital scanning electron microscope with an energy dispersive detector (EDS) provided by Oxford Instruments.

RESULTS AND DISCUSSION

The investigation of using an immersion test to measure the product consistency of the MWF was driven by the need to have a means of assessing product consistency of the as-cast waste form. Preliminary tests using a different, and unsatisfactory, sample preparation method (liquid nitrogen treatment followed by an impact mortar) to obtain crushed samples were undertaken and initiated the sample preparation methods discussed here. The reason that the liquid nitrogen sample preparation method was deemed to be unsatisfactory was that it damaged the microstructure to an extent that was judged unrepresentative of the original material.

The crushed MWF surrogate material was subjected to testing as described above in the experimental section. The results are presented in Tables I and II. The primary focus of the investigation was to examine the release of the major constituents present in every anticipated metal waste product to be produced. Thus the following analytes were chosen: Fe, Cr, Ni, Mo, Mn, and Zr. Table I. contains the results of the elemental analysis of the test solutions from the 14 day immersion test. Table II contains the results of the elemental analysis of the acid strip solutions from the vessels. Since this paper presents the first application of this protocol to the MWF the applicability of the acid strip was being evaluated and thus was not performed on all runs. The acid strip was shown to be absolutely required for this procedure because the majority of material released to solution was collected in the acid strip.

Table I. The elemental results from the crushed sample immersion test performed on a 316 SS/15 Zr alloy prepared using the Drill bit/Micro-mill (DBMM) and the Bandsaw/Micro-mill (BSMM) methods. Blank solutions were run in vessels for 14 days simultaneously and analyzed. The tests were run in several batches so the blank for that representative batch is immediately following the samples in the table below. All values are in μg/L (ppb). Each entry represents the results of a single run. The analytical uncertainty for each element is listed in the table by each entry in parenthesis. The pH and leachant used in this test are indicated in the table.

Sample Id.	Fe	Cr	Ni	Mo	Mn	Zr	Leach Sol.	pH
DBMM-1	<48	<2.4	10(±15%)	1100(±10%)	6.9(±10%)	5.5(±10%)	DW	7.4
DBMM-2	<48	<2.4	7(±20%)	1100(±10%)	6.5(±10%)	5.1(±10%)	DW	7.5
DBMM-1 Blank-1/2	<5	<0.24	<0.14	0.06(±30%)	<0.05	0.05(±15%)	DW	6.3
DBMM-3	<28	1.7(±30%)	<2.1	130(±10%)	1.23(±20%)	0.14(±50%)	DW	8.4
DBMM-4	<28	0.64(±90%)	10.5(±20%)	710(±10%)	9.2(±10%)	<0.07	DW	7.4
DBMM-5	<28	0.54(±95%)	11.5(±20%)	780(±10%)	15(±10%)	<0.07	DW	7.2
Blank-3/5	<28	<0.53	<2.1	0.5(±20%)	<0.23	0.25(±10%)	DW	6.5
DBMM-6	<28	24(±10%)	2.7(±80%)	1500(±10%)	0.25(±90%)	<0.07	SGW	8.6
DBMM-7	<28	25(±10%)	3.5(±60%)	1400(±10%)	0.29(±80%)	0.12(±60%)	SGW	8.7
DBMM-8	<28	34(±10%)	3.5(±60%)	1500(±10%)	0.25(±90%)	0.50(±15%)	SGW	8.8
Blank-6/8	<28	<0.5	<2.1	0.86	<0.23	<0.07	SGW	8.8
BSMM-1	<48	15(±20%)	6.9(±20%)	550(±10%)	4.5(±15%)	0.67(±10%)	DW	8.3
BSMM-2	<48	4.7(±45%)	7.1(±20%)	730(±10%)	5.1(±10%)	0.46(±15%)	DW	8.4
BSMM-1 Blank-1/2	<5	<0.24	<0.14	0.05(±35%)	<0.05	0.01(±45%)	DW	6.8

To investigate the material actually used for the test SEM was employed. This analysis evaluates several important aspects: 1) the rough geometry of the material to discern the validity of using the spherical approximation to calculate the surface area, 2) the condition of the microstructure following sample preparation and 3) whether the sample utilized was representative of the whole. Figure 1 is a low magnification back scattered electron (BSE) SEM micrograph showing general characteristics of the material. The material displayed in Fig. 1 is prepared pre-test material which has a similar appearance using either sample preparation method described in the experimental section. Figure 2 represents a long axis view of a representative particle from the material featured in Fig. 1. For comparison purposes Fig. 3 displays the microstructure of a sample from an actual metal waste form ingot produced from cladding hulls of irradiated fuel. This cross section indicates that the fundamental microstructure is unchanged during preparation and that the particle is a representative sample of the bulk. The SEM analysis shows that very little distortion of the microstructure occurs.

Table II. The elemental results from the acid strip analysis. The samples correspond to those in Table I. All values are in µg/L (ppb). Each entry represents the results of a single run. The analytical uncertainty for each element is listed in the table by each entry in parenthesis. Acid strips were not performed on all vessels.

Sample Id.	Fe	Cr	Ni	Mo	Mn	Zr
DBMM-3	18000(±20%)	100(±10%)	3300(±10%)	71(±40%)	1200(±10%)	250(±10%)
DBMM-4	12000(±20%)	62(±10%)	2000(±10%)	150(±20%)	220(±10%)	45(±10%)
DBMM-5	19000(±20%)	78(±10%)	3700(±10%)	220(±15%)	330(±10%)	54(±10%)
Blank-3/5	30(±10%)	0.96(±15%)	<0.39	<0.06	0.21(±10%)	8.0(±10%)
DBMM-6	1200(±10%)	55(±10%)	330(±10%)	32(±10%)	18(±10%)	240(±10%)
DBMM-7	1000(±10%)	52(±10%)	270(±10%)	25(±10%)	14(±10%)	69(±10%)
DBMM-8	850(±10%)	57(±10%)	230(±10%)	20(±10%)	13(±10%)	83(±10%)
Blank-6/8	120(±10%)	0.81(±15%)	4.0(±10%)	0.57(±10%)	1.4(±10%)	4.0(±10%)

Observations

1) The sample preparation was fairly quick providing drill shavings are available. The equipment involved is operable in a remote environment with little to no modification.
2) The samples of crushed material appear to be very similar to glass samples prepared in our laboratory for this type of test. The general geometry is very similar and the estimated surface area is therefore thought to be accurate.
3) The microstructure of the material is such that no phase segregation occurs during sample preparation.
4) The use of DW results in more Fe, Mo and Mn being released whereas SGW results in more Ni being released. The amount of Cr and Zr released is roughly equivalent in each case.
5) The Mo is present in the leachate test solution more so than the acid strip. This is always true and is more accentuated in the tests performed with simulated well water.
6) The Fe, Cr, Ni, Mn and Zr are more concentrated in the acid strips for all tests. This is consistent with their being present on the vessel either as material adhered to the vessel walls or as corrosion products that were not successfully rinsed out prior to the acid strip step. A very small stain was present on the vessel that was not removed with the water rinses prior to the acid strip step. This stain is consistent with iron corrosion products.
7) The results of the elemental analysis of the test and the acid strip solutions must be examined as two parts of the same test. The leachants used, DW or SGW, makes some small difference in the test results as one would expect. This is most likely caused by the anion content of the SGW since it contains chloride at the ~10 ppm level.
8) The precision of the test results is fairly good for Fe, Cr, Ni, and Mo, i.e., percent relative standard deviation of 30% or less. The Mn and Zr results appear to have either large standard deviation or incorporate spurious data. More results would decide the matter definitively in the future.

Figure 1. The BSE SEM micrograph of the material prepared for the crushed sample immersion test. The material has been crushed and sorted for the test.

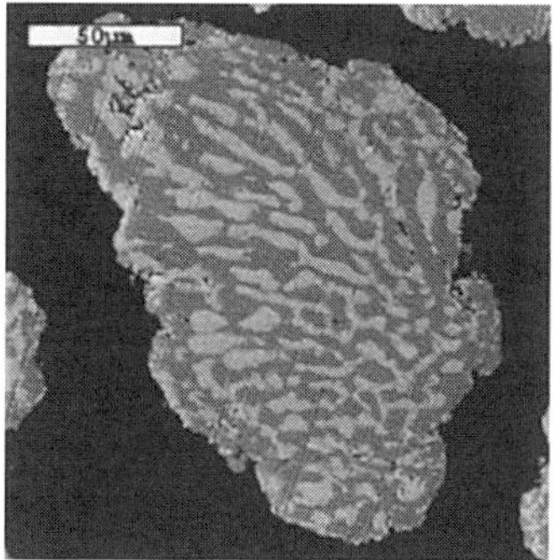

Figure 2. A Cross Section of a Representative Particle Produced Using the Method described in the Experimental Section. The lighter contrast phase is the Fe_2Zr intermetallic phase whereas the dark contrast phase is the iron solid solution phase.

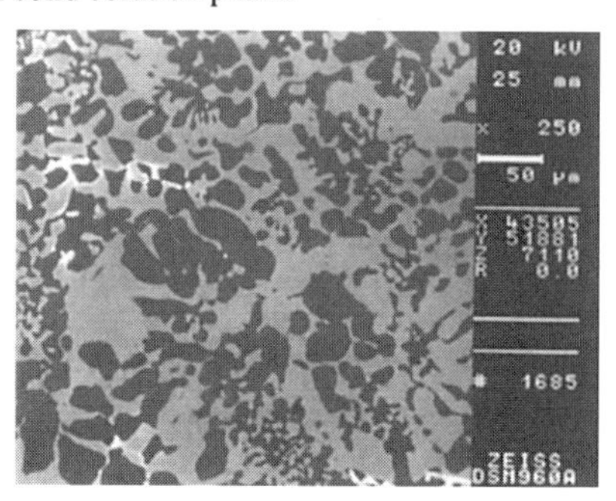

Figure 3. Microstructure of a sample of an actual radioactive metal waste form ingot produced from cladding hulls. The phases are identical to Fig. 2.

CONCLUSIONS

As stated in the introduction the main premise of this paper was to present the results of a candidate method for assessing product consistency for the purpose of exercising process control of the MWF production process. It is important to consider our criteria for such a test. It must: 1) detect an out-of-control process, 2) it must be operational in a remote (hot cell) environment and 3) it should be scalable to accommodate larger metal waste form ingots planned in the future.

The crushed sample immersion test has the potential to meet all the criteria mentioned above. It is scalable to larger ingots in that a small sample can be easily obtained using the method described here, the drill bit/micro-mill method in particular. It is hot cell compatible in that all equipment described here is currently operational in a hot cell environment at Argonne National Laboratory. It is a sensitive method that has the potential to detect compositional and phase differences in the MWF product, although the results presented here are those for an ingot of the target composition. Future plans include the testing of ingots that have compositions outside the specifications for the MWF to establish specific elemental release limits.

REFERENCES

[1] McDeavitt, S. M., Abraham, D. P., Park, J. Y., J. Nucl. Mat., **257**, 21 (1998).

[2] Keiser, D. D., Westphal, B. W., ANS Proceedings "Third Topical Meeting DOE Spent Nuclear Fuel and Fissile Materials Management", Charleston, SC, 668 (1998).

[3] Office of Civilian Radioactive Waste Management, Waste Acceptance System Requirements Document, E00000000-00811-1708-00001, rev. 2 (1996).

[4] Marra, J. C., Marra, S. L., Bibler, N. E., ANS Proceedings "Third Topical Meeting DOE Spent Nuclear Fuel and Fissile Materials Management", Charleston, SC, 731 (1998).

[5] C1220-97, "Determining Chemical Durability of Nuclear, Hazardous, and Mixed Waste Glasses: The Product Consistency Test (PCT)", ASTM, Philadelphia (1999).

[6] Johnson, S. G., Keiser, D. D., Noy, M., O'Holleran, T. P., Frank, S. M., MRS proceeding Sci. Basis Waste Mang. 1998, in press (1999).

BEHAVIOR OF ACTINIDE IONS DURING SLUDGE WASHING OF ALKALINE RADIOACTIVE WASTES

ANDREW H. BOND,* KENNETH L. NASH,* ARTEM V. GELIS,* MARK P. JENSEN,*
JAMES C. SULLIVAN,* LINFENG RAO**
*Chemistry Division, Argonne National Laboratory, Argonne, IL, 60439 USA
**MS 70A-1150, Lawrence Berkeley National Laboratory, Berkeley, CA 94720 USA

ABSTRACT

It is difficult to accurately predict actinide behavior during the alkaline leaching of Hanford's radioactive sludges due to the diverse chemical and radiolytic conditions existing in these wastes. The results of Pu dissolution during experimental washing of sludge simulants from the $BiPO_4$, Redox, and PUREX processes shows that ≤ 2.1% Pu is dissolved during contact with alkaline media, but up to 65.5% Pu may be dissolved in acidic media. The dissolution of Cr, Fe, Nd, and Mn has also been observed, and the results of solid state, radioanalytical, and spectroscopic investigations are detailed.

INTRODUCTION

Various processes were used at Hanford for the large-scale purification of Pu for defense uses. When production of ^{239}Pu commenced at Hanford in 1944, Pu(IV) was coprecipitated with $BiPO_4$ while U(VI) remained in solution as a SO_4^{2-} complex. After solid-liquid separation, the Pu(IV) in the $BiPO_4$ precipitate was oxidized to soluble Pu(VI) by BiO_3^-, MnO_4^-, or $Cr_2O_7^{2-}$ prior to further purifications that included LaF_3 coprecipitation [1]. This solid-liquid-based separation was replaced eight years later by the Redox process in which Pu(VI) and U(VI) were partitioned into methyl isobutyl ketone from concentrated aqueous $Al(NO_3)_3$. Separation of Pu(VI) from U(VI) was accomplished by addition of Fe(II), causing reduction to Pu(III) that reports to the aqueous phase [2]. After only a few years in operation, the Redox process was replaced by PUREX solvent extraction in which U(VI) and Pu(IV) are extracted from 3-4 M HNO_3 into 30% (v/v) tri-*n*-butyl phosphate in an aliphatic hydrocarbon diluent. Recovery of Pu again involves reduction to the trivalent state, principally by U(IV) or Fe(II) [2].

An understanding of the processing history and chemistry unique to each Pu purification process is now important because wastes from these activities must be remediated. Acidic production waste effluents were adjusted to above pH 9 with NaOH prior to tank storage. The results of such a drastic pH change are the formation of hydrolytic polymers and large quantities of $NaNO_3$ that have now stratified into sludge, supernatant, and saltcake layers. Vitrification is the preferred radioactive waste immobilization strategy, but the borosilicate glass formulations are sensitive to the presence of the P, Al, and Cr that reside in the sludge materials. Hanford's alkaline radioactive waste remediation strategy proposes a high temperature leach of the residual sludge materials with 3 M NaOH [3] to remove the P, Al, and Cr. Based on the available experimental data, it has been assumed that actinide ions will remain in the sludge during leaching procedures. Unfortunately, this assumption does not account for the diverse redox and solution chemistry of the actinides that is readily perturbed by hydrolysis, complexation, solubility, disproportionation, or redox reactions involving matrix ions [4]. Because sludge washing operations will encounter a heterogeneous mixture of solids whose thermodynamic and kinetic behavior with respect to actinide and matrix ion dissolution is only poorly understood, the

Mat. Res. Soc. Symp. Proc. Vol. 608 © 2000 Materials Research Society

present work investigates how the mobilization of Pu is influenced by the concurrent dissolution of Cr, Mn, and Fe from $BiPO_4$ (Cycle 3), Redox, and PUREX sludge simulants.

EXPERIMENTAL

Details of the synthesis, sludge washing, and spectroscopic analyses appear elsewhere [5]. Modifications to the published [6] synthesis of the $BiPO_4$ (Cycle 3) sludge simulant were made: 100% La was substituted with 33.3% each of La, Nd, and Eu. Al, Ti, and Fe are present in actual sludge samples and were added to the simulant referred to as $BiPO_4$ (Modified). Scanning electron microscopy (SEM) and energy dispersive spectroscopy (EDS) were performed on $BiPO_4$, $BiPO_4$ (Modified), and PUREX simulant sludges that had been dried at 120 °C. All SEM examinations were performed on a JEOL 6400 instrument. EDS data collection employed a Noran detector with data reduction by the Vantage software package. Consecutive sludge washing experiments generally involved sorption of ^{238}Pu(III/IV) or ^{238}Pu(VI) from a pH = 9.5(5) solution onto the desired sludge simulant followed by contact with a 0.10 M Na_2CO_3 + 0.50 M NaOH + 1.0 M $NaNO_3$ tank waste supernatant simulant. The remaining sludge washing operations involved consecutive contact of each ^{238}Pu-containing sludge simulant with 0.01 M NaOH + 0.01 M $NaNO_2$, 3.0 M NaOH, H_2O, 0.05 M glycolic acid + 0.10 M NaOH, 0.10 M HNO_3, 2.0 M HNO_3, and 0.50 M 1-hydroxyethane-1,1-diphosphonic acid (HEDPA). All Pu percentages are relative to the activity of Pu initially added to the sludge simulant slurry.

RESULTS

The influence of thermal and radiolytic "aging" of mixtures of actual sludge materials is expected to impact many of their chemical properties, and the laboratory investigations of individual sludge simulants described here provide baseline information in the absence of mixing and "aging" effects. One thousand-fold magnification (Figure 1) of the $BiPO_4$ (Modified) simulant reveals the apparent formation of μm-sized crystallites with well defined edges, whereas the PUREX surfaces appear less ordered. A variety of P-based (e.g., $BiPO_4$, $FePO_4$, $AlPO_4$, $La_4(P_2O_7)_3$, etc.) and Cr-based (e.g., Cr(O)OH, $Bi_{38}CrO_{60}$, $Fe(Cr/Fe)_2O_4$, etc.) crystalline phases have been identified in Hanford sludge samples [3, 7].

Despite the sampling limitations of EDS, the principal components in the $BiPO_4$ and PUREX sludge simulants are readily identified (Figure 1). The compositions of the actual $BiPO_4$, Redox, and PUREX sludge materials and the simulants prepared here are reported in Table I. In general the analyses compare favorably, except that all three simulants analyzed by EDS contain > 3-5 times the Mn found in the actual wastes.

Figure 2 shows negligible dissolution of Pu by the tank waste supernatant simulant or by 0.01 M NaOH + 0.01 M $NaNO_2$ (the sluice liquid in Hanford sludge pretreatment). Dissolution of Pu(III/IV) (1.4%) and Pu(VI) (2.1%) from the PUREX simulant by 3.0 M NaOH is low but potentially significant. Plutonium is not effectively removed from the sludge simulants by H_2O or 0.05 M glycolic acid + 0.10 M NaOH. Contact of the Redox sludge simulant with 0.10 M HNO_3 shows release of 19.5% Pu(III/IV) and 31.6% Pu(VI) in unfiltered samples [5], underscoring the importance of transport by suspended solids. Sludge washing with 2.0 M HNO_3 effects the greatest dissolution of Pu(III/IV) and Pu(VI) from all four sludge simulants. The $BiPO_4$ (Modified) simulant releases the most Pu activity: Pu(III/IV) = 65.5% and Pu(VI) = 19.7%. The remaining three simulants all release ≤ 9.9% Pu. Sludge washing of $BiPO_4$ and $BiPO_4$ (Modified) simulants by the strong chelating agent HEDPA (derivatives of which

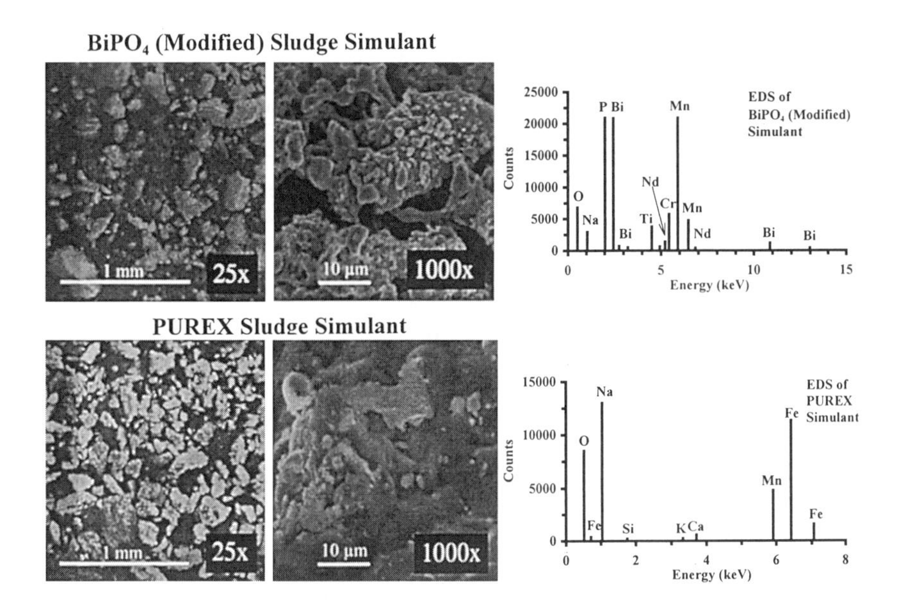

Figure 1: SEM and EDS analyses of BiPO4 (Modified) and PUREX sludge simulants dried at 120 ˚C.

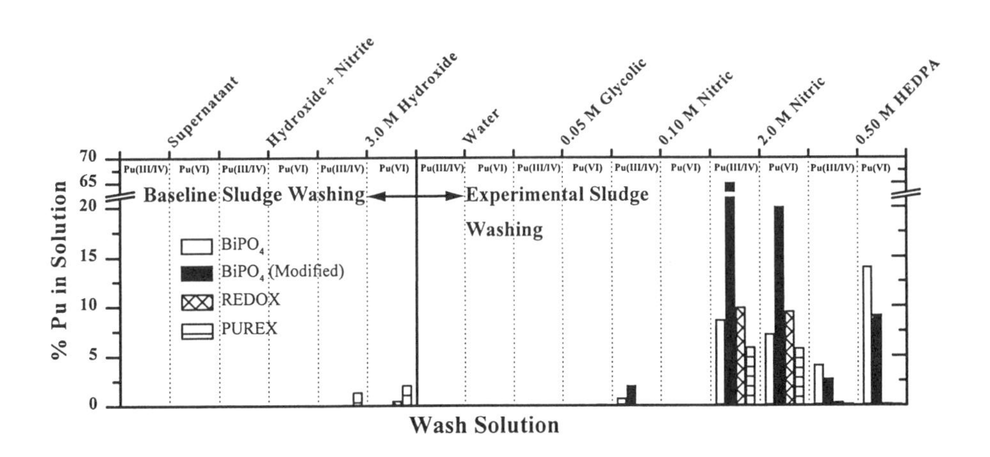

Figure 2: % Pu dissolved in wash solutions after contact with the BiPO4, BiPO4 (Modified), Redox, or PUREX sludge simulants at 23(2) ˚C. Solution concentrations are reported in the Experimental section.

TABLE I

Analyses of Bulk Constituents in Radioactive Waste Sludges and Simulant Sludge Materials

	$BiPO_4$	$BiPO_4$ [a]	$BiPO_4$ (Modified) [a]
Component	Waste Analysis [b] %	Simulant Analysis [c] %	Simulant Analysis [c] %
Al + Fe	1.4	0.0	3.9
Bi	23.6	5.5	22.2
Cr	5.3	7.6	6.7
La	10.5	16.9 (La + Nd + Eu)	7.5 (La + Nd + Eu)
Mn	7.2	33.1	24.3
Na	11.8	1.6	3.3
Ti	1.7	0.0	2.6
Anions	12.0	13.8 (P)	8.9 (P)
Volatiles	14.3	NM [d]	NM
O_2	12.2	21.5	20.6

Redox [e]		PUREX		
Component	Waste Analysis [b] %	Component	Waste Analysis [b] %	Simulant Analysis [c] %
Al	25.4	Al	7.4	NM
Ca	1.5	Ca	1.6	0.7
Cr	3.8	Fe	23.6	29.9
Fe	5.1	K	0.0	0.2
Mn	1.1	Mn	2.2	11.6
Na	15.8	Na	12.2	28.2
Si + Ti	6.2	Si + Ti + Zr	4.9	0.4
Anions	0.5	Anions	1.7	NM
Volatiles	21.0	Volatiles	25.3	NM
O_2	19.6	O_2	21.1	28.8

[a] Modifications to the $BiPO_4$ sludge syntheses are described in the Experimental section. [b] Values are in % (w/w) for sludges dried to 120 °C. "Volatiles" represents organics and water lost by firing to 750 °C. O_2 calculated by difference [8]. [c] Elemental analyses by EDS. [d] NM = not measured. [e] SEM and EDS on the Redox simulant could not be performed due to restrictions on the analysis of radioactive (U) samples in the instrument.

have been proposed as final radioactive waste forms [9]) causes dissolution of ≈ 3.4 times more Pu(VI) than Pu(III/IV).

Chromium removal from the solids targeted for vitrification is a primary objective of sludge pretreatment at Hanford [3]. Contact of the $BiPO_4$, $BiPO_4$ (Modified), and Redox simulants with 3.0 M NaOH afford visible absorption spectra that closely match the 2 x 10^{-4} M Na_2CrO_4 standard spectrum (Figure 3). The comparatively low absorbance for the $BiPO_4$ simulant may be an artifact of Cr removal by the preceding wash [5]. The $BiPO_4$ (Modified) and Redox wash solutions may be calculated (ε from [10]) to yield CrO_4^{2-} concentrations of 1.3 x 10^{-4} and 9.5 x 10^{-5} M, respectively, in 3.0 M NaOH. These values are in agreement with the decreasing Cr contents of the sludge simulants (Table I).

Absorbances for the 2.0 M HNO_3 Redox and PUREX wash solutions are negligible because Fe is almost quantitatively removed from these simulants by the preceding 0.10 M HNO_3 wash. However, dissolution of Cr(III) from the $BiPO_4$ simulant affords a broad band, while the $BiPO_4$ (Modified) sample yields a convoluted spectrum with contributions from Cr(III) and Fe(III).

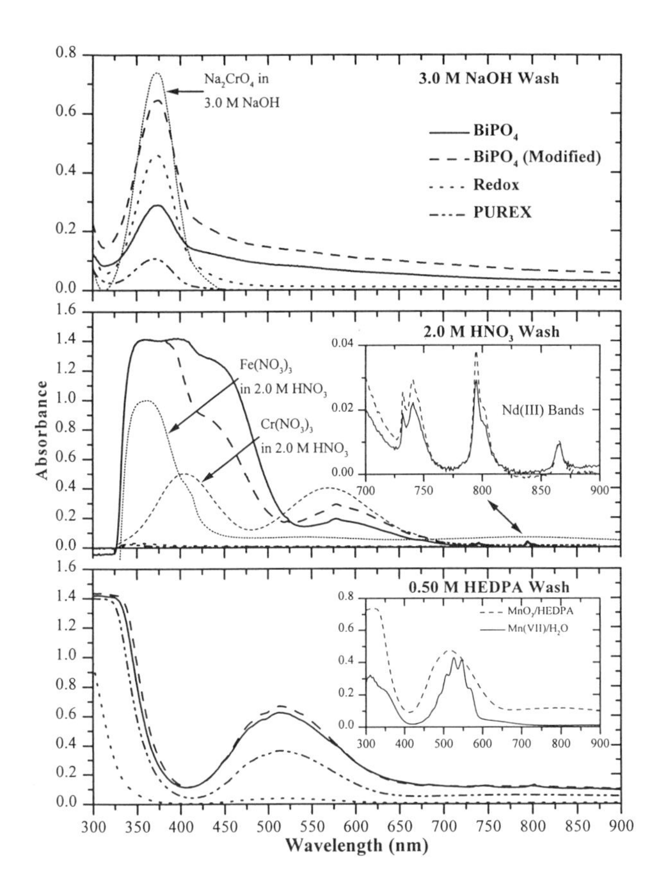

Figure 3: Absorption spectra of 3.0 M NaOH, 2.0 M HNO_3, and 0.50 M HEDPA wash solutions after contact with the respective sludge simulants.

An examination of the ≈ 700-900 nm wavelength region (inset) in the 2.0 M HNO_3 wash solutions after contact with the $BiPO_4$ simulants shows very weak bands that are characteristic for Nd(III). The Nd(III) concentration, and by inference that of La and Eu, is 3.5×10^{-3} M [11].

Contact of the $BiPO_4$, $BiPO_4$ (Modified), and PUREX sludge simulants with 0.50 M HEDPA generates wash solutions purple in color. The visible spectra of these solutions show almost featureless broad bands centered around 513 nm. There is no absorption in this region in the wash solution from the Redox sludge, as it contains the least Mn (Table I). Contact of pristine MnO_2 with 0.50 M HEDPA affords a spectrum (inset) similar to those from the $BiPO_4$, $BiPO_4$ (Modified), and PUREX sludge simulants. Examination of standard spectral data [12] for aqueous Mn(III) shows λ_{max} values of ≈ 500-550 nm with an intense high energy band below 400 nm. Based on these and other observations [5], it can be surmised that a Mn(III)/HEDPA complex is responsible for the broad bands observed in the 0.50 M HEDPA wash solutions from the $BiPO_4$, $BiPO_4$ (Modified), and PUREX sludge simulants.

CONCLUSIONS

The radioanalytical and spectroscopic data indicate that ≤ 2.1% Pu is liberated during CrO_4^{2-} dissolution by 3.0 M NaOH. Experimental sludge washing by 2.0 M HNO_3 causes dissolution of Pu(III/IV), Pu(VI), Ln(III) (Ln = La, Nd, and Eu), and Fe(III) (where present) from the two $BiPO_4$ simulants. Manganese leaching by 0.50 M HEDPA is accompanied by dissolution of up to 4.1% Pu(III/IV) and 14.3% Pu(VI) from the $BiPO_4$ and $BiPO_4$ (Modified) sludge simulants. Future work will involve γ-irradiation of sludge simulants to gain insight into those properties, including actinide oxidation state and chemical speciation, that are influenced by radiolytic effects.

ACKNOWLEDGMENTS

This work is funded by the Environmental Management Sciences Program, Offices of Energy Research and Environmental Management, USDOE, under contract number W-31-109-ENG-38.

REFERENCES

1. F. Weigel, J. J. Katz, and G. T. Seaborg, in *The Chemistry of the Actinide Elements, Second Edition*, edited by J. J. Katz, G. T. Seaborg, and L. R. Morss (Chapman and Hall, London, 1986), p. 499.
2. C. Musikas and W. W. Schulz, in *Principles and Practices of Solvent Extraction*, edited by J. Rydberg, C. Musikas, and G. R. Choppin (Marcel Dekker, New York, 1992), p. 413.
3. G. J. Lumetta, B. M. Rapko, J. Liu, and D. J. Temer, in *Science and Technology for Disposal of Radioactive Tank Wastes*, edited by W. W. Schulz and N. J. Lombardo (Plenum, New York, 1998), p. 203.
4. G. R. Choppin, A. H. Bond, and P. M. Hromadka, J. Radioanal. Nucl. Chem. **219**, 203 (1997).
5. A. H. Bond, K. L. Nash, A. V. Gelis, J. C. Sullivan, M. P. Jensen, and L. Rao, Sep. Sci. Technol. submitted (1999).
6. M. J. Kupfer, *Preparation of Nonradioactive Substitutes for Radioactive Wastes*, Rockwell Hanford Operations, Report DOE/ET/41900-8, 1981.
7. J. L. Krumhansl, P. V. Brady, P. C. Zhang, S. Arthur, and J. Liu, in *Book of Abstracts, 218th ACS National Meeting* (American Chemical Society, Washington, DC, 1999), p. NUCL 55.
8. W. W. Schulz, M. M. Beary, S. A. Gallagher, B. A. Higley, R. G. Johnston, F. M. Jungfleisch, M. J. Kupfer, R. A. Palmer, R. A. Watrous, and G. A. Wolf, *Preliminary Evaluation of Alternative Forms for Immobilization of Hanford High-Level Defense Wastes*, Rockwell Hanford Operations, Report RHO-ST-32, 1980.
9. E. P. Horwitz, H. Diamond, R. C. Gatrone, K. L. Nash, and P. G. Rickert, Process Metall. **7A**, 357 (1992).
10. C. H. Delegard, N. N. Krot, V. P. Shilov, A. M. Fedoseev, N. A. Budantseva, M. V. Nikonov, A. B. Yusov, A. Y. Garnov, I. A. Charushnikova, V. P. Perminov, L. N. Astafurova, T. S. Lapitskaya, and V. I. Makarenkov, *Development of Methods for Dissolving Some Cr(III) Compounds Present in Hanford Site Tank Sludges*, Pacific Northwest National Laboratory, 1999.
11. M. P. Jensen, *unpublished results*, Argonne National Laboratory, 1999.
12. IUPAC, in *Tables of Spectrophotometric Absorption Data of Compounds Used for the Colorimetric Determination of Elements* (Butterworths, London, 1963), p. 320.

CESIUM REMOVAL FROM THE FUEL STORAGE WATER AT THE SAVANNAH RIVER SITE R-BUILDING DISASSEMBLY BASIN USING 3M EMPORE®-MEMBRANE FILTER TECHNOLOGY

L. N. Oji, M. C. Thompson, Kurt Peterson
Westinghouse Savannah River Company, Savannah River Site
Aiken, SC 29808; Thomas M. Kafka 3M Center, 3M Company
St. Paul, MN 55144-1000

ABSTRACT

This report describes results from a seven-day demonstration of the use of a 3M Empore® membrane filter loaded with ion exchange material (potassium cobalt hexacyanoferrate (CoHex)) for radioactive cesium uptake from the R-Disassembly Basin at the Savannah River Site. The goal of the demonstration was to evaluate the efficacy of the Process Absorber Development Unit (PADU), a water pre-filtration/CoHex configuration on a skid, to remove cesium from R-Disassembly basin water at a linear processing flow rate of 22.7 liters per minute (1,200 liters/minute/m^2). During the seven-day demonstration, over 210,000 liters of R-Disassembly basin water was processed through the PADU without cesium breakthrough.

INTRODUCTION

The R-Reactor Disassembly Basin at the Savannah River Site (SRS) was used for over fourteen years mainly for temporary storage of irradiated fuel to remove radioactive decay heat from the fuel assemblies. The Basin currently contains no fuel, target assemblies or other nuclear reactor components. The Basin holds an estimated volume of nineteen million liters of tritiated water and 2-3 inches of radioactive sludge on the Basin floor. The sludge primarily consists of corrosion products from aluminum-clad fuel assemblies and debris from underwater machining operations. The Basin water pH ranges from 7.06 to 7.66 (1). The principal radionuclides of environmental concern in the Basin are summarized in Table 1.

Table 1. Average concentration of radionuclides and competing ions of interest in R-Reactor Basin water (1).

Radionuclide	Activity*, pCi/mL	Competing ions	Concentration, mg/L
Tritium	41,000	Sodium	15.22
Cesium-137	8.25	Potassium	27.57
Strontium-90	23.5	Calcium	12.11

* Total beta/gamma activity = 243 dpm/mL

One basin stabilization option (removal of traces of radionuclides) considers ion exchange-based decontamination technologies to remove cesium-137 and strontium-90. This demonstration project evaluated technology that allows ion-specific separations to concentrate low levels of cesium-137 while generating minimum secondary waste.

The 3M empore® technology, which was evaluated for this cesium removal project, was developed with support from the Department of Energy's Efficient Separations and Processing (ESP) Crosscutting Program. Previous low-volume flow rate (≈ 1.0 liter/minute) demonstrations with this technology had been carried out at the Pacific Northwest National Laboratory (PNNL) (2) and 3M at Idaho National Engineering Environmental Laboratory (3).

Mat. Res. Soc. Symp. Proc. Vol. 608 © 2000 Materials Research Society

The objectives of this ion exchange treatment project were:
(1) To demonstrate the use of 3M Empore membrane/ion exchange filter integrated system to remove cesium from the basin at a flow rate up to 22.7 liters per minute, (2) To determine the effectiveness of a new configuration for the (spiral-wound) web-like 3M membrane in a cartridge, and (3) To provide field data for the scale-up of the technology to more than 100 liters per minute flow rate.

EXPERIMENTAL

MEMBRANE DESIGN

The Empore®web-like membrane developed by 3M is an inert matrix of polytetrafluoroethylene fibrils, which can be loaded with element-selective ion exchange materials. Potassium cobalt hexacyanoferrate (CoHex) was selected for cesium removal. The membrane is configured as a cartridge and inserted into a commercial filter housing. It is in this cartridge form that the CoHex was used in the R-Disassembly basin cesium removal demonstration. An effective pre-filtration system was used upstream of the empore cartridges to remove R-Disassembly Basin particulates and to prevent the clogging of the Empore cartridges during lengthy unattended operation periods.

PADU DESIGN

The PADU consists of a portable skid unit containing pumps, piping, pre-filter system, and 3M Empore membrane/CoHex cartridges (Figure 1). The Basin water intake orifice was about 1.6 meters into the basin. This orifice was guarded with an inlet strainer made of stainless steel. The strainer prevented large particles from going into the pre-filter units. The two positive displacement pumps, in parallel, can deliver water at either 3.8 or 22.7 liters per minute. In this demonstration the PADU was operated in the 22.7 liters per minute mode. Each of the two pre-filter units (coarse and fine assemblies) was housed in a high-pressure stainless steel vessel. Both the coarse (0.45 microns) and fine (0.1 microns) filter assemblies contained six filter cartridges. The six filter cartridges were arranged in parallel inside each vessel and the two vessels were connected in series. The coarse filter housing was connected directly to the displacement pumps and to the Basin water intake line via a 3.1-meter long stainless steel pipe. Each pre-filter unit was equipped with a pressure gauge, a sampling port and a pressure release/water drain valve. The coarse and fine filters could be replaced independently.

The high-pressure switch, located on the intake side of the PADU pre-filter system, was designed to initiate a shut-off of the PADU if the pressure reading on any of the pre-filters went above 100 psi due to clogging. Another switch, a low-pressure one, located after the fine guard pre-filter assembly initiated the PADU shut-off on occasions when the PADU pressure was lower than 15 psi due to a leak in the PADU.

The 3M Empore/CoHex membrane was housed in a high-pressure stainless steel vessel. This unit trapped radioactive cesium and thus was lowered about three meters into the Basin water for shielding and radioactive heat dissipation purposes. This vessel was also equipped with pressure gauge/pressure release valve and was connected to the rest of the PADU unit via a totalizing flow meter. A 10-meter long copper line (effluent line) from the 3M Empore/CoHex membrane vessel contained a valve for obtaining samples before the treated Basin water was returned into the Basin.

The PADU was equipped with a 24-hour clock wired to indicate a self shut-off time, which by design was expected to take place when the pressure readings on the pre-filters was either above 100 psi or below15 psi.

PADU OPERATION

To test the system, more than 6,800 liters of distilled water was pumped through the PADU. During these trial runs the PADU was checked for leaks, sampling procedures, and auto shut down responses resulting from system pressure greater than 100 psi and lower than 15 psi. The last two conditions were simulated by valve restrictions and intentionally inducing leaks in the PADU. During operations, the PADU was checked for leaks during sampling periods. No leaks were observed during the demonstration. The stainless steel housing containing the six CoHex cartridges was lowered into the basin to provide radioactive shielding. Processed water was returned to the basin about 30 meters from the intake orifice to prevent agitation of the basin water.

Samples were collected from three sampling locations on the PADU (Figure 1): pre-filtration samples (samples collected before entering pre-filter housings), post pre-filtration samples (after leaving filter housing) and post-cesium decontamination through 3M Empore filter samples. Each sampling port was fitted with a toggle switch, which could be flushed before sample collection. Daily samples were collected from each port at intervals of two hours beginning at 0900 hours. A total of 12 samples were collected on a typical 24 hour time period. 200-mL samples were collected in 250-mL high-density polyethylene bottles from each specific sampling port. Radiation on the outside of the stainless steel pre-filter housings was measured during each sampling period. After approximately 74 hours of continuous running, the PADU coarse pre-filter was changed because the filter clogged with particulate from the basin.

A sodium iodide (NaI (Tl)) scintillation detector, housed in a lead-shielded cavity to eliminate background influence from the disassembly basin and interfaced to a computer, was used in the field to obtain the approximate activity for cesium-137 in all influent and effluent samples collected (gamma ray spectrum for Cs-137 at $\approx$ 662 KeV). A one-hour time delay in the estimation of cesium activity in the samples using the field detector was employed to minimize the effect of barium-137 (daughter products of cesium-137 with a half-life of 30 minutes). At the end of each day of operation, all samples collected were carried back to the laboratory for final analytical quantitation for cesium.

RESULTS AND DISCUSSION

In this demonstration, the position of the intake orifice relative to the depth of the basin (17 meters) ensured that the sludge materials at the Basin bottom (less than 3% by volume) was not agitated. However, during the actual Basin clean up operation the pre-filtration system would have to be engineered or modified to incorporate a filtration system capable of "backflusing" to clear the clogged membrane filters. This would be necessary especially as the pumping of the water approaches the Basin bottom.

At the end of the seven-day demonstration, the external gamma radiation dose rate on the 11-gauge stainless steel housing (CoHex cartridge housing) was 70 mR/hr at 30 cm. External radiation measurements for the stainless steel pre-filter housings were never greater than 400 dpm/mL β/γ at 30 cm. Field estimates of cesium activity in all samples collected, using the portable-sodium iodide scintillation detector, showed that the CoHex treated effluent samples had no detectable cesium activity. The analytically determined average cesium concentration in the basin water (influent) was 83.4 $\pm$ 2.4 pCi/mL. This value is consistent with a value of 82.5 pCi/mL documented in previous reports on the R-disassembly basin water chemistry (1). The average cesium concentration in the Basin water samples coming through the CoHex filter (effluent), after processing more than 210,000 liters of basin water, was 0.052 pCi/mL. This cesium analysis in the effluent was based on a gamma scan counting time of 10,000 seconds with average mean activity uncertainty of 28%.

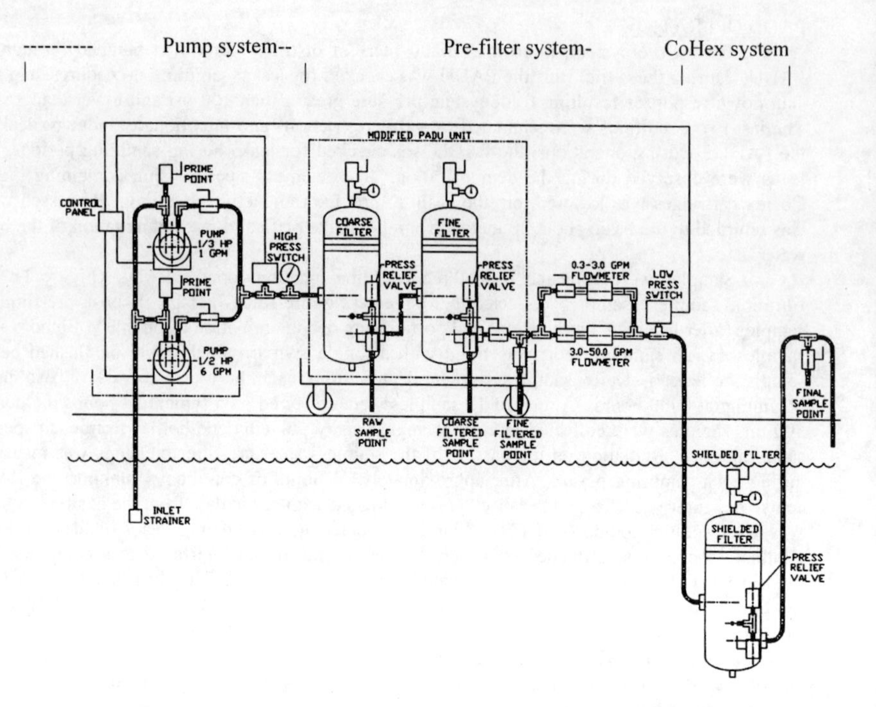

Figure 1. The R-Reactor Disassembly Basin PADU engineering design schematic, consisting of pump, pre-filter and CoHex ion exchange systems.

The cesium decontamination factor, D_f (initial cesium concentration in the basin water/final cesium concentration in the effluent), based on the above numbers is 1604 (83.4 pCi/mL/0.052 pCi/mL). When the effluent samples were analyzed based on a counting time of only 1000 seconds, with average mean activity uncertainty of 4%, the cesium concentration in the effluent is 0.23 pCi/mL. A new D_f value of 363 is obtained.

The Department of Energy (DOE) release-limit for cesium-137 is 3 pCi/mL and with an average Basin water cesium-137 concentration of 83.4 pCi/mL, the required D_f would be 27.8 (83.4 pCi/mL / 3 pCi/mL). With the PADU/CoHex set up it was possible in this demonstration to obtain a D_f better than 1600, which is 58 times better than the required D_f. Even a conservative D_f value of 363, obtained with limited counting time, is still 13 times better than the required D_f. The magnitude of these D_f values shows that the cesium uptake efficiency by CoHex, in the presence of competing ions like sodium and potassium, is significantly large for this single stage decontamination.

Pressure readings on the coarse pre-filter housing increased by about 10.4 psi every 24 hours. The fine pre-filter pressure increases were significantly lower at about 1.0 psi per 24 hours. These increases in pressure, due to the clogging of primarily the coarse pre-filters, can be attributed to the presence of particulate matter in the Basin water. A plot of the PADU continuous operation pressure changes across the coarse and fine pre-filter cartridges versus time

Figure 2, shows that the pressure changes across the pre-filter system was about 10.4 ± 1 psi every 24 hours.

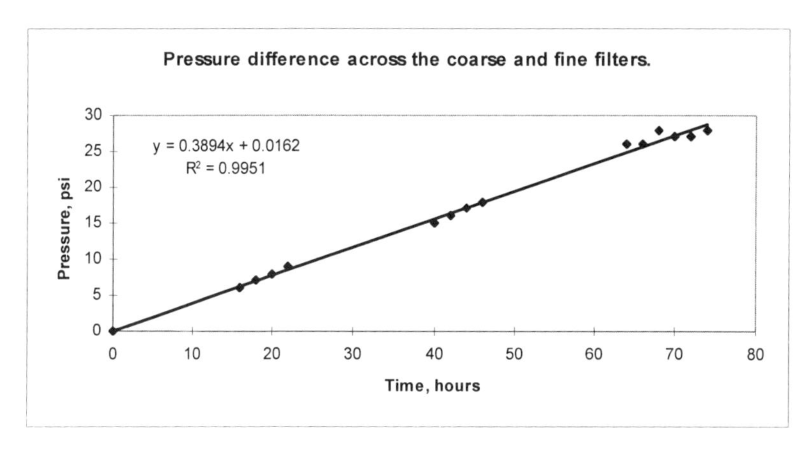

Figure 2. Pressure difference across a set of coarse and fine pre-filters as a function of time. Pressure changes across the pre-filter system averaged 10.4 ± 1.0 psi every 24 hours.

The Basin water flow rate across the PADU, determined from the totalizing flow meter and the slope of a plot of total volume of water processed with time, averaged 22.7 liters per minute. The calculated CoHex cartridge intake surface area is 0.019 m^2 (six cartridges). Based on a linear flow rate of 22.7 liters per minute, the approximate flow rate per unit area is 1,200 liters/minute/m^2, suggesting a very short contact time between the CoHex and the Basin water.

Secondary waste generated from this demonstration (clogged filters and cesium loaded CoHex cartridges) are not sufficiently radioactive and would be disposed off as low or intermediate level radioactive waste.

In the second phase of this PADU/CoHex demonstration, modifications will be made to the unit in order to attain a Basin water-processing rate of 100-150 liters per minute. Also, in this second phase of the project technical questions related to the long term radiation stability (Ten ^{137}Cs half-life)/radiolytic degradation products of the proprietary 3M CoHex membrane and filter units, including ways to limit the frequent changing of the pre-filter cartridges would be addressed.

CONCLUSIONS

The linear flow rate, approximately 22.7 liters of basin water per minute, is sufficiently low to ensure that there is sufficient contact time for ion exchange reactions in the spiral membrane/CoHex arrangement. The estimated minimum cesium-137 decontamination factor (D_f) based on a minimum cesium detection limit of 0.052 pCi/mL, is 1604. This D_f value, at the basin cesium-137 concentration of 83.4 pCi/mL, is 58 times better than the DOE requirement.

After processing over 210,000 liters of the R-disassembly basin water there was no cesium-137 breakthrough. The calculated total amount of cesium-137 adsorbed on the CoHex in a cartridge unit is 17,500 μCi, or about 2,900 μCi per cartridge. The pressure reading change across the PADU coarse and fine pre-filter units was 10.4 ± 1.0 psi per 24 hours time period.

Because of the clogging of the coarse and fine pre-filter units, they will need to be changed every 6.7 days and 70 days, respectively, for continuous operation of the PADU at 22.7 liters of basin water per minute.

The high cesium-137 decontamination factor obtained at 22.7 liters per minute flow rate, shows that a modified PADU, one capable of processing over 100 liters per minute, can be safely operated to meet DOE requirements for cesium-137 decontamination of water in the R-Disassembly Basin. However, at this processing flow rate the pre-filter system would have to be engineered or modified to minimize the clogging and frequent change-outs observed in this demonstration.

ACKNOWLEDGEMENTS

This work was performed under the auspices of the U.S. Department of Energy by the Savannah River Technology Center. The Department of Energy's Efficient Separations and Processing (ESP) Crosscutting program provided funding for this project.

REFERENCES

1. J. B. Pickett, Analytical Results of the 1997 R-Disassembly Basin Sampling Program (U), FDD-ENG.-98-0029, Westinghouse Savannah River Company, Savannah River Site, Aiken, SC.
2. Bechtel Hanford, Inc. Richland, WA, "Demonstration of Radionuclides Removal at the 105-N Basin using the 3M System", Report BHI-00759, Rev. 0.
3. Herbst, R. S., K. N. Brewer, T. A. Todd (Lockheed Idaho Technologies Company), L. A. Bray (PNNL), T. Kafka, R. L. White (3M). " Decontamination of TAN Injection Well Water using 3M Web technology'. Report, November 1995, INEL-95\0589, Lockheed Idaho Technologies Company, Idaho, Falls, ID.

CHARACTERIZATION OF AND WASTE ACCEPTANCE RADIONUCLIDES TO BE REPORTED FOR THE SECOND MACRO-BATCH OF HIGH-LEVEL WASTE SLUDGE BEING VITRIFIED IN THE DWPF MELTER

T.L. FELLINGER, N.E. BIBLER, W.T. BOYCE, and J.J. OLSON
Savannah River Technology Center, Westinghouse Savannah River Company, Aiken, S.C. 29808

ABSTRACT

The Defense Waste Processing Facility (DWPF), at the Savannah River Site (SRS), is currently processing the second million gallon batch (Macro-Batch 2) of radioactive sludge slurry into a durable borosilicate glass for permanent geological disposal. To meet the reporting requirements as specified in the Department of Energy's Waste Acceptance Product Specifications (WAPS), for the final glass product, the nonradioactive and radioactive compositions must be provided for a Macro-Batch of material [1]. In order to meet this requirement, sludge slurry samples from Macro-Batch 2 were analyzed in the Shielded Cells Facility of the Savannah River Technology Center (SRTC). This information is used to complete the necessary Production Records at DWPF so that the final glass product, resulting from Macro Batch 2, may be disposed of at a Federal Repository. This paper describes the results obtained from the analyses of the sludge slurry samples taken from Macro-Batch 2 to meet the reporting requirements of the WAPS. Twenty eight elements were identified for the nonradioactive composition and thirty one for the radioactive composition. The reportable radioisotopes range from C-14 to Cm-246.

INTRODUCTION

High level liquid waste generated, by F and H Canyon operations, over many years is stored at SRS in double walled underground steel tanks as a caustic slurry. The DWPF is processing and immobilizing this high level liquid waste into a durable borosilicate glass for permanent geological disposal. The current radioactive sludge slurry batch that DWPF is processing is called Macro-Batch 2. Macro-Batch 2 consists primarily of Tank 42 sludge slurry that was transferred to and mixed with the remaining small heel of the first Macro-Batch.

Six samples (~80 mL each) were taken of Macro-Batch 2 and sent to SRTC Shielded Cells for full chemical and radionuclide analysis. The sludge slurry samples were combined, dissolved, and analyzed for nonradioactive elemental composition by Inductively Coupled Plasma- Emission Spectroscopy (ICP-ES) and Atomic Adsorption (AA). Twenty eight nonradioactive elements for Macro-Batch 2 were identified. The radioactive elemental composition for the sludge slurry was determined by Inductively Coupled Plasma – Mass Spectroscopy (ICP-MS) and counting techniques. Special separation techniques were used to detect C-14, I-129, Cm-246 and Cm-247 in the sludge slurry. This is the first time these radionuclides have been detected and measured in radioactive sludge slurry at the SRS. The WAPS states that all radioisotopes that have a half-life longer than 10 years and contribute greater than 0.05% of the total radioactivity (Becquerels or Curie basis) during the first 1100 years after production must be reported [1]. The DWPF has extended the criteria to include all radioisotopes that have a half-life longer than 10 years and contribute greater than 0.01% of the total radioactivity during the first 1100 years after glass production. The radioisotopes from these analyses that met the WAPS criteria of having a half life longer than 10 years were decayed by a computer program from time zero (May 1999) to 2000 years (May 3999). This was completed in order to determine the radioactivity of the glass 1100 years after production in the DWPF as required by the WAPS. Thirty one radioisotopes were identified as reportable for Macro Batch 2.

Mat. Res. Soc. Symp. Proc. Vol. 608

EXPERIMENTAL

Combination of the Sludge Slurry Samples

Six samples of Macro-Batch 2 sludge slurry from SRS Tank 51 were received at SRTC in 80 mL stainless steel bottles. These sludge slurry samples were placed into the SRTC Shielded Cells and then combined into a one-liter container. To ensure all of the sludge slurry solids had been removed from each stainless steel bottle, each bottle was carefully rinsed. The final volume of the combined sludge slurry was ~454 mL with about 19 wt. % solids.

Analytical Methods

Dissolution of the Macro-Batch 2 sludge slurry sample was performed remotely in the Shielded Cells of SRTC. A portion of the sludge slurry sample was dissolved in quadruplicate by two separate dissolution methods. Details of these dissolution methods have been published [2]. Prior to dissolution, the sludge slurry was dried at 115°C in a drying oven. A standard glass with a composition similar to the DWPF glass, Analytical Reference Glass-1 (ARG-1) [3], was also dissolved and analyzed concurrently with the sludge slurry samples. This was done to confirm that the dissolutions were complete and the analytical procedures were performed correctly.

The dissolved samples were diluted so that only a small potion of the radioactivity was removed from the Shielded Cells. The resulting solutions were then analyzed using ICP-ES, AA, ICP-MS and counting techniques.

Identification of Reportable Radionuclides for Macro Batch 2

In order to determine the reportable radionuclides for Macro-Batch 2 all radioactive U-235 fission products and all radioactive activation products that could be in the SRS HLW were considered.

It was evident that some of the radioisotopes in the sludge slurry could not be measured because of their low concentration in the sludge slurry. For these isotopes, an estimate of the concentration was made by calculating their concentration based on their U-235 fission yields. An estimate of the concentrations for Se-79, Rb-87, Pd-107, In-115, Sn-121m and Sn-126 was done in this way. This was accomplished by dividing the fission yield (a known yield value from the fission of U-235) for each isotope by the average Fission Yield Scaling Factor (FYSF). The average FYSF is determined by calculating a FYSF (known fission yield / elemental wt. %) for appropriate low and high mass fission products and averaging the results. Details concerning the calculation of the FYSF and the fission products that are appropriate have been published [4]. Figures 1 and 2 present the measured and calculated concentrations for the low mass and the high mass U-235 fission products that could be detected for Macro-Batch 2. Explanations for those concentrations that deviate from the calculated concentrations have been published [4]. For example, results at masses 88,107,110, and 140 can be assigned to natural Sr, Ag, Cd, and Ce isotopes in the waste.

After determining and estimating the concentrations for all of the radioactive U-235 fission products and all of the radioactive activation products that could be present in the sludge slurry, a list of radioisotopes considered for the decay calculations was generated.

The initial activities (Becquerels/kg of dried sludge) for the radioisotopes were calculated by an Excel spreadsheet using the weight percent reported for each measured or calculated isotope and its specific activity. These initial activities were then entered into a computer program and decayed over a time period starting with time zero (May 1999) to 2000 years (May 3999) for each of the isotopes of concern. Excel spreadsheets were then used to calculate the total activity (Ci/kg of dried sludge slurry) at each time and the percent of the activity that each of the radionuclides contributed. After these calculations were complete, each radionuclide was plotted as percent activity versus time in years. A total of nine plots were generated. A check of the decay calculations was performed using another computer program. The results of both programs were in good agreement, indicating the calculation of activity over time was correct for the isotopes of concern.

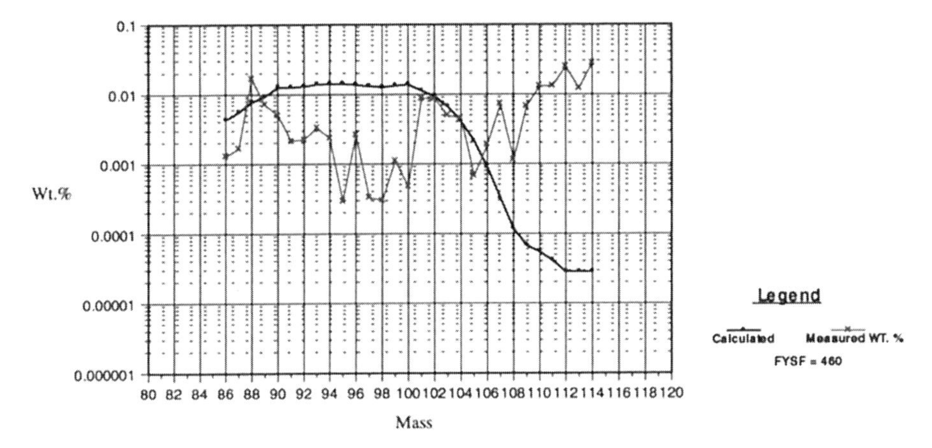

Figure 1.Measured and Calculated Concentrations of U-235 Low Mass Fission Products in Macro-Batch 2

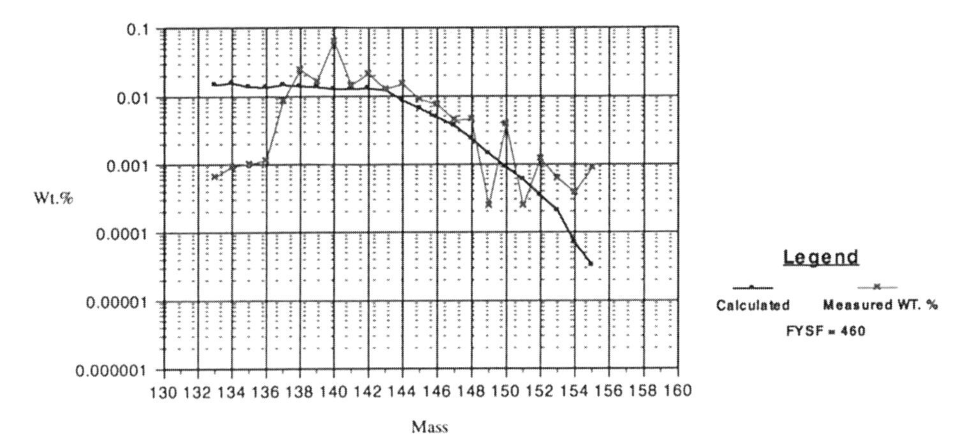

Figure 2. Measured and Calculated Concentrations of U-235 High Mass Fission Products in Macro-Batch 2

RESULTS

Nonradioactive and Radioactive Composition of Macro-Batch 2

In order to comply with the requirements of the WAPS for this Macro-Batch of sludge slurry, DWPF is required to provide the nonradioactive and radioactive composition of the waste form. Table I provides the nonradioactive composition of the sludge slurry for Macro-Batch 2. Table II presents the radioactive composition of Macro-Batch 2 used in the decay calculations. The standard deviation and the percent Relative Standard Deviation (RSD) for both tables are provided in the parenthesis. The absence of a standard deviation and percent RSD, in the tables, indicates that only one value was obtained.

Table I. Nonradioactive Composition for Macro-Batch 2 Sludge Slurry

Element	Weight Percent *, a	Element	Weight Percent *, a
Ag[b]	1.85E-02 (± 8.8E-04, 4.8)	Mg	1.15E00 (± 3.9E-02, 3.9)
Al	7.66E00 (± 2.2E-01, 2.9)	Mn	3.30E00 (± 1.2E-01, 3.5)
B[b]	5.49E-03 (± 1.7E-03, 30)	Mo[b]	4.61E-03 (± 3.5E-04, 7.6)
Ba[b]	4.61E-02 (± 1.3E-03, 2.7)	Na[b]	5.78E00 (± 2.4E-01, 4.2)
Ca	2.18E00 (± 1.1E-01, 5.0)	Ni	3.44E-01 (± 1.4E-02, 4.0)
Cd	1.10E-01 (± 3.7E-03, 3.4)	P	6.55E-01 (± 1.6E-01, 24)
Cr	1.32E-01 (± 5.4E-03, 4.1)	Pb[b]	7.48E-02 (± 3.8E-03, 5.1)
Cu	2.97E-02 (± 1.1E-03, 3.8)	Si[b]	1.34E00 (± 1.2E-02, 0.93)
Co[b]	7.64E-03 (± 6.3E-04, 8.2)	Sn[b]	9.60E-03 (± 7.0E-04, 7.3)
Fe	2.15E01 (± 6.9E-01, 3.2)	Sr[b]	1.93E-02 (± 7.1E-04, 3.7)
Hg[c, d]	8.61E-01 (± 4.2E-02, 4.9)	Ti[b]	1.70E-02 (± 6.1E-04, 3.6)
K[d, e]	<5.0E-02	V[b]	7.84E-03 (± 6.6E-04, 8.4)
La[b]	2.13E-02 (± 6.3E-04, 3.0)	Zn[b]	3.60E-02 (± 2.1E-03, 5.8)
Li[b]	6.57E-03 (± 3.7E-03, 5.7)	Zr[b]	2.13E-02 (± 2.2E-03, 11)

* The sludge slurry sample was dried overnight at 115°C in a drying oven. Results are present on a dry total solids basis.
[a] Majority of the results are determined by ICP-ES unless otherwise indicated and are the average of eight sample results.
[b] Average of four results only.
[c] Average of three results only.
[d] Results determined by AA method.
[e] Detection Limit of the Instrument.

Table II. Initial Radioactive Composition for Macro-Batch 2 Sludge Slurry used in Decay Calculations

Element	Results	Units*	Element	Results	Units*
C-14[a]	9.91E-08 (± 5.7E-08, 57)	Wt. %	Th-232[d]	3.10E-01 (± 1.4E-02, 4.6)	Wt. %
Ni-59[a]	8.10E-05	Wt. %	U-233[d]	4.52E-04 (± 1.8E-05, 4.0)	Wt. %
Co-60[b]	6.19E-08 (± 4.4E-09, 7.1)	Wt. %	U-234[d]	4.65E-04 (± 4.3E-05, 9.2)	Wt. %
Ni-63[a]	6.0E-06	Wt. %	U-235[d]	1.30E-02 (± 4.8E-04, 3.7)	Wt. %
Se-79[c]	1.22E-04	Wt. %	U-236[d]	1.14E-03 (± 8.4E-05, 7.4)	Wt. %
Rb-87[c]	5.54E-03	Wt. %	Np-237[d]	1.83E-03 (± 9.6E-05, 5.3)	Wt. %
Sr-90[a]	3.10E-03 (± 7.9E-05, 2.5)	Wt. %	U-238[d]	1.92E00 (± 7.9E-02, 4.1)	Wt. %
Zr-93[d]	3.27E-03 (± 9.9E-04, 30)	Wt. %	Pu-238[e]	5.81E-04 (± 4.8E-05, 8.3)	Wt. %
Tc-99[d]	1.10E-03 (± 6.7E-05, 6.1)	Wt. %	Pu-239[d]	7.87E-03 (± 2.6E-04, 3.3)	Wt. %
Pd-107[c]	3.70E-04	Wt. %	Pu-240[d]	7.82E-04 (± 7.5E-05, 9.5)	Wt. %
Cd-113[d]	1.20E-02 (± 7.3E-04, 6.0)	Wt. %	Pu-241[a]	2.75E-05 (± 2.9E-06, 11)	Wt. %
In-115[c]	2.39E-05	Wt. %	Pu-242[d]	9.41E-05 (± 1.7E-05, 18)	Wt. %
Sn-121m[c]	3.91E-05	Wt. %	Am-241[b]	1.94E-04 (± 1.1E-05, 5.1)	Wt. %
Sn-126[c]	1.20E-04	Wt. %	Am-243[a]	2.84E-05 (± 2.4E-06, 8.6)	Wt. %
I-129[a]	6.16E-04 (± 2.1E-04, 34)	Wt.%	Cm-243[a]	2.72E-07(± 1.0E-08, 3.8)	Wt. %
Cs-135[d]	6.54E-04	Wt. %	Cm-244[a]	5.46E-06 (± 3.3E-07, 6.1)	Wt. %
Cs-137[b]	1.84E-04 (± 5.6E-06, 3.0)	Wt. %	Cm-246[a]	1.32E-05	Wt. %
Sm-151[d]	2.40E-04 (± 2.2E-05, 9.0)	Wt. %	Cm-247[a]	2.48E-06	Wt. %
Eu-154[b]	3.89E-06 (± 1.2E-07, 3.2)	Wt. %	Cf-250[c]	2.60E-08	Wt. %

* The sludge slurry was dried overnight at 115°C in a drying oven. All results presented on a dry total solids basis.
[a] Special Separation Technique
[b] Gamma Scan
[c] Calculated Value for wt.%
[d] ICP-MS results
[e] Calculation based on Total Alpha and ICP-MS results.

Examples of Plots (Percent Activity vs. Time) Generated from Decay Calculations

The computer program used to calculate the decay of the radionuclides also calculated the daughters of the parent radionuclide which contribute to the total activity. Some of these daughters have very short half-lives and are in secular equilibrium with the parent's activity or in some cases the granddaughters and great granddaughters are in secular equilibrium with the daughter's activity. An example of the daughter being in secular equilibrium with the parent can be seen in Figure 3 for Sr-90 and Y-90. After some time, these two curves overlap one another and the two activities become equal. In Figure 3 it can be seen that the only radionuclide that is not reportable is Rb-87 because it remains less than 0.01%.

An example of the granddaughter being in secular equilibrium with the daughter can be seen in Figure 4. Figure 4 is the decay chain for U-233. In this case the daughter of U-233, Th-229, has a long half-life (7.3E4 years) and alpha decays to Ra-225 which has a ~14 day half-life. The Th-229 and Ra-225 activities therefore overlap one another. In this case, no initial activity was assumed for Th-229 and Ra-225 because of the long half-live of U-233 (1.6E5 years). Note that after 400 years Th-229 and Ra-225 contribute greater than 0.01 % of the total activity. However Ra-225 is not reportable because of its short half-life ($t_{1/2}$ = 14.8 days). The same arguments apply to Ac-225 ($t_{1/2}$ = 10.0 days) and Fr-221 ($t_{1/2}$ = 4.8 minutes) in this Figure.

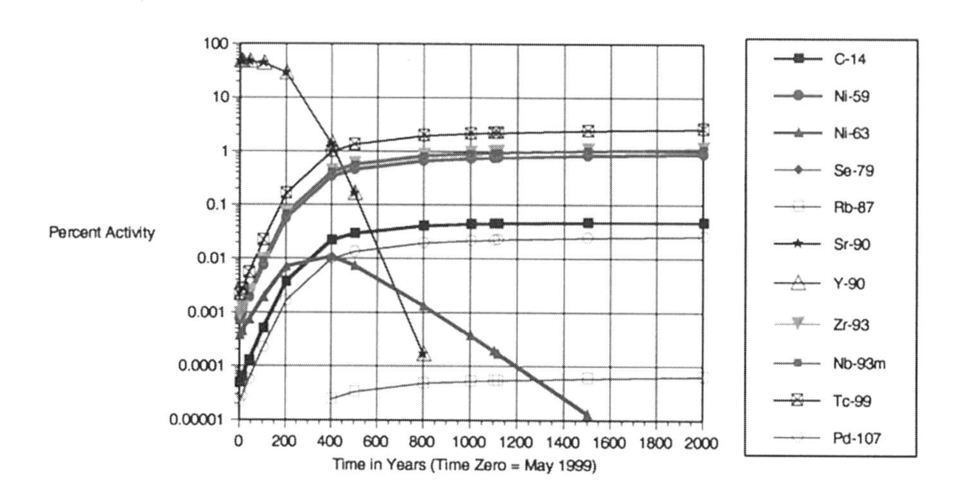

Figure 3. Calculated Percent Activity for Selected Fission Products over Time

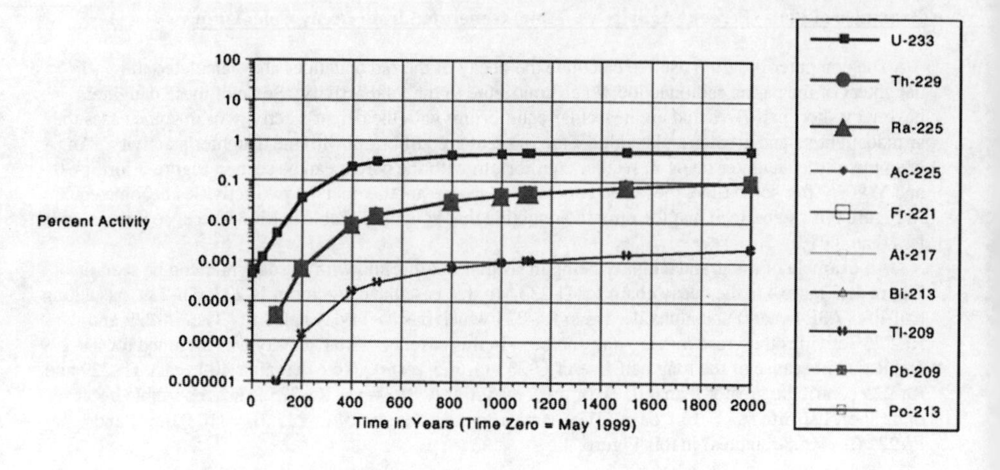

Figure 4. Calculated Percent Activity for U-233 over Time

CONCLUSIONS

1. The nonradioactive and radioactive composition of the DWPF Macro-Batch 2 has been determined.

2. Thirty-one radioisotopes were identified as reportable for the DWPF Macro-Batch 2 sludge slurry. The radioisotopes are listed below:
   C-14, Ni-59, Ni-63, Se-79, Sr-90, Zr-93, Nb-93m, Tc-99, Pd-107, Sn-126, I-129, Cs-135, Cs-137, Sm-151,Th-229, Th-230, U-233, U-234, U-235, U-236, Np-237, U-238, Pu-238, Pu-239, Pu-240, Pu-241, Am-241, Pu-242, Am-243, Cm-244, and Cm-246

REFERENCES

1. Office of Environmental Restoration and Waste Management, *Waste Acceptance Product Specifications for Vitrified High-Level Waste Forms*, Revision 2, USDOE Document EM-0093, U.S. Department of Energy, Germantown, MD, 1996.

2. W.F. Kinard, N.E. Bibler, C.J. Coleman, and R.A. Dewberry, J. Radioanal. Nucl. Chem., Vol. 219, pp.197, 1997.

3. G.L. Smith, *Characterization of Analytical Reference Glass 1 (ARG-1)*, PNNL-8992, Pacific Northwest Laboratory Report, (12/93).

4. N.E. Bibler, W.F. Kinard, W.T. Boyce and C.J. Coleman, J. Radioanal. Nucl. Chem., Vol. 234, pp.159-163, 1998.

*

For further information, contact:

Terri L. Fellinger
Westinghouse Savannah River Company
Aiken, S.C. 29808
Telephone: (803) 725-7745
e-mail: terri.fellinger@srs.gov

MATRIX-ASSISTED INFRARED-LASER DESORPTION-IONIZATION MASS SPECTROMETRY OF ORGANIC MOLECULES ON $NaNO_3$

D. R. ERMER, M. BALTZ-KNORR, D. NAKAZAWA*, M. R. PAPANTONAKIS and R. F. HAGLUND, JR.
Department of Physics and Astronomy and W. M. Keck Foundation Free-Electron Laser Center, Vanderbilt University, Nashville TN 37235 richard.haglund@vanderbilt.edu
*Physics Department, Reed College, 3203 SE Woodstock Boulevard, Portland, OR 97202-8199

ABSTRACT

We demonstrate sensitive mass identification of model organic compounds directly from sodium nitrate by infrared matrix-assisted laser desorption and ionization mass spectrometry. Sensitivity limits of order 10^{-6} are achieved for resonant desorption, with a linear calibration verified for crown ether molecules over almost three decades in concentration. We also have observed that crown ether molecules can abstract atomic sodium from the sample when irradiated at the resonant frequency of the NO_3 stretching vibrations around 7.1 μm. A model for this process is proposed which suggests great promise for this technique on untreated tank waste.

INTRODUCTION

High-sensitivity, rapid and economical characterization of high-level tank wastes is a critical unmet need in the Department of Energy's environmental management program [1]. We are developing mass spectrometry protocols for molecular constituents of high-level tank wastes in solid, liquid and slurry forms using tunable, ultrashort-pulse, infrared lasers by investigating: the effects of laser wavelength, intensity and fluence on desorption and ionization of atomic and molecular species from nitrate, phosphate, sulfate, carbonate and ferrocyanide samples; the correlation between extinction coefficient, deposited laser energy and desorption-ionization thresholds in solid, liquid and slurry materials using photoacoustic, surface-plasmon and FTIR spectrometry; and the systematic effects of laser parameters and chemical reactivity between matrix and analyte on our ability to quantitate the desorption and ionization of desorbed species, including the formation of adduct complexes on analyte ions. In this work, we particularly seek to capitalize on the high efficiency of desorption, bond-breaking and ionization in solids, liquids, and slurries when initiated by resonant vibrational excitation with picosecond laser pulses.

In this paper, we present some of the principal findings emerging from our initial studies of desorption of organic molecules from sodium nitrate, including: (1) the capacity of a tunable infrared laser, in conjunction with time-of-flight mass spectrometry, to identify organics on solid and water slurries of sodium nitrate; (2) initial calibration curves for ion yields of eighteen-crown ether (specifically, 18C6) as a function of concentration down to 1 ppm; and (3) demonstration that Na^+ adduct ions are formed during in the desorption of 18C6 from sodium nitrate. The fact that the most efficient desorption occurs at the resonant frequency for the NO_3 stretching vibration permits us to develop a tentative model for this adduct formation.

This work suggests that IR-MALDI can be a useful quantitative analytical tool in three areas where current methods are either insufficient to meet DOE needs or simply do not exist: (1) real-time, quantitative, high-sensitivity molecular as well as atomic speciation for on-line monitoring of waste process streams; (2) measurement of chemical energy content of tank waste; and (3) rapid analysis of surface soils and subsurface contaminated sediments present in DOE sites. One of the biggest advantages of IR-MALDI is the fact that it is possible to desorb and ionize molecules from liquid matrices, including water, because water has strong fundamental and combination bands throughout the mid-IR spectrum.

Mat. Res. Soc. Symp. Proc. Vol. 608 © 2000 Materials Research Society

EXPERIMENTAL METHOD AND APPARATUS

Matrix-assisted laser desorption-ionization (MALDI) coupled with time-of-flight mass spectrometry (MS) is a sensitive method for detecting molecules at concentrations down to attomoles and with nanogram sample masses. In MALDI-MS, analyte molecules at low concentrations are prepared in an absorbing matrix which is irradiated by laser pulses at moderate intensity (~1-100 $MW \cdot cm^{-2}$). The total ion yield is proportional to the specific energy deposition:

$$\text{Yield} \propto \frac{\text{Energy}}{\text{Volume}} \cong \frac{\text{Fluence}}{\text{Optical Absorption Depth}} = F \cdot \left[\alpha_o(\omega) + \beta I\right] \tag{1}$$

where F is the laser fluence ($J \cdot cm^2$); α_o is the linear optical absorption coefficient (cm^{-1}) which is generally a strong function of the frequency ω in the infrared; β is the nonlinear absorption coefficient ($cm \cdot W^{-1}$); and I is the laser intensity ($W \cdot cm^{-2}$). Since the ion yield is proportional to both laser frequency through α_o and laser intensity through βI, the advantages of high-intensity, tunable, ultrashort-pulse lasers are self-evident. Moreover, at laser pulse durations shorter than 2 ps or so, the energy is deposited much faster than thermal diffusion times, guaranteeing that the desired, spatially localized electronic or vibrational excitation is not dissipated by mixing with the delocalized, harmonic modes of the phonon bath. Unlike laser ablation mass spectrometry [2], MALDI routinely produces not only atomic-ion but also intact molecular ion-ion spectra.

Our experiments used a picosecond free-electron laser (FEL) in which the wavelength λ_{FEL} of the FEL is related to the electron-beam energy and the wiggler parameters by [3]

$$\lambda_{FEL} = \frac{\lambda_W}{2\gamma^2}\left(1 + \kappa^2\right) \tag{2}$$

where λ_W is the wiggler spatial period; γ is the ratio of the kinetic and rest energies of the electrons, and κ is a magnetic-field-dependent quantity of order unity. The FEL wavelength can be varied continuously by changing either the wiggler field, the electron-beam energy, or both. The Vanderbilt FEL delivers a pulse train comprising some $2 \cdot 10^4$ 0.7-1.0-ps micropulses in a 4-μs macropulse, with an interpulse spacing of 350 ps [4]. Typical micropulse energies are a few μJ, yielding macropulse energies of order 30-150 mJ in a diffraction- and transform-limited TEM_{00} mode. A broadband HgCdTe electro-optic switch (II-VI, Inc., Saxonburg PA) can be used to slice a short pulse train (≥50 ns) from the macropulse with a contrast ratio of order 160:1 [5].

A Comstock (Oak Ridge, TN, USA) model TOF-101-MS, one-meter linear time-of-flight (TOF) mass spectrometer was used with a specially modified ion source with conical electrodes; ions were detected with a chevron-type dual microchannel plate (Galileo, Inc.). The measurements on $NaNO_3$ slurries were carried out using a cryogenically cooled sample stage in the linear section of a two-meter Comstock reflectron time-of-flight mass spectrometer, likewise equipped with a dual chevron microchannel-plate detector. Base vacuum in both systems is of order $5 \cdot 10^{-7}$ Torr.

Our model tank-waste substrate, sodium nitrate, is isoelectronic to calcium carbonate, with the same crystal structure and a similar response to electronic excitation [6]. It is also a primary constituent of the "salt cake" which forms when high-level radioactive waste is diluted with sodium hydroxide and stored in mild-steel storage tanks [7]. Like $CaCO_3$, sodium nitrate has a broad absorption resonance around 7.1 μm which is created by the superposition of the six different asymmetric stretching modes of the NO_3 molecular subgroup.

EXPERIMENTAL RESULTS

Previous studies of infrared free-electron-laser-induced desorption and ionization on the desorption and ionization of binary and ternary oxides have demonstrated that:

- Infrared laser ablation from $NaNO_3$, $CaCO_3$ and SiO_2 exhibits lower shock velocities and planar material removal than for visible or ultraviolet laser ablation [8]; and
- MALDI measurements of EDTA in succinic acid show substantially greater efficiency with the ps pulse train of the FEL than with the ns pulse from an Er:YAG laser [9].
- MALDI wavelength thresholds do not correlate exactly with IR absorption bands of either the matrix or the analyte for ps pulses, but seem to be slightly blue-shifted [9].

Recent results using picosecond *tunable* IR-FEL radiation in the 2 - 10 micron wavelength range show [10] that desorption and ionization by excitation of O-H, H_2O, C=O and CO_3 vibrational modes are possible in solid matrix microcrystallites, thin films and macroscopic crystals.

Our recent FEL-MALDI mass spectrometry experiments have shown that it is possible to identify toluene, benzene and carbon tetrachloride adsorbed on the surface of sodium nitrate crystals even in relatively small quantities. In the spectra shown below, the delayed extraction feature on our mass spectrometer was unused, so the mass resolution is degraded from its optimum value by a factor of ~15. The sodium nitrate samples used for these studies were grown from supersaturated solution, forming clear crystals several mm on a side.

Figure 1 shows three mass spectra taken from these $NaNO_3$ samples after dosing with toluene and benzene under the following conditions: Run 3, $\lambda = 3.2$ µm, first shot; Run 4, $\lambda = 3.5$ µm, first shot; Run 5, $\lambda = 3.5$ µm, fortieth shot. The mass peaks from toluene ($m/z = 92$) and for the NO_3 radical ($m/z = 62$) are evident in every shot. However, the benzene peak ($m/z = 78$) is seen only in the last spectrum, possibly indicating its dependence on surface conditioning. At an FEL wavelength of 3.5 µm, the $NaNO_3$ is nearly transparent, with the only absorption coming from the overtone of the NO_3 stretching vibrations together with a small contribution from hydrated water in the crystal. Hence the MALDI process in this case is not very surface specific, but may involve laser-absorption processes occurring deeper in the crystal.

In Figure 2, taken near the resonant wavelength of the ν_2- ν_4 asymmetric stretching vibration at 7.0 µm, the absorption of the FEL light is much stronger, and the toluene and benzene peaks are clearly evident in Run 8 — taken on the 200th shot after the initial dosing. After sitting without laser irradiation in a vacuum of $5 \cdot 10^{-7}$ torr (Run 9), the toluene peak is still evident, as is

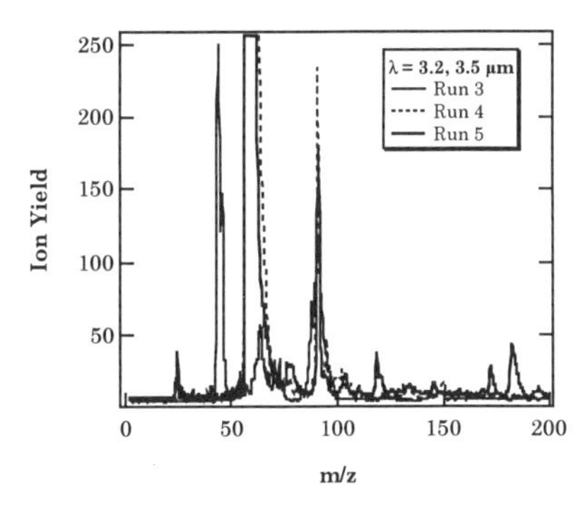

Figure 1. Ion spectra from FEL irradiation of a crystalline $NaNO_3$ sample dosed with toluene and benzene, at wavelengths around 3 µm (near the OH stretch). The conditions of Runs 3, 4 and 5 are explained in the text.

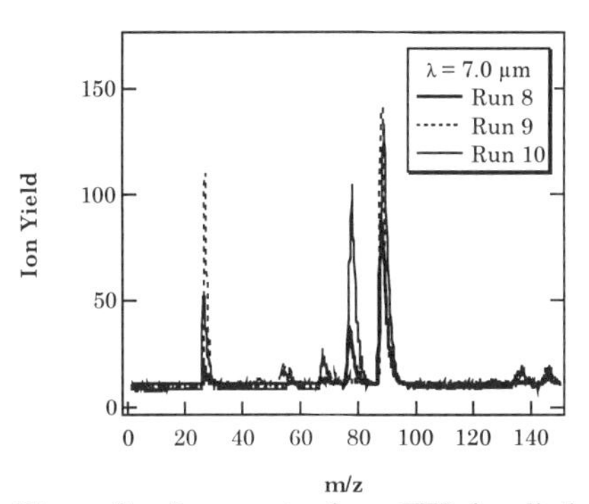

Figure 2. Ion spectra from FEL irradiation of a $NaNO_3$ crystal dosed with toluene and benzene, near the asymmetric stretch of the nitrate group at 7 µm. The conditions of Runs 8, 9 and 10 are explained in the text.

the N_2 ion peak (*m/z* = 28); however, the benzene peak is missing. Nevertheless, the benzene and toluene peaks are both evident in Run 10, taken after irradiating the sample for 30 s at the maximum FEL pulse-repetition rate of 30 Hz. Apparently the burst of laser irradiation after the incubation in vacuum brings benzene from the interior of the crystal to the surface. Thus

- Tuning to the correct wavelength *and* being able to deposit energy in picosecond bursts is an important element of seeing all the key constituents of the sample.
- Wavelength tuning permits some discrimination between surface adsorbates and the constituents of the bulk sample (as in the 3.5 μm irradiation experiment).

Since the phosphates, sulfates and carbonates all have vibrational bands like those of the nitrates, similar conclusions should apply to major constituents of the tank waste solids. Moreover, it may be possible to study rates of surface-to-surface and bulk-to-surface diffusion of molecular impurities in the nitrates using ultrashort-pulse infrared LDI.

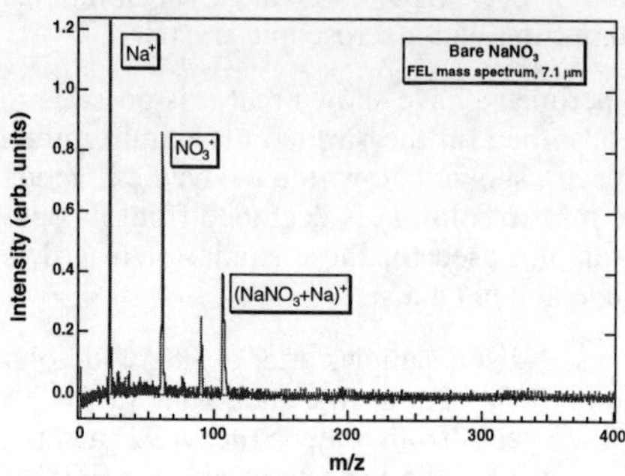

Figure 3(a). 18C6 on $NaNO_3$ slurry.

In another set of experiments, slurries of H_2O and $NaNO_3$ (in a 1:6 ratio by mass) dosed with 18-crown-6 $(C_2H_4O)_6$ molecular weight 264.3 – were frozen on a cryogenically cooled sample stage and irradiated at various FEL wavelengths using the full macropulse duratoin of 4 μs. When bare $NaNO_3$ is irradiated on resonance, one sees Na^+, NO_3^+ and $(NaNO_3+Na)^+$ in the mass spectrum, indicating that the crystal is broken into its constituent subgroups. [Figure 3(a)] When dosed with the crown ether (hereinafter referred to as 18C6), at a concentration of 1:100 by volume, quite distinctive behaviors are observed on- and off-resonance. At 3.4 μm, where $NaNO_3$ still has some residual absorption from the first overtone of the NO_3 stretching vibration, only the Na^+ ion is seen with any significant yield, although there is the faintest suggestion of a desorption-ionization event in the spectrum at the place where 18C6 should appear [Figure 3(b)]. When the FEL is tuned in to the nitrate resonance, however, a clear signal of $(18C6+Na)^+$ is seen, indicating that the crown ether has abstracted a sodium ion from the crystal. The signal is detectable above background down to concentrations as low as a few ppm. The signals from the $NaNO_3$ crystal itself are similar to those seen from the bare crystal, as shown in Figure 3(c).

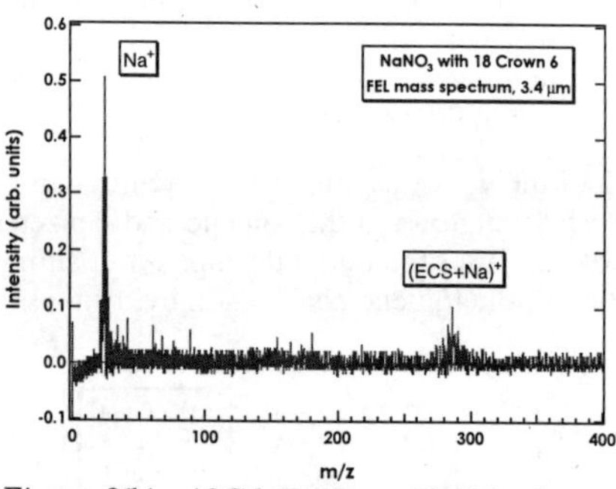

Figure 3(b). 18C6 (ECS) on $NaNO_3$ slurry.

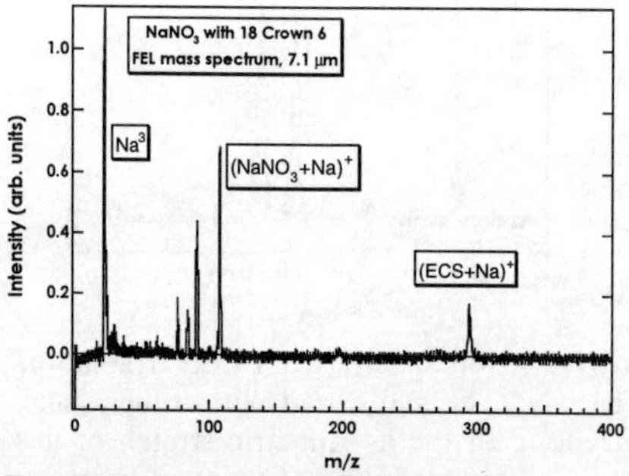

Figure 3(c). 18C6 (ECS) on $NaNO_3$ slurry.

Given the crystal structure of $NaNO_3$, in which the nitrate groups are located in close proximity to the Na ions, we propose the following mechanism for the creation of the crown-ether ions with Na adducts. On resonance, the absorption depth is extremely short, of order ~100-200 nm. This is much shorter than the thermal diffusion length, and hence leads to a condition of explosive vaporization or "phase explosion." In this condition, the short absorption depth leads to heating of the material above the critical thermodynamic temperature at such a rate that no equilibration with the surrounding material is possible and the absorption zone makes an extremely rapid phase transition from a solid to explosive melting and vaporization. When the $NaNO_3$ is irradiated on-resonance, the nitrogen ion in all six asymmetric stretch modes vibrates opposite to one or more of the oxygen ions. The local dipole moment of the NO_3 group shifts spatially at an extremely rapid rate, alternately becoming more and less attractive to the nearest-neighbor Na^+ ions. The crown ether has a large π-electron concentration in its crown, and thus is highly electronegative, making it straightforward for the 18C6 to pick up the adduct ion. This picture presumes, of course, that the 18C6 exists as a neutral molecular interstitial in the sodium nitrate matrix, probably a reasonable assumption given the ionicity of $NaNO_3$.

Calibration of ion yields remains a significant problem in both UV- and IR-MALDI [11]. We have found that in the 18C6 case, a linear calibration over several decades of concentration should make it possible to make quantitative measurements with IR-MALDI on $NaNO_3$ substrates. 18C6 concentration measurements for mass spectra taken at a laser wavelength of 7.1 μm were made by integrating under the $Na(NO_3)_2^+$ and the $(18C6+Na)^+$ positive-ion peaks in the TOF data at different as-prepared concentrations of 18C6. This technique seems to work well in giving quantitative information. Even though the TOF signal intensity varies greatly from one laser pulse to the next, the ratio of the two peaks remains relatively constant and can be used to calibrate the concentration of 18C6. Concentrations from $1{:}10^6$ to 1:30 were used and the resulting graph is shown in Figure 4. The curve can be fit to an exponential of the form $R = A\ exp(\alpha\ C)$, where R is the ratio of the mass peaks, C is the concentration of 18C6, A = 0.245 and α=164. The lowest concentration in this figure was prepared in a ratio of $10^{-6}{:}1$ in an $NaNO_3$ slurry, indicating that the sensitivity does not come at the cost of the linearity of the measurement.

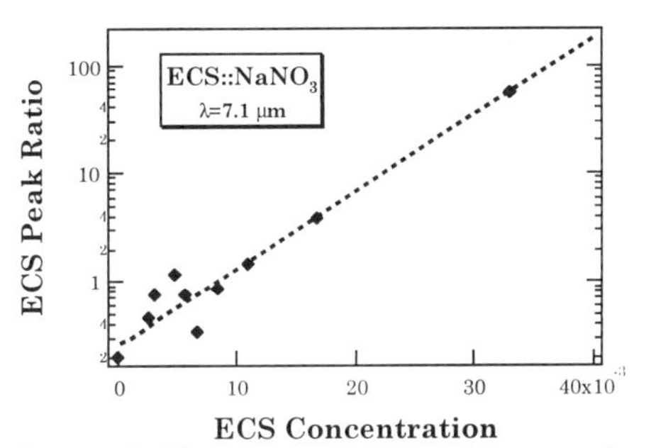

Figure 4. Experimental calibration curve for 18C6, showing peak area ratio, as defined in the text, as a function of 18C6 concentration.

It is not at present known whether organics, such as those used in these early experiments, are adsorbed on the surface of tank solids; it is far more likely, given the heterogeneous character of the tank solids, that there are both surface-bound and bulk-bound species. Since other experimental evidence shows that resonant desorption and ablation of $NaNO_3$ produces a quite uniform vaporization over a short absorption depth, this may make little difference. This remains a subject for future investigations.

CONCLUSIONS

Characterization of the molecular species present in heterogeneous wastes is needed throughout all phases of mixed waste remediation. The current off-line methods for identifying chemical species are laboratory-based, slow, expensive, produce secondary waste streams, and do not provide critical characterization data needed to make decisions in a timely manner. Advances in spectroscopy – such as those demonstrated here – could have significant impact in all phases of environmental management of mixed high-level wastes. Infrared MALDI mass spectrometry with tunable, ultrashort-pulse lasers has great potential because it

- can identify molecular (*e.g.*, organics) and atomic species (*e.g.*, Cs) embedded in or adsorbed on complex nitrates, phosphates, sulfates or ferrocyanide substrates; and
- is sensitive enough to require extremely low masses of materials for analysis, specific enough to identify species unambiguously and yet general enough to work for the heterogeneous mix of tank-waste materials.

While the major thrust of the proposed research involves the use of a free-electron laser which today is found only in a few state-of-the-art laboratories, the current directions of technology development in ultrafast infrared laser development point toward the medium-term deployability of the mass spectrometry protocols described herein as potentially field-capable techniques. One can envisage, for example, a special purpose optical parametric amplifier with the requisite pulse energies [a few μJ] at appropriate wavelengths to volatilize efficiently and to ionize the analytes in nanoliter samples of tank waste. Experiments with such an OPA are in progress at Vanderbilt.

ACKNOWLEDGEMENTS

This research is supported by the United States Department of Energy, Office of Science (Grant Number DE-FG07-98ER62710). MB-K acknowledges support from a Molecular Biophysics Training Grant funded by the National Institutes of Health, [2T32GM08320-91]. DN was supported by the Research Experience for Undergraduates program of the National Science Foundation [NSF99-104352]. We thank Wayne Hess and Michael Alexander of the Pacific Northwest National Laboratories for helpful discussions. Last but not least, we thank the staff of the Keck Foundation Free-Electron Laser Center for expert operation of the FEL, which is supported by the Office of Naval Research through the Medical Free-Electron Laser Program [N00014-1-94-1023].

REFERENCES

1 D. M. Camaioni *et al.*, *Organic Tanks Safety Program FY95 Waste Aging Studies*, PNL-10161, Pacific Northwest Laboratory, Richland, Washington (1995).

2 R. E. Russo and X. Mao, "Chemical Analysis by Laser Ablation," in *Laser Ablation and Desorption*, eds. J. C. Miller and R. F. Haglund, Jr. (Boston: Academic Press, 1998).

3. C. A. Brau, *Free-Electron Lasers* (New York: Academic Press, 1990).

4. G. S. Edwards *et al.*, IEEE J. Sel. Top. Quantum Electron. **2**, 810-817 (1996).

5 K. Becker, B. R. Johnson and G. S. Edwards, Rev. Sci. Instrum. **37**, 592 (1993).

6 R. A. Bradley, Jr., E. Lanzendorf, M. I. McCarthy, T. M. Orlando and W. P. Hess, J. Phys. Chem. **99**, 11715-11721 (1997). M. I. McCarthy, K. A. Peterson and W. P. Hess, J. Phys. Chem. **100**, 6708 (1996).

7 J. A. Campbell *et al.*, Anal. Chem. **66**, 1208A (1994).

8 O. Yavas, E. L. Maddocks, M. R. Papantonakis and R. F. Haglund, Jr., Appl. Phys. Lett. **71**, 1287 (1997). See also Appl. Surf. Sci. **129**, 335 (1998).

9 W. P. Hess, H. K. Park, O. Yavas and R. F. Haglund, Jr. Appl. Surf. Sci. **129**, 615 (1998).

10. R. Cramer, R. F. Haglund, Jr. and F. Hillenkamp, J. Am. Soc. Mass Spectrom.**7**, 1187 (1996). Hee K. Park and R. F. Haglund, Jr., Appl. Phys. A **64**, 431-438 (1997). R. Cramer, R. F. Haglund, Jr. and F. Hillenkamp, Int. J. Mass Spectrom. Ion Proc **169/170**, 51 (1997).

11 A. I. Gusev, W. R. Wilkinson, A. Proctor and D. M. Hercules, Anal. Chem. **67**, 23 (1995).

EVALUATION OF PHASE- AND ELEMENT DISTRIBUTION AFTER NON-TRADITIONAL *IN-SITU* VITRIFICATION (NTISV) AT LOS ALAMOS NATIONAL LABORATORY ON A SIMULATED ADSORPTION BED

Thomas HARTMANN
Environmental Science and Waste Technology Division, Los Alamos National Laboratory, Los Alamos, NM 87545, USA, hartmann@lanl.gov

ABSTRACT

A cold test demonstration of the Geosafe® Non-Traditional *In-Situ* Vitrification technology (NTISV) was performed at a simulated adsorption bed by using commercial-scale equipment. A successful demonstration of NTISV at Los Alamos National Laboratory (LANL) is expected to provide a useful technology for remediation needs of EM-40 in the U.S. Department of Energy (DOT) complex on legacy buried waste containing organics and radionuclides in geological environments. Therefore, a simulated test bed including surrogates was prepared to evaluate the impact this technology has on immobilizing radionuclides. The total volume of the cold demonstration vitrification was 3.1 m x 4.6 m x 4.9 m resulting in 176,667 kg of glass or glassy constituents. A drill core sample was removed from the glass monolith and divided into 13 samples to gain spatial (vertical) information about (a) vitrification progress, (b) phase distribution, and (c) element distribution. Based upon optical microscopy, least square and Rietveld refinement of the X-ray diffraction data, the phase constitution was determined and a temperature profile estimated. The glass compositions as well as the concentrations of surrogates and trace metals were analyzed by using X-ray Fluorescence Analysis, Electron-beam Microprobe Analysis, and Scanning Electron Microscopy. In the main melting zone, in the depth range 3.7 m to 6.3 m, NTISV successfully vitrified the geological formation (Bandelier Tuff, Unit 2/3 of Tshirege) and convert it mainly into 98-99 wt.-% silicate-based glass and 1-2 wt.-% iron-based metal inclusions. Within this zone the glass compositions have been found to be nearly constant and the surrogate elements (cerium and cesium) are distributed homogeneously.

INTRODUCTION

A cold test demonstration of the Geosafe Non-Traditional in-situ Vitrification technology (NTISV) was performed at a simulated adsorption bed by using Geosafe's commercial-scale equipment. Between 1945 and 1961, three former adsorption beds received effluent from a nuclear laundry and are now contaminated with a number of radionuclides. A successful demonstration of NTISV at LANL is expected to provide a useful technology for remediation needs on legacy buried waste containing organics and radionuclides in geological environments. Therefore, a simulated adsorption bed including surrogates was prepared to evaluate the impact of this technology on immobilizing radionuclides. The surrogates (6.5-kg Cs_2CO_3 and 30-kg CeO_2) were placed on the floor of the adsorption bed. The total volume of the vitrified tuff after cold demonstration was estimated to be 3.1 m x 4.6 m x 4.9 m (10 ft x 15 ft x 16 ft) resulting in 67.95 m^3 or 176,667 kg of glass or glassy constituents. If NTISV is capable of providing perfect homogeneous element distribution within this estimated glass monolith, the surrogate inventory should increase the background concentrations up to 29 mg/kg in Cs and 138 mg/kg in Ce within the glass. The background element concentrations have been found to be 2 to 3 mg/kg Cs and 111 mg/kg Ce [1]. In this work, results of phase distribution and element distribution as a function of depth are presented.

Mat. Res. Soc. Symp. Proc. Vol. 608

EXPERIMENTAL

A drill core was removed from the glass monolith, which represents the vitrified material and the amount of glass-forming oxides, as well as the surrogate charge. This drill core was divided into 13 samples (Table 1) which were analyzed to gain spatial (vertical) information about (a) vitrification progress, (b) phase distribution, and (c) element distribution by following methods:

- X-ray diffraction (XRD, Scintag XDS 2000) and least square refinement (Rietveld refinement [2]) to determine the phase constitution, as well as the vitrification progress.
- Optical polarization microscopy (Zeiss) to determine homogeneity, and residual phases.
- X-ray Fluorescence (XRF, Rigaku 3064) to analyze glass composition, as well as the distribution of surrogates and trace metals.
- Electron-beam Microprobe Analysis (EMPA, Cameca SX50 operated at 15 kV, 30 nA) to analyze glass matrices, and metal inclusions with high spatial resolution.
- Scanning Electron Microscopy (SEM, Noran Instruments ADEM) to characterize the nature of metal inclusions.

RESULTS AND DISCUSSION

As a result of studies by Rietveld refinement on X-ray diffraction data, and optical microscopy, three different phase constitutional zones were distinguished as a function of depth (Table 1, Fig.1).

Table 1: Samples analyzed by XRD, XRF, EMPA, SEM, and optical microscopy

Upper Zone		**Main Melting Zone**		**Lower Zone**	
Sample Name	**Depth Below Surface**	**Sample Name**	**Depth Below Surface**	**Sample Name**	**Depth Below Surface**
TH1-10 **6608**	**1.8 m to 3.05 m** **6 ft to 10 ft**	**TH4-15.9** **6611**	**4.7 m to 4.85 m** **15.5 ft to 15.9 ft**	**TH11-20.5** **6618**	**6.25 m** **20.5 ft**
TH2-12 **6609**	**3.2 m to 3.66 m** **10.5 ft to 12 ft**	**TH5-16** **6612**	**4.85 m to 4.9 m** **15.9 ft to 16.0 ft**	**TH12-21.5** **6619**	**6.55 m** **21.5 ft**
TH3-15 **6610**	**3.66 m to 4.6 m** **12 ft to 15 ft**	**TH6-16.8** **6613**	**4.9 m to 5.1 m** **16.0 ft to 16.8 ft**	**TH13-22.5** **6620**	**6.55 m to 7.0 m** **21.5 ft to 22.8 ft**
		TH7-17 **6614**	**4.9 m to 5.2 m** **16.8 ft to 17.0 ft**		
		TH8-18 **6615**	**5.18 m to 5.5 m** **17.0 ft to 18.0 ft**		
		TH9-18.9 **6616**	**5.5 m to 5.8 m** **18.0 ft to 18.9 ft**		
		TH10-20 **6617**	**5.8 m to 6.25 m** **18.9 ft to 20.5 ft**		

In the depth range between 1.8 m and 4.6 m (upper zone) vitrification of the simulated adsorption bed progresses from no significant vitrification (TH1-10) to nearly complete vitrification (TH3-15). In the main melting zone between 4.7 m and 6.25 m temperatures of 1700 °C and higher were achieved. As a consequence, the simulated adsorption bed is fully vitrified. Unfortunately, redox conditions caused by the technological set-up (graphite electrodes, graphite starter plates, no oxidizing agent) are not in favor to reach optimum solubilities of transition metals within the silicate-based glass melt. As a result, a liquid-liquid miscibility gap between the silicate-based melt and an iron-metal-based melt occurs. The metal–based melt tend to accumulate to bigger aggregates in order to minimize its surface energy, and in a second stage to segregate and to deposit at the bottom of the melt which is caused by its about 3 times higher density. Fortunately, the high viscosity of the melt caused by high SiO_2 content prevents significant segregation of the metal inclusions within the temperature and time range applied in the cold test experiment. In the depth range between 5.2 m and 6.25 m small amounts (1-2 wt.-%) of the high-temperature modification of cristobalite (SiO_2) were detected. The presence of this SiO_2 polymorph indicates a temperature regime of 1470 °C to 1700 °C. Below 6.25 m (lower zone) the temperature decreases rapidly to about 1470 °C to 870 °C. This temperature range is not sufficient for a complete vitrification of the simulated adsorption bed. As a result, the glassy content decreases to about 30 wt.-% and no vitrification products could be found in depths of 6.6 m or below. At a 7-m depth, residual low-cristobalite, as a frozen leftover from the devitrification in the geological formation (Bandelier tuff) is present, which is a good indication that the temperature did not exceed 800 °C.

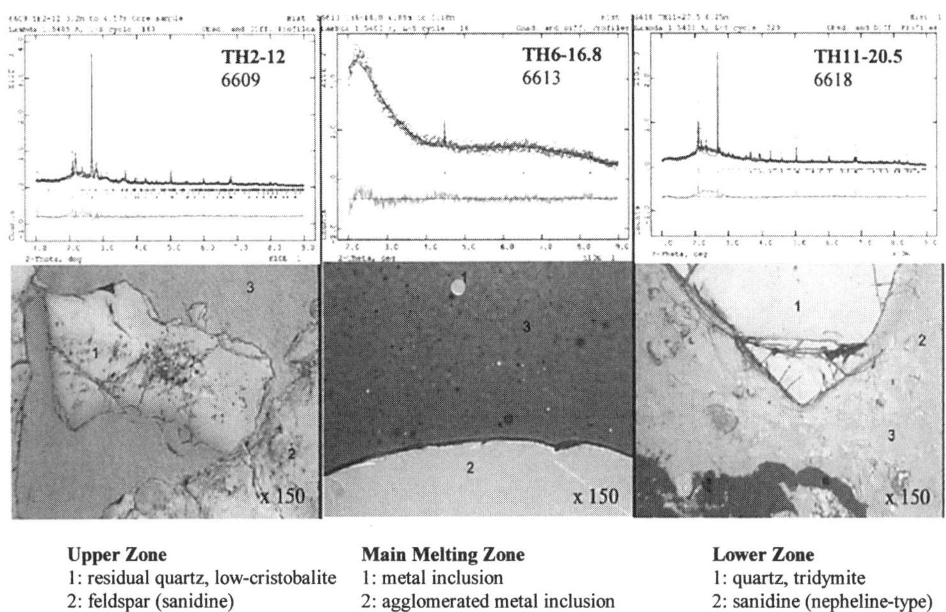

Figure 1: Rietveld refinement and optical microscopy of a representative sample from each zone

Based upon least square refinement of the X-ray diffraction data, the phase constitution can be determined and a temperature profile estimated (Fig.2). X-ray fluorescence analysis (XRF) was used to analyze the major oxide content of the samples (Fig. 3), and the distribution of the surrogates (Ce, Cs) (Fig. 5). XRF was also used to determine the concentrations of trace metals (Ba, Zr, Sr, Rb, Zn, Nb, Y, V, Cr, and Ni) within the drill core samples. Electron-beam microprobe analysis (EMPA) was used for quantitatively analyzing glass matrices (Fig. 4) and metal inclusions.

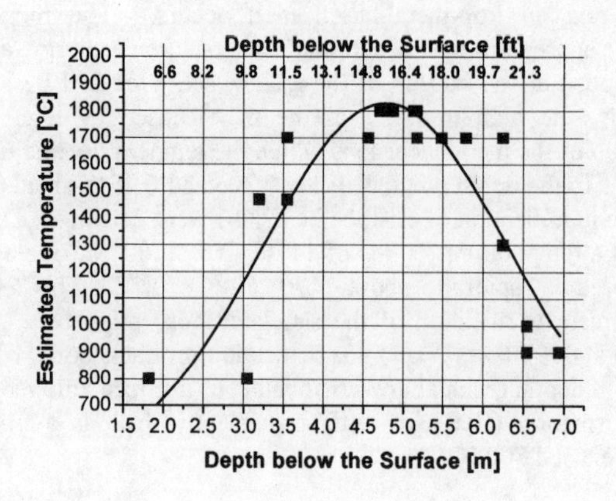

Figure 2: Estimated temperature profile as a function of depth based on the determined phase constitutions

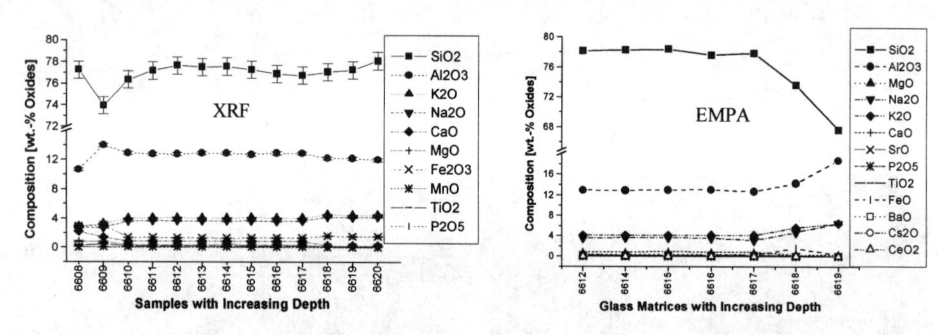

Figure 3: Sample composition measured by XRF

Figure 4: Glass compositions measured by EMPA

Glass compositions in the main melting zone (6610 to 6617) were found to be similar. The amount of the main glass-former SiO_2 is very consistent and varies between 76.4 (±0.8) wt.-% and 76.8 (±0.8) wt.-%. This high and homogeneous content in SiO_2 provides the glass monolith with excellent properties as an actinide host. Furthermore, the low (10 wt.-%) content of pure network modifier (K_2O, Na_2O, CaO, MgO) is allowing a perfectly linkage between the SiO_2 tetrahedron within the glass structure and in all three dimensions. This strong linkage between the structural elements within the glass structure will result in good corrosion resistance under severe weathering condition. Surface-related samples (e.g. 6609) are influenced by the infiltration of weathering products. The sample composition of 6608 is affected by the heat cover (sand, gravel, cobble) and the SiO_2 content is shifted towards higher amounts, from estimated 70-71 wt.-% to 77.3 (±0.8) wt.-% SiO_2. Excluding impacts of the experimental set-up, higher surface concentrations of Al_2O_3, CaO, MgO, Fe_2O_3, TiO_2, and P_2O_5 were measured, while the SiO_2, K_2O, and Na_2O contents were lower. The NTISV technology is capable to provide appropriate convection and intermixing within the melt, and the resulting glass compositions were found to be between tuff composition and the composition oft the near surface area. Below 6.25 m vitrification in the NTISV cold demonstration was not complete, which is indicated by lower SiO_2 concentrations in the glass phase (6618, 6619). The temperatures in depths below 20.5 ft were too low to completely melt the tuff, and residual SiO_2-modifications remained.

In the group of trace metals (Ba, Zr, Sr, Rb, Zn, Nb, Y, V, Cr, Ni) the Zn concentration does reflect the behavior of volatile or semi-volatile metals in this cold test vitrification experiment (Fig.5). Zinc is volatile under the temperature condition of the main melting zone and is not becoming part of the glass structure. Zinc oxide easily reduces under the redox conditions in the experiment and metallic Zn is transported apart from the heat source and condensed in a lower temperature regime of about 907 °C. The colder area surrounding the heat source is highly enriched with zinc, indicating how the other volatile metals of this group (Cd, Hg) will act.

The distributions of cerium and cesium in the cold test experiment were strongly related (Fig.6). The surrogates placed at 1.8-m depth are enriched in the main melting zone at constant concentrations. The homogeneous surrogate distributions within the main melting zone demonstrate that intermixing caused by convection is appropriate. In the cold demonstration experiment the cerium background concentration (tuff) was increased from 95(±7) mg/kg to 212(±7) mg/kg in the main melting zone. The cesium concentration ranged from 18(±6) mg/kg (tuff) to 47(±6) mg/kg in the main melting zone. Because of the semi-volatile character of cesium, the Cs-concentration in the tuff formation close to the melting zone was enriched (18 mg/kg) and above the expected cesium background concentration of 2-3 mg/kg. In a first approximation, the cesium concentration profile apart from the heat source can be estimated and a cesium background concentration of 3(±1) mg/kg will be reached in a 7.3-m to 7.35-m depth, about 1.30 m below the main melting zone. The lateral spread of cesium will certainly exceed this 1.30-m vertical spread, since volatile metals try to move to lower surrounding pressure.

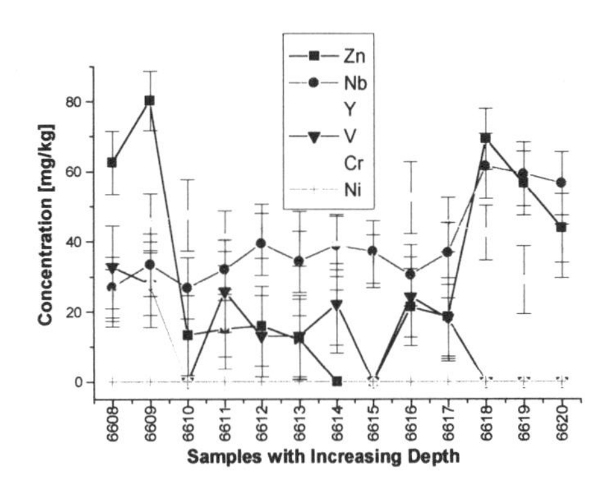

Figure 5: Zinc distribution within the experiment among other trace metals

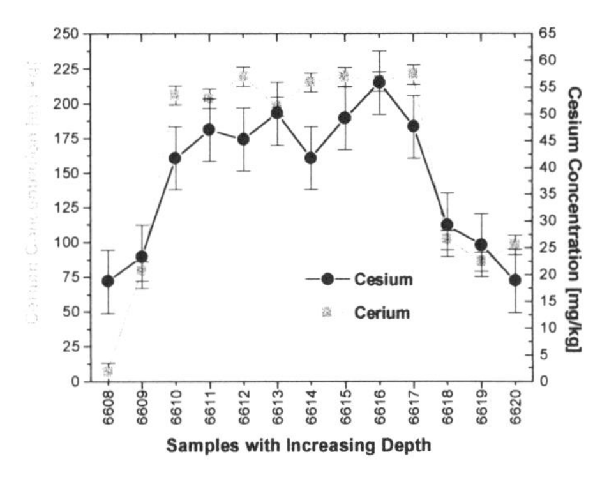

Figure 6: Cerium and Cesium accumulation in the main melting zone

Element compositions of metal inclusions were under investigation by using electron-beam microprobe analysis (EMPA) as well as scanning electron microscopy (SEM). The metal inclusions mainly consist of iron, and a α-Fe(Si) phase ($Fe_{94}Si_5P_1$) with 2-3 wt.-% silicon, and a phosphorus enriched iron-based phases ($Fe_{96}P_4$ to $Fe_{80}P_{20}$) were identified. The metal inclusions are ideal host for non-ferrous metal as opposed to rare earth elements, and enrichments with niobium, titanium, vanadium, and nickel were measured. The surrogate concentrations in metal inclusions are similar or below the concentrations in the glass matrices. As far as the complete system is understood, the formation of metal inclusion causes only marginal negative effects on the hydrodynamic behavior of the glass monolith under weathering conditions.

CONCLUSION

The deployment of Geosafe's Non Traditional *In-Situ* Vitrification (NTISV) on Bandelier tuff (Unit 2/3 of the Tshirege member) was successful for fully vitrifying the geological formation in a depth range of 3.7 m to 6.25 m. The energy transfer into this main melting zone was sufficient to achieve temperatures of 1700 °C and higher and to initiate appropriate material flow and convection for obtaining a very homogeneous product. As a result, the amount of primary glass-former SiO_2 is consistently high. Since the content of pure network modifier (K_2O, Na_2O, CaO, MgO) is low (about 10 wt.-%), a perfect three-dimensional linkage of the SiO_2 tetrahedron in the glass structure will be achieved. These high ratios of glass former to network modifier allow high solubilities of radionuclides within the glass structure and causes high corrosion resistance under severe weathering conditions. Unfortunately, the redox conditions in this experiment, caused by the technological set-up do not favor optimum solubilities of transition metals within the silicate-based glass melt. Since in this experiment the iron inventory is low (1-2 wt.-%), the melt in the technological process was never in danger of freezing. However, there is a potential risk for freezing the vitrification process if the total iron-metal inventory exceeds 10-15 wt.-%. In these cases oxidizing agents should be introduced. Metal inclusions are not suitable as actinide hosts, and the concentrations of the surrogates were similar or below the concentrations within the glass matrices. The hydrodynamic performance of the monolith will be slightly diminished by the presence of these metal inclusions, but will still be appropriate to stabilize the expected radioactive content in the material disposal areas (MDAs) for a long period of time. The surrogates are nearly entirely enriched in the main melting zone exhibiting constant concentrations. The enrichment of these surrogate metals in the main melting zone strongly supports the deployment of the NTISV technology in plutonium contaminated sites, since the thermodynamic data for cerium and plutonium are very similar. Plutonium will also be enriched in the main melting zone where glass-forming conditions are optimal. The cesium distribution indicated that minor amounts of semi-volatile radionuclides and transition metals will be driven off the heat source and will migrate into surrounding non-vitrified rock formations. This potential relocation of hazardous material can not be avoided by changing glass composition or redox conditions and should be further investigated and better understood.

ACKNOWLEDGMENTS

This research was sponsored by the US Department of Energy under the Program Code MR8A 0403 A186.

REFERENCES

[1] Geosafe Corporation, *Non-traditional in-situ vitrification at the Los Alamos National Laboratory- Cold demonstration report*, GSC report 363, 64 pages, 1999.

[2] A.C. Larson, R.B. Von Dreele, *General Structure Analysis System*, LA-UR-86-748, LANL report, Los Alamos National Laboratory, Los Alamos, NM, USA, 1986.

CONCEPTS FOR DRY PROCESSING OF SPENT NUCLEAR FUEL FOR RECYCLING TO LIGHT-WATER REACTORS

Jerry Christian, James Sterbentz, David Abbott, K.R. Czerwinski* and R. Cacciapouti**
Idaho National Engineering and Environmental Laboratory, National SNF Program;
*Massachusetts Institute of Technology, Dept of Nuclear Engineering, Cambridge, MA
**Duke Engineering Services, Marlborough, MA.

ABSTRACT

We have initiated a study of improved methods for implementing dry (AIROX) processing of commercial spent nuclear fuels for recycling back into light-water reactors. In this proliferation-resistant recycle, the spent fuel is converted to a powder, blended with fresh medium-enriched uranium powder, and refabricated into fuel elements. Evaluations of neutronic characteristics show that it will be necessary to remove a substantial portion of the neutron-absorbing fission products in the spent fuel, especially lanthanides and rhodium, in order to achieve efficient utilization of the spent fuel. We have already modeled oxidative vaporization of selected fission products from the powder at 1000 °C. In addition to permanent gases and fission products that are vaporized during pellet sintering, this can remove Tc, Mo, and some Ru and improve the neutronics. A number of approaches are being evaluated for removing lanthanides and rhodium, initially by thermodynamic modeling and review of literature. The lanthanides exist in solid solution with UO_2 so separations methods will require conversion to fine powder; rhodium is present as a metallic inclusion in the epsilon phase. Chlorination of finely powdered oxide at 1100-1200 °C would vaporize substantial portions of Nd, Eu, Gd, and Rh. A fraction of the uranium would also vaporize; if significant, it could be recovered for recycle. Magnetic and electrostatic methods were evaluated for separation of lanthanide from spent fuel. They are not likely to be practical. However, static separation techniques may be applicable for removing rhodium in the fine powder metallic inclusions. These technical considerations provide the basis for a suggested experimental program.

INTRODUCTION

In AIROX treatment of fuel, the primary concept is to heat the fuel that is removed from the cladding to 900–1,000°C in oxygen to remove the semivolatile fission products before subsequent processing. The size of the particles that are heated will play an important role in the ability to achieve volatilization. This may entail oxidizing the fuel pellets at *ca.* 400°C to convert them to U_3O_8; in the cladding puncture approach to decladding, this will be automatically done as part of the decladding operation. If other means were used, such as laser cutting of the cladding away from the fuel, the oxidation would follow removal of the pellets.

At this point, the U_3O_8 powder may or may not be of a form to release the semivolatiles when heated. Therefore, alternative paths of further particle size reduction may need to be evaluated. These are milling the fuel to physically size-reduce the powder and reducing it in H_2 to achieve the size reduction. An alternative approach might be to comminute the fuel to small particles of UO_2, heat them to 900–1,000°C, and then introduce oxygen.

Neutronic analyses have demonstrated that the first dry recycle fuel composition is a viable nuclear fuel. This fuel composition is approximately 75% by volume spent nuclear fuel (SNF) and the complement fresh feed material, i.e. 17% U-235 enriched uranium oxide. The fresh feed represents a significant amount of material and a relatively high enrichment. A second recycle

Mat. Res. Soc. Symp. Proc. Vol. 608

fuel composition would require even more fresh feed (50% by volume at 17% enrichment). As long as the uranium enrichment remains below 20%, it is considered proliferation resistant. In order to make dry recycle more efficient and economical, a reasonable goal would be to reduce the feed enrichment and/or the feed volume percent.

Reduction of the feed enrichment or volume percent requires a reduction in the spent nuclear fuel poison concentrations or the total negative reactivity. In essence, this is the additionally neutron capture attributed to the presence of fission products. In 20-year-old spent PWR fuel, the largest elemental negative reactivity contributors are the rare earth fission product elements. These include Sm, Gd, Nd, Eu. The elements contribute 22.8%, 13.2%, 12.7%, and 4.6% of the total spent fuel negative fission product reactivity, respectively, or 54.2% of the total. The next largest contributor of negative reactivity in the spent fuel is from the metallic inclusions composed of Mo, Ru, Rh, Tc, and Pd. The stable rhodium isotope (Rh-103) alone accounts for 11.8% of the total negative reactivity.

The following sections look at the potential use of magnetic and electrostatic separation techniques to extract both rare earth fission products and the metallic inclusions from the spent fuel based on their differences in magnetic susceptibility and electrical conductivity relative to UO_2. Standard industrial magnetic and electrostatic separation techniques anticipated for these separations are discussed elsewhere [1,2]. We survey the material properties of the rare earth oxides and metallic inclusions and evaluate separation applicability.

Separation Analysis

Magnetic Separation of Rare Earth Elements

The rare earth oxides are unfortunately distributed more or less uniformly as a solid solution throughout the spent nuclear fuel urania matrix. In addition, the rare earth oxides and urania are all paramagnetic materials. Separation using standard magnetic separation techniques would be virtually impossible, despite the fact that the magnetic susceptibilities of the rare earth oxides are significantly higher than that of UO_2. The rare earth oxides susceptibilities range from a factor of 5-25 greater than the uranium oxide. Despite these significant differences in susceptibilities, the rare earth oxides mixed uniformly throughout the urania make magnetic separation ineffectual.

However, recent developments [3] in the oxidation of UO_2 to U_3O_8 followed by further heating at higher temperatures (1000 to 1600 C) can lead to a rare earth-rich fluorite phase and a rare earth-poor phase in the U_3O_8. Use of this heat treatment technique may allow for the application of a magnetic separation technique. Further investigation is required.

Magnetic Separation of Rhodium

Rhodium does not form an oxide like the rare earth elements, but rather has an affinity to accumulate with other noble metals in intermetallic inclusions or precipitates. These metallic inclusions or "white inclusions", as they are referred to because they appear as white beads or spots on micrographs of spent fuel, are solid metal ingots composed of several noble metal elements. These metallic inclusions basically form separate entities within the UO_2 fuel matrix with particle sizes ranging up to 10 microns in diameter [4].

The metallic inclusions are composed mainly of Mo (55 wt%), Tc (22wt%), Ru (22 wt%), Rh (6 wt%), and Pd (<1 wt%) atoms [5]. Note these weight percentages are in good agreement with their respective fission yields, which suggests that the majority of these noble-metal fission product atoms have agglomerated into the inclusions.

Rhodium and constituent noble metals in the spent fuel metallic inclusions may lend themselves to magnetic separation techniques due the observed differences in their magnetic susceptibilities relative to that of uranium oxide (UO_2). The metallic inclusion constituents have lower magnetic susceptibilities relative to UO_2. The composite magnetic susceptibility of the metal inclusions is less than that of the bulk uranium oxide by a factor of 22. This difference may be sufficient for a reasonable magnetic separation efficiency. Particle size and magnetic field strengths are important parameters related to separation efficiency and require further investigation.

Electrostatic Separation of Rhodium

Perhaps the most promising technique for the separation of the metallic inclusions from the spent nuclear fuel matrix is electrostatic separation. Electrostatic separation exploits the differences in the electrical properties between the relatively high conductivity metallic inclusions and the poor electrical conductivity of the bulk uranium oxide (UO_2). Uranium oxide is basically an insulator or dielectric material. References 1 and 2 discuss the principles of electrostatic separation and present a variety of electrostatic separation techniques.

The table below gives the electrical conductivity of the five metallic constituents in the metallic inclusions along with UO_2 for comparison.

Table I. Electrical conductivities of the metallic inclusion elements (Ref. 6,7,8).

Element	Temperature (C)	Electrical Resistivity (microhm-cm)	Electrical Conductivity (Sieverts/m)
Mo	0	5.0	2.00E+7
	0	5.2	1.92E+7
Tc	100	22.6	4.42E+6
Ru	0	7.1	1.41E+7
Rh	0	4.30	2.32E+7
	20	4.51	2.22E+7
Pd	0	10.0	1.00E+7
	20	10.8	9.26E+6
UO_2	20	3.8E+10	2.63E-3

A composite electrical conductivity for the metallic inclusions is estimated to be approximately 1.58E+7 S/m. The bulk UO_2 electrical conductivity is approximately 2.63E-3 S/m. This huge difference in electrical conductivity (factor of 6.6E+9) between these inclusions and the bulk UO_2 leads to the possibility of an electrostatic separation.

Although a large difference in the electrical conductivities between the metallic inclusions and the bulk uranium oxide exists, further investigation needs to be performed to select the most efficient and economical charging technique associated with the electrostatic separation technique. Particle sizes in the range of 5-10 microns (fine powder) also need to be considered in the selection process. Reference 2 gives a description of a fine metal powder electrostatic separator known as an electrostatic sieve. This type of separator would be a good starting point for the development of dry recycle in-line electrostatic separator for the metallic inclusions.

In conclusion, the electrostatic separation technique appears to be quite promising for the separation of the rhodium (Rh-103) isotope and the other noble metal poisons in the spent fuel metallic inclusions from the bulk uranium oxide matrix. In order to incorporate the electrostatic separator in the dry recycle process, it is anticipated that the spent fuel would first need to be milled to a particle size in the range of 5-10 microns. The fine oxide powder would then be sent through the electrostatic separator before the oxidation-reduction steps. In this way, the rhodium and other noble metal poisons (Mo, Tc, Ru, and Pd) could be removed together.

Neutronic Analysis

The removal of fission product nuclides with the AIROX process results in a gain of 1.2%Δρ (total worth) in the initial reactivity of the AIROX recycle assemblies. The total initial fission product nuclide worth is 6.9%Δρ, resulting in 5.7%Δρ in remaining fission product worth. In order to overcome this remaining fission product worth, a significant initial fissile inventory is required in the assembly design to meet cycle length requirements or additional fission products need to be removed.

To evaluate the worth of the fission products, the CASMO-3 [9] and ORIGEN-S [10] codes were used. CASMO-3 is an integral transport lattice code used to generate cross section data as a function of burnup for a variety of conditions. The CASMO-3 model is a two-dimensional, one-eighth assembly infinite lattice representation. The fuel, gap, and cladding for each individual pin are modeled explicitly in CASMO-3, as are the moderator, guide tubes and control rods. CASMO-3 calculates assembly average reactivity, two-group constants, discontinuity factors, xenon and samarium reaction rate and reactivity data, and kinetics parameters. For AIROX-removed nuclides that are not explicitly represented in CASMO-3, ORIGEN-S is used to calculate the reactivity worth associated with these nuclides. The ORIGEN-S code calculates time-dependent concentrations and source terms for the large number of isotopes associated with neutron transmutation, fission and radioactive decay.

Table II. CASMO-3 and ORIGEN-S Fission Product Nuclide Worth. AIROX Case Assembly with Assumed AIROX Nuclides Removed. Full Power, Equilibrium Conditions at 0 MWd/kg Exposure, Westinghouse 3411 MWth PWR

Fission Product Nuclides	CASMO-3 Total Worth (%Δρ)	Assumed AIROX Removal (%)	CASMO-3 Removed Worth (%Δρ)
Am-241,242,243	1.99	0	0.00
Gd-155	0.95	0	0.00
Nd-143,145	0.77	0	0.00
Rh-103,105	0.67	0	0.00
Xe-131	0.49	100	0.49
Cs-133,134,135	0.48	90	0.43
Eu-153,154,155	0.31	0	0.00
Ag-109	0.11	0	0.00
Kr-83	0.04	100	0.04
Cm-242,243,244,245,246	0.003	0	0.00
LFP1 (lumped, non-saturating)	0.41	0	0.00
LFP2 (lumped, slowly saturating)	0.64	30**	0.19
Total	6.86		1.15

The reactivity worth for the various isotopes are provided in Table II for a 3411 MWth Westinghouse PWR. The table represents the full power, equilibrium conditions at an exposure of MWd/kg. A 0.736 weight fraction of AIROX material and 0.264 weight fraction of fresh UO_2 at 17.0 w/o U-235. AIROX material from 4.00 w/o U-235 depleted assemblies at 40 MWd/kg exposure, following a 20 year decay time. A 30% removal at start of irradiation approximates the worth of the following AIROX removed nuclides which are not explicitly modeled in CASMO-3, based on the following worth from ORIGEN-S (Table III).

Table III. Worths evaluated with CASMO-S

Nuclides	Total Worth (%$\Delta\rho$)	Removed %	Removed Worth %$\Delta\rho$
Remaining Kr	0.002	100	0.002
Remaining I	0.014	100	0.014
Remaining Xe	0.012	100	0.012
All Ru	0.172	90	0.155
All Cd	0.015	75	0.011
All Te	0.003	75	0.002
All In	0.006	75	0.005

CONCLUSIONS

An experimental program is needed to determine the optimum paths to follow that will provide fuel particle properties suitable for oxidative vaporization of the semivolatile fission products and for the final preparation of the powder for blending with feed material and pellet fabrication. Other considerations for treatment options will include throughput rate, capital and operating costs, and simplicity of operations and maintenance. These apply to the decladding approach as well.

The key to successful semivolatile removal will be to form very small particles to enable ready diffusion of oxygen to the metal inclusions and of the oxidized fission product vapors out of the matrix. UO_2 requires high temperatures to sinter. However, U_3O_8 can begin to sinter at temperatures above 450°C [11]. An experimental study would investigate the oxidative removal of fission products at approximately 900 to 1,000°C in oxygen from the powder in the form of U_3O_8 and, also, in the form of UO_2. The former would be at a higher oxidation potential, while the latter would retain the small particle size longer. But the UO_2 would oxidize during the process. The dynamic processes will affect the release of the semivolatile oxides. A research program would characterize these dynamic processes, including the effects of oxygen partial pressure over the UO_2 on fission product release rate and UO_2 oxidation rate.

The final demonstration of fission product volatility behavior must be done with actual irradiated LWR fuel that is processed via the selected conditions. Preliminary work, however, can be done with unirradiated fuel. Because the chemical conditions of the fission products affect the volatility behavior, it is important to properly simulate the irradiated fuel. This means incorporating the various fission products in the proper phase(s), and, if possible, the same sizes as exist in spent fuel. Atomic Energy of Canada, Limited (AECL) has developed and characterized a synthetic irradiated nuclear fuel called SIMFUEL [12] that simulates the composition and microstructure of high-burnup UO_2 nuclear fuels. Microscopic analysis shows the microstructure to be very similar to that of spent fuel as characterized by Kleykamp. In addition SIMFUEL can incorporate technetium, an important fission product for neutronic considerations [13].

REFERENCES

1. T.E. Carleson, N.A. Chipman, C.M. Wai, “Separation Techniques in Nuclear Waste Management”, CRC Press, 1996.

2. D.W. Green (editor), “Perry’s Chemical Engineers’ Handbook”, sixth edition, McGraw-Hill Book Company, 1973.

3. P. Taylor and R.J. McEachern, “Process to remove rare earths from spent nuclear fuel”, Atomic Energy of Canada Limited, US Patent No. 5,597,538, January 28, 1997.

4. H. Kleykamp, “The Chemical State of the Fission Products in Oxide Fuels”, Journal of Nuclear Materials 131 (1985) 221-246.

5. D.R. Olander, “Fundamental Aspects of Nuclear Reactor Fuel Elements”, TID-26711-P1, prepared for the Division of Reactor Development and Demonstration, Energy Research and Development Administration, published by Technical Information Center, U.S. Department of Energy, 1976.

6. R.C. Weast, editor, “Handbook of Chemistry and Physics”, page F-140, 51th edition, The Chemical Rubber Co., 18901 Cranwood Parkway, Cleveland, Ohio, 44128.

7. John A. Dean, editor, “Lange’s Handbook of Chemistry”, pages 3-2 to 3-6, copyright 1979, McGraw-Hill Book Company.

8. D.R. Lide, editor-in-chief, “CRC Handbook of Chemistry and Physics”, page **12**-119, 71st edition 1990-1991, The Chemical Rubber Co., 18901 Cranwood Parkway

9. M. Edenius and B. Forssen, *CASMO-3 - A Fuel Assembly Burnup Program*, Version 4.4 and Amendment for Version 4.7 (Rev. 2), Studsvik/NFA-89/3, Original Release in November 1989, Rev. 3 Release in June 1993.

10. O. Hermann and R. Westfall, *ORIGEN-S: SCALE System Module to Calculate Fuel Depletion, Actinide Transmutation, Fission Product Buildup and Decay, and Associated Radiation Source Terms*, NUREG/CR-0200, Revision 5, Volume 2, Section F7, ORNL/NUREG/CSD-2/V2/R5, Draft User Documentation for Release of SCALE-4.3, Oak Ridge National Laboratory, September 1995.
11. R. C. Hoyt, L. F. Grantham, R. G. Clark, and P. W. Twitchell, *AIROX Dry Reprocessing of Uranium Oxide Fuels,* Report ESG-DOE-13276, July 12, 1979.

12. P. G. Lucuta, R. A. Verrall, Hj. Matzke, and B. J. Palmer, "Microstructural feature of SIMFUEL Simulated high-burnup UO2-based nuclear fuel," *J. Nucl. Mater.,* **178**, 48-60 (1991).

13. P. G. Lucuta, Aceram, Ontario, Canada, personal communication, January 1999.

POWERFUL GELS FOR POWER PLANT DECONTAMINATION

D. CHEUNG*, J.L. PASCAL*, S. BARGUES** and F. FAVIER*
* LAMMI, ESA 5072 CNRS, University Montpellier 2, c.c.015, F-34095 Montpellier Cedex 05, pasfav@univ-montp2.fr
** STMI, ZAC de Courcelles, 1 route de la Noue, F-91196 Gif-sur-Yvette, sbargues@stmi.fr

ABSTRACT

Decontamination for on-site maintenance during operation and for decommissioning of end-of-operation plants can be achieved using thixotropic corrosive gels to dissolve the radioactive deposit trapped at the surface of stainless steels. The dissolving agent is a corrosive solution of a Ce(IV) salt stabilized in concentrated nitric acid. Rheological properties of the gels are adjusted to give optimal spraying characteristics and surface adhesion using silica of small particle size and non-ionic surfactants. The formulation of the gels is such that they allow easy decontamination processing, and minimize the liquid and solid waste generated.

INTRODUCTION

Among the main requirements and objectives of modern decontamination processes are maximization of decontamination factors, low cost, low hazard and exposure risks, ease of processing, little generated waste, etc. Chemical processes meet some of these requirements by removing the contaminated micrometric layer of oxide from the surface of steel alloys used in nuclear facilities. Most methods of chemical decontamination operate by immersion of the contaminated components in concentrated and aggressive solutions, which necessitates the dismantling of plant segments and relatively large decontaminating baths. Solutions have also been projected onto plant components. In that case, no dismantling is required but the contact time with the surface is too short. In addition, as for the immersion method, high volumes of liquid radioactive waste are generated. The use of foaming agents is particularly useful for hollow and complex systems but generates a large amount of organic waste.

With the aim of avoiding these disadvantages while keeping the high efficiency of chemical methods, gels can be used instead of solutions or foams. In principle, due to their thixotropic behavior, gels can be sprayed directly onto the plant components (pools, floors, ceilings, pipes, etc) which do not have to be dismantled, and the efficiency of the process is ensured by the prolonged contact of the corrosive medium with the surface. The rinsing protocol and the method for treatment of the generated waste must also be established.

In this work, we have developed a decontamination process using corrosive thixotropic gels for the treatment of metallic surfaces [1]. Active corrosive species are studied in acidic solution, and inorganic and organic additives that control the solution viscosity have been chosen on the basis of the minimum solids content associated with satisfactory rheological properties. The final formulation of these gels was based on quantitative and qualitative measurements of their corrosivity. The study of the chemical behavior of the various species in the gel and after rinsing allows us to define a treatment technology for the waste generated.

The decontamination process using such gels has already been proven on several nuclear plants and decommissioned sites. In 1997 and 1998, principal French plants and a few foreign sites were treated, including: decontamination of 'BR' or 'BK' pools from several EDF plants (Chinon, Cruas, Dampierre, Blayais, Cattenom, Saint-Laurent-des-Eaux, etc), tanks and glove bags from Cogema at La Hague, various parts of a moderately contaminated room in a Japanese

Mat. Res. Soc. Symp. Proc. Vol. 608 © 2000 Materials Research Society

plant, 304L stainless steel rooms at the National Institute of Radioelements in Belgium, and others. The production of these gels amounts to several tons per year.

EXPERIMENTAL

Ammonium cerium (IV) nitrate, $(NH_4)_2Ce(NO_3)_6$ (99.99%), was purchased from Aldrich, HNO_3 (65%, for analysis) from Carlo Erba, trioxyethylene glycol-n-decyl ether ($C_{10}E_3$) from SEPIC and, dioxyethylene glycol-n-hexyl-ether (C_6E_2) and tetraoxyethylene glycol-n-dodecyl ether ($C_{12}E_4$), from SIGMA. Voltametric measurements were conducted with an EG&G 273A potensiostat. A rotating platinum electrode (2500 rd/min, $3.14.10^{-2}$ cm^2) was used as working electrode. Reference and counter electrodes were also made of platinum. Surface excesses [2] were measured by the Wilhelmy plate method using a Prolabo tensiometer. A TOC5000 Shimadzu instrument was used to establish surfactant-solid adsorption isotherms in water and nitric acid solution by measurement of the equilibrium concentration in solution of organic carbon. Rheogram curves at various shear rates were obtained on a Paar Physica USD200 using a Kel-F Z3 cell. Quantitative corrosion properties were determined by weighing the mass loss from 316L and 304L stainless steel plates (100 cm^2) after contact of the various gels (1 $kg.m^{-2}$). Micrographs were obtained with a Cambridge S360 microscope and surface microanalysis was done with a LINK analyzer coupled to the SEM apparatus. Raman spectra of solutions were recorded with a Dilor spectrometer using a Spectra Physics argon laser (5145 Å). Samples were contained in 4 mm Pyrex tubes. IR spectra of gaseous by-products were recorded using a BOMEM DA8 FTIR spectrometer. The sampling technique has been reported elsewhere [3].

RESULTS

Formulation of Thixotropic Corrosive Gels

The dissolving agent is a corrosive cerium (IV) solution, $(NH_4)_2Ce(NO_3)_6$ (hereafter CAN), stabilized in a concentrated nitric acid medium. To establish the phenomenological characteristics of the corrosion process in such media and then to determine the optimal concentrations of CAN and HNO_3, electrochemical analysis of these solutions was done using a differential pulse voltametric technique. The redox potential of the Ce^{4+}/Ce^{3+} couple is highest in 2 M to 3 M nitric acid [4]. At these molarities the solubility of CAN is limited to 710 $g.dm^{-3}$ (Fig.1 left). Fig.1 (right) shows the relative concentration of the various ceric species as a function of CAN molarity, established on the basis of current/potential curves obtained by

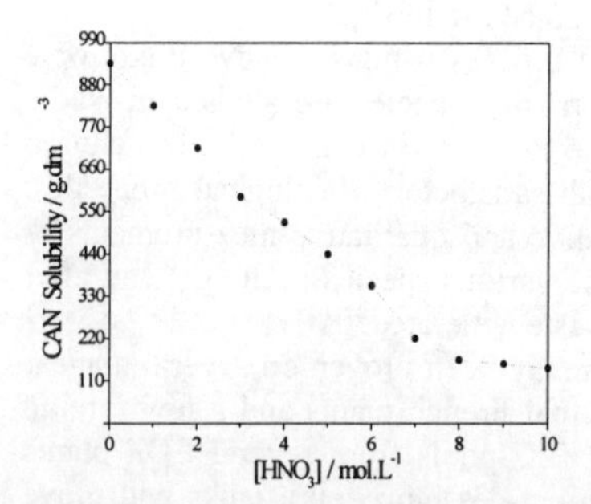

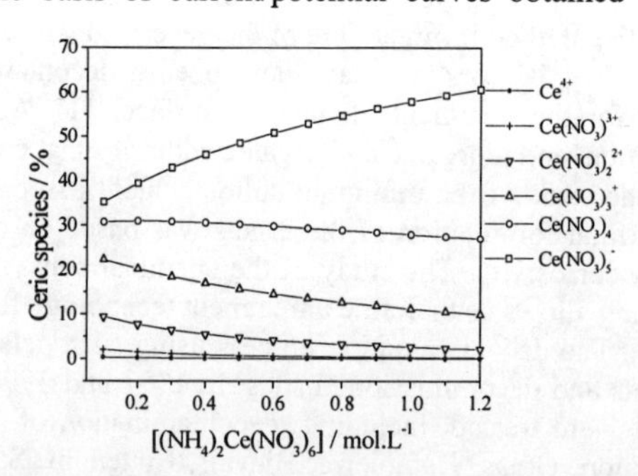

Figure 1. **left**: solubility of CAN in HNO_3; **right**: Distribution of ceric species in 3 M HNO_3 solutions.

varying CAN and HNO_3 concentrations, and using the equilibrium constants of Ce(IV) inner-sphere nitrate complexes in solution [5]. The most corrosive behavior of the solution is reached when the CAN molarity is around 1 M in HNO_3 of 2-3 M.

To obtain a gel from this solution, the first approach was to use an inorganic support, such as small particles of inorganic oxides, with high chemical stability and that can provide good viscosity properties. Fumed silica is highly stable even at very low pH. On the basis of simple rheological tests (10 s^{-1} for 1 minute on 10 and 20% SiO_2 suspensions in HNO_3 (3 M)), an acidic silica, Cabosil M5 (Cabot), was chosen among other silicas selected for their hydrophobic-hydrophilic balance, acidic properties, BET surface area, particle size, etc. The size of the primary particles of Cabosil M5 is between 7 and 15 nm, and the BET surface area is 210 $m^2.g^{-1}$.

To reduce the volume of solid waste generated, the weight percentage of silica was limited to less than 5%. Under these conditions, the gel is highly soluble in water, and thus allows for efficient rinsing at low pressure. However, with 5% of Cabosil M5, the gel is still liquid and its viscosity remains too low (100 mPa.s at 10 s^{-1}). For a sufficiently high viscosity, the silica content should be increased to 20%, which would be a severe drawback for the reasons outlined above. Alternatively a co-additive can be used to achieve the desired viscosity properties.

Surfactants generally can be used to vary the rheological properties of colloidal solutions of silica [6]. Non-ionic surfactants are relatively stable chemically, especially polyoxyethylene glycol ethers. In neutral colloidal solutions of silica, they produce gels [7,8] but their adsorptive properties and physical behavior in highly acidic, saline and oxidative media are unknown. They are highly soluble in water and we have shown that the critical micelar concentration (CMC) of $C_{10}E_3$ is reduced by a factor of 10 in a solution 1 M in CAN and 2.8 M in HNO_3 from that in water [9]. At the same time, the size of the polar oxyethylene glycol part is increased by a factor of 2. Adsorption isotherms on silica show that in water, for 1 m^2 of SiO_2, the coverage ratio increases from 0.6 m^2 to 1.6 m^2 and 3.3 m^2 from $C_{12}E_4$ to $C_{10}E_3$ and C_6E_2. In HNO_3 (2.8 M), these coverage ratios are doubled. In such media, several layers of adsorbed surfactant molecules build up at the Cabosil M5 surface. Such modification of the surface physicochemical characteristics will influence the interparticle interactions and therefore the rheological properties. The chemical stability of these surfactants in such media is satisfactory.

Rheology

The main interest in gels lies in their thixotropic properties [10,11]: liquid during spraying, and solid when stationary, allowing strong adherence to surfaces. Thixotropy is also characterized by a re-build time when the gel structure evolves from breakdown to recovery states. The re-build time can vary from tenths of seconds up to one second. Figure 2 shows schematic rheograms of various gels of Cabosil M5 and summarizes the effects of the formulation on the rheological properties. After shearing for 60 seconds at a shear rate of 700 s^{-1} (time sufficiently long to reach break-down equilibrium and simulating the spraying conditions), a final shear rate of 10 s^{-1} (representing a stationary state and adherence to the surface) is imposed and a partial build-up occurs after an equilibrium time (re-build time). The equilibrium viscosity at 10 s^{-1} and the re-build time vary drastically depending on the amount of silica, the acid concentration, the nature and concentration of the non-ionic surfactants, and the salinity and electrolytic properties of the medium. The viscosity at 10 s^{-1} and re-build time attain satisfactory values, up to 2 Pa.s and less than 5 s, when a polyoxyethylene glycol ether surfactant is added. Further gain is reached in CAN/HNO_3 solutions. The highest viscosity at 10 s^{-1}, 15.5 Pa.s, and the shortest re-build time, less than 1 s, are obtained using 1wt% of C_6E_2 surfactant. The amount of Cabosil M5, for optimal spraying characteristics and adherence to the surface without leaking is then the desired 5%.

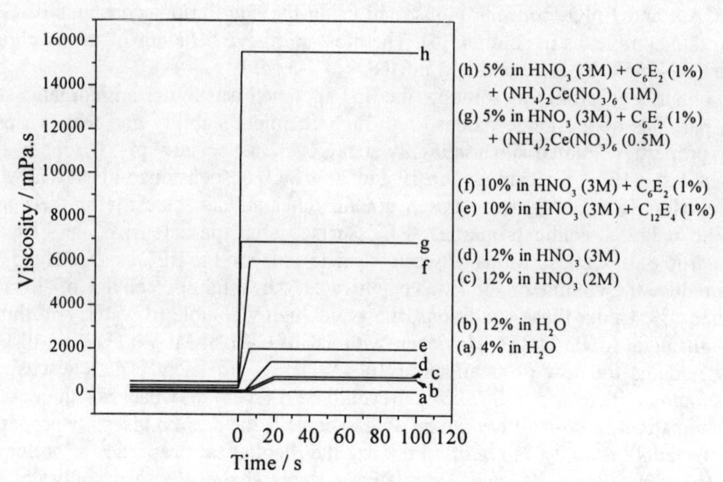

Figure 2. Schematic rheograms of various gels of Cabosil M5. Shearing for 60 s at 700 s^{-1} followed by a shear rate of 10 s^{-1} for 100 s.

Corrosion and Decontamination Results

The cerium (IV) content and the contact time with the surface govern the ability of the gels to corrode stainless steels. The thickness of the layer removed reaches 1.5 μm at the highest CAN molarity (1.2 M) in HNO_3 (3 M). The depth removed is directly proportional to the number of gel applications (Table I). After treatment, scanning electron micrographs (Fig.3) show no marked differences in the surface and cross sectional morphologies. The depth of localized corrosion remains close to that of the reference sample. The Ra parameters (average difference of the measured profile compared with the average roughness), measured by the microroughness diamond tip method (five points), are systematically lower after gel contact than in the reference sample. Corrosion therefore proceeds by smoothing the surface of metallic grains rather than by attack of the inter-granular space. X-ray fluorescence surface analysis confirms that the surface elemental composition remains unchanged after treatment. After six hours of contact, the gels no longer corrode the steel (Fig.4).

Table I. Microroughness parameter (Ra), depth of corrosion, and localized corrosion at the grain boundaries as a function of the CAN molarity and number of applications on 304L samples.

CAN / mol.L^{-1}	Treatment duration / h	Ra	Depth of corrosion /μm	Localized corrosion / μm	
				average	maximum
0.8	6	0.29	1.04	<2	4
	2 x 6	0.53	2.08	2	5
1.0	6	0.33	1.23	<2	4
	2 x 6	0.59	2.47	3	7
1.2	6	0.40	1.48	<2	5
	2 x 6	0.80	2.93	3 to 4	10
Reference sample		0.94	-	<2	4

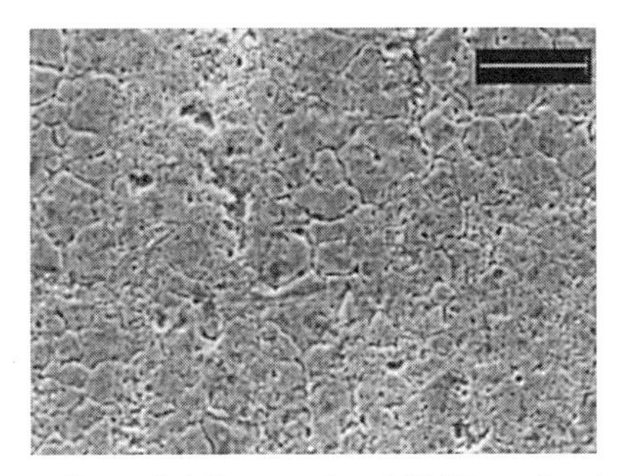

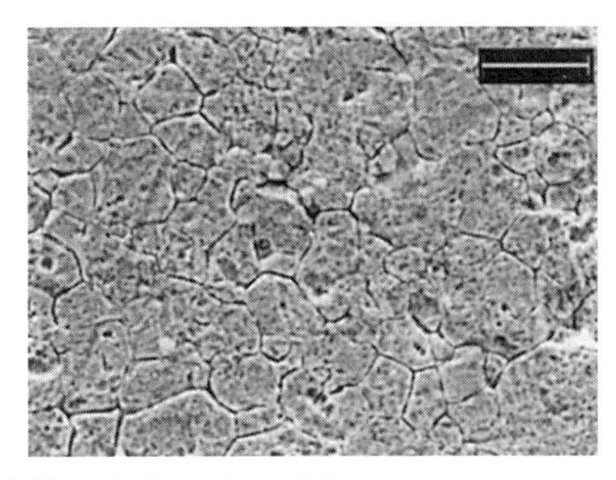

Figure 3. Micrographs of 316 L surface before (left) and after (right) 5 hour contact with a gel ([CAN] = 1 M, [HNO_3] = 3 M, Cabosil M5 (5%), and C_6E_2 (1%)). Bars represent 50 μm.

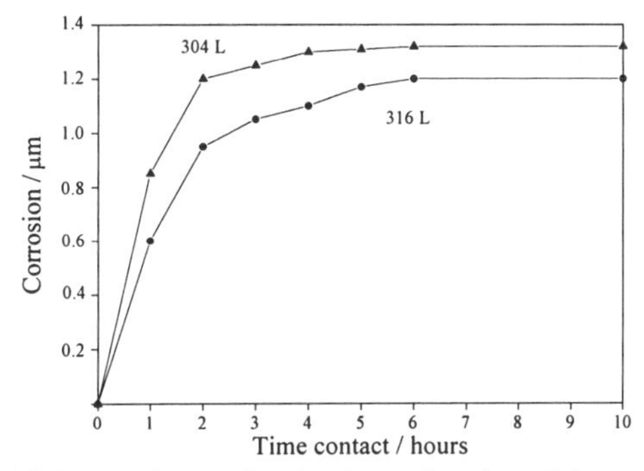

Figure 4. Thickness of corrosion (μm) as a function of the contact time for a gel with [CAN] = 1 M, [HNO_3] = 3 M, Cabosil M5 (5%), and C_6E_2 (1%).

Decontamination results [12] from several plant items made of 316 L stainless steel (^{60}Co contamination) show that decontamination factors (DF) vary from 2 to more than 1900, depending on the initial contamination level (as measured by the radiation rate) and the complexity of the components to be treated. The best results are obtained with simple systems with an initial contamination level higher than 1500 $Bq.m^{-2}$ which, after 3 applications of 6 hours each, decreased to 0.8 $Bq.m^{-2}$, leading to an overall Df greater than 1900. Some items have a DF greater than 40 after only a single treatment.

By-products and Generated Waste

After application, the gel is easily rinsed with water under pressure, and the generated waste is recovered for treatment. Simultaneously the excess of Ce(IV) salt is reduced and the organic material is degraded by addition of H_2O_2. The pH is adjusted to 7 with NaOH. At this pH, the Ce(III) salt, $(NH_4)_3Ce_2(NO_3)_9$, is reoxidized and the resulting CeO_2, $2H_2O$ precipitates. Filtration or evaporation then recovers the precipitate. Vibrational spectroscopy shows that the chemical nature of the gels evolves during the corrosion process [9]. Polyoxyethylene glycol ethers are progressively oxidized to heavy carboxylic acids, aldehydes, and esters and finally CO_2. Note that (i) no organic nitrates (high hazard risk) are generated in these media; (ii) the oxidation of the surfactant to give CO_2 limits the Chemical Oxygen Demand (COD) in the waste.

CONCLUSION

Decontamination using a gel is perfectly suited to on-site maintenance of operational plants and decommissioning of end-of-operation installations. It consists of application of the gel by spraying it on the items to be treated. Using this process, exposure is minimized since the process equipment works at a distance. The gel adheres to any surface (reverse, vertical or complex) and operates by dissolving the radioactive deposit. No subsequent damage of the materials treated is observed. A few hours after application (typically less than 6 hours), the gel is rinsed with water under high pressure and the waste generated is recovered for treatment and filtration. This ease of cleaning generates limited waste, which is then captured in solid form or in solution, the subsequent treatment of which is known and straightforward.

ACKNOWLEDGMENTS

The Centre National de Recherche Scientifique (CNRS) and the Société des Techniques en Milieu Ionisant (STMI) are gratefully acknowledged for financial support.

REFERENCES

1. S. Barguès, F. Favier, J.L. Pascal, J.P. Lecourt, and F. Damerval, French Patent No. FR9603517 extended to nuclear countries (May 29 1998).

2. A.W. Adamson, *Physical Chemistry of surfaces*, 5th ed. (J. Wiley & sons, New York, 1990), pp. 64-75.

3. J.L. Pascal, Thesis, Contribution to the study of Halogen Oxides. Structure and Reactivity of Chlorine Trioxide. Anhydrous Perchlorates. Ozon Oxidation of Bromine. Bromine Oxides (Br(I) to Br(V))Montpellier (1978).

4. A.A. Noyes and C.S. Garner, J. Am. Chem. Soc., **58**, 1265 (1936).

5. V.S. Smelov and Yu I. Vereshchagin, Russ. J. Inorg. Chem., 9, **12**, 1494 (1964).

6. Y. Otsubo, Adv. in Colloid and Interface Science, **53**, 1 (1994).

7. F. Podo, A. Ray, and G. Nemethy, J. Am. Chem. Soc., **95**, 6164 (1973).

8. C. Tandford, *The Hydrophobic Effect* (J. Wiley & sons, New York, 1967).

9. S. Barguès, Thesis, Thixotropic Corrosive Gels for Nuclear Decontamination, University of Montpellier (1998).

10. F. Moore, Trans. Brit. Ceram. Soc., **58**, 470 (1959).

11. J. Sestak, Prog. Trends Rheol. II, 18 (1988).

12. STMI (private communication, February 1998).

DEVELOPMENT AND TESTING OF A NEW POROUS CRYSTALLINE MATRIX (GUBKA) FOR STABILIZING ACTINIDE SOLUTIONS

Albert S. Aloy,* A. G. Anshits, ** A. A. Tretyakov, *** D. A. Knecht,**** T. J. Tranter**** and Y. Macheret****
*V. G. Khlopin Radium Institute, 2-nd Murinskiy Ave., St. Petersburg, 194021, Russia;
** Krasnoyarsk Scientific Center of Siberian Branch of Russian Academy of Sciences (KSC RAS), 42 K. Marx St., Krasnoyarsk 660049, Russia;
*** Federal State Unitary Enterprise "Mining and Chemical Combine" (FSUE MCh C), 53 Lenin St., Zheleznogorsk, Krasnoyarsk Region, 660033, Russia;
**** Idaho National Engineering and Environmental Laboratory, P. O. Box 1625, Idaho Falls, ID 83415

ABSTRACT

This paper describes the results of a joint research program of the Russian institutes at St. Petersburg, Krasnoyarsk and Zheleznogorsk with the Idaho National Engineering and Environmental Laboratory. A new "Gubka" ("sponge" in Russian) material was used to sorb and stabilize surrogate problematic actinide solutions, which contained lanthanide mixtures in nitric acid and tracer americium–241, by using repeated saturation-drying-calcining cycles. These tests resulted in maximum loading up to about 45 wt.% nitrate salts after drying and 33 wt.% oxides after calcination. The rates of americium–241 recovery were measured in 6 M nitric acid at 60 °C. Gubka samples loaded with cerium and neodymium oxides were hot pressed at 29 MPa and 20-1000 °C, resulting in a 35 % volume reduction.

INTRODUCTION

U. S. Department of Energy (DOE) sites, such as the Savannah River Site (SRS) and the Hanford site, store problematic actinide solutions, which are residues of past weapons production and which require stabilization under the U. S. Defense Nuclear Facilities Safety Board (DNFSB) Recommendation 94-1. These solutions include isotopes of americium and curium (Am/Cm) at SRS and plutonium nitrate solutions at Hanford. The "Gubka" ("sponge" in Russian) material, was developed by the Institute of Chemistry and Chemical Technology (Siberian Branch of the Russian Academy of Sciences) from coal power plant fly ash and consists of sintered glass-ceramic microspheres, or cenospheres, of mainly calcia-silica-alumina compositions. The Gubka material was previously tested for high-temperature catalysis applications [1]. Preliminary studies have shown the feasibility of using Gubka matrices to stabilize such rare earth elements as well as long-lived radionuclides technetium–99, zirconium–95, neptunium and plutonium. The process consists of repeated saturation-drying cycles at below boiling temperatures. The final waste form is a stable material, suitable for safe, long-term storage or transportation. The trapped isotopes can be recovered by dissolution in acid or immobilized further, such as by vitrification or hot pressing.

Inorganic materials were previously studied as a primary matrix for radioactive waste immobilization in porous glasses (P. Macedo, Catholic University, U.S.) [2, 3]; in silica gel (A. Nardova, Institute of Chemical Technology, Russia) [4, 5]; in foam corundum (V. Zakharov, Institute of Physical Chemistry, Russia) [6, 7, 8, 9] and in porous fireclay (chamotte) and diatomite, which were studied by scientists at V. G. Khlopin Radium Institute over many years (1975–1985). However, unlike the new Gubka material described in this paper, the drying process with the previous materials required boiling the solution.

Mat. Res. Soc. Symp. Proc. Vol. 608 © 2000 Materials Research Society

This paper describes tests with Gubka samples to stabilize surrogate SRS Am/Cm solutions. The purpose of the tests is to demonstrate technical feasibility of the Gubka technology to stabilize a DNFSB 94-1 problematic actinide solution composition. Future tests are also planned to demonstrate the feasibility of using Gubka to stabilize other problematic solution compositions.

EXPERIMENT

Gubka Properties

Preparation of the Gubka material involves recovery and separation of hollow glass crystal microspheres (cenospheres) of desired composition from Kuznetskii coal fly ash produced by the Tom-Usinskaya and Novosibirskaya power plants. The cenospheres consist of calcia-alumina-silica microspheres with and without ferric phases. The cenospheres are then consolidated with a binder to form a porous crystalline matrix in blocks of different sizes and configurations. Representative characteristics of the Gubka samples used in this study are shown in Table I. The Gubka samples typically sorb water in 30 seconds to about 60 % of the equilibrium value at room temperature. The drying time to constant mass after such water sorption was determined to be 1.5 – 3 hours at 60 – 40 °C, respectively, and one hour at 100 °C.

Table I. Representative Gubka sample characteristics.

Mass, g	Volume, cm^3	Bulk Density, g/cm^3	Water Absorption, Volume %	Porosity, %	Phase Composition	Acid Dissolution,* %
1.4	2.3	0.6	87	51	Amorphous + α–quartz	< 1

*Acid dissolution in 3 –12 M HNO_3 at 20, 40 and 60 °C for 3 hours.

Loading Tests Using Lanthanide Solution

The composition of lanthanide solutions, similar to the Am/Cm tank solution at SRS, is shown in Tables II and III. Both a concentrated and dilute solution was tested to determine loading behavior on Gubka. Table II shows the composition of the concentrated solution in 3.25 M nitric acid, and Table III shows the composition of the dilute solution in 0.5 M nitric acid.

Table II. Composition of concentrated lanthanide and Am–241 solution in 3.25 M HNO_3

Salts	$La(NO_3)_3$	$Ce(NO_3)_3$	$Pr(NO_3)_3$	$Nd(NO_3)_3$	$Sm(NO_3)_3$	$Eu(NO_3)_3$	$Gd(NO_3)_3$	$Am(NO_3)_3$
g/L	25	28	28	59	13.5	2.7	6.6	29kBq/mL

Table III. Composition of dilute lanthanide and Am-241 solution in 0.5 M HNO_3

Salts	$La(NO_3)_3$	$Ce(NO_3)_3$	$Pr(NO_3)_3$	$Nd(NO_3)_3$	$Sm(NO_3)_3$	$Eu(NO_3)_3$	$Gd(NO_3)_3$	$Am(NO_3)_3$
g/L	2.5	2.8	1.8	5.9	1.35	0.27	0.66	4.2kBq/mL
Salts	$Al(NO_3)_3$	$Mn(NO_3)_2$	KNO_3	$Zn(NO_3)_2$	$Ca(NO_3)_2$	$Fe(NO_3)_3$	$Na(NO_3)$	$Zr(NO_3)_2$
g/L	8.6	0.79	0.56	0.06	0.27	24.1	2.0	0.01

The results of representative loading tests with the lanthanide solutions and Gubka samples are described in Table IV. The samples were loaded in 10 or 20 cycles for the concentrated solution and in 15 cycles for the dilute solution and dried at 130 °C after each cycle. After the desired number of loading cycles were completed, the samples were calcined at a temperature of 400 or 800 °C. Under the drying conditions at 130 °C, lanthanide nitrate loading in the range up to 49 and 30 wt.% was achieved for the concentrated and dilute lanthanide solution, respectively. By calcining at 400 or 800 °C, a lanthanide oxide loading of up to 33 or 20 wt.% was obtained for the concentrated or dilute solution, respectively.

Table IV. Characteristics of Gubka samples loaded with concentrated lanthanide solution (Samples 647, 648, and 524) and dilute lanthanide solution (Sample 240)

Sample No.	Mass, g	Volume, cm^3	Density, g/cm^3	Number of cycles	Drying T, °C	Drying Time, Min	Loading, wt.%
647	1.29	2.48	0.52	10	130 400	60 30	45 30
648	1.25	2.48	0.50	20	130 800	60 30	46 31
524	1.71	2.66	0.60	20	130 400	60 30	49 33
240	1.36	2.66	0.51	15	130 400	60 30	30 20

Figure 1 shows a scanning electron micrograph of the Gubka sample saturated with lanthanide nitrate salts after drying at 130°C. In the micrograph, the lanthanide nitrate solid can be seen forming in the iterstitial volume between the Gubka cenospheres of the Gubka structure.

Recovery Tests Using 6 M Nitric Acid

The rate of Am–241 recovery from lanthanide nitrate saturated Gubka samples was measured by placing an approximately 2.5 cm^3 sample in 100 mL of 6 M HNO_3 at 60 °C and periodically sampling and analyzing the solution for Am-241 using gamma spectrometry. The study was performed until an equilibrium concentration of americium was established in the solution. The rate of Am–241 recovery shown in Figure 2 results in nearly complete recovery in approximately 60 min from Gubka samples loaded for 10 cycles with lanthanide nitrates and calcined at 400 or 800°C to produce lanthanide oxides. In some cases, the shape of the porous matrix cylinder was observed to change during the Am–241 recovery tests. Based on gamma-spectrometry measurement of the Gubka sample after completion of the recovery test, the residual Am-241 remaining on the sample was about 3 % of the original sample loading.

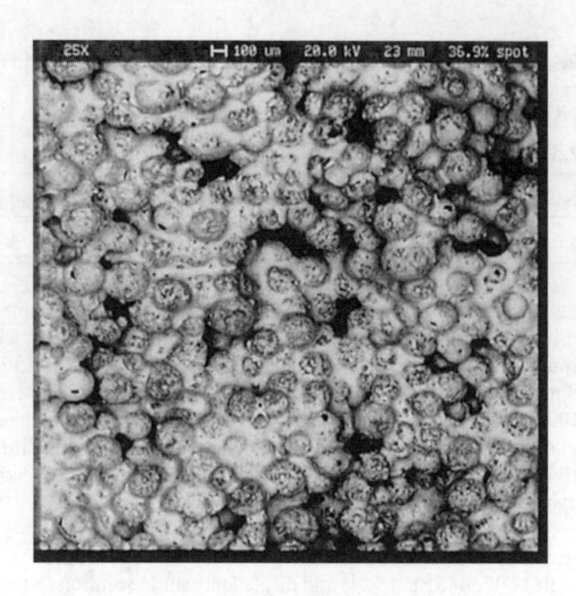

Figure 1. Scanning electron micrograph of Gubka sample saturated with lanthanide nitrate salts after drying at 130°C.

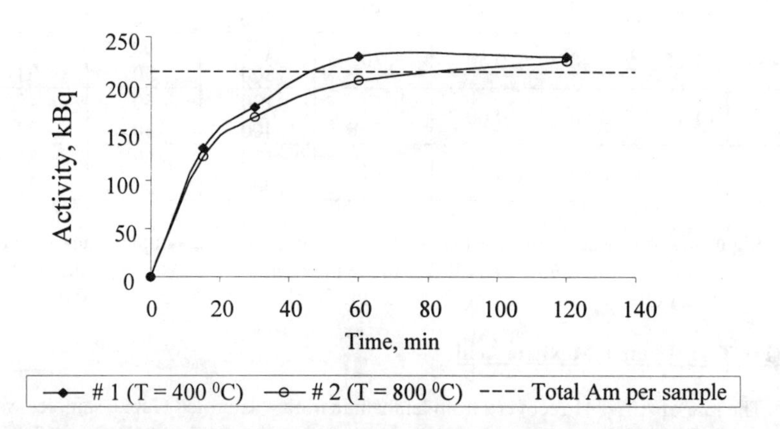

Figure 2. Rate of Am–241 recovery from Gubka samples at 60 °C in 6M HNO_3 (10 cycles saturation–drying-calcination)

Compaction of Gubka by Uniaxial Hot Pressing

Compaction of the porous Gubka matrix was evaluated using hot uniaxial pressing (HUP) with press molds of high–quality graphite type AG–1500. The linear shrinkage, shown in Figure 3, of the samples was measured in air at 20 to 1000 °C under a constant HUP pressure of 300 kg/cm^2 (29.4 MPa). Samples tested include the following:

- Unloaded Gubka
- Gubka, loaded with 12 wt.% CeO_2 after10 cycles of sorption–drying–calcination
- Gubka, loaded with 22 wt.% Nd_2O_3 after 10 cycles of sorption–drying–calcination

The linear shrinkage has similar characteristics and is independent of Gubka loading with CeO_2 or Nd_2O_3. Most of the shrinkage occurs at 775 to 875 °C and is practically complete at 900°C. At the test completion, the compacted Gubka material was attached to the graphite molds, which indicated a liquid–phase sintering mechanism. The phase compositions and characteristics of the resulting hot-pressed products are shown in Table V. The amorphous phase is the main phase in all cases.

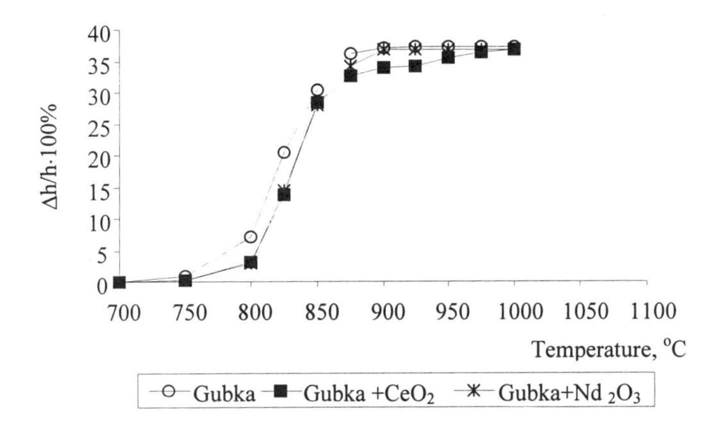

Figure 3. Densification in air at 29.4 MPa as a function of temperature for Gubka samples loaded with cerium or neodymium oxides.

Table V. Characteristics of Gubka hot-pressed products

Hot-Pressed Products	Density, g/cm^3	Porosity, %	Phase Composition
Unloaded Gubka	2.03	0.95	Amorphous phase
Gubka + CeO_2	2.56	2.36	Amorphous phase, CeO_2 ,$Ce_2Si_2O_7$
Gubka + Nd_2O_3	2.57	1.00	Amorphous phase, Nd_2O_3, $Nd_2Si_2O_7$, $Ca_2Nd_8(SiO_4)_6O_2$

CONCLUSIONS

Gubka consists of an amorphous phase and α–quartz with a high porosity of about 50 % and a low bulk density of about 0.6 g/cm^3 with less 1 % mass loss in concentrated nitric acid at 60 °C and 3 hr. Gubka samples were loaded by sorption of surrogate problematic actinide solutions containing tracer amercium-241 by using multiple loading–drying cycles, with a sorption time of about 30 seconds and one hour drying at 100 °C for each cycle. Using a concentrated lanthanide solution in 3.25 M nitric acid, similar to the composition of an SRS Am/Cm tank solution, a maximum loading of about 49 wt.% nitrate salts was obtained after 10-20 saturation–drying cycles at 130 °C and about 33 wt.% after calcining at 400 or 800 °C. Using a dilute lanthanide solution with about one-tenth the total concentration, lower loadings of 30 and 20 wt.% were obtained for the nitrates and oxides, respectively. Greater than approximately 97 % of the americium could be recovered in 6 M nitric acid at temperature 60 °C from Gubka samples, which were calcined at 400 or 800 °C. Hot uniaxial press compaction of Gubka samples, either unloaded or loaded with cerium and neodymium oxide resulted in a linear shrinkage of about 35 % at a constant pressure of 29.4 MPa in a temperature range of 750 – 850 °C. As a result of these tests, the Gubka material appears feasible for stabilizing a surrogate composition of the SRS Am/Cm solution composition. Further tests are planned to demonstrate the technical feasibility of using the Gubka process to stabilize other problematic actinide solutions in the DOE complex.

REFERENCES

1. E. V. Fomenko, E. V. Kondratenko, A. N. Salanov, O. A. Bajukov, A. A. Talyshev, N. G. Maksimov, V. A. Nizov, and A. G. Anshits, Catalysis Today, **42**, pp. 267-272 (1998).
2. J. H. Simmons, P. B. Macedo, Aaron Barkatt, and T. A. Litovitz, Nature, **278**, p. 729 (1979).
3. P. B. Macedo, D. C. Tran, J. H. Simmons, M. Saleh, A. Barkatt, C. J. Simmons, N. Lagakos, and E. Dewitt, in *Ceramics in Nuclear Waste Management*, edited by T. D. Chikalla and J. E. Mendel, (Technical Information Center **US DOE CONF–790420)**, Cincinnati, OH, 1979, pp. 321-326.
4. A. K. Nardova and O. S. Tumanova, in *Proceedings of Int. Topical Mtg. On Nuclear and Hazardous Waste Management Spectrum'96*, August 18–23, 1996, Seattle, Washington, (ANS, 1996, pp. 2154-2160).
5. A. K. Nardova, E. A. Filippov, and G. F. Egorov in *Proceedings of Int. Topical Mtg. On Nuclear and Hazardous Waste Management Spectrum'96*, August 18–23, 1996, Seattle, Washington, (ANS, 1996, pp. 2120-2122).
6. M. A. Zaharov, T. I. Potemkina, A. A. Kozar', Inorganic Materials, **29**, #3, pp. 379-380 (1993).
7. K. I. Portnoy, V. I. Fadeeva, and N. I. Timofeeva, Atom Energy, **14**, #6, pp. 559–562 (1963).
8. A. S. Nikiforov, M. A. Zaharov, and □. □. Kozar', Atom Energy, **70**, #3, pp. 188–191 (1991).
9. M. A. Zaharov, □. □. Kozar', and A. S. Nikiforov, Reports of USSR Academy of Science, **314**, #6, pp. 1441–1444 (1990).

CHARACTERIZING TRANSPORT AND SORPTION IN ION-SPECIFIC RESIN COLUMNS USING NUCLEAR MAGNETIC RESONANCE (NMR) IMAGING

D.F. CAPUTO*, D.G CORY*, M. DRAYE**, AND K.R. CZERWINSKI*
*Department of Nuclear Engineering, NW13-219, Massachusetts Institute of Technology, Cambridge, MA 02139, kczer@mit.edu
**Ecole National Supeior de Chemie, 11 rue Pierre et Marie Curie 75005 PARIS, France

ABSTRACT

The goal of this work is to assess the physical transport properties of Gd through an ion exchange column while determining the sorption properties of the resin. By coupling the physical transport with the chemical sorption, further insight into the behavior of the ion exchange resin can be gained. NMR imaging provides a powerful, non-destructive, means to extract spatial information from complex systems on a near real-time basis. An important example is liquid flow through granular media. With the use of a chemically reactive NMR contrast agent, the chemical speciation can be traced along the physical flow path of the granular media. In this study, trivalent gadolinium (Gd^{3+}) was selected based on its chemical similarity to typical high-level waste components, ^{241}Am and ^{244}Cm, and for its paramagnetic contrasting abilities in NMR experiments. NMR imaging results of flow experiments are provided showing a characteristic flow phenomena and resin column loading profiles. ICP-AES data are provided to show resin ion exchange capacities (IECs) and breakthrough curves. The use of NMR imaging with a Gd^{3+} tracer will lead to a better understanding of the transport and sorption properties of these ion-specific resins. This technique can be applied to other complex flow systems such as environmental transport.

INTRODUCTION

Ion-specific resins have been developed to partition similar inorganic chemical species, such as Cs from Na, from waste streams in radioactive waste reprocessing operations. The resins used in these studies are synthetic organic structures with phenol functional groups. The resins are made selective by incorporation of a chelating compound within the resin structure. These types of resins have been shown to form stable complexes with a variety of radionuclides [1]. Due to the recent development of these synthetic resins, there is a general lack of information on the behavior of the resins under operational conditions. Many analytical methods exist to provide information on the bulk characteristics of a packed resin column, however, few methods exist to extract real-time spatial information from the columns under operational conditions. The method chosen in this study is nuclear magnetic resonance (NMR) imaging with the use of Gd tracer to examine new ion-specific resins for the separation of trivalent metal ions.

NMR imaging is a non-invasive technique to probe the spatial structure of a variety of complex systems [2-5]. By using a paramagnetic ion in a water-based solution, the behavior of that ion can be traced as it moves through the NMR sampling volume, since the paramagnetic ion reduces the NMR relaxation time of neighboring water spins. The ability to use an ion as an NMR tracer provides a means to assess the physical structure of porous media while identifying the chemical behavior along the flow path. Specifically, the chemical speciation (sorption, complexation, colloid formation, and precipitation) can be traced along a flow path, which leads to a greater understanding of the ultimate fate of the ion. This NMR tracer method can be

Mat. Res. Soc. Symp. Proc. Vol. 608

applied to a variety of systems where chemical behavior must be coupled to physical transport in order to predict the change in ion concentration along a flow path.

EXPERIMENTAL THEORY AND SETUP

Theory

The work presented in this paper uses the trivalent lanthanide, Gd^{3+}, as a chemical analog for trivalent actinides. Lanthanides and actinides have long been used as chemical analogs for each other based primarily on the gradual filling of the 4f and 5f subshells by the lanthanides and actinides, respectively, which result in similar relative orbital energies [6-7]. In particular, the trivalent lanthanides and actinides have similar ionic radii, which results in similar ligand-metal orbital interactions as seen in the resultant absorption spectra and stability constants [8-9].

The Gd^{3+} ion has the additional physical property of being paramagnetic, since it has a half-filled 4f subshell, providing seven unpaired electrons [10]. This electron configuration creates a highly localized increased magnetic moment around the Gd^{3+} ion, and protons in the immediate vicinity the ion will experience increased relaxation rates. The influence of the Gd^{3+} ion on the proton relaxation is highly localized. The NMR experiments used in this paper make use of the increased spin-lattice, T_1, relaxation rate of protons near the Gd^{3+} ion to track the physical and chemical behavior of the Gd^{3+} ions as they move through a porous structure.

Setup

The NMR system consists of an Oxford horizontal (15cm bore) 3T magnet equipped with an integrated gradient set coupled to a Bruker AMX spectrometer for gradient control, pulse application, and data acquisition. The NMR programs used were 2- and 3-D slice selective, spin echo imaging sequences with an initial inversion recovery step to separate the fast and slow relaxing components.

The flow system consists of an Edwards high-vacuum pump, a Nold 6L Water De-aerator [11], a Varian stepper motor/piston pump, a source of compressed air, and a set of pressure reservoirs with associated tubing, valves, gauges, and NMR sample tube. The flow system is able to evacuate the test specimen and tubing, fill the sample media with deaerated water (or another liquid), and then apply a positive pressure to assure saturation, thus eliminating air bubbles and subsequent susceptibility effects. All of these components are linked in a closed loop flow system that permits complete control of fluid movement through the test specimen. The vacuum pump is used to draw air out of solution in the water deaerating unit and to evacuate the piston chamber, specimen tube, flow tubing, and pressure reservoirs down to a vacuum of 0.1 MPa (~30" of Hg). The water deaerating unit is operated in a batch mode and can produce up to 6 L of water with an air content of less than 0.6 ppm dissolved oxygen, DO (9 ppm DO ambient level)[12]. The deaerated water is then used to saturate the test specimen via the piston pump or the pressure reservoirs. The piston pump is used to control the flow rate and pressurization of the system. This system is capable of displacing up to 375 ml of fluid per stroke at a rate of up to 5 ml/sec at pressures in excess of 6.0 MPa (~60 atm). By over-pressuring the liquid in the reservoirs at the inlet end and displacing the piston at the outlet end of the sample tube, the liquid can be moved through the system in plug-flow fashion. In this experiment, the system was under 0.42 MPa of pressure during the entire flow experiment. The flow rate was held constant at 0.1 ml/sec or approximately 0.5 cm/sec through the 10.0 cm long sand and resin filled column.

The phenolic ion-exchange resins used in these experiments are described and were prepared according to the procedures in ref. [13-14]. The NMR sample contained 0.2 g (dry) of the resorcinol-formaldehyde (RF) resin. The resin was sieved with sand (high purity, ignited) into a 0.8 cm inner diameter NMR sample tube to produce a homogenous resin/sand layer approximately 2.0 cm long. This homogenized mixture allowed for good flow pathways through the sample and provided maximum available resin surface area for ion exchange. Pure sand was placed on each end of the resin/sand mixture to ensure good dispersion and flow development of the inlet solution. The inlet solution was 1.0 mM Gd^{3+} at pH 5.4, and thus the Gd ion was not hydrolyzed under experimental conditions.

The ICP-AES system used for this work is the Spectroflame ICP-D model Inductively Coupled Plasma - Atomic Emission Spectroscopy. The samples were extracted from the column in a batch process of between 10 and 50 ml. The samples were then filtered through 0.45 um filters to remove any particulates which could clog the ICP-AES system. Three measurements we taken for all samples, and the average and standard deviations were calculated. A calibration curve was generated using Aldrich ICP Gd Calibration solution diluted over the range of expected concentrations. A lower limit of detection of 0.7 μM was established at two standard deviations above the background intensity using de-ionized water.

RESULTS

The series of 2-D NMR images in Figure I shows the ingress of a 1.0 mM gadolinium (Gd) solution into a 0.8 cm diameter column packed with a homogenous mixture of 2.0 g sand and 0.4 g (wet) resorcinol-formaldehyde (RF) ion specific resins. The RF resin/sand mixture was saturated with water before Gd ingress. The images were generated using a two-dimensional, slice-selective, spin echo imaging sequence with T_1 weighting. The paramagnetic nature of the Gd^{3+} ion causes a decrease in the T_1 of neighboring protons thus allowing for a separation of the NMR signal from the free water and Gd solution. The image size is 256 by 256 pixels corresponding to 0.8 cm in height and 1.25 cm in width. The slice thickness is approximately 0.1 cm. The image sequence (Fig. I, Image 2) clearly shows the development of a characteristic parabolic flow, which eventually leads to a fingering flow phenomena (Fig. I, Image 5). The fingering flow phenomena is characteristic in inhomogeneous flow environments with varying hydraulic conductivity.

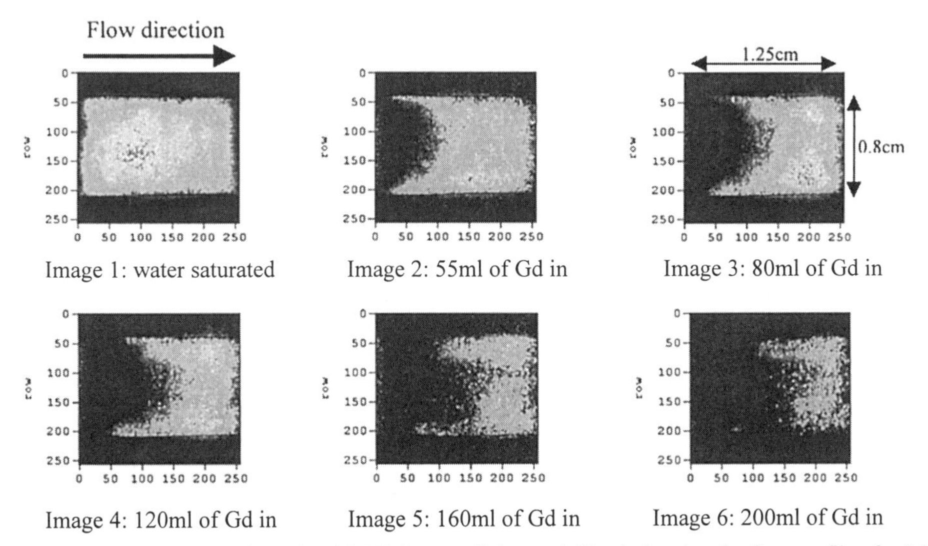

Figure I: Sequence of 2-dimensional NMR images (0.8cm x 1.25cm) showing the flow profile of a 1.0 mM Gd solution entering a RF ion exchange resin column. Image 5 clearly shows fingering flow phenomena which can be coupled to effluent data as shown in the following breakthrough curve.

The plot in Figure II is a typical breakthrough curve of the ratio of the effluent Gd concentration to the initial Gd concentration versus the total volume of fluid through the column. The initial Gd concentration was 1.0 mM, and the resin reached saturation after approximately 760 ml of the Gd solution. This corresponds to an ion exchange capacity of 3.8 milliequivalents per gram of dry resin (meq/g), and a proton exchange capacity of 11.4 meq/g, which correlates well with the literature values for this resin of 11.5 meq/g [14]. The plot also shows the high affinity of the RF resin's functional groups for Gd^{3+} by the immediate reduction of the effluent concentration to zero upon flushing with water.

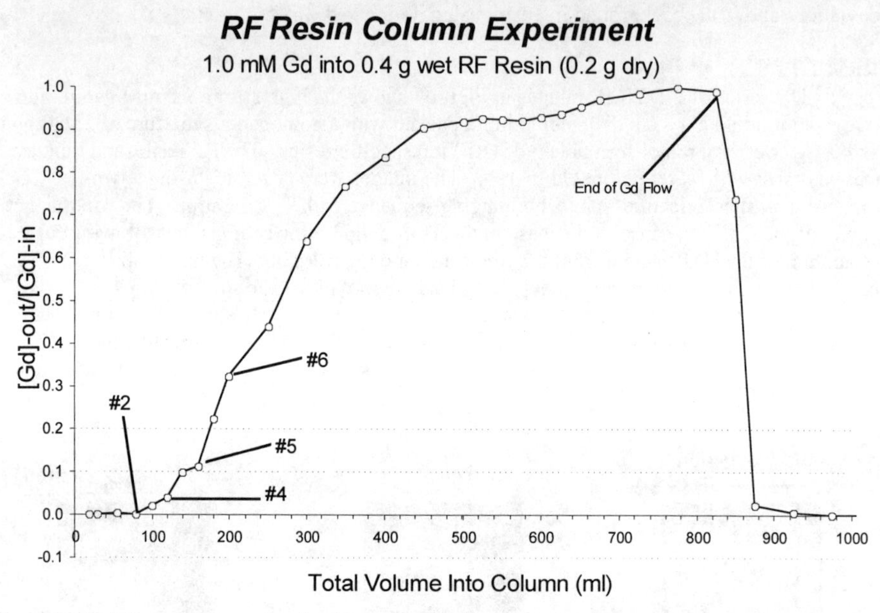

Figure II: Breakthrough curve showing the ratio of the effluent Gd concentration to the initial Gd concentration versus the total volume of fluid through the column. The numbers represent selected NMR images in Figure I.

The 3-D NMR image in Figure III shows the physical structure of the sand/resin column and the sorption behavior of the Gd^{3+} ions. The image size is 0.8 cm in diameter by 1.0 cm long with a minimum spatial resolution of approximately 80 um (128 x 128 x 128 pixels). The resin is completely loaded with Gd^{3+} ions, and the higher loading capacities are visible in the dark areas to the rear of the image. The white areas in the bulk of the image show uniform loading with the exception of a few voids spaces, which are visible at the near end of the image. The heterogeneous voids are due to areas of low sorption, which is a result of sand layering or resins with low Gd loading capacity. The homogenous (spherical) voids are most likely the result of large sand grain sizes or agglomerated resins, since great care was taken to exclude all air from the sample. The 3-D image clearly shows the heterogeneous nature of ion exchange in granular resin columns.

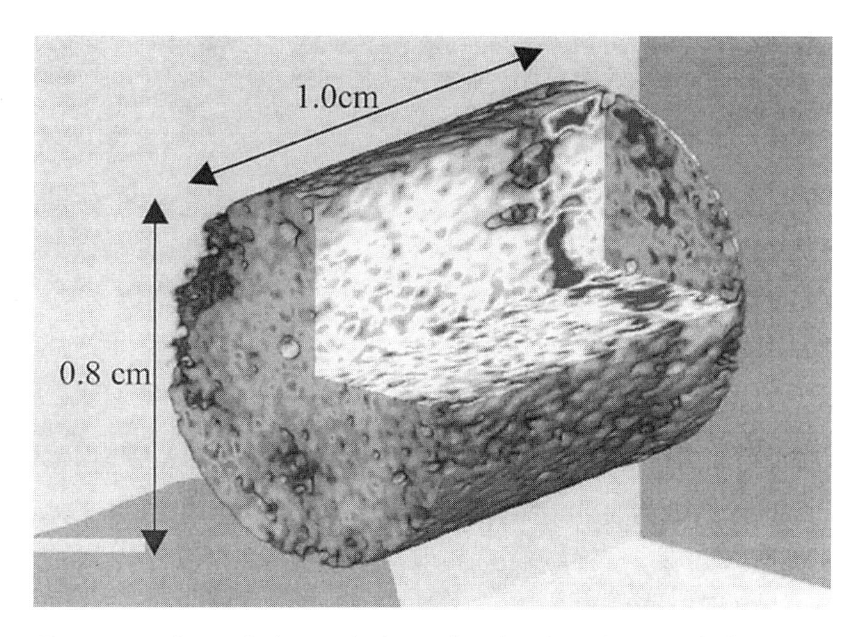

Figure III: 3-D NMR image of a homogenized RF resin and sand sample, 0.8 cm in diameter by 1.0 cm in length. The resin in completely loaded with Gd^{3+} ions, and the heterogeneous nature of the loading is easily visualized. Voids in the sample are also readily apparent.

CONCLUSIONS

NMR imaging of Gd flow is a viable technique to assist in the understanding of physical transport and chemical behavior of trivalent metal ions in a variety of porous media flow systems. The flow of a 1.0 mM Gd^{3+} ion solution into an RF resin and sand column was imaged in both two and three dimensions. The 2-D images were able to correlate the shape of the break-through curve with the shape of the flow front. These correlated images could be beneficial to explain anomalies in column test results, such as those arising from fingering and preferential flow phenomena. The 3-D imaging provides good qualitative information on the heterogeneous nature of ion exchange in homogeneously packed granular resin columns. In particular, 3-D imaging has the ability to map out flow pathways and sorption sites along those pathways, which can prove to be extremely useful in understanding metal ion flow in hard packed structures.

The flow system and chemical analysis were well suited to this type of study. The flow system's deaeration techniques were able to remove virtually all air from the sample to ensure complete liquid saturation. The flow rate control and sample collection methods ensured a high degree of consistency between batch samples. The ICP-AES analysis of the effluent was highly reproducible with low standard deviation over the range of concentrations measured. The calculated gadolinium exchange capacity for this flow experiment of 11.4 meq/g correlated well with results of previous ion exchange studies.

The Gd ion is a highly effective NMR contrast agent for use in imaging experiments of ion exchange resin columns and other granular media systems. The use of Gd^{3+} as an NMR tracer shows great promise as an analytical method to increase the understanding of physical fluid transport and coupled chemical reactivity in heterogeneous systems. Through a greater understanding of the chemical reactivity coupled to the physical transport, better models can be developed which will lead to better prediction of the fate of trivalent lanthanides and actinides in the environment.

REFERENCES

1. N. Dumont, A. Favre-Reguillon, B.Dunjic, M. Lemaire, Sep. Sci. Tech., 31(7), 1001-1010 (1996).

2. J.D. Seymour and P.T. Callaghan, AIChE Journal, **43 No.8**, 2096-2111 (August 1997).

3. A. Caprihan and E. Fukushima, Physics Reports, **198 No. 4**, 198-235 (1990).

4. S. Chen, F. Qin, K. Kim, and A.T. Watson, AIChE Journal, **39 No. 6**, 925-934 (June 1993).

5. A. Feinaur, S.A. Altobelli, and E. Fukushima, Mag. Res. Imaging, **15 No. 4**, 479 (1997).

6. G.T. Seaborg, Actinides Rev., 1, 3 (1967).

7. J.E. Huheey, *Inorganic Chemistry*, 3rd ed. (Harper Row Publishers, New York, 1983), p.803.

8. P.R. Fields and T. Moeller, Advances in Chemistry Series, No. 71, American Chemical Society, Washington, DC, (1967).

9. W. Stumm and J.J. Morgan, *Aquatic Chemistry: Chemical Equilibria and Rates in Natural Waters*, 3rd ed. (John Wiley and Sons Inc., New York, 1996).

10. P.T. Callaghan, "Principles of Nuclear Magnetic Resonance Microscopy," Clarendon Press, Oxford (1991).

11. W. Nold, "The Nold DeAerator Manual," Revision 18, Walter Nold Company, Natick, MA.

12. W. Nold, The Nold DeAerator, Design News, 68, (17 Aug 81).

13. K.R. Czerwinski, M. Draye, J. Foos, and A. Guy: Ion Selective Resins: Development and Applications for Nuclear Waste Management. MRS Scientific Basis for Nuclear Waste Management XX. In press.

14. M. Draye, K.R. Czerwinski, A. Favre-Reguillon, J. Foos, A. Guy, and M. Lemaire, Sep. Sci. Tech., In press.

Volatilization of Fission Products from Metallic Melts in the Melt-Dilute Treatment Technology Development for Al-Based DOE Spent Nuclear Fuels

Thad M. Adams, Andrew J. Duncan, and Harold B. Peacock, Jr.
Westinghouse Savannah River Company
Savannah River Technology Center
Aiken, SC 9808

Abstract

The melt-dilute treatment technology is being developed to facilitate the ultimate disposition of highly enriched Al-Base DOE spent nuclear fuels in a geologic repository such as that proposed for Yucca Mountain. Currently, approximately 28 MTHM is expected to be returned to the Savannah River Site from domestic and foreign research reactors. The melt-dilute treatment technology will melt the fuel assemblies to reduce their volume and alloys them with depleted uranium to isotopically dilute the 235U concentration and reduce the potential for criticality and proliferation concerns. A critical technology element in the development of the melt-dilute process is the development of an offgas system. Experimental tests using both cesium surrogates and radioactive cesium have shown that zeolite 4A is an effective gaseous cesium trap and as a result a preliminary offgas system concept has been developed employing dry zeolite 4A absorber beds as the primary cesium trapping medium. Final, validation of this offgas concept will occur during pilot-scale irradiated testing.

Introduction

The melt-dilute treatment technology is being developed to facilitate the ultimate disposition of highly enriched Al-Base DOE spent nuclear fuels in a geologic repository such as that proposed for Yucca Mountain. Currently, approximately 28 MTHM is expected to be returned to the Savannah River Site from domestic and foreign research reactors. The melt-dilute treatment technology will melt the fuel assemblies to reduce their volume and alloys them with depleted uranium to isotopically dilute the ^{235}U concentration. The resulting alloy is cast into a form for long term geologic repository storage. Benefits accrued from the melt-dilute process include the potential for significant volume reduction; reduced criticality potential, and proliferation concerns

A critical technology element in the development of the melt-dilute process is the development of offgas system requirements. The volatilization of radioactive species during the melting stage of the process primarily constitutes the offgas in this process. Several of the species present following irradiation of a fuel assembly have been shown to be volatile or semi-volatile under reactor core melt-down conditions. Some of the key species that have previously been studied are krypton, iodine, and cesium. All of these species have been shown to volatilize during melting experiments however, the degree to which they are released is highly dependent upon atmosphere, fuel burnup, temperature, and fuel composition. With this in mind an analytical and experimental program has been undertaken to assess the volatility and capture of species under the melt-dilute operating conditions.

Melt-Dilute Program Development

A typical irradiated research and test reactor fuel assembly contains more than 100 chemical species that include fission products, actinides, and light elements. Table 1 displays a listing of the key radiological and non-radiological elements that have been analytically predicted to volatilize during treatment of DOE SNF using the melt-dilute technology. From this listing it can be seen

Mat. Res. Soc. Symp. Proc. Vol. 608 © 2000 Materials Research Society

that of all the radiological elements predicted to be volatile the Cs-137 and Kr-85 results in the greatest mass and activity.

Conventional techniques for trapping krypton gas are commercially available. Currently, there are no commercially available techniques or devices for trapping volatile cesium. Hence, developing a process for trapping Cs is important and needs development.

Previous work [1] on the melt-refining of uranium metal EBR-II assemblies performed at ANL in the mid-60's has provided guidance with respect to the trapping of volatile cesium from metallic melts. In these experiments, sodium bonded uranium metal EBR-II assemblies were melted at temperatures on the order of 1200°C. Like U-Al fuels, the major volatile component of these EBR-II fuels was cesium. As a result, bench-scale experiments were conducted to assess a suitable means of trapping cesium vapor and particulate. Several packed granular beds consisting of various media including activated alumina, activated carbon, molecular sieves, magnesia, zirconia, and silica gel were tested at temperatures from 500-800°C. From these experiments, three of the media–activated alumina, activated carbon, and molecular sieves—appeared suitable for preventing cesium breakthrough of the packed bed. Thus, these cesium absorbing media will serve as the basis for the fundamental experimental melt-dilute offgas development effort.

Table 1 Listing of the Key Radiological and Non-Radiological Volatile Elements

Element	Mass (g)	Curies
Antimony	0.0002	0.042
Barium	0.011	<E-11
Bromine	0.176	0
Cesium	19.3	817
Cadmium	0.134	0.06
Europium	0.0009	0.0193
Gadolinium	0.00004	<E-09
Iodine	1.28	0.0002
Krypton	3.64	72
Magnesium (NR)	60	0
Manganese (NR)	29	0
Rubidium	3.54	<E-07
Samarium	0.0012	0.0008
Strontium	0.006	5.04
Tellurium	0.456	0.43
Xenon	0.422	<E-21
Zinc (NR)	5	0
NR= Non-Radiological--i.e., Cladding Additions		

Experimental Program

The approach to the development and design of the melt-dilute offgas system has been incremental in nature as shown in Figure 1. The development program began with bench-scale non-radioactive surrogates. Much of the cesium was observed to remain in the crucible, bulk metal, or in the oxide layer on the surface of the metallic melt [2,3]. These tests provided screening of different absorber media and absorber bed configurations on a sample size approximately 1/10th that of a full-scale SNF assembly. Based on the output of the bench-scale surrogate tests, a full-scale surrogate offgas system and furnace were developed for further testing and validation. This apparatus was capable of testing full-scale surrogate SNF assemblies containing representative concentration of total cesium. Simultaneous with these full-scale surrogate tests, a bench-scale irradiated SNF coupon test was conducted to assess the impact of irradiation on the offgas behavior. These tests were performed at ANL with coupons of irradiated U-Al RERTR fuels in a furnace and offgas system similar to our bench-scale surrogate apparatus. This paper will provide the results from the full-scale surrogate and irradiated SNF coupon tests.

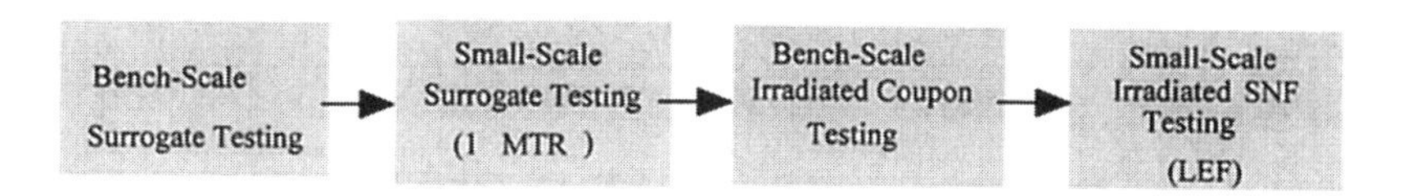

Figure 1 Development Path for the Melt-Dilute Treatment Technology Offgas System

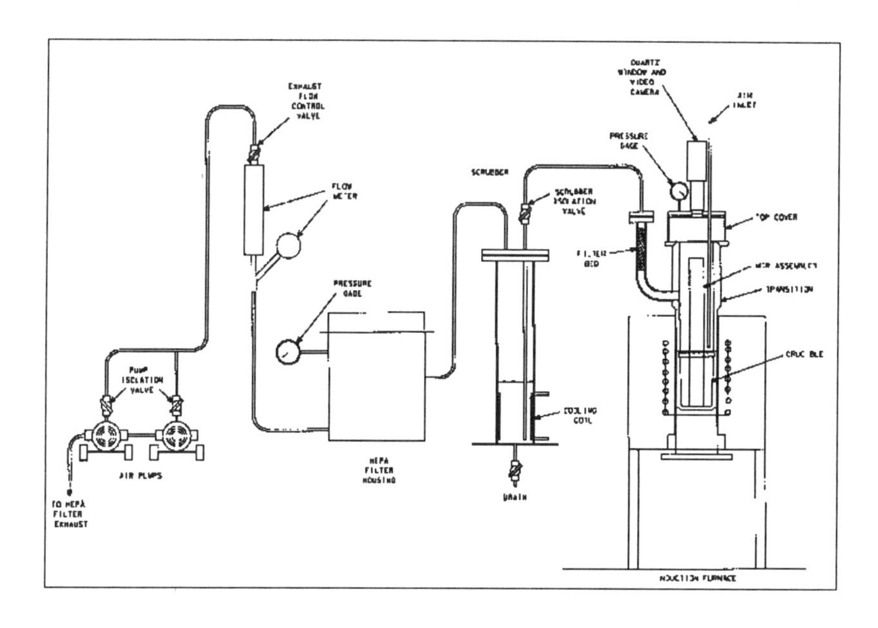

Figure 2 Integrated Melt-Dilute Treatment Apparatus

Experimental Methods

SRTC Small Scale Tests

Following the bench-scale screening tests, an integrated melt-dilute treatment apparatus was developed as shown in
Figure 2. This apparatus consisted of a 125kW induction furnace and stirring unit with an offgas system made up of an absorber bed, water scrubber, silica gel dryer, a HEPA filter, and two vacuum pumps. This system was designed to be versatile in nature in that it is capable of accommodating several different absorber bed designs and positions.

The approach to the experiments conducted on this full-scale system was similar to the bench-scale system with the ultimate goal being to develop an absorber bed that has a performance efficiency of 100%--i.e., allows zero break-through. Surrogate SNF assemblies doped with cesium carbonate as the cesium surrogate and were fabricated in the Materials Laboratory at SRTC. These mock SNF assemblies were melted at 850 °C in the full-scale apparatus and held at temperature for times on the order of 1 hour. Following this treatment cycle, the melt was allowed to solidify in the crucible and then the entire system was dismantled and each piece washed with either deionized water or dilute nitric acid. These wash samples were analyzed for cesium content using atomic absorption spectroscopy.

Irradiated Testing

Bench-Scale irradiated coupon tests have been performed at the Alpha-Gamma Hot Cells at ANL-E. These tests were performed as indicators to the behavior of irradiated materials with respect to both process melting and offgas behavior. The relatively small sample size and fission product loading makes these tests valuable only to general behavioral trends and not as relative absolutes. An apparatus was designed by SRTC and ANL-E personnel to simulate as close as possible the melt-dilute treatment conditions as tested in the full-scale surrogate experiments. One major difference, however, was the absence of any stirring—mechanical or induction – of the melt-pool in the irradiated coupon test. The apparatus contained a graphite crucible for melting the SNF coupons and a zeolite 4A-absorber bed approximately 1 inch in diameter and 5 inches long. The airflow in the test assembly was held constant at 10 scfh. At the outlet side of the zeolite absorber bed a water bubbler was inserted to monitor for bed break-through.

Results and Discussion

SRTC Small Scale Testing:

From the series of experiments conducted at850 °C in the full-scale apparatus and held at temperature for times on the order of 1 hour the results for the system efficiency were established. The performance efficiency values were determined by examining the percentage difference between the cesium found in the wet scrubber and the tubing connecting the bed and scrubber compared to the initial cesium loading of the surrogate MTR. These relative efficiency values are reported in Table 2. Once again as in the bench-scale results the uncertainty for these measurements is approximately 0.01 % based on the sensitivity of the atomic absorption cesium analysis of the liquid wash samples. The results of these experiments indicate that the offgas system with the concentric ring bed design is the most consistently effective at containing cesium. Additionally, the release fractions from these melt-experiments are higher than the bench-scale tests most likely due to stirring, however are they are still within the range of 5% reported by Taleyarkhan [4].

Irradiated Testing

Bench-Scale irradiated coupon tests have been performed at the Alpha-Gamma Hot Cells at ANL-E. These tests were performed as indicators to the behavior of irradiated materials with respect to both process melting and offgas behavior. The gamma scan for the test assembly is shown in Figure 3. The scan starts below the specimen chamber and continues up past the filter chamber. The top curve shows the gross activity and the bottom shows the Cs-137 activity; there is some background signal from the hot cell that accounts for the approximately 20,000 counts seen as the baseline of the gross activity curve. Both curves show two peaks, the larger peak is associated with the specimen chamber. There is a slightly elevated section just above the specimen chamber (from the 5-8 in. mark), which is probably due to some Cs-137 plating out on the stainless steel tubing during the cool-down. In the absorber bed chamber, there are two Cs peaks present a small peak followed by a larger peak. The smaller peak may be associated with Cs fines that were trapped in the steel screen, located at the bottom of the filter to retain the filter media. No activity is seen above the absorber bed.

Table 2 Full-Scale Surrogate Testing Offgas Efficiency Results

Bed Design	Test Conditions	System Efficiency %	% Volatilized	Mass Balance %
Straight Pass	Side Exit, 1scfm	99.17	3.13	26.97
Straight Pass	Side Exit, 1scfm	97.92	5.21	37.92
Straight Pass	Side Exit, 1scfm	97.65	6.40	54.89
Straight Pass	Side Exit, 1scfm	97.86	5.19	53.68
Straight Pass	Side Exit, 6scfm	98.31	3.93	61.05
Straight Pass	Top Exit, 1scfm	97.55	3.71	50.50
Straight Pass	Top Exit, 1scfm	98.45	3.35	54.38
Radial Plate	1scfm	99.34	2.35	59.36
Radial Plate	1scfm, Cs metal	94.34	27.65	52.80
Concentric Ring	1scfm	99.89	2.35	26.10
Concentric Ring	1scfm	99.99	0.51	13.68

About 78% of the Cs-137 remained in the specimen chamber, 16% was trapped in the filter and approximately 6% was plated out on the stainless steel tube just above the specimen chamber. The total release fraction for this experiment is approximately 22%. This value is greater than the values reported by Taleyarkhan and for most of our SRTC test, however, this value is consistent with previous SNF coupon melt studies performed at Hanford.

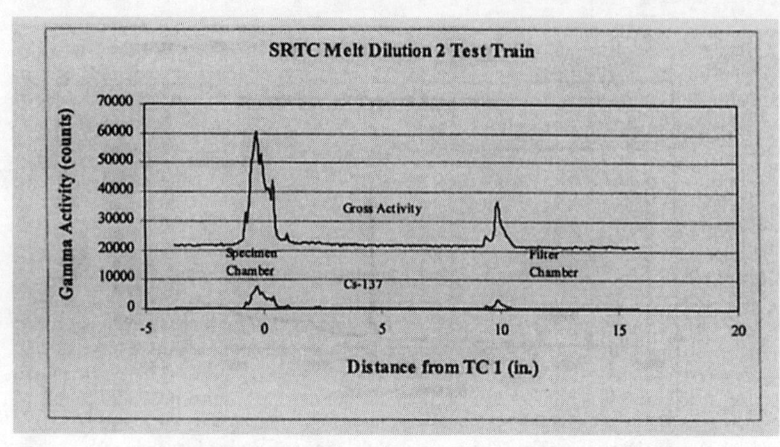

Figure 3 **Gamma Scan for the entire Bench Scale Irradiated Testing Apparatus**

Conclusions

The major findings from the full-scale surrogate experiments were that of the three-absorber bed designs tested the baffled concentric ring design provided the highest and most reproducible efficiencies. However, the design as tested did not provide zero breakthrough. As a result, the absorber bed that is being designed for the pilot-scale irradiated facility will be of the same basic design with a 50% increase in bed volume to attempt to achieve zero breakthrough.

The results from the bench-scale irradiated SNF coupon have provided further confirmation as to the effectiveness of a zeolite absorber bed to trap volatile cesium. No evidence of bed break-through was detected. Additionally, these results provide further validation to fundamental analytical calculations for predicting volatile species from the melt-dilute process. According to these calculations cesium would be the major volatile of concern and in the bench-scale irradiated coupon test the only radionuclide detected in any part of the test assembly using gamma scanning was cesium. Furthermore, although more cesium was released from this irradiated coupon melt than in previous bench and full scale surrogate tests, the amount volatilized was still below the maximum cesium release reported in the literature. Finally, the behavior documented from this irradiated coupon melt fully supported/validated all of the surrogate melt studies.

Lastly, the final step in the development and design of the melt-dilute offgas system will culminate with a full-scale irradiated melt-dilute facility at the Savannah River Site. This facility will be capable of melting full-scale irradiated SNF assemblies and will possess a prototypical offgas system that has been initially designed based on the bench and full-scale surrogate and bench-scale irradiated offgas experiments.

Acknowledgements

This work was supported by the U. S. Department of Energy under contract No. DE-AC09-96SR18500. The authors would also like to thank Adam Cohen and his staff at the Alpha-Gamma Hot Cells at Argonne National Laboratory for conducting the irradiated SNF coupon test.

References

1 J. Wolkoff and A. Chilenskas, "The Melt Refining of Irradiated Uranium: Application to EBR-II Fast Fuel, IX, " Sorption and Retention of Sodium and Cesium Vapor on Stationary Beds at Elevated temperature," Nuclear Science and Engineering, 9, pp 71-77 (1961).

2 T. M. Adams et al, The Melt-Dilute Treatment Technology Offgas Development Status Report, WSRC-TR-99-00145, April, 1999.

3 M. E. Hodges and M. L.Hyder, "Offgas-Studies for the Melt-Dilute Program", ANS Third Topical Meeting, DOE Spent Nuclear Fuel and Fissile Materials Management, Charleston, SC, September 8-11, 1998.

4 R. P. Taleyarkhan, "Analysis and Modeling of Fission Product Release from Various Uranium-Aluminum Plate-Type Reactor Fuels," Nuclear Safety, Vol. 33, No. 1, January-March 1992.

Corrosion and Characterization of Glass Wasteforms

THE EFFECT OF ADDING CRYSTALLINE SILICOTITANATE ON THE DURABILITY, LIQUIDUS, AND VISCOSITY OF SIMULATED HIGH-LEVEL WASTE GLASSES AT SAVANNAH RIVER SITE

J. R. HARBOUR, T. B. EDWARDS, R. J. WORKMAN
Westinghouse Savannah River Company, Aiken, SC 29808, john.harbour@srs.gov

ABSTRACT

This report provides a summary of the results obtained for a limited variability study for glasses containing Crystalline Silicotitanate (CST), Monosodium Titanate (MST), and either simulated Purex or HM sludge. Twenty-two glasses containing Purex sludge and three glasses containing HM sludge were fabricated and tested. The fabricated glasses were tested for durability using the 7-day Product Consistency Test (PCT) and characterized by measuring the viscosity at 1150°C and by determining an approximate, bounding liquidus temperature. The current models used by Defense Waste Processing Facility (DWPF) for predicting durability, viscosity, and liquidus temperature were applied to all 25 glasses. The goal of this work was to identify any major problems from a glass perspective, within the scope of this effort, which could potentially preclude the use of CST at DWPF.

INTRODUCTION

One of the alternative salt disposition flowsheets being considered by Savannah River Site (SRS) would require that the Defense Waste Processing Facility (DWPF) vitrify a coupled feed containing High Level Waste (HLW) and Crystalline Silicotitanate (CST). A glass variability study was therefore conducted to explore the processibility and product quality of the glass composition region for this alternative to the In-Tank Precipitation (ITP) Process. The objective of this study was to obtain information on the feasibility of incorporating anticipated levels of CST into DWPF glass with and without doubling the nominal levels of Monosodium Titanate (MST).

The glasses for this study were selected from a set of candidate glasses that involved Purex and HM sludge types, covered sludge loadings (in the glass) of 22, 26, and 30 oxide weight percent (wt%), utilized CST loadings (in the glass) of 3, 6, and 9 oxide wt%, and included MST concentrations (in the glass) at 1.25 and 2.5 wt%. For each composition, the remainder of the glass consisted of a proprietary glass former composition developed by M. K. Andrews of Savannah River Technology Center (SRTC).

The selection of Purex sludge for this study was based on the knowledge that this sludge type has historically been the most difficult sludge to incorporate into glass. Depleted uranium was introduced into the simulated Purex sludge to represent the uranium content (~ 9 wt% oxide) of actual sludge.

One of the major elements of concern for this study was titanium, which the DWPF currently restricts to a value of less than 1 wt% TiO_2 in glass. The introduction of CST and MST in this study results in TiO_2 levels in the glasses that significantly exceed the current limit. Furthermore, CST introduces proprietary components with unknown impact on glass quality and processing properties.

Mat. Res. Soc. Symp. Proc. Vol. 608 © 2000 Materials Research Society

The primary property of interest in this study was the durability (as measured by the 7-day Product Consistency Test, PCT [1]) of the test glasses. The PCT is the recognized standard for determining the durability of vitrified HLW, and the Environmental Assessment (EA) glass is the reference standard for assessing acceptable durability determined using the PCT. For a glass to demonstrate acceptable durability its PCT leach rate must be 2-sigma better than the PCT leach rate of the EA glass.

Processing properties of interest for these glasses included viscosity at 1150°C and liquidus temperature. Viscosity was measured at SRTC using a Harrop viscometer. To gain insight into the liquidus property for the CST glasses, an effort was made to obtain an upper bound on the T_L's using isothermal holds at 900, 950, 1000, and 1050°C and non-quantitative XRD evaluations with a sensitivity of ~ 0.7 to 1 wt% for crystalline Trevorite in the glass.

RESULTS

Measurements of the compositions of the test glasses were conducted by SRTC. Standards were included with the test glasses and bias-corrections were conducted for many of these measurements. The sum of oxides for all of these glasses fell within the interval of 95 to 105 wt%, a measure of the quality of the analytical results.

Glass Durability: Product Consistency Test

All of the CST glasses, after being batched and fabricated, were subjected to the 7-day PCT as an assessment of their durabilities [1]. More specifically, Method A of PCT (ASTM C1285) was used for these measurements. Durability is the critical product quality metric for vitrified nuclear waste. The PCT responses (for four elements: boron, silicon, sodium, and lithium) were normalized to the elemental concentrations in glass, and reported with units of grams-per-liter (g/L) using the measured, measured bias-corrected, and target compositions. In addition to the CST test glasses, the EA glass and an Approved Reference Material (ARM) standard glass were subjected to the PCT for validation to determine glass acceptability.

All 25 glasses were very durable as measured by the PCT. The PCT values clustered within the interval from 0.64 to 0.91 g/L for boron for all of the Purex glasses except one and ranged from 0.37 to 0.43 g/L for boron for the HM glasses. For comparison, the EA glass has a boron rate of 16.7 g/L [1].

Figure 1 is a plot of the DWPF model that relates the common logarithm of the normalized PCT (in this case for B) to a linear function of a free energy of hydration term (ΔG_p, kcal/100g glass) derived from the glass (measured and bias-corrected) compositions [2]. Prediction limits (at 95% confidence) for individual PCT results are also plotted around this linear fit. The PCT results for EA (shown as a diamond), ARM (shown as a "**z**"), and the CST glasses (shown as open squares) are presented on this plot. Note that the CST results reveal acceptable PCT values although they are not well predicted by the current DWPF durability model. Almost all of the PCT values for the CST glasses are slightly higher than the upper prediction limit. The behaviors of Si, Na, and Li release are similar to that of B: acceptable but unpredictable durabilities.

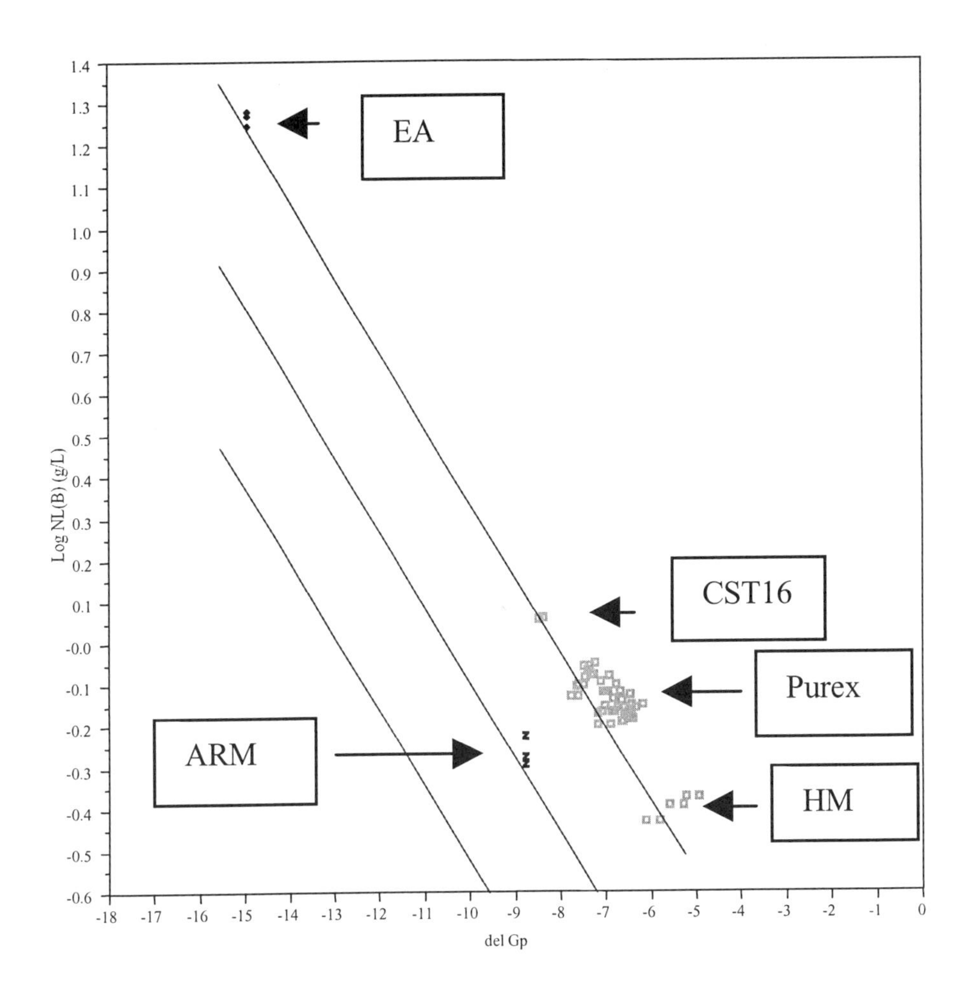

Figure 1
Log NL(B) (g/L) by del Gp(Using Measured & Bias-corrected CST Glass Compositions & EA and ARM reference compositions)

Figure 1 reveals that the PCT results for these CST glasses are consistently underpredicted by the current DWPF durability model. Possible reasons for this behavior include: (1) proprietary elements in CST not being adequately addressed by the current durability models, (2) the CST glasses may be phase separated thus violating one of the prerequisites for the use of the durability models and (3) a tendency for the durability models to underpredict glasses with ΔG_p values greater than –8.5 kcal/100g glass.

Viscosity at 1150°C.

Viscosity measurements were made on these CST glasses at SRTC using a Harrop, high-temperature viscometer. The viscosity, η, (in Poise) of each of these glasses at 1150°C was estimated ($\hat{\eta}$) from a Fulcher equation fitted to a set of viscosity measurements taken over an appropriate range of temperatures. The functional form of the (three-parameter) Fulcher equation (expressed in Poise) used to fit these data is given by equation (1):

$$\ln \hat{\eta} = A + \frac{B}{(T - C)} \tag{1}$$

where A, B, and C represent the parameters of the model that were determined from the available viscosity measurements at various temperatures (represented by T, expressed in °C). The fitted model was then used to predict the viscosity of the given glass at 1150°C.

Figure 2 provides a plot of the measured viscosities versus the viscosities predicted from the DWPF model using target, measured, and bias-corrected compositions [3]. A 45-degree (diagonal) line is also shown for reference.

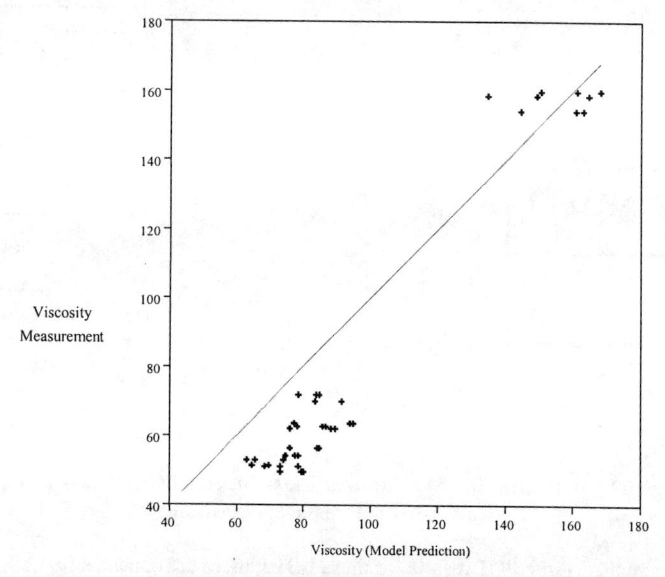

Figure 2: Viscosity Measurements versus Property Predictions in Poise

The melt viscosities at 1150°C for CST glasses fabricated using Purex are all well within the operating range for DWPF (20 to 100 Poise). Therefore, processing these glasses in the DWPF would not appear to be a problem from a viscosity perspective. The measured viscosities for all three of the HM glasses are well above the upper limit and, thus, outside of the operating window for DWPF. Although no Blend sludge (a sludge produced from blending of all the tank waste) glasses were fabricated, viscosity predictions for these glasses suggest that viscosity values may be close to 100 Poise, the upper limit for DWPF operations.

There are several interesting trends observed in the data. Whether one uses the target, measured, or bias-corrected measured compositions to predict viscosities using the current model, the predicted viscosities are always higher than the measured viscosities for the glasses batched using Purex sludge. This may not be unexpected since some of the elements in CST and MST are not considered in the viscosity model. This overprediction does not occur for the glasses made using HM sludge. However, these viscosities are beyond DWPF's operating window. Further work may be required to include the additional elements introduced by CST into the viscosity model.

Liquidus Temperature (T_L)

Twenty-four hour isothermal holds at 50°C intervals were used to estimate the liquidus temperature. XRD was selected as the method of detection for crystal formation in the glasses after each isothermal hold. It is estimated that the sensitivity of XRD (non-quantitative) is ~ 0.7 to 1 wt% for a crystalline phase (in this case, Trevorite [4]). Therefore, for this type of measurement, absence of detection of a crystalline phase was evidence that the liquidus temperature is less than the temperature of that isothermal hold. On the other hand, detection of Trevorite (or any other primary crystalline phase) indicates that the liquidus temperature is higher than the temperature of the isothermal hold. The bounds on the liquidus temperatures for these CST glasses, estimated to the detection capabilities of XRD, are plotted against model predictions [3] in Figure 3. The points on the plot represent the lowest temperatures without crystal detection. A 45-degree (diagonal) line is also shown for reference.

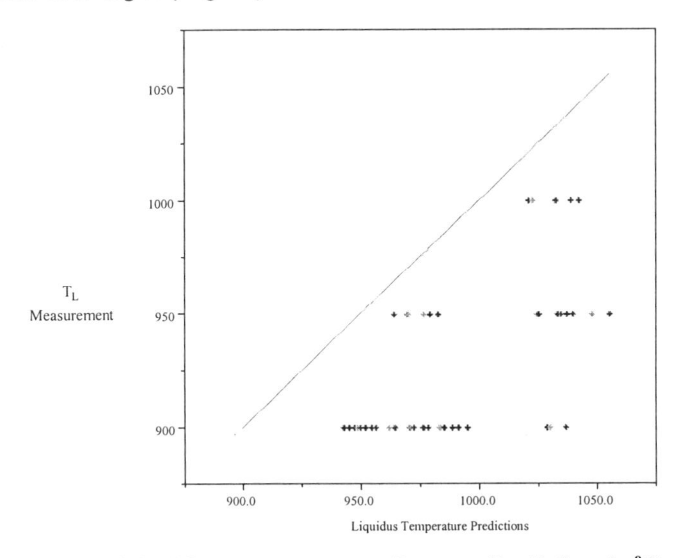

Figure 3: T_L Measurements versus Property Predictions in °C

The bounded estimates of T_L suggest that the liquidus temperatures of these glasses are within the DWPF operating window for this property. The model predictions versus the bounding estimates suggest that the model may be conservative for these glasses. Therefore, it may be necessary to modify the liquidus temperature model to include elements of CST and prevent unnecessarily conservative constraints on this property.

CONCLUSIONS

All 25 glasses were very durable as measured by the PCT. The PCT values clustered within the interval from 0.64 to 0.91 g/L for boron for all of the Purex glasses except one and ranged from 0.37 to 0.43 g/L for boron for the HM glasses. For comparison, the reference EA glass had a boron PCT release of 16.7 g/L [1]. A remarkable finding from this study was the highly clustered nature of the results. The 22 Purex-loaded glasses clustered tightly in one region, whereas the HM glasses clustered at an even lower value for boron release. These measured boron releases were generally greater than the upper 95% prediction limit of the model for these glasses. This type of behavior has been observed before for a range of glasses predicted to be very durable. The highly clustered nature of the results suggests that model revisions could be made to ensure predictability of the glasses.

For the 22 wt% Purex glasses, no crystals were detected in the bulk at 900°C or at the top surface of the glasses. For the 26 wt% Purex glasses, only two of the six glasses had bulk crystals after 24 hours at 900°C, and crystallization was no longer evident after the 24 hour hold at 950°C. For the 30 wt% Purex glasses, crystals were evident at higher temperatures but below the XRD detection limit at 1000°C.

For the Purex containing glasses, all viscosities were well within the DWPF range of 20 to 100 poise. The viscosity model, in general, overpredicted the measured viscosities for these glasses. This is not surprising given the fact that the model was not developed for glasses incorporating CST elements. On the other hand, the HM sludge-containing glasses had, as predicted, viscosities at 1150°C (~160 Poise) that were far above the 100 Poise limit. Thus, the HM sludge-containing glasses fabricated would not be acceptable for processing in the DWPF.

These conclusions are limited by the scope of this task. One limitation of this scope was the absence of any investigation of kinetic effects. Thus, one can not rule out that amorphous phase separation occurs with centerline cooling, for example, and this could have a deleterious effect on the durability of the glass. A second limitation was the restriction on independent variation of chemical constituents. In a major variability study, ranges are established for each element, and a statistically designed set of glasses identified which not only covers a larger region of compositional space, but also provides the potential for revealing (or confirming) relationships between the properties and the glass compositions. A third limitation was that only approximate and bounding measurements of the liquidus temperatures were made.

REFERENCES

[1] ASTM C1285-97, "Standard Test Methods for Determining Chemical Durability of Nuclear Waste Glasses: The Product Consistency Test (PCT)," 1997.

[2] C. M. Jantzen, J. B. Pickett, K. G. Brown, T. B. Edwards, and D. C. Beam, "Process/Product Models for the Defense Waste Processing Facility (DWPF): Part I. Predicting Glass Durability from Composition Using a Thermodynamic Hydration Energy Reaction Model (THERMO) (U)," WSRC-TR-93-672, Rev. 1, September 28, 1995.

[3] K. G. Brown and R. L. Postles, "SME Acceptability Determination for DWPF Process Control (U)," WSRC-TR-95-0364, Revision 3, February 21, 1996.

[4] C. A. Cicero, S. L. Marra, and M. K. Andrews, "Phase Stability Determinations of DWPF Waste Glasses (U)," WSRC-TR-93-227, Revision 0, 1993.

CONDUCTIVITY – CHEMISTRY RELATIONSHIP IN SIMULATED NUCLEAR WASTE GLASS MELTS

S. K. Sundaram and Elvis Q. Le[1]
Pacific Northwest National Laboratory
Richland, WA 99352, sk.sundaram@pnl.gov

ABSTRACT

Electrical conductivity of three simulated nuclear waste glass melts was measured over a wide range of temperatures, using a high-accuracy, calibration-free, coaxial-cylinders technique. Glass chemistries representing both low-activity and high-level wastes at Hanford and Savannah River sites were used. The temperature dependence of the conductivity of these glasses was presented. The conductivity values measured at 1150°C were compared to the conductivity values predicted from first-order mixture model.

INTRODUCTION

Electric melting is a well-established mode of production of nuclear waste glass melts due to high thermal efficiency, ease of operation, yield of homogenous glass with the required properties, lower volatilization losses, and remote operation [1]. Electrical conductivity of glass melts is one of the most important processing parameters in the joule-heated melters used for vitrification of nuclear wastes. It should be in the range 10 – 100 S/m for successful melter operation. A lower conductivity at the melting temperature would require a higher voltage across the electrodes resulting in conduction within the melter refractory and could also cause melter start-up difficulties unless undesirably large electrical power system are supplied. A higher conductivity would result in current exceeding the recommended maximum density for the melter electrodes.

Two different electrical conductivity models for prediction of properties as a function of glass composition were reported in the literature. The approach taken by the first principles process-product model by Jantzen [2] was based on glass structural considerations, expressed as a calculated non-bridging oxygen (NBO) term. Empirical first- and second-order mixture models were fit using a composition variation study (CVS) data (developed by the Pacific Northwest National Laboratory (PNNL) from 1989-1994)[3] to relate conductivity and other properties to glass composition. A model consisting of the Arrhenius equation with its coefficients (A and B) expanded in the forms of first-order mixture models accounted for 96 % of the variation in the CVS measured electrical conductivity data in the 950-1250°C range. For all CVS electrical conductivity testing, a probe with two platinum-10% rhodium blades was inserted into the glass to a known depth and the resistance of the glass between the blades was determined at discrete frequencies. For modeling, the resistance values at 1-kilohertz was used for consistency. Hrma and coworkers [4] developed mixture models for the logarithm of viscosity and that of electrical conductivity at 1150°C. First-order models within the temperature range from 950 to 1250°C fitted the experimental data fairly well, accounting for roughly 98 and 96% of the variability in

---

[1] Department of Mechanical Engineering, Washington State University, Tri-Cities Campus, Richland, WA 99352

Mat. Res. Soc. Symp. Proc. Vol. 608 © 2000 Materials Research Society

the data, respectively. The authors reported the effects of glass components on the electrical conductivity of melts at 1150°C in the form of a series for decreasing electrical conductivity: $Li_2O >> Na_2O >>$ Others $> Fe_2O_3 \approx B_2O_3 > Al_2O_3 \approx ZrO_2 \approx CaO \approx MgO > SiO_2$.

The measurement techniques used in the above modeling works, two-wire or crucible techniques, were generally considered to be low-accuracy techniques [5]. Therefore, a high-accuracy, calibration-free coaxial cylinders technique developed by Schiefelbein and Sadoway[6] was selected because accurate electrical conductivity measurements could be made using this technique in high temperature, relatively conductive, and highly corrosive melts like molten oxides in radioactive waste glass melts. The authors had elaborated this technique and presented their data for two melts in the CaO-MgO-SiO_2 system. Schiefelbein[7] applied this technique to a glass melt (Corning 7059). In the present investigation, the coaxial-cylinders technique was applied to select nuclear waste glass melts and the conductivity measured was compared to the predicted conductivity from the first-order mixture CVS model.

EXPERIMENTAL

Electrochemical impedance spectroscopy (Solartron 1260, Impedance/Gain-phase Analyzer) was used for measurements. Electrical conductivity probe design, measurement procedure, and data analysis used in the present investigation are identical to the one used in the reported works [5-7]. A brief description of the theoretical background, data collection and analysis, and conductivity calculation is presented below.

The electrical conductivity of a melt is derived from its impedance, $\tilde{Z}_{sol'n}$ (the tilde denotes a complex number). An actual impedance measurement contains conributions from other sources, for example, electrodes ($\tilde{Z}_{electrode}$) and leadwires ($\tilde{Z}_{leads}$), so that $\tilde{Z}_{meas} \neq \tilde{Z}_{sol'n}$. Assuming a simple equivalent circuit of the coaxial-cylinders cell [5], the impedances in series can be summed as follows:

$$\tilde{Z}_{meas} = \tilde{Z}_{leads} + \tilde{Z}_{electrode} + \tilde{Z}_{sol'n}$$

Because the conductivity κ derives solely from $\tilde{Z}_{sol'n}$, the other contributions must be eliminated from $\tilde{Z}_{meas}$. This is accomplished by measuring the impedance of the system with the electrodes shorted, then subtracting this from $\tilde{Z}_{meas}$. In addition, the purely resistive part of $\tilde{Z}_{sol'n}$, denoted $(Z^{real}_{sol'n})^*$, must be isolated. The following are the two steps involved in accomplishing the isolation:

1) Impedance measurements are first taken over a wide frequency range at many successive depths of immersion.
2) Using a short circuit correction, $\tilde{Z}_{sol'n}$ at each immersion is isolated from $\tilde{Z}_{meas}$.

After the isolation, the value of $(Z^{real}_{sol'n})^*$ at each immersion is found by choosing the value of $Z^{real}_{sol'n}$ at which $(-Z^{im}_{sol'n})$ is a minimum. Then $1/(Z^{real}_{sol'n})^*$ is plotted vs the corresponding relative immersion, ξ. The slope of this straight line, $[d(1/(Z^{real}_{sol'n})^*)]/d\xi$, is multiplied by the constant, $[\ln(b/a)]/2\pi$, to yield the electrical conductivity of the melt, κ. The variables a and b represent the diameters of the inner and outer electrodes, respectively.

Three glass compositions, a low-activity waste (LAW) and a high-level waste (HLW) glass, representing Hanford wastes and a Defense Waste Process Facility (DWPF) HLW glass were chosen for the present investigation. The target compositions are presented in Table I. The LAW and HLW glasses were melted at 1300°C and 1150°C, respectively. The glass was melted for 2 hours at the processing temperature and quenched by pouring on stainless steel plate. Each glass was melted in two batches (500-grams each) first and then melted the two portions together again for homogeneity. The glass was then crushed into powder using a tungsten mill. The glass powder was used for electrical measurements over a temperature range of 1000-1500°C in 100-150°C increments. The electrical conductivity was measured at four temperatures for five immersion depths for each glass.

Table I. Target Glass Compositions (weight %)

Constituent	LAW Glass (Hanford)	HLW Glass (Hanford)	HLW Glass (DWPF)
Al_2O_3	12.00	6.02	5.08
B_2O_3	9.00	7.03	8.64
Fe_2O_3	0.84	14.56	11.59
Li_2O	---	---	3.55
K_2O	0.26	---	---
Na_2O	20.00	14.92	12.86
SiO_2	56.78	45.14	50.89
ZrO_2	0.40	3.01	0.22
Others	0.72	9.32	7.17

RESULTS AND DISCUSSION

As all of the data can not be shown, an example is shown here for the Hanford HLW composition at 1000°C. The raw impedance data ($\tilde{Z}_{meas}$) for the first run, as well as the short data[2], ($\tilde{Z}_{short}$), for the first run and the short run are shown in Figure 1. The impedance of the short was subtracted from the measured impedance, point by point over the entire frequency range at each immersion. The result is also shown in Figure 1. The value of $(Z^{real}_{sol'n})^*$ at each immersion was then determined by choosing the value of $Z^{real}_{sol'n}$ at which ($-Z^{im}_{sol'n}$) was a minimum. Then $1/(Z^{real}_{sol'n})^*$ was plotted vs. the corresponding relative immersion, ξ, as shown in Figure 2. The slope of the straight line was multiplied by the constant, $[\ln (b/a)]/2\pi$, to yield the electrical conductvity of the glass melt, 14.22 S/m for the melt at 1000°C.

The temperature dependence of conductivity (Arrhenius fit) of the melts is presented in Figure 3. The activation energies of the melts are comparable. High R^2 values (> 0.97) indicate good fit to data. The conductivity range, 11-56 S/m, meets the specification for vitrification.

From the first-order mixture model [3], the ε_{1150} values estimated from the individual Arrhenius fits can be used as the dependent variable in the following form:

$$\ln \varepsilon_{1150} = \sum_{j=1}^{10} b_i x_i$$

The estimated coefficients of the model reported [3] have been used to estimate the conductivity at 1150°C and are shown along with corresponding measured values in Table 1.

---

[2] Impedance measurements taken with the electrodes shorted

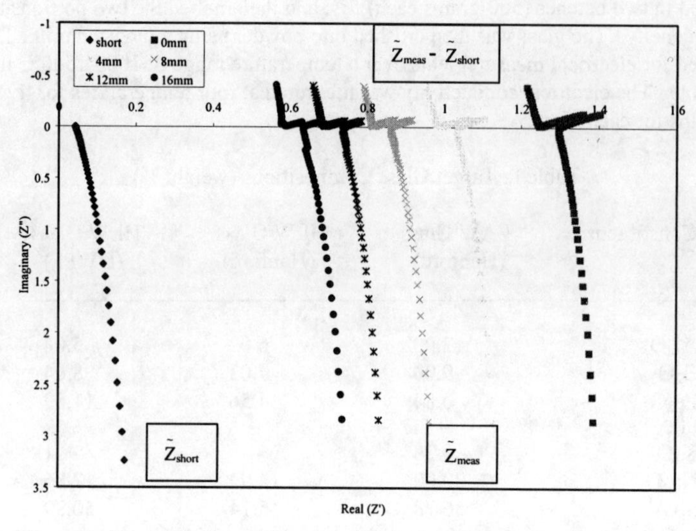

Figure 1. Raw, Short, and Corrected Data for Different Immersion Depths (0, 4, 8,12,16 mm from the surface of the melt)

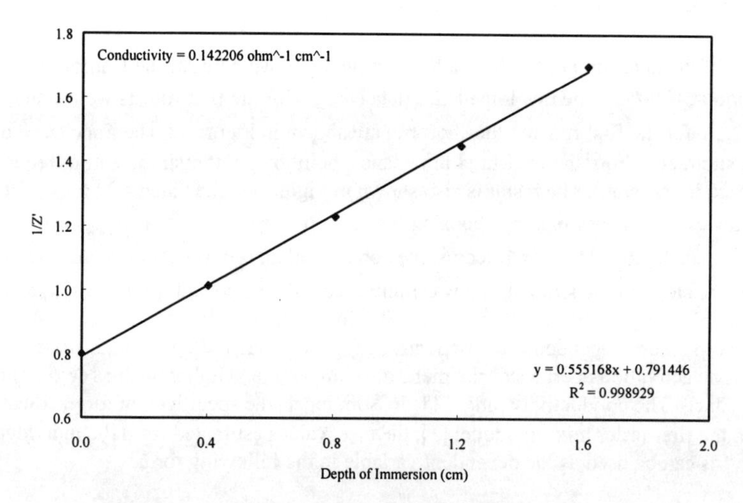

Figure 2. $1/(Z^{real}_{sol'n})^*$ vs. Relative Immersion, ξ

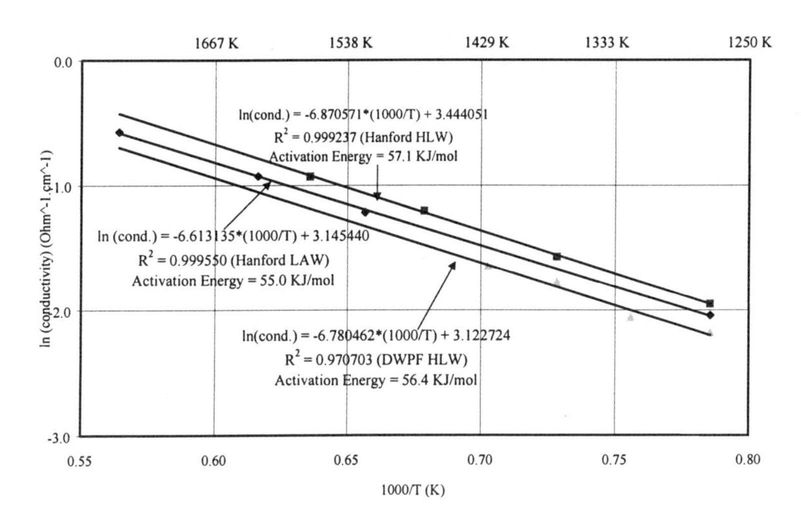

Figure 3. Temperature Dependence of Electrical Conductivity

In the case of Hanford LAW and Hanford HLW glasses, the first-order mixture model has predicted the conductivity values close to the corresponding measured values as shown in Table 1, indicating application of the model to these compositions. In the case of DWPF HLW glass, the model overestimated the conductivity. The deviation can not be explained due to use of only one glass composition in this study. The prediction will be refined with more DWPF glasses in future.

Table 1. Experimental vs. First-Order Mixture Model Data

Glass	Conductivity, S/m	
	Coaxial-Cylinders Technique	First-Order Mixture Model
LAW (Hanford)	22.28	22.20
HLW (Hanford)	25.07	20.06
HLW (DWPF)	19.37	33.02

CONCLUSIONS

The data show that the coaxial-cylinders technique is a useful tool in accurately measuring the electrical conductivity of the nuclear waste glasses. There is a good agreement between the experimental and predicted (from the first-order mixture model) data for Hanford LAW and HLW glasses. A difference between the measured and predicted data for DWPF HLW glass observed needs further evaluation.

ACKNOWLEDGEMENTS

The authors gratefully acknowledge the useful discussions with Dr. Susan L. Schifelbein, Corning Inc. and Professor Donald R. Sadoway, Department of Materials Science and Engineering, Massachusetts Institute of Technology, Cambridge, MA. The authors also acknowledge the technical review by Dr. Pavel Hrma at PNNL. The Environment Management Science Program (EMSP) of the DOE supported the research and preparation of this paper.

REFERENCES

1. J. Stanek, "Electrical Properties of Glass," pp. 15-88 in Electric Melting of Glass, Elsevier Scientific Publishing Company, Amsterdam, The Netherlands, 1977.
2. C. M. Jantzen, "First Principles Process-Product Models for Vitrification of Nuclear Waste: Relationship of Glass Composition to Glass Viscosity, Resistivity, Liquidus Temperature, and Durability," pp. 37-51 in Nuclear Waste Management IV, Editor: G. G. Wicks, D. F. Bickford, L. Roy Bunnell, Ceramic Transactions, Volume 23, The American Ceramic Society, Westerville, Ohio, USA, 1991.
3. P. R. Hrma, G. F. Piepel, M. J. Schweiger, D. E. Smith, D.-S. Kim, P. E. Redgate, J. D. Vienna, C. A. LoPresti, D. B. Peeler, M. H. Langowski in Property/Composition Relationships for Hanford High-Level Waste Glasses Melting at 1150°C, PNL-10359, Vol. 1 & 2, December 1994.
4. P. R. Hrma, G. F. Piepel, D. E. Smith, P. E. Redgate, and M. J. Schweiger, "Effect of Composition and Temperature on Viscosity and Electrical Conductivity of Borosilicate Glasses for Hanford Nuclear Waste Immobilization," pp. 151-158 in Environmental and Waste Management Issues in the Ceramic Industry, Editor: G. B. Mellinger, Ceramic Transactions, Volume 39, The American Ceramic Society, Westerville, Ohio, USA, 1994.
5. S. L. Schiefelbein, "A New Technique to Measure the Electrical Properties of Molten Oxides," Ph. D. Thesis, Massachusetts Institute of Technology, 1996.
6. S. L. Schiefelbein and D. R. Sadoway, "A High-Accuracy, Calibration-Free Technique for Measuring the Electrical Conductivity of Molten Oxides," Metallurgical and Materials Transactions B, 28B 1141-1149 (1997).
7. S. L. Schiefelbein, "A High-Accuracy, Calibration-Free Technique for Measuring the Electrical Conductivity of Glass Melts," pp. 99-113 in Electrochemistry of Glass and Ceramics, Editors: S. K. Sundaram, D. F. Bickford, and E. J Hornyak, Jr., Ceramic Transactions, Volume 32, The American Ceramic Society, Westerville, Ohio, USA, 1999.

Liquidus Temperature of High-Level Waste Borosilicate Glasses with Spinel Primary Phase

Pavel Hrma, John Vienna, Jarrod Crum, and Greg Piepel
Pacific Northwest National Laboratory, Box 999, Richland, WA
Martin Mika
Institute of Chemical Technology, Technicka 5, 16628 Prague, Czech Republic

Abstract

Liquidus temperatures (T_L) were measured for high-level waste (HLW) borosilicate glasses covering a Savannah River composition region. The primary crystallization phase for most glasses was spinel, a solid solution of trevorite ($NiFe_2O_4$) with other oxides (FeO, MnO, and Cr_2O_3). The T_L values ranged from 859 to 1310°C. Component additions increased the T_L (per mass%) as Cr_2O_3 261°C, NiO 85°C, TiO_2 42°C, MgO 33°C, Al_2O_3 18°C, and Fe_2O_3 18°C and decreased the T_L (per mass%) as Na_2O -29°C, Li_2O -28°C, K_2O -20°C, and B_2O_3 -8°C. Other oxides (CaO, MnO, SiO_2, and U_3O_8) had little effect. The effect of RuO_2 is not clear.

Introduction

Spinel, a solid solution of trevorite ($NiFe_2O_4$) with other oxides (FeO, MnO, and Cr_2O_3), forms in glasses with high concentrations of Cr, Ni, and Fe [1-6]. Spinel may have an adverse impact on melter performance, causing sludge accumulation, pouring difficulties, cold-cap freezing, and foam stabilization. To reduce spinel precipitation in the Joule-heated melter, the liquidus temperature (T_L) is required to be 100°C below the nominal melter operating temperature. For many high-level waste (HLW) streams, this constraint has a huge economic impact because it lowers the waste loading to a value below that which other properties would allow. Changing the melter design to enable a higher waste loading is considered risky because the current melters have been demonstrated as reliable. However, with more precise relationships between T_L and glass composition, glass formulations can be optimized [7-10] to balance the risk of melter failure against the economic loss from reduced waste loading. Mika et al. [3] measured T_L over a broad HLW glass composition region by adding components to a baseline glass. In this study, we use a layered statistical design [11] to cover a Savannah River composition region. By fitting linear approximation functions to data, we obtain partial specific liquidus temperatures (T_{Li}) for major glass components. These values can be used for optimized formulation of HLW glasses within the corresponding composition regions [12].

Experimental

To distinguish between Mika's [3] and the present databases, we labeled the former as SP and the latter as SG. The SG and SP composition regions are summarized in Table I. The SG test matrix contains 53 glasses (on the boundary, in the interior, and at the center of the SG region). Their compositions are listed elsewhere [13].

Each 450-g batch was made by blending oxides, alkali and calcium carbonates, boric acid, and $Ru(NO)(NO_3)_3$ in an agate milling chamber for 6 min. Batches were melted for 1 h in a covered Pt-10%-Rh crucible. The melting temperature was $T_5 + \Delta T$, where T_5 is the temperature at which the calculated melt viscosity was 5 Pa·s and $\Delta T \geq 100$°C. The T_5 value was estimated from approximation functions fitted to the Hanford viscosity database [2]. This unusually high melting temperature was necessary to dissolve some of the RuO_2 that if not dissolved would

Mat. Res. Soc. Symp. Proc. Vol. 608 © 2000 Materials Research Society

interfere with the detection of spinel crystals (SG glasses contained 0.09 mass% RuO_2, an oxide that is virtually insoluble in glass at 1150°C [14]). The glasses were milled in a tungsten carbide chamber for 8 min and remelted for 1 h to ensure homogeneity.

To measure T_L, approximately 2.5-g glass samples in covered Pt-10%-Rh crucibles were heat treated at a constant temperature (T) and checked by optical microscopy for the presence of crystals. The heat-treatment time varied to ensure redox equilibrium with air and to minimize volatilization. When the difference in T between heat-treated glass samples with and without crystals was ≤10°C, the T_L value within this interval was estimated according to the size, shape, and number of crystals in the sample. We believe that this is within ±5°C. We used a standard glass with T_L = 1040°C to make a correction for the temperature difference between the measuring thermocouple and the sample. Crystalline phases in some samples were identified by X-ray diffraction and scanning electron microscopy (SEM). Crystal composition was estimated using SEM energy dispersive spectroscopy (EDS).

RESULTS

The primary crystallization phase was spinel in 44 glasses and clinopyroxene in 7 glasses. The T_L values ranged from 865°C to 1310°C for glasses with the spinel primary phase and from 793°C to 996°C for glasses with clinopyroxene. The nearly uniform size of spinel crystals ranged from 1 to 12 μm and made them easily distinguishable from much smaller crystals (<1 μm) that occasionally precipitated during quenching. Figure 1A shows the morphology of spinel crystals that were isolated by dissolving the glass in acid (crystals are not of a uniform size in this special case). Star-shaped spinel crystals formed in some glasses (Figure 1B).

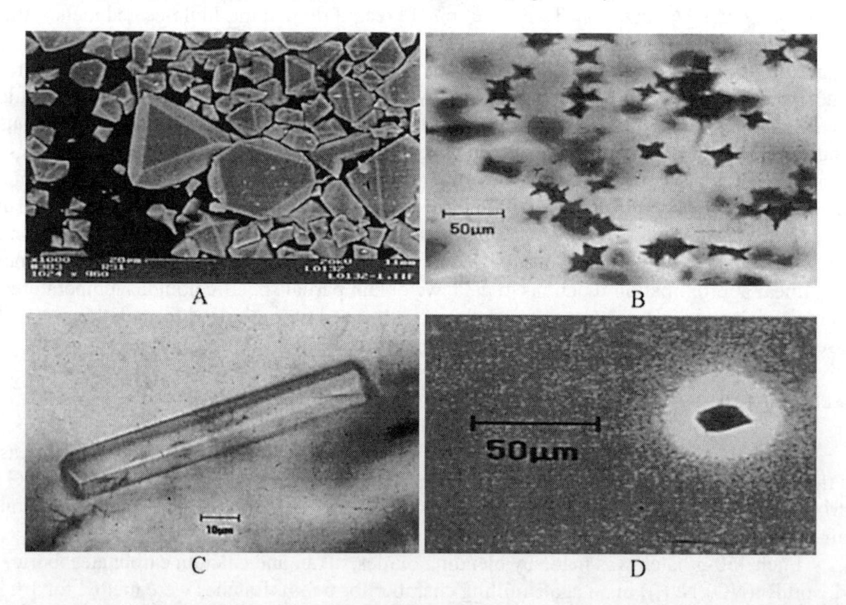

Figure 1. A Spinel crystals chemically separated from a HLW glass (SEM); B star shaped spinel crystals (optical microscope, transmitted light); C a clinopyroxene crystal (optical microscope); D a halo around a spinel crystal (SEM)

From EDS analyses, Ni and Fe were major components of spinel. The Cr content in spinel decreased with decreasing temperature and was higher at the interior of crystals than on the surface. This indicates that at temperatures just below T_L, Cr_2O_3 reacts with NiO and Fe_2O_3, forming a solid solution of nichromite ($NiCr_2O_4$) and chromite ($FeCr_2O_4$). At lower temperatures, spinel is increasingly enriched in trevorite ($NiFe_2O_4$) and magnetite (Fe_3O_4). Other spinel components, Mn, Al, and Mg, were detected at 1 to 2 cation%. Small fractions of Si and Na in some crystals were probably inclusions. Clinopyroxene (Figure 1C) did not appear in glasses with <50 mass% SiO_2; its composition, in cation%, was 8.5 to 9.1 Na, 8.3 to 9.5 Mg, 10.6 to 11.7 Ca, 0.3 to 0.7 Ni, 1.2 to 1.4 Mn, 12.3 to 14.8 Fe, 0.5 to 0.9 Cr, and 55.0 to 55.6 Si.

An interesting phenomenon was the presence of a colorless halo around spinel crystals in some brown glasses after a short heat treatment (Figure 1D). The glass composition within and outside the halo was identical within the accuracy of the quantitative EDS. The halo was probably caused by a difference in glass redox.

We calculated T_{Li} values by fitting the equation

$$T_L = \sum_{i=1}^{N} T_{Li} g_i \qquad (1)$$

to experimental data. Here g_i is the i-th component mass fraction from the as-batched composition and N is the number of components. T_{Li} values for the SG and SP databases and their combination (SG&SP) are listed in Table I. Partial specific properties are generally functions of composition. The T_{Li} values in Table I are approximations as constants.

Table I also shows R^2 and s values for the overall fits to the three databases. R^2 is the fraction of the variation in T_L database values accounted for by Equation (1). The quantity s is the root mean square error and estimates the experimental standard deviation if Equation (1) adequately represents the true T_L-composition relationship. The R^2 values in Table I suggest that Equation (1) provides reasonable approximations to the $T_L(g_i)$ relationships for each of the three databases. Based on glasses measured multiple times [13], the imprecision standard deviation is 4°C. Considering 2 to 3 standard deviations, most glasses are expected to have T_L within the range of ±8°C or ±12°C of the measured values. The s values (Table I) are larger, from 23°C to 27°C, suggesting that Equation (1) is not a perfect approximation of the true $T_L(g_i)$ relationship.

As seen in Table I, the SP component ranges are broader than the SG component ranges for some components (SiO_2, B_2O_3, MgO, MnO, NiO, Fe_2O_3, Cr_2O_3, RuO_2), partially overlap for other components (Al_2O_3, Na_2O), and do not overlap at all for one component (Li_2O). Further, the SG study varies several components (K_2O, CaO, TiO_2, U_3O_8) not varied in the SP study. These differences are probably the cause of a constant difference of roughly 150°C that was seen between measured T_L values in the SG data set and calculated T_L values based on the SP data set (and vice versa). However, as Figure 2 demonstrates, the combined SG&SP data set shows consistency, suggesting that the corresponding T_{Li} values can be used for glasses in both SP and SG composition regions.

The effects of glass components on T_L can be judged by the T_{Li} values or by other measures. For example, in a glass with T_L = 1050°C, adding a component with T_{Li} > 1050°C will increase T_L, whereas an addition of a component with T_{Li} < 1050°C will decrease T_L. Alternatively, the slope

$$S_i = (T_{Li} - T_{L,ref})/(1 - g_{i,ref}) \qquad (2)$$

estimates the change in T_L per unit addition a component addition to a reference glass.

Table I. Composition regions, partial specific liquidus temperatures, and component effects

	Composition Region		T_{Li} (°C)			S_i (°C/mass%)			T'_{Li} (°C/mass%)		
	SG	SP	SG	SP	SG&SP	SG	SP	SG&SP	SG	SP	SG&SP
Al_2O_3	.025-.080	.040-.160	2777	3330	2980	18±2	22±2	20±2	18±2	25±2	20±2
B_2O_3	.050-.100	.000-.120	346	363	304	-8±2	-9±2	-8±2	-7±2	-5±2	-7±2
CaO	.003-.020		2215		2162	12±6		11±6	12±6		12±6
Cr_2O_3	.001-.003	.000-.012	27108	19242	21038	261±53	181±20	200±21	261±53	184±20	200±21
Fe_2O_3	.060-.150	.060-.150	2700	2605	2730	18±1	15±3	19±1	17±1	18±3	17±1
K_2O	.015-.038		-885		-711	-20±4		-18±4	-19±4		-17±4
Li_2O	.030-.060	.000-.030	-1639	-1599	-1587	-28±3	-30±7	-28±3	-26±3	-24±7	-26±3
MgO	.005-.025	.004-.060	4345	2900	3592	33±5	17±4	26±4	33±5	21±4	26±4
MnO	.010-.030	.000-.040	507	1962	958	-6±5	8±6	-1±4	-5±5	11±6	-0.3±4
Na_2O	.060-.110	.080-.200	-1565	-1851	-1654	-29±2	-34±3	-30±2	-26±2	-27±2	-26±2
NiO	.001-.020	.000-.030	9445	8329	9231	85±5	72±9	83±5	84±5	75±9	82±5
RuO_2	.0009	.0003	(d)	31192	15461	(d)	300±224	144±225	(d)	304±224	145±224
SiO_2	.430-.590	.380-.600	1004	838	993	-1±1	-8±2	-1±1	n.a.	n.a.	n.a.
TiO_2	.002-.006		5262		5361	42±23	34±11	43±23	43±23		44±22
U_3O_8	.000-.055		1429		1444	4±2		4±2	4±2		5±2
Others[(a)]		.045-.070		4383	3209		33.98	21±4		35±11	22±4
N [(c)]			53	43	96						
R^2			0.97	0.94	0.96						
s (°C)			26.8	23.5	27.0						

(a) Others for SP glasses were a mixture of 22 minor components with ZrO_2, Nd_2O_3, La_2O_3, CdO, MoO_3, F, and $SO_3 > 3$ mass% of the Others mix. SG glasses did not contain any other components than those listed.
(b) S_i was calculated by Equation (2) using SG05 as the reference glass (see [13]). T'_{Li} was calculated by Equation (3). Both were multiplied by 0.01 to express a change in T_L per 1 mass% change in a component. The ± values represent the standard deviations of the S_i and T'_{Li} values.
(c) N is the number of data points used to fit Eq. (1). SP RuO_2 data and replicate data were used in fitting Eq. (1), so the resulting T_{Li} and T'_{Li} values in this table differ slightly from those reported in [13].
(d) Because the level of RuO_2 in SG glasses was fixed, it was normalized out of the SG glass compositions used to model spinel T_L.

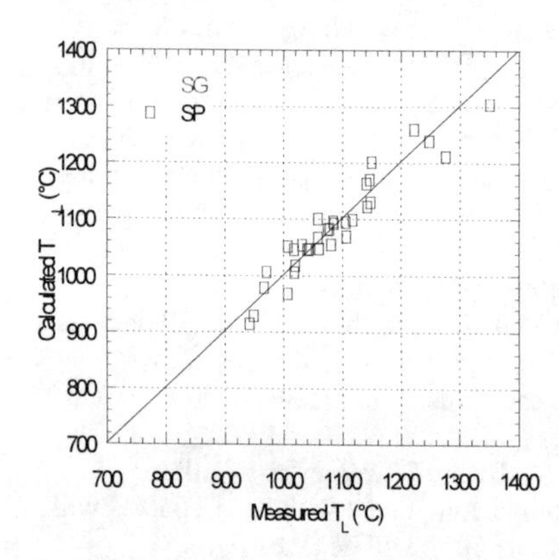

Figure 2. Calculated (using SG&SP T_{Li} coefficients) versus measured T_L values for SG and SP glasses

Finally, the difference $T'_{Li} = T_{Li} - T_{Ln,}$, called the i-th component relative partial specific liquidus temperature, compares the i-th component with the n-th component. SiO_2 is a good choice for the n-th component because it is the most abundant component in the glass. The S_i and T'_{Li} values multiplied by 0.01 (to express a change in T_L per 1 mass% change in component mass fraction) are listed in Table I.

DISCUSSION

Apart from being of general interest, the knowledge of component effects is important for glass formulation to optimize waste loading while keeping T_L low enough for safe processing. If crystallization limits waste loading, T_L of the glass would be close to 1050°C, a conventional limit for Joule-heated melters. According to their effects on T_L, glass components for both the SG and SP databases can be arranged into four groups: (Cr_2O_3, NiO) >> (MgO, TiO_2, Al_2O_3, Fe_2O_3) > (U_3O_8, MnO, CaO, B_2O_3, SiO_2) > (K_2O, Li_2O, Na_2O). We shall discuss these groups in terms of T'_{Li} coefficients because they vary somewhat less among the databases than the slopes.

Components in the first group (Cr_2O_3 and NiO) strongly increase T_L. Replacing 1 mass% SiO_2 with 1 mass% Cr_2O_3 results in a $T'_{L,Cr2O3} \approx 200$°C/mass%. Nickel oxide also strongly increases T_L: $T'_{L,NiO} \approx 80$°C/mass%.

Components in the second group (MgO, TiO_2, Al_2O_3, and Fe_2O_3) moderately increase T_L. Fe_2O_3, the major spinel component, increases T_L less than Cr_2O_3, NiO, MgO, and possibly TiO_2 ($T'_{L,Fe2O3} \approx 17$°C/mass%). Cr_2O_3 and NiO initiate spinel formation as the glass cools. Although magnetite can precipitate from iron-rich glass in the absence of Cr_2O_3 and NiO, maghemite (Fe_2O_3) is more likely to form [15]. The relatively strong effect of MgO on T_L within the spinel primary field of HLW glasses ($T'_{L,MnO} \approx 20$ to 35°C/mass%) is well known [4]. Another oxide that increases T_L within the spinel field is Al_2O_3 ($T'_{L,Al2O3} \approx 18$ to 25°C). MgO and Al_2O_3 are minor components in spinel that precipitates from HLW glass below 1300°C. Considering the extremely narrow range of TiO_2 concentration (0.15 to 0.6 mass% in SG), its effect ($T'_{L,TiO2} \approx 44$ to 40°C/mass%) is less certain than that of other glass components (note the high standard deviation of S_{TiO2} and $T'_{L,TiO2}$ values in Table I).

Components in the third group (U_3O_8, MnO, CaO, B_2O_3 and SiO_2) have little effect on T_L ($T'_{Li} \leq 10$°C/mass%). It is uncertain to which group RuO_2 belongs. The RuO_2 concentration was not varied in SG and SP glasses. The RuO_2 effect, shown in Table I, was estimated from additional data (to be reported elsewhere). RuO_2 is even less soluble in glass than Cr_2O_3. RuO_2 solubility increases at higher temperatures. As the glass cools, RuO_2 partly precipitates as needle-like crystals and partly becomes a component in spinel [16]. RuO_2 concentration in spinel was too low to be detectable by SEM-EDS, but it was detected by direct chemical analysis of spinel crystals separated from glass.

Components in the fourth group, consisting of alkali oxides (K_2O, Li_2O, and Na_2O), decrease T_L ($T'_{L,Na2O} \approx -26$°C/mass%, $T'_{L,Li2O} \approx -26$°C/mass%, and $T'_{L,K2O} \approx -18$°C/mass%).

CONCLUSIONS

According to their effects on T_L, glass components fall into four groups. Cr_2O_3 and NiO strongly increase T_L, MgO, TiO_2, Al_2O_3, and Fe_2O_3 moderately increase T_L, U_3O_8, MnO, CaO, B_2O_3, and SiO_2 have little effect, and K_2O, Li_2O, and Na_2O decrease T_L. Within the composition region of HLW glasses, the effects of glass components on T_L have been expressed in term of T_{Li} coefficients (the partial specific T_Ls). These coefficients are useful tools for HLW glass formulations that optimize waste loading while reducing the risk of melter failure due to an accumulation of spinel. The present study did not investigate the effect of glass redox on T_L. This effect and the nonlinear blending effects of components will be pursued in a future study.

ACKNOWLEDGEMENTS

Funding for this task was provided by the Tanks Focus Area and the U.S. Department of Energy Office of Science and Technology. Tommy Edwards from the Savannah River Technology Center developed the SG test matrix. Mike Schweiger trained and advised two of the authors in their technical work. Ron Sanders and Meiling Gong analyzed the glasses and crystals. Jim Young performed SEM-EDS analyses and David McCready conducted a part of the x-ray diffraction work. The authors thank Denis Strachan and Wayne Cosby for helpful suggestions. Martin Mika is grateful to Associated Western Universities for his fellowship at Pacific Northwest National Laboratory (PNNL). PNNL is operated for the U.S. Department of Energy by Battelle under Contract DE-AC06-76RLO 1830.

REFERENCES

1. P. Hrma, G. F. Piepel, M. J. Schweiger, D. E. Smith, D.-S. Kim, P. E. Redgate, J. D. Vienna, C. A. LoPresti, D. B. Simpson, D. K. Peeler, and M. H. Langowski. 1994. *Property / Composition Relationships for Hanford High-Level Waste Glasses Melting at 1150°C,* PNL-10359, Pacific Northwest Laboratory, Richland, Washington
2. P. Hrma, G. F. Piepel, P. E. Redgate, D. E. Smith, M. J. Schweiger, J. D. Vienna, and D.-S. Kim. 1995. *Ceram. Trans*. 61, 505-513.
3. M. Mika, M. J. Schweiger, and P. Hrma. 1997. "Liquidus Temperature of Spinel Precipitating High-Level Waste Glass," *Scientific Basis for Nuclear Waste Management* (Editors W. J. Gray and I. R. Triay), Vol. 465, p. 71-78, Material Research Society, Pittsburgh, Pennsylvania.
4. K.-S. Kim and P. Hrma. 1994. *Ceram. Trans*. 45, 327-337.
5. J. G. Reynolds and P. Hrma. 1997. *Scientific Basis for Nuclear Waste Management* (Editors W. J. Gray and I. R. Triay), Vol. 465, p. 65-70, Material Research Society, Pittsburgh, Pennsylvania.
6. J. D. Vienna, P. Hrma, D. S. Kim, M. J. Schweiger, and D. E. Smith. 1996.*Ceram. Trans*. 72, 427-436.
7. P. Hrma and P. A. Smith. 1994. "The Effect of Vitrification Technology on Waste Loading," *Proc. Int. Top. Meeting Nucl. Hazard. Waste Manag. Spectrum '94*, Vol. 2, pp. 862-867.
8. P. Hrma. 1994. *Ceram. Trans.* 45, 391-401.
9. P. Hrma, J. D. Vienna, and M. J. Schweiger. 1996. *Ceram. Trans*. 72, 449-456.
10. P. Hrma and R. J. Robertus. 1993. *Ceram. Eng. Sci. Proc.* 14 [11-12] 187-203.
11. G. F. Piepel, C. M. Anderson, and P. E. Redgate. 1993. *1993 Proceedings of the Section on Physical and Engineering Sciences*, 205-227, American Statistical Association, Alexandria, Virginia.
12. P. Hrma. 1998. *Ceram. Trans*. 87, 245-252.
13. P. Hrma, J. D. Vienna, M. Mika, J. V. Crum, and G. F. Piepel: *Liquidus Temperature Data for DWPF Glass*, PNNL-1170, Pacific Northwest National Laboratory. Richland, Washington, 1999.
14. H. D. Schreiber, F. A. Settle, P. L. Jamison, J. P. Eckenrode, and G. W. Headley. 1986. *J. Less-Common Metals*, 115, 145-154.
15. D-S. Kim, P. Hrma, D. E. Smith, and M. J. Schweiger. 1994. *Ceram. Trans*. 39, 179- 189.
16. C. J. Capobianco and M. J. Drake. 1990. *Geochem. Cosmochim. Acta* 54, 869-874.

LIQUIDUS TEMPERATURE OF RARE EARTH-ALUMINO-BOROSILICATE GLASSES FOR TREATMENT OF AMERICIUM AND CURIUM

Brian J. Riley, John D. Vienna, and Michael J. Schweiger, *Pacific Northwest National Laboratory, Richland, WA 99352*

David K. Peeler and Irene A. Reamer, *Westinghouse Savannah River Company, Aiken, SC 29808*

ABSTRACT

The liquidus temperatures, T_L, of rare earth-alumino-borosilicate glasses were measured as a function of glass composition. The T_L values ranged from 1153°C to 1405°C. Three primary crystalline phases were identified in the study. The most frequently encountered was a rare earth silicate phase in which the T_L values ranged from 1153°C to 1405°C. Al_2O_3 was encountered in glasses with an Al_2O_3:SiO_2 mass ratio greater than 1, with T_L values ranging from 1242°C to 1305°C. Alumino-silicate crystals were encountered as the primary phase in glasses with less than 43 mass% of mixed rare earth oxides (Ln_2O_3) with T_L values between 1164°C and 1255°C. A linear relationship between total mixed rare earth oxide concentration and T_L was found within all three primary phase fields. In the rare earth silicate primary phase field, normalizing the Ln_2O_3 concentration with its mean ionic radius enhanced this linear relationship.

INTRODUCTION

Approximately 11,000 liters (3,600 gallons) of solution containing isotopes of Am and Cm are stored in Tank 17.1 in the Savannah River Site's (SRS) F-Canyon. These isotopes were recovered during plutonium-242 production campaigns in the mid and late 1970s. The continued storage of these isotopes was identified as an item of primary concern in the Defense Nuclear Facility Safety Board's (DNFSB) Recommendation 94-1. The process selected to treat the contents of this tank includes (Marra et al. 1999):

- pretreating the tank contents in existing canyon vessels to separate actinides and lanthanides from other impurities (primarily iron, aluminum, and sodium)
- vitrifying the actinide/lanthanide fraction of the tank contents in the Multi-Purpose Processing Facility (MPPF) in the F-Canyon
- shipping the actinide containing glass to Oak Ridge National Laboratory (ORNL) for storage and separation of actinides from glass
- irradiating the separated Cm to form heavy elements such as Cf for civilian and military applications.

Peeler and colleagues developed a glass composition specifically for the treatment of this material (Fellinger et al. 1998, Peeler 1998, Peeler and Edwards 1999, and Peeler et al. 1999). The 25SrABS glass frit (33.68 mass% SiO_2, 25.00 La_2O_3, 24.87 Al_2O_3, 13.54 B_2O_3, and 2.91 SrO), when combined with 20 to 50 mass% mixed lanthanide/actinide oxides, produces a glass with adequate properties for the process. A two-phase study was performed to quantify the impact of variation in glass composition on the properties of this glass. In the Phase I of this study, the concentration of mixed rare earth oxides, Ln_2O_3, was systematically varied between 38 and 54 mass% along with its composition (Peeler 1998 and Peeler and Edwards 1999). Phase II was designed to systematically vary the concentration and composition of Ln_2O_3 and the concentrations of SiO_2, Al_2O_3, B_2O_3, and SrO in glass (Peeler and Edwards 1999). The concentration of Ln_2O_3 varied between 48.9 and 64.1 mass percent. The ratios of SiO_2, Al_2O_3, B_2O_3, and SrO remained constant and equal to those in the 25SrABS frit in 27 of the 55 glasses.

The properties, including liquidus temperature (T_L), normalized release from glass subjected to the product consistency test, viscosity-temperature relationship, density, and lanthanide recoverability, were characterized for the 87 test glasses (Peeler et al. 1999b and Vienna et al. 1999). In this paper, we discuss the impact of composition on the T_L of test glasses.

Mat. Res. Soc. Symp. Proc. Vol. 608 © 2000 Materials Research Society

PROCEDURE

Eighty-seven glasses were fabricated by mixing boric acid, $SrCO_3$, and oxides of rare earth elements, silicon, aluminum, and other minor chemical additives. The targeted and measured compositions of these glasses are listed in Peeler et al. (1999b). Table I lists the measured composition of test glasses. The chemical batches were melted in Pt/Rh crucibles with lids. Phase I glasses were melted at 1350°C for ½ h and quenched on a steel plate. Phase II glasses were melted at 1450°C for 1 h, quenched on a steel plate, ground in a tungsten carbide mill, and remelted at 1450°C for 1 h.

Roughly 30 g of each glass, used for T_L measurements, was broken from the quenched glass patty, crushed in a tungsten-carbide milling chamber, and sieved to capture particles between 0.5 mm and 4 mm. The sized glass was cleaned in an ultrasonic bath of ethanol for 2 min and dried at 200°C for 2 h. Glass samples were placed in labeled 1 cm^3 Pt/Rh boxes with tight fitting lids and heat treated in uniform temperature furnaces for 24 ± 2 h. For each glass, a sample was heat-treated at 1200°C and analyzed using low magnification optical microscopy (OM). Additional samples were then prepared and heat-treated at higher temperatures if the previous sample contained crystals or lower temperatures if the previous sample didn't contain crystals. Once the temperature range around T_L was narrowed to roughly ± 25°C, samples were analyzed by OM at higher magnifications (up to 1000×) and in many cases scanning electron microscopy (SEM). This process continued until the T_L was narrowed to within ± 5°C. Roughly 15 tests were performed on each glass. The primary crystalline phase for each glass was identified by a combination of techniques including OM, SEM with energy dispersive spectroscopy (EDS), and x-ray diffractometry (XRD).

RESULTS AND DISCUSSION

The T_L of the 87 Phase I & II glasses are listed in Table I along with measured glass compositions. The T_L values range between 1153°C (AC2-52) and 1405°C (AC2-04). Only two glasses had T_L above 1350°C, AC2-04 and AC2-18. Three crystal types, aluminosilicates (AlS), rare earth silicates (RES) and aluminum oxide (Al_2O_3), were identified as the primary phases in the test glasses. In Phase I, 22 glasses were in the RES primary phase field and 10 were in the AlS field. RES was the primary phase in 47 Phase II glasses and Al_2O_3 was the primary phase in 8 of the glasses.

Figure 1 shows the $Al_2O_3 \cdot Ln_2O_3 \cdot SiO_2$ ternary submixture with test glasses plotted using different symbols representing their primary phase. Primary phase field boundaries are apparent from this figure. The boundary between AlS and RES phase fields is at roughly 48 normalized mass% Ln_2O_3. The boundary between RES and Al_2O_3 phase fields is at roughly the Al_2O_3:SiO_2 ratio of 1.0.

Table I. Measured glass composition (mass%), T_L (°C), and primary phase (p) for test glasses

Glass	p(1)	T_L	Al_2O_3	B_2O_3	Ce_2O_3	Er_2O_3	Eu_2O_3	Gd_2O_3	La_2O_3	Nd_2O_3	Pr_2O_3	SiO_2	Sm_2O_3	SrO
AC1-01	+	1272	16.56	7.35	4.59	4.75	0.67	1.66	20.14	12.57	4.60	21.85	3.61	1.65
AC1-02	S	1255	20.58	10.00	3.01	2.88	0.38	0.71	23.02	5.06	2.74	27.96	1.40	2.26
AC1-03	S	1246	21.56	10.34	2.94	2.91	0.36	0.74	21.49	5.93	2.57	27.78	1.08	2.30
AC1-04	S	1255	21.82	10.71	2.46	2.34	0.33	0.87	21.96	6.36	2.15	27.56	1.10	2.34
AC1-05	S	1164	21.34	9.93	3.62	3.15	0.41	1.00	20.38	8.42	2.31	26.16	1.09	2.19
AC1-06	S	1230	21.31	9.42	2.56	2.91	0.43	1.17	21.58	6.50	3.15	26.63	2.27	2.07
AC1-07	S	1187	20.07	9.95	3.03	2.95	0.43	1.04	21.70	8.16	2.74	26.04	1.64	2.25
AC1-08	+	1184	19.48	8.64	3.73	4.21	0.46	1.43	22.23	7.81	2.85	24.58	2.62	1.96
AC1-09	+	1212	20.08	8.96	2.98	2.76	0.47	0.90	20.67	10.29	3.78	25.39	1.76	1.96
AC1-10	+	1216	19.43	8.65	3.73	3.57	0.52	1.19	21.79	9.25	3.38	24.51	2.03	1.95
AC1-11	+	1267	16.72	8.28	3.86	4.06	0.59	1.91	19.18	14.52	4.74	21.54	2.97	1.63
AC1-12	+	1238	19.09	8.13	4.78	4.50	0.64	1.52	21.58	8.40	4.60	23.29	1.69	1.78
AC1-13	+	1241	18.46	7.83	4.05	3.97	0.61	1.53	20.08	10.45	3.81	23.42	4.05	1.74
AC1-14	+	1240	17.49	8.56	3.98	3.74	0.54	1.73	19.44	13.41	3.67	22.42	3.26	1.76
AC1-15	+	1254	17.08	7.66	5.60	4.63	0.62	1.55	21.64	12.27	3.91	21.56	1.81	1.67

Glass	p[1]	T_L	Al_2O_3	B_2O_3	Ce_2O_3	Er_2O_3	Eu_2O_3	Gd_2O_3	La_2O_3	Nd_2O_3	Pr_2O_3	SiO_2	Sm_2O_3	SrO
AC1-16	+	1195	19.42	8.71	3.90	3.62	0.54	1.37	20.44	9.86	3.49	23.81	2.93	1.91
AC1-17	S	1246	21.75	10.09	2.48	2.40	0.36	0.83	22.78	6.40	2.33	26.86	1.40	2.32
AC1-18	S	1197	20.87	9.30	3.14	2.81	0.43	0.95	21.55	7.76	2.84	26.73	1.59	2.03
AC1-19	+	1246	17.34	7.67	4.82	4.49	0.66	1.49	21.11	12.17	4.39	21.85	2.26	1.75
AC1-19b	+	1165	18.25	8.60	4.30	3.76	0.52	1.28	22.10	9.17	3.43	24.38	2.22	1.99
AC1-20	+	1247	18.96	7.96	4.21	4.16	0.61	1.36	21.08	10.76	4.02	23.19	1.84	1.85
AC1-21	+	1180	19.96	9.15	3.84	3.59	0.52	1.26	20.32	8.99	3.45	24.92	2.13	1.87
AC1-22	S	1224	21.66	10.59	2.52	2.38	0.34	0.78	21.87	6.61	2.36	27.51	1.22	2.16
AC1-23	S	1201	20.16	9.81	3.11	2.98	0.43	1.03	21.67	8.07	2.85	25.96	1.84	2.09
AC1-24	+	1163	19.98	8.93	3.56	3.41	0.49	1.12	20.20	9.39	3.37	25.51	2.17	1.87
AC1-25	+	1244	17.79	8.13	4.36	4.07	0.60	1.44	20.55	11.37	4.03	23.01	2.78	1.87
AC1-26	+	1298	18.22	7.96	4.78	4.61	0.63	1.54	19.70	11.87	4.21	22.52	2.28	1.68
AC1-27	+	1168	19.98	8.85	3.88	3.61	1.22	0.61	20.47	8.89	3.19	25.73	1.67	1.90
AC1-28	+	1208	19.65	8.48	3.80	3.64	1.31	0.62	20.17	9.83	3.58	24.34	2.76	1.82
AC1-29	+	1259	16.95	8.28	4.64	4.22	1.50	0.63	21.43	11.08	4.29	22.90	2.14	1.94
AC1-30	+	1242	17.74	8.11	4.55	4.26	1.51	0.63	21.14	11.03	4.16	23.10	1.98	1.79
AC1-31	+	1186	20.28	9.39	3.85	3.59	1.19	0.64	20.76	8.69	3.24	25.18	1.23	1.96
AC2-01	+	1308	11.93	6.08	8.60	7.53	0.47	1.15	23.13	8.39	7.58	21.80	1.95	1.39
AC2-02	A	1257	16.68	9.10	8.34	3.01	0.52	2.91	23.38	8.87	7.45	16.43	1.90	1.40
AC2-03	+	1313	11.99	7.97	8.85	3.01	1.26	1.18	23.75	18.22	3.71	16.41	1.71	1.93
AC2-04	+	1405	11.93	5.73	3.53	3.05	0.51	2.92	24.80	17.91	6.82	16.34	4.49	1.98
AC2-05	+	1241	17.07	6.18	3.53	2.99	1.28	2.92	23.13	10.32	3.04	22.72	4.74	2.07
AC2-06	+	1309	11.97	9.29	3.49	7.99	0.58	1.16	22.45	17.59	3.12	16.22	4.79	1.35
AC2-07	+	1209	17.48	6.23	3.31	7.67	1.23	1.05	22.13	10.46	3.02	23.98	1.97	1.45
AC2-08	+	1242	16.54	6.43	8.37	3.05	1.26	1.23	17.93	8.33	7.12	22.97	4.76	2.01
AC2-09	+	1323	11.98	6.18	8.62	7.52	1.26	2.83	18.98	12.73	7.64	15.92	4.95	1.38
AC2-10	+	1255	11.97	8.89	8.34	7.47	0.48	2.75	18.34	8.10	3.73	22.85	5.00	2.07
AC2-11	A	1292	16.90	6.10	8.30	3.01	0.52	1.05	18.51	18.85	3.79	16.51	5.12	1.33
AC2-12	+	1333	11.80	6.51	3.43	3.44	0.59	1.16	18.80	20.14	6.62	23.69	1.94	1.89
AC2-13	+	1181	17.13	9.12	3.41	3.08	0.53	1.15	18.82	10.37	6.71	23.42	4.80	1.46
AC2-14	A	1250	17.86	8.76	3.49	7.51	1.30	1.08	18.19	14.62	6.85	16.24	2.00	2.10
AC2-15	+	1279	12.03	9.43	3.38	3.04	1.27	2.86	18.36	19.33	2.97	23.88	1.98	1.46
AC2-16	A	1298	17.25	6.30	3.47	7.85	0.57	2.80	18.59	19.83	3.15	16.27	1.86	2.06
AC2-17	+	1273	14.50	7.85	5.80	5.27	0.93	1.98	20.57	12.87	5.18	19.97	3.38	1.72
AC2-18	+	1366	11.91	6.09	9.58	2.21	1.52	0.91	16.82	23.78	3.25	16.18	5.60	2.15
AC2-19	+	1249	12.22	8.30	9.96	8.64	0.26	0.88	23.84	6.07	3.61	22.91	1.27	2.03
AC2-20	+	1343	12.25	5.97	9.39	2.30	1.45	3.24	23.98	13.43	8.65	16.41	1.45	1.48
AC2-21	A	1242	14.69	8.89	6.20	5.51	0.32	2.02	20.71	12.45	5.46	16.08	5.60	2.04
AC2-22	+	1284	12.17	9.17	2.64	2.40	1.53	3.23	23.76	6.54	7.49	23.23	6.04	1.79
AC2-23	A	1282	17.47	6.29	2.57	2.40	0.34	0.73	24.44	17.81	7.58	16.74	1.52	2.10
AC2-24	+	1319	12.21	9.37	2.56	3.55	0.41	3.24	24.16	22.83	2.45	16.34	1.51	1.38
AC2-25	A	1305	17.36	5.85	2.59	8.83	1.47	3.27	24.53	10.06	2.39	16.40	5.32	1.93
AC2-26	+	1208	17.50	5.57	2.89	8.64	1.44	0.76	24.06	6.09	2.43	23.64	5.58	1.41
AC2-27	+	1242	16.29	6.20	9.49	2.29	0.33	3.32	16.94	6.26	8.51	23.35	5.58	1.43
AC2-28	A	1260	16.98	8.88	9.78	8.09	1.45	3.31	16.85	6.35	8.69	16.13	1.48	1.99
AC2-29	+	1300	12.15	9.64	2.64	9.06	0.41	0.85	17.30	17.26	7.96	16.47	4.93	1.34
AC2-30*	+	1227	17.51	9.20	2.49	2.31	1.51	0.69	16.94	21.04	2.36	23.03	1.56	1.36
AC2-31	+	1305	12.21	6.23	2.55	8.88	0.30	3.19	17.28	20.53	2.40	22.93	1.49	2.02
AC2-32	+	1264	14.57	7.30	6.33	5.48	0.86	2.08	21.11	12.24	5.34	19.78	3.28	1.63
AC2-33	+	1271	14.90	7.62	5.55	4.98	0.81	1.87	20.73	13.48	4.95	19.98	3.34	1.78
AC2-34	+	1232	16.20	7.59	5.33	4.65	0.74	1.78	21.58	11.98	4.69	21.25	2.39	1.82
AC2-35	+	1289	14.36	7.27	5.79	5.31	0.85	1.93	20.91	14.09	5.09	19.19	3.52	1.70

Glass	p(1)	T_L	Al_2O_3	B_2O_3	Ce_2O_3	Er_2O_3	Eu_2O_3	Gd_2O_3	La_2O_3	Nd_2O_3	Pr_2O_3	SiO_2	Sm_2O_3	SrO
AC2-36	+	1294	14.09	7.58	5.98	5.55	0.82	1.90	20.35	14.24	5.30	19.14	3.41	1.65
AC2-37	+	1305	13.49	7.28	6.09	5.73	0.91	2.08	21.02	14.29	5.46	18.52	3.53	1.60
AC2-38	+	1308	13.23	7.07	6.23	5.80	0.94	2.04	20.81	14.99	5.66	17.99	3.64	1.60
AC2-39	+	1318	13.09	6.75	6.58	5.81	0.92	2.10	20.83	15.29	5.92	17.42	3.82	1.46
AC2-40	+	1325	12.87	6.57	6.35	6.04	0.97	2.08	20.55	15.72	6.08	17.53	3.84	1.41
AC2-41	+	1324	12.67	6.86	6.53	6.24	0.95	2.17	20.66	15.66	5.95	16.97	3.83	1.51
AC2-42	+	1180	17.33	8.72	4.26	3.97	0.62	1.32	21.92	10.27	3.85	23.18	2.54	2.02
AC2-43	+	1227	15.98	8.29	4.98	4.47	0.71	1.60	21.67	11.95	4.35	21.22	2.90	1.88
AC2-44	+	1305	13.45	6.96	6.33	5.85	0.94	1.99	20.48	14.87	5.69	18.07	3.78	1.58
AC2-45	+	1259	14.93	7.55	10.62	4.94	0.63	1.51	19.51	10.89	4.91	20.12	2.74	1.65
AC2-46	+	1333	12.36	6.32	6.88	6.45	0.97	2.22	20.03	16.69	6.30	16.35	4.03	1.39
AC2-47	+	1272	14.71	7.30	5.81	5.04	0.79	1.90	21.30	13.47	5.20	19.36	3.38	1.73
AC2-48	+	1181	17.28	9.38	4.29	3.90	0.67	1.40	21.66	10.34	3.86	22.67	2.65	1.92
AC2-49	+	1220	15.87	8.35	5.39	4.31	0.74	1.68	20.36	12.61	4.50	21.04	2.85	1.84
AC2-50	+	1272	14.46	7.77	5.60	5.15	0.85	1.85	20.56	13.42	5.00	19.86	3.28	1.59
AC2-51	+	1305	13.20	6.92	6.41	5.60	0.92	2.13	20.12	15.32	5.67	17.94	3.57	1.56
AC2-52	+	1153	17.44	9.35	0.25	4.16	0.67	1.53	22.55	11.36	3.46	24.02	2.50	2.17
AC2-53*	+	1192	16.36	8.30	4.89	4.41	0.71	1.52	20.94	11.79	4.38	21.50	2.74	1.83
AC2-54	+	1326	12.54	6.42	6.92	6.12	1.02	2.30	19.44	17.01	5.88	16.57	3.80	1.44

(1)Primary crystalline phase, "+"-RES, "A"-Al_2O_3, and "S"-AlS

A strong linear trend between T_L and concentration of Ln_2O_3 in glass is shown in Figure 2. This trend is negative in the AlS primary phase field (≤43 mass% Ln_2O_3) and positive in both the RES and Al_2O_3 phase fields. The effect of Ln_2O_3 on T_L was quantified using the relationship:

$$T_L = \sum T_i g_i, \quad (1)$$

where T_i and g_i are the partial specific T_L and the mass fraction of the ith component in the glass, respectively. The partial specific T_L values (T_i) for each of the primary phase fields are listed in Table II. With sufficient data in the RES phase field, Equation 1 was also fitted with i=Ln_2O_3, Al_2O_3, B_2O_3, SiO_2, and SrO. The predictability of this model is low with an R^2 value of only 0.640. In an attempt to improve the predictability of T_L in the RES primary phase field, the average ionic radius of Ln was taken into account. Shelby and Kohli (1983) showed that properties of rare earth-aluminosilicate glasses scaled linearly with ionic radii and Mika et al. (1999) showed that the influence of cations on the T_L of waste glasses was a linear function of their ionic radii. Using the crystal radii of Shannon (1976), the mass averaged rare earth radius (r_{Ln}) was calculated and used in fit T_L data according to the relationship:

$$T_L = a + b g_{Ln} r_{Ln} \quad (2)$$

where a and b were found to be 808°C, and 723°C/Å, respectively, with an R^2 value of 0.727. This is a considerable improvement in predictability without any additional fitting parameters.

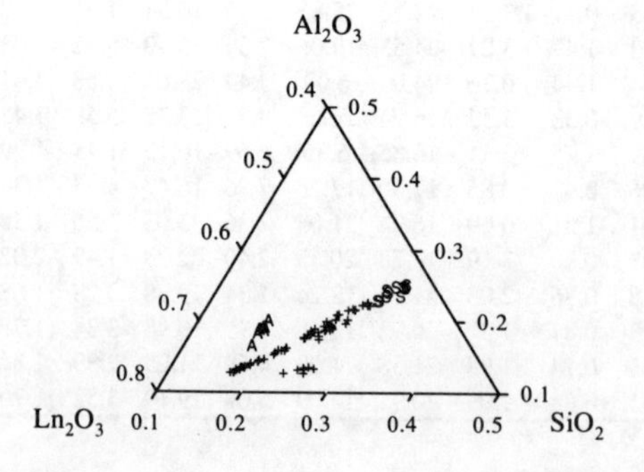

Figure 1. Al_2O_3·SiO_2·Ln_2O_3 ternary submixture with test glasses plotted with different symbols for each primary phase: "A"-Aluminum Oxide (Al_2O_3), "S"-Aluminosilicates (AlS), and "+"-Rare Earth Silicates (RES).

Table II. Partial specific T_L values (°C)

Phase	Ln_2O_3	Al_2O_3	B_2O_3	SiO_2	SrO	glass[1]	R^2
Al_2O_3	1859	-	-	-	-	501	0.868
AIS	352	-	-	-	-	1793	0.528
RES	1569	-	-	-	-	896	0.637
RES	1555	766	955	930	1837	-	0.640

[1] glass represents all other components not specifically fitted

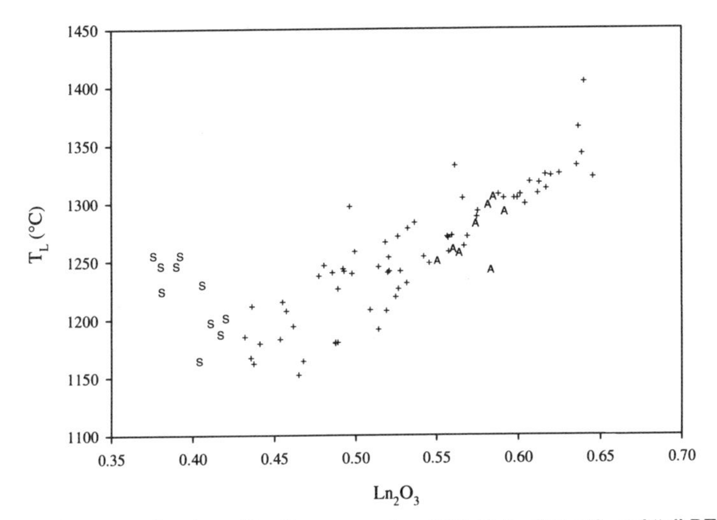

Figure 2. T_L as a function of Ln_2O_3 concentration. "A"-Al_2O_3, "S"-AlS, and "+"-RES

ACKNOWLEDGEMENTS

The authors would like to acknowledge TB Edwards for test matrix development and many helpful discussions, JV Crum for assistance in T_L measurements and XRD analyses, JS Young and JE Coleman for SEM analyses, and PR Hrma for helpful discussions. The research reported in this paper was partially funded by the U.S. Department of Energy's (DOE) office of Environmental Management through the Savannah River Sites Americium/Curium Treatment Project. Battelle operates Pacific Northwest National Laboratory for the DOE under contract DE-AC06-76RLO 1830. BJR acknowledges partial support from the Associated Western Universities.

REFERENCES

1. A. P. Fellinger, M. A. Baich, B. J. Hardy, G. T. Jannik, T. M. Jones, J. E. Marra, C. B. Miller, D. H. Miller, D. K. Peeler, T. K. Synder, M. E. Stone, and D. C. Witt, *Americium-Curium Vitrification Process Development,* WSRC-MS-98-00864, Westinghouse Savannah River Company, Aiken, South Carolina (1998).

2. D. K. Peeler, *Am-Cm: SrABS Glass Variability Study, Technical Task and QA Plan*, WSRC-RP-98-01144, Rev. 0, Westinghouse Savannah River Company, Aiken, South Carolina, (1998).

3. D. K. Peeler, and T. B. Edwards, *Phase 2 Am/Cm Glass Variability Study, Technical Task and QA Plan*. WSRC-RP-99-00272, Rev. 0, Westinghouse Savannah River Company, Aiken, South Carolina, (1999).

4. D. K. Peeler, T. B. Edwards, I. A. Reamer, J. D. Vienna, D. E. Smith, M. J. Schweiger, B. J. Riley, and J. V. Crum, *Composition/Property Relationships for the Phase 1 Am/Cm Glass Variability Study (U)*, WSRC-TR-99-00055, Westinghouse Savannah River Company, Aiken, South Carolina (1999).

5. D. K. Peeler, T. B. Edwards, I. A. Reamer. J. D. Vienna, D. E. Smith, M. J. Schweiger, B. J. Riley, and J. V. Crum, "Composition/Property Relationships for the Phase 1 Am/Cm Glass Variability Study," to be published (2000) in the *Proceedings of the International Symposium on Waste Management Technologies in Ceramic and Nuclear Industries,* Ceramic Transaction series.

6. D.K. Peeler, T.B. Edwards, T.S. Rudisill, I.A, Reamer, J.D. Vienna, D.E. Smith, M.J. Schweiger, and B.J. Riley *Composition / Property Relationships for the Phase 2 Am/Cm Glass Variability Study (U)*, WSRC-TR-99-00393, Rev. 0, Westinhouse Savannah River Company, Aiken, South Carolina (1999b).

7. J. D. Vienna, B. J. Riley, M. J. Schweiger, D. E. Smith, D. K. Peeler, and I. A. Reamer, *Property Data for Simulated Americium/Curium Glasses*, PNNL-13009, Pacific Northwest National Laboratory, Richland, Washington (1999).

8. J. E. Marra, M. A. Baich, A. P. Fellinger, B. J. Hardy, G. T. Jannik, T. M. Jones, C. B. Miller, D. H. Miller, D. K. Peeler, T. K. Synder, M. E. Stone, and D. C. Witt, "A Batch Process for Vitrification of Americium/Curium Solution (U)," to be published in the *ACS proceedings* (1999).

9. J. E. Shelby, and J. T. Kohli, "Rare-Earth Aluminosilicate Glasses," *J. Am. Ceram. Soc.*, **73** [1], 39-42, (1983).

10. M. Mika, J. V. Crum, and P. Hrma, "Spinel Precipitation In High-Level Waste Glass," in *Proceedings of the 5th ESG Conference,* Prague, Czech Republic, (1999).

11. R. D. Shannon, "Revised Effective Ionic Radii and Systematic Studies of Interatomic Distances in Halides and Chalcogenides," in *Acta Cryst.* **A32**, 751-767, (1976).

GADOLINIUM AND HAFNIUM ALUMINO-BOROSILICATE GLASSES: Gd AND Hf SOLUBILITIES

Donggao Zhao*, L.L. Davis**, Liyu Li**, C.S. Palenik*, L.M. Wang*, D.M. Strachan**, R.C. Ewing*

*Department of Nuclear Engineering and Radiological Sciences and Department of Geological Sciences, University of Michigan, Ann Arbor, Michigan 48109-2104, dzhao@umich.edu

**Pacific Northwest National Laboratory, P.O. Box 999 (K6-24), Richland, Washington 99352

ABSTRACT

The solubilities of Hf and Gd in sodium alumino-borosilicate glasses based on the target compositions were examined and confirmed by electron microprobe analysis. The measured compositions of essentially crystal-free glasses are generally homogeneous and close to the target compositions. Therefore, the solubilities of Gd and Hf in sodium alumino-borosilicate glasses based on the target glass compositions are valid. However, for glasses containing precipitates (crystals grown from the melt) and undissolved HfO_2 with overgrowths, the chemical compositions are often heterogeneous and may be significantly different from the target compositions. Precipitated crystalline phases include a rare earth silicate with the apatite structure ($NaGd_9(Si_{5.25}B)O_{26}$) in a gadolinium sodium alumino-borosilicate glass and a HfO_2 phase in hafnium sodium alumino-borosilicate glasses.

INTRODUCTION

Alumino-borosilicate glasses are proposed waste forms for the immobilization of plutonium-containing waste (this does not include the excess weapons plutonium that is to be converted to a crystalline ceramic) and miscellaneous spent nuclear fuels [1-4]. In the present solubility study, new glass compositions in the four-component system (Na_2O-B_2O_3-Al_2O_3-SiO_2) with neutron absorbers (Gd and Hf) have been synthesized to study the effect of variation of the bulk composition on the solubility of Gd and Hf [5-7]. Gadolinium and Hf solubilities in glasses are investigated because 1) Gd and Hf have high neutron capture cross-sections and are effective neutron absorbers that can be used to prevent criticality events during the storage; 2) Gd^{3+} is a surrogate element for Pu^{3+} [6]; and 3) Hf^{4+} is a surrogate element for Pu^{4+} [7]. Experiments have been performed to evaluate the effects of elements such as Na, Al and B on the solubility of Gd and Hf in the borosilicate glasses [6, 7].

The solubility limit is defined as the highest concentration of an element (e.g., Gd or Hf) in the glass above which crystallization or phase separation occurs [6]. Chemical heterogeneity and mass losses during glass synthesis may affect solubility determination. Therefore, it is necessary to directly determine distributions and contents of each individual element in the glasses and associated precipitated crystals. Electron microbeam techniques, such as electron microprobe analysis (EMPA), backscattered electron (BSE) imaging, and energy dispersive X-ray spectroscopy (EDS) were used. We present the compositional features of representative glasses and precipitated crystals as determined by EMPA.

ANALYTICAL METHODS

The glass synthesis method is described in [6] and [7]. Care was taken to minimize mass loss due to volatilization. Thirteen Gd- or Hf-bearing glass samples, with or without precipitated phase, were studied (Table I). Precipitated crystals in some samples are not homogeneously distributed. Therefore, although optically and by XRD, crystals are known to be present, the shard of glass prepared for EMPA may be crystal-free. In addition, some samples examined here

Mat. Res. Soc. Symp. Proc. Vol. 608 © 2000 Materials Research Society

Table I. Gd- and Hf-bearing sodium alumino-borosilicate glass samples.

Sample #	Description
Al15Gd18	Clear Gd glass. No observable crystals
B15Gd42	Clear Gd glass. No observable crystals
B15Gd48	Clear Gd glass with crystalline phase. Elements in crystals: Si, Al, Na, O and Gd
Na10Gd20	Clear Gd glass. No observable crystals
B15Hf30	Clear Hf glass. No observable crystals
B15Hf31	Clear Hf glass. No observable crystals
Na30Hf30	Clear Hf glass. No observable crystals
Na30Hf34	Clear Hf glass. No observable crystals
Na30Hf35a	Clear Hf glass with no crystals
Na30Hf35b	Different Hf glass from sample Na30Hf35a. Clear glass with euhedral HfO_2 crystals (up to tens of μm in size). Elements confirmed in crystals: Hf and O
PL0.35Hf8a	Clear Hf glass with bladed crystals radiating outward from undissolved HfO_2 particles. Heat-treated for one hour at 1560°C and one hour at 1450°C. Elements confirmed in crystals: Hf and O
PL0.35Hf8b	Hafnium glass with tiny crystals. Heat treated for 30 minutes at 1400°C after initially melted at 1560°C for one hour and one hour at 1450°C. Elements confirmed in crystals: Hf and O
PL0.85Hf32	Clear Hf glass with well-developed, hexagonal crystals. Heat-treated for one hour at 1560°C and three hours at 1350°C. Elements confirmed in crystals: Hf and O

were not formed under the same conditions as the samples used in the solubility study. For example, the Hf-bearing glasses, in particular, were melted for longer times in order to grow the large HfO_2 crystals. The samples were prepared by mounting the glass grains in epoxy resin discs that were polished for determination of chemical composition and crystal morphology using BSE, EDS and EMPA. The size of the glass grains ranged from a few millimeters to up to one centimeter.

Backscattered electron imaging and EDS spectra were used to examine the homogeneity of the glass matrix and precipitated crystals, and to characterize the morphology of precipitated crystals. A four-spectrometer Cameca CAMEBAX electron microprobe analyzer (wavelength dispersive system) was used to determine chemical compositions on the polished surfaces. The Cameca PAP correction routine was used in data reduction. The directly analyzed elements were Si, Al, Na, and Gd/Hf with O obtained by stoichiometry and B obtained by difference. Other elements were excluded because the glass and crystals were synthesized with known elemental contents. During each analytical session, spectrometers were verified for position, and each standard was recalibrated and analyzed prior to analyzing the glass samples. A focused beam in spot mode was not used because the glass samples may be damaged by the electron beam, due to the presence of the light elements, B and Na. Standards were SiO_2 for Si, andalusite (Al_2SiO_5) for Al, albite ($NaAlSi_3O_8$) or jadeite ($NaAlSi_2O_6$) for Na, Gd phosphate ($GdPO_4$) for Gd, and metallic Hf for Hf. Glass standards were not used due to alkali loss.

Different EMPA procedures were tested to search for a proper protocol for analyzing the glass samples. To evaluate potential loss of Na, time scans (counts vs. time plot) were conducted using a beam size of 6 x 6 μm^2, accelerating voltage of 15 kV, beam current of 6 nA, and peak and background counting times of 10 and 5 seconds, respectively. The counts vs. time plots show that the counts from NaKα decrease with time, whereas, counts of AlKα, SiKα and GdLα peaks do not change with time, suggesting that Na loss is a problem at these operating conditions. This was confirmed by analyzing the same spot three times consecutively at the same conditions. The Na_2O content decreases 0.5 wt % each time, with the first analysis closest to the target composition. Thus, beam size was increased to 15 x 15 μm^2. To eliminate or reduce the effect of count time on alkali loss at the increased beam size, Na was counted for 5 periods and then extrapolated to zero time. However, the correction is negligible. For example, for Al15Gd18, NaKα count rates were 566.64, 576.16, 546.10, 542.59, and 562.13 counts/second for counting period #1 to # 5, with an extrapolated zero time count of 568.43 counts/second. This confirms that Na concentration changed just slightly or not at all with time at these conditions. In summary, in order to minimize volatilization of sodium in glass, enlarged beam sizes, lower beam current and shorter count times were used. The optimized procedure is beam size 15 x 15 μm^2, beam current 6 nA, accelerating voltage 15 kV, peak and background counting times of 10 and 5 seconds, B by difference, and O by charge balance. All the analyses of glass matrices were obtained using this procedure. Chemical compositions of precipitated crystals were obtained under the similar conditions [8].

RESULTS

Characteristics of samples and morphology of precipitated crystals

Gd-bearing glasses. The BSE images show that Gd-bearing glass samples Al15Gd18, B15Gd42 and Na10Gd20 are homogeneous in composition and that there are no crystalline phases present in these samples. However, one Gd-bearing glass, sample B15Gd48, contained abundant precipitated crystals in the glass matrix [8]. The shape of the precipitated phase in B15Gd48 is elongate, acicular, prismatic or dendritic with hexagonal cross-sections, indicating that the phase is crystalline, not amorphous. The precipitated crystals are up to 200 μm in length [8]. In a BSE image of sample B15Gd48 (Figure 1A), the glass matrix in the upper left area is darker than those in the central and lower right areas, suggesting that the glass matrix is not homogeneous. The darker area of the BSE image is enriched in Si, Al and Na, whereas the brighter area is enriched in Gd.

Hf-bearing glasses. The BSE images show that samples B15Hf30, B15Hf31, Na30Hf30, Na30Hf34 and Na30Hf35a do not contain crystalline phases and that the glass matrices appear to be homogeneous in composition. Although the first piece of sample Na30Hf35 (labeled as Na30Hf35a) does not contain crystals, a second glass chip (Na30Hf35b) does contain crystals, suggesting that the precipitated crystals are not homogeneously distributed. The precipitated crystals are 20 μm (Figure 1B), and they surround a bubble in the glass (Figure 1C). The EDS spectra indicate that the crystalline phase in Na30Hf35b contains only Hf and O. In PL0.35Hf8a, crystals were found, but were not common. Almost all the crystals are elongated or needle-like (Figure 1D), probably representing the end-on sections of the platy squares of HfO_2 identified optically and by XRD. In PL0.35Hf8b, crystals are common. In contrast to the crystals in PL0.35Hf8a, the crystals in this grain have two different shapes: one is elongated or needle-like, the same as those in PL0.35Hf8a (not common); the other is micron-sized, undissolved HfO_2 with overgrowths (common) (Figure 1E). Sample PL0.35Hf8b was heated for additional 30 minutes at 1400°C after initially melted at 1560°C for one hour and at 1450°C for one hour for PL0.35Hf8a. The differences in size and crystal morphology are consistent with the different melting times (Table I). From EDS spectra, the precipitated crystals in PL0.35Hf8a and PL0.35Hf8b are composed of Hf and O. Precipitated crystals are also common in PL0.85Hf32, but they are bladed and relatively large as compared with the crystals in PL0.35Hf8a and PL0.35Hf8b (Figure 1F). The precipitated crystals in PL0.85Hf32 also consist of Hf and O.

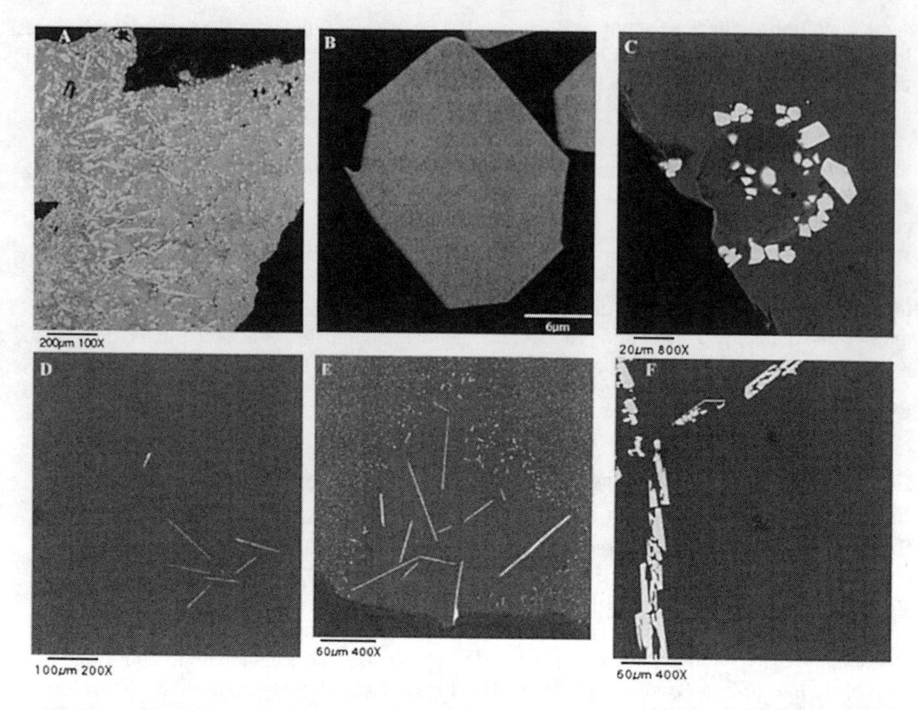

Figure 1. Backscattered electron images of the glass samples containing the precipitated crystalline phases. A) B15Gd48 with elongated, acicular, prismatic or dendritic Gd silicate apatite crystals; cross sections are often hexagonal; glass matrix in the upper left area slightly darker than those in the central and lower right areas, indicating chemical heterogeneity of the matrix. B) Na30Hf35b with euhedral HfO_2 crystals. C) Na30Hf35b with HfO_2 crystals surrounding a bubble in the glass. D) PL0.35Hf8a showing the edges of platy HfO_2 crystals (not common). E) PL0.35Hf8b showing smaller micron-sized, undissolved HfO_2 with overgrowths and edges of larger platy HfO_2 crystals. F) PL0.85Hf32 with bladed HfO_2 crystals.

Chemical compositions of glass matrix

More than one analysis was obtained for each sample. To examine the chemical homogeneity of samples, a few EMPA profiles were completed. Average compositions of sodium alumino-borosilicate glasses are given in Table II.

Homogeneity of glass matrix. The crystal-free Gd-bearing glasses (Al15Gd18, B15Gd42 and Na10Gd20) are homogeneous in chemical composition. In sample B15Gd48, which contains precipitated Gd-rich silicate crystals, two compositional domains were identified in the glass matrix. The glass matrix from the darker upper left area of the BSE image (Figure 1A) is enriched in Si, Al and Na and depleted in Gd, as compared with the glass matrix from the brighter area (Table II; Figure 2). The heterogeneity of the glass matrix in sample B15Gd48 may be caused by crystallization of Gd crystals that take Gd from the glass matrix. Most Hf-bearing glasses are homogeneous in composition. However, the distributions of Hf across the glass grains showed variations. For example, HfO_2 varies significantly in B15Hf31, while Al_2O_3, Na_2O and SiO_2 remain generally constant across the glass grain (Figure 3). Chemical

Table II. Electron microprobe analyses (wt %) of Gd and Hf alumino-borosilicate glasses.

Sample	*Al15Gd18*		*B15Gd42*		*B15Gd48*			*Na10Gd20*		*B15Hf30*	
Point	Ave (7)	Target	Ave (10)	Target	Ave (5)	Ave (4)	Target	Ave (8)	Target	Ave (33)	Target
SiO_2	38.10	39.84	30.94	32.68	28.80	34.04	29.30	47.72	49.90	37.03	39.44
Al_2O_3	16.31	16.91	4.43	4.62	4.30	5.89	4.14	6.86	7.06	5.54	5.58
Na_2O	12.72	13.70	10.90	11.23	10.75	14.02	10.07	7.49	8.58	12.15	13.56
Gd_2O_3	18.90	18.00	42.57	42.00	45.39	31.13	48.00	20.83	20.00	32.02*	30.00*
B_2O_3	13.97	11.55	11.16	9.47	10.75	14.91	8.49	17.10	14.46	13.26	11.42

* Content of HfO_2.

Sample	*B15Hf31*		*Na30Hf30*		*Na30Hf34*		*Na30Hf35*			*PL0.35Hf8*		*PL0.85Hf32*
Point	Ave (35)	Target	Ave (37)	Target	Ave (78)	Target	a Ave (22)	b Ave (41)	Target	a Ave (24)	b Ave (4)	Ave (11)
SiO_2	37.16	38.88	33.67	35.96	30.12	33.90	31.06	30.93	33.39	44.53	47.38	38.75
Al_2O_3	5.53	5.50	5.10	5.08	4.77	4.79	4.73	4.84	4.72	21.53	22.18	6.38
Na_2O	11.66	13.37	17.16	18.54	16.36	17.48	15.74	15.74	17.22	7.21	7.53	11.93
HfO_2	33.05	31.00	32.58	30.00	36.33	34.00	37.79	36.92	35.00	8.40	4.98	29.44
B_2O_3	12.61	11.25	11.49	10.42	12.42	9.83	10.67	11.57	9.67	18.33	17.93	13.50

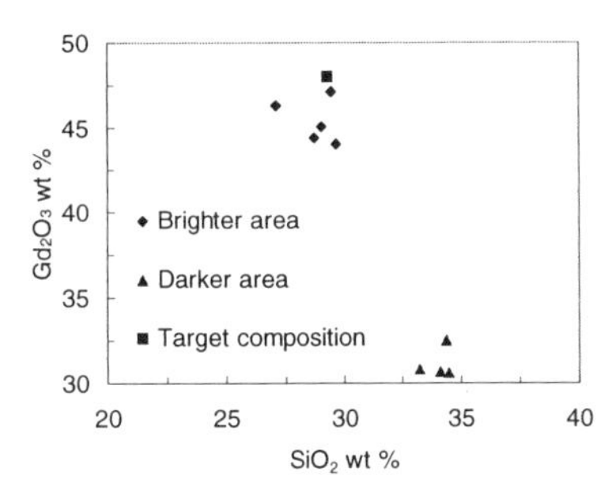

Figure 2. Distributions of SiO_2 and Gd_2O_3 in the glass matrix for B15Gd48 that contains precipitated crystals. Low Gd_2O_3 (31.13 wt %) and high SiO_2 (34.04 wt %) spots are in the darker area of the BSE image (the upper left corner of Figure 1A). High Gd_2O_3 (45.39 wt %) and low SiO_2 (28.80 wt %) spots are in the brighter area (the lower right area of Figure 1A)

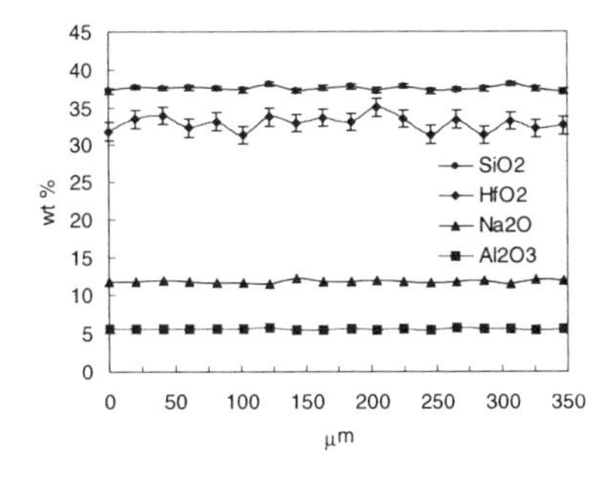

Figure 3. Oxide variations in Hf glass B15Hf31. HfO_2 varies significantly while other oxides remain constant. The profile steps are 20 μm.

compositions of two grains of Na30Hf35 (above the solubility limit) are the same and are close to the target compositions, although one grain contains the crystalline phase HfO_2 (Table II). This suggests that crystallization of small amount of crystals in glass matrix does not change the composition of glass matrix significantly.

Differences between target and measured compositions. For the crystal-free Gd-bearing glasses, such as Al15Gd18, B15Gd42 and Na10Gd20, the measured contents of SiO_2 Al_2O_3 and Na_2O are the same as or slightly lower than those for the target composition; whereas, Gd_2O_3 is slightly higher. The 1σ uncertainties of measurement are from 0.80 to 1.30 wt % for Gd_2O_3, 0.35 wt % for SiO_2, from 0.10 to 0.20 wt % for Al_2O_3, and from 0.20 to 0.45 wt % for Na_2O. Most differences between target and measured compositions are less than 1 wt %, indicating that the measured and the target compositions are essentially the same (Table II). However, for B15Gd48, which contains precipitated Gd silicate crystals, the differences between the target and the measured compositions are significant. The glass matrix has 31.13 wt % Gd_2O_3 in the darker area and 45.39 wt % Gd_2O_3 in the relatively bright areas (Figure 1A), both different from the 48.00 wt % Gd_2O_3 of the target composition. The composition of the brighter areas is closer to the target composition (Table II). For Hf-bearing glasses without or with a small amount of precipitated HfO_2 crystals, the measured Al_2O_3, SiO_2 and Na_2O are the same as, lower, or slightly lower than those for the target composition. The measured HfO_2 is approximately 2 wt % higher than HfO_2 in the target composition. In general, the measured compositions of Hf-bearing glasses are close to the target compositions, if the glass matrix contains no or only a small amount of precipitated HfO_2 crystals. However, if there are abundant precipitated crystals, the compositions of the glass matrix may vary significantly. For example, in two grains of PL0.35Hf8, the glass matrix with fewer HfO_2 crystals (PL0.35Hf8a) contains 8.40 wt % HfO_2; whereas, the glass matrix with more HfO_2 crystals (PL0.35Hf8b) contains 4.98 wt % HfO_2 (Table II). Measured compositions demonstrate that composition of the glass matrix is the same as or close to the target composition if there are no precipitated crystals or only a small amount of precipitated crystals.

Chemical compositions of precipitated crystals

Precipitated crystals identified in gadolinium alumino-borosilicate glasses (B15Gd48) were described in [8]. The phase is hexagonal, probably a rare earth silicate with the apatite structure $A_{4-x}REE_{6+x}(SiO_4)_{6-y}(PO_4)_y(F,OH,O)_2$ (where A = Li, Na, Mg, Ca, Sr, Ba, Pb and Cd, and REE = La, Ce, Pr, Nd, Pm, Sm, Eu and Gd). The empirical formula is $NaGd_9(Si_{5.25}B)O_{26}$. The presence of B in the precipitated crystals was detected from wavelength dispersive spectrum (WDS) and its contents were obtained by the differences between 100 and the measured totals [8]. For Hf-bearing glasses, the precipitated crystals are HfO_2, as identified by EDS spectra, XRD, and electron microprobe analysis. Due to the small size of the precipitated crystals, detected Si, Al and Na are most likely from the glass matrix. However, the possible existence of these elements and compositional zonings in the HfO_2 crystals are being examined.

CONCLUSIONS

1. Direct measurements of the glass compositions by EMPA demonstrate that glass matrices are homogeneous in chemical composition if there are no or few precipitated crystals. Our EMPA data also show that the measured glass compositions are in general the same as, or close to, the target compositions if there are no or few precipitated crystals.
2. Glass samples with crystals that grew from the melt or abundant undissolved HfO_2 powder are often heterogeneous in chemical composition; and their actual chemical compositions may be significantly different from the target compositions.
3. The precipitated crystalline phase identified in a gadolinium alumino-borosilicate glass is a rare earth silicate apatite; and the precipitated crystals identified in hafnium alumino-borosilicate glasses are HfO_2.
4. Therefore, electron microprobe data confirm that the solubilities of Gd and Hf in sodium alumino-borosilicate glasses based on the target glass compositions are valid if there are no or few precipitated crystals in glass matrices.

ACKNOWLEDGMENT

This study is supported by the Environmental Management Science Program, US Department of Energy through grant # DE-FG07-97-ER45672. The electron microprobe was acquired under Grant # EAR-82-12764 from the National Science Foundation. Liyu Li and Linda L. Davis are grateful to Associated Western Universities for facilitating their postdoctoral appointments at Pacific Northwest National Laboratory.

REFERENCES

1. H. Matzke, J. van Geel, In Disposal of Weapons Plutonium (eds. E.R. Merz and C.E. Walter), Kluwer Academic Publishers, 1996, 93-105.
2. R.C. Ewing, W.J. Weber, W. Lutze, In Disposal of Weapons Plutonium (eds. E.R. Merz and C.E. Walter), Kluwer Academic Publishers, 1996, 65-83.
3. Technical Summary Report for Surplus Weapons-Usable Plutonium Disposition, Report DOE/MD-0003, Office of Fissile Materials Disposition, U.S. Department of Energy, July 17, 1996.
4. L.W. Gray and T.H. Gould, Immobilization Technology Down-Selection Radiation Barrier Approach, Report UCRL-ID-127320, Lawrence Livermore National Laboratory, Livermore, CA, 1997.
5. X. Feng, H. Li, L.L. Davis, L. Li, J.G. Darab, M.J. Schweiger, J.D. Viena, and B.C. Bunker, Distribution and Solubility of Radionuclides in Waste Forms for Distribution of Pu and Spent Nuclear Fuels, Ceramic Transactions, 93, 409-419 (1999).
6. L. Li, D.M. Strachan, H. Li, L.L. Davis, M. Qian, Peraluminous and peralkaline effects on Gd_2O_3 and La_2O_3 solubilities in sodium-alumino-borosilicate glasses, Ceramic Transaction (in press).
7. L. Davis, L. Li, J.G. Darab, H. Li, D. Strachan, P.G. Allen, J.J. Bucher, I.M. Craig, N.M. Edelstein, D.K. Shuh, Scientific Basis for Nuclear Waste Management, XXII, Materials Research Society, Pittsburgh (in press).
8. Donggao Zhao, L.M. Wang, R.C. Ewing, Liyu Li, L.L. Davis, D.M. Strachan, Gd-Apatite Precipitates in a Sodium Gadolinium Alumino-Borosilicate Glass, Journal of Crystal Growth (submitted).

JOINT VITRIFICATION OF VARIOUS MIXED WASTES

O.I. KIRYANOVA, T.N. LASHTCHENOVA, F.A. LIFANOV, S.V. STEFANOVSKY, O.V. TOLSTOVA
SIA Radon, 7th Rostovskii per. 2/14 Moscow 119121 RUSSIA, itbstef@cityline.ru

ABSTRACT

Mixed wastes involve radioactive constituents and hazardous components and must be conditioned to be disposed. Joint vitrification of low and intermediate level radioactive wastes (LILW) from nuclear power plants, together with spent catalysts, sorbents, and cathode ray tube (CRT) glass is proposed as a suitable means to condition these wastes. Preliminary experiments on vitrification in crucibles in a laboratory-scale resistive furnace were carried out that demonstrate possibility of 40 wt.% LILW salt loading in a batch. Homogeneous silicate- and borosilicate-based glasses have been obtained and characterized. A phase separation problem at high sulfate and chloride content caused a "yellow phase" to be formed. To prevent this, sulfates and chlorides must be separated from the LILW. The most promising method of mixed waste vitrification is inductive melting in a cold crucible (IMCC) because of its high throughput rate, high achievable temperatures, and long lifetime of the apparatus. Bench-scale tests have been carried out. Preliminary testing of IMCC of spent CRT glass and a combined batch of this glass and LILW salt surrogate was also conducted. LILW salt loading in the batch reached 30 wt.%. The starting melt was formed by heating electrically conductive silicon carbide rods in a high frequency electromagnetic field. The CRT glass melt has low electric conductivity and high viscosity at temperatures of about 1200 °C, which does not allow the required IMCC starting melt to be formed. Addition of up to 30 wt.% of surrogate RW salts to the batch increases the electric conductivity and reduces the viscosity of the melt to values that support IMCC. The batch is preferably fed as a calcine.

INTRODUCTION

To protect the environment, radioactive and other hazardous wastes must be treated and disposed. From an economic point of view, joint treatment of various wastes is preferable to treating each waste alone. When radioactive and highly toxic wastes are processed, the principal requirements are a maximum waste volume reduction factor, process safety and reliability, and final waste form durability. The most efficient method of radioactive and hazardous waste conditioning is vitrification. At the present time, vitrification is considered as the most promising technology for conditioning of high-level radioactive wastes and LILW, as well as mixed and hazardous wastes containing various toxic agents. In Russia such wastes are collected and disposed at special centralized sites [1].

LILW contains major alkali metal salts, predominantly as nitrates, minor boron, aluminum, iron, calcium, and magnesium compounds, and sulfates and chlorides. Inorganic hazardous wastes include silicon, aluminum, lead, and transition metal oxides and phosphates, which can be glass formers or intermediates. Spent cathode ray tube (CRT) glass is a hazardous inorganic waste and needs to be transformed into a safe form for disposal. Smelting of LILW salts and spent CRT breakage yields glassy materials with good chemical durability suitable for long-term storage. Preliminary experiments [2, 3] have demonstrated the possibility of joint vitrification of LILW and CRT glass in both a resistive furnace and a cold crucible to produce chemically durable glass. This paper describes new SIA Radon achievements in this area, including research on the appli-

Mat. Res. Soc. Symp. Proc. Vol. 608 © 2000 Materials Research Society

cation of CRT glass as a glass forming agent in the vitrification of Nuclear Power Plant (NPP) wastes.

EXPERIMENTAL

Typical institutional and NPP LILW compositions are shown in Table I. In all the wastes, sodium and nitrate ions are predominant. NPP wastes from VVER type reactors (Russian analog of PWR) also contain potassium and boron. Occasionally, rather high amounts of iron, sulfate and chloride ions also may be present. The major radioactive constituent is ^{137}Cs. Volume activities of ^{134}Cs and corrosion products are lower by 1-2 orders of magnitude. Minor actinides are also present.

Table I. LILW compositions

Parameters		Institutional waste	NPP wastes from RBMK			NPP wastes from VVER		
	Ions		Leningrad	Kursk	Chernobyl	Kalinin	Kola	Novo-Voronezh
Chemical composition, $kg \cdot dm^{-3}$	Na^+	50-150	71.0	82.0	15.0	104-130	70	20-30
	K^+	1-10	3.1	2.4	1.5	7-30	48	5-15
	NH_4^+	n.m.*	2.1	-	n.m.	0.1	0.1	n.m
	Ca^{2+}	5-20	<0.1	0.9	n.m.	0.6	-	-
	Mg^{2+}	1-10	-	0.4	-	-	-	-
	Al^{3+}	0.5-2	0.3	-	-	-	-	-
	Fe^{3+}	3-15	<0.1	0.9	5-10	-	4.2	<0.1
	NO_3^-	100-350	150-180	195.0	35.0	100-200	155	80-100
	Cl^-	5-20	28-32	11.8	-	2-9	5.5	<0.1-5
	SO_4^{2-}	3-35	3-6	9.2	n.m.	0.5-10	26	n.m.
	PO_4^{3-}	1-5	0.1	0.7	n.m.	-	-	-
	BO_3^{3-}	<0.1-2	0.5	-	-	70.8	74	40-70
	MnO_2	<0.1-0.5	-	-	3.5	-	-	-
Total, $kg \cdot dm^{-3}$		250-600	300-350	332	60-65	450-500	520	150-230
Radio-nuclide composition, $Bq \cdot m^{-3}$	^{137}Cs	10^7-10^{10}	$5 \cdot 10^6$	$4.8 \cdot 10^9$	$3.7 \cdot 10^8$	$1.9 \cdot 10^{10}$	$7.5 \cdot 10^{10}$	$1.2 \cdot 10^{10}$
	^{134}Cs	10^6-10^9	-	$6.8 \cdot 10^7$	$1.0 \cdot 10^7$	$6.5 \cdot 10^9$	$8.0 \cdot 10^9$	$9.0 \cdot 10^8$
	^{60}Co	10^5-10^7	$2.8 \cdot 10^7$	$2.7 \cdot 10^7$	n.m.	$2.7 \cdot 10^8$	$1.5 \cdot 10^9$	$2.6 \cdot 10^7$
	^{54}Mn	10^5-10^6	n.m.	n.m.	n.m.	n.m.	$1.5 \cdot 10^8$	n.m.
	^{239}Pu	10^3-10^5	n.m.	$5.0 \cdot 10^6$	n.m.	$1.5 \cdot 10^6$	n.m.	n.m.
pH		9-12	12.9	12.0	~7	9-11	7-8	10-11.5

* n.m. – not measured. RBMK – Russian fast neutron channel-type reactor.

The composition of NPP waste surrogates used in the experiments are shown in Table II. We have chosen two compositions, typical for wastes from RBMK and VVER type reactors.

Table II. Compositions NPP waste surrogates (in wt.%)

NPP type	$NaNO_3$	KNO_3	NH_4NO_3	NaCl	Na_2SO_4	H_3BO_3
RBMK	70.7	3.1	4.0	19.3	7.4	-
VVER	62.3	16.3	-	1.0	0.6	19.8

The CRT used consists of screens and tubes (Table III) in a weight fraction ratio of approximately 2:1. CRT glass may be also used as an additive for vitrification of spent inorganic ion-exchanger, incinerator ash, and other solid wastes.

On a laboratory scale, glassy samples were produced from mixtures of ball-milled CRT glass and waste surrogates in alumina crucibles placed in a resistive furnace at 1200-1400 °C. followed by pouring the melt onto a metal plate. Bench-scale tests were carried out using the Radon cold crucible based unit (Figure 1) at the same temperatures.

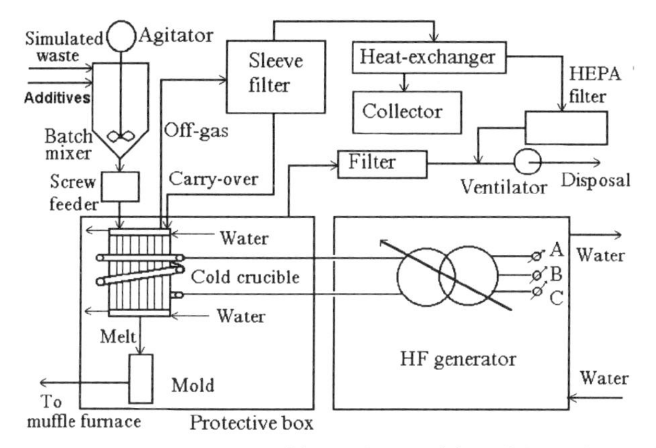

The samples obtained were examined with X-ray diffraction (XRD, DRON-4 diffractometer, Cu K_{α} radiation), and replica electron microscopy (EMV-100L). Leach rates of Na and Pb from products were measured using the IAEA test [4].

Figure 1. Block diagram of the Radon bench-scale cold crucible system.

Table III. Composition of CRT glasses (in wt.%)

Glass	Li_2O	Na_2O	K_2O	MgO	CaO	BaO	PbO	Al_2O_3	Sb_2O_3	SiO_2	F	Total
Screen	0.65	7.0	7.0	-	-	12.0	-	5.0	-	67.5	0.65	99.8
Tube	0.5	5.5	5.5	2.6	9.8	2.3	13.5	3.4	0.2	56.4	0.5	100.2
Mix	0.6	4.5	4.5	1.0	4.0	8.1	5.4	4.4	0.1	66.8	0.6	100.0

RESULTS

Lab-scale tests of joint vitrification of surrogate NPP wastes and CRT glass

All the melts produced from surrogate VVER waste and CRT glass were fully homogeneous after melting at 1200 °C. No crystalline phases have been found by either visual inspection or XRD. Leach rates of sodium and lead ions are also quite low (Table IV). For all the melting tests, actual weight losses exceeded calculated weight losses by 1.2 – 1.4 wt.% due to volatilization of some components. The most volatile elements in the melts are known to be boron, sodium, cesium, ruthenium, and lead [5,6]. In the given case the major contributors to the loss are sodium, boron, and lead.

The melts produced from surrogate RBMK waste and CRT at 1200 °C were phase-separated. One of the phases was glass melt, the second one was a sulfate–chloride-containing phase ("yellow phase"). From the XRD data, this is presumed to be composed primarily of sodium sulfate and sodium chloride, with a minor sodium metasilicate component. Glass melting at higher temperatures (>1250 °C) did not produce this phase separation. It is hypothesized that disappearance of the "yellow phase," was probably due to vaporization of its constituents. This is confirmed by the larger difference between actual and calculated weight losses, which can reach almost 10 wt.%, at the higher temperature (Table IV). The major components of the "yellow phase" would obviously lead to "yellow phase" vaporization, but at high waste oxide content losses of the other volatile species also occurred. Leach rates of sodium ions from RBMK waste glasses are higher than from VVER waste glasses.

Notwithstanding the sulfate and chloride volatility, high alkali RBMK waste glasses dissolve a

Glass	Composition, wt.%		Weight loss, wt.%			Viscosity, Pa·s (T °C)	Glass density, g/cm³	Leach rate, g/(cm²·day)	
	Waste oxide	CRT glass	Calculated	Actual	Difference (volatility)			Na^+	Pb^{2+}
V-1	10	90	11.6	12.8	1.2	7.2 (1200)	2.74	$2 \cdot 10^{-6}$	10^{-8}
V-2	15	85	16.5	17.7	1.2	5.8 (1200)	2.73	$3 \cdot 10^{-6}$	10^{-8}
V-3	20	80	20.8	22.1	1.3	4.6 (1200)	2.67	$6 \cdot 10^{-6}$	10^{-7}
V-4	25	75	24.7	26.1	1.4	3.8 (1200)	2.59	$9 \cdot 10^{-6}$	10^{-7}
V-5	30	70	28.3	29.5	1.2	3.3 (1200)	2.47	$2.6 \cdot 10^{-5}$	$1 \cdot 10^{-6}$
R-1	10	90	8.6	13.7	5.1	>10 (1250)	2.75	$4 \cdot 10^{-6}$	10^{-7}
R-2	15	85	12.4	20.1	7.7	8.7 (1250)	2.75	$6 \cdot 10^{-6}$	10^{-7}
R-3	20	80	15.9	23.1	7.2	5.7 (1250)	2.75	$1.3 \cdot 10^{-5}$	$2 \cdot 10^{-6}$
R-4	25	75	19.1	23.1	4.0	4.2 (1250)	2.76	$2.8 \cdot 10^{-5}$	$2 \cdot 10^{-6}$
R-5	30	70	22.1	31.9	9.8	3.7 (1250)	2.77	$5.2 \cdot 10^{-5}$	$3 \cdot 10^{-6}$

Table IV. Results on preparation of glasses from NPP waste and CRT glass

more sulfate ions than VVER waste glasses and elevated sodium leaching occurs from the sodium-sulfate–rich constituent, whereas Pb^{2+} ions incorporated in the glass network are leached significantly more slowly.

Most of the glass melts produced from NPP waste surrogates and CRT glass have a viscosity suitable for electric melting both in a Joule heated ceramic melter and a cold crucible.

Bench-scale tests on joint vitrification of surrogate NPP wastes and CRT glass

Tests on joint vitrification of VVER surrogate waste and CRT glass were performed using the bench-scale cold crucible system. The cold crucible (inside diameter 135 mm, surrounded by a copper inductor with an internal diameter of 195 mm) was placed in a process box. The unit was energized from a high frequency generator with 60 kW and an operating frequency of 1.76 MHz.

In the first test, the crucible was filled with CRT glass breakage and silicon carbide rods inserted to heat the glass and initiate melting. To reduce the melting temperature and electric resistivity, VVER waste surrogate was admixed in portions until homogeneous melt formation was achieved; this was followed by removal of the rods and continuous batch feeding. The calculated glass composition was as follows: waste oxide – 25 wt.%, CRT glass – 75 wt.%.

Melt separation, a major problem, occurred because of the difference between the specific gravity of the surrogate waste salts and that of the CRT glass breakage used. To eliminate this effect, several methods of batch preparation such as moisturizing, compaction of the mixture of salts and finely ground CRT breakage, and calcination were tested. The best results were achieved using a calcined batch. No melt separation was observed. The average specific melt throughput rate reached 6 kg/(dm²·h).

Lab-scale tests on joint vitrification of incinerator ash and CRT glass

Properties of the glassy slags produced by smelting of CRT glass and actinide-containing incinerator ash generated by the Radon flame incinerator [7] at 1350 °C are given in Table V.

The glassy slags contained minor crystalline phases. In the slags with high ash content (~90 wt.%) the composition of the major crystalline phase was close to that of silicocarnotite $Ca_5(PO_4)_2SiO_4$. Major diffraction peaks were (in Å): 4.00 (20%), 3.82 (40%), 3.55 (20 %), 3.29 (60 %), 3.16 (60 %), 3.10 (50 %), 2.98 (50 %), 2.91 (70 %), 2.80 (100 %), 2.61 (80 %), 2.56 (70

Table V. Properties of glassy slags from CRT glass and incinerator ash

Slag	Composition, wt.%		Loss, %		Density, g/cm^3	Leach rate, $g/(cm^2 \cdot day)$ on 28th day			
	CRT glass	Ash	^{137}Cs	α-emitters		^{137}Cs	^{90}Sr	α-emitters	Pb^{2+}
S-1	10	90	6.2	0.2	2.77	$2.7 \cdot 10^{-5}$	10^{-7}	10^{-8}	10^{-7}
S-2	20	80	4.3	0.1	2.74	$1.2 \cdot 10^{-5}$	10^{-7}	10^{-8}	10^{-7}
S-3	30	70	2.5	0.1	2.72	$7 \cdot 10^{-6}$	10^{-7}	10^{-8}	10^{-7}
S-4	40	60	2.0	0.1	2.69	$4 \cdot 10^{-6}$	10^{-8}	10^{-8}	$1 \cdot 10^{-6}$
S-5	50	50	1.6	<0.1	2.67	$2 \cdot 10^{-6}$	10^{-8}	$<10^{-8}$	$1 \cdot 10^{-6}$

%), among others (Figure 2). A second crystalline phase present was an aluminosilicate with a set of diffraction peaks close to those of kalsilite $KAlSiO_4$. There are deviations of the peak positions for both phases from the JCPDS database for these compounds due to the presence of more complex compositions. The composition of the silicophosphate phase in Samples S-3 and S-4, with lower ash content, is closer to nagelshmidtite $Ca_7(PO_4)_2(SiO_4)_2$ rather than silicocarnotite (Figure 2). These silicophosphate phases are hosts for uranium, plutonium and other actinides. Major isomorphic substitution schemes are $2Ca^{2+} = An^{4+}$ + vacancy and $Ca^{2+} + P^{5+} = (An,Ln)^{3+} + Si^{4+}$. Other phases that occasionally occurred in the slags were spinels and aluminosilicates, which can be host phases for iron group elements and cesium and strontium, respectively. The slags have low leachability (Table V), demonstrating their capability to be used as waste forms for disposal of LILW.

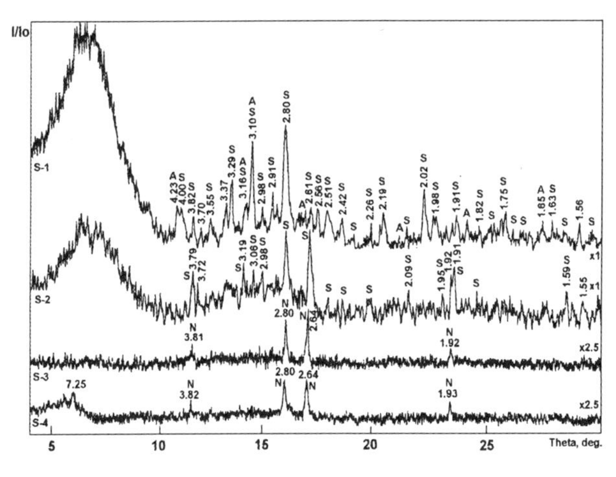

Figure 2. XRD patterns of the glassy slags.
A – aluminosilicate, N – nagelschmidtite, S – silicocarnotite.

Lab-scale tests on joint vitrification of VVER waste, incinerator ash and CRT glass

Joint vitrification of liquid and solid LILW would be expedient for NPPs having both vitrification and incineration facilities. Preliminary tests using CRT glass as a glass forming agent added to attain better homogeneity of the melt and final waste form, were also carried out. As seen from Table VI, durable glasses containing up to 30 wt% VVER waste oxides and 60 wt.% ash oxides may be obtained.

Table VI. Compositions (in wt.%) and properties of glassy products from VVER waste surrogate, incinerator ash, and CRT glass.

Nos	VVER waste oxides	Ash oxides	CRT glass	Density, g/cm^3	Leach rates, $g/(cm^2 \cdot day)$		
					^{137}Cs	^{90}Sr	α-emitters
G-1	30	60	10	2.81	$5.3 \cdot 10^{-6}$	$5 \cdot 10^{-7}$	10^{-8}
G-2	25	65	10	2.79	$3.0 \cdot 10^{-6}$	$3 \cdot 10^{-7}$	10^{-8}
G-3	20	70	10	2.76	$1.3 \cdot 10^{-6}$	$1 \cdot 10^{-7}$	10^{-8}

XRD has shown that the products from combined VVER waste surrogate, incinerator ash, and CRT glass are predominantly vitreous, but traces of crystalline phases with a silico-phosphate composition were also found. XRD patterns of samples G-1, G-2, and G-3 are similar to the XRD patterns of the glassy slags S-3 and S-4, where only strongest peaks due to a nagelschmidtite-type phase appeared. Electron micro-photographs demonstrate the phase-separated structure of the glassy products (Figure 3).

One of the phases is more resistant to acid attack than the other. At high VVER waste content, the "framework" is more durable than the "drops", whereas at lower VVER oxides content and higher ash oxides content in the glass, the "framework" becomes more leachable. The more highly leachable phase is enriched with alkali elements and probably concentrates cesium and, to a lesser extent, strontium.

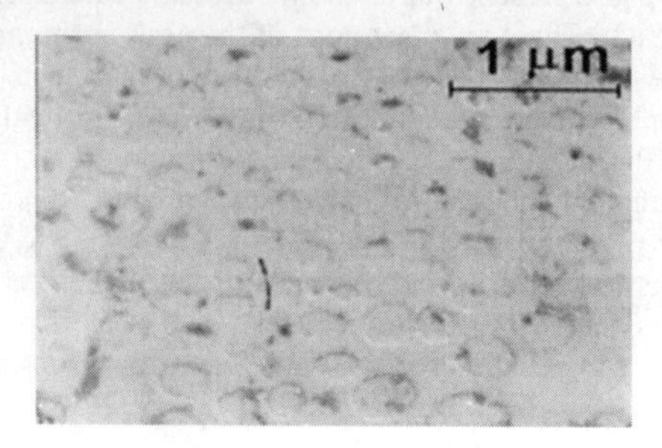

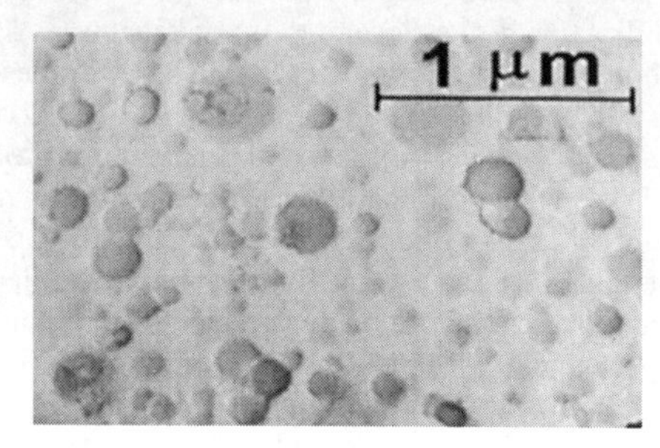

Figure 3. Replica electron microphotographs of the glasses G-1 (left) and G-3 (right). Breaches were etched with 0.1 M HCl.

CONCLUSION

Mixtures of radioactive and hazardous wastes, such as NPP (VVER and RBMK types) wastes, incinerator ash, and CRT glass in widely varying proportions may be converted to durable glass. The possibility of mixed waste vitrification using cold crucible melting has been demonstrated.

This work was performed under financial support of International Science and Technology Center (Project #1131).

REFERENCES

1. I.A. Sobolev, Management of Institutional Radioactive Wastes. Experience of Scientific and Industrial Association "Radon" (SIA "Radon", Moscow, 1995).
2. I.A. Sobolev, G.V. Makartchenko, S.V. Stefanovsky, F.A. Lifanov, Glass and Ceramics (Russ.) [3], 8 (1991).
3. S.A. Dmitriev, F.A. Lifanov, S.V. Stefanovsky, A.P. Kobelev, V.I. Kornev, A.E. Savkin, O.A. Knyazev, T.N. Lashtchenova, M.I. Ojovan, N.N. Buravtchenko, S. Merlin, P. Roux. In *Waste Management '96. Proceedings* (Tucson, AZ, 1996) Rep. 27-3. CD ROM.
4. ISO 6961-82, Vienna, 1984.
5. *Glass Technology* (Russ.), edited by I.I. Kitaigorodsky (Moscow, 1967).
6. A.S. Nikiforov, V.V. Kulitchenko, M.I. Zhikharev, *Treatment of Liquid Radioactive Wastes* (Russ., Energoatomizdat, Moscow, 1984).
7. S.A. Dmitriev, I.A. Knyazev, F.A. Lifanov, A.E. Savkin, S.V. Stefanovsky, I.D. Tolstov, in *1996 International Conference on Incineration and Thermal Treatment Technologies*. Proceedings (Savannah, GA, 1996), pp. 247-251.

Tc-99 AND Cs-137 VOLATILITY FROM THE DWPF PRODUCTION MELTER DURING VITRIFICATION OF THE FIRST MACROBATCH OF HLW SLUDGE AT THE SAVANNAH RIVER SITE

N. E. BIBLER, T. L. FELLINGER, S. L. MARRA, R. J. O'DRISCOLL, J. W. RAY, AND W. T. BOYCE

Savannah River Technology Center and the Defense Waste Processing Facility, Westinghouse Savannah River Company, Aiken, S.C. 29808

ABSTRACT

Technetium-99 and cesium-137 are two radionuclides in high level waste (HLW) that can volatilize from high temperature melters during the immobilization of the HLW into a borosilicate glass. At Savannah River Site (SRS) we have obtained data that indicate that this volatilization is small from the full scale production melter in the Defense Waste Processing Facility (DWPF). These data were obtained during the vitrification of the first HLW macrobatch at SRS. This campaign lasted ~2.5 years and produced ~9 hundred metric tons of glass from ~1.6 million liters of HLW. Losses of Tc-99 and Cs-137 were determined by comparing their measured concentrations in the glass with their respective predicted concentrations based on the composition of the HLW being vitrified. For three glass samples taken during the campaign, the measured and predicted concentrations agreed within 7% or better indicating a small loss of either radionuclide. The DWPF melter operates with a cold cap on the surface of the melt. This cold cap could enhance the radionuclide retention, especially Tc-99.

INTRODUCTION

The DWPF at the SRS has completed vitrifying the first macrobatch of radioactive HLW sludge slurry into a stable borosilicate glass. This HLW is composed of U-235 fission products and radioactive actinide isotopes produced in the reactors at SRS during production of special nuclear material for the U.S. Government. However the major constituents of the HLW are nonradioactive elements used in the processes at SRS to purify the special nuclear material. The DWPF process converts the caustic HLW slurries into a borosilicate glass by mixing the pretreated slurry with a glass forming frit and then feeding this mixture to a joule-heated melter at 1150°C. The molten glass is then poured into stainless steel canisters that are 3m tall and 0.6m in diameter. The canister is then sealed by welding a plug in the top of the canister. It is then stored prior to shipment to a Federal Repository. During the vitrification process there are two U-235 fission products, Tc-99 and Cs-137, which have the possibility of volatilizing from the melter. For example, Tc-99 forms an oxide (Tc_2O_7) that has a boiling point of approximately 311°C. [1] An extensive study by Vida [2] has shown that substantial amounts of Tc-99 can be volatilized during vitrification in laboratory melters and in a large scale melter. Further, when both Tc and Cs are in the waste, it has been shown nominally 60% of the Tc and 10% of the Cs were volatilized in laboratory scale melter tests. [3]

During the processing of Macrobatch One in the DWPF, data were obtained that allow a determination of the losses of Tc-99 and Cs-137 from the DWPF production melter. These determinations were made by comparing the measured concentrations of Tc-99 and Cs-137 in three glass samples taken at different times throughout the campaign with predicted concentrations based on measurements of the Tc-99 and Cs-137 concentrations in the HLW being vitrified. The results of the determinations are presented in this paper. Direct quanitative measurement of the Tc-99 and Cs-137 losses by analyzing the melter offgas samples in the DWPF would be extremely difficult due to the equipment configuration in the DWPF.

The Macrobatch One campaign in the DWPF lasted ~30 months from 3/96 to 9/98. Details of the campaign are summarized in Table I. The DWPF production melter holds ~6.6E03kg (~2.5E03L) of

Mat. Res. Soc. Symp. Proc. Vol. 608

molten glass. The top surface area of the melt is $2.6m^2$ and is 95% covered by a cold cap while the waste-glass frit slurry (~60 wt% water) is being fed to the melter. The feed rate is nominally 2L/min. Note in Table I that there was approximately a 21 month period between the first two and the third glass sample. These samples were taken from the pour stream of the melter using a special removable sampler that fit into the top of a canister and collected ~40 grams of glass. The glass samples were sent from the DWPF to the Savannah River Technology Center (SRTC) where they were analyzed. Samples of the HLW sludge slurry taken from the 3.7E06L DWPF feed-tank (Tank 51) to the DWPF were also analyzed at SRTC.

Table I. Properties of the Macrobatch One Campaign in the DWPF

Volume of Sludge Slurry Vitrified	~1.6 Million Liters
Amount of Glass Produced	~0.9 Million Kilograms
Number of Canisters Filled	495
First Glass Sample Taken During Pouring of Canister 50	Taken on 9/2/96
Second Glass Sample Taken During Pouring of Canister 61	Taken on 9/23/96
Third Glass Sample Taken During Pouring of Canister 409	Taken on 6/30/98

EXPERIMENTAL

The sludge slurry and glass samples had to be dissolved and analyzed in order to determine their composition. These dissolutions were performed remotely in the Shielded Cells of SRTC using procedures developed and tested at SRTC. The resulting solutions were diluted and removed from the Shielded Cells for analyses by various techniques. For the HLW sludge slurry, a large sample was thoroughly mixed and portion taken for dissolution. This sample was dried at 115°C and four samples were taken from it for dissolution. Each of these samples was dissolved by two separate dissolution methods. These methods were an aqua regia method at 115°C in sealed Teflon vessels and a sodium peroxide fusion followed by a HNO3 dissolution. For the glass dissolutions, the three glass samples listed in Table I were crushed and sieved prior to dissolution. The portion passing through a 200 mesh sieve was used for dissolution of each glass. Quadruplicate samples of each were dissolved by two separate dissolution methods. The first used a mixture of HF/HNO3/HCl/H3BO3 solutions at 115°C in sealed Teflon vessels. The second was a sodium peroxide fusion. The solutions from the acid dissolutions in sealed Teflon vessels were used for the determination of Tc-99 or Cs-137, since no losses of these radionuclides could occur during these dissolution procedures. A standard glass with a composition similar to the DWPF glass, (Analytical Reference Glass-1)[4] was dissolved and analyzed each time a set of sludge slurry or glass samples were dissolved in order to check if the dissolutions were complete and the analytical procedures were performed correctly. For each set of dissolutions, the results confirmed that this was the case.

The major nonradioactive waste elements (excluding oxygen) in the dissolved HLW sludge slurry and the glass samples were measured by analyzing the diluted solutions from the Shielded Cells by inductively coupled plasma emission spectroscopy (ICP-ES). Cs-137 was determined by gamma counting the Ba-137m gamma ray that is in secular equilibrium with Cs-137. Tc-99 was determined by inductively coupled plasma mass spectroscopy (ICP-MS). It is proven below that the response at mass 99 in these solutions was indeed due to Tc-99 and not any other isotope such as natural Ru-99 that could have been in the waste.

The analytical method ICP-MS gives the concentrations of analytes as a function of mass. For radioactive HLW solutions, it has been shown that specific isotopes can be assigned to these masses by considering the natural isotopes in the waste and the U-235 fission products at that respective mass.[5] At mass 99, there are two possibilities – natural Ru-99 and the U-235 fission product Tc-99. It can be proven that the response at mass 99 is from Tc-99 and not Ru-99 by considering the concentrations measured at masses 101, 102, and 104. These masses can only be due to natural or fission product Ru. All the radioactive U-235 fission products with these masses have half lives too short to still be present in the sludge at a reasonable concentration. There are no other natural isotopes at these masses except

Pd-104 at mass 104. However Pd-104 is blocked in the isobaric beta decay chain by stable Ru-104. Table II gives the measured concentrations at masses 101, 102, and 104 in the dried HLW sludge slurry. The natural abundances and fission yields of Ru-101, 102, and 104 are also given. The distributions based on natural abundance and fission yields are compared to the measured distribution in the last three columns of Table II.

Table II. Proving that the Isotope Measured at Mass 99 is not Natural Ru-99

Isotope	Natural Abundance (%)	U-235 Fission Yield	Measured Concentration Wt.%	Distribution Based on Abundance	Distribution Based on Fission Yield	Measured Distribution
Ru-101	12.1	5.18	1.3E-03	0.19	0.46	0.46
Ru-102	31.6	4.29	1.1E-03	0.51	0.38	0.39
Ru-104	18.6	1.88	4.0E-04	0.30	0.16	0.14

The measured distribution clearly fits the distribution of fission yields indicating the isotopes measured at these masses are fission product Ru and not natural Ru. The response at mass 99 must then be due to Tc-99 which, with its long half life (2E05 years), blocks the Ru-99 in the beta decay of the isobaric chain at mass 99.

RESULTS AND DISCUSSION

Pertinent Concentrations in the Dried Sludge Slurry and in the Three Glass Samples

The radioactive and nonradioactive composition of the HLW sludge slurry that was vitrified as Macrobatch One was determined at SRTC prior to processing this material in the DWPF. An example of this composition for Tank 51 dried sludge slurry is given in Reference 5. The elements Fe, Al, Mn, and Ca, are the major nonradioactive elements (excluding oxygen) in the HLW that are solidified in the glass. They comprise ~88 elemental percent of the dried sludge slurry. The other major nonradioactive elements are Na and Mg (8%), but these are also present in the glass forming frit and thus their concentrations are not specific to the HLW in the final glass. Since Fe, Al, Ca, and Mn are the major elements in the HLW, the concentrations of these were used in the method to determine the losses of Tc-99 and Cs-137 from the DWPF production melter. Table III shows the measured concentrations of Fe, Al, Ca, and Mn in the dried sludge slurry and in the three glass samples taken during processing of Macrobatch One. The measured concentrations of Tc-99 and Cs-137 are also presented. Standard deviations are given for the replicate samples dissolved and analyzed.

Table III. Concentrations (Weight Percent) of Major HLW Elements and Tc-99 and Cs-137 in the Dried HLW Sludge Slurry and Three DWPF Glass Samples from the Macrobatch One Campaign[a]

Species	Sludge Slurry	Canister 50	Canister 61	Canister 409
Fe	25.8±0.7	9.11±0.21	8.44±0.37	8.78±0.31
Al	6.36±0.30	2.34±0.16	2.13±0.13	2.26±0.10
Ca	2.39±0.12	0.88±0.05	0.79±0.04	0.93±0.03
Mn	2.61±0.12	0.88±0.01	0.81±0.03	0.86±0.03
Tc-99	(1.28±0.06)E-03	(4.34±0.20)E-04	(4.08±0.28)E-04	(4.67±0.26)E-04
Cs-137	(7.17±0.32)E-05	(2.62±0.03)E-05	(2.49±0.10E-05)	(2.45±0.09)E-05

[a] For the dried sludge slurry, the result is the average and standard deviation for three samples. For the glasses, eight samples were analyzed except for Tc-99 where only four samples were analyzed.

The concentrations measured in the three glass samples are in good agreement as expected indicating that the composition of the HLW did not change significantly during processing. Also note the concentrations in the glasses are lower than those in the dried HLW sludge slurry. This lowering is due to the addition of the nonradioactive glass forming frit that dilutes the HLW elements. This dilution is one of the bases for predicting what the concentrations of Tc-99 and Cs-137 should be in the glass if they are not volatilized from the melter.

Method for Predicting Tc-99 and Cs-137 Concentrations in the Glass

If Tc-99 and Cs-137 are not volatilized from the melter, they should be diluted by the same factor that causes the dilution of the major HLW elements. This presumes that Fe, Al, Ca, and Mn are not lost in the vitrification process. This has been shown to be the case by comparing the compositions of vitrified samples from two melter feed tank batches (measured in the DWPF process control analytical laboratory) with the respective compositions of the glass being poured (Canisters 50 and 61) while those batches were in the melter feed tank.[6] The agreement was excellent for the major nonradioactive elements and for Cs-137 (Tc-99 could not be measured in the DWPF process control analytical laboratory.). This waste dilution factor (WDF) is the concentration of an element in the glass divided by its respective concentration in the dried sludge slurry. Table IV shows the WDF calculated in each glass based on each of the major elements in the HLW. The average, standard deviation, and percent relative standard deviation for each glass sample are also given in Table IV.

Table IV. Waste Dilution Factors (WDF) Based on Major Waste Elements in Three DWPF Glass Samples in the Macrobatch One Campaign

Element	**Canister 50**	**Canister 61**	**Canister 409**
Fe	2.83	3.06	2.94
Al	2.71	2.98	2.81
Ca	2.71	3.02	2.58
Mn	2.96	3.21	3.05
Average	**2.80**	**3.07**	**2.84**
Std. Dev.	0.12	0.10	0.20
Percent RSD	4.1	3.2	7.2

The final averages for the three glasses are in good agreement with the relative precision of each average being better than 8% based on the four elements. The predicted concentration of Tc-99 and Cs-137 in each of the glasses is then its respective concentration in the dried HLW sludge slurry divided by the appropriate WDF. These results are presented in the next section.

Comparison of Measured and Predicted Tc-99 and Cs-137 Concentrations in the Glass

Table V shows the measured and predicted and concentrations of Tc-99 and Cs-137 in each of the three glass samples taken during processing of Macrobatch One. The percent difference for each was calculated from the formula

$$\% \text{ Diff.} = ((\text{Pred. Conc.} - \text{Meas. Conc.}) / \text{Pred. Conc.}) \text{ X } 100 \qquad (1)$$

It can be seen that in all cases the predicted and measured concentrations are in close agreement suggesting little volatilization of the Tc-99 or Cs-137. In those cases where a negative difference is indicated, the measured concentration was larger than the predicted. This is due to experimental errors in measuring the respective concentrations.

Table V. Measured and Predicted Concentrations (Wt. %) of Tc-99 and Cs-137 in Three DWPF Glass Samples in Macrobatch One Campaign

	Canister 50			Canister 61			Canister 409		
Isotope	**Meas.**	**Pred.**	**Diff.**	**Meas.**	**Pred.**	**Diff.**	**Meas.**	**Pred.**	**Diff.**
Tc-99	4.34E-04	4.51E-04	3.8%	4.08E-04	4.16E-04	1.8	4.67E-04	4.53E-04	-3.2%
Cs-137	2.62E-05	2.53E-05	-3.4%	2.49E-05	2.34E-05	-6.5%	2.45E-05	2.54E-05	3.7%

It should be mentioned that the results in Table V may be prone to a large relative error because we are trying to determine a small difference between two relatively large numbers. For example, for the glass in Canister 409, the minimum concentration predicted for Tc-99 in the glass is 4.01E-04 wt.% while the maximum is 5.08E-04 wt.%. This is a difference of 20-25% and is due to just the measured precision of the average concentration of Tc-99 in the waste and the precision of the average WDF presented in Table IV for Canister 409. On this basis it is difficult to assign an exact number to the possible losses of Tc-99 and Cs-137; however, based on these data, it appears that these losses are small. As stated earlier, a much better estimate could be made if it were possible to get a direct measurement the Tc-99 and Cs-137 being captured by the DWPF offgas system.

As mentioned before, during feeding of the DWPF melter, the melt pool is 95% covered with a cold cap due primarily by heat transfer from the molten glass to the water from the slurry. The water is evaporated and collected by the offgas condensate tank in the DWPF. This cold cap would clearly decrease the volatility of the Tc-99 and Cs-137. The cold cap is burned off during those times that the melter is not being fed; however, these times were short compared to the times that the melter was being fed. Another possibility for decreasing the volatility of Tc-99 is that the Tc-99 may be in the +4 valance state due to formic acid being added to the sludge slurry in the DWPF as a process chemical in a pretreatment step. The +4 state forms a less volatile oxide that the +7 state. [2]

At SRTC we had the opportunity to make these same type of measurements during a demonstration of the DWPF vitrification process using a remote slurry fed research melter in the Shielded Cells.[7] This demonstration was with the HLW from Tank 51 prior to its washing to be sent to the DWPF. The melt pool of this melter contained only 10kg of glass and the feed rate was so slow that a cold cap never formed.[7] As with Macrobatch One we measured the concentrations of Fe, Al, Ca, and Mn, in the waste and in a sample of the final glass. In this case a WDF of 3.07±0.21 was obtained.[5] The measured concentrations of Tc-99 and Cs-137 in the sludge and glass along with their predicted concentrations are presented in Table VI.

Table VI. Measured and Predicted Concentrations (Wt. %) of Tc-99 and Cs-137 in Tank 51 Glass Prepared in the Small Slurry Fed Research Melter in SRTC

	Sludge	Glass		
Isotope	**Measured.**	**Measured**	**Predicted**	**Difference**
Tc-99	1.7E-03	3.8E-04	5.5E-04	31%
Cs-137	6.7E-05	2.2E-05	2.3E-05	4.3%

In this campaign, the results indicate that 31% of the Tc-99 was lost while very little Cs-137 was lost. In this case, loss of the Tc-99 may be due to the absence of a cold cap.

CONCLUSIONS

- The data presented in this paper indicate that there is very little loss of Tc-99 and Cs-137 from the DWPF production melter during the vitrification of SRS HLW sludge slurries. This may be due to the fact that the DWPF melter operates with the ~95% of the surface of the melt covered by a cold cap.

- In a small research slurry fed melter, nominally 30% of the Tc-99 was lost while very little of the Cs-137 was lost. This melter did not operate with a cold cap supporting the conclusion that a cold cap enhances retention of the Tc-99.

ACKNOWLEDGEMENT

This paper was prepared in connection with work done under Contract No. DE-AC09-96SR18500 with the U. S. Department of Energy.

REFERENCES

1. J. A. Rard, *Critical Review of the Chemistry and Thermodynamics of Technetium and Some of Its Inorganic Compounds and Aqueous Species,* UCRL-53440, Lawrence Livermore National Laboratory, Livermore, CA, 1983.

2. J. Vida, *The Chemical Behavior of Technetium During the Treatment of High-Level Radioactive Waste,* KfK 4642, Translated by J. R. Jewett, PNL-TR-497, Westinghouse Hanford Co., 6/23/1994.

3. H. Lammertz, E. Merz, and ST. Halaszovich, *Technetium Volatilization during HLLW Vitrification,* Scientific Basis for Nuclear Waste Management VIII, Mat. Res. Soc. Symp. Vol 44, p. 823, Materials Research Society, Pittsburgh, PA, 1985.

4. G.L. Smith, *Characterization of Analytical Reference Glass 1 (ARG-1)*, PNNL-8992, Pacific Northwest Laboratory Report, (1993).

5. N.E. Bibler, W.F. Kinard, W. T. Boyce, and C.J. Coleman, J. Radioanal. Nucl. Chem., **234**, p. 159-163 (1998).

6. N. E. Bibler, J. W. Ray, T. L. Fellinger, O. B. Hodoh, R. S. Beck, and O. G. Lien, *Characterization of the Radioactive Glass Currently Being Produced by the Defense Waste Processing Facility at the Savannah River Site,* Proceedings – Waste Management '98, CD-ROM Session 14 (1998).

7. M. K. Andrews and N. E. Bibler, Ceramic Transactions, 39, The American Ceramic Soc., Westerville, OH, 1994, p. 205.

AMERICIUM/CURIUM VITRIFICATION PROCESS DEVELOPMENT PART II

Andrew P. Fellinger, Mark A. Baich, Jon W. Duvall, Timothy M. Jones, John E. Marra, Carey B. Miller, Donald H. Miller, David K. Peeler, Theresa K. Snyder, Michael E. Stone and Douglas C. Witt, Westinghouse Savannah River Company, Savannah River Site, Aiken, SC 29808

ABSTRACT

At the Savannah River Site (SRS) we are currently finalizing the design for a multi-system vitrification process that will be installed in the F-Canyon Multi-Purpose Process Facility (MPPF), an existing highly shielded, remotely operated facility. Authorization to proceed beyond the preliminary design based on the recommendation of a Formal Design Review Board was requested in May of 1999.

The Savannah River Technology Center (SRTC) Process Development Group has been conducting research and developing a process to identify equipment design bases and process operating parameters since 1996. The goal of the project is to stabilize a tank of ~11,000 liters of nitric acid solution containing valuable isotopes of americium (Am) and curium (Cm). Vitrification has been selected as the most attractive alternative for stabilization and provides the opportunity for recovery and eventual reuse of the actinides. The final glass form will be placed in interim storage awaiting a disposition by the Department of Energy. This paper presents a brief history of the stabilization program and an overview of the entire Am/Cm stabilization process. This paper also provides details of a specific processing issue related to drain tube pluggage (devitrification) that was encountered during the development of the baseline batch vitrification process, and the remedy employed to reduce the potential for further drain tube pluggage.

INTRODUCTION

The SRTC Process Development Group has been developing a process to stabilize ~11,000 liters of nitric acid solution containing valuable isotopes of americium (Am) and curium (Cm) at the request of the Defense Nuclear Facility Safety Board since 1996. This solution also contains relatively high concentrations of rare-earth fission products formed during production of the Am and Cm in SRS reactors. The originally proposed process entailed an application of resistance heating to a slightly modified, commercially available platinum-rhodium slab-type bushing melter. Challenges with off-gas system pluggage, feed system pluggage and melter durability associated with efforts to directly vitrify the nitric acid feed solution in the slab-type bushing melter were encountered through four developmental melter designs between 1996 and 1998. Technical issues associated with processing and difficulties of the continuous feed Am/Cm bushing melter pilot system have been documented by Smith, et.al. [1]

In January of 1998 an alternative batch process was proposed. The background for changing to a batch vitrification process and the simplification of the melter system as compared to the bushing melter are described by Marra, et.al.[2] As a new induction heated, cylindrical melter was being installed, the development of vitrification experiments with the product slurry from the new pretreatment step were carried out in a resistance heated platinum melter to validate the technical feasibility of the new process flowsheet. The success and technical feasibility of the batch flowsheet scoping studies in early 1998 have been described in Part I of this paper by Fellinger, et.al.[3] The radically altered operating scenario (batch versus continuous-feed), the modified flowsheet (addition of an oxalic acid precipitation step and slurry feed) and the new full-scale induction melter system were successfully demonstrated in four integrated runs in late 1998. The success of the process and equipment changes described above provided the confidence to specify the design basis for a preliminary design of the vitrification system racks in May of 1999. However, sporadic drain tube pluggage due to devitrification was encountered during subsequent process optimization and upset testing which required further investigation. This paper presents the results of that investigation and how the drain tube pluggage was eliminated.

Am/Cm BATCH VITRIFICATION PROCESS OVERVIEW

A multi-step feed pretreatment phase will be conducted in existing canyon tanks prior to introduction of the concentrated Am/Cm material to the Multi-Purpose Processing Facility (MPPF). The pretreated and concentrated Am/Cm solution (1,100 liters) will then be transferred in 48-liter batches to the MPPF Liquid Feed Tank for the final stabilization phase. The final stabilization phase inside the MPPF is fundamentally a two-step process. The first step, precipitation, entails a further de-nitration of the feed by precipitating the actinides and lanthanides as oxalates and separating the solids from the remaining aqueous fraction. The second step, vitrification, entails combining the lanthanide and actinide oxalate solids with glass formers and vitrifying at a nominal temperature of 1450°C.

The first step inside the MPPF, the precipitation process, is comprised of an oxalate precipitation performed in a 25 liter, temperature controlled reaction vessel (Precipitator Vessel) located above, but off-center of the melter vessel. The precipitation is accomplished by adding a controlled volume of 7 liters of feed from the Liquid Feed Tank to the Precipitator Vessel, and then adding 8 wt% oxalic acid to the one-molar nitric acid feed at a 1.74:1 volume ratio while mixing (motor operated, shaft driven, four blade impeller). Following the settling of the insoluble oxalate solids and decanting of the supernate, a 0.1 molar oxalic acid wash is performed as the batch is again mixed. A second settling and decant follows leaving a volume of approximately 1.9 liters of oxalate solids and approximately 0.6 liters of free oxalic wash. The Precipitator Vessel is then positioned over the melter, the contents mixed and gravity dropped into the melter vessel onto a bed of strontium-alumina-borosilicate glass cullet (25SrABS). The drop is quickly followed by a de-ionized water rinse to remove any residual solids from the walls of the Precipitator Vessel.

The second step inside the MPPF, the vitrification process, is performed in a 12.7 cm (5") diameter Pt/Rh cylindrical melter, consisting of three independent induction heating zones. The vessel wall, conical section and drain tube heating zones are shown in Figure 1. The heat profile of the melter at various steps of the vitrification phase (drying, calcination and vitrification) has proven critical in ensuring the glass and process integrity. Initially, only the melter wall zone is heated to approximately 200°C to dry the precipitated solids (~100°C in the bed) by evaporating the free aqueous fraction. The application of excessive power to the bed and/or the application of the lower conical heating zone during drying will result in unacceptable entrainment of the melter contents into the off-gas system, or in the worst instance an eructation of solids; either of which would be undesirable with the highly radioactive material. Once drying is complete, the segregated batch (glass cullet and dried oxalates) is heated to 1450°C at various controlled rates, ranging between 5 and 10°C per minute through a programmed automatic sequencing of the melter vessel and conical section heating zones. The dried oxalates are converted to oxides through calcination as the batch is heated through 900°C. The conversion of oxalates to oxides liberates CO and CO_2 from the bed that is allowed to burn and vent from the top of the melter into a decoupled off-gas hood. Vitrification occurs as the glass cullet softens and begins to incorporate the feed

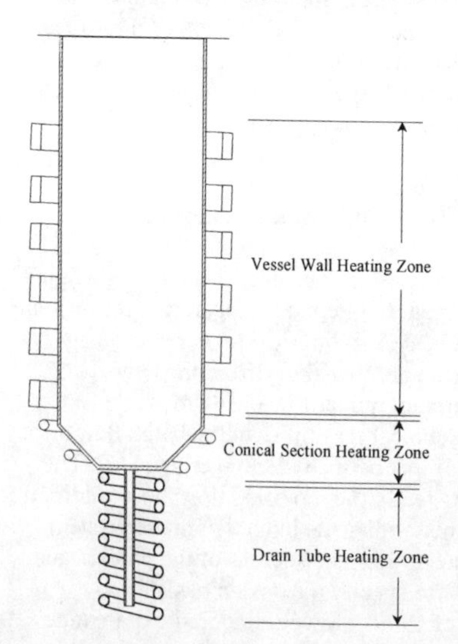

Figure 1 – CIM Vessel Heating Zones

oxides into a low viscosity glass matrix. As the bed heats through 1140°C, oxygen is liberated from the thermal reduction of CeO_2 producing bubbles that are allowed to diffuse through the glass as the viscosity decreases at higher temperatures.

After reaching the target setpoint temperature in the molten glass pool, the drain tube heating zone is heated at a rate of approximately 15°C per minute to 1450°C, while the molten glass pool is held at approximately 1450°C. Upon completion of the mixing process, the batch is drained into a canister. The molten glass is drained through a 0.53 cm (0.21 inch) diameter by 15 cm (6 inch) long platinum-rhodium drain tube by turning off the cooling air that maintains a glass plug at the tip of the drain tube. Pour initiation using this freeze-valve application is typically within 20-30 seconds and continues with a continuous pour stream. The pour is terminated by reapplication of the cooling air to leave the drain tube with a glass plug to allow subsequent processing. It was found during testing that the drain tube was prone to pluggage due to crystal formation in the glass. As a result of this study, pluggage was eliminated by inserting a Pt/Rh bubbler tube along the centerline of the melter to about 2.5 cm (1 inch) from the bottom of the vessel. Argon gas is bubbled through the low viscosity glass at approximately 0.71 SLPM (1.5 SCFH) during the hold period to promote homogenization and ensure an evenly distributed glass loading.

RESULTS AND DISCUSSION

Initial process development runs at SRTC had targeted a total lanthanide (Ln) loading of 60 wt% using a non-radioactive surrogate feed. The composition of the surrogate feed is given in reference 3. The process demonstration runs using a melter of the exact size that will be used in MPPF, were meeting the requirement for incorporating all of the surrogate feed materials into the glass matrix; however, testing was exhibiting evidence of a non-homogenous melt as attested by a rather large fluctuation in glass pour rates during the drain cycle. Further, on multiple occasions, the drain tube had plugged after inadvertently cooling during the drain cycle. During a pour cycle, "grab" samples were taken every nine seconds and weighed to determine pour rate. The samples were then submitted for compositional analysis. The total measured lanthanide content of the consecutive glass samples displayed a correlation to the pour rates recorded (shown in Figure 2). The "grab" sample analysis proved that a significant high to low distribution of lanthanide loading

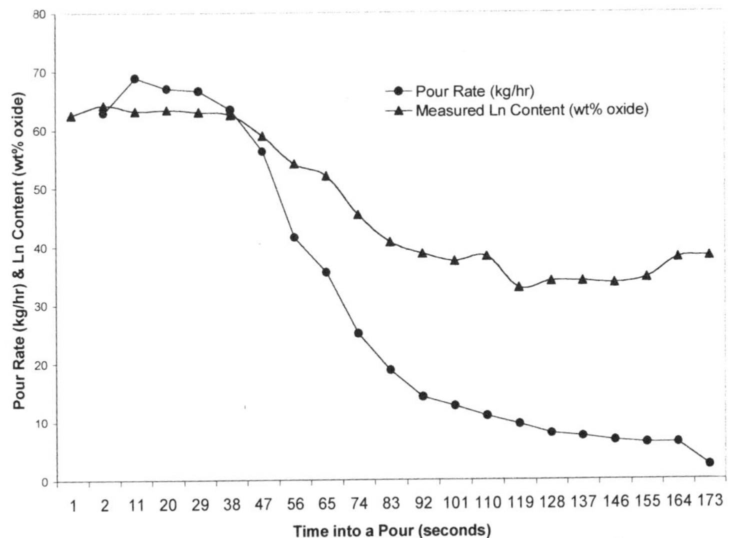

Figure 2 –Results of lanthanide loading and glass pour rates based on pour stream samples measured in a test without mixing of the melt pool.

existed from the bottom to the top of the glass pool in the melter. This difference in loading was affecting the viscosity of the glass over the drain cycle producing the large fluctuations in glass pour rate. Laboratory studies by Peeler, et.al. [4&5] provided data revealing that variations in lanthanide content have a dramatic effect on the viscosity of the glass system further supporting the pour rate differences recorded. The laboratory crucible studies also provided data suggesting the mechanism responsible for the drain tube pluggage.

A glass composition variability study conducted by Peeler, et. al.[5] has identified a crystalline phase field that would cause the drain tube pluggage experienced in the pilot system. The variability study identified a rare earth silicate (RES) phase for surrogate loadings greater than 62 wt% lanthanides with liquidus temperatures as high as ~1400°C. Further, the kinetics of devitrification at temperatures below liquidus increases as the lanthanide content of the glass increases, fully supporting the rapid tube pluggage experienced in the pilot system. Scanning electron microscopy (SEM) and energy dispersive spectroscopy (EDS) showed Si, La, Nd, and O as major components and Ce, Pr, Sm, Gd, and Er as the minor components of the RES crystals [5]. Although the crystals were observed in four distinct morphologies, chemically the crystals were the same. Two examples of the RES crystal morphologies identified in the variability study are shown in Figures 3 & 4.

Figure 3 – SEM Micrograph of Needle-Like RES Crystals (AC2-04 with Dense, Settled RES Crystals)

Figure 4 - SEM Micrograph of Rectangular Capped RES Crystals on the Surface (AC2-46 at 1321°C)

Pilot system testing and the identification of the RES crystals linked the drain tube pluggage to the conversion of the cullet / oxalate mixture to a glass product that is highly influenced by the reactions occurring at glass former – feed oxide interface. As a result of the interaction between the 25SrABS cullet and surrogate oxides, a more dense glass is initially formed relative to the cullet that begins to settle to the bottom of the melt pool. As the more dense glass settles and additional oxides are incorporated into the glass matrix, the amount of oxides to be incorporated into the glass matrix decreases until too few oxides remain to be incorporated into the excess cullet. This "lightly" loaded glass is less dense than the highly loaded glass at the bottom of the melt pool. As the melter process transitions into the draining cycle, without forced mixing, a high initial glass flow rate results (i.e. the higher the lanthanide content the lower the glass viscosity).

Drain tube pluggage was only encountered if the glass in the drain tube was allowed to cool after draining was initiated. High lanthanide loading tests (>~60 wt% Ln) at the pilot facility proved that if the higher loaded glass were allowed to fill the volume of the drain tube, and the drain tube was subsequently cooled to a temperature below ~1300°C, the rare earths would be expected to precipitate to a silicate crystalline phase (e.g., RES crystals shown above) effectively plugging the drain tube. Initial glass flow rates in the range of 60-70 kg/hr correlating to 62-68 weight percent (lanthanide basis) loaded glasses have been recorded. As the lanthanide loading decreases during the pour, the viscosity of the glass increased, resulting in a lower flow rate. While the chemical inhomogeneity would be acceptable from a glass durability and recovery perspective, in August of 1999, an argon bubbler was added to the design basis to promote the mixing of the glass during the hold period. The forced mixing of the glass pool would be expected to distribute the surrogate oxides more evenly over the entire melt pool ultimately resulting in a more uniform product in terms of composition throughout the entire draining cycle. Coupled with the more evenly distributed loading, the targeted glass loading on the total lanthanide basis was decreased to 49 wt%. Therefore, with a more homogenous glass composition achieved due to forced mixing, the total lanthanide content of the glass within the drain tube (during an inadvertent cooling) is within an acceptable range for very limited devitrification, mitigating the potential for pluggage. This has been demonstrated successfully through several tests at the pilot system. Figure 5 depicts the results of the pour rate to lanthanide loading correlation at a targeted 49 wt% in a test after the argon bubbler was emplaced.

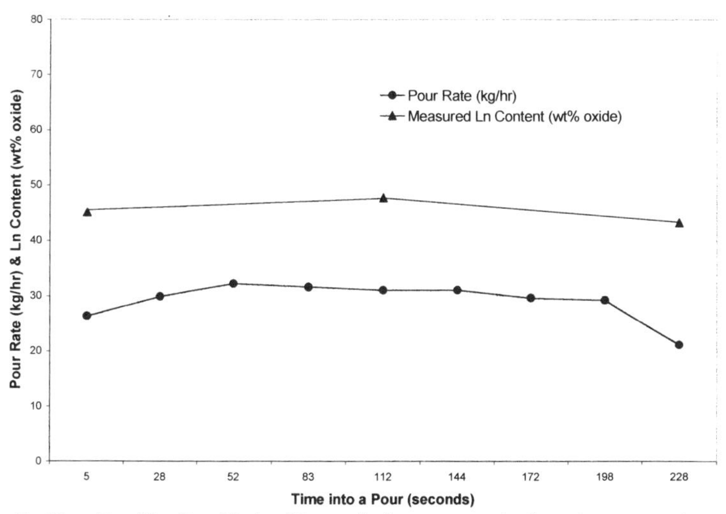

Figure 5 – Results of lanthanide loading and glass pour rates based on pour stream samples measured in a test after an argon bubbler was emplaced to promote mixing of the melt pool.

CONCLUSION

Difficulties with rare-earth silicate crystal pluggage of the drain tube due to devitrification during process optimization and upset testing at the melter pilot facility have been mitigated by mixing the molten glass using an argon bubbler system and reducing the targeted lanthanide loading from 60 wt% to 49 wt%. The pilot facility equipment continues to prove to be considerably more robust and less intricate than the originally proposed slab-type bushing melter system, and has translated into a preliminary design that offers operational flexibility and complete remotability. Additional operational studies to investigate and generate a thorough understanding of operational variability, process upsets and recovery methods are underway.

ACKNOWLEDGEMENT

This paper was prepared in conjunction with work done under Contract No. DE-AC09-96SR18500 with the U.S. Department of Energy. By acceptance of this paper, the publisher and/or recipient acknowledges the U.S. Government's right to retain a nonexclusive, royalty-free license in and to any copyright covering this paper, along with the right to reproduce and to authorize others to reproduce all or part of the copyrighted paper.

REFERENCES

1. M.E. Smith, A.P. Fellinger, T.M. Jones, C.B. Miller, D.H. Miller, T.K. Snyder, M.E. Stone, and D.C. Witt, "Americium/Curium Melter 2A Pilot Tests (U)," published in the *Environmental Issues and Waste Management Technologies in Ceramic and Nuclear Industries IV, Volume 93*, The American Ceramic Society, 1999, pp. 381-389.

2. J.E. Marra, M.A. Baich, A.P. Fellinger, B.J. Hardy, T.M. Jones, C.B. Miller, D.H. Miller, D.K. Peeler, T.K. Snyder, M.E. Stone, J.C. Whitehouse, and D.C. Witt, "Americium-Curium Vitrification Pilot Tests – Part II," published in the *Environmental Issues and Waste Management Technologies in Ceramic and Nuclear Industries IV, Volume 93*, The American Ceramic Society, 1999, pp. 391-398.

3. A.P. Fellinger, M.A. Baich, B.J. Hardy, G.T. Jannik, T.M. Jones, J.E. Marra, C.B. Miller, D.H. Miller, D.K. Peeler, T.K. Snyder, M.E. Stone, and D.C. Witt, "Americium-Curium Vitrification Process Development (U)," published in the *Proceedings of the Symposium on the Scientific Basis for Nuclear Waste Management XXII, Volume 556*, The Materials Research Society, 1999, pp. 367-373.

4. D.K. Peeler, T.B. Edwards, and I.A. Reamer, "Composition / Property Relationships for the Phase 1 Am/Cm Glass Variability Study (U)," Report Number WSRC-TR-99-00055, January 1999.

5. D.K. Peeler, T.B. Edwards, and I.A. Reamer, "Composition / Property Relationships for the Phase 2 Am-Cm Glass Variability Study (U)," Report Number WSRC-TR-99-00393, May 2000.

LIQUIDUS TEMPERATURE AND PRIMARY CRYSTALLIZATION PHASES IN HIGH-ZIRCONIA HIGH-LEVEL WASTE BOROSILICATE GLASSES

TREVOR PLAISTED, PAVEL HRMA, JOHN VIENNA, AND ANTONIN JIRICKA
Pacific Northwest National Laboratory, Box 999, Richland, WA 99352

ABSTRACT

Liquidus temperature (T_L) studies of high-Zr high-level waste (HLW) borosilicate glasses have identified three primary phases: baddelyite (ZrO_2), zircon ($ZrSiO_4$), and alkali-zirconium silicates, such as parakeldyshite ($Na_2ZrSi_2O_7$). Using published T_L data for HLW glasses with these primary phases, we have computed partial specific T_Ls for major glass components. On the Na_2O-SiO_2-ZrO_2 submixture, we have determined approximate positions of the boundaries between the baddelyite, zircon, and parakeldyshite primary phase fields. The maximum that can dissolve at 1150°C in a borosilicate HLW glass subjected to common processability and acceptability constraints appears to be 16.5 mass% ZrO_2.

INTRODUCTION

Several published composition variation studies (CVS) [1-8] provide databases linking the liquidus temperature (T_L) associated with Zr-containing primary phases to high-level waste (HLW) glass composition. Three Zr-containing phases were identified. The zircon ($ZrSiO_4$) primary phase field occupied most of the region covered by the published data. The rest was divided between baddelyite (ZrO_2) and alkali zirconium silicates, mainly parakeldyshite ($Na_2ZrSi_2O_7$). Studies [2-5] express T_L as a function of glass composition by the equation

$$T_L = \sum_{i=1}^{N} T_{Li} g_i \tag{1}$$

where T_{Li} is the i-th component partial specific T_L (also called i-th component coefficient), g_i is the i-th component mass fraction in glass, and N is the number of components (usually limited to N-1 major components plus Others, the sum of all minor components; see the Table 1a footnote). Equation (1) represents a nearly planar liquidus surface. This is a satisfactory approximation of the real T_L surface on a constraint region within a multicomponent mixture. Such regions have narrow ranges of mass fractions of components. In this paper, we evaluated the following databases: Hanford CVS [1,2], simulated transuranic (TRU) waste CVS [3-5], and Idaho National Engineering Environmental Laboratory (INEEL) CVS [6-8]. To obtain T_{Li} values, we sorted the glasses into three groups according to the primary phase (zircon, baddeleyite, and parakeldyshite) and fitted Equation (1) to T_L versus g_i data for each group.

T_L VERSUS COMPOSITION DATA

This section summarizes the published T_L databases [1-8] that we used to compute the T_{Li} values. In all studies [1-8], the primary phases were identified by x-ray diffraction and scanning electron microscopy. T_L was measured by the gradient furnace method only in the earliest study, the Hanford CVS [1], that covers the glass-composition region envisaged for Hanford before 1995 [9]. A disadvantage of this method is that at $T > 850$°C, the surface-tension gradient induced by volatilization may generate melt convection that shifts the crystallization front. In later studies [3-8], this method was replaced by the uniform-temperature heat-treatment method.

Mat. Res. Soc. Symp. Proc. Vol. 608

Out of more than 100 glasses tested in Hanford CVS [1], 25 precipitated a Zr-containing primary phase: 18 precipitated zircon, 1 precipitated baddeleyite, and 6 formed an alkali-zirconium silicate, probably parakeldyshite. Although the Hanford CVS [1] did not focus solely on high-Zr glasses, it allowed the authors to determine a set of 10 component coefficients [2] for glasses with Zr-containing primary phases (all three phases taken together).

The TRU CVS [3-5] was designed for high-Zr Hanford HLW glasses. The initial study [3] tested 36 compositions and varied 9 components, one at a time. Zircon was the primary phase in 27 glasses, baddeleyite in 7 glasses, and 2 glasses precipitated CeO_2. T_{Li} values were determined only for glasses with a zircon primary phase. The following study [4] evaluated data from [1] and [3] and developed several sets of T_{Li} values. The final TRU CVS [5] added 10 compositions derived from the same baseline as in [3] except that two or three components were varied at a time. Several sets of T_{Li} values and second-order terms were developed.

Studies [6-8] extended the high-Zr composition region to incorporate INEEL high-Zr HLW. The first of these studies [6] developed a glass (marked B1-2 in Table Ia) with as high Zr content as possible while maintaining acceptable processing properties. This glass formed parakeldyshite as a primary phase. The following study [7] covered a broad composition region for INEEL wastes. Out of 44 glasses, 5 precipitated lithium, sodium, and sodium-lithium-zirconium-silicates as primary phases. The final study [8] generated 30 glasses with the same calculated viscosity value (6 Pa·s) at 1150°C. The primary phases were zircon in 17 glasses, baddeleyite in 6 glasses, and parakeldyshite in 7 glasses.

Space limitations do not allow us to list all T_L and g_i data. Table Ia shows only data that are not readily available and Table Ib lists the reference glasses.

Table Ia. Compositions (in Mass Fractions) and T_L Values for HLW High-Zr Glasses that Precipitate Alkali-Zirconium-Silicate as the Primary Phase ([6] and [7] data only)

ID	Al_2O_3	B_2O_3	Bi_2O_3	CaO	CeO_2	Fe_2O_3	Li_2O	MgO	Na_2O	P_2O_5	SiO_2	ZrO_2	Others*	T_L °C
IG1-10	.0000	.0653	.0000	.0000	.0000	.0005	.0900	.0000	.1560	.0000	.5439	.1400	.0043	981
IG1-12	.0317	.1500	.0000	.0000	.0000	.0005	.0000	.0000	.1530	.0000	.4205	.1400	.1043	1075
IG1-25	.0750	.0750	.0000	.0000	.0000	.0005	.0510	.0000	.1273	.0362	.4744	.1050	.0556	991
IG1-33	.0000	.0500	.0000	.0000	.0000	.0000	.0116	.0000	.2000	.0000	.4936	.1400	.1048	1310
IG1-39	.0751	.1250	.0000	.0000	.0000	.0000	.0675	.0000	.0883	.0125	.4469	.1049	.0798	855
Bl-2	.0169	.0500	.0000	.0009	.0000	.0005	.0700	.0000	.1687	.0056	.5220	.1503	.0151	1077

* Others contain 7 minor components: BaO, Cs_2O, CuO, Gd_2O_3, K_2O, PbO, and SrO.

Table Ib. Compositions (in Mass Fractions) and T_L Values for Reference Glasses

ID	Phase	Ref.	Al_2O_3	B_2O_3	Bi_2O_3	CaO	CeO_2	Fe_2O_3	Li_2O	MgO	Na_2O	P_2O_5	SiO_2	ZrO_2	Others	T_L °C
Zr-1	ZS	[3]	.0450	.1000	.0000	.0009	.0002	.0000	.0750	.0000	.1100	.0021	.5114	.1500	.0054	1064
IG1-25	PK	[7]	.0750	.0750	.0000	.0000	.0000	.0005	.0510	.0000	.1273	.0362	.4744	.1050	.0556	991
CVS2-88	Z	[1]	.1050	.0600	.0000	.0400	.0000	.0250	.0700	.0050	.1403	.0000	.4597	.0750	.0200	982

Primary phases: ZS zircon, Z baddeleyite, PK parakeldyshite

Partial Specific Liquidus Temperatures

Table II summarizes the T_{Li} values computed by fitting Equation (1) to the T_L values of glasses within the three primary phase fields using data from studies [1–8]. The effect of the i-th component on T_L can be judged from the difference between T_{Li} and the computed T_L value of a reference glass ($T_{L,ref}$). Table III shows the effect (S_i) of adding 1 mass% of the i-th component to a reference glass shown in Table Ib for each primary phase. The S_i values were calculated by the formula $S_i = 0.01(T_{Li} - T_{L.ref})/(1 - g_{i,ref})$, where subscript *ref* stands for the reference glass. The $T_{L.ref}$ values were obtained from Equation (1) and the T_{Li} values listed in Table II.

Zircon (ZS), the most common primary phase, occurs in $n = 70$ out of 103 compositions in the combined database (n is the number of data points within a primary phase field). We computed several sets of T_{Li} coefficients for ZS. The largest set ($N = 13$) is composed of 12 major oxides (Al_2O_3, B_2O_3, Bi_2O_3, CaO, CeO_2, Fe_2O_3, Li_2O, MgO, Na_2O, P_2O_5, SiO_2, and ZrO_2) plus Others. Note that the number and kind of components in Others may change from set to set and varies from study to study. Interestingly, moving three components, Bi_2O_3, CaO, and CeO_2, that are absent or minor in most glasses to Others (reducing the number of coefficients to $N = 10$), reduced S_i for Others from 5.3 to 3.3°C/mass%. Adding Fe_2O_3, a component with the smallest effect ($S_i = 0.6$°C/mass%), to Others decreased the number of coefficients to $N = 9$. We also reduced the component set to $N = 8$ by removing from the database 4 glasses in which P_2O_5 was the single component varied in the TRU CVS [3]. To compare the coefficients for ZS with those for baddeleyite (Z) and alkali-zirconium-silicates (labeled as PK), we moved MgO to Others ($N = 7$). However, we do not recommend this set for estimating T_L because the effect of MgO on T_L is one of the largest (approximately 30°C/mass%).

Table II. T_{Li} (in °C) for Glasses with Zircon (ZS), Baddeleyite (Z), and Parakeldyshite (PK) as Primary Phases

	ZS					Z	PK
N	13	10	9	8	7	7	7
Al_2O_3	1570	1535	1570	1593	1494	5372	1688
B_2O_3	554	527	568	550	707	-686	168
Bi_2O_3	1274						
CaO	855						
CeO_2	1843						
Fe_2O_3	1353	1119					
Li_2O	-1448	-1275	-1300	-1257	-436	-4463	-1085
MgO	3973	4083	3974	3914			
Na_2O	-2130	-1935	-1907	-1884	-1423	-1523	2612
P_2O_5	2062	2256	2166				
SiO_2	1154	1116	1078	1052	1143	974	568
ZrO_2	4510	4470	4561	4602	3463	6244	2346
Others	1593	1395	1355	1404	1376	1701	970
n	70	70	70	66	66	14	19
R^2	0.857	0.842	0.840	0.837	0.752	0.548	0.736
s [°C]	32	32	32	34	41	93	85

Table III. Predicted Change in T_L (in °C/mass%) from Adding 1 Mass% of a Component to a Reference Glass

	ZS					Z	PK
N	13	10	9	8	7	7	7
Al_2O_3	5.3	4.9	5.3	5.6	4.5	49.4	7.6
B_2O_3	-5.7	-6.0	-5.5	-5.7	-3.9	-17.4	-8.8
Bi_2O_3	2.1						
CaO	-2.1						
CeO_2	7.8						
Fe_2O_3	2.9	0.5					
Li_2O	-27.2	-25.3	-25.6	-25.1	-16.2	-58.2	-21.8
MgO	29.1	30.2	29.1	28.5			
Na_2O	-35.9	-33.7	-33.4	-33.1	-27.9	-28.7	18.6
P_2O_5	10.0	11.9	11.0				
SiO_2	1.8	1.0	0.2	-0.2	1.7	0.5	-8.0
ZrO_2	40.5	40.0	41.1	41.7	28.3	57.3	15.2
Others	5.3	3.3	2.9	3.4	3.2	7.7	-0.2

In Table II, R^2 is the fraction of the variation in T_L database values accounted for by Equation (1), and the quantity s is the root mean square error that estimates the experimental error standard deviation if Equation (1) does not have a significant lack of fit. For ZS with $N = 8$

to 10, $R^2 = 0.84$ suggests that Equation (1) provides a reasonable approximation to the relationships between glass composition and T_L. The s value of approximately 32°C for N > 8 includes the measurement error of 10 to 15°C and the model lack of fit. Removing MgO from the set of major components resulted in a lower R^2 and higher s. Generally, the R^2 and s values are such that the models have some lack of fit, and thus the S_i values in Table III should be considered to have moderate uncertainties. The measured versus calculated T_L values, shown for the $N = 13$ (SZ) set in Figure 1, indicate that Equation (1) fits the data without a significant bias.

The baddeleyite (Z) and PK primary phase fields were supported by smaller numbers of data points than that of ZS ($n = 14$ for Z and 19 for PK). This limited the number of T_{Li} values to $N = 7$ (i = Al_2O_3, B_2O_3, Li_2O, Na_2O, SiO_2, ZrO_2, and Others). Both sets have large values of s (85°C for PK and 93°C for Z) and low R^2 values, though the R^2 value for PK is comparable to the $N = 7$ set for ZS (Table II). The very low R^2 value for Z indicates that the corresponding T_{Li} values provide a rather rough estimate of T_L. This is also evident from extremely large positive and negative values of coefficients and a large T_{Li} value for Others. Clearly, the T_L database for baddeleyite is insufficient to support the model.

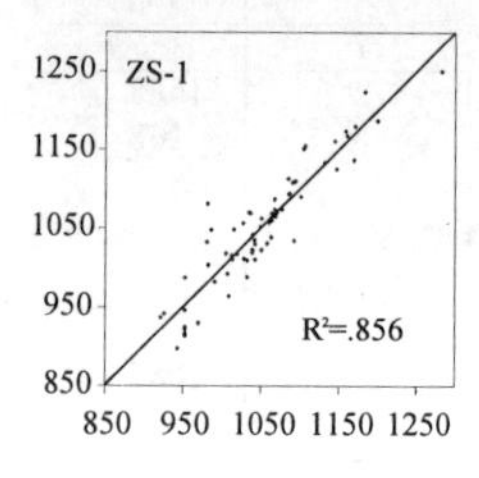

Figure 1. Measured Versus Calculated T_L (ZS-13 indicates zircon as the primary phase with $N = 13$)

Not unexpectedly, ZrO_2 has the strongest effect on T_L (approximately 41°C/mass% for ZS). Within the ZS and Z fields, alkali oxides decrease T_L (Na_2O by 34°C/mass% and Li_2O by 25°C/mass% for ZS). Not surprisingly, Na_2O increases T_L for glasses with PK as the primary phase, but the strong negative effect of Li_2O is unexpected, considering that Li_2O is also found in alkali-zirconium-silicates. Interestingly, P_2O_5 increases T_L for ZS by 11°C/mass%. Zirconium phosphate was not detected as a crystalline phase in the databases analyzed. Al_2O_3 somewhat increases (by 5°C/mass%) and B_2O_3 decreases (by 6°C/mass%) T_L for glasses with the ZS primary phase, but the T_L for Z is predicted to be strongly increased by Al_2O_3. SiO_2 has virtually no effect except for glasses with PK as the primary phase.

PRIMARY PHASE FIELDS: GRAPHICAL REPRESENTATION

For a graphical representation of the primary phase fields, we calculated component coefficients on the Na_2O-SiO_2-ZrO_2 submixture, a composition subspace, on which mass fractions are defined as $g_i' = g_i/(1 - g_{others})$, (i = Na_2O, SiO_2, ZrO_2), where the subscript *others* stands for all remaining glass components. Ternary submixtures have previously been used to illustrate phase behavior of HLW glasses. For example, the $(Na,Li)_2O$-B_2O_3-SiO_2 submixture displays phase separation [10], and the Na_2O-Al_2O_3-SiO_2 submixture indicates formation of nepheline [11]. Crum et al. [3] used the Na_2O-ZrO_2-SiO_2 submixture to delineate boundaries between primary phase fields for Zr-containing HLW glasses. They observed that baddeleyite was likely to form if $g'_{ZrO_2} > 0.22$ zircon was likely to form if $0.08 < g'_{ZrO_2} < 0.22$ and $g'_{Na_2O} < 0.22$, and alkali-zirconium-silicates and rare-earth-zirconates were likely to form if $0.08 < g'_{ZrO_2} < 0.22$ and $g'_{Na_2O} > 0.22$. Other than Zr-containing primary phases were likely to precipitate if $g'_{ZrO_2} < 0.08$.

We applied the Na_2O-ZrO_2-SiO_2 submixture approach to our database. In addition to drawing boundaries between primary phase fields, we also depicted T_L-isotherms using a modified form of Equation (1), i.e., $T_L^p = T_{L0}^p + T_{Lz}^p g_z' + T_{Ln}^p g_n'$, where the superscript p indicates the primary phase field, and the subscripts z and n stand for ZrO_2 and Na_2O. On the boundary between p and r primary phase fields, $T_L^p = T_L^r$, and thus $T_{L0}^p + T_{Lz}^p g_z' + T_{Ln}^p g_n' = T_{L0}^r + T_{Lz}^r g_z' + T_{Ln}^r g_n'$, or, after rearrangement

$$g_n' = k_{pr} + l_{pr} g_z' \tag{2}$$

where $k_{pr} = \dfrac{T_{L0}^r - T_{L0}^p}{T_{Ln}^p - T_{Ln}^r}$ and $l_{pr} = \dfrac{T_{Lz}^r - T_{Lz}^p}{T_{Ln}^p - T_{Ln}^r}$. Equation (2) represents the phase field boundary and the k_{pr} and l_{pr} coefficients relate the boundary slopes and intercepts to those of isotherms.

Attempts to compute the isotherms from the T_L database were successful only for ZS as the primary phase. We estimated the boundaries between primary phase fields by drawing lines that separate the data points for individual phases (Figure 2). Thus we obtained six values for k_{pr} and l_{pr} coefficients (two for each boundary). These values are shown in Table IV. Then we forced the isotherms for PK to intersect with the isotherms for SZ on the ZS-PK boundary. Finally, we obtained T_{Li} coefficients for Z by connecting intersects of ZS and PK isotherms with Z-ZS and Z-PK boundaries. The T_{Li} values are listed in Table V.

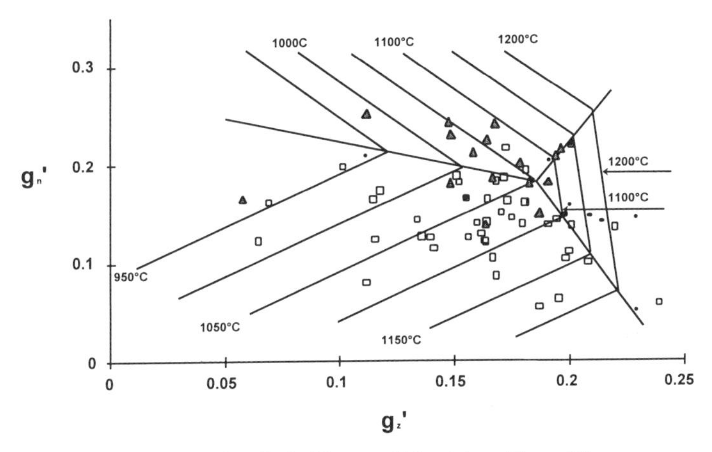

Figure 2. Na_2O-SiO_2-ZrO_2 Ternary Submixture Phase Diagram
(□—zircon, ▲—parakeldyshite, •—baddeleyite)

Figure 2 clearly indicates the opposite effect of Na_2O on T_L for ZS and PK fields whereas Z is little affected by Na_2O. It also shows that while T_L increases with ZrO_2 content in all fields, the slope is much steeper in the baddeleyite field. The primary-phase boundaries reasonably agree with those previously estimated by Crum et al. [3].

As the low R^2 values in Table V indicate, the ternary-phase diagram is a useful aid for rough orientation when formulating high-Zr HLW glasses, but cannot be used for estimating T_L. For example, we can estimate glass composition with a maximum ZrO_2 content that meets the constraint $T_L \leq 1050°C$ (needed to restrict solid phase settling in Joule-heated melters [1]). The 1050°C isotherm passes through the common intersection of the field boundaries, corresponding to 18.6 mass% ZrO_2 in the submixture. Since the submixture is only a fraction of the overall composition, the actual ZrO_2 content in a HLW glass with $T_L \leq 1050°C$ will be lower. The maximum ZrO_2 content experimentally verified in a glass with $T_L \leq 1050°C$ was 16.5 mass% [6-8].

Table IV. Coefficients for Boundaries between the Primary Phase Fields for Na_2O-SiO_2-ZrO_2 Ternary Submixture

	k_{pr}	l_{pr}
ZS-PK	0.270	-0.482
ZS-Z	0.758	-3.109
Z-PK	-0.356	2.891

Table V. T_L Component Coefficients (in °C) for Na_2O-SiO_2-ZrO_2 Ternary Submixture

	T^p_{L0}	T^p_{Lz}	T^p_{Ln}	R^2
ZS	1033	1070	-1006	0.598
Z	25	5202	324	0.104
PK	393	2210	1358	0.630

CONCLUSIONS

Available data on T_L for HLW glasses with Zr-containing primary crystallization phases provide T_{Li} values for all major HLW glass components. These values are reasonably accurate for HLW glasses that precipitate zircon as their primary phase and should be used with caution for HLW glasses with alkali-zirconium silicates as their primary phase. The current T_L database for HLW glasses with baddeleyite as the primary phase is insufficient to provide dependable T_{Li} values. Because T_L changes strongly with the ZrO_2 content in the baddeleyite field, we do not recommend formulating HLW glasses that form baddelyite as the primary phase.

ACKNOWLEDGMENTS

The authors thank Greg Piepel, Linda Davis, and Wayne Cosby for helpful suggestions. Trevor Plaisted is grateful to Associated Western Universities for his fellowship at Pacific Northwest National Laboratory (PNNL). Funding for this task was provided by the U.S. Department of Energy Office of Science and Technology. PNNL is operated for the U.S. Department of Energy by Battelle under Contract DE-AC06-76RLO 1830.

REFERENCES

1. P. Hrma, G. F. Piepel, M. J. Schweiger, D. E. Smith, D.-S. Kim, P. E. Redgate, J. D. Vienna, C. A. LoPresti, D. B. Simpson, D. K. Peeler, and M. H. Langowski, *Property/Composition Relationships for Hanford High-Level Waste Glasses Melting at 1150°C*, PNL-10359, Pacific Northwest Laboratory, Richland, Washington (1994).
2. D-S. Kim and P. Hrma, *Ceram. Trans.* 45, 327-337 (1994).
3. J. V. Crum, M. J. Schweiger, P. Hrma, J. D. Vienna. *Mat. Res. Soc. Symp. Proc.* 465, 79-85 (1997).
4. Q. Rao, G. F. Piepel, P. Hrma, and J. V. Crum. *J. Non-Cryst. Solids* 220, 17-29 (1997).
5. J. V. Crum, P. Hrma, M. J. Schweiger, and G. F. Piepel, *Ceram. Trans.* 87, 271-277 (1998).
6. D. K. Peeler, I. A. Reamer, J. D. Vienna, and J. V. Crum, *Preliminary Glass Formulation Report for INEEL HAW*, Savannah River Technology Center, Aiken, SC (1998).
7. B. A. Staples, D. K. Peeler, J. D. Vienna B. A. Scholes, and C. A. Musick, *The Preparation and Characterization of INTEC HAW*, INEEL/EXT-98-00971, Lockheed Martin Idaho Technologies Co, Idaho Falls, Idaho (1999).
8. J. D. Vienna, D. K. Peeler, T. J. Plaisted, R. L. Plaisted, I. A. Reamer, and J. V. Crum. *Ceram. Trans.* 107, in press.
9. G. F. Piepel, P. E. Redgate, and P. Hrma. *Ceram. Trans.* 61, 489-496 (1995).
10. D. K. Peeler and P. Hrma. *Ceram. Trans.* 45, 219-229 (1994).
11. H. Li, J. D. Vienna, P. Hrma, D. E. Smith, and M. J. Schweiger. *Mat. Res. Soc. Proc.* 465, 261-268, (1997).

ASSESSMENT OF NEPHELINE PRECIPITATION IN NUCLEAR WASTE GLASS VIA THERMOCHEMICAL MODELING

T. M. BESMANN*, K. E. SPEAR**, AND E. C. BEAHM*
*Oak Ridge National Laboratory, Oak Ridge, TN 37831-6063, tmb@ornl.gov
**Materials Science and Engineering Department, Pennsylvania State University, University Park, PA 16802-5005

ABSTRACT

A thermochemical representation of the Na-Al-Si-B-O system relevant for nuclear waste glass has been developed based on the associate species approach for the glass solution phase. Thermochemical data were assessed and associate species data determined for binary and ternary subsystems in the Na_2O-Al_2O_3-B_2O_3-SiO_2 system. Computed binary and ternary phase diagrams were compared to published diagrams during this process, with adjustments in data made as necessary to obtain consistent thermodynamic values. The resulting representation for the four oxide system was used to help understand the problem of nepheline precipitation in certain waste glass formulations.

INTRODUCTION

High-level nuclear and transuranic wastes are currently foreseen as being incorporated in a host glass for permanent disposal. A large number of glasses have been explored, with borosilicate glass as the typical base composition. Glass compositions are under development at Pacific Northwest National Laboratory (PNNL) and Savannah River Laboratory that will allow dissolution of the waste species in a glass matrix. Issues of glass stability are important in that the glass must remain mechanically intact and retain a low leach rate on exposure to moisture. A somewhat opposing goal is to maximize waste loading of the glass, with a significant economic gain associated with incremental increases in waste content.

A problem identified at PNNL is the precipitation of a nepheline phase ($Na_2O \cdot Al_2O_3 \cdot 2SiO_2$) within certain compositions during the cooling of glass, which weakens the network structure by removing the glass formers Al_2O_3 and SiO_2. The result is that nepheline precipitation in high-level waste glass limits waste species loading. It has been observed that compositions rich in Al_2O_3 and Na_2O are particularly prone to precipitating nepheline[1].

The rapid kinetics of nepheline formation in the cooling glass suggests that an equilibrium thermodynamic model may provide useful insights with regard to the compositional parameters governing its precipitation [1]. Currently, there are limited thermochemical models for complex glass compositions, with much of the guidance for glass chemistry based on empirical or semi-empirical approaches. In the work reported in this paper, a thermochemical model for the Na_2O-Al_2O_3-B_2O_3-SiO_2 crystalline and glass system has been developed and applied to the problem of nepheline precipitation.

SOLUTION MODEL

An associate model developed in the 1980s was used to represent the thermochemical behavior of liquid oxide solutions. With an accurate model of the liquid solution, the supercooled liquid therefore represents the chemically complex nuclear waste glass. The model was initially utilized for complex solutions by Hastie, Bonnell, and co-workers [2-5]. Other thermodynamic models have been used to represent nuclear waste glass, most notably the

Mat. Res. Soc. Symp. Proc. Vol. 608

modified quasichemical model of Pelton, Blander and co-workers [6-7] who initially used it to represent molten slag phases, and later, nuclear materials, including waste glass [8]. The associate model, however, is substantially easier to understand and use, and yet the model still accurately represents the limiting thermodynamic activities of components in these metastable glass phases.

Energies of interaction between end-member component oxides beyond those of ideal mixing often exist and their use in complex systems results in a significant multiplication of terms. These energies are included in the associate model by adding "associate species," with their respective formation energies, to the solution. Thus, the complex interaction terms are simply embodied in additional species in the solution. For example, in using the associate model for the Na_2O-Al_2O_3 binary oxide system, an ideal liquid solution phase was created from the liquid components $Na_2O(l)$ and $Al_2O_3(l)$, along with a $NaAlO_2(l)$ associate liquid species. In calculating the equilibrium state of the liquid phase containing the three species, the minimization of the total free energy determines the relative mole fractions of the species. This is conveniently accomplished using the thermochemical computational software ChemSage™ [9]. Since the system that includes the associate species is treated as an ideal solution, the activities are by definition equivalent to the species mole fraction. A more detailed description of the approach can be found in a recent review paper [10].

Na_2O-Al_2O_3-B_2O_3-SiO_2 MODEL

In developing the thermodynamic data file for the quaternary Na_2O-Al_2O_3-B_2O_3-SiO_2 system, the thermodynamic and phase diagram data for six binary (Na_2O-Al_2O_3, Na_2O-B_2O_3, Na_2O-SiO_2, Al_2O_3-B_2O_3, Al_2O_3-SiO_2, and B_2O_3-SiO_2) and four ternary (Na_2O-Al_2O_3-B_2O_3, Na_2O-Al_2O_3-SiO_2, Na_2O-B_2O_3-SiO_2, and Al_2O_3-B_2O_3-SiO_2) subsystems were assessed and optimized. The approach involved adjusting thermodynamic data so as to reproduce equilibrium phase diagrams as a means of testing and generating thermodynamic information for glass forming oxide systems. Since a phase diagram graphically depicts the equilibrium chemistry and thermodynamic properties of a system, the diagram can be calculated if the thermodynamic properties are known for all chemical species/phases that can form in the system. A large fraction of the needed thermochemical information has not been measured or reported, so a set of procedures for estimating or calculating the information was developed [10]. In all cases, the complete set of thermochemical information for a system is refined and tested to give reasonable thermodynamic and phase diagram information over wide ranges of temperature and composition.

ChemSage™ [9] was the primary tool for developing an assessed, internally consistent thermodynamic database, and for subsequent calculations of the equilibrium chemical behavior of the glass systems. The needed thermodynamic data are obtained from the literature and sources such as the assessed SGTE substance database associated with ChemSage™, our estimates, and simultaneously comparing and optimizing sets of phase equilibria and thermodynamic data. A primary source of phase diagram information is the set of volumes Phase Diagrams for Ceramists [11], plus literature that includes reports of previously optimized thermochemical data.

Several liquid associate species were used in modeling the quaternary Na_2O–Al_2O_3–B_2O_3–SiO_2 system. As a means of providing equal weighting to each liquid associate species, each species was adjusted to have a total of two non-oxygen atoms in its formula. A listing of the species is given below:

Single oxide liquid species

Na_2O	Al_2O_3	B_2O_3	Si_2O_4

Binary oxide liquid species

$(1/4)Al_6Si_2O_{13}$	$NaAlO_2$		
$(1/3)Na_4B_2O_5$	$NaBO_2$	$(1/3)Na_2B_4O_7$	$(1/5)Na_2B_8O_{13}$
$(2/5)Na_4SiO_4$	$(2/3)Na_2SiO_3$	$(1/2)Na_2Si_2O_5$	

Ternary oxide liquid species

$(2/3)NaAlSiO_4$	$(1/2)NaAlSi_2O_6$

An example of the computed phase relations compared with that experimentally determined [12] can be seen in the pseudo-binary diagrams for the nepheline/carnegeite-albite-silica system (Fig. 1). Although the homogeneity ranges for the crystalline phases were not modeled, the computed phase diagram reproduces the observed phase relations remarkably well given the simplicity of the model.

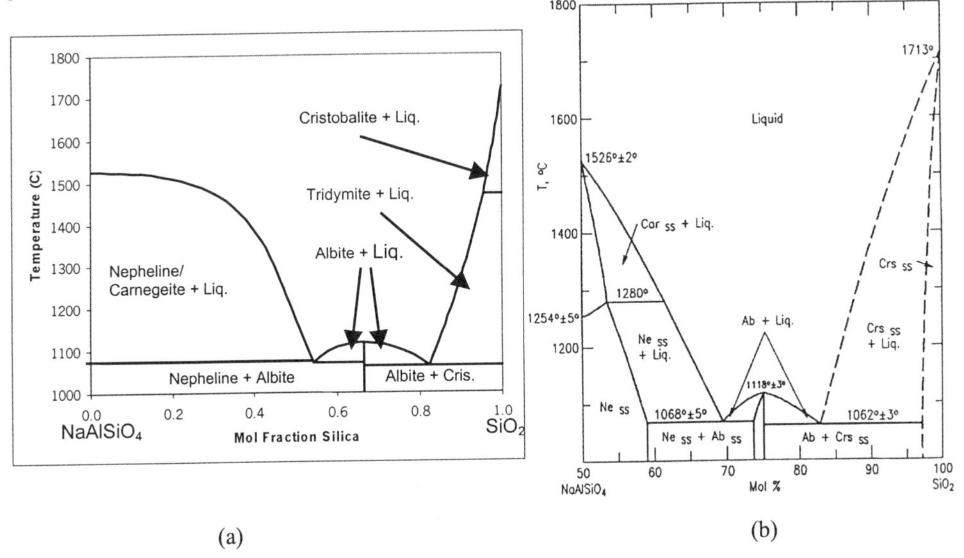

Fig. 1. (a) Computed and (b) published pseudo-binary phase diagram for the nepheline/carnegeite-albite-silica system (No. 10013 in ref. [12]).

NEPHELINE-GLASS PSEUDO-EQUILIBRIUM

As has been observed, the precipitation of nepheline occurs during glass cooling in the absence of other crystalline phases [1]. It is thus possible to model this behavior utilizing pseudo-equilibrium calculations in which nepheline and the glass/liquid phase are the only phases allowed to form, with all other crystalline phases prevented from being present. The thermochemical computations, therefore, would indicate either the glass alone is present, or the glass is present in equilibrium with nepheline.

Stability Region

Utilizing the ChemSage™ software and the thermochemical data file for the Na_2O-Al_2O_3-B_2O_3-SiO_2 system, the composition space of the ternary oxide Na_2O-Al_2O_3-SiO_2 system was explored at 800°C with no boria present and with 30 wt.% boria. The results can be seen in the ternary diagram of Fig. 2 with the binary oxides and the nepheline composition indicated.

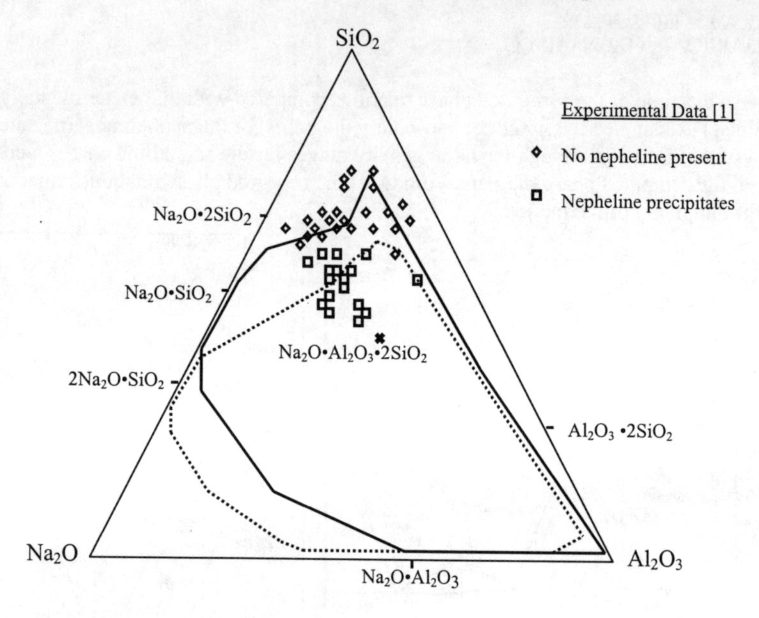

Fig. 2. Ternary Na_2O-Al_2O_3-SiO_2 phase space (wt. %) showing the computed stability region for nepheline plus the glass phase at 800°C with no boria (——) and 30 wt.% boria (.....) along with experimentally determined precipitation data [1].

Apparent from the calculational results is the wide compositional range over which nepheline is stable. The maximum silica composition lies along the SiO_2-$Na_2O{\bullet}Al_2O_3$ join. Important for the selection of waste compositions is the observation that the stability region decreases to lower silica content with increasing boria. Experimental results are also shown on the diagram, and agree reasonably with the results of the calculations. In the experimental work a variety of boria contents were used which span the 0-30 wt.% range, however, the Na_2O-Al_2O_3-B_2O_3-SiO_2 system also contained other components representative of practical waste compositions such as Li_2O, K_2O, CaO, and Fe_2O_3. Similar calculations were performed for the system at 600°C, and in that case the stability range for nepheline moved to higher silica contents, having a maximum on the ternary diagram of 90 wt.% silica without boria and 75 wt.% silica with 30 wt.% boria.

Formation Temperature

To aid in controlling nepheline precipitation it is useful to know the temperature at which the phase will form. Figure 3 is a plot of the calculated formation temperatures in the presence of the glass phase over the nepheline-albite ($Na_2O \bullet Al_2O_3 \bullet 6SiO_2$) compositional range. The calculations were performed with no boria present and with 30 wt% boria, with and without other phases allowed to form. The boria content can be seen to substantially reduce the formation temperature, with the potential presence of other phases not having a significant effect.

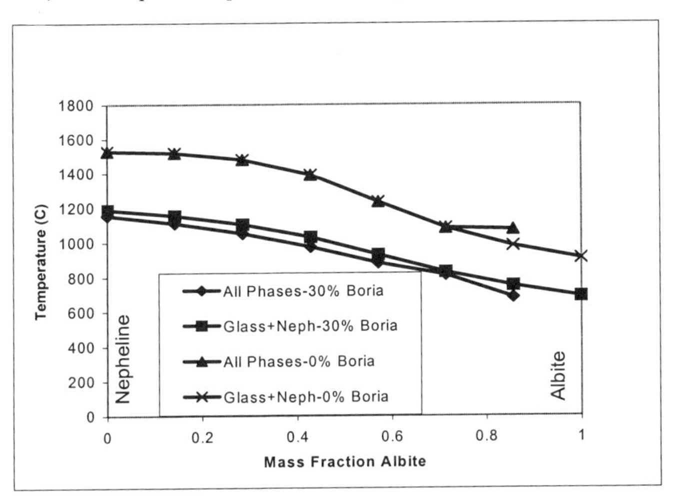

Fig. 3. Formation temperature for nepheline with no boria and 30 wt.% boria, assuming all phases can form or only the glass and nepheline.

The amount of nepheline expected to form under the pseudo-equilibrium conditions also varies with composition. Utilizing the thermochemical computations the amount of the glass and nepheline phases predicted can be determined under any set of conditions. As expected from the phase relations, both higher boria and higher silica results in lower nepheline amounts.

CONCLUSIONS

A relatively simple thermochemical model for the Na_2O-Al_2O_3-B_2O_3-SiO_2 system has been developed and has been shown to be accurate and predictive. The model can be easily expanded to include other chemical components. Other waste constituents will be added in the future to allow further modeling of high-level nuclear waste forms.

The predicted stability region for the formation of nepheline can have a significant influence on the choice of waste glass compositions. From the current work it is apparent that nepheline formation can be avoided by utilizing compositions high in silica and boria. In addition, even within the nepheline formation region, the amount of nepheline can be suppressed by higher silica and boria contents. These effects are directly related to lowering the activity of Na_2O in the glass phase, restricting the sodium available to form nepheline.

Reducing the liquidus temperature or the temperature at which the nepheline phase is computed to form can also aid in controlling nepheline formation and content. Compositions with lower nepheline formation temperatures are less likely to actually form the phase during cooling of the glass since at lower temperatures kinetics and mass transport will be slower.

These results are in general agreement with experimental observations [1]. They should provide guidance in the development of waste formulations that can avoid or minimize the detrimental formation of crystalline nepheline inclusions in glass that can occur during waste canister cooling.

ACKNOWLEDGMENTS

The authors would like to thank J. M. Vitek and D. F. Wilson for their constructive reviews. The research was supported by the DOE Environmental Management Science Program funded by the Office of Environmental Management's Office of Science and Technology, and administered jointly with the Office of Energy Research under contract DE-AC05-96OR22464 with Lockheed Martin Energy Research Corporation.

REFERENCES

1. H. Li, J. D. Vienna, P. Hrma, D. E. Smith, and M. J. Schweiger in *Scientific Basis for Nuclear Waste Management XX*, edited by W. J. Gray and I. R. Triay (Mater. Res. Soc. Proc. **465**, Pittsburgh, PA 1997), p. 261-276.

2. J.W. Hastie and D.W. Bonnell, High Temp. Sci. **19**, p. 275 (1985).

3. J.W. Hastie, Pure and Appl. Chem. **56**, p. 1583 (1984).

4. J.W. Hastie, E.R. Plante, and D.W. Bonnell, *Vaporization of Simulated Nuclear Waste Glass, NBSIR 83-2731*, NIST, Gaithersburg, MD (1983).

5. D.W. Bonnell and J.W. Hastie, High Temp. Sci. **26**, p. 313 (1990).

6. A.D. Pelton and M. Blander, Met. Trans. B, **17B**, p. 805 (1986).

7. M. Blander and A.D. Pelton, Geochim. Cosmochim. Acta, **51**, p. 85 (1987).

8. A. D. Pelton, Pure and Applied Chem. **69**(11), p. 2245 (1997).

9. G. Eriksson and K. Hack, Met. Trans B **21B**, p. 1,013 (1990).

10. K. E. Spear, T. M.Besmann, and E. C.Beahm, MRS Bulletin **24**(4), p. 37 (1999).

11. *Phase Diagrams for Ceramists*, Volumes 1-12, (1964-1996) The American Ceramic Society, Westerville, OH.

12. A. E. McHale and R. S. Roth, *Phase Equilibrium Diagrams*, *Vol. XII*, American Ceramic Society, Columbus, OH, 1996, p. 172.

CHARACTERISATION AND RADIATION RESISTANCE OF A MIXED-ALKALI BOROSILICATE GLASS FOR HIGH-LEVEL WASTE VITRIFICATION

J.M. RODERICK *, D. HOLLAND *, C.R. SCALES **
*Physics Department, Warwick University, Coventry, CV4 7AL, UK, phrxi@csv.warwick.ac.uk
**British Nuclear Fuels Ltd, Sellafield, Seascale, Cumbria, UK

ABSTRACT

Glasses related to those used for the vitrification of high-level waste (HLW) have been produced from the sodium-lithium borosilicate system. Their thermal and structural characteristics have been measured for a wide range of boron oxide contents (NBS series) and for a range of alkali oxide to boron oxide ratios, R, for several fixed silica to boron oxide ratios, K (ABS series). The NBS series of glasses was seen to exhibit a maximum in the glass transition temperature, T_g, and a minimum in the fraction of 4-coordinated borons, N_4, at 22 mol% and 29 mol% boron oxide, respectively. With variation of K and R, initial results indicate an increase in T_g to a maximum before a gradual decrease. Density measurements show a general increase before remaining within experimental error at larger R. Initial α-particle irradiation tests have been carried out on several base glass samples and show evidence of B(α,n) reactions, the occurrence of which could have important consequences for the future viability of wasteforms designed for reprocessing of higher burn-up fuel.

INTRODUCTION

Borosilicate glass is the current medium of choice for the immobilisation of radioactive high-level waste (HLW) in the UK and a concern that has been raised is the interaction of this particular system with certain types of commerical wastes anticipated over the next 10-15 years. Reprocessing of higher burn-up fuels with higher actinide contents can increase the alpha radiation dose to the storage matrix, with repercussions for its stability and consequent viability. The current work is based on a quaternary borosilicate and in order to ensure that the matrix is sufficiently resistant to the cumulative effects of long-term α-radiation exposure, laboratory simulation of these effects is used, albeit on a much shorter timescale. This particular study has concentrated on the interaction of individual matrix components with α-radiation. Theory [1] predicts the occurrence of B(α,n)N reactions, leading to chemical and ballistic consequences for the whole wasteform [2,3].

Studies of the Li_2O-B_2O_3-SiO_2 and Na_2O-B_2O_3-SiO_2 systems have been carried out by a number of workers [4-7] but less data is available on this mixed alkali system. In order to optimise this system for future use, thermal and structural characteristics have been investigated for a range of boron oxide contents.

EXPERIMENT

The raw batch materials were Wacomsil® quartz (SiO_2), sodium tetraborate ($Na_2B_4O_7$), boric acid (H_3BO_3), sodium carbonate (Na_2CO_3) and lithium carbonate (Li_2CO_3), all reagent grade. The powders were weighed and mixed thoroughly before being transferred to a 90Pt/10Rh crucible and heated at 300°C min^{-1} to the required melting temperature, which was maintained for approximately 20 minutes. The melts were quenched in distilled water and allowed to dry. The glasses were remelted at the same temperature for 10 minutes and either poured directly into moulds or splat-quenched, depending on the rate of devitrification of each

Mat. Res. Soc. Symp. Proc. Vol. 608 © 2000 Materials Research Society

sample. Two separate series of glasses were prepared. In the NBS series, the amount of B_2O_3 in the current composition used by BNFL was varied whilst the ratios of the other components remained the same. In the ABS series, the SiO_2/B_2O_3 ratio (*K*) was kept at fixed values and the alkali/B_2O_3 ratio (*R*) was varied. The NBS series of glasses were annealed at 450°C for 1 hour. The various values of *K* and *R* used are shown in Fig I.

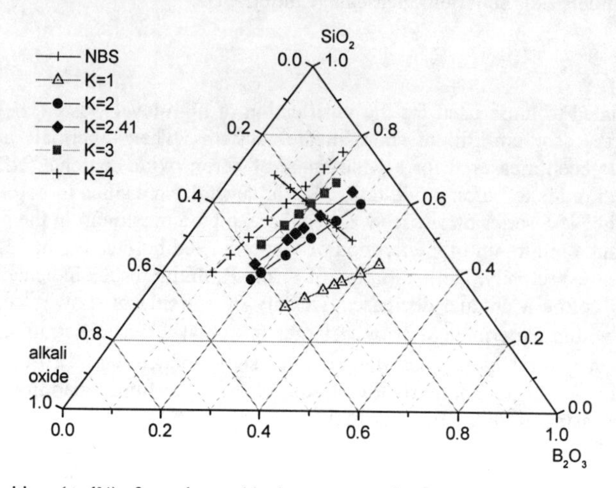

Fig I Batch compositions (mol%) of samples used in the present work. **O** represents samples of insufficient thermal stability or chemical durability.

Preliminary chemical analysis of the boron content of the samples has been carried out using quantitative ^{11}B NMR (Table 1). Boron can be present in the network in 3-fold coordination to oxygen to give trigonal planar [BO_3] units (B_3) or 4-fold coordination to oxygen to give tetrahedral [BO_4] units (B_4). By comparing the combined intensities of the B_3 and B_4 peaks with those of a reference material (Pyrex®) of known boron content, an estimate of the boron content of each sample was made. The experimental details of physical and structural characterisation using DTA, NMR and density measurement have been presented previously [8].

RESULTS

The nominal and analysed boron compositions of the glass samples are listed in Table I. Most samples were formed easily and produced transparent, amorphous glasses. Approaching the glass-making limit for this system, several compositions proved to be unsuitable for use and are highlighted with circles in Fig I. The variations in calculated density of each sample, as a function of K and R, are shown in Fig.II. It can be seen that there is a general trend toward a significant increase in measured density at low R, before a levelling off, within experimental error, at larger R. This is in fairly close agreement with the results obtained by Tang & Jiang [9] for the Na_2O-B_2O_3-SiO_2 system, where the value of R at which the density begins to level off increases with K. Fig IIb illustrates the similarity of this behaviour with that observed by Budhwani and Feller [10] for the Na- and Li-borosilicate systems. The Li-Na-borosilicate dependence on density lies between that of the individual alkali ternary systems, with a tendency to follow that of the Na-borosilicate.

Table I.
Nominal and analysed boron composition of prepared samples. Boron content analysed via MAS NMR.

Sample	K	R	Nominal B_2O_3 (mol %)	Analysed B_2O_3 (mol %) (± 5% of value)	Sample	K	R	Nominal B_2O_3 (mol %)	Analysed B_2O_3 (mol %) (± 5% of value)
NBS7	-	-	8.57	-	2B	2	0.3	30	30.7
NBS6	-	-	13.57	-	2C	2	0.63	27.5	29.1
MW	-	-	18.57	17.7	2D	2	1	25	25.7
NBS8	-	-	23.57	19.8	2E	2	1.44	22.5	25.1
NSB14	-	-	25.57	22.6	2F	2	2	20	24.8
NSB9	-	-	28.57	27.8	2G	2	2.26	19	23.6
NSB10	-	-	33.57	31.5	3B	3	0.54	22	20.1
ABS01	2.41	0.38	26.38	24.8	3C	3	1	20	21.68
ABS02	2.41	0.85	23.46	-	3D	3	1.26	19	20.34
ABS03	2.41	1.14	21.99	22.2	3E	3	2.25	16	17.82
ABS04	2.41	1.29	21.26	21.7	3F	3	0.75	21.05	19.34
ABS05	2.41	1.46	20.52	21.3	3G	3	1.6	17.85	-
ABS06	2.41	2.27	17.59	-	4B	4	0.33	18.75	17.12
1H	1	0.35	42.5	-	4C	4	0.71	17.5	17.32
1B	1	0.5	40	40.7	4D	4	1.15	16.25	16.44
1F	1	0.67	37.5	42.5	4E	4	1.67	15	-
1G	1	0.75	36.5	37.1	4F	4	2.27	13.75	-
1C	1	0.86	35	37.2	4G	4	3.00	12.5	-
1E	1	-	32.5	Outside glass	4H	4	3.88	11.25	-
1D	1	-	30	formation region.	4I	4	-	10	Outside g.f.r

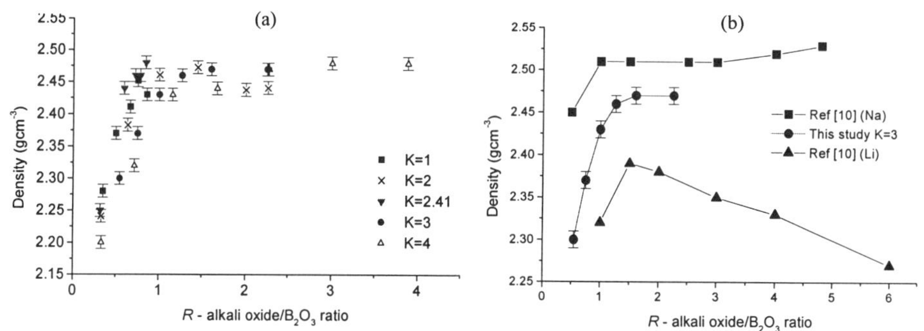

Fig II. (a) The dependence of density on R, for constant K, in the Na_2O-Li_2O-B_2O_3-SiO_2 glass system. (b) Comparison of K=3 results with Li-borosilicate and Na-borosilicate glasses, taken from [10]. Lines have been drawn to guide the eye.

The coordination of boron in the samples has been analysed using ^{11}B NMR and the results are summarised in Fig. III. These indicate behaviour similar to that observed in the ternary alkali borosilicate system. There exists a critical value of R, R_o, beyond which the fraction of 4-coordinated boron atoms, N_4, decreases. The value of R_o is a function of K, as is the maximum in N_4 at R_o.

Fig. IIIb also helps to illustrate the similarity of the behaviour of this system to that of the Na- and Li-borosilicate systems [4,6]. Initial addition of alkali increases T_g for all values of K up to a value of R, R_m, at which point T_g begins to decrease with increasing R. The rate of decrease

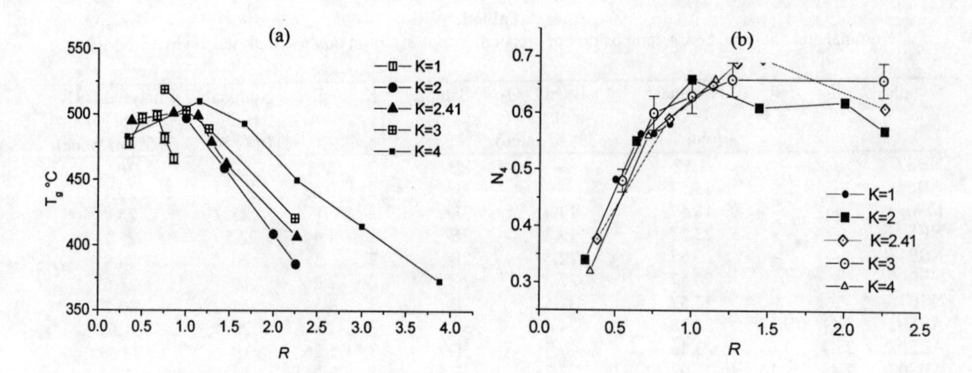

Fig. III. The dependence of (a) glass transition temperature (T_g) and (b) boron coordination on R for constant K in the Na_2O-Li_2O-B_2O_3-SiO_2 glass system. Lines have been drawn to guide the eye

is dependent on K and those samples with large K retain the highest value of T_g at any particular R. The values of R_o and R_m do show some agreement with the results reported by Boekenhauer [4] as shown in Table 2. It should be stressed that these figures are based on the results obtained up to the point of writing and that data from further samples will produce a more detailed and reliable picture. This will then be used in conjunction with structural models that have already been suggested [5,6,10] for the ternary borosilicate systems, to create a model for the effect of composition on structural evolution in the quaternary borosilicate system. As an additional point, although the relevant data has not been presented here explicitly, thermal analysis of the glasses has shown convincing evidence for the presence of phase separation in many of the samples used in the study. As discussed previously [11], in two-phase glasses, each glass must transform from the supercooled liquid state to the glassy state within its own transformation range. Hence, a two-phase glass must exhibit two different T_gs. This has been observed in the current study and further results will be published in the near future.

Table II.
Comparison of values of R which maximise T_g and N_4 for different K
(a) For Li-borosilicate system, from [4]. (b) For present work.

K	The value of R which maximises:			
	T_g		N_4	
	(a)	(b)	(a)	(b)
1	0.6	0.625	1.1	-
2	1.0	-	1.5	1
2.41	-	0.76	-	1.5
3	1.5	-	1.8	1.75
4	-	1.25	-	-

The boron coordination and glass transition temperature of the modified BNFL base glass (NBS series) were also investigated and the results are shown in Fig IV. During the initial stage of radioactive decay the glass will be exposed to elevated temperatures for prolonged periods. It is thus desirable that T_g be as high as possible without compromising the cost of processing or heavy element solubility. Fig IV illustrates how the addition of B_2O_3 leads to a measureable increase in T_g before decreasing beyond ~24 mol% B_2O_3 content. It was observed that there was a corresponding decrease in processing temperature with increased boron content which could mean that a glass containing a greater B_2O_3 content than that currently used could provide a more stable and cost effective wasteform. Changes in boron speciation are consistent with the removal of the extra oxygen needed to form B_4 as alkali oxide and SiO_2 are replaced by B_2O_3.

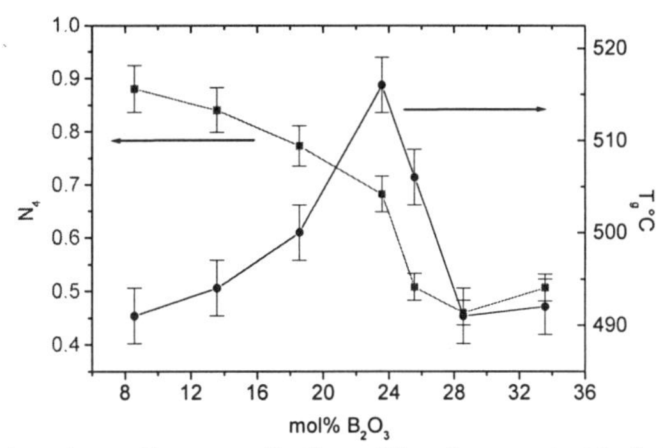

Fig. IV. The dependence of boron coordination and T_g on boron content for the NBS samples. Lines have been drawn to guide the eye.

Preliminary investigation of the behaviour of the NBS glasses under irradiation has produced evidence of interactions between boron and α-particles according to the reaction,

$$^{10}B + \alpha \rightarrow ^{13*}N + n \quad (1)$$

This has been concluded from the results of monitoring the irradiated sample for a period of time with a Ge-scintillator γ-ray detector. The excited ^{13}N nucleus releases a γ-ray via positron annihilation with a characteristic half-life of 10 minutes. This was observed as shown in Fig V. The presence of this reaction implies the possibility of interaction between the boron and α-particles generated by the decay of actinides present in the waste. This could compromise the critical properties, such as strength and chemical durability, that the glass composition has been tailored to produce. Further irradiation testing of the NBS series glasses will be carried out utilising a moderated He-3 neutron detector. Similar work has been carried out previously [1], and has established the propensity for boron to interact with α-particles

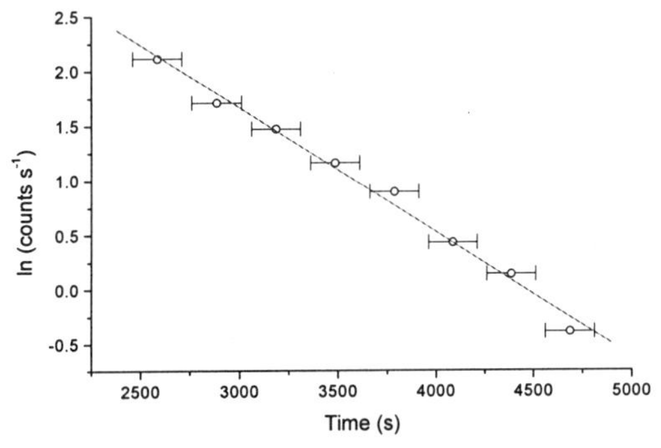

Fig V. Decay characteristics of radioactive species present in base glass after irradiation. ($\tau_{1/2}$ 10.1 minutes)

CONCLUSIONS

Changing the relative proportions of B_2O_3 in the current waste vitrification glass used by BNFL (the NBS series) changes physical properties such as T_g by a significant amount. Preliminary characterisation of the ABS glass samples has provided evidence of a change in the boron coordination and glass transition temperature with the amount of alkali oxide in the glass, similar to that observed in the single-alkali systems. Together with information on density variations, this will enable a determination of the likely structural species present in the glasses and more specifically, how the roles attributed to the alkali species are apportioned between the Li and Na. Further investigation of the ABS series samples using thermal expansion measurements and ^{29}Si MAS NMR will provide a more detailed picture of the short-range order present in the system over a large range of compositions. A suitable composition has been chosen and several samples have been produced to study the mixed-alkali effect in this particular system, the results of which will be reported at a later date. Preliminary investigation of the interaction of alpha radiation with the NBS glass has shown evidence of a reaction with ^{10}B, which will be further studied using neutron and γ-ray spectrometry.

ACKNOWLEDGEMENTS

The authors would like to express their gratitude to Dr. N.M. Clarke of Birmingham University for his advice and collaboration on use of their Radial Ridge Cyclotron Facility. This research has been supported by British Nuclear Fuels Ltd. and funded by an EPSRC Studentship award.

REFERENCES

1. G.J.H. Jacobs & H. Liskien, *Ann. Nucl. Energy* **10** No.10 (1983), 541.
2. W.S. Snyder and J. Neufeld, *Physical Review* **97** No.6 (1955), 1636.
3. R.C. Ewing and W.J. Weber, *J. Mater. Res.* **12** No.8 (1997), 1946.
4. R. Boekenhauer et al, *J. Non-Cryst. Solids* **175** (1994), 137.
5. W.J. Dell & P.J. Bray, *J. Non-Cryst. Solids* **58** (1983) 1.
6. M.P. Brungs and E.R. McCartney, *Phys. Chem. Glasses* **16** (1975), 48.
7. J. Xhong and X. Wu, *J. Non-Cryst. Solids* **107** (1988), 81.
8. J.M. Roderick, D. Holland and C.R. Scales, *Proceedings of the Third Conference on Borate Glasses, Crystals and Melts,* Phys. Chem. Glasses, in press.
9. Y. Tang & Z. Jiang, *J. Non-Cryst Solids* **189** (1995), 251.
10. K. Budhwani & S. Feller, *Phys. Chem. Glasses* **36** No.4 (1975), 183.
11. O.V. Mazurin, M.V. Streltsina & A.S. Totesh, *Phys. Chem. Glasses* **10** No.2 (1969), 63.

THE BEHAVIOR OF SILICON AND BORON IN THE SURFACE OF CORRODED NUCLEAR WASTE GLASSES: AN EFTEM STUDY

E. C. Buck*, K. L. Smith**, and M. G. Blackford**
*Argonne National Laboratory
**Australian Nuclear Science and Technology Organization

ABSTRACT

Using electron energy-loss filtered transmission electron microscopy (EFTEM), we have observed the formation of silicon-rich zones on the corroded surface of a West Valley (WV6) glass. This layer is approximately 100-200 nm thick and is directly underneath a precipitated smectite clay layer. Under conventional (C)TEM illumination, this layer is invisible; indeed, more commonly used analytical techniques, such as x-ray energy dispersive spectroscopy (EDS), have failed to describe fully the localized changes in the boron and silicon contents across this region. Similar silicon-rich and boron-depleted zones were not found on corroded Savannah River Laboratory (SRL) borosilicate glasses, including SRL-EA and SRL-51, although they possessed similar-looking clay layers. This study demonstrates a new tool for examining the corroded surfaces of materials.

INTRODUCTION

Following the initial high dissolution rate of waste glasses (forward rate), there is a significant drop in the dissolution rate as the reaction progresses. This behavior is interpreted and modeled by assuming silicic acid (H_4SiO_4) saturation with respect to the dissolving solid, an affinity reaction term [1]. However, Bourcier [2] has shown that regression analysis of release rate data cannot discriminate between either the Grambow reaction affinity model or a surface transportation model (i.e., diffusion term). Gin et al. [3] maintain that a "gel" layer forms on the corroded glass surface, and this layer acts as a diffusion barrier to the continued dissolution of the glass. Until now the direct evidence for "gel" layers has remained elusive. In this study, we present direct evidence of a silica-rich layer with (EFTEM), which may be the "gel" described by others [3]; however, this layer has only been found in a limited set of simulant waste glasses.

Reacted borosilicate glasses commonly exhibit a thin smectite clay layer that is clearly discernible by conventional transmission electron microscopy (CTEM). At the stage of glass reaction when the dissolution rate starts to drop, this layer is typically between 50 and 200 nm thick and consists of a fine bundles of smectite clays oriented perpendicular to the surface. This layer is probably too porous to act as a diffusion barrier (see Figure 1).

Electron Energy-Loss Spectroscopy (EELS) is a well-established TEM technique that is based on the fact that the energy electrons lose as they pass through TEM specimens is determined by the crystal chemistry of the specimen (see Figure 2). Energy filtered imaging (EFI or EFTEM) is a relatively new technique wherein images and compositional maps of TEM specimens are formed using electrons from a limited sections of full EELS spectra [4]. Energy filtered imaging is quantitative, which means that the contrast seen in the images is directly related to the number of atoms per unit area [4]. Furthermore, the spatial resolution is better than that obtained with EDS [5] or secondary ion mass spectrometry [6], two techniques that have previously been used to profile elements through surfaces. Using EFI, we were able to directly observe the distribution of the major glass components, Si, B, and O, in various glasses. In some borosilicate glasses, direct evidence was found of a silica-rich layer, which may be a diffusion barrier.

Mat. Res. Soc. Symp. Proc. Vol. 608

a

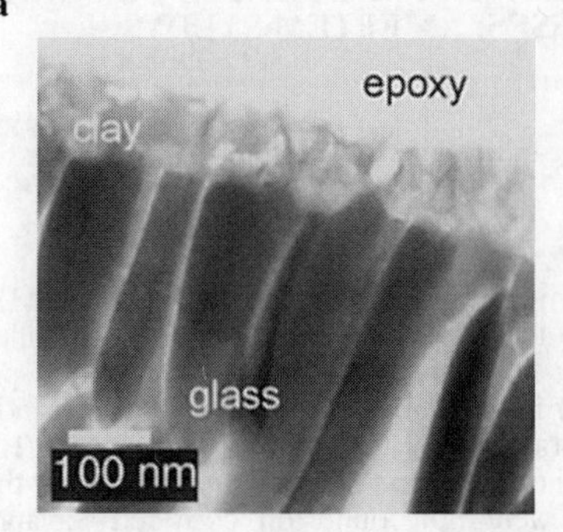

b

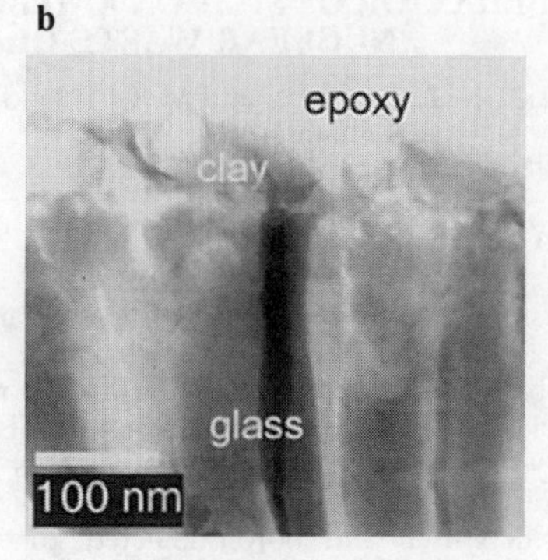

Figure 1. Thin clay layer visible from bright field TEM images of cross-sectioned (**a**) WV6 and (**b**) SRL-51S borosilicate glasses after 182 d reaction. With CTEM imaging the most prominent feature on the corroded surface was the smectite clay layer. The glass underneath displayed almost uniform contrast, indicating no compositional or structural changes. The samples were prepared by embedding corroded glass in epoxy and thin sectioning with an ultramicrotome. The glass shatters during this process, producing shards of glass which can be seen in the images.

a　　b

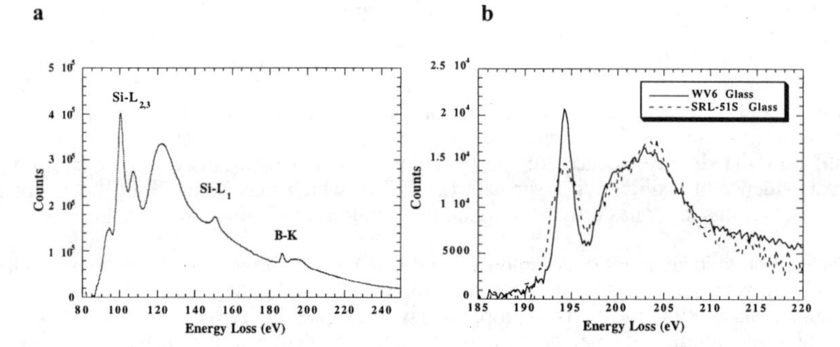

Figure 2. (**a**) Electron energy-loss spectroscopy of WV6 glass showing Si-$L_{2,3}$ and B-K edges. The B-K-edges for (**b**) WV6 glass and SRL-51S glass indicate that the WV6 glass has a higher level of trigonal boron compared to the SRL-51S glass.

EXPERIMENTAL PROCEDURE

We examined TEM specimens of two borosilicate glasses after they had been corrosion tested for various times. The glasses were a non-radioactive West Valley glass (WV6) and simulated SRL Tank 51 glass (SRL-51S) that were prepared at the Catholic University of America and Argonne National Laboratory (ANL), respectively. The nominal compositions of the glasses are listed in Tables 1 and 2 (excluding elements not analyzed with EDS). These glasses are part of a suite of nuclear waste glasses that were reacted under extended Product Consistency Test-B (PCT-B) conditions [7] as follows. The glass was crushed and sieved into ~75-150 μm particles, then equal masses of sieved glass and EJ-13 simulated groundwater were placed in stainless steel Parr™ vessels. The initial glass surface area to leachant volume (S/V) ratio was 20,000 m^{-1}. Tests were maintained at 90°C for periods of 14 to 364 d. The solution analyses from these tests have been published in detail elsewhere [7]. After the corrosion tests were terminated, selected particles of the reacted glass were extracted, embedded in an epoxy resin, and thin-sectioned with an ultramicrotome. The thin-sections were examined at the JEOL2010F-GIF2000 facility of the Australian Nuclear Science and Technology Organisation (ANSTO) and at the JEOL2000FXII-GIF200 facility of ANL.

Compositional maps were calculated using the "3-window method" [4]. Two pre-edge images were used to model and strip the background from under the post-edge region. Images were produced from the $Si\text{-}L_{2,3}$, B-K, and O-K edges. The EFTEM images were obtained at a microscope magnification of 1200x using 0.1 eV/channel dispersion, 2x binning, and exposure times ranging from 1 s to 10 s. The boron near edge structure describes the proportion of trigonal ($^{[3]}B$) and tetrahedral ($^{[4]}B$) boron in the glass [8,9].

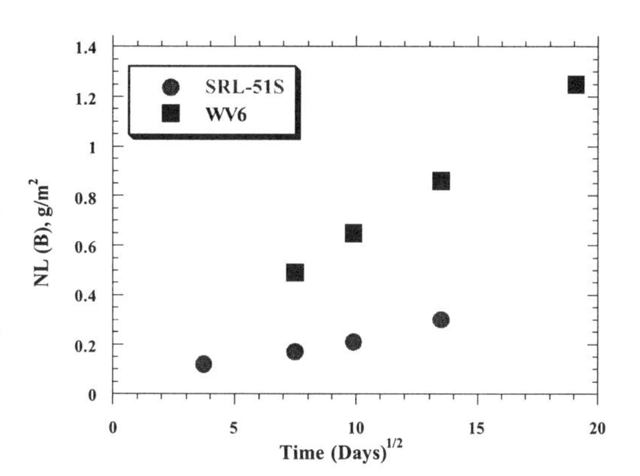

Figure 3. Plot showing release of boron from SRl-51S and WV6 glasses at an S/V of 20 000m^{-1} at 90°C versus root time. Adapted from Bates et al. [7]].

RESULTS

Corrosion Testing Results

As stated above, detailed solution analyses from all the corrosion tests that were conducted on the glasses in this study are published elsewhere [7]. Pertinent to this study is the fact that the reported release rates of boron from the WV6 glass are higher than those from the SRL-51S glass (see Figure 3). For example, after 182 d the normalized release of boron from the SRL-51S and WV6 glasses were 0.3 and 0.85 g/m^2 respectively, whereas the measured Si concentrations, were 398 and 110 mg/L, and the pHs were 11.4 and 9.

Electron Energy-Loss Spectroscopy and Energy Filtered Imaging Results

The B-K core-loss edge can be used to probe the chemistry around B, providing information on the coordination environment and on the relative amounts of trigonal ($^{[3]}B$) and tetrahedral ($^{[4]}B$) boron in a sample [8,9]. The boron edge is characterized by two strong features; a sharp peak at ~194 eV (the edge feature of $^{[3]}B$) and the broader peak at 198-200 eV (the edge feature of $^{[4]}B$). The spectrum in Fig. 2b suggests that the WV6 glass possesses a higher level of $^{[3]}B$ than SRL-51S, based on the relative intensities of these features.

Energy filtered imaging revealed the presence of silicon-rich zones on the corroded surface of WV6-45 sample directly underneath layers of precipitated smectite clay. The thickness of the silicon-rich layer on samples reacted at 20 000 m^{-1} varies from ~100 nm after 182 d to 150 nm thick after 364 d. Figure 4 shows EFTEM maps from a WV6 sample reacted at 20 000 m^{-1} for 182 d.

An extremely thin zone enriched in silicon was also seen on a WV6 glass specimen that had been reacted at 2 000 m^{-1} for 30 days. Conventional TEM showed that the clay layer on this latter specimen was <10 nm thick and patchy and EFI silicon maps showed silicon enrichment to a depth of ~10 nm. These silicon-rich zones in all the specimens examined were not readily identifiable using EDS. Energy filtered mapping of the SRL-51S glass reacted at 20 000 m^{-1} for 182 d (see Figure 5) shows no surface enrichment of silicon even though the clay layer has a similar thickness to the clay layer observed on the WV6 glass. The boron map showed the absence of boron in the clay layer.

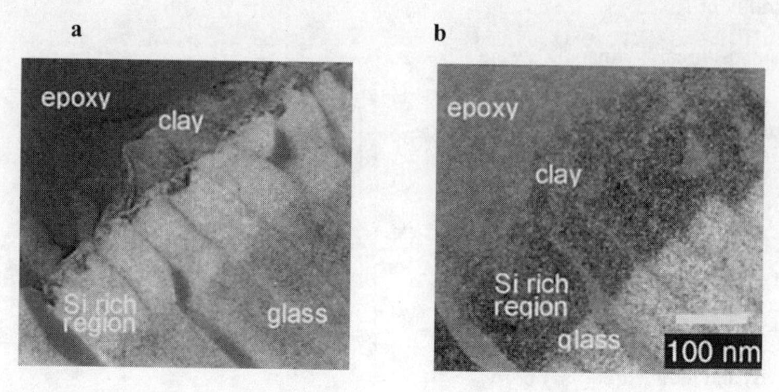

Figure 4. West Valley (WV6) glass reacted at 20 000 m^{-1} for 182 d. The oxygen map showed both the glass and the thin clay layer. The silicon map (**a**) showed the presence of an enriched region just underneath the clay layer. The boron map (**b**) revealed the complete absence of boron in the 150 nm thick layer underneath the clay layer.

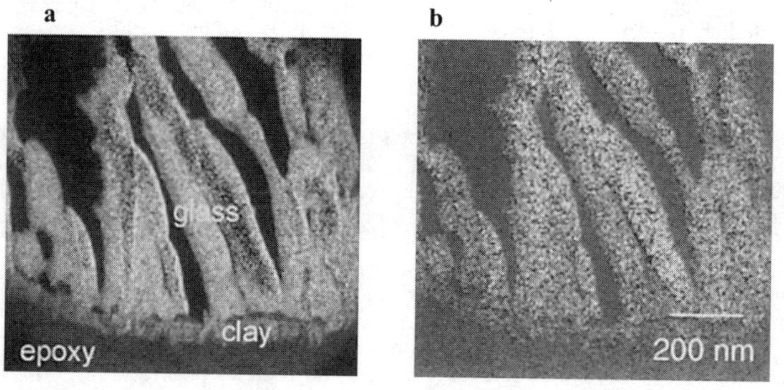

Figure 5. Savannah River Laboratory (SRL) Tank 51S simulated sludge glass reacted at 20 000 m^{-1} for 182 d. The silicon map (**a**) showed no enrichment of silicon even though the clay layer is similar in thickness to the layer observed in Figure 4. The boron map (**b**) indicated the absence of boron in the clay layer.

Table 1. Analyses[1] of WV6 Glass Reacted at an S/V of 20 000 m^{-1} for 182 d (n= number of measurements, s =standard deviation).

Oxide (wt%)	Clay Layer[2] (n=5)		Altered Si-Rich Region[2] (n=5)		Unreacted Glass[2] (n=5)		Reported Glass Composition[3,4]
	mean	s	mean	s	mean	s	
Na_2O[5]	2.38	0.34	0.55	0.16	0.43	0.44	10.42
MgO	4.68	0.18	0.37	0.10	0.75	0.17	1.16
Al_2O_3	13.79	0.63	10.86	0.32	9.55	0.26	7.82
SiO_2	51.86	0.94	61.14	2.51	63.20	1.75	53.43
P_2O_5	0.80	0.64	3.73	1.16	3.33	0.61	ND[6]
SO_3	0.74	0.24	0.43	0.05	0.39	0.13	ND[6]
K_2O	1.44	0.23	0.75	0.33	0.66	0.55	6.52
CaO	0.79	0.32	0.60	0.19	0.52	0.14	0.63
TiO_2	0.76	0.06	1.23	0.09	1.23	0.07	1.04
MnO	5.01	0.12	0.35	0.09	0.78	0.15	1.30
Fe_2O_3	13.45	0.23	12.22	0.39	12.03	0.86	15.64
NiO	0.21	0.07	ND[6]	--	ND[6]	–	0.33
ZrO_2	0.84	0.20	1.87	0.27	1.74	0.30	1.72
ThO_2	1.63	0.43	5.32	0.34	4.62	0.23	ND[6]
UO_2	ND	--	0.58	0.30	0.78	0.28	0.77
Total	98.40	--	100	--	100.00	–	100.00

Table 2. Analyses[1] of SRL51S Glass Reacted at an S/V of 20 000 m^{-1} for 182 d.

Oxide (wt%)	Clay Layer[2] (n=2)		Unreacted Glass[2] (n=6)		Reported Glass Composition[4]
	mean	s	mean	s	
Na_2O[5]	3.06	1.16	0.59	0.54	10.44
MgO	3.92	0.64	2.17	0.16	2.23
Al_2O_3	7.88	0.43	6.98	0.24	5.67
SiO_2	56.75	0.18	69.06	1.40	61.20
P_2O_5	1.36	0.19	1.14	0.10	0.61
SO_3	0.76	0.08	0.11	0.06	ND[6]
K_2O	0.44	0.27	0.33	0.31	1.51
CaO	0.39	0.16	1.54	0.17	1.51
Cr_2O_3	0.58	0.18	0.59	0.05	0.48
MnO	1.97	0.28	1.32	0.09	1.54
Fe_2O_3	18.94	0.79	14.48	0.34	13.32
NiO	0.57	0.13	0.37	0.04	0.30
UO_2	0.74	0.40	1.24	0.29	1.19
Total	97.32	–	99.90	–	100

[1] Data are normalized and exclude the major elements B and Li, as well as the minor elements, including rare earths and noble metals.

[2]Quantification performed with NIST DTSA software [see G. R. Lumpkin, K. L. Smith, and R. Gieré, *Micron* **28** (1994) 57-68]. Errors were 5% for most elements.

[3]W. L. Ebert, *Ceramic Trans.* **61** (1996) pg. 473.

[4]Dissolved and analyzed glass by mass spectrometry [6]. Reported uncertainty in measurement was approximately 15% for most elements.

[5]The focused electron probes in scanning and transmission electron microscopes cause alkali (in particular Na^+) migration in alkali aluminosilicate glasses and minerals, making chemical analyses of these phases difficult [see G. B. Morgan and D. London, *Amer. Mineral.* **81** (1996) 1176-1185].

[6]ND = no data available for this element.

DISCUSSION

The clays from the tests of WV6 and SRL-51S were enriched in Mg, Mn, and Ni relative to the glass and depleted in U (see Tables 1 and 2). These analyses of clay agree with previous tests with similar borosilicate glasses [5,6,10]. There was significant enrichment of Th, Zr, and P in the Si-rich layer in the WV6 glass. The compositional variations in the WV6 and SRL-51S glass may account for the variation in boron coordination. The energetics of the charge balance mechanisms in borosilicate glass dictate that $NaAl^{3+}$ should form before $NaFe^{3+}$ or NaB^{3+}; Na is used to charge balance Al^{3+} and once all the Al is accounted for, NaB^{3+} will start to emerge [11]. This might explain the higher level of $^{[3]}B$ in the WV6 glass.

Although the WV6 boron release data (see Figure 3) follow a $t^{1/2}$ relationship, this does not prove that the observed Si-rich layer is a diffusion barrier. However, supporting evidence comes from EFTEM images obtained at 364 d where a 100-200 nm thick Si-enriched layer was also found. The SRL-51S boron release data do not appear to follow the $t^{1/2}$ relationship, and no Si-rich layer was observed with EFI. As the compositional maps reflect the actual density of atoms [4], the Si-rich layer on the corroded WV6 glass has a higher amount of silicon per unit area than the underlying uncorroded glass, even though the EDS result indicates that the Si level is lower in the layer. This suggests that restructuring of the glass occurred as boron was leached from the surface. Changes in the deformation produced by the microtome in this region observed in some CTEM images also support this contention. The Si-rich layer found in the corroded WV6 glass does agree with the models of glass corrosion and, therefore, this layer might be the real "gel' layer [3].

CONCLUSION

We have a) demonstrated differences in B coordination between two types of borosilicate glass, b) proven the occurrence of Si-rich zones in corroded WV6 glass, c) shown that the level of boron within the Si-rich layer is low, and d) shown that the boundary between the Si-layer and glass underneath is very sharp. This study demonstrates that EELS and EFTEM are a powerful tools for characterizing and assessing the corrosion behavior of materials.

ACKNOWLEDGEMENTS

Tests were run by W. L. Ebert, J. K. Bates, and J. E. Emery. One of us (ECB) would like to thank the Australian Nuclear Science and Technology Organisation for providing financial support for a visit to their JEOL2010F-GIF2000 Microscopy Facility in Sydney, Australia.

REFERENCES

[1] D. M. Strachan, W. L. Bourcier, and B. P. McGrail, *Rad. Waste Mgmt. Environ. Res.* **19** (1994) 129-145.
[2] W. L. Bourcier, *Mater. Res. Soc. Symp. Proc.* **333** (1994) 69-82.
[3] S. Gin, E. Vernaz, C. Jégou, and F. Larche, Proc. 18th Inter. Congress on Glass (1998), Section **B10** pp. 64-69; G. Berger, C. Claparols, C. Guy, and V. Daux, *Geochim. Cosmo. Acta.* **58** (1994)4875-4886.
[4] F. Hofer, W. Grogger, G. Kothleitner, and P. Warbichler, *Ultramicroscopy*, **67** (1997) 83-103.
[5] J. K. Bates, D. L. Lam, and M. J. Steindler, *Mater. Res. Soc. Symp. Proc.* **15** (1983) 183-190.
[6] B. W. Biwer, J. K. Bates, T. A. Abrajano, and J. P. Bradley, *Mater. Res. Soc. Symp. Proc.* **176** (1990) 255-263
[7] J. K. Bates et al., ANL Technical Support Program for DOE Office of Environmental Management, Annual Report October 1994- September 1995 ANL-96/11 (1996).
[8] L. A. J. Garvie, P. R. Buseck, and A. J. Craven, *Amer. Miner.* **80** (1995) 1132-1144.
[9] M. E. Fleet and S. Muthupari, *J. Non Cryst. Sol.* **255** (1999) 233-241.
[10] W. L. Ebert, J. K. Bates, and W. L. Bourcier, *Waste Mgmt.* **11** (1994) 205-221.
[11] J. J. De Yoreo and A. Navrotsky, *J. Am. Ceram. Soc.* **73** (1990) 2068-2072.

CORROSION OF GLASS-BONDED SODALITE AS A FUNCTION OF pH AND TEMPERATURE

L. R. MORSS, M. L. STANLEY, C. D. TATKO, AND W. L. EBERT
Chemical Technology Division, Argonne National Laboratory, Argonne, IL 60439

ABSTRACT

This paper reports the results of corrosion tests with glass-bonded sodalite, a ceramic waste form (CWF) that is being developed to immobilize radioactive electrorefiner salt used to condition spent sodium-bonded nuclear fuel, and with sodalite and binder glass, the two major components of the CWF. These tests were performed with dilute pH-buffered solutions in the pH range of 5-10 at temperatures of 70 and 90□C to determine the pH dependences of the forward dissolution rates of the CWF and its components. The tests show that the pH dependences of the dissolution rates of sodalite, binder glass, and glass-bonded sodalite are similar to the pH dependence of dissolution rate of borosilicate nuclear waste glasses, with a negative pH dependence in the acidic region and a positive pH dependence in the basic region. The dissolution rates are higher at 90□C than at 70□C. Our results on the forward dissolution rates and their temperature and pH dependences will be used as components of a waste form degradation model to predict the long-term behavior of the CWF in a nuclear waste repository.

INTRODUCTION

The Electrometallurgical Treatment Program at Argonne National Laboratory (ANL) is developing a conditioning process for treatment of some of the U.S. Department of Energy spent sodium-bonded nuclear fuels that may not be suitable for direct geological disposal. In this process, uranium is electrorefined from spent fuel in a molten LiCl-KCl electrolyte, and fission products (Rb, Cs, Sr, Ba, and rare earths) and actinides accumulate in the molten salt. To make the salt acceptable for disposition in a repository, the salt is blended with dried zeolite 4A, $Na_{12}(AlSiO_4)_{12}$, at 500□C. The salt-loaded zeolite is mixed with a glass binder in a 3:1 mass ratio of salt-loaded zeolite to glass. The mixture is processed at high temperature and pressure to convert the zeolite into sodalite, $Na_8(AlSiO_4)_6Cl_2$, forming glass-bonded sodalite, the ceramic waste form (CWF) [1]. This paper reports the dissolution rates of the sodalite and glass binder components of the CWF, as well as the CWF itself, as a function of temperature and solution pH. These rates will be used to provide parameters for the ceramic waste form model.

Several studies of the effects of pH and solution conditions on dissolution have been carried out on glasses and aluminosilicate minerals [2-7]. The equation for the dissolution rate of aluminosilicate minerals is based upon the model of Aagaard and Helgeson [8], who postulated the formation of an activated complex at the mineral surface, the decomposition of which is the rate-controlling step in dissolution:

$$\text{rate} = S\ \{ k_0 \bullet 10^{(\eta \bullet pH)} \bullet e^{(-E_a/RT)} \bullet (1 - Q/[H_4SiO_4]_{sat})\} \qquad (1)$$

where

rate = dissolution rate of material ($g\bullet d^{-1}$)
S = surface area of material (m^2)
k_0 = rate constant ($g\bullet m^{-2}\bullet d^{-1}$)
η = pH dependence factor
E_a = activation energy ($kJ\bullet mol^{-1}$), reflecting temperature dependence

Mat. Res. Soc. Symp. Proc. Vol. 608 © 2000 Materials Research Society

R = gas constant (8.314×10^{-3} kJ•mol^{-1}•K^{-1})
T = absolute temperature (K)
Q = concentration of H_4SiO_4 in solution (mg/L)
$[H_4SiO_4]_{sat}$ = saturation concentration of H_4SiO_4 in solution (mg/L)

Our study has determined values of η, the parameter that represents the pH dependence of dissolution rate, from short-term corrosion tests of glass binder, sodalite, and CWF in dilute buffer solutions at pH values between 5 and 10 at 70 and 90□C.

EXPERIMENT

Samples of borosilicate glass binder (Pemco Corp, Baltimore, MD) synthetic sodalite with a clay binder, and CWF were prepared by hot isostatic pressing. Monoliths of each material were cored and cut into wafers nominally 10 mm in diameter and 1 mm thick. The wafers were polished with abrasives to a 600-grit finish. Scanning and transmission electron microscopies have shown that the CWF has two major phases (glass and sodalite) as well as minor phases (nepheline, halite, oxides, rare earth silicates, and clay). Preparation details and sample characterizations have been published [1].

The polished pellets were tested according to the MCC-1 procedure [9] in Teflon containers with buffered leachant solutions having the compositions and pH values shown in Table I. To achieve the ratio of 10 m^{-1} for the specimen surface area to leachant volume (S/V), a typical wafer with geometric surface area of 2.00 cm^2 was placed in buffer solution of volume 20.0 mL. The sealed vessels were placed in a constant-temperature oven at 70 or 90□C for 1 to 10 days. The pH was measured before and after each test using an Orion Ross combination semi-micro electrode, calibrated with reference buffer solutions at 70□C or 90□C. All pH measurements were carried out with the test or reference solution in a constant-temperature bath.

TABLE I. Buffer Compositions Used in pH Buffer Tests and Measured Buffer pH Values

Buffer Composition	pH, 25°C	pH, 70°C	pH, 90°C
0.0095m KHph[a] + 0.0027m LiOH	4.85	5.01	5.07
0.0038m KHph[a] + 00031m LiOH	5.82	5.97	6.01
0.0263m TRIS[b] + 0.010m HNO_3	8.47	7.25	6.15
0.064m H_3BO_3 + 0.010m LiOH	8.39	8.10	7.94
0.012m H_3BO_3 + 0.010m LiOH	9.81	9.35	9.15
0.00098m HNO_3 + 0.0117m LiOH	11.96	10.66	10.23

[a] KHph: Potassium hydrogen phthalate.
[b] TRIS: Tris(hydroxymethyl)aminomethane.

The MCC-1 tests have been conducted at 90□C for 1, 2, 3, 5, 7, and 10 days, or at 70□C for 3, 5, 7, and 10 days. Tests at 40□C are in progress. The test durations were selected to be short enough that the rate of corrosion would be as close as possible to the forward rate. Results of MCC-1 tests with an alkali borosilicate glass at 90□C had previously shown that 3, 5, 7, and 10-day tests yield the forward rate [10]; tests at shorter durations may be affected by surface roughness, while tests at longer durations show the effect of the affinity term ($1-Q/[H_4SiO_4]_{sat}$). Blanks were run to measure background concentrations.

Termination of the tests required taking aliquots for pH measurement at the testing temperature and at 25□C, and after 0.45-μm filtration and acidification for inductively coupled plasma-atomic emission spectroscopy or for inductively coupled plasma-mass spectroscopy of matrix and minor elements. The vessels were then subjected to acid stripping with 1% HNO_3 solution; cation concentrations in the acid strip solutions were negligible.

RESULTS

The concentration of silicon in solution after test termination provides the best measure of matrix dissolution of glass-bonded sodalite, since both glass binder and sodalite have high silicon concentrations. The measured Si concentrations in solutions from buffered MCC-1 tests with each material (glass binder, sodalite, and CWF) were used to calculate the normalized mass losses, $NL(Si) = (m - m_b)/f_{Si} \bullet S$, where m is the mass of silicon in the test solution, m_b is the mass of silicon in the experimental blank, f_{Si} is the mass fraction of silicon in the material, and S is the sample surface area. At each temperature, the NL(Si) from tests at each buffer pH increased rapidly during the first three days, then increased more slowly to 10 days. The 90□C NL(Si) data for CWF are plotted in Fig. 1.

The normalized dissolution rates, $NR(Si) = d[NL(Si,t)]/dt$, were calculated by linear regression of the 3- to 10-day releases into solution for glass, sodalite, and CWF at 70□C and 1- to 5-day releases at 90°C; the release rates are listed in Table II. The logarithms of the NR(Si) data were plotted as a function of pH from tests at each temperature, as shown in Fig. 2. The logarithmic form of eq. (1) shows that the values of η are the slopes of the regression lines in Fig. 2.

$$\text{Log (rate/S)} = \log k_0 + \eta \bullet \text{pH} - \log [\exp(E_a/RT)] + \log (1 - Q/[H_4SiO_4]_{sat}) \qquad (2)$$

The values of η in the acidic and basic regions are listed in Table III.

TABLE II. Normalized Dissolution Rates NR(Si) ($g \bullet m^{-2} \bullet d^{-1}$) of Sodalite, Binder Glass, and Ceramic Waste Form (CWF) as a Function of pH at 70 and 90°C [a]

pH	Sodalite, 90°C [b]	Glass, 70°C	Glass, 90°C	CWF, 70°C	CWF, 90°C
5.1	2.6±0.6	0.025±0.002	0.088±0.020	1.4±0.2	1.8±0.4
6.0	0.64±0.15	0.0093±0.0019	0.056±0.012	0.48±0.02	0.67±0.18
7.0	0.38±0.24		0.056±0.006		0.69±0.11
7.3		0.016±0.002		0.19±0.03	
7.9	0.98±0.29		0.93±0.21		1.3±0.1
8.3		0.22±0.01		0.40±0.03	
9.2	1.2±0.3		1.5±0.4		1.5±0.2
9.6		0.50±0.02		0.50±0.03	
10.2	2.6±1.2		5.3±1.3		3.3±0.8

[a] Uncertainties are standard errors.

[b] Measurements at 70□C are in progress.

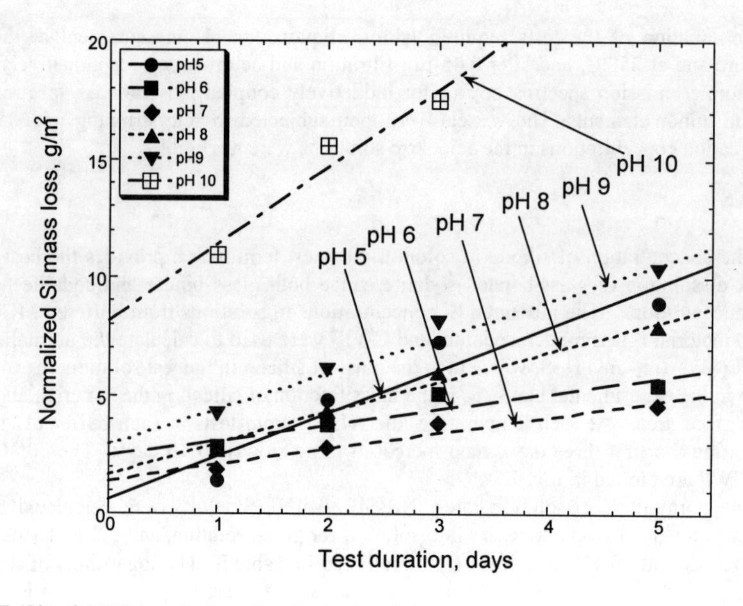

Fig. 1. NL(Si) of CWF in solution after 1-, 2-, 3-, and 5-day tests in buffer solutions at 90□C.

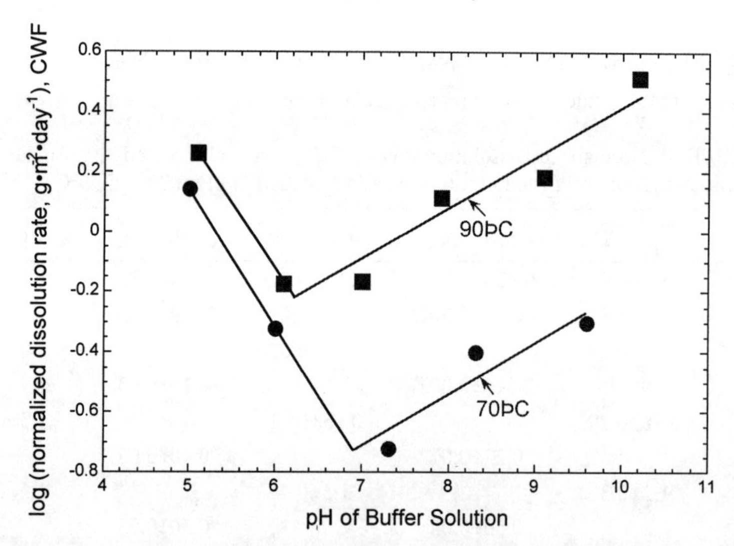

Fig. 2. Log (silicon dissolution rate) from CWF as Function of pH in Buffered MCC-1 Tests at 70 and 90°C.

TABLE III. Values of η for Binder Glass, Sodalite, and CWF at 70 and 90□C (uncertainties of η are standard errors)

Material	η (acidic)	η (basic)
Glass (70□C)	-0.54*	0.63±0.24
Glass (90□C)	-0.20*	0.55±0.15
Sodalite (70□C)	In progress	In progress
Sodalite (90□C)	-0.47±0.12	0.23±0.05
CWF (70□C)	-0.46*	0.18±0.07
CWF (90□C)	-0.44*	0.17±0.03

*2-point fit; standard error estimated as ±0.2

DISCUSSION

Controlled-pH MCC-1 tests in the pH 5-10 region show normalized silicon mass losses that increase with test duration, rapidly in the first few days (Fig. 1) and then more slowly during 5-10 day test durations. Tentative dissolution rates have been calculated in the acidic and basic regions at 70 and 90□C. The dissolution rates are lower at 70□C than at 90□C at all pH values. The dissolution rates have a V-shaped behavior with minima at neutral solution pH values and increase as solution pH becomes more acidic or more basic (Fig. 2).

The dissolution rates are consistent with dissolution rates for nuclear waste glasses and aluminosilicate minerals. The pH dependence of the dissolution of glass, salt-loaded sodalite, and CWF display the V-shape behavior reported for nuclear waste glass [4]. For the CWF as well as its components, the rate of dissolution is at a minimum near the point of neutral pH, which is 7.00 at 25□C, decreasing to 6.40 at 70□C and 6.21 at 90□C [11]. The results for salt-loaded sodalite are in fair agreement with dissolution tests of natural sodalite as a function of pH at 50□C: η(acidic) = -0.6 and η(basic) = 0.06 [12]. Other studies found a "flattened V" pH dependence. For example, Knauss and Wolery reported the dissolution rate of albite to be independent of pH over the range pH 3 to 8 [2] and the dissolution rate of muscovite to be pH independent between pH 5 and 7 [3]. Other aluminosilicate mineral dissolution rates are pH dependent over a wider pH range [5].

The dissolution rate of the CWF is expected to be a linear combination of the dissolution rates of its two components, weighted in terms of the CWF composition of 75 mass % salt-loaded sodalite and 25 mass % glass, in the absence of reaction between glass and sodalite during the hot isostatic pressing of the CWF, and in the absence of synergism (interactive effects) between glass and sodalite dissolution. The pH dependence of the dissolution rate, η, is also expected to be a weighted linear combination of the glass and sodalite rates. The weighted results of Table III at 90□C yield calculated η values for CWF of –0.40 in the acidic region and 0.31 in the basic region. Inconsistencies between calculated and measured dissolution rates and pH dependences of the dissolution rates indicate that there may be reaction between glass and sodalite during the hot isostatic pressing of the CWF.

Additional buffered MCC-1 tests are being carried out at lower and higher pH values to better define the η values. Additional tests are also in progress at 40□C. Values of η at three temperatures will permit the calculation of E_a, which reflects the temperature dependence of the dissolution rate.

ACKNOWLEDGMENTS

The authors acknowledge timely and thorough analytical support by Doris Huff, Kevin Quigley, Yifen Tsai, and Stephen Wolf. Research supported by U.S. Department of Energy, Nuclear Energy Research and Development Program, under contract W-31-109-ENG-38.

REFERENCES

1. ANL-NT-119, *Ceramic Waste Form Handbook,* L. R. Morss, compiler, 1999.
2. K. G. Knauss and T. J. Wolery, "Dependence of Albite Dissolution Kinetics of pH and Time at 25° and 70□C," Geochim. Cosmochim. Acta **50**, 2481-2497 (1986).
3. K. G. Knauss and T. J. Wolery, "Muscovite Dissolution as a Function of pH and Time at 70□C," Geochim. Cosmochim. Acta **53**, 1493-1501 (1989).
4. K. G. Knauss, W. L. Bourcier, K. D. McKeegan, C. I. Merzbacher, S. N. Nguyen, F. J. Ryerson, D. K. Smith, H. C. Weed, "Dissolution Kinetics of a Simple Analogue Nuclear Waste Glass as a Function of pH, Time, and Temperature," Mat. Res. Soc. Symp. Proc., **176**, 371-381 (1990).
5. E. H. Oelkers, J. Schott, J. Devidal, "The Effect of Aluminum, pH, and Chemical Affinity on the Rates of Aluminosilicate Dissolution Reactions. Geochim. Cosmochim. Acta **58**, 2011-2024 (1994).
6. P. K. Abraitis, D. J. Vaughan, F. R.Livens, J. Monteith, D. P. Trivedi, J. S. Small, "Dissolution of a Complex Borosilicate Glass at 60°C: The Influence of pH and Proton Absorption on the Congruence of Short-Term Leaching," Mat. Res. Soc. Symp. Proc. **506**, 47-54 (1998).
7. W. Bourcier, "Affinity Functions for Modeling Glass Dissolution Rates," UCRL-JC-131186 (1998).
8. P. Aagaard and H. C. Helgeson, "Thermodynamic and Kinetic Constraints on Reaction Rates among Mineral and Aqueous Solutions, I. Theoretical Considerations," Am. J. Science, **282**, 237-285 (1982).
9. American Society for Testing and Materials, "Annual Book of ASTM Standards," **12.01**, Standard Test Method for Static Leaching of Monolithic Waste Forms for Disposal of Radioactive Waste, C1220-98, pp. 1-16 (1998).
10. B. P. McGrail, W. L. Ebert, A. J. Bakel, D. K. Peeler, "Measurement of Kinetic Rate Law Parameters on a Na-Ca-Al Borosilicate Glass for Low-Activity Waste," J. Nucl. Materials **249**, 1765-189 (1997).
11. H. S. Harned and B. B. Owen, *The Physical Chemistry of Electrolytic Solutions,* 3rd ed. (Reinhold, New York, 1958), pp. 643-649.
12. K. Montgomery, "The Synthesis and Dissolution of Sodalite: Implications for Nuclear Waste Disposal," M. Sc. Thesis, Dept. of Geology, University of Alberta, Canada (1986).

Plutonium Silicate Alteration Phases Produced by Aqueous Corrosion of Borosilicate Glass

J. A. Fortner, C. J. Mertz, A. J. Bakel, R. J. Finch, and D. B. Chamberlain

Chemical Technology Division, Argonne National Laboratory, Argonne, IL 60439

ABSTRACT

Borosilicate glasses loaded with ~ 10 wt % plutonium were found to produce plutonium-silicate alteration phases upon aqueous corrosion under a range of conditions. The phases observed were generally rich in lanthanide (Ln) elements and were related to the lanthanide orthosilicate phases of the monoclinic Ln_2SiO_5 type. The composition of the phases was variable regarding [Ln]/[Pu] ratio, depending upon type of corrosion test and on the location within the alteration layer. The formation of these phases likely has implications for the incorporation of plutonium into silicate alteration phases during corrosion of titanate ceramics, high-level waste glasses, and spent nuclear fuel.

INTRODUCTION

The lanthanide borosilicate (LaBS) glass is chemically durable and can dissolve substantial amounts of plutonium as well as the neutron absorbers gadolinium and hafnium [1,2,3]. The prototype LaBS formulation tested in this work, however, contained no hafnium but did contain some zirconium, which is expected to be similar chemically. Although a titanate ceramic has been chosen over the LaBS glass as a plutonium immobilization form [4], the corrosion mechanisms observed in testing of the LaBS glass may provide insight into the behavior of plutonium and lanthanide elements during alteration of other waste forms. Specifically, silicon-rich groundwater may be available to alter waste forms in the proposed Yucca Mountain repository, a site located in a hydrologically unsaturated zone and composed of welded and devitrified tuff. The local groundwater from the nearby USGS J-13 well contains ~ 40 ppm dissolved silica. Silicon-rich groundwater has been found to cause low-temperature alteration of natural zirconolites [5], an observation with direct bearing on the long-term behavior of the titanate phases in the target plutonium immobilization ceramic. Plutonium is also an important constituent of spent nuclear fuel and is a trace element in high-level waste (HLW) glass. Corrosion tests of HLW glass indicate that released plutonium is associated with colloidal particles, which are dominated by smectite-clays (a silicate) [6]. Silicon-based colloids have been implicated in the migration of spent weapons plutonium from the Nevada Test Site [7], although Kersting *et al.* attributed the plutonium incorporation to sorption rather than co-precipitation.

EXPERIMENT

Monolithic samples of LaBS glass loaded with 10 weight % plutonium were reacted with water vapor at 100% relative humidity and 200°C for periods of 14 to 56 days. The test procedure is described in detail elsewhere [1, 8]. Parallel corrosion tests, whereby crushed LaBS glass was immersed in deionized water at 90 °C (a modified ASTMC-1285 Product Consistency Test or PCT), were conducted for extended periods of time (> 98 days). The surfaces of the glass samples, along with alteration phases, were examined with a transmission electron microscope (TEM) and scanning electron microscope (SEM) to determine the characteristic alteration products. Vapor alteration of the LaBS glass for 14 days produced macroscopic crystallites of a plutonium-lanthanide silicate material, which is, to our knowledge, the first plutonium-based alteration phase observed in glass-water reaction. An extensive alteration layer was found on the glass surface containing amorphous aluminosilicate layered with bands of a cryptocrystalline plutonium-lanthanide silicate that was similar to the surface crystals but relatively depleted in lanthanides. Of particular interest was evidence of size selection among the lanthanide elements lanthanum, neodymium, and gadolinium, as well as separation of the lighter lanthanides from plutonium. There is little evidence that gadolinium, an important neutron absorber, is separated from the plutonium.

Mat. Res. Soc. Symp. Proc. Vol. 608 © 2000 Materials Research Society

After vapor hydration at 200°C and 100% relative humidity for 14 days, the surface of the LaBS glass monolith was speckled with minute (~ 5 μm) crystallites that appeared white against the dark glass. The surface of the glass itself had a slight crust, or alteration film, which appeared less advanced than what would typically be observed from a similar test on a HLW glass. Samples of the white surface crystallites were embedded in epoxy resin and thin sectioned with a Riechert ultramicrotome. The thin-sectioned material was examined using a JEOL 2000FX II transmission electron microscope operating at 200 keV and equipped with x-ray (EDS) detectors and an electron energy loss spectrometer (EELS). Cross sections of the altered surface film on the glass were taken by scoring the specimen with a diamond scribe, and embedding fractured material for thin sectioning in a manner similar to the surface particles. Intact glass surfaces were also examined with an SEM and light microscopy. Samples of LaBS glass from crushed glass immersion tests at 90°C were also examined. The LaBS glass reacted very slowly under these conditions, and little evidence of an alteration layer was observed until at least 98 days of testing. Longer-term tests have not been examined in detail owing to a shift in program focus to the titanate ceramic waste forms [4]. In each case where lanthanide-silicate alteration products were observed, the details of their structure and composition were dependent upon test type, test duration, and even location within or upon the altered glass surface.

RESULTS AND DISCUSSION

The SEM examination of glass monolith surfaces from vapor hydration testing revealed copious particles of a lanthanide (La, Nd, and Gd) plutonium silicate composition, distinct from the LaBS glass (Figure 1). These particles typically had a rosette or "onion-skin" structure, and were loosely attached to the glass surface. These particles are the white microcrystals observed visually and sampled for TEM as described above.

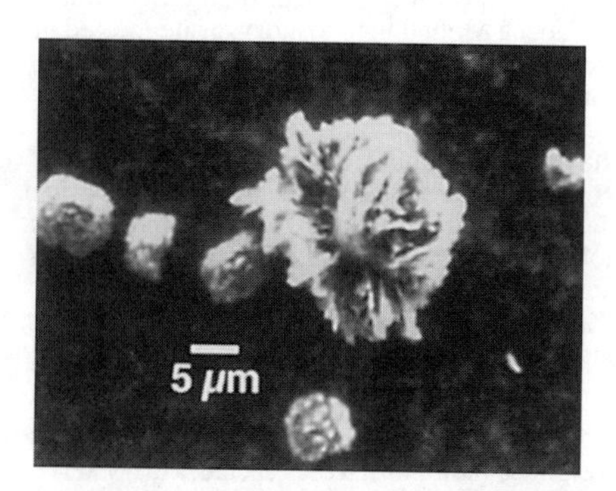

Figure 1. SEM backscatter micrograph of a plutonium silicate "rose" from LaBS vapor alteration at 200°C.

The TEM revealed a sharp, crystalline diffraction from the surface particles clearly distinct from that of Ln oxides, Ln oxyhydroxides, Ln-uranium oxides, or $PuSiO_4$ (where Ln = lanthanide). The experimentally-derived *d*-spacings are compared with JCPDS-ICDD diffraction data for two Ln_2SiO_5 structures in Table 1 [9]. Both EDS (Figure 2) and EELS (Figure 3) suggest a composition enriched in the larger lanthanides (La > Nd > Gd), depleted in silicon and nearly devoid of aluminum relative to the LaBS glass. The low energy EELS signal from boron, readily detected in the glass, was not observable from the crystals. Interestingly, the ratio of the plutonium to gadolinium remained about the same from the glass to the alteration crystal for both EDS and EELS measurements (Figures 2 and 3). Along with the diffraction data, we are led to propose a lanthanide-silicate type structure, with plutonium and gadolinium occupying a similar site in the structure. The mechanism driving size selectivity favoring the larger lanthanum and neodymium ions in this structure is unknown.

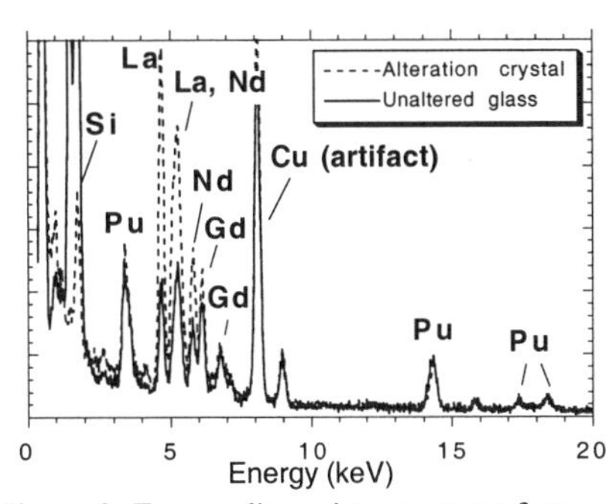

Figure 2. Energy dispersive spectrum from the large Pu-lanthanide-silicate crystals from the 14-day vapor hydration test. The EDS spectrum from unaltered LaBS glass is shown for comparison, normalized to the Pu signal.

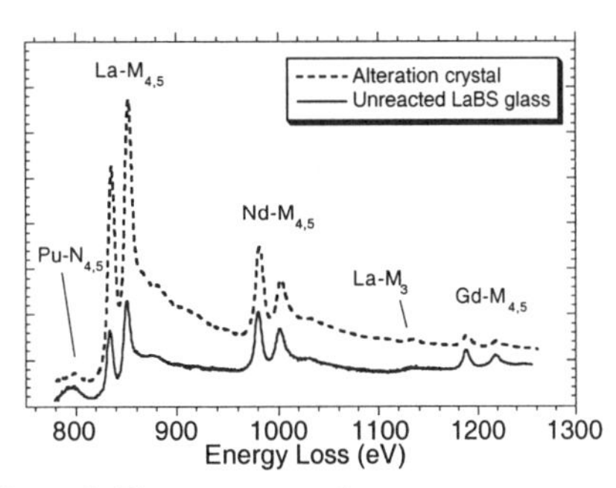

Figure 3. Electron energy loss spectra (EELS) of surface material from the 14-day vapor hydration of LaBS glass. The spectra reveal different partitioning of the rare earth elements (La, Nd, and Gd) in the crystalline alteration phase than in the unaltered glass.

Table 1. Electron diffraction results from the surface crystal from a 14-day vapor hydrated LaBS. Reference JCPDS-ICDD [9] data reported for $I/I_0 \geq 0.3$ or for match with experiment.

Experimental d-spacings (nm)	Nd_2SiO_5 JCPDS-ICDD 40-284 (nm)	La_2SiO_5 JCPDS-ICDD 40-234 (nm)
	0.5591	0.572
0.489	0.4859	
	0.4382	0.4418
0.387	0.3891	0.3805
0.351	0.3413	0.3511
	0.3180, 0.3128	
0.286	0.2937, 0.2853	0.2992, 0.2945
	0.2788	0.2850, 0.2831
0.260	0.2604	0.2652
0.245	0.2467	0.2450
0.224	0.2215, 0.2264	0.2202, 0.2192
0.202	0.20046	0.2035, 0.1984
	0.19357	
0.187	0.18678	0.1877
0.176	0.17535	0.17699
0.1652	0.16580	0.16555
0.1549	0.15450	0.15641
0.1487	0.14892	0.14729

Table 2. Electron diffraction results from intralayer crystallites from the 14-day vapor hydrated LaBS. Reference JCPDS-ICDD [9] data reported for $I/I_0 \geq 0.3$ or for match with experiment.

Experimental d-spacings C18500 (nm)	Nd_2SiO_5 JCPDS-ICDD 40-284 (nm)
	0.5591
	0.4859
	0.4382
	0.3891
0.339	0.3413
0.310, 0.307	0.3180, 0.3128
	0.2937, 0.2853
0.275	0.2788
	0.2604
0.243	0.2467
	0.2215, 0.2264
0.196	0.20046
0.192	0.19357
	0.18678
0.170	0.17535
0.166, 0.163	0.16580

Cross sections of the surface film were then examined by TEM, revealing a complex alteration structure tenaciously bonded to underlying LaBS glass (Figure 4). Within the alteration layer were very small crystallites having a similar lanthanide plutonium silicate composition. The crystallites were imbedded in a matrix of a fibrous aluminosilicate phase, which appeared similar to the mineral *imogolite* ($Al_2SiO_3(OH)_4$)[10]. Electron diffraction indicated that the intralayer lanthanide plutonium silicate crystals were less ordered than the surface crystals (perhaps cryptocrystalline), with only a few *d*-spacings identifiable. These are tabulated in Table 2 for several diffraction measurements. Notably, EDS indicated that these intralayer crystals were less enriched in the lighter lanthanides than were the surface crystals. This is reflected by a systematic shift in the observed d-spacings to higher values, owing to the larger ionic size of the lighter lanthanides. This is illustrated in Figure 5, where *d*-spacings are plotted against Ln ions and experimental values [9, 11].

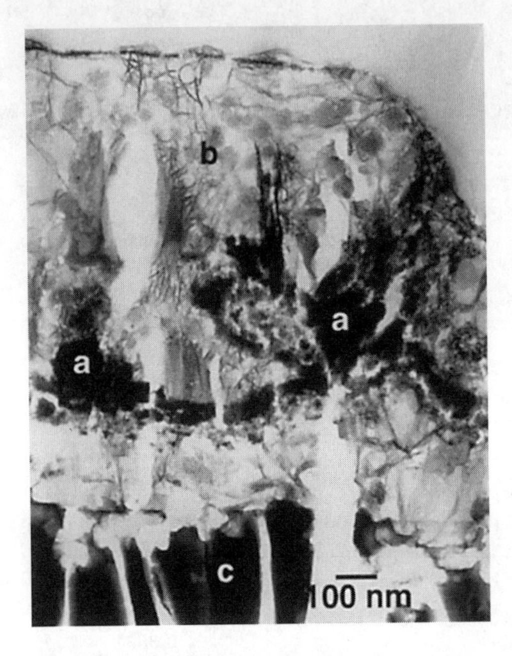

Figure 4. Transmission electron micrograph of the surface alteration layer on LaBS glass from a 14-day vapor hydration test showing (a) crystalline Pu silicates (b) aluminosilicate regions, and (c) unaltered LaBS glass. The "picket fence" appearance of the glass is an artifact of the microtoming.

Samples of the modified (98-day) PCT-reacted samples were examined by TEM (Figure 6). Generally, the LaBS glass reacted very slowly under these conditions relative to HLW-type glasses, likely owing to the absence of alkali oxides in the LaBS [1,2]. Only a slight hint of surface alteration could be observed, along with occasional minute (< 100 nm) alteration products of plutonium lanthanide silicate. These alteration products produced no crystalline diffraction pattern, and were likely amorphous. Nonetheless, their chemical composition strongly suggests that they are structurally related to the larger crystals observed in the more aggressive vapor hydration conditions.

The gadolinium orthosilicate phase of the monoclinic Ln_2SiO_5 type has been characterized by Smolin and Tkachev [11]. These phases have a layered structure similar to the synthetic titanites. In the proposed structure (figure 7), the Ln (or plutonium, in the present case) occupies two distinct sites, the six-coordinated octahedral sites and the seven-fold decahedral sites. The SiO_2 tetrahedra share three oxygen atoms with the octahedral lanthanide and one oxygen with the decahedral lanthanide. To our knowledge, these phases have not previously been synthesized except at high temperature from melts. However, the presence of a long Si—O bond in the decahedral site [11], which can be replaced by Si—OH in hydrous minerals, makes the low temperature synthesis of this phase plausible.

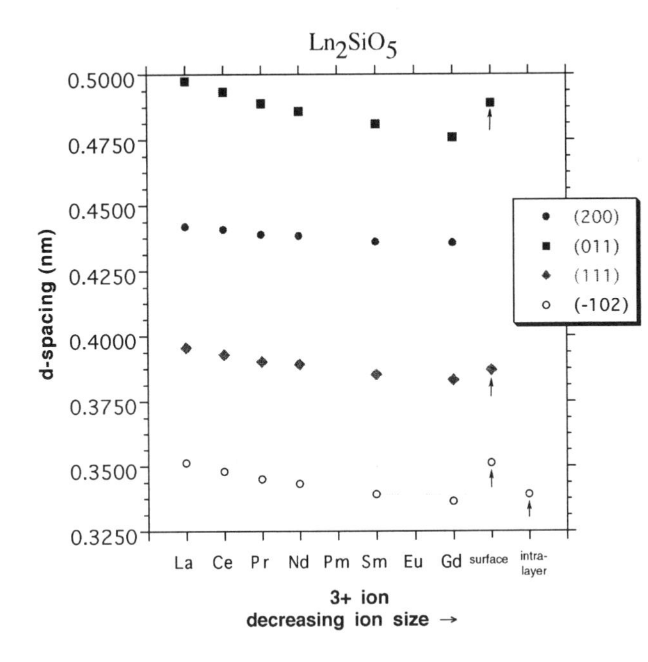

Figure 5. Experimental d-spacings (horizontal lines with data from this study marked by arrows) for surface and intralayer crystals from the 14-day vapor hydration test. These are compared with literature values [9, 11] for Ln_2SiO_5 structures as a function of 3+ ion size.

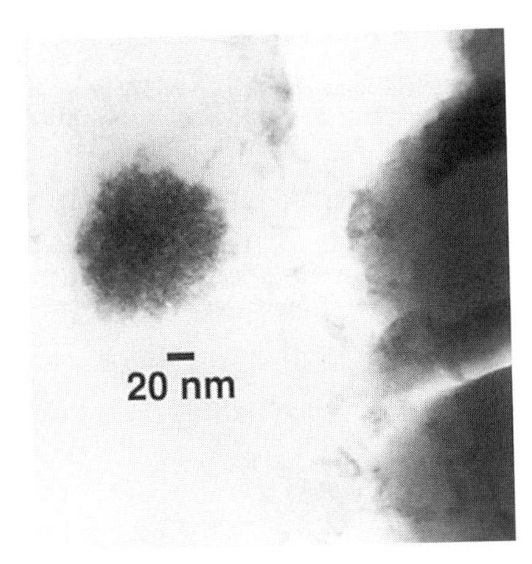

Figure 6. A TEM micrograph of a plutonium silicate alteration phase from a 98-day immersion test of LaBS glass at 90°C. This phase appears as a small clump at left. The glass (right) is accompanied by wisps of aluminum-rich debris from the surface.

Figure 7. Depiction of the gadolinium orthosilicate phase of the monoclinic Ln_2SiO_5 type originally described by Smolin and Tkachev [11]. The Ln occupies two distinct sites, six-fold octahedral sites, rendered as the light polyhedra, and seven-fold decahedral sites, which appear as the central ions with wire frame oxygen polyhedra. The SiO_2 tetrahedra are rendered as dark gray, and share three oxygen atoms with the octahedral Ln and one oxygen with the decahedral Ln.

REFERENCES

1. A. J. Bakel, C. J. Mertz, D. B. Chamberlain, E. C. Buck, S. F. Wolf, and J. A. Fortner, *Proceedings of the 18th International Congress on Glass*, San Francisco, July 5-10 1988.
2. N. E. Bibbler, W. G. Ramsey, T. F. Meeker, and J. M. Pareizs, *Mat. Res. Symp. Proc.* **412**, 65 (1996).
3. T. F. Meaker, D. K. Peeler, J. C. Marra, J. M. Pareizs, and W. G. Ramsey, *Mat. Res. Symp. Proc.* **465**, 1281 (1997).
4. Department of Energy, *Record of Decision for the Storage and Disposition of Weapons-Useable Fissile Materials Final Programmatic Environmental Impact Statement* 62 FR3014, Office of the Federal Register, Washington D. C., January 14, 1997.
5. G. R. Lumpkin, K. L. Smith, M. G. Blackford, K. P. Hart, P. McGlinn, R. Gieré, and C. T. Williams, *Proceedings of the 9th Pacific Basin Nuclear Conference*, (Sydney, Australia, May 1-6, 1994).
6. See, for instance, E. C. Buck and J. K. Bates, *Applied Geochemistry,* **14**, 635 (1999); J.A. Fortner, S F. Wolf, E.C. Buck, C.J. Mertz, and J.K. Bates, *Mater. Res. Soc. Symp. Proc.* **465**, 165 (1997).
7. A. B. Kersting, D. W. Eford, D. L. Finnegan, D. J. Rokop, D. K. Smith, and J. L. Thompson, *Nature* **397**, 56 (1999).
8. D. J. Wronkiewicz, *et al., Mat. Res. Symp. Proc.* **294**, 183 (1993); D. J. Wronkiewicz, *et al., Mat. Res. Symp. Proc.* **333**, 259 (1994).
9. International Centre for Diffraction Data, Newtown Square, PA 19073-3273, U.S.A.
10. J. A. Speer and P. H. Ribbe, in *Orthosilicates*, Reviews in Mineralogy Vol. 5, H. Ribbe, ed. (Mineralogical Society of America, Washington, D. C., 1982) pp. 429-450.
11. Yu. I. Smolin and S. P. Tkachev, *Sov. Phys.-Crystal.* **14** 14 (1969).

THE ROLE OF ALTERATION PHASES IN INFLUENCING THE KINETICS OF GLASS DISSOLUTION

D.J. Wronkiewicz and K.A. Arbesman
Department of Geology and Geophysics, University of Missouri-Rolla, Rolla, MO 65409

ABSTRACT

The potential effect of alteration phases on the kinetics of glass corrosion has been examined in a preliminary series of Product Consistency Tests (0.5, 1, 3, 7, 35, and 91 days). Crushed samples of a relatively simple Li-Na-Ca-K-Al-B-Si glass were reacted in the presence of a relatively high ionic strength fluid, to which various alteration phases (analcime, adularia, chabazite, or Na-montmorillonite) were added as "seed-crystals". The release of boron and lithium were used to monitor the corrosion rate of the glass. In general, corrosion rates varied only slightly between the tests with different seed-crystals types. Boron and lithium contents in tests with analcime or adularia were slightly higher than tests with Na-montmorillonite or chabazite present. Silicon concentrations did not display any consistent variation over the testing interval, remaining relatively similar to the starting leachant value of 3.5×10^{-2} M. The concentration of aluminum, however, decreased significantly during the first 35 days of testing and could be inversely correlated to boron and lithium concentrations. The concentration of aluminum then increased between 35 and 91 days, whereas boron and lithium concentrations remained relatively static. The noted correlation between aluminum and boron (or lithium) suggests a coupling of the rate of glass corrosion with aluminum concentration.

INTRODUCTION

The successful deployment of a waste form for the disposal of radioactive materials will require an assessment of its radionuclide retentive properties under a range of potential storage conditions. Part of this assessment will require the establishment of a link between the short-term performance of the waste form, as measured in standardized durability tests, with its potential long-term behavior in an actual repository setting.

Studies examining alteration processes of borosilicate glasses suggest that alteration phases may have a profound influence on the kinetics of glass corrosion and radionuclide release rates [1]. The two dominating effects can be correlated to alteration phase controls on the activities of dissolved glass (or crystal) components and the sorption of radionuclides. The activity of specific aqueous species (e.g., H_4SiO_4) may, in turn, influence the dissolution rates of glass and/or other crystalline solids [2-5]. The aqueous solubilities of both silicon and aluminum are specie dependent, with speciation often being influenced by pH [6]. Alteration phases may also moderate the activities of H^+ or OH^- ions by ion exchange and/or the formation of various hydroxide-bearing solid phases. The long-term dissolution rates of nuclear waste glasses may therefore be influenced in a variety of ways by the formation of alteration phases.

The purpose of this study is to experimentally examine the effect of various alteration phases on the kinetics of glass corrosion. All of the seed crystals used in this study have previously been noted, through studies of experimentally altered borosilicate glass and the weathering of naturally occurring volcanic glasses, to be important alteration phases that occur during glass reaction paragenesis [1, 7-11].

Mat. Res. Soc. Symp. Proc. Vol. 608 © 2000 Materials Research Society

EXPERIMENTAL PROCEDURE

Alteration phases that have been previously been noted to occur during the corrosion of natural and simulated borosilicate nuclear waste glasses [1, 7-11] were chosen as "seed crystals" for the tests. These phases include adularia ($KAlSi_3O_8$), analcime ($NaAlSi_2O_6 \cdot H_2O$), chabazite ($Ca_2Al_4Si_8O_{24} \cdot 12H_2O$), and Na-montmorillonite ($(Na_{0.7}(Al_{3.3}Mg_{0.7})[Si_8O_{20}](OH)_4 \cdot nH_2O$). Adularia, analcime, and chabazite crystals were obtained from the University of Missouri-Rolla mineral museum. The samples were crushed, passed under a magnet to remove metal filings from the crushing apparatus, and then hand picked under a microscope to remove and isolate the phase of interest. The Na-montmorillonite sample was obtained as a standard from the Source Clay Minerals Repository, Department of Geology, University of Missouri-Columbia.

The leachant solution was prepared by reacting a mixture of powdered chemical oxides in deionized water at 90°C for a period of nine months. The leachant was then passed through a 0.45 μm filter and stored in a Teflon vessel at 90°C prior to testing. The powdered oxides used in the preparation of the leachant were mixed in a proportion that matched the composition of the glass used in the tests (Table I), except that boron and lithium were omitted. The elimination of boron and lithium from the leachant recipe allowed us to use them to monitor the rate of glass dissolution. This leachant preparation resulted in a fluid composition approximating that expected following long-term glass reaction and therefore would be close to equilibrium with respect to the various alteration phases that may occur. This leachant fluid had the following composition: Na, 1.19×10^{-1}M; Si, 3.51×10^{-2}M; K, 5.59×10^{-3}M; Ca, 2.36×10^{-5}M; Li, 4.4×10^{-6}M; Al, 3.5×10^{-6}M; Mg, 2.8×10^{-6}M; Sr, 6.4×10^{-7}M; and pH = 10.86.

The glass used in the tests was of a relatively simple Li-Na-Ca-K-Al-B-Si composition and represents a potential binder material to be utilized in a glass-ceramic waste form [12]. The glass was provided in a powdered form by Argonne National Laboratory. The glass powder was prepared for testing by melting into beads, then hand crushed and sieving the glass to collect a -100 to +200 mesh (74 to 149 μm) size fraction. The composition of the glass is given in Table I. Lithium was not reported as a constituent of the glass by the manufacturer [12], but its presence is indicated by its release to the leachate during testing.

The experiments with glass plus seed-crystals were conducted in a product consistency test (PCT) format using screw-top Teflon® vessels with an internal volume of 60 cm^3. Two grams of crushed glass were placed directly on the vessel bottom along with 0.1 grams of one of the four seed-crystal types. A fifth set of tests was initiated with just the crushed glass (without

Table I. Composition of Oxide Mixture Used to Prepare Leachant Solution and Composition of the Binder Glass (in oxide wt.%).

Component	Oxide Mixture Used to Prepare Leachant	[a]Binder Glass Composition
SiO_2	79.63	65.97
Al_2O_3	9.35	7.80
Na_2O	8.77	7.19
CaO	1.46	1.27
K_2O	0.78	0.66
SrO	0.01	0.01
B_2O_3	not added	17.10
Total	100.00	100.00

a, calculated from data of Simpson et al. [12]. Lithium was not reported as a component, however, its presence in the glass is recorded by its release into the leachate solutions.

seed crystals). Twenty grams of the leachant solution were then added to each vessel prior to testing. Experiments were run in a batch mode at a temperature of 90 ± 2°C, while test intervals ranged from 0.5, 1, 3, 7, 35, and 91 days. After completion of the prescribed test interval, the leachate solutions were filtered through a 0.45 μm filter and acidified with nitric acid prior to analysis. Solutions were analyzed by inductively coupled plasma/optical emission spectroscopy (ICP/OES), using a facility available at the University of Missouri-Columbia. Analytical accuracy determined for standards ICV-1A and ICV-1B were better than +/- 10% for Si, B, K, Ca, Li, and Mg, while Al values were within a +/- 13% error limit. The release of relatively soluble B and Li components of the glass were used to monitor the extent of glass corrosion.

RESULTS AND DISCUSSION

Solution Results

The concentration of boron in the leachate solution progressively increased for all PCTs throughout the first 35 days of testing (Fig. 1a). Between 35 and 91 days, the rate of boron release flattened to a level that did not display any perceptible change within the analytical uncertainty error of +/- 10%. Boron concentrations varied only slightly between tests using different types of seed-crystals. This pattern indicates that the presence of the alteration phases had only a minimal impact on glass reaction rates. The tests with adularia or analcime crystals displayed slightly higher boron concentrations than tests with Na-montmorillonite or chabazite. Concentration differences between tests were most pronounced between 0.5 to seven days, whereas differences for the 35- and 91-day tests were minimal.

Many of the leachate solutions from tests containing seed crystals had slightly lower boron concentrations than the control tests without seed crystals present. This trend was most notable for all of the 0.5-day tests and all tests with chabazite for up to seven days in length (Fig. 1a). Boron concentrations were generally the lowest for tests with chabazite, followed by those with Na-montmorillonite. Both phases tend to have high cation exchange capacities and thus may act as adsorbents for cations in solution. Lithium concentrations also closely parallel those of boron (Fig. 1b). It is possible that the boron and lithium may have been removed from the leachate *via* a process of ion exchange with the seed crystals. Such a process could have important implications for the assessment of glass performance, as boron is often used as a monitor of glass reaction rates (a premise of the current study). Current glass reaction models could underestimate the rate of glass corrosion if empirical data used in the models were based from tests where boron adsorption was occurring. A comparison of the boron and lithium trends may provide some enlightenment towards these observations. Ion exchange processes generally favor the incorporation of higher valence charge and larger radii ions into solid phases. Because lithium occurs in solution as a univalent Li^+ ion with a relatively small ionic radius, it may play a rather limited role in ion exchange processes. Since the lithium and boron release trends are so similar (Figs. 1a and b) it may be expected that the release trends observed for both elements are a function of corrosion processes affecting the glass samples rather than any ion exchange reactions.

Although silicon concentrations decreased slightly during the first 0.5 days, the concentration quickly rebounded to a value that was higher than the starting leachant after one day of testing (Fig. 1c). These changes may be correlated to solution pH values, that also increase from ~10.5 to 10.9 between 0.5 and one day of testing. Although silicon concentrations fluctuated over the remainder of the test intervals, no consistent change in concentration patterns were noted, despite a moderately strong decrease in pH values that

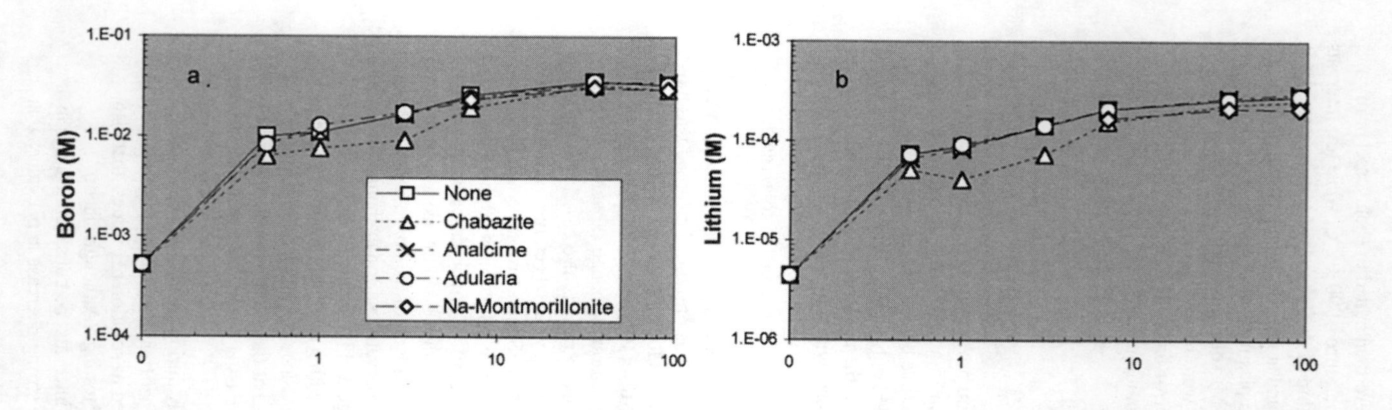

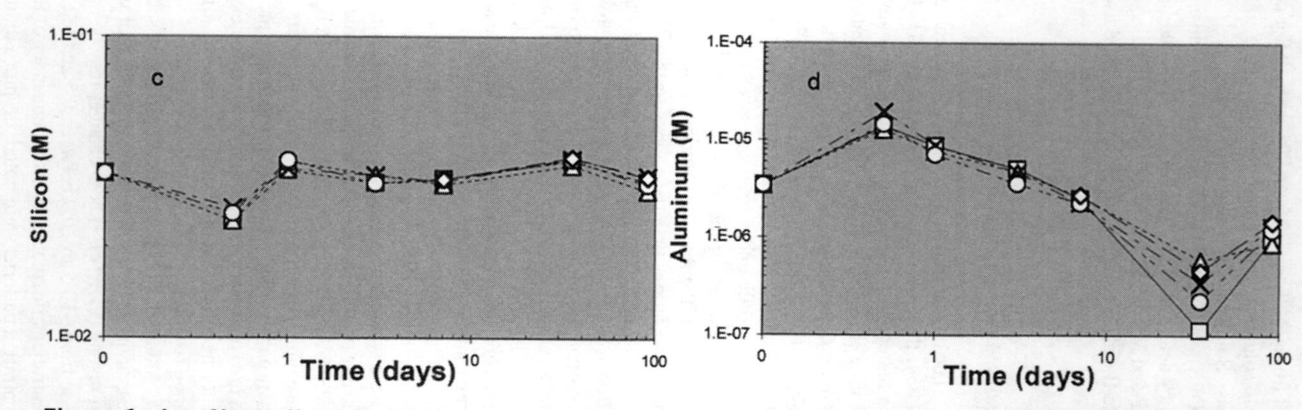

Figure 1. Log Normalized Cation Concentrations from Product Consistency Tests. Pattern key given in figure a notes the various types of seed crystals added to the tests.

occurred between three and seven days of testing (~10.9 to 10.2). Silicon contents also did not vary significantly between tests with different types of seed crystals. These concentrations thus remained similar to the starting leachant value (3.5×10^{-2} M) for the remainder of the testing intervals. This pattern suggests that the silicon concentration may be controlled by its saturation with respect to an alteration phase that formed in the tests.

Aluminum displays a consistent and progressive decrease in concentration during the first 35 days of testing, then an increase up to the 91-day PCTs (Fig 1d). Such a trend suggests that the aluminum was initially being consumed either by its incorporation into alteration phases or surface adsorption during the first 35 days of testing. An aluminum-rich phase would have to dissolve between 35 and 91 days to account for the increase observed for aluminum, without any corresponding increase in boron or lithium (Figs. 1a, b, and d). These patterns also suggest a coupling between the concentration of aluminum and the rate of glass reaction. An identification for the aluminum phase cannot be postulated at this time because none of the other cations present in the seed crystals (Ca, K, Na, and Si) display any correlative changes. Significant changes in aluminum concentration were not observed between tests with different types of seed crystals, nor do the aluminum trends display any correlation with pH values.

Solids Analysis

Experiments seeded with analcime or adularia crystals were characterized by a relatively large degree of sample grain clumping following testing. The process of clumping and grain attachment in these tests can usually be attributed to the precipitation of secondary mineral phases. In contrast, samples with chabazite displayed only minor evidence of grain clumping, whereas the samples with Na-montmorillonite and the control tests conducted without any seed crystals did not display any noticeable clumping of particles.

Glass fragments from the 91-day PCT runs with seed crystals were sectioned and polished for examination using scanning electron microscopy/energy dispersive x-ray spectrometry techniques (SEM/EDS; Fig. 2). A small percentage of the altered glass grains display the presence of one, or in some regions, two leached layers of variable thicknesses. The thicker layers were notably depleted in sodium (from approximately four elemental weight% in the core region, to 3.3% in the inner layer, and 2% in the outer layer). No changes in the distribution of other major glass elements (Si, Al, K, and Ca) were noted in the alteration layers except for a slight residual enrichment in silicon. The depletion of sodium from the glass during alteration is consistent with its enrichment in the leachate solution. Boron and lithium cannot be detected using the SEM/EDS equipment; thus no determination of their concentration profiles could be made. It is expected that these two elements would display a leaching profile similar to sodium.

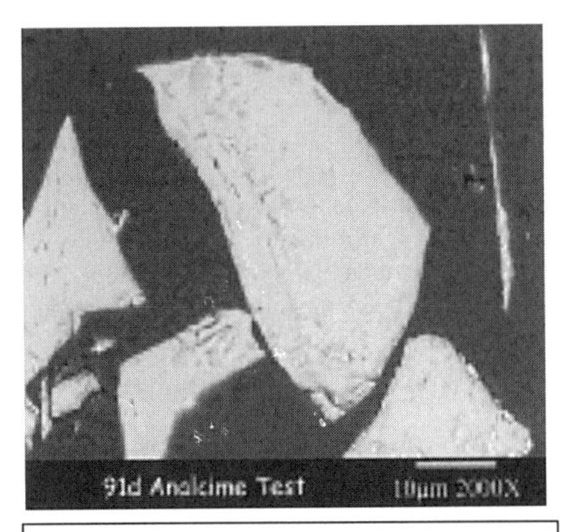

Figure 2. Scanning Electron Microscopy Image of Reacted Glass Particles After 91-Day Product Consistency Test. Thin reaction layer occurs around particle edges.

CONCLUSIONS

- The presence of various alteration phases (analcime, adularia, Na-montmorillonite, or chabazite) was found to have only a limited effect on the kinetics of glass corrosion in high-ionic strength leachant solutions. The experiments could be ranked in the following order of decreasing rate of glass reaction (based on boron release):

  adularia > analcime > Na-montmorillonite > chabazite.
- Silicon concentrations were relatively invariant over time and did not display any correlation with boron or lithium release.
- Aluminum concentrations decreased progressively during the first 35 days of testing and were inversely correlated to the release of boron and lithium. Aluminum concentrations increased between 35 and 91 days, while boron and lithium remained relatively static. These correlations suggest that glass reaction rates are coupled to aluminum concentration in relatively high ionic strength fluids.

ACKNOWLEDGEMENTS

The authors would like to acknowledge Carol Nabelek, University of Missouri-Columbia for assistance with the ICP-OES analyses. Sara Kaps provided assistance with the tedious task of separating seed crystals for testing. Funding for this research was provided through Argonne National Laboratory, under contract 980292401. The following Argonne personnel also provided technical assistance throughout this project: Lester Morss, Lin Simpson, Michele Lewis, and William Ebert.

REFERENCES

1. J.K. Bates, L.J. Jardine, and M.J. Steindler, Science, **218**, 51-54 (1982).
2. B. Grambow and D.M. Strachan, in *Scientific Basis for Nuclear Waste Management XI,* edited by M.J. Apted and R.E. Westerman (Mater. Res. Soc. Proc. **112**, Pittsburgh, PA 1988), p. 713-724.
3. A.C. LaSaga and G.V. Gibbs, Amer. J. Sci. **290**, 263-295 (1990).
4. E.Y. Vernaz and J.L. Dussossoy, Appl. Geochem. **1**, 13-22 (1992).
5. G. Berger, C. Claparols, C. Guy, and V. Daux, Geochim. Cosmochim. Acta, **58**, 4875-4886 (1994).
6. K.V. Ragnarsdottir, Geochim. Cosmochim. Acta, **57**, 2439-2449 (1993).
7. A. Iijima in *Natural Zeolites Occurrence, Properties, Use*, edited by L.B. Sand and F.A. Mumpton, Pergamon Press, Oxford, 1978, p. 135-143.
8. R.L. Hay and S.G. Guldman, Clays and Clay Minerals, **35**, 449-457 (1987).
9. J. Caurel, E. Vernaz, and D. Beaufort in *Scientific Basis for Nuclear Waste Management XIII,* edited by V.M. Oversby and P.W. Brown (Mater. Res. Soc. Proc. **176**, Pittsburgh, PA 1990), p. 309-318.
10. E.Y. Vernaz, A. Loida, G. Malow, J.A.C. Marples, and H.J. Matzke, H.J. Third EC Conf. on Radioactive Waste Management and Disposal, Luxembourg, Sept. 17-21, 1990 EUR-13389, p. 302-315 (1991).
11. D.J. Wronkiewicz, J.K. Bates, E.C. Buck, J. Hoh, J. Emery, and L.M. Wang, Argonne National Laboratory Report, ANL-97/15, 238 p. (1997).
12. Simpson, L.J., Wronkiewicz, D.J., and J.A. Fortner, "Development of test acceptance standards for qualification of the glass-bonded zeolite waste form," Argonne National Laboratory Report, ANL-NT-51, 121 p. (1997).

ESTIMATING MODEL PARAMETER VALUES FOR TOTAL SYSTEM PERFORMANCE ASSESSMENT

WILLIAM L. EBERT, VLADISLAV N. ZYRYANOV, JAMES C. CUNNANE
CMT Division, Argonne National Laboratory, Argonne, IL 60439, ebert@cmt.anl.gov

ABSTRACT

The intrinsic dissolution rates of nine borosilicate waste glasses were extracted from the results of MCC-1 tests conducted for durations long enough that the solution pH reached a nearly constant value but short enough that the buildup of dissolved species did not affect the dissolution rate. The effects of the pH and temperature on the measured rates were deconvoluted to determine the sensitivity of the rate to the glass composition. The intrinsic dissolution rates were similar for all of these glasses and were not correlated with the glass composition. The mean and standard deviation of the intrinsic dissolution rates of these glasses are log $\{k_0/[g/(m^2 \bullet d)]\} = 8.2 \pm 0.2$.

INTRODUCTION

The results of total system performance assessment (TSPA) calculations will play an important role in the design and licensing of the federal repository for high-level radioactive waste disposal. These calculations will be used to evaluate the long-term containment of radionuclides by the waste forms and the engineered barrier systems and to ensure that regulatory requirements will be met throughout the service life of the disposal system. A mechanistic rate expression has been developed for borosilicate waste glasses and was included in the TSPA-Viability Assessment report [1]. That expression contains terms for the forward dissolution rate (k_f), which depends on the glass composition, pH, and temperature, and for the reaction affinity term, which quantifies the feedback effect of solute species on the glass dissolution. The rate expression can be written as:

$$\text{rate} = S \bullet k_f \bullet (1 - Q/K) \quad (1)$$

where S is the surface area. The value of the affinity term, which is the term in parentheses, depends on the saturation index of the solution, Q/K, where K is a quasi-thermodynamic constant. The forward rate can be written as:

$$k_f = k_0 \bullet 10^{h \cdot pH} \bullet e^{(-E_a/RT)} \quad (2)$$

where k_0, h, and E_a are parameters for the effects of the glass composition, pH, and temperature on the rate. These parameters values must be measured experimentally. Previous tests have shown the values of h and E_a can be assumed to be independent of the glass composition [2-5]. In this paper, we present the results of tests conducted to determine the values of the intrinsic dissolution rate (k_0) for glass compositions that are representative of likely high-level waste glasses.

Mat. Res. Soc. Symp. Proc. Vol. 608 © 2000 Materials Research Society

EXPERIMENTAL METHOD

The glasses used in these tests include reference compositions for the Defense Waste Processing Facility (DWPF), the West Valley Demonstration Project (WVDP), and Hanford tank wastes. The glass compositions are given in Table 1. The SRL 51S glass is a nonradioactive homologue of glass that was made at DWPF with sludge from Tank 51 [6]. The SRL 202U, SRL 165U, and SRL 131U glasses represent possible DWPF waste glasses. The WV6 glass is a nonradioactive homologue of the glasses made at the WVDP. A glass formulated to represent a potential waste glass made with Hanford tank wastes [7] is also being tested; we refer to that glass as Hanford-D glass. The intrinsic rates of three other glasses were measured to evaluate the effects of high and low aluminum contents: Hanford-L, LD6-5412, and PNL 7668. The glass we refer to as Hanford-L was formulated at BNFL, Inc., as a reference glass for low-activity waste forms. The LD6-5412 glass is a reference low-activity waste glass for Hanford tank wastes that contains a much higher concentration of aluminum than typical high-level waste glasses. The PNL 7668 glass was included as a composition without aluminum. These glasses were tested to provide added insight regarding the effect of composition on the intrinsic dissolution rate.

The nine glasses used in this study provide a wide range in the concentrations of key glass components, including Al_2O_3 from 0 to 13 mass %, B_2O_3 from 5.3 to 13 mass %, Na_2O from 9 to 20 mass %, and SiO_2 from 31 to 59 mass %. While the compositions of high-level waste glasses have not been finalized for Hanford tank wastes or for high-level wastes in Idaho, it is likely that the concentrations of the key glass-forming components of the waste form will be within these ranges, and that the intrinsic dissolution rate measured in the present study will provide an upper bound to the intrinsic dissolution rates of waste glasses developed in the future.

The MCC-1 tests [8] were conducted with disk-shaped monolithic specimens nominally 10 mm in diameter and 1 mm thick. The faces of the samples were polished to a 600-grit final finish and cleaned to remove fines. Samples were placed flat on perforated Type 304L stainless steel or titanium stands in Type 304L stainless steel vessels. Enough demineralized water was added to the vessel so that the geometric glass surface area/water volume ratio was about 10 m^{-1}. The vessel was sealed with a Teflon gasket and closure fitting, then placed in a convection oven set at 90°C for periods ranging from 1 to 33 days. Blank tests with only demineralized water added to the vessel were conducted for similar durations. At the end of the test, an aliquot of the solution was analyzed for pH with a combination electrode. The remaining solution was acidified with ultrapure, concentrated nitric acid and analyzed with inductively coupled plasma-atomic emission spectroscopy (ICP-AES).

RESULTS AND DISCUSSION

The results of the short-term MCC-1 tests are plotted in Figs. 1 and 2 as the normalized elemental mass loss based on the solution concentrations of B, Na, and Si against the test duration. The normalized elemental mass loss, NL(i), was calculated using the following expression

$$NL(i) = C(i) / \{(S/V) \bullet f(i)\} \quad (3)$$

Table 1. Composition of Reference Glasses, in oxide mass %

Oxide	SRL 51S	SRL 202U	SRL 165U	SRL 131A	WV6[a]	Hanford-D[b]	Hanford-L[c]	LD6-5412	PNL 76-68[d]
Al_2O_3	5.27	3.84	4.08	3.27	6.00	10.13	11.97	12.89	-
B_2O_3	7.41	7.97	6.76	9.65	12.89	6.99	8.85	5.34	8.98
CaO	1.39	1.2	1.62	1.23	0.48	4.04	-	3.90	2.54
Cr_2O_3	0.44	0.08	<0.01	0.13	0.14	0.04	0.02	-	0.43
Cs_2O	-	-	-	-	0.08	-	-	-	1.06
Fe_2O_3[e]	12.2	11.4	11.74	12.7	12.02	22.95	5.77	0.124	9.16
K_2O	1.39	3.71	-	3.86	5.00	0.86	3.10	1.37	-
La_2O_3	0.52	-	<0.05	-	0.04	-	-	-	4.11
Li_2O	4.54	4.23	4.18	3.0	3.71	3.00	-	-	-
MgO	1.79	1.32	0.7	1.31	0.89	0.11	1.99	0.035	-
MnO_2	1.41	2.21	2.79	2.43	1.01	0.41	-	0.0047	-
MoO_3	-	0.02	<0.01	-	0.04	-	-	-	1.85
Na_2O	9.60	8.92	10.85	12.1	8.00	15.74	20.0	20.23	14.2
Nd_2O_3	-	-	<0.05	-	0.14	-	-	-	1.40
NiO	0.26	0.82	0.85	1.24	0.25	0.10	-	-	0.19
P_2O_5	0.58	-	0.023	-	1.20	0.72	0.080	0.119	0.64
SiO_2	56.3	48.9	52.86	43.8	40.98	30.19	38.25	58.91	42.3
SrO	-	0.03	0.11	0.01	0.02	4.16	-	-	0.40
ThO_2	0.014	0.26	-	-	3.56	-	-	-	-
TiO_2	-	0.91	0.14	0.65	0.80	0.19	2.49	0.84	2.96
UO_2	1.1	1.93	0.92	2.73	0.59	-	-	-	-
ZnO	0.13	0.02	0.04	0.02	0.02	0.01	4.27		3.27
ZrO_2	-	0.1	0.66	0.22	1.32	-	2.49	0.0068	1.76

[a] Provided by The Catholic University of America.
[b] As-formulated composition [7].
[c] As-formulated composition for LAWA33 glass provided by PNNL.
[d] Composition from [9].
[e] All Fe is represented as Fe(III).

where C(i) is the measured solution concentration, S/V is the glass surface area/solution volume ratio, and f(i) is the mass fraction of element i in the glass. (The concentrations of B, Na, and Si in the blank tests were below detection limit and were neglected in the calculations.) Linear regression lines are drawn through the results in Figs. 1 and 2 to provide graphical measures of the normalized dissolution rates NR(B), NR(Na), and NR(Si). These were used as estimates of the forward dissolution rate (k_f); the origin was not included in the regression fits for any of the glasses. The results of tests conducted for less than two days were not included in the regression fits because (1) they are strongly affected by disturbance of the outer surface by the surface preparation and (2) the solution pH increased rapidly over the first two days due to the consumption of protons during the initial dealkalization of the surface. While high-energy sites generated during surface preparation are present on all samples, their contribution to the total amount of glass dissolved is expected to become constant for a given glass after the outer surface dissolves. For tests with each glass, the pH values became nearly constant after the first few days.

The results of tests conducted for longer than about 10 days were excluded from the regression because the dissolution rate was likely affected by the buildup of dissolved glass components. The results of these tests lie below the regression lines drawn through the results of tests conducted for between about two and ten days. The test results for most glasses are well fit by linear regression over the interval of two to ten days; the regression coefficients are $R^2 > 0.92$. An exception is the results of tests with Hanford-D glass, which show more scatter.

In the plots in Figs. 1 and 2, results shown by solid symbols were included in the linear regression to determine the slope, which is the dissolution rate, while results shown by open symbols were excluded from the regression. The slopes based on the releases of B and Si and the average solution pH values for the tests included in the regression analyses were used to calculate the intrinsic dissolution rates. The release of sodium is presumed to occur by ion exchange rather than by matrix dissolution, even though the release rates of B, Na, and Si are similar. For convenience, we express Eq. 2 in logarithmic form and solved for log k_0:

$$\log k_0 = \log k_f - h \bullet \mathrm{pH} - \log \{\exp(-E_a/RT)\} \tag{4}$$

Although the pH and temperature dependencies have only been measured for a few glasses [2-5], the values of h and E_a are similar for these glasses. For the present analysis, we used h = 0.4 and E_a = 80 kJ/mol. The solution pH was measured at room temperature. The pH at the test temperature of 90°C was assumed to be 1.6 units lower than the pH measured at room temperature due to the difference in the equilibrium constants of water at 25 and 90°C. This adjustment to the pH has the effect of increasing the value of log k_0 by 0.64 and the value of k_0 by a factor of 4.4 for all glasses. The temperature-adjusted pH values were used to calculate the value of the pH term (i.e., h•pH) at 90°C, where h = 0.4. Since all tests were conducted at 90°C, the value of the temperature term with E_a = 80 kJ/mol is –11.51 for all tests. The intrinsic dissolution rates calculated by using Eq. 4 and the normalized dissolution rates based on B and Si are summarized in Table 2.

The average intrinsic dissolution rate of the reference high-level waste glasses is almost the same as the average for all nine glasses. If the higher value of log k_0(B) and log k_0(Si) for each glass is used, the average for all glasses is log $\{k_0/[g/(m^2 \bullet d)]\} = 8.26 \pm 0.15$ and the percent relative standard deviation is 1.5%. The percent relative standard deviations in the forward rates (for all glasses) are 42% for NR(B) and 23% for NR(Si). This indicates that most of the

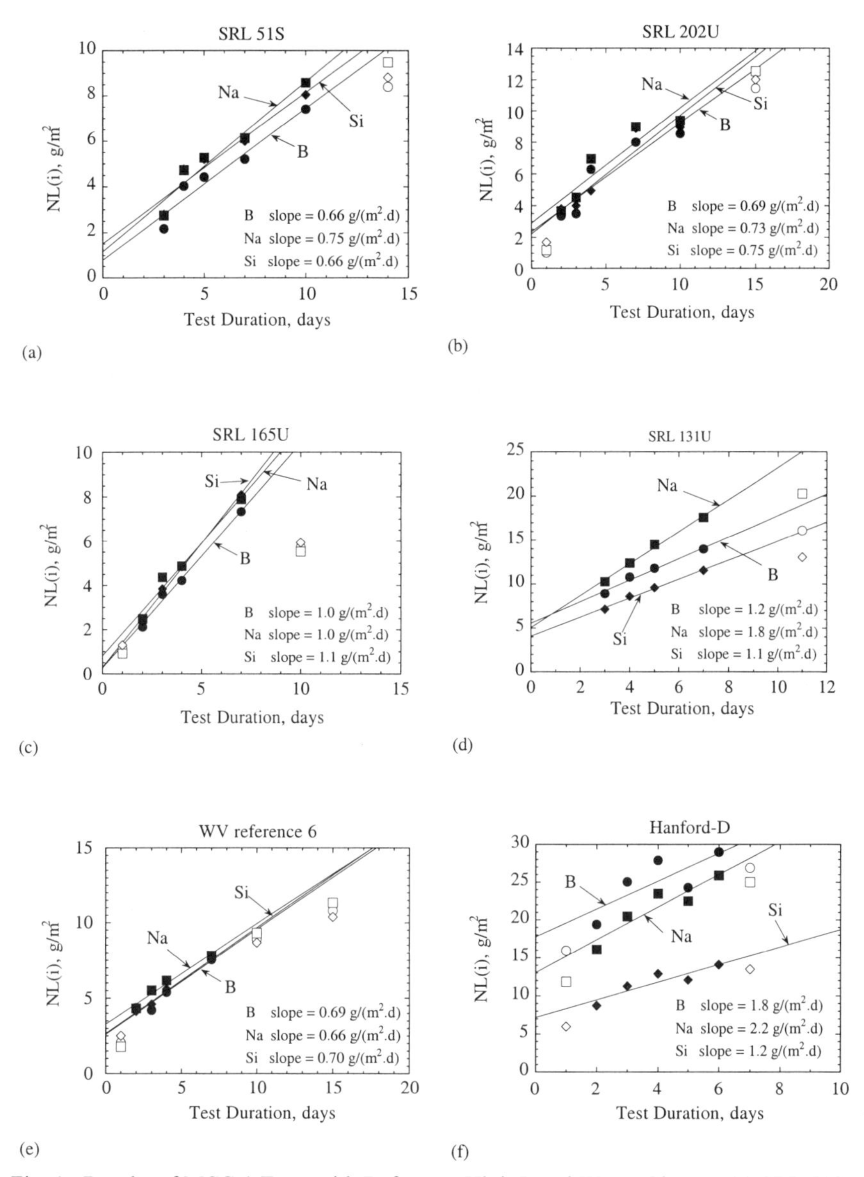

Fig. 1. Results of MCC-1 Tests with Reference High-Level Waste Glasses: (a) SRL 51S, (b) SRL 202U, (c) SRL 165U, (d) SRL 131U, (e) WV reference 6, (f) Hanford-D. Results shown by open symbols were excluded from regression analysis.

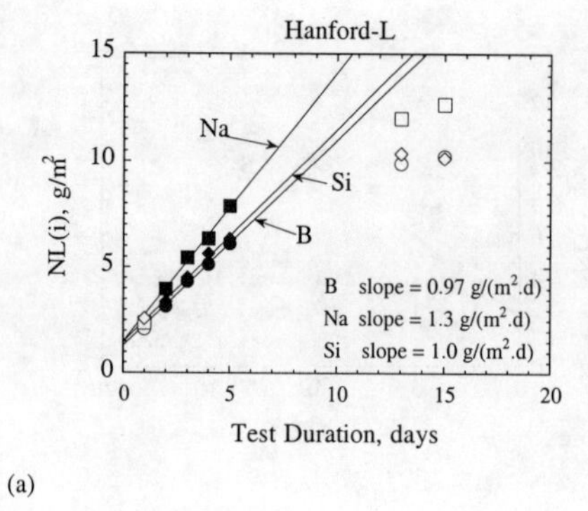

(a)

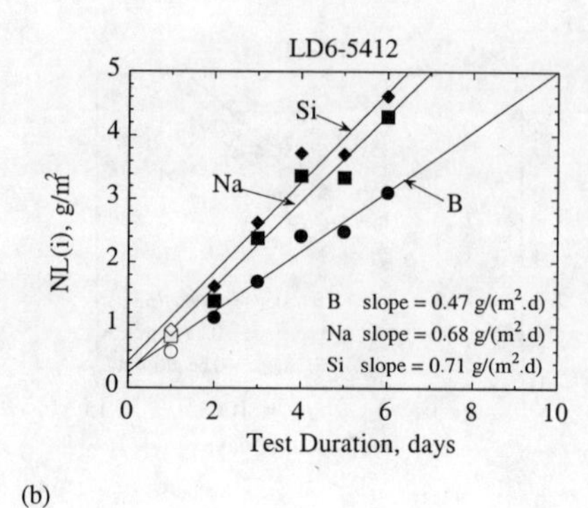

(b)

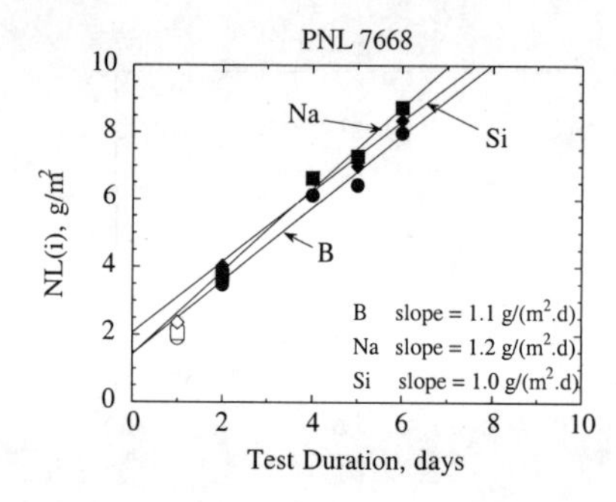

(c)

Fig. 2. Results of MCC-1 Tests with (a) Hanford-L, (b) LD6-5412, and (c) PNL 7668 Glasses. (●) NL(B), (■) NL(Na), and (◆) NL(Si). Results shown by open symbols were excluded from regression analysis.

difference in the dissolution rates of these glasses in short-term MCC-1 tests is due to the small differences in the solution pH values. A conservative upper bound for the intrinsic dissolution rate is the mean plus two standard deviations, which is log $\{k_0/[g/(m^2\bullet d)]\} = 8.5$ if the higher of the values based on boron or silicon is used. The forward dissolution rate at any temperature and pH can be calculated by using Eq. 1. This is the maximum possible dissolution rate of a waste glass under particular temperature and pH conditions. It is also the most conservative bound to the long-term corrosion rate of disposed high-level waste glasses. The similarity in the intrinsic dissolution rates of glasses having significantly different compositions is probably an indication that the same rate limiting reaction controls the dissolution rate for each glass. Secondary effects of the glass composition may become important as corrosion proceeds, such as the effect on the solution pH and the affinity term. These effects can dominant the glass response in other tests.

These results indicate that the intrinsic dissolution rate cannot be used to discriminate between glasses that have different chemical durabilities under anticipated disposal conditions. Instead, the response in a test that is sensitive to solution feedback effects must be used. We are currently evaluating the use of the product consistency test (PCT) for this purpose.

Table 2. Summary of Normalized Dissolution Rates (NR) and Intrinsic Rate Constants

Glass	Average pH[a] (25°C)	NR, $g/(m^2 \bullet d)$		$\log\{k_0/[g/(m^2 \bullet d)]\}$	
		B	Si	B	Si
SRL 51S [b]	9.9	0.66	0.66	8.01	8.01
SRL 202U [b]	9.8	0.69	0.75	8.07	8.11
SRL 165U [b]	9.6	1.0	1.1	8.31	8.35
SRL 131U [b]	9.8	1.2	1.1	8.31	8.27
WV6 [b]	9.5	0.69	0.70	8.19	8.20
Hanford-D [b]	10.5	1.8	1.2	8.57	8.27
Hanford-L	9.5	0.97	1.0	8.34	8.35
LD6-5412	9.3	0.47	0.71	8.10	8.28
PNL 7668	9.2	1.1	1.0	8.51	8.47
6 HLW glasses mean±s		1.01 ± 0.44	0.92 ± 0.24	8.18 ± 0.12	8.16 ±0.14
All 9 glasses mean ±s		0.95 ± 0.40	0.91 ± 0.21	8.23 ± 0.16	8.23 ±0.16

[a] Average pH value measured at room temperature in tests included in regression.
[b] Reference high-level waste glass.

CONCLUSIONS

The values of the intrinsic dissolution rates of nine borosilicate glasses having a wide range of compositions were estimated from the results of short-term MCC-1 tests by deconvoluting the effects of temperature and solution pH. The intrinsic dissolution rates extracted from the results of tests were similar: $\log\{k_0/[g/(m^2 \bullet d)]\} = 8.2 \pm 0.2$. The very small variation in the determined rates indicates that the same intrinsic dissolution rate can be used for high-level waste glasses in performance assessment calculations with little added uncertainty. An upper bound for the intrinsic dissolution rates of borosilicate waste glasses can be assumed to be $\log\{k_0/[g/(m^2 \bullet d)]\} = 8.6$ at the 95% confidence limit.

ACKNOWLEDGMENT

Laboratory assistance provided by Jeffrey W. Emery, Lohman Hafenrichter, and Michael K. Nole. This task was performed under the guidance of the Yucca Mountain Site Characterization Project (YMP) and is part of activity D-20-28 in the YMP/Lawrence Livermore National Laboratory Spent Fuel Scientific Investigation Plan. This work was supported by the U. S. Department of Energy under contract W-31-109-ENG-38.

REFERENCES

1. "Total System Performance Assessment - Viability Assessment (TSPA-VA) Analyses Technical Basis Document: Waste Form Degradation," Chapter 6, Radionuclide Mobilization Preliminary, and Transport Through the Engineered Barrier System. Civilian Radioactive Waste Management System report B00000000-01717-4301-00004 REV 01, Las Vegas, Nevada, 1998.

2. K. G. Knauss, W. L. Bourcier, K. D. McKeegan, C. I. Merzbacher, S. N. Nguyen, F. J. Ryerson, D. K. Smith, and H. C. Weed, in *Scientific Basis for Nuclear Waste Management XIV*, edited by V.M. Oversby and P.W. Brown (Mater. Res. Soc. Proc., **176**, Pittsburgh, PA, 1990) pp. 371-381.

3. T. Advocat, J. L. Crovisier, E. Vernaz, G. Ehret, and H. Charpentier, in *Scientific Basis for Nuclear Waste Management XIV*, edited by T. Abrajano, Jr. and L.H. Johnson (Mater. Res. Soc. Proc., **212**, Pittsburgh, PA, 1991) pp. 57-64.

4. B. P. McGrail, W. L. Ebert, A. J. Bakel, and D. K. Peeler, J. Nucl. Mat., 249, 175 (1997).

5. P. K. Abraitis, D. J. Vaughan, F. R. Livens, L. Monteith, D. P. Trivedi, J. S. Small, in *Scientific Basis for Nuclear Waste Management XXI*, edited by I. G. McKinley and C. McCombie (Mater. Res. Soc. Proc., **509**, Pittsburgh, PA, 1998) pp. 47-54.

6. M.K. Andrews and N.E. Bibler, Ceram. Trans., Vol. 39, 205 (1993).

7. C. L. Crawford, D. M. Ferrara, B. C. Ha, and N. E. Bibler, in *Proceedings of Spectrum '98, International Conference on Decommissioning and Decontamination and on Nuclear and Hazardous Waste Management, Denver, CO, September 13-18*, La Grange Park, IL, 1998, pp. 581-588.

8. *Standard Test Method for Static Leaching of Monolithic Waste Forms for Disposal of Radioactive Waste*, Standard C1220-98, American Society for Testing and Materials, West Conshohocken, PA. (1998).

9. C. Q. Buckwalter, L. R. Pederson, and G. L. McVay, J. Non-Cryst. Solids, 49, 397-412 (1982).

AUTHOR INDEX

SUBJECT INDEX